METHODS IN ENZYMOLOGY

VOLUME III

METHODS IN ENZYMOLOGY

METHODS IN ENZYMOLOGY

Edited by

SIDNEY P. COLOWICK and NATHAN O. KAPLAN

McCollum-Pratt Institute
The Johns Hopkins University, Baltimore, Maryland

VOLUME III

ACADEMIC PRESS

A Subsidiary of Harcourt Brace Jovanovich, Publishers

New York   London   Toronto   Sydney   San Francisco

ACADEMIC PRESS, INC.
111 Fifth Avenue, New York, New York 10003

United Kingdom Edition published by
ACADEMIC PRESS, INC. (LONDON) LTD.
24/28 Oval Road, London NW1

LIBRARY OF CONGRESS CATALOG CARD NUMBER: 54-9110

Printed and bound in the United Kingdom
Transferred to Digital Printing, 2011

PREFACE

The first two volumes of this treatise were devoted to methods for the preparation and assay of enzymes. In the present volume are described the synthesis, isolation, qualitative detection, quantitative determination, and properties of the corresponding substrates and coenzymes.

As pointed out in the Preface to Volume I, the information on substrates and coenzymes was not included with that on the corresponding enzymes; this was done in order to avoid duplication, since a single compound so often serves as substrate or coenzyme for numerous enzyme systems. In order to help minimize any inconvenience resulting from this separation, the organization of the present volume has been designed to correspond as closely as is feasible to that for the enzymes in Volumes I and II. Furthermore, cross-references are included in order to expedite correlation of material in Volumes I and II with that in Volume III.

Obviously, the material in the present volume, while presented nominally for use in connection with enzyme studies, has much broader application. The analytical procedures and methods of preparation will, of course, be of value in any biochemical studies involving these substances. From this point of view, the present volume may be used independently of the preceding ones. On the other hand, the present volume is essential for maximal usefulness of Volumes I and II.

Since the original outlines for these volumes were prepared, important advances have of course been made in various areas. An attempt has been made to include in this volume some of the more important recently discovered metabolites, such as carbamyl phosphate and the various nucleoside di- and triphosphates now known to be involved in nucleic acid synthesis. In the case of such additions to Volume III, there are of necessity no sections on the corresponding enzymes in the preceding volumes.

We wish to express our appreciation to Mrs. Maryda Colowick for her invaluable assistance in the preparation of the subject indexes for Volumes I, II, and III, and to Dr. John B. Wolff for his translation of article 36 in this volume.

We wish again to express our gratitude to all of the investigators who have contributed to these volumes, as well as to Volume IV on Special Techniques, for which all solicited manuscripts have already been received.

It is also our sad duty to record here that four of the contributors to this treatise have died during the past year: Walter Christian, George H.

Hogeboom, Seymour Korkes, and Carl Neuberg. We wish to express our own deep sense of personal loss in the passing of Seymour Korkes, who was for many years a dear and inspiring friend to both of us.

Baltimore, October 9, 1956   SIDNEY P. COLOWICK
                                                                                        NATHAN O. KAPLAN

Contributors to Volume III

Article numbers are shown in parentheses following the names of contributors.
Affiliations listed are current.

S. ABRAHAM (7), *University of California, Berkeley, California*

SHIRO AKABORI (44), *Osaka University, Osaka, Japan*

ELIZABETH P. ANDERSON (143A, part II), *National Institutes of Health, Bethesda, Maryland*

REGINALD M. ARCHIBALD (75), *The Rockefeller Institute for Medical Research, New York, New York*

C. ARTOM (59), *Bowman-Gray School of Medicine, Wake Forest College, Winston-Salem, North Carolina*

GILBERT ASHWELL (12), *National Institutes of Health, Bethesda, Maryland*

V. H. AUERBACH (89), *Harvard Medical School, Boston, Massachusetts*

ROBERT BALLENTINE (145, 148), *The Johns Hopkins University, Baltimore, Maryland*

H. A. BARKER (53, 61), *University of California, Berkeley, California*

S. B. BARKER (42), *University of Alabama, Birmingham, Alabama*

WILLIAM S. BECK (32), *New York University, New York, New York*

AARON BENDICH (106), *Sloan-Kettering Institute, New York, New York*

A. A. BENSON (15), *Pennsylvania State University, State College, Pennsylvania*

RONALD BENTLEY (41), *University of Pittsburgh, Pittsburgh, Pennsylvania*

ARIEH BERGER (81, parts II and III), *Weizmann Institute of Science, Rehovoth, Israel*

LOUIS BERGER (117, 118, 119), *Sigma Chemical Company, St. Louis, Missouri*

SAMUEL P. BESSMAN (67, 83A), *University of Maryland, Baltimore, Maryland*

P. K. BHATTACHARYYA (96), *National Chemical Laboratory, Poona, India*

SIMON BLACK (47), *National Institutes of Health, Bethesda, Maryland*

KONRAD BLOCH (86), *Harvard University, Cambridge, Massachusetts*

DAVID H. BROWN (21), *Washington University, St. Louis, Missouri*

ERNEST BUEDING (60), *Louisiana State University, New Orleans, Louisiana*

HELEN B. BURCH (139, 141), *Washington University, St. Louis, Missouri*

DOROTHY D. BURFORD (148), *The Johns Hopkins University, Baltimore, Maryland*

ROBERT MAIN BURTON (43), *National Institutes of Health, Bethesda, Maryland*

HARRIS BUSCH (70), *University of Illinois, Chicago, Illinois*

WILLIAM L. BYRNE (24, part II), *Duke University, Durham, North Carolina*

G. L. CANTONI (85), *National Institutes of Health, Bethesda, Maryland*

CARLOS E. CARDINI (114, 115), *Instituto de Investigaciones Bioquimicas, Fundácion Campomar, Buenos Aires, Argentina*

H. E. CARTER (96), *University of Illinois, Urbana, Illinois*

WALTER CHRISTIAN* (125), *C. F. Boehringer and Soehne, Mannheim-Waldhof, Germany*

J. ANTHONY CIFONELLI (4), *University of Chicago, Chicago, Illinois*

MARGARET M. CIOTTI (45, 128), *The Johns Hopkins University, Baltimore, Maryland*

PHILIP P. COHEN (94), *University of Wisconsin, Madison, Wisconsin*

WALDO E. COHN (107, 120), *Oak Ridge National Laboratories, Oak Ridge, Tennessee*

S. P. COLOWICK (127), *The Johns Hopkins University, Baltimore, Maryland*

GERTY T. CORI (8), *Washington University, St. Louis, Missouri*

* Deceased.

vii

R. D. DeMoss (40), *University of Illinois, Urbana, Illinois*

George de Stevens (6), *Ciba Pharmaceutical Products, Inc., Summit, New Jersey*

Daniel H. Deutsch (68), *California Foundation for Biochemical Research, Los Angeles, California*

Albert Dorfman (4), *University of Chicago, Chicago, Illinois*

J. W. Dubnoff (91), *California Institute of Technology, Pasadena, California*

Edward L. Duggan (78), *University of California, Berkeley, California*

A. H. Ennor (116), *The John Curten School of Medical Research, Australian National University, Canberra, A.C.T., Australia*

Cecil Entenman (55, 56), *U. S. Naval Radiological Defense Laboratory, San Francisco, California*

S. P. Felton (140), *University of Washington, Seattle, Washington*

William H. Fishman (9), *Tufts Medical College, Boston, Massachusetts*

G. S. Fraenkel (97), *University of Illinois, Urbana, Illinois*

Dexter French (3), *Iowa State College, Ames, Iowa*

Theodore E. Friedemann (66), *Fitzsimons Army Hospital, Denver, Colorado*

M. Friedkin (26), *Washington University, St. Louis, Missouri*

Howard Gest (52), *Western Reserve University, Cleveland, Ohio*

Jesse P. Greenstein (82), *National Institutes of Health, Bethesda, Maryland*

Isidor Greenwald (37), *Bellevue Medical Center, New York University, New York, New York*

I. C. Gunsalus (138, 142), *University of Illinois, Urbana, Illinois*

W. Z. Hassid (7, 16B), *University of California, Berkeley, California*

E. Hoff-Jørgensen (110), *Biochemical Institute, Copenhagen, Denmark*

B. L. Horecker (13, 19, 23, 28, 29, 29A, 30, 124, part II), *National Institutes of Health, Bethesda, Maryland*

Rollin D. Hotchkiss (102, 105), *The Rockefeller Institute for Medical Research, New York, New York*

F. M. Huennekens (140), *University of Washington, Seattle, Washington*

Robert B. Hurlbert (111), *M. D. Anderson Hospital, Houston, Texas*

Jerard Hurwitz (29A), *Washington University, St. Louis, Missouri*

M. E. Jones (94A), *Massachusetts General Hospital, Boston, Massachusetts*

H. M. Kalckar (26, 121, 143A), *National Institutes of Health, Bethesda, Maryland*

Nathan O. Kaplan (14, 45, 123, 128, 129), *The Johns Hopkins University, Baltimore, Maryland*

Ephraim Katchalski (81), *Weizmann Institute of Science, Rehovoth, Israel*

Z. I. Kertesz (5), *New York State Agricultural Experiment Station, Cornell University, Geneva, New York*

Hans Klenow (25), *University of Copenhagen, Copenhagen, Denmark*

C. A. Knight (100), *University of California, Berkeley, California*

W. E. Knox (89), *Harvard Medical School, Boston, Massachusetts*

Seymour Korkes* (71), *Duke University, Durham, North Carolina*

Arthur Kornberg (124, 131), *Washington University, St. Louis, Missouri*

L. O. Krampitz (50), *Western Reserve University, Cleveland, Ohio*

J. O. Lampen (27), *The Squibb Institute, New Brunswick, New Jersey*

Judith Lange (146), *University of California, Berkeley, California*

Henry A. Lardy (24, 31, 33), *University of Wisconsin, Madison, Wisconsin*

Helge Larsen (72), *Norway Institute of Technology, Trondheim, Norway*

Ennis Layne (73), *University of Maryland, Baltimore, Maryland*

Marjorie B. Lees (57), *McLean Hospital, Waverley, Massachusetts*

Albert L. Lehninger (51, 126), *The Johns Hopkins University, Baltimore, Maryland*

* Deceased.

Luis F. Leloir (17, 114, 115, 143), *Instituto de Investigaciones Bioquimicas, Fundácion Campomar, Buenos Aires, Argentina*

M. Levy (74), *New York University, New York, New York*

Katharine F. Lewis (49), *Lankenau Research Institute, Philadelphia, Pennsylvania*

F. Lipmann (94A), *Massachusetts General Hospital, Boston, Massachusetts*

D. L. MacDonald (88), *University of California, Berkeley, California*

R. M. McCready (16B), *U. S. Dept. of Agriculture, Albany, California*

William D. McElroy (122), *The Johns Hopkins University, Baltimore, Maryland*

Ines Mandl (22, 34, 35), *Columbia University, New York, New York.*

Kingsley M. Mann (24, parts III and IV), *University of Wisconsin, Madison, Wisconsin*

Roy Markham (108, 112), *Cambridge University, Cambridge, England*

Elizabeth S. Maxwell (143A, part I), *National Institutes of Health, Bethesda, Maryland*

Alton Meister (65), *Tufts University, Boston, Massachusetts*

Agnete Munch-Peterson (121), *University of Copenhagen, Copenhagen, Denmark*

Victor A. Najjar (75A), *The Johns Hopkins University, Baltimore, Maryland*

Alvin Nason (144), *The Johns Hopkins University, Baltimore, Maryland*

Erwin Negelein (36), *Institut für Medizin und Biologie der Deutschen Akademie der Wissenschaften, Berlin-Buch, Germany*

A. C. Neish (46), *Prarie Regional Laboratory, Saskatoon, Saskatchewan, Canada*

Carl Neuberg* (22, 34, 35), *New York Medical College, New York, New York*

D. J. D. Nicholas (144, 149), *University of Bristol Research Station, Long Ashton, Bristol, England*

* Deceased.

G. David Novelli (132, 134), *Oak Ridge National Laboratories, Oak Ridge, Tennessee*

Evelyn L. Oginsky (92), *Merck Institute of Therapeutic Research, Rahway, New Jersey*

Alejandro C. Paladini (17, 143), *Instituto de Investigaciones Bioquimicas, Fundácion Campomar, Buenos Aires, Argentina*

Edmond S. Perry (62), *Distillation Products Industries, Rochester, New York*

Ralph E. Peterson (87, part III), *National Institutes of Health, Bethesda, Maryland*

Robert E. Phillips (68), *California Foundation for Biochemical Research, Los Angeles, California*

Paul E. Plesner (25), *University of Copenhagen, Copenhagen, Denmark*

Theodore Posternak (16A, 18), *University of Geneva, Geneva, Switzerland*

Edison W. Putman (10, 11), *University of California, Berkeley, California*

E. Racker (54), *Public Health Research Institute of the City of New York, New York, New York*

Gale W. Rafter (127), *The Johns Hopkins University, Baltimore, Maryland*

S. Ratner (93), *Public Health Research Institute of the City of New York, New York, New York*

W. E. Razzell (138), *University of Illinois, Urbana, Illinois*

Warwick Sakami (95), *Western Reserve University, Cleveland, Ohio*

Gerhard Schmidt (38, 58, 98, 101, 109, 109A), *The Boston Dispensary, Boston, Massachusetts*

Walter C. Schneider (99), *National Institutes of Health, Bethesda, Maryland*

Thomas John Schoch (2), *Corn Products, Inc., Argo, Illinois*

Milton W. Slein (20), *Camp Detrick, Frederick, Maryland*

Emil L. Smith (80), *University of Utah, Salt Lake City, Utah*

ROBERTS A. SMITH (142), *University of Illinois, Urbana, Illinois*

ESMOND E. SNELL (77, 133), *University of California, Berkeley, California*

JOHN E. SNOKE (86), *University of California, Los Angeles, California*

MICHAEL SOMOGYI (1), *Jewish Hospital, St. Louis, Missouri*

MORRIS SOODAK (48), *Massachusetts General Hospital, Boston, Massachusetts*

L. SPECTOR (94A), *Massachusetts General Hospital, Boston, Massachusetts*

JOSEPH R. SPIES (76), *U. S. Department of Agriculture, Washington, D. C.*

E. R. STADTMAN (39, 131, 137), *National Institutes of Health, Bethesda, Maryland*

THRESSA C. STADTMAN (63), *National Institutes of Health, Bethesda, Maryland*

R. Y. STANIER (88), *University of California, Berkeley, California*

JAKOB A. STEKOL (84), *The Institute for Cancer Research, Philadelphia, Pennsylvania*

WILLIAM STEPKA (79), *Medical College of Virginia, Richmond, Virginia*

JOSEPH R. STERN (69), *Western Reserve University, Cleveland, Ohio*

FRANCIS E. STOLZENBACH (129), *The Johns Hopkins University, Baltimore, Maryland*

BERNARD L. STREHLER (122), *Gerontology Section, National Institutes of Health, Baltimore, Maryland*

J. L. STROMINGER (143A, part) I, *Washington University, St. Louis, Missouri*

P. K. STUMPF (29A), *University of California, Berkeley, California*

HERBERT TABOR (90), *National Institutes of Health, Bethesda, Maryland*

HAROLD TARVER (146), *University of California, Berkeley, California*

C. A. THOMAS, JR. (104), *Eli Lilly and Co., Indianapolis, Indiana*

ALEXANDER R. TODD (113), *Cambridge University, Cambridge, England*

EZRA L. TOTTON (24, part I), *University of Wisconsin, Madison, Wisconsin*

SIDNEY UDENFRIEND (87), *National Institutes of Health, Bethesda, Maryland*

KIHACHIRO UEHARA (44), *Osaka University, Osaka, Japan*

J. E. VARNER (64), *Ohio State University, Columbus, Ohio*

HEINRICH WAELSCH (83), *Columbia University, New York, New York*

KURT WALLENFELS (125), *Chemisches Laboratorium der Universität, Freiburg, Germany*

T. P. WANG (130, 135, 136), *Institute of Physiology and Biochemistry, Shanghai, China*

SIDNEY WEINHOUSE (49), *The Lankenau Hospital Research Institute, Philadelphia, Pennsylvania*

EUGENE L. WITTLE (133), *University of Texas, Austin, Texas*

HARLAND G. WOOD (52), *Western Reserve University, Cleveland, Ohio*

W. A. WOOD (19), *University of Illinois, Urbana, Illinois*

STEPHEN ZAMENHOF (103), *Columbia University, New York, New York*

Outline of Organization

VOLUME III
PREPARATION AND ASSAY OF SUBSTRATES

VOLUME III
TABLE OF CONTENTS

Section I. Carbohydrates

Section V. Nucleic Acids and Derivatives

Section VI. Coenzymes and Related Phosphate Compounds

Outline of Volumes I, II, and IV

VOLUME I
PREPARATION AND ASSAY OF ENZYMES

Section I. General Preparative Procedures

A. Tissue Slice Technique. B. Tissue Homogenates. C. Fractionation of Cellular Components. D. Methods of Extraction of Enzymes. E. Protein Fractionation. F. Preparation of Buffers.

Section II. Enzymes of Carbohydrate Metabolism

A. Polysaccharide Cleavage and Synthesis. B. Disaccharide, Hexoside and Glucuronide Metabolism. C. Metabolism of Hexoses. D. Metabolism of Pentoses. E. Metabolism of Three-Carbon Compounds. F. Reactions of Two-Carbon Compounds. G. Reactions of Formate.

Section III. Enzymes of Lipid Metabolism

A. Fatty Acid Oxidation. B. Acyl Activation and Transfer. C. Lipases and Esterases. D. Phospholipid and Steroid Enzymes.

Section IV. Enzymes of Citric Acid Cycle

VOLUME II
PREPARATION AND ASSAY OF ENZYMES

Section I. Enzymes of Protein Metabolism

A. Protein Hydrolyzing Enzymes. B. Enzymes in Amino Acid Metabolism (General). C. Specific Amino Acid Enzymes. D. Peptide Bond Synthesis. E. Enzymes in Urea Synthesis. F. Ammonia Liberating Enzymes. G. Nitrate Metabolism.

Section II. Enzymes of Nucleic Acid Metabolism

A. Nucleases. B. Nucleosidases. C. Deaminases. D. Oxidases. E. Nucleotide Synthesis

Section III. Enzymes in Phosphate Metabolism

A. Phosphomonoesterases. B. Phosphodiesterases. C. Inorganic Pyro- and Polyphosphatases. D. ATPases. E. Phosphate-Transferring Systems

Section IV. Enzymes in Coenzyme and Vitamin Metabolism

A. Synthesis and Degradation of Vitamins. B. Phosphorylation of Vitamins. C. Coenzyme Synthesis and Breakdown.

Section V. Respiratory Enzymes

A. Pyridine Nucleotide-Linked, Including Flavoproteins. B. Iron-Porphyrins. C. Copper Enzymes. D. Unclassified.

VOLUME IV
SPECIAL TECHNIQUES FOR THE ENZYMOLOGIST

Section I. Techniques for Characterization of Proteins (Procedures and Interpretations)

A. Electrophoresis; Macro and Micro. **B.** Ultracentrifugation and Related Techniques (Diffusion, Viscosity) for Molecular Size and Shape. **C.** Infra-red Spectrophotometry. **D.** X-ray Diffraction. **E.** Light Scattering Measurements. **F.** Flow Birefringence. **G.** Fluorescence Polarization and Other Fluorescence Techniques. **H.** The Solubility Method for Protein Purity. **I.** Determination of Amino Acid Sequence in Proteins. **J.** Determination of Essential Groups for Enzyme Activity.

Section II. Techniques for Metabolic Studies

A. Measurement of Rapid Reaction Rates; Techniques and Applications, Including Determination of Spectra of Cytochromes and Other Electron Carriers in Respiring Cells. **B.** Use of Artificial Electron Acceptors in the Study of Dehydrogenases. **C.** Use of Percolation Technique for the Study of the Metabolism of Soil Microorganisms. **D.** Methods for Study of the Hill Reaction. **E.** Methods for Measurement of Nitrogen Fixation. **F.** Cytochemistry.

Section III. Techniques for Isotope Studies

A. The Measurement of Isotopes. **B.** The Synthesis and Degradation of Labeled Compounds (Including Application to Metabolic Studies): Monosaccharides and Polysaccharides; Citric Acid Cycle Intermediates; Glycolic, Glyoxylic and Oxalic Acids; Purines and Pyrimidines; Porphyrins; Amino Acids and Proteins; Steroids; Methylated Compounds and Derivatives; Sulfur Compounds; Fatty Acids; Phospholipids; Coenzymes; Iodinated Compounds; Intermediates of Photosynthesis; O^{18}-Labeled Phosphorus Compounds.

Errata for Volume I

P. 324, line 4: should read "0.25 M" instead of "0.04 M."
P. 324, line 8: after "TPN," insert "0.05 ml. of glucose-6-phosphate."
P. 443, running head: should read "Lactic Dehydrogenase of Muscle."
P. 504, chapter heading: should read "*Leuconostoc*" instead of "*Leuconostic*."
P. 610, line 9: after the first sentence of *Step 3*, insert the following: "After centrifugation, the precipitate is discarded and 193 ml. of acetone are added to the supernatant fluid."

Errata for Volume II

P. 183, line 5: should read "Yeast extract (Difco) 10 g."
P. 188, section on Aspartic Acid Decarboxylase: for β-alanine, read α-alanine; for α-carboxyl group, read β-carboxyl group.
P. 513, line 1: for 112 mg., read 11.2 mg.

Section I

Carbohydrates

[1] Preparation of Glycogen, Nitrogen- and Phosphorus-Free, from Liver

By MICHAEL SOMOGYI

This is a modification of the classic Pflüger method, the essential change consisting in the use of low alcohol concentration for precipitation. Efforts of many workers to purify glycogen by numerous reprecipitations failed because impurities always adhered to the glycogen when alcohol concentrations prescribed by Pflüger were employed. Fractional precipitations with variable alcohol concentrations to obtain pure glycogen are applicable also to methods other than Pflüger's, as, for instance, to aqueous extracts made from liver. The procedure described in 1934[1] is briefly as follows:

Introduce the ground liver into a flask, add 2 ml. of concentrated (50 to 60%) NaOH for each gram of liver, and heat, immersed in boiling water, for 3 hours, with occasional agitation during the first hour. On cooling without disturbance, a cake of soap congeals on the surface, occluding all solid impurities, and the perfectly clear, almost gelatinous fluid can be poured off. (Hydrolysis with 60% KOH, followed by salting out with NaCl, leads to the same result.) To the soap in the flask add water about equal in volume to the decanted fluid, heat until the soap is evenly distributed, then add NaCl until it again separates out, cool, and filter. Unite the two fluids, measure the volume, and introduce with continuous agitation 50 ml. of 95% alcohol for each 100 ml. of alkaline fluid. This results in an alcohol concentration of about 33%. Allow the precipitate to settle overnight at room temperature, syphon off the clear layer of the supernatant fluid, and from the rest separate the glycogen by centrifugation or filtration. Prepare a washing solution by adding 1 vol. of alcohol to 2 vol. of a 20% NaOH solution; wash on the filter or in the centrifuge tube until the washing fluid is colorless or nearly so, and perform a last washing with 95% alcohol. (The use of the centrifuge in this operation ensures speed and efficacy, since the precipitate can be thoroughly stirred and mixed with the washing fluid.)

Dissolve the impure glycogen in water while still moist with alcohol, and remove the insoluble particles by filtration. To the filtrate add 2 N HCl until it gives a distinct but not maximum acid reaction (brownish-blue color) with Congo paper. Some precipitate separates out. Measure the volume of the solution, and add 50 ml. of 95% alcohol for each 100 ml.

[1] M. Somogyi, *J. Biol. Chem.* **104**, 245 (1934).

in order to bring about a better flocculation of the precipitate. Filter, returning the first fractions of the filtrate to the filter paper. Add to the filtrate more 95% alcohol so as to increase its total amount to 80 ml. for each 100 ml. of the original acid aqueous solution. This results in an alcohol concentration of approximately 45%, sufficient to effect the complete precipitation of glycogen in the acid medium. Allow the precipitate to settle for a few hours, or overnight if convenient, separate the precipitate by centrifugation or filtration, and wash consecutively twice with 45%, twice with 95% alcohol, and finally with ether. The strong alcohol and the ether are designed mainly for dehydration; and to further this effect, stir with.a glass rod in order to enhance the contact of the glycogen with the fluids. After the last washing, place the centrifuge tube at a slant, mouth down, to allow the ether to evaporate. The glycogen soon can be removed from the tube in the form of a fine white powder and should be left exposed to air for further drying.

The product thus obtained still contains an appreciable amount of NaCl, and as a rule two precipitations with 45% alcohol are required for its removal. If in the final precipitation the alcohol causes only milky opacity but no flocculation, owing to the absence of electrolytes, addition of 0.3 ml. of 0.1 N HCl per 100 ml. of fluid suffices to effect immediate precipitation, and the glycogen may be separated and dried as described.

It is advisable to use redistilled alcohol and ether, as impurities of the commercial products might be adsorbed by glycogen in the course of precipitation and washing. (One worker, for example, observed gradual increase of the phosphorus content of glycogen in the course of repeated reprecipitation.)

Analysis shows such preparations to be free of phosphorus and nitrogen and virtually ash-free. The aqueous solutions are perfectly clear, transparent and nearly colorless in transmitted light, and opalescent, but never milky, in reflected light even in concentrations as high as from 5 to 10%. On hydrolysis with acid no trace of yellow color and none of the familiar flimsy precipitate appears, either in the acid state or after neutralization.

[2] Preparation of Starch and the Starch Fractions

By THOMAS JOHN SCHOCH

Preparation of Starch

Starches for scientific studies should be chosen with considerable discrimination. When a commercial starch is employed, the manufacturer can usually supply top-quality unmodified starch of high viscosity and with a history of minimum chemical treatment. In general, packaged laundry or household starches should not be employed, since these are frequently modified to render them more suitable for their intended use. Where a starch must be isolated from natural sources, three general procedures are recommended: (1) for tuber starches such as potato, (2) for dry cereal grains such as corn, (3) for ground cereal flours such as wheat. These methods are designed to produce a good yield of high-quality starch, without attempting to achieve complete recovery.

Potato Starch. Clean, sound potatoes are washed, peeled, sliced, and ground in the Waring Blendor for 5 minutes with 2 to 3 vol. of distilled water. To avoid damage to the starch granules, the Blendor should be run at 85 to 90 volts from a variable transformer. The slurry is screened through No. 9 (97-mesh per inch) silk or nylon bolting cloth. The cloth is continuously scraped with a spoon, and the starch is washed through with a jet of distilled water. The pulp is returned to the Blendor and ground for an additional 5 minutes with sufficient water to give a thin slurry. This is rescreened and the pulp discarded. The combined starch suspensions are then passed through No. 12 (125-mesh) bolting cloth into a tall glass jar or cylinder and allowed to settle for approximately 1 hour, or until the starch deposits as a firm dense cake. The top-quality starch sediments most rapidly, followed by small-granule and soft-granule starch and by any residual fiber. Hence the supernatant should be decanted when it is still somewhat cloudy with these undesired fractions. If the sedimentation has proceeded too far, the soft top surface of the settled starch cake may be rinsed with a small amount of distilled water. At this point, the starch may advantageously be screened again through 125-mesh bolting cloth to remove persistent traces of fiber. It is then stirred up with 20 to 30 vol. of water and allowed to settle in a glass cylinder. The supernatant is decanted, and the process of resuspension in fresh water and sedimentation is repeated until the starch settles cleanly to a dense firm cake with no evidence of a cloudy supernatant or a soft top surface. The starch cake is then suspended in methanol, filtered on a Büchner, washed with methanol, and dried at 40 to 50°.

Corn Starch. One kilogram of clean, dry seed corn is placed in a wide-mouthed glass bottle, covered with 2 l. of 0.45% sodium metabisulfite solution ($Na_2S_2O_5$), and the bottle stoppered and placed in a thermostated water bath at 50° for 24 hours. The swollen kernels should then be completely softened; if not, additional steeping is required. The steep liquor is drained off, and the corn is ground in successive small portions with 2 to 3 vol. of distilled water in the Waring Blendor. To avoid damaging the starch granules, the Blendor is operated at 85 to 90 volts, and the grinding time should not exceed 5 minutes. The slurry is then screened and washed through 97-mesh bolting cloth, and the pulp squeezed as dry as possible. The pulp is reground for 5 minutes in the Blendor with additional water and again screened. The combined starch suspension is then passed through 125-mesh bolting cloth. In most instances, sedimentation of this suspension does not give a clean-cut "break" between starch and gluten. Consequently, 20 to 25% by volume of Pentasol[1] is added to swell and float the gluten (unpublished method of S. A. Watson and C. W. Stewart). The mixture is stirred for 15 minutes, then centrifuged in a Sharples continuous-flow supercentrifuge; the starch is deposited in the bowl, and the gluten and Pentasol discharged in the overflow. Equally good separation is obtained in a bottle centrifuge operating at 2000 r.p.m.; the starch deposits cleanly, and the gluten and oil pass into the upper Pentasol layer. In either case, the sedimented starch is resuspended in fresh distilled water, 20% of Pentasol stirred in, and the mixture centrifuged. This process is repeated once or twice again, or until the intermediate water layer is perfectly clear and there is no visible gluten in the upper Pentasol layer. The starch is then suspended in methanol, screened through No. 25 (200-mesh) bolting cloth, filtered on a Büchner, washed with alcohol, and dried at 40 to 50°.

Wheat Starch. Sufficient water is worked into wheat flour to give a strong cohesive dough. This is placed in a wet cloth bag (preferably made from fine bolting cloth), and the doughy mass is hand-kneaded under water until all the starch is expelled from the rubbery gluten. The starch suspension is then screened and repeatedly sedimented as described for potato starch.

Defatting Procedure for Starches. All starches of cereal origin should be exhaustively extracted with alcohol to remove associated fatty acids.[2] Presence of the latter seriously impedes enzyme conversion of the linear fraction, giving rise to insoluble residues erroneously termed "amylohemicellulose" or "amylocellulose." Small amounts of starch can be con-

---

[1] Trade-marked commercial mixture of primary amyl alcohols marketed by Sharples Solvents.

[2] T. J. Schoch, *J. Am. Chem. Soc.* **64,** 2954 (1942).

veniently defatted by 24 to 48 hours' Soxhlet extraction with 95% ethyl alcohol. For larger quantities, 1 kg. of the dry starch is suspended in 3 l. of 85% methanol in a three-necked 5-l. flask equipped with reflux condenser and motor-driven stirrer. A short piece of closely fitting glass or stainless steel tubing through a rubber stopper in the center neck of the flask provides an adequately sealed bearing for the stirrer shaft. The stirred suspension is heated for 1 hour at gentle reflux, either by a mantle heater or by immersion in a hot water bath. The starch is then filtered hot on a Büchner and washed with 85% methanol. The starch cake is extracted twice further in similar fashion with 85% methanol, then filtered and dried at 50°.

Fractionation of Starch

The common starches contain both linear and branched types of polysaccharides, frequently designated as "amylose" and "amylopectin," respectively. According to the preferred technique of Lansky et al.,[3] the linear fraction is precipitated as a microcrystalline complex by cooling a cooked starch solution in contact with excess Pentasol.[1] Normal primary amyl alcohol is second choice as a precipitant. The starch may be fractionated directly without preliminary defatting.

Primary Separation. In a 5-gallon Pyrex bottle, 300 g. of starch is suspended in a mixture of 15 l. of distilled water and 1 l. of Pentasol; 8.2 g. of anhydrous KH_2PO_4 and 1.8 g. of anhydrous K_2HPO_4 are added to buffer the pH at 6.2 to 6.3. The bottle is fitted with a reflux condenser and a high-speed propeller agitator, the propeller shaft passing through a bearing in the rubber stopper of the bottle. The bottle is heated by immersion in a suitable boiling water bath. The contents are first stirred in the cold to effect uniform suspension of the starch, then heated to the boiling point (92°) with continuous agitation. The mixture is gently refluxed for 3 hours, allowed to cool overnight, and refrigerated for 24 hours; continuous stirring is maintained throughout these operations. The linear fraction complex crystallizes as minute spherules or as needle clusters. Better crystal formation is obtained if the bottle is wrapped with cloth as insulation during the cooling period. The product is best separated if the refrigerated mixture is passed at a flow rate of 250 to 300 ml./min. through a continuous-flow Sharples supercentrifuge fitted with a clarifier bowl and operating at 50,000 r.p.m. During centrifugation, the discharged supernatant should be repeatedly examined under the microscope; the appearance of crystals of the linear fraction indicates that the bowl is completely filled. The precipitated material in the centrifuge bowl should be dense and firm, with no evidence of slime. The col-

[3] S. Lansky, M. Kooi, and T. J. Schoch, *J. Am. Chem. Soc.* **71**, 4066 (1949).

lected supernatant is again centrifuged to remove any residual linear fraction. To recover the branched fraction, the supernatant is treated with an equal amount of methanol, and the mixture refrigerated overnight. The precipitated mass is dehydrated by being ground for 2 to 3 minutes in the Waring Blendor with fresh methanol, then filtered on a Büchner and dried at 50°. Depending on the variety of starch processed, this branched fraction contains 0 to 3 % of linear material. This cannot be removed by any known means.

Recrystallization of Linear Fraction. The linear fraction is purified by two recrystallizations, with *n*-butyl alcohol as precipitant rather than Pentasol. A mixture of 15 l. of distilled water and 1 l. of *n*-butyl alcohol is brought to a boil in an unstoppered 5-gallon Pyrex bottle immersed in a boiling water bath. The moist linear fraction from the centrifuge bowl is gradually added with very vigorous mechanical agitation to effect rapid dispersion and solution of the added material. Heating and agitation are continued for 20 to 30 minutes, or until substantially complete solution is effected. The hot solution is then passed through the Sharples supercentrifuge to remove any dirt, cellular material, and protein carried over from the original starch. This should not exceed 0.5 % of the weight of the starch. An excessive deposit in the bowl of the centrifuge indicates incomplete solution of linear fraction, owing either to too low a temperature or to inadequate stirring during solution of the crude material. This can usually be recovered by dissolving separately in boiling water and recentrifuging. The total centrifuged solution is reheated to boiling in the water bath, with additional butyl alcohol if necessary to saturate the system, and cooled overnight with continuous agitation. Refrigeration for an additional 24 hours is advantageous. The recrystallized linear fraction is collected in the supercentrifuge and for optimum purity should be recrystallized a second time in similar fashion. Purity is determined by iodine affinity. In a typical instance, the crude Pentasol-precipitated linear fraction from corn starch adsorbed 16.5 % iodine by this test; successive recrystallizations with butyl alcohol raised this value to 18.4 %, 18.8 %, 18.8 %, and 19.0 %. Additional crystallizations effected no further change. Purity and molecular size of the two starch fractions are adequately characterized by iodine affinity and intrinsic viscosity; typical values are listed in the table. The moist linear fraction complex cannot be dried directly without causing retrogradation to a permanently insoluble state. To avoid this, the material is first dehydrated by being stirred for several hours in 10 vol. of cold *n*-butyl alcohol and then filtered. This treatment is repeated several times, and the final product dried in the vacuum oven at 75° to give a fluffy white powder. Several precautions should be observed in dissolving this dry linear fraction, in order to avoid retrogradation. The requisite volume of water is brought to a vigorous boil over a free flame while being stirred with a mechanical agitator to create a deep vortex. The dry linear fraction is sifted into this vortex at such a rate that it is immediately dispersed and dissolved without lumping. The solution is gently boiled for several minutes to drive off adsorbed butyl alcohol. Any trace of retrograded linear material can be

CHARACTERISTICS OF THE PENTASOL-SEPARATED FRACTIONS FROM VARIOUS STARCHES
(Linear fraction repeatedly recrystallized with n-butyl alcohol)

	Iodine affinity			Intrinsic viscosity	
	Parent starch	Linear fraction	Branched fraction	Linear fraction	Branched fraction
Wheat	5.21	19.9	0.5	1.54	1.18
Corn	5.30	19.0	0.4–0.6	1.27	1.25–1.36
Sago (unbleached)	5.10	18.5	0.2	1.13	0.81
Potato, Maine	4.13	19.9	0.4	1.85	1.45
Potato, Idaho	4.65	—	0.15	2.32	1.52
Tapioca	3.27	18.6	0.0	2.25	1.26

removed by hot filtration through a wad of Pyrex glass wool. At concentrations higher than 2%, solutions of the linear fraction tend to set up to irreversible gels when cooled to room temperature. If a supercentrifuge is not available, small quantities of starch can be fractionated with a mixture of n-butyl and isoamyl alcohols and the linear fraction separated with an ordinary bottle centrifuge, according to the method of Wilson, Jr., et al.[4]

Preparation of Pure Branched Starch Substance

The branched starch fraction as obtained by fractionation with amyl alcohol usually contains 2 to 3% of unremoved linear material. Similarly, the ordinary waxy maize starch of commerce contains 5 to 6% of normal corn starch granules as a result of cross-pollination. These can be readily distinguished by staining them lightly with iodine and examining them under the microscope; the normal corn starch granules appear deep blue or black, the waxy granules light red. When branched starch substance of utmost purity is required, either of two alternative methods may be employed.

[4] E. J. Wilson, Jr., T. J. Schoch, and C. S. Hudson, J. Am. Chem. Soc. 65, 1380 (1943).

From Genetically Pure Waxy Maize Seed. The dealers in hybrid seed corn can frequently furnish small amounts of waxy maize seed which has been hand-picked to remove normal blue-staining corn kernels. The Bear Hybrid Corn Company (Decatur, Illinois) has supplied certified waxy seed with less than one part per thousand of blue-staining corn kernels. This waxy seed is steeped in 0.45% metabisulfite solution and the granular starch isolated as previously described for corn starch. This starch is free from any significant amount of linear starch fraction (i.e., less than 0.03%).

From Commercial Waxy Maize Starch. If blue-staining corn starch is gelatinized and cooked in water saturated with cyclohexanol, the granules undergo only a restricted swelling and do not dissolve even on prolonged boiling. Under similar circumstances, waxy starch granules swell freely and eventually dissolve. This difference in behavior may be employed to remove the contaminating 5 to 6% of blue-staining corn starch from commercial waxy maize starch. A three-necked 3-l. flask is fitted with a reflux condenser and a high-speed propeller agitator. An adequate seal for the propeller shaft is provided by a bearing through a rubber stopper in the center neck of the flask. Into the flask are placed 50 g. of commercial waxy maize starch, 2 l. of distilled water, and 200 ml. of cyclohexanol. The system is buffered at pH 6.1 to 6.3 by the addition of 1.64 g. of KH_2PO_4 and 0.36 g. of K_2HPO_4. The cold mixture is stirred for 5 to 10 minutes to suspend the starch and dissolve the cyclohexanol, and the flask then placed in the vigorously boiling water bath, agitation being continued throughout the heating period. Mantle heaters are not recommended for aqueous starch pastes. The mixture is heated to the boiling point (98°) and gently refluxed for 1 to 1.5 hours. The starch gelatinizes to a viscous paste which progressively thins out as the swollen granules dissolve. If a sample of the resulting solution is lightly stained with iodine and examined under the microscope, the normal corn starch granules will appear only slightly swollen. The hot solution is then passed twice through the Sharples continuous-flow supercentrifuge to remove this normal corn starch; any slime which drains from the bowl of the centrifuge is rejected. As a less desirable alternative, the hot solution may be spun in a bottle centrifuge, in which case the separated supernatant should be carefully examined under the microscope to ensure complete removal of blue-staining starch granules. Recheck of pH should show no change. The branched starch material is precipitated with methanol, dehydrated, and dried as described under starch fractionation. The product should give a red or violet-red color with iodine. Unlike the genetically pure waxy maize starch, this preparation contains no granular starch, and consequently it forms a paste in cold water.

Preparation of Phytoglycogen from Sweet Corn

The soluble polysaccharide in certain varieties of sweet corn appears to be very similar or even identical in molecular structure to animal glycogen. This so-called phytoglycogen can be readily isolated in a relatively pure and undegraded state. In particular, the protein content can be reduced to a very low level without resorting to treatment with such agents as trichloroacetic acid or hot caustic alkali, either of which may cause degradation. Any of the common golden hybrid sweet corn varieties (e.g., Golden Bantam) appears to be a suitable source. Fresh corn may be used in season, and deep-frozen ears can usually be obtained throughout the year. Canned corn or ears which have been subjected to any heat treatment cannot be used. Commercial yellow dent corn (feed or field corn) contains little or no glycogen. Better yields of glycogen are obtained if the corn is fully ripe and the kernels completely developed. In fact, it is preferable to use corn which is past the usual eating or canning stage.

The ears are husked and the kernels sliced off close to the cob with a sharp knife. A dry weight should be run on a sample of the cut kernels if glycogen yield is to be calculated. The kernels are ground for 5 minutes in the Waring Blendor with an equal weight of water, and the slurry screened through No. 9 (97-mesh) silk or nylon bolting cloth to remove coarse fiber, the magma being squeezed as dry as possible. The press cake is ground a second time in the Blendor with fresh water, then screened and the residue discarded. The combined extracts are screened and washed through No. 17 (163-mesh) bolting cloth to remove additional fiber. This progressive screening is more rapid than can be achieved by screening the original slurry through No. 17 bolting cloth. The resulting starchy milk is then passed twice through a Sharples supercentrifuge fitted with a clarifier bowl and operating at 50,000 r.p.m. The first pass is made at a relatively rapid flow rate (500 ml./min.); the centrifuge is then cleaned out, and the second pass made at a slower flow rate (150 to 200 ml./min.). This centrifugation completely removes the starch and any fine fiber, together with insoluble gluten and a portion of entrained oil. The supernatant contains the phytoglycogen and soluble protein, together with some oil and pigment. This solution is then heated in a boiling water bath with occasional stirring. At approximately 75°, the soluble protein coagulates as a soft yellow curd. The solution is maintained at 95° for 15 minutes, then supercentrifuged to remove the coagulated protein. A small amount of slimy material may likewise be deposited in the centrifuge bowl; this is discarded.

The supernatant from the protein separation contains the phytoglycogen. It should exhibit the typical blue-white haze of glycogen; any yellowish color indicates the presence of protein, probably due to insufficient coagulation. The glycogen is precipitated by the addition of 3 vol. of methanol, and the mixture is boiled for 15 to 30 minutes in the hot water bath with continuous agitation to prevent bumping. It is then removed from the bath, the glycogen allowed to settle out, and the supernatant decanted. Three volumes of fresh methanol is added, and the glycogen again digested in the hot water bath for 30 minutes. This rigorous dehydration is necessary to avoid subsequent gumming on filtration or on drying. The crude glycogen is filtered on a Büchner and dried to constant weight in the vacuum oven at 70°. At this stage, it should be a fluffy white or off-white powder; hard crusts indicate incomplete dehydration with methanol. The product still contains some removable protein which has become insolubilized during the drying operation. Consequently, it is redissolved by being sifted into vigorously stirred boiling water, digested for 30 minutes in the boiling water bath with continuous agitation, and the solution passed at a slow flow rate (150 ml./min.) through the supercentrifuge. Any lumps of undissolved glycogen should be separately dissolved in boiling water and centrifuged. A further small amount of protein is deposited in the centrifuge bowl. The purified glycogen is precipitated and dehydrated by successive treatments with boiling methanol, as previously described. Owing to low electrolyte content, it may be necessary to add a pinch of sodium chloride to effect flocculation by the methanol. After drying to constant weight in the vacuum oven at 70°, the product should be a fluffy pure-white powder. The protein content of various batches has ranged from 0.040 to 0.19%. The yield of purified glycogen is about 30% of the dry weight of the cut corn kernels.

The pH of the mashed corn normally runs 7.0 to 7.1. Hence no precautions are observed to buffer the system to prevent hydrolytic degradation. Intrinsic viscosities of various samples in 1 M KOH have ranged from 0.081 to 0.092, depending on the type and maturity of the corn. All samples gave red to red-brown colorations with iodine, indicating the complete absence of linear starch material. Although sweet corn appears to be devoid of α-amylase, it seems advisable to carry through the initial stages of preparation rapidly and without interruption. Depending on the quantity processed, 2 to 4 hours' time is required to reach the first precipitation with methanol. A bottle centrifuge operating at 2000 r.p.m. may be used for small quantities, but the preparation is much more tedious, and starch and protein removal is more difficult. The dried seed of sweet corn contains a high proportion of phytoglycogen, but the protein appears to lose its heat-coagulable character during field-drying of the corn and hence cannot be removed without chemical treatment.

Determination of Iodine Affinity

The linear starch fraction binds iodine to give a stable blue complex, whereas the branched fraction has only a weak affinity for iodine. Bates *et al.*[5] originally observed that potentiometric titration of the starch substance with iodine provided quantitative differentiation between the fractions. A somewhat simpler and more convenient modification[3] of this method is described below.

Apparatus. The apparatus required includes a Leeds and Northrup Type K potentiometer and No. 2420-C galvanometer, a Beckman saturated calomel electrode, a bright platinum electrode, and a water bath thermostated at 30.0°. For less accurate work, a sensitive pH meter calibrated in millivolts may be substituted for the potentiometer and galvanometer. Provision must be made for stirring the solution mechanically during titration.

Reagents

Stock iodine solution, 0.5 M with respect to KCl, 0.5 M to KI, and containing exactly 2.000 g. of iodine per liter. Standardize against sodium arsenite, and adjust if necessary. If stored in a glass-stoppered brown or Pyrex actinic red bottle, this solution maintains its value indefinitely.

Iodine solution for titration. Stock iodine solution is quantitatively diluted tenfold. This diluted solution is unstable and must be prepared daily.

0.5 M KI solution. Store in a brown or Pyrex actinic red bottle.

Also, 1.0 M KOH and 0.5 M HCl.

Calibration. A calibration chart must first be prepared to relate the e.m.f. reading to the amount of free iodine in solution, under conditions identical with those employed for the starch titration. For this purpose, 373 mg. of KCl and 830 mg. of KI are dissolved in exactly 100.0 ml. of water (at 30°) in a 250-ml. beaker. The latter is placed in the 30° bath, the stirrer and electrode assembly adjusted, and the solution titration potentiometrically with the tenfold dilution of stock iodine. Since the solution and the iodine reagent are both 0.05 M with respect to KCl and to KI, there is no change in salt concentration during the titration. From the carefully plotted curve of this titration, a chart is prepared giving the free iodine in solution corresponding to each half-millivolt reading from 230 to 290 mv.

Titration of Starch Sample. Before evaluation of iodine affinity, it is essential that the sample be exhaustively defatted by Soxhlet extraction

[5] F. L. Bates, D. French, and R. E. Rundle, *J. Am. Chem. Soc.* **65**, 142 (1943).

with ethyl alcohol, then dried and pulverized to pass a 60-mesh sieve. A clean, dry 250-ml. beaker is weighed to 0.1 g. on a sensitive torsion balance. An appropriate amount of the starch sample (i.e., approximately 40 mg. of the linear fraction, 100 mg. of whole starch, or 200 mg. of the branched fraction) is weighed on the analytical balance to an accuracy of 0.1 mg. and transferred to the beaker. Five milliliters of 1 M KOH is then added by pipet, and the sample immediately dispersed in the alkali by

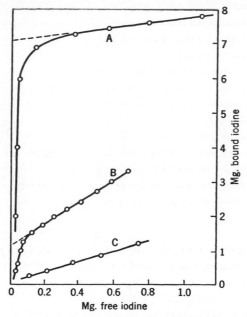

FIG. 1. Method for plotting potentiometric evaluation of iodine affinity, showing titration of (A) 38.4 mg. of linear fraction, (B) 182 mg. of branched fraction, and (C) 190 mg. of phytoglycogen from sweet corn.

grinding with a stirring rod, with care to avoid the formation of diffi-cultly soluble lumps or clots. The mixture is allowed to stand with occa-sional stirring for 1 hour, or until a clear solution is effected. It is then neutralized to methyl orange with 0.5 M HCl. At this point, the solution should be perfectly clear, with no evidence of lumps or undissolved particles. With some granular starches and with some branched fractions, the sample will not completely dissolve in the alkali. In such cases, the sample is neutralized, diluted with 25 to 50 ml. of water, and heated care-fully to incipient boiling over a low flame. The beaker (cooled if neces-sary) is placed on the torsion balance, and 10 ml. of 0.5 M KI is added by pipet, followed by sufficient water to give a total weight of 100.9 g. over the weight of the empty beaker (i.e., weight of 100.0 ml. of water + 373 mg. of KCl + 830 mg. of KI + sample). Except for the presence of starchy material, this solution is identical with that titrated in the foregoing calibration. It is placed in the 30° bath and titrated in similar fashion, and the e.m.f. determined at 8 to 10 points between 230 and 280 mv. After each addition of iodine, the system is allowed to equilibrate for 2 minutes before the voltage is read. For each point in the titration, the free iodine in solution is calculated from the corresponding e.m.f. of the calibration chart, and this amount is subtracted from the total amount of iodine added at that point (i.e., milliliters of iodine added × 0.200 mg.) to give the bound iodine. Free iodine is then plotted against bound iodine as shown in Fig. 1. The upper linear segment of this curve is extrapolated to intersect the axis, and from this amount of bound iodine is calculated the per cent iodine affinity. Results must be converted to a dry basis by separate moisture determination on the sample (viz., 4 hours *in vacuo* at 70°).

As an indication of precision, an average deviation from the mean on a large number of duplicate analyses was ±0.08% iodine affinity. Typical values are listed in the table. It is strongly recommended that the linear characteristics of various starch substances be expressed merely as iodine affinity and not calculated as percentage of linear substance in the sample.

Intrinsic Viscosity

The intrinsic viscosity of various starch materials provides a reliable index of molecular dimensions. It is capable of detecting minor hydrolytic changes too small to be measured by any reducing value method. Likewise, it permits valid comparison between substances of widely different molecular size. An optimal solvent for viscosity evaluation is 1 M KOH, which effects rapid and complete dispersion of most starchy substances. Although atmospheric oxygen causes a slow progressive degradation of alkaline starch systems, the intrinsic viscosity can be determined within 3 hours' time, and hence no precautions need be observed to exclude air. The linear and branched starch fractions require certain variations in mode of solution,[3] as described below.

Materials

1 M KOH, standardized and adjusted between 0.99 and 1.01 M.
5 M KOH, adjusted between 4.95 and 5.05 M.
Cannon-Fenske capillary viscometer, size 100 only.
Constant-temperature bath thermostated at 30.0 ± 0.2°.

Linear Starch Substances. 5.00 g. of the linear starch fraction (calculated on a dry basis by separate moisture determination) is placed in a clean, dry 600-ml. beaker. Approximately 200 ml. of distilled water is quickly added, and the mixture *immediately* and vigorously stirred with a glass propeller stirrer, to give a suspension of small gel particles with no large lumps. Exactly 100 ml. of 5 M KOH is added by pipet, and the mixture is stirred for 30 minutes, which usually suffices to give complete solution. The latter is then transferred and rinsed into a 500-ml. volumetric flask and made up to mark with water. This gives a 1.00% concentration of the linear fraction in 1.00 M KOH. The solution should be filtered through a wad of Pyrex glass wool to remove any trace of fiber or insoluble material which might impede capillary flow. Accurate dilutions are made to some five to six concentrations between 0.1 to 1.0%, with 1 M KOH for diluting. Viscosity is determined at each concentration with the Cannon-Fenske No. 100 pipet maintained in the 30° bath. Flow time is determined to a precision of ± 0.1 second (average deviation from the mean), the viscometer being filled at least twice and three flow times being run for each filling. Flow time is likewise determined for the 1 M KOH, and specific viscosity at each dilution is calculated as

$$\frac{\text{Flow time of solution} - \text{Flow time of 1 } M \text{ KOH}}{\text{Flow time of 1 } M \text{ KOH}}$$

The function (specific viscosity divided by per cent concentration) is plotted against per cent concentration, and the lower linear segment of this curve extrapolated to zero concentration. This intercept is the intrinsic viscosity.

Branched Starch Substances. 2.50 g. of the branched fraction (calculated to a dry basis) is quantitatively sifted into 200 to 300 ml. of cold water in a 600-ml. beaker, with vigorous stirring to avoid lumping. The beaker is heated in a boiling water bath for 15 to 30 minutes, then allowed to cool to room temperature, stirring being continued throughout the heating and cooling periods. Surface evaporation may cause insoluble skins on the side of the beaker; this is minimized by stirring through a perforated watch-glass cover during the heating and cooling periods. Then 100.0 ml. of 5 M KOH is added, the solution made up to 500 ml. with water, and viscosities determined as described for linear starch materials.

The viscosity curve of undegraded linear fraction begins to curve upward at about 1% concentration, owing to mutual interference between molecules. Hence, for purposes of extrapolation, the useful portion of this curve is the linear segment below this point. Similarly, viscosity of the undegraded branched fraction must be measured between 0.1 and 0.5%. However, hydrolyzed starch fractions and various enzymatic dextrins may advantageously be run at considerably higher concentrations, without an upswing in the viscosity curve. For example, the phytoglycogen from sweet corn gives a curve which is substantially linear between 1 and 5%. However, no entirely satisfactory method has been found for evaluating the intrinsic viscosities of unmodified granular starches, apparently owing to the difficulty in dissociating the granule structure. If necessary, these can be run by the procedure recommended for the branched starch fraction, but the curves are poor and accuracy of measurement is low. Values for retrograded and nonretrograded linear fraction are identical. Precision of the intrinsic viscosity is better than ±2%, representing the average deviation from the mean on a wide variety of duplicate runs. The No. 100 viscometer should be employed exclusively, to avoid slight differences due to energy characteristics of the viscometer.

[3] Preparation of Schardinger Dextrins

By DEXTER FRENCH

General Principles

An enzyme preparation from cultures of *Bacillus macerans* has the ability to convert starch through a nonhydrolytic breakdown into the Schardinger dextrins.[1] These nonreducing Schardinger dextrins have the ability to complex with iodine and various organic solvents to form crystalline complexes, often of very low solubility in water. By selective complex formation during or after the enzymolysis it is possible to separate[2] crude Schardinger dextrin mixtures into three distinct compounds: cyclohexaamylose (α-dextrin), cycloheptaamylose (β-dextrin), and cyclo

öctaamylose (γ-dextrin).

Enzyme Preparation[1,3]

Bacillus macerans (American Type Culture Collection No. 7069) is grown on a sterilized medium of 10% sliced old potatoes[4] and 2% $CaCO_3$ in distilled water for 1 to 2 weeks at 40°. The chilled culture fluid is centrifuged and/or filtered through a clarifying filter (diatomaceous earth may be used) and preserved in the refrigerator under toluene.

[1] E. B. Tilden and C. S. Hudson, *J. Am. Chem. Soc.* **61**, 2900 (1939).
[2] D. French, M. L. Levine, J. H. Pazur, and E. Norberg, *J. Am. Chem. Soc.* **71**, 353 (1949).
[3] E. B. Tilden and C. S. Hudson, *J. Bacteriol.* **43**, 527 (1942).
[4] New potatoes cannot be used; if old potatoes are not available, substitute 4% rolled oats in the medium.

Such crude enzyme solutions have been kept for many months under refrigeration.

Enzyme Assay[3]

For approximate work the Tilden assay is convenient and sufficiently accurate. One milliliter of the enzyme solution is incubated with 2 ml. of a 3% soluble starch solution at 40°. At intervals 3 drops of the digest are transferred to a spot plate and mixed with 1 drop of 0.1 N iodine in 0.1 M potassium iodide. A droplet of the mixed solution is streaked on a microscope slide and examined during and after evaporation. Initially a blue iodine color and blue-black crystals are seen, but, as the iodine color changes to brown-violet, dichroic needles are seen to spread toward the center of the evaporating droplet. The stage at which these characteristic needles are first seen is the *end point*. The *unit of enzyme activity* is the amount of enzyme which will convert 1 mg. of soluble starch to the end point at 40° in 1 minute; thus in this assay the activity per milliliter is 60 divided by the time required to the end point (in minutes). A *conversion period* is defined[2] as the time of reaction which under the conditions used would be just sufficient to convert soluble starch to the Tilden end point.

α-Dextrin. Enzymolysis of a 3 to 5% potato starch paste[5] for two to six conversion periods in the absence of a dextrin precipitant (i.e., the reaction mixture should be sterile or protected against microorganisms by a poison such as thymol) yields a mixture from which α-dextrin can be readily obtained by concentration to about 15% solids, precipitation with trichloroethylene, and centrifugation of the crystalline complex.[6] The crude precipitate is suspended in water, boiled to remove trichloroethylene, treated with carbon, filtered, and cooled. After dilution to about 2% solids concentration, bromobenzene is added to effect precipitation of any β- or γ-dextrin which may be present. The mixture is best stirred with a mechanical stirrer or shaken overnight to ensure complete equilibration with the bromobenzene. The resulting suspension is filtered with suction, the filtrate boiled down to about 40% of the original volume, and the bromobenzene precipitate, which is usually small, worked up with β-γ crudes. The solution of crude α-dextrin is then treated with trichloroethylene or tetrachloroethane by being stirred or shaken overnight and the crystalline complex separated by suction filtration. The precipitate is air-dried, boiled with about 5 parts of water until the vapors are free from precipitant, clarified by filtration with carbon if necessary, treated with 1.5 vol. of normal propyl alcohol, and allowed to cool. The crystalline propyl alcohol complex is filtered off after several days and

[5] For preparation of starch pastes see Schoch Vol. III [2].
[6] The Sharples supercentrifuge is very useful for this type of separation.

recrystallized by dissolving the air-dry material in 4 parts of hot water and adding 1.5 vol. of propyl alcohol. The final crystallization may be carried out by dissolving the pure propanol complex in water, boiling off the propanol, concentrating the solution to about 35% solids, and allowing it to crystallize by evaporation at room temperature for a day or two. (Alcohol-free solutions of the dextrins should be guarded against mold growth by being covered while hot.) The crystals are filtered off and allowed to air-dry. On drying in the vacuum oven the crystals lose about 10% moisture; the specific rotation of the anhydrous material is $[\alpha]_D = +150.5 \pm 0.5°$.

β-Dextrin. Enzymolysis of 3 to 5% paste of potato or waxy corn starch is carried out without a precipitant for two to three conversion periods. The digest, which is very fluid, is clarified by filtration through earth, fresh enzyme is added, a layer of toluene is added, and the enzymolysis is allowed to proceed for thirty to fifty conversion periods. β-Dextrin-toluene complex separates during the digestion, and to secure a maximum yield of dextrin the digestion vessel should be agitated from time to time to promote contact between the toluene and the enzymolysis mixture. After the conversion the precipitate is most conveniently separated by a supercentrifuge. The crude material is suspended in boiling water, toluene is removed by boiling until there is no odor of toluene in the vapor, and the boiling hot solution is treated with carbon and filtered with pressure through a clarifying filter.[7] The concentration at this point should not exceed 20 to 25% solids; higher concentrations give trouble through crystallization of the β-dextrin during filtration. If it is necessary to use suction filtration, this may be carried out at a concentration of 10% solids, and the filtrate should be concentrated by boiling to the 20 to 25% level. The filtrate should be perfectly clear; if not, it is filtered hot through successively finer filters, with carbon and filter aid if necessary. The resulting solution is allowed to cool, whereupon large crystals of β-dextrin form. After standing overnight in the cold, the crystals are filtered off, washed, and purified by repeated crystallization from 4 to 5 parts of boiling water. In this way is obtained a product which is ash-free, is completely soluble in water to give a clear solution, and contains about 14% water of crystallization. After the product has dried to constant weight at 70° in the vacuum oven, the rotation is $[\alpha]_D = +162.5 \pm 0.5°$.

γ-Dextrin. The enzymolysis with potato or waxy maize starch is carried out in the same manner as described above for β-dextrin, except that after the initial brief conversion and clarification the reaction is allowed to proceed in the absence of a precipitant for thirty to fifty conversion

[7] Horm laboratory pressure filter, F. R. Horman and Co., Inc., 17 Stone Street, Newark 4, New Jersey.

periods. γ-Dextrin is scarcely detectable in the early phases of the enzymolysis when the α-dextrin concentration is at a maximum. If the reaction mixture is not sterile, thymol may be used to inhibit molds. When the conversion is complete, the mixture is concentrated to about one-fourth the original volume, stirred with trichloroethylene, and the precipitate separated by centrifugation.[4] The precipitate is dissolved in boiling water, clarified, concentrated to about 25% solids, and allowed to stand at room temperature a day or two. Seeding with β-dextrin may be advisable to promote fairly complete removal of this component. The precipitate is removed, and the clear filtrate is diluted to about 3% solids. This solution is stirred overnight with bromobenzene, and the resulting suspension is filtered with suction. The precipitate is washed, dissolved in boiling water and boiled to remove the bromobenzene, concentrated to about 20% solids, and treated with 1.5 vol. of normal propyl alcohol. After standing for a few days at room temperature, the crystals of the γ-dextrin–propanol complex are removed by suction filtration and allowed to air-dry. γ-Dextrin is best purified by dissolving the air-dry complex in 4 parts of boiling water, filtering with carbon if the solution is colored or turbid, and treating the resulting solution with 1.5 vol. of n-propyl alcohol. Recrystallization should be repeated until there is no evidence for α- or β-dextrins in the 60% propanol mother liquors. Generally two or three recrystallizations are required. The propanol cannot be removed from this complex by either air or vacuum drying, but it is readily removed by boiling in water. By concentrating a boiled aqueous solution to 40 to 50% solids and letting it stand at room temperature for a few days, large, clear crystals of γ-dextrin hydrate are formed. The crystals effloresce in air; the air-dry material loses 8.3% of its weight on drying at 70° in the vacuum oven. The specific rotation of the anhydrous material is $[\alpha]_D = +177.4 \pm 0.5°$.

[4] Mucopolysaccharides

By ALBERT DORFMAN and J. ANTHONY CIFONELLI

General Considerations

Although a number of methods have been devised for the preparation of acid mucopolysaccharides, none of these is completely satisfactory. An ideal method should yield, by a relatively simple procedure, a product of theoretical composition as closely similar as possible to the appropriate compound in its natural state.

Since the methods to be described are intended primarily for the preparation of substrates for hyaluronidase, they will be limited to procedures for the preparation of hyaluronic acid (HA) and chondroitinsulfuric acid of cartilage (CSA-A). It should be pointed out that there are at least three other acid mucopolysaccharides in mammalian tissues: chondroitinsulfuric acid B (CSA-B)[1] (probably identical with β-heparin,[2] polysaccharide B,[3] and a fraction isolated from rat and rabbit skin[4]), keratosulfate,[5] and chondroitin.[6] CSA-B and keratosulfate are resistant to the action of both testicular and bacterial hyaluronidases, and chondroitin is said to resemble HA in that it is hydrolyzed by both types of enzyme. CSA-A is hydrolyzed by testicular hyaluronidase but not by the bacterial enzyme.[7,8]

HA has been prepared from a wide variety of sources.[9] Methods of preparation satisfactory for a given source do not necessarily yield similar results when applied to other starting materials. The preparation of material of high analytical purity has only rarely been achieved, and in no case has the existence of a monodisperse preparation been established.

The various procedures employed in the preparation of acid mucopolysaccharides involve the following basic steps: (1) extraction, (2) removal of protein, (3) precipitation, and (4) final purification.

Efficiency of extraction of acid mucopolysaccharides varies with both the polysaccharide in question and the starting material. In general, HA is much more readily extracted than is CSA-A and can usually be extracted in good yield by neutral aqueous solvents. Alkali or high salt concentrations are necessary to extract CSA-A in good yield, particularly from cartilage. This poses considerable difficulty, since extraction by alkali, although resulting in a large yield, produces a product of low molecular weight.[10-12] Milder extraction methods result in lower yields of material which appear to be combined with protein.[13]

[1] K. Meyer and M. M. Rapport, *Science* **113**, 596 (1951).
[2] R. Marbet and A. Winterstein, *Helv. Chim. Acta* **34**, 2311 (1951).
[3] H. Smith and R. C. Gallop, *Biochem. J.* **53**, 666 (1953).
[4] S. Schiller, M. B. Mathews, H. Jefferson, J. Ludowieg, and A. Dorfman, *J. Biol. Chem.* **211**, 717 (1954).
[5] K. Meyer, A. Linker, E. A. Davidson, and B. Weissmann, *J. Biol. Chem.* **205**, 611 (1953).
[6] E. A. Davidson and K. Meyer, *J. Biol. Chem.* **211**, 605 (1954).
[7] M. B. Mathews, S. Roseman, and A. Dorfman, *J. Biol. Chem.* **188**, 327 (1951).
[8] K. Meyer and M. M. Rapport, *Arch. Biochem.* **27**, 287 (1950).
[9] K. Meyer and M. M. Rapport, *Advances in Enzymol.* **13**, 199 (1950).
[10] J. E. Jorpes, *Biochem. Z.* **204**, 354 (1929).
[11] G. Blix and O. Snellman, *Arkiv Kemi Mineral. Geol.* **19A**, 32 (1945).
[12] M. B. Mathews and A. Dorfman, *Arch. Biochem. and Biophys.* **42**, 41 (1953).
[13] M. B. Mathews, *Federation Proc.* **14**, 252 (1955).

The removal of proteins has been attempted by a wide variety of methods. Detailed references to most of the methods that have been employed are given by Tolksdorf.[14] The methods have included the use of proteolytic enzymes, adsorption on $Zn(OH)_2$, Lloyd's reagent, kaolin, bentonite, and Magnesol. Proteins have also been removed by heating, by the use of protein precipitants, and by shaking with organic solvents (Sevag procedure). Glycogen, which contaminates mucopolysaccharides from some sources, may be removed by treatment with amylase.[15]

After removal of protein, the acid mucopolysaccharides can be precipitated by the addition of ethyl alcohol or acetone. The best results are usually obtained by precipitation of the salt from a slightly alkaline medium containing an alkali metal salt. Fractional precipitation of salts of mixtures of mucopolysaccharides has been used for separation of acid mucopolysaccharides.[2,16]

Several new methods have recently been employed for the final purification of acid mucopolysaccharides. Excellent separation of mixtures of HA and CSA-A have been obtained by Gardell et al.[17] and Schiller et al.[4] by slab electrophoresis on Celite or potato starch.

Separations on paper have also been reported by Reinits[18] and by Kerby.[19] Such electrophoretic separations do not result in separation of CSA-A and CSA-B. Recently, Davidson and Meyer[6] showed that a mixture of polysaccharides obtained from cornea could be separated by chromatography on Dowex 1 (1% cross-linked).

Below are described in detail the procedures for the preparation of HA and CSA-A. It should be again emphasized[20] that the use of such substrates for assay of hyaluronidase requires careful attention to all details of the assay procedure. Comparison to standard enzyme should be made on the same batch of substrate, since in some cases the substrate may be contaminated with inhibitors. Bacterial enzymes may give different assay curves than the testicular hyaluronidase on a given substrate.

Preparation of HA

This method, which has recently been developed in this laboratory, was chosen for detailed presentation because of its simplicity for preparation of HA useful for most purposes. The starting material selected is

[14] S. Tolksdorf, in "Methods of Biochemical Analysis" (Glick, ed.), Interscience Publishers, New York, 1954.
[15] K. Meyer, Physiol. Revs. 27, 335 (1947).
[16] K. Meyer and E. Chaffee, J. Biol. Chem. 138, 491 (1941).
[17] S. Gardell, A. H. Gordon, and S. Aquist, Acta Chem. Scand. 4, 907 (1950).
[18] K. G. Reinits, Biochem. J. 53, 79 (1953).
[19] G. P. Kerby, Proc. Soc. Exptl. Biol. Med. 83, 263 (1953).
[20] A. Dorfman, Vol. I [19].

human umbilical cord, since this source is most readily available. The
method has also been used for isolation of HA from bovine vitreous humor
and from supernatants of streptococci cultures. Although vitreous humor
gives an excellent product free of contamination by sulfated polysaccharides, the yield is low. Filtrates from HA-producing strains of streptococci are an excellent source of HA but are available only to laboratories
experienced in growing organisms under appropriate conditions for
maximal HA production. Synovial fluid, although a good source of HA, is
not readily available.

Reagents

0.9% NaCl.
Darco G-60.
Cellulose powder (Whatman ashless, cellulose powder, coarse grade,
    W & R Balston, Ltd., England).

Procedure. Fresh human umbilical cords are washed copiously in water
to remove as much blood as possible. They are then stored in acetone,
and when sufficient material is collected the cords are cut into 1-inch sections and ground in a mechanical meat grinder. The resultant material
is washed several times with fresh acetone and air-dried.

One hundred grams of acetone-dried, ground umbilical cords is mixed
with 2 l. of 0.9% NaCl, and the mixture, to which toluene is added as
preservative, is shaken on a mechanical shaker overnight. After centrifugation for 10 minutes the viscous supernatant fluid is decanted, the
residue mixed with 1.5 l. of 0.9% saline, and the mixture shaken once more
overnight. The combined extracts are then ready for passage over a carbon column (9.5 × 15 cm.) to remove protein and nucleic acid impurities. The column, consisting of equal parts by weight of Darco G-60
and cellulose powder is washed with several volumes of distilled H_2O prior
to use. The progress of purification is followed by examining the effluent
at intervals for absorption at 260 to 280 mμ. Saturation of the carbon by
impurities from the extract is indicated by the appearance in the effluent
of material showing ultraviolet absorption.

Some HA is retained on the carbon column, and mucopolysaccharide is
not detected in the effluent until approximately 2 vol. of extract has passed
through. After the extract has passed through the column, 1 vol. of H_2O is
necessary to wash all mucopolysaccharide from the column. A rate of flow
of 3 to 6 ml./min. has been found satisfactory with the size column
indicated. To maintain this rate, weak suction is usually necessary, since
the column tends to slow down with usage.

After concentration *in vacuo* at 35 to 40° to a volume of 80 to 100 ml.,

3 to 4 vol. of ethanol is added to precipitate the mucopolysaccharide. Removal of salts is accomplished by redissolving the precipitate and dialyzing against distilled water. A slight turbidity may develop at this point, and clarification is accomplished by centrifugation or filtration. The free acid may be obtained by passage of the dialyzed preparation over a cation exchange resin.

The yield of mucopolysaccharide from 100 g. of dried cord is 3.0 to 3.5 g. The ratios of hexosamine/uronic acid/nitrogen are $1:1:1$, within the experimental error of the analytical methods utilized. In an experiment utilizing paper electrophoresis, it was established that this preparation of HA is contaminated with less than 2% of sulfated polysaccharides.

Preparation of CSA-A

The method described produces a CSA-A preparation of high analytical purity but of relatively low molecular weight.[12] It differs from previously published methods employing strong alkali primarily in the use of phosphomolybdic acid as a protein-precipitating agent. This modification was developed in this laboratory by Dr. Saul Roseman.

Reagents

Acetone-dried, powdered bovine nasal septa.
2% NaOH.
6 N HCl.
Phosphomolybdic acid, reagent grade, 10% solution.
Amberlite IR-400 converted to its hydroxyl form.
Amberlite IR-120 converted to its sodium form.

Procedure. *Step 1. Extraction.* Five hundred grams of cartilage is stirred for 2 days with 3 l. of 2% NaOH at 2°. During this operation, the mixture is covered with a thin layer of toluene or chloroform to prevent bacterial growth. The mixture is centrifuged at about 3000 r.p.m., and the residue is re-extracted with 1 l. of the alkali under the same conditions. The supernatant is adjusted to pH 7 to 8 and kept at 2°. After the second extraction is complete, the mixture is again centrifuged, and the supernatant is combined with the first extract. The total extract is adjusted to pH 7 to 8 with HCl.

NOTES (1) All steps should be carried out in the cold room, if possible. (2) The extraction is performed by stirring the tissue vigorously with the alkali. (3) The supernatants are extremely turbid, but no attempt is made to clarify them at this point.

Step 2. Removal of Protein. The protein is removed with phosphomolybdic acid. This substance is readily reduced by most metals to yield a blue color so that it is preferable to work in a glass or enamel system until the molybdate has been removed. The molybdate is removed by passage of the solution over Amberlite IR-400 (OH⁻) which readily removes the excess phosphomolybdic acid without retaining any significant quantities of CSA.

To the combined supernatants at 2°, 10% phosphomolybdic acid is added rapidly with vigorous stirring. Sufficient phosphomolybdic acid is added so that the final pH is 3 to 3.5 (the exact quantity necessary can be determined on an aliquot using more dilute phosphomolybdic acid); 600 to 800 ml. of 10% phosphomolybdic acid is usually required. The mixture is centrifuged in the cold at high speed and the supernatant solution separated from the residual protein. If the supernatant is turbid, it is filtered through Super-Cel.

Step 3. Removal of Phosphomolybdic Acid and Conversion to Sodium Chondroitin Sulfate. The clear filtrate (if any reduced molybdate is present, it may have a bluish or greenish tinge) is then stirred vigorously with approximately 200 g. of well-washed Amberlite IR-400 (OH⁻) for about 20 minutes. The resin is removed by filtration and the filtrate is tested for phosphomolybdate.

Test for Phosphomolybdate. A small amount of powdered zinc is added to 3 ml. of the filtrate. If the solutions turn blue before or after acidification with acetic acid, molybdate is present. It is most convenient to carry out this test in a small centrifuge tube so that the zinc will not interfere with the observation of the color. If any molybdate is present in the filtrate, it is again treated with resin.

The clear, colorless solution is dialyzed against running tap water for 1 day and distilled water for 2 days (toluene is placed in the dialysis bags). The dialyzed solution is stirred with Amberlite IR-120, with Na⁺, and finally passed over the regenerated resin twice to assure complete conversion to the sodium salt.

The sodium chondroitin sulfate is precipitated by pouring the aqueous solution into a 4 vol. of 95% ethyl alcohol, half-saturated with NaCl. After being allowed to settle, the supernatant alcohol is removed by means of a sintered-glass filter stick, and the precipitate is centrifuged, washed with ethyl alcohol and ether, and placed in a vacuum desiccator. The yield from 500 g. of the tissue (acetone-dried) is 29 to 35 g.

Analyses

Considerable confusion with regard to the identification of mucopolysaccharides results from difficulties in performing adequate analyses. Space does not permit inclusion of detailed directions for all the appropriate analyses, but the following notes may be of value to the investigator interested in characterizing mucopolysaccharide preparations.

1. *Moisture.* The hygroscopic nature of these compounds makes exact moisture determination difficult. It is not always possible clearly to distinguish conditions responsible for actual water loss and loss due to destruction. For this reason the expression of individual analyses in terms of percentage dry weight is usually inadequate, and more meaningful information is obtained by means of ratios of different components obtained from analyses performed on samples of identical moisture content.

2. *Hexosamine Analyses.* Since the original publication of the Morgan-Elson method for the determination of hexosamines by the color reaction originally described by Zuckerkandl and Meissner, a great number of modifications of this method have been published. It is apparent from the many studies of this method that it is subject to many errors and must be utilized with great care to obtain satisfactory results. The recent modification of Boas,[21] which depends on the separation of hexosamines from interfering substances by the use of an ion exchange resin, seems to give most satisfactory results. It is now reasonably well established that, if acetylation is carried out with acetyl acetone, equivalent color densities are obtained with glucosamine and galactosamine. It has been shown by Aminoff et al.,[22] and confirmed by Roseman and Moses,[23] that the acetyl hexosamines as obtained by acetylation by acetic anhydride give different color equivalents. Leskowitz and Kabat[24] have attempted to accomplish the same purpose by the colorimetric determination of the reduced dinitrophenyl derivatives of hexosamines after chromatographic separation.

Under the best conditions the determination of hexosamines can rarely be performed with less than a 10% error, and the results obtained after hydrolysis of polysaccharides are apparently somewhat low.

3. *Uronic Acid.* Uronic acid can be determined with considerable accuracy by decarboxylation in acid according to various modifications of the Tollens-Lefevre reaction, particularly that of Burkhart et al.[25] Although the accuracy of the latter method is excellent, it is limited by the large amount of material necessary, an objection that has been partially overcome by the more micro modification of Tracey.[26] A possible limitation of the decarboxylation methods for uronic acid analyses is the report by Blix,[27] which indicates that the so-far-uncharacterized com-

[21] N. F. Boas, *J. Biol. Chem.* **204**, 553 (1953).

[22] D. Aminoff, W. T. J. Morgan, and W. M. Watkins, *Biochem. J.* **51**, 379 (1952).

[23] S. Roseman and F. E. Moses, unpublished results.

[24] S. Leskowitz and E. A. Kabat, *J. Am. Chem. Soc.* **76**, 4887 (1954).

[25] B. Burkhart, L. Baur, and K. P. Link, *J. Biol. Chem.* **104**, 171 (1934).

[26] M. V. Tracey, *Biochem. J.* **43**, 185 (1948).

[27] G. Blix, *Acta Physiol. Scand.* **1**, 29 (1940).

ponent "scialic acid" also yields carbon dioxide on acid decarboxylation. A valuable method for uronic acid analysis is the carbazole method of Dische,[28] which cannot be used as a fundamental method of uronic acid analysis in view of the differences of color production from different polysaccharides but is of great value in the analysis of different preparations of known mucopolysaccharides. The marked differences by this method in the color equivalent of different CSA's is to be particularly noted.

4. *Sulfate Analyses.* Sulfate analyses are best performed after Carius oxidation, but such methods can be utilized only in the established absence of protein. In the presence of other forms of sulfur, sulfate analysis is carried out subsequent to acid hydrolysis of the ester sulfate and can be performed on a microscale by the method of Anderson.[29]

5. *N-Acetyl Analyses.* N-Acetyl analyses are carried out by the conventional Kuhn-Roth procedure with *p*-toluenesulfonic acid or mineral acids. However, the procedure of Weissenberger,[30] utilizing chromic acid, is probably somewhat safer (in the absence of compounds containing methyl groups).

6. *Other Analyses.* No special comment need be made regarding nitrogen, ash, or titrimetric analyses which can be done by conventional methods.

* * * *

The authors are greatly indebted to Dr. Martin B. Mathews, Dr. Saul Roseman, and Dr. Sara Schiller who have, at various times, participated in the testing and development of analytical and preparative methods in this laboratory.

[28] Z. Dische, *J. Biol. Chem.* **167**, 189 (1947).
[29] L. Anderson, *Acta Chem. Scand.* **7**, 689 (1953).
[30] E. Weissenberger, *Mikrochim. Acta* **33**, 51 (1948).

[5] Preparation and Determination of Pectic Substances

By Z. I. KERTESZ

Three compounds or groups of compounds may be listed as substrates of pectic enzymes.[1] The first one of these, protopectin, is the substrate of protopectinase. Our knowledge of this enzyme and of protopectin itself is so limited[2] that both will be overlooked in this discussion. The second

[1] See Vol. I [18], where a brief definition of the various pectic substances discussed here may also be found.
[2] Z. I. Kertesz, "The Pectic Substances," Interscience Publishers, New York, 1951.

group is the infinite number of possible pectins (pectinic acids) which act as substrates of some polygalacturonases (pectinase, pectin-polygalacturonase) and of pectinesterase (pectase, pectin-methylesterase). The third is pectic acid, which serves as substrate of some polygalacturonidases.

Preparation of Purified Pectin (Pectinic Acid)

Pectin used in enzyme research is hardly ever prepared in the laboratory from plant tissue. Some commercial pectin preparations of high purity are now easily available, and these may be used in enzyme work without any further purification. These preparations are usually sold for pharmaceutical purposes, and their purity requirements and other characteristics are clearly defined.[3]

When such "pure" (undiluted) pectins are not available, almost any make of good-quality dry (powdered) pectin is suitable, after appropriate purification. This consists mostly in the removal of various compounds (sugars, buffers, calcium salts) deliberately added by the manufacturer for the purposes of standardization and improving the utility of pectins in the making of jams, jellies, and similar fruit products.[2]

For purification, such products may be dissolved in hot water to give about a 1% solution, cooled, made about 0.05 N HCl by the addition of concentrated HCl, and precipitated by slowly pouring into the vigorously stirred solution a twofold volume of 95% ethanol. After a few hours' standing, the pectin is filtered off (nylon cloth is very suitable), redispersed, and soaked in (unacidified) 95% ethanol overnight. Then it is filtered off again, soaked first in an ethanol-ether mixture, then in a little ether, squeezed as dry as possible, and rubbed dry in a large hot mortar. A few minutes' drying over a hot radiator helps to remove the last traces of ethanol and ether. Often, particularly in high-grade pectin preparations, the removal of the undesired added components can be easily accomplished by extraction of the finely powdered pectin with cold 80% ethanol containing 0.05 N HCl, followed by several washings with 95% ethanol and drying as noted above.

Properties

Pectin used as a substrate in enzyme research should give in a 1% concentration highly colloidal but practically colorless solutions. The dry pectin should contain not less than 80% anhydrogalacturonic acid and over 7% methoxyl, be free of starch, and show not more than 4% ash on ignition. However, it should be free of heavy metals, and its content

[3] See the various editions of the *National Formulary*, published by The National Formulary Committee, Washington, D. C., for these purity definitions. Several salves, etc., prepared with pectin are listed in the *N.F.* since its seventh edition.

of polyvalent ions like Ca and Mg should be low. For further criteria of purity, see already quoted references.[2,3]

Preparation of Pectic Acid

There is no pectic acid of sufficient purity for enzyme research now available commercially. Although pectic acid occurs as such in some plant tissues,[2] it is almost invariably prepared from purified pectins in the laboratory. Usually careful de-esterification with cold dilute alkali is applied rather than enzymic methods. Since it is known that in addition to saponification the alkali causes some other yet little understood degradative changes, the alkali treatment should be kept as mild and brief as possible.

If the starting material is not a sufficiently purified pectin, it is first extracted or precipitated in the manner described above. However, in such cases it is not necessary to dry the pectin; it can be dissolved without complete removal of the (neutral) ethanol. A cold, about 1% solution of the pectin is treated under constant mechanical stirring with small increments of dilute NaOH until pH 8.5 is reached. This pH is maintained with the further addition of alkali until it does not change during a 15-minute period. Then 0.5 N HCl is added slowly and under constant stirring until the pectic acid is precipitated. The latter is separated by filtration, washed with a little hot water, and then redispersed in water with just the minimum amount of alkali sufficient to dissolve the pectic acid. The solution is put through beds of cation and anion exchange resins[4] and precipitated with double the volume of 95% ethanol containing 0.01% HCl. The precipitate is filtered off, washed with pure 95% ethanol, an ethanol-ether mixture, and finally with a little ether, and then dried in a hot mortar under constant rubbing. If no ion exchange resins are available, the pectic acid is precipitated as stated above but washed with several portions of the acid ethanol before proceeding to the pure 95% ethanol and subsequent operations.

Properties

Pectic acid is easily precipitated by traces of polyvalent ions as calcium, for instance, and therefore a very low ash content and the virtually complete absence of polyvalent ions is important. Similarly, all the usual precautions must be exercised to assure that the distilled water and the various chemicals are sufficiently free of metallic ions and polyvalent ions of the alkali earth metal group, in particular.

In making up solutions of pectic acid, often some alkali has to be applied. This is no major detriment, since in most cases some alkali will

[4] K. T. Williams and C. M. Johnson, *Ind. Eng. Chem., Anal. Ed.* **16**, 23 (1944).

324,000 to 480,000 for cotton cellulose as determined by the viscosity method. Standard cellulose is hydrolyzed by strong mineral acids but is insoluble in 17.5% alkali.

Enzymatic Properties. The hydrolytic effect of various cellulase preparations on cellulose has been outlined in detail in Vol. I [20].

II. Hemicelluloses

Preparation

Principle. Hemicellulose has been defined by Norman[4] as "those polysaccharides extractable from plant tissues by treatment with dilute alkalis, either hot or cold, but not with water, and which may be hydrolyzed to constituent sugars and sugar acids by dilute mineral acids." However, this description has been found by Mitchell and Ritter[5] to be somewhat specific. In the course of their investigations, four different fractions of hemicellulose have been isolated.

Reagents

Holocellulose from maple wood.[6]
2% sodium carbonate.
4% sodium hydroxide solution.
10% sodium hydroxide solution.

Procedure. FRACTION I. A water-soluble fraction of hemicellulose was prepared by extracting the holocellulose with 30 parts of boiling water for

ANALYSIS OF HEMICELLULOSE FRACTIONS[a]

No.	Uronic acid anhydride	Xylan	Methoxyl	Acetyl	Hexosan	$[\alpha]_D$
IA	17.1	46.1	2.7	9.3	24.7	−38
IB	15.8	48.7	2.3	9.2	24.0	−25
II	28.9	54.7	2.6	—	13.7	−34
III	12.2	79.2	2.1	—	6.5	−70
IV	9.3	80.9	2.3	—	7.5	−83

[a] R. L. Mitchell and G. J. Ritter, *J. Am. Chem. Soc.* **62,** 1958 (1940).

1 hour. The filtrate was concentrated, and the hemicelluloses were precipitated by slowly adding 95% ethyl alcohol with stirring, after which they were filtered off and designated as fraction IB (see the table). This frac-

[4] A. G. Norman, "The Biochemistry of Cellulose, Hemicelluloses, Polyuronides and Lignins," p. 37, Oxford University Press, 1937.
[5] R. L. Mitchell and G. J. Ritter, *J. Am. Chem. Soc.* **62,** 1958 (1940).
[6] E. F. Kurth and G. J. Ritter, *J. Am. Chem. Soc.* **56,** 2720 (1934).

tion constituted 3.0% of the holocellulose. The precipitate obtained by adding acetone to the filtrate of IB is designated as fraction IA. It constituted 1.6% of the holocellulose.

FRACTION II. The hemicellulose fraction was isolated by treating the insoluble residue from extraction I with 10 parts of cold 2% sodium carbonate solution for 48 hours. The dissolved material was precipitated with alcohol, filtered, and freed from salts by suspending it in water, acidifying with hydrochloric acid, adding alcohol and acetone, and filtering. It constituted 2.2% of the holocellulose.

FRACTION III. A fraction soluble in 4.0% sodium hydroxide solution was prepared from the insoluble residue remaining after the removal of fraction II. It was extracted at room temperature by using 10 parts of alkali solution to 1 part of the insoluble residue and recovered by precipitating with alcohol, filtering, suspending the material in water, acidifying, precipitating with alcohol and acetone, and filtering. It constituted 14.9% of the holocellulose.

FRACTION IV. The residue from extraction III was treated with 10 parts of boiling 10% sodium hydroxide solution for 1 hour. The fraction was recovered from the alkaline solution and purified in the same manner as fraction III. It constituted 3.5% of the holocellulose.

Chemical and Physical Properties

The hemicellulose fractions were washed with alcohol and ether, dried at 80° in a vacuum oven, and then analyzed. The table presents the results of those analyses.

III. Cellobiose

Preparation

Principle. The procedure outlined is based on the hydrolysis of cellobiose octaacetate according to the methods described by Braun[7] and Hudson *et al.*[8]

Reagents

    Cellobiose octaacetate.
    Barium methylate.

Procedure. To a suspension of 3.0 g. of finely powdered cellobiose octaacetate (α and β forms, m.p. 180 to 185°) in 150 ml. of methyl alcohol is added a molecular equivalent of barium methylate, and the mixture is

[7] G. Braun, *Org. Syntheses* **17**, 34–36 (1937).

[8] W. T. Haskins, R. M. Hann, and C. S. Hudson, *J. Am. Chem. Soc.* **64**, 1289 (1942).

stirred vigorously for 1 hour at room temperature. The cellobiose is obtained in the form of prisms in quantitative yield.

Chemical Properties

β-Cellobiose melts at 225 to 226° with decomposition. An aqueous solution of the substance shows initial and final rotations of 16.2° and 34.9°, respectively, with a mutarotation rate of 0.0043 at 20°.

[7] Chemical Procedures for Analysis of Polysaccharides

By W. Z. HASSID and S. ABRAHAM

I. Determination of Glycogen and Starch

Determination of Glycogen by Modified Pflüger Method

Principle. The method originally used by Pflüger[1] and modified by Good *et al.*[2] consists in the digestion of the tissue in hot concentrated KOH, precipitation of the glycogen with ethanol, hydrolysis of the glycogen with acid, and determination of the glucose in the hydrolyzate as reducing sugar.

Reagents

30% KOH.
95% ethanol.
60% ethanol.
0.6 N HCl or H_2SO_4.
Saturated sodium sulfate.
Reagents for determination of reducing sugar, depending on the method.[3-8]

Procedure. Approximately 1 g. of animal tissue is dropped into a previously weighed 15-ml. Pyrex centrifuge tube containing 3 ml. of 30%

[1] E. F. W. Pflüger, "Das Glycogen," p. 53, Bonn, 1905.
[2] C. A. Good, H. Kramer, and M. Somogyi, *J. Biol. Chem.* **100**, 485 (1933).
[3] M. Somogyi, *J. Biol. Chem.* **117**, 771 (1937); **160**, 61, 69 (1945); **195**, 19 (1952).
[4] N. Nelson, *J. Biol. Chem.* **153**, 375 (1944).
[5] W. Z. Hassid, *Ind. Eng. Chem., Anal. Ed.* **9**, 228 (1937).
[6] S. M. Horvath and C. A. Knehr, *J. Biol. Chem.* **140**, 869 (1941).
[7] J. T. Park and M. J. Johnson, *J. Biol. Chem.* **181**, 149 (1949).
[8] B. Mendel and P. L. Hoogland, *Lancet* 16 (1950).

potassium hydroxide solution. After delivery of the sample, the tube and contents are reweighed and the weight of sample is determined by difference. The tissue is then digested by heating the tube in a boiling water bath for about 20 to 30 minutes. When the tissue is dissolved, 0.5 ml. of saturated sodium sulfate is added and the glycogen is precipitated by the addition of 1.1 to 1.2 vol. of 95% ethanol. The contents are stirred with a stirring rod, and the rod washed with a small quantity of 60% ethanol. The tube and contents are heated again until the mixture begins to boil, then cooled and centrifuged at 3000 r.p.m. The mother liquor is decanted, and the test tube is allowed to drain. The remaining adhering alcohol may be expelled by heating the tube in a boiling water bath. The precipitated glycogen is redissolved in 2 ml. of distilled water and reprecipitated with 2.5 ml. of 95% ethanol, the alcoholic supernatant liquid decanted, and the tube drained as before.

The purified glycogen is hydrolyzed as follows: 6 ml. of 0.6 N HCl or H_2SO_4 is introduced into a test tube provided with an air-cooled condenser or covered with a glass bulb, and the glycogen is hydrolyzed by heating for 2 to 2.5 hours in a boiling water bath. The solution is cooled, neutralized with 0.5 N NaOH, with phenol red as an indicator, transferred to a volumetric flask, and diluted to an appropriate volume, depending on the amount of sugar present and the method chosen for determination of glucose.[3-8] Good *et al.* claim that for the sugar determination oxidation with copper reagents[2] is preferable to oxidation with alkaline ferricyanide. In the experience of other investigators, however, determination of reducing sugar by oxidation with ferricyanide[5-7] gives equally satisfactory results.

In calculating the amount of glycogen from the glucose determined in the hydrolyzed glycogen sample, a conversion factor of 0.93 is taken.

Determination of Glycogen with Anthrone Reagent

Principle. The tissue sample containing the glycogen is digested with 30% KOH as described in the above method. The glycogen is treated with the anthrone reagent and determined colorimetrically as glucose.[9,10]

Reagents

30% KOH.
95% ethanol.
60% ethanol.
95% H_2SO_4.

[9] S. Seifter, S. Seymour, B. Novic, and E. Muntwyler, *Arch. Biochem.* **25**, 191 (1950).
[10] J. Fong, F. L. Schaffer, and P. K. Kirk, *Arch. Biochem. and Biophys.* **45**, 319 (1953).

0.2% anthrone solution. This reagent is made by dissolving 0.2 g. of anthrone[11] in 100 ml. of 95% sulfuric acid. The reagent is not stable in solution and should be kept in the refrigerator. It should be prepared fresh every 2 days.

A standard glucose solution containing 20 γ of glucose per milliliter. Photoelectric colorimeter.

Procedure. Seifter *et al.*[9] showed that in tissues such as liver where the glycogen content is high (approximately from 1.5 to 9%) the glycogen can be determined in the presence of proteins. The method is comparatively brief, since the necessity for glycogen precipitation and hydrolysis is eliminated. The glycogen is directly determined colorimetrically with anthrone reagent.

A 1-g. sample is digested in a 15-ml. test tube with 30% KOH as described in the previous method. The digest is cooled, transferred quantitatively to a 50-ml. volumetric flask, and diluted to the mark with water. The contents of the flask are thoroughly mixed, and a measured aliquot is then further diluted with water in a second volumetric flask so as to yield a solution of glycogen concentration of approximately 3 to 30 γ/ml. The determination is then carried out as follows.

A 5-ml. aliquot of the solution containing an amount of carbohydrate equivalent to 15 to 150 γ of glucose is transferred to a colorimetric tube. Into a second tube is introduced 5 ml. of the glucose standard containing ·100 γ of hexose. To a third tube is added 5 ml. of distilled water, which serves as a blank. The tubes are submerged in cold water, 10 ml. of the anthrone reagent is added to each test tube from a fast-flowing pipet or buret, and the reactants are mixed by swirling the tubes. The cold tubes are covered with glass marbles and heated for 10 minutes in a boiling water bath. They are then immediately cooled in a bath containing cold water and read in the colorimeter at 620 mμ after the galvanometer has been set at 100 with the blank. With an Evelyn colorimeter, the amount of glycogen in the aliquot used is calculated from the following equation:

$$\gamma \text{ of glycogen in aliquot} = \frac{100 \times U}{1.11 \times S}$$

where U = the optical density of the unknown test solution.

S = the optical density of the 100-γ glucose standard.

1.11 = the factor determined by Morris[12] for the conversion of glucose to glycogen, with this equation.

---

[11] Anthrone can be prepared according to directions given in *Org. Synthesis* 1, 52 (1932). A summary of the synthesis is given by Morris.[12] Anthrone can be obtained commercially from a number of chemical companies.

[12] D. L. Morris, *Science* 107, 254 (1948).

Determination of Glycogen in Tissues of Low Glycogen Content (Less Than 1%)

A 25- to 100-mg. sample is weighed on a microtorsion balance and placed in a graduated 12-ml. centrifuge tube containing 1 ml. of 30% KOH. The tube is placed in a boiling water bath for 20 minutes, the digest cooled, and 1.25 ml. of 95% ethanol added. The contents are mixed with a stirring rod, and the rod is washed with a small quantity of 60% ethanol. The contents of the tube are gently brought to a boil in a hot water bath, again cooled, and centrifuged for 15 minutes at 3000 r.p.m. The supernatant liquid is decanted, and the tube is allowed to drain on filter paper for a few minutes. The precipitated glycogen is redissolved in 1 ml. of distilled water, reprecipitated with 1.25 ml. of 95% ethanol, and the tube drained as before

The glycogen is hydrolyzed by the addition of 1 ml. of 0.6 N HCl to the centrifuge tube, which is provided with an air reflux condenser, and heated on the water bath for 2 to 2.5 hours. The solution is cooled, neutralized with 0.5 N NaOH, with phenol red as an indicator, and diluted to 5 ml. The sugar is determined by any of the standard methods for microdetermination of sugars.[2-4,6-8]

The glycogen can also be determined with the anthrone reagent.[9,10] Hydrolysis of the polysaccharide is not required; and the determination can be carried out directly on the precipitated glycogen. The sedimented glycogen in the above procedure is dissolved in exactly 5 ml. of water and reacted with the anthrone reagent as previously described for the determination of larger amounts of glycogen. The reaction mixture is transferred to an appropriate colorimeter tube and read in the colorimeter. The glycogen content is then calculated from the equation given for the anthrone method.

Assay of C¹⁴-Labeled Glycogen

Principle. Glycogen isolated by the previously described method may be contaminated with small amounts of amino acids. When this polysaccharide is determined by the usual analytical procedures, these contaminants do not interfere. In dealing with C¹⁴-labeled glycogen, however, where its radioactivity is determined, the amino acids, which may possess a high specific activity, can cause a considerable error and should be eliminated.

Procedure. Approximately 500 mg. of liver tissue is placed in 2 ml. of 30% potassium hydroxide and heated on a steam bath for 50 minutes. After precipitation of the glycogen with alcohol in the presence of sodium sulfate, the polysaccharide is dissolved in water and dialyzed against distilled water for 36 hours. This removes the amino acids and any other dialyzable substances that may be present. The dialyzed solution is filtered into a volumetric flask and diluted to volume with water. Aliquots of this solution are plated directly on aluminum disks and assayed for their C^{14} content in the usual manner.

Determination of Starch

Principle. The method described below is essentially that developed by Pucher *et al.*[13] Starch is extracted from dried plant tissue with perchloric acid, precipitated with iodine, and recovered as starch which is hydrolyzed with acid and determined as glucose.

Reagents

Ethanol-water. 1680 ml. of 95% ethanol is diluted with water to make 2 l. of 80% ethanol.

Perchloric acid, 52%. 270 ml. of 72% perchloric acid is diluted with 100 ml. of water. The solution should be stored in glass-stoppered containers.

Iodine-potassium iodide. 7.5 g. of iodine and 7.5 g. of potassium iodide are ground with 150 ml. of water, diluted to 250 ml., and filtered through a No. 3 Whatman paper with suction.

Alcoholic sodium chloride. 350 ml. of ethanol, 80 ml. of water, and 50 ml. of 20% aqueous sodium chloride solution are diluted to 500 ml. with water.

Alcoholic sodium hydroxide, 0.25 N. 350 ml. of ethanol, 100 ml. of water, and 25 ml. of 5 N sodium hydroxide are diluted to 500 ml. with water and filtered through a No. 3 Whatman paper with suction.

Hydrochloric acid, 0.7 N. 60 ml. of concentrated hydrochloric acid is diluted to 1 l.

Reagents for determination of reducing sugar, depending on the method chosen.[3-8]

In addition, 20% aqueous sodium chloride, 0.05% phenol red indicator solution, 0.5 N sodium hydroxide, and 0.1 N oxalic acid are required.

Procedure. If the determination is to be carried out on fresh plant material it should be rapidly dried in a vacuum oven at 90°. A ventilated oven, provided with a fan to circulate the air over the tissue at 70 to 80°, can be conveniently used. The dried material is then ground to pass a 50- to 80-mesh screen. A 50- to 250-mg. sample (chosen according to the

[13] W. G. Pucher, C. S. Leavenworth, and H. B. Vickery, *Anal. Chem.* 20 850(1948).

expected starch content) is placed in a 50-ml. centrifuge tube, a few drops of 80% ethanol are introduced to wet the ground material, so as to prevent clumping, 5 ml. of water is added, and the mixture is thoroughly stirred. The soluble sugars are then extracted as follows: 25 ml. of hot 80% ethanol is added, and the mixture stirred and centrifuged after standing for 5 minutes. The alcoholic solution is decanted and discarded. The residue is extracted again with 30 ml. of fresh hot 80% ethanol, stirred, centrifuged, and the alcoholic solution discarded as before. The extraction is repeated twice more for a total of four times in order to completely remove the soluble sugars.

The sugar-free residue is then treated with 5 ml. of water and stirred with a glass rod while 6.5 ml. of 52% perchloric acid is added. Stirring is continued for about 5 minutes and thereafter occasionally during 15 minutes; 20 ml. of water is then added and the mixture centrifuged. The aqueous starch solution is decanted into a 100-ml. volumetric flask, and the extraction is repeated by adding 5 ml. of water to the residue and stirring while adding 6.5 ml. of the diluted perchloric acid reagent. Stirring is continued for about 5 minutes and then occasionally during 20 minutes. The contents of the tube are then washed into the 100-ml. flask containing the first extract. The combined extracts and insoluble material are diluted to mark and filtered on a dry filter paper; the first 5 ml. of solution is discarded. Immediate analysis is to be preferred, but the solutions may be stored in the refrigerator for as long as 48 hours if necessary.

An aliquot of the starch extract of 1 to 10 ml., depending on the starch content, is transferred to a test tube calibrated at 10, 15, and 20 ml.; 5 ml. of 20% sodium chloride and 2 ml. of iodine-potassium iodide reagent are added, and the solution is mixed. After standing for at least 20 minutes, the tube is centrifuged and the supernatant liquid is decanted with extreme care to avoid loss of precipitate. The precipitated starch-iodine complex is then suspended in 5 ml. of alcoholic sodium chloride wash solution by gently tapping the tube, centrifuged, and the fluid decanted. To the packed precipitate 2 ml. of 0.25 N alcoholic sodium hydroxide is added, and the tube is gently shaken and tapped until all the blue color is discharged. A stirring rod must not be used, and ample time for decomposition of the complex must be allowed. The liberated starch is then centrifuged and washed with 5 ml. of alcoholic sodium chloride as before.

Two milliliters of 0.7 N hydrochloric acid is added to the precipitate. The tube is covered with a glass bulb and heated for 2.5 hours in a constant-level water bath provided with a cover which contains holes to accommodate the tubes. The bath is maintained at vigorous ebullition. It is important that holes not occupied by a tube be covered. The tube is cooled to room temperature, a few drops of 0.05% phenol red are

2,3-Di-O-methyl-D-glucose, prepared according to Irvine and
  Scott.[19]
Whatman No. 1 filter paper (large sheets).
Methyl ethyl ketone saturated with water.[20]
p-Anisidine hydrochloride solution (saturated solution in n-
  butanol).
0.1 N iodine solution.
0.2 M sodium bicarbonate-0.2 M sodium carbonate mixture (pH
  10.6).
2 N sulfuric acid.
0.01 N sodium thiosulfate solution.
Starch indicator.

Acetylation of Amylopectin or Glycogen.[21] Starch or glycogen is more
readily acetylated when it is freshly precipitated from aqueous solution.
The procedure is carried out as follows: 20 g. of amylopectin or glycogen
is dissolved in 1 l. of hot water and precipitated by the addition of 2.5 l.
of 95% ethanol. The precipitated product is collected on a Büchner
funnel and washed with ethanol and ether; the slightly moist polysaccharide is transferred to a 2-l. round-bottomed flask and stirred with
300 ml. of formamide until dissolved. Five hundred milliliters of pyridine
is added slowly with continuous mechanical stirring. This is followed by
the addition of 400 ml. of acetic anhydride in small portions over a period
of 1 hour. Stirring is continued for several hours, and the mixture is
allowed to stand overnight at room temperature. The solution is then
filtered through a fine cotton or nylon cloth on a Büchner funnel and
poured into a large excess of water to precipitate the acetylated product.
If the solution is too viscous to filter, it can be diluted with glacial acetic
acid. The precipitated acetylated product is collected on a sintered Pyrex
glass filter, washed with water until the filtrate gives a neutral reaction,
and then dried *in vacuo* at 80°.

Starch or glycogen is not completely acetylated by one such procedure.
Analysis of several amylopectins after the first acetylation showed an
acetyl content of approximately 40% (calculated $COCH_3$ content for the
triacetate, $(C_6H_7O_5(CH_3CO)_3)_n$, 44.8%). This process is repeated by dissolving the partially acetylated product in 500 ml. of pyridine and
acetylating with 400 ml. of acetic anhydride as before. After the second

[19] J. C. Irvine and J. C. Scott, *J. Chem. Soc.* **103**, 571, 575, 582 (1913).
[20] L. Boggs, L. S. Cuendet, I. Ehrenthal, R. Koch, and F. Smith, *Nature* **166**, 520
  (1950).
[21] For a summary on the preparation and properties of starch esters, see R. L. Whistler,
  Advances in Carbohydrate Chem. **1**, 279 (1945).

acetylation at room temperature, the polysaccharide is completely acetylated, giving the theoretical acetyl value. The values for the specific rotations in chloroform obtained for amylopectin triacetates prepared by this mild treatment range from $[\alpha]_D + 163°$ to $+175°$. The specific rotation of acetylated glycogen in chloroform is approximately $[\alpha]_D + 170°$.

The acetyl content of starch acetate or glycogen acetate can be determined by a variety of methods.[22-24]

Acetylation of Amylose. The amylose is reprecipitated as follows: 20 g. of the product is dissolved in 700 ml. of 3% potassium hydroxide, neutralized with dilute acetic acid, and precipitated by the addition of an equal volume of 95% ethanol. The precipitate is filtered and washed with ethanol and ether. The freshly precipitated amylose is suspended in formamide and acetylated in the presence of pyridine with acetic anhydride at room temperature as described for the acetylation of amylopectin. Two acetylations are required to completely acetylate the amylose. The specific rotations of various acetylated amyloses in chloroform range from $[\alpha]_D + 170°$ to $+178°$.

Methylation of Starch, Starch Fractions, or Glycogen.[18,25-27] The starch or glycogen triacetate is simultaneously deacetylated and methylated as follows: 16 g. of the acetylated product is dissolved in 200 ml. of acetone, warmed to 55°, and treated with 100 ml. of dimethyl sulfate and 300 ml. of 30% sodium hydroxide, these reagents being added in ten equal portions at 10-minute intervals with vigorous mechanical stirring. At the end of this operation 200 ml. of boiling water is added, and the reaction mixture is heated on a steam bath. As the acetone is removed by evaporation, the partially methylated material separates out. This product is separated out from the liquid phase, dissolved in acetone, and remethylated with the same quantities of reagents as described above. After a total of eight such methylations, the product is partially purified by suspending the material in approximately 200 ml. of hot water and boiling for a few minutes. Inasmuch as methylated starch or glycogen is fairly insoluble in boiling water, the loss of material is negligible by this operation. The water is decanted while hot, and the material dried and dissolved in chloroform. The solution is filtered through a Büchner funnel containing a layer of Celite or talc, and the chloroform evaporated to a small volume. The methylated product is then precipitated by the addition of petroleum

[22] E. P. Clark, *Ind. Eng. Chem., Anal. Ed.* **8**, 487 (1936); **9**, 539 (1937).
[23] L. B. Genung and R. C. Mallatt, *Ind. Eng. Chem., Anal. Ed.* **13**, 369 (1941).
[24] R. L. Whistler and A. Jeans, *Ind. Eng. Chem., Anal. Ed.* **15**, 317 (1943).
[25] W. N. Haworth and E. G. V. Percival, *J. Chem. Soc.* **1932**, 2277.
[26] W. N. Haworth, E. L. Hirst, and M. D. Woolgar, *J. Chem. Soc.* **1935**, 177.
[27] W. N. Haworth, E. L. Hirst, and F. A. Isherwood, *J. Chem. Soc.* **1937**, 577.

ether (b.p. 30 to 60°) and dried *in vacuo* at 95°. A product (starch or glycogen) having a specific rotation in chloroform of approximately $[\alpha]_D + 215°$ and a methoxyl content of about 44% is usually obtained by this procedure.

The methoxyl content of the methylated product is conveniently determined by the method of Clark.[28]

Hydrolysis of Methylated Polysaccharide.[16] Approximately 50 mg. of the methylated product is placed in a 12 × 0.5-cm. glass tube, and 1 ml. of 4% methanolic hydrogen chloride is added. The tube is sealed and immersed in a boiling water bath for 6 hours. During this methanolysis reaction the polysaccharide is degraded to its constituent monosaccharide with the simultaneous production of methyl glucosides. After methanolysis the tube and the contents are cooled, the tube is cautiously opened, and the methanol boiled off. An aqueous solution of 5 ml. of 4% aqueous hydrochloric acid (approximately 1 N) is then added, the tube sealed again, and the methyl glucosides hydrolyzed at 100° by immersion in a boiling water bath for 7 hours. The tube and contents are cooled, opened, and the solution neutralized with solid silver carbonate. The silver chloride is removed by filtration, the solution is passed through an ion exchange column containing 5 ml. of Dowex 50 and 5 ml. of Duolite A-4, and the neutral effluent evaporated to a thin sirup (about 10% solids).

Paper Chromatographic Separation of an Artificial Mixture of 2,3,4,6-Tetra-, 2,3,6-Tri-, and 2,3-Di-O-methyl-D-Glucose.[16,29] In order to establish the R_f values of the methylated glucose derivatives and the length of time required for their chromatographic separation on paper, a preliminary run should be made with a mixture of known methylated derivatives expected to be present in the hydrolysis products of methylated starch or glycogen.

With the aid of a fine capillary pipet an aliquot of approximately 0.01 ml. of an artificial mixture containing 2,3,4,6-tetra-, 2,3,6-tri-, and 2,3-di-O-methyl-D-glucose, each in approximately a 1% solution, is placed on the starting line of a 15 × 50-cm. Whatman No. 1 filter paper. At a distance of about 3 cm. from this spot is placed a similar spot of a 1% aqueous solution of 2,3,4,6-tetra-O-methyl-D-glucose alone. At a similar distance another spot containing the methylated starch or glycogen hydrolyzate is placed. The filter paper is put in a closed cabinet and the chromatogram developed with a water-saturated methyl ethyl ketone solution.[20] The chromatogram is allowed to proceed until the solvent has moved approximately 40 cm. from the origin. The paper is then removed,

[28] E. P. Clark, *J. Assoc. Offic. Agr. Chemists* **15**, 136 (1932); **22**, 622 (1939); see also E. P. Clark, "Semimicro Qualitative Organic Analysis," p. 68, Academic Press, New York, 1943.

[29] A. E. Flood, E. L. Hirst, and J. K. N. Jones, *J. Chem. Soc.* **1948**, 1679.

dried, and sprayed with a solution containing p-anisidine hydrochloride in n-butanol. On heating the paper chromatogram at approximately 100°, the positions of the separated sugars are indicated by reddish spots. The R_f values of the separated sugars are determined by taking the ratio between the distance from the starting line to the center of the spot and the distance through which the tetra-O-methyl-D-glucose has moved.

Paper Chromatographic Separation of the Hydrolysis Products of Methylated Starch or Glycogen.[16,29] Two pieces of filter paper are cut to suitable dimensions as described above, and spots of the sirup obtained from the hydrolysis products of the methylated starch or glycogen are placed about 5 mm. apart along the starting line with a fine glass capillary. The spots are of the order of 0.002 ml. and should be approximately uniform in size. They should not overlap, and the two end spots should not be too near the edge of the paper. Each of the end spots should be well separated from the next so that after chromatographic development narrow strips containing these spots can be cut from the paper sheet. The place at which the cuts are to be made is indicated by small pencil marks. Two paper sheets are prepared in this way, placed in a closed airtight cabinet, and the chromatogram developed by descending chromatography with water-saturated methyl ethyl ketone. A third filter paper of the same dimensions, which serves as a blank, is placed in the same or similar cabinet and developed by descending chromatography with the same solvent as the papers containing the methylated sugars.

After the chromatogram has run until the solvent front has reached the lower edge of the filter paper, the sheets are removed and dried in the air or in an oven at approximately 50°. The two narrow strips on either side of the paper are cut from each paper sheet at the places previously marked. These strips containing single sample spots are sprayed with the solution of p-anisidine hydrochloride and heated at 100° or higher to bring out the color of the spots.

From the color patterns of the chromatogram obtained on these narrow strips, the position of the methylated glucose derivatives can be located and the degree of separation of the methylated hexoses detected. By matching these strips alongside the major sheet from which they were cut, bands of paper can be cut, each containing a single methylated sugar component. If caution is taken and the sheets are hung vertically in the cabinets, it is found that the separated sugars lie on parallel horizontal lines. Finally paper strips which are equal in size to each of the methylated sugar-containing areas are cut from the non-sugar-containing paper to serve as paper blanks. Since the filter papers are well washed by the solvent during the development, the blanks appear to be consistently low and the papers do not require special treatment prior to use.

The methylated glucose derivatives are quantitatively removed from the cut-out paper strips as follows: One marginal end of each paper strip is cut to a pointed tip, and the other end is placed in a trough containing water and set in a small chromatography cabinet. The methylated sugars are eluted from the strips by allowing the water to drip off the pointed end by capillary descent into 22 × 3-cm. test tubes fitted with ground-glass stoppers. It has been found that when about 1 ml. of the eluate is collected all the methylated sugar is removed from the strip. The paper blanks are similarly eluted with water.

Quantitative Determination of the Methylated Glucose Derivatives with Sodium Hypoiodite.[16] The eluted methylated sugar samples and blanks are diluted in the test tubes to 5 ml., 1 ml. of 0.1 N solution of sodium hypoiodite is accurately pipetted into each tube, 2 ml. of a mixture of 0.2 M sodium bicarbonate and 0.2 M sodium carbonate (pH 10.6) is added, and the test tube stoppered. The solutions should not contain more than 2.5 mg. of sugar; otherwise the oxidation will not be complete. The stoppers are moistened with a little 10% potassium iodide solution in order to prevent loss of iodine due to evaporation. After 2 to $2\frac{1}{2}$ hours the stoppers are washed, and the reaction mixtures are diluted to 25 ml. and acidified by carefully running 2 N sulfuric acid down the side of the tube to avoid vigorous evolution of carbon dioxide. The tubes are then stoppered pending titration of the liberated iodine with 0.01 N sodium thiosulfate solution.

Standard aqueous solutions of pure samples of 2,3,4,6-tetra-O-methyl-D-glucose, 2,3,6-tri-O-methyl-D-glucose, and 2,3-di-O-methyl-D-glucose are analyzed simultaneously, their concentrations being adjusted so that 5 ml. will contain approximately the same quantity of methylated sugar as in the hydrolyzate to be analyzed. With these reducing values as reference, the amount of each methyl glucose derivative in the unknown sample can be estimated.

A considerable amount of 3,6-di-O-methyl-D-glucose, together with a small quantity of monomethyl glucose, is usually found in the hydrolysis mixture of the methylated polysaccharides. Their presence can be attributed to the partial hydrolysis of 2,3,6-tri-O-methyl-D-glucose. The reducing values of these two methylated derivatives can be assumed to be the same as that of the 2,3-di-O-methyl-D-glucose. In calculating the results, the smaller molecular weight of the monomethyl glucose must be taken into consideration.

The R_f values of 2,3,4,6-tetra-O-methyl-D-glucose, 2,3,6-tri-O-methyl-D-glucose, and 2,3-di-O-methyl-D-glucose in methyl ethyl ketone are 0.82, 0.56, and 0.30, respectively.

Once the amounts of the various methylated derivatives have been obtained, the average chain length of the polysaccharide can be calculated from the proportion of the 2,3,4,6-tetra-O-methyl-D-glucose (end group) as follows:

$$\text{Average chain length} = \frac{\dfrac{\text{mg. tetra}}{236} + \dfrac{\text{mg. tri}}{222} + \dfrac{\text{mg. di}}{208} + \dfrac{\text{mg. mono}}{194}}{\dfrac{\text{mg. tetra}}{236}}$$

By this procedure and with n-butanol-ethanol-water-ammonia mixture as a developing solvent, Hirst et al.[16] obtained the following results of end-group analysis of waxy maize starch.

R_f in n-butanol-ethanol-water-ammonia	Methylated glucose derivative	Mg. found	Composition, %
1.00	2,3,4,6-Tetra-O-methyl	0.28	4.2
0.81	2,3,6-Tri-O-methyl	5.37	80.0
0.58	2,3-Di-O-methyl	0.27	4.0
0.51	3,6-Di-O-methyl	0.70	10.4
0.26	Mono-O-methyl	0.09	1.3

End-Group Determination of Starch or Glycogen by Periodate Oxidation

Principle. This procedure is based on the observation that when hexopyranosides are attacked by periodate the ring structure is disrupted with the removal of the third carbon atom of the hexose as formic acid. Since amylopectin consists of a multitude of branched chains made up of 1,4-linked glucose units, the terminal glucose unit of each chain, like that in the nonreducing end of maltose, contains three hydroxyls on contiguous carbon atoms 2, 3, and 4, and yields 1 mole of formic acid on oxidation with periodate. Theoretically, only one reducing glucose unit should be present in an undegraded amylopectin or glycogen molecule, and the negligible amount of formic acid produced from this reducing end group may then be ignored. Thus, a quantitative determination of the formic acid produced by oxidation of amylopectin or glycogen with periodate will give a measure of the number of nonreducing end glucose units from which the average chain length of the branches in the polysaccharide molecule can be calculated.

On oxidation of amylose, formic acid will be liberated from both ends of the long straight-chain molecule, the nonreducing glucose unit producing 1 mole of formic acid, and the reducing glucose unit at the other end giving rise to 2 moles of formic acid. In calculating the end group of

for titrating the maltose samples) either with bromocresol purple to pH 6 or potentiometrically with a glass electrode.

Blank determinations are obtained by titrating solutions of similar carbohydrate samples containing sodium metaperiodate solution which has been reduced by ethylene glycol. The blanks do not have to be kept in the cold. They are titrated immediately after addition of the reagents.

In the event that the amount of formic acid obtained from the maltose is less than that of the theoretical, either the time of oxidation for the amylose is increased or a proper correction is made in the calculation of the chain length of the amylose from the formic acid.

The chain length of amylose, in terms of anhydroglucose units, is obtained by calculating the grams of amylose per 3 moles of formic acid formed and dividing the result by 162 (molecular weight of anhydroglucose). Thus, for example, a titration of 2.21 ml. of 0.01 N sodium hydroxide obtained from oxidation of 0.5 g. of amylose gives a chain length of 420 glucose units.

[8] Enzymatic Procedures for the Analysis of Glycogen and Amylopectin

By GERTY T. CORI

There are two enzymatic procedures available for the study of the structure of glycogens and amylopectins. The first one consists in the determination of the percentage of branch points.[1,2] The results are in good agreement with those obtained by chemical methods (methylation or periodate oxidation). In the second and more complete one[3] the polysaccharide is degraded stepwise, tier by tier, which permits the determination of chain length in each tier and the characterization of successive limit dextrins. Up to nine-tenths of the macromolecules has actually been peeled off by the latter procedure. What remained still had a molecular weight of about 300,000. The structure near the only reducing end group of these polysacchardies has not yet been studied. No chemical methods are available for stepwise degradation.

Two enzymes are used: (1) Muscle phosphorylase, which in the presence of inorganic phosphate starts its action at the nonreducing end groups and forms glucose-1-phosphate. It acts only on the 1:4 linked

[1] G. T. Cori and J. Larner, *J. Biol. Chem.* **188**, 17 (1951).

[2] B. Illingworth, J. Larner, and G. T. Cori, *J. Biol. Chem.* **199**, 631 (1952).

[3] J. Larner, B. Illingworth, G. T. Cori, and C. F. Cori, *J. Biol. Chem.* **199**, 641 (1952).

glucose residues, and its activity is stopped when it approaches the outermost tier of branch points. (2) Amylo-1,6-glucosidase, which by hydrolytic action splits off the glucose residue which is in 1:6 linkage at the branch points. It acts only after phosphorylase has removed the glucose residues in 1:4 linkage which cover the residue in 1:6 linkage.[4]

Determination of Percentage of Branch Points

The aim is to degrade the polysaccharide completely by the simultaneous action of the two enzymes and to determine in the reaction mixture glucose-1-phosphate (and hexose-6-phosphate if the glucosidase is contaminated with phosphoglucomutase and phosphohexoisomerase) and free glucose. From the ratio of these end products of enzymatic action, the percentage of branch points is calculated, a value which is practically identical with the end group value, since branch points are related to end groups as $(n - 1)$ to n, n being a large figure. Glycogens prepared with glacial acetic acid sometimes contain acetyl groups and have to be saponified previous to the use of the enzymatic method because otherwise they are incompletely degraded.[2]

Procedure. Polysaccharide, 10 to 12 mg., is dissolved in 1.2 ml. of H_2O; 0.1 ml. is used for hydrolysis (3 hours at 100°) in 1 N HCl (total volume, 0.5 ml.). After hydrolysis the solution is neutralized, carefully transferred with repeated washings to a 10-ml. volumetric flask, and reduction is determined in 1.0 ml. with reagent No. 60 of Shaffer and Somogyi. The same reagent, combined with the colorimetic method of Nelson, is also used in other determinations of reducing values.

Reaction Mixture

Polysaccharide, 1 ml.
KPO$_4$, pH 7, 1.0 M, 1 ml.
Versene[5], pH 7, 0.005 M, 1 ml.
Phosphorylase, 20 to 50 γ (third or later crystals[6]).
Glucosidase, 0.3 to 0.5 mg. (purified to stage 3).[7]
H_2O to 12 ml., incubation at 30°.

The reaction mixture is cooled to 0° before addition of enzymes which are also at 0°. Aliquots of 0.5 ml. are removed immediately and after 1 and 2 hours for precipitation with Schenck reagent and determination of

---

[4] β-Amylase cannot be substituted for phosphorylase, since glucosidase has little if any effect on the β-amylase limit dextrin.
[5] Presence of cysteine in the reaction mixture is to be avoided because it increases the blank values.
[6] G. T. Cori, B. Illingworth, and P. J. Keller, Vol. I [23].
[7] G. T. Cori, Vol. I [25].

glucose plus phosphorylated sugars. Five milliliters is removed immediately and after 2 hours for precipitation with West reagent and determination of free glucose. Zero-hour blanks should be very low for both methods. They are determined to be sure that no contamination with reducing substances has taken place.

Determination of Sum of Free and Phosphorylated Sugars. The reaction mixture (0.5 ml.) is pipetted into 1.0 ml. of Schenk reagent (2.5% $HgCl_2$ in 0.5 N HCl), and 0.5 ml. of H_2O is added. The filtrate is treated with H_2S, aerated, and placed in a boiling water bath for 5 minutes (HCl is about 0.3 N). Then 0.5 ml. is pipetted into Klett tubes and neutralized. Reducing sugar is then determined (boiling time 30 minutes).

The semipurified glucosidase contains phosphoglucomutase and phosphohexoisomerase. The phosphorylated sugars are glucose-1-phosphate, glucose-6-phosphate, and fructose-6-phosphate. Glucose-1-phosphate is split in 5 minutes at 100° in 0.3 N HCl. The other two esters gives their full reduction equivalent under the conditions used.

Completeness of digestion of polysaccharide is checked by comparing the enzymatically obtained values with those obtained by acid hydrolysis.

Determination of Free Glucose. The reaction mixture (5.0 ml.) is delivered into 0.7 ml. of West reagent (28% $FeSO_4$ + 34% $HgSO_4$ in 1.5 N H_2SO_4). Solid $BaCO_3$ (analytical grade) is added under vigorous shaking until the reaction is neutral to bromothymol blue paper. The filtrate obtained by suction is acidified with a trace of H_2SO_4 until it is acid to Congo paper. After treatment with H_2S it is filtered, and the filtrate is aerated and neutralized. It should remain clear and colorless after neutralization. Reduction is determined in 1.0-ml. aliquots as just described. Absence of amylase activity should be established by comparing the values obtained by reduction of alkaline copper solution with values obtained by a specific method for glucose, in which maltose and other oligosaccharides (which are present when there is contamination with amylase) do not react. A microenzymatic method has been proposed in which purified yeast hexokinase, ATP, Zwischenferment and TPN are added to the West filtrate and the reduction of TPN is determined in the spectrophotometer.[1,2] Only 10 to 20 γ of glucose is needed for this procedure; for the colorimetric method at least 40 γ should be present.

Calculation of per cent branch points:

$$\frac{\text{Free glucose}}{\text{Free glucose} + \text{phosphorylated hexose}} \times 100 = \text{Per cent branch points}$$

Analysis by Stepwise Enzymatic Degradation

For this method many-times-recrystallized phosphorylase, freed of the last traces of glucosidase, is needed. Glucosidase of relative crudity may be used; contamination with phosphorylase is immaterial, since no phosphate is added while it acts. In conjunction with the branch point percentage it is often sufficient for the characterization of a glycogen or amylopectin to determine only the length of the outer chains by submitting the polysaccharide to one exhaustive phosphorylase action and determining the glucose-1-phosphate in the reaction mixture.[8] For a more complete characterization the limit dextrin$_1$ (LD$_1$) is isolated. Then, glucosidase is allowed to act on it, and free glucose is determined in the reaction mixture. Subsequently the glucosidase is destroyed, and inorganic phosphate and phosphorylase are added again in order to prepare LD$_2$. The glucose-1-phosphate which is liberated is again determined. These alternating enzyme actions can be repeated many times.

The successive phosphorylase LD's were found to be similar in structure, as shown by the fact that β-amylase removed from each two maltose units (per main chain) and that the per cent end group remained fairly constant. The per cent of total branch points present in each tier became progressively less as the reducing end group was approached. Only one kind of model could be made to fit the experimental data, namely, one which presents these polysaccharides as multibranched, treelike structures.

Procedure. At least 700 mg. of polysaccharide is required to permit digestion to the LD$_4$ level and characterization of each LD. A solution in H$_2$O is prepared, and the concentration is again determined by acid hydrolysis of an aliquot. KPO$_4$ solution (pH 6.7 to 7.0) is added so that the polysaccharide is present in about 1% solution, the phosphate is about 0.3 M, and Versene 5×10^{-4} M. Phosphorylase (seventh to eleventh crystals), about 0.5 mg. per 100 mg. of polysaccharide, is added.

The reaction mixture (0.1 or 0.2 ml.) is analyzed for glucose-1-phosphate after 0, 2, and 3 hours of incubation at 30°. At 0 hours the solution has to be at 0°. The samples are diluted to 1 ml., and an equal volume of Schenk solution is added, filtered, and the filtrate treated as described in the previous section. (Polysaccharide is removed with the HgS precipitate.) Glucose-1-phosphate is completely split during 5 minutes at 100°, (the solution contains about 0.3 N HCl), and glucose is determined, after neutralization, by the Somogyi-Nelson procedure. Phosphorylase action should be exhaustive after 2 hours. The third-hour sample is taken merely to ascertain that no further degradation has taken place. If the phosphorylase is contaminated with glucosidase, degradation slowly continues. The average length of the outer chains in different glycogens

---

[8] β-Amylase can be substituted for phosphorylase if only the length of the outer chains is to be determined. The β-amylase LD is smaller than the phosphorylase LD (see below).

varies considerably, so that about 30 to 45% of the glucose molecules in glycogen is liberated in this first degradation.

From the bulk of the reaction mixture the LD_1 is isolated by precipitation with an equal volume of ethanol. The precipitate is dissolved in H_2O, a small amount of insoluble material (protein) is removed by centrifugation, and the pH adjusted to 4 to 5 (first blue to Congo paper) by cautious addition of dilute HCl (about 0.05 N). A second precipitation in 50% ethanol follows; this precipitation at acid pH should be carried out once more to be sure that all phosphates have been removed. Finally the LD is precipitated at neutral reaction in 50% ethanol and dehydrated with 95 and 100% ethanol and anhydrous ether and dried at room temperature. It is weighed so that the yield, which should be 85 to 95% of the calculated, is known. The percentage of branch points of the LD_1 is determined in a 10- to 12-mg. aliquot.

The rest (about 400 mg. when the starting weight was 700 mg.) is dissolved in water, an aliquot removed for acid hydrolysis, and Versene added, so that final concentrations are about 1% LD_1 and 5 × 10^{-4} M Versene, pH 7. Semipurified glucosidase is added (about 1 mg. per 100 mg. of polysaccharide). Aliquots of 0.2 ml. are withdrawn after 0, 2, and 3 hours, diluted to 1 ml., and precipitated with 1 ml. of Schenk solution. In the final neutralized filtrate (no hydrolysis is carried out) glucose is determined by the Park-Johnson[9] method, which allows accurate determination of as little as 10 γ of glucose. It is very desirable to check the values with those obtained with the microenzymatic method which is specific for glucose. The glucosidase action should be complete after 2 hours; the third-hour sample serves to show that there is no further liberation of glucose.

The bulk of the reaction mixture after 3 hours of incubation is placed in a boiling water bath for 3 minutes, cooled, and the denatured protein filtered off. LD_2 is prepared by phosphorylase action, according to the procedure described for the preparation of LD_1. The alternating enzyme actions are repeated as desired or until most of the polysaccharide has been digested.

[9] J. T. Park and M. J. Johnson, *J. Biol. Chem.* **181**, 149 (1949).

[9] Preparation and Assay of Substrates for β-Glucuronidase

By WILLIAM H. FISHMAN

The requirements of a good substrate for assaying β-glucuronidase may be listed as follows: (1) The glucuronide or a suitable derivative should be a crystalline product which can be efficiently purified. (2) It should be readily hydrolyzed by the enzyme. (3) The amount of aglucurone liberated should be accurately measured in microgram amounts. (4) The rate of hydrolysis should be linear during the course of the incubation. (5) Interference by tissue constituents and by unhydrolyzed substrate should be negligible. (6) The assay procedure should be convenient and rapid, requiring only the usual laboratory apparatus. The disadvantages of relying on the determination of increase in reducing power of the digests have been cited.

The substrate phenolphthalein glucuronic acid satisfies these requirements. Its use in the assay of β-glucuronidase is now widespread, and the results may be accepted with confidence.

Preparation of the Substrate Phenolphthalein Glucuronic Acid[1-3]

Principle. Rabbits injected with sodium phenolphthalein phosphate excrete phenolphthalein glucuronic acid in the urine which is isolated as the cinchonidine derivative. From this pure product, a 0.01 M solution of phenolphthalein glucuronic acid is prepared to provide the standard substrate solution.

Sodium Phenolphthalein Phosphate. First dry chloroform (40 ml.) and then a mixture of (redistilled) phosphorous oxychloride (50 ml.) and chloroform (50 ml.) was added with mechanical stirring to 50 g. of phenolphthalein in a 1-l. round-bottomed flask cooled in an ice bath. To this was added dropwise dry pyridine (40 ml.) with stirring continued for a total of 3 to 5 hours. The next day, 150 ml. of distilled water was added in small quantities to the ice-cooled reaction flask, followed by an excess of 40% sodium hydroxide solution (approximately 300 ml.). On cooling, needle-like crystals (sodium phosphate) appeared and were removed by filtration. When an excess of concentrated hydrochloric acid (Congo Red paper) was added to the aqueous phase, phenolphthalein diphosphoric acid precipitated as a gum which could be conveniently separated from the mixture with a glass stirring rod. This gum was warmed in a porcelain

[1] W. H. Fishman, B. Springer, and R. Brunetti, *J. Biol. Chem.* **173,** 449 (1948).
[2] P. Talalay, W. H. Fishman, and C. Huggins, *J. Biol. Chem.* **166,** 757 (1946).
[3] A. A. Di Somma, *J. Biol. Chem.* **133,** 277 (1940).

dish on a boiling water bath, and 20 ml. of concentrated sodium hydroxide (100 g. plus 100 ml. of distilled water) was added gradually with stirring. Solution of the gum was completed with distilled water (final volume 200 ml.).

Cinchonidine Derivative of Phenolphthalein Glucuronic Acid. Six milliliters of this phenolphthalein phosphate was brought to neutrality with weak alkali and diluted to 20 ml. with distilled water. This amount was injected subcutaneously daily at two widely separated sites on the skin of each rabbit for 6 days. The animals received carrots and cabbage in addition to pellets, the water intake being limited. The daily urine collections (under toluene) were completed on the morning of the eighth day.

The filtered toluene-free urine was acidified to Congo Red paper with 6 *N* HCl and shaken vigorously. Then 600-ml. portions of urine were extracted with four successive 170-ml. portions of ethyl acetate. The ethyl acetate phase was centrifuged and dried by decantation through cotton.

The ethyl acetate extract was then slowly saturated with cinchonidine, a gram or two in a cotton gauze bag being agitated in the solution. This was repeated several times. A fine white crystalline precipitate appeared which is the cinchonidine derivative of phenolphthalein glucuronic acid. The separation of the product was completed overnight in the cold. The product was separated on filter paper, dried, and dissolved in the minimum quantity of hot methyl alcohol to which was added 4 vol. of hot ethyl acetate. A rather pure white crystalline product is thus obtained. Additional amounts of the conjugate may be obtained by extracting the gum which forms on the outside of the gauze bag with methyl alcohol and crystallizing the product by the addition of ethyl acetate and also by concentrating the mother liquors to small volume at 50°. Cinchonidine salt many times recrystallized melts at 210 to 211°. However, products melting at 200° were found to be satisfactory for assay purposes. The yield varies between 0.5 g. and 1.0 g. per rabbit per day.

Phenolphthalein Glucuronic Acid, 0.01 M. 0.788 g. of the pure cinchonidine salt of phenolphthalein monoglucuronide (molecular weight 788) was mixed with an excess of approximately 2 *N* hydrochloric acid (20 ml.) and 20 ml. of ethyl acetate. The mixture was transferred quantitatively to a separatory funnel with washings of ethyl acetate. The ethyl acetate layer was collected and decanted through a cotton plug in a funnel into a 500-ml. Erlenmeyer flask. The aqueous layer was then extracted with six successive portions (20 ml.) of ethyl acetate, the ethyl acetate layers being collected in the Erlenmeyer flask after decantation through the same cotton plug. The solvent was evaporated by drawing a current of dry clean air into the flask. The gummy colorless residue is the glucuronide. To this was added 20 ml. of hot distilled water and sufficient 0.1 N sodium hydroxide solution to bring the solution to pH 5.0 (pH paper). The mixture was transferred quantitatively with water washings to a 100-ml. volumetric flask, made up to the mark, and well shaken. If turbidity forms, a small amount of charcoal is stirred in, and the mixture is filtered.

Assay of Glucuronidase Activity Utilizing the Substrate Phenolphthalein Glucuronic Acid

Conditions have been published for the assay of β-glucuronidase in solutions of the purified enzyme (Vol. I [31]), in tissue extracts and blood serum,[1,4] in leucocytes,[5] in spinal fluid,[6] and in vaginal fluid,[7] with phenolphthalein glucuronic acid as the substrate.

Among the sources of error which arise are (1) failure to maintain a pH of 10.2 to 10.4 in the alkalinized digest prior to reading, (2) adsorption of phenolphthalein on protein if it is precipitated,[4] and (3) unnecessary purification of the aqueous homogenate of tissue before assay, which introduces a loss of enzyme.

The use of other substrates and assay procedures for β-glucuronidase is referred to in a recent review article.[8]

[4] W. H. Fishman, *in* "The Enzymes" (Sumner and Myrbäck, eds.), Vol. 1, Part 1, p. 635, Academic Press, New York, 1950.
[5] A. J. Anlyan, J. Gamble, and H. A. Hoster, *Cancer* **3**, 116 (1950).
[6] A. J. Anlyan and A. Starr, *Cancer* **5**, 578 (1952).
[7] W. H. Fishman, S. C. Kasdon, and F. Homburger, *J. Am. Med. Assoc.* **143**, 350 (1950).
[8] W. H. Fishman, *in* "Advances in Enzymology" (Nord, ed.), Vol. 16, p. 361. Interscience Press, Inc., New York, 1955.

[10] Column Chromatography of Sugars

By EDISON W. PUTMAN

Charcoal Columns

Principle. Oligosaccharides in aqueous solution are adsorbed chromatographically on charcoal and selectively desorbed with dilute solutions of ethanol according to the method of Whistler and Durso.[1] This method is particularly useful for the isolation of sufficient quantities

[1] R. L. Whistler and D. F. Durso, *J. Am. Chem. Soc.* **72**, 677 (1950).

of enzymically synthesized disaccharides to permit determinations of structure.[2]

Materials Required

Activated carbon, Darco G-60 (Atlas Powder Co., New York, N. Y.).

Celite No. 535 (Johns-Manville, New York, N. Y.).

Chromatography tube. A Pyrex glass tube, 50 × 5 cm., is constricted at one end and sealed to a 10-cm. length of 10-mm. O.D. Pyrex glass tubing. Larger or smaller tubes possessing the same relative dimensions may be used, depending on the amount of adsorbent required.

Column Preparation. Sixty grams of Darco G-60 is mixed intimately with 120 g. of Celite by shaking vigorously in a dry 2-l. Erlenmeyer flask. Water is added, and the mixture is stirred until it is a homogeneous thick slurry. A plug of glass wool is placed at the bottom of the chromatography tube, and an aqueous suspension of Celite is poured in until a bed about 1 cm. thick is built up over the glass wool. The tube, supported at its base by a 1-l. filter flask bearing a Neoprene filter adaptor, is clamped to a ring stand in a vertical position. The charcoal slurry is then poured into the tube and allowed to pack by gravity. The sides of the tube are washed down with water so that the top of the column is covered with approximately 5 cm. of water as it settles. The upper end of the tube is then closed by a one-holed rubber stopper bearing a short length of glass tubing through which water from a reservoir placed 2 to 3 feet above the column is admitted by means of a flexible siphon. The column is then washed with about 1.5 l. of water at a flow rate of 10 to 20 ml./min. maintained by partial evacuation of the filter flask. A column prepared in this manner with a 30-cm. bed has a hold-up volume of about 400 ml. and will adsorb up to 4 g. of oligosaccharides.

Sugar Separation. An aqueous solution of the sugars to be separated, containing not more than 4 g. of oligosaccharide, is deproteinized and adjusted to pH 7.0. The water in the tube is drained to the level of the charcoal bed, the receiver is emptied, and the sugar solution is transferred to the column. The solution is allowed to drain into the bed, and the sides of the tube are washed down successively with two 25-ml. portions of water allowed to drain to bed level after which 100 ml. of water is added and the siphon delivery from the reservoir is connected to the top of the tube.

[2] E. W. Putman, C. Fitting Litt, and W. Z. Hassid, *J. Am. Chem. Soc.* **77**, 4351 (1955).

The first 300 ml. of effluent may be discarded after which separate fractions of 50 ml. are collected until a positive qualitative test for reducing sugars is obtained. The monosaccharides and salts travel with the solution front and are removed completely from the column with an additional 1.5 to 2 l. of water. The progress of elution is followed by testing the effluent after successive collections of about 250 ml.

After complete removal of the monosaccharide component the column is drained to bed level, and the water in the reservoir is replaced by a 5% solution of ethanol. The sides of the tube are washed down with small portions of 5% ethanol which are allowed to drain to the level of the charcoal bed. About 100 ml. of 5% ethanol is added to the column, and the siphon delivery from the reservoir is attached to the tube. The course of desorption and elution of a disaccharide with 5% ethanol follows that of the elution of a monosaccharide with water. The disaccharides move with the solvent front and appear in the effluent after about 400 ml. has been collected. A total of approximately 2 l. of 5% ethanol is required for the complete elution of disaccharides.

When the effluent no longer shows traces of disaccharide, the concentration of alcohol in the eluent is increased to 15% in order to desorb trisaccharide if present. The sequence of events is the same as in the elution of the disaccharide.

After the desired separation has been achieved, the column is stripped by washing with about 2 l. of 50% alcohol and prepared for reuse by washing with 2 l. of water. When the column is not in use the efflux tube is closed with a rubber stopper.

The alcoholic solutions of the di- and trisaccharide fractions are concentrated by vacuum distillation, filtered, and reduced by sirups by further concentration in a vacuum desiccator. Crystallization of the sirups is induced by the addition of hot absolute ethanol and trituration of the mixtures.

If it is desirable to isolate the sugar in the monosaccharide fraction, the solution is deionized with suitable ion exchange columns prior to concentration and ultimate crystallization. In cases where mixtures of monosaccharides are present and difficulties are encountered in fractional crystallization, the mixture may be resolved by partition chromatography on a cellulose column.

Cellulose Columns

Principle. The separation of sugars by cellulose column chromatography, originated by Hough *et al.*,[3] is an application of the principle of partition chromatography that the components of a solute mixture may

[3] L. Hough, J. K. N. Jones, and W. H. Wadman, *J. Chem. Soc.* **1949**, 2511.

be separated by a continuous countercurrent extraction of the solution with a second solvent immiscible with the first, provided the components possess different partition coefficients in the two-phase system. The cellulose column supports an immobile aqueous phase through which a partially miscible organic solvent is permitted to flow. The components of a sugar mixture placed at the top of the column are successively eluted in the same sequence that their partition coefficients favor the mobile phase.

Materials Required

Whatman standard-grade ashless powdered cellulose.

Chromatography tube. A Pyrex glass tube, 50×3.8 cm., constricted at one end and fitted with a No-Lub stopcock (Scientific Glass Apparatus Co., Bloomfield, N. J.). A tube this size will accommodate sufficient cellulose to effect the separation of 1 g. of a sugar mixture.

Fraction collector. A variety of fraction-collecting devices are available commercially. A siphon-actuated collector will require a 5-ml. siphon.

Equipment for paper chromatography (see Vol. III [11]).

Column Preparation. One hundred grams of Whatman powdered cellulose placed in a 2 l. beaker is covered with acetone and stirred vigorously with a mechanical stirrer. Sufficient acetone is used so that a thin homogeneous slurry results. A plug of glass wool is placed in the bottom of the chromatography tube, and the stopcock is opened. As much as possible of the cellulose suspension is poured into the tube. Stirring of the suspension in the beaker is continued, and the rest of the slurry is added before the acetone in the tube has drained to the level of the cellulose bed. The column is first washed with 150 ml. of acetone, then successively with 150-ml. portions of the following mixture: 80% acetone and 20% water saturated butanol, 60% acetone and 40% water-saturated butanol, 40% acetone and 60% water-saturated butanol, 20% acetone and 80% water-saturated butanol, and finally water-saturated butanol, the developer employed in effecting the sugar separations. The solvents are admitted by siphon through a stopper at the top of the chromatography tube.

Sugar Separation. The sugar mixture to be separated is deionized with suitable ion exchange resins (Dowex 50 and Duolite A4) and concentrated to a thin sirup. Preliminary examination of the mixture by two-dimensional paper chromatography will disclose the number of components in the mixture, and their approximate R_f values in a butanol-acetic acid-water mixture will give an indication of the efficiency with which they can be separated on a cellulose column. If the R_f's differ by more than 20%, fairly clean fractions can be cut; when the difference is less than 20%, there is considerable overlap.

The solvent in the chromatography tube is drained to the level of the cellulose bed, and the sugar mixture is pipetted onto the top of the column. Enough dry powdered cellulose to absorb the sirup is added to the column and covered with a disk of Whatman No. 3 filter paper cut to the inside dimensions of the tube. Gentle pressure is applied to seat the disk and level the cellulose. Ten milliliters of water-saturated butanol is pipetted onto the disk, and the stopcock at the bottom of the column is adjusted so as to give a flow rate of about 1 ml./min. Two more 10 ml. portions of solvent are added to the top of the column and allowed to drain to bed level, after which 50 ml. of solvent is added and the siphon delivery from the solvent reservoir is connected to the top of the tube. The fraction collector is set into operation, and 5-ml. fractions are collected.

After about 500 ml. of eluate has been collected, an analysis of the fractions is commenced while the column remains in operation. By means of a capillary pipet aliquots of about 0.02 ml. from each fraction are spotted in order of collection on a large sheet of Whatman No. 1 filter paper, five rows of twenty spots. The paper is sprayed with a saturated solution of p-anisidine hydrochloride and heated to develop the colors indicative of the presence of reducing sugars. In the ideal situation there will be gaps between series of fractions which react with the indicator.

Additional aliquots of the fractions reacting positively with the indicating reagent are spotted in order 2 cm. apart on the starting lines of filter paper sheet chromatograms. Specimens of the original mixture are included as reference markers. The chromatograms are developed with a butanol-acetic acid-water mixture, sprayed with the indicating reagent, and heated to reveal the distribution of the sugars among the fractions collected. The collection and analysis of fractions is continued until the last component of the mixture has been eluted from the column.

It is now possible to select and combine the fractions which are chromatographically pure with respect to the same component. The solvent is removed by vacuum distillation. The residue is dissolved in water, treated with charcoal, filtered, and reduced to a sirup by concentration in a vacuum desiccator. The sirup can usually be crystallized by adding absolute ethanol and triturating the mixture.

2-inch tab which projects into the solvent trough is held in position by placing the anchor rod on the half-inch fold. The troughs are filled with the appropriate solvents (about 150 ml.), and the lids are clamped into position. If only one or two papers are to be developed in a large cabinet, it is advisable to include extra sheets of filter paper in adjacent troughs on either side as "saturation sheets" to minimize solvent evaporation from the developing chromatograms.

The time required for development with the butanol solvents is about 18 hours, whereas the papers developed with phenol require from 24 to 30 hours.

When the solvents have descended to the lower edges of the filter paper sheets the chromatograms are removed from the cabinets. Manipulation of the wet papers can be carried out most easily by clipping them to the removable glass support rods with stainless-steel spring clips.[3] The chromatograms can then be readily removed from the cabinets by means of the glass rods which now function as supports during the drying process.

After the papers have been dried in the hood they are sprayed with a p-anisidine solution and heated to develop the colored spots which indicate the positions of the various sugars on the chromatograms. The chromatograms can be analyzed more easily if the spots are outlined with a pencil. It is now possible to estimate the number of components in the test solution by counting the colored spots on the chromatograms. The major components in the unknown mixture will be visible where an application of 0.01 ml. of the solution was made on the chromatogram; however, additional spots due to minor constituents, if present, will be visible where applications of 0.02 and 0.04 ml. were made.

The area and intensity of the spots are indicative of relative sugar concentrations; however, two sugars with similar mobilities may overlap and thus appear in a one-dimensional chromatogram as a single sugar of intermediate mobility. To avoid this error in interpretation, two-dimensional chromatograms of the unknown mixture should be prepared and developed simultaneously with the one-dimensional sheets.

Two-Dimensional Analysis. The filter paper sheet is folded along the 24-inch edge as previously described. In addition to the parallel line drawn 1 inch from the support fold, a second line is drawn 3 inches from, and parallel to, the 18-inch edge of the sheet. The intersection of these two lines is the origin of the two-dimensional chromatogram. At least two such chromatograms should be prepared. A 0.01-ml. aliquot of the unknown solution is applied to the origin of the first paper, and four 0.01-ml. aliquots to the origin of the second, allowing time for the spots to dry between applications.

After these chromatograms have been developed with phenol and dried until the bulk of the solvent has been evaporated, the initial folds are smoothed out and new folds are made along the 18-inch edge next to the origin. The chromatograms are then developed in the second dimension with the butanol-acetic acid-water mixture. When this second development has been completed and the papers have been dried, they are sprayed with p-anisidine and heated to reveal the positions of the sugars.

APPROXIMATE R_f VALUES OF FREE SUGARS IN AQUEOUS PHENOL (A) AND BUTANOL-ETHANOL-WATER (B)

Compound	A	B	Compound	A	B
Methylglyoxal	0.91	0.85	Tagatose	0.45	0.21
Dihydroxyacetone	0.90	0.48	Galactose	0.42	0.14
Glyceraldehyde	0.53	0.52	Glucose	0.36	0.15
Erythrose	0.69	0.44	Mannose	0.43	0.21
2-Deoxyribose	0.78	0.46	Mannoheptulose	0.36	0.15
Ribulose	0.65	0.33	Sedoheptulose	0.41	0.20
Xylulose	0.59	0.37	Cellobiose	0.27	0.04
Arabinose	0.53	0.22	Lactose	0.31	0.02
Lyxose	0.49	0.27	Maltose	0.32	0.06
Ribose	0.61	0.29	Isomaltose	0.20	0.03
Xylose	0.44	0.25	Melibiose	0.25	0.04
Hydroxymethylfurfural	0.92	0.80	Sucrose	0.36	0.09
Fucose	0.62	0.30	Turanose	0.41	0.08
Rhamnose	0.61	0.36	Melizitose	0.30	0.03
Fructose	0.51	0.21	Raffinose	0.25	0.02
Sorbose	0.41	0.18			

The two-dimensional chromatograms give more reliable information as to the number of components present. It is relatively easy to identify a spot on the two-dimensional chromatograms with that produced on each of the unidimensional runs by orienting the origins and directions of solvent flow of the one-dimensional chromatograms with those of the two-dimensional chromatograms and comparing similarities in mobility, color, size, and intensity of the spots produced. When the positions of the unknown components are compared with the positions of the known sugars in the standard mixtures, approximate R_f values can be assigned to the unknown sugars for the two solvent systems employed. Comparison of these assigned values with the values listed in the table will result in most cases in a tentative identification of the unknown sugar components. Approximate R_f values of the sugars in the systems butanol-acetic acid-water and ethyl acetate-acetic acid-water have not been listed in the table, since the relative mobilities of the various sugars in these solvents are essentially the same as in butanol-ethanol-water.

The R_f values of most of the sugars are considerably lower in the butanol solvents than they are in phenol. However, the butanol spot pattern can be extended by increasing the time of development. As a result the solvent will drip from the lower edge of the papers. In order to obtain a uniform spot pattern, it is essential to serrate the lower edges of the chromatograms so that the dripping will be uniform. When the travel of the sugars is increased by extended development, it is often possible to separate cleanly compounds that have low but different R_f's. The disaccharides have extremely low R_f values in the butanol-ethanol-water mixture. Thus it is possible to make exceptionally wide separations of glucose, fructose, and sucrose by extending the developing time with this solvent. Butanol-acetic acid-water is a useful "general solvent" in which the R_f values of the sugars have a wide range. Although the sugars also have relatively low R_f values in the ethyl-acetate-acetic acid-water mixture, the solvent travels rapidly. Thus, it is quite an effective solvent for the separation of di- and trisaccharides under conditions of extended development; however, this solvent must be freshly prepared, as the ester tends to hydrolyze.

Isolation and Identification

In order to verify the identification of unknown sugars, it is desirable to have milligram quantities of the individual components. Such quantities can be acquired by isolations from band chromatograms. A filter paper sheet prepared for unidimensional chromatographic development is marked at 1.5-cm. intervals along the starting line. The test solution, which has previously been concentrated to 5 to 10% free sugar, is applied in aliquots of 0.01 ml. to the marks so as to form a band of sugar along the starting line. The selection of a suitable solvent to separate the mixture is determined by the results of the previous two-dimensional analysis.

After the chromatogram has been developed and dried, strips approximately 2 cm. in width are cut from both of the margins and the center of the chromatogram. The strips are sprayed to indicate the positions of the sugars, and after the spots on the strips have been outlined the chromatogram is reassembled. Straight lines connecting the upper and lower limits of the equivalent separated spots are drawn across the body of the chromatogram in order to delineate the positions of the sugars. These bands containing the separated sugars are labeled as to their tentative identification and cut out from the filter paper sheet.

Elution. One end of each strip is cut to a point and the other end is placed between two microscope slides which are held together with rubber bands. The slides holding the strips of paper are placed in the troughs in the eluting cabinet, and the troughs are filled with water. As soon as the strips assume a vertical position, 5-ml. beakers are placed beneath them, and the shelf is positioned so that the pointed tips dip into the beakers. The bottom of the cabinet and the inside of the door are sprayed with water from a wash bottle, and the door is clamped shut. The elution of a 5-inch strip requires about 2 hours in order to collect 4 to 5 drops of eluate which contain the bulk of the sugar present on the strip (2 to 3 mg.).

If some of the eluates are known to contain two sugars, because the individual components could not be cleanly separated in the solvent system employed, it is necessary to subject these eluates to a second chromatographic band development in a solvent that will effect a separation.

This paper chromatographic method of isolation may be used to separate as much as 200 mg. of a three-component system on eight chromatograms. If a larger amount is to be separated, it is more practical to employ cellulose column chromatography. The method can, however, be used most effectively in the isolation and purification of high-specific-activity C^{14}-labeled sugars produced photosynthetically.[13] The recoveries are essentially quantitative, since the position of the bands of sugar can be located exactly by placing an X-ray film in contact with the chromatogram. Four repetitions of the chromatographic isolation, with two different solvents, yielded 99.8% chromatographically pure glucose.

After the several components of the tentatively identified mixture have been isolated, it is necessary to chromatograph them unidimensionally, side by side with authentic specimens of the suspected sugars, in both the phenol and the butanol solvents. If it can be demonstrated that a component has the same R_f values in both solvents as those possessed by the authentic specimens and that both sugars undergo exactly the same color reactions when sprayed with the various indicating reagents, it can be assumed that the sugars are identical. If sufficient material is available, biochemical tests should be applied as further confirmation.

Quantitative Procedure

After the identities of the components in a mixture have been established, it is often possible to quantize them by applying specific analytical procedures to the original solution. However, if one of the components present interferes with the determination of another, the sugars must be isolated as previously described and determined separately.

Aliquots of 0.01 ml. of the solution to be analyzed (approximately 1% with respect to each component) are spotted at 1.5-cm. intervals along the starting line of a one-dimensional sheet chromatogram which is

[13] E. W. Putman and W. Z. Hassid, *J. Biol. Chem.* **196**, 749 (1952).

developed with a solvent that will effect complete separation of the sugars present. A clean sheet of filter paper to be used in the preparation of paper blanks is developed at the same time. At the end of the run, four parallel ¾-inch guide strips are cut from the chromatogram, one from each edge, and one 5 inches in from each of the remaining edges, leaving three 5-inch sections of the chromatogram intact.

The positions of the separated sugars on the guide strips are revealed by the indicating reagent, and the corresponding areas are excised from the intact sections of the chromatogram with parallel cuts. Areas identical in size and position are cut from the blank chromatogram.

The resultant 5-inch strips are labeled with respect to the sugar they contain and the section from which they were cut. Prior to eluting a strip, it is cut on the long diagonal to yield two right triangles which are placed between the microscope slides so that their right angles are contiguous one with the other at the center of the slides. The result is an isosceles triangle formed by two pieces of paper held together between the slides at its base. If the paper strips are too wide to be held conveniently by microscope slides, strips of Plexiglas the length of the elution trough may be used. After the papers have been positioned in the trough it is filled with water and the elution is continued until approximately 1 ml. of eluate is collected in the 5-ml. beakers placed under each strip. The eluates are transferred quantitatively to 10-ml. volumetric flasks and made up to volume. These dilute sugar solutions, about 100 γ/ml., are analyzed by conventional colorimetric methods.

Analytical Methods

Ketohexoses as well as di- and trisaccharides containing ketohexoses may be determined by Roe's method.[14] A 2.0-ml. aliquot of the sugar solution containing 10 to 100 γ of ketohexose is placed in a colorimeter tube. Two milliliters of 0.1% resorcinol in 95% ethanol and 6 ml. of 10 N HCl containing 0.001% $FeCl_3$ are added. The contents of the tube are thoroughly mixed by shaking, and it is heated for 8 minutes in a water bath at 80° to develop the color. The tube is cooled and compared with standards at 490 mμ.

Pentoses and ketoheptoses are determined by the Mejbaum orcinol method.[15] The sample, 4 to 40 γ of pentose, is diluted to 3 ml. in a colorimeter tube, and 3 ml. of a 1% solution of orcinol in 12 N HCl containing 0.1% $FeCl_3$ is added. After thorough mixing, the tube is heated for 40 minutes in a boiling water bath, cooled, and compared with standards at 660 mμ. The orcinol reagent must be freshly prepared. A stock solution

[14] J. H. Roe, *J. Biol. Chem.* **107**, 15 (1934).
[15] W. Z. Mejbaum, *Z. physiol. Chem.* **258**, 117 (1939).

of $FeCl_3$ in concentrated HCl may be kept on hand, but the orcinol should not be added until the determination is to be made.

Aldohexoses as well as di- and trisaccharides can be determined by the anthrone method.[16] A 1.0-ml. aliquot of the solution containing 20 to 200 γ of sugar is placed in a colorimeter tube, and 5 ml. of a 0.1% solution of anthrone in 72% H_2SO_4 is added. After a thorough mixing, the tube is placed in a boiling water bath for 12 minutes. The tube is then cooled and compared with standards at 620 mμ. Although the anthrone method is relatively insensitive toward pentoses, it may be used satisfactorily in the determinations of a variety of other sugars.[17]

The ratio of the sugars present in the paper strips cut from the same section of the chromatogram represents the ratio existing in the original mixture. If a component which can be determined directly in the original mixture, e.g., a pentose or a ketohexose, is present, this sugar serves as a reference sugar. By expressing the concentration ratios in terms of the known concentration of the reference sugar, the absolute concentrations of the various sugars in the original mixture may be calculated. If such a sugar is not present in the system, a reference sugar having an R_f value different from any component in the mixture may be weighed into a known volume of the test solution before chromatography.[2] The results obtained from the three sections of the chromatogram should check within 5% of the average values.

Alternative methods for the quantitative analysis of sugars separated on paper chromatograms have been described by Isherwood.[18]

Quantitative Analysis of Reducing Oligosaccharides

The method of Wadman et al.[19] has proved useful in the analysis of oligosaccharides obtained during the course of polysaccharide hydrolysis.

A known volume of the solution to be analyzed, containing 0.1 to 4% sugar, is mixed with an equal volume of a 10% solution of N-(1-naphthyl)-ethylenediamine dihydrochloride in a solvent consisting of 50% triethylamine, 40% ethanol, and 10% water by volume. Solutions of known amounts of authentic oligosaccharides (maltose, maltotriose, etc.) are treated similarly in order to provide calibration values.

Accurately measured aliquots, 5 to 10 μl., of the known and test solutions as well as a 1:1 dilution of the naphthylethylenediamine reagent are applied 3 inches apart along the starting line of a Whatman No. 1 filter paper sheet. The paper is then heated at 100° for 30 minutes in

[16] N. J. Fairbairn, *Chemistry & Industry* **1953**, 86.
[17] L. H. Koehler, *Anal. Chem.* **24**, 1576 (1952).
[18] F. A. Isherwood, *Brit. Med. Bull.* **10**, 202 (1954).
[19] W. H. Wadman, Gwen J. Thomas, and A. B. Pardee, *Anal. Chem.* **26**, 1192 (1954).

order to effect a coupling of the sugar with the amine. Under these conditions about 60% of the sugar reacts.

The chromatograms are developed by the descending method with a solvent consisting of 40 parts of n-butanol, 12 parts of ethanol, 16 parts of water, and 1 part of concentrated NH_4OH by volume. The time required for development is approximately 18 hours. The chromatograms are dried at room temperature and examined under ultraviolet light.

The uncombined amine is found immediately behind the solvent front, and the positions of the sugar derivatives, which are easily located because of their fluorescence, are arranged in order of increasing molecular weight as well-separated spots behind the amine. The R_f of the maltose derivative is approximately 0.6. Dextrins containing up to eight sugar units have been separated from acid hydrolyzates of amylose by this method.

After the fluorescent areas on the chromatograms are outlined with pencil, they are excised along with the minimum excess of paper that will allow one end of the spot to be held between microscope slides during elution and the other end to be tapered. Blanks are provided by cutting areas of identical size, shape, and distance from the origin from the chromatogram below the point of application of the amine solution which was not mixed with sugar. The excised strips of paper are then eluted chromatographically with a 1% solution of $Na_3PO_4 \cdot 12H_2O$. All fluorescent material is removed in the first milliliter of eluate collected. The eluates are transferred to 5-ml. volumetric flasks, diluted to volume, and analyzed for fluorescence by means of a Beckman DU spectrophotometer equipped with a fluorescence accessory attachment. An exciting wavelength of 355 mμ is used in these determinations which show a straight-line relationship between per cent transmittance and maltose concentration over the range of 5 to 40 γ of maltose applied to the chromatogram.

Determinations may also be made by reading the optical density at 340 mμ in the conventional manner. Although less sensitivity is obtained than when fluorescence is determined and the blank is higher, 20 to 200 γ of maltose can be estimated readily.

[12] Colorimetric Analysis of Sugars

By GILBERT ASHWELL

I. General Remarks

The failure of the classical methods of organic chemistry to cope with the problems of identification and determination of small quantities of carbohydrates obtained from biological materials has necessitated the development of highly sensitive and specific techniques capable of measuring microgram quantities of sugars in the presence of various tissue constituents. In general, such tests have been devised by heating aqueous solutions of the sugar with a strong acid, thereby converting it to furfural or a derivative of furfural. A color is then produced by the addition of an organic developer such as indole, orcinol, diphenylamine, or carbazole. Under appropriate conditions, these tests can be made to yield proportional and reproducible results.

In the use of any color test for an unknown material of biological origin there are two considerations of importance to the individual investigator. First, there is the problem of specificity. It must be emphasized that no single colorimetric assay is absolutely specific. Therefore, identification of an unknown sugar becomes credible only when concurring results are obtained in at least two independent assay procedures. It would seem highly improbable that two completely different sugars should give identical absorption in the same range in two different reactions. Second, there is the problem of control blanks. Many of the organic substances present in tissues and tissue extracts react with strong acids, especially with concentrated H_2SO_4, to form compounds with a marked absorption in the blue and ultraviolet region. To compensate for this it is necessary to run with every sample of the unknown to which developer is added another tube containing all the additions of the above save the developer. The former is read against a water blank with developer, and the latter against a water blank without developer. The difference between these curves is then specifically a measure of the reaction between the sugar and the color-producing reagent.

Color reactions may be described as either qualitative or quantitative. Qualitative tests are extremely useful in determining the presence of a new or unsuspected sugar in a reaction mixture. They are also of value in the systematic identification of an unknown compound. In this case careful attention to the rate and quality of color development, the effect of an altered heating time, and the spectral curve of the reaction products may yield valuable information on the nature of the unknown. Quantitative tests, on the other hand, are carried out on known sugars under rigorously prescribed conditions where the possibility of interfering substances has been carefully eliminated and the only variable is the concentration of the substance being measured. Under these conditions, and with appropriate blanks, the determination may be performed with a reasonable measure of confidence in the validity of the results so obtained.

Apart from the two commonly used methods described in the section on total hexoses, procedures based on the reducing properties of sugars will not be considered here. These methods, although of great usefulness under certain clearly defined conditions, are completely nonspecific and rarely appropriate for the investigation of complex systems. In addition, there are many excellent compilations of these methods which are readily available.[1]

II. Determination of Hexoses

Many of the reagents commonly employed for the determination of a specific class of sugars are suitable, with slight modifications, to measure total carbohydrates. Thus, orcinol, carbazole, indole, and diphenylamine have all been used for this purpose. Although these reactions logically require a separate classification, they will be discussed individually in order to prevent unnecessary duplication.

A. Aldohexoses

There are, unfortunately, no simple colorimetric procedures capable of distinguishing aldohexoses in the presence of ketohexoses. This is understandable, since the reactive component, which is presumed to be hydroxymethylfurfural, is formed more readily from fructose than from glucose. However, since the keto sugars are readily determined by a variety of techniques, the value so obtained may be deducted from the total hexose concentration, thus providing an approximate measure of the aldohexoses. With this procedure it is essential to know the identity of the sugar being studied, since the extinction coefficients obtained from the individual hexoses vary appreciably in any given assay.

Several reactions have been reported wherein the color produced by the aldo sugars differs from that of the ketoses, thereby providing a simple qualitative means of identification. α-Naphthol,[2] tryptophan,[3]

[1] F. J. Bates and associates, *Natl. Bur. Standards (U.S.) Circ.* **C440** (1942); C. A. Browne and F. W. Zerban, "Physical and Chemical Methods of Sugar Analysis," John Wiley & Sons, New York, 1941.

[2] Z. Dische, *Mikrochemie* **7**, 33 (1929).

[3] M. R. Shetler, J. V. Foster, and M. R. Everett, *Proc. Soc. Exptl. Biol. Med.* **67**, 125 (1948).

and aminoguanidine,[4] under appropriate conditions, yield characteristically different spectra, depending on the nature of the hexose present. The usefulness of these reactions is limited by the fact that they cannot be applied for quantitative purposes in mixtures. In addition, since they are not specific, a small amount of aldohexose will be completely masked in the presence of an excess or the ketose.

Recently, the well-known oxidation of aldo sugars by hypobromite has been adapted for microdeterminations. Here, 10 to 100 γ of the aldose is incubated with an excess of Br_2 (either as a saturated Br_2-water or a 0.5 M alcoholic solution) at room temperature for 30 minutes. Solid $BaCO_3$ or phosphate buffer at pH 7.5 is added to prevent excess acidity. At the end of the incubation period, the unreacted Br_2 is removed by aeration with He. The aldoses are completely destroyed under these conditions; the ketoses are largely unaffected. Total hexose determination before and after oxidation gives a quantitative measure of the aldose concentration. In addition to an internal standard, it is essential to run controls in which the sugar is added after the unreacted Br_2 is removed, since the presence of Br^- may affect the assay.

This procedure was found by Horecker et al.[5] to be applicable to the qualitative determination of ribulose[6] in the presence of ribose where the sugars were assayed by the orcinol method. Subsequently, Slein and Schnell[7] showed that it could be applied quantitatively to mixtures of glucose and fructose or heptose and heptulose when assayed by the cysteine-H_2SO_4 method of Dische. In principle, this is a simple and accurate method, when properly controlled, and the only precise way of determining aldohexose in the presence of ketohexose.

B. Ketohexoses

Resorcinol Method of Roe

This method, a modification of the Seliwanoff reaction, was described by Roe[8] in 1934. Recent improvements in sensitivity and proportionality have been reported in adapting the test for inulin determination. The later modification is given here.[9]

Reagents. The developing reagent is prepared by dissolving 0.1 g. of resorcinol and 0.25 g. of thiourea in 100 ml. of glacial acetic acid. This

[4] H. Tauber, *Anal. Chem.* **25**, 826 (1953).

[5] B. L. Horecker, P. Z. Smyrniotis, and J. F. Seegmiller, *J. Biol. Chem.* **193**, 383 (1951).

[6] For a more sensitive and specific test for ribulose, see comments under the cysteine-carbazole test for ketohexoses.

[7] M. W. Slein and G. W. Schnell, *Proc. Soc. Exptl. Biol. Med.* **82**, 734 (1953).

[8] J. H. Roe, *J. Biol. Chem.* **107**, 15 (1934).

[9] J. H. Roe, J. H. Epstein, and N. P. Goldstein, *J. Biol. Chem.* **178**, 839 (1949)

pentose, and tetrose will interfere with the detection and determination of fructose by this reaction. On the other hand, it is possible to detect the presence of these substances in the unknown by their carbazole reaction if the presence of ketoses and trioses can be excluded or accounted for. For this purpose, a combination of the carbazole and resorcinol methods may be of advantage.

Diphenylamine Reaction

This reaction has been used extensively for the determination of inulin.[16-18] The test is based on the Ihl-Pechmann reaction for fructose and is not to be confused with the diphenylamine test for DNA which is described in the section on deoxy sugars. The procedure of Rothenfusser[19] for the determination of fructose and its esters was modified by Dische in 1929.[2] It was reported at that time that a 3-minute heating was sufficient to develop the color of fructose and that a 30-minute heating would permit a quantitative determination of the aldoses as well. A more recent modification is given here.[20]

Reagents. The reagent is prepared by mixing 10 ml. of a 10% solution of diphenylamine in absolute alcohol with 90 ml. of glacial acetic acid and adding 100 ml. of concentrated HCl. The solution is stable indefinitely. Commercial preparations of diphenylamine are best recrystallized from 70% alcohol.

Procedure. To 1 part of the solution containing 10 to 50 γ of fructose, 2 parts of the reagent is added, and the mixture is heated in a boiling water bath for 10 minutes. The tubes are cooled in tap water and read at 635 mμ. The color is stable for many hours.

Comments. The diphenylamine reaction produces colored compounds with all carbohydrates, although the intensity varies markedly with the time of heating. Under the above conditions, the color intensities produced by fructose and glucose are about 30:1. In this, as in the carbazole reaction, phosphorylated fructose esters give the same value per micromole as does pure fructose.[21]

When the reaction mixture is heated for 3 minutes instead of 10, the ratio of color production for fructose/glucose is much higher and can be used profitably as a qualitative test for the presence of ketohexose in the presence of a great excess of total hexoses. However, all hexoses produce

[16] P. Kruhøffer, *Acta Physiol. Scand.* **11**, 1 (1946).
[17] A. S. Alving, J. Rubin, and B. F. Miller, *J. Biol. Chem.* **127**, 609 (1939).
[18] A. C. Corcoran and I. H. Page, *J. Biol. Chem.* **127**, 601 (1939).
[19] S. Rothenfusser, *Chem. Zentr.* **80**, II, 934 (1909).
[20] Z. Dische, *J. Biol. Chem.* **204**, 983 (1953).
[21] Z. Dische, *Mikrochemie* **10**, 129 (1931).

a blue color with absorption maxima at 635 and 520 mμ and a minimum at 560 mμ. Heptoses give a violet color with a single sharp maximum at 560 mμ. Thus, the shape of the absorption curve between 520 and 635 mμ provides a means of tentative identification of the seven carbon sugars in mixtures with hexoses. Ketopentoses give approximately 10% of the color produced by fructose at 635 mμ. Aldopentoses and trioses give a greenish-yellow color; hexuronic acids appear brown-red with an absorption maximum at 520 mμ and a second maximum in the violet part of the spectrum. The 2-deoxypentoses yield a blue color with peaks at 600 and 520 mμ.

Indole Reaction

The early investigations of Weehuizen[22] and Fleig[23] on the colorimetric determination of sugars by means of skatole and indole were re-examined by Dische and Popper in 1926[24] and by Dische in 1929.[2] It was found that the violet color produced by skatole was unsatisfactory for quantitative determinations, but the orange color developed by indole was both sensitive and proportional to extremely small concentrations of sugars. As indicated earlier, this reaction is readily modified so as to provide a means of determining total carbohydrates. Consequently both procedures are described here.

Reagents. The 75% acid reagent is prepared by adding 70 vol. of concentrated H_2SO_4 to 30 vol. of water. Indole is made up as a 1% solution in absolute alcohol.

Procedure. FOR FRUCTOSE. To 1 ml. of a solution containing 5 to 25 γ of fructose, 0.1 ml. of the indole solution is added, followed by 4 ml. of the 75% H_2SO_4. The tube is shaken and allowed to stand at room temperature. The samples are read at 480 mμ after 2 hours.

FOR TOTAL CARBOHYDRATES. To 1 ml. of a solution containing 5 to 25 γ of sugar is added 2 ml. of concentrated (98%) H_2SO_4. The tubes should be kept in ice water and shaken while the acid is being added to prevent excessive heating. Then 0.1 ml. of the indole solution is added, and the tubes are put in a water bath at 100° for 10 minutes. After cooling they are read at 480 mμ. The color is stable.

Comments. This assay procedure has not been widely used, and there is little information concerning its application in the recent literature. It is included here, however, because of its extreme sensitivity for fructose in the presence of glucose. According to our observations, the color intensity increases continually after addition of the acid, and a 2-hour

[22] F. Weehuizen, *Chem. Zentr.* **78**, I, 134 (1907).
[23] C. Fleig, *Chem. Zentr.* **79**, II, 1954 (1908).
[24] Z. Dische and H. Popper, *Biochem. Z.* **175**, 371 (1926).

reading time was chosen as an arbitrary compromise between color development and specificity. After 24 hours, the color intensity is doubled and the ratio of fructose/glucose color production is halved. Under the conditions described here, the reactivity of fructose as compared to glucose is better than 125:1.

When the assay is carried out as described for total carbohydrates, the marked selectivity towards ketohexoses disappears and the reaction serves as a sensitive tool for the determination of small amounts of carbohydrates. Pentoses, hexoses, and heptoses all give the reaction with a maximum around 480 mμ. The color development with tetroses is negligible. In both modifications, the color lacks proportionality when amounts less than 5 γ are measured.

C. Total Hexoses

It should be pointed out that there are three categories of procedures treated in this section. These include the carbazole and cysteine-H_2SO_4 reactions which permit the identification of individual hexoses in mixtures, and the anthrone method in which the hexoses are indistinguishable from each other. The third group, reducing sugar methods, is nonspecific and can be used only under carefully defined conditions.

Carbazole Reaction for Hexoses

The method presented here is a modification by Gurin and Hood[25] of the reaction described by Dische[13,26] in which it was shown that the various hexoses reacted differently in the color produced when heated with sulfuric acid and carbazole. Seibert and Atno[27] subsequently used this procedure for the determination of total carbohydrates in serum polysaccharides.[28]

Reagents. The acid reagent is prepared by adding 8 vol. of cold concentrated H_2SO_4 to 1 vol. of cold water. The acid should be free of nitrogen. Carbazole is used as a 0.5% solution in absolute alcohol and is stable for one to two months. Resublimed commercial carbazole has been found satisfactory.[27]

Procedure. Nine milliliters of the H_2SO_4 reagent is measured into a widemouthed test tube and chilled in an ice bath. One milliliter containing 50 to 200 γ of sugar is carefully layered above the acid and chilled again for several minutes. The tube is then shaken while still immersed

---

[25] S. Gurin and D. B. Hood, *J. Biol. Chem.* **131**, 211 (1939).

[26] Z. Dische, *Mikrochemie* **8**, 4 (1930).

[27] F. B. Seibert and J. Atno, *J. Biol. Chem.* **163**, 511 (1946).

[28] For determination of the total carbohydrates in serum polysaccharides, a modification of the tryptophan reaction has been reported recently by J. Bodin, C. Jackson, and M. Schubert, *Proc. Soc. Exptl. Biol. Med.* **84**, 288 (1953).

in the ice bath. Insufficient care in chilling often produces irregular results. The tube is then removed, 0.3 ml. of the carbazole added, and the contents rapidly mixed in a vigorously boiling water bath for 10 minutes. After cooling, the tubes are read at 540 and 440 mμ. The color is stable for hours. As little as 25 γ may be estimated safely.

Comments. Inasmuch as the slope of the absorption curve from 440 to 540 mμ is different for each of the four common hexoses, this ratio is calculated in order to identify the hexose or hexose mixture present. The absorption at 540 mμ is then compared with calibration curves established with varying concentrations of the appropriate sugars or sugar mixture. This procedure can be used to differentiate simple solutions of the hexoses, although complex mixtures obviously cannot be determined in this fashion. The ratios $D_{540}:D_{440}$, as given by Seibert and Atno[27] for the various hexoses, are as follows: glucose 4.89; galactose 1.90; and mannose 0.81. However, it is strongly advised that these figures be followed only as a guide and that the appropriate standards be run with every determination, since slight alterations in temperature or time of heating will change them somewhat.

Holzman *et al.*[29] have systematically examined such variables as time of heating, concentration of acid, and amount of carbazole added in an effort to increase the sensitivity of the assay. However, the greater sensitivity with their modification is gained at the expense of specificity, and under their conditions the colors obtained with the various hexoses are not sufficiently distinctive to permit ready differentiation by spectrophotometric measurements.

The carbazole reaction, it must be emphasized, is given by all classes of carbohydrates and is not unique for hexoses. Arabinose and DNA both give the peak at 540 mμ, the former yielding about one-half and the latter about twice the color intensity produced by glucose. Aldo- and ketoheptoses give an orange color with a maximum at 480 mμ.[20] Amino acids do not interefere except for tryptophan, where as little as 50 γ introduces a serious error. The method has been employed successfully in studies on a number of neutral polysaccharides, ovomucoid, serum mucoid, plasma proteins, and casein, but lipids, sulfhydryl compounds, and hexuronic acids will produce interferences.

Cysteine-H$_2$SO$_4$ Reaction for Hexoses

This method, reported by Dische *et al.* in 1949,[30] is a specific application of the so-called basic cysteine reaction of sugars.[31] It is of consider-

[29] G. Holzman, R. V. MacAllister, and C. Niemann, *J. Biol. Chem.* **171,** 27 (1947).
[30] Z. Dische, L. B. Shettles, and M. Osnos, *Arch. Biochem.* **22,** 169 (1949).
[31] Z. Dische, *J. Biol. Chem.* **181,** 379 (1949).

able value, since two reaction products are formed, both of which can be employed for the quantitative determination of hexoses. In addition, under certain conditions,[20] the reaction serves as an extremely sensitive and quantitative assay for heptoses and methylpentoses.

Reagents. The acid reagent is prepared by adding 6 vol. of concentrated H_2SO_4 to 1 vol. of water. The cysteine hydrochloride is made up as a 3% solution in water and must be freshly prepared for each determination.

Procedure. 4.5 ml. of the acid is placed in a widemouthed test tube immersed in ice water. After several minutes, 1 ml. of the solution containing 10 to 100 γ of hexose is added with shaking and continuous cooling in ice water. After a few minutes the tube containing the reaction mixture is replaced in tap water to bring it to room temperature and then kept for exactly 3 minutes in a boiling water bath. After cooling, 0.1 ml. of the cysteine solution is added and the mixture is shaken. In the presence of hexoses, a yellow color appears a few minutes after the addition of the cysteine with a maximum absorption at 415 mμ. The difference in absorption at 415 and 380 mμ ($\Delta D_{415-380}$) is a quantitative measure of hexose concentration.

Comments. The curves for all four hexoses investigated show the same peak, differing only in height. At a concentration of 50 γ the relative intensities are: glucose 1.0; fructose 1.04; galactose 0.58; mannose 0.45. The phosphorylated hexoses react similarly.[10] Methylpentoses give a faint yellow color with a maximum at 400 mμ. The absorption due to these compounds is equal at 380 and 415 mμ and slight for pentoses, deoxypentoses, and hexuronic acids. The $\Delta D_{415-380}$ is, therefore, zero for methylpentoses and negligible with all the above-mentioned classes of sugars except hexoses, for which it is positive. Ten micrograms of hexoses per milliliter can be detected safely by this difference. For quantitative purposes, the test should be read 2 hours after addition of the cysteine.

On standing at room temperature for 48 hours, the primary reaction products are unstable and the absorption maximum at 415 mμ disappears. At the same time, other color reactions appear which are different for the different hexoses. Galactose and sorbose give blue, glucose and fructose green, and mannose yellow colors. Apparently this represents the mixture of two reaction products, one with a maximum at 600 mμ, and one with an intensive absorption between 380 and 430 mμ with no characteristic maximum and appearing yellow. This reaction serves as a convenient and qualitative test for galactose and sorbose. For quantitative use, the reaction is run in the presence of 0.1 mg. of mannose which shifts the peak of the maximum to 605 mμ and increases the proportionality among all the hexoses. Aldo- and ketoheptoses yield an orange color which slowly turns pink with a sharp peak at 506 to 510 mμ. This color can be read after 18 hours at room temperature and is stable for days.[20]

Cysteine-H$_2$SO$_4$ Reaction for Mannose

This reaction[30] differs from the previously described method in the concentration of acid, the time of heating, and the addition of cysteine. It can be applied as a qualitative test for all the common hexoses and as a quantitative test for mannose.

Reagents. The acid reagent is prepared by mixing 70 vol. of cold concentrated H$_2$SO$_4$ with 30 vol. of cold water. The cysteine hydrochloride is freshly prepared as a 3% aqueous solution.

Procedure. To 0.9 ml. of solution containing 50 to 100 γ of hexose is added 0.1 ml. of the cysteine followed by 5 ml. of the H$_2$SO$_4$ reagent. The tube is heated for 10 minutes in a boiling water bath and cooled in tap water. The samples are read at 370 and 400 mμ, 48 hours after heating.

Comments. All the common hexoses react strongly under the above conditions to yield absorption curves with characteristic properties in the range 350 to 400 mμ. Advantage can be taken of the geometrical distribution of these curves to formulate an expression which serves to identify three of these sugars. Thus $\Delta D_{375-400} - \Delta D_{350-375}$ is zero for galactose, negative for glucose, and both positive and proportional for mannose. It is substantially zero for methylpentoses and DNA. Pentoses and hexuronic acids give slightly negative values. Furthermore, since in this, as in the previously described reaction, galactose forms a blue color with a maximum at 600 mμ, the ratio $D_{600} : D_{350-375}$ provides a convenient means for the identification of this sugar. The ratio is 4.5 for galactose, and 1.4 or less for glucose, mannose, and fructose.

For the quantitative determination of 20 to 100 γ of mannose, in the presence of other classes of sugars, the expression $\Delta D_{370-400}$ is used. This value is zero for deoxyribose and barely significant for methylpentoses and hexuronic acids. It is proportional to the concentration of all hexoses and additive in mixtures. Consequently, in the determination of mannose by this procedure, the presence of other hexoses must be ascertained and their contribution to the delta reading subtracted out. Although the presence of free pentose interferes seriously with this method, when the pentose is substituted in position 5, as in the nucleotides and coenzymes, the effect on the assay is slight. The presence of free pentoses in the unknown may be readily determined by measuring the $\Delta D_{320-340}$ at 2 and 48 hours after the heating. Any increase in the value on standing indicates the presence of significant amounts of pentoses and hexuronic acids.

The two modifications of the cysteine-H$_2$SO$_4$ method for hexoses described here yield three reactions, the products of which differ con-

Procedure. Blood filtrates may be prepared for analysis as follows: 1 vol. of blood is mixed with 15 vol. of water, and 2 vol. of $Ba(OH)_2$ is added with stirring. After the solution has turned brown, 2 vol. of $ZnSO_4$ is added and again stirred. The reaction mixture is allowed to stand for several minutes before filtering.

One milliliter of the filtrate containing 5 to 300 γ of glucose is pipetted into a narrow test tube graduated at 25 ml. One milliliter of a mixture (prepared the same day) of 25 parts of copper reagent A to 1 part of B is added. Blanks and standards are treated the same way. The solutions are then mixed and heated for 20 minutes in a boiling water bath. After cooling, 1 ml. of the arsenomolybdate reagent is added. The samples are mixed thoroughly and diluted to the mark. A stable blue color quickly appears and is read at 500 to 520 mμ.

Comments. The sensitivity of the reaction can be increased fourfold by reading at 660 mμ which is the point of maximum absorption. The lower wavelength indicated here is chosen as a compromise between the sensitivity lost and the advantages gained by reducing the effect of variation in such factors as blank due to reagent and reoxidation of cuprous oxide. The large amount of Na_2SO_4 present gives adequate protection against reoxidation for most purposes so that Folin-Wu or covered tubes are not essential. However, if high accuracy with quantities of glucose below 5 γ is needed, these precautions should be taken. In general, the deviation in duplicate samples should not be more than 3% and in most cases will be less than 1.5%.

In addition to blood glucose determinations, this procedure has been used for tissue sugars, glycogen, urine sugars, maltose, and glucuronic acid.

Ferricyanide Method of Park and Johnson

The determination of blood sugar by means of ferricyanide reduction in alkaline solution has been used for many years.[40] The present modification[41] offers the advantage of extreme sensitivity.

Reagents. The ferricyanide solution is prepared by dissolving 0.5 g. of potassium ferricyanide in 1 l. of water. It is stored in a brown bottle. The carbonate-cyanide reagent consists of 5.3 g. of sodium carbonate and 0.65 g. of KCN per liter of solution. The ferric iron reagent is made up to contain 1.5 g. of ferric ammonium sulfate and 1 g. of Duponol[42] in 1 l. of 0.05 N H_2SO_4.

[40] H. C. Hagdorn and B. N. Jensen, *Biochem. Z.* **135,** 46 (1923).
[41] J. T. Park and M. J. Johnson, *J. Biol. Chem.* **181,** 149 (1949).
[42] Duponol (sodium monolauryl sulfate) may be obtained from E. I. du Pont Co., Inc., Wilmington, Delaware. The purpose of this addition is to keep the Prussian blue in suspension. When present in excess, it causes air bubbles to remain suspended in the solution.

Procedure. The sample, deproteinized and neutralized when necessary, containing 1 to 9 γ of glucose is placed in an 18-mm. test tube and diluted to 1 to 3 ml. as desired. One milliliter each of the ferricyanide and carbonate-cyanide reagent is added. After mixing, the tube is heated in a boiling water bath for 15 minutes and cooled. Five milliliters of the ferric iron solution is then added, and the mixture shaken. A blue color develops which can be read in 15 minutes at 690 mμ. The precision is ± 0.2 γ.

Comments. In carrying out this test on acid filtrates or strongly buffered solutions, proper control of the pH is essential. Incomplete reduction may occur during the heating step when the pH of the reaction mixture is less than 10.5. Similarly, turbidities may develop on addition of the ferric iron if the pH of the final mixture is above 2.

The widespread use of this procedure is a tribute to its sensitivity, accuracy, and simplicity. However, it must not be employed indiscriminately. As with all reducing methods, it is not specific for sugars and cannot be used for quantitative purposes in complex mixtures.

III. Determination of Pentoses

In addition to the methods outlined here for the determination of pentoses, there are several alternate procedures which may be of value under special circumstances. The more important of these seem to be the methods of Roe and Rice[43] and of Tracey,[44] employing *p*-bromoaniline and aniline, respectively, as the color-producing agent. These reactions are appreciably less affected by the presence of hexoses and uronic acids than is the case with orcinol. The carbazole reaction[45] is moderately sensitive but lacks specificity. In it, as in the cysteine reaction, xylose gives twice the color of the other three aldopentoses. A recent modification of the phloroglucinol test for pentoses has appeared[46] which is quite elaborate and less sensitive than the methods described below.

The Orcinol Reaction

This method, first proposed by Bial,[47] has been extensively studied in a number of laboratories.[48–51] Sørensen and Haugaard[52] have described

[43] J. H. Roe and E. W. Rice, *J. Biol. Chem.* **173,** 507 (1948).
[44] M. V. Tracey, *Biochem. J.* **47,** 433 (1950).
[45] S. Gurin and D. B. Hood, *J. Biol. Chem.* **139,** 775 (1941).
[46] H. von Euler and L. Hahn, *Svensk Kem. Tidskr.* **58,** 251 (1946) [*Chem. Abstr.* **41,** 2108 (1947)].
[47] M. Bial, *Deut. med. Woch.* **28,** 253 (1902); **29,** 477 (1903).
[48] G. Embden and M. Lenhartz, *Z. physiol. Chem.* **201,** 149 (1931).
[49] Z. Dische and K. Schwartz, *Mikrochim. Acta* **2,** 13 (1937).
[50] H. K. Baranscheen and A. Peham, *Z. physiol. Chem.* **272,** 81 (1942).
[51] H. G. Albaum and W. W. Umbreit, *J. Biol. Chem.* **167,** 369 (1947).
[52] M. Sørensen and G. Haugaard, *Biochem. Z.* **260,** 247 (1933).

conditions under which it reacts with all classes of sugars and can be employed for the determination of total carbohydrates. The most widely used modification for pentoses is that of Mejbaum,[53] and it is the basis for the following procedure.

Reagents. Ferric chloride is dissolved in concentrated HCl to a final concentration of 0.1 % $FeCl_3$. Orcinol is made up in 95 % alcohol, 100 mg./ml. This solution is unstable and should be prepared fresh daily. The commercial reagent is frequently unsatisfactory and should be recrystallized from benzene.

Procedure. To 3 ml. of the unknown containing 10 to 30 γ of pentose, 3 ml. of the HCl-$FeCl_3$ reagent is added, followed by 0.3 ml. of the alcoholic orcinol solution. The tubes are then heated for 40 minutes in a boiling water bath, cooled, and read at 670 mμ. Evaporation may be minimized by employing glass-stoppered tubes or by covering the mouths of the tubes with carefully cleaned marbles.

Comments. Although the original procedure described a 20-minute heating period, it was shown by Albaum and Umbreit[51] that the color development with ribose-3-P and free pentoses was only 87 to 88 % completed after 25 minutes at 100°. Horecker *et al.*[5] found that a 40-minute heating time was required for maximum color production. The increased time of heating, however, has a marked disadvantage in that it increases the interference of other sugars. Thus, glucose has a small extinction coefficient for the shorter boiling periods which grows progressively larger and continues to increase even after 60 minutes at 100°. This difficulty is less pronounced in the Dische modification[20] which calls for a higher concentration of acid and a shorter time of heating. This procedure is carried out as follows.

To 1.5 ml. of the unknown is added 3.0 ml. of an acid reagent prepared by adding 0.5 ml. of a 10 % $FeCl_3 \cdot 6H_2O$ solution to 100 ml. of concentrated HCl. This is followed by 0.2 ml. of a freshly prepared 6 % solution of orcinol in 95 % alcohol. This solution is heated in a boiling water bath for 3 minutes for qualitative or for 20 minutes for quantitative determination of pentoses. Under these conditions, both ribose and ribose-3-P are completely developed in 20 minutes. Maximum absorption is at 665 mμ.

In all modifications of the Bial reaction, extreme care must be taken to exclude the presence of sugars other than pentoses before the relatively simple uncompensated procedures described above can be used as a quantitative measure of pentose concentration. Brown[54] and Drury[55]

[53] W. Mejbaum, *Z. physiol. Chem.* **258,** 117 (1939).
[54] A. H. Brown, *Arch. Biochem.* **11,** 269 (1946).
[55] H. F. Drury, *Arch. Biochem.* **19,** 455 (1948).

have described methods to be employed when glucose is present in the unknown solution. A simple alternative procedure[56] is the determination of a wavelength at which the absorption due to glucose is the same as that at 670 mμ. On a known glucose sample this is usually found at about 580 mμ. The difference $D_{670} - D_{580}$ is then a true measure of the pentose present. Since this difference is only about 25% lower than the reading at 670 mμ, this does not involve a large decrease in the sensitivity of the reaction. In the presence of even small amounts of fructose or tetrose, the determination of pentose by the above methods becomes impossible.

Sedoheptulose was shown by Horecker and Smyrniotis[57] to give a brown color in the orcinol test with a broad maximum around 580 mμ which was satisfactory for quantitative measurements. All eight possible heptuloses give the same maximum although with varying sensitivity.[58] Aldoheptoses, on the other hand, show widely differing maxima. α-Gluco-, manno-, and guloheptose give a green color with a peak at 665 mμ.[20] Ribulose forms a second absorption band at 540 mμ which essentially disappears when the ribulose is phosphorylated.[5,14] Other interfering compounds of importance in biological preparations are 2-deoxyribose, DNA, methylpentoses, and hexuronic acids, all of which absorb in the region of 670 mμ.[59]

A simple and convenient method for the determination of free or phosphorylated pentoses in the presence of adenosine, AMP, ADP, or ATP consists in a preliminary alkaline hydrolysis.[60] When the solution is made 0.1 N with NaOH and heated to 100° for 5 minutes, the free and phosphorylated pentoses are completely destroyed but the purine bound sugars are unaffected. Therefore, by determining the orcinol value before and after alkali treatment, both constituents of the mixture may be calculated. The orcinol reaction may also be used to advantage for a study of the coenzymes found in tissues. After a 20-minute heating (Dische modification) the extinction coefficient for ATP, DPN, TPN, and CoA is about the same, whereas after 3 minutes the adenylic acid moiety of DPN and TPN reacts only one-half as strongly as that of ATP.[56]

[56] Z. Dische, G. Ehrlich, C. Munoz, and L. Von Sallmann, *Am. J. Ophthalmol.* **36,** 54 (1953).

[57] B. L. Horecker and P. Z. Smyrniotis, *J. Am. Chem. Soc.* **74,** 2123 (1952).

[58] The heptuloses were generously provided by Drs. J. W. Pratt and N. K. Richtmyer of the National Institutes of Health.

[59] Z. Dische, *in* "The Nucleic Acids" (Chargaff and Davidson, eds.), Academic Press, New York, 1954. The reader is referred to this paper for a detailed study of the methods employed in nucleic acid determination and a critical analysis of the reaction mechanisms involved. The author is grateful to Dr. Dische for the opportunity to examine the article while still in manuscript form.

[60] Z. Dische, *Naturwissenschaften* **26,** 252 (1938).

compounds, the special techniques will be treated separately in the appropriate sections.

IV. Determination of Tetroses

There is at present no colorimetric procedure available for the determination of tetroses. Since these compounds are unable to form furfural or its derivatives, which are known to be the active intermediaries in color production, their lack of reactivity is understandable. However, they will interfere in certain assays. Thus, the determination of pentoses by orcinol is invalidated in their presence. Both erythrose and erythrulose give the nonspecific maximum at 430 mμ in this reaction with appreciable absorption at 670 mμ.[15] Similarly, the measurement of trioses by cysteine-carbazole is complicated when they are present. In general, the lack of a peak at 430 mμ in the orcinol reaction is presumptive evidence for their absence.

V. Determination of Uronic Acids

The quantitative measurement of uronic acids may be carried out with great precision by several of the modifications[62,63] of the procedure of Lefèvre and Tollens.[64] Here the CO_2 liberated from the hexuronic acid is trapped and determined gravimetrically. This is the method of choice where milligram quantities of the pure uronide or its polymers are available. It is, however, of little value for microdeterminations in the presence of other sugars or proteins. For this purpose two colorimetric procedures are available and are presented here.

Naphthoresorcinol Method

The use of naphthoresorcinol was proposed by Tollens in 1907 for the specific colorimetric determination of glucuronic acid in urine.[65] Until recently, it has been the only procedure available and has been extensively studied in an attempt to apply it quantitatively to biological materials. A later modification is reported here.[66]

Reagents. The naphthoresorcinol is prepared as a 10% solution in 95% alcohol. The reagent is centrifuged or filtered; a clear, colorless solution is essential. It is stable for several days when stored in a cool dark place. For preparation of the HCl reagent, concentrated HCl is diluted so as to contain 19 g. per 100 ml.

Procedure. Five milliliters of a solution containing 40 to 200 γ of uronic acid are pipetted into a 50-ml. glass-stoppered cylinder, 0.2 ml. of the

[62] B. Burkhart, L. Baur, and K. P. Link, *J. Biol. Chem.* **104**, 171 (1934).

[63] K. Freudenberg, H. Gudjons, and G. Dumpert, *Ber.* **74**, 245 (1941).

[64] K. U. Lefèvre and B. Tollens, *Ber.* **40**, 4513 (1907).

[65] B. Tollens, *Ber.* **41**, 1788 (1908).

[66] W. Deichmann, *J. Lab. Clin. Med.* **28**, 770 (1943).

19% HCl is added, and the mixture placed in a water bath at 75° for 45 minutes. Five milliliters of concentrated HCl and 1 ml. of naphthoresorcinol are added, and the cylinder is allowed to stand for 90 minutes in a water bath at 50°. After cooling, the solution is extracted with 10 ml. of ether. When the two layers have separated, sufficient ether is added to bring the ether volume to exactly 15 ml., and the solution is mixed by gentle swirling of the cylinder. The purple color of the ethereal layer is stable for about 1 hour. Maximum absorption is at 570 mμ.

Comments. This procedure was especially designed for the analysis of glucuronides in urine, and the modifications are well adapted to reduce most of the naturally occurring interfering substances to a minimum. The quantitative recovery of glucuronic acid or menthol and borneol glucuronides was found to be uninfluenced by appreciable concentrations of hexoses, pentoses, proteins, and ascorbic acid. This is largely due to the lower temperature employed (most modifications heat to 100° for 15 to 30 minutes)[67] and the great excess of naphthoresorcinol present. In those cases where impurities cause difficulty with the assay, treatment with lead acetate and subsequent recovery of the precipitated hexuronic acids is frequently helpful. Under these conditions, care must be taken, since an excess of the lead salt leads to low recoveries of the uronic acid. Mucic acid also gives the reaction, but the amount of this compound usually present in urine will give no trouble.

The analysis of the polyuronides is less successful. Jorpes and Bergström[68] found that heparin fails to react at all with this reagent. Meyer *et al.*[69] state that none of the modifications of the naphthoresorcinol test gives accurate values for uronic acids in the native fluids with the possible exception of urine and suggest that the use of this reagent be limited to the analysis of pure uronides. For the determination of polyuronides, these authors recommend methanolysis rather than hydrolysis, since the latter tends to decarboxylate the uronic acid. In addition, they suggest the use of benzol rather than ether as the solvent of choice for the extraction of the colored reaction product.

In short, this procedure is satisfactory for the analysis of glucuronic acid and hexuronides in urine, but for more complex mixtures of polyuronides the carbazole reaction is preferable.

Carbazole Reaction for Uronic Acids

The use of carbazole for the determination of uronic acids was described by Dische in 1947.[70] Subsequently, a modification of this reaction

[67] G. B. Maugham, K. A. Evelyn, and J. S. L. Browne, *J. Biol. Chem.* **126,** 567 (1938).
[68] E. Jorpes and S. Bergström, *J. Biol. Chem.* **118,** 447 (1937).
[69] K. Meyer, H. S. Bloch, and E. Chaffee, *Federation Proc.* **1,** 125 (1942).
[70] Z. Dische, *J. Biol. Chem.* **167,** 189 (1947).

was published in which the concentration of acid, the temperature, and the time of heating were altered.[71] Under these conditions, the ratio of color development of galacturonic/glucuronic acid is 30:1, although the over-all sensitivity is appreciably less. However, by determining the ratio of color produced in the two reactions, a constant is obtained which is characteristic for many of the known polyuronides and which is of considerable value in the identification of these compounds. In addition to the above, reactions with thioglycolic acid[72] and with cysteine[73] have been reported which seem specific for glucuronic and galacturonic acid, respectively, but are less sensitive.

Reagents. The carbazole solution is prepared as a 0.1% solution in absolute alcohol and is stable indefinitely in the cold. Commercial carbazole is readily purified by sublimation.

Procedure. To 1 ml. of the solution containing 5 to 100 γ of uronic acid, 6 ml. of concentrated H_2SO_4 is added, and the tube is shaken while immersed in tap water. The mixture is then heated for 20 minutes in a vigorously boiling water bath, cooled, and 0.2 ml. of the carbazole added. After a few minutes, a pink color appears with a maximum at 530 mμ. The tubes are allowed to stand at room temperature for 2 hours before reading.

Comments. In contrast to the naphthoresorcinol test which is essentially a group reaction for hydroxycarbonyl acids, the carbazole reaction is far more specific. Thus, ascorbic, mucic and hydroxypyruvic acid, all of which are positive with naphthoresorcinol, fail to react with carbazole. Pentoses and hexoses, in high concentrations, will interfere with the assay by forming a yellow and red-brown color, respectively. For qualitative purposes, these sugars may readily be distinguished from the uronic acids by their characteristic behavior on dilution with water. The addition of 3.8 ml. of water to the reaction mixture causes an intense violet color to appear which grows rapidly in intensity, whereas with uronic acids the solution becomes colorless in a few minutes.

For quantitative determination of uronic acid in the presence of appreciable amounts of true sugars, it is necessary to know the nature and the approximate concentration of the latter. Since the intensity of color development is both additive and proportional, the absorption due to the sugars can readily be deducted to ascertain the value contributed by the uronic acid. For this purpose, it is essential to run a sugar stand-

[71] Z. Dische, *J. Biol. Chem.* **183**, 489 (1950).

[72] Z. Dische, *J. Biol. Chem.* **171**, 725 (1947).

[73] Z. Dische, *Arch. Biochem.* **16**, 409 (1948). It should be pointed out that there is a confusing misprint in this paper. In determining the absorption increment between two wavelengths as a test for galacturonic acid, the expression $E_{5400} - E_{4800}$ is given. It should read $E_{5400} - E_{6000}$.

ard with every determination of uronic acid in body fluids and tissue extracts.

The presence of serum proteins will not usually interfere with the determination, since they exert a negligible effect even when present in a 20-fold excess. Should their concentration rise to 0.1%, however, the color of the uronic acid is depressed about 20%. Sulfhydryl compounds, on the other hand, will interfere and must be removed from the assay sample. This can be done by making the solution alkaline with solid Na_2CO_3 and allowing the mixture to stand at room temperature for 30 minutes.

The carbazole reaction cannot be used for an accurate determination of the absolute values of the hexuronic acid content of polysaccharides, since the chromogenic values of the several polyuronides deviate from the values obtained from an equivalent amount of the free hexuronic acids. Hyaluronic and pectic acid as well as type I and III of the pneumococcus polysaccharides give values 5 to 15% higher and chondroitin sulfuric acid about 10% lower than the corresponding free acids. Heparin and alginic acid deviate markedly from the majority of the polyuronides. The former yields about 60% more and the latter about 80% less color than would be expected on the basis of their uronic acid content. Free galacturonic acid is approximately 20% less sensitive than free glucuronic acid. However, in spite of these drawbacks, the reaction has proved useful as a guide in the isolation of these compounds and for comparative studies on the content of polysaccharides in tissues. Furthermore, in cases where the polyuronides are merely suspected, a negative carbazole reaction is strong evidence for their absence, since there is no known member of this group which fails to give the characteristic color.

VI. Determination of Hexosamines

In addition to the two reactions described here for the determination of hexosamines, a new method has recently been reported.[74] This procedure utilizes the ability of the amino sugars to liberate the whole of their nitrogen content when heated for a few minutes at 100° with 1 N alkali. The nitrogen is then determined by nesslerization. The assay may be used to estimate 50 to 500 γ of material.

Method of Elson and Morgan

The earliest practical method for the determination of hexosamine was developed by Elson and Morgan in 1933.[75] Since that time numerous modifications have been designed to increase the sensitivity and to reduce the interference caused by true sugars and other naturally occurring

[74] M. V. Tracey, *Biochem. J.* **52**, 265 (1952).
[75] L. A. Elson and W. T. J. Morgan, *Biochem. J.* **27**, 1824 (1933).

the determination of methylpentoses in the presence of an excess of other types of carbohydrates.

Reagents. The H_2SO_4 reagent is prepared by adding 6 vol. of cold concentrated H_2SO_4 to 1 vol. of cold water. Cysteine hydrochloride is freshly prepared as a 3% aqueous solution.

Procedure. One milliliter of a solution containing 2 to 10 γ of methylpentose is placed in a widemouthed test tube and chilled in an ice bath. Then 4.5 ml. of the acid reagent is added, and the mixture shaken while still in the ice bath. The mixture is warmed to room temperature for a few minutes, placed in an actively boiling water bath for either 3 or 10 minutes, and then cooled in tap water. To the cool solution, 0.1 ml. of the cysteine is added with shaking. A very faint yellowish color appears. The difference in the optical density at 396 and 427 mμ ($\Delta D_{396-427}$) is determined 2 hours after addition of the cysteine.

Comments. Since the color developed in the cysteine reaction depends on the time of heating, the two reactions may be described as CyR3 and CyR10, respectively. The latter reaction is more specific for methylpentoses and therefore used in quantitative measurements, although the former possesses the advantage that hexoses can be determined simultaneously.

Methylpentoses absorb in the region of 360 to 430 mμ with a sharp maximum at 400 mμ. Hexoses show a maximum at 415 mμ but, owing to their distribution around this peak, their absorption at 427 mμ is equal to that at 396 mμ. The equivalence of these readings for hexoses is best established with each determination on a known hexose, since minor variations in carrying out the assay will alter the latter value slightly. At 396 mμ, methylpentoses exhibit almost maximum absorption, whereas their reading at 427 mμ is very small. In the range 396 to 427 mμ, pentoses, 2-deoxypentoses, and hexuronic acids give essentially horizontal curves; the value of $\Delta D_{396-427}$ is practically zero for all the above classes of sugars, whereas it is highly positive for methylpentoses. The determination of these sugars is unaffected by a fourfold excess of hexoses and an eightfold excess of pentoses. The appearance of a pink color indicates the presence of pentoses in appreciable amounts, and the contribution of this constituent to the delta reading should be determined on the basis of the orcinol or basic cysteine reaction.

$\Delta D_{396-427}$ is strictly proportional to the concentration of methylpentose. Since the assay is not affected by a considerable excess of hexoses or pentoses, it is possible to determine the methylpentose content of polysaccharides by comparing the absorption with a standard solution of fucose or rhamnose.

The presence of proteins in moderate amounts does not seriously interfere with the assay. A concentration of 0.1% serum albumin produces only a 3% depression in the measurement of a 0.001% solution of fucose. However, the use of an internal standard is indicated in all assays where interference by unknown tissue contaminants may play a role. Furthermore, in this as in all color reactions involving concentrated H_2SO_4, the unknown samples should be run with and without cysteine to blank out nonspecific reactions.[108]

The investigation of the specificity of the cysteine reaction for methylpentoses has been confined to the examination of known substances occurring in nature. Since the possibility exists that rarely found branched or anhydro sugars may react in a similar fashion, the value of this test for unknown sugars lies chiefly in its ability to eliminate their presence. However, their tentative identification can be confirmed if the unknown is found to react characteristically in a second and independent assay. Such an assay, utilizing the unique sensitivity of the reaction product to dilution with water, has been described.[109]

[108] For details, see introductory remarks.
[109] Z. Dische and L. B. Shettles, *J. Biol. Chem.* **192**, 579 (1951).

[13] The Orcinol Reaction for Mixtures of Pentose and Heptulose

By B. L. HORECKER

Principle. The general applications of the orcinol reaction are discussed elsewhere.[1] With this reagent pentoses yield a product with absorption maximum at 670 mμ; with heptuloses the absorption maximum occurs at about 580 mμ. The concentration of these substances in a mixture can be calculated from density measurements at 670 mμ and 580 mμ.

Reagents

Ferric chloride-hydrochloric acid. One-tenth per cent $FeCl_3 \cdot 6H_2O$ in concentrated hydrochloric acid (38%).

Orcinol. One hundred milligrams per milliliter in absolute alcohol. Commercial orcinol was recrystallized from a mixture consisting of 2 parts of chloroform and 3 parts of benzene.

Arabinose standard. The stock solution is made up to contain 3 mg./ml. (20 μM./ml.). This is diluted 1:20 for use in the assay.

[1] See Vol. III [12].

has been used to follow reactions involving free glucose. Slein *et al.*[10] have applied the method to determine the affinity for glucose for various hexokinases; it has also been of value in following the action of amyloglucosidases.[11] It would appear that the spectrophotometric method is more sensitive for the microassay of glucose than the manometric technique described above for the glucose oxidase. A limitation of the method is that the yeast hexokinase is not specific, as it will promote the phosphorylation of fructose and other sugars as well as glucose. The affinity of the enzyme for glucose is quite high, however, whereas that for fructose quite low. If the enzyme preparations are free of the hexose isomerases, then the system would be specific for glucose, since the Zwischenferment will react only with glucose-6-phosphate.

[10] M. W. Slein, G. T. Cori, and C. F. Cori, *J. Biol. Chem.* **186**, 763 (1950).
[11] G. T. Cori, Vol. III [8].

[15] Sugar Phosphates, Paper and Column Chromatography[1]

By A. A. BENSON

The impetus given by the application of paper and ion exchange chromatography to the identification and purification of phosphorylated metabolic intermediates can scarcely be overestimated. Standards of purity have been re-evaluated, and the identification of unanticipated sugar phosphates has exposed new pathways for biological synthesis and degradation.

These two complementary techniques allow rapid separation of a great variety of phosphate esters on practically any scale. Trace amounts of compounds of high specific radioactivity may be separated as well as multigram amounts of nucleic acid degradation products. This new range of sensitivity demands the establishment of new criteria of purity, a factor which has often received too little attention in commercially available products. We may expect that sugar phosphates of reliable purity will become increasingly available as commercial products.

Both paper and column chromatography of sugar phosphates are natural extensions of the original applications of these techniques. In each case the initial difficulties have been overcome, and their use is expanding rapidly as their value is comprehended.

[1] The work described in this paper was sponsored by the U.S. Atomic Energy Commission.

I. Filter Paper Chromatography

Chromatography on pure paper may be considered an establishment of a dynamic equilibrium or partition between solution (adsorption) in a fixed semi-aqueous layer upon the cellulose fiber and solution in the moving organic phase. Although the role of adsorption is certainly greater for sugar phosphates than for the free sugars, the general principles of relative solubilities may be applied in predicting their movement in a particular solvent system. The phosphate group determines much of the solubility properties of the compound. This requires that the chromatographic method be particularly effective in order to separate compounds by virtue of the less-polar carbohydrate moiety.

The hydrophilic nature of the phosphate esters results in relatively low R_f values with solvents normally used for amino acids and sugars. A number of stronger solvents have been developed, either by increasing water content, or by increasing acidity. The picric acid ($pK = 0.38$) solvent developed by Hanes and Isherwood[2] successfully represses the dissociation of the phosphate group ($pK = 2.12$) to such an extent that it offers almost no retarding influence on the movement of the sugar.

Limitations of the Method. Salt toleration is a serious handicap in the application of the method to biochemical preparations, which are rarely salt-free. The paper may not be overloaded with any solute, salts, fats, or phosphates without serious distortion. Salts may be avoided in the preparation of the sample, or preliminary barium precipitation or anion resin adsorption may be used to effect suitable purification. Barium salts are readily decationized with a cation exchange resin such as 400-mesh Dowex 50. Some degree of salt toleration has been obtained by using solvents saturated with the salt.

Readily hydrolyzable esters are labile in many of the acidic or basic solvents commonly employed. Reduced temperature or faster development with a less acidic solvent should be considered.

Paper Purification. Calcium and magnesium salts in paper cause irreversible adsorption of phosphates which cannot be tolerated in their chromatography. Whatman paper Nos. 1, 4, and 52 and Schleicher and Schuell No. 589 have been used with suitable prewashing.

Cations have been removed by washing with N hydrochloric acid and 0.02% aqueous Versene (Versenes, Inc., formerly Bersworth Chemical Co., Framingham, Massachusetts).[3] Oxalic acid solutions (1%) have been successful in reducing phosphate adsorption and yield an "acidic" paper

[2] C. S. Hanes and F. A. Isherwood, *Nature* **164,** 1107 (1949).

[3] D. C. Mortimer, *Can. J. Chem.* **30,** 653 (1952).

Chromatographic Solvents

The solvent chosen should be determined by the nature of the mixture to be separated and the impurities present. There appears to be no ideal formulation for universal separation. Both acid and basic solvents have been successfully used. There seems to be no particular advantage to those miscible with water or those of the single-phase type. All of them require an aqueous component, except a novel formamide solvent developed by Mortimer.[3]

Pre-equilibration of the paper in the solvent vapor is advisable in most instances and generally requires 2 to 4 hours. The advisability of pre-equilibration is dependent on solvent composition.

Ascending Chromatography of Phosphate Esters.[4] The concentrated sample of decationized (Dowex 50) phosphates is dried in a 0.5-cm.-diameter origin on a 28-cm. square of Schleicher and Schuell No. 589 blue ribbon filter paper. The paper is then stapled into a cylindrical form, with contact of the adjacent edges avoided. A cylindrical jar, 15 cm. in diameter and 30 cm. tall, containing 100 ml. of the acid solvent and covered by a glass plate serves as the chamber. The cylindrical paper with a stainless wire through it is temporarily held to the glass cover by a strong permanent magnet during the 2-hour equilibration period. The magnet is then removed, and the solvent development begun. After this the paper is dried and re-equilibrated and developed in the alkaline solvent.

Acid Solvent. Methanol 80 vol., formic acid (88%) 15 vol., and water 5 vol. Develops 28 cm. in 6 to 6.5 hours at 2°.

Basic Solvent. Methanol 60 vol., ammonium hydroxide (specific gravity 0.9015) 10 vol., and water 30 vol. Develops 28 cm. in 12 to 15 hours at 2°.

Separation of the glycolytic intermediates is shown in Fig. 2. This technique has found application in several laboratories. It was used for separation of phosphorylated compounds formed by incorporation of radiophosphate by pea seed meal and extracts after preliminary barium precipitation.[8]

Picric Acid Solvent (Modified Hanes-Isherwood Solvent). Picric acid is sufficiently strong to repress the ionization of phosphate groups. The esters then have good R_f values and separate well in a solvent containing less water than required to move ionic phosphates. Disadvantages of the picric acid solvent are that it is sensitive to the degree of hydration of the paper, the picric acid is nonvolatile and must be removed before further chemical operations, the yellow picric acid front moves slower

[8] B. Axelrod, R. S. Bandurski, and P. Saltman, *Federation Proc.* **10**, 158 (1951).

than the butanol front, and monophosphates move as rapidly as the diphosphorylated compounds (i.e., F-6-P vs. FDP).

Procedure. A solution of 2 g. of picric acid in 80 ml. of *t*-butanol and 20 g. of water is used for descending chromatography on oxalic acid-washed Whatman No. 4 paper without pre-equilibration. Development of 55 cm. requires 20 hours whereupon inorganic phosphate has moved 35 cm.

Phenol Solvent. Phenol is distilled at atmospheric pressure into tared bottles, and 0.39 part of decationized water is added to give a single

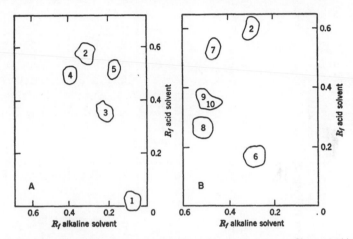

Fig. 2. Two-dimensional chromatograms of glycolytic intermediates.[4] (A) Barium-insoluble fraction: 1, ATP; 2, orthophosphate; 3, FDP; 4, 3-PGA; 5, 2-PGA. (B) Barium-soluble fraction: 2, orthophosphate; 6, adenosine-3-P; 7, P-pyruvate; 8, G-1-P; 9, F-6-P; 10, G-6-P.

phase. The solvent is stored in the dark at 5° to prevent oxidation. The impurities in the phenol distort the chromatography of hydroxy acids and sugar phosphates. Although the R_f values of sugar phosphates in this solvent are low, the solvent may be allowed to flow past the serrated edge of the paper to achieve improved separations. Organic dyes (croceine scarlet, ponceau 4R, tropeolin 000) placed near the origin may be used to observe the rate of solvent flow after the paper is completely wet.

R_f Values and Position Constants of Phosphate Esters

The measurement and comparison of absolute R_f values is hardly practical for the identification of compounds in complex systems. Variations in paper hydration, atmosphere saturation, and temperature all affect the absolute R_f values obtained. Only rarely, however, do they affect the relative positions of phosphorylated compounds. It is always better to

TABLE I

R_f Values[a] and Position Constants[b] for Phosphate Esters in Chromatographic Solvents[c]

Solvent	Ethyl acetate 3, Acetic acid 3, Water 1	Methyl Cellosolve 7, Methyl ethyl ketone 2, 3 N NH_4OH 3	Ethyl acetate 1, Formamide 2, Pyridine 1	t-Butanol 80, Picric acid 2 g., Water 20	Isopropyl ether 90, 90% Formic acid 60	Phenol 72 g., Water 28	Butanol 100, Propionic acid 50, Water 70	Methanol 80, 88% Formic acid 15, Water 5	Methanol 60, 28% NH_4OH 10, H_2O 30
Temperature	4°	26°	26°	22°	20°	22°	22°	2°	2°
Paper	Whatman No. 1	Whatman No. 1	Whatman No. 1	Whatman No. 1	Whatman No. 1	Whatman No. 4	Whatman No. 1	Schleicher and Schuell No. 589	Schleicher and Schuell No. 589
Reference	Mortimer[d]	Mortimer[d]	Mortimer[d]	Wilson[e]	Hanes and Isherwood[f]			Bandurski and Axelrod[g]	Bandurski and Axelrod[g]
Orthophosphate	*33* 100	*21* 100	*50* 100	100	100	*22* 100	100	*63* 100	*28* 100
Phosphoglycolaldehyde	—	—	—	—	—	~170	~73	—	—
Phosphoglycolic acid	—	*39* 192	*54* 114	106	—	*23* 102	75	—	—
Glycerol-1-phosphate	*26* 79	*19* 90	—	—	—	—	—	*46* 13	*18* 64
Glyceraldehyde-3-phosphate	*7* 22	—	—	—	—	—	—	—	—
Dihydroxyacetone phosphate	—	—	—	—	—	170	59	—	—
2-Phosphoglyceric acid	*27* 81	*41* 200	*23* 47	—	—	—	—	—	—
3-Phosphoglyceric acid	*23* 71	*22* 116	*28* 57	85	59	*22* 100	65	*50* 79	*35* 125
2,3-Diphosphoglyceric acid	*11* 35	*7* 36	*15* 30	—	—	—	—	—	—
Phosphopyruvic acid	—	—	—	120	—	*26* 110	92	*52* 82	*46* 165
Phosphoerythronic acid	—	—	—	—	—	74	51	—	—
Ribose-1-phosphate	*15* 45	*40* 197	*50* 110	—	—	—	—	—	—

	C1	C2	C3	C4	C5	C6	C7	C8	C9	C10	C11	C12	C13	C14	C15
Ribose-5-phosphate	*14*	—	—	—	—	—	—	—	—	139	49	—	—	—	—
Ribulose-5-phosphate	*12*	—	—	—	—	—	—	—	—	147	53	—	—	—	—
Ribulose-1,5-diphosphate	—	—	—	—	—	—	58	18	*8*	26	22	—	—	—	—
Glucose-1-phosphate	—	37	*36*	170	89	*44*	—	18	*27*	—	—	*27*	43	*60*	215
Glucose-6-phosphate	—	29	*29*	140	100	*50*	40	—	*26*	113	40	*38*	60	*48*	170
Uridine diphosphate glucose	—	—	—	—	—	—	dec.	—	*26*	111	15	—	—	—	—
Glucose-1,2-cyclic phosphate	—	—	—	—	—	—	54	—	—	170	49	—	—	—	—
Glucose-1,6-diphosphate	—	—	—	—	—	—	—	—	—	26	22	—	—	—	—
Fructose-1-phosphate	—	—	—	—	—	—	—	—	—	135	46	—	—	—	—
Fructose-6-phosphate	*17*	48	*36*	171	108	*54*	61	28	*29*	125	46	*34*	54	*44*	156
Fructose-1,6-diphosphate	*8*	25	*8*	37	26	*13*	—	15	—	26	22	*40*	63	*24*	86
Galactose-1,2-cyclic phosphate	—	—	—	—	—	—	45	—	—	170	49	—	—	—	—
Mannose-6-phosphate	—	—	—	—	—	—	52	—	*29*	125	46	—	—	—	—
Mannoheptulose phosphate	—	—	—	—	—	—	—	—	—	113	40	—	—	—	—
Sedoheptulose-7-phosphate	—	—	—	—	—	—	52	—	*27*	113	40	—	—	—	—
Sedoheptulose diphosphate	—	—	—	—	—	—	—	—	—	26	22	—	—	—	—
Sucrose phosphate	—	—	—	—	—	—	32	—	—	113	40	—	—	—	—

a Italicized numbers give R_f values of per cent of solvent travel.

b P constants relative to orthophosphate.

c A. A. Benson, in "Modern Methods of Plant Analysis" (M. V. Tracey and K. Paech, eds.), Springer Verlag, Berlin, 1954.

d D. C. Mortimer, Can. J. Chem. 30, 653 (1952).

e A. T. Wilson, Thesis, University of California, 1954.

f C. S. Hanes and F. A. Isherwood, Nature 164, 1107 (1949).

g R. S. Bandurski and B. Axelrod, J. Biol. Chem. 193, 405 (1952).

compare the position of an unknown compound to that of a similar authentic material rather than to depend on R_f measurements. For this reason, Mortimer[3] has used the term "position constant" in which the position of a substance is defined relative to that of inorganic phosphate. It might be better to compare even more similar compounds, however, in order to establish more reliable comparisons.

Table I includes published data for position constants of a variety of compounds and conditions. It must be borne in mind that these values are useful but not unequivocal in identifying an unknown. Successful cochromatography does not necessarily demonstrate identity of the two substances. Chemical evidence is always invaluable in identifying an unknown. One or more observations of chemical transformation is more likely to provide conclusive identification than would chromatography in several solvent systems. When the chemical properties of the suspected compound are known, it is often possible to use a chromatographic identification for the product of the chemical transformation.

Chromatographic Separation of 2- and 3-PGA on Molybdate Paper (R. W. Cowgill)

Several methods for separation of 2- and 3-PGA have been reported in the literature and found by other workers to have limitations. The method of Cowgill (unpublished) is based on the well-known interaction of 3-PGA and molybdate and has proved very successful.

Procedure. The mixture of acids (25 to 50 γ each) is chromatographed on Whatman No. 52 or Schleicher and Schuell 602 ED paper which has previously been wet with 0.5% sodium molybdate solution and dried. Whatman No. 1 paper may be used, although the spots are not so well formed as with the denser papers. The solvent is 1 part 88 to 90% formic acid, 29 parts water, and 70 parts 95% ethanol freshly prepared before use. After 20 hours of descending development, the paper is dried. 2-PGA is found to have traveled 2.5 times as far as the more strongly complexed 3-PGA and 0.48 times as far as inorganic phosphate.

Application of Borate Complex Formation. Certain mixtures of sugar phosphates are readily separable by virtue of their ability to complex with borate ion (cf. use of borate in column chromatography). The highly dissociated complexes, being more hydrophilic, have lower R_f values. Separation of the ribose phosphates in borate solvent has been reported by Cohen and Scott.[9] Extension of this method with dilute borate buffers should be helpful in separating mixtures where differences in borate complexing is indicated.

[9] S. S. Cohen and D. B. McNair Scott, *Science.* 111, 543 (1950).

Method. A solution of 0.64% boric acid in 80% ethanol is used for descending development on Sleicher and Schuell No. 597 paper. Arabinose-5-P (R_f 0.25), xylose-5-P (R_f 0.25), ribose-3-P (R_f 0.19), and glyceraldehyde-3-P (R_f 0.19) migrate readily; ribose-5-P and hexose phosphates fail to move.

II. Exchange Resin Chromatography of Sugar Phosphates

Strong base anion exchange resins have been effectively used in separating large quantities of phosphate esters, with good recovery and little hydrolysis of acid- or alkali-labile compounds. Separation is effected by taking advantage of differences in affinity of the anionic phosphate and carboxyl groups for the resin in which the following equilibrium is established:

$$[\text{Rosin } \overset{+}{\text{N}}\text{R}_3\text{Cl}^-] + \text{sugar-OPO}_0\text{H}^- \rightleftarrows [\text{Resin-}\overset{+}{\text{N}}\text{R}_3 \ \overset{-}{\text{H}}\text{O}_3\text{PO-sugar}] + \text{Cl}^-$$

The equilibrium is, of course, dependent on pH of the medium and acid strength of the weak acid involved.

One may compare the affinities of a variety of compounds for a unit mass of resin and predict relative elution rates. These can be determined by analysis of an equilibrated mixture of resin and adsorbate[10] where

$$\frac{\text{Fraction of solute on resin}}{\text{Fraction of solute in solution}} = \text{Constant}$$

The mobility of the sugar phosphate differs from that of chloride or other anion, and as the eluting solution flows over the column the phosphate ester will be steadily displaced. Our understanding of the affinities of sugar phosphates for resins has been clarified by Khym *et al.*[11] to a point where it may be possible to predict elution characteristics of known compounds. In fact, exchange resin elution characteristics have confirmed structure assignments for unknown esters.

Factors Affecting Acidity of Phosphate Groups. The sugar phosphates vary in affinity for anion resins for several reasons. The differences in acidity of sugar monophosphates arises by virtue of possible internal hydrogen bonding of the phosphate group with adjacent hydroxyl groups.[11,12] When there is a possibility of hydrogen bonding between a phosphate —OH group and the ring oxygen (as in G-1-P or R-5-P and R-1-P), the degree of dissociation of the phosphate hydrogen will be diminished, especially when a six-membered ring rather than a seven-

[10] R. M. Wheaton and W. C. Bauman, *Ind. Eng. Chem.* **43**, 1088 (1951).
[11] J. X. Khym, D. G. Doherty, and W. E. Cohn, *J. Am. Chem. Soc.* **76**, 5523 (1954).
[12] W. D. Kumler and J. J. Eiler, *J. Am. Chem. Soc.* **65**, 2355 (1943).

membered ring results. Those structures in which the phosphate oxygen can form a hydrogen bond with an adjacent —OH group lead to increased dissociation. With two adjacent —OH groups the acid-strengthening

"Acid-weakening" H-bonding   "Acid-strengthening" H-bonding

effect is still more pronounced (i.e., ribopyranose-3-P). These predictions were borne out by ion exchange separations of the ribose phosphates.[11]

The diphosphates are more strongly adsorbed on the anion resin. In order to be moved on the column, both anionic groups must be displaced simultaneously by the eluting agent. Therefore, a stronger or more concentrated eluting agent is required to move the diphosphates. Phosphorylated carboxylic acids likewise are difficultly eluted from the resin at pH values where the carboxyl group is dissociated. They would be only difficultly separable from the diphosphates were it not possible to repress the dissociation of the carboxyl group by including a stronger acid in the eluting solution. In acid solution, then, PGA and 6-P-G behave as if they were simple monophosphate esters.

Borate Complexing. The formation of borate complexes (reviewed by Böeseken[13]) by *cis*-hydroxyl groups was applied successfully by Khym and Cohn[14] for separation of monophosphate esters. Borate ion exists only in alkaline solutions where phosphate esters are completely dissociated and hence more strongly bound by the anion resin. Elution with borate ion alone involves excessive concentrations of eluant. Although the borate form of an anion resin can be used,[15] the interference of the excess borate presents an unnecessary difficulty when phosphates are to be recovered. Low concentrations (10^{-2} to 10^{-5} M) of borate are effective in complex formation[14] in sodium sulfate or ammonium chloride-ammonium hydroxide eluting solutions and hence affect the ion exchange affinities of the sugar phosphates. Furanose forms like F-6-P and R-5-P, on the other hand, exhibit very strong affinity for the exchanger with but small concentrations of borate ion (0.0005 M). The dilute borate solution also allows a more effective recovery of phosphate ester.[14]

Three generalizations concerning the reactions of borate with sugars

[13] J. Böeseken, *Advances in Carbohydrate Chem.* **4**, 189 (1949).
[14] J. X. Khym and W. E. Cohn, *J. Am. Chem. Soc.* **75**, 1153 (1953).
[15] M. Goodman, A. A. Benson, and M. Calvin, *J. Am. Chem. Soc.* **77**, 4257 (1955).

have been set forth:[11] (1) *cis-α*-glycols are normally strongly complexed; (2) pyranose ring systems are not complexed compared to furanose systems; (3) the stronger the complex, the greater the ionization and affinity for anion exchangers. The efficacy of borate in separating a given group of phosphates may be deduced from these principles and is demonstrated in the accompanying examples. The pyranose structures of ribopyranose-2-P, ribopyranose-3-P, and R-4-P are not rigid-ring forms, and their vicinal hydroxyl groups are spread apart and are less readily complexed.[16,17] The strength of the borate complex is generally a function of the number of possible forms. When a *cis-α*-glycol system occurs only in an α or β form of the sugar, the borate complex is accordingly weaker, i.e., ribose-3-P.

Recovery of Separated Phosphates after Borate Elution. Khym *et al.*[11] removed borate from column eluates in the following manner: The eluate, containing sugar phosphate, sulfate, and borate, is absorbed on a small anion column (IRA-400-acetate, 8 cm.^{2} × 4 cm.) after removal of sodium ion with cation resin and sulfate with excess barium acetate. The IRA-400-acetate column adsorbs sugar phosphates and inorganic phosphate but not borate because at pH 8.0 to 8.5 the phosphates are divalent and strongly bound and the high acetate concentration displaces the small amount of borate. The phosphates are then eluted with 125 ml. of 1 M ammonium acetate, decationized with Dowex 50-H$^+$, and the acetic acid removed by evaporation *in vacuo*.

Separation Methods

Separation of G-1-P, G-6-P, F-6-P, R-5-P, F-1,6-diP, P$_i$, 2-PGA, and the Adenosine Phosphates by Dilute Borate Elution (Khym and Cohn[14]). Separation of the monophosphates in this mixture requires the additional factor of borate complex formation. Adjustment of pH allows separation of the dibasic acids (ADP, 2-PGA) from the monobasic ones.

Ion Exchanger. A 200- to 400-mesh trimethylammonium polystyrene resin (Dowex 1) is freed of fines by decantation of a stirred suspension or by passing a slow stream of water through a vertical column of resin suspension until the finer particles have passed out in the overflow. The resin is slurried into a column 0.86 cm.^{2} × 12 cm. The resin is converted to the chloride form by washing with 1 N HCl and washed thoroughly with water to remove excess acid.

Preparation of Phosphates. Dowex 50 in the acid form is prepared by acidifying a batch in a Büchner funnel with excess 1 N HCl and thor-

---

[16] J. X. Khym and L. P. Zill, *J. Am. Chem. Soc.* **74**, 2090 (1952).
[17] H. T. MacPherson and E. G. V. Percival, *J. Chem. Soc.* **1937**, 1920.

oughly washing for 1 to 2 hours with distilled water. Excess water is removed by suction, and the damp resin may be stored for use. Solutions of phosphate salts are treated with several batches of the acid Dowex 50 which is removed by filtration or centrifugation. The usual capacity of Dowex 50 is 5 to 10 meq./g. of resin, but an excess is advisable.

Procedure. The solution of acidic esters in dilute ammonia (pH 8.5) is passed through the column at a rate of 3.5 ml./min. Free sugars are removed with 100 ml. of 0.001 M ammonium hydroxide. A succession of eluting agents, in the order described by the elution curve of Fig. 3, is passed through the column to selectively desorb the components of the mixture. Assay of the effluent samples is described by the data of Table II.

Separation of the Ribose Phosphates. A classical example of the efficacy of the anion resin in separating similar structures was reported by Khym *et al.*[11] They explain the elution succession of the isomeric ribose monophosphates on the basis of acidity of the phosphate groups in the five positions and the stability of borate complexes of R-1-P and R-5-P and the α and β forms of R-3-P and R-5-P.

Method. A pH 8.5 ammoniacal solution of 3 to 10 mg. each of the ribose esters is absorbed on a 300-mesh Dowex 1 (sulfate) column, 0.86 cm.^{2} $\times$ 12 cm., and washed with water to remove nonanionic materials. The column is eluted successively with the solutions described in Fig. 4 at about 3.5 ml./min., and the eluates are analyzed by the orcinol method.[18]

It is seen that R-2-P and R-4-P are unaffected by borate, and that R-3-P and R-5-P form complexes, with R-5-P the more strongly affected. R-1-P is strongly complexed by borate, but its phosphate acidity is diminished by the hydrogen-bonding effect.

Separation of Polybasic Esters, PGA, P-Glycolic Acid, FDP, and Ribulose Diphosphate.[15] *Principle.* Addition of dilute strong acid to the eluting agent represses dissociation of the carboxylic acids and consequently their affinity for the anion resin.

MATERIALS CHROMATOGRAPHED: 9.5 mg. of barium 3-phosphoglycerate; 10 mg. of barium fructose diphosphate; 250,000 c.p.m. of ribulose diphosphate eluted from a stripe chromatogram of labeled photosynthetic products from *Scenedesmus.*

ELUTING AGENT: 0.15 N NaCl + 0.05 N HCl.

ELUTING RATE: 0.11 ml./mm.

FRACTION VOLUME: 0.4 ml. calculated from elution rate and time interval of sample changer.

[18] E. Volkin and W. E. Cohn, *in* "Methods in Biochemical Analysis" (D. Glick, ed.), Vol. 1, pp. 298–299. Interscience Publishers, New York, 1954.

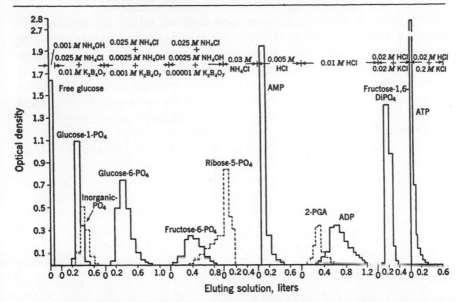

FIG. 3. Ion exchange separation of sugar phosphates, inorganic phosphate, adenosine phosphate, and phosphoglyceric acid in the amounts given in Table II. Exchanger, Dowex 1 (chloride), 300 mesh, 0.86 cm.² × 10 cm. Flow rate 3.5 ml./min.[14]

TABLE II[a]

ANALYTICAL DATA FOR SEPARATION DEMONSTRATED IN FIG. 3

Compound	Assay method	Wavelength, mμ	Approximate amount added,[b] ~mg.	Recovered, %
Glucose	Anthrone	620	5	101
Glucose-1-PO₄	Anthrone	620	10	99
Glucose-6-PO₄	Anthrone	620	10	93
Fructose-6-PO₄	Anthrone	620	5	92
Fructose-1,6-DiPO₄	Anthrone	620	10	95
Inorganic PO₄(K₂HPO₄)	Phosphate	660	2	105
2-PGA	Phosphate	660	4	95
Ribose-5-PO₄	Orcinol	660	5	90
AMP	Ultraviolet absorption	260	8	95
ADP	Ultraviolet absorption	260	5	102
ATP	Ultraviolet absorption	260	6	100

[a] J. X. Khym and W. E. Cohn, *J. Am. Chem. Soc.* **75**, 1153 (1953).

[b] The milligram quantities given for the sugar phosphate represent the free sugar content of these substances. The quantities given for inorganic phosphate and 2-PGA are calculated as total phosphorus present. The amount of each adenosine derivative was calculated from extinction coefficients.

Procedure. A 0.6-cm. diameter × 28-cm. column of 200 to 400 mesh Dowex 2 anion resin (from which fines have been removed by decantation) is prepared by pouring a slurry of the resin in distilled water into the column. After the resin has settled evenly and most of the water has passed through the column, more suspended resin is added. It should be emphasized that the water level should never be allowed to go below the

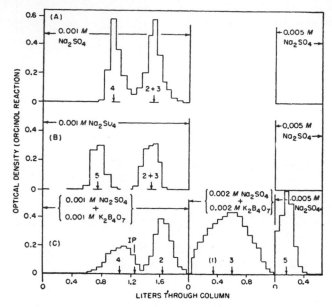

Fig. 4. Ion exchange separation of ribose phosphates in the presence and absence of borate (Khym *et al.*[11]). (*A*) Hydrolyzate of 60 mg. of commercial adenylic acids heated with 6 ml. of Dowex 50-H$^+$ in 6 ml. of H$_2$O for 3 hours at 100° to yield ribose-2-, -3-, and -4-phosphates eluted with 2 l. of 0.001 M Na$_2$SO$_4$ followed by 700 ml. of 0.005 M Na$_2$SO$_4$. (*B*) 8 mg. of ribose-5-phosphate plus 4 mg. each of ribose-2- and -3-phosphates eluted as in *A*. (*C*) 3 to 10 mg. each of ribose-2-, -3-, -4-, and -5-phosphates, eluted successively with 2 l. of 0.001 M Na$_2$SO$_4$ plus 0.001 M K$_2$B$_4$O$_7$, 1.1 l. of 0.002 M Na$_2$SO$_4$ plus 0.002 M K$_2$B$_4$O$_7$, and 0.005 M Na$_2$SO$_4$. (The probable position of the ribose-1-P peak, from a separate experiment, is shown in parentheses.)

resin level. It is not satisfactory to apply suction to the bottom of the column, as uneven packing and air pockets may result.

The column is washed with excess 1 N HCl and then with distilled water until the acidity is no longer detectable.

A solution of the two barium salts and the labeled ribulose diphosphate in 50 ml. of water is passed onto the column and thoroughly washed to remove the cations. The eluting agent is passed through the column from a motor-driven greased syringe through a syringe needle and rubber serum bottle stopper sealing the top of the column.

The results of this separation are described by the curve of Fig. 5A. Radioactivity is determined by preparing "infinitely thin" plates on aluminum disks from each fraction and counting under a thin-window G.-M. tube. Phosphorus analysis of each fraction gives the block curve. It is seen that the added PGA and a certain amount of the radioactivity coincide and that there is a considerable peak of P-glycolic acid (identified by cochromatography on paper) lying between the FDP and PGA peaks. Ribulose diphosphate is more strongly adsorbed on the resin than FDP.

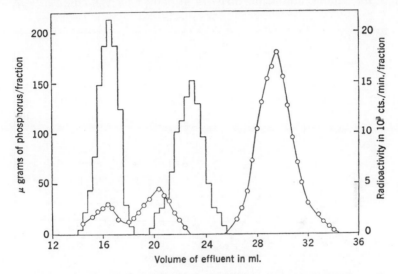

FIG. 5A. Acid elution of phosphoglyceric acid, fructose diphosphate, and radioactive ribulose diphosphate from a Dowex 2 (chloride) column (28 × 0.6 cm.). Smooth curve denotes C^{14}. Block curve represents phosphorus analyses.

Another experiment with no added carrier phosphates gave an almost identical elution curve (Fig. 5B). There is apparently little distortion of the radioactivity elution curve by the added carriers. The quantity of labeled ribulose diphosphate added in this experiment was ca. 0.1 micromole. The amounts of PGA and P-glycolic, therefore, were much less. Even at this low concentration, very little irreversible adsorption is observed.

Isolation of Sedoheptulose Phosphate (Horecker et al. and Klenow[19]). The mixture obtained from the action of transketolase on 703 micromoles of R-5-P containing 235 micromoles of sedoheptulose-7-phosphate and residual pentose phosphates was chromatographed on a Dowex 1 (for-

[19] B. L. Horecker, P. Z. Smyrniotis, and H. Klenow, *J. Biol. Chem.* **205**, 661 (1953); see also Vol. III [30].

mate) column. The elution order—sedoheptulose-7-P, R-5-P, ribulose-5-P—may be explained by the acid-weakening effect of the ring oxygen adjacent to the 7-phosphate group, the same effect in R-5-P, and the acid-strengthening effect of the adjacent —OH group of ribulose-5-P.

Procedure. A 2.5-cm. diameter × 13-cm. column of 300-mesh Dowex 1 (10% cross-linked) is converted to the formate form by passing 2 *M* sodium formate through it until the effluent gives a negligible silver chloride precipitate on addition of silver ion. After the column is washed with water, the equilibration mixture is placed on the column, and the

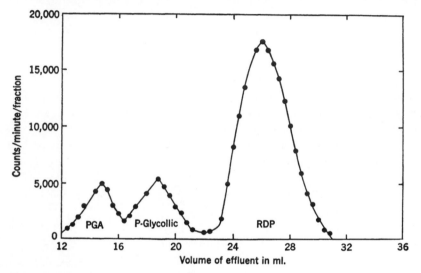

Fig. 5*B*. Acid elution of radioactive ribulose diphosphate from a Dowex 2 chloride column (28 × 0.6 cm.).[15] Eluting agent; 0.15 *N* NaCl + 0.05 *N* HCl; eluting rate 0.11 ml./min.; fraction volume: 0.4 ml.

column is again washed with 50 ml. of water. The mixture is eluted with 0.2 *N* formic acid containing 0.03 *N* sodium formate at the rate of 2 ml./min., and 25 ml. of effluent fractions is collected. Heptulose assay by the orcinol method on 0.02-ml. samples showed that sedoheptulose phosphate emerged between fractions 24 and 29, and the pentose phosphates followed in fractions 35 to 50, exhibiting two peaks. The first at fraction 38 contained R-5-P; the last at fraction 46 contained ribulose-5-phosphate. The sedoheptulose-7-P in fractions 24 to 29 was adjusted to pH 6.2 with 5.0 ml. of 4 *N* NaOH and 21.0 ml. of saturated barium hydroxide solution. The precipitate obtained on addition of 4 vol. of ethanol was collected, washed with 10 ml. of 80% ethanol, and dried *in vacuo.* The yield was 117 mg. of barium sedoheptulose-7-P (73% pure).

"**Gradient Elution**" **Technique.** Elution peaks obtained in column chromatography often have satisfactorily sharp fronts but suffer from trailing. When this leads to overlapping of peaks, it may be alleviated by an innovation in elution technique developed by Busch et al.[20] for separation of carboxylic acids on the anion resin.

To obtain a gradual increase in acidity, concentrated acid from a reservoir is passed into a mixing flask filled with distilled water and mixed constantly by a magnetic stirrer. The solution from the mixing flask is passed into the 1-cm.-diameter chromatographic column. This technique avoids the discontinuity of changing eluant solutions and produces sharp symmetrical elution peaks.

Method for Sedoheptulose Mono- and Diphosphates (Horecker et al.[21]). A transaldolase reaction mixture prepared from 250 micromoles of sedoheptulose-7-P[32] and HDP is placed on a Dowex 1 (formate) column, 11.5 cm. $\times$ 3.1 cm.², and eluted with 0.2 N formic acid containing 0.5 N sodium formate after it is passed through a stirred 400-ml. water reservoir at 1 ml./min. Fractions of 8 ml. were collected and analyzed with the orcinol reaction. Fractions 15 to 25 contained sedoheptulose-7-P, and fractions 75 to 84 contained sedoheptulose di-P with very sharp peaks at 20 and 77 and negligible trailing.

[20] H. Busch, R. B. Hurlbert, and V. R. Potter, *J. Biol. Chem.* **196,** 717 (1952).
[21] B. L. Horecker, P. Z. Smyrniotis, H. Hiatt, and P. Marks, *J. Biol. Chem.* **212,** 827 (1955); see also Vol. III [30].

[16A] Chemical Syntheses of Aldohexose-1-phosphates

By THEODORE POSTERNAK

I. α-Aldohexose-1-phosphates

Principles of the Syntheses

a. According to the procedure of Cori, Colowick, and Cori,[1-3] an α-acetobromoaldose is treated with trisilver phosphate. The condensation product, which is mainly a triester, is partially hydrolyzed by acid in order to remove two sugar residues and then deacetylated. This method can be used when the α-acetobromoaldose is readily available in great quantities.

[1] C. F. Cori, S. P. Colowick, and G. T. Cori, *J. Biol. Chem.* **121,** 465 (1937).
[2] S. P. Colowick, *J. Biol. Chem.* **124,** 557 (1938).
[3] M. E. Krahl and C. F. Cori, *Biochem. Preparations* **1,** 33 (1949).

b. An α-acetohalogenoaldose is treated with silver diphenylphosphate. After removal of the phenyl groups by catalytic hydrogenation and of the acetyl groups by alkali, an α-aldose-1-phosphate is obtained as main product.[4] As this synthesis does not involve the loss of two-thirds of the starting sugar, the yields are much better than by the first method. This second procedure is therefore to be preferred when only small amounts of α-acetohalogenoaldoses are available.

Preparation of the α-Acetohalogenoaldoses

β-D-Glucose Pentaacetate.[5] Twenty-five grams of dry, finely powdered glucose and 12 g. of finely powdered anhydrous sodium acetate are ground and mixed in a mortar. The mixture is placed in a 0.5-l. flask with 125 g. of reagent-grade acetic anhydride and heated on a steam bath with vigorous agitation until all the material is dissolved; this usually requires 30 minutes. The heating is then continued for 2 hours more. The solution is cooled and poured, with mechanical agitation, into 1 l. of ice water, and the stirring is continued for 2 hours. The precipitate is filtered with suction and resuspended for 3 hours in 1 l. of ice water. It is then filtered again with suction, and all possible water is pressed out. The compound is finally recrystallized from 125 ml. of alcohol. Yield, 35 to 40 g.; melting point, 126 to 130°. The product is of sufficient purity for further use. Melting point of the pure compound, 133°; $[\alpha]_D^{20} = +4.9°$ (methanol, $c = 1$).

α-Tetraacetylglucosylbromide.[5] Twenty-five grams of dry, finely powdered β-glucose pentaacetate is placed in a 100-ml. glass-stoppered Erlenmeyer flask and cooled to 5°. Fifty grams of a solution of dry hydrobromic acid saturated in acetic acid at 0° is added. The stoppered flask is shaken gently until all the solid is dissolved, and the mixture is allowed to stand at room temperature for 2 hours. The solution is then diluted with 100 ml. of chloroform and poured with stirring into 350 ml. of ice water. The chloroform layer is separated, and the water solution is extracted with 25 ml. of chloroform. The combined chloroform solutions are shaken with 200 ml. of ice water and dried by agitation for 10 minutes with fused calcium chloride. The clear solution is concentrated by distillation *in vacuo* to a volume of about 40 ml. Petroleum ether is then added slowly with vigorous stirring. The compound separates in long needles which can be recrystallized in the same way. Yield, 15 g.; melting point, 88 to 89°; $[\alpha]_D^{19} = +199°$ (chloroform, $c = 4.5$). The pure compound is stable for years if kept over $CaCl_2$ or P_2O_5 and soda lime.

[4] T. Posternak, *J. Am. Chem. Soc.* **72**, 4824 (1950).

[5] E. Fischer, *Ber.* **49**, 584 (1916).

β-D-*Galactose Pentaacetate and α-Tetraacetylgalactosylbromide.* The above described procedures can be used for the preparation of *β*-galactose pentaacetate, melting point 142°, $[\alpha]_D = +7.5°$ (benzene, $c = 6$), and of α-acetobromogalactose, melting point 82 to 83°, $[\alpha]_D^{20} = +236°$ (benzene, $c = 8$).

β-D-*Mannose Pentaacetate.*[6] Ten grams of dry, finely powdered D-mannose is added to a mixture, which has been precooled to 0°, of 67 g. of dry pyridine and 50 g. of acetic anhydride. The suspension is shaken at 0° until all the solid dissolves. The mixture is allowed to stand at 0° for 2 days more. The solution is then poured slowly with stirring into 250 ml. of ice water. The separated oil becomes crystalline on being rubbed under water with a glass rod. The mannose pentaacetate is recrystallized three times from 96% ethyl alcohol. Yield, 13.5 g.; melting point, 117 to 118°; $[\alpha]_D^{21} = +24.1°$ (chloroform, $c = 2$).

α-Tetraacetylmannosyl Chloride.[7] Ten grams of dry *β*-D-mannose pentaacetate is refluxed for 4 hours with exclusion of moisture with a mixture of 45 g. of dry chloroform and 4.9 g. of titanium tetrachloride. At first a canary-yellow precipitate is formed which dissolves after a short while. The solution is poured with stirring into 2 vol. of ice water. The chloroform layer is separated and washed twice with 30 ml. of ice water; it is then dried for 2 hours on fused calcium chloride. The solvent is then evaporated *in vacuo* (bath temperature, 30°). When the remaining sirup is covered with anhydrous ether and rubbed with a glass rod, crystallization begins. The compound is purified by dissolution in the minimum amount of anhydrous ether followed by careful addition of petroleum ether. Yield, 8 g.; melting point, 81°; $[\alpha]_D^{20} = +91°$ (chloroform; $c = 3$); α-tetraacetylmannosyl-1-chloride crystallizes much more easily than the corresponding bromide.

Preparation of α-Aldose-1-phosphates by the Trisilver Phosphate Procedure

Trisilver Phosphate.[3] 11.9 g. of Na_2HPO_4 and 42.7 g. of $AgNO_3$ are each dissolved in 1 l. of water. With mechanical stirring, the solution of the former is poured slowly into that of the latter. The mixture is then stirred for 30 minutes more. The canary-yellow precipitate is filtered with suction, washed successively with water, absolute alcohol, and anhydrous ether, and dried *in vacuo* in a brown desiccator. All these operations should be carried out under reduced illuminations. For further use, it is recommended to take freshly prepared trisilver phosphate.

α-Glucose-1-phosphate.[3] 17.5 g. of dry, finely powdered trisilver phos-

---

[6] E. Fischer and R. Oetker, *Ber.* **46**, 4029 (1913).
[7] E. Pacsu, *Ber.* **61**, 1508 (1928).

phate, 50 g. of α-tetraacetylglucosylbromide, and 150 ml. of dry benzene are placed in a three-necked 500-ml. flask fitted with a mercury seal, a mechanical stirrer, and a reflux condenser equipped with a calcium chloride tube. The mixture is refluxed with vigorous stirring under reduced illumination. After 1 hour, 8 g. of trisilver phosphate is added and the refluxing is continued for ½ hour more. The silver salts are separated by centrifugation and washed with benzene. The combined benzene solutions are evaporated to dryness under reduced pressure (bath temperature 30°). The last traces of solvent are removed after 5 to 6 hours in a high vacuum at room temperature.

The condensation product is then dissolved in 800 ml. of a mixture of 14.0 ml. of concentrated hydrochloric acid and 850 ml. of absolute methyl alcohol. The filtered solution is allowed to stand at 20 to 25°. At hourly intervals 0.01- to 0.02-ml. aliquots are withdrawn and analyzed for inorganic phosphates. When the inorganic phosphate has risen to 20% of the total phosphate (after 8 to 12 hours), the solution is diluted with an equal volume of methyl alcohol and brought to pH 8.0 to 8.5 by addition of 0.3 N barium hydroxide. The reaction mixture is left overnight at 4°, and the pH is brought back to 8.0 to 8.5 with barium hydroxide. The precipitate of barium salts is then collected by centrifugation and washed with methyl alcohol and ether and dried *in vacuo*. It is then extracted four times with 40 ml. of cold water, vigorous shaking for 10 to 15 minutes during each extraction, then centrifuged. The liquids are pooled, and the barium salt is reprecipitated by addition of 2 vol. of ethanol; it is collected by centrifugation and washed successively with alcohol and ether and dried *in vacuo*. Yield, 3 to 6 g.

For conversion into the crystalline potassium salts, the barium salt is dissolved in 10 parts of water. An equivalent amount of 10% K_2SO_4 is added. The barium sulfate is removed by centrifugation. By slow addition of 1.7 vol. of ethanol at 0°, potassium glucose-1-phosphate crystallizes in fine needles.

The same prescription can be used for the preparation of α-mannose-1-phosphate[2] and α-galactose-1-phosphate.[2,8] The former is isolated as barium salt, the latter as crystalline potassium salt.

Preparation of α-Aldose-1-phosphates by the Silver Diphenylphosphate Procedure

Diphenyl Chlorophosphonate.[9] 250 g. of phosphorus oxychloride and 250 g. of phenol are placed in a two-necked flask fitted with a reflux condenser and a thermometer. The condenser is equipped with a calcium

[8] H. W. Kosterlitz, *Biochem. J.* **33**, 1087 (1939).
[9] P. Brigl and H. Müller, *Ber.* **72**, 2121 (1939).

chloride tube at the top. In the course of 5½ hours the temperature of the mixture is raised to 180°. When the evolution of hydrogen chloride has become slow, the temperature is raised in 30 minutes to 225°. It is then raised quickly to 260° and maintained at this temperature for 1 minute. The reaction mixture is fractioned by distillation *in vacuo*. The liquid distilling at 191 to 194° (12 mm.) is refractioned. Yield, 190 g. of pure diphenyl chlorophosphonate.

Silver Diphenylphosphate.[10] Five milliliters (6.35 g.) of diphenyl chlorophosphonate is heated on a steam bath with 20 ml. of 2.5 N sodium hydroxide until all the material is dissolved. After neutralization to phenolphthalein with concentrated nitric acid, 50 ml. of water and 85 ml. of 10% silver nitrate are added. The mixture is heated to boiling and filtered hot in order to remove the silver chloride. After cooling in ice, the fine needles (yield, 4.6 g.) are filtered and recrystallized from hot water. Before use the compound is dried at 100°.

α-Glucose-1-phosphate.[4] Six-tenths gram of α-acetobromoglucose is dissolved in 2 ml. of dry benzene. 0.52 g. of dry finely powdered silver diphenylphosphate is added, and the mixture is refluxed for 30 minutes with exclusion of moisture. After the addition of 0.25 g. of silver diphenylphosphate, the mixture is refluxed again for 30 minutes. After centrifugation and washing of the silver salts with dry benzene, the solvent is evaporated under reduced pressure. The residue is dried in a high vacuum and then dissolved in 8 ml. of absolute ethanol. The filtered solution is exhaustively hydrogenated at room temperature and at atmospheric pressure in the presence of 100 mg. of platinum oxide; 8 moles of H_2 per mole of substance is consumed in about 2 hours. Depending on the quality of the catalyst, the uptake is sometimes too slow; it is then necessary to add more catalyst, or the platinum can be reactivated by shaking it with air for 5 to 10 minutes. After the hydrogenation, 1.0 ml. of 10 N sodium hydroxide is added. The mixture is heated to boiling for 3 minutes. A gummy precipitate sometimes separates, which may be dissolved by the addition of about 10 ml. of water. The milky solution is heated again to boiling for 2 minutes. After cooling, it is neutralized with glacial acetic acid. Ethanol and cyclohexane are then removed by distillation *in vacuo*, and a concentrated aqueous solution of 500 mg. of barium acetate is added. The precipitate (mainly barium phosphate) is removed by centrifugation and washed with water. By the addition of 3 vol. of ethanol to the combined water solutions, barium glucose-1-phosphate is precipitated. It is centrifuged down, washed with ethanol and ethyl ether, and dried *in vacuo*. The barium salt (350 mg.) is converted to the crystalline potassium salt (200 mg.) as described above (see trisilver phosphate procedure).

[10] T. Posternak, *J. Biol. Chem.* **180**, 1269 (1949).

α-D-*Galactose-1-phosphate*[4] and α-D-*Mannose-1-phosphate*.[11] By the silver diphenylphosphate procedure, these two sugar phosphates can be prepared exactly as described above, from α-acetobromogalactose and α-acetochloromannose. It is recommended to purify them as crystalline brucine salts. First, 1.05 g. of brucine sulfate is added to a water solution of 500 mg. of the barium salt. The suspension is heated on a steam bath; barium sulfate is removed by centrifugation from the hot mixture and well washed with water. The combined supernatant fluids are evaporated to dryness *in vacuo*. The residue is crystallized from 2 parts of hot water or by addition of acetone to a more dilute water solution. The brucine salt is finally suspended in 3 ml. of water, and 10% potassium hydroxide is added dropwise until a permanent pink color is obtained with phenolphthalein. Brucine is removed by exhaustive extraction with chloroform.

Potassium α-D-galactose-1-phosphate crystallizes in needles during slow addition of ethanol at 0° to the water solution.

The potassium salt of α-mannose-1-phosphate could not be obtained in crystalline state. This sugar phosphate is therefore precipitated as barium salt by addition of a small excess of barium acetate and of 2 to 3 volumes of ethanol.

II. β-Aldohexose-1-phosphates

Principles of the Synthesis

α-Acetobromoglucose and α-acetobromogalactose condense with silver dibenzylphosphate with Walden inversion at carbon 1.[12] After removal of the benzyl groups by catalytic hydrogenation and deacetylation with alkali, a β-aldose-1-phosphate is formed.[13]

The same compounds can also be prepared by condensation of the α-acetobromoaldoses with "monosilver phosphate" followed by deacetylation with alkali.[14]

Preparation of β-Glucose-1-phosphate and β-Galactose-1-phosphate

Silver Dibenzylphosphate.[15] A mixture of 12 g. of benzyl chloride, 12 g. of trisilver phosphate, and 100 ml. of anhydrous ether is refluxed for

[11] T. Posternak and J. P. Rosselet, *Helv. Chim. Acta* **36**, 1614 (1953).

[12] α-Acetobromomannose condenses, however, with silver dibenzylphosphate or with "monosilver phosphate" without inversion. After removal of the benzyl groups and deacetylation, α-mannose-1-phosphate is therefore formed.[11] These two procedures can be used for the preparation.

[13] M. L. Wolfrom, C. S. Smith, D. E. Pletcher, and A. E. Brown, *J. Am. Chem. Soc.* **64**, 23 (1942).

[14] F. J. Reithel, *J. Am. Chem. Soc.* **67**, 1056 (1945).

[15] W. Lossen and A. Köhler, *Ann.* **262**, 212 (1891).

2 hours with exclusion of moisture and with mechanical stirring. The operation is carried out under reduced illumination. The insoluble silver salts are filtered with suction and washed with ether. After evaporation of the solvent on a water bath, the partly crystalline residue is kept *in vacuo* at 30° for a few hours. An oily impurity is removed by pressing the crystals of tribenzylphosphate on a porous plate. Melting point, 60°.

Five grams of tribenzylphosphate are refluxed for 10 hours with 25 ml. of 10% methyl alcoholic potassium hydroxide. Most of the solvent is then evaporated *in vacuo*. The residue is dissolved in water. By acidification, dibenzyl hydrogen phosphate is precipitated; it is recrystallized from chloroform-petroleum ether; melting point, 79 to 80°. The silver salt is precipitated by adding an alcoholic solution of silver nitrate to an alcoholic solution of the acid. Silver dibenzylphosphate can be recrystallized from hot water; for further use, it is dried at 100°; melting point, 222°.

Tetraacetyl-β-glucose-1-dibenzylphosphate.[13] In a flask fitted with a mechanical stirrer and a condenser, 14 g. of dry, finely powdered silver dibenzylphosphate, 10 g. of α-acetobromoglucose, 5 g. of powdered Drierite (anhydrous calcium sulfate), and 60 ml. of dry benzene are warmed at 60° for 30 minutes with mechanical agitation and exclusion of moisture. The mixture is then refluxed for 1½ hour, mechanical stirring being maintained throughout. After cooling, the mixture is filtered through a precoat of decolorizing charcoal. After complete evaporation of the solvent *in vacuo*, the residue can be crystallized from a concentrated ether solution by dropwise addition of petroleum ether. Yield, 10 g.; melting point of the pure material, 78 to 79°; $[\alpha]_D^{25} = -9°$ (chloroform, $c = 3.5$). The crude compound, however, is of sufficient purity for further use.

β-Glucose-1-phosphate.[13] One gram of the preceding compound is dissolved in 8 ml. of absolute ethanol. The filtered solution is hydrogenated in the presence of 0.2 g. of palladous oxide. Two moles of hydrogen per mole of substance is consumed in 30 to 40 minutes. One milliliter of 10 N sodium hydroxide is added. A gelatinous precipitate is formed; water is added in sufficient amount in order to dissolve it. After neutralization with acetic acid, ethanol is removed by distillation *in vacuo* and a concentrated solution of 500 mg. of barium acetate is added. The precipitate is removed by centrifugation. The barium salt of the sugar phosphate (500 mg.) is precipitated by addition of 2 to 3 vol. of ethanol and converted as described above to the brucine salt which crystallizes from hot water as decahydrate. The anhydrous form is obtained by heating over P_2O_5 at 100° *in vacuo;* melting point, 162 to 166°. The brucine salt is transformed in the usual way to the potassium or sodium salts, which in

SUMMARY OF SOME PROPERTIES OF SYNTHETIC ALDOHEXOSE-1-PHOSPHATES

Substance	Total P and acid-labile P, %	Aldohexose by hydrolysis, %	$[\alpha]_D$ (water)	Hydrolysis constants, $K = 2.3/t \log_{10} (A/A - x)$
α-Glucose-1-phosphate $C_6H_{11}O_9PK_2 \cdot 2H_2O$	8.33	48.4	+78° (25°; c = 4)[a]	0.00299 (0.25 N HCl; 37°)[b] 0.0115 (N HCl; 33°)[c]
$C_6H_{13}O_9P$ (free acid)	—	—	+120° (25°; c = 4)[b]	0.00265 (0.95 N H_2SO_4; 30°)[d]
α-Galactose-1-phosphate $C_6H_{11}O_9PK_2 \cdot 2H_2O$	8.33	48.4	+98° (18°; c = 2.375)[e] +100° (32°; c = 1.57)[f] +143° (25°)[g]	0.0059 (0.25 N HCl; 37°)[e]
$C_6H_{13}O_9P$ (free acid)			+148.5° (18°, c = 1.678)[e]	
α-Mannose-1-phosphate $C_6H_{11}O_9PBa$			+36° (25°)[g] +33.7° (23°; c = 3.0)[d]	0.0019 (0.95 N H_2SO_4; 30°)[d]
$C_6H_{11}O_9PBa \cdot 3H_2O$[h]	6.90	40.1		
$C_6H_{11}O_9PBa \cdot 2H_2O$[i]	7.18	41.7	+58° (25°)[g]	
$C_6H_{13}O_9P$ (free acid)				
β-Glucose-1-phosphate Brucine salt (decahydrate) $C_6H_{11}O_9P$ Na_2			−17° (29°; c = 1.7)[c] +13° (20°; c = 3.83 calculated as free acid)	0.0345 (N HCl; 33°)[c]
β-Galactose-1-phosphate $C_6H_{11}O_9PBa \cdot 3H_2O$[j]	6.90	40.1	+31.2° (30°; c = 1.2)[k]	0.0059 (0.25 N HCl; 37°)[k]

[a] M. L. Wolfrom and D. E. Pletcher, *J. Am. Chem. Soc.* **63,** 1050 (1941).
[b] C. F. Cori, S. P. Colowick, and G. T. Cori, *J. Biol. Chem.* **121,** 465 (1937).
[c] M. L. Wolfrom, C. S. Smith, D. E. Pletcher, and A. E. Brown, *J. Am. Chem. Soc.* **64,** 23 (1942).
[d] T. Posternak and J. P. Rosselet, *Helv. Chim. Acta* **36,** 1614 (1953).
[e] H. W. Kosterlitz, *Biochem. J.* **33,** 1087 (1939).
[f] T. Posternak, *J. Am. Chem. Soc.* **72,** 4824 (1950).
[g] S. P. Colowick, *J. Biol. Chem.* **124,** 557 (1938).
[h] Dried at room temperature, *in vacuo.*[g]
[i] Dried at 110°, *in vacuo,* over P_2O_5.[d]
[j] Dried at room temperature, *in vacuo,* over P_2O_5.[k]
[k] F. J. Reithel, *J. Am. Chem. Soc.* **67,** 1056 (1945).

their turn can be converted to the barium salt. β-Galactose-1-phosphate has been prepared by the same procedure.[14]

III. Identification and Estimation of Degree of Purity

The purity of the synthetic samples of aldohexose-1-phosphates is determined on the basis of analytical data, optical rotation, and lability to acid hydrolysis (see the table). They have no reducing power, but reducing properties appear after a short acid hydrolysis. Synthetic samples can also be characterized by the degree of conversion by phosphoglucomutase. α-Glucose-1-phosphate is transformed by the enzyme to the extent of 95% to the acid-stable 6-phosphate. α-Mannose-1-phosphate and α-galactose-1-phosphate are converted to a similar extent, but much more slowly, by the muscle enzyme. All these transformations require the presence of catalytic amounts of α-glucose-1,6-diphosphate.[16] The enzyme has no detectable action on β-aldohexose-1-phosphates.

It has been shown[16] that samples of synthetic α-glucose-1-phosphate contain variable amounts of glucose-1,6-diphosphate (from practically 0 to 0.03%) which it is difficult to remove by crystallization of the potassium salt.

[16] E. W. Sutherland, M. Cohn, T. Posternak, and C. F. Cori, *J. Biol. Chem.* **180**, 1285 (1949).

[16B] Preparation of α-D-Glucose-1-phosphate[1] by Means of Potato Phosphorylase

By R. M. McCready and W. Z. Hassid

Assay Method

Principle. α-D-Glucose-1-phosphate is formed by phosphorolysis of starch in the presence of the enzyme phosphorylase and orthophosphate by the following reaction:

$$\text{Starch} + \text{inorganic phosphate} \underset{\substack{\text{potato} \\ \text{phosphorylase}}}{\rightleftharpoons} \alpha\text{-D-Glucose-1-phosphate}$$

The starch is phosphorylized to α-D-glucose-1-phosphate which can be separated from the unreacted starch, inorganic phosphate, and the impurities introduced by the potato juice (source of enzyme). At pH 6.5 an

[1] A chemical method for the preparation of α-D-glucose-1-phosphate is described by M. E. Krahl and C. F. Cori in *Biochem. Preparations* **1**, 33 (1949). The present enzymatic method has the advantage in that the yields are high, the starting materials inexpensive, and the procedure by far less laborious. This procedure has also been described in *Biochem. Preparations* **4**, 63 (1955).

equilibrium is established with a ratio of 4 parts of orthophosphate to 1 part of α-D-glucose-1-phosphate.

The orthophosphate is removed as the insoluble magnesium-ammonium salt. The solution containing the α-D-glucose-1-phosphate is passed through a strong-acid cation exchange resin and then through a medium-basic anion exchange resin. The properties of the strongly acidic α-D-glucose-1-phosphoric acid are such that it is readily absorbed by the anion exchange resin in the presence of very high concentrations of weaker acids. The resin containing the absorbed α-D-glucose-1-phosphoric acid is washed thoroughly with water to remove traces of starch and other nonionic impurities. The ester is then eluted from the washed resin with potassium hydroxide solution and crystallized from methanol as its dipotassium dihydrate salt.

Reagents

Phosphate buffer, pH 6.7. Thirty-five grams (0.2 mole) of dipotassium hydrogen phosphate (K_2HPO_4) and 27 g. (0.2 mole) of potassium dihydrogen phosphate (KH_2PO_4) are dissolved in 500 ml. of water.

Starch solution. Twenty grams of dry starch (Lintner's soluble starch or a similar soluble starch product[2]) is added to 500 ml. of boiling water and stirred vigorously. The solution is then permitted to cool to room temperature.

Enzyme preparation. Two medium-sized potatoes[3] are peeled and sliced into 1-cm. shoestring-sized strips. About 300 g. of these slices is blended in an electric blender with 300 ml. of water at high speed for about 2 minutes, the contents poured into a Nylon bag, and pressed by hand to remove the pulp. After about 5 minutes, when the starch granules have settled and the cloudy solution has been decanted from the residue, about 500 ml. of 1:1 potato juice is obtained.

Ion exchange resins. A strongly acidic high-capacity cation exchange resin, such as Dowex 50, is regenerated with hydrochloric acid and washed with water according to the directions given by the manufacturer.[4] A medium-basic high-capacity exchange resin

---

[2] Raw potato, wheat, cassava, or corn starch may also be used, but boiling should be continued for about 30 minutes to aid in dispersing the starch.

[3] Idaho russet, Burbank, White Rose, and other varieties of potatoes may be used.

[4] A most effective resin for this purpose is found to be a strongly acidic, high-capacity cation exchange resin such as Amberlite IR 112, Permutit Q, and Dowex 50. The Amberlite resins are manufactured by the Rohm and Haas Co., Philadelphia, Pa.; Dowex resins are obtained from the Dow Chemical Co., Midland, Mich.; and the Permutit resins from the Permutit Corp., 330 West 42nd St., New York 18, N. Y.

such as Permutit A (1953)[5] is regenerated to its basic form and washed as recommended by the manufacturer.

Procedure

1. Preparation of the Digest.[6] The 1:1 potato juice, the 500 ml. of starch solution, and the 500 ml. of buffer solution are mixed thoroughly, and about 0.25 g. of phenylmercuric nitrate is added as a preservative. The reaction mixture is permitted to stand for at least 16 hours (see Table I). The solution is then heated rapidly to 95° to coagulate proteins

TABLE I

PRODUCTION OF 10-MINUTE HYDROLYZABLE PHOSPHATE (α-D-GLUCOSE-1-PHOSPHATE)
IN A POTATO JUICE-STARCH-BUFFER MIXTURE

Time of reaction, hr.	Phosphorus, γ/ml. of digest		
	Inorganic	Inorganic + 10-min. hydrolysis	10-min. hydrolysis
0	25	50	25
2	25	275	250
16	25	900	875
24	25	950	925
40	25	1050	1025
60	25	1100	1075

and to stop further enzyme action. After the solution has cooled, the insoluble material is removed by filtration with suction on an 18.5-cm. filter paper precoated with diatomaceous silica filter aid. To the clear solution is added 86 g. (0.4 mole) of magnesium acetate hydrate (MgAc$_2$· 4H$_2$O). The mixture is stirred to dissolve the salt and the pH adjusted to 8.5 with 14% ammonia (concentrated NH$_4$OH, diluted 1:1). After removal of the precipitated magnesium ammonium phosphate by filtration, the solution is ready for isolation of the α-D-glucose-1-phosphate.

---

[5] The anion exchange resin best suited for the purpose of removing the α-D-glucose-1-phosphoric acid from dilute solutions containing large amounts of weak (acetic) acid is one having weak- or medium-basic groups and a high absorption capacity. Amberlite IR 4B, Permutit S (before 1951), or Permutit A (1953) may be used. Owens, Goodban, and Stark [*Anal. Chem.* **25**, 1507 (1953)] note "—that resin S obtained from the Permutit Corp., New York, in 1951 was similar in base strength and high capacity to the product now manufactured under the name resin A." Resin S as now offered is a bead-type, strong-base exchange resin of lower capacity and was unsuitable for their experiments and also for the isolation of α-D-glucose-1-phosphate reported here.

[6] The procedure is similar to that described earlier by R. M. McCready and W. Z. Hassid [*J. Am. Chem. Soc.* **66**, 560 (1944)].

2. Determination of the Course of Starch Phosphorolysis. The formation of the α-D-glucose-1-phosphate during the course of the reaction can be determined by analysis of the 10-minute hydrolyzable phosphorus (α-D-glucose-1-phosphate) is the digest during several time intervals. The inorganic phosphate resulting from the hydrolysis of the ester is conveniently determined colorimetrically by the method of Allen.[7] The determination is carried out as follows:

To a 1-ml. aliquot of digest in a 25-ml. volumetric flask, 15 ml. of water, 2.5 ml. of magnesium ammonium chloride solution (55 g. of magnesium chloride hexahydrate and 70 g. of ammonium chloride dissolved in 650 ml. of water and added to 350 ml. of 10% ammonium hydroxide), and 1.5 ml. of ammonium hydroxide solution (25% w/v aqueous) are added, and the solution in the flask is diluted to the mark with water. The precipitate of magnesium ammonium phosphate formed from the free phosphate in the solution is removed by gravity filtration through a dry, fluted filter paper (Whatman No. 12, 18.5 cm.). A 1-ml. aliquot of this inorganic phosphate-free solution is measured into a 25-ml. volumetric flask and put aside as a control. Another 1-ml. aliquot is placed into a 25-ml. volumetric flask containing 18 ml. of water and 2 ml. of 60% w/v perchloric acid. The flask and contents (about 1 N acid concentration) are immersed in a boiling water bath for 10 minutes and then cooled to 20°. Two milliliters of amidol reagent and 1 ml. of ammonium molybdate reagent are added, diluted with water to the 25-ml. mark, and the contents mixed.

The 1-ml. control sample of the phosphate-free diluted digest, previously set aside in the 25-ml. flask, is treated with the same reagents (18 ml. of water, 2 ml. of perchloric acid, 2.5 ml. of amidol reagent, and 1 ml. of ammonium molybdate reagent), but the ester is not hydrolyzed by heating the mixture. The density of the color of the two samples is determined colorimetrically.[7] Typical results showing the course of the phosphorolysis of starch in this digest are shown in Table I.

[7] R. J. L. Allen, *Biochem. J.* **34B**, 858 (1940).

Reagents

     Perchloric acid, 60 % w/v.

     Amidol reagent. 2 g. amidol (2,4-diaminophenoldihydrochloride) and 40 g. of sodium bisulfite dissolved in water to 200 ml.

     Ammonium molybdate. 8.3 g. of ammonium molybdate dissolved in 100 ml. of water containing 1 drop of ammonia.

     Standard phosphate solution, 1.0967 g. of potassium dihydrogen phosphate dissolved in 250 ml. of water. This solution contains 1 mg. of P per milliliter and is diluted as a standard. Phosphate is determined with the range of 5 to 70 γ of phosphorus per determination.

The data in this table show that 875 γ of inorganic phosphate was incorporated into glucose-1-phosphate per milliliter of digest during a period of 16 hours. The amount of dipotassium α-D-glucose-1-phosphate dihydrate is therefore

$$\frac{0.000875 \times 1500 \times 100}{8.33} = 15.7 \text{ g.}$$

3. Preparation and Use of Ion Exchange Resins for the Recovery of α-D-Glucose-1-phosphate from Dilute Solutions. CATION EXCHANGE RESIN. The Dowex 50 as obtained from the manufacturer must be washed and regenerated to the H-form. Operation of the ion exchange columns is essentially the same as previously described.[6]

A cylindrical glass column is used, 5.0 cm. in diameter by 60 cm. in length, of approximately 1-l. capacity, open at the top, and with a tubulation at the bottom. A small wad of glass wool is packed into the bottom of the tube, and then 600 ml. of wet Dowex 50 is added. Two liters of 5% hydrochloric acid is passed through the resin at a rate of about 3 l. (5 bed vol.) per hour. The acid is washed out of the regenerated resin bed with about 4 l. of distilled water at the same rate. At that point the pH of the effluent is about 3.5 and the solution does not give any titratable acidity.

The digest is then passed through the column at a rate of about 5 bed vol. per hour, and finally the resin bed is washed once with 600 ml. of water. This quantity of resin will permit the total volume of the digest to pass through without exceeding its exchange capacity. As the digest passes the column, the pH of the effluent changes from an initial value of 3.5 to a final pH of about 1.8. This effluent (pH 1.6 to 2.0) is now ready to be passed through the anion exchange column.

ANION EXCHANGE RESIN. The column used is similar in design to that previously mentioned but of different dimensions, 2.5 cm. in diameter and 45 cm. in length, with a capacity of approximately 100 ml. Fifty milliliters of wet Permutit A resin is placed into the column, and 500 ml. of 5% sodium hydroxide is passed through it at a flow rate of about 5 bed vol. per hour. The alkali is washed out with 2 l. of water at a rate of 10 bed vol. per hour. When the pH of the effluent is about 8.9 to 9.0, the column is ready to be used for absorption of the α-D-glucose-1-phosphoric acid.

The acidic effluent of the starch digest which has been passed through the Dowex 50 column is now passed through the column of Permutit A at a rate of about 10 bed vol. per hour. This amount of resin is more than sufficient to remove the α-D-glucose-1-phosphate in the digest. Since the color of the resin darkens as the α-D-glucose-1-phosphate is absorbed, the extent of its saturation can be readily observed. The column of resin is washed with 2 l. of water at a rate of 10 bed vol. per hour to remove traces of starch, sugars, and other impurities. The α-D-glucose-1-phosphate is then removed by passing a solution of 5% potassium hydroxide through the column. Table II shows the results of the displacement of the α-D-glucose-1-phosphate from the resin.

TABLE II

COURSE OF ELUTION OF α-D-GLUCOSE-1-PHOSPHATE FROM A COLUMN CONTAINING 30 G. OF MOIST (50% WATER) PERMUTIT A BY SUCCESSIVE 50-ML. PORTIONS OF 5% POTASSIUM HYDROXIDE SOLUTION

Elution fraction	pH[a]	Glucose-1-phosphate in the eluate as per cent of original digest[b]
1	5.7	4.2
2	5.6	57.0
3	12.0	25.8
4	12.2	4.0
5	12.6	1.0
6	13.0	0.3

[a] pH measurements made using a Beckman Model G pH meter with sodium electrodes.
[b] Total 10-minute hydrolyzable P in the original digest is 15.7 g.

When 3 vol. of methanol is slowly added with stirring to the first 200 ml. of the alkali effluent, the α-D-glucose-1-phosphate crystallizes spontaneously as its dipotassium dihydrate salt. After cooling, the crystalline potassium salt of the ester is filtered off and dissolved in 100 ml. of water. The product is then recrystallized by the addition of 1 ml. of 1% potassium hydroxide and 3 vol. of methanol. The crystals are collected by filtration and dried *in vacuo* over calcium chloride. A yield of 14.3 g. of α-D-glucose-1-phosphate dipotassium dihydrate (91% of the theory) is obtained.

ALTERNATIVE PROCEDURE FOR THE ISOLATION OF α-D-GLUCOSE-1-PHOSPHATE. The acidic effluent obtained after the digest has passed through the cation exchange column (1600 ml., pH 1.8) is treated with 30 g. of air-dry (50% moisture content) Permutit A resin, and the mixture is stirred for 30 minutes. The resin on which the α-D-glucose-1-phosphate was absorbed is collected by filtration and washed on the filter paper with a large quantity of water to remove impurities. The moist resin is placed in a beaker and treated with 100 ml. of 5% potassium hydroxide solution which liberates the α-D-glucose-1-phosphate. After standing for 5 minutes the alkaline solution (pH 12 to 12.5) containing the ester is filtered, and the resin is washed twice with 50-ml. portions of water. The solution and washings are combined, 3 vol. of methanol is added, and the product isolated as previously described. After two crystallizations about 14 g. of α-D-glucose-1-phosphate dipotassium dihydrate (89% of the theory) is obtained.

Identification of the α-D-Glucose-1-phosphate

Elementary analysis[1] indicates this product to be $C_6H_{11}O_5PO_4K_2 \cdot 2H_2O$. The ester is completely hydrolyzed to glucose and orthophosphate by heating at 100° for 30 minutes with 1 N sulfuric acid. An osazone can be prepared from the hydrolysis products which is shown to be glucosazone. The glucose is also identified by paper chromatography. The P content of the ester agrees with that of the theoretical, 8.33%.

In the presence of potato phosphorylase the product forms a starch-like polysaccharide which stains blue with iodine.

The specific rotation of the dipotassium dihydrate salt is $[\alpha]_D^{20} = +78°$ (4% solution in water).

[17] Isolation of Glucose-1,6-diphosphate

By LUIS F. LELOIR and ALEJANDRO C. PALADINI

Assay Method

Principle. The rate of the reaction catalyzed by phosphoglucomutase,

$$\text{G-1-P} + \text{glucose-1,6-diphosphate} \rightleftarrows \text{Glucose-1,6-diphosphate} + \text{G-6-P}$$

is, under certain conditions, proportional to the concentration of glucose-1,6-diphosphate.[1] It can be measured by the disappearance of labile phosphate (7 minutes at 100° in 1 N acid) or by the reducing power of the G-6-P formed. The latter procedure will be described. With this method it is possible to measure amounts of glucose-1,6-diphosphate ranging from 0.2 to 0.8 × 10^{-3} micromole with an error of about 10%.

Reagents

G-1-P, 10 micromoles/ml. Sodium or potassium salt.

0.1 M magnesium sulfate.

Glucose-1,6-diphosphate, 0.2 micromole/ml. For use it is diluted ten times. Both solutions can be kept for months in the frozen state.

[1] C. E. Cardini, A. C. Paladini, R. Caputto, L. F. Leloir, and R. E. Trucco, *Arch. Biochem.* **22,** 87 (1949).

Enzyme. Dried brewer's yeast is suspended in 3 vol. of water, incubated for 2 hours at 37°, and then centrifuged. The supernatant is then diluted with 4 vol. of water. The activity remains unchanged on storage in the frozen state for several months. For the preparation of the purified phosphoglucomutase, see Vol. I [36].

Procedure. Mix the following components: 0.2 ml. of G-1-P, 0.01 ml. of magnesium sulfate, 0.02 ml. of enzyme, and variable amounts of the glucose-1,6-diphosphate solution to a total volume of 0.3 ml. Usually two or three different amounts of the unknown are compared with a curve obtained in the same series with two or three samples of the standard solution. The reaction is started by adding the enzyme, and after 10 minutes at 37° it is interrupted by the addition of 1.5 ml. of Somogyi's sugar reagent. The tubes are heated for 10 minutes in a boiling water bath, cooled, and 1.5 ml. of Nelson's reagent and water to complete the volume to 7.5 ml. are added (see Vol. III [12]).

Remarks. Since no buffer is used, all the samples should be adjusted to pH 7.4 to 7.5. Blanks without G-1-P are run at the same time. These blanks are necessary because crude solutions may reduce initially or may develop reducing power during incubation. For instance, samples containing sucrose give a high blank due to the invertase present in the yeast extract. If blanks are high, it is convenient to purify partially by precipitation with lead acetate. The most frequent cause of error is the presence of salts which inhibit the enzyme.[2] Sometimes a precipitation with lead acetate is useful in this case also.

Substances with SH groups hardly affect the activity with the yeast preparation, and, moreover, they can be destroyed by heating in alkaline solution.

Purification Procedure

1. Preparation of the Yeast Extract. The technique of Neuberg and Lustig for preparing fructose diphosphate (see Vol. III [22]) is used with fresh baker's yeast, saccharose, and ether. The glucose-1,6-diphosphate content increases during incubation at 30° up to 24 hours, and slowly decreases thereafter.

Hours	0	2	5	24
Glucose-1,6-diphosphate content (μM.)	11	185	415	445

The substitution of saccharose for glucose or starch does not improve the yield. After incubation, the proteins are coagulated by heating, and the mixture is filtered through fluted paper.

[2] G. T. Cori, S. P. Colowick, and C. F. Cori, *J. Biol. Chem.* **124**, 543 (1938).

2. Precipitation with Lead Acetate. To the filtrate of step 1, glacial acetic acid is added (2 ml. per 100 ml.), and then 25% lead acetate until the supernatant gives no more precipitate on addition of more reagent. The suspension is filtered on Büchner funnels through Celite Super-Cel. The cake is suspended in water and decomposed with H_2S. After filtration and aeration, the liquid is made alkaline to phenolphthalein with NaOH.

COMMENTS. The treatment with H_2S should be carried out rapidly, since glucose-1,6-diphosphate is acid-labile. The omission of step 2 does not give good results.

3. Destruction of Fructose Diphosphate. The liquid from step 2 is treated with 0.05 vol. of 5 N NaOH and then heated in a large boiling water bath. Half an hour after the thermometer has reached 90°, the liquid is cooled. Acetic acid is added to pH 7, followed by an excess of magnesium acetate. The mixture is left overnight at 5°. The $Mg_3(PO_4)_2$ is filtered off, and NH_4OH is added until phenolphthalein gives a rose color. After filtration, the absence of inorganic phosphate in the filtrate is checked analytically. Excess lead acetate is then added. The suspension, which is slightly alkaline to litmus, is filtered, and the precipitate is decomposed with H_2S, filtered, and aerated.

COMMENTS. During the alkaline treatment the liquid becomes dark brown, and nearly all the organic phosphate is hydrolyzed. The Seliwanoff reaction for fructose[3] becomes negative. The amount of alkali can be increased up to 1 N and the heating up to 3 hours without appreciable destruction of glucose-1,6-diphosphate.

The elimination of the inorganic phosphate with magnesia mixture in the usual manner results sometimes in a loss of about 40% of glucose-1,6-diphosphate by coprecipitation. Excess NH_4OH should be avoided, since it interferes in the subsequent lead precipitation.

4. Barium Fractionation. Excess barium acetate and 0.2 vol. of alcohol are added, and the liquid is filtered through Celite Super-Cel. The dry precipitate (step 4a, Summary of Purification Procedure) is suspended in 10 vol. of water, cooled in ice, and adjusted to pH 3.5 (bromophenol blue) with HCl. The precipitate is discarded, and the liquid is treated with 3 vol. of 96% ethyl alcohol. The acid barium salt is separated with hardened filter paper, dissolved in water, adjusted with $Ba(OH)_2$ to pH 8, and 0.2 vol. of alcohol is added. The precipitate is then centrifuged and dried with alcohol and ether (step 4b, Summary of Purification Procedure).

COMMENTS. This fractionation has been described by MacFarlane[4]

[3] J. H. Roe, *J. Biol. Chem.* **107,** 15 (1934).
[4] M. G. MacFarlane, *Biochem. J.* **33,** 565 (1939).

for the purification of FDP. In some cases the procedure can be repeated; in others the acid barium salt can be dried and treated directly as described in step 5. The latter procedure is to be preferred.

After this step it can be considered that all the acid-labile phosphate corresponds to the glucose-1,6-diphosphate, since all the FDP has been destroyed and G-1-P is separated as the barium soluble salt.

SUMMARY OF PURIFICATION PROCEDURE

Step	Amount of extract	Micromoles of glucose-1,6-diphosphate		Relation P total/ P acid-labile
1. Filtered yeast extract	1700 ml.	650[a]		226[b]
2. After Pb and H_2S	1300 ml.	680[a]		130[b]
3. Alkali, Mg^{++}, Pb, H_2S	580 ml.	—		15[b]
4a. First Ba salt	5.06 g.	545[a]	680	10
4b. Fractionated Ba salt	1.85 g.		450	6.5
5a. Acetone precipitation	—	F_1	50	2.2
		F_2	128	2.1
		F_3	107	2.5
		F_4	28	3.8
5b. Final Ba salt ($F_1 + F_2$)	—		140	2.07

[a] Estimations by enzymatic method.
[b] P total/micromoles of glucose-1,6-diphosphate.

5. *Precipitation with Acetone.* The barium salt is suspended in water as a thick paste, cooled, and 5 N H_2SO_4 added until thymol blue gives a rose color. After the suspension has been checked for the absence of barium ions, it is filtered on a Büchner funnel through Celite and a small amount of Norit. The filtrate is treated with 10 vol. of acetone and centrifuged. The oily precipitate is dissolved in a small volume of water and stored in ice (fraction 1). To the supernatant a drop of concentrated NH_4OH is added followed by centrifugation. Three or four fractions are collected in this manner and analyzed separately for inorganic, acid-labile, and total phosphate (step 5a, Summary of Purification Procedure). The fractions which show a relation total phosphate/phosphate acid-labile near 2 are mixed, acidified with acetic acid, and treated with $BaCl_2$. The $BaSO_4$ is centrifuged off and washed. The combined supernatant and washings are adjusted to pH 8 with filtered saturated $Ba(OH)_2$, 0.2 vol. of alcohol added, and centrifuged. The precipitate is washed with 20% alcohol until the supernatant gives no reaction for chloride and then dried with alcohol-ether (step 5b). The yield and purification in different steps are shown in the table. One of the barium salts obtained in this manner

and dried *in vacuo* over $CaCl_2$ had an acid-labile phosphate content of 3.48% and 8.5% of total phosphate. For an anhydrous dibarium hexose diphosphate the theoretical total phosphate is 10.15%. The Seliwanoff reaction for fructose gives values which are equal to those given by equivalent amounts of glucose. In some preparations the values were slightly higher but could be lowered by a second treatment with alkali.

Properties

The barium and lead salts of glucose-1,6-diphosphate are insoluble in water. In general, the solubility of the salts of glucose-1,6-diphosphate is very similar to that of the corresponding salts of FDP. Nevertheless, they differ remarkably in their behavior toward alkali; the glucose diphosphate molecule resists 3 hours of heating in 1 N NaOH at 90° without being destroyed, whereas FDP is completely degraded by a much milder treatment.

Heating in 0.1 N acid at 100° leads to complete liberation of the labile phosphate of glucose-1,6-diphosphate in 9 to 10 minutes. This property can be used analytically as a method of estimation, provided that other labile phosphorus-containing compounds are absent.

[18] Chemical Syntheses of Glucose-1,6-diphosphates

By THEODORE POSTERNAK

Principle

α-Glucose-1,6-diphosphate can be synthesized by both general methods which have been indicated for the synthesis of α-aldohexose-1-phosphates (see Vol. III [16A]). The silver diphenylphosphate procedure affords a better yield of α-glucose-1,6-diphosphate than the trisilver phosphate procedure; β-glucose-1,6-diphosphate is formed as a by-product in the latter synthesis. The starting material in both procedures is 2,3,4-triacetyl-α-glucosylbromide-6-diphenylphosphate.

Preparation of 2,3,4-Triacetyl-α-glucosylbromide-6-diphenylphosphate

6-Trityl-β-D-glucose-1,2,3,4-tetraacetate.[1,2] A mixture of 12 g. of anhydrous glucose, 19.3 g. of trityl chloride, and 50 ml. of anhydrous pyridine is heated on the steam bath with exclusion of moisture and with stirring by hand until solution is complete. Without cooling, 36 ml. of

[1] B. Helferich and W. Klein, *Ann.* **450**, 219 (1926).
[2] D. Reynolds and W. L. Evans, *Organic Syntheses* **22**, 56 (1942).

acetic anhydride is added in one portion. After standing for 12 hours, the solution is poured slowly, with mechanical stirring, into a mixture of 1 l. of ice water and 50 ml. of acetic acid. The resulting suspension is vigorously stirred mechanically for 2 hours. After filtration, the precipitate is suspended again in 1 l. of ice water and stirred for 15 minutes. The precipitate is filtered, washed well with cold water, and dried in a vacuum desiccator over H_2SO_4. The solid is then digested with 50 ml. of ether in order to dissolve the α isomer. The fraction which remains insoluble or which separates (β isomer) is dissolved in approximately 300 ml. of 95% ethanol, and the solution is filtered while hot. Fine needles of 6-trityl-β-D-glucose tetraacetate separate on cooling. Yield, 17 g.; melting point, 158 to 162°. The product is of sufficient purity for further use. Recrystallization from 95% ethanol raises the melting point to 166 to 167°; $[\alpha]_D^{19} = +44.8°$ (pyridine).

β-D-*Glucose-1,2,3,4-tetraacetate*.[1,2] 4.6 g. of 6-trityl-β-D-glucose-1,2,3,4-tetraacetate is dissolved on the steam bath in 20 ml. of acetic acid. After cooling to approximately 10°, 1.8 ml. of a saturated solution of dry hydrogen bromide in acetic acid is added, and the reaction mixture is shaken for about 45 seconds. The precipitate of trityl bromide is removed *at once* by filtration with suction on a sintered glass funnel and washed with some acetic acid. The filtrate is poured *immediately* into 100 ml. of cold water. The tetraacetate is extracted with 25 ml. of chloroform. The extract is washed four times with ice water and dried over anhydrous sodium sulfate. The chloroform is evaporated *in vacuo* at room temperature. When the remaining sirup is covered with 10 ml. of anhydrous ether and rubbed with a glass rod, crystallization takes place immediately. The product is recrystallized by dissolving it in the minimum amount of chloroform and adding anhydrous ether until crystallization begins. Yield, 1.5 g.; melting point, 128 to 129°; $[\alpha]_D^{20} = +12°$ (chloroform).

1,2,3,4-Tetraacetyl-β-D-glucose-6-diphenylphosphate.[3] Six grams (4.72 ml.) of diphenyl chlorophosphonate[4] is added, dropwise, from a buret, with continuous shaking and cooling in an ice bath, to a precooled solution of 7.1 g. of 1,2,3,4-tetraacetyl-β-D-glucose in 20 ml. of anhydrous pyridine. The mixture is then kept for 15 minutes at 0° and for 18 hours at 10°. One-half milliliter of water is added and, after ½ hour, the mixture is poured slowly, with stirring, into 600 ml. of ice water. When the precipitate becomes granular, it is filtered and shaken for 1 hour in the cold with 200 ml. of ice water. After filtration, the product is dissolved in 100 ml. of chloroform. The solution is washed once with cold 5% HCl[5] and twice

---

[3] H. A. Lardy and H. O. L. Fischer, *J. Biol. Chem.* **164**, 513 (1946).
[4] See Vol. III [16A].
[5] The reaction of the water layer must remain acid to Congo red paper.

with ice water; it is then dried over anhydrous sodium or calcium sulfate and evaporated *in vacuo* to a sirup. Crystals are obtained by careful addition of petroleum ether. Yield, 10.9 g.; melting point, 64 to 66°. For recrystallization, the substance is dissolved in hot acetic ether and petroleum ether is added. Melting point, 68°; $[\alpha]_D^{22} = +16.5°$ (anhydrous pyridine, $c = 1.37$).

2,3,4-Triacetyl-α-glucosylbromide-6-diphenylphosphate.[6] 5.5 g. of 1,2,3,4-tetraacetylglucose-6-diphenylphosphate is suspended in 6 ml. of glacial acetic acid saturated at 0° with dry hydrobromic acid. The mixture is left for 2½ hours at room temperature. If the suspension is frequently shaken, the compound will dissolve in ½ hour. The solution is then diluted with 12 ml. of ice-cold chloroform and shaken with 40 ml. of ice water. After removal of the chloroform layer, the water solution is extracted with 6 ml. of chloroform. The combined chloroform extracts are washed several times with small amounts of ice-cold water until the reaction is only faintly acid to Congo red paper. The combined chloroform extracts are dried by shaking them for a few minutes with anhydrous calcium chloride. Most of the solvent is then evaporated by distillation *in vacuo* at room temperature. Crystals are obtained by slow addition at 0° of petroleum ether to the concentrated solution. The compound is recrystallized in the same way from chloroform-petroleum ether. Yield, 5 g.; the substance melts first at 74 to 75°, solidifies again, and then melts at 87 to 88°; $[\alpha]_D^{26} = +149°$ (chloroform, $c = 1.49$).

Preparation of α- and β-Glucose-1,6-diphosphates by Condensation with Trisilver Phosphate[6]

The preceding bromo compound (4.7 g.) is refluxed, with exclusion of moisture, with 14 ml. of dry benzene and 1.2 g. of freshly prepared, dry, finely powdered trisilver phosphate.[4] Two additional portions of 1.2 g. of trisilver phosphate are added, one after ½ hour and the other after 1 hour, and the mixture is then refluxed for 1 hour more. The silver salts are filtered with suction and washed with dry benzene. The combined solutions are evaporated to dryness under reduced pressure, and the last traces of solvent are removed after 5 to 6 hours in a high vacuum. The residue is then dissolved in the cold in 30 ml. of dry methanol. After filtration, the solution is hydrogenated at room temperature and atmospheric pressure in the presence of 500 mg. of platinum oxide. If the uptake of hydrogen becomes too slow, it is necessary after a while to add 250 mg. more of catalyst. The theoretical amount (8 moles H_2 per mole of sugar) is usually absorbed in 8 hours.

The filtered solution which does not contain any inorganic phosphate

[6] T. Posternak, *J. Biol. Chem.* **180**, 1269 (1949).

is diluted to 100 ml. with methanol, and 1.6 ml. of concentrated hydrochloric acid is added. The mixture is allowed to stand at room temperature. At hourly intervals, 0.1-ml. samples are withdrawn and analyzed for inorganic phosphate. As soon as the inorganic phosphate has risen to about 20% of the labile phosphate, the solution is neutralized to phenolphthalein with normal sodium hydroxide. Most of the methyl alcohol is removed by distillation *in vacuo*. The inorganic phosphate is then precipitated by an excess of magnesia mixture prepared with magnesium chloride. After neutralization with acetic acid, the solution is diluted with water to 200 ml., a concentrated solution of 3.5 g. of barium acetate is added, and the mixture is heated to boiling.

The flocculent precipitate of barium glucose-1,6-diphosphate is filtered with suction while hot. An additional amount is precipitated by adding 0.25 vol. of alcohol to the filtrate. Total yield, 0.7 g. For purification the compound is dissolved at 0° in the minimum amount of very dilute hydrochloric acid. After weak alkalinization with ammonia, the mixture is heated to boiling, and the precipitate is filtered hot and washed with hot water, alcohol, and ether. The compound is a mixture, in variable proportions, of the barium salts of α- and β-glucose-1,6-diphosphates.

Separation and Properties of α- and β-Glucose-1,6-diphosphates. The separation can be carried out by fractional crystallization of the brucine salt, the salt of the β form being less soluble in water than the salt of the α form. Two hundred milligrams (containing 18.0 mg. P) of the finely powdered barium salt of a mixture of the α and β isomers is suspended in 3 ml. of water, 590 mg. of brucine sulfate is added, and the mixture is heated, with constant stirring, gradually to 100°. It is centrifuged hot, and the barium sulfate is washed exhaustively with water. Excess barium and sulfate ions must be removed by addition of the minimum required amount of brucine sulfate or barium chloride. The solution (volume, 10 ml.) is left overnight at 25°. The crystals (200 mg.) consisting mainly of the brucine salt of the β form are recrystallized repeatedly from 100 times their weight of water. $C_6H_{14}O_{12}P_2$ $(C_{23}H_{26}O_4N_2)_4$ dried in vacuo, at 110°, over P_2O_5. $[\alpha]_D^{25} = -30.4° \pm 1.6°$ (water, $c = 0.64$).

The mother liquor of the first crystallization is evaporated to dryness *in vacuo*, and the crystalline residue is washed with 2 ml. of ice water. It contains mainly the salt of the α form contaminated with some of the β form. In order to remove the latter, the crystals are dissolved in 5 ml. of water and the solid which crystallizes at room temperature is removed. The purity of the soluble fraction can be checked polarimetrically; if the specific rotation is more negative than $-18°$, the fractionation has to be repeated. The purified brucine salt of the α form dried in vacuo, at 110°, over P_2O_5 has a specific rotation: $[\alpha]_D^{26} = -16.3° \pm 1.3$ (water, $c = 0.8$).

In order to transform the brucine salts into the sodium salts, sodium hydroxide is added to a suspension in water of the brucine salt, until a permanent pink color is obtained with phenolphthalein. The solution is then exhaustively extracted with chloroform in order to remove brucine and the pH is adjusted to 8.0.

For conversion into the barium salt, excess barium acetate is added to the solution of the sodium salt. The mixture is then diluted with 1 vol. of ethanol, in order to precipitate the barium salt completely; it is centrifuged down and washed with 50% ethanol, absolute ethanol, and ether. In a dry state it can be stored without decomposition for years. The analysis of the salt of the α form dried at 110°, *in vacuo*, over P_2O_5 corresponds to the formula $C_6H_{10}O_{12}P_2Ba \cdot H_2O$. These barium salts are only slightly soluble in cold water and still less soluble in hot water.

For enzymatic assays, the barium salts are converted to the sodium or potassium salts in the following way. They are suspended in ice-cold water and dissolved with a minimum amount of hydrochloric acid followed by the addition of the calculated amount of sodium or potassium sulfate. After neutralization with sodium or potassium hydroxide, the barium sulfate is removed by centrifugation. All these operations should be carried out at 4°. α Form (sodium salt): $[\alpha]_D^{28} = +83° \pm 4°$ (water, $c = 0.229$ calculated as anhydrous free acid). β Form (sodium salt): $[\alpha]_D^{26} = -19° \pm 2°$ (water, $c = 0.373$) calculated as anhydrous free acid.

Pure samples of the glucose-1,6-diphosphates have no reducing power, but reducing properties appear after a short acid hydrolysis; one-half of the total phosphorus is split after 5 minutes of heating at 100° in normal sulfuric acid. Hydrolysis constants in N H_2SO_4, at 30°: α form 7.8 $\times$ 10^{-4}; β form 3.13 $\times$ 10^{-3}. α-Glucose-1,6-diphosphate activates phosphoglucomutase, whereas the β form is enzymatically inactive. If the enzyme and the glucose-1-phosphate preparations used are coenzyme-free, the concentration of α-glucose-1,6-diphosphate which gives half-maximal velocity is 5 $\times$ 10^{-7} M in the usual reaction mixture containing 5 $\times$ 10^{-3} M G-1-P, 1.5 $\times$ 10^{-3} M magnesium sulfate, and 2.5 $\times$ 10^{-2} M cysteine. Maximal activation requires a concentration of about 4 $\times$ 10^{-6} M.[7]

Preparation of α-Glucose-1,6-diphosphate by Condensation with Silver Diphenyl Phosphate[6]

A mixture of 0.6 g. of 2,3,4-triacetyl-α-glucosylbromide-6-diphenylphosphate, 0.36 g. of silver diphenylphosphate,[4] and 2 ml. of dry benzene is refluxed for 15 minutes with exclusion of moisture. Two-tenths gram of silver diphenylphosphate is then added, and the refluxing is continued

[7] E. W. Sutherland, M. Cohn, T. Posternak, and C. F. Cori, *J. Biol. Chem.* **180**, 1285 (1949).

for 30 minutes. The silver salts are removed by centrifugation and washed with benzene. After distillation of the solvent *in vacuo*, the residue of the benzene solutions is dissolved in a small amount of chloroform, reprecipitated with petroleum ether, dried *in vacuo*, and redissolved in 10 ml. of anhydrous methanol. The filtered solution is hydrogenated in the presence of 200 mg. of platinum oxide. Usually the theoretical amount of hydrogen (16 moles per mole of sugar) is absorbed in 1 hour; if the gas uptake becomes too slow, it is necessary to add 100 mg. more of catalyst. After filtration, normal sodium hydroxide is added to the cooled solution, until a permanent pink color is obtained with phenolphthalein. A gummy precipitate separates which is dissolved by addition of water. The solution is then concentrated *in vacuo* in order to remove methanol and cyclohexane. The inorganic phosphate which represents about 30% of the total phosphate is precipitated with an excess of magnesia mixture prepared from magnesium chloride. The pH of the filtered solution is adjusted to 8.5 with acetic acid. By addition of 0.5 g. of barium acetate and heating to boiling for 2 minutes, the barium salt is precipitated; it is centrifuged while hot and washed with hot water, alcohol, and ether. The compound (115 mg.), which already consists mainly of α-glucose-1,6-diphosphate, is purified as brucine salt as indicated above. Sodium salt: $[\alpha]_D^{25} = +80.3° \pm 2.4°$ (water, $c = 0.406$, calculated as anhydrous free acid in water).

When ethanol instead of methanol is the solvent for the hydrogenation, the amount of inorganic phosphate formed is about twice as small and the final yield is twice as great. However the isolated glucose-1,6-diphosphate contains about 25% of the β form, and the separation of the components has to be carried out by fractional crystallization of the brucine salts as described above.

[19] D-Glucose-6-phosphate

By B. L. HORECKER and W. A. WOOD

Preparation

Principle. Fructose-6-phosphate is converted to glucose-6-phosphate by the action of hexose phosphate isomerase. In the presence of barium ion the equilibrium

$$\text{D-Fructose-6-phosphate} \leftrightarrows \text{D-Glucose-6-phosphate}$$

is displaced by the precipitation of a highly insoluble barium glucose-6-phosphate heptahydrate, and the reaction proceeds essentially to completion.

Reagents. Fructose-6-phosphate may be obtained by acid hydrolysis[1] of fructose-1,6-diphosphate.[2] Commercial barium fructose-6-phosphate, which contains about 40% of barium fructose-6-phosphate and 15% of barium glucose-6-phosphate, is an inexpensive starting material. Active preparations of hexose phosphate isomerase may be prepared from rat or rabbit skeletal muscle or from bacterial extracts. The enzyme preparations must be freed from sulfate and phosphate ions by dialysis.

A crude hexose phosphate isomerase from muscle is adequate for the preparation (for the preparation of the purified enzyme, see Vol. I [37]). Fresh rat skeletal muscle is ground with a small hand meat grinder, extracted with an equal weight of cold water, and strained through several layers of cheesecloth. The residue is again extracted with one-half the original quantity of water. The combined extracts are dialyzed against cold distilled water until free of inorganic phosphate.[3] The dialyzed solution is centrifuged for 10 minutes at 18,000 × g, and the precipitate is discarded.

Procedure. Two grams of commercial barium fructose-6-phosphate is dissolved in 20 ml. of water, and the insoluble residue is removed by centrifugation.[4] The supernatant solution, containing approximately 0.68 millimole of glucose-6-phosphate and 1.74 millimoles of fructose-6-phosphate, is treated with 4.0 ml. of the dialyzed isomerase preparation. Several seed crystals are added, and the reaction is allowed to proceed at room temperature. After 16 hours the mixture is chilled to 0° and filtered. The crystalline precipitate is washed with a small volume of cold water and dried in air. The product weighs 1.0 g. and contains 1.92 millimoles of barium glucose-6-phosphate heptahydrate (79% of theory). It may be dissolved in hydrochloric acid and recrystallized by neutralizing the solution.

Properties

The properties of glucose-6-phosphate and several methods of analysis have been described previously.[5]

Assay. A sensitive and highly specific method for the determination of glucose-6-phosphate is the reduction of triphosphopyridine nucleotide

[1] C. Neuberg, H. Lustig, and M. A. Rothenberg, *Arch. Biochem.* **3,** 33 (1943); see Vol. III [22].
[2] H. Z. Sable, *Biochem. Preparations* **2,** 52 (1952); see Vol. III [22].
[3] An 8-hour dialysis period is satisfactory.
[4] Some commercial batches of fructose-6-phosphate give a deeply colored yellow-orange solution and, in this procedure, yield appreciable quantities of an amorphous product which retains the chromogen. Such batches may be treated with activated carbon and filtered while hot to give clear solutions which yield only the crystalline product.
[5] H. A. Lardy, *Biochem. Preparations* **2,** 39 (1952).

taken at 30 minutes, 1 hour, and 3 hours and treated, as was the initial sample in order to follow the decrease in nonphosphorylated reducing sugar. The aliquots showed that 50% of the mannose had been phosphorylated in 1 hour, and 80% in 3 hours.

After 3 hours, 3 ml. of 1 M barium acetate, a few drops of phenolphthalein, and about 0.6 ml. of 2 N NaOH were added to give a slightly pink color. The suspension was left in the cold overnight, and the water-insoluble barium salts were centrifuged off and washed with 10 ml. of cold 0.05 M barium acetate adjusted to about pH 8.5. The supernatant fluids were combined, treated with 4 vol. of ethanol, and left in the cold overnight. The alcohol-insoluble barium salts were centrifuged off, washed twice with 12-ml. portions of cold 80% ethanol, and drained well. The precipitate was stirred up with 40 ml. of water and centrifuged. The dark gray residue was washed with 2 ml. of water, and the washing was added to the main solution of clear, colorless water-soluble salts. Phenol red was added to this solution, and a trace of 1 M acetic acid to adjust the pH to about 7.5. One-half milliliter of 0.5 M silver nitrate was added, and then a trace of NaOH to restore the pink color of phenol red. This procedure removes any traces of ATP or its derivatives, which are discarded after centrifugation.[8] A slight excess of chloride ion (0.2 ml. of 1 N HCl) was added to remove silver ion, and the clear supernatant fluid was made up to 50 ml. with water and 0.1 ml. of 2 N NaOH to give a purplish color. Then 0.2 ml. of 1 M barium acetate and 200 ml. of ethanol were added, and the suspension was cooled for 30 minutes in an ice bath before centrifugation. The precipitate was washed with 15 ml. of cold 80% ethanol and dried in a vacuum desiccator at room temperature. The grayish cake was dissolved with 9 ml. of water, and a trace of insoluble residue was centrifuged off. Barium was removed by the addition of a slight excess of 0.5 M sulfuric acid (about 1.2 ml.). The supernatant liquid was adjusted to pH 7 with NaOH and made up to 15 ml., which gave a solution of about 0.04 M M-6-P. This solution keeps well when frozen.

Assay Method

Phosphorus analysis, by the method of Fiske and SubbaRow,[9] indicated negligible inorganic orthophosphate. After wet-ashing with sulfuric acid and hydrogen peroxide, the organic phosphorus concentration was found to be 0.043 M.

---

[8] S. E. Kerr and K. Seraidarian, *J. Biol. Chem.* **159**, 211 (1945). It is important that the treatment with silver nitrate be done after precipitation of the alcohol-insoluble barium salts; otherwise a voluminous precipitation of the silver salt of Tris would occur.

[9] C. H. Fiske and Y. SubbaRow, *J. Biol. Chem.* **66**, 375 (1925); see Vol. III [114].

The resorcinol color reaction of Roe[10] indicated the presence of some F-6-P which resulted from a trace of PMI activity in the heat-treated yeast extract. Since aldohexoses give the resorcinol color reaction to a much smaller degree than fructose, part of the color may have been due to the mannose moiety. An enzymatic assay showed the presence of about 2% hexosemonophosphates other than M-6-P. The following test system was made up in a cell having a 1-cm. light path:

> 1.0 ml. 0.1 M Tris, pH 7.5
> 1.25 ml. H_2O
> 0.2 ml. 0.001 M TPN [11]
> 0.2 ml. 0.1 M $MgCl_2$
> 0.05 ml. phosphoglucose isomerase[12]
> 0.2 ml. G-6-P dehydrogenase[13]
> ———
> 2.90 ml.

One micromole of the M-6-P preparation was added to this mixture in a spectrophotometer set at 340 mμ. The reduction of TPN [14] became maximal in about 5 minutes and showed the presence of about 2% F-6-P plus G-6-P.

The same test system, with the further addition of PMI,[15] was then used to assay total hexose-6-P in the preparation, using a limiting amount (0.1 micromole) of organic P. Reduction of TPN was complete in about 15 minutes and indicated the presence of 0.1 micromole total hexose-6-P. The M-6-P preparation was, therefore, 98% pure so far as phosphorus compounds were concerned.

[10] J. H. Roe, *J. Biol. Chem.* **107,** 15 (1934); see Vol. III [12].

[11] TPN is available commercially. For methods of isolation and purification, see Vol. III [124].

[12] A fraction free of PMI should be used; see Vol. I [37]. Sufficient enzyme should be added so that the rate of reaction of trace amounts of F-6-P will be rapid enough to assure that an end point is reached in a conveniently short time (e.g., 5 to 10 minutes). 0.25 mg. of a fraction having about 12,000 units/mg. is satisfactory.

[13] Prepared as described by A. Kornberg, *J. Biol. Chem.* **182,** 805 (1950); see Vol. I [43]. Sufficient enzyme (about 0.1 unit) should be used to assure rapid attainment of an end point. Since such preparations may also contain a substance which catalyzes the reoxidation of TPNH, it is important not to add a large excess of the enzyme. Reoxidation of TPNH will be indicated by a slow decrease in spectrophotometer readings after a maximal reading has been reached.

[14] For each micromole of G-6-P present, or formed enzymatically from other hexosemonophosphates, 1 micromole of TPN will be reduced. This causes an increase in optical density (log I_0/I) of 2.07 units when the volume is 3 ml. and the light path is 1 cm.; see Vol. I [43].

[15] About 50 units (F-6-P color assay) were used; see Vol. I [37].

[21] Preparation and Analysis of D-Glucosamine-6-phosphate

By David H. Brown

Preparation

Principle. D(+)-Glucosamine (2-Amino-2-deoxy-D-glucose) has been shown to be rapidly phosphorylated by ATP in the presence of crystalline yeast hexokinase to give D-glucosamine-6-phosphate.[1] Brain hexokinase also catalyzes this phosphorylation,[2,3] the product presumably being the same as that in the case of the yeast enzyme. The phosphate ester can be isolated from an enzymatic incubation mixture by precipitating it as the ethanol-insoluble barium salt and purifying it by reprecipitation. In the procedure nucleotides are removed by mercury precipitation.

Enzymatic Synthesis. Although crystalline yeast hexokinase can, of course, be used (see Vol. I [32]), a procedure is given below in which a yeast protein fraction containing 20% hexokinase is used. Hexokinase of this purity is rather readily obtainable.

Reagents

> 0.20 M Tris buffer, pH 8.0.
> 0.10 M ATP, sodium salt (Sigma or Pabst), pH 6.7.
> 0.25 M $MgCl_2$.
> D-Glucosamine hydrochloride (Pfanstiehl), solid.
> Serum albumin, crystalline, bovine, solid.
> 0.001 M EDTA (ethylenediaminetetraacetic acid), sodium salt, pH 6.6.
> Yeast hexokinase, 20% pure.[4]

Procedure. A reaction mixture is prepared by mixing 80 ml. of Tris, 20 ml. of ATP, 12 ml. of $MgCl_2$, 1.29 g. of D-glucosamine hydrochloride, 10 mg. of serum albumin, 23 ml. of water, and 60 ml. of EDTA. The pH of the solution is adjusted to 8.0 by the addition of about 4.2 millimoles of NaOH added as a 2 M solution of NaOH (preferably Baker's C.P. stick form, "Low in PO_4"). The volume of the mixture must be measured accurately if it is desired to determine by change in 10-minute-labile P (100°, 1 N H+) the extent of enzymatic phosphorylation and, hence, the

---

[1] D. H. Brown, *Biochim. et Biophys. Acta* **7**, 487 (1951).

[2] R. P. Harpur and J. H. Quastel, *Nature* **164**, 693 (1949).

[3] A. Sols and R. K. Crane, *Federation Proc.* **12**, 271 (1953).

[4] L. Berger, M. W. Slein, S. P. Colowick, and C. F. Cori, *J. Gen. Physiol.* **29**, 379 (1946), see Vol. I [32].

amount of glucosamine-6-phosphate formed. The solution is brought to 30° in a water bath, and 2 ml. of a yeast protein fraction is added containing about 3 mg. protein per milliliter, of which 20% is hexokinase. Serum albumin is added to the solution to increase the total protein concentration which should be about 80 γ/ml. The initial concentration of ATP is 0.01 M, and that of glucosamine 0.03 M.

The solution is incubated at 30° for 30 minutes. With the amount of hexokinase used above, the reaction is substantially complete (formation of 0.01 M glucosamine-6-phosphate) in less than 15 minutes. The final pH will be about 7.4. Either the solution may be frozen and stored overnight, or the purification of the ester may be begun immediately.

Purification Procedure

The pH of the solution is adjusted to 8.2 by the addition of about 1.2 millimoles of NaOH (care should be taken not to overalkalinize). Twenty-five milliliters of 0.92 M barium acetate are added (or more if precipitation is not complete), and the mixture allowed to stand in ice for 30 minutes. The water-insoluble barium salts are then centrifuged in the cold, and the supernatant fluid is filtered to free it from traces of suspended solid. The precipitate may be discarded, since less than 10% of the total glucosamine-6-phosphate formed can be recovered from it. The volume of the filtrate of water-soluble barium salts is measured, and 4 vol. of cold 95% ethanol is added to it. The mixture should be allowed to stand overnight in the cold.

The heavy flocculent precipitate is removed by centrifugation in the cold, and, after the supernatant fluid has been discarded, the barium salts are washed once with twice their volume of cold absolute ethanol and the washings discarded. The precipitate is largely dissolved in 40 ml. of water with the aid of about 6.5 ml. of 1 M HNO_3. In all, enough HNO_3 is added to make the pH 2.0 $\pm$ 0.1. Without removing any suspended solid, 1.5 ml. of 15% $Hg(NO_3)_2$ in 0.5 M HNO_3 is added to precipitate nucleotides. The mixture is kept in ice for 30 minutes and then centrifuged. The supernatant fluid is tested for completeness of precipitation with the 15% $Hg(NO_3)_2$ reagent. As much as another 0.5 ml. may be required. Care should be taken not to add more $Hg(NO_3)_2$ than is necessary for complete precipitation, since otherwise large amounts (up to 30%) of glucosamine-6-phosphate may be precipitated. The voluminous white precipitate is centrifuged and discarded. The supernatant fluid is decanted through filter paper, cooled in ice, and treated with H_2S to precipitate HgS. The mixture is then filtered and the filtrate aerated briefly (5 to 10 minutes) to remove excess H_2S. The odor of polysulfides remains.

To the clear solution 0.5 ml. of 0.92 M barium acetate is added, followed by enough 2 M NaOH to bring the pH to 8.2 (about 6 millimoles required). After standing for 30 minutes in ice, the mixture is filtered in the cold and the precipitate discarded. To the filtrate four times its volume of cold 95% ethanol is added, and the mixture is allowed to stand overnight in the cold.

The precipitate of barium glucosamine-6-phosphate obtained by centrifugation is dissolved in the minimum amount of cold water (about 50 ml.), filtered to remove some granular precipitate, and reprecipitated with 4 vol. of cold ethanol. Solution in the minimum amount of cold water and reprecipitation in the cold is repeated once more, and the final precipitate is dissolved in no more than 60 ml. of cold water and filtered. It is recommended that this solution be analyzed for total P (H_2SO_4-H_2O_2 digestion) in order to determine its glucosamine-6-phosphate content. A calculated amount of standard 0.50 M H_2SO_4 then is added to remove Ba^{++} and to acidify the solution. The calculation is made on the basis of one Ba^{++} being present for each P atom found by analysis. The resulting mixture has a pH between 2.5 and 3.5, depending on the concentration of the ester. $BaSO_4$ is removed from it by centrifugation, and the supernatant fluid (test for complete removal of Ba^{++}) is stored in the deep-freeze. The solution contains 0.01 to 0.02 M glucosamine-6-phosphate. The yield is about 50% based on the ATP taken (yields vary from 40 to 70%).[5] Glucosamine-6-phosphate is not entirely stable (see Properties), and keeping it at neutrality or in weakly alkaline solution should be avoided. Although its barium salt can be isolated in the dry state ($C_6H_{12}O_8NPBa\cdot2\frac{1}{4}H_2O$) by washing it with absolute ethanol and dry ethyl ether, followed by drying at room temperature *in vacuo* over P_2O_5,[1] the salt becomes colored and has altered properties. Therefore, it is desirable to store the ester as the free acid frozen in solution.

Properties

Glucosamine-6-phosphate is stable to acid hydrolysis (1 N H^+, 100°); about 1.5% of the ester is hydrolyzed in 30 minutes in H_2SO_4. It reduces alkaline copper reagents, and 1 mole of the ester consumes the expected 4 moles of $NaIO_4$ at room temperature in an unbuffered solution of initial pH 4.5.[1] Solutions of the ester invariably have an absorption spectrum with a maximum at 273 mμ. There is a minimum at 241 mμ,

---

[5] The use of cationic exchange resins to remove Ba^{++} from the final solution has been investigated. Duolite C-3 in its hydrogen form can be used, but it removes considerable amounts of the glucosamine-6-phosphate as well as the Ba^{++} and seems to catalyze a reaction by which a part of the reducing power of the ester is lost (see Properties). The use of ion exchange resins in this preparation is not recommended.

and sometimes an ill-defined shoulder near 300 mμ. The spectrum is believed to be that of a substance of unknown structure produced from the ester by standing in solution. The absorption at 273 mμ increases with time; this increase is much more rapid in neutral and alkaline solutions than in acid solution and appears to be due to an irreversible change probably involving carbon atom 1 of the glucosamine molecule, since the reducing power of old solutions may decline. A suggestion has been made concerning the possible type of compound responsible for the observed absorption spectrum.[1]

The second and third ionization constants of the ester have been determined[6] to be pK_2' = 6.08 and pK_3' = 8.10 at 26°. The specific rotation of the dipolar ion of the ester is $[\alpha]_D^{24}$ = +48.5° (c, 0.51% in H$_2$O; pH 2.50).[6]

Analysis

Glucosamine-6-phosphate may be determined by making use of its reduction of reagent 60 of Shaffer and Somogyi[7] and subsequently employing the arsenomolybdate reagent of Nelson[8] for color development. Glucosamine hydrochloride should be used as a standard. Both substances give linear curves which do not pass through the origin, owing to the fact that a constant small amount of the amino sugar is destroyed in the alkaline reagent more rapidly than it can reduce the Cu^{++}. Standard curves must always be determined. For analysis of 0.05 micromole or less of glucosamine-6-phosphate, the ferricyanide oxidation method of Park and Johnson[9] may be used,[6] but the only suitable standard is a pure solution of the ester itself, since it reduces much less than does glucosamine on a molar basis. Neutral salts have a marked effect on this method.[6]

The ester can also be determined by means of total P after H$_2$SO$_4$-H$_2$O$_2$ digestion (method of Fiske and SubbaRow).[10]

Glucosamine-6-phosphate gives about 90% as much color on a molar basis as glucosamine in an Elson-Morgan type of test[11] in which the sample, in a volume of 2 ml., is heated for 20 minutes at 100° with 1 ml. of 2% (in 0.5 N Na$_2$CO$_3$) 2,4-pentanedione; the cooled solution is treated with 1 ml. of a solution of 6 N HCl in 50% ethanol containing 1.3% p-dimethylaminobenzaldehyde. After dilution with ethanol, the color is developed for 30 minutes at 37° and read at 540 mμ.

[6] D. H. Brown, *J. Biol. Chem.* **204,** 877 (1953).
[7] P. A. Shaffer and M. Somogyi, *J. Biol. Chem.* **100,** 695 (1933).
[8] N. Nelson, *J. Biol. Chem.* **153,** 375 (1944).
[9] J. T. Park and M. J. Johnson, *J. Biol. Chem.* **181,** 149 (1949).
[10] C. H. Fiske and Y. SubbaRow, *J. Biol. Chem.* **66,** 375 (1925).
[11] L. A. Elson and W. T. J. Morgan, *Biochem. J.* **27,** 1824 (1933).

Enzymatic Reaction

Phosphoglucomutase from rabbit muscle has been found to catalyze the conversion of glucosamine-6-phosphate to glucosamine-1-phosphate.[6] Added α-glucose-1,6-diphosphate accelerates the reaction. An apparent equilibrium is established at pH 7.53 and 30° at which 19.2% of the 6-ester is converted to 1-ester. Glucosamine-1-phosphate is more acid-labile than glucosamine-6-phosphate but yields all its labile P only after 30 minutes of hydrolysis at 100° in 1 N HClO$_4$.

[22] Ketohexose Phosphates

By INES MANDL and CARL NEUBERG

I. Fructose-1,6-diphosphate

Fructose-1,6-diphosphate (FDP) is most easily prepared from commercial baker's yeast.[1] This eliminates the necessity of washing and pressing brewer's yeast and yields satisfactory results with most brands. Fresh Atlantic, Federal, Blue Ribbon, and National Grain have been used successfully, but fresh Fleischmann's yeast cannot be used. Excellent results, however, were obtained with dried Fleischmann's as well as dried National Grain and Federal yeasts. Procedures for the use of both fresh or dried preparations are given by Neuberg and Lustig.[1]

Preparation of Crude FDP with Fresh Baker's Yeast[1]

Fresh baker's yeast (450 g.) and 150 ml. of a plasmolytic agent (ether or CCl$_4$) are added to a solution of 200 g. of sucrose, 42 g. of NaH$_2$PO$_4$·2H$_2$O, and 11 g. of NaHCO$_3$ in 1 l. of tap water. The mixture is shaken until homogeneous, the bottle stoppered in such a way that CO$_2$ gas can escape, and phosphorylation allowed to proceed in an incubator at 37°. The time required for complete phosphorylation differs with the yeast and with temperature. With Atlantic yeast at 37° it takes about $4\frac{1}{2}$ hours. The completeness of the process can be tested on a filtered aliquot of the fermentation mixture. To 2 ml. of filtrate are added 3 ml. of 2.5% NH$_4$OH, 1 ml. of 10% NH$_4$Cl, and 3 ml. of the usual Mg mixture. Immediate precipitation will not occur if phosphorylation is complete.

At this stage proteins are removed by coagulating through heating on a boiling water bath with addition of 10 ml. of a solution of 10% octyl alcohol in ethanol or by adding 50 ml. of a 20% solution of picric acid in hot ethanol and leaving it in a refrigerator for 2 hours. In either case the

[1] C. Neuberg and H. Lustig, *J. Am. Chem. Soc.* **64**, 2722 (1942).

clear filtrate is neutralized with 4 N NaOH until it is just alkaline to phenolphthalein, and the product is precipitated by adding 55 g. of $CaCl_2$ in 100 ml. of water. After heating on the boiling water bath for a short time, the precipitated Ca FDP is filtered off by suction while still warm. It is washed with hot water, and also with alcohol if picric acid deproteinization has been used. The yield is 22 to 24 g.

Ba FDP can be prepared in an analogous manner, but the picric acid method is not recommended for the Ba salt because of the relative insolubility of Ba picrate.

Preparation with Dry Yeast[1]

A mixture of 300 g. of yeast that has been well dried at room temperature or commercially available dried yeast, 150 ml. of CCl_4, and 1 l. of the sugar-phosphate solution as above are incubated at 37°. With Fleischmann's dried yeast phosphorylation is usually complete in $1\frac{1}{2}$ hours, and the product is worked up as before.

Purification by Reprecipitation from Acid Solution with Alkali[1,2]

Some purification is obtained by dissolving the moist, crude Ca FDP in 250 ml. of 2 N acetic acid, adding 125 ml. of H_2O, filtering if necessary, and adding 2 N NaOH to the clear solution until it is just alkaline to phenolphthalein.[1] Alternatively, 10 g. of Ca FDP is dissolved in 64 ml. of 2 N HCl + 50 ml. of ice-cold H_2O, and 6.5 ml. of 2 N NaOH is added dropwise.[2] The precipitate is filtered off after heating on the boiling water bath a short time and washed with hot water, in which Ca FDP is difficultly soluble. The yield is about 80% of the crude material.

Conversion of Ca FDP to Mg FDP[2,3]

The easily water-soluble (solubility 30%) and stable Mg salt is frequently more desirable as starting material for further preparation or tests than the difficultly soluble Ca, Ba, or the gummy alkali derivatives. It can be prepared by suspending Ca FDP in H_2O and shaking it with the equivalent amount of oxalic acid solution until the oxalate ions disappear. The Ca oxalate is filtered off, and the free FDP is neutralized by shaking with pure MgO. Then CO_2 is introduced, $MgCO_3$ is filtered off, and the clear filtrate is poured into four times its volume of alcohol. After a few hours the precipitate is filtered off by suction, washed with alcohol, dissolved in ice water, filtered again, and reprecipitated with alcohol.

[2] C. Neuberg and S. Sabatay, *Biochem. Z.* **161**, 240 (1925).
[3] M. Kobel and C. Neuberg, *in* "Handbuch der Pflanzenanalyse," Vol. 2, p. 556, G. Klein, ed., Springer Verlag, Vienna, 1932.

Purification via the Acid Strychnine Salt[4]

To prepare completely pure FDP from the crude preparation of commercially available material, the method of Neuberg, *et al.*[4] is recommended. It is based on the fact that the acid strychnine salt, unlike the alkaline earth salts and other FDP salts, crystallizes well and is very stable.

Neutral Ba or Ca FDP is converted into the acid strychnine salt by the procedure of Neuberg and Dalmer.[5] Strychnine sulfate dissolved in water is shaken on a shaking machine for 5 hours with one and one-half times the calculated amount of FDP salt. After this period no free SO_4^{--} should be left. The precipitate, consisting of $BaSO_4$, excess FDP, and free strychnine, is filtered or centrifuged off, and the clear solution is concentrated *in vacuo* until crystals begin to form. It is then boiled with alcohol to redissolve all but a small flocculent precipitation of unchanged Ba FDP which is filtered off. The alcoholic solution is again concentrated *in vacuo* and left to crystallize if necessary with the addition of some ethyl acetate. The resulting crystals are difficultly soluble in absolute alcohol, even when hot, more easily in aqueous alcohol, and almost insoluble in ethyl acetate, acetone, or ether. With warm water some dissociation occurs. The salts are recrystallized by dissolving in 20 parts of 85% alcohol and adding 15 parts of ethyl acetate to the hot solution. This is repeated four times. The pure substance has the formula

$$C_6H_{10}O_4(PO_4H_2)_2(C_{21}H_{22}N_2O_2)_2$$

Then 52.2 g. of the recrystallized acid salt (in glittering small needles) is dissolved in twenty times its weight of 90% methanol. After the solution has cooled but before strychnine salt crystallizes out, 199 ml. of a clear filtered solution of $Ba(OH)_2 \cdot H_2O$ in methanol containing 94.75 g. of pure $Ba(OH)_2 \cdot H_2O$ in 1 l. of water-free CH_3OH is added with vigorous shaking. The tetrabasic Ba FDP formed, $C_6H_{10}O_4(PO_4Ba)_2$, is quantitatively precipitated with 100 ml. of acetone, but it still contains some free strychnine. This is extracted after the precipitate is centrifuged off by vigorous shakings first with a mixture of 1 part of acetone and 9 parts of methanol, then with ethanol, and finally with chloroform containing 5% CH_3OH, until the extract no longer gives a violet to cherry red strychnine spot test with the Mandelin reagent consisting of 1% NH_4 vanadate in concentrated H_2SO_4. The product obtained is a fluffy white powder completely free of inorganic P or N impurities. It easily dissolves in solutions of neutral NH_4 salts in accord with the purity test of Neuberg and

---

[4] C. Neuberg, H. Lustig, and M. A. Rothenberg, *Arch. Biochem.* **3**, 33 (1943).

[5] C. Neuberg and O. Dalmer, *Biochem. Z.* **131**, 188 (1922).

Sabatay.[2] The α_D of the free FDP is determined by dissolving the Ba salt in the requisite amount of 2 N HBr to give $BaBr_2$ and making up with water: $[\alpha]_D^{17} = +4.04°$ to $+4.15°$.

Purification via the Acid Ba Salt, $C_6H_{10}O_4$ $(PO_3H)_2Ba^4$

Crude FDP containing inorganic P or mixed with monoesters may be converted to the corresponding acid salt and precipitated by methyl or ethyl alcohol. Any acid-forming alcohol-soluble alkaline earth salts are suitable—e.g., HBr, $HClO_4$, or $CCl_3 \cdot COOH$. To 6.29 g. of Ba FDP are added at 3° 50 g. of chopped ice and 50 ml. of 0.4 N HBr. The resulting clear solution is immediately added dropwise with constant stirring to 600 ml. of EtOH. After 5 minutes at 3°, the precipitate is filtered off by suction and washed with alcohol. This eliminates $BaBr_2$, which is alcohol-soluble, as well as inorganic P, F-6-P, etc., which remain in solution. After washing with alcohol and other, the acid salt may be reprecipitated from 70 ml. of cold water with 400 ml. of cold alcohol, quickly filtered, and washed with pure EtOH and water-free acetone. It can be reconverted to the normal tetrabasic salt by precipitation from aqueous solution with $Ba(OH)_2$. The resulting Ba FDP is pure. The intermediary acid salt gives satisfactory analytical data only if the starting material is Ba FDP already purified via the strychnine compound.

Analysis. Besides the possibility of estimating organic P after acid digestion or hydrolysis with phosphomonoesterase, spectrophotometric procedures have been successfully applied. Vishniac and Ochoa[6] measure the reduction of DPN in a system containing aldolase,[7] phosphotriose,[8] isomerase,[9] phosphoglyceraldehyde dehydrogenase,[10] Bücher's transphosphorylating enzyme,[11] and 0.16 μM. of DPN, 0.5 μM. of ADP, 1.5 μM. of $MgSO_4$, 400 μM. of glycine, 150 μM. of K phosphate buffer, pH 7, in a total volume of 3 ml. (FDP → two phosphotrioses); each micromole of DPN reduced corresponds to 0.5 μM. of FDP. PGA does not interfere with this method.

A similar principle forms the basis of the somewhat simpler method of Slater.[12] Making use of the rabbit muscle fraction A of Racker,[13] which contains aldolase, triosephosphate isomerase, and glycerol phosphate dehydrogenase, he too takes the change in optical density at 340 mμ due

[6] W. Vishniac and S. Ochoa, *J. Biol. Chem.* **195**, 75 (1952).

[7] Vol. I [39].

[8] Vol. III [31].

[9] Vol. I [37].

[10] Vol. I [60].

[11] Vol. I [63].

[12] E. C. Slater, *Biochem. J.* **53**, 157 (1953).

[13] E. Racker, *J. Biol. Chem.* **167**, 843 (1947).

to the disappearance of reduced DPN as a measure of the concentration of FDP, according to the equation

$$\text{FDP} \xrightarrow{\text{aldolase}} \text{dihydroxyacetone P} + \text{glyceraldehyde P}$$

$$\xrightarrow[\text{isomerase}]{\text{triosephosphate}} 2 \text{ dihydroxyacetone P}$$

$$\Big\downarrow \begin{array}{l} \text{glycerol P} \\ \text{dehydrogenase} + \\ 2 \text{ DPNH} \end{array}$$

$$2 \text{ glycerol P} + 2 \text{ DPN}$$

This method measures the sum of FDP and the triose phosphates. Sorbose-1-P would react in the same way. Phosphopyruvate or phosphoglycerate and also oxaloacetate interfere, since they react as ½ FDP. Large amounts of inorganic P have no effect and need not be separated. If the enzyme preparation contains phosphohexokinase, this should be removed by filtering through Super-Cel. In practice, optical densities at 340 mμ of a control solution of 0.3 ml. of buffer (0.25 M glycylglycine or trihydroxymethylaminoethane, pH 7.6), 0.3 ml. of reduced DPN (approximately 8×10^{-4} M), and 0.3 ml. of rabbit muscle fraction in a total volume of 2.5 ml. and unknown solutions which in addition contain 0.03 to 0.08 μM. of FDP are compared at intervals. The reference solution contains 0.3 ml. of buffer only.

Mention may also be made of a paper chromatographic method for the separation, identification, and semiquantitative analysis of FDP and related compounds. Hanes and ·Isherwood[14] use descending chromatograms on Whatman Nos. 1 and 4 papers with solvent systems such as: (1) *tert*-butanol (80 ml.)-water (20 ml.)-picric acid (4 g.); (2) *n*-propanol (60 ml.)-water (10 ml.)-concentrated NH$_4$OH(30 ml.); (3) *tert*-amyl alcohol (90 ml.)-90% formic acid (30 ml.); (4) ethyl acetate (100 ml.)-water (100 ml.)-pyridine (45 ml.). The paper is sprayed with an acid molybdate solution of 5 ml. of 60% perchloric acid, 10 ml. of N HCl, and 25 ml. of 4% NH$_4$-molybdate in a total volume of 100 ml. of water, then heated to 85° for 7 minutes to hydrolyze the esters. A phosphomolybdate complex is formed which gives a blue spot on a buff background of Mo$_2$S$_3$ when developed in a jar of H$_2$S gas. The authors prefer the NH$_4$ salts of the esters and get good separations of FDP from other hexose phosphates and 3-PGA except in solvent 1, where FDP and F-6-P give the same R_f value. Preliminary washing of the paper with 2 N acetic acid, then water, is recommended. For quantitative estimations the spots are cut out, ashed with 0.5 ml. of 3 vol. concentrated H$_2$SO$_4$ + 2 vol. 60% perchloric acid. The inorganic P resulting is determined by the method of Berenblum

[14] C. S. Hanes and F. A. Isherwood, *Nature* **164,** 1107 (1949), see Vol. III [15].

and Chain.[15] Five to ten micrograms of P in 4 to 8 cm.^{2} of paper can be determined with an error of $\pm 1.5\%$; for 1 γ of P the error is $\pm 3\%$. Several modifications of this method have been published. For references see footnotes 29–34 at the end of this section.

II. Preparation and Identification of Fructose-6-phosphate

This compound (F-6-P) is prepared by partial hydrolysis of Ca or Ba FDP based on the observation[16] that the P group in position 1 is split off by acids more readily than that in position 6. The one disadvantage of the original method[16] is that at 100° the acid causes destruction of some F-6-P. This disadvantage is overcome[4] by carrying out the hydrolysis at 35° for a prolonged period of about 5 days instead of 30 to 60 minutes at 100°. After that time about half the weight of the original FDP salt is converted to the corresponding F 6 P compound, and if the starting material was pure the F-6-P is equally pure.

Procedure According to Neuberg, Lustig, and Rothenberg[4]

Thirteen grams of Ca FDP prepared and purified as described in the previous section is dissolved in 150 ml. of N HCl, and the solution is incubated at 35°. After 5 days the solution is cooled and neutralized first with $CaCO_3$ until it is congo red, and then with a fine suspension of $Ca(OH)_2$ until it is just alkaline to phenolphthalein. Immediately thereafter CO_2 is bubbled through the solution, and the mixture is heated for 3 minutes in a vigorously boiling water bath and filtered by suction while still warm. The precipitate is washed with hot water; both filtrate and washings are concentrated *in vacuo* at 35° to approximately 90 ml. This concentrate is filtered and added slowly with vigorous stirring in a thin stream to four times its volume of 95% alcohol. The precipitate is allowed to flocculate in the refrigerator overnight, then filtered off by suction, washed with 95% alcohol, and dried in a vacuum desiccator. It can be reprecipitated by dissolving it in 70 ml. of H_2O and dropping it into 275 ml. of 95% alcohol. The material is free of chlorine.

If any FDP is still present, a test sample of the filtered solution containing 5% of the salt becomes turbid on boiling. In this case the entire product is dissolved in 150 ml. of H_2O, and 10 ml. of absolute alcohol is added. On boiling on a water bath FDP precipitates and can be filtered off while hot. The remaining solution is then concentrated *in vacuo* and again worked up as described above. The yield is 6 to 6.5 g.

Ba F-6-P is prepared analogously, with N HBr for the partial hydrolysis. Since $BaBr_2$ is soluble in alcohol, the F-6-P salt is obtained free

[15] I. Berenblum and E. Chain, *Biochem. J.* **32**, 295 (1938).
[16] C. Neuberg, *Biochem. Z.* **88**, 432 (1918).

of halogen. The Ba salt sediments even more readily than the Ca salt. Analysis of the Ba salt: $[\alpha]_D^{19}$ (in water) = $+3.58°$. After double transformation with the exact quantity of K_2SO_4, the reducing power of a 5.3% solution of F-6-P amounts to 82% of that of an equivalent solution of fructose. Neither the aqueous solution of the Ba salt nor its solution in an equivalent amount of 0.1 N HBr uses more than traces of Br_2 after standing for 7 days at room temperature.

Analysis. Besides the usual determinations of organic P, colorimetric, spectrophotometric, and paper chromatographic methods of analysis have been used.

Slein[17] found that the colorimetric method of Roe[18] for the estimation of fructose by heating with the resorcinol reagent for 10 minutes at 80° may be applied directly to F-6-P, which gives 79% of the color given by free fructose. The same author[17] measures F-6-P concentration spectrophotometrically by the change in optical density at 340 mμ due to TPN reduction in a system containing Lohmann's isomerase[19] from rabbit muscle extract and Zwischenferment[20] from brewer's yeast in accordance with the equation

$$\text{F-6-P} \xrightarrow[\text{isomerase}]{} \text{G-6-P} \xrightarrow[\text{TPN}]{\text{Zwischenferment}} \text{phosphoglyceric acid}$$

The sum of F-6-P and G-6-P and possibly mannose-6-phosphate is measured.

Another spectrophotometric method is that of Slater,[12] who measured the sum of FDP and F-6-P (plus G-1-P or G-6-P, but *not* F-1-P) by the disappearance of DPN when $\sim$P is in excess.

$$\text{FDP} + 2 \text{ reduced DPN} + Mg^{++} + \text{ATP excess} + \text{F-6-P(G-1-P, G-6-P)} \xrightarrow[\text{fractions A + B}]{\text{rabbit muscle}} 2 \text{ glycerol P} + 2 \text{ DPN}$$

Racker's[13] enzyme fraction A is the same as that used for the determination of FDP; his fraction B is mainly phosphohexokinase. In the absence of the glucose phosphate, F-6-P can be calculated by the difference between this method and that for FDP only.[12] F-1-P is not measured by this procedure.

Paper chromatography may be applied to F-6-P by the method described for FDP[14] (see p. 166), except that no separation of F-6-P and F-1-P was obtained by the authors. Separation of fructose monophosphates from glucose phosphates and FDP is possible by this procedure.

[17] M. W. Slein, *J. Biol. Chem.* **186**, 753 (1950).
[18] J. H. Roe, *J. Biol. Chem.* **107**, 15 (1934).
[19] K. Lohmann, *Biochem. Z.* **262**, 137 (1933); see Vol. I [37].
[20] E. Negelein and W. Gerischer, *Biochem. Z.* **284**, 289 (1936); see Vol. I [42].

III. Preparation and Assay of Fructose-1-phosphate

Fructose-1-phosphate (F-1-P) can be prepared by two different methods, both enzymatic, with FDP as the starting material.

1. Hydrolysis of FDP by bone phosphatase[21] gives a mixture of F-1-P, F-6-P, and aldose monophosphates. The latter can be removed by oxidation of the Ba salts with Br_2; the remaining fructose monophosphates are converted into their brucine salts and separated by fractional crystallization from water.[22]

2. Enzymatic condensation of dihydroxyacetone-1-phosphate and D-glyceraldehyde by muscle or yeast aldolase, which should theoretically give a mixture of F-1-P and sorbose-1-phosphate, actually yields F-1-P only, almost quantitatively and irreversibly.[23] In the presence of the enzyme, FDP, used as starting material, is converted to dihydroxyacetone phosphate and glyceraldehyde phosphate; the latter is in turn converted to dihydroxyacetone phosphate.

FDP $\rightarrow$ dihydroxyacetone P + glyceraldehyde P $\rightarrow$ dihydroxyacetone P

Procedure According to Tankó and Robison[22]

FDP is prepared as described on p. 162 or obtained commercially from Schwarz Laboratories (Mt. Vernon, New York) as Ca and Mg salts. Bone phosphatase may be prepared by the method of Martland and Robison.[24] Purified phosphatase yields a product of stronger levorotation, but this is probably due to the absence of aldose phosphates produced by the phosphohexokinase present in the crude extract rather than higher F-1-P content. Crude bone extract is prepared by Tankó and Robison[22] for the partial hydrolysis of FDP yielding Ba hexose monophosphates. They contain about 15% of the total P of FDP, of which 30% is aldose monophosphates and 50% is F-1-P. Hydrolysis is carried out at 25° in the presence of chloroform and stopped by adding H_2SO_4 to lower the pH to 5, then precipitating organic and inorganic P by basic Pb acetate. This precipitate is decomposed with H_2SO_4 and converted to the Ba salts with $Ba(OH)_2$. The inorganic P precipitates as insoluble $Ba_3(PO_4)_2$ and is filtered off.

Oxidation of Aldose Phosphates. Fourteen grams of Ba hexose monophosphates is dissolved in 240 ml. of H_2O and treated at room tempera-

[21] M. Mcleod and R. Robison, *Biochem. J.* **27**, 286 (1933).
[22] B. Tankó and R. Robison, *Biochem. J.* **29**, 961 (1935).
[23] O. Meyerhof, K. Lohmann, and P. Schuster, *Biochem. Z.* **286**, 319 (1936); see Vol. I [39 and 40].
[24] M. Martland and R. Robison, *Biochem. J.* **23**, 237 (1929); see Vol. II [82].

ture with 1.5 ml. of Br_2; 70 ml. of cold saturated solution of $Ba(OH)_2$ (0.47 N) is then added in 10-ml. portions at intervals over a period of 20 minutes with constant shaking. (This is 200% excess Br_2 and 50% excess $Ba(OH)_2$ over that calculated.) After 5 minutes of standing the solution is acidified with 12 ml. of N HCl, and excess Br_2 is removed by aerating for 1 hour. The pH, now about 4.2, is raised to 8 with $Ba(OH)_2$, and the precipitated Ba phosphohexonate is filtered off. The fructose phosphates are recovered from the filtrate by precipitation with alcohol and freed from the phosphohexonate by repeated solution in small quantities of H_2O and fractional precipitation with alcohol. The fraction finally precipitated between 18 and 66% alcohol has 7.81% P; $[\alpha]_{5461} = -27°$

Fractional Crystallization of Brucine Salts

The fructose monophosphates are separated by first converting them to their brucine salts according to Robison and King.[25] This procedure consists in dissolving the Ba salts in five times their weight of H_2O and then treating them with the amount of 5 N H_2SO_4 required to precipitate all the barium. $BaSO_4$ is removed by centrifugation and washed once with H_2O and three times with 96% alcohol. The acid solution is treated with slight excess of brucine (2.2 g. for each milliliter of 5 N H_2SO_4) of which a little less than half is added as a solid before centrifugation. The remaining alkaloid is added as a solution in hot 50% alcohol after the final washing is completed. After about a week at 0°, the precipitate is filtered off and the salts recrystallized by dissolving them in seven or eight times their weight of 20% EtOH at 40°, leaving them at 0°, filtering, and repeating. A third recrystallization is made from boiling 95% MeOH.

The salts may also be recrystallized[22] by dissolving them in 80% alcohol, cooling them to $-15°$, and leaving them for 12 hours at 0° after crystallization has begun. The purification of the product is effected by threefold redissolution in small volumes of 40% alcohol at 30° and addition of absolute alcohol at 90% concentration. The salts are dissolved in 40% alcohol and crystallized at 0°.

To separate the F-1-P the purified brucine salts are recrystallized from H_2O in which F-1-P brucine salt is less soluble then in dilute alcohol. Solution is effected at 40°, and crystallization is allowed to take place at room temperature. The brucine salt so obtained has 2.95% P; $[\alpha]_{5461} = 52.1°$. The Ba salt tested at $c = 6.1\%$ shows $[\alpha]_{5461} = -39°$; the free ester $[\alpha]_{5461} = -64.2°$ ($c = 11.3\%$).

The Ba salt is further purified by repeated solution in 10 parts of H_2O and precipitation with alcohol. Traces of sparingly soluble substances

[25] R. Robison and E. J. King, *Biochem. J.* **25**, 323 (1931).

left are filtered off, and the salt is then obtained by fractional precipitation with alcohol. The fraction between 20 and 70% alcohol is filtered off, washed with absolute alcohol, and dried.

Procedure According to Meyerhof, Lohmann and Schuster[23]

To 100 ml. of Na FDP[26] equivalent to 750 mg. of P_2O_5 (prepared from the commercial Ca salt with the calculated amount of oxalic acid and subsequent neutralization with NaOH) and 100 ml. of 1.5% D-glyceraldehyde is added 100 ml. of rabbit muscle extract previously dialyzed against H_2O for 4 hours and mixed with one-ninth its volume of 2.6% $NaHCO_3$. After standing for 30 minutes at 40°, the solution is deproteinized with 20 ml. of 40% TCA, the residue washed with 50 ml. of 4% TCA, and the filtrate neutralized with NaOH. To this solution 15 ml. of 25% Ba acetate and alcohol to a concentration of 20% are added. The precipitate is discarded, and alcohol to a concentration of 60 to 65% is added to the supernatant. This precipitate is twice redissolved in 10% alcohol and reprecipitated. The yield of Ba salt obtained after alcohol and ether washing was 3.69 g.

To 3.3 g. of this Ba salt 100 ml. of 60% alcohol is added, and any glyceraldehyde that might adhere to the salt is brought into solution by shaking for 1 hour. The residue is then filtered off, washed with 60% alcohol, and dissolved in 80 ml. of H_2O. The insoluble part is removed, and the Ba salt precipitated with alcohol.

Assay. Optical rotation of the free acid, $[\alpha]_D^{20} = -52.5°$; $[\alpha]_{5461} = -64.3°$. Hydrolysis for 5 minutes in N HCl at 100° splits off 55.0% of the total P (Tankó[22]); F-1-P $\rightarrow$ 55.8%. Fructose-1-phosphate is not fermented by fresh yeast but by a solution of zymase which can be made by a simple method[27] from commercially available baker's yeast. The pure substance should not take up any Br_2 even on standing at room temperature for 1 week with about $N/10$ bromine water containing 2% KBr.[28] Intestinal rat phosphatase[23] splits off D-fructose which may be tested for in the usual way (see Vol. III [12]).

Colorimetric and paper chromatographic methods described for F-6-P (see Vol. III [12, 15]) are also applicable to F-1-P. Paper chromatographic methods published since the work of Hanes and Isherwood[14] may sometimes be better suited for the separation or identification of the

---

[26] The readily soluble Mg FDP may be preferable to the gumlike and unstable alkali salts; Mg FDP is a stable solid substance of 30% water solubility and may be easily prepared according to Neuberg and Kobel as described on p. 163. It is also commercially available at Schwarz Laboratories, Mount Vernon, New York.

[27] C. Neuberg and I. S. Forrest, *Z. physiol. Chem.* **295,** 110 (1953).

[28] C. Neuberg and H. Collatz, *Biochem. Z.* **223,** 494 (1930).

ketohexosephosphates. Bandurski and Axelrod[29] separate the free acids, i.e., FDP and F-6-P (obtained by shaking the salts with the cation exchange resin Dowex 50) by ascending two dimensional chromatography in a basic and an acidic solvent. They recommend CH_3OH—NH_4OH—H_2O and CH_3OH—$HCOOH$—H_2O as the solvent systems and then spray with essentially the same reagent mixture as Hanes and Isherwood.[14] Benson et al.[30] get good separations of radioactive compounds formed during photosynthesis by radiograms. Phenol and BuOH—CH_3COOH—H_2O systems for one-dimensional and BuOH—CH_3CH_2COOH—H_2O, then phenol for two-dimensional are recommended. These authors also separate FDP and F-6-P by elution from Dowex A1 anion exchange resin with 0.1 M NaCl. F-6-P emerges first, FDP second. The method of Burrows et al.[31] is similar to that of Bandurski and Axelrod,[29] modified for dipping instead of spraying by replacing water with acetone. Mortimer[32] gets excellent results with the solvent systems (a) ethyl acetate-acetic acid-water (3:3:1), (b) ethyl acetate-pyridine-formamide (6:4:1 or 1:2:1) and (c) methyl cellosolve-methyl ethyl ketone-3 N NH_4OH (7:2:3) or two-dimensional with (a) or (c) followed by (b). Paper electrophoresis in veronal buffer pH 8.6 at 200 V $3\frac{1}{2}$ hr. has been used by Schild and Bottenbruch[33] to identify and follow the purification of FDP and F-6-P. Using P^{32} in the preparation of the esters, radioactivity can be measured along the paper strip and separate peaks distinguished. Wade and Morgan[34] combine paper ionophoresis and chromatography.

[29] R. S. Bandurski and B. Axelrod, J. Biol. Chem. **193**, 405 (1951).
[30] A. A. Benson, J. A. Bassham, M. Calvin, T. C. Goodale, V. A. Haas, and W. Stepka, J. Am. Chem. Soc. **72**, 1710 (1950). See Vol. I [15].
[31] S. Burrows, F. S. M. Gryllis, and J. S. Harrison, Nature **170**, 800 (1952).
[32] D. C. Mortimer, Canad. J. Chem. **30**, 653 (1952).
[33] K. T. Schild and L. Bottenbruch, Z. physiol. Chem. **292**, 1 (1953).
[34] H. E. Wade and D. M. Morgan, Biochem. J. **60**, 264 (1955).

[23] Preparation and Analysis of 6-Phosphogluconate

By B. L. HORECKER

Preparation

Principle. Previous methods for the oxidation of glucose-6-phosphate to 6-phosphogluconate with bromine involve the control of pH with solid barium carbonate.[1-3] However, barium glucose-6-phosphate now

[1] R. Robinson and E. J. King, Biochem. J. **25**, 323 (1931).
[2] O. Warburg and W. Christian, Biochem. Z. **292**, 287 (1937).
[3] J. E. Seegmiller and B. L. Horecker, J. Biol. Chem. **192**, 175 (1951).

precipitates as a crystalline heptahydrate which has a very low solubility in water, and under these circumstances the reaction proceeds very slowly. In the procedure outlined here the oxidation is carried out in the absence of barium ions and the pH is maintained by the addition of sodium hydroxide.

Reagents

Potassium sulfate, 0.57 M in water.

Barium hydroxide, saturated solution at room temperature.

Procedure. Barium glucose-6-phosphate heptahydrate[4] (2.0 g., 3.84 millimoles) is dissolved in 20 ml. of water with the aid of 0.40 ml. of concentrated hydrochloric acid and treated with 8.0 ml. of 0.57 M potassium sulfate. The precipitated barium sulfate is digested in a boiling water bath for 10 minutes, cooled, and centrifuged. The supernatant solution is adjusted to pH 5.4 with 0.2 ml. of 4 N NaOH and treated with 0.5 ml. of bromine. During the following 20 minutes the pH is maintained between 4.8 and 5.9 by the addition of 3.0 ml. of 4 N NaOH. At the end of this time excess bromine is removed by aeration with nitrogen for 60 minutes. Concentrated hydrochloric acid (0.5 ml.) is added until the solution yields a blue color with congo red paper. Sulfate is precipitated with 1.0 ml. of 1 M barium acetate and removed by centrifugation. The supernatant solution is gassed with nitrogen and adjusted to a permanent pink end point by the addition of 21 ml. of saturated barium hydroxide solution. The suspension is treated with 15 ml. of ethanol and filtered with suction. The precipitate is resuspended in 50 ml. of 20% ethanol, filtered, and dried overnight *in vacuo* over KOH and $CaCl_2$. The yield is 1.79 g. of white powder.

Properties

Analysis. For analysis 30.0 mg. of product is dissolved in 3.0 ml. of 0.07 M hydrochloric acid. To a 1.0-cm. absorption cell are added 0.75 ml. of water, 0.15 ml. of 0.25 M glycylglycine buffer, pH 7.5, 0.1 ml. of 0.1 M $MgCl_2$, 0.10 ml. of TPN (1.4 μM./ml.), and 0.01 ml. of phosphogluconic dehydrogenase[5] (30 units). Density is determined at 340 mμ and 0.03 ml. of 6-phosphogluconate solution, diluted twenty times, and the density followed at 340 mμ until a constant value is reached. The amount of 6-phosphogluconate is calculated from the equation

$$6\text{-PG (micromoles)} = \frac{d_f - d_0}{6.22} \times 1.14$$

where d_f and d_0 are the final and initial density readings, respectively.

[4] See Vol. III [19].

[5] See Vol. I [42].

Purity. By the enzymatic assay the purity of the preparation is about 95%, uncorrected for moisture content. It contains no glucose-6-phosphate. The yield from glucose-6-phosphate is 93%.

[24] Synthetic Ketohexose Phosphates

I. D-Tagatose-6-phosphate

By Ezra L. Totton and Henry A. Lardy

$$
\begin{array}{l}
\text{H}_2\text{C}-\text{OH} \\
\quad | \\
\text{HOC}-\\
\quad | \\
\text{HOCH} \\
\quad | \\
\text{HOCH} \\
\quad | \\
\text{HCO}-\\
\quad | \\
\text{H}_2\text{COPO}_3\text{Ba}
\end{array}
$$

D-Tagatose[1] was converted to 1,2;3,4-diisopropylidene-D-tagatose by the procedure of Reichstein and Bosshard.[2]

1,2;3,4-Diisopropylidene-D-tagatose-6-diphenyl Phosphate

To 1.5 g. (0.0057 mole) of 1,2;3,4-diisopropylidene-D-tagatose, in 14 ml. of cold dry pyridine, 1.5 ml. (0.007 mole) of diphenylchlorophosphonate was added dropwise. After the reaction mixture had remained at 0° for 30 minutes, it was placed in the cold room at 5° for 24 hours. The reaction mixture was poured onto 200 ml. of finely cracked ice while being rapidly stirred. The stirring was continued for 30 minutes, and the mixture extracted with 125 ml. of chloroform. The chloroform solution was washed first with 25 ml. of dilute HCl and four times with a total of 100 ml. of distilled water. It was dried over Na_2SO_4 (anhydrous) and concentrated under reduced pressure to a dry sirup. The sirup was taken up in 200 ml. of absolute ethanol; distilled water was added to turbidity, and, after scratching, the mixture was allowed to remain at 5° for 24 hours. The product crystallized in long silky needles. After two such recrystallizations, the product weighed 2.5 g. (86% of theory); melting

[1] E. L. Totton and H. A. Lardy, *J. Am. Chem. Soc.* **71**, 3076 (1949).
[2] T. Reichstein and W. Bosshard, *Helv. Chim. Acta* **17**, 757 (1934).

point 94 to 95°; $[\alpha]_D^{25} = 32°$ ($c = 2$ in chloroform). Further recrystallization did not change the melting point or the rotation.

Analysis—$C_{24}H_{29}O_9P$ *(492.5)*

Calculated: C 58.51, H 5.93, P 6.28
Found: C 58.03, H 6.12, P 6.21
            C 58.34, H 6.16, P 6.21

1,2;3,4-Diisopropylidene-D-tagatose-6-phosphoric Acid

A solution of 1.3 g. (0.0026 mole) of 1,2;3,4-diisopropylidene-D tagatose-6-diphenyl phosphate in 25 ml. of absolute ethanol was shaken with 0.12 g. of platinum catalyst and hydrogen at a pressure slightly greater than atmospheric. The reaction stopped after the theoretical amount of hydrogen had been consumed (8 moles). The reaction was complete in 5 hours. After removal of the catalyst by centrifugation, the alcoholic solution was concentrated to a sirup under reduced pressure. The product was obtained in crystalline form from 100 ml. of pentane. After remaining at 5° for 24 hours, the crystals were collected, washed with pentane, and dried over paraffin and calcium chloride; weight 0.300 g. (33.4%); melting point 92°, $[\alpha]_D^{25} = 43.6°$ ($c = 0.5$ in absolute ethanol).

Analysis—$C_{12}H_{21}O_9P$ *(340.3)*

Calculated: C 42.33, H 6.22, P 9.11
Found: C 42.71, H 6.32, P 9.17
                      P 9.17

Barium-D-tagatose-6-phosphate

1,2;3,4-Diisopropylidene-D-tagatose-6-diphenyl phosphate (2.5 g.) was cleaved by reduction with hydrogen and platinum as above. The sirup was crystallized from pentane and the pentane decanted, but no further attempt was made to purify the product. The crystal sludge gave a negative Fehling's test. Fifteen milliliters of distilled water was added to the crystals, and the solution warmed on a steam bath for 3 minutes. The solution then gave a strongly positive Fehling's test. Barium hydroxide solution (CO_2-free) was added dropwise until the pH reached 10.2 and, after filtering, the solution (60 ml.) was poured into 240 ml. of 95% ethanol. The mixture was placed in the cold room at 5° for 24 hours. The mother liquor was decanted, and the white amorphous precipitate was collected on the centrifuge. It was washed once with absolute ethanol; once with each of the following absolute ethanol-ether mixtures, 80:20, 50:50, 20:80; and twice with anhydrous ethyl ether. The dried product weighed 1.5 g. (75%); $[\alpha]_D^{25} = 5.65°$ ($c = 1.06$ in water).

Analysis—$C_6H_{11}O_9PBa$ (395.5)

Calculated: C 18.22, H 2.81, P 7.83
Found: C 17.58, H 3.11, P 7.83
                        P 7.84

Tagatose-6-phosphate is converted to tagatose-1,6-diphosphate by the phosphohexokinase of beef brain[3] or by the highly purified phosphohexokinase of rabbit muscle.[4] The tagatose-1,6-diphosphate so produced is cleaved by crystalline muscle aldolase although at a rate much slower than FDP is cleaved.

II. Fructose-1-phosphate

By WILLIAM L. BYRNE and HENRY A. LARDY

$$
\begin{array}{l}
H_2C\text{—}OPO_3Ba \\
|\\
HO\text{—}C\text{———}\\
|\\
HOCH\\
|\\
HCOH\\
|\\
HC\text{—}O\text{—}\\
|\\
H_2COH
\end{array}
$$

The reaction is carried out in a 2-necked flask equipped with a dropping funnel and a vent through a calcium chloride tube. The flask is cooled in an ice bath. To 10 g. of 2,3;4,5-diisopropylidene-D-fructose[5] dissolved in 20 ml. of ice-cold pyridine, 12.1 g. (1 M equivalent) of diphenylchlorophosphonate[6] is added in portions of about 1 ml. over a period of 30 minutes. Three hours after the final addition, the reaction mixture is poured into 500 ml. of well-stirred ice water. The mixture is extracted repeatedly with ether, and the combined ether layers are washed twice with each of the following: 1 N H_2SO_4, 0.5 N NaOH, and water. The ether solution is dried overnight over K_2CO_3 and is then evaporated to a syrup.

The crystalline 1-diphenylphosphate-2,3;4,5-diisopropylidene-D-fructose (melting point 52.5°) described by Brigl and Müller[7] is obtained by crystallization from a large volume of petroleum ether. Seed crystals may

[3] E. L. Totton and H. A. Lardy, *J. Biol. Chem.* **181**, 701 (1949).
[4] K.-H. Ling and W. L. Byrne, *Fed. Proc.* **13**, 253 (1954).
[5] H. Ohle and I. Koller, *Ber.* **57**, 1566 (1924); D. Bell, *J. Chem. Soc.* **1947**, 1461.
[6] E. Baer, *Biochem. Preparations* **1**, 51 (1949); **2**, 97 (1952).
[7] P. Brigl and H. Müller, *Ber.* **72**, 2121 (1939).

be obtained by taking up a small portion in hot petroleum ether, allowing it to cool, and decanting; after this procedure is repeated a few times, the compound crystallizes on scratching. Yield, 16.6 g. (75% of theory).

Hydrogenation of 13 g. of 1-diphenylphosphate-2,3;4,5-diisopropylidene-D-fructose in methanol, hydrolysis of isopropylidene groups, and isolation of the barium salt of F-1-P are accomplished by the procedure described for L-sorbose-1-phosphate. The yield of barium salt was 68%.

The synthetic compound is free of FDP and other hexose phosphates as determined by an ion exchange separation procedure. F-1-P is cleaved by muscle aldolase, and the dihydroxyacetonephosphate liberated can be determined by its oxidation of DPNH using α-glycerophosphate dehydrogenase.[8] For further information on fructose-1-phosphate, see Vol. III [22].

III. L-Sorbose-1-phosphate

By KINGSLEY M. MANN and HENRY A. LARDY

$$H_2COPO_3Ba \text{ (or HK)}$$
$$|$$
$$C\text{—OH}$$
$$|$$
$$HOCH$$
$$|$$
$$HCOH$$
$$|$$
$$OCH$$
$$|$$
$$H_2COH$$

2,3;4,6-Diisopropylidene-L-sorbose-1-diphenyl Phosphate

To 5 g. of 2,3;4,6-diisopropylidene-L-sorbose[9] dissolved in cold (0°) dry pyridine, 7.25 g. of diphenylchlorophosphonate[6] was added dropwise with stirring. Near the end of the addition white crystals of pyridine hydrochloride appeared. The reaction mixture was left overnight at 4° and then decomposed by pouring onto 500 ml. of ice and water. The oil which settled out was extracted with three 100-ml. portions of chloroform. The latter solution was then washed successively with ice-cold solutions of dilute hydrochloric acid and sodium bicarbonate. After a final water wash the chloroform solution was dried with anhydrous sodium sulfate and the solvent completely removed under a vacuum,

[8] T. Baranowski and T. R. Niederland, *J. Biol. Chem.* **180,** 543 (1949).
[9] S. Maruyama, *Sci. Pap. Inst. Phys. Chem. Research (Tokyo)* **27,** 59 (1935). The compound is an intermediate in the manufacture of vitamin C and may be obtained from commercial sources.

leaving a viscous residue, 7.8 g. (83% of theory); $[\alpha]_D^{22} = -11.7°$ ($c = 5.0$ in chloroform).

Analysis—$C_{24}H_{29}O_9P$ (*492.4*)

Calculated: C 58.52, H 5.93, P 6.29
Found: C 58.47, H 5.81, P 6.36
          C 58.27, H 5.97, P 6.45

Barium L-Sorbose-1-phosphate

Five grams of 2,3;4,6-diisopropylidene-L-sorbose-1-diphenyl phosphate was dissolved in 25 ml. of redistilled methanol together with 1 ml. of water and reduced in the presence of 0.5 g. of platinum oxide. After 6 hours the reduction stopped when 9.5 moles of hydrogen had been consumed, indicating partial removal and hydrogenation of the isopropylidene residues. The catalyst was removed by filtration and the solvents by distillation (bath temperature 50°) under a vacuum (10 mm.). Then the residual oil was taken up in 20 ml. of water and heated on a steam bath for 35 minutes to hydrolyze the remaining isopropylidene groups. After neutralization of the solution to the pink color of phenolphthalein with saturated barium hydroxide, the inorganic phosphate was separated by centrifugation or filtration through a diatomaceous earth pad (Johns-Manville, Celite). On addition of 2 vol. of cold acetone, the barium salt of L-sorbose-1-phosphate precipitated. Purification was achieved by dissolving the salt in 50 ml. of water, removing the insoluble material, and reprecipitating with acetone. The barium salt was collected by centrifugation and washed several times with acetone and then with peroxide-free ether. By drying in a vacuum desiccator over sulfuric acid a powdery white amorphous solid was obtained; yield, 1.5 g.; $[\alpha]_D^{28} = -7.2°$ ($c = 2.5$ in 0.104 N HCl).

Analysis—$C_6H_{11}O_9PBa\cdot2H_2O$ (*431.6*)

Calculated: C 16.69, H 3.50, P 7.19, Ba 31.91
Found: C 16.83, H 3.62, P 7.31, Ba 31.5

Crystalline Potassium Hydrogen L-Sorbose-1-phosphate

The barium salt of L-sorbose-1-phosphate (1.5 g.) was passed slowly through a repeatedly acid- and base-washed column of Amberlite IR-100 in the potassium cycle. After evaporation of the barium-free effluent to 15 ml. under a vacuum (bath temperature, 40 to 50°), the solution was adjusted to pH 3 with glacial acetic acid. Ethanol was then added until a faint turbidity persisted, and the solution was allowed to stand at room temperature until crystals formed. Small periodic additions of alcohol were made until no more crystals separated. They were collected on a sintered glass funnel, washed with absolute ethanol and then with ether, and dried *in vacuo* over sulfuric acid; yield, 0.55 g. Once dried in this manner the compound appeared quite stable, and, since it is readily soluble in water, it is most convenient for study in biological systems. $[\alpha]_D^{30} = -16.5°$ ($c = 2.0$ in water).

Analysis—$C_6H_{12}O_9PK$ (298.3)

Calculated: C 24.15, H 4.06, P 10.40, K 13.11
Found: C 24.01, H 4.20, P 10.3, K 13.3

L-Sorbose-1-phosphate is a strong inhibitor of hexokinase.[10] It is cleaved by crystalline muscle aldolase to dihydroxyacetonephosphate and L-glyceraldehyde, although the cleavage can be detected only by employing an assay system which removes one of the products.[11] The compound is acid-labile, as is shown in the following chart.

HYDROLYSIS OF BARIUM L-SORBOSE-1-PHOSPHATE IN 1 N HCl AT 98.5°

Time, min.	5	10	20	30	60
Hydrolysis	35.8	58.9	83.4	90.8	95.0
k, min.$^{-1}$	0.089	0.089	0.090	0.080	

IV. L-Sorbose-6-phosphate

By KINGSLEY M. MANN and HENRY A. LARDY

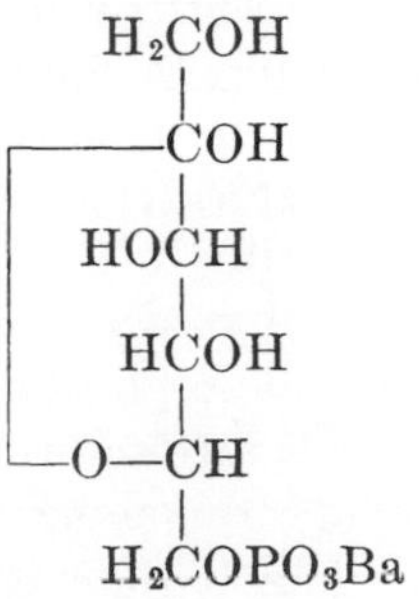

$$
\begin{array}{c}
H_2COH \\
|\\
COH \\
|\\
HOCH \\
|\\
HCOH \\
|\\
O{-}CH \\
|\\
H_2COPO_3Ba
\end{array}
$$

2,3-Isopropylidene-L-sorbose-1,6-bis(diphenyl Phosphate)

Five grams of 2,3-isopropylidene-L-sorbose,[12] dissolved in 30 ml. of dry pyridine, was esterified by the dropwise addition, with stirring, of 12.9 g. (2.1 moles) of diphenylchlorophosphonate. As the last of the acid

[10] H. A. Lardy, V. D. Wiebelhaus, and K. M. Mann, *J. Biol. Chem.* **187**, 325 (1950).
[11] T. Tung, K.-H. Ling, W. L. Byrne, and H. A. Lardy, *Biochim. Biophys. Acta* **14**, 488 (1954).
[12] H. H. Schlubach and P. Olters, *Ann.* **550**, 140 (1942).

chloride was added, a white precipitate of pyridine hydrochloride settled out. The reaction mixture was left overnight at 4° and then poured, in a thin stream, into a vigorously stirred mixture of ice and water. The resulting pale-yellow oil gradually solidified and was removed by filtration. Washing by repeated suspension in water and decantation yielded a solid, 13.2 g., melting point 79 to 86°, which still contained traces of pyridine. Repeated crystallization from ethanol alone or ethanol and water raised the melting point to 100.5 to 101.5°; yield, 8.5 g.; $[\alpha]_D^{23} = -4.2°$ ($c = 5$ in chloroform).

Analysis—$C_{33}H_{34}O_{12}P_2$ (684.6)

$$\text{Calculated: C 57.88, H 5.00, P 9.05}$$
$$\text{Found:} \quad \text{C 58.00, H 5.01, P 8.95}$$
$$\text{C 58.00, H 4.80, P 9.04}$$

Barium L-Sorbose-6-phosphate

The reduction of the tetraphenyl ester (5.0 g.) described above was carried out in 25 ml. of glacial acetic acid or methanol with 0.5 g. of platinum oxide to catalyze the reaction. After the theoretical amount of hydrogen for cleavage and hydrogenation of the phenyl groups had been consumed the catalyst was removed and the solvent evaporated under a vacuum. By repeated solution of the oily residue in absolute ethanol and removal of the solvent at 50° and 8 mm., the last traces of acetic acid could be removed. Deacetonation and hydrolytic removal of the 1-phosphate were achieved by dissolving the viscous material in 20 ml. of water and heating on the steam bath for 45 minutes. On adjustment of the solution to the pink color of phenolphthalein with saturated barium hydroxide, a copious white precipitate of the barium salts of sorbose diphosphate[13] and inorganic phosphate appeared. After removal of the precipitate by centrifugation, 2 vol. of acetone was added to the supernatant solution. The barium L-sorbose-6-phosphate was freed of all traces of the diphosphate by taking the salt into solution with water, adjusting the pH to 10 with barium hydroxide, removing the insoluble material, and reprecipitating with acetone. The barium salt was dried in the same manner as the 1-phosphate. $[\alpha]_D^{29} = -12.0°$ ($c = 2.0$ in water).

Analysis—$C_6H_{11}O_9PBa$ (395.5)

$$\text{Calculated: C 18.23, H 2.80, P 7.83, Ba 34.7}$$

Analysis—$C_6H_{11}O_9PBa \cdot H_2O$ (413.5)

$$\text{Calculated: C 17.43, H 3.17, P 7.49, Ba 33.2}$$
$$\text{Found:} \quad \text{C 17.16, H 3.17, P 7.57, Ba 33.3}$$

[13] Partially purified sorbose diphosphate has been prepared from this mixed precipitate.

HYDROLYSIS OF BARIUM L-SORBOSE-6-PHOSPHATE IN 1 N HCl AT 99°

Time, min.	5	20	45	120	180	360
Hydrolysis	5.3	20.6	39.6	73.4	85.5	96.2
k, min.$^{-1}$	0.0116	0.0115	0.0112	0.0110	0.0107	0.00905

[25] Preparation of Ribose-1-phosphate

By PAUL E. PLESNER and HANS KLENOW

Principle. Ribose-1-phosphate (R-1-P) is formed by the phosphorolytic cleavage described by Kalckar:[1]

Inosine + orthophosphate $\rightleftarrows$ R-1-P + hypoxanthine

The reaction is catalyzed by nucleoside phosphorylase and pulled by oxidizing the hypoxanthine formed with xanthine oxidase. The reaction may be followed by determining the amount of acid-labile phosphate formed, using the Fiske-SubbaRow and Lowry-Lopez reactions (see Vol. III [115]).

Enzymes. Xanthine oxidase is prepared according to the method of Ball,[2] with the modifications described by Kalckar *et al.*[3] A rather unfractionated enzyme may prove satisfactory if care is taken that the preparation does not contain any phosphatase. One enzyme unit is defined as the amount of enzyme which causes an increase in extinction of 0.001 per minute at 293 mμ in a 3-ml. cuvette (light path 1 cm.) under the following standard conditions: phosphate buffer, 0.05 M, pH 7.4, 2.6 ml.; Aypoxanthine, 0.005 M, 0.3 ml.; enzyme + H_2O to 3.0 ml.

Nucleoside phosphorylase is prepared according to Price, *et al.*[4] one unit being defined as that of xanthine oxidase (see above) in the following system: phosphate buffer, 0.05 M, pH 7.4, 2.6 ml.; inosine, 0.005 M, 0.2 ml.; excess xanthine oxidase, nucleoside phosphorylase + H_2O to 3.0 ml.

Procedure

Preparation of the Barium Salt of Ribose-1-phosphate. The following mixture is incubated at room temperature under an oxygen atmosphere and shaken vigorously: phosphate buffer, 0.2 M, pH 7.4, 8.95 ml. (1790

[1] H. M. Kalckar, *J. Biol. Chem.* **167**, 477 (1947).
[2] E. G. Ball, *J. Biol. Chem.* **128**, 51 (1939).
[3] H. M. Kalckar, N. O. Kjeldgaard, and H. Klenow, *Biochim. et Biophys. Acta* **5**, 575 (1950).
[4] V. E. Price, M. C. Otey, and P. Plesner, Vol. II [64].

μM.); tris(hydroxymethyl)aminomethane-formic buffer, 0.6 M, pH 7.6, 32.2 ml. (19.3 μM.); nucleoside phosphorylase, 5.0 ml. (89,000 units); xanthine oxidase, 7.16 ml. (130,000 units); inosine, 384 mg. (1430 micromoles).

After 6 hours of incubation (1×10^6 counts per minute of P^{32}-labeled orthophosphate is added), the mixture is poured on a Dowex 1 column (6×10 cm.) in the formate form. The column is eluted with 0.1 M Na formate-formic acid, pH 5.0 (7 l. of 0.1 M Na formate + 1.3 ml. of concentrated formic acid). The R-1-P appears immediately before the labeled phosphate, the appearance of which signals the completion of the elution (4 to 5 resin bed volumes at a flow rate of 5 ml./min.). To the pooled fractions containing acid-labile phosphate (700 ml.) are added 2 ml. of 2 M barium acetate, enough NaOH to bring the pH to 8, and 4 vol. of 96% ethanol. The mixture is left in the cold overnight. The precipitate is centrifuged off, washed with ethanol and ether, and dried *in vacuo*. The yield is 795 μM. This preparation is free of coenzyme activity toward phosphoglucomutase and phosphoribomutase.

Preparation of Cyclohexylamine Salt. Fifty-four micromoles of barium R-1-P is dissolved in 0.225 ml. of water; 14.8 mg. of cyclohexylamine sulfate dissolved in 0.2 ml. of water is added (following the procedure described by Friedkin[5]). The barium sulfate formed is removed by centrifugation and washed once with 0.2 ml. of water; to the pooled supernatants is added 7 vol. of butanol, and the mixture is centrifuged to remove the amorphous precipitate. To the supernatant is added 1 vol. of ether, whereby fine crystals are formed. After 2 to 3 hours at 0°, the crystals are collected by centrifugation, washed in ether, and dried. The precipitate shows a yield of 66.7% of the barium salt.

Yields and Purity of R-1-P

The yields and purity at the different steps of the preparation calculated from the initial amount of inosine are:

	Yield	Purity
After 3¼ hours of incubation	47%	
After 5 hours of incubation	70%	
After Dowex chromatography	68%	
As barium salt	55.7%	80%
As cyclohexylamine salt	37.2%	99%

A purity of 99% of the weight of the R-1-P cyclohexylamine salt could be accounted for as acid-labile phosphate in the Fiske-SubbaRow reaction.[6] Under the microscope the salt appeared as needle-shaped crystals.

[5] M. Friedkin, *J. Biol. Chem.* **184**, 449 (1950).
[6] C. H. Fiske and T. SubbaRow, *J. Biol. Chem.* **66**, 375 (1925).

[26] Preparation of Deoxyribose-1-phosphate

By M. FRIEDKIN and H. M. KALCKAR

I. Preparation of Barium Salt after Action of Purine Nucleoside Phosphorylase[1]

The phosphorolysis of guanine deoxyriboside was carried out by incubating the following mixture at 24° for 24 hours: 100 mg. of guanine deoxyriboside, 600 mg. of Na_2HPO_4, 24 ml. of 0.1 M tris(hydroxymethyl)-aminomethane-HCl buffer, pH 8.5, and 3.6 ml. of calf liver nucleoside phosphorylase (see Vol. II [64]). After incubation the mixture was cooled in an ice-water bath and then centrifuged (the guanase present in the enzyme preparation converts guanine to xanthine which precipitates during the incubation period). The supernatant was extracted three times with 85-ml. portions of n-butanol saturated with water. The layers were separated by centrifugation. Denatured protein appeared at the interface. To the aqueous layer were added 14.4 ml. of 0.5 M $MgCl_2$—5.0 M NH_4Cl and 1.2 ml. of 15 N NH_3. After 3 hours in a refrigerator, the $MgNH_4PO_4$ precipitate was removed by filtration through a pad of cotton and paper pulp. To the clear filtrate were added 3.6 ml. of ammoniacal barium acetate, 2.0 g. of $Ba(C_2H_3O_2)_2 \cdot H_2O$, 10 ml. of H_2O, and 0.6 ml. of 15 N NH_3. The mixture was placed in a refrigerator. The precipitate which formed overnight was spun down and discarded. To the clear supernatant were added 4 vol. of ethanol and 0.4 ml. of 15 N NH_3. After 10 hours in the refrigerator, the barium salt which had formed was centrifuged, washed two times by suspension in ethanol, and finally freed of alcohol by suspension in diethyl ether. The salt was dried *in vacuo* at room temperature.

The crude barium salt of deoxyribose-1-phosphate was reprecipitated as follows: 95 mg. of the crude salt was extracted three times with 2.0-ml. portions of cold water. After three such extractions an insoluble residue was left after centrifugation and was discarded. To the water extracts was added 15 vol. of dry butanol, which on stirring yielded a one-phase system due to the slight solubility of water in butanol. The barium salt which precipitated on the addition of butanol was spun down, washed three times with diethyl ether, and dried *in vacuo* at room temperature. The final yield of purified barium salt, starting with 100 mg. of guanine deoxyriboside, was 15 mg.

[1] M. Friedkin, *J. Biol. Chem.* **184,** 449 (1950).

II. Preparation of the Cyclohexylamine Salt from the Barium Salt[1]

First 35.7 mg. of barium salt of deoxyribose-1-phosphate was dissolved in 400 μl. of H_2O. Then 23.5 mg. of cyclohexylamine sulfate[2] in 100 μl. of H_2O was added. The precipitate of $BaSO_4$ was spun down. The clear supernatant was pipetted into 7.5 ml. of absolute n-butanol. This mixture was vigorously stirred until all the water had dissolved in the organic solvent, and then placed in a refrigerator overnight. A small amount of precipitate was spun down and discarded. To the clear supernatant was added approximately an equal volume of diethyl ether. After standing at room temperature for 15 to 20 minutes, the butanol–ether–water mixture yielded a mass of fine long needles. The crystals were collected, washed with ether, and redissolved in 0.2 ml. of water. Then 3.0 ml. of dry n-butanol was added. The butanol-water mixture was filtered through a sintered glass disk. One hundred milliliters of ether was then added gradually until the solution was faintly cloudy. After standing at room temperature for 1 hour, the solution was filled with a mass of fine long needles. The cyclohexylamine salt was collected, washed with ether, and dried $in\ vacuo$ at room temperature. Seven milligrams of recrystallized salt was obtained. Qualitative tests indicated the presence of deoxyribose (Dische cysteine-H_2SO_4 color reaction, Vol. III [12]) and inorganic P and the absence of sulfate. The salt sinters and starts to decompose at 152°.

$C_{17}H_{37}O_7N_2P$ (molecular weight 412.46)   Found: P, 7.34

Calculated: P, 7.51

(P was determined as inorganic P at pH 1 by the method of Gomori.[3])

III. Direct Preparation of the Cyclohexylamine Salt after Action of Thymidine Phosphorylase[4]

Dicyclohexylammonium hydrogen phosphate was prepared by addition of 27.2 ml. of 90% H_3PO_4 in 250 ml. of 95% ethanol to 100 ml. of cyclohexylamine in 500 ml. of ethanol. The precipitated salt thus formed was recrystallized from hot water. Incubation mixtures consisted of 2.42 g. of thymidine (10 mM), 2.96 g. of dicyclohexylammonium hydrogen phosphate (10 mM), and 35 ml. of thymidine phosphorylase (700 mg. of protein; activity 3.5). After 7 hours at 38°, the enzymatic reaction was

---

[2] Cyclohexylamine sulfate $(C_6H_{11}NH_2)_2 \cdot H_2SO_4$, was prepared as follows: To 1.0 ml. of cyclohexylamine dissolved in 5 ml. of ethanol was added 3.7 ml. of alcoholic H_2SO_4 (1.0 ml. of 10 N H_2SO_4 + 9.0 ml. of absolute ethanol). A crystalline precipitate of cyclohexylamine sulfate formed immediately. The salt was washed with ethanol and dried $in\ vacuo$ at room temperature.

[3] G. Gomori, $J.\ Lab.\ Clin.\ Med.$ **27**, 955 (1942).

[4] M. Friedkin and D. Roberts, $J.\ Biol.\ Chem.$ **207**, 257 (1954).

terminated by the addition of 15 vol. (525 ml.) of n-butanol. After 5 minutes of vigorous agitation, 525 ml. of diethyl ether was added, and the mixture was chilled in an ice bath for 20 minutes and filtered for the removal of protein and unreacted dicyclohexylammonium hydrogen phosphate. On the addition of 1050 ml. of diethyl ether, crystalline dicyclohexylammonium deoxyribose-1-phosphate was obtained. (The unreacted thymidine is soluble at this point and can be easily recovered by evaporation of the mother liquor.) The salt was recrystallized by redissolving in H_2O (17 mg./ml.) and then adding 15 vol. of n-butanol and 45 vol. of diethyl ether. (Final yield, 9 to 12.5%, based on thymidine initially present, and 36 to 50%, based on thymidine reacted.)

Preparation of Thymidine Phosphorylase[5]

Two hundred and fifty gram of horse liver powder[6] was shaken for 15 minutes with 2.5 l. of chilled redistilled water. After standing for 30 to 60 minutes at 3 to 4°, the dispersion was spun in a refrigerated International centrifuge in 250-ml. cups at 2500 r.p.m. for 15 minutes. The slightly turbid supernatant fluid was then filtered through fluted filter paper (fraction I).

To fraction I, obtained from four generations as described above (6700 ml.), was added 23 g. of $(NH_4)_2SO_4$ per 100 ml., and, after complete solution of the salt, the pH was adjusted to 7.5 to 8.0 by the cautious addition of 15 M NH_4OH (approximately 12 to 15 ml.).

The precipitated protein was allowed to settle out overnight. Most of the clear supernatant fluid was siphoned off the next morning, and the remaining bottom layer was spun in a Servall centrifuge at 8000 r.p.m. The precipitate, 76 ml., was taken up with cold water to a total volume of 250 ml. and centrifuged, yielding 238 ml. of clear supernatant fluid (fraction II).

To fraction II was added 45 ml. of ammoniacal saturated $(NH_4)_2SO_4$ (250 ml. of saturated $(NH_4)_2SO_4$ at 25° plus 1.7 ml. of 15 M NH_4OH). The protein was spun down, and this procedure was repeated by the further addition of 45 ml. of ammoniacal saturated $(NH_4)_2SO_4$. The first and second protein precipitates were combined, dissolved by the addition of cold water to a volume of 200 ml., and dialyzed overnight against redistilled water (fraction III).

The resulting material is suitable for use in the preparation of deoxy-

---

[5] M. Friedkin and D. Roberts, *J. Biol. Chem.* **207**, 245 (1954).

[6] The source of this material is "liver powder 37°," a product of the Viobin Corporation, Monticello, Ill. Fresh horse liver is defatted by treatment with ethylene dichloride and dehydrated by azeotropic distillation of the solvent. One gram of powder is equivalent to 4.5 g. of fresh tissue.

[28] Preparation and Analysis of Ribose-5-phosphate

By B. L. HORECKER

Preparation

Principle. Two methods of preparation of ribose-5-phosphate are given. One of these is based on the hydrolysis of adenylic acid;[1] the other involves the esterification of free ribose with ATP in the presence of ribokinase and is useful for the preparation of ribose-5-phosphate labeled with C^{14} or P^{32}. Ribokinases have been described by Sable[2] and by Cohen *et al.*;[3] a satisfactory preparation is also obtained from a strain of *Aerobacter aerogenes.*[4]

Reagents

Adenylic acid (commercial adenosine-5-phosphoric acid).
ATP (commercial preparations of the sodium salt).

Procedure I. Hydrolysis of Adenylic Acid. One gram of adenosine-5-phosphoric acid (2.9 millimoles) is dissolved in 10 ml. of 1 N hydrochloric acid and heated for 30 minutes in a boiling water bath. The solution is cooled, adjusted to pH 6.5 with 30 ml. of saturated barium hydroxide, and centrifuged. From the supernatant solution the barium salt of ribose phosphate is precipitated by the addition of 250 ml. of ethanol. The precipitate is washed with 10 ml. of absolute ethanol and extracted with three 10-ml. portions of water. The residue (mainly barium phosphate) is discarded. The precipitate which forms when the extracts are combined is dissolved by the addition of water, and the solution is treated with 4 vol. (180 ml.) of ethanol. The precipitate is collected by centrifugation and dried *in vacuo*. The yield is 500 mg. of the barium salt (1.38 millimoles, 48%).

Analysis. 50.2 mg. of the barium salt, dissolved in 0.02 N HCl, was found to contain:

Pentose (orcinol)	143 micromoles
Phosphate (total)	139 micromoles
Adenine (280-mμ absorption)	2 micromoles
Theoretical (molecular weight = 365)	138 micromoles

[1] G. A. LePage and W. W. Umbreit *J. Biol. Chem.* **148,** 255 (1943).
[2] H. Z. Sable, *Proc. Soc. Exptl. Biol. Med.* **75,** 215 (1950); see also Vol. I [49].
[3] S. S. Cohen, D. B. M. Scott, and M. Lanning, *Federation Proc.* **10,** 173 (1951).
[4] B. L. Horecker, M. Gibbs, H. Klenow, and P. Z. Smyrniotis, *J. Biol. Chem.* **207,** 393 (1954).

A solution of barium salt in 0.02 N HCl, containing 120 micromoles of pentose per milliliter, had a rotation of $+0.632°$. For ribose-5-phosphoric acid, $[\alpha]_D^{20} = +22.8°$.

Procedure II. Preparation of Ribokinase. An organism isolated by enrichment culture on a ribose-mineral salt medium has been identified as a strain of *Aerobacter aerogenes.*[4] The American Type Culture Collection strain also yields active enzyme preparations. The medium contains 0.15% of K_2HPO_4, 0.05% of KH_2PO_4, 0.02% of $MgSO_4$, 0.10% of $(NH_4)_2SO_4$, and 0.10% of D-ribose. One liter of medium is innoculated with 50 ml. of a 24-hour culture grown at 37°, incubated at this temperature for 24 hours, and harvested with the Sharples supercentrifuge. Two grams of wet cells, derived from 4 l. of medium, is ground for 5 minutes at 0° with 6 g. of alumina. The mixture is suspended in 15 ml. of 0.05 M 2-amino-2-hydroxymethyl-1,3-propanediol (Tris) buffer, pH 7.5, and centrifuged at $18,000 \times g$. The extract containing the ribokinase activity is frozen and stored at $-12°$. For purification 40 ml. of extract is treated with 8.0 ml. of protamine sulfate solution (salmine, Lilly, 10 mg./ml.), kept for 3 minutes at 0°, and centrifuged. The precipitate is homogenized with 10.0 ml. of 0.05 M K_2HPO_4 and centrifuged. The supernatant solution is treated with 0.18 ml. of 0.5 M KH_2PO_4 and adjusted to pH 5.8 with 1 N acetic acid (0.43 ml.). This solution is heated to 60° in a 70° water bath, kept for 5 minutes at 60°, and cooled in ice to 2°. The heated solution is adjusted to pH 7.4 with 1.0 ml. of 2 N ammonium hydroxide and centrifuged. The supernatant solution containing the ribokinase can be stored indefinitely at $-16°$. It contains some pentose phosphate isomerase but is free of transketolase.

Ribokinase activity is assayed in a mixture containing 0.05 ml. of 0.02 M D-ribose, 0.05 ml. of 0.10 M $MgCl_2$, 0.05 ml. of 0.045 M ATP, 0.50 ml. of 0.05 M Tris buffer, pH 7.5, and 0.05 to 0.20 ml. of ribokinase, incubated at 25°. Aliquots of 0.1-ml. samples are added to 0.5 ml. of ethanol containing 0.01 ml. of 2 M barium acetate. The precipitated phosphate esters are removed by centrifugation, and 0.1 ml. of the supernatant solution is assayed for free pentose by the orcinol method.[5] A unit of activity is defined as the amount of enzyme required to esterify 1 micromole of ribose in 1 hour, under the conditions of the test.

For the preparation of ribose-5-phosphate the incubation mixture contains 2.5 ml. of 0.03 M D-ribose, 4.7 ml. of 0.1 M $MgCl_2$, 4.4 ml. of 0.045 M ATP, 24.0 ml. of 0.05 M Tris buffer, pH 7.5, and 9.5 ml. of ribokinase (90 units). After 60 minutes at room temperature esterification is essentially complete. The incubation mixture is centrifuged and the supernatant solution passed through a Dowex 1 (formate) column

[5] See Vol. III [13].

1.1 cm. × 20 cm. The column is washed with 10 ml. of water and the phosphate esters eluted with 0.2 N formic acid containing 0.03 N sodium formate. Fractions of 5.0 ml. are collected, and 0.03-ml. aliquots assayed for pentose. Fractions 68 to 77, containing the pentose esters, are partly neutralized with 1.8 ml. of 4 N sodium hydroxide and adjusted to pH 6.5 with 5.0 ml. of saturated barium hydroxide. The barium salts are precipitated with 220 ml. of ethanol, collected by centrifugation, washed with 10 ml. of 80% ethanol, and dried *in vacuo*. The yield is 24.4 mg., containing 56 micromoles of pentose ester. The purity is about 85%, and the recovery 63%. The product is mainly ribose-5-phosphate, with small amounts (approximately 10%) of ribulose-5-phosphate. For the preparation of P^{32}-labeled pentose phosphate it is desirable to use a smaller excess of ATP; however, at lower levels of ATP the reaction rate is greatly reduced.

[29] Preparation and Analysis of Ribulose-5-phosphate

By B. L. HORECKER

Preparation

Principle. In the oxidation of 6-phosphogluconate by purified phosphogluconic dehydrogenase preparations from yeast, the first product to accumulate is ribulose-5-phosphate which is then slowly converted, by traces of an isomerase in the enzyme preparation, to ribose-5-phosphate.[1] Thus maximum yields of the ribulose ester are obtained only if the incubation is not prolonged beyond the time necessary to accomplish the oxidative decarboxylation. The reaction is carried out with catalytic amounts of TPN in the presence of pyruvate and lactic dehydrogenase to reoxidize the TPNH formed:[2]

6-Phosphogluconate + TPN⁺ → Ribulose-5-phosphate + CO_2
+ TPNH + H⁺

Pyruvate + TPNH + H⁺ → Lactate + TPN⁺

Reagents

6-Phosphogluconate. One gram of the barium salt[3] is dissolved in 8.0 ml. of water with the aid of 0.7 ml. of glacial acetic acid.

[1] B. L. Horecker, P. Z. Smyrniotis, and J. E. Seegmiller, *J. Biol. Chem.* **193**, 383 (1951); see Vol. I [42].

[2] A. M. Mehler, A. Kornberg, S. Grisolia, and S. Ochoa, *J. Biol. Chem.* **174**, 961 (1948); see Vol. I [67] or [69].

[3] See Vol. III [23].

Barium ion is removed by the addition of 2.37 ml. of 2.2 N sulfuric acid and centrifugation. The supernatant solution is adjusted to pH 7.4 with 3.55 ml. of 5 N potassium hydroxide and diluted to 25.0 ml. with water.

Sodium pyruvate, 72.0 mg. of commercial sodium pyruvate[4] in 3.0 ml. of water.

TPN, 31.2 mg., 86% pure, in 1.0 ml. of water.

Phospholguconic dehydrogenase, 376 units/ml.; 32.7 mg./ml.

Lactic dehydrogenase, 13.7 mg./ml.

Procedure. The reaction mixture contains 1.3 ml. of water, 5.2 ml. of 6-phosphogluconate (350 micromoles), 2.0 ml. of sodium pyruvate (440 micromoles), 0.2 ml. of TPN (7.2 micromoles), 0.8 ml. of phosphogluconic dehydrogenase (300 units), and 0.5 ml. of lactic dehydrogenase (6.8 mg. of protein). After 10 minutes at 25° the reaction mixture is treated with 1.0 ml. of 50% trichloroacetic acid, cooled to 0°, diluted with 20.0 ml. of water, and centrifuged. The supernatant solution is adjusted to pH 6.4 with 5.7 ml. of saturated $Ba(OH)_2$, and after 1 hour at 0° the precipitated barium sulfate is centrifuged, washed with 6.0 ml. of water, and discarded. The supernatant solution and washing are combined and treated with 0.5 ml. of 1 M barium acetate and 4 vol. (180 ml.) of absolute ethanol. After 1 hour at 0°, the precipitate is collected by centrifugation, washed with 80% ethanol, and dried *in vacuo* over $CaCl_2$ and KOH. The product, 100 to 130 mg., contains a mixture of ribulose-5-phosphate and ribose-5-phosphate, with the latter comprising 15 to 35% of the total pentose. Approximately 70% of phosphate is present as pentose phosphate.

Chromatography. Further separation of ribulose-5-phosphate and ribose-5-phosphate is accomplished by ion exchange chromatography. The barium salt (145 mg.) is dissolved in 10 ml. of water and adsorbed on a Dowex 1 (formate) column,[5] 2.5 × 15 cm., and the column washed with 15 ml. of water. Elution is carried out at the rate of 2.5 ml./min. with a formate buffer containing 0.1 M formic acid and 0.03 M sodium formate. Fractions of 25 ml. are collected and analyzed for pentose by the orcinol method.[6] Two peaks are observed, one with maximum in fraction 65 (ribose-5-phosphate) and the second with maximum in fraction 72 (ribulose-5-phosphate), with some overlapping. For the precipitation of the ribulose-5-phosphate, tubes 68 to 80 (320 ml.) are pooled and brought to pH 6.3 by the addition of 64 ml. of saturated $Ba(OH)_2$. The barium salt is precipitated with 1540 ml. of ethanol, collected by cen-

---

[4] Nutritional Biochemicals Co.

[5] For the preparation of the resin see DPN preparation, Vol. III [124].

[6] See Vol. III [13].

Procedure

Formation of Equilibrium Mixture. The ribose-5-phosphate solution (42.0 ml., 4200 μM.) is adjusted to pH 7.5 with 1.0 ml. of 2 N KOH and treated with 1 ml. of isomerase (30,000 units, 12 mg.) and 5 ml. of epimerase (800 units, 45 mg.). The mixture is incubated at 38°, and the reaction followed by measurement of ribose-5-phosphate with the phloroglucinol assay.[7] The reaction is essentially complete in about 5 minutes. After 1 hour at 38° the solution is treated with 10 ml. of 10% trichloroacetic acid followed by 15 ml. of 1 M barium acetate. The precipitated barium sulfate is removed by centrifugation. The supernatant solution is adjusted to pH 5.5 with 0.5 ml. of saturated NaOH and treated with 4 vol. of absolute ethanol. The precipitate is allowed to flocculate at 0° for 2 hours, collected by centrifugation, and washed with 100 ml. of 80% ethanol. Dried at high vacuum, the product weighs 2.0 g. (crude barium salt mixture).

Formation of Ribulose-1,5-diphosphate. The barium salt mixture is dissolved in 25 ml. of 0.02 N acetic acid and decomposed with a slight excess (10 ml.) of 0.6 M K_2SO_4. Barium sulfate is removed by centrifugation. The supernatant solution is treated with 10 ml. of 0.35 M ATP, 1.5 ml. of 1 M $MgCl_2$, and 2.5 ml. of 0.2 M glutathione. The pH is adjusted to 7.8, and 1.5 ml. of phosphoribulokinase (3000 units, 18 mg.) is added. The solution is stirred continuously, and the pH maintained at 7.8 by the addition of 1 N NaOH. The reaction is complete after about 55 minutes, when little further change in pH is noted. At this time, 20 ml. of 10% trichloroacetic acid is added, together with sufficient 6 N HCl (0.1 ml.) to bring the pH to 2.0. This solution is treated with 20 g. of charcoal and filtered with suction. On the basis of its absorption at 260 mμ, the filtrate should be essentially free of adenine nucleotides. The charcoal mass is washed by suspending it in 100 ml. of water and filtering. After three washings the combined filtrates (525 ml.) are adjusted to pH 6.0 with 3.0 ml. of 5 N NaOH. This solution is treated with 10 ml. of 1 M barium acetate and centrifuged to remove barium sulfate. The supernatant solution (525 ml.) is treated with 175 ml. of ethanol and the precipitate allowed to flocculate overnight at 0°. The solution is centrifuged and the precipitate washed with 100 ml. of 80% ethanol and dried *in vacuo* over KOH. This is the ribulose diphosphate fraction; yield = 1.0 g.; purity = 73%.

The 25% ethanol supernatant solution is treated with 1900 ml. of absolute ethanol, cooled at 0° for 2 hours, and centrifuged. The precipitate is washed and dried as before. This is the xylulose-5-phosphate fraction; yield = 0.7 g.; purity = 68%.

[7] Z. Dische, unpublished procedure.

Analysis

Ribulose-1,5-diphosphate. 58.5 mg. of the barium salt dissolved in 2.0 ml. of 0.1 N HCl was found to contain: ribulose (orcinol),[8] 73.4 μM.; phosphate (organic), 152.6 μM.; phosphate (inorganic), 33.4 μM.; adenine (280-mμ absorption), 0 μM.; theoretical (molecular weight = 580), 101.0 μM.

Xylulose-5-phosphate. 38.5 mg. of the barium salt dissolved in 2.0 ml. of 0.1 N HCl was found to contain: xylulose (orcinol),[9] 72.0 μM.; phosphate (organic), 70.2 μM.; phosphate (inorganic), 27.8 μM.; adenine 0 μM.; theoretical (molecular weight = 360) 105.0 μM.

[8] In this test both ribulose diphosphate and xylulose-5-phosphate yield absorption at 670 mμ which is 0.57 times that given by equal amounts of ribose or arabinose.
[9] Between 1 and 2% of the ketopentose is ribulose.

[30] Preparation and Analysis of Heptulose Phosphates

By B. L. HORECKER

Preparation

Principle. Sedoheptulose-7-phosphate is prepared from ribose-5-phosphate by the action of pentose phosphate isomerase and transketolase. A purified transketolase preparation from spinach contains both enzymes.[1]

Ribose-5-phosphate + ribulose-5-phosphate
$\rightleftarrows$ Sedoheptulose-7-phosphate + glyceraldehyde-3-phosphate

Sedoheptulose-1,7-diphosphate is prepared from sedoheptulose-7-phosphate and fructose-1,6-diphosphate by the action of aldolase and transaldolase.[2] Transaldolase is prepared from yeast,[3] aldolase is prepared from rabbit muscle.[4] The products are isolated by ion exchange chromatography.

Sedoheptulose-7-phosphate + glyceraldehyde-3-phosphate
$\rightleftarrows$ Fructose-6-phosphate + tetrose phosphate

[1] B. L. Horecker, P. Z. Smyrniotis, and H. Klenow, *J. Biol. Chem.* **205,** 661 (1953); see also Vol. I [53].
[2] B. L. Horecker, P. Z. Smyrniotis, Howard H. Hiatt, and Paul A. Marks, *J. Biol. Chem.* **212,** 827 (1955).
[3] See Vol. I [55].
[4] J. F. Taylor, A. A. Green, and G. T. Cori, *J. Biol. Chem.* **173,** 591 (1948); see Vol. I [39].

minutes by 1 N H_2SO_4 at 100°; the remaining organic phosphate shows the same acid stability as sedoheptulose-7-phosphate.

Enzymatic Properties. Sedoheptulose-7-phosphate yields active glycolaldehyde with transketolase[1] and is a dihydroxyacetone donor with transaldolase.[8] Sedoheptulose diphosphate is cleaved by aldolase about one-half as rapidly as is fructose diphosphate. The K_m for sedoheptulose diphosphate with aldolase is 0.1×10^{-3} M., compared with 0.06×10^{-3} M. for fructose diphosphate.

[8] B. L. Horecker and P. Z. Smyrniotis, *J. Am. Chem. Soc.* **75**, 2021 (1953)

[31] DL-Glyceraldehyde-3-phosphate

By HENRY A. LARDY

$$
\begin{array}{l}
\text{CHO} \\
| \\
\text{CHOH} \\
| \\
\text{CH}_2\text{OPO}_3{}^{--}
\end{array}
$$

D-Glyceraldehyde-3-phosphate may be prepared enzymatically[1] from hexosediphosphate or chemically by the recently developed procedure of Ballou and Fischer.[2] More commonly the substrate is used in the DL form after synthesis by the procedure of Baer and Fischer.[3,4] Their method is described here.

The starting material, DL-glyceraldehyde dimer, may be prepared in the laboratory[5] and is also available commercially. Baer[4] has found it advantageous to dry the dimer by refluxing with acetone and to disperse it by passing through a 200-mesh sieve.

DL-Glyceraldehyde-1,3-bisdiphenylphosphate Dimer

To a solution of 26 g. of diphenylchlorophosphonate in 9 ml. of ice-cold, dry pyridine, 2.5 g. of DL-glyceraldehyde is added and the mixture is stirred mechanically in a sealed flask immersed in an ice-salt bath near 0°. The pasty mixture must be stirred sufficiently vigorously to prevent the temperature from rising above 10°. After stirring for 4 to 6 hours the mixture may be kept overnight in the refrigerator. If the tem-

[1] O. Meyerhof and R. Junowicz-Kocholaty, *J. Biol. Chem.* **149**, 71 (1943).
[2] C. E. Ballou and H. O. L. Fischer, *J. Am. Chem. Soc.* **77**, 3329 (1955).
[3] E. Baer and H. O. L. Fischer, *J. Biol. Chem.* **150**, 223 (1943).
[4] E. Baer, *Biochem. Preparations* **1**, 50 (1949).
[5] *Org. Syntheses, Coll. Vol.* **2**, 305 (1943).

perature is allowed to rise during the reaction, the mixture darkens and may even become black. A satisfactory product is obtained nevertheless, but the yield may be greatly decreased.

The cold reaction mixture is stirred up with 60 ml. of cold 95% ethanol, and when the paste is homogeneous it is filtered with suction. The residue is washed with small portions of ice-cold ethanol until it is colorless. The product is then dissolved in 25 ml. of warm dioxane and is filtered. Another 15 to 20 ml. of dioxane may be used in small portions to rinse the original container and funnel to effect a quantitative transfer. The residue which does not dissolve in dioxane is unreacted glyceraldehyde and may amount to as much as 600 mg.

Occasionally the product may crystallize from dioxane after filtration. If this occurs it is best to redissolve the compound and allow crystallization to occur more slowly. Petroleum ether is added slowly to the dioxane solution with scratching to induce crystallization. As crystallization proceeds, more petroleum ether is added. After a total of 180 ml. has been added, precipitation is rapidly completed and the compound may be collected by filtration within a few hours. The yield varies from 5.8 to 8.2 g. The compound is usually obtained in its pure form with a melting point of 110° to 111°.

Dimeric DL-Glyceraldehyde-1-bromide-3-diphenylphosphate

The phosphate groups at the hemiacetal positions are cleaved by treatment of the 1,3-bisdiphenylphosphate dimer with 30% hydrobromic acid in glacial acetic acid.

Commercially available HBr in glacial acetic acid is cooled in an ice bath, and 33 ml. is added to 9.4 g. of DL-glyceraldehyde-1,3-bisdiphenylphosphate dimer in a glass-stoppered flask. The mixture is shaken occasionally for a few minutes until it dissolves and is then allowed to stand at room temperature for 24 hours. The crystals, which begin to form within ½ hour, are collected by filtration and washed with successive quantities of ice water. Alternatively, the crystalline paste may be poured into 250 ml. of ice and water which is thoroughly stirred for several minutes. The crystals are then collected, washed with ice water, and dried *in vacuo* over P_2O_5 or calcium chloride. The crude product weighs about 6.0 to 6.6 g. It is recrystallized from the minimum volume of purified[4] ethyl acetate (100 to 150 ml.). When dry the compound melts at 161° to 162°.

Dimeric DL-Glyceraldehyde-1-bromide-3-phosphoric Acid

The phenyl groups are removed from the above ester by reductive cleavage with hydrogen in the presence of Adam's platinum oxide catalyst. Occasionally this step results in the formation of an acid-stable compound which is not the desired one. It appears that the acid-stable compound is formed more frequently if the highly active commercially available Pt_2O is used. Better results are obtained with catalyst prepared in the laboratory.[6]

Two and one-half grams of dimeric glyceraldehyde-1-bromide-3-diphenylphosphate is suspended in 50 ml. of purified and dry glacial acetic acid. Commercial glacial acetic may be purified by refluxing with CrO_3, distilling, and storing over granular anhydrous calcium sulfate. The suspension is shaken with 0.6 g. of Pt_2O catalyst in an atmosphere of dry hydrogen. The apparatus should be equipped with a simple gas buret, such as an inverted graduate cylinder attached to a leveling bulb, to permit quantitative measurement of hydrogen uptake. As reduction proceeds, the poorly soluble diphenyl ester continues to dissolve, and the solution becomes entirely clear when reduction is near completion. A total of 16 mole equivalents of hydrogen should be consumed. If hydrogen uptake stops before the theoretical amount has been consumed, the apparatus may be flushed with nitrogen, a small quantity of fresh catalyst added, and hydrogenation may be continued to complete the reduction. Fresh catalyst must never be added in the presence of hydrogen. It ignites!

When the reduction is completed, the catalyst is separated by filtration and washed with small portions of dry acetic acid. The clear solution is evaporated to dryness under a vacuum and at a bath temperature not exceeding 25°. The residue may be stored in the distilling flask, in a vacuum desiccator over NaOH flakes. The dried crude product can be removed as needed, dissolved, and neutralized carefully for enzyme experimentation. A pure and more stable form of the compound may be obtained as the dioxane addition compound described below.

Dioxane Oxonium Salt of Dimeric DL-Glyceraldehyde-1-bromide-3-phosphoric Acid

The dried residue from the reduction is dissolved in 5 vol. of tributyl phosphate, filtered or centrifuged if necessary, and 10 vol. of purified dioxane is added. The solution is allowed to stand at room temperature for a day. The pure dioxane addition compound (molecular weight 642; $P = 9.65\%$) crystallizes out in good yield. It is collected by filtration and washed with small portions of cold dioxane. It may be stored for a considerable time in a vacuum desiccator, especially in the cold, without appreciable decomposition. The compound should contain little or no inorganic phosphate.

Solutions of the dioxane addition compound hydrolyze spontaneously

[6] R. Adams, V. Voorhees, and R. L. Shriner, *Org. Syntheses, Coll. Vol.* **1**, 463 (1941).

when dissolved in water. They should be neutralized carefully, for the glyceraldehyde-3-phosphate is readily hydrolyzed by alkali. The bromide ion and dioxane which are liberated do not interfere with most enzyme systems. If desired, the compound may be converted to the calcium salt as described in the original paper of Baer and Fischer.[3]

[32] Assay of Triose Phosphates

By WILLIAM S. BECK

The two triose phosphates, D-glyceraldehyde-3-phosphate (G-3-P) and dihydroxyacetone phosphate (DHA-P), are both involved in the reactions of aldolase, triosephosphate isomerase, and certain phosphatases, and, in addition, each is acted on by a number of enzymes specific for one or the other of the two isomers. Thus, G-3-P is oxidized to phosphoglyceric acid by triosephosphate dehydrogenase and DPN and reacts specifically with transketolase, DR-aldolase, and triokinase; DHA-P alone participates in the reactions of certain aldolases and is reduced by α-glycerophosphate dehydrogenase and reduced DPN.

Chemically, the compounds are aldose-ketose isomers containing alkali- and acid-labile phosphate ester linkages, both of which shared properties—the labile phosphate bond and the carbonyl group—may be exploited in assaying total triose phosphate. Exposure to 2 N NaOH for 20 minutes at room temperature completely hydrolyzes both esters to orthophosphate and lactic acid. Acid hydrolysis yields orthophosphate and methyl glyoxal; in 1 N HCl at 100°, both esters are hydrolyzed 44% at 7 minutes and 90% at 30 minutes.

Among the properties which permit the estimation of one triose phosphate in the presence of the other are the enzyme specificities mentioned above, the susceptibility of G-3-P and the resistance of DHA-P to oxidation by iodine in weakly alkaline solution, and the relatively slower rate of 2,4-dinitrophenylosazone formation by the aldotriose G-3-P as compared to the ketotriose DHA-P.

All these properties have been used with varying success in assaying the triose phosphates and the enzymes with which they interact.

Estimation of Total Triose Phosphate

General Considerations. The procedures most frequently used are based on (1) the determination of total alkali-labile phosphorus;[1] (2) the assay

[1] O. Meyerhof and K. Lohmann, *Biochem. Z.* **275**, 89 (1934).

of the methyl glyoxal formed by heating triose phosphates in acid by the colorimetric method of Ariyama,[2] or by the procedure of Dounce and Beyer[3] in which methyl glyoxal is further converted to acetaldehyde which is then assayed by a modification of the lactic acid method of Barker and Summerson;[4] (3) optical changes in systems containing excess triosephosphate isomerase and either α-glycerophosphate dehydrogenase and DPNH (in the absence of phosphate or arsenate)[5] or triosephosphate dehydrogenase, DPN, and arsenate;[6,7] and (4) the spectrophotometric measurement of colored 2,4-dinitrophenylhydrazine derivatives.[8] The 2,4-dinitrophenylhydrazine derivatives have also been measured gravimetrically,[9] although the procedure is rather cumbersome.

The need to measure *total triose phosphate* arises most commonly in connection with the estimation of aldolase activity.

The Determination of Aldolase.[10] When aldolase is to be measured in preparations known to be free of triosephosphate dehydrogenase and isomerase, the enzymatic method of Warburg and Christian[6] is satisfactory. In this procedure, the rate of DPNH production is observed spectrophotometrically in a system containing excess triosephosphate dehydrogenase, DPN, and arsenate. Since only one of the two triose phosphates formed in the aldolase reaction is being measured, it is necessary to assume that this represents half of the total triose phosphate. Directions for this procedure are given in Vol. I [39, 40]. It should be noted that the enzymatic methods are useful only in purified systems and require reagent enzymes that are demonstrably free of interfering enzymes.

Total triose phosphate produced by aldolase activity in crude tissue extracts, homogenates, and sera can be measured accurately by the use of

[2] N. Ariyama, *J. Biol. Chem.* **77**, 359 (1938).
[3] A. L. Dounce and G. T. Beyer, *J. Biol. Chem.* **173**, 159 (1948).
[4] S. B. Barker and W. H. Summerson, *J. Biol. Chem.* **138**, 535 (1941).
[5] E. Racker, *J. Biol. Chem.* **167**, 843 (1947).
[6] O. Warburg and W. Christian, *Biochem. Z.* **314**, 149 (1943).
[7] O. Meyerhof and P. Oesper, *J. Biol. Chem.* **170**, 1 (1947).
[8] J. Sibley and A. Lehninger, *J. Biol. Chem.* **177**, 859 (1949).
[9] E. Baer and H. O. L. Fischer, *J. Biol. Chem.* **150**, 213 (1943).
[10] The comments which follow are concerned chiefly with matters relating to the determination of triose phosphates. Further discussion of the enzymological features of the aldolase assay may be found in Vol. I [39, 40] and in the papers of Meyerhof,[11] Cook and Dounce,[12] Sibley and Lehninger,[8] Bruns,[13] and Beck.[14]
[11] O. Meyerhof, *in* "The Enzymes" (Sumner and Myrback, eds.), Vol. II, Part 1, pp. 162–182. Academic Press, New York, 1951.
[12] J. L. Cook and A. L Dounce, *Proc. Soc. Expt. Biol. Med.* **87**, 349 (1954).
[13] F. Bruns, *Biochem. Z.* **325**, 429 (1954).
[14] W. S. Beck, *J. Biol. Chem.* **212**, 847 (1955).

carbonyl trapping agents such as hydrazine and cyanide. These fix the triose phosphates quantitatively, thus preventing further enzymatic transformation, and, at the same time, serve to displace the reaction equilibrium. The triosephosphate hydrazones can then be assayed in deproteinized filtrates as total alkali-labile phosphorus (as described in Vol. I [39]) or by one or another of the colorimetric methods. Of these, the methyl glyoxal procedures are rarely used because they yield high blank values. The best procedure is probably that of Sibley and Lehninger,[8] a modification of which[14] is presented here.

Reagents

FDP solution (0.05 M). Commercial preparations of the barium salt of fructose-1,6-diphosphate are satisfactory. Remove barium by the addition of an equivalent amount of Na_2SO_4 followed by filtration. Adjust the pH to 8.6 with NaOH. Solutions may be stored in the freezer.[15]

Hydrazine sulfate (0.22 M). Analytical reagent-grade preparations are used. Adjust the solution to pH 8.6 with NaOH.

Tris buffer (0.1 M, pH 8.6). Dissolve 24.2 g. of tris(hydroxymethyl)-aminomethane in water and dilute to 500 ml. Adjust the pH to 8.6 with HCl, and dilute to 1 l. The choice of buffer may influence the results, although reports of various authors have differed on the extent and explanation of this effect. Aldolase activity at pH 8.6 tends to be higher in Tris than in collidine buffer, whereas at 7.4 the reverse seems to be the case. A recent report[12] states that at 7.4 *serum* aldolase activity (unlike that of tissue preparations) is higher in Tris than in collidine.

10% Trichloroacetic acid.

0.75 N NaOH.

2,4-DNPH solution. Accurately weigh out 1.0 g. of 2,4-dinitro-phenylhydrazine, dissolve, and dilute to 1 l. with 2 N HCl.

Enzyme. Dilute samples to be tested (enzymes, extracts, or homogenates) to contain in a 0.2-ml. aliquot an amount of aldolase which will split 0.1 to 0.4 μM. of FDP in 15 minutes. One milliliter of normal serum contains about this amount of aldolase; therefore in assaying serum aldolase a 1-ml. sample should be used with a correspondingly smaller volume of water.

Procedure. To a tube containing 1.0 ml. of buffer, 0.25 ml. of FDP, 0.25 ml. of hydrazine, and 0.8 ml. of water (omitted in serum assays), · add 0.2 ml. of enzyme (or 1.0 ml. of serum). Incubate for 15 minutes at

[15] Methods of isolation and purification of FDP are given in Vol. III [22].

38°, then terminate the reaction by adding 2.0 ml. of 10% TCA, and centrifuge. The control incubation is identical except that FDP is not added until after the TCA. Transfer 1.0 ml. of the protein-free filtrate to a cuvette, and add 1.0 ml. of 0.75 N NaOH with mixing. Allow the tubes to stand for 10 minutes at room temperature; then add 1.0 ml. of 2,4-DNPH solution, and incubate for 60 minutes at 38°. Add 7.0 ml. of NaOH and, after 10 minutes, read the optical density at 550 mμ. The control tube may ordinarily be used as the blank, although it may be desired to distinguish between a sample blank and reagent blank, in which case a tube may be carried in which water is substituted for enzyme.

Calculation of Results. Meyerhof and Beck[16] defined a unit of aldolase activity as that amount of enzyme which will produce 1 mg. of alkali-labile phosphorus per minute at 38°. The aldolase unit proposed by Sibley and Lehninger,[8] however, was based on the Warburg notation and was defined as the amount of enzyme which will split 1 mm.^{3} of FDP in 1 hour at 38°. Although such definitions serve adequately in expressing comparative data (as in purification schemata or clinical determinations), it seems to the writer that they complicate life unnecessarily by imposing the need always to remember which of several alternative definitions is being used, without adding any compensatory clarity. The use of volumetric units to express rates of substrate disappearance seems particularly unfortunate in view of the increasing obsolescence of the manometric method for measuring glycolysis. Accordingly, it is recommended that aldolase activity be expressed as micromoles of FDP split (or triose phosphate formed) per whatever time unit was used in the assay procedure (in this case 15 minutes) per tissue or enzyme reference unit. This procedure is equally satisfactory for expressing comparative data and, in addition, conveys information of possible interest to those concerned with actual or potential velocity rates of component enzymes in multi-enzyme systems.

Calibration Procedures. A problem common to all methods for estimating triose phosphates is the difficulty in obtaining pure primary reference standards. Until recently, synthetic triose phosphates were difficult to prepare and were unstable even when stored dry at −20°. Commercial preparations have almost invariably been grossly impure. Recently, however, Ballou and Fischer[17] have devised newer methods for the synthesis of triose phosphates which yield cyclohexylammonium salts of the dialkyl acetal derivatives of extreme purity and stability. In the experience of the writer, samples of these compounds, when converted to the corresponding free acids, have been entirely satisfactory reference

[16] O. Meyerhof and L. V. Beck, *J. Biol. Chem.* **156**, 109 (1944).
[17] C. E. Ballou and H. O. L. Fischer, *J. Am. Chem. Soc.* **77**, 3329 (1955). See Vol. III [31].

standards, displaying essentially theoretical enzymatic and alkali-labile phosphorus analyses on a weight basis.

Until these compounds are generally available, total triose phosphate methods may be calibrated by the determination of total alkali-labile phosphorus in filtrates on which 2,4-DNPH colors have been separately measured. This calibration method is relatively accurate, although the large number of necessary determinations, blanks, controls, and standards makes it both cumbersome and susceptible to technical error. In addition, it cannot be used in conjunction with serum aldolase determinations because of interference by serum inorganic phosphate. In this case, it is necessary to use serial dilutions of an aldolase preparation for the calibration procedure. The burden thus created for the clinical laboratory has now been partially eased by the advent of acceptable aldolase preparations on the commercial market (Worthington, Nutritional, and others). The details of the alkali-labile P calibration procedure are given by Sibley and Lehninger.[8]

Further studies[14] on the Sibley and Lehninger procedure have shown that pure glyceraldehyde may be used as a secondary standard. It was found that the chromogens derived from the triosephosphate hydrazones on addition of 2,4-DNPH and NaOH are identical for G-3-P and DHA-P and consist essentially of a mixture of methylglyoxal-2,-4-dinitrophenylosazone and pyruvic 2,4-dinitrophenylhydrazone. The same chromogenic compounds are formed from the trioses glyceraldehyde and dihydroxyacetone, but in different proportions, so that the triose chromogen extinction values differ from those of the triosephosphate chromogens. However, if the chromogen development reaction is continued for 60 minutes, the color value of the triosephosphate chromogen consistently equals 1.6 to 1.9 times the color value of the triose chromogen. Therefore, if used in conjunction with a correction factor based on a prior alkali-labile phosphorus calibration, glyceraldehyde may be used as a secondary standard.

Glyceraldehyde solutions should be allowed to depolymerize in aqueous solution at room temperature for 2 to 4 days. Solutions should be stored in the refrigerator and prepared fresh every two weeks, preferably with boiled water. A satisfactory standard may be prepared by diluting 50 mg. of DL-glyceraldehyde to 10.0 ml. with water. Prior to use, a 1:25 dilution is made, yielding a solution containing 2.22 μM./ml. Tubes containing buffer, hydrazine, 0.5- to 1.0 ml.-aliquots of diluted glyceraldehyde solution, and water to a volume of 2.5 ml. are carried through incubation, TCA treatment, and chromogen development, in parallel with the aldolase incubations.

The use of dihydroxyacetone as a secondary standard has also been reported by Lowry and associates.[18] These authors also made the useful suggestion that any tendency for 2,4-DNPH chromogen to precipitate (particularly in microassays) may be combated by adding methyl Cellosolve to the chromogen development reaction.

Estimation of Individual Triose Phosphates in Mixtures

General Considerations. The principles used to determine G-3-P and DHA-P in each other's presence were referred to above. Their chief application has been in the assay of triosephosphate isomerase.

In the classical method of Meyerhof,[16,19,20] G-3-P is selectively oxidized to phosphoglyceric acid by alkaline iodine and assayed as alkali-labile phosphate disappearance or, more accurately, by the increase in optical rotation in the presence of added molybdate (the polarimetric method for phosphoglyceric acid). It has been pointed out, however, that, unless conditions are precisely controlled, incomplete aldotriose and partial ketotriose oxidation frequently occur.

The enzymatic methods—some of the several variations of which have been described by Oesper and Meyerhof,[21] Warburg and Christian,[22] and Racker[5]—are all subject to the strictures given above about interfering enzymes. If G-3-P or DHA-P is measured as produced by optical methods using DPN-linked dehydrogenases, reaction mixtures must be rigidly freed of aldolase and triosephosphate isomerase. Similarly, if residual or newly formed triose phosphate is measured after termination of the incubation by heat, acid, or removable protein precipitants such as perchloric acid, reagent enzymes must be pure and isomerase-free (see Schade[23]).

It has recently been reported[24] that triose phosphates in mixtures and triosephosphate isomerase activity may be assayed on the basis of the difference in rate of formation of the chromogenic 2,4-dinitrophenylosazone derivatives of G-3-P and DHA-P. By means of carefully timed chromogen development reactions, it is possible to determine the G-3-P/DHA-P ratio in filtrates and, from parallel determinations of total triose phosphate as alkali-labile phosphorus, the concentrations of the individual triose phosphates and hence their rate of conversion can be calculated. The method is relatively simple but requires pure primary reference com-

[18] O. H. Lowry, N. R. Roberts, M. Wu, W. Hixon, and E. Crawford, *J. Biol. Chem.* **207**, 19 (1954).
[19] O. Meyerhof and W. Kiessling, *Biochem. Z.* **279**, 40 (1935).
[20] O. Meyerhof and R. Junowicz-Kocholaty, *J. Biol. Chem.* **147**, 71 (1943).
[21] P. Oesper and O. Meyerhof, *Arch. Biochem.* **27**, 223 (1950).
[22] O. Warburg and W. Christian, *Biochem. Z.* **314**, 399 (1943).
[23] A. Schade, *Biochem. et Biophys. Acta* **12**, 163 (1953).
[24] W. S. Beck, *Arch. Biochem. and Biophys.* in press.

pounds and repeated standardization. In addition, the method can only yield linear activity vs. enzyme concentration curves over a limited range of enzyme concentrations, owing to the limitation set by the magnitude of the ratio between the ϵ_M values of the DHA-P and G-3-P chromogens.

Separation of Triose Phosphates

In tissue fractionation, triose phosphates appear in the barium-soluble–alcohol-precipitable fraction. In preparing substrates for enzyme experiments, some of the earlier workers used this method and another method involving hydrazine interception in large-scale aldolase incubations, followed by benzaldehyde and ether extraction. More recently, a variety of chromatographic procedures have appeared in the literature, references to which may be found in reviews by Block et al.[25] and Isherwood.[26]

[25] R. Block, E. Durrum, and G. Zweig, "A Manual of Paper Chromatography and Paper Electrophoresis," 2nd ed. Academic Press, New York, 1955.
[26] F. A. Isherwood, Brit. Med. Bull. **10**, 202 (1954).

[33] L-α-Glycerophosphate

By HENRY LARDY

$$H_2COH$$
$$|$$
$$HOCH$$
$$|$$
$$H_2COPO_3Ba$$

The synthesis of the natural isomer of α-glycerophosphate is accomplished by the procedure of Baer and Fischer.[1] The reaction is carried out in a two-necked flask fitted with a dropping funnel and a vent through a calcium chloride tube. The flask is cooled at all times by a salt-ice bath at −15°.

A solution of 8.75 g. of D-isopropylidene-glycerol[2] in 24 ml. of pure, dry quinoline is added dropwise with shaking during a period of 20 minutes to a solution of 10.5 g. of phosphorus oxychloride in 30 ml. of quinoline.[3] After 1 hour the flask may be removed from the cold bath and

[1] E. Baer and H. O. L. Fischer, *J. Biol. Chem.* **128**, 491 (1939).
[2] E. Baer and H. O. L. Fischer, *J. Biol. Chem.* **128**, 463 (1939); E. Baer, *Biochem. Preparations* **2**, 31 (1952).
[3] This solution is prepared by adding phosphorus oxychloride dropwise to precooled quinoline.

permitted to come to room temperature to complete the reaction and to facilitate handling the mixture which is extremely viscous at the low temperature of the bath. The reaction mixture is then poured slowly into a vigorously stirred mixture of 30 ml. of H_2SO_4, 28 g. of Ag_2CO_3, and 700 ml. of ice and water. After 30 minutes of stirring the solution is filtered through a layer of charcoal on a precoated funnel and silver ion is removed from the solution by treatment with H_2S. The clear filtrate is heated in a boiling water bath for 30 minutes to free it of H_2S and to remove the isopropylidene group. The solution is cooled to 0° and, with stirring, finely pulverized $Ba(OH)_2 \cdot 8H_2O$ (about 100 g. is required) is added slowly until the solution becomes alkaline to phenolphthalein. Excess barium ion is removed with carbon dioxide and, after filtration, the aqueous solution is concentrated by distillation in the vacuum of a water pump and with a bath at 40°. When the volume of the solution has been reduced to about 60 ml., a slight turbidity is removed by filtration. The barium salt of α-glycerophosphate is precipitated from solution by heating in a boiling water bath for about 20 minutes and collected, while hot, on a hard paper or sintered glass Büchner funnel. After drying *in vacuo* over P_2O_5, the compound has the composition of the anhydrous barium salt (molecular weight 307.5). Yields of 30 to 37% of theoretical are obtained.

For biological work the barium salt is dissolved in water with the aid of a little acid. Sodium or potassium sulfate is then added, and after removal of the precipitated $BaSO_4$ the solution is brought to the desired pH and volume.

The compound is extremely resistant to hydrolysis by the standard analytical procedure (1 N HCl, 100°). In hot alkaline solution the phosphate group migrates to the β position and optical activity is lost.

[34] Preparation of D(−)-3-Phosphoglyceric Acid

By INES MANDL and CARL NEUBERG

Natural 3-PGA may be prepared by three different methods:

1. It may be prepared biologically from yeast and sugar phosphate solution in the presence of acetaldehyde or another hydrogen acceptor and fluoride.[1-4] Such substances prevent a simultaneous formation of

[1] C. Neuberg and M. Kobel, *Biochem. Z.* **263**, 219 (1933).
[2] C. Neuberg and M. Kobel, *Biochem. Z.* **264**, 456 (1933).
[3] C. Neuberg and M. Kobel, *Angew. Chem.* **46**, 711 (1933).
[4] C. Neuberg and H. Lustig, *Arch. Biochem.* **1**, 311 (1942).

α-glycerophosphate, and the fluoride inhibits the transformation of the PGA by yeast or maceration juice. In the original method[1] the starting material was Na FDP, but it was soon found[2,3] that with phosphorylating yeast the FDP need not be isolated but could be replaced by sucrose or glucose, thus considerably simplifying the procedure. The method described below[4] has the further advantage of using commercial baker's yeast instead of the more difficultly available washed and pressed bottom yeast.

2. Racemic 2-PGA prepared from Na β-glycerophosphate[5] can be converted by muscle extract to D(−)-3-PGA, which can be isolated in pure form, owing to the low solubility of the Ba salt of PGA compared to that of 2-PGA.[6]

3. Total synthesis has been accomplished from free D-glyceric acid and "metaphosphoric acid" ester[7] by an adaptation of the method originally devised for the racemic compound.[8,9] D(−)-3-PGA is the main product and is separated as the acid Ba salt from the D(+)-2-PGA formed at the same time.

Preparation According to Neuberg and Lustig[4]

Sucrose (120 g.), 22.2 g. of $NaH_2PO_4·H_2O$ and 6.6 g. of $NaHCO_3$ in 600 ml. of water together with 300 g. of fresh National Grain yeast and 90 ml. of CCl_4 are kept at room temperature until phosphorylation is complete, i.e., about 25 hours or less, until Mg mixture no longer precipitates phosphate in the clearly filtered solution. At that time 840 ml. of 2% acetaldehyde, 140 ml. of 0.2 M NaF, 1200 ml. of 10% glucose, 380 g. of fresh yeast, and 20 ml. of CCl_4 are added. After thorough mixing and shaking, the mixture is again kept at room temperature for 24 hours. To deproteinize, the mixture is heated for ½ hour with 5 ml. of glacial acetic acid on the water bath and centrifuged. To each 100 ml. of clear filtrate 5 ml. of glacial acetic acid and 7 ml. of 50% Ba acetate are then added, the slight precipitate filtered off immediately, and the clear filtrate kept in the refrigerator for 48 hours. On scratching, crystallization sets in readily and is complete after 48 hours. The yield of crude product filtered off with suction and washed with water amounts to 16.8 g.

It may be mentioned here that other yeasts can be used, but National Grain gave most consistent results. Other plasmolytic agents may replace CCl_4 except CCl_3H. CCl_4 and ether were found to be more effective than the generally used toluene.

[5] W. Kiessling, *Ber.* **68**, 243 (1935).
[6] O. Meyerhof and W. Kiessling, *Biochem. Z.* **276**, 239 (1935).
[7] C. Neuberg, *Arch. Biochem.* **3**, 105 (1943).
[8] C. Neuberg, F. Weinmann, and M. Vogt, *Biochem. Z.* **199**, 248 (1928).
[9] M. Vogt, *Biochem. Z.* **211**, 1 (1929).

Acetaldehyde can be substituted by other reducible substances, e.g., methylene blue and in particular furfural, which is more convenient and at least as effective as acetaldehyde. Five per cent furfural (300 ml.) is used with the quantities cited. Phosphorylation at 37° gives lower yields, which are probably due to some hexose monoester formation. With dried yeasts phosphorylation is much faster but the extra step is seldom justifiable. If phosphorylation is not absolutely complete, pure PGA may still be obtained, since any inorganic P present is removed in the purification process of the acid Ba salt. Special precipitation with Mg acetate is not necessary. KF or NH₄F may replace NaF. Although the acid Ba salt is slightly soluble, the yield is not increased by higher concentration of acetaldehyde and fluoride. The addition of free glucose with the fluoride may be omitted.

Purification by Precipitation from Acid Solution.[4] Four grams of crude Ba salt is dissolved in 280 ml. of 0.05 N HCl, warmed, and filtered. The clear filtrate is precipitated with 560 ml. of alcohol. The milky precipitate changes to silky crystal when stirred. It is left in the refrigerator for some time, then filtered by suction and washed with dilute alcohol until free from chloride. The yield of pure acid BaPGA is 3.5 g.

Purification by Precipitation from NH₄ Salt Solution.[10] This method is based on the observation that the presence of NH₄ salts prevents the crystallization of the acid Ba salt of D(−)-3-PGA.[7] Any NH₄ salt which is itself soluble in the precipitating alcohol may be used; the amount needed may vary according to the purity of the Ba salt. One gram of acid Ba-PGA is heated with 25 ml. of 25% NH₄Cl. The Ba salt dissolves, and the adhering impurities can be filtered off. Then 100 to 125 ml. of CH₃OH is added to the clear filtrate which becomes opaque and condenses into a mass of crystals. These are filtered off, washed first with 75% EtOH, and then with pure EtOH or MeOH. The recovered Ba salt, $C_3H_5O_8PBa·2H_2O$, is free from NH₄⁺. $[\alpha]_D^{20} = -13.8°$.

Recovery of 3-PGA from the Mother Liquor.[4] Usually it is not worth the large amounts of alcohol required to work up the mother liquor. This should be done, however, if the yields are less than 12 g. for the amounts indicated. An equal volume of alcohol is added to the filtrate from the acetic acid-Ba acetate precipitation, and the mixture is left in the icebox for 24 hours. The precipitate is filtered off, washed with 50% alcohol, and suspended in 440 ml. of H₂O, to which 2 N HCl is then added until the solution is 0.05 N. Most of the precipitate goes into solution. The solution is filtered with fuller's earth, and the acid Ba salt is precipitated in the clear filtrate with 900 ml. of alcohol. After stirring, characteristic crystals appear; they are filtered off and washed and dried. The additional yield usually amounts to 5.6 g.

[10] I. S. Forrest and C. Neuberg, *Biochim. et Biophys. Acta* **11**, 588 (1953).

Preparation According to Meyerhof and Kiessling[6] via Synthetic Racemic 2-PGA[5]

Twenty grams of Na β-glycerophosphate, containing not more than 1 to 2% of the α compound, and 25 g. of Ba acetate are dissolved in 150 ml. of water, and 15 g. of Br_2 is added. To neutralize the HBr formed, 3 N NaOH is added dropwise, with stirring sufficient to keep the pH just acid to litmus. After 20 minutes almost all the Br_2 is used up and amorphous Ba 2-PGA precipitates. The precipitate is allowed to settle for 1 hour on ice, then is centrifuged off and washed two to three times with water. It is dissolved in dilute HCl, Ba^{++} precipitated with H_2SO_4, and inorganic P precipitated with Mg mixture in ammoniacal solution (see Vol. III [114]). To the remaining solution NaOH and alcohol are added to obtain three fractions as follows: (1) At a pH just acid to congo with 1 vol. of alcohol, 1 to 2 g. of 65 to 75% pure salt is separated. (2) At more alkaline pH but still weakly acid to litmus, 1½ vol. of alcohol yields the main fraction of 4 to 6 g. of 100% pure compound. (3) With more alcohol at neutral pH a larger amount of less pure material is obtained. (The purity can be determined by observing the carboxylatic split with yeast maceration juice.) For analysis the amorphous Ba salt is converted to a crystalline Ag salt by dissolving it in dilute HNO_3, precipitating Ba^{++} with H_2SO_4, and adding excess 25% $AgNO_3$ while neutralizing with NH_4OH. The Ag salt which first separates amorphous is redissolved in hot dilute HNO_3, mixed with 0.33 vol. of hot alcohol, and neutralized by dropwise addition of 10% NH_4OH until turbidity persists. After being left overnight at 0°, crystals separate.

Formation of D(−)-*3-PGA from* DL-*2-PGA.*[6] Forty-one milliliters of racemic 2-PGA prepared from 1.5 g. of the Ba salt is added to 100 ml. of muscle extract dialyzed for 4 hours as well as 2.8 ml. of 4.2% NaF and 14.5 ml. of 2.6% $NaHCO_3$ saturated with 5% CO_2 in N_2. The mixture is left at 28° for 2 hours, then deproteinized with 30 ml. of 40% TCA. Two grams of Ba acetate is added to the clear filtrate which is then almost neutralized with NaOH and precipitated with 2 parts of alcohol. The precipitate is dissolved in dilute HCl, the insoluble residue filtered off, Ba^{++} precipitated with H_2SO_4, and inorganic P removed with Mg in ammoniacal solution. Excess 25% Ba acetate is then added to the supernatant which is acidified to congo. The $BaSO_4$ is centrifuged off, and the weakly acid solution is reprecipitated with alcohol. The precipitate is redissolved in HCl, and alcohol is added dropwise until a crystalline precipitate begins to separate. After settling on ice, the product is separated and recrystallized twice in the same way. Large plates of the crystalline salt are pure D(−)-3-PGA. The yield is 170 mg., fermentable to 95%. Unfortunately the method is cumbersome and gives poor yields.

Preparation by Total Synthesis According to Neuberg[7]

D-Glyceric acid (26.5 g., 0.25 mole) is mixed with 41 g. (0.375 mole) of "metaphosphoric acid ethyl ester" (a mixture of 90% iso- and 10% tetraethyltetrametaphosphoric acid ester).[11] As the reaction proceeds, the clear mixture becomes cloudy and heat develops. The reaction is maintained at room temperature by occasionally dipping the flask into cold water for about 1 hour, then terminated by heating for 2 hours on a water bath, excluding moisture. The inhomogeneous mixture is then shaken several times with glass beads and chloroform to extract unused "metaphosphoric acid ester." The residue is dissolved in 300 ml. of water, and NH_4OH is added until the mixture is just acid to litmus. After standing another hour at room temperature, it is boiled for 45 minutes to saponify the ethyl ester present. More NH_4OH is added to definite alkalinity, and boiling is resumed for 15 minutes. After cooling, inorganic P is precipitated with 20% Mg acetate and filtered off after being kept in the refrigerator overnight. Since the presence of NH_4 salts prevents crystallization of the acid Ba 3-PGA, NH_3 is expelled by nascent $BaCO_3$; $Ba(OH)_2$ is added, and a slow stream of CO_2 is bubbled through the solution warmed *in vacuo* to 35°. The volume is kept constant by replacing the water evaporated off, and the process is continued as long as NH_3 escapes from the mixture. The amount of $Ba(OH)_2$ and the time required to free all NH_3 is best determined on a small aliquot. The amount of $Ba(OH_2)$ needed is about 110% of that calculated from the alkaline equivalent of the Mg acetate, glyceric acid, and ethyl metaphosphate employed. When all NH_3 has been removed, 4 N HBr is added dropwise until a clear solution is obtained. This is warmed and precipitated with hot saturated $Ba(OH)_2$ until the pH is alkaline to phenolphthalein, followed by 50 ml. of CH_3OH. The mixture is brought to a boil, filtered by suction while still hot, and the precipitate washed with hot 10% methanol. The Ba salts are stirred with a little water and just enough 2 N HBr to bring them into solution. The resulting pH should be just acid to congo and may be reduced by adding Ba acetate if necessary. Acid Ba D(−)-3-PGA is then reprecipitated with 95% ethanol added until a slight cloudiness appears. The mixture is stirred, the walls of the container scratched, and the container left in the refrigerator for 3 days. The characteristic crystals which form are filtered off by suction, washed first with ice water, and then with alcohol. The yield is 18.9 g. of almost pure substance. After two recrystallizations from water-alcohol contain-

[11] R. Rätz and E. Thilo, *Ann.* **572**, 173 (1951).

ing a small amount of HBr, $[\alpha]_D^{18} = -13.27°$. The synthetic compound is identical with the natural D(−)-3-PGA and completely fermentable. Further purification may be effected by precipitation from NH_4 salt solution as described above.[10] The Ba salt can be converted to the free acid with H_2SO_4 in the usual way.

The acid Ba salt dried over $CaCl_2$ at 20 to 22° shows the composition $C_3H_5O_8PBa \cdot 2H_2O$. Products with higher or lower H_2O content are objectionable. Differing optical rotations indicate either impurities or excessive or insufficient drying. A product free from crystal water may be prepared by drying in the Abderhalden drying apparatus in a high vacuum (1 to 2 mm.) at 80° for 48 hours, with benzene as the heating bath.[10]

Analysis. Quantitative polarimetric determinations are based on the fact that, in the presence of molybdate, complexes with high optical rotations form. Meyerhof and Schulz[12] add 1 ml. of 25% NH_4 molybdate to 2 ml. of a neutral solution containing 0.2 to 6 mg. of PGA. The exact amount of D(−)-3-PGA is calculated from the difference in rotation before and after addition of the molybdate solution. The values for the complex are $[\alpha]_D^{20} = -745°$; $[\alpha]_{5461}^{20} = -950°$.

Other substances which show similarly increased rotations in the presence of molybdate interfere with the determination and should be absent. These include malic acid, lactic acid, isocitric acid, carbohydrates, and the corresponding alcohols and acids as well as D(+)-2-PGA.

Rapoport[13] claims that Eegriwe's[14] observation that glyceric acid gives a deep blue color when heated with 2 ml. of 0.1% naphthoresorcinol in concentrated H_2SO_4 is equally applicable to PGA. This allows colorimetric estimations of PGA, provided the PGA is pure. Unfortunately a large number of substances, including F-6-P, FDP, and glycerophosphate, give similar colors and interfere with the procedure.[15]

The paper chromatographic method of Hanes and Isherwood[16] described for FDP (see Vol. III [15, 22]) is also applicable to 3-PGA. The same is true for the modifications of Bandurski and Axelrod,[17] Mortimer[18] and Wade and Morgan[19] (see Vol. III [22]). In addition Aronoff and Vernon[20] separate PGA and related compounds from the reaction mixture

[12] O. Meyerhof and W. Schulz, *Biochem. Z.* **297**, 60 (1938).
[13] S. Rapoport, *Biochem. Z.* **291**, 429 (1937).
[14] E. Eegriwe, *Z. anal. Chem.* **95**, 323 (1933).
[15] C. Neuberg and H. Lustig, *Exptl. Med. and Surg.* **1**, 14 (1943).
[16] C. S. Hanes and F. A. Isherwood, *Nature* **164**, 1107 (1949).
[17] R. S. Bandurski and B. Axelrod, *J. Biol. Chem.* **193**, 405 (1951).
[18] D. C. Mortimer, *Canad. J. Chem.* **30**, 653 (1952).
[19] H. E. Wade and D. M. Morgan, *Biochem. J.* **60**, 264 (1955).
[20] S. Aronoff and L. Vernon, *Arch. Biochem. Biophys.* **28**, 424 (1950); **32**, 237 (1951).

by ion exchange resins, then chromatograph with phenol-water or butyric acid-butanol-water (2:1:1) and measure radioactivity where applicable or spray with a modified molybdate reagent (20 ml. 10% NH$_4$ molybdate added to 3 ml. conc. HCl, followed by 5 ml. NH$_4$Cl).

Radiograms and regular chromatograms are also used by Cohen and Oosterbaan[21] with the solvent systems diisopropyl ether-formic acid (3:2) and tert. or isoamyl alcohol-water (2:1)-p-toluenesulfonic acid (1 g./45 ml.). These authors detect PGA by the radioactive P^{32} or an NH$_4$ molybdate-formic acid-perchloric acid spray, followed by heating to 85° for 7 min. and reduction with H$_2$S to yield blue spots.

Fletcher and Malpress[22] besides substituting acetic for formic acid in the Bandurski and Axelrod[17] solvent system, recommend enzymatic hydrolysis with alkaline phosphomonoesterase (Vol. II [80–82]) to liberate o-phosphate from acid-stable compounds, especially 3-PGA.

[21] J. A. Cohen and R. A. Oosterbaan, *Chem. Weekblad* **49**, 308 (1953).
[22] E. Fletcher and F. H. Malpress, *Nature* **171**, 838 (1953).

[35] Preparation of D(+)-2-Phosphoglyceric Acid

By INES MANDL and CARL NEUBERG

2-Phosphoglyceric acid (2-PGA) may be prepared by two distinct methods; in both of them 2-PGA is isolated from the mother liquor remaining after precipitation of the less-soluble acid Ba salt of 3-PGA. Meyerhof and Kiessling[1] obtained the compound from yeast maceration juice, when sugar fermented in the presence of fluoride. Neuberg[2] accomplished total synthesis from D-glyceric acid and metaphosphoric acid ethyl ester.

Procedure According to Meyerhof and Kiessling[1]

Separation from the Mother Liquor of 3-PGA.[3] To 100 g. of dry yeast or 100 ml. of maceration juice are added 17 ml. of 4.2% NaF, 80 ml. of 2.6% NaHCO$_3$, 5 ml. of 5% Na FDP, 30 ml. of 10% glucose, 40 ml. of 4% acetaldehyde, and 130 ml. of H$_2$O; CO$_2$ is added until saturation. After 4 hours at 28° the solution is deproteinized with 400 ml. of 40% TCA and filtered. Acid Ba 3(−)-PGA is then precipitated with 50 ml. of 25% Ba acetate, after the addition of concentrated NaOH till only just

[1] O. Meyerhof and W. Kiessling, *Biochem. Z.* **276**, 239 (1935).
[2] C. Neuberg, *Arch. Biochem.* **3**, 105 (1943).
[3] O. Meyerhof and W. Kiessling, *Biochem. Z.* **267**, 313 (1933).

acid to litmus and 1 vol. of alcohol. The precipitate can be purified as described in the previous section.

Isolation of Pure D-2-PGA.[1] After 15 g. of the Ba salt has been separated three times, an equal volume of alcohol is added and the new precipitate consisting of 3-PGA as well as inorganic P, D(+)-2-PGA, phosphoric acid, and phosphorus-free impurities is centrifuged off. It is dissolved in dilute HCl until just acid to congo, with excess which might lead to racemization avoided. A further 0.25 vol. of alcohol is then added with stirring, and the crystalline precipitate consisting of a mixture of D(−)-3-PGA and D(+)-2-PGA is separated. This alcohol precipitation is repeated several times until the ratio of the optical rotation of the mixture no longer increases to the fermentation value—i.e., until all 3-PGA has been separated. The final precipitate is then redissolved in the calculated amount of dilute HNO_3 and any phosphopyruvic acid is destroyed by adding a few milliliters of 5% $HgCl_2$ and letting the solution stand for 20 minutes at room temperature. Hg is then precipitated with H_2S, inorganic P with Mg mixture, and the precipitates filtered off. By the addition of Ba acetate and 3 parts of alcohol in weakly acid solution, the D(+)-2-PGA is precipitated. This is redissolved in dilute HCl and reprecipitated with 2 vol. of alcohol. On stirring, the amorphous precipitate slowly becomes crystalline. Reprecipitation yields the optically pure salt; $[\alpha]_D$ (in HCl) $= +24.28°$. The yield is 1 to 1.5 g. from 100 g. of yeast.[3] This separation of 2-D-PGA from the mother liquor is applicable by whatever method 3-D-PGA has been precipitated (see Vol. III [34]).

Procedure According to Neuberg[2]

Reaction of 26.5 g. of glyceric acid and 41 g. of "metaphosphoric acid ethyl ester" as described in Vol. III [34] yields D(−)-3-PGA as the chief product together with a smaller amount of D(+)-2-PGA in a ratio of about 10:1. Details for the separation of 3-PGA as its acid Ba salt are given there. The isolation of 2-PGA from the remaining filtrate is based on the finding[4] that the Sr salt of 3-PGA is much less soluble in hot than in cold water. The entire filtrate is thus freed from Ba^{++} by addition of Na_2SO_4, neutralized with NaOH, concentrated to 35 ml., and mixed with sufficient hot saturated $Sr(OH)_2·8H_2O$ solution to give no further precipitate. The Sr salt formed is centrifuged off, washed first with ice water, then with methanol and dried *in vacuo*. It is then stirred with 40 ml. of H_2O, and 0.5 ml. of M $HClO_4$ is added dropwise until the salt is dissolved. On addition of 2 vol. of alcohol the solution becomes cloudy, and on boil-

[4] C. Neuberg and M. Kobel, *Biochem. Z.* **272**, 462 (1934).

ing a microcrystalline precipitate of acid Sr D(−)-3-PGA separates. This is filtered off, and 15% basic lead acetate is added to the filtrate until it is just alkaline to litmus. Subsequently 30% normal Pb acetate is added until no further precipitate forms. The precipitate is centrifuged, washed three times with 10% alcohol, and ground with H_2O. The aqueous suspension is decomposed with H_2S, and the filtrate is freed from H_2S by introducing CO_2. After the dissolved CO_2 is eliminated in turn, the solution of free PGA is treated with hot saturated $Ba(OH)_2$ at a pH alkaline to phenolphthalein until precipitation is complete. The neutral Ba salt obtained is converted to the acid salt with HBr and then precipitated with alcohol. It is filtered off by suction, washed with dilute alcohol, and dried in a desiccator. The salt dissolves almost completely in a large quantity of boiling H_2O. When the filtered solution is concentrated on a water bath, the first crystals appearing on the edges still include some D(−)-3-PGA salt and are filtered off. The remaining filtrate is further concentrated until a crystal fuzz forms, then allowed to cool slowly. The salt formed in this way consists almost exclusively of the D(+)-2 compound. Some amorphous particles are still present, but three recrystallizations, with the first crystals discarded each time, yield 1.8 g. of the acid Ba salt of D(+)-2-PGA; $[\alpha]_D^{22}$ (in N HCl) $= +23.2°$. It is completely fermentable and identical with the natural product.

Analysis. The methods described for 3-PGA (see Vol. III [34]) are also applicable to 2-PGA. The optical rotation of the molybdate complex is $[\alpha]_D^{20} = -68°$.

[36] Synthesis, Determination, Analysis, and Properties of 1,3-Diphosphoglyceric Acid[1]

By ERWIN NEGELEIN

Synthesis

Principle. Glyceraldehyde 3-phosphate dehydrogenase catalyzes the following reversible reaction:[2]

$$GAP + \text{phosphate} + DPN^+ \rightleftarrows DPGA + DPNH + H^+ \qquad (1)$$

---

[1] For further details consult, E. Negelein and H. Brömel, *Biochem. Z.* **301**, 135 (1939); **303**, 132 (1939).

[2] Abbreviations used: DPGA = 1,3-diphosphoglyceric acid; GAP = glyceraldehyde phosphate; 3-PGA = 3-phosphoglyceric acid.

The reducing enzyme of fermentation (alcohol dehydrogenase) reoxidizes the resultant DPNH to DPN in a reaction which is also reversible:

$$\text{DPNH} + \text{H}^+ + \text{acetaldehyde} \rightleftarrows \text{DPN}^+ + \text{ethanol} \qquad (2)$$

The sum of reactions 1 and 2 is

$$\text{GAP} + \text{phosphate} + \text{acetaldehyde} \rightleftarrows \text{DPGA} + \text{ethanol} \qquad (3)$$

Here DPN acts as a catalyst. Under the conditions stated below, reaction 3 goes almost to completion from left to right. DPGA is isolated as the crystalline, sparingly soluble tetrastrychnine salt.

It is essential that both enzyme preparations be pure. DPGA cannot be prepared with impure protein fractions because, on the one hand, DPGA decomposes as soon as it is formed, and on the other, added GAP may be used up by conversion to dihydroxyacetone phosphate.

Reagents

GAP, 0.070 M solution of DL-GAP synthesized according to Fischer and Baer.[3] Only the naturally occurring D-component of the racemic mixture reacts.

DPN,[4] 0.004 M solution.

Glyceraldehyde 3-phosphate dehydrogenase, 0.75% solution of crystalline enzyme prepared according to Warburg and Christian.[5]

Alcohol dehydrogenase, 0.60% solution of crystalline enzyme, prepared according to Negelein and Wulff.[6]

Phosphate, 0.50 M solution (9.5 moles Na_2HPO_4 : 0.5 mole KH_2PO_4).

Acetaldehyde, 1.0 M solution.

Strychnine chloride, saturated solution (about 0.1 M).

Procedure. GAP solution (12.6 ml.) is neutralized with N NaOH to pH 7.5. To this are added 5.2 ml. of phosphate solution and 3.4 ml. of DPN solution, and the volume is brought to 36 ml. with water. The mixture is maintained at 18° in a water bath, and 0.34 ml. of acetaldehyde solution, 0.60 ml. of glyceraldehyde 3-phosphate dehydrogenase solution, and 0.60 ml. of alcohol dehydrogenase solution are added. These enzyme concentrations suffice to effect a rapid reaction rate. After 2 minutes,

[3] H. O. L. Fischer and E. Baer, *Ber.* **65,** 337 (1932); see also C. E. Ballou and H. O. L. Fisher, *J. Am. Chem. Soc.* **77,** 3329 (1955). See Vol. III [31].

[4] See Volume III [124, 125].

[5] O. Warburg and W. Christian, *Biochem. Z.* **301,** 221 (1939); **303,** 40 (1939). See Vol. I [61].

[6] E. Negelein and H. J. Wulff, *Biochem. Z.* **289,** 436 (1937); **290,** 445 (1937); **293,** 351 (1937). See Vol. I [79].

and again after 4 minutes, a further 0.34 ml. of acetaldehyde solution is added. The pH is 7.6.

The reaction is complete in 25 minutes at 18°. The mixture is then brought to pH 2.10 with about 6 ml. of N H$_2$SO$_4$ and poured *immediately* into 445 ml. (10 vol.) of cold acetone. Most of the inorganic phosphate remains in solution; DPGA is precipitated as the acid salt. The precipitate is centrifuged in the cold, washed once with cold acetone, and dried rapidly in a desiccator under high vacuum. DPGA is unstable in this state and should be further processed as soon as possible. It is taken up in about 28 ml. of cold water. The enzymes which were denatured by the acidification step remain as a residue and are filtered off. The clear solution is neutralized with about 1.3 ml. of N NaOH. One milliliter of this solution contains about 1.24×10^{-5} mole of DPGA. The total DPGA is about 80 to 85 % of the reactive portion of GAP used as starting material.

To 29 ml. of neutral DPGA solution is added 29 ml. of strychnine chloride solution. Crystallization of the strychnine salt begins shortly afterwards. After about 15 hours at 0°, the crystals are suction-filtered in the cold and washed with a little cold water. For recrystallization, the material is dissolved in 25 ml. of cold water with dropwise addition of the minimum necessary amount of N HCl. To the clear solution 0.2 N NaOH is added dropwise until crystallization begins. The solution is allowed to stand for a short time at 0°; then the dropwise addition of 0.2 N NaOH is continued until the solution becomes neutral to litmus paper. After standing for a few hours at 0°, the recrystallized strychnine salt is suction-filtered and washed in the cold, as above.

To remove the strychnine, the recrystallized material is placed with about 10 ml. of cold water into a separatory funnel and extracted several times with cold chloroform, with enough 0.2 N NaOH added each time to keep the aqueous phase weakly alkaline (pH about 7.6). The strychnine-free DPGA solution is placed at 0° in a vacuum desiccator over silica gel in order to remove dissolved chloroform and to concentrate the solution somewhat. This procedure yields 10 ml. of a solution containing 2.52×10^{-4} mole of DPGA, or 57 % of the reactive portion of starting GAP.

Assay Method

Principle. DPGA is determined by the reverse of reaction 1 which proceeds sufficiently far to completion from right to left when solutions free of inorganic orthophosphate are used. In the reaction an amount of DPNH is oxidized which is equivalent to the DPGA, thus decreasing the absorbency in the near ultraviolet. The change in absorbency is measured spectrophotometrically in the customary way.

Reagents

Pyrophosphate, 0.10 M solution pH 7.9 (0.10 mole $Na_4P_2O_7$ + 0.060 mole HCl).

DPNH,[7] 2.2×10^{-3} M.

Glyceraldehyde 3-phosphate dehydrogenase, 0.1% solution of crystalline enzyme.[5]

Procedure. The quartz cuvette for measurement of absorbency has a volume of 3.0 ml. and light path $d = 0.60$ cm. The cuvette contains 1.00 ml. of pyrophosphate, 0.20 ml. of DPNH, 1.57 ml. of water, and 0.20 ml. of DPGA solution to be assayed (not more than 0.40×10^{-6} mole). The total liquid volume is 2.97 ml.

The absorbency ($\ln I_0/I$) is determined at 334 mμ (Hg line). Then 0.03 ml. of glyceraldehyde 3-phosphate dehydrogenase solution is added and the decrease in absorbency ($\Delta \ln I_0/I$) is measured, taking into consideration the dilution (1%) resulting from the addition of 0.03 ml. of enzyme solution. The amount, m, of DPGA added is calculated as follows:

$$m = 3 \times \frac{1}{\Delta\beta} \times \frac{1}{d} \times \Delta \ln \frac{I_0}{I} \text{ [moles]}$$

where $\Delta\beta$ is the difference in specific extinctions between DPNH and DPN at $\lambda = 334$ mμ. $\Delta\beta_{334 m\mu} = 12.6 \times 10^6$ [cm.²/mole].[1] The presence of strychnine does not interfere in the spectrophotometric titration of DPGA, so that the concentration of a solution of the strychnine salt can also be determined by this method.

Analysis

The recrystallized strychnine salt, dried *in vacuo* at 60°, had the following composition:[1]

Calculated for $C_{87}H_{96}N_8P_2O_{18}$ (tetrastrychnine salt):

C, 65.13; H, 6.04; N, 6.99; P, 3.87

Found:

C, 64.85; H, 6.13; N, 7.08; P, 3.89

64.69   6.12   6.99   3.79

In the colorimetric estimation of P according to Briggs as modified by

[7] See Volume III [126, 127].

Martland and Robison,[8] the labile P is split off and measured as in-organic P.

Properties

Stability. DPGA decomposes spontaneously into 3-PGA and inorganic phosphate. This process occurs so readily that no state is known in which the compound will remain unchanged for any length of time. In aqueous solution the course of the decomposition follows the equation $-dc/dt = k \times c$. At 38° and pH 7.2, $k = 0.026$ min.$^{-1}$.[1] Hence 2.6% was hydrolyzed per minute, or 50% in 27 minutes.

The rate of hydrolysis is least between pH 7 and 9; at 0° 6% of DPGA decomposes in 24 hours. The material is more stable when frozen in weakly alkaline solution; the loss in 24 hours is then only 3%. Drying a neutral solution of DPGA in a desiccator caused a 77% loss, whereas the dry preparation obtained by acetone precipitation at pH 2.1 lost only 19% in 24 hours in a desiccator. The crystalline strychnine salt and an amorphous calcium salt were also unstable in the dry state.

Asymmetric Carbon Atom. A neutral solution of DPGA causes very little rotation of polarized light. For PGA resulting from the spontaneous decomposition of DPGA, $[\alpha]_D^{20°} = -675°$ in 8.3% ammonium molybdate solution.[1] Under the same conditions 3-PGA gave $[\alpha]_D^{20°} = -682°$.

Ultraviolet Absorption Spectrum. A neutral aqueous solution of DPGA has a characteristic absorption band at 215 mμ that disappears on decomposition of DPGA. The specific extinction is

$$\beta_{215\,m\mu} = 2.11 \times 10^5 \;[\text{cm.}^2\,\text{mole}^{-1}]$$

Adjacent at 200 mμ is the rising portion of a second band which does not disappear on decomposition of DPGA but is shifted by about 5 mμ to longer wavelengths. Both bands of DPGA lie so close together that no minimum can be seen because of the overlapping of the two bands. A solution of the split products of DPGA and a solution of the equivalent mixture of 3-PGA and phosphate exhibit the same absorption spectrum.

[8] M. Martland and R. Robison, *Biochem. J.* **20**, 847 (1926).

[37] Preparation and Analysis of Barium 2,3-diphosphoglycerates

By Isidor Greenwald

$$Ba_5(C_3H_3O_{10}P_2)_2 \text{ and } Ba_3(C_3H_5O_{10}P_2)_2$$

Preparation

The original procedure[1] has been modified to include only one precipitation with lead acetate. Defibrinated pig blood obtained at an abattoir has been the most convenient source. Horse blood or erythrocytes or *fresh* human erythrocytes may also be used. "Outdated" human blood yields no appreciable quantity.

Steps 1 and 2. Removal of Protein and Precipitation of Inorganic Phosphate, Preformed or Derived from Labile Organic Phosphoric Acids. Dilute the blood with 3 vol. of water. Add 1 vol. of 20% (w/v) trichloroacetic acid solution. Dilute the packed erythrocytes with 4 or 5 vol. of water, and add 2 vol. of trichloroacetic acid. Shake well, and filter on fluted papers. Allow to drain overnight. To each liter of filtrate, add 1 g. of $Mg(NO_3)_2 \cdot 6H_2O$ and then make slightly alkaline to phenolphthalein with concentrated ammonia water. After several hours, add 50 ml. of concentrated ammonia water to each liter. Allow to stand overnight. Filter, but do not wash.

Step 3. Precipitation and Decomposition of Lead Diphosphoglycerate. Neutralize the filtrate with nitric acid, and add a concentrated solution of lead acetate until trial of a portion of the supernatant with *dilute* lead acetate solution produces no precipitate. Avoid a large excess of lead acetate. Allow to stand overnight.

Filter, and wash thoroughly. It is best to remove the precipitate from the filter, suspend it in water, and then filter it again. Repeat this procedure until the magnesium has been removed. (Precipitate the lead in a small quantity of the filtrate with hydrogen sulfide, filter, and test the filtrate with sodium hydroxide.) It may be necessary to add a few drops of lead acetate solution to the later wash liquids. Finally, press out the precipitate between filter papers, or suck it dry on a suction filter. Suspend it in water, and decompose with hydrogen sulfide. Filter out the lead sulfide, and wash with water.

Step 4. Precipitation of the Pentabarium Salt. Aerate the combined filtrate and wash water for 30 minutes to remove the hydrogen sulfide.

[1] I. Greenwald, *J. Biol. Chem.* **63**, 339 (1925).

Add a drop of a 1% solution of phenolphthalein and then a hot concentrated solution of barium hydroxide until a pink color is obtained. Allow to stand for a few hours. Filter out the precipitate, and press it between papers. Dissolve it in a convenient volume of water by adding dilute hydrochloric acid until the mixture is distinctly acid to Congo red paper. Filter if necessary, and reprecipitate with *filtered* barium hydroxide solution. After a few hours filter on suction or on fluted papers and press out between papers. When air-dried, the precipitate has the composition $Ba_5(C_3H_3O_{10}P_2)_2 \cdot 3H_2O$.

Step 5. Preparation of the Tribarium Salt. Suspend the preferably still moist pentabarium salt in water, and dissolve it by adding dilute hydrochloric acid until Congo red paper is only slightly blued. Add 2 vol. of 95% alcohol. After a few hours, filter out the precipitate with suction or on fluted papers, and press it out between papers. Redissolve as before, and reprecipitate with alcohol. After a few hours, filter, wash with alcohol, dry in air and then at 110°. The dried material has the composition $Ba_3(C_3H_5O_{10}P_2)_2$.

If too much hydrochloric acid has been used to dissolve either barium precipitate, more alcohol will be needed to secure complete precipitation and the material will then contain less barium and more phosphorus than is indicated by the formula. However, the composition always is that of a diphosphoglyceric acid in which between two and three hydrogens have been replaced by equivalents of barium. Yield, from 1 to 1.3 g. of tribarium salt, or 125 to 160 mg. of phosphorus per liter of blood.

Analysis

Diphosphoglyceric acid is extremely resistant to hydrolysis. The only certain method is complete oxidation. Any of the usual methods for analyzing for barium and phosphorus may be employed. The following have been used.

Determination of Barium. Dissolve approximately 0.1 g. of either of the barium salts in about 200 ml. of water and 10 ml. 2 M hydrochloric acid. Add 10 ml. of M sulfuric acid, drop by drop. Heat, just below boiling, for several hours. Filter on a Gooch crucible, and wash thoroughly. Dry, ignite, and weigh as $BaSO_4$.

Determination of Phosphorus. To the combined filtrate and washings from the barium sulfate, add 2 or 3 ml. of concentrated sulfuric acid. Evaporate to small volume, and transfer to a 100-ml. Kjehldahl flask. Boil the liquid until it begins to char. Allow to cool, add a drop or two of concentrated nitric acid, and heat again. If necessary, add more nitric acid. The liquid must remain clear and colorless after it has been heated until dense white fumes fill the neck of the flask. Allow to cool. Transfer to a beaker with about 200 ml. of water. Add a solution containing about 0.2 g. of $MgCl_2 \cdot 6H_2O$ or of $Mg(NO_3)_2 \cdot 6H_2O$, a drop of 1% phenolphthalein solution and concentrated ammonia water until slightly pink. After at least 2 hours, add 20 ml. of concentrated ammonia water and allow to stand overnight. Filter on a Gooch crucible, wash with a mixture of 1 vol. of concentrated ammonia water, 3 vol. of alcohol, and 7 vol. of water. Allow to dry in air, and weigh as $MgNH_4PO_4 \cdot 6H_2O$.[2]

[2] W. Jones and M. E. Perkins, *J. Biol. Chem.* **55**, 343 (1923).

[38] Preparation of Phosphopyruvic Acid

By GERHARD SCHMIDT

The crystallized silver barium salt of phosphoenolpyruvic acid ($C_3H_2O_6PBaAg \cdot 2H_2O$), molecular weight 446.35, which was discovered by Lohmann and Meyerhof,[1] is the most satisfactory compound for the isolation of phosphopyruvic acid from biological material as well as from the reaction mixture obtained during its laboratory synthesis, which offers by far the most convenient methods for the preparation of this metabolite. The first synthesis of phosphoenolpyruvate was accomplished by Kiessling,[2,3] and his procedure is the basis of the improved techniques described more recently by Schmidt and Thannhauser[4] and by Ohlmeyer.[5] Either of these two modifications of Kiessling's techniques is suitable for the preparation of the substance, but the original procedure of Kiessling is frequently unsuccessful, owing to the insufficient amount of quinoline recommended in Kiessling's directions.

Another convenient procedure has been developed by Baer and Fischer[6,7] on the basis of the phosphorylation of β-chlorolactate with phosphorus oxychloride in dimethylaniline to phosphoryl-β-chlorolactate and its subsequent transformation to phosphoenolpyruvate by treatment with an alcoholic solution of potassium hydroxide at room temperature. The starting material, β-chlorolactic acid, can easily be prepared[6] by oxidation with nitric acid of glycerol monochlorohydrin or epichlorohydrin, both of which are commercially available and inexpensive.

[1] K. Lohmann and O. Meyerhof, *Biochem. Z.* **273**, 60 (1934).
[2] W. Kiessling, *Ber.* **68**, 597 (1935).
[3] W. Kiessling, *Ber.* **69**, 2331 (1936).
[4] G. Schmidt and S. J. Thannhauser, *J. Biol. Chem.* **149**, 369 (1943).
[5] P. Ohlmeyer, *J. Biol. Chem.* **190**, 21 (1951).
[6] E. Baer and H. O. L. Fischer, *J. Biol. Chem.* **180**, 145 (1949).
[7] E. Baer, *Biochem. Preparations* **2**, 25 (1952).

β-Chlorolactic acid is now listed in the catalogues of several chemical supply firms. No reports concerning the suitability of the commercial substance for the preparation of phosphopyruvic acid are available as yet. Baer and Fischer call attention to the possible contamination of β-chlorolactic acid preparations with oxalic acid and describe a procedure for its removal. The final isolation of silver barium phosphopyruvate in the Baer-Fischer procedure involves steps similar to those of the modified Kiessling synthesis. Since the details of the Baer-Fischer synthesis are easily available,[6,7] the method described here is limited to the modifications of Kiessling's synthesis.

Procedure. One hundred grams of pyruvic acid (Eastman) and 400 g. of quinoline (Eastman) are distilled under reduced pressure on a water pump. It is not necessary to distill these substances over dehydrating agents; it is sufficient to discard the fraction of the pyruvic acid and the first turbid drops of the distillate from quinoline which contain most of the moisture.

The phosphorylation should be carried out as soon as possible after the distillation, preferably not later than on the following day. Forty grams of pyruvic acid is dissolved in 120 ml. of quinoline in a 2-l. beaker. To this mixture, a solution of 40 g. of phosphorus oxychloride in 40 ml. of quinoline is added from a dropping funnel equipped with a soda lime tube, within 20 minutes, and with continuous stirring. The temperature of the mixture rises soon to 70° to 80° during the whole reaction. If it rises higher, the beaker is immersed in a dish of ice water. The acidity is frequently checked with Congo paper, and more quinoline is added if the color turns blue. The reaction mixture darkens very soon. In the later stages its consistency changes into that of a thick paste, owing to the crystallization of quinoline chloride. After the addition of the total amount of phosphorus oxychloride, the stirring of the paste is continued until the temperature falls to about 50° (usually within 20 to 30 minutes).

Approximately 1 l. of a mixture of ice cubes and water is added after completion of the phosphorylation. The reaction product is extracted from the solid cake by slowly adding a 33% solution of sodium hydroxide until the aqueous phase reacts neutral to litmus paper. The appearance of a white, cloudy opalescence caused by the separation of quinoline on addition of sodium hydroxide is a convenient test for judging the progress of the neutralization. At the end of the neutralization, the cake obtained by the phosphorylation is completely dissolved.

The mixture is transferred to a separatory funnel and kept for half an hour. The aqueous layer is separated from the quinoline, brought to pH 9 by the addition of sodium hydroxide, and precipitated with 2 vol. of alcohol. After half an hour the slimy precipitate is filtered off over a thin layer of Hyflo filter aid (Johns-Manville). A 25% aqueous solution of barium acetate is added to the filtrate until the precipitation is complete. The precipitate is centrifuged off, washed once with 66% alcohol, and suspended in 500 ml. of water. The barium salts are decomposed by the addition of 25% sulfuric acid. After the removal of the barium sulfate by centrifugation, the combined supernatant and washings are made alkaline by the addition of ammonia. After the addition of magnesia mixture (55 g. of crystallized magnesium chloride and 105 g. ammonium chloride are dissolved in water and brought to a volume of 1 l.), the solution is kept in the refrigerator for half an hour and filtered over a Büchner funnel. The filtrate is neutralized against litmus with acetic acid, and an excess of a 25% solution of barium acetate is added. The precipitation is completed by adding an equal volume of alcohol. The barium salts are centrifuged off and washed once with 50% alcohol. The precipitate is suspended in water and extracted with a small excess of hydrochloric acid. The barium sulfate is centrifuged off, and the reddish brown supernatant is shaken with a few grams of Norit. *It is essential for the success of the preparation that complete and prompt decoloration be accomplished at this stage.* We found that some samples of Norit are without sufficient effect. In this case, other samples of charcoal should be tried until the solution is *colorless*.

After centrifugation, the supernatant is neutralized with sodium hydroxide until it is weakly acid against litmus paper. The reprecipitation of the barium salts is completed by adding an equal volume of alcohol. The barium salts are centrifuged off and washed once with 50% alcohol. They are dissolved in $N/5$ nitric acid and decomposed by the addition of the necessary amount of sulfuric acid. A *small* excess of sulfuric acid does not interfere with the procedure. After the removal of the barium sulfate by centrifugation, an excess of a concentrated solution of silver nitrate is added to the combined supernatant and washings. The silver chloride is removed by filtration on a Büchner funnel (no filter aid should be used). The clear filtrate is neutralized with ammonia until the color of blue litmus turns only faintly red. The silver phosphopyruvate which precipitates during the neutralization is filtered off on a small Büchner funnel and thoroughly washed with water. (It is advisable to test the completeness of the precipitation by adding silver nitrate to a sample of the filtrate.)

The silver salt is dissolved in a measured volume of $N/5$ nitric acid. For every 40 ml. of nitric acid, 3 ml. of a 25% solution of barium acetate is added. Frequently, a precipitate forms on the addition of the barium acetate. In this case, an additional measured volume of $N/5$ nitric acid is added which dissolves the precipitate. More barium acetate solution is added (3 ml. per 40 ml. of nitric acid), and the alternating addition of nitric acid and barium acetate is continued until the solution remains clear (or becomes only faintly opalescent) after the last addition of barium acetate. An amount of $N/5$ ammonia solution equivalent to the total amount of nitric acid used is heated to 90°, and the solution of phosphopyruvic acid is slowly poured into the hot ammonia solution. The copious precipitate of silver barium phosphopyruvate which forms immediately is partially crystalline and partially amorphous. On standing for a short time, the amorphous part becomes completely crystalline. After cooling at room temperature in a dark place, the suspension is placed in the refrigerator and filtered on a Büchner funnel after 24 hours. The crystals are washed with ice water and consist of practically pure silver barium phosphopyruvate. The substance can be recrystallized by dissolving it in $N/5$ nitric acid and by adding the solution to the equivalent amount of the hot $N/5$ ammonia as described above.

Four grams of the silver barium salt is usually obtained from the phosphorylation of 40 g. of pyruvic acid.

Preparation of Silver Barium Phosphopyruvate According to Ohlmeyer. Ohlmeyer's procedure differs from the method just described mainly in three points: (1) A smaller excess of quinoline is used. (2) The removal of inorganic phosphate by precipitation with magnesia mixture is omitted. (3) The decoloration with charcoal prior to the reprecipitation of the barium salts is omitted.

The phosphorylation is carried out by adding dropwise to an ice-cold solution of 30 g. of freshly distilled pyruvic acid in 90 g. of freshly distilled quinoline an ice-cold mixture of 30 g. of phosphorus oxychloride and 30 g. of freshly distilled quinoline. After completion of the reaction the sticky phosphorylation product is mixed under stirring with 150 ml. of a 25% aqueous solution of sodium hydroxide, which is added dropwise, and with 150 ml. of ice water. After all solid material has been dissolved, the mixture is brought to pH 8 by further additions of a 25% solution of sodium hydroxide. The mixture is centrifuged, and the dark aqueous solution (approximately 320 ml.) is separated from the dark layer of quinoline.

The solution of the phosphate compounds is mixed with 600 ml. of methanol, centrifuged, and the supernatant solution precipitated by the addition of 40 ml. of a saturated aqueous solution of barium acetate. The centrifuged precipitate is washed twice with 600 ml. of a 2:1 mixture of methanol and water and dissolved in the minimal necessary amount of 0.3 N nitric acid at a pH between 2.8 and 3.2. Twenty-five milliliters of a 25% solution of silver nitrate is added. After removal of the precipitate by centrifugation, the pH of the supernatant liquid is adjusted to a value between 5.0 and 5.2 by the addition of about 450 ml. of 0.2 N ammonia. The suspension is kept overnight in the refrigerator. The precipitate, which may already contain crystals of the silver barium phosphopyruvate, is separated by centrifugation, and the supernatant is kept in the refrigerator for further crystallization. The precipitate is dissolved in an amount of 0.3 N nitric acid sufficient to bring the pH to 3 (approximately 70 ml. of 0.3 N nitric acid); any insoluble material is removed by centrifugation. The solution is brought to pH 5 with ammonia and the precipitate is centrifuged off, and the supernatant solution is placed in the refrigerator. This reprecipitation is repeated three times. Finally, the crystals from all supernatant solutions are collected and recrystallized in the following way. The precipitate is dissolved in 0.3 N nitric acid, and the solution is diluted with an equal volume of water. After addition of 0.3 ml. of saturated barium acetate solution, the mixture is filtered and the precipitate is washed with water. Ammonia is added to the filtrate until a turbidity appears. When the suspension is shaken, crystals appear. The mixture is brought to pH 6 with ammonia, placed in the refrigerator for an hour, and filtered. The crystals are separated from the supernatant solution, washed once with a 67% aqueous methanol solution, twice with methanol, and twice with ether.

The supernatant solution is mixed with 5 ml. of a saturated solution of barium acetate and placed in the refrigerator for several days. An additional crop of approximately 1 g. of crystallized silver barium phosphopyruvate can be recovered. Total yield: 4.9 g. of silver barium salt from 30 g. of redistilled pyruvic acid.

Some Properties of Phosphopyruvic Acid Which Are Useful for Its Analytical Determination. The amounts of phosphopyruvate at various stages of the isolation procedures can be easily assayed on the basis of the rapid conversion of the phosphoryl group to inorganic phosphate by hypoiodites. According to Lohmann and Meyerhof,[1] approximately 3 ml. of a suitably diluted aqueous solution of phosphopyruvate is mixed with 1 ml. of 0.1 N iodine solution and 1 ml. of 0.1 N sodium hydroxide. After standing for 10 to 20 minutes, the mixture is acidified with 1 ml. of N hydrochloric acid, and the inorganic phosphate is determined according to the method of Fiske and Subbarow.

The instantaneous and quantitative liberation of inorganic phosphate from phosphopyruvate in the presence of mercuric salts is likewise suitable for the analytical control of the isolation procedure.[1]

The unimolecular reaction constant during the hydrolysis of phosphopyruvic acid in N hydrochloric acid at 100° is 3.5×10^{-3}. The hydrolysis is complete within 60 minutes under these conditions (degrees of hydrolysis: 93% after 30 minutes, 56% after 10 minutes).

Bandurski and Axelrod[8] described conditions for the identification of phosphopyruvic acid by two-dimensional paper chromatography.

Utter and Kurahashi[9] found that phosphopyruvic acid may be purified efficiently by a passage through a column of Dowex 1 resin. They obtained by this step the acid prepared from the first barium precipitate of the Baer-Fischer synthesis[6] in 50% yield and in a state of 90% purity.

Stability. Silver barium phosphopyruvate is stable for years at room temperature in brown glass containers. A slight discoloration after long storage may be disregarded if the salt is being used for the preparation of phosphopyruvic acid.

Drying Conditions.[1] At 78°, silver barium phosphopyruvate loses, per mole, between 0.9 and 1 mole of its 2 moles of crystal water within 1 hour in high vacuum over P_2O_5. Drying at 100° in a high vacuum results in discoloration and larger loss of weight.

[8] R. S. Bandurski and B. Axelrod, *J. Biol. Chem.* **193**, 405 (1951).
[9] M. F. Utter and K. Kurahashi, *J. Biol Chem.* **207**, 821 (1954).

[39] Preparation and Assay of Acetyl Phosphate

By E. R. STADTMAN

Preparation of Acetyl Phosphate (Acetyl∼P) by Reaction of Orthophosphate with Isopropenyl Acetate (Procedure A)

Principle. This method, which has been previously described,[1] is based on the fact that orthophosphoric acid is readily acetylated by isopropenyl acetate to give monoacetyl dihydrogen phosphate. The product is isolated as the dilithium salt by fractional crystallization from dilute alcoholic solution.

Reagents

85% sirupy phosphoric acid.
Isopropenyl acetate (freshly distilled).
Concentrated sulfuric acid.
Lithium hydroxide, 4 N.

Procedure. Two hundred milliliters of isopropenyl acetate in a 1-l. flask is cooled in an ice bath. Then, with stirring, 25 ml. of 85% sirupy phosphoric acid is added dropwise, and finally 1.0 ml. of concentrated sulfuric acid is added as catalyst. The reaction mixture is placed in a water bath at 30° and is stirred constantly to prevent localizing heating.

[1] E. R. Stadtman and F. Lipmann, *J. Biol. Chem.* **185**, 549 (1950).

After 30 minutes, or when the reaction mixture becomes pale yellow in color, it is cooled in an ice bath and 100 g. of ice is added. The mixture is neutralized to pH 5.0 with cold 4 N lithium hydroxide (about 300 ml.). During neutralization, the mixture may separate into two phases, since the isopropenyl acetate which is present in excess is relatively insoluble in water. It is therefore necessary to shake the mixture vigorously in an ice bath during the addition of lithium hydroxide so as to extract the acetyl phosphoric acid from the organic phase. The neutralization should be carried out as rapidly as possible without, however, allowing the temperature to rise unduly. Once the pH is adjusted to 5.0, the acetyl phosphate is relatively stable and subsequent operations may be carried out more leisurely.

Finally, cold water is added to bring the aqueous layer to 500 ml. in volume and the mixture is extracted three times with 500-ml. amounts of ethyl ether to remove the excess isopropenyl acetate. The aqueous portion is carefully adjusted to pH 8.0 with 4.0 N lithium hydroxide, and the turbid solution is centrifuged to remove the insoluble lithium phosphate. The clear supernatant solution is placed in an ice bath, and, with stirring, 5 vol. of cold ethyl alcohol is added slowly. The dilithium monoacetyl-phosphate which crystallizes out is filtered on a Büchner funnel, washed with absolute ethanol and finally with ether, and then dried in a vacuum desiccator over P_2O_5 and paraffin.

By this procedure 20 to 45 g. of dilithium acetyl~P is obtained. Sometimes the product first obtained is only 80 to 90% pure. To obtain an analytically pure product the crude material is further purified by fractional precipitation from a 10% solution by addition of ethanol. The precipitate formed with 0.5 vol. of ethanol contains most of the impurities and is discarded. The pure compound is then precipitated by the slow further addition of 5 vol. of cold ethanol and is collected and dried as described above.

Properties. When kept dry and cold (0° to −10°) the lithium acetyl~P may be stored for years without appreciable decomposition. Neutral solutions of acetyl~P may be stored for months at −10° without appreciable decomposition. Solutions of the compound are very unstable at low and high pH and at elevated temperatures. Complete hydrolysis occurs in 5 minutes at 100°.

Preparation of Acetyl~P by Reaction of K_2HPO_4 with Acetic Anhydride (Procedure B)

This method was previously described by Avison[2] and is based on the fact that acetic anhydride will readily acetylate orthophosphate at neutral to slightly alkaline pH in aqueous pyridine solution. The acetyl

[2] A. W. D. Avison, *J. Chem. Soc.* **1955**, 732.

phosphate is isolated as the dilithium salt by precipitation from dilute alcoholic solution.

Reagents

Pyridine.
Dipotassium hydrogen phosphate, 0.25 M.
Lithium hydroxide, 4 N.
Ethyl alcohol.

Procedure. Ninety-five grams of pyridine (1.2 moles) and 200 ml. of 0.25 M dipotassium hydrogen phosphate solution are stirred together at 0° to give a clear homogeneous solution. Then 9.6 ml. of acetic anhydride (0.1 mole) is added dropwise to the vigorously stirred solution over a 5-minute interval during which time the temperature may rise to 3° to 4°. After 35 minutes 38 ml. of 4 N lithium hydroxide is added cautiously. The final pH (after tenfold dilution with water) should be about 7.6. Cold ethanol (2300 ml.) is then added slowly, and after standing for 2 hours at 0° the precipitate is filtered off, washed with absolute ethanol and then with ether, and finally dried in a vacuum desiccator over P_2O_5 and paraffin. By this procedure 6 to 7 g. of dilithium acetyl$\sim$P of about 70% purity is obtained. A product of higher purity may be obtained by recrystallization of the dilithium salt as described in procedure A above.

Comment. Since the acetylation of orthophosphate by procedure B is essentially quantitative, it has, with slight modification, been used for the synthesis of P^{32}-labeled acetyl$\sim$P.[3]

Assay Method

Principle. This method was previously described by Lipmann and Tuttle[4] and is based on the fact that acyl phosphates react rapidly with hydroxylamine at pH 6.5 to 7.0 to form hydroxamic acids, which in the

---

[3] Details of the procedure as modified by Kornberg for the synthesis of P^{32}-labeled acetyl$\sim$P are as follows: pyridine (0.95 ml.), K_2HPO_4 (1 M, 0.5 ml., containing P^{32}-labeled phosphate as desired), and water (1.5 ml.) are mixed in a 30-ml. Erlenmeyer flask. The mixture is called to 0° or less, then kept in ice. Acetic anhydride in slight excess (0.11 ml.) is added over a 3-minute period to the mixture which is constantly shaken. LiOH (4 N) is added 2 minutes later to adjust the pH to about 7.5 (about 4.5 ml. is required). Ethanol (23 ml. at −15°) is added slowly with agitation of the reaction mixture. After 1 hour at 0°, the precipitate is collected by centrifugation, washed twice with cold ethanol, and dried *in vacuo* over $CaCl_2$ and KOH. About 400 micromoles of acetyl$\sim$P is recovered, a yield of about 80% based on phosphate. The purity is about 60%. No inorganic orthophosphate is normally detected in the crude product. See A. Kornberg, S. R. Kornberg, and E. S. Simms, *Biochim. et Biophys. Acta* **20**, 215 (1956).

[4] F. Lipmann and L. C. Tuttle, *J. Biol. Chem.* **159**, 21 (1945).

presence of ferric salts produce red to violet complexes that may be quantitated by colorimetric analysis.

Reagents

Neutralized, hydroxylamine solution. Prepare just before use by mixing equal volumes of 28% solution of $NH_2OH \cdot HCl$ (4.0 M) and 14% NaOH (3.5 M).

Ferric chloride solution. Prepare by mixing equal volumes of 5% $FeCl_3$ (in $N/10$ HCl), 12% TCA, and 3 N HCl.

Hydroxamic acid reference standards: (1) Dissolve 30 mg. of acetohydroxamic acid in 100 ml. of H_2O (2.0 ml. of this solution = 4.0 micromoles). (2) Alternatively a hydroxamic acid standard solution may be prepared from succinic anhydride as follows: Dissolve 1.0 g. of succinic anhydride in 40 ml. of the neutralized hydroxylamine solution, and dilute to 100 ml. When 1 ml. of this stock solution is diluted to 40 ml. with water, 2.0 ml. of the resulting solution gives a color equivalent to 4.0 micromoles of acetohydroxamic acid.

Procedure. To 2.0 ml. of test solution containing 0.5 to 5 micromoles of acetyl~P, add 1.0 ml. of the neutralized hydroxylamine reagent. After 10 minutes at room temperature, add 3.0 ml. of the ferric chloride reagent. If protein is present, remove it by centrifugation, and read the colored supernatant solution immediately in a colorimeter at 540 mμ. The optical density is proportional to the concentration of acetohydroxamic acid over the range of 0 to 6 micromole per 6.0 ml. of final solution.

Specificity. Under the mild conditions described the method will measure acid anhydrides, acid chlorides, thiol esters,[5] acyl phosphates and N-acyl imidazoles. All these compounds give essentially equivalent color values. Ordinary esters, aldehydes, amides, and ketones do not interfere even at relatively high concentrations.

Comments. High concentrations of various anions depress color development. Thus, fluoride in concentrations of 100, 50, and 25 micromoles per 6 ml. of final reaction mixture depresses the color development 33, 7, and 0%, respectively. With phosphate, 100 and 50 micromoles depress the color 13 and 4%, respectively, and depression with 200 micromoles of sulfate ion is 8%. Citrate in comparable concentrations is without effect. Other anions have not been tested.

[5] Thio esters of β-keto acids do not react with hydroxylamine to give hydroxamic acids. Instead, pyrazolones are formed which give much less intense colors with the iron reagent.

[40] Preparation and Determination of Gluconic, 2-Ketogluconic, and 5-Ketogluconic Acids

By R. D. DeMoss

I. Gluconic Acid

Preparation

Reagents

0.1 M Na_2CO_3 (106 mg. per 10 ml.).
0.1 M I_2 (254 mg. per 10 ml.) in 0.1 M KI (166 mg. per 10 ml.).

Procedure. To 100 mg. of D-glucose dissolved in 10 ml. of 0.1 M Na_2CO_3, add 10 ml. of a solution of 0.1 M in 0.1 M KI. Mix, and incubate at 30° for 30 minutes. Bubble CO_2-free nitrogen gas through the mixture for 1 hour at 30° to remove free I_2. If the experimental conditions are not sensitive to the presence of iodide and iodate ions, as is often the case, the solution of gluconic acid may be used without further treatment, after adjustment to the desired pH. Iodide and iodate ions may be removed by passing the solution through an anion exchange resin such as Dowex 50-H+. Gluconic acid is eluted from the column with 0.05 N H_2SO_4.

Determination

D-Gluconic acid may be determined by measuring the change in rotation due to formation of the molybdate complex,[1] or enzymatically by measuring the extent of DPN reduction coupled with the conversion of the gluconic acid to 6-phosphogluconic acid. Although both methods are described here, the enzymatic method is recommended. In addition, manometric methods have also been described.

Optical Rotation Method

Reagents

Saturated ammonium molybdate solution.
Glacial acetic acid.

Procedure. Measure the optical rotation of a solution containing 5 to 20 mg. of D-gluconic acid per milliliter. Add 0.5 vol. of saturated ammonium molybdate solution and 0.2 vol. of glacial acetic acid. Mix the components, and allow to stand in the dark at room temperature for 3 hours. Measure the optical rotation of the supernatant solution. The quantity

[1] T. A. Bennet-Clark, *Biochem. J.* **28**, 45 (1934).

of D-gluconic acid present in the original solution is calculated from a standard curve previously prepared with known amounts of gluconic acid. If a salt of gluconic acid is used for determination, the mixture should be acidified with a slight excess of HCl before the polarimetric measurements are made.

Enzymatic Method

In the presence of the appropriate enzymes and cofactors, D-gluconate may be converted to 6-phosphogluconate (6-PG) and subsequently oxidized with concomitant reduction of DPN or TPN according to equations 1 and 2.

$$\text{D-Gluconate} + \text{ATP} \xrightarrow{\text{Mg}^{++}} \text{6-PG} + \text{ADP} \qquad (1)$$

$$\text{6-PG} + \begin{bmatrix} \text{DPN} \\ \text{or} \\ \text{TPN} \end{bmatrix} \rightarrow \begin{bmatrix} \text{DPNH} \\ \text{or} \\ \text{TPNH} \end{bmatrix} + \text{H}^+ + \text{CO}_2 + \text{ribulose-5-P} \qquad (2)$$

Gluconokinase catalyzes reaction 1 and may be prepared from yeast.[2] Reaction 2 is catalyzed by 6-PG dehydrogenase which may be prepared from yeast (TPN-linked)[3] or *Leuconostoc mesenteroides* (DPN-linked).[4]

Reagents

Gluconokinase.
6-PG dehydrogenase.
ATP, 0.1 M, pH 7.4.
MgCl$_2$, 0.1 M.
DPN or TPN, 0.0027 M (molecular weight 664 or 744).
Glycylglycine, 0.04 M, pH 7.4.

Procedure. To a cuvette with a 1-cm. light path, containing 5 units of gluconokinase (1 unit of yeast gluconokinase is defined here as the amount of enzyme which will catalyze the formation of 1 micromole of CO$_2$ per 10 minutes in the standard manometric assay conditions described by Cohen[5]), 5 units of 6-PG dehydrogenase, 50 μM. of MgCl$_2$, 5μM. of ATP, 1.35 μM. of DPN or TPN (according to the source of 6-PG dehydrogenase), and 60 μM. of glycylglycine in a total volume of 2.8 ml., add 0.2 ml. of a sample containing 20 to 100 γ of D-gluconate. To a second cuvette containing the same components, add 0.2 ml. of distilled water in place of the gluconate sample. Allow the cuvettes to stand at room temperature until the reaction is complete (no further increase in absorp-

[2] H. Z. Sable and A. J. Guarino, *J. Biol. Chem.* **196,** 395 (1952).
[3] B. L. Horecker and P. Z. Smyrniotis, Vol. I [42].
[4] R. D. DeMoss, Vol. I [43].
[5] S. S. Cohen and R. Raff, *J. Biol. Chem.* **188,** 501 (1951).

tion at 340 mμ). The difference in absorption at 340 mμ is proportional to the quantity of D-gluconate added (in the 0.2-ml. sample) and may be calculated according to equation 3.

$$E_{340} \times 0.478 = \mu M. \text{ of D-gluconate (in the 0.2-ml. sample)} \quad (3)$$

Comments. Since the enzyme preparations used for the determination are only partially purified, the method is subject to error in so far as other substances which are capable of giving a false reaction may be present in the gluconate sample. This possibility may be controlled partially by using both the DPN- and TPN-linked 6-PG dehydrogenases in separate experiments. If the same result is obtained in both cases, the gluconate sample may be presumed to be free of interfering substances.

Physical Constants. Free acid:[6] $[\alpha]_D^{20} = +12$ (in H$_2$O, $c = 2.841$). Ca salt:[7] $[\alpha]_D^{20} = +7.5$ (in H$_2$O, $c = 1$). Brucine salt:[8] melting point 155 to 157 (anhydrous). Phenylhydrazide:[8] melting point 200 to 201. Methyl ester:[9] melting point 174 to 175.

Paper Chromatogram Procedures. SOLVENTS: (1) Ethanol:methanol: H$_2$O:formic acid (15:15:9:2, v/v), R_f 0.76. (2) Ethanol:methanol:H$_2$O (9:9:2, v/v),[10] R_f 0.22. (3) Butanol:pyridine:H$_2$O (6:4:3, v/v)[11] R_f 0.15.

SPRAY REAGENTS: (1) 0.1 N AgNO$_3$ in 5 N NH$_4$OH.[10] (2) 0.05% bromophenol blue.[11]

II. 2-Ketogluconic Acid

Preparation

There does not appear to be a suitable chemical method for the preparation of 2-ketogluconate in good yield. Therefore, a biological method is described which involves the isolation of 2-ketogluconate from a complex medium.[12] Several strains of the *Pseudomonas fluorescens* group of bacteria catalyze the conversion of glucose to 2-ketogluconate.[12]

Reagents

Culture: *Pseudomonas fluorescens.*

Medium contains, per 100 ml., glucose, 10.0 g.; KH$_2$PO$_4$, 0.06 g.; MgSO$_4$·7H$_2$O, 0.025 g.; yeast extract (Difco), 0.5 g. Adjust to pH 7.2, dispense 100-ml. quantities into 500-ml. Erlenmeyer flasks, and sterilize at 120°.

20% urea solution, sterilize at 120°.

[6] K. Rehorst, *Ber.* **61**, 163 (1928).
[7] A. Dyfverman, B. Lindberg, and D. Wood, *Acta Chem. Scand.* **5**, 253 (1951).
[8] F. W. Jensen and F. W. Upson, *J. Am. Chem. Soc.* **47**, 3019 (1925).
[9] H. Ohle, *Ber.* **70B**, 2153 (1937).
[10] F. N. Stokes and J. J. R. Campbell, *Arch. Biochem.* **30**, 121 (1951).
[11] H. J. Koepsell, F. H. Stodola, and E. S. Sharpe, *J. Am. Chem. Soc.* **74**, 5142 (1952).
[12] L. B. Lockwood, B. Tabenkin, and G. E. Ward, *J. Bacteriol.* **42**, 51 (1941).

Dry CaCO₃. Dispense 2.5 g. per tube, and sterilize in a dry oven at 165° for 1 hour.

Procedure. Prior to inoculation, to each 100 ml. of medium add 1.0 ml. of 20% urea solution and 2.5 g. of dry CaCO₃. Inoculate each 100 ml. of fresh medium with 1.0 ml. of a 24-hour culture which has been grown in the same medium. Incubate for 8 days on a shaking apparatus at 30°. The course of the biological oxidation may be followed quantitatively by measuring the optical rotation of aliquots of the culture supernatant. Maximum 2-ketogluconate concentration is attained when the maximum negative rotation is observed.

Centrifuge the culture, and evaporate the supernatant *in vacuo* at 25 to 30° to approximately one-fourth the original volume. Let stand overnight at 0°, filter off the crystalline material, and recrystallize from the minimal amount of water to obtain the pure calcium salt of 2-keto-gluconate. The yield should fall within the range of 70 to 90%, depending on the amount of glucose consumed.

Suitable conditions for preparing small quantities of 2-ketogluconate have not been established, although it is highly probable that a combination of resting cell suspension and ion exchange column procedures would achieve this end.

Determination

A nonspecific chemical method for the determination of 2-keto-gluconate, utilizing the Shaffer-Hartman copper reagent, has been employed by Stubbs *et al.*[13] The somewhat more specific method of Lanning and Cohen[14] is described below.

Reagent

o-Phenylenediamine dihydrochloride, 2.5 g. per 100 ml. of distilled water (freshly prepared).

Procedure. To a 2.0-ml. neutral aqueous sample containing 10 to 100 γ of 2-ketogluconate, add 1.0 ml. of the *o*-phenylenediamine reagent. Mix, heat in a boiling water bath for 30 minutes, and cool to room temperature. Measure the optical density of the reaction mixture at 360 and 330 mμ. For 2-ketohexonic acids, the ratio of optical density at 330 mμ to that at 360 mμ is approximately 1.51. The optical density of the reaction mixture at 330 mμ is proportional of the 2-ketogluconate concentration and may be estimated from a standard curve prepared with authentic 2-ketogluconate.

[13] J. J. Stubbs, L. B. Lockwood, E. T. Roe, B. Tabenkin, and G. E. Ward, *Ind. Eng. Chem.* **32**, 1626 (1940).
[14] M. C. Lanning and S. S. Cohen, *J. Biol. Chem.* **189**, 109 (1951).

Specificity of Methods. The data of the table are taken from Lanning and Cohen[14] and show the relative reactivities of several related compounds under the standard assay conditions.

RELATIVE REACTIVITIES OF SEVERAL RELATED COMPOUNDS UNDER STANDARD ASSAY CONDITIONS[a]

	mM. $\times$ 10⁴/OD (330 mμ)
2-Keto-D-gluconate	4.02
2-Keto-D-galactonate	4.36
2-Keto-D-gulonate	3.32
2-Keto-D-glucoheptonate	8.63
2-Keto-D-galactoheptonate	3.96
Pyruvate	17.0
5-Keto-D-gluconate	67.8
Dehydroisoascorbate	2.77
Dihydroxyacetone	21.9
D-Glucose	66.0
D-Ribose	51.3
D-Gluconate	2000

[a] Data from Lanning and Cohen.[14] The values given represent the calculated amounts of the respective compounds required to produce one unit of optical density under the assay conditions described.

Physical Constants. Free acid:[15] $[\alpha]_D^0 = -99.62$ (in dilute HCl). Brucine salt:[15] melting point 166 (decomp.). Methyl ester:[16] melting point (decomp.).

Paper Chromatogram Procedures. SOLVENTS: (1) Ethanol:methanol:water (9:9:2, v/v),[10] R_f 0.39. (2) Ethanol:water (9:1, v/v),[17] R_f 0.33. (3) Butanol:pyridine:water (6:4:3, v/v),[11] R_f 0.15.

SPRAY REAGENTS: (1) α-Napthylamine; 2%, in *n*-butanol:methanol (1:1, v/v) containing 5% trichloroacetic acid.[18] (2) *o*-Phenylenediamine; 2% of the dihydrochloride in 80% ethanol.[14] (3) Aniline hydrogen phthalate; 0.93 g. of aniline and 1.66 g. of phthalic acid anhydride in 100 ml. of water–saturated *n*-butanol. (4) 3,5-Dinitrosalicylic acid; 0.5% in N NaOH.[11]

III. 5-Ketogluconic Acid

This acid is a product of glucose oxidation by several members of the genus *Acetobacter*. Stubbs *et al.*[13] have studied the conditions for 5-ketogluconate production from glucose by *Acetobacter suboxydans* on a commercial scale. The conditions established by these authors have been

[15] H. Ohle and G. Berend, *Ber.* **60**, 1159 (1927).
[16] H. Ohle and R. Wolter, *Ber.* **63**, 843 (1930).
[17] F. C. Norris and J. J. R. Campbell, *Can. J. Research* **C27**, 253 (1949).
[18] L. Hough, J. K. N. Jones, and W. H. Wadman, *J. Chem. Soc.* **1950**, 1702.

adapted for the method described below. A chemical procedure involving the oxidation of glucose to 5-ketogluconate by hypobromous acid has been published,[19] but is not recommended because of the low yield and the high degree of gluconic acid impurity (see also Barch[20]).

Preparation

Reagents

Culture: *Acetobacter suboxydans.*

Medium contains, per 100 ml., glucose, 10.0 g.; yeast extract (Difco), 0.5 g. Dispense in 100-ml. quantities in 500-ml. Erlenmeyer flasks and sterilize at 120°.

Dry $CaCO_3$. Dispense in 2.7-g. quantities in test tubes and sterilize at 165° for 1 hour.

Procedure. Prior to inoculation add 2.7 g. of sterile dry $CaCO_3$ to each 100 ml. of medium. Inoculate each 100 ml. of fresh medium with 5 ml. of a 24-hour culture grown in the same medium. Incubate the cultures at 25° for approximately 33 hours on a shaking device. The amount of $CaCO_3$ used represents slightly less than the theoretical amount required to combine with the organic acid formed. When all the glucose has been consumed, the $CaCO_3$ dissolves, owing to the formation of soluble calcium gluconate (the intermediate product). Subsequently, 5-ketogluconate is produced, and crystals of the calcium salt begin to appear in the medium.

To recover the calcium salt of 5-ketogluconate, filter the culture through a medium-porosity sintered-glass filter. Wash the residue with small amounts of cold water, ethanol, and ether, and dry at 100°. The salt may be recrystallized from water.

Determination

Stubbs *et al.*[13] applied the Shaffer-Hartmann copper reagent to the determination of 5-ketogluconate. Perlman[21] has described relatively specific conditions using the nonspecific reducing method of Nelson.[22] The latter procedure is described here.

Reagent

Arsenomolybdate. Dissolve 25 g. of ammonium molybdate in 450 ml. of distilled water, add 21 ml. of concentrated H_2SO_4 (specific gravity 1.84), and mix. Add 3 g. of $Na_2HAsO_4 \cdot 7H_2O$ dissolved in 25 ml. of distilled water, mix, and incubate at 37°

[19] E. W. Cook and R. T. Major, *J. Am. Chem. Soc.* **57**, 773 (1935).
[20] W. E. Barch, *J. Am. Chem. Soc.* **55**, 3653 (1933).
[21] D. Perlman, *J. Biol. Chem.* **215**, 353 (1955).
[22] N. Nelson, *J. Biol. Chem.* **153**, 375 (1944).

for 24 to 48 hours. Dilute 1 part of solution with 2 parts of distilled water, and store in a glass-stoppered brown bottle.

Procedure. To a 1.0-ml. sample containing 5 to 80 γ of 5-ketogluconate, add 5 ml. of diluted arsenomolybdate reagent. Mix, and incubate at 50° for 2 hours. Measure the optical density of the mixture at 660 mμ. The quantity of 5-ketogluconate is estimated from a standard curve prepared with authentic 5-ketogluconate.

Specificity. The following compounds do not interfere under the standard assay conditions at levels of up to 800 γ per sample:[21] glucose, mannose, sorbose, galactose, maltose, gluconic acid, glucuronic acid, and galacturonic acid. The reducing power of fructose and sucrose is 8% and 3%, respectively, of that of 5-ketogluconate under the standard assay condition.

5-Ketogluconate also gives colored products on reaction with naphthoresorcinol or resorcinol.

Physical Constants. Free acid:[23] $[\alpha]_D^{20} = -14.5$ (in H_2O, $c = 2$). Ca salt solubility:[13] 0.2 g. per 100 ml. of water. Drying at 85° to 90° results in no loss in weight[19] (distinguishes from Ca saccharate). Brucine salt:[7] melting point 172 to 173.

Paper Chromatogram Procedures. SOLVENTS: (1) Ethanol:methanol: water (9:9:2, v/v),[10] R_f 0.24.

SPRAYS: (1) *p*-Anisidine·HCl; 3% in *n*-butanol.[18] (2) Anthraquinone; 3% in alcoholic 5% HCl.[18] (3) Urea; 3% in alcoholic 5% HCl.[18] (4) Diphenylamine; 2% in *n*-butanol:methanol (1:1, v/v) containing 5% trichloroacetic acid.[18] (5) α-Naphthylamine; 2% in No. 4 solvent.[18] (6) Aniline hydrogen phthalate; 0.93 g. of aniline and 1.66 g. of phthalic acid anhydride in water-saturated *n*-butanol. (7) Resorcinol, 3% in alcoholic HCl (20:1, v/v).[7] (8) Ammoniacal AgNO₃.[7] (9) *o*-Phenylenediamine; 2% of the dihydrochloride in 80% ethanol.

[23] M. L. Boutroux, *Ann. chim. phys.* **21**, 565 (1890).

[41] Preparation and Analysis of Kojic Acid

By RONALD BENTLEY

Preparation

Principle. Kojic acid (5-hydroxy-2-hydroxymethyl-1,4-pyrone) has been prepared synthetically[1,2] but is most readily obtained by the action

[1] K. Maurer, *Ber.* **63B**, 25 (1930).

[2] M. Stacey and L. M. Turton, *J. Chem. Soc.* **1946**, 661.

of fungi, particularly of the genus *Aspergillus*, on carbohydrates. It is often available commercially.

Mold Cultures. Kojic acid formation is most abundant with some strains of the *Aspergillus flavus-oryzae* group. Suitable strains of *Aspergillus flavus* maintained in the culture collection of the Fermentation Division, Northern Regional Research Laboratory (U. S. Department of Agriculture) are the following: NRRL 484,[3] NRRL 625,[4] and NRRL 2405 (the latter is a strain used in a recent study of kojic acid biosynthesis;[5] it was originally referred to as *A. flavus-oryzae* and was not previously maintained in the NRRL collection). Subcultures of the molds are kept on Czapek-Dox agar slants by standard methods (for full details, the "Manual of the Aspergilli" should be consulted[6]).

Culture Medium. A modified Czapek-Dox liquid medium has been found to give consistently good production of kojic acid.[5] It has the following composition (w/v); $NaNO_3$, 0.2%; K_2HPO_4, 0.1%; KCl, 0.05%; $MgSO_4 \cdot 7H_2O$, 0.05%; $FeSO_4 \cdot 7H_2O$, 0.001%; Difco yeast extract, 0.1%; glucose, 10%. First, 200-ml. portions of this medium in 1-l. conical flasks are sterilized (15 p.s.i. for 20 minutes) and inoculated by transferring a loop of spores from an agar subculture. The flasks are allowed to stand for 10 days at 28 to 30°, during which time a surface mycelium develops. On more prolonged standing kojic acid disappears from the solution. After sterilization of the flask and its contents (15 p.s.i. for 20 minutes) the medium is decanted through a filter and the remaining mycelium is washed by swirling with water (2 × 25 ml.). The combined medium and washings are extracted continuously with ether for 24 hours. Evaporation of the ether extract yields crude crystalline kojic acid. The yield is between 1.5 and 2.0 g. per 200 ml. of medium.

Purification. Pure kojic acid forms prismatic needles, melting point 152°. Kojic acid may be recrystallized from water, and if the sample is discolored a preliminary treatment with charcoal should be included. However, owing to the solubility of kojic acid in water (3.95 g. per 100 ml. at 20°), this purification is attended by some loss. For small amounts, it is better to dissolve the crude kojic acid in the minimum volume of warm methanol and, after charcoal treatment if appropriate, add ether slowly to the point of incipient crystallization. Kojic acid crystallizes on standing. Kojic acid has been separated from other biochemically important

[3] O. E. May, A. J. Moyer, P. A. Wells, and H. T. Herrick, *J. Am. Chem. Soc.* **53,** 774 (1931).

[4] H. N. Barham and B. L. Smits, *Ind. Eng. Chem.* **28,** 567 (1936).

[5] H. R. V. Arnstein and R. Bentley, *Biochem. J.* **54,** 493 (1953).

[6] C. Thom and K. B. Raper, "A Manual of the Aspergilli," The Williams and Wilkins Co., Baltimore, 1945.

organic acids by partition chromatography on Celite 545.[7] Vacuum sublimation onto a cold finger condenser has also proved useful in the preliminary purification of kojic acid isolated in tracer studies.[5]

Analysis

Principle. The methods available for kojic acid assay have been reviewed by Foster.[8] In practice the most convenient is based on the intense red color produced by kojic acid with dilute solutions of ferric chloride.[9] Since similar colorations are given by other compounds (e.g., α-hydroxy acids, phenolic compounds), this is not a specific reagent. It is, however, more sensitive than any of the other methods. Further, it is generally true that organisms producing kojic acid do so to the virtual exclusion of other acids or phenols liable to produce a red color with ferric chloride. Only one example of the joint production of kojic acid and substantial amounts of another acid is known. In this case an unidentified *Aspergillus* with yellow to ochraceous conidia produced both kojic and itaconic acids, the proportion depending on the temperature of incubation.[10] Under the assay conditions to be described it has been found that 25 to 400 γ of kojic acid can be determined without interference in the presence of 0 to 200 γ of itaconic acid.

The following modification of the method used by Arnstein and Bentley[11] is useful over the range 25 to 400 γ.

Reagents

Kojic acid standard solution, 0.5 g. of recrystallized kojic acid in 500 ml. of water; dilute 1:10 for calibration.

Ferric chloride solution, 1.0 g. of $FeCl_3 \cdot 6H_2O$ is dissolved in 100 ml. of 0.1 N HCl.

Procedure. A suitable aliquot of culture medium or other sample is treated with 4 ml. of 1% ferric chloride and diluted to 10 ml. The color develops immediately and is stable for some hours. It is read in a Klett-Summerson colorimeter with the 540-mμ filter. (The complex has absorption maximum at 505 mμ.) A blank is prepared from the same volume of uninoculated culture medium and 4 ml. of ferric chloride. Calibration is carried out in the same way with aliquots of the stock kojic acid solution.

Kojic acid containing culture media from *Aspergillus* species do not

[7] E. F. Phares, E. H. Mosbach, F. W. Denison, and S. F. Carson, *Anal. Chem.* **24,** 660 (1952).
[8] J. W. Foster, "Chemical Activities of Fungi," Academic Press, New York, 1949.
[9] K. Saito, *Botan. Mag.* (*Tokyo*) **21,** 7 (1907).
[10] J. L. Yuill, *Nature* **161,** 397 (1948).
[11] H. R. V. Arnstein and R. Bentley, *Biochem. J.* **54,** 508 (1953).

need to be deproteinized before estimation. If deproteinization of a sample is necessary, it should first be carried out with an appropriate concentration of TCA. Concentrations of up to 1% TCA in the final solution have no effect on the color. For solutions containing higher TCA concentrations the calibration should be carried out at the appropriate TCA concentration. A larger quantity of ferric chloride might be necessary.

[42] Preparation and Colorimetric Determination of Lactic Acid

By S. B. BARKER

Analytical Method[1]

Principle. Lactic acid is converted quantitatively into acetaldehyde on being heated with concentrated sulfuric acid. The acetaldehyde is determined by measurement of the purple color formed with p-hydroxydiphenyl.

Reagents

20% solution of $CuSO_4 \cdot 5H_2O$.
4% solution of $CuSO_4 \cdot 5H_2O$.
Solid $Ca(OH)_2$, powdered.
Sulfuric acid, concentrated, specific gravity 1.84.
1.5% solution of p-hydroxydiphenyl in 0.5% of NaOH. Dissolve 1.5 g. of solid p-hydroxydiphenyl in 100 ml. of 0.5% NaOH.
Lactate standards.

Procedure

Step 1. Removal of Protein. Any of the common procedures may be employed, such as trichloroacetic acid, tungstic acid, zinc hydroxide, or cadmium hydroxide. It is preferable to remove the precipitated protein by centrifugation, but filtration is possible with a high-grade paper tested to make certain that no contaminant is extracted.

Step 2. Treatment with Copper and Calcium. This removes glucose and other interfering or reacting substances. One to five milliliters of protein-free filtrate, containing 20 to 100 γ of lactic acid, is added to 1 ml. of 20% $CuSO_4$ solution, and water is added to a total volume of 10.0 ml. Approximately 1 g. of powdered $Ca(OH)_2$ is added, the test tube capped

[1] S. B. Barker and W. H. Summerson, *J. Biol. Chem.* **138,** 535 (1941).

characteristic violet of acetaldehyde. Propylene glycol would probably be present only if added. The free amino acid methionine reduces the amount of color yielded by lactate and acetaldehyde, probably by formation of a compound which does not react with p-hydroxydiphenyl. Other interfering substances must be eliminated by experimental trial.

SUBSTANCES INTERFERING WITH LACTATE DETERMINATION

Compound	Color obtained	Maximum removed by Ca-Cu, γ/ml. final solution	Color yielded by 1 γ in terms of γ of lactic acid
Removed at Low Concentration			
Dihydroxyacetone	Similar to lactate	1	0.02
Glyceric aldehyde	Similar to lactate	10	0.005
p-Hydroxyphenyl-lactic acid	Similar to lactate	15	0.005
Malic acid	Similar to lactate	5	0.01
Pyruvic acid	Similar to lactate	5	0.025
Rhamnose	Similar to lactate	5	0.02
Not Removed at All			
Acetaldehyde	Similar to lactate		2.05
Acrolein	Similar to lactate		0.10
Methylglyoxal	Similar to lactate		0.80
Propylene glycol	Similar to lactate		0.16
Djenkolic acid	Blue-green		
Formaldehyde	Blue-green		
α-Hydroxybutyric acid	Blue-gray		
Methionine	(See text)		

From Barker and Summerson, *J. Biol. Chem.* **138**, 535 (1941), by permission of the publisher.

Analytical Precautions

The principal source of error is contamination by lactic acid itself, in sweat, saliva, or other biological material. Great care must be exercised to avoid handling test-tube lips, pipet tips, etc. Glassware must be scrupulously clean, preferably by means of soap and detergent; if chromic acid cleaning mixture is employed, traces of chromate should be removed by passing all glassware through an alkaline wash solution. After thorough rinsing with tap and distilled water, drying should be done in an oven to minimize exposure to room dust, and articles should be stored in a protected place.

Any of the solutions may become contaminated with lactate, in which case it must be replaced. The only reagent requiring special precautions is the concentrated sulfuric acid. This should be highest quality reagent grade, but each lot number must be tested, since some contain unknown substances interfering with the reaction which are not revealed by the manufacturer's analyses. Although nitrate is known to interfere, some unsatisfactory batches contain practically no nitrogen. Grasselli acid has proved most consistently satisfactory. The acid is best added from some type of all-glass automatic pipetting device; if a buret is used, the stopcock must be entirely free from grease. Chilling the sulfuric acid before it is added to the lactate solution at 0° will eliminate occasional variations caused by local overheating.

Eastman Kodak p-hydroxydiphenyl may be used without further purification. If the solution is stored in a dark bottle, it is stable at room temperature for several weeks or indefinitely in a refrigerator. When discoloration or precipitation is prominent, a fresh solution should be prepared. A pipet delivering 0.1 ml. in 2 drops may be inserted in the stopper; the reagent should be dropped directly on the surface of the acid rather than run down the inside wall of the test tube. The subsequent mixing should be thorough for uniform dispersion of the p-hydroxydiphenyl. The color reaction is not significantly affected by a slight excess of either the reagent or copper ions. Color development is complete in 30 minutes at 30°, and longer standing at this stage does not alter the results. The curve relating color development to temperature is almost flat between 20 and 30°. Final color intensity falls off rapidly at 35°, owing to rapid destruction of p-hydroxydiphenyl. Thus, the acid solution must be well cooled after the 5-minute heating period before addition of the color reagent, or low results will be obtained.

Heating the solution for 90 seconds, after color development is complete, is to destroy excess reagent and to stabilize the color. Although not critical between 1 and 2 minutes, a longer heating period changes the spectrat characteristics of the color. After the heated solution is cooled, the color diminishes about 5% after 3 hours at room temperature. It should also be noted that the earlier heating in step 3 may be varied from 3 to 10 minutes without significant changes. The 4% $CuSO_4$ solution may be added to the copper-calcium filtrate at any stage in the procedure before addition of the p-hydroxydiphenyl. The special role of the cupric ions is discussed in the original publication.[1]

In general, although elasticity of the procedure has been discussed, it is recommended that conditions be held as constant as possible for maximum reproducibility.

Preparation of Lactate

For standards, lithium lactate is preferred, since it is easily prepared in anhydrous form, whereas zinc lactate contains water of crystallization.

ture, and determine the optical density at 570 mμ. The unknown glycerol samples should always be analyzed simultaneously with blank and standard glycerol samples.

The chromatographic separation of glycerol from other polyols is effected as follows: Wet 3 g. of the washed Celite thoroughly with 2.5 ml. of water. Slurry the wet Celite with ethyl acetate, and use it to prepare a chromatographic column. Slurry 0.5 g. of dry Celite with ethyl acetate, and place on top of the wet Celite in the column. Discard the excess ethyl acetate. Carefully add the glycerol sample, containing 1 to 5 mg. of glycerol, in 0.5 ml. of water to the top of the dry Celite layer, and cover it with 2 ml. of ethyl acetate. Pass 50 ml. of ethyl acetate through the column to elute acetoin in the first 20 ml. of eluate and 2,3-butanediol in the next 30 ml. After the ethyl acetate level is below that of the top of the Celite in the column, add benzene:n-butanol and collect the eluted fractions. The first 40 ml. of the benzene:n-butanol eluted contains ethanediol and 1,2-propanediol; the next 50 ml. contains the glycerol. The 5- and 6-carbon sugars and sugar alcohols remain on the column. The rate of elution is considerably increased by the use of a pressure bulb (Davol No. 1556 Cautery Set; double bulb with net). Mix the glycerol fraction with 50 ml. of water, and concentrate on a steam bath. Dilute the 10 to 15 ml. of the glycerol solution remaining after evaporation to an appropriate volume, and analyze a suitable aliquot by the periodic acid procedure.

Discussion

Periodic acid oxidation of any terminal vicinal hydroxyl groups, as in sugar alcohols, will yield formaldehyde and will be determined as glycerol. For this reason, crystalline mannitol is a good primary standard for this procedure. When glycerol occurs in a mixture of polyols, it is possible to separate the glycerol by means of a Celite chromatographic column. Pentoses, hexoses, and sugar alcohols remain on the column; ethanediol, 1,2-propanediol, 2,3-butanediol, erythritol, acetoin, and glycerol can be quantitatively resolved. Dihydroxyacetone, if present, will be in the glycerol fraction.

II. Dihydroxyacetone[3]

Assay Method

Principle. Ketoses, but not aldoses, react with resorcinol in concentrated hydrochloric acid to give a colored compound. The intensity of the color development is a function of the ketose concentration.

---

[3] Based on the Seliwanoff test for ketoses, T. Seliwanoff, *Ber.* **20**, 181 (1887), and modified by C. Neuberg, *Z. physiol. Chem.* **31**, 564 (1901), and R. E. Asnis and A. F. Brodie, *J. Biol. Chem.* **203**, 153 (1953).

Reagents

0.001 M dihydroxyacetone.

2% resorcinol in concentrated hydrochloric acid.

Procedure. The ketose sample is diluted, so that 1 ml. contains between 0.1 and 0.2 μM. of dihydroxyacetone. Prepare a reagent blank and standards containing 0.05, 0.1, and 0.2 μM. of dihydroxyacetone per milliliter. Add 4 ml. of 2% resorcinol in hydrochloric acid to 1 ml. of the ketose solutions. Mix the solutions thoroughly, and allow them to incubate at room temperature (25°) for about 8 to 12 hours. Determine the color intensity with a Klett 54 filter or at 490 mμ with a spectrophotometer. The dihydroxyacetone concentration is estimated by comparison with the standard curve.

Discussion

The color developed by dihydroxyacetone is interfered with by any other ketose, but not by aldoses or polyols. Thus, dihydroxyacetone may be estimated in the presence of glycerol and glyceraldehyde. Dihydroxyacetone may be separated from hexoses, pentoses, tetroses, and acetoin by means of the Celite column described for the determination of glycerol.

[44] Preparation and Analysis of Dihydroxymaleate and Hydroxypyruvate

By SHIRO AKABORI and KIHACHIRO UEHARA

I. Preparation of Dihydroxymaleic Acid

Dihydroxymaleic acid is prepared by oxidation of tartaric acid according to a modification of the method described by Nef.[1] To a hot solution of 50 g. of *d*-tartaric acid in 30 ml. of water is added 1 g. of ferrous sulfate ($FeSO_4 \cdot 7H_2O$) and 1.2 g. of rochelle salts in 20 ml. of water. The mixture is cooled at once to 0° in a salt ice bath, and to it is added 320 ml. of 3.5% hydrogen peroxide solution in portions over a period of about 80 minutes, with continuous stirring, at a temperature of 0° to −5°. Great care must be taken not to add an excess of the peroxide. The mixture is cooled and stirred at this temperature for 2 hours after the addition of peroxide is complete. As the oxidation proceeds, the solution becomes a slight violet.

[1] J. U. Nef, *Ann.* **357**, 291 (1907).

At the end of the specified time the solution is cooled and, with vigorous stirring, a mixture of 51 ml. of concentrated sulfuric acid (ca. 95%) and 102 ml. of fuming sulfuric acid containing as much as 50% of sulfur trioxide is added at a temperature below $-5°$. The slight yellow solution is kept in the refrigerator for about 10 days. The crystalline dihydroxymaleic acid hydrate is collected, washed very thoroughly with a small amount of cold water to remove all traces of sulfuric acid, and dried over sulfuric acid in a vacuum desiccator. The yield of pure, anhydrous material is 15 g.

If recrystallization is necessary, 20 g. of the product is dissolved in 160 ml. of absolute alcohol at room temperature and the solution is filtered. The filtrate is cooled in an ice bath and gently stirred during the dropwise addition of 150 ml. of water; the crystalline dihydroxymaleic acid is collected, washed, and dried as described above.

Properties and Determination

The dehydrated substance decomposes, without melting, at 159°. Crystalline dihydroxymaleic acid dihydrate forms in shining leaflets with a pearly luster, which give off water when placed in a vacuum over sulfuric acid, leaving a white amorphous powder. The substance is very sparingly soluble in cold water and ether but is more soluble in ethyl or methyl alcohol.

In aqueous solution the substance slowly decomposes, with evolution of carbon dioxide, at 50 to 60°, but it is very stable in ethyl alcohol solution.

Dihydroxymaleic acid rapidly reduces silver and cupric salts, as well as potassium permanganate. Ferric salt gives a blackish color, changing to beautiful violet on the addition of caustic alkali. Titanium chloride gives a red color on addition of sulfuric acid. The latter may be used also as a quantitative colorimetric method. With mercuric nitrate dihydroxymaleic acid forms a very unstable precipitate of the dimercuric salt, which is decomposed by water at 70° into tartronic acid, carbon dioxide, and mercury. With excess of mercuric nitrate the products of decomposition are mercury, carbon dioxide, and nitric acid, but no tartronic acid is formed. The quantitative determination may be made by estimating any one of these products or the excess of mercuric nitrate.

The titrimetric method for the quantitative estimation of dihydroxymaleic acid consists in decomposing dihydroxymaleic acid by boiling the aqueous solution and determining the decrease in acidity by titration to phenolphthalein.[2]

The use of 2,6-dichlorophenolindophenol to determine dihydroxy-

[2] H. Schmalfuss and H. Barthmeyer, Z. physiol. Chem. 160, 196 (1926).

maleic acid is based on the fact that the colored indicator is quantitatively and rapidly reduced by dihydroxymaleic acid in acid solution to a colorless compound. Since the indophenol method depends on the reduction of the reagent by dihydroxymaleic acid, any substance, such as vitamin C, having a reducing potential lower than the dye is a possible source of interference.

By heating dihydroxymaleic acid with a solution of p-nitrophenylhydrazine and acetic acid in a boiling water bath, the p-nitrophenylhydrazone of tartronic acid semialdehyde (decomposes at 240°) is obtained. On addition of p-nitrophenylhydrazine and acetic acid to an aqueous solution of dihydroxymaleic acid which has been previously heated at 60° for 15 minutes, the p-nitrophenylosazone of glyoxal (decomposes at 310°) is obtained.

II. Hydroxypyruvic Acid

Hydroxypyruvic acid is prepared by hydrolysis of bromopyruvic acid according to the procedure of Sprinson and Chargaff.[3]

Preparation of Bromopyruvic Acid

Pyruvic acid (3.5 g.) is heated to 45 to 50°, and 6.5 g. of bromine, previously dried by shaking with concentrated sulfuric acid, is added by drops, so that the temperature is kept at 45 to 50° throughout the reaction, with gentle stirring. The flask is cooled, if necessary, in a pan of cold water (the reaction is exothermic). After the addition of bromine is complete (about 2 hours) the stirring is continued for half an hour longer. The thick, fuming, sirupy reaction mixture is poured into a large crystallizing dish, the flask is washed with hot benzene, and the washing is added to the reaction mixture. The mixture is placed in a vacuum desiccator over sodium hydroxide pellets, and the solvent is removed by suction.

The mixture is kept in vacuo for 2 or 3 days, with occasional stirring, until no more fumes of hydrogen bromide are given off, and crystalline bromopyruvic acid, colored a slight yellow, is obtained.

The product is recrystallized from dry chloroform to obtain colorless crystals of bromopyruvic acid, melting at 58 to 62°. The yield is 5.1 g.

Identification of Bromopyruvic Acid

Bromopyruvic acid 2,4-dinitrophenylhydrazone melting at 180° and the quinoxaline derivative (I) melting at 235°, prepared by the following procedure, are useful for the identification of bromopyruvic acid.

[3] D. B. Sprinson and E. Chargaff, J. Biol. Chem. 164, 417 (1946).

$$
\begin{array}{c}
\text{NH} \\
\text{benzene ring fused} \quad \text{CO} \\
\text{C—CH}_2\text{Br} \\
\text{N}
\end{array}
$$

(I)

Preparation of Quinoxaline Derivative

On addition of 0.5 ml. of 2 N hydrochloride and 110 mg. of o-phenyl-ene-diamine to 167 mg. of bromopyruvic acid in 1.1 ml. of water, the quinoxaline derivative immediately crystallizes. The mixture is allowed to stand in the refrigerator. Then the crystals are collected, and the derivative is recrystallized from water. The substance is a slight yellow, and its aqueous solution shows a violet fluorescence.

Preparation of Hydroxypyruvic Acid

To a solution of 835 mg. (5 mM.) in 5 ml. of water is added 50.6 ml. of 0.196 N sodium hydroxide (2 equivalents) by drops, to keep the pH below 8.5, with stirring, at room temperature.

Although the solution of hydroxypyruvate so prepared is contaminated by a small amount of its isomer, bromopyruvic acid, and sodium bromide, it seems to be useful for most experiments. If the contamination of bromides is undesirable, the barium salt of hydroxypyruvic acid may be prepared by the following procedure.

To 50 ml. of the hydroxypyruvic acid solution, prepared as above, is added 1.0 ml. of concentrated hydrochloric acid. The mixture is concentrated under reduced pressure to a heavy sirup in a current of nitrogen gas, and the residual sirup is extracted three times with 1 ml. of absolute alcohol each time. The evaporation residue of the extract is dissolved in 1 ml. of water and neutralized with saturated barium hydroxide. Then ethyl alcohol is added to precipitate barium hydroxypyruvate. For the purification of this substance, it is dissolved in as small a quantity of water as possible with addition of a few drops of 0.1 N hydrochloric acid and precipitated by addition of methyl alcohol. The barium salt of hydroxypyruvic acid prepared in this manner, although contaminated with a small amount of the isomer, is free from bromopyruvic acid.

Properties

Solutions of hydroxypyruvic acid reduce silver nitrate at room temperature. Ferric salt gives no distinct color reaction, unless alkali is added, resulting in a deep-violet solution. Hydroxypyruvic acid 2,4-dinitrophenylhydrazone is orange and melts at 162°.

Hydroxypyruvic acid consumes 1 mole of periodic acid per mole and yields 0.9 mole of formaldehyde. Oxidation with ceric sulfate in 2 N sulfuric acid produces approximately 1 mole of carbon dioxide per mole.

Hydroxypyruvic acid is considerably stable in acidic solution, whereas it undergoes easily tautomeric change in a weakly alkaline medium.

Descending paper chromatography of the solution of hydroxypyruvic acid, prepared by hydrolysis of bromopyruvic acid, is carried out in the solvent system ethyl alcohol-water-acetic acid (100:25:1), and the spots are developed with an ammoniacal solution of silver hydroxide. Bromopyruvic acid, hydroxypyruvic acid, and its isomer are found to have R_f values of 0.66, 0.40, and 0.24, respectively.

[45] Enzymatic Determination of Ethanol

By NATHAN O. KAPLAN and MARGARET M. CIOTTI

Assay Method

Principle. The method is based on the reduction of DPN by ethanol in the presence of alcohol dehydrogenase. Bonnichsen and Theorell[1] have applied the procedure for the determination of ethanol in blood. These authors used the crystalline horse liver alcohol dehydrogenase[2] as the source of enzyme. The liver enzyme has a much higher affinity for ethanol than the yeast alcohol dehydrogenase[3] and is more suitable for the determination of small levels of ethanol. From 0.07 to 1.4 μmoles of ethanol can be estimated by this procedure.

Procedure. Bonnichsen and Theorell[3] carry out the assay as follows: 800 γ of DPN (80% purity) and the alcohol solution (not to exceed 1 ml.) are added to cuvettes. The volume is made up to 3.2 ml. by adding a buffer semicarbazide solution of the following composition: 3 vol. of 0.1 N NaOH + 7 vol. of 0.1 N glycine in 0.1 N NaCl + 1 vol. of semicarbazide solution (1.12 g. of semicarbazide in 100 ml. of 0.1 N NaOH). The resulting pH is about 9.2. The light absorption at 340 mμ is read, and 0.04 ml. of pure liver alcohol dehydroganase[2] (0.8 mg./ml.) is added. After 30 minutes at room temperature the mixtures are again read at 340 mμ. A standard curve is obtained by using the following concentrations of ethanol: 0.205 μmoles, 0.411 μmoles, 0.686 μmoles, and 1.32 μmoles. A blank (reagents without ethanol) is also run. Standard curves are essential

[1] R. K. Bonnichsen and H. Theorell, *Scand. J. Clin. Lab. Invest.* **3**, 58 (1951).
[2] See Vol. I [78].
[3] H. Theorell and R. K. Bonnichsen, *Acta Chem. Scand.* **5**, 1105 (1951).

with every analysis because of variation in temperature and enzyme activity.

Discussion

In order to determine blood ethanol the samples are distilled and then determined. The method can be used successfully to determine ethanol production in fermentations. Some higher aliphatic alcohols will also react, but at a slower rate. Methanol does not interfere with the ethanol analyses.

Recently we have used the yeast alcohol dehydrogenase and the acetyl pyridine analog of DPN (APDPN)[4] to estimate ethanol. APDPN is of

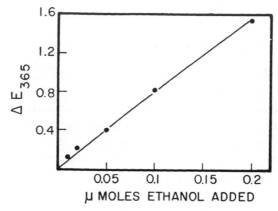

Fig. 1.

advantage because it has a considerably more positive potential than DPN, and hence its reduction is much more favored.[5] The method is perhaps twenty times as sensitive as the procedure described by Bonnichsen and Theorell.[3] Although no extensive use of the method has as yet been made, we believe it of value to report the procedure at present because of its sensitivity. The reaction is carried out as follows: 1 mg. of APDPN,[6] 0.01 ml. of a dialyzed crystalline yeast alcohol dehydrogenase preparation[7] (dialyzed against 0.01 M KCl for 19 hours at 4°), and the sample containing alcohol, diluted to 0.5 ml. with 0.1 M pyrophosphate buffer (pH 9.5). The reaction mixture is incubated for 30 minutes at 37°. The samples are diluted to 1 ml. with H_2O, and the optical density at 365 mμ taken. A blank without ethanol is always included, since the

[4] N. O. Kaplan and M. M. Ciotti, *J. Am. Chem. Soc.* **76**, 1713 (1954).
[5] N. O. Kaplan, M. M Ciotti, and F. E. Stolzenbach, *J. Biol. Chem.* **221**, 833 (1956).
[6] See Vol. III [129].
[7] Yeast alcohol dehydrogenase can now be obtained commercially.

reagents contain small amounts of ethanol. One of our standard curves is reproduced in Fig. 1. The values in the curve have been corrected for the blank reading. As can be seen from the figure, the method can determine less than 0.01 of a micromole of ethanol.[8]

Another method which we have used is based on the following reaction:

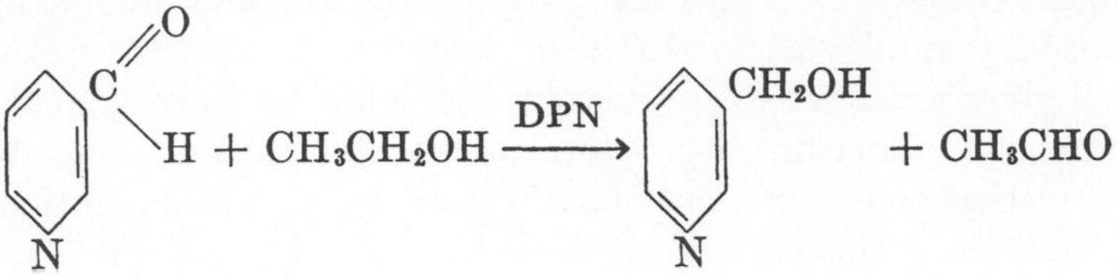

$$\underset{N}{\overset{}{\bigcirc}}\!\!-\!\!C\!\!\overset{O}{\underset{H}{\diagdown}} + CH_3CH_2OH \xrightarrow{\text{DPN}} \underset{N}{\overset{CH_2OH}{\bigcirc}} + CH_3CHO$$

This reaction is catalyzed by the liver alcohol dehydrogenase and not by the yeast enzyme.[8] The equilibrium is far in favor of the formation of acetaldehyde and the pyridyl-3-carbinol. As a result, ethanol is almost quantitatively oxidized. The reaction can be followed spectrophotometrically because the pyridyl-3-aldehyde has a much greater extinction coefficient at 270 mμ than the pyridyl carbinol. The decrease has been found to be proportional to the level of ethanol. Through this method, we have been able to follow yeast fermentation of glucose with 100 γ of whole yeast, directly in the spectrophotometer. It is of interest that DPN is essential for the reaction, although only a trace of the nucleotide is required.[8]

[8] N. O. Kaplan and M. M. Ciotti, in preparation.

[46] Chemical Procedures for Separation and Determination of Alcohols

By A. C. NEISH

General Principles and Selection of Method

Ethanol and related compounds can be determined with good precision by oxidation with acid dichromate solutions. An accurately measured amount of dichromate is added, and the amount remaining at the end of the reaction is measured iodometrically.[1] This method has been used and modified by numerous workers. Ethanol is oxidized stoichiometrically to acetic acid, 4 equivalents of dichromate being consumed per mole of ethanol.

[1] B. Kuriloff, Ber. **30**, 741 (1897).

$$2K_2Cr_2O_7 + 8H_2SO_4 + 3C_2H_5OH \rightarrow 3CH_3COOH + 2Cr_2(SO_4)_3$$
$$+ 2K_2SO_4 + 11H_2O$$

Methanol, isopropanol, and 2,3-butanediol can be oxidized stoichiometrically to carbon dioxide, acetone, and acetic acid (2 moles), respectively. Acetic acid is quite resistant to further oxidation by dichromate. All the other fatty acids are attacked by dichromate, and, under the conditions usually employed, mixtures of acids are formed. For example, n-butanol gives a mixture of n-butyric, propionic, and acetic acids, and n-propanol gives a mixture of propionic and acetic acids. Such oxidations are not stoichiometric and must be calibrated by standard solutions.

Although the estimation of alcohols by dichromate is precise, especially when a stoichiometric reaction occurs, it is extremely unspecific. The majority of organic compounds of interest to biochemists are readily attacked. For this reason it is essential to separate the alcohol from the other oxidizable components of the mixture before applying the oxidation. This can be done by distillation or by microdiffusion. Distillation is easy to use if only a few samples are to be analyzed, but microdiffusion methods are preferable when only small amounts are available or when a large number of determinations are to be carried out. A mixture of alcohols may be separated by partition chromatography on Celite columns and estimated separately. Positive identification of an alcohol can be made by preparation of the crystalline 3,5-dinitrobenzoate. These esters are suitable for radioactivity measurements and permit use of the isotope dilution technique. Details for these various procedures are given below.

Determination of Ethanol by Microdiffusion

Principle. The principles of microdiffusion have been thoroughly treated by Conway.[2] The method described below is based on that of Winnick.[3] The ethanol is separated from interfering substances by diffusion into an acid dichromate solution, where it is oxidized irreversibly to acetic acid, thus causing the diffusion to go to completion. The amount of ethanol is determined by measuring the amount of dichromate reduced.

Reagents

    Potassium dichromate (0.05 N) in 10 N sulfuric acid.

    Potassium iodide. Dissolve 25 g. of KI in water, and adjust the volume to 50 ml.

    Sodium thiosulfate (0.1 N).

    Soluble starch solution (1%).

[2] E. J. Conway, "Microdiffusion Analysis and Volumetric Errors," Crosby Lockwood & Son, London, 1947.

[3] T. H. Winnick, *Ind. Eng. Chem. Anal. Ed.* **14**, 532 (1942).

Procedure. Pipet exactly 1 ml. of the acid dichromate solution into the center well of a Conway No. 1 microdiffusion unit. Grease the lid with Vaseline and place it lightly in position, leaving an opening at one side. Add 1 ml. of the solution to be analyzed (0.1 to 0.4 mg. of ethanol) to the outer well, and seal the unit immediately. Let stand overnight (18 hours or more), and then add 1 to 2 ml. of water to the center well and 0.5 ml. of the KI solution. Titrate the iodine formed (at once) with 0.1 N sodium thiosulfate solution, from a micrometer syringe, or a microburet of suitable capacity, until the solution is a pale yellow. Add 2 drops of the starch solution, and continue the titration until the blue color disappears. A blank containing everything but ethanol is run with each set of determinations. The difference between the milliequivalents of thiosulfate required by the blank and by the sample, multiplied by 11.5, equals the milligrams of ethanol in the sample.

Limitations and Modifications. Any volatile organic compound is likely to interfere. If the solution is at pH 6 to 8, interference from volatile acids will be eliminated. Acetone is not oxidized readily enough to interfere. 2,3-Butanediol diffuses too slowly to cause appreciable error. A recent modification[4] reduces the elapsed time of the determination to 1 hour. This is accomplished by the use of special units which permit the diffusion to be carried on at a higher temperature and a lower pressure.

Determination of Ethanol in Distillates

Principle. Proteins and other substances likely to cause frothing are removed by coprecipitation with zinc hydroxide as described by Somogyi.[5] The neutral aqueous solution is then distilled to half its volume without reflux. All the simple alcohols pass into the distillate along with a reproducible fraction of certain less-volatile materials, such as acetoin.[6] If only one volatile alcohol is present, as is often the case, it can be determined directly by dichromate oxidation of the distillate. Corrections can be applied for interfering substances, such as acetoin, if they can be determined by an independent method.

Reagents

Phenol red (0.05%). Grind 100 mg. of phenol red powder in a mortar with 28.5 ml. of 0.01 N NaOH, and dilute the mixture to 200 ml. with water.

Sodium hydroxide (N).

[4] W. B. McConnell, *Am. J. Clin. Pathol.* **22,** 1223 (1952).
[5] M. Somogyi, *J. Biol. Chem.* **86,** 655 (1930).
[6] A. F. Langlykke and W. H. Peterson, *Ind. Eng. Chem. Anal. Ed.* **9,** 163 (1937).

Zinc sulfate (25%). Dissolve 250 g. of $ZnSO_4 \cdot 7H_2O$ in 1 l. of water. Determine the strength of this solution in terms of N NaOH by diluting 10 ml. to 100 ml. with water and titrating to the first pink color, using phenolphthalein as the indicator.

Oxidizing reagent. Dissolve 1.96 g. of potassium dichromate in 550 ml. of water, and add enough concentrated sulfuric to give a volume of 1 l. when cooled to room temperature. This is 0.04 N dichromate in approximately 18 N sulfuric acid.

Potassium iodide. Reagent-grade crystals.

Sodium thiosulfate (N), accurately standardized.

Soluble starch solution (1%).

Preparation of Sample. The oxidation described below should be used only on distillates. Some samples can be distilled without the zinc hydroxide treatment, but they should be neutralized first to prevent distillation of volatile acids. Solutions containing proteins or other macromolecular substances are neutralized to pH 7 to 8 with phenol red indicator or a glass electrode. The neutral solution is treated with $\frac{1}{15}$ vol. of the zinc sulfate solution, and then an equivalent amount of sodium hydroxide is added, with good mixing. The mixture is adjusted to a definite volume (about twice the original volume), centrifuged, and the clear supernatant liquid decanted. An aliquot of this clarified solution can usually be distilled without frothing or bumping.

The distillation may be carried out in any simple distillation apparatus. Good results can be obtained with 10 ml. of the clarified solution in a 50-ml. boiling flask and distilling exactly 5 ml. into a calibrated receiver. The distillate contains all the simple alcohols (ethanol, isopropanol, n-butanol, etc.) as well as volatile ketones such as acetone and diacetyl. If the apparatus is insulated to prevent reflux, the distillate will also contain a definite, reproducible fraction of moderately volatile compounds such as acetoin.[6] The apparatus should be checked to make sure that alcohol is being condensed quantitatively.

Procedure. An accurately measured 1- to 2-ml. aliquot of a distillate containing 1 to 4 mg. of ethanol, or its equivalent, is pipetted into a 250-ml. glass-stoppered Erlenmeyer flask. The flask should be free of any organic matter, such as dust or grease. An accurately measured 10-ml. portion of the oxidizing reagent is added, preferably by a syringe equipped with mechanical stop. The flask is stoppered, and the contents mixed thoroughly. After standing for 30 minutes at room temperature, 100 ml. of distilled water is added and then 0.5 to 1 g. of KI crystals. After the iodide is dissolved, the iodine is titrated using N sodium thiosulfate in a micrometer syringe, or suitably diluted thiosulfate in an ordinary micro- or macroburet. About 0.5 ml. of the starch solution is added near the end point, and the titration is continued until the deep-blue color of the starch-iodine complex changes to the pale-blue color of the trivalent chromium ion.

A blank is run with each set of determinations. The difference between the thiosulfate required by the blank and by the sample gives the milli-equivalents of dichromate reduced by the alcohol in the sample. One milliequivalent of dichromate is reduced by 11.5 mg. of ethanol, 22 mg. of acetoin, or 30 mg. of isopropanol. These compounds are oxidized stoichiometrically. Diacetyl and *n*-butanol are oxidized empirically, about 30 and 16 mg. being required, respectively, for reduction of 1 milli-equivalent of dichromate. If *n*-butanol is being determined, the method should be calibrated against a standard solution. Corrections can be applied for diacetyl and acetoin if they are estimated by specific methods on another aliquot of the distillate (see Vol. III [50]). A correction for acetone is not necessary, since it is not oxidized appreciably under the conditions outlined.

Limitations and Modifications. There are numerous modifications of this general procedure. Many of them use a lower concentration of sulfuric acid and thus require a heating period to complete the reaction in a reasonable time. The chief limitation of the method is its low degree of specificity.

Chromatographic Analysis of Mixtures of Alcohols

Principle. The simple aliphatic alcohols can be separated by partition chromatography on Celite-water columns with carbon tetrachloride and chloroform as developing solvents.[7] The technique of packing the column allows the sample to be put on as an aqueous solution. Since the develop-ing solvents are not oxidized by dichromate, it is possible to determine the quantity of alcohol in each fraction by the same general method given for distillates.

Reagents

> Purified Celite. Slurry Celite 535 with concentrated hydrochloric acid, and allow the mixture to stand at room temperature for 20 to 22 hours. Wash with distilled water, twice by decantation, and then in a Büchner funnel until free of acid. Wash the acid-free product with ethanol, then ether, dry in air, and then at 150 to 170° overnight.
>
> Carbon tetrachloride. Wash reagent-grade carbon tetrachloride (400 ml.) in a separatory funnel with two 400-ml. portions of distilled water, and then filter it through a dry filter paper.

[7] A. C. Neish, *Can. J. Research* **B28,** 535 (1950); *Can. J. Chem.* **29,** 552 (1951).

Chloroform. Wash reagent-grade chloroform (400 ml.) six times in a separatory funnel with an equal volume of water each time, and then filter it through a dry filter paper. This exhaustive washing is necessary to remove the alcohol added as a preservative. Since washed chloroform is not very stable, it should be prepared the day it is used.

Chloroform-carbon tetrachloride (50%). Mix equal volumes of washed chloroform and washed carbon tetrachloride, and filter the mixture through a dry paper. Prepare each day. Do not mix chloroform and carbon tetrachloride and then wash the mixture, since a solvent with different properties will be obtained.

Oxidizing reagent, thiosulfate, KI, and starch solution, as used for determination of alcohol in distillates.

Preparation of Sample. The procedure should be applied to distillates. These may be prepared and analyzed as described on pp. 257–259. If the concentration of alcohols is too low, the distillates (20 ml.) may be concentrated tenfold by redistillation through a column (2 × 20 cm.) of glass helices (⅛ inch) at a 1:1 reflux ratio. The first 2-ml. portion of the solution to distill contains all the alcohols. The correct amount of reflux can be obtained by a cold finger cooled by an adjustable air stream.

Procedure. Mix 3 g. of the Celite thoroughly with 2.5 ml. of distilled water with a stainless steel spatula. Slurry the mixture in washed carbon tetrachloride and pour it into a 300 × 15-mm. o.d. Pyrex chromatogram tube fitted with a perforated porcelain plate and a tight-fitting filter paper disk. Allow the Celite to settle first by gravity, and then compress it to a column about 8 cm. long by inserting another paper disk and forcing it down with a snug-fitting glass ramrod. Slurry 0.5 g. of dry Celite in the carbon tetrachloride, and compress it on top of the Celite-water column with a third paper disk. Decant any excess solvent, and pipet 0.5 ml. of an aqueous solution of the alcohol mixture (containing about 1 to 5 mg.) on top of the column, taking care to prevent any from getting on the walls of the tube except for 2 to 3 mm. above the top of the packing. While the sample is being absorbed into the layer of "dry" Celite, change the receiver to an empty 50-ml. graduated cylinder. Wash the sample in with two 1-ml. portions of carbon tetrachloride, and then start the development with this solvent. During the development, force the solvent through the column by gentle air pressure at the rate of 4 to 5 drops per second. Do not allow any air to get into the packing. When changing solvents, pour off any solvent left in the tube before adding the new solvent. Collect five successive fractions as follows:

1. 20 ml. of CCl_4; contains all the n-butanol, higher alcohols, diacetyl, and acetone.
2. 22 ml. of CCl_4-$CHCl_3$ (50%); contains n-propanol, isopropanol, and acetoin.
3. 25 ml. of $CHCl_3$; contains all the ethanol.
4. 15 ml. of $CHCl_3$; free of alcohols.
5. 40 ml. of $CHCl_3$; contains all the methanol.

The last two fractions may be omitted if methanol is absent.

Each fraction is transferred to a 250-ml. glass-stoppered Erlenmeyer flask. Five to ten milliliters of carbon tetrachloride is used to rinse out the graduate. Exactly 10 ml. of the oxidizing reagent is added, preferably from a syringe, and the flask is allowed to stand for 30 minutes at room temperature (20 to 25°). During this period it should be shaken vigorously, at least five times, for about 10 seconds each time. When the oxidation is finished, add 100 ml. of water and 1 to 2 g. of KI to the flask, and titrate the iodine with thiosulfate, adding 0.5 ml. of the starch solution when near the end point. The end point is reached when vigorous shaking of the flask for 15 to 20 seconds no longer results in appearance of the blue starch-iodine complex in the aqueous phase. The number of milliequivalents of dichromate reduced by the alcohol in any fraction is calculated from the difference between its titer and that of a blank containing an equal amount of the same developing solvent. Isopropanol, ethanol, and methanol reduce 2, 4, and 6 equivalents of oxidant per mole, as expected for oxidation to acetone, acetic acid, and carbon dioxide, respectively. The corresponding factor for n-butanol (about 4.5) is empirical and should be checked by a standard solution oxidized in parallel with the sample. Corrections can be made for diacetyl and acetoin if they are determined by more specific methods on an aliquot of the distillate. Acetone does not interfere.

Limitations and Modifications. The method is useful for obtaining information about the composition of mixtures of volatile alcohols, especially when the amount is too small to permit fractional distillation. It is too tedious for large numbers of samples. The ability to employ aqueous samples permits extension to the analysis of glycols, glycerol and other polyols.[7] It is possible to separate n-butanol from isoamyl alcohol by the procedure described above, although there is only 1 to 2 ml. of alcohol-free effluent between the two fractions. The columns may be used for three successive analyses, if the upper layer (0.5 g.) of Celite is dug out and replaced each time. A recent modification[8] employs a Celite-silicic

[8] S. Dal Nogare, *Anal. Chem.* **25**, 1874 (1953).

acid mixture which is stated to give better results than Celite alone. The procedure described above has been in use over a period of four years in the author's laboratory and has always given good results. Anyone following the method for the first time should check the separations and retention volumes against standard mixtures by collecting and analyzing a number of successive 2-ml. samples. Good separations depend on having the proper Celite/water ratio, and it is possible that this may vary between different batches of Celite. The Celite sold as "Analytical Filter-Aid" is useless for this type of chromatography.

Identification of Alcohols as the 3,5-Dinitrobenzoates

Principle. Positive identification of an alcohol can be made by a mixed melting point of a suitable crystalline derivative with an authentic sample. The 3,5-dinitrobenzoates have good properties for this purpose. They can be prepared from dilute aqueous solutions of the alcohols,[9] and, if a mixture is encountered, the esters can be separated by adsorption chromatography.[10] The dinitrobenzoates should be useful derivatives for the determination of alcohols by isotope dilution, the ethyl ester, at least, being readily obtained in a form suitable for radioactivity determinations.[11] The following procedure outlines a simple method for obtaining the ethanol derivative. The more-involved procedure of Lipscomb and Baker,[9] which has been tried on all the lower alcohols, may give better yields.

Reagents

3,5-Dinitrobenzoyl chloride, obtainable from Eastman Kodak Co.
Sodium hydroxide (40%). Dissolve 40 g. of NaOH pellets in water, adjust the volume to 100 ml., and cool in a refrigerator.

Procedure. Mix 1 ml. of aqueous distillate (containing about 50 mg. of ethanol) with 0.5 ml. of the sodium hydroxide solution in a 20-ml. beaker. Cool the mixture in an ice bath, and then add 0.4 g. of 3,5-dinitrobenzoyl chloride. Triturate with a stirring rod vigorously for 1 minute and then occasionally for 5 to 7 minutes. Dilute the mixture with an equal part of cold water, filter with suction, wash thoroughly with cold water, and dry in air or a vacuum desiccator. The yield is about 100 mg. of ester melting at about 91 to 93° (lit. 93°). This may be purified (and obtained in a physical condition suitable for mounting by filtration) by solution in hot pyridine (1 ml.) followed by addition of 20 ml. of water.

[9] W. N. Lipscomb and R. H. Baker, *J. Am. Chem. Soc.* **64**, 179 (1942).
[10] J. W. White, Jr., and E. C. Dryden, *Anal. Chem.* **20**, 853 (1948).
[11] A. C. Neish, *Can. J. Botany* **31**, 265 (1953).

After cooling to 3 to 5° and standing for several hours, the fine crystalline precipitate is filtered out, washed with cold water, and dried as above.

[47] Microanalytical Method for Acetic and Other Volatile Acids

By Simon Black

Principle. Volatile acids such as acetic diffuse from an acidified sample in the main chamber of a microdiffusion vessel to an alkaline solution in the center cup. After liberation of CO_2 from the cup by an acid buffer, a microtitration is made. The latter measures volatile acid plus the excess of acid buffer over initial alkali. A blank titration must therefore be subtracted from each determination to obtain the volatile acid value.

Although the method is nonspecific, some distinction between the several volatile fatty acids can be made on the basis of their distribution behavior between ether and water. Extraction of the acid into ether prior to its determination also serves to eliminate many interfering substances.

Special Equipment

Glass-stoppered centrifuge tubes, 30-ml. capacity. These may be obtained from the Wilkins-Anderson Co., 111 N. Canal St., Chicago, Ill., as Maizel-Gerson reaction vessels.

A 10-ml. pipet with its tip bent 90° with respect to the pipet axis. The curved tip facilitates withdrawal of supernatant ether, leaving an aqueous layer undisturbed.

Glass-stoppered Erlenmeyers, 50-ml. capacity.

Center cups. These are made by cutting the lower 15-mm. portion from shell vials (15 mm. o.d.) which are obtainable from most scientific supply companies. They should be discarded after they have been used once or twice. For greater precision steel cups may be used.[1]

A microburet which delivers with an accuracy of ± 0.1 μl.

A magnetic stirrer and a stirring bar made by sealing a 4-mm. length of iron wire in glass.

Reagents

Alkaline trapping solution. Dissolve 0.124 g. of $K_2CO_3 \cdot 1.5$ H_2O plus 0.25 g. of KNO_3 (to increase hygroscopicity) in distilled water, and dilute to 50 ml.

[1] S. Black, *Arch. Biochem.* **23**, 347 (1949).

80% H_2SO_4. Add 400 ml. of concentrated H_2SO_4 (c.p. or reagent) to 100 ml. of distilled water.

0.1 M monosodium maleate. Dilute 5 ml. of 1.0 N NaOH and 1.5 ml. of isoamyl alcohol (to reduce surface tension) plus 0.58 g. of maleic acid with distilled water to 50 ml.

0.1 N standard NaOH.

Indicator solution. Dilute 1 ml. of phenolphthalein indicator, 0.1% in 95% alcohol, to 25 ml. with CO_2-free (boiled) distilled water. This solution should be made up on the day of use.

5 N H_2SO_4.

Alkaline Na_2SO_4. Dissolve 20 g. of anhydrous Na_2SO_4 (c.p. or reagent) plus 10 ml. of 1 N NaOH and dilute to 100 ml. Sodium hydroxide solutions without Na_2SO_4 do not remove fatty acids quantitatively from ether.

Ethyl ether. This reagent should be of good quality, substantially free of peroxides.

10% sodium tungstate.

Procedure. One milliliter of a copper-lime filtrate,[2] containing not more than 2 microequivalents of volatile acid, is pipetted into a glass-stoppered Erlenmeyer. One milliliter of 80% H_2SO_4 plus about 10 mg. of Ag_2SO_4 (to retain chloride) are added, and the flask is shaken gently to ensure mixing. A center cup, containing 100.0 μl. of the alkaline trapping solution, is placed centrally on the floor of the flask with the aid of forceps. The cup stands directly in the acidified sample. The glass stopper is put loosely in place (do not twist or "lock"), and the flask set in a 100 to 105° oven. After 14 to 16 hours the center cup is removed and 40.0 μl. of monosodium maleate solution is added. The cup is then tilted gently, if necessary, to ensure distribution of this reagent over its bottom. After 10 to 20 minutes, during which CO_2 escapes, about 0.3 ml. of indicator solution and a magnetic stirrer are added. A titration is then made in the cup, with stirring, with 0.1 N NaOH. The amount of volatile acid is calculated by substrating a blank titration value.

When the amount of interfering materials make necessary a preliminary separation from the volatile acid, the following ether extraction procedure may be used: Three milliliters of a protein-free filtrate is placed in a 30-ml. glass-stoppered centrifuge tube. Two-tenths milliliter of 5 N H_2SO_4 is added, 15 ml. of ether run in, and the tube shaken vigorously for 2 minutes. Ten milliliters of the supernatant ether is transferred with the special pipet to a second stoppered tube containing 1.2 ml. of the alkaline Na_2SO_4 solution. This tube is shaken for 1 minute. One milliliter of the

[2] A. L. Lehninger and S. W. Smith, *J. Biol. Chem.* **173,** 773 (1948).

aqueous layer is then transferred with a fine-tipped pipet to a 50-ml. Erlenmeyer flask, and the analysis carried out as described above by the microdiffusion procedure. To calculate the amount of volatile acid present in the original 3 ml. of filtrate, the quantity found is divided by the factor 0.55 multiplied by whatever fraction of the acid in question the table indicates should be extracted by this procedure.

EXTRACTION OF VOLATILE ACIDS FROM WATER WITH ETHYL ETHER
I. 15 ml. of ether used to extract 3.2 ml. of aqueous solution.[a]
II. 10 ml. of ether used to extract 10 ml. of aqueous solution.

| | Per cent of acid found in ether layer | |
Acid extracted	I	II
Acetic	71	32
Propionic	87	60
Butyric	95	83
Valeric	98	93
Caproic	100	98
Caprylic	100	100

[a] The small amount of calcium in copper-lime filtrates causes additional retention in the aqueous layer of about 0.1 micromole of volatile acid; this amount is independent of the total volatile acid in the sample. Addition of 0.1 ml. of 10% Na_2WO_4 prior to extraction causes precipitation of the calcium (which need not be removed) and elimination of this source of error.

Specificity. The method is nonspecific for volatile acids. However, the different distribution characteristics between water and ether (see the table) can be used as an aid in identification. It is also possible to replace the titration procedure with a colorimetric test specific for acetic acid.[3] Numerous sources of interference with this colorimetric method are eliminated by prior use of the microdiffusion procedure. A method for acetic acid which employs paper chromatography is also available.[4]

Some interference is also given in the titrimetric procedure by lactic, β-hydroxybutyric, and pyruvic acids. Diffusion of the latter can be eliminated by $HgSO_4$ (about 10 mg.). When $HgSO_4$ is used in conjunction with the extraction procedure, ether should first be eliminated by drying the flask in an oven; 1 ml. of water plus solid $HgSO_4$ are then added before the 80% H_2SO_4.

Precision. Known quantities of acetic acid can be recovered from tissue samples, by means of the ether extraction and microdiffusion procedures, with an average deviation of 0.03 μM.

[3] J. O. Hutchens and B. M. Kass, *J. Biol. Chem.* **177,** 571 (1949).
[4] E. P. Kennedy and H. A. Barker, *Anal. Chem.* **23,** 1033 (1951).

Advantages and Disadvantages. The advantages of this method, compared to the Duclaux procedure,[5] are in the possibility of analyzing many samples simultaneously with simple apparatus, and in its suitability for small quantities of sample. Its low specificity, however, limits its usefulness in experiments involving crude extracts, or in other cases where the samples contain potential sources of interference.

[5] T. E. Friedeman, *J. Biol. Chem.* **123**, 161 (1938). See Vol. III [61].

[48] An Enzymatic Micromethod for the Determination of Acetate

By MORRIS SOODAK

Acetate + ATP + CoA + Sulfanilamide →
        Acetyl-Sulfanilamide + AMP + CoA + Pyrophosphate

Assay Method

Principle. In the enzymatic acetylation of sulfanilamide by a partially purified pigeon liver extract preparation,[1,2] acetate may be made the limiting factor.[3] This system has been adapted for the quantitative determination of small amounts of acetate. If proper care is taken to ensure acetate-free reactants, the system permits the estimation of 0.025 to 0.20 micromole of acetate in 0.24 ml. In this range the acetylation of sulfanilamide measured as a disappearance of sulfanilamide (Bratton and Marshal procedure[4]) bears a linear or near-linear relationship to the acetate concentration.

Reagents

    0.02 *M* sulfanilamide.
    0.1 *N* crystalline Na ATP.
    0.1 *M* MgCl$_2$.
    1.0 *M* K citrate.
    CoA solution, 500 units/ml.
    Assay mixture. The above solutions are mixed in the following proportions, dispensed in approximately 3-ml. portions, and stored in the deep-freeze: 2 ml. of sulfanilamide, 3 ml. of ATP,

[1] N. O. Kaplan and F. Lipmann, *J. Biol. Chem.* **174**, 37 (1948).
[2] N. O. Kaplan and F. Lipmann, *Federation Proc.* **6**, 266 (1947).
[3] M. Soodak and F. Lipmann, *Federation Proc.* **7**, 190 (1948).
[4] A. C. Bratton and E. K. Marshall, Jr., *J. Biol. Chem.* **128**, 537 (1939).

0.5 ml. of $MgCl_2$, 2 ml. of K citrate; 2 ml. of CoA, and 0.5 ml. of water.

1.0 M Tris, pH 8.0.

0.5 M cysteine·HCl.

0.005 M standard acetic acid solution.

Enzyme. Partially purified preparation of extract of pigeon liver acetone powder.

5% TCA.

4.0 N HCl.

0.1% sodium nitrite (prepared fresh daily).

0.5% ammonium sulfamate.

0.1% N-(1-naphthyl)ethylenediamine dihydrochloride (in dark bottle in icebox).

Procedure. The reaction mixture, final volume 0.5 ml., is set up in 13 × 100-mm. test tubes as follows: distilled water, 0.05 ml. of Tris, exactly 0.05 ml. of assay mixture (containing 0.4 micromole of sulfanilamide, 1.5 micromoles of ATP, 0.5 micromole of $MgCl_2$, 10 micromoles of K citrate, and 5 units of CoA), 0.01 ml. of cysteine/HCl, to 0.24 ml. of neutralized (pH 7) acetate solution (0.025 to 0.2 micromole), and 0.15 ml. of enzyme. Standard acetic acid samples (0.0, 0.05, 0.1, 0.15, and 0.2 micromole of acetate) are set up simultaneously to establish the standard curve. One tube containing only the assay mixture and Tris serves as a base line for unreacted sulfanilamide. The tubes are gassed with nitrogen for 20 seconds, stoppered with rubber stoppers, and incubated for 1 hour at 37°.

At the end of the incubation period 2 ml. of 5% TCA is added to stop the reaction. After standing for 10 minutes, the protein is centrifuged down, and the supernatants are poured over into clean test tubes. The sulfanilamide remaining is determined by the method of Bratton and Marshall. One-milliliter aliquots are transferred to large tubes (20 × 150 mm.), and 9 ml. of water is added to each. This is followed by addition of 0.5 ml. of 4 N HCl. One milliliter of 0.1% sodium nitrite is added, mixed well, and the mixture is allowed to stand for 3 minutes. Then 1.0 ml. of 0.5% ammonium sulfamate is added, the tubes are well mixed and rotated at an angle to be sure to destroy any nitrite remaining on the walls of the tube, and allowed to stand for 2 minutes. Next 1.0 ml. of 0.1% N-(1-naphthyl)ethylenediamine solution is added. After mixing and standing a few minutes the color is read in a Klett colorimeter with a No. 54 filter. The decrease in color is a measure of the sulfanilamide acetylated, and the unknown acetate concentration is obtained by comparison with the standard acetate curve. The Klett reading for the unreacted sulfanilamide is about 170. The 0.0 acetate tube should give a delta of about 30 with acetate-free reagents. This blank reading may be due to acetate present in the enzyme and to traces of the citrate-ATP-CoA enzyme.[5] The delta for 0.05 micromole of acetate ranges from 20 to 30 on the Klett.

The assay may also be run in 1-ml. final volume. This permits the use of unknown acetate samples to 0.48 ml. In this case all the reagents are doubled as well as the enzyme, the incubation time increased to 2 hours, and the reaction stopped with 4 ml. of TCA.

Preparation of Assay Enzyme. The pigeon liver acetone powder as well as the aged extracts are prepared according to the method of Kaplan and Lipmann,[2] which is described in Vol. III [132], except that a 5% extract (5 g. of acetone powder is extracted with 100 ml. of ice-cold 0.02 M KHCO$_3$ solution) is used. The 5% extract ensures a lower acetate blank in the final enzyme preparation.

Ice-cooled aged extract (50 to 85 ml.) is made 70% saturated by adding solid ammonium sulfate (65 g. of ammonium sulfate per 100 ml. being taken as 100% saturation). After addition of the ammonium sulfate and standing at room temperature for $\frac{1}{2}$ hour, the precipitate is centrifuged at 3000 to 4000 r.p.m. for $\frac{1}{2}$ hour in a cold centrifuge. The resulting supernatant is discarded, and the residue is dissolved by adding exactly 30 ml. of ice-cold 0.02 M KHCO$_3$. The increase in volume is taken to be the volume of the 70% residue. The ammonium sulfate content of the residue is taken into account in the next step. More 0.02 M KHCO$_3$ is added to bring the volume to that of the original aged extract taken for fractionation. The solution is then brought to 40% saturation with solid ammonium sulfate. It stands at room temperature for $\frac{1}{2}$ hour and is then centrifuged as above. The precipitate is discarded. The supernatant is raised to 70% saturation and treated as above. The 40 to 70% fraction is dissolved in the minimum quantity of ice-cold 0.02 M KHCO$_3$ and dialyzed overnight in the cold against 10 l. of dialyzing fluid containing 50 g. of KCl, 20 g. of KHCO$_3$, and 2 g. of cysteine·HCl. This dialyzed 40 to 70% fraction is the enzyme used in the assay. The entire procedure must be followed if one is to obtain an enzyme essentially free of acetate. The enzyme is stable for months when kept at $-20°$.

Interfering Substances. When 10 μmoles of the following compounds are tested in the assay system, the resulting interference in terms of acetate equivalents in micromoles is: formate, 0; propionate, 0.07; butyrate, 0.13; valerate, 0.04; pyruvate, 0; and ethanol, 0. Acetoacetate presents a special problem. Some of the assay enzymes contain an aceto-

[5] P. A. Srere and F. Lipmann, *J. Am. Chem. Soc.* **75**, 4875 (1953).

acetate-activating enzyme, whereas others do not. The interference caused by 0.5 micromole of acetoacetate will vary from 0 to 0.6 micromole of acetate equivalent according to the assay enzyme used. The interference by acetoacetate may be eliminated in the following way: the solution to be assayed for acetate is made 0.1 N with respect to HCl, incubated at 37° for 16 hours, and then neutralized. This treatment breaks down the acetoacetic acid to acetone without the formation of any acetate.

The pigeon liver system is not very sensitive to sodium ions,[6] which do not inhibit strongly, as in the case of the heart system of Von Korff.[7]

The assay enzyme contains an esterase for acetylsalicylic acid.

Preparation of Sample for Assays. Since most biological materials do not interfere with the assay, it is usually sufficient to stop the enzymatic activity of the solution under investigation and take a proper aliquot at pH 7 to 8 for assay. Enzymic reactions are stopped by placing the tubes in a boiling water bath for 1 minute. In Warburg experiments with whole microorganisms it is only necessary to centrifuge out the cells. Neutralized acetic acid distillates may, of course, be used.

Comments

Von Korff[7] has described a specific enzymatic micromethod for acetate (0.05 to 0.4 micromole) using a preparation derived from rabbit heart ventricle mitochondria, in which DPN reduction is followed. Rose[8] has applied the specific acetokinase reaction of *E. coli* for the determination of acetate in the range 0.5 to 2.5 micromoles. Acetohydroxamic acid is determined.

[6] M. E. Jones, personal communication.
[7] R. W. Von Korff, *J. Biol. Chem.* **210,** 539 (1954).
[8] I. A. Rose, see Vol. I [97].

[49] Determination of Glycolic, Glyoxylic, and Oxalic Acids

By KATHARINE F. LEWIS and SIDNEY WEINHOUSE

I. Oxalic Acid

Oxalic acid crystallizes with two molecules of water in monoclinic prisms having a melting point of 101.5°. After fusion, water is lost and the anhydrous acid sublimes at 150 to 160°; in a sealed tube it melts at 189.5°. It is soluble in 10.5 parts of water at 15°, easily soluble in ethanol, and only slightly soluble in ether (1.5 g. per 100 ml. at 25°). All the salts except those of the alkali metals or magnesium are insoluble in water; the highly insoluble calcium salt (6.7 mg./l. at 13°), is also insoluble in aqueous solutions of acetic acid, which differentiates it from most other organic acids. Heated with concentrated sulfuric acid or acetic anhydride, the acid or its salts yield both CO and CO_2. It decolorizes acid (but not basic) solutions of potassium permanganate.[1] Despite its low solubility in ether, oxalic acid may be removed from aqueous solutions by continuous ether extraction. With an extractor of the type in which ether droplets are continuously passed upward through the aqueous solution by dispersion through a coarse sintered glass disk, oxalic acid in 0.01 M solution in 1.0 M sulfuric acid can be extracted completely in 24 hours. Under the same conditions 70% of glycolic acid can be extracted in 24 hours and 94% in 48 hours. However, glyoxylic acid is not readily extractable with ether; only 26% is extracted in 24 hours and 45% in 48 hours under the same conditions.

A most conveniently prepared derivative of oxalic acid is the S-benzylthiuronium salt[2] which crystallizes with high purity. A concentrated solution of the sodium or potassium salt in aqueous alcohol is added rapidly, with stirring, to a slight excess of a 15% solution of S-benzylthiuronium chloride in hot ethanol. On cooling, di-(S-benzylthiuronium) oxalate crystallizes; melting point 193°. For other derivatives refer to Huntress and Mulliken.[1]

Identification by Color Test

This test depends on the formation of aniline blue, a triphenylmethane dyestuff, formed by melting together oxalic acid and diphenylamine; it is also formed when insoluble oxalates are warmed with diphenylamine and sirupy phosphoric acid.[3] The reaction is quite specific; it is not given by formic, acetic, propionic, tartaric, citric, succinic, dihydroxymaleic, benzoic, phthalic, tricarballylic, glycolic, or glyoxylic acids. The limit of identification is 5 γ with a concentration limit of 1 to 10,000, or 0.1 mg./ml.

Method. A crystal of the substance (a solution must be evaporated to dryness) is melted with a little solid diphenylamine in a micro test tube over a free flame. On cooling, the melt is taken up in a drop of alcohol; a blue color indicates the presence of oxalic acid; a blank test remains colorless. To detect oxalic acid in a mixture containing other anions precipitated by calcium, the solution is acidified with acetic acid and a solution of

[1] E. H. Huntress and S. P. Mulliken, "Identification of Pure Organic Compounds," p. 99, John Wiley & Sons, New York, 1941.
[2] J. J. Donleavy, *J. Am. Chem. Soc.* **58**, 1005 (1936).
[3] F. Feigl and O. Frehden, *Mikrochemie* **18**, 272 (1935).

CaCl₂ is added. The precipitate is collected by filtration or centrifugation, and the color test is conducted directly on the solid. The sensitivity of the test can be increased by adding a few drops of water and extracting the dye from the aqueous layer with ether; as the layers separate, the blue dye collects at the junction of the two layers.

Quantitative Determination of Oxalic Acid in Biological Materials

Oxalic acid may be extracted from tissues by standard methods of maceration or homogenization with dilute hydrochloric acid solution. In virtually all methods of oxalate determination, precipitation with calcium ions is the method of choice. The highly insoluble calcium oxalate may be determined directly by weighing, either as such, or after conversion to calcium oxide, sulfate, or carbonate by ignition. Most convenient, however, is solution in dilute mineral acid and titration with potassium permanganate. The solution is acidified strongly with sulfuric acid, heated to 60°, and directly titrated to a permanent pink with 0.1 N KMnO₄ (1 ml. = 4.5 mg. of oxalic acid). Where the amounts of oxalate available or its concentration are low, it can be concentrated by continuous extraction with ether. The following procedure of Powers and Levatin,[4] designed for oxalate determination in urine, will serve as a guide in determining small quantities of oxalate in tissues and biological fluids. In this method about 10 ml. of solution containing approximately 0.025 mg./ml. of oxalic acid is acidified with 0.4 ml. of concentrated hydrochloric acid and submitted to continuous ether extraction for 6 hours. One milliliter of 2% acetic acid solution is added to the ether extract, and ether is removed by evaporation at 70°. The residue is transferred by pipet or syphon to a centrifuge tube, and the extraction flask washed with a total of 4 ml. of 95% ethanol, which in turn is transferred to the centrifuge tube.

Calcium oxalate is precipitated by adding 0.6 ml. of 10% CaCl₂ solution; the mixture is stirred by means of an air current introduced through a fine capillary. The solution is covered *without stirring* with 2 ml. of acid-alcohol solution (60 ml. of 95% ethanol, 10 ml. of 2% acetic acid, and 20 ml. of water). After standing overnight, the tube is centrifuged for 30 minutes at 2000 r.p.m., the supernatant solution decanted, and the tube inverted to drain for 2 minutes. The sides of the tube are washed with 2 ml. of acid-alcohol solution, and the precipitate broken up with a stirring rod, which is removed and washed with 2 ml. of acid-alcohol solution. The precipitate is again centrifuged for 30 minutes, and the decanting and draining repeated. The tube is then heated for a few minutes in a water bath to ensure the removal of all the alcohol.

One milliliter of 2% H₂SO₄ is blown from a pipet directly on the pre-

4 H. H. Powers and P. Levatin, *J. Biol. Chem.* **154,** 207 (1944).

crystals, melting point 98°, which deliquesce to a sirup on exposure to air. It forms metallic salts which are quite stable and insoluble in water (silver salt, 0.007 g. per 100 ml.; barium salt, 0.005 g. per 100 ml.).[8] Solutions of glyoxylic acid in dilute mineral acid are stable indefinitely in the cold; the acid also can be quantitatively recovered from solution and stored as the bisulfite addition product (vide infra).

Identification

Because of its reactive aldehyde group, glyoxylic acid forms many characteristic carbonyl derivatives in essentially quantitative yields. Glyoxylic phenylhydrazone is prepared by treating a solution of the acid with a saturated solution of phenylhydrazine in 1 N HCl saturated with ammonium chloride. The phenylhydrazone precipitates from the aqueous solution as yellow needles, difficultly soluble in cold water, but easily soluble in alcohol, melting point 143 to 145°.[9]

The semicarbazone can be prepared in essentially quantitative yield as follows. An aqueous solution of glyoxylic acid is heated for a few minutes on a water bath with an excess of a solution of 1 g. of semicarbazide hydrochloride and 1.5 g. of sodium acetate dissolved in 10 ml. of water. The highly insoluble semicarbazone which separates on cooling is recrystallized from aqueous ethanol, melting point 202 to 203°.[10]

Glyoxylic 2,4-dinitrophenylhydrazone is prepared by mixing a solution of the acid with an 0.1% solution of 2,4-dinitrophenylhydrazine in 2 N HCl. The yellow precipitate, which forms immediately if the concentration of glyoxylic acid is sufficiently high, is recrystallized from aqueous ethanol. It exists in cis-trans forms, one melting at 190°, the other at 202°.[11,12] The 2,4-dinitrophenylhydrazone has an absorption spectrum which differs sufficiently from those of other natural carbonyl compounds to make possible its identification in the presence of interfering materials by the method of Friedemann and Haugen.[13] The spectrum of its sodium salt, together with those of two carbonyl compounds commonly present in biological materials, pyruvic and α-ketoglutaric acids, is shown in Fig. 1.

Paper chromatography of the 2,4-dinitrophenylhydrazone using 0.05 N NaOH as the developing agent gives a characteristic brown spot with an R_f value of 0.57. The 2,4-dinitrophenylhydrazones of pyruvic

[8] W. H. Hatcher and G. W. Holden, Trans. Roy. Soc. Canada, Sec. III [3] 19, 11 (1925).
[9] S. P. Mulliken, "A Method of Identification of Pure Organic Compounds," Vol. 2, p. 60, John Wiley & Sons, New York, 1916.
[10] Beilstein's Handb. organ. Chem. 3, Suppl. 1, 209 (1929).
[11] C. Neuberg and K. Kobel, Biochem. Z. 256, 475 (1932).
[12] S. Ratner, V. Nocito, and D. E. Green, J. Biol. Chem. 152, 119 (1944).
[13] T. E. Friedemann and G. E. Haugen, J. Biol. Chem. 147, 415 (1943).

acid give two yellow spots in this medium, with R_f values of 0.58 and 0.41, and α-ketoglutaric acid gives a single yellow spot with $R_f = 0.67$.

The hydrazones can also be chromatographed in butanol saturated with 3% ammonia (v/v).[14] In this medium the R_f value of glyoxylic 2,4-dinitrophenylhydrazone is 0.24; for pyruvic, 0.34; for α-ketoglutaric, 0.05. The lower limit of identification is approximately 25 γ for both methods.

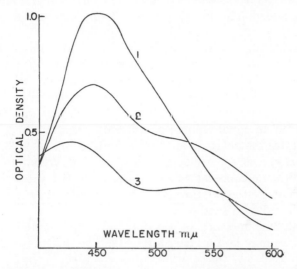

Fig. 1. Absorption spectrum of 2,4-dinitrophenylhydrazone of glyoxylic acid (1) compared with pyruvic acid (2) and α-ketoglutaric acid (3), 0.001% solution of hydrazones in N NaOH.

Quantitative Determination

Bisulfite Binding. An excess of 0.3 N sodium bisulfite is added to an aliquot of the solution and the pH brought to 2.5 to 3.0 with dilute HCl. After standing for 15 minutes, the excess bisulfite is titrated with standardized 0.05 N iodine solution. The bisulfite addition product is broken up by the addition of 3 ml. of a saturated solution of sodium carbonate, and the released bisulfite is titrated with iodine.

$$1 \text{ ml. } 0.05 \ N \ I_2 = 1.85 \text{ mg. glyoxylic acid}$$

Colorimetric Determination of 2,4-Dinitrophenylhydrazone. See Vol. III [66].

Glyoxylic acid in microgram quantities may be determined by the method of Friedmann and Haugen for the determination of keto acids.[13]

14 D. Cavallini, N. Frontali, and G. Toschi, *Nature* **163,** 568 (1949).

Synthesis of Glyoxylic Acid

Glyoxylic acid, which is unavailable commercially, may be synthesized for biochemical work from oxalic acid by the method of Weinhouse and Friedmann[15] based on the procedure of Benedict.[16]

Ten millimoles of oxalic acid is dissolved in 10 ml. of 1 N HCl, and the solution is placed in an ice-salt bath (on cooling, part of the oxalic acid crystallizes out); 5 drops of a saturated aqueous mercuric chloride solution are then added. This is followed by 500 mg. (20 millimoles) of magnesium turnings of the type used in the Grignard reaction, added in small portions, with constant stirring, over a period of 45 minutes. If the magnesium is added too rapidly, the temperature rises and the yield is decreased. The pH of the solution is tested periodically with pH paper, and if it rises above 3.0 a drop or two of concentrated HCl is added. The gray paste of magnesium salts is filtered off with suction. To the clear filtrate is added an equal volume of a freshly prepared 20% solution of sodium bisulfite, the solution is stirred, and 10 ml. of 95% ethanol is carefully layered over the solution without mixing. When crystals of the bisulfite addition compound begin to form at the interface (usually within 30 minutes), the mixture is vigorously stirred, refrigerated overnight, and filtered with suction.

The crystalline mixture of bisulfite addition product and excess sodium bisulfite may be purified as follows. The mixture is dissolved in 5.0 ml. of 1 N HCl, and the solution is concentrated on the steam bath to 2.5 ml. and cooled. Precipitation with 2.5 ml. of 20% sodium bisulfite solution and 2.5 ml. of ethanol is repeated as above. The bisulfite addition product may be stored as such or may be decomposed and the glyoxylic acid kept in a dilute aqueous HCl solution in which it is indefinitely stable. The addition product is decomposed by dissolving it in 5 ml. of 1 N HCl, evaporating the solution to 2.5 ml., adding 2.5 ml. of water, and repeating the evaporation to ensure complete removal of SO_2. The solution is assayed for glyoxylate by the bisulfite binding method and stored in the refrigerator. The product, thus obtained in a yield of 20 to 30%, contains no oxalate, and all the carbon is accounted for as glyoxylate by bisulfite binding, or by 2,4-dinitrophenylhydrazone or semicarbazone formation.

[15] S. Weinhouse and B. Friedmann, *J. Biol. Chem.* **191**, 707 (1951).
[16] S. R. Benedict, *J. Biol. Chem.* **6**, 51 (1909).

[50] Preparation and Determination of Acetoin, Diacetyl, and Acetolactate

By L. O. KRAMPITZ

I. Preparation and Determination of Acetoin

Purification of the Dimer of Racemic Acetoin

Racemic acetoin is commercially available; however, it usually exists as a mixture of monomeric and dimeric forms. The dimeric form is a white crystalline material melting at 94° which readily dissolves in water in the monomeric form. Conversion of acetoin to the dimeric form is facilitated by the addition of a few granules of zinc and incubation at 0° with frequent shaking (Kling[1]). The white crystalline precipitate is filtered, washed with ether, and dried *in vacuo* at room temperature. If further purification is required, the crystalline material can be distilled under nitrogen, and the distillate collected between 72° and 75° and treated again with zinc granules at 0° (Berl and Bueding[2]). The latter authors report a melting point at 97.5° to 98.5° for repeatedly distilled and crystallized material. Observed freezing-point depression in aqueous solution showed that the solute was monomeric.

Preparation of (−)Acetoin

Silverman and Werkman[3] and Juni[4] have described an enzyme system in *Aerobacter aerogenes* which catalyzes the conversion of pyruvate to (−)acetoin and carbon dioxide. Berl and Bueding[2] have employed an acetone powder of the organism for the preparation of (−)acetoin.

Growth of Organism and Preparation of Acetone Powder. Ten liters of a medium containing 1% peptone, 0.5% beef extract, 0.5% yeast extract, and 0.5% NaCl (pH adjusted to 7.7 with NaOH) is inoculated with 100 ml. of a 24-hour culture of *A. aerogenes* grown in the same medium. The 10 l. of culture is aerated at 37° for 18 hours, and the cells are harvested with a Sharples centrifuge. The cell paste is suspended in 50 vol. of cold (−18°) acetone for an hour with occasional shaking. The acetone is decanted, and the cells are dried by vacuum desiccation. Under cool and dry conditions, the powder retains activity for several months.

Preparation of (−)Acetoin. One hundred milligrams of acetone powder

[1] A. Kling, *Ann. chim. et phys.* **5**, 550 (1905).
[2] S. Berl and E. Bueding, *J. Biol. Chem.* **191**, 401 (1951).
[3] M. Silverman and C. H. Werkman, *J. Biol. Chem.* **138**, 35 (1941).
[4] E. Juni, *J. Biol. Chem.* **195**, 715 (1952).

is mixed with 50 ml. of M sodium phosphate buffer, pH 5.6. This suspension is mixed with 10 g. of sodium pyruvate and sufficient mixture of $5 \times 10^{-5} M$ $MnCl_2$ and 0.05 M sodium phosphate buffer (pH 5.6) to bring the total volume to 1000 ml. The mixture is incubated at 35° with constant shaking for 5 hours. The pH is adjusted hourly to approximately 5.6 with 4 N HCl. The proteins are precipitated with 800 ml. of 12.5% $ZnSO_4 \cdot 7H_2O$ (in 0.3 N H_2SO_4) and 235 ml. of 3 N NaOH (8 ml. of the zinc sulfate solution is neutralized by 2.35 ml. of the NaOH solution). The precipitate is removed by centrifugation at 0°, washed with water, and the ..ashings added to the supernatant portion. After adjustment to pH 7.2 the solution is distilled to dryness under a stream of nitrogen (32° to 35°; 30 to 33 mm. Hg). For concentration of the acetoin in the distillate, 50 g. of $NaHSO_3$ is added, and the solution is distilled to dryness under a stream of nitrogen as above. The residue is dissolved in 150 ml. of water, and solid $NaHCO_3$ is added until no further effervescence occurs. The pH is brought to 8.8 with 5 N NaOH. The solution is distilled under nitrogen as above. The distillate contains approximately 7 to 8 mg. of (−)acetoin per milliliter; $[\alpha]_D = -80°$.

Determination of Acetoin

Principle. Acetoin is oxidized under conditions of the test to diacetyl which reacts with the guanidino group of creatine to form a red color. The chemical reactions involved in the development of the red color are not fully understood. The test is based on the principle of the Voges-Proskauer[5] reaction and developed quantitatively by Westerfeld.[6]

Reagents

0.5% creatine.

5.0% α-naphthol in 2.5 N NaOH.

The α-naphthol is redistilled under nitrogen and the solution prepared *immediately* before using.

Procedure. To 5 ml. of solution containing 1 to 12 γ of acetoin or diacetyl are added consecutively 1 ml. of 0.5% creatine and 1 ml. of the 5% α-naphthol solution. The color is allowed to develop for 10 minutes for diacetyl and 1 hour for acetoin. The peak of absorption is between 520 and 540 mμ. Readings should be taken immediately after the times indicated above. A standard curve is prepared for both acetoin and diacetyl, since the color intensity on a molar basis is not exactly identical. Straight-line relationships are obtained between 0 and 12 γ.

Specificity. The test is nearly specific for acetoin and diacetyl. The related five- and six-carbon ketols interfere; however, their occurrence

[5] O. Voges and B. Proskauer, *Z. Hyg. u. Infektionskr.* **28**, 20 (1898).

[6] W. W. Westerfeld, *J. Biol. Chem.* **161**, 495 (1945).

in biological systems is uncommon. Ninety micrograms of methylglyoxal gives as much color as 1 γ of diacetyl. When glucose is heated in acid solutions (hydrolysis conditions), diacetyl is formed.

Application of Determination to Biological Systems. Unless interfering materials are present, acetoin may be determined on deproteinated filtrates. Most of the common deproteinating reagents may be employed, providing the filtrate is adjusted to approximately neutrality prior to addition of the assay reagents. The alkaline conditions employed in the test are important.

Oxidation of Acetoin to Diacetyl. When interfering substances are present in deproteinated biological solutions, acetoin is oxidized by $FeCl_3$-$FeSO_4$ reagent to diacetyl and the latter distilled. Aliquots of the distillate are employed for the determination, using the creatine and α-naphthol reagents.

Reagents

$FeSO_4 \cdot 7H_2O$ (solid).
50% $FeCl_3 \cdot 6H_2O$; 500 g. of $FeCl_3 \cdot 6H_2O$ dissolved in water and diluted to 1 l.
10 N H_2SO_4.

The oxidizing reagent is prepared by dissolving 10 g. of $FeSO_4 \cdot 7H_2O$ in 30 ml. of 50% $FeCl_3$. Two milliliters of this solution and 1 ml. of 10 N H_2SO_4 are added to 7 ml. of the acetoin-containing solution. The tube is stoppered lightly and placed in a boiling-water bath for 30 seconds. Then the stopper is inserted tightly, and the tube is heated for 30 minutes in the boiling-water bath. The tube is cooled, and the contents transferred to a distillation apparatus which will accommodate 10 ml. of liquid without any carryover of the material to the distillate. Five milliliters is directly distilled into a 10-ml. graduate cylinder, and the contents diluted, depending on the amount of diacetyl present. The color reaction described above is carried out with this solution. If the oxidizing mixture of $FeCl_3$-$FeSO_4$ is present when solutions containing glucose are heated in the presence of acid, glucose does not give rise to any large amount of diacetyl. Large amounts of pyruvic and lactic acids give rise to small amounts of diacetyl under the above oxidizing conditions; however, these quantities of pyruvic and lactic acid are rarely encountered in biological solutions.

II. Preparation and Determination of Diacetyl

Purification of Diacetyl

Diacetyl is available commercially and can be purified by distillation, collecting the fraction which boils at 86 to 87° Aqueous solutions may be prepared by oxidation of acetoin with the $FeCl_3$-$FeSO_4$ solutions as described in the above procedure for the determination of acetoin.

Determination of Diacetyl

Principle. This assay is based on the condensation of diacetylmonoxime with urea for the determination of the latter. For the determination of diacetyl the converse is employed; i.e., urea and hydroxylamine are reagents used for the determination of diacetyl.[7]

Reagents

Diacetyl standard, a solution of 100 mg. of diacetyl in 100 ml. of H_2O. This solution should be kept in the icebox when not in use and, owing to the volatility of diacetyl, should be freshly prepared about every 2 weeks. For assay this solution is diluted 1:10 to give a concentration of 100 γ/ml.

Urea solution, 3% solution in water.

Hydroxylamine, solution containing 10 mg./ml. in water.

Sulfuric-phosphoric acid mixture, 1 vol. of concentrated H_2SO_4 and 3 vol. of sirupy H_3PO_4.

Procedure. Various dilutions of a trichloroacetic or metaphosphoric acid filtrate of the tissue or bacterial juice are made in such a manner that 1 ml. contains approximately 100 γ or less of diacetyl. One milliliter of the dilution is then transferred to a test tube, and 1 ml. of the hydroxylamine solution, 1 ml. of urea solution, and 2 ml. of the H_2SO_4-H_3PO_4 mixture are added. The volume is then adjusted to 6 ml. with water, and the tube is rotated rapidly to mix. A diacetyl standard is run with concentrations of 100, 80, 60, 40, and 20 γ of diacetyl treated in the same manner as the filtrates above.

After mixing, the tubes are placed in a boiling-water bath for 35 to 40 minutes, cooled, and the optical density measured on a photometer with a 470-mμ filter. A reagent blank is used to set the instrument scale at zero. A standard curve is plotted from the known diacetyl readings, and the diacetyl concentration of the unknown is calculated by interpolation. Acetoin does not give a color with the reagents but may be determined with the method after oxidation and distillation with $FeCl_3$-$FeSO_4$ mixture described for the conversion of acetoin to diacetyl.

Prill and Hammer[8] have described a method for the determination of diacetyl based on the formation of the intensely colored ammonoferrous

[7] A. G. C. White, L. O. Krampitz, and C. H. Werkman, *Arch. Biochem.* **9,** 229 (1946).
[8] E. A. Prill and B. W. Hammer, *Iowa State Coll. J. Sci.* **12,** 385 (1938).

dimethylglyoximate. The method requires distillation; however, it has the advantage that few substances interfere.

III. Preparation and Determination of α-Acetolactic Acid

Preparation

Principle. α-Acetolactic acid is prepared by hydrolysis of methylacetoxyethylacetoacetate with NaOH.[9] The methylacetoxyethylacetoacetate is prepared by the oxidation with lead tetraacetate of the labile hydrogen of methyl substituted ethylacetoacetate.

$$2CH_3 \cdot CO \cdot C(CH_3)H \cdot COOEt + Pb(OAc)_4 \rightarrow$$
$$2CH_3 \cdot CO \cdot C(CH_3)(OAc)COOEt + Pb(OAc)_2$$
$$CH_3 \cdot CO \cdot C(CH_3)(OAc)COOEt + 2NaOH \rightarrow$$
$$CH_3 \cdot CO \cdot C(CH_3)(OH)COONa + CH_3COONa + C_2H_5OH$$

Reagents

Methyl-substituted ethylacetoacetate.
500 ml. of absolute ethyl alcohol.
23 g. (1 g. atom) of metallic sodium.
130 g. (1 mole) of ethylacetoacetate.
150 g. (∼1 mole) of methyl iodide
Absolute benzene (thiophene-free).
Three-necked 2-l. flask fitted with mechanical stirrer, reflux condenser, and dropping funnel.

Procedure

METHYL-SUBSTITUTED ETHYLACETOACETATE. A 2-l. three-necked flask is fitted with a mechanical stirrer, reflux condenser, and dropping funnel. Open ends of apparatus are fitted with $CaCl_2$ tubes for anhydrous conditions.

To 500 ml. of absolute alcohol, 23 g. of sodium is slowly added until complete solution is affected. Then 130 g. of ethylacetoacetate is added, and the mixture is brought to a gentle boil with stirring. The pH is alkaline. Then 150 g. of CH_3I is slowly added, with continued boiling and stirring until the reaction is neutral.

A fractionating column and condenser are fitted to the flask, and the alcohol is distilled off under reduced pressure. When NaI precipitates in large quantity, the remaining solution is decanted and the salt is washed with 100 ml. of benzene. The benzene is added to the solution containing the product. The benzene and remaining alcohol are distilled off under reduced pressure, and the methyl-substituted ethylacetoacetate, which boils at approximately 30° at 6 mm., is collected.

[9] L. O. Krampitz, *Arch. Biochem.* **17**, 81 (1948).

METHYLACETOXYETHYLACETOACETATE. With vigorous mechanical stirring under anhydrous conditions, 260 g. of Pb(OAc)₄ is added in small portions to a mixture of 280 g. of thiophene-free anhydrous benzene and 86.4 g. (0.6 mole) of methyl-substituted ethylacetoacetate. During the addition of the Pb(OAc)₄ the temperature is not allowed to rise above 35°, after which the mixture is heated to 40° for 5 hours with vigorous stirring. The mixture is allowed to stand an additional 24 hours at room temperature, filtered, and the precipitate washed five times with 100-ml. portions of benzene. The washings are combined with the original solution. To remove any acetic acid present in the benzene, it is washed with 100-ml. portions of water until the water is neutral to bromothymol blue. The benzene solution is dried over MgSO₄, filtered, and the benzene distilled off under reduced pressure. The remaining ester is distilled at a pressure of 1 to 10 mm. A fraction is obtained at 38° to 43° which contains mostly unreacted methyl-substituted ester. Another fraction (48 g.) is obtained at 79° to 83°. A fraction remains which is a highly polymerized yellow oil.

The fraction collected at 79° to 83° (1 to 10 mm.) is relatively pure methylacetoxyethylacetoacetate. It may be purified further by redistillation under reduced pressure.

This ester may be hydrolyzed to sodium α-acetolactate by 2 equivalents of NaOH when incubated at 25° for 40 minutes. Sodium acetate and ethyl alcohol, the other products of hydrolysis, are also present. For most metabolic experiments these products do not interfere. Separation of α-acetolactic acid from the hydrolytic mixture is difficult because of the lability of the acid.

Determination of α-Acetolactic Acid

Principle. α-Acetolactic acid is easily decarboxylated by a variety of methods to acetoin and carbon dioxide. Colormetric determination of acetoin can then be performed as described above, and/or the volume of CO_2 resulting from the decarboxylation can be measured.

$$CH_3 \cdot CO \cdot C(CH_3)OH \cdot COOH \rightarrow CH_3 \cdot CHOH \cdot CO \cdot CH_3 + CO_2$$

Colorimetric Determination. *Reagents.* Same as for acetoin.

Procedure. Since the reagents for the determination of acetoin (creatine-α-naphthol) do not react with α-acetolactic acid, the latter is decarboxylated to acetoin. The most convenient method is to acidify (Congo red) an aliquot of the deproteinated mixture containing 2 to 18 γ of α-acetolactic acid with 5 N H₂SO₄ followed by heating the solution for 15 minutes in a boiling-water bath. After neutralization of the solution, the acetoin is determined with the creatine-α-naphthol reagents as described in the section on determination of acetoin.

Volumetric Analysis of α-Acetolactic Acid. *Principle.* Reagents (aniline-citrate;[10] aluminum sulfate-potassium acid phthalate[11]) which decarboxylate β-keto acids may be used for the volumetric determination of the CO_2 formed during the decarboxylation of α-acetolactic acid.

Procedure. Follow the procedure outlined in Vol. III [67] for the determination of oxalacetic acid.

Enzymatic Determination. *Principle.* A crude preparation of α-acetolactic decarboxylase is easily prepared and can be used to determine α-acetolactic acid by measuring either the volume of CO_2 or the amount of acetoin formed. The enzyme decarboxylates the dextrorotary isomer at a rate about twenty times that for the levorotary isomer.

Procedure. The crude enzyme is prepared from cells of *Aerobacter aerogenes*. A complete description is given of this preparation by Juni in Vol. I [72]. Proceed through step 2. The preparation at this stage forms acetoin from pyruvate; however, this activity can be destroyed by heating the preparation at 70° for 2 minutes. The denatured proteins are removed by centrifugation, and the clear supernatant is used for the α-acetolactic decarboxylase. A neutralized solution containing the α-acetolactic acid (an amount which can be conveniently measured on Warburg apparatus) is placed in the main chamber of a Warburg vessel. Phosphate buffer, pH 6.0, final concentration 0.1 *M*, is added. Then 0.2 ml. of the heated *Aerobacter aerogenes* extract is added from the side arm after temperature equilibration. The enzyme may be stored for several months in a frozen state without losing appreciable activity.

[10] H. L. Edson, *Biochem. J.* **29**, 2082 (1935).
[11] H. A. Krebs and L. V. Eggleston, *Biochem. J.* **39**, 408 (1945).

[51] Colorimetric Determination of Acetoacetate

By ALBERT L. LEHNINGER

Principle.[1] Acetoacetate is decarboxylated in the presence of aniline hydrochloride catalyst. The acetone formed is then determined colorimetrically as the 2,4-dinitrophenylhydrazone.[2]

Procedure. A 3.0-ml. sample in 3% trichloroacetic acid, containing not more than 1.5 micromoles of acetoacetate, is incubated with 4.0 ml. of 4 *M* aniline hydrochloride, freshly filtered and extracted (see Notes),

[1] S. S. Barkulis and A. L. Lehninger, *J. Biol. Chem.* **190**, 339 (1951).
[2] L. A. Greenberg and D. Lester, *J. Biol. Chem.* **154**, 177 (1944).

and 1.0 ml. of 1 M acetate buffer (pH 5.0) in a 50-ml. glass-stoppered centrifuge tube at 30° for 90 minutes to decarboxylate the acetoacetate. Then 5.0 ml. of 0.1% 2,4-dinitrophenylhydrazine in 2 N HCl and 10 ml. of carbon tetrachloride are added. The tubes are stoppered tightly and shaken vigorously in a shaking machine. The aqueous layer is then removed as completely as possible by aspiration through a capillary siphon and discarded. The CCl_4 extract is then washed in the tube with two 20-ml. portions of H_2O and a 10-ml. portion of 0.5 N NaOH. Each time, the aqueous layer is removed as completely as possible by aspiration, by means of a capillary tube connected to a water aspirator.

After transfer of the CCl_4 layers to colorimeter tubes the optical density is read at 420 mμ against a reagent blank provided by carrying 3.0 ml. of 3% trichloroacetic acid through the entire procedure. For standards, acetone solutions made up in 3% trichloroacetic acid are used. At least two levels are employed: 0.5 micromole and 1.0 micromole per 3.0 ml. of 3% trichloroacetic acid. Such standard solutions may be kept in the cold for at least 2 weeks without change. The optical density is linear with concentration up to about 1.5 micromoles.

Notes on Method. (1) The aniline hydrochloride solution employed above must first be freed of reactive chromogens causing a high blank value by several extractions with equal volumes of chloroform followed by filtration through paper. The aniline hydrochloride solutions so prepared may be deeply colored, particularly if commercial aniline hydrochloride (Eastman) is employed. After extraction with $CHCl_3$, however, they will not contain materials capable of yielding a yellow color in the analysis. On standing at room temperature for a week or longer, the 4 M aniline hydrochloride may again acquire an appreciable blank. Reextraction with $CHCl_3$ and filtration will suffice to remove the chromogen.

2. The time of extraction of the hydrazone into CCl_4 should be determined for the particular shaking apparatus employed.

3. The method as described does not distinguish between acetoacetate and acetone. In addition it is evident that any neutral carbonyl compound capable of forming a hydrazone soluble in CCl_4 will also react, such as acetaldehyde and formaldehyde.

Acidic carbonyl compounds such as pyruvic, oxalacetic, and α-ketoglutaric acids do not give a color in the reaction in concentrations likely to be present in tissues or enzyme reaction media. Their hydrazones are extracted from CCl_4 solution by 0.5 N NaOH.

Other Procedures. Acetoacetate may also be determined manometrically, with aniline citrate catalyst;[3] however, this method requires much

[3] P. P. Cohen, *in* "Manometric Techniques" (Umbreit, Burris, and Stauffer, eds.), p. 175. Burgess Publishing Co., Minneapolis, 1949.

larger samples than the colorimetric procedure described above. Aceto-acetate may be also determined in microgram quantities colorimetrically[4] or fluorometrically.[5] In addition, many reports deal with determination of acetone, as the Denigès complex (cf. ref. 6), but this procedure is much more laborious and offers no advantages.

Determination of β-Hydroxybutyrate. None of the published methods appears to be completely satisfactory, since they depend on oxidation of β-hydroxybutyrate to acetone by dichromate, a reaction which is neither specific for β-hydroxybutyrate (cf. ref. 7) nor complete. The methods described by Weichselbaum and Somogyi[6] and Greenberg and Lester[2] appear to be most useful.

[4] P. G. Walker, *Biochem. J.* **58,** 699 (1954).
[5] G. Leonhardi and I. Glasenapp, *Z. physiol. Chem.* **286,** 145 (1951).
[6] T. E. Weichselbaum and M. Somogyi, *J. Biol. Chem.* **140,** 5 (1941).
[7] N. L. Edson and L. F. Leloir, *Biochem. J.* **30,** 2319 (1936).

[52] Determination of Formate

By HARLAND G. WOOD and HOWARD GEST

Separation

Distillation. Formic acid is usually separated from other components of a reaction mixture by steam distillation. Its pK_a is 3.67, and the solution to be distilled is adjusted to about pH 2.0 (blue to Congo red paper) with dilute sulfuric acid. During steam distillation at constant volume approximately 50% of the formic acid which is present is volatilized in each 2 vol. of distillate (see the table). Thus 50% is obtained in 100 ml. of distillate from 50 ml. or in 50 ml. of distillate from 25 ml., and in the succeeding 2 vol. approximately 25% more is recovered. In order to obtain substantially quantitative recovery approximately 12 vol. of distillate is required.[1] To recover the formic acid in a relatively small volume, the solution is distilled directly to a small volume and then an additional 12 vol. is collected by steam distillation at constant volume. If the solution contains sugar, heating under alkaline conditions must be avoided, since formic acid is a decomposition product of the sugar.

Chromatographic Separation. Formic acid may be separated from other volatile acids by chromatography on phosphate-Celite columns, but the recovery is not quantitative (see Vol. III [60]). Essentially quantita-

[1] Sixty-five per cent of acetic acid is volatilized in 2 vol. of distillate, and recovery is practically complete in 8 vol. of distillate.

tive recovery can be obtained, however, with sulfuric acid-Celite or water-Celite columns.[2,3]

STEAM DISTILLATION OF FORMIC ACID AT CONSTANT VOLUME FROM DIFFERENT VOLUMES OF SOLUTION

Ml. of solution distilled[a]	Ml. of distillate	Multiple	Recovery in fraction, %	Total recovery, %
50	100	2	52.6	52.6
	200	4	25.9	78.5
	300	6	12.2	90.7
	400	8	5.6	96.3
	600	12	3.5	99.8
	800	16	0.8	100.6
25	50	2	47.9	47.9
	100	4	26.6	74.5
	150	6	13.7	88.2
	200	8	6.6	94.8
	300	12	3.9	98.8
	400	16	0.6	99.4
10	25	2.5	52.0	52.0
	50	5.0	26.7	78.7
	75	7.5	11.8	90.5
	100	10	4.5	95.0
	125	12.5	1.4	96.4

[a] 0.252 N formic acid.

Macromethods of Determination

Formic acid is oxidized to CO_2 by mercuric ions, and this forms the basis of rather specific methods for its determination.

Oxidation with Mercuric Oxide.[4] A solution (100 to 500 ml.) containing 0.2 to 2 millimoles of the free acid is placed in a three-necked round-bottomed flask fitted with an aeration tube and a dropping funnel and connected to a reflux condenser which is attached to an anhydrone drying train. The solution is boiled and aerated for 10 minutes to free it of CO_2. It is then cooled, and 3 to 5 g. of mercuric oxide is quickly added to the flask. The CO_2 collector is attached to the drying train. CO_2-free air is

[2] E. F. Phares, E. H. Mosbach, F. W. Denison, Jr., and S. F. Carson, *Anal. Chem.* **24,** 660 (1952).

[3] M. H. Peterson and M. J. Johnson, *J. Biol. Chem.* **174,** 775 (1948).

[4] O. L. Osburn, H. G. Wood, and C. H. Werkman, *Ind. Eng. Chem. Anal. Ed.* **5,** 247 (1933).

passed through the apparatus while the mixture is gently boiled for 20 minutes. Ten milliliters of 25% phosphoric acid is then added through the funnel, and the boiling and aeration are continued for 15 minutes. The CO_2 is determined gravimetrically. Other acidic volatile compounds including pyruvic acid were not found to interfere.[4] Best success has been obtained when the formic acid is not neutralized before oxidation.

The method is very helpful in tracer studies, since the CO_2 may be collected in alkali and plated directly as $BaCO_3$. The other volatile acids, freed of formic acid, may then be recovered by acid distillation of the filtered residue of oxidation.

Oxidation with Mercuric Acetate.[5] This method is somewhat more convenient than the mercuric oxide method and is preferred if the other volatile acids are not to be recovered from the residue of oxidation. The equipment is the same as for the mercuric oxide oxidation.

Five milliliters of 1 N acetic acid is added, and the solution is boiled and aerated to remove dissolved CO_2. Then the CO_2 collector is attached, and 20 ml. of 0.3 M mercuric acetate is added slowly. The solution is boiled and aerated for 20 minutes, and the liberated CO_2 is determined gravimetrically.

Pyruvate has not been found to yield CO_2 by this method, although it is stated that it does interfere.[6]

Micromethods of Determination

Manometric Determination with Ceric Sulfate and Palladium.[6] Formic acid may be determined manometrically in the Warburg apparatus by measurement of the CO_2 produced on oxidation with ceric sulfate and palladium. Ceric acid is not as specific an oxidizing reagent as mercuric salts; interference due to other oxidizable compounds can be minimized, however, if the formic acid is first distilled, neutralized, and obtained as the sodium salt by evaporation to dryness. Ten micromoles of formic acid may be determined with ±3% error in the presence of 5 micromoles of lactic acid or 10 micromoles of pyruvic acid. The original article should be consulted for details.

Colorimetric Microdetermination of Formate.[7] The following method (a modification of a procedure suggested by Peabody[8]) is based on reduction of formate to formaldehyde with Mg as described by Grant,[9] followed

[5] H. D. Weike and B. Jacobs, *Ind. Eng. Chem. Anal. Ed.* **8,** 44 (1936).
[6] M. J. Pickett, H. L. Ley, and N. S. Zygmuntowicz, *J. Biol. Chem.* **156,** 303 (1944).
[7] The authors are indebted to Miss Marion A. Koser for making numerous tests of the procedure described.
[8] R. A. Peabody, private communication.
[9] W. M. Grant, *Anal. Chem.* **20,** 267 (1948).

by reaction of the formaldehyde with acetylacetone to form a colored complex under the conditions specified by Nash.[10]

REDUCTION. A 0.5-ml. sample containing not more than 1.1 micromoles of formate is placed in a 13 × 100-mm. test tube, and the tube is cooled in an ice bath. An 80-mg. coil of Mg (made by rolling a strip 10 cm. long × 3 mm. wide and stored in a desiccator containing NaOH pellets) is briefly dipped into 0.01 N HCl, rinsed in distilled water, shaken, and dropped into the sample. Ten drops of cold concentrated HCl (total volume, 0.5 ml.) are added, one drop at a time, at 1-minute intervals (the tube contents are mixed by brief agitation after each addition).

COLOR DEVELOPMENT. One minute after the last drop of HCl is added, 3 ml. of cold 1 N NaOH is introduced into the tube, and the contents are thoroughly mixed with a stirring rod. The tube is then centrifuged (room temperature) to sediment the precipitate of $Mg(OH)_2$. A 2-ml. aliquot of the supernatant liquid is added to 2 ml. of Nash's reagent B (2 ml. of redistilled acetylacetone + 3 ml. of glacial acetic acid + 150 g. of ammonium acetate made up to 1 l. with water). After mixing, the tube is incubated at 37° for 45 minutes and the sample read in a Klett-Summerson photometer with the 420-mμ filter (or in the Beckman spectrophotometer at 412 mμ) in the usual manner. When 0.5-ml. samples containing 0.3, 0.6, and 0.9 micromole of formate are treated as described above, typical Klett readings observed are 28, 56, and 84, respectively. The relationship between photometer readings and amount of formate is approximately linear over the range 0 to 1.1 micromoles, but for unknown reasons the slope of the curve shows some variation from day to day.

An alternative determination is based on color development with chromotropic acid as described by Grant.[9] In this method, the tube is removed from the ice bath one minute after the last addition of HCl (see Reduction), and 1.5 ml. of chromotropic acid reagent (0.6 g. of chromotropic acid in 20 ml. of water added to 180 ml. of concentrated H_2SO_4) is added. After mixing, the tube is placed in a boiling water bath for 30 minutes (with a marble on top) and then centrifuged to remove the white precipitate. The color of the supernatant liquid is measured in a colorimeter or a spectrophotometer at 570 mμ. The relationship between log galvanometer readings and amount of formic acid is reported to be linear over the range 0 to 10 γ.

Although Grant implies that the reduction procedure (see above) is not satisfactory when the 0.5-ml. sample contains more than 15 γ of formic acid (as tested in the chromotropic acid method), satisfactory

[10] T. Nash, *Biochem. J.* **55**, 416 (1953).

results have been obtained with as much as 1.1 to 1.2 micromoles of formate in the acetylacetone determination. In comparing the two alternative procedures it should be noted that the color complex developed with chromotropic acid is not as stable as that obtained with acetylacetone. With either method, blanks and HCOONa standards (in 0.01 N HCl) are carried through the procedures indicated.

It is reported that few substances other than formate yield formaldehyde on reduction by Mg under the conditions described. Since CO_2 may interfere in this respect, all solutions are acidified prior to analysis. It is evident that preformed formaldehyde and labile compounds or polymers of formaldehyde will interfere with the determination of formate. In connection with the chromotropic acid method, Grant[9] describes a procedure for removal of formaldehyde by reaction with phenylhydrazine. Both the chromotropic reaction and the relatively mild acetylacetone reaction appear to have a high degree of specificity for formaldehyde. Relatively large concentrations of certain substances (e.g., acetaldehyde), however, will interfere; papers describing the detailed testing of the color development reactions should be consulted in this regard.[10,11]

Biological Methods

Conversion to CO_2 and H_2. Resting cell suspensions of most strains of *Escherichia coli* and of certain other colon-aerogenes bacteria (obtained by anaerobic growth in media containing glucose and complex supplements such as yeast extract) decompose formate according to the following equation (for detailed discussion, see Gest[12]):

$$HCOOH \rightarrow CO_2 + H_2$$

Under an atmosphere of N_2 or He, the decomposition is complete and can be readily measured by conventional Warburg techniques. A survey of the available data indicates that best results are obtained with moderately concentrated, fresh cell suspensions in phosphate buffer of pH about 7 at a temperature of 37°. It should be noted that the pH showing optimal *rate* is influenced by formate concentration.[13] The minimum quantity of formate which can be estimated is determined by the inherent limitations of the Warburg apparatus. Since 2 moles of gas are produced per mole of formate, it appears that in ordinary practice the lower limit would be 0.5 to 1.0 micromole of formate. Possible complications which must be considered in applying this procedure are: (1) CO_2 and H_2 are also produced from pyruvate or compounds which the organism can con-

[11] D. A. MacFadyen, *J. Biol. Chem.* **158**, 107 (1945).
[12] H. Gest, *Bacteriol. Revs.* **18**, 43 (1954); see also Vol. I [88].
[13] M. Stephenson, *Ergeb. Enzymforsch.* **6**, 139 (1937).

vert to pyruvate, and a (2) number of compounds inhibit the reaction markedly—these include formaldehyde, nitrate, nitrite, cyanide, and fluoride. Fumarate (or compounds convertible to fumarate, such as malate or aspartate) may also interfere by acting as a hydrogen acceptor. These potential sources of difficulty can, of course, be minimized by preliminary steam distillation of the formate.

The enzyme system which catalyzes formate degradation to CO_2 and H_2 (hydrogenlyase) can be readily prepared in the form of a stable cell-free extract as follows.[14] *Escherichia coli* strain Crookes (American Type Culture Collection 8739) is grown in deep stationary culture at 37° in a medium of the composition 1% glucose, 1% yeast extract, 1% tryptone. After 10 to 12 hours of growth (cell yield-1.25 g./l.), the cells are harvested in the Sharples centrifuge and ground for several minutes in a large mortar with Alumina A 301 (62.5 g. per 25 g. of wet cell paste). The ground mixture is extracted with 1.5 ml. of cold water for each gram of wet cell paste, and the suspension is centrifuged for 10 minutes at 16,000 to 20,000 × g to remove alumina and residual intact cells. This step is followed by a 1-hour centrifugation (16,000 to 20,000 × g) which sediments most of the other cell debris. The viscous supernatant fluid is dispensed in tubes, which are flushed with N_2 or He, stoppered, and stored in the deep-freeze. Suitable conditions for the manometric estimation of formate (0.5 to 1.0 micromole or more) are as follows: 0.25 ml. of extract in 0.065 M phosphate buffer, pH 6, at 30° under an atmosphere of helium (final volume, 1.2 ml.). Both H_2 and CO_2 are also produced, at a relatively slow rate, from pyruvate under the foregoing conditions; smaller quantities of extract (e.g., approximately 0.05 ml. in a typical instance) in the same final volume, however, decompose formate rapidly but do not evolve gas from pyruvate. The activity of the enzyme system is subject to inhibition by most of the compounds already noted in connection with the assay using intact cells. In addition, inhibitions by DPN, TPN, ATP, and a variety of other compounds have been noted, particularly with the more dilute extracts. Although extensive tests have not been made, preliminary observations indicate that H_2 production by the extract is specific for formate under the conditions described.

Oxidation to CO_2 and H_2O. Formate is oxidized to CO_2 and H_2O by a DPN-linked dehydrogenase system in peas and beans, according to the reaction

$$HCOOH + \tfrac{1}{2}O_2 \rightarrow CO_2 + H_2O$$

The system has been partially purified by Mathews and Vennesland[15]

---

[14] H. Gest, "Phosphorus Metabolism" (McElroy and Glass, eds.), Vol. 2, p. 522, The Johns Hopkins Press, Baltimore, 1952.

[15] M. B. Mathews and B. Vennesland, *J. Biol. Chem.* **186,** 667 (1950); see Vol. I [87].

and by Davison.[16] Purification of approximately twentyfold was achieved by Davison by the following procedure.

Sifted bean meal (150 g.) was extracted with 450 ml. of 0.1 M Na$_2$-HPO$_4$ for 1 hour at room temperature with occasional stirring. The homogenate was squeezed through muslin, centrifuged for 15 minutes at 2800 r.p.m., and the supernatant, No. 1, treated with solid $(NH_4)_2SO_4$ to 0.8 saturation. The precipitate was filtered off on kieselguhr, and the cake rubbed up with 75 ml. of 0.03 M Na$_2$HPO$_4$. This suspension was dialyzed against running tap water overnight. The dialyzate was spun for 30 minutes at 3600 r.p.m., and a large amount of inert protein removed. The supernatant, No. 2, about 200 ml., was used for fractionation with calcium phosphate gel (30 mg. dry weight per milliliter). Calcium phosphate gel (20 ml. per 100 ml. of supernatant) was added to fraction 2, and the suspension mixed and allowed to stand at room temperature for 30 minutes. The gel was spun off, and the supernatant, No. 3, treated again with 20 ml. of calcium phosphate per 100 ml. of supernatant. The supernatant, No. 4, from this adsorption was then treated with 30 ml. of calcium phosphate per 100 ml. The supernatant, No. 5, after spinning, was treated a second time with 30 ml. of calcium phosphate per 100 ml. of supernatant. The supernatant, No. 6, from this adsorption was then treated with 40 ml. of calcium phosphate per 100 ml., the enzyme being brought down on the gel at this stage. The supernatant was discarded, and the gel eluted with three successive lots of 25 ml. of 0.1 M Na$_2$HPO$_4$ for 15 minutes each time. This preparation, No. 7, was neutralized to pH 7.0 with 0.3 M KH$_2$PO$_4$ and dialyzed against distilled water at 4° overnight. The dialyzed preparation, No. 8, may be stored at $-10°$ for several weeks without loss of activity or, alternatively, freeze-dried and the dry powder stored.

With 50 micromoles of formate, quantitative oxidation was observed according to the foregoing equation, under the following conditions: 1.8 ml. of enzyme preparation (27 mg. of freeze-dried fraction, No. 8); 0.6 ml. of 0.3 M phosphate buffer, pH 7.0; 0.3 ml. of CoI (1.5 mg.); 0.1 ml. of heart diaphorase, containing 1 γ of bound riboflavin; 0.1 ml. of 0.005 M methylene blue. Final volume, 3 ml.; temperature, 38°. With regard to specificity of the reaction, it may be noted that the purified system does not oxidize acetate, oxalate, lactate, pyruvate, formaldehyde, or acetaldehyde. Unfortunately, the range of concentration over which formate is completely oxidized has not been reported. The dehydrogenase is inhibited by cyanide, azide, Fe^{+++}, and Cu^{++}.

Intact cells of aerobically grown *Escherichia coli* and particulate preparations from such cells can oxidize formate quantitatively to CO_2

[16] D. C. Davison, *Biochem. J.* **49**, 520 (1951).

For example, intermediates isolated as barium salts which are not completely freed of barium before being tested in a phosphate medium; in test systems which contain Mn^{++} and phosphate, gradually developing turbidity can be encountered even at low concentrations of reagents.

Assay Method for Triose Phosphates

Principle. Determination of alkali-labile P and colorimetric procedures are used for the determination of triose phosphate (Vol. III [32]). If differentiation between glyceraldehyde-3-phosphate and dihydroxyacetone phosphate is required, more specific enzymatic tests can be used. The method described below is based on the reduction of DPN (measured at 340 mμ) in the presence of triose phosphates and the appropriate enzymes.

Reagents

0.06 *M* pyrophosphate buffer, pH 8.5.

0.15 *M* sodium arsenate.

1% solution of DPN.

1% solution of glyceraldehyde-3-phosphate dehydrogenase (Vol. I [60]), prepared from rabbit muscle in the presence of Versene[1] and recrystallized seven to eight times until free of triosephosphate isomerase and α-glycerophosphate dehydrogenase.

2% solution of triosephosphate isomerase.[2] As a source of this enzyme lyophilized preparations of step 4 of glyoxalase I (Vol. I [70]) can be used. Crystalline isomerase preparations from muscle (Vol. I [57]) should be even more suitable, provided they are free of α-glycerophosphate dehydrogenase and aldolase.

Procedure. Place 1.2 ml. of H_2O in a quartz cell having a 1-cm. light path, then add 1.5 ml. of pyrophosphate buffer, 0.1 ml. of arsenate, 0.1 ml. of DPN, and 0.05 ml. of unknown sample. Place the same reagents into the control cell, but substitute 0.2 ml. of H_2O for DPN and arsenate. After initial readings are taken, add 0.05 ml. of triosephosphate dehydrogenase to both cells and allow the reaction to go to completion. Take readings, and correct for initial absorption. Calculate glyceraldehyde-3-phosphate content of sample (0.1 micromole of DPNH gives a density reading of 0.210). To determine dihydroxyacetone phosphate, add 0.01 ml. of triosephosphate isomerase to both cells and again allow the reaction to go to completion. Correct for initial absorption, and calculate dihydroxyacetone phosphate content as above.

[1] I. Krimsky and E. Racker, *J. Biol. Chem.* **198**, 721 (1952).

[2] I. Krimsky and E. Racker, unpublished data.

Specificity. Although the enzyme reacts with glyceraldehyde and several other nonphosphorylated aldehydes,[3] the test is quite suitable for the determination of triose phosphates at low substrate concentration, because of the low affinity of the enzyme to nonphosphorylated substrates.

As an alternative method dihydroxyacetone phosphate and glyceraldehyde-3-phosphate can be measured by a combination of α-glycerophosphate dehydrogenase (Vol. I [58]) and triosephosphate isomerase (Vol. I [57]) in the presence of DPNH, but unless the enzymes are free of aldolase, fructose-1,6-diphosphate is included in the estimations.

Assay Method for Nonphosphorylated Aldehydes

Principle. Aldehydes can be assayed spectrophotometrically with DPN-linked enzymes measuring the absorption of DPNH at 340 mμ. Either (1) alcohol dehydrogenase and DPNH or (2) aldehyde dehydrogenase and DPN are used.

Reagents 1

1 M potassium phosphate buffer, 7.2.

0.003 M DPNH.

0.1% solution of alcohol dehydrogenase, in 0.1% bovine serum albumin and 0.01 M potassium phosphate, pH 7.2 (Vol. I [79]).

Procedure 1. Place 2.75 ml. of distilled water, 0.05 ml. of buffer, 0.1 ml. of DPNH, and 0.05 ml. of alcohol dehydrogenase in a quartz cell having a 1-cm. light path. Take initial readings, and add 0.05 ml. of aldehyde sample to the experimental quartz cell as well as to the control cell which contains all the reagents except DPNH. When reduction of the aldehyde in the sample (which should contain between 0.02 and 0.2 micromole of acetaldehyde) is completed, as indicated by constancy of density readings, subtract the final density value from the initial reading, after correcting for dilution of DPNH absorption due to addition of the sample. (0.1 micromole of acetaldehyde results in a density change of 0.21.)

Reagents 2

0.1 M pyrophosphate buffer, pH 9.3.

0.01 M DPN.

Aldehyde dehydrogenase solution containing 2000 to 5000 units/ml. (Vol. I [83]).

Procedure 2. Place 2.5 ml. of distilled water, 0.3 ml. of pyrophosphate buffer, 0.05 ml. of DPN, and 0.1 ml. of aldehyde dehydrogenase in a quartz cell having a 1 cm.-light path. The control cells contain the same

[3] E. Racker and I. Krimsky, *J. Biol. Chem.* **198**, 731 (1952).

solution, except for DPN. After addition of the sample, which may contain between 0.02 and 0.4 micromole of acetaldehyde, density readings are taken until the reaction ceases. Calculations are made as described for the other tests.

Specificity. Since both alcohol dehydrogenase and aldehyde dehydrogenase act on several aldehydes, the tests lack specificity. This is true particularly of the aldehyde dehydrogenase, which reacts with a large variety of aliphatic and aromatic aldehydes. On the other hand, alcohol dehydrogenase has a narrower range of substrates (cf. Vol. I [79]) when used at low enzyme concentrations.

Assay Method for Ketoaldehydes

Principle.[4] A specific test for methylglyoxal and other ketoaldehydes is available owing to the fact that glyoxalase I catalyzes the condensation between ketoaldehydes and glutathione, resulting in the formation of thiol esters which can be determined spectrophotometrically at 240 mμ or colorimetrically with hydroxylamine.[5]

Reagents

1 M potassium phosphate buffer, pH 6.6.
2% solution of neutralized glutathione.
0.5% solution of glyoxalase I (cf. Vol. I [70]).

Procedure. Place 2.7 ml. of distilled water in a quartz cell having a 1-cm. light path, and then add 0.1 ml. of buffer, 0.05 ml. of glutathione, and 0.1 ml. of glyoxalase I. The control cell contains the same reagents, except the enzyme, in a final volume of 2.95 ml. Take initial readings at 240 mμ, and then add 0.05 ml. of sample (containing 0.05 to 1 micromole of methylglyoxal) to both cells. On completion of the reaction, take final readings. After corrections due to initial readings and dilution by the sample have been made, the amount of methylglyoxal can be calculated. (1 micromole of methylglyoxal gives a density reading of 1.130.)

Specificity. This method is specific for ketoaldehydes. No other compound has been found to date to react with glyoxalase I. Owing to its own pronounced absorption of 240 mμ, phenylglyoxal is more conveniently measured by the hydroxylamine reaction[5] after being completely transformed into the thiol ester by glyoxalase I.

[4] E. Racker, *J. Biol. Chem.* **190**, 685 (1951).
[5] F. Lipmann and L. C. Tuttle, *J. Biol. Chem.* **159**, 21 (1945); see also Vol. III [39].

Section II

Lipids and Steroids

[55] General Procedures for Separating Lipid Components of Tissue

By CECIL ENTENMAN

For the rapid and complete removal of lipids from tissues, three things must occur: (1) The tissue must be subdivided under conditions that do not favor breakdown of the lipids. (2) The solvent used must be capable of penetrating the divided tissue and breaking the protein-lipid bond. (3) The tissue must be washed completely free of the lipid by repeated treatment with lipid-free solvent. Some of the difficulties involved in meeting these conditions have been discussed by Bloor.[1] The techniques of tissue preparation and the techniques of solvent extraction are so interrelated that it is extremely difficult to discuss them separately. Consequently, in the present section, a number of *complete procedures* are given whereby lipids may be separated from tissues. However, it should be realized that other combinations may be just as useful. The choice of one procedure over another is very greatly dependent on the equipment present in the laboratory, the experimental design, and the nature of the determination to be made on the extracts obtained. From the variety of procedures presented it should not be difficult to select one which meets most requirements.

Solvents

Only high-quality solvents should be used in the separation of lipids from tissues. Commercially available solvents are adequate for most purposes. If desired, the solvents can be purified by redistillation over the appropriate chemical.[2] If gravimetric or oxidative methods of analysis are to be used, it is advisable to purify all solvents.

Ethyl Ether. Only peroxide-free ethyl ether should be used. Ether prepared for anesthetic purposes and stored in tin-lined cans is ready for use, but after the container has been opened peroxides will form in a few days.

Petroleum Ether. Petroleum ether with a boiling point from 30 to 68° should be used. For many lipid determinations it is not necessary to purify the solvent. If oxidative or gravimetric determinations are to be used, however, it is advisable to allow the petroleum ether to stand for a week or more over concentrated sulfuric acid, reflux the petroleum ether

[1] W. R. Bloor, "Biochemistry of the Fatty Acids," pp. 37–46, Reinhold Publishing Corp., New York, 1943.

[2] I. L. Chaikoff and A. Kaplan, *J. Biol. Chem.* **106,** 267 (1934).

over KOH pellets, and distill slowly.[2] Petroleum ether prepared in this manner will not leave an oxidizable residue when a portion is evaporated. Merck's petroleum ether "for fat determination" does not require further purification.

Acetone. Anhydrous acetone of high purity can be obtained commercially. Water, if present for any reason, can be removed by allowing the acetone to stand over calcium chloride and then distilling.

Ethyl Alcohol. Only 95 to 100% ethyl alcohol should be used. If lipids are to be determined by gravimetric or oxidative techniques, the alcohol should be refluxed over KOH and then distilled.

Chloroform. Analytical reagent-grade.

Methyl Alcohol. Analytical reagent-grade.

Special Treatment for Certain Tissues

Cholesterol and fatty acids can be successfully removed from most tissues by the procedures described below. For the complete removal of *all* lipids, however, certain procedures have advantages over others. For some tissues, certain treatments are essential. The following is an attempt to recommend the best method for a particular tissue and to mention additional pertinent facts.

Whole Blood, Serum, or Plasma.[3] The hot alcohol-ether extraction is recommended (procedure 12).

Red Blood Cells.[4] Washed red blood cells are hemolyzed by addition of an equal volume of water. Lipids are then extracted with *cold* alcohol-ether as described in procedure 12.

Brain. Lipids in this tissue can be removed by treatment with hot alcohol-ether and then extracted in the Soxhlet apparatus (procedure 7) with chloroform instead of ethyl ether. Changus *et al.*[5] used ethyl ether and petroleum ether, but the author considers chloroform to be the better solvent for lipids of this tissue. The best method for this tissue, however, is described in procedure 10, which is a modification of the techniques described by Folch *et al.*[6] and Sperry.[7,8]

Skin. For large samples, snip the tissue into fine bits with scissors and extract by procedure 7. For small samples, procedure 5 has been found to be satisfactory. It is essential to keep the temperature at 0° at all times

[3] E. M. Boyd, *J. Biol. Chem.* **114,** 223 (1936).

[4] E. M. Boyd, *J. Biol. Chem.* **115,** 37 (1936).

[5] G. W. Changus, I. L. Chaikoff, and S. Ruben, *J. Biol. Chem.* **126,** 493 (1938).

[6] J. Folch, I. Ascoli, M. Lees, J. A. Meath, and F. N. LeBaron, *J. Biol. Chem.* **191,** 833 (1951).

[7] W. M. Sperry, *J. Biol. Chem.* **209,** 377 (1954).

[8] W. M. Sperry, "Methods of Biochemical Analysis," pp. 83–111, Interscience Publishers, New York, 1955.

during the homogenization and centrifugation in the presence of tri-chloroacetic acid; otherwise lipid will be lost in the supernatant acid.[9]

Muscle. Procedure 6 is recommended, although if the tissue is finely divided most of the other methods are satisfactory.

Artery. Procedure 6 is recommended. Procedure 5 can be used.[10]

Carcass. Rat[11] and mouse carcass lipids can be extracted as described in procedure 7.

Bone. Bone lipids can be extracted by procedure 6, or if the bones are cracked open with bone shears or scissors the lipids can be removed conveniently as described in procedure 7.

Lymph. The hot alcohol-ether method (procedure 12) should be used, extracting twice with hot alcohol-ether instead of once. Lipids are redissolved with ethyl ether instead of petroleum ether from a concentrate of the alcohol-ether extract.

Procedure 1. *Tissue Preparation:* Homogenization in Alcohol-Ether.
Lipid Extraction: Homogenizing Tubes.

Comment. This procedure is useful when it is desired to obtain all the lipids from relatively small amounts of tissue and quantitatively to study certain constituents in the residue remaining. The procedure is not adaptable to tissues difficult to disintegrate because the heat generated tends to boil off the solvent.

Tissue Preparation. Up to 1 g. of fresh tissue is placed in a Tenbroeck tissue-grinding tube of about 7-ml. capacity. Five milliliters of alcohol-ether (3:1) is added, and the tissue is completely homogenized. This can be done easily and safely with a mechanical apparatus commercially available.[12] The pestle is removed from the tube and washed down with alcohol-ether.

Lipid Extraction. The tube with contents is heated in a tube heater[13] or water bath for 1 hour at 60°. The tube is removed from the heater, centrifuged for 10 minutes at 2000 r.p.m., and the alcohol-ether extract

---

[9] W. Lee, A. K. Davis, C. Entenman, and G. E. Sheline, *Arch. Biochem. and Biophys.* **54**, 146 (1955).

[10] S. Chernick, P. A. Srere, and I. L. Chaikoff, *J. Biol. Chem.* **179**, 113 (1949).

[11] E. M. Boyd, M. L. Connell, and H. D. McEwen, *Can. J. Med. Sci.* **30**, 471 (1952).

[12] This apparatus consists of a cone-driven motor which turns a pestle with a ground-glass or Teflon surface while the homogenizing tube is mechanically raised and lowered. It is useful in reducing tissues to a fine state of subdivision, especially such tissues as skin or muscle which are hard to homogenize by other means (Hallikainen Instruments, Berkeley, California).

[13] An electrically operated, thermostatically controlled heating unit containing holes into which tubes can be placed for extraction processes requiring controlled heat (Hallikainen Instruments, Berkeley, California).

decanted into a side-arm Erlenmeyer flask.[14] The tissue residue is treated in a similar fashion with a second portion of alcohol-ether, and the extract added to the side-arm flask. The residue is then washed with two to three portions of ethyl ether, and after centrifugation the ether extracts are combined with the alcohol-ether extracts. The combined extracts are concentrated almost to dryness, the last 10 to 20 ml. being removed under an atmosphere of nitrogen or carbon dioxide. Petroleum ether is added to the side-arm flask, heated to boiling on a steam bath or hot plate, mixed thoroughly with a swirling motion,[15] and after cooling decanted into a volumetric flask. The aqueous phase, which is caught in the side arm of the Erlenmeyer flask, is extracted with two to three additional portions of petroleum ether in a similar manner. The combined petroleum ether extracts contain all the lipids in the tissue.

Procedure 2. *Tissue Preparation:* Tissue Disintegrator.
                Lipid Extraction: Centrifuge Tubes.

Comment. The tissue is placed in a 50-ml. round-bottomed centrifuge tube, finely divided with a tissue disintegrator[16] in the presence of alcohol-ether, and all the lipids are extracted. This procedure is rapid and quantitative, since time is not lost in the transfer of tissue from the disintegrating apparatus to an extraction vessel; the tubes may be conveniently handled in large numbers in a tube heater and in the centrifuge; losses of tissue are not incurred; and excess solvent is not used in transferring the tissue to a separate flask. Furthermore, the sample may be placed in the tube, frozen with liquid air, and stored in a freezer until needed; or it can be disintegrated first in the presence of alcohol[17] and then stored until needed.

---

[14] Useful in extracting the lipids present in a small amount of one liquid phase by means of larger amounts of an immiscible, volatile solvent of lower specific gravity. The vessel consists of an Erlenmeyer flask with a side arm made by sealing a glass tube into the flask at an angle of 45° near the top of the flask. The capacity of the sidearm is approximately 1.5 ml. for a 125-ml. flask and about 5 ml. for a 250-ml. flask [C. Entenman, A. Taurog, and I. L. Chaikoff, *J. Biol. Chem.* **155**, 13 (1944)].

[15] Conveniently done with a rotary shaker which swirls the liquid in the flask rapidly and safely (New Brunswick Scientific Co., New Brunswick, New Jersey).

[16] Consists of a device for holding a tube firmly and a motor-driven shaft to the end of which are attached cutting blades. When used in conjunction with a centrifuge, both the subdivision of the tissue and the extraction of the lipids can be carried out in a single 50-ml. centrifuge tube (Hallikainen Instruments, Berkeley, California).

[17] If the tissue is not to be extracted immediately, it should be disintegrated in alcohol instead of alcohol-ether. Then, at the time of extraction, ether is added to make the alcohol-ether ratio 3:1. For best results the tissue should be disintegrated as soon as possible after placing it in the alcohol-ether. Otherwise it may harden to such an extent that fine subdivision is not possible by this method.

Tissue Preparation. Up to 4 g. of tissue is placed in a 50-ml. round-bottomed centrifuge tube. Approximately 3 vol. of alcohol-ether[17] is added to the tube, and the tissue is finely divided by means of a tissue disintegrator.[16] The cutting blades and shaft holder are washed down with a little alcohol-ether, and the tube is removed from the apparatus.

Lipid Extraction. More solvent is added to make the ratio of alcohol-ether to tissue approximately 10:1, and the tube is placed in a tube heater[13] or water bath for 1 hour at 60°. The tube is removed from the tube heater, centrifuged, and the supernatant decanted into a 125-ml. Erlenmeyer flask with a side arm.[14] The residue is extracted with a second portion of hot alcohol-ether as above, and finally the residue is washed three times with 10-ml. portions of ethyl ether.[18] All extracts are combined in the Erlenmeyer side-arm flask. The flask is heated on a steam bath or hot plate until 1 ml. or less of fluid remains. Fifteen to twenty milliliters of petroleum ether is added and brought to boiling, and the flask is thoroughly shaken with a swirling motion.[15] The petroleum ether is decanted into an Erlenmeyer flask. The aqueous phase is extracted two to three more times with hot petroleum ether, all extracts being combined in the Erlenmeyer flask. The petroleum ether is evaporated to the correct volume for quantitative transfer to a volumetric flask of the desired size.

Procedure 3. *Tissue Preparation:* Grind with Sodium Sulfate.
Lipid Extraction: Chloroform (Soxhlet).[19]

Comment. This procedure is applicable to soft tissues such as the liver. The sodium sulfate is used to dehydrate the tissue in order that the chloroform may then dissolve the lipids.[20]

Tissue Preparation. A weighed portion of tissue is placed in a dry mortar and ground thoroughly with approximately 8 g. of anhydrous sodium sulfate per gram of tissue.

Lipid Extraction. The powder is transferred to a filter paper, placed in a Soxhlet apparatus,[21] and extracted with chloroform for 48 hours. The

---

[18] If the tissue is not very finely divided, it should be transferred to a filter paper or a sintered-glass filter after the second hot alcohol-ether extraction and the residue extracted with ethyl ether in a Soxhlet apparatus for 6 hours or longer.

[19] R. Weil and D. Stetten, *J. Biol. Chem.* **168,** 129 (1947).

[20] If only a few small tissue samples are to be extracted, this procedure has some advantage in that the equipment is simple and the operation easy. For larger samples and for rapid analysis, however, the grinding with sodium sulfate is very tedious and the total preparation is extremely bulky. One is cautioned against using the chloroform extract without further purification, by redissolving the lipids in petroleum ether or by adding methanol and purifying by procedure 10, since considerable nonlipid materials are picked up from the tissue by chloroform.

[21] A screw-cap vial has been used instead of the Soxhlet apparatus for the chloroform

chloroform extract is evaporated to dryness under an atmosphere of nitrogen, and the lipid re-extracted with hot petroleum ether.[22] The extract contains all the tissue lipids.

Procedure 4. *Tissue Preparation:* Waring Blendor.
          Extraction Solvent: Alcohol-Ether (3:1) with and without Soxhlet.[23]

Comment. The Waring blendor and similar apparatus offer a convenient means of reducing tissue to a fine state of subdivision after which lipids can be removed. If the blendors of this general design are used, however, the blended mixture has to be transferred to another vessel for extraction, a procedure which frequently requires an excess of solvent in order to attain a quantitative transfer. In addition large pieces of tissue frequently remain even after blending for some time.

Tissue Preparation. The fresh tissue[24] is placed in the container of the Waring blendor (the microcontainer is used if the tissue sample is 15 g. or less). Approximately 3 ml. of alcohol is added for each gram of tissue. The container is covered, and the apparatus turned on for 3 minutes or longer in order to get fine subdivision. The contents of the vessel are transferred quantitatively to an Erlenmeyer flask using alcohol in the transfer.

Lipid Extraction. Sufficient alcohol and ether are then added to make the alcohol—ether ratio 3:1 and the solvent mixture—tissue ratio at least 10:1. A glass bubble stopper is placed on the flask to retard the loss of solvent, and the flask is heated in a water bath at 60° for 6 to 18 hours. (For liver, 6 hours is sufficient.[25]) The extract is decanted through fat-

---

extraction [R. W. Payne, *Endocrinology* **45**, 305 (1949)]. The powdered preparation is transferred from the mortar to a clean, dry, 50-ml. screw-cap vial, the cap of which is lined with tinfoil and seated in cork. Thirty milliliters of chloroform are added, and the screw cap sealed (0.5 g. of liver is used with 4 g. of sodium sulfate). The tube is turned end over end at 50 to 60 r.p.m. for 16 hours by means of a mechanical device or allowed to stand for 72 hours with intermittent shaking by hand. It is essential that the preparation be free of water during the extraction with chloroform. Excess water can be detected by placing a pinch of anhydrous copper sulfate in the solution. A change of color to blue indicates water contamination.

[22] D. Koch-Weser, P. B. Szanto, E. Farber, and H. Popper, *J. Lab. Clin. Med.* **36**, 694 (1950).

[23] D. Jensen, I. L. Chaikoff, and H. Tarver, *J. Biol. Chem.* **192**, 395 (1951).

[24] With large tissues it is sometimes advantageous to blend the tissue thoroughly in a large container, then transfer a weighed portion to the microcontainer, add solvent, and blend thoroughly again.

[25] C. B. Shaffer and F. H. Critchfield [*J. Biol. Chem.* **174**, 489 (1948)], after grinding rat liver in a Waring blendor, extracted it with at least 20 ml. of alcohol-ether (3:1) per gram of tissue *without the use of heat.* The extraction was carried out in a volu-

free filter paper into an Erlenmeyer flask with a side arm[14] or into a Kjeldahl concentrating flask with a side arm.[26] The tissue is then also transferred to the filter paper, and the residue, after draining, is washed with six portions of ethyl ether.[27] Sufficient time is allowed to elapse between successive washes such that draining is complete each time. The volume of the alcohol-ether extract is reduced to about 1 ml. by heating the flask on a steam bath or the Kjeldahl flask in a water bath at 60° under reduced pressure, the last 20 to 100 ml. of solvent being removed under an atmosphere of nitrogen or carbon dioxide. The lipids are extracted with three to four portions of hot petroleum ether, the aqueous phase being caught each time in the side arm of the flask. The volume of petroleum ether is then reduced by evaporation and transferred to a volumetric flask of appropriate size.

Procedure 5. *Tissue Preparation:* Homogenize in Trichloroacetic Acid. *Extraction Solvent:* Alcohol-Ether.

Comment. This procedure is useful where it is desired to remove the acid-soluble components (as contaminants in tracer experiments or for study) and to determine the lipid components on the same sample. The procedure is of particular value in the determination of lipids from such tissues as skin[9] or artery[10] where it is so difficult to disintegrate the tissues that the heat generated during the process volatilizes solvents such as alcohol.

Tissue Preparation. A weighed portion of tissue (up to 1 g.) is placed in a 7-ml. all-glass homogenizing tube (Tenbroeck tissue grinder), and to the tube are added 5 ml. of 10% ice-cold trichloroacetic acid (TCA). The tissue is ground to a fine pulp by mechanical means.[12] The entire operation is carried out at 0° by placing ice in the container around the

---

metric flask, and, after several hours, aliquots were removed for analysis. It has been the author's experience that the lipids of liver, after treatment in the Waring blendor, are not all consistently removed unless heat is used. The Waring blendor frequently does not reduce the liver to particles sufficiently fine for cold solvent to leach out the lipid. If heat is used, however, the techniques described by Shaffer and Critchfield are applicable to a number of tissues in addition to the liver.

[26] A Kjeldahl flask with a ground-glass joint and a side arm (as described above) about 1 inch below the joint. The ground-glass cap for the flask has a capillary tube through which nitrogen or carbon dioxide can pass into the flask, and a second tubing of larger bore which can be connected to a vacuum line. The capillary tube extends to within a short distance of the bottom of the flask.

[27] If desired, the ether washes may be omitted and the tissue extracted for 6 to 24 hours with ethyl ether in a Soxhlet apparatus. This should be done, in most instances, where the tissue has been stored in alcohol for some time or if the treatment in the blendor did not produce finely subdivided particles.

homogenizing tube while the tissue is being ground. The material is then centrifuged[28] at 0°, and the supernatant containing acid-soluble components is decanted. The residue is washed three times with 5-ml. portions of TCA at 0° in order to remove completely the acid-soluble components.

Lipid Extraction. The residue is then extracted three to five times with alcohol-ether (3:1)[29] and twice with ether. The alcohol-ether extractions are carried out at 60° for 1 hour. All extracts are decanted, after centrifugation, into an Erlenmeyer flask with a side arm.[14] The extracts are concentrated to a small volume on a steam bath in an atmosphere of nitrogen, and the lipids redissolved in warm petroleum ether (for details see Procedure 2).

Procedure 6. *Tissue Preparation:* Freeze and Pulverize.

Extraction Solvent: Alcohol-Ether, then Chloroform (Soxhlet).

Alcohol-Ether, then Ether (Soxhlet).

Methanol-Chloroform (Soxhlet).

Comment. By freezing and pulverizing it is possible to get fairly representative finely subdivided samples from large and tough tissues or even from the whole bodies of small animals from which lipids may be extracted. In addition, changes in lipids that might occur during the sampling of tissue[30] may be prevented by such treatment.

Tissue Preparation. Tissues can be frozen and pulverized for extraction by one of several techniques. McKibbin and Taylor[31] froze tissue with carbon dioxide snow and then ground it with dry ice in a mortar. Graeser *et al.*,[32] using a dry ice-cooled tissue crusher, pulverized 1 to 10 g. of tissue frozen with liquid air. Entenman,[33] after freezing large tissue samples or a whole rat in liquid air, finely pulverized them in a steel

---

[28] By mixing a filtering aid (Celite) with the homogenized liver pulp, C. A. Olmsted [Master's thesis, University of California, Berkeley, 1954] was able to remove the acid-soluble components by using a sintered-glass funnel and washing with cold trichloroacetic acid (this procedure will not work satisfactorily for fatty tissues unless done at 0°). The residue on the funnel, after drying with alcohol, was extracted for 18 hours with ether in a Soxhlet apparatus. The combined alcohol-ether extracts contained all the tissue lipids.

[29] C. Artom and M. A. Swanson [*J. Biol. Chem.* **193,** 473 (1951)] dehydrated the residue by allowing it to stand overnight in acetone and then extracted the remaining lipids with boiling ethanol and ethanol-ether (2:1). The extracts obtained were used for the determination of total lipid and lipid phosphorus.

[30] D. Fairbairn, *J. Biol. Chem.* **157,** 645 (1945).

[31] J. M. McKibbin, and W. E. Taylor, *J. Biol. Chem.* **178,** 17 (1949).

[32] J. B. Graeser, J. E. Ginsberg, and T. E. Friedmann, *J. Biol. Chem.* **104,** 149 (1934).

[33] C. Entenman, unpublished observations.

mortar, and Teague et al.[34] dried tissues from the frozen state under a high vacuum and then powdered them in a ball mill.

Lipid Extraction. When the tissue has been pulverized by one of the above-mentioned methods, the lipids may be extracted with alcohol-ether according to procedures described by Kaucher et al.[35] or after the manner described in procedures 7 and 8. Methanol-chloroform (1:1) used in a Soxhlet apparatus (18 hours) is also effective, especially if the tissue is prepared by the method of Teague et al.[34] McKibbin and Taylor[31] refluxed the pulverized tissue with alcohol-ether (3:1), filtered the extract through a sintered filter, dried the residue at 50°, and extracted it for 6 hours with chloroform in a Soxhlet apparatus. Tests showed that only negligible amounts of lipid remained in the residue after such treatment.

Procedure 7. *Tissue Preparation:* Meat Grinder.
          Extraction Solvent: Alcohol-Ether, then Ether (Soxhlet).[36,37]

Comment. This procedure is particularly useful in the extraction of all lipids from large tissue samples and from the whole bodies of small animals. In many instances it is desirable to obtain a representative sampling of a large tissue, and this aim can be achieved by grinding the tissue several times in a meat grinder of the shearing type.

Tissue Preparation. The fresh tissue is ground three times in a shearing-type meat grinder, and all the tissue or a weighed portion thereof is transferred to an Erlenmeyer flask (widemouthed flasks for large samples). If all the tissue is to be extracted, the tissue adhering to the grinder should be transferred to the extraction flask by means of a spatula and alcohol from a washing bottle. Five to ten milliliters of alcohol-ether mixture (3:1) per gram of tissue is added to the flask, and the tissue and solvent are thoroughly mixed as quickly as possible in order to prevent clumping. This can be done by stoppering the flask tightly with a tinfoil-covered cork or stopper and shaking the flask violently. The tissue adhering to the sides of the flask can be removed with a steel spatula.

Lipid Extraction. The flask is placed in a water bath for 2 hours or more at 60°. The mixture is agitated frequently by means of a stirring rod. While still warm, the alcohol-ether extract is decanted through fat-free filter paper into a Kjeldahl concentrating flask[26] or an Erlenmeyer

[34] D. M. Teague, H. Galbraith, F. C. Hummel, H. H. Williams, and I. G. Macy, *J. Lab. Clin. Med.* **28,** 343 (1942).
[35] M. Kaucher, H. Galbraith, V. Button, and H. H. Williams, *Arch. Biochem.* **3,** 203 (1943).
[36] A. Kaplan and I. L. Chaikoff, *J. Biol. Chem.* **108,** 201 (1935).
[37] F. W. Lorenz, C. Entenman, and I. L. Chaikoff, *J. Biol. Chem.* **123,** 527 (1938).

flask with sidearm.[14] The residue is extracted a second time for 2 hours[38] at 60° with 5 ml. of alcohol-ether (3:1) per gram of fresh tissue, and the extract decanted through the same filter paper. The residue is then transferred to the filter paper,[39] using ethyl ether and a spatula to make the transfer quantitative. The residue is washed thoroughly with ethyl ether, allowed to drain, transferred to the Soxhlet chamber, and extracted with ethyl ether for 6 to 24 hours. After combining the ether extract with the alcohol-ether extract the combined mixture may be made to a convenient volume for lipid analysis or it may be reduced to a low volume and the lipids extracted with ethyl or petroleum ether (see procedure 2 or 4 for details).

Procedure 8. *Tissue Preparation:* Grind with Sand.
         Extraction Solvent: Alcohol-Ether (Soxhlet).[40]

Comment. This procedure is useful in the separation of lipids from either fresh tissue or from tissue that has hardened by long standing in a dehydrating solvent. Grinding with sand is an effective though tedious means of subdividing a tissue sample so that the lipids can be completely extracted. The method is widely used, since the apparatus needed is available in most laboratories.

Tissue Preparation. Fresh tissue or tissue that has been stored in alcohol is transferred to a mortar, and a small amount of lipid-free fine sand is added.[41] When the tissue has been thoroughly ground, the alcohol in which the tissue had been stored is poured into the mortar and the resulting mixture is transferred to the extraction flask through a wide-

---

[38] For small samples of finely divided tissue a second extraction of 1 hour is sufficient.

[39] If a large portion of tissue is being extracted, the tissue residue should not be transferred to the filter paper, but should be covered with ethyl ether and allowed to stand for 18 to 24 hours. A piece of fat-free cotton is placed over the siphon outlet of a large Soxhlet chamber, the residue is transferred directly to the Soxhlet chamber, the filter paper from the alcohol-ether decantations is folded and placed on top, and all are covered with another piece of fat-free cotton. After continuous extraction for 18 to 24 hours, the ether extract is combined with the previously obtained alcohol-ether extracts in the Kjeldahl concentrating flask (the volume should be reduced before the ether extract is added). The flask is placed in a water bath at 60°, and the extracts concentrated under reduced pressure in an atmosphere of nitrogen until it is apparent that most of the remaining liquid is water as evidenced by beads of moisture on the neck of the flask. The concentrate is extracted five or six times with large volumes of ethyl ether, and the volume of the combined ether extracts reduced by evaporation and transferred to a volumetric flask of appropriate size. Ethyl ether, not petroleum ether, is recommended when the aqueous phase is large.

[40] A. Taurog, C. Entenman, B. A. Fries, and I. L. Chaikoff, *J. Biol. Chem.* **155,** 19 (1944).

[41] If a large portion of tissue is to be ground, it is best to grind it a little at a time.

stemmed funnel.[42] A steel spatula is used, and first alcohol, then ether, from washing the bottles, to make the transfer quantitative. When all the tissue has been ground and transferred to the extraction flask, more alcohol is added, if necessary, until 6 to 10 ml. of alcohol per gram of tissue is present.

Lipid Extraction. The flask is covered with a glass bubble stopper or watch glass to retard evaporation and is heated in a water bath at 60° for 2 hours. The extract is decanted, while still warm, through fat-free filter paper into a Kjeldahl concentrating flask[26] or Erlenmeyer. flask with a side arm.[14] The residue is heated for 1 hour with a second portion of alcohol, and the alcohol and tissue residue transferred quantitatively to the filter paper. After the residue and filter paper are extracted with ether in a Soxhlet apparatus for 18 hours, the combined alcohol-ether extracts are made to a convenient volume for study or concentrated and the lipids redissolved in petroleum ether as described in procedure 2 or 4.

Procedure 9. *Tissue Preparation:* Procedures 2, 4, 6, 7, and 8.
                Extraction Sclvent: Hot Alcohol.[1,43,44]

Comment. The use of hot alcohol as a solvent for lipids has been discussed at some length by Bloor.[1] The method as described by Sperry[43] is applicable to large samples of tissue. The use of hot alcohol only is less efficient, but if continued over a long period of time all the lipids can be removed.[33,44]

Tissue Preparation. The tissues may be subdivided in preparation for extraction by any one of several methods already presented (procedures 2, 4, 6, 7, and 8).

Lipid Extraction. The tissue is then placed in a modified Clark extractor[43] or in a glass siphon extractor cup[45] which can be hung below a condenser in the extraction flask, or in a Soxhlet apparatus designed for hot extractions.[46] The siphoning action ensures repeated changes of solvent around the tissue, and the alcohol vapors warm the contents in extractor cup, thus promoting more efficient removal of the lipids. Extraction is complete in 24 to 48 hours. The alcohol extract can be made to

---

[42] Some sizes of wide-stemmed funnels are available commercially. Such funnels are easily made by cutting the bottom and lip from Erlenmeyer flasks of appropriate size.

[43] W. M. Sperry, *J. Biol. Chem.* **68,** 357 (1926).

[44] L. L. Reed, F. Yamaguchi, W. E. Anderson, and L. B. Mendel, *J. Biol. Chem.* **87,** 147 (1930).

[45] A Bailey-Walker siphon extraction cup or an extraction siphon cup described in *Ind. Eng. Chem.* **6,** 78 (1914).

[46] Obtainable from Kopp Laboratory Supplies, Inc., New York, New York.

volume for study or evaporated and the lipids redissolved in petroleum ether by procedures already described (procedure 2 or 4).

Procedure 10. *Tissue Preparation:* Homogenization with Chloroform-Methanol (2:1).
Lipid Extraction: Chloroform-Methanol (2:1).

Comment. This method is particularly useful in the extraction of lipids from brain[6,7] and in obtaining a purified extract for the determination of tissue total lipids.[6-8] Folch did not use heat in the extraction, but Sperry heated just to boiling to coagulate the proteins and facilitate filtration.

Tissue Preparation and Lipid Extraction. A tissue sample of approximately 1 g., together with chloroform-methanol (2:1), is added to a large Tenbroeck or Potter-Elvehjem tissue grinder and homogenized. Complete homogenization is usually achieved in 3 minutes. The homogenate is quantitatively transferred to a volumetric flask (25 ml. or larger) using chloroform-methanol (2:1). After making to volume the contents are mixed and then filtered through fat-free filter paper into a glass-stoppered flask.

Purification of Extract.[47] A small beaker is placed in a second beaker ten times as large and nine-tenths full of water. An aliquot of the chloroform-methanol extract is delivered by pipet into the smaller beaker, care being taken to avoid turbulence and backflow into the pipet. (Sperry[8] pipetted the extract into a small vial, covered the extract with water, and lowered the vial into a beaker of water.) The beaker is covered and allowed to stand overnight, during which time methanol and water-soluble impurities diffuse away from the chloroform. A "fluff" is formed at the interface between the water—methanol and chloroform. The upper phase is almost completely removed by means of suction. A layer of water only 3 to 4 mm. thick should remain. One-fourth the volume of methanol originally present in the aliquot of extract is added to the small beaker to dissolve the fluff. If a clear solution is not obtained by stirring the mixture, more methanol is added dropwise until the solution is crystal clear. The mixture may be transferred to a volumetric flask and made to volume with chloroform—methanol (2:1), or, if it is desired that the protein moiety be removed from the proteolipids present, a beaker containing an aliquot or all the extract may be placed in a vacuum oven at

---

[47] An alternative method of purification, suggested in a personal communication to the author, is as follows: Pipet an aliquot of the chloroform-methanol extract into a glass-stoppered centrifuge tube. Add 4 vol. of water, stopper, and mix gently by inversion. Centrifuge for 10 minutes, and withdraw the upper phase by means of suction. Wash once or twice with the upper phase obtained by shaking 4 vol. of water with 1 vol. of chloroform-methanol (2:1). Remove the protein as described in procedure 10.

60° and the extract evaporated to dryness in an atmosphere of nitrogen. The residue is extracted twice with hot chloroform-methanol (2:1), and the extracts are filtered through fat-free filter paper into a volumetric flask and made to volume with chloroform—methanol.

Procedure 11. *Tissue Preparation:* Digestion and Saponification with KOH.
Extraction Solvent: Petroleum Ether, Ethyl Ether, Chloroform.

Comment. In the procedures to be described below the tissue lipids are liberated by treatment with strong alkali, and the freed lipids are dissolved in petroleum ether, ethyl ether, or chloroform. In some of the procedures the lipid is completely removed from the alkaline digest, whereas in other procedures the lipid is uniformly dispersed in a known volume of lipid solvent and, by analysis of an aliquot, the total amount present can be calculated. Only total cholesterol and total fatty acids can be obtained by these techniques, owing to the strongly destructive action of the alkali on the ester linkages of the lipids.

Tissue Preparation. A weighed sample of tissue is placed in a suitable container (an Erlenmeyer flask, a side-arm flask,[14] the chamber of a liquid-liquid Soxhlet,[48] a screw-cap culture tube,[49] or a beaker). Two milliliters of aqueous 30% KOH per gram of tissue is added, and the container heated by a hot plate, tube heater,[13] steam bath, or infrared lamps at 80 to 100° to digest the tissue.[50] An equal volume of alcohol is added to make the alcohol concentration approximately 50%.[51] Saponification of the lipids is completed by heating the alcoholic KOH digest[52] at 80 to

---

[48] Liquid-liquid soxhlets are useful in removing lipids from relatively large volumes of liquid phase by means of a volatile, immiscible solvent of lower specific gravity. For illustrations see W. W. Umbreit, R. H. Burns, and J. F. Stauffer, "Manometric Techniques and Tissues Metabolism," 2nd ed., p. 158, Burgess Publishing Co., Minneapolis, Minnesota, 1949; S. Paulkin, A. G. Murray, and H. R. Watkins, *Ind. Eng. Chem.* **17**, 612 (1925).

[49] Screw-cap culture tubes with Teflon-lined caps are used in preference to glass-stoppered tubes or flasks, since a very tight seal is formed without becoming "frozen" as does a ground joint under the action of strong alkali.

[50] For large tissues it is convenient to transfer the aqueous digest, as soon as the tissue is completely dissolved, to a volumetric flask, using alcohol in the transfer and ending with a final concentration of about 50 to 60% alcohol. Aliquots can then be taken for complete saponification and extraction of lipids.

[51] Alcoholic KOH is more efficient than aqueous KOH for saponification because it keeps the fats more uniformly dispersed so that the alkali can act. This mixture is recommended for large samples and for fatty tissues.

[52] Saponification can also be carried to completion with aqueous KOH digests, but it may take as long as several days if the tissue is large and fatty. Nonfatty tissues may be completely saponfied in several hours at 80 to 100°.

cholesterol with ether or petroleum ether and separatory funnels. The alkaline-alcohol mixture, including the KOH and water washes of the ether extracts, are transferred to an Erlenmeyer flask and heated until free of alcohol (Sperry did not completely remove the alcohol), acidified to the bromocresol green end point with 10 N sulfuric acid, and allowed to stand for at least 1 hour at 4 to 10° to permit fatty acid aggregates to form. The precipitated salts together with the fatty acids are filtered on a Büchner funnel with Whatman No. 1 filter paper. The filtrate should be crystal clear; if the first portion is cloudy, it is returned to the funnel. The filter paper is sucked as dry as possible, and the clear filtrate is discarded. The fatty acids are removed from the salt and filter paper by washing several times with acetone, which is allowed to remain in contact with the salt for some time before applying suction. The salt remaining on the filter paper is washed with petroleum ether back into the Erlenmeyer flask and extracted with several portions of petroleum ether. The acetone extracts are concentrated almost to dryness on a steam bath under an atmosphere of carbon dioxide, and the remaining aqueous solution is transferred to a separatory funnel. The flask is washed with the petroleum ether used in washing the salt, and the washings are transferred to the separatory funnel. The aqueous solution is washed three times with petroleum ether; the combined extracts are washed with water until the water washes are neutral, dried over sodium sulfate, and filtered into a volumetric flask.

Extraction of Fatty Acids and Cholesterol. (*H*) *With Petroleum Ether and Separatory Funnels.* The alkaline-alcohol solution remaining after saponification as described above (under Tissue Preparation) is transferred to a separatory funnel maintaining the alcohol concentration at about 50%. The mixture is acidified with 10 N sulfuric acid. Petroleum ether is added (5 vol. per volume of alkaline-alcohol digest) and shaken vigorously, and after standing the petroleum ether is drained out of the funnel. After four more similar treatments the combined ether extracts are evaporated to a low volume and washed with water until the water washes are neutral. The remaining petroleum ether extract is transferred to a volumetric flask.

I. *With Petroleum Ether and Liquid-Liquid Soxhlets.*[33,48] The alkaline-alcohol mixture after digestion and saponification as described above (Tissue Preparation) is added to the extraction chamber of a liquid-liquid Soxhlet,[48] acidified with 18 N sulfuric acid, and the fatty acids and cholesterol removed by continuous extractions for 48 hours with petroleum ether. If ethyl ether is used for the extraction, the alcohol should be evaporated from the digest prior to acidification. Ethyl ether extracts cholesterol with greater efficiency than does petroleum ether.

J. With Petroleum Ether and Side-arm Flasks.[14,33] The alkaline-alcohol digest of whole tissues or aliquots of large tissue digests obtained as described above may be completely digested and saponified in a side-arm flask and extracted with petroleum ether or ethyl ether. The flasks containing the alkaline-alcohol digests are heated at 80 to 100° for several hours, and then the mixture is allowed to go almost to dryness. After acidification with 18 *N* sulfuric acid and addition of 1 to 2 ml. of water to give a sufficient aqueous phase, the lipids are extracted by vigorous shaking with warm petroleum ether. The solvent is decanted while still warm, and the aqueous phase extracted three additional times in a similar manner. The combined ether extracts are reduced to a low volume and made to volume.

K. With Petroleum Ether and Screw-Cap Tubes.[33] The alkaline-50% alcohol-tissue digest equivalent to less than 2 g. of fresh tissue is pipetted into a screw-cap culture tube and heated in a tube heater[13] at 80° for 1 hour to complete the saponification. After cooling, the mixture is acidified with 18 *N* sulfuric acid, an exact amount of petroleum ether is added (15 to 30 ml.), and the tube tightly capped and shaken vigorously with a mechanical shaker for 5 to 10 minutes. The tube is centrifuged, and aliquots are removed for cholesterol and fatty acid determinations.

L. With Chloroform and Screw-Cap Tubes.[57] The method is the same in all respects as procedure D described above for removing cholesterol only with chloroform and screw-cap tubes, except that after saponification the mixture is acidified with 5 *N* sulfuric acid. When a measured volume of chloroform is added and the mixture treated as directed, the tissue fatty acids and cholesterol are uniformly dispersed in the chloroform subnatant and aliquots can be removed for analysis.

M. With Chloroform in Erlenmeyer Flasks.[57] Tissues are digested and saponified with alcoholic KOH in an Erlenmeyer flask as described above, or aliquots of tissues after saponification may be transferred to Erlenmeyer flasks. The alcohol is removed by evaporation on a steam bath, 2 drops of 0.04% bromocresol green is added, and the mixture is acidified by the addition of 5 *N* sulfuric acid. An exact volume of chloroform is added (5 to 20 ml./g. of fresh tissue), and the flask is tightly stopped and shaken vigorously for several minutes. The emulsified mixture is transferred to a centrifuge tube and centrifuged for 5 to 10 minutes. The tissue lipids are uniformly dispersed in the subnatant chloroform phase. With a syringe and a long needle an aliquot can be removed for analysis for total fatty acids or total cholesterol. If desired, an aliquot of the chloroform extract may be purified by evaporation just to dryness under nitrogen, and after addition of a few milliliters of water the lipids can be redissolved in petroleum ether.

Procedure 12. *Extraction of Body Fluids with Alcohol-Ether.*

Hot Extraction. Alcohol-ether (3:1) has been found to be an excellent solvent for lipids from plasma and other body fluids.[58,59] The use of heat is recommended in extracting lipids from body fluids with the alcohol-ether mixture. This makes it possible completely to remove the lipids from the body fluids with less solvent, and body fluids of widely varying lipid content can be adequately treated.

An exact amount of whole blood, serum, plasma, lymph, or other body fluid is pipetted into a graduated screw-cap culture tube.[60] With a syringe, alcohol-ether[61] (3:1) is added forcefully through a long wide-bore needle. The mixture is made to volume with alcohol-ether (3:1). The ratio of solvent to body fluid should be not less than 15:1, and preferably 20:1 or more. The tube is capped tightly, shaken thoroughly, and placed in a tube heater[13] or water bath at 60° for 1 hour.[5] After centrifugation, the supernatant is decanted into another screw-cap tube[60] which is capped tightly until needed for analysis. If large amounts of body fluids are to be extracted, volumetric flasks may be used. If preferred, the hot extraction can be carried out by the same techniques described for cold extraction with, if desired, 15 to 20 vol. of alcohol-ether instead of 25 to 30 vol.

Cold Extraction. Cold extraction is adequate for many body fluids when the lipid concentration is not high. Lymph is very difficult to extract completely with cold alcohol-ether. A greater amount of solvent is required for cold extraction than for hot extraction.[31,59]

An exact amount of body fluid is pipetted into a suitable vessel[60,62] and to it is added forcefully, as described above, 20 to 30 vol.[59] of alcohol-ether (3:1). After 15 minutes at room temperature (or more if convenient or desired), the mixture may be made to volume, filtered through fat-free filter paper, and aliquots taken for analysis. Or the mixture, without being made to volume, may be filtered into a concentrating flask,[26] and the residue and filter paper washed thoroughly with ether. The concen-

---

[58] W. R. Bloor, *J. Biol. Chem.* **17**, 377 (1914); **22**, 133 (1915).

[59] E. M. Boyd, *J. Biol. Chem.* **114**, 223 (1936).

[60] A graduated teflon-lined screw-cap culture tube is used because it can be made to volume immediately, capped tightly, conveniently heated in a tube heater[13] or water bath at 60° without loss of solvent, and centrifuged while still warm. Large numbers of samples can be handled with much less effort, equipment, and supplies than is possible if the extract is obtained by filtration. The seal provided by the screw cap, being much superior to that of glass stoppers, is also desirable during times when the room temperature may be elevated.

[61] If the samples are to be stored for longer than a day or two, only alcohol should be used at this time, the proper amount of ether being added at the time of extraction.

[62] A volumetric flask is very convenient for this extraction.

trating flask[63] is placed in a water bath at 60°, and the solvent almost completely removed under reduced pressure and in an atmosphere of nitrogen. Petroleum ether is added, and the mixture heated to facilitate removal of the lipids from the aqueous phase. After decanting, the treatment with petroleum ether is repeated with several small portions to effect complete extraction of the lipids.

Procedure 13. *Extraction of Body Fluids with Alcohol-Acetone.*

Alcohol-acetone (1:1) ranks well with alcohol-ether (3:1) as a mixture for the extraction of lipids from body fluids, particularly from serum and plasma. Acetone-alcohol has been most widely used to extract cholesterol from plasma, since it not only extracts the cholesterol efficiently but also provides a better medium from which to precipitate quantitatively the free cholesterol.[33,64] Although the mixture has not been widely used as a solvent for all lipids in body fluids, it is capable of extracting them quantitatively if 20 vol. or more of the mixture is used per volume of body fluid and if heat is applied.[33] The techniques used in the extraction of lipids with hot acetone-alcohol are the same as for hot extraction with alcohol-ether and need not be repeated here.

[63] A side-arm flask[14] may be used instead of the concentrating flask,[26] and the solvent can be removed by heating on a steam bath. A stream of nitrogen is directed into the flask when almost all the solvent has been evaporated.

[64] R. Schoenheimer and W. M. Sperry, *J. Biol. Chem.* **106**, 745 (1934).

[56] Preparation and Determination of Higher Fatty Acids

By CECIL ENTENMAN

The methods presented below are useful in the measurement of the long-chain fatty acids in the extracts prepared as described in Vol. III [55]. Included are descriptions of fatty acid determination by oxidation-titration, oxidation-colorimetry and alkali-titration, and in addition procedures are presented for the estimation of esterified and unesterified fatty acids. The manometric method for the determination of fatty acid carbon[1] is not presented. The choice of method depends on the kind, amount, and nature of lipids in the extract. It should be noted that the oxidative procedures give a good estimate of the *amount* of fatty acids present and are less affected by variations in chain length. The results are, however, influenced by the presence of any non-fatty acid-oxidizable

[1] D. D. Van Slyke and J. Folch, *J. Biol. Chem.* **136**, 509 (1940).

materials in the final extract. On the other hand, the other methods presented give a good estimate of the *milliequivalents* of fatty acids, but the amounts of fatty acids are difficult to determine unless the chain length is known. The titrimetric methods are the most suitable for microanalysis of fatty acids.

Colorimetric Determination of Fatty Acids after Oxidation with Sulfuric Acid-Dichromate Mixture

Principle. The method is based on the color change produced by the reducing action of fatty acids on a sulfuric acid-dichromate mixture.[2] The fatty acids are liberated by saponification, redissolved after acidification, the solvent removed, and the lipids oxidized. A correction has to be made for cholesterol, if present. The method has been adapted to the determination of fatty acids by Bloor,[3,4] and by Kibrick and Skupp,[5] and for total lipids by Bragdon.[6] The procedure given below is a modification of the Bragdon method.

Reagents and Apparatus

Sodium ethylate, 1 *M*, prepared immediately before use by adding 0.5 g. of metallic sodium to 20 ml. of 95% ethanol, cooling during the reaction.

Alcoholic KOH solution, made immediately before using by diluting 5 ml. of aqueous KOH solution (10 g. of reagent-grade KOH added to 20 ml. of distilled water) to 75 ml. with 95% ethanol.

5 *N* sulfuric acid.

Alcoholic sulfuric acid, equal parts of 5 *N* sulfuric acid and 95% ethanol.

Carborundum crystals.[7]

Potassium dichromate oxidation mixture. Dissolve 15 g. of reagent-grade potassium dichromate in 1000 ml. of reagent-grade concentrated sulfuric acid. Heat in a covered vessel for 2 hours at about 100° on a steam bath or hot plate. Do not heat above 120°. Store in glass-stoppered bottle.

Palmitic acid standard. Dissolve 100 mg. of palmitic acid, recrystallized and dried to constant weight, in ethyl alcohol, and dilute to 100 ml. Pipet 1-ml. aliquots into a series of 16 × 150-mm Teflon-

---

[2] I. Bang, *Biochem. Z.* **91**, 86, 235 (1918).

[3] W. R. Bloor, *J. Biol. Chem.* **77**, 53 (1928).

[4] W. R. Bloor, *J. Biol. Chem.* **170**, 671 (1947).

[5] A. C. Kibrick and S. J. Skupp, *Arch. Biochem. and Biophys.* **44**, 134 (1953).

[6] J. H. Bragdon, *J. Biol. Chem.* **190**, 513 (1951).

[7] Available from Central Scientific Co., Chicago, Illinois. The crystals, not exceeding 1 mm. in diameter, should be freed of oxidizable material by treating at 100° for 2 hours with the sulfuric acid-dichromate mixture and washed with distilled water.

lined screw-cap culture tubes,[8] and evaporate the solvent by
placing the tubes in a 100° tube heater. Pass a gentle stream of
nitrogen into the tubes to remove the last traces of solvent. Cap,
and store until needed. Each tube should contain 1 mg. of palmitic
acid.

Cholesterol standard. Dissolve 100 mg. of cholesterol, which has
been recrystallized four times from alcohol, in ethyl alcohol,
and make to 100 ml. Pipet 1-ml., aliquots into 16 × 150-ml.
screw-cap culture tubes,[8] and remove the solvent as described
above. Store the tubes until needed. Each tube should contain
1 mg. of cholesterol.

All solvents used anywhere in the procedure should be purified (see
Vol. III [55]). Reflux the ethyl alcohol over KOH, and then distill.
The ethyl ether should be peroxide-free.[9] Purify the petroleum ether
(boiling point, 30 to 60°) by allowing it to stand for several weeks
over c.p. concentrated sulfuric acid. Use about 100 ml. of sulfuric
acid per liter of petroleum. Shake frequently, decant into a distillation flask containing KOH pellets, and distill slowly. (The
sulfuric acid can be used for three treatments of the petroleum
ether, and then it should be discarded.) Merck's petroleum ether
"for fat extraction" is satisfactory without further purification.

Test tubes and Teflon-lined screw-cap culture tubes,[8] 16 × 150 mm.
Tube heater[10] to heat the 16 × 150-mm. tubes at 100°.
Glass bubble stoppers[8] to cover the 16 × 150-mm. test tubes.
Flat-tipped stirring rods, 6 mm. in diameter, 8 inches long.

Saponification. Pipet an aliquot of lipid extract, obtained by procedures previously described, into a 16 × 150-mm. screw-cap Teflon-
lined culture tube. The aliquot should not exceed 10 ml. and should contain 1 to 6 mg. of fatty acids. (If a petroleum ether extract or an extract
containing acetone is used, the solvent should be almost completely
removed by evaporation and 5 ml. of 95% ethanol added.) Add a clean
Carborundum crystal[7] to the tube, and set the tube in the tube heater[10]
at 100°. After 10 minutes, add 2 ml. of 1 M sodium ethylate or alcoholic
KOH. Evaporate the mixture almost to dryness by heating in the tube

[8] Available from Hallikainen Instruments, Berkeley, California. Cleaned by heating
at 100° for 2 hours with the oxidation mixture. Wash in distilled water, dry, and
store in a place free from solvent vapors.
[9] Peroxide-free ether can be obtained by adding a little hydroxylamine hydrochloride
to the stock ether bottle prior to distillation of the ether. The ether used in the
procedure should be freshly distilled.
[10] An electrically operated, thermostatically controlled heating unit containing holes
into which tubes can be placed for reactions requiring controlled heat (Hallikainen
Instruments, Berkeley, California).

heater over a period of 1 to 1½ hours. Almost all the alcohol should be evaporated, but do not let the tube go to dryness. A gentle stream of nitrogen aids in removing the excess alcohol. Remove the tube from the heater, cool somewhat, and acidify by addition of 1 ml. of the alcoholic sulfuric acid solution. Heat the tube briefly (20 seconds) in the tube heater, then rotate it in an almost horizontal position in order that all soaps react with the acid. Place the tube in a 20° water bath, and add exactly 15 ml. of petroleum ether.[11] Immediately cap the tube tightly with a clean Teflon-lined cap, shake vigorously for 30 seconds or longer, and return it to the 20° water bath for 30 minutes or longer in order for the fine particles to settle out.

Fatty Acid Oxidation. Pipet aliquots of the petroleum ether phase, containing 0.5 to 2 mg. of fatty acids, into scrupulously clean 16 × 150-mm. test tubes.[8] Place in a rack two tubes containing 1 mg. of *palmitic acid standard*, two tubes containing 1 mg. of *cholesterol standard*, and one clean empty tube to serve as a *reagent blank*, and add approximately 5 ml. of petroleum ether to each tube. Then place all tubes in the tube heater at 100°, and evaporate the petroleum ether completely. Through a capillary tube extending about half-way down the tube, pass in a stream of nitrogen for 2 minutes to remove all traces of solvent. (This step is accomplished conveniently with a glass manifold.) Remove the tubes from the heater, and cool. With a syringe-pipet[11] add 7 ml. of the potassium dichromate oxidizing mixture. Agitate the mixture by lateral motion, cover with glass stoppers, and place in the tube heater at 100°. After 1 hour remove the tubes and place them in a cold-water bath. When cool, add 7 ml. of distilled water to each tube by means of a syringe-pipet. Add the water down the sides of the tubes in such a manner that little mixing occurs. Stir contents cautiously with a flat-tipped glass rod to avoid excessive heating and loss of water vapors. When cool, transfer the solutions to cuvettes, set the reading of the reagent blank to zero at 600 mμ in the spectrophotometer, and determine the readings of the unknowns and standards.

Calculation

Mg. fatty acids in aliquot oxidized =

$$\frac{[\text{Optical density of unknown} - (\text{optical density of 1 mg. cholesterol standard} \times \text{mg. cholesterol in aliquot}^{12})]}{\text{Optical density of the 1 mg. palmitic acid standard}}$$

[11] A syringe-pipet is convenient for repetitive, accurate delivery of fixed volumes (available from Hallikainen Instruments, Berkeley, California).

[12] Cholesterol, if present in the extract, must be determined by procedures described elsewhere in order to apply this correction.

Silver Dichromate Oxidation—Thiosulfate Titration

Principle. The basis of the method is the change produced by the reducing action of fatty acids on a sulfuric acid-dichromate mixture,[2] the excess dichromate being determined by titration with thiosulfate solution. These techniques have been used for the determination of cholesterol,[3,13,14] fatty acid,[3,14] phospholipids,[14,15] and total lipids.[3] For the determination of fatty acids, saponification of the lipids is required.

Reagents and Apparatus

Sodium ethylate. Prepared fresh by adding 0.5 g. of metallic sodium to 20 ml. of ethanol. Keep the mixture cool while it is reacting.

Dilute sulfuric acid, 1 part concentrated sulfuric acid and 9 parts of water.

Starch solution. Boil 1000 ml. of water, and, while it is actively boiling, stir in a slurry made with 5 g. of soluble starch and a little cold water. Boil for an additional 1 minute.

Potassium dichromate solution, 1 *N*. Dissolve 49 g. of reagent-grade potassium dichromate in distilled water, and dilute to 1 l.

Nicloux reagent. Dissolve 5 g. of silver nitrate in 25 ml. of distilled water in a 100-ml. beaker. To this add 5 g. of potassium dichromate dissolved in 50 ml. of water. Transfer the mixture to 50-ml. centrifuge tubes, using a minimum of water in the transfer. Separate the precipitated silver dichromate by centrifugation, wash twice with water by centrifugation to remove nitric acid, and dissolve the cake of precipitate, without drying, in 500 ml. of reagent-grade concentrated sulfuric acid. Heat the mixture for several hours at 80 to 100°, cool, and store in glass-stoppered bottles.

Sodium thiosulfate, 0.1 *N*.

Potassium iodide, 10%.

Carbon dioxide. The ordinary commercial product from the tank delivered through a tube containing cotton as a filter.

Organic Solvents. Alcohol, ethyl ether, and petroleum ether purified as described in Vol. III [55].

Electric oven.[16] The oven should be capable of maintaining the

---

[13] R. Okey, *J. Biol. Chem.* **88,** 367 (1930).

[14] E. M. Boyd, *Am. J. Clin. Pathol.* **8,** 77 (1938).

[15] W. R. Bloor, *J. Biol. Chem.* **82,** 273 (1929).

[16] A tube heater[10] or a pressure cooker[5] may be substituted for the electric oven, the reaction mixture being heated in tubes and transferred to Erlenmeyer flasks for titration with thiosulfate.

chemical in water and diluting to 100 ml. Solution 2 is 3.5 N sodium hydroxide made by dissolving 14.0 g. of NaOH in water and diluting to 100 ml.

Ferric chloride, 0.37 M in 0.1 HCl. Dissolve 8.9 g. of $FeCl_3 \cdot 6H_2O$ in 0.1 N HCl, and dilute to 100 ml. with 0.1 N HCl.

Hydrochloric acid, approximately 4 N, made by adding 1 part of concentrated HCl to 2 parts of water.

Alcohol-ether mixture (3:1), 3 parts of 95% ethanol and 1 part of ethyl ether (peroxide-free[9]).

Ethyl ether, peroxide-free.[9]

Trimyristin standard. Dissolve 200 mg. of recrystallized trimyristin in 200 ml. of chloroform at 20°. Pipet 3-ml. aliquots (containing 3 mg. of trimyristin) into culture tubes with Teflon-lined screw-caps.[8] Evaporate the chloroform at 60° in a stream of nitrogen, cap, and store until needed. When the standard is needed, add 15 ml. of alcohol-ether mixture (3:1), cap tightly, heat for 2 minutes at 60°, shake, and cool to 20°. Each milliliter contains 0.2 mg. of trimyristin.

Procedure. Pipet aliquots of an alcohol-ether (3:1) extract of tissue or body fluids (see procedures in Vol. III [55]) into 16 × 150-mm. culture tubes with Teflon-lined screw caps.[23] The aliquot should contain 0.3 to 0.8 mg. of fatty acids. For the reagent blank pipet into a screw-cap culture tube 3 ml. of alcohol-ether mixture (3:1), and for the fatty acid standard pipet duplicate 3-ml. aliquots of the alcohol-ether (3:1) solution of the trimyristin standard into two culture tubes.

To each tube add 1.0 ml. of the hydroxylamine-sodium hydroxide solution, cap, mix with a lateral motion, and let stand for 20 minutes at room temperature. Add 0.6 ml. of 4 N HCl, and mix. Add 1 ml. of peroxide-free ethyl ether, and mix. Set the reagent blank to zero at 520 mμ in the spectrophotometer, and read the standards and unknowns.

Calculations

Mg. esterified fatty acid in 3-ml. aliquot =

$$\text{Optical density of unknown} \times \frac{(\text{mol. wt. of fatty acid in unknown})}{(\text{optical density of standard})} \tag{1}$$
$$\times 3 \frac{(\text{mg. in aliquot of standard})}{(\text{mol. wt. of standard})}$$

[23] If tissue extracts are used whose final solvent mixture is not alcohol-ether (3:1), the solvent must be removed by evaporation under a stream of nitrogen and the lipids dissolved in alcohol-ether (3:1). Extracts containing *acetone* must be avoided, or, if used, the acetone must be completely removed under heat and a stream of nitrogen before proceeding. Acetone vapors must not be present while the procedure is being carried out.

Meq. esterified fatty acid in 3-ml. aliquot =

$$\frac{\text{Optical density of unknown} \times \text{meq. of standard}}{\text{Optical density of standard}} \quad (2)$$

Unesterified Fatty Acids

Principle. The method described, a modification of the technique of Davis[24] as reported by Seldon and Westphal,[25] involves treatment of the sample with NaCl and HCl followed by extraction with ethyl ether, and titration of the acids with sodium hydroxide. The method is applicable to tissues as well as to sera. Tissue extracts can be prepared as described in Procedure 6, Vol. III [55] and purified, if desired by the method of Fairbairn.[26] A simple method for unesterified fatty acids in sera was reported by Grossman *et al.*[27]

Reagents

Nile blue, 0.1% in ethanol.
Sodium hydroxide 0.02 N.
Petroleum ether (purified).
Alcohol (purified).
Sodium chloride, 1%.
Hydrochloric acid, 1 N.

Procedure. If serum is used, add the sample (15 ml.) to 10 ml. of 1% NaCl solution and 4 ml. of 1 N HCl in a 50-ml. centrifuge tube, and extract once with 8 ml. and then six times with 4 ml. each of ether, using a glass plunger for gentle mixing.[28] Then cover the tube with tinfoil, centrifuge for 10 minutes at approximately $800 \times g$, and pipet the supernatant ether phase into a cylindrical separatory funnel. Wash the combined ether extracts with 5-ml. portions of 1% NaCl solution to litmus-neutrality of the aqueous phase (generally eight times).

[24] B. D. Davis, *Arch. Biochem.* **15**, 351 (1947).
[25] G. L. Seldon and U. F. Westphal, Army Medical Research Laboratory, Fort Knox, Kentucky, Report No. 182, March 1955.
[26] D. Fairbairn, *J. Biol. Chem.* **157**, 633, 645 (1945).
[27] M. I. Grossman, L. Palm, G. H. Becker, and H. C. Moeller, Medical Nutrition Laboratory Fort Fitzsimons Army Hospital, Denver, Colorado, Report No. 131, July 1954. Two milliliters of serum or plasma was added to a mixture of 0.2 ml. of 0.2 M phosphate buffer (pH 6.0) and 2 ml. of 9% ethanol. This mixture was extracted three times with 3-ml. portions of petroleum ether (boiling point, 30 to 60°) by shaking for 1 minute, centrifuging, and removing the supernatant with a syringe and needle. The combined petroleum ether extracts were evaporated to dryness in a bath at 70°, and the residue was taken up in 2 ml. of alcohol, heated, and titrated with 0.02 N aqueous sodium hydroxide, with thymol blue as an indicator.
[28] Efficient emulsification without splattering is produced by rapidly raising and lowering a long glass rod, with a flat tip flattened to a disk fitting the tube like a piston.[24]

Transfer the washed ether solution or a tissue extract, prepared as described in Vol. III [55], A[29] in small portions to a small titration vessel, evaporating the solvent almost to dryness after each addition, and finally to complete dryness under a stream of nitrogen. Avoid overheating. Dissolve the residue in 0.5 ml. of ethanol by brief boiling, cool, add 1 drop of 0.1% Nile blue[30] in ethanol, and bubble nitrogen through for 10 minutes. Titrate the solution with aqueous 0.02 N sodium hydroxide to a just-pink end point.

Carry out a blank experiment with each set of samples, starting with 1 ml. of 1% NaCl solution instead of serum. An alcoholic solution of pure fatty acid can be used as a standard. The results are expressed as milliequivalents of fatty acid.

Titrimetric Determination of Fatty Acids

Principle. The titrimetric method described below is a slight modification of the method of Page and Michaud[31] which was derived from those of Bloor et al.,[32] Stoddard and Drury,[33] Smith and Kik,[34] and Schmidt-Nielsen.[35] An extract containing lipid is saponified with potassium hydroxide with heat under an atmosphere of nitrogen. The fatty acids are liberated with hydrochloric acid and dissolved in hot benzene (cf. ref.[36-38]). An aliquot of the benzene extract, after cooling, is evaporated to dryness, and the fatty acids dissolved in alcoholic thymol blue and titrated with tetramethyl ammonium hydroxide, using a micrometer buret. The same procedure is applicable to the determination of larger quantities of fatty acids if adjustments are made in the amounts of reagents and type of apparatus used.

[29] Lipid extracts of tissues or body fluids, prepared as described in Vol. III [55] (procedures 6 and 12), with due precautions being taken to prevent autolysis,[26] can be used after the extracts have been taken to dryness under a stream of nitrogen and redissolved in ethyl ether. Phospholipids, if present, are acidic, and should be removed with acetone and magnesium chloride as described by Fairbairn.[26]

[30] Thymol blue has been used, but the end point is not so good, according to Seldon and Westphal.[25]

[31] E. Page and L. Michaud, *Can. J. Med. Sci.* **29**, 239 (1951).

[32] W. R. Bloor, K. F. Pelkan, and D. M. Allen, *J. Biol. Chem.* **52**, 191 (1922).

[33] V. L. Stoddard and P. E. Drury, *J. Biol. Chem.* **84**, 741 (1929).

[34] M. E. Smith and M. C. Kik, *J. Biol. Chem.* **103**, 391 (1933).

[35] K. Schmidt-Nielsen, *Compt. rend. trav. lab. Carlsberg Sér. chim.* **24**, 233 (1942).

[36] Instead of extracting the liberated fatty acids with benzene, some investigators[32,33,36,37] have preferred to separate the fatty acids after saponification by precipitation in an acid solution, filter, wash, with 5% NaCl to remove mineral acid, dissolve in alcohol, and titrate.[34,35] The technique is well described by Kaiser and Kagan.[38] Although this technique permits the isolation of pure fatty acids, errors may arise owing to loss while filtering, especially of the unsaturated fatty acids.

[37] E. B. Man and E. F. Gildea, *J. Biol. Chem.* **99**, 43 (1932–1933).

[38] E. Kaiser and B. M. Kagan, *Anal. Chem.* **23**, 1879 (1951).

Reagents and Apparatus

Potassium hydroxide, 30% in water.

Hydrochloric acid, 25%.

Thymol blue, 0.01% in absolute alcohol. The solution is partly neutralized with 0.02 N tetramethyl ammonium hydroxide $(N(CH_3)_4OH)$, so that complete neutralization is reached with the further addition of about 0.8 μl. of the base per milliliter of alcoholic thymol.

Absolute alcohol.

Petroleum ether, 30 to 60°.

Pure benzene.

Tetramethyl ammonium hydroxide,[39] 0.02 N in absolute alcohol. This solution is kept in a Pyrex bottle equipped with a siphon and a CO_2 absorption tube to prevent contamination with CO_2.

Micrometer buret,[40] capacity 0.1 ml., each scale corresponding to 0.0001 ml.

Saponification. Pipet an aliquot of lipid extract obtained by procedures described in Vol. III [55] into a 16 × 150-mm. screw-cap Teflon-lined culture tube[8] containing a small Carborundum chip.[7] The aliquot should not exceed 10 ml. and should contain 0.3 to 1 mg. of fatty acids.[41] (If a petroleum ether extract or an extract containing acetone is used, the solvent should be evaporated and 95% ethanol added to make a total volume of about 4 ml.) After adding exactly 0.1 ml. of potassium hydroxide, place the tube, in a tube heater[10] at 100° and heat under a stream of nitrogen (delivered through a glass manifold if a large number of samples are treated simultaneously) until the last trace of alcohol has been evaporated. The solution should not be evaporated to complete dryness.

Extraction of the Fatty Acids.[36] After saponification, cool the tube slightly and add exactly 0.2 ml. of hydrochloric acid. Warm the tube in the tube heater; then swirl it in such a manner that all soaps and excess alkali react with the acid. After cooling, add exactly 5.0 ml. of benzene at 20°, cap the tube tightly, warm to about 60° (no hotter), and shake vigorously. Allow the tube to stand for 30 minutes or longer, or centrifuge for 3 minutes to free the walls from water.

[39] Used in preference to sodium or potassium hydroxide because of their instability and because potassium carbonate is more insoluble in alcohol [N. Kretchmer, *J. Am. Oil Chemists' Soc.* **25,** 404 (1948)].

[40] Available from the Emil Greiner Co., New York, New York.

[41] Page and Michaud[31] used about 80 μl. of plasma (determined by weighing in a melting point tube), extracted the lipids with alcohol-acetone (1:1), then saponified and extracted the fatty acids with benzene in a 5-ml. volumetric flask. An aliquot of the benzene was taken for titration of the fatty acids.

Titration. After cooling to 20°, pipet a 4-ml. aliquot of the benzene extract into a 5-ml. test tube, and evaporate the benzene to dryness in a tube heater or on a steam bath under a stream of nitrogen. Add exactly 1 ml. of alcoholic thymol blue, taking care to wash down the walls of the tube. By means of a fine capillary, bubble a fine stream of nitrogen or CO_2-free air through the solution for several minutes in order to remove any carbon dioxide. Then titrate the fatty acids with tetramethyl ammonium hydroxide from a micrometer buret whose tip extends below the surface of the liquid. The end point is reached when the indicator changes from a bright yellow to a persistent yellowish-green tinge. Place a stoppered control tube containing 1 ml. of the acid alcoholic thymol blue in permanence next to the solution being titrated. A reagent blank is run with each set of determinations, the titration value being about 1 μl.

Calculations. V = net titration value in microliters; N = normality of $N(CH_3)_4OH$; and W = body fluid in milligrams or microliters.

$$\text{Meq. of fatty acid in aliquot titrated} = \frac{V \times N}{1000} \quad (1)$$

Fatty acids in body fluids, usually expressed as milliequivalents per liter, are calculated as follows:

$$\text{Meq./l.} = \frac{V \times N \times 1000}{W} \times \frac{5}{4} \quad (2)$$

Normal plasma values vary from 6 to 20 meq./l. To convert to milligrams of fatty acids per 100 ml., multiply the milliequivalents per liter by 27.72, the average molecular weight of blood fatty acids, according to Stoddard and Drury.[33]

[57] Preparation and Analysis of Phosphatides[1]

By MARJORIE B. LEES

Assay Methods

The quantitative analysis for total and individual phosphatides presents numerous problems associated with the completeness of their extraction, the removal of nonlipid contaminants, hydrolysis procedures, and the specificity of the methods. The complete extraction of the phos-

---

[1] The excellent books by H. Witcoff, "The Phosphatides" (Reinhold Publishing Corp., New York, 1951) and H. J. Deuel, Jr., "The Lipids" Volume I (Interscience Publishers, Inc., New York, 1951) are recommended for further references and discussion of much of the material presented.

phatides and the removal of nonlipid contaminants are prerequisites for any method for the determination of total phosphatides (Vol. III [55]). The most commonly used method is the analysis of an adequately extracted, dialyzed sample for phosphorus. This is as accurate as and much simpler than the oxidative method of Bloor[2] or the determination of fatty acids recommended by Artom.[3] Analyses for lipid phosphorus can be carried out by the method of Sperry[4] or by any other suitable procedure. The phosphorus value obtained is converted to phosphatides by means of a factor based on the average phosphorus content of the phosphatides. Although various investigators have used experimentally determined factors, multiplying the lipid P by 25 (based on an average P content of 4%) is adequate for most purposes.

The best methods for the analysis of individual phosphatides are those based on a study of their hydrolysis products. Methods dependent on the solubility of the lipids are completely inadequate, since the solubility of the individual lipids is influenced by the presence of other lipids in a mixture. The cephalin fraction, for example, is insoluble in alcohol, whereas the phosphatidyl ethanolamine isolated therefrom is completely alcohol-soluble. After hydrolysis of the phosphatides, a combination of several analytical techniques can be used to determine the amounts of the main components of the mixture. Table I shows the distribution of various chemical components among the principal phosphatides. Phosphatides such as cardiolipin, acetal phosphatides, and phosphatidic acids have been omitted from this discussion, since they either have been isolated only from special tissues or have not been sufficiently characterized to warrant their inclusion here. Brief descriptions of their preparation may be found in Witcoff's book.[1] The following relationships can be seen from Table I: (1) lecithin and sphingomyelin are the only known choline-containing lipids; (2) the amount of the lecithin plus cephalin fraction is represented by the amount of glycerol present; (3) the total amino nitrogen value gives the sum of the phosphatidyl ethanolamine and phosphatidyl serine fractions. Glycerol is determined by the method of Blix;[5] inositol can be determined microbiologically or chemically;[6] methods for the quantitative determination of choline, ethanolamine, serine, and sphingosine are described in Vol. III [59]. In evaluating the analytical results, however, the effect of the minor phosphatides should not be overlooked.

[2] W. R. Bloor, *J. Biol. Chem.* **82**, 273 (1929).

[3] C. Artom, *Bull. soc. chim. biol.* **14**, 1386 (1932).

[4] W. M. Sperry, *Ind. Eng. Chem. Anal. Ed.* **80**, 46 (1938).

[5] G. Blix, *Mikrochim. Acta* **1**, 75 (1937).

[6] D. W. Woolley, *J. Biol. Chem.* **140**, 453 (1941); B. S. Platt and G. E. Glock, *Biochem. J.* **37**, 709 (1943).

Methods based on preferential hydrolysis constitute a slightly different approach to the analysis of individual phosphatides. Schmidt *et al.*[7] have shown that the P of monoaminophosphatides becomes acid-soluble under conditions where sphingomyelin remains intact. This has been used as the basis for the assay of sphingomyelin described below. Hack[8] has developed a similar method for blood phosphatides based on the fact that the choline from lecithin can be hydrolyzed whereas that from sphingomyelin remains intact. After acidification and removal of the unreacted

TABLE I

CHEMICAL COMPOSITION OF THE PRINCIPAL PHOSPHATIDES

Lipid	P	N	α-NH$_2$-N^a	NH$_2$-N^b	Choline	Glycerol
	Atom grams				Moles	
Phosphatidyl choline (lecithin)	1	1	0	0	1	1
Cephalins						
Phosphatidyl ethanolamine	1	1	0	1	0	1
Phosphatidyl serine	1	1	1	1	0	1
Diphosphoinositide	2	0	0	0	0	1
Sphingomyelin	1	2	0	0	1	0

[a] By the ninhydrin-CO$_2$ method of D. D. Van Slyke, R. T. Dillon, D. A. Mac-Fadyen, and P. Hamilton, *J. Biol. Chem.* **141**, 627 (1941).

[b] By the nitrous acid manometric method of D. D. Van Slyke, *J. Biol. Chem.* **83**, 425 (1929).

sphingomyelin, the filtrate is analyzed for P and choline. From these values and the total P, values for lecithin, cephalin, and sphingomyelin can be calculated. It should be noted that these hydrolysis procedures are not necessarily applicable to all tissues. Brain lipids in particular often behave differently from those of other tissues. For example, Sperry and Brand's hydrolysis procedure for sphingomyelin could not be applied satisfactorily to brain lipids.[9]

Criteria for Purity of Phosphatides. The development of adequate criteria for the determination of the degree of purity of phosphatide preparations still awaits future research. Recent studies using countercurrent distribution and chromatographic techniques have made a beginning in that direction. It is known that the individual phosphatides isolated thus far are not single chemical compounds but are mixtures of closely related substances differing at least in their fatty acid residues.

[7] G. Schmidt, J. Benotti, B. Hershman, and S. J. Thannhauser, *J. Biol. Chem.* **166**, 505 (1946).

[8] M. H. Hack, *J. Biol. Chem.* **169**, 137 (1947).

[9] J. Folch and W. M. Sperry, *Ann. Rev. Biochem.* **17**, 147 (1948).

It is significant that, as yet, no synthetic phosphatide has been prepared which is identical with the substances isolated from tissues. Since we are not dealing with a homogeneous chemical compound, the determination of physical constants is not sufficiently precise to be of value in a consideration of the purity of a phosphatide sample. Melting point and solubility, for example, are affected by the chain length and degree of unsaturation of the fatty acid residues.

At the present state of knowledge the best approach to a consideration of the degree of purity of a phosphatide preparation is by means of chemical analyses for the amounts of impurities present. The following points should be noted: (1) The complete removal of nonlipid contaminants should be ensured by adequate dialysis of the preparation. It is known that lipids can carry impurities with them through lengthy fractionation procedures. (2) The elemental chemical analysis and the amounts of the hydrolysis products obtained should be consistent with the accepted formula for that phosphatide. (3) No hydrolysis products other than those required by the accepted formula should be present; phosphatide samples should be free of carbohydrates and cholesterol. (4) Hydrolysis products should be present in large enough amounts to exclude the possibility of their being contaminants. Proof that a constituent is combined chemically can best be obtained by a study of the kinetics of its liberation by several agents. (5) the $N:P$ ratio, although widely used in the literature, is not a particularly sensitive criterion for purity, since large amounts of contaminants may be present without affecting the ratio significantly. (6) All the nitrogen (or phosphorus) should exist in the form required by the accepted formula for that phosphatide. With the exception of sphingomyelin, which contains both sphingosine and choline, the nitrogen should be present as a single chemical constituent. Some of the above points will be discussed under the individual phosphatides.

Preparation of Lecithin (Phosphatidyl Choline)

Most of the methods for the preparation of lecithin depend on the precipitation of its cadmium salt from an alcoholic solution. The method of Pangborn,[10] described below for the preparation of egg lecithin, is a simplification of that of Levene and his co-workers.[11,12] In Pangborn's method the lecithin-$CdCl_2$ complex is broken with a specific combination of organic solvents, after which the cadmium is removed with dilute alcohol (Fig. 1).

[10] M. Pangborn, *J. Biol. Chem.* **188**, 471 (1950).
[11] P. A. Levene and H. S. Simms, *J. Biol. Chem.* **48**, 185 (1921).
[12] P. A. Levene and I. P. Rolf, *J. Biol. Chem.* **72**, 587 (1927).

Step 1. Extraction and Preparation of Lecithin-CdCl₂. The yolks of twelve fresh eggs are stirred for a few seconds in a Waring blendor; 200 ml. of acetone is added, and the mixture is stirred for about 30 seconds, transferred to a beaker, stirred thoroughly with 400 ml. more of acetone, and

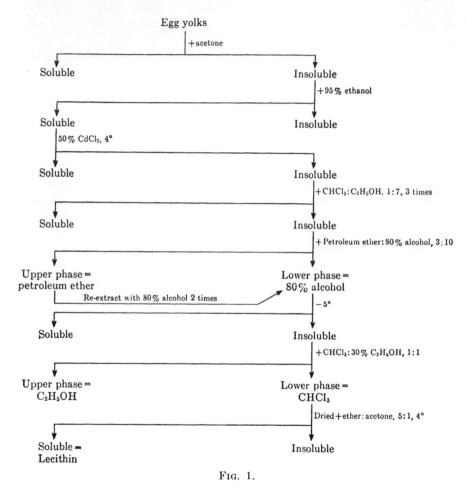

Fig. 1.

filtered by suction. The filter cake is extracted five times in the blendor with 200-ml. portions of acetone. The creamy white yolk powder thus obtained is extracted for 30 minutes in a mechanical shaker with 800 ml. of 95% ethanol and filtered by suction. This alcoholic extract is precipitated with a slight excess of 50% aqueous CdCl₂ (approximately 15 ml.). After standing for 1 hour at 4°, the mixture is filtered by suction and the precipitate is washed twice with acetone while still on the filter.

Step 2. Purification of Lecithin-CdCl₂. The precipitate is dissolved in 100 ml. of chloroform. The slightly cloudy, faintly brown solution is poured with constant mixing into 700 ml. of ethanol containing 10 ml. of 50% aqueous $CdCl_2$ to prevent dissociation of the lecithin salt. After intermittent shaking for 10 minutes at room temperature, the mixture is filtered by suction. The precipitate is redissolved in chloroform and treated with ethanol and $CdCl_2$ as before, except that it is allowed to stand for 30 minutes with frequent shaking. A third chloroform-alcohol precipitation is carried out as above.

The cadmium salt is then suspended in 150 ml. of petroleum ether, and to it is added 500 ml. of 80% alcohol previously saturated with petroleum ether and containing 0.1% $CdCl_2$. After vigorous shaking in a separatory funnel, the alcohol layer is drawn off and the petroleum ether layer is re-extracted twice with a total of 100 ml. of the above 80% alcohol mixture per gram of material remaining in the petroleum ether layer (as determined by drying an aliquot). The combined alcohol extracts are concentrated to two-thirds the original volume to remove the petroleum ether. The concentrated lecithin-CdCl₂ mixture is left at about −5° overnight, after which it is filtered by suction.

Step 3. Removal of Cadmium. The lecithin-CdCl₂ precipitate is dissolved in 150 ml. of chloroform, an equal volume of 30% ethanol is added, and the mixture is shaken vigorously for 5 minutes. Under these conditions the lecithin double salt dissociates, and the $CdCl_2$ is washed out in the dilute alcohol. Four extractions with 30% alcohol should suffice to remove all the cadmium. An aliquot of the aqueous-alcoholic layer may be tested for the presence of cadmium ions with a drop of 5% $AgNO_3$. One extra extraction should be carried out to ensure complete removal of the cadmium.

Step 4. Final Purification. The chloroform solution is dried by vacuum distillation of the solvent. The lecithin is dissolved in 100 ml. of anhydrous ether, and to it is added 20 ml. of acetone. The mixture is placed at 4° overnight, and the finely flocculent precipitate which forms is removed by filtration over a Büchner funnel. The clear filtrate is evaporated to dryness *in vacuo,* and the purified lecithin is dissolved in absolute alcohol. Yield, 7 grams; P, 4.03%; N, 1.98%; NH₂-N, 0.02%; N:P, 1.06:1.

Modifications for Tissues

Hanahan and Jayko[12a] have isolated an individual, completely unsaturated lecithin from baker's yeast by a method involving a final purification of the material on an aluminum oxide column. Since their

[12a] D. J. Hanahan and M. E. Jayko, *J. Am. Chem. Soc.* **74**, 5070 (1952).

preparation has been well characterized, it should prove particularly useful for enzyme studies.

The preparation of lecithin from tissue extracts is more difficult, and each material presents special problems. A summary of Pangborn's modification for the preparation of beef heart lecithin is as follows.

Step 1. Removal of Cardiolipin.[13] 6.2 kg. of fresh, defatted beef hearts is ground in a meat grinder. The minced tissue is extracted twice at room temperature with 1.2 ml. of acetone per gram of fresh tissue. The acetone is removed by suction; the residue from the second extraction is dried before a fan until the acetone odor is gone. It is ground to a fine powder which is then extracted three times with 2 l. of 95% methanol for each 300 g. of powder, each extraction being carried out for a week at room temperature with frequent shaking. After the final extraction, the residue is washed with methanol and discarded. As each extract is collected, it is precipitated with 20% aqueous $BaCl_2$ and stored at 3 to 6°. The pooled extracts are centrifuged; the precipitate can be processed for cardiolipin while lecithin is prepared from the supernatant fluid.

Step 2. Precipitation of Lecithin-CdCl₂ and Removal of Cadmium. The procedure is then essentially as described above. Lecithin is precipitated from its alcoholic solution by the addition of excess 50% aqueous $CdCl_2$; the precipitate is taken up in 200 ml. of chloroform and precipitated with 1400 ml. of ethanol a total of three times, after which the cadmium is removed. The cadmium-free chloroform solution is evaporated to dryness and taken up in 180 ml. of anhydrous ether. To the slightly cloudy solution is added 36 ml. of acetone. The precipitate which forms when the mixture is refrigerated overnight is removed by filtration, and the clear filtrate is evaporated to dryness and redissolved in a small amount of alcohol. Yield, 18 g.; NH_2-N, 0.25%.

Step 3. Further Purification. The alcoholic solution is made alkaline to phenolphthalein with 10 ml. of saturated aqueous $Ba(OH)_2$, the mixture is immediately neutralized with CO_2, and 2 ml. of saturated NaCl is added with vigorous shaking to flocculate the precipitate. The precipitate is removed by gravity filtration, and the clear filtrate is reprecipitated with $CdCl_2$. The cadmium salt is precipitated twice by pouring its chloroform solution into alcohol and then fractionated once by the petroleum ether-80% alcohol method described for egg lecithin. The cadmium is then removed, and the lecithin is dissolved in ether to which 20% of its volume of acetone is added. After refrigeration at −5° overnight, the small amount of precipitate is removed by filtration, and the filtrate is dried and dissolved in a small amount of alcohol. Yield, 10.5 g.; P, 4.21%; N, 1.94%; N:P, 1.01; NH_2-N, 0.01%; iodine number, 83.7.

13 M. Pangborn, *J. Biol. Chem.* **161**, 71 (1945).

Step 4. Purification for Serological Tests.[14] The above preparation can be separated chromatographically into fractions of different serological activity (in the Wasserman test for syphilis). A glass column (50 × 250 mm.) is packed with Magnesol-Celite. After impregnation of the column with benzene, 500 ml. of lecithin in 15 ml. of benzene is added at the top of the column, which is then developed with 500 ml. of 2% *tert*-butyl alcohol in benzene. The column is extruded, and a narrow vertical band is streaked with 1% potassium permanganate in 2.5 N NaOH. The zones which thus show are then eluted with absolute ethanol. This procedure can be applied successfully to commercial lecithin preparations or to lecithin-CdCl$_2$ complexes. Residual impurities can also be removed with aluminum oxide columns.[15]

Properties

Pure lecithin is a whitish, paraffinlike substance which is extremely hygroscopic. When dry it can be ground to a powder, but on taking up water it becomes a waxy, sticky mass. On exposure to light and air the material darkens and develops a disagreeable odor owing to the instability of the fatty acids. No definite melting point can be determined.

Lecithin is readily soluble in methanol, ethanol, benzene, ether, petroleum ether, chloroform, carbon tetrachloride, and carbon disulfide. It is also soluble in pyridine, glycerol, and acetic acid. It is, however, insoluble in methyl acetate and acetone. Lecithin may be precipitated from its water emulsion by the addition of acetone, acids, or inorganic salts. Lecithin containing completely saturated fatty acids (hydrolecithin) is insoluble in ether and is found as a contaminant in the sphingomyelin fraction unless special precautions are taken to remove it.

Lecithin is easily hydrolyzed by either acid or alkali. Hack,[8] using normal KOH at 37° for 16 hours, found glycerophosphate, choline, and fatty acids among the hydrolysis products. Since migration of the phosphoric acid radical occurs during hydrolysis, the configuration of the naturally occurring substance cannot be determined from the configuration of the isolated glycerophosphoric acid.[16] The best evidence is that most, if not all, of the lecithin occurs naturally in the α configuration.[17]

Lecithin probably exists in the form of a zwitterion. Some sort of internal neutralization is to be expected because of the presence of a strongly acidic and a strongly basic group. Lecithin in solution is essentially neutral and has almost no buffering power in the physiological range. The

[14] F. A. H. Rice and A. G. Osler, *J. Biol. Chem.* **189**, 115 (1951).
[15] M. Faure and J. Legault-Demare, *Bull. soc. chim. biol.* **32**, 509 (1950).
[16] J. Folch, *J. Biol. Chem.* **146**, 31 (1942).
[17] E. Baer and M. Kates, *J. Biol. Chem.* **175**, 79 (1948).

isoelectric point is rather difficult to determine, but the best values place it in the neighborhood of 6.4 [18] (theory = 7.5).[19]

Purity. The most sensitive criterion for the purity of a lecithin preparation is the absence of α-amino nitrogen.[20] No base other than choline should be present. Lecithin preparations should have a N:P ratio of 1:1; all the phosphorus should be saponifiable under the conditions of the procedure of Schmidt *et al.;* and theoretical amounts of choline should be obtained after adequate hydrolysis.

Preparation of Cephalin Phosphatides

The multiple nature of the cephalin fraction was first recognized by Folch and Schneider.[21] In the method of Folch[22] described below, the crude cephalin mixture from brain is separated into three different fractions on the basis of their differing solubilities in chloroform-ethanol mixtures. Increasing amounts of ethanol are added to a chloroform solution of cephalin, and the precipitates are collected at various concentrations of ethanol. In this manner three fractions can be isolated: (1) diphosphoinositide, which shows relatively low solubility in alcohol; (2) phosphatidyl serine; and (3) phosphatidyl ethanolamine, which is freely soluble in alcohol. The success of the separation of the crude cephalin mixture is dependent on the fact that phosphatidyl ethanolamine is present as the free ampholyte whereas the other phosphatides are in the salt form. In previous procedures the use of HCl resulted in their conversion to the free acid form which is evidently more difficult to separate. Aside from the preparation described below, probably the only pure phosphatidyl ethanolamine preparation has been that of Rudy and Page.[23] Although analysis shows the presence of phosphatidyl serine in various tissues, it has never been prepared in a pure state from any tissue except brain. Inositol phosphatides have been reported from a variety of sources. Woolley[24] isolated an inositol-containing phosphatide from soybeans which he named lipositol. No formula has been proposed for this compound, although galactose, ethanolamine, tartaric acid, saturated and unsaturated fatty acids, phosphoric acid, and inositol have been identified among its hydrolysis products. It is possible that lipositol is a mixture of closely related lipids.

[18] H. B. Bull and V. L. Frampton, *J. Am. Chem. Soc.* **58,** 594 (1936).

[19] H. Fischgold and E. Chain, *Biochem. J.* **28,** 2044 (1934).

[20] D. D. Van Slyke, R. T. Dillon, D. A. MacFadyen, and P. Hamilton, *J. Biol. Chem.* **141,** 627 (1941).

[21] J. Folch and H. A. Schneider, *J. Biol. Chem.* **137,** 51 (1941).

[22] J. Folch, *J. Biol. Chem.* **146,** 35 (1942); **177,** 497 (1949).

[23] H. Rudy and I. H. Page, *Z. physiol. Chem.* **193,** 251 (1930).

[24] D. W. Woolley, *J. Biol. Chem.* **147,** 581 (1943).

Step 1. Preparation of Crude Brain Cephalin. Fresh ox brains are freed of their membranes, and 100-g. portions of tissue are ground in a Waring blendor with 300 ml. of acetone for 2 minutes. The portions are combined, and acetone is added until there is at least 3.8 ml./g. of tissue. After a few minutes the acetone is removed by suction and discarded. The precipitate is extracted once more with acetone, once with alcohol, and twice with petroleum ether, 4 ml. of each solvent being used per gram of original tissue. The two petroleum ether extracts are combined and dried completely by vacuum distillation. This dried material is then suspended in 200 ml. of ethyl ether for each kilogram of original tissue and allowed to stand in the refrigerator for a day or two. If a clear supernatant does not separate within three days, water is added to the extent of 1% of the volume of ether. If this does not result in the appearance of a clear upper fraction, the ether suspension should be diluted one-fold with ether. Once a supernatant solution amounting to about one-fourth of the total volume is obtained, the suspension is centrifuged and the precipitate is washed twice with cold ether.

The combined ether extracts are dried by vacuum distillation at room temperature, and the residue is dissolved in 50 ml. of ether for each kilogram of initial brain tissue. The ethereal solution is placed in the refrigerator overnight, and any precipitate which forms is discarded. The solution is diluted with an equal volume of ether, and 5 vol. of alcohol is added with stirring to precipitate the cephalin. After the mixture stands at room temperature for about 1 hour, a clear supernatant is formed. The precipitate is collected on a Büchner filter and suspended in 100 ml. of acetone per kilogram of original brain tissue. The suspension is shaken for 40 minutes to remove acetone-soluble impurities. After removal of the supernatant fluid by suction, acetone is again added to the precipitate and the procedure is repeated. Yield, 15 g. of a tan cephalin powder per kilogram of initial tissue; C, 55.2%; P, 4.13%; N, 1.59%; NH_2-N, 1.51%; α-NH_2-N, 0.76%; NH_2-N:N, 0.96; P:N, 1.17.

Step 2. Fractionation of Brain Cephalin by Chloroform-Alcohol Method. One gram of the crude cephalin preparation described above is dissolved in 8 ml. of chloroform, and to it is added 1.45 times as much alcohol as chloroform by volume. A turbidity develops and, after standing for about 1 hour, the mixture is centrifuged and resolves itself into a viscous underlayer (fractions I and II, inositol phosphatide fraction) and a clear supernatant. The supernatant is decanted, and to it is added 25 ml. of alcohol. The precipitate (fraction III, phosphatidyl serine) which appears is collected on a Büchner filter and dried. The filtrate is concentrated *in vacuo* to half its volume and allowed to stand in the refrigerator for 2 or 3 days. The precipitate (fraction IV) is removed by filtration in the cold and dried. The filtrate is concentrated to 1 ml., and to it is added 5 ml. of acetone. After standing for 1 day in the refrigerator, the acetone-insoluble precipitate (fraction V, phosphatidyl ethanolamine) is removed by filtration and dried. Table II gives an analysis of the above fractions.

TABLE II[a]

ANALYSIS OF FRACTIONS ISOLATED FROM BRAIN CEPHALIN BY CHLOROFORM-ALCOHOL METHOD

Components	Fraction I (inositol phosphatide), %	Fraction II (inositol phosphatide), %	Fraction III (phosphatidyl serine), %	Fraction IV, %	Fraction V (phosphatidyl ethanolamine), %
C	55.0	59.0	60.2	63.0	66.1
P	4.25	3.86	3.58	3.60	3.65
N	1.15	1.36	1.62	1.75	1.78
Amino N[b]	1.15	1.36	1.64	1.60	1.50
α-Amino N[c]	0.70	0.80	1.47	0.60	<0.02
Inositol	6.8	3.4	<0.20	<0.20	<0.20
Iodine No.	65.0		39.8		78.0
Ash	16.7		12.8		2.5
Yield, g. per 100 g. cephalin	22.0	10.0	27.0	8.0	15.0

[a] From J. Folch, *J. Biol. Chem.* **146**, 35 (1942).

[b] By the nitrous acid manometric method of D. D. Van Slyke, *J. Biol. Chem.* **83**, 425 (1929).

[c] By the ninhydrin-CO_2 method of D. D. Van Slyke, R. T. Dillon, D. A. Mac-Fadyen, and P. Hamilton, *J. Biol. Chem.* **141**, 627 (1941).

Step 3. Purification of Inositol Phosphatide by Chloroform-Methanol Method. The viscous underlayer (fractions I and II) is treated with ethanol, and the solid precipitate which forms is collected on a Büchner filter and dried. It is then processed at 4° as follows: One gram of crude inositol phosphatide is dissolved in 12 ml. of chloroform, and to the clear solution is added 22 ml. of methanol. The mixture is shaken for 30 minutes in a shaking machine, after which the precipitate is collected by centrifugation. It is then redissolved in 12 ml. of chloroform, 22 ml. of methanol is added to the solution, and the procedure is repeated as before as long as the supernatant solution contains material with less than 4.5% P. If at any point the precipitate fails to dissolve completely in chloroform, the insoluble material can be removed by precipitation with 2 vol. of ether. If the material has not been previously dialyzed, twelve successive precipitations usually suffice, but as many as thirty precipitations may be required on a dialyzed preparation.

The crude diphosphinositide so obtained contains large amounts of water-soluble contaminants. These may be removed by dialysis as described below. Table III gives the analysis of the fractions obtained before and after dialysis.

TABLE III
ANALYSIS OF BRAIN DIPHOSPHOINOSITIDE[a]

Component	Before dialysis, %	After dialysis, %
C	31.5	46.0
N	0.79	0.4–0.6
P	11.08	7.0–7.3
Inorganic P	6.10	<0.05
α-NH$_2$-N	0.10	<0.02
NH$_2$-N (after acid hydrolysis)		0.1–0.4
Carbohydrate (as galactose)	1.7	0.4–1.8
Glycerol	4.9	9.0
Inositol		21.0
K	8.05	$\left\{ <0.01 \right.$
Na	3.25	
Ca	0.30	0.64
Mg	1.30	2.88
Yield, g./kg. fresh tissue	1.0	0.5–0.6

[a] Data from J. Folch, *J. Biol. Chem.* **177**, 505 (1949).

Step 4. Dialysis of Fractions. A 3% aqueous emulsion of each of the fractions is prepared by shaking 1 g. of material with 30 ml. of water until homogeneous. The emulsion is dialyzed at 4° for 4 days against distilled water with several changes of the outside liquid. The undialyzable fraction is then lyophilized, and the fluffy white powder which is obtained is suspended in acetone, collected on a Büchner filter, and dried.

Step 5. Purification of Phosphatidyl Serine.[25] The above method results in phosphatidyl serine preparations which are 85 to 90% pure; i.e., between 85 and 90% of the total nitrogen is present as α-amino nitrogen. If lower values are obtained or if slightly greater purity is required, the following procedure is recommended: One gram of the preparation is dissolved in 10 ml. of chloroform, and to it is added 16.5 ml. of absolute ethanol. A turbidity develops which, on standing or by centrifugation, resolves itself into a viscous underlayer and a clear supernatant solution. The supernatant solution is decanted, and to it is added 30 ml. of absolute

[25] J. Folch, *J. Biol. Chem.* **174**, 439 (1948).

ethanol. The precipitate which separates is collected and dried. It contains a higher concentration of α-amino nitrogen than does the mother substance.

Phosphatidyl serine preparations of greater than 92% purity are not obtained consistently. The main contaminant appears to be phosphatidyl ethanolamine, since all the nitrogen can be accounted for as amino nitrogen.[26] It is essentially free of cerebrosides (<0.1% carbohydrate), lecithin and sphingomyelin (<0.1% choline), and cholesterol (<0.1%).

Properties of Phosphatidyl Ethanolamine

Freshly prepared phosphatidyl ethanolamine is a slightly sticky white powder which becomes dark brown and stickier on standing in the dark in a vacuum desiccator. This change in physical appearance is evidently not accompanied by any observable change in elementary composition. The material contains 1.7% water, which can be removed at 80° *in vacuo*. However, it returns to its original weight on storage in a desiccator over $CaCl_2$. As isolated from brain by neutral solvents, it is ash-free.

Phosphatidyl ethanolamine, unlike the cephalin mixture from which it is derived, is freely soluble in alcohol. It is also soluble in chloroform, petroleum ether, carbon disulfide, benzene, hot acetic acid, and wet ether. It is insoluble in anhydrous ether and acetone and forms an emulsion with water. It is hydrolyzed relatively easily by either acids or alkali. The iodine number of the product isolated by Folch is 78, indicating the presence of two double bonds for each P atom. The few physical measurements available have been carried out on impure preparations and therefore do not warrant discussion.

Purity. All the nitrogen of phosphatidyl ethanolamine should be present as amino nitrogen as determined by the nitrous acid procedure of Van Slyke.[26] The presence of serine can be excluded by the absence of α-amino nitrogen. A phosphatidyl ethanolamine preparation should be free of choline and inositol; all the phosphorus should be saponifiable under the conditions of the procedure of Schmidt *et al.*;[7] and it should show a N:P ratio of 1:1.

Properties of Phosphatidyl Serine

Phosphatidyl serine is a free-flowing white powder with a slight tendency to darken. After long periods of storage *in vacuo* in the dark at room temperature there is no change in its elementary composition but there is some sort of molecular rearrangement involving a sharp drop of its α-amino nitrogen and primary amino nitrogen content. Storage in chloroform solution in a dry icebox seems to be more satisfactory. Phosphatidyl

[26] D. D. Van Slyke, *J. Biol. Chem.* **83**, 425 (1929).

serine, like phosphatidyl ethanolamine, contains a small percentage of water. It is freely soluble in chloroform, ethyl ether, and petroleum ether, but insoluble in ethanol, methanol, and acetone. In the presence of base, phosphatidyl serine forms a stable, concentrated emulsion with water. However, acidification of the emulsion to pH 1.5 results in a complete precipitation of base-free phosphatidyl serine. The process may be reversed by the addition of theoretical amounts of alkali.

Phosphatidyl serine is a strongly acidic compound, having one basic group (the —NH$_2$ group of serine) and two acidic groups (the —COOH group of serine and one group from phosphoric acid). Therefore, at physiological pH it binds one equivalent of base for each atom of P. This is shown by the fact that phosphatidyl serine isolated from brain by the use of neutral solvents and freed of water-soluble impurities by dialysis contains Na and K. Preparations obtained by the method described above contain much more K than Na, but Ca and Mg are absent. The same preparations also give theoretical values for an equimolar mixture of oleic and stearic acids.

Glycerophosphoric acids, L-serine, and fatty acids have been isolated as hydrolysis products of phosphatidyl serine in approximately 1:1:2 proportions. Intact phosphatidyl serine reacts with ninhydrin and with nitrous acid in the same way as does an α-amino acid, showing that both the carboxyl and the amino groups are free. Its inability to react with periodic acid indicates that the hydroxyl group must be combined.

Purity. All the nitrogen of a phosphatidyl serine preparation should be present as free α-amino nitrogen; the presence of any other type of nitrogen indicates the existence of impurities. This is the only quantitative method for differentiating phosphatidyl serine from phosphatidyl ethanolamine, as other criteria listed under phosphatidyl ethanolamine apply to both substances.

Properties of Diphosphoinositide

Diphosphoinositide prepared from brain by neutral solvents is an acidic phosphatide which is obtained as a salt of calcium and magnesium.[27] It is a white, gritty powder which is not emulsified in water and is insoluble in most organic solvents except wet chloroform. It can be converted to the potassium salt which forms a stable emulsion with water but is insoluble in the usual organic solvents. It is possible to obtain a solution of this material in an organic solvent by diluting a 5% emulsion with methanol or 2:1 chloroform-methanol. The base can be removed by treatment with dilute acid, indicating that the bases are present in a saltlike linkage. The base-free diphosphoinositide is readily soluble in

[27] J. Folch, *J. Biol. Chem.* **177**, 505 (1949).

water, forming an acid solution which, on titration, uses one equivalent of base per atom of P present.

Brain diphosphoinositide is both alkali- and acid-labile. Among the products of hydrolysis, fatty acids, glycerol, and inositol metadiphosphate have been isolated in approximately equimolar proportions. The chemical structure of inositol metadiphosphate has been established by Folch by (1) elementary analysis, (2) isolation of the theoretical amount of inositol, (3) titration to pH 8.7, which shows the presence of two free acid groups for each phosphoryl radical, and (4) a study of the reaction products with HIO_4, which shows that the two phosphoryl groups are in the meta position on the inositol. The attachment of the remaining constituents in the molecule is as yet unknown. The nitrogen and carbohydrate are probably present as impurities. The following formula has been postulated on the basis of the experimental evidence, where R and R' stand for unknown radicals:

$$
\begin{array}{c}
\text{OR} \\
\diagup \\
\text{HC·O·PO} \\
\diagup \qquad \diagdown \\
\qquad \qquad \text{OH} \\
\text{HCOH} \qquad \text{HCOH} \\
| \qquad\qquad | \qquad \text{OR'} \\
\qquad\qquad\qquad \diagup \\
\text{HCOH} \qquad \text{HC·O·PO} \\
\diagdown \qquad \diagup \qquad\qquad \diagdown \\
\qquad \qquad\qquad\qquad \text{OH} \\
\text{HCOH}
\end{array}
$$

Purity. Diphosphoinositide preparations should be free of nitrogen and choline; phosphorus and inositol should be present in a 2:1 ratio.

Sphingomyelin (Monoacylsphingosylphosphoryl Choline)

Assay Method

Principle. The older methods for the determination of sphingomyelin depended either on its solubility,[28] which gave low results, or on its precipitation as a reineckate,[29] which gave high results. The analysis for sphingomyelin developed by Schmidt et al.[7] is based on the fact that these compounds are much more resistant to alkali than are the glycerophosphatides. The conditions of the procedure are such that the selective saponification of the glycerophosphatides is brought about. Since sphingomyelin is represented by the difference between the total and the saponi-

---

[28] E. Kirk, *J. Biol. Chem.* **123**, 623 (1938).

[29] B. N. Erickson, I. Avrin, D. M. Teague, and H. H. Williams, *J. Biol. Chem.* **135**, 671 (1940).

fiable phosphorus, the values can only be considered approximate in cases where the sphingomyelin content of the tissue is small. On the other hand, it is very useful for detecting the presence of small amounts of hydrolecithin which often contaminates sphingomyelin samples.

Procedure. Suitable aliquots of the lipid samples to be tested are placed in wide Pyrex test tubes and evaporated to dryness on a water bath. Each sample is emulsified with 5 ml. of N potassium hydroxide at 37° for 24 hours. Schmidt recommends continual shaking of the emulsion, but we have not found this to be necessary. The emulsion is precipitated by the addition of 1 ml. of 5 N hydrochloric acid and 5 ml. of 10% trichloro-acetic acid. After standing for 1 hour at room temperature (to hydrolyze acetal phosphatides), the samples are centrifuged and filtered. The phos-phorus in the original lipid solution and in the clear trichloroacetic acid filtrate is determined. The difference between the two values represents the nonsaponifiable or sphingomyelin phosphorus.

Preparation

Sphingomyelin prepared by the usual methods[30,31] is contaminated with a large proportion of hydrolecithin. The observation that sphingo-myelin is resistant to N alkali at 37°, in contrast to hydrolecithin which is completely hydrolyzed under those conditions (see above), made it possible to follow the separation of the two substances in the course of their preparation. In the following procedure developed by Thannhauser and his collaborators,[32] the crude ether-insoluble phosphatide mixture is precipitated from glacial acetic acid with a large volume of acetone. After alkaline hydrolysis to destroy the hydrolecithin, sphingomyelin can be prepared in pure form from the precipitate. Lung tissue is used as the starting material since, in addition to containing large amounts of sphin-gomyelin, it contains only small amounts of cerebrosides. The same procedure may be used for spleen. The large amount of cerebrosides in brain makes the fractionation of the ether-insoluble material from that tissue much more difficult. The modifications necessary for the isola-tion of brain sphingomyelin have been described by Thannhauser and Boncoddo.[33]

Step 1. Preparation of Crude Sphingomyelin. Fifty pounds of beef lung is minced and washed twice with acetone, filtered, and dried *in vacuo* at 60°. The tissue is then ground to a powder and extracted continuously

[30] P. A. Levene, *J. Biol. Chem.* **24,** 69 (1916).
[31] S. J. Thannhauser and P. Setz, *J. Biol. Chem.* **116,** 527 (1936).
[32] S. J. Thannhauser, J. Benotti, and N. F. Boncoddo, *J. Biol. Chem.* **166,** 669, 677 (1946).
[33] S. J. Thannhauser and N. F. Boncoddo, *J. Biol. Chem.* **172,** 141 (1948).

with ether for 3 days. After standing in the refrigerator overnight, the ether extract is filtered over Hyflo filter aid (Johns-Manville). The precipitate is re-extracted with ether in a Soxhlet apparatus for several days to remove any further ether-soluble impurities. The residue is taken up in 1 l. of 9:1 petroleum ether-methanol and filtered. The filtrate is concentrated to a thin sirupy liquid which is precipitated with 1 to 2 l. of acetone. After standing in the refrigerator overnight, the precipitate can be collected by filtration. The precipitate is mainly a mixture of sphingomyelin and hydrolecithin. Yield, 20 to 40 g.; P, 3.8 to 4.0%, of which 40 to 50% is saponifiable.

Step 2. Separation of Sphingomyelin from Bulk of Hydrolecithin. The crude ether-insoluble phosphatide mixture is warmed slightly with 10 vol. of glacial acetic acid. After standing overnight at room temperature, the solution is filtered and the residue is re-extracted with 10 vol. of acetic acid. The pooled filtrates are concentrated to a small volume and then precipitated with 1000 to 1500 ml. of acetone. After standing overnight in the refrigerator the suspension is filtered. The precipitate contains sphingomyelin and hydrolecithin in the proportion 2:1 and is completely free of unsaturated monoaminophosphatides.

Step 3. Preparation of Sphingomyelin by Alkaline Hydrolysis. Ten grams of the above material is suspended in a small amount of water and ground to a paste. A total of 200 ml. of 0.25 N NaOH is added, and the suspension is shaken at 37° for 4 or 5 days. It is then acidified with glacial acetic acid, refrigerated overnight, and filtered over Hyflo filter aid. The precipitate is washed with acetone, followed by ether, and then extracted in a Soxhlet apparatus for 2 to 3 days with ether. The contents of the thimble are dialyzed for 24 hours against running water to remove inorganic contaminants; the dialyzed suspension is filtered over Hyflo filter aid and washed with acetone. The residue is taken up in 9:1 petroleum ether-methanol and filtered. The filtrate is concentrated to a small volume and precipitated with 1000 to 1500 ml. of acetone. Yield of precipitate, 3 to 4 g.

Step 4. Removal of Traces of Cerebrosides. The precipitate is taken up in a small amount of 9:1 petroleum ether-methanol and run through an Al_2O_3 chromatographic column for the selective absorption of cerebrosides. The sphingomyelin is recovered by precipitation of the effluent fluid with acid and recrystallization from hot ethyl acetate. N, 3.4%; P, 3.58%; N:P, 2:1; and less than 1% of the total P is saponifiable.

Properties

Lung sphingomyelin so obtained is a white crystalline substance which is relatively stable to light and air and is nonhygroscopic. It is soluble in benzene, chloroform, warm alcohol, and hot ethyl acetate but insoluble in ether and acetone. It can be crystallized from hot ethyl acetate and it emulsifies in water easily. Its melting point is 209°. Sphingomyelin contains two asymmetric carbon atoms and is therefore optically active. Thannhauser et al.[32] report $[\alpha]_D^{29} = +6.25$ (4% solution in 1:1 chloroform-methanol). This is slightly lower than the values reported by Levene,[30] but the latter's preparations were undoubtedly contaminated with hydrolecithin. Sphingomyelin is capable of forming complexes with metallic compounds such as cadmium chloride and chloroplatinic acid. The insoluble compound with Reinecke acid has been used for analytical purposes. However, Hack[34] has shown that the sphingomyelin reineckate formed is simply an adsorption complex and not a stoichiometric chemical compound and is therefore unsuitable as an analytical method. The iodine number of the sphingomyelin isolated by Thannhauser is 30 (theoretical − 36), indicating the presence of one double bond. Since sphingosine contains a double bond, only saturated fatty acids can be present. Approximately equal amounts of palmitic and lignoceric acids were found in sphingomyelin prepared from lung or spleen. On the other hand, the component fatty acids of brain sphingomyelin are stearic, nervonic, and lignoceric acids.

The acid-base relationships of sphingomyelin are very much like those of lecithin, since they both have a strong acid and a strong basic group. The basic character of sphingosine need not be considered, since its amino group is bound in amide linkage. Sphingomyelin exists as a zwitterion. It shows no buffering power over a wide range of pH's and shows no combination with either chloride or sodium ions.[35] Chain and Kemp[36] determined the isoelectric point of sphingomyelin to be 6.0. This somewhat low value was attributed to the presence of an acidic impurity.

Purity. Sphingomyelin preparations should be free of glycerol, free amino nitrogen and saponifiable phosphorus. All the phosphorus of sphingomyelin is resistant to the action of alkali under the conditions described above by Schmidt et al.[7] Nitrogen and phosphorus are present in a 2:1 ratio, in contrast to lecithin, phosphatidyl ethanolamine, and phosphatidyl serine, which have 1:1 ratios. Choline and sphingosine should be the only nitrogen-containing residues present.

[34] M. H. Hack, *J. Biol. Chem.* **166**, 455 (1946).
[35] H. N. Christensen and A. B. Hastings, *J. Biol. Chem.* **136**, 387 (1940).
[36] E. Chain and I. Kemp, *Biochem. J.* **28**, 2052 (1934).

[58] Preparation of O-(L-α-Glycerylphosphoryl)choline, Phosphorylcholine, O-(L-α-Glycerylphosphoryl)- ethanolamine, and Phosphorylethanolamine

By GERHARD SCHMIDT

I. O-(L-α-Glycerylphosphoryl)choline (L-α-GPC) (Molecular Weight 257.2)

$$CH_2OH \cdot CHOH \cdot CH_2O—P(OH)OOCH_2CH_2N(CH_3)_3$$

The procedure described by Schmidt *et al.*[1] for the isolation of L-α-glyc-erylphosphorylcholine from autolyzed pancreas and that described by Baer and Kates[2] for its organic synthesis have now been superseded for the purpose of preparing the pure ester by the excellent method of Tattrie and McArthur,[3] which is based on the catalytic hydrolysis of lecithin by mercuric chloride. This reagent, which was introduced into lipid chemistry by Feulgen[4] as a catalyst of the hydrolysis of acetal-phospholipids, was applied by Dawson[5] for the cleavage of cephalin and permits the hydrolysis of lecithin without secondary isomerization of its glycerylphosphorylcholine moiety.

Some Properties of L-α-GPC. The ester linkages of choline and ethanolamine in their glycerylphosphoryl esters are very labile toward acids, even at room temperature. At approximately neutral reaction, however, aqueous solutions of these compounds are stable for several days in the refrigerator. The choline group of L-α-GPC is quantitatively liberated in the form of free choline by heating its aqueous solutions in 1 N hydrochloric acid at 100° for 20 minutes.[6] No inorganic phosphoric acid is formed under these conditions. The phosphoryl groups are transformed to a mixture of α- and β-glycerophosphoric acids. L-α-GPC is insoluble in acetone and ether, but considerably soluble in 95% ethanol, and very easily soluble in water. No water-insoluble compounds of L-α-GPC are known. In concentrated alcoholic solutions, L-α-GPC forms precipitates on addition of an alcoholic solution of ammonium reineckate. It also forms alcohol-insoluble complexes with cadmium chloride. These complexes can be stored. Solutions of free L-α-GPC can be conveniently prepared by shaking an aqueous solution of the cadmium chloride complex

[1] G. Schmidt, B. Hershman, and S. J. Tannhauser, *J. Biol. Chem.* **161**, 523 (1945).
[2] E. Baer and M. Kates, *J. Am. Chem. Soc.* **70**, 1394 (1948).
[3] N. H. Tattrie and C. S. McArthur, *Can. J. Biochem. Physiol.* **33**, 761 (1955).
[4] R. Feulgen and K. Imhäuser, *Biochem. Z.* **181**, 30 (1927).
[5] R. M. C. Dawson, *Biochem. J.* **59**, 5 (1955).
[6] G. Schmidt, L. N. Greenbaum, P. Fallot, A. C. Walker, and S. J. Thannhauser, *J. Biol. Chem.* **212**, 887 (1955).

with solid silver carbonate at room temperature. For enzyme experiments, it might be preferable to achieve the decomposition of the cadmium chloride complexes by ion exchangers according to Tattrie and McArthur[3] (see below).

L-α-GPC consumes per mole 1 mole of periodate. The reaction is complete within a few minutes at room temperature.

Behavior against Enzymes. L-α-GPC is completely resistant toward acid prostatic phosphatase and purified alkaline intestinal phosphatase. However, when the incubation with these enzymes is preceded by a 20-minute hydrolysis with N hydrochloric acid at 100° and by subsequent adjustment of the pH to the activity optimum of the respective phosphatase, quantitative enzymatic hydrolysis to inorganic phosphate takes place.

Preparation[3] of L α-GPC-Cadmium Chloride Complex, $(C_8H_{22}O_7NP)_2$-$(CdCl_2)_3$, (Molecular Weight 1100)

One hundred grams of crude egg phospholipids containing 70% of lecithin is homogenized in a Waring blendor with 2000 ml. of water. The emulsion is mixed with an equal volume of 99% ethanol containing 200 g. of mercuric chloride. The mixture is refluxed for 26 hours, cooled, and extracted four times with 600-ml. portions of ether. The solution is treated with hydrogen sulfide in the presence of barium carbonate, and the mercurous sulfide is filtered off. The clear filtrate is concentrated under reduced pressure at a bath temperature of 45° until the ether and most of the alcohol are removed. The yellow solution is passed through a mixture of 2 vol. of Amberlite IR-45 base and 1 vol. of Amberlite IRC-50 acid. The clear, colorless effluent is further concentrated at 45° bath temperature under reduced pressure to a sirup which is dissolved in 500 ml. of 99% ethanol. A saturated solution of cadmium chloride in 95% ethanol is added until the precipitation of the GPC-cadmium chloride complex is complete. After 3 hours of standing in the refrigerator, the dense white precipitate is filtered over a Büchner funnel, washed with cold 99% alcohol and with ether, and dried over phosphorus pentoxide in a vacuum desiccator. The yield is approximately 34 g. (65 to 70% of the theoretical yield). The choline content of the amorphous cadmium complex is almost theoretical.

Preparation of the Crystalline L-α-GPC-Cadmium Chloride Complex, $(C_8H_{22}O_7NP)CdCl_2$ (Molecular Weight 494.7), According to Baer and Kates[2]

A solution of 12.4 g. of the amorphous cadmium chloride complex in 150 ml. of water is gradually diluted with 600 ml. of 99% ethanol. The clear solution is kept for 12 hours at room temperature, and for another

12 hours in the refrigerator at 5°. The crystalline precipitate is filtered off and washed with a small volume of cold 80% ethanol. The air-dried crystals weigh approximately 7 g. (yield 60%).

After drying *in vacuo* over phosphorus pentoxide at 56°, a weight loss of 10.92% occurs, corresponding to a content of 3 moles of water. The dried substance analyzes correctly for a compound of 1 mole of L-α-GPC with 1 mole of $CdCl_2$ (molecular weight 440.6); $\alpha_D^{25} = -1.33°$ in water (C_1 10.16).

Preparation of Crystallized L-α-GPC

Crystallization of L-α-GPC was achieved for the first time by Tattrie and McArthur.[3] A 2% solution of the crystallized cadmium complex is passed through the above-mentioned mixture of Amberlites. The clear, colorless filtrate is concentrated under reduced pressure at 45° bath temperature to a sirup which is further dried at 56° *in vacuo* over phosphorus pentoxide for 48 hours. Three grams of the vitreous product is dissolved in 20 ml. of 99% ethanol at 60°. The clear solution is placed in the refrigerator at 5° for 3 hours, and subsequently at 15° for 3 hours. The crystals are washed with cold 99% ethanol and finally with ether. Yield, 2 g. (dried *in vacuo* at 56° over phosphorus pentoxide).

II. Preparation of Phosphorylcholine as Crystallized Calcium Phosphorylcholine Chloride (Molecular Weight 275.7)

$$C_5H_{13}O_4NPCaCl \cdot 4H_2O$$

Phosphorylcholine was synthesized by Plimmer and Burch[7] by esterification of choline chloride on a boiling water bath with a mixture of phosphoric acid and phosphorus pentoxide. The efficiency of this method was improved by Riley,[8] who described a procedure for the preparation of P^{32}-labeled (or C^{14}-labeled) phosphorylcholine. This procedure differs from the original technique of Plimmer and Burch mainly by the higher reaction temperature and by more elaborate precautions against the presence of moisture. Since Riley's procedure is almost as simple as that described by Plimmer and Burch, it is advisable to use his technique also for the preparation of the nonlabeled ester.

Some Properties of Phosphorylcholine. Phosphorylcholine is very stable toward acid and alkaline hydrolyzing reagents. It is rapidly and quantitatively hydrolyzed by prostatic and intestinal phosphatases. In

[7] R. H. A. Plimmer and W. J. N. Burch, *Biochem. J.* **31**, 398 (1937).
[8] R. F. Riley, *J. Am. Chem. Soc.* **66**, 512 (1944).

contrast to the glycerylphosphoryl ester, phosphorylcholine chloride forms well-crystallized calcium and barium salts, and a crystalline mercuric chloride compound. In concentrated ethanolic solution it forms a difficultly soluble reineckate. In contrast to choline, phosphorylcholine is not precipitated by ammonium reineckate in aqueous solutions.

Preparation of Phosphorylcholine

A solution of 150 mg. of P^{32}-labeled phosphoric acid is placed in a pear-shaped flask of 35-ml. capacity which is fitted with a standard taper-24 stopper and two outlets with glass stopcocks. The phosphoric acid is brought to 100% concentration by heating it in an oil bath the temperature of which is gradually increased to 180°. During the evaporation, a stream of dry air is passed over the solution. After cooling, the phosphoric acid is mixed with 214 mg. of dry choline chloride. A Soxhlet thimble containing a mixture of 1 g. of phosphorus pentoxide with asbestos is supported in the space over the reaction mixture. The flask is evacuated and kept at 165° for 12 hours under vacuum. On cooling, the product is dissolved in a few milliliters of water, 100 mg. of calcium chloride is added, and the solution is made alkaline with phenolphthalein as indicator by adding a saturated solution of calcium hydroxide. The solution obtained after filtration from calcium phosphate is concentrated further and filtered from calcium carbonate. The filtrate is just acidified with hydrochloric acid and brought again to weakly alkaline reaction with the calcium hydroxide solution. On addition of an equal volume of alcohol the calcium salt of phosphorylcholine chloride crystallizes as platelets. The crystals are washed with 50% alcohol and then with absolute alcohol. It is recrystallized by dissolving the crystals in water and by adding an equal volume of alcohol to the solution.

The crystallized barium salt of phosphorylcholine chloride may be obtained in a similar way.

The yield of the calcium salt before recrystallization is approximately 60% of the theory. The method described can be used for the preparation of larger batches.

III. Preparation of O-(L-α-Glycerylphosphoryl)ethanolamine (L-α-GPE) + H₂O (Molecular Weight 233.2)

$$CH_2OH \cdot CHOH \cdot CH_2O—P(OH)OOCH_2 \cdot CH_2NH_2$$

L-α-GPE was the first glycerophosphoryl ester obtained in pure crystalline form. It was discovered by Feulgen and Bersin,[9] who obtained the

---

[9] R. Feulgen and Th. Bersin, *Z. physiol. Chem.* **260**, 217 (1939).

substance during the hydrolysis of crystalline acetalphospholipids from muscle in 1939. Thannhauser et al.[10] isolated it in 1951 in a similar manner from crystallized acetalphospholipids of brain. Its preparation by organic synthesis was described by Baer and Stancer.[11]

Each of the procedures mentioned yields L-α-GPE in pure crystalline form, but each method is very time-consuming and expensive. In all probability, a reasonably convenient procedure for the preparation of L-α-GPE from natural phospholipids could be developed by replacing the crystalline acetalphospholipids as starting material with a more accessible phospholipid fraction containing L-α-GPE as exclusive nitrogenous component. This prerequisite appears to be essential at present, since all attempts to crystallize the substance from a mixture of water-soluble phosphodiesters have so far been unsuccessful. L-α-GPE is readily crystallized when no other phosphodiesters are present. For this reason, the most promising approach to the problem seems to be the preparation from the phosphatidylethanolamine fraction of Folch.[11a] This fraction can be obtained with a reasonable expenditure of time and effort in sufficient quantities to provide starting material for the preparation of L-α-GPE. There is no reason why the mercuric chloride hydrolysis described by Dawson[5] and applied successfully by Tattrie and McArthur[3] for the preparation of L-α-GPC should not be adaptable for the preparation of L-α-GPE.

Since the preparation of L-α-GPE from phosphatidylethanolamine has not been practically explored as yet, this possible approach can be offered only as a suggestion. Although the preparation of L-α-GPE from crystallized acetalphospholipids is easy once the starting material is available, the fact must be considered that processing of 40 pounds of beef brain yields approximately 1 g. of crystallized acetalphospholipids, and that 0.7 g. of L-α-GPE was obtained by Thannhauser et al.[10] from 3.5 g. of acetalphospholipid. For this reason, the preparation of L-α-GPE from natural sources will not be described here and the directions given will be limited to those devised by Baer and Stancer[11] for its organic synthesis from D-acetoneglycerol. Figure 1 summarizes the reactions in this synthesis.

Starting Materials. D-*Acetoneglycerol* is prepared by oxidation of 1,2,5,6-diacetone D-mannitol with lead tetraacetate to D-acetoneglyceraldehyde and subsequent reduction of the latter with Raney nickel.[11b]

Phenylphosphorus oxychloride (*boiling point 244°*) is best prepared

[10] S. J. Thannhauser, N. F. Boncoddo, and G. Schmidt, *J. Biol. Chem.* **188**, 417 (1951)
[11] E. Baer and H. C. Stancer, *J. Am. Chem. Soc.* **75**, 4510 (1953).
[11a] J Folch, *J. Biol. Chem.* **146**, 35 (1942).
[11b] E. Baer, *Biochem. Preparations* **2**, 31 (1952).

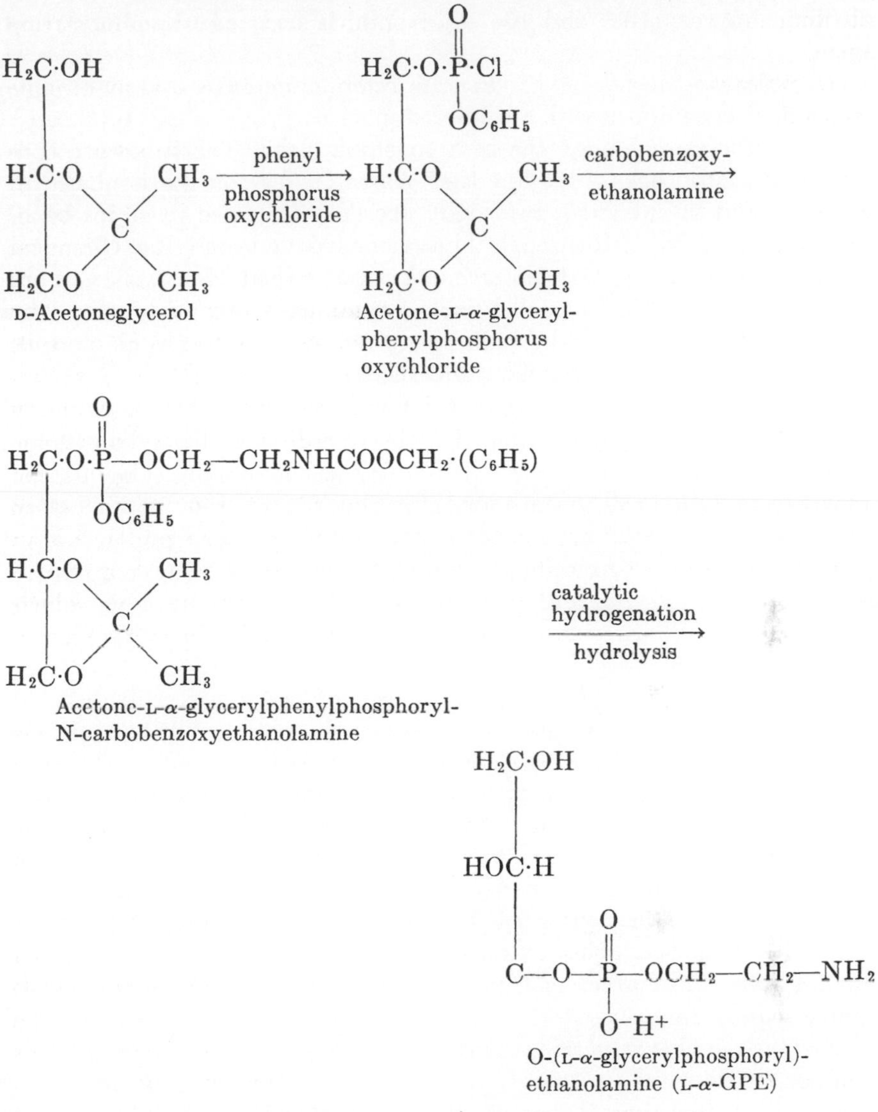

$$
\begin{array}{ccc}
\text{H}_2\text{C·OH} & & \overset{\displaystyle\text{O}}{\underset{\displaystyle\text{OC}_6\text{H}_5}{\text{H}_2\text{C·O·P·Cl}}} \\[2em]
\text{H·C·O}\quad\text{CH}_3 & \xrightarrow[\substack{\text{phosphorus}\\\text{oxychloride}}]{\text{phenyl}} & \text{H·C·O}\quad\text{CH}_3 \\
\qquad\text{C} & & \qquad\text{C} \\
\text{H}_2\text{C·O}\quad\text{CH}_3 & & \text{H}_2\text{C·O}\quad\text{CH}_3
\end{array}
$$

D-Acetoneglycerol — Acetone-L-α-glyceryl-phenylphosphorus oxychloride — carbobenzoxy-ethanolamine

$$\text{H}_2\text{C·O·P}-\text{OCH}_2-\text{CH}_2\text{NHCOOCH}_2\cdot(\text{C}_6\text{H}_5)$$

$$\underset{\displaystyle\text{OC}_6\text{H}_5}{|}$$

$$
\begin{array}{c}
\text{H·C·O}\quad\text{CH}_3 \\
\qquad\text{C} \\
\text{H}_2\text{C·O}\quad\text{CH}_3
\end{array}
\qquad \xrightarrow[\text{hydrolysis}]{\substack{\text{catalytic}\\\text{hydrogenation}}}
$$

Acetone-L-α-glycerylphenylphosphoryl-
N-carbobenzoxyethanolamine

$$
\begin{array}{c}
\text{H}_2\text{C·OH} \\
| \\
\text{HOC·H} \\
| \qquad\quad \overset{\displaystyle\text{O}}{\|} \\
\text{C}-\text{O}-\text{P}-\text{OCH}_2-\text{CH}_2-\text{NH}_2 \\
\underset{\displaystyle\text{O}^-\text{H}^+}{|}
\end{array}
$$

O-(L-α-glycerylphosphoryl)-
ethanolamine (L-α-GPE)

FIG. 1. Synthesis of L-α-GPE from D-Acetoneglycerol.

according to Zenftman, McGillivray, and Imperial Industries, Ltd.,[12] by refluxing 58.4 g. of phenol with 220 g. of phosphorus oxychloride in the presence of 8 g. of iron filings for 4 hours. The reaction product is fractionated by distillation under reduced pressure, and the fraction

[12] H. Zenftman, R. McGillivray, and Imperial Industries, Ltd., British Patent 651,656 (April 4, 1951) [C.A. **45**, 9081 (1951)].

distilling between 103° and 106° at 9 mm. is used as phosphorylating agent.

Carbobenzoxychloride[13,14,14a] (*Benzyl Chloroformate*). A 3-l. round-bottomed flask is equipped with a rubber stopper carrying an exit tube and a delivery tube extending to the bottom of the flask. The tubes are connected with stopcocks. Five hundred grams of dry toluene is placed in the flask, and the weight is recorded. The flask is cooled in an ice bath, and phosgene is bubbled through the toluene from a tank (Ohio Chemical Company) until 109 g. (1.1 moles) has been absorbed. The exit gases are passed through a 2-l., ice-cooled flask containing a liter of toluene (the inlet should almost reach the bottom) to remove any phosgene, and, in addition, through a calcium chloride tube.[14]

The connection to the phosgene tank is then replaced by a separatory funnel from which 108 g. (104 ml., 1 mole) of redistilled benzyl alcohol is added rapidly to the phosgene solution under gentle shaking. The mixture is allowed to stand in an ice bath for half an hour and at room temperature for 2 hours. The solution is concentrated under reduced pressure at a bath temperature not exceeding 60° for the purpose of removing hydrogen chloride, phosgene, and most of the toluene. The residue, which weighs 200 to 220 g., contains 155 to 160 g. of carbobenzoxychloride (approximately 90% yield).

Carbobenzoxyethanolamine.[15–17] A solution of 24.4 g. of freshly distilled ethanolamine in 50 ml. of water is placed in a three-necked flask equipped with a stirrer. The flask is immersed in an ice bath, and 68 g. of carbobenzoxychloride and 100 ml. of sodium hydroxide are added during the course of 40 minutes. This stirring is continued for half an hour in the ice bath and for half an hour at room temperature. The reaction mixture is just made acid to Congo paper and extracted with ether. The ether layer is washed twice with 0.5 N hydrochloric acid and with water, dried with magnesium sulfate, and left overnight at −30°. The filtered crystals have the correct melting point between 60° and 62.5°. Yield, approximately 35 g.

*Acetone-*L-*α-glycerylphenylphosphoryl-N-carbobenzoxyethanolamine.* It is important to maintain strictly anhydrous conditions during the whole procedure of phosphorylation. A 1-l., three-necked, thick-walled round flask with ground-glass joints is equipped with an oil-sealed, motor-driven

[13] N. Bergmann and L. Zervas, *Ber.* **65,** 1192 (1932).

[14] H. E. Carter, R. L. Frank, and H. W. Johnston, *Org. Syntheses* **23,** 13 (1943).

[14a] Carbobenzoxychloride may be purchased from Mann Research Laboratories, Inc., 136 Liberty Street, New York 6, New York.

[15] *Org. Syntheses* **14,** 2; *Coll. Vol.* **2,** 4 (1943).

[16] E. Chargaff, *J. Biol. Chem.* **118,** 417 (1937).

[17] W. G. Rose, *J. Am. Chem. Soc.* **69,** 1384 (1947).

stirrer, a calcium chloride tube, and a dropping funnel. Then 63.3 g. (0.3 mole) of monophenylphosphoryl dichloride and 100 ml. of glass beads of 6- to 7-mm. diameter are placed in the flask. The flask is immersed in a cold bath (−10°), and a mixture of 39.6 g. (0.3 mole) of freshly prepared D-acetone-glycerol, whose specific optical rotation must not be lower than +13.5°, and 39.0 ml. (0.33 mole) of dry quinoline is added dropwise within 10 minutes to the vigorously stirred contents of the flask. The rates of the addition of the acetoneglycerol solution are rapid at the beginning and slower during the later phases of phosphorylation. Five minutes after the addition of the acetoneglycerol solution has been completed, the cold bath is removed and the reaction mixture is allowed to attain room temperature. To the hard, white reaction mixture containing acetone-L-α-glycerylphenylphosphorylchloride as the principal product is added immediately 100 ml. of anhydrous pyridine (refluxed over barium oxide and redistilled under exclusion of moisture). The mixture is stirred until a fine suspension has formed and immersed in a water bath of a temperature between 10° and 15°. To the vigorously stirred mixture is added a solution of 58.5 g. of carbobenzoxyethanolamine in 120 ml. of anhydrous pyridine which has been brought to room temperature. The stirring is continued for 2 hours.

The pyridine is removed as completely as possible from the reaction mixture by distilling the mixture from the reaction vessel at 35° to 40° bath temperature at 8- to 10-mm. pressure. The viscous residue is stirred with 200 ml. of petroleum ether (boiling point 35° to 60°) for a few minutes, and the layers are allowed to separate. The petroleum ether extract is decanted, and the extraction is repeated with another 200-ml. portion of petroleum ether.

The petroleum ether-insoluble residue is extracted in a similar manner four times with 200-ml. portions of diethyl ether. The ether extracts are filtered and washed in succession as rapidly as possible with two 150-ml. portions of ice-cold 5 N sulfuric acid, 150 ml. of water, 150 ml. of a half-saturated solution of sodium bicarbonate, and finally with 150 ml. of water. The ether solution is dried with anhydrous sodium sulfate and concentrated under reduced pressure at a bath temperature of 35° to 40°. The residue is kept under a vacuum of 0.2 mm. at a bath temperature of 40° for 5 hours. The colorless oil weighs approximately 110 g., corresponding to a yield of almost 80%.

Removal of the Protecting Groups. The hydrogenolysis of the carbobenzoxy group, the deacetonation, and the removal of the sulfuric acid must be carried out on the same day because of the instability of L-α-GPE at acid reaction. The final solution of L-α-GPE may be stored overnight in the cold.

Preparation of Palladium Black.[14,18] Seventeen grams of palladium is dissolved in aqua regia in a 2-l. Kjeldahl flask, and the solution is evaporated twice to dryness. The residue is taken up in a small amount of hydrochloric acid. Then 1.5 l. of boiling water and 5 ml. of formic acid of the specific gravity of 1.22 are added to the contents of the flask. The solution is made just alkaline with a solution of potassium hydroxide. To the faintly alkaline suspension, more formic acid is added from a buret under vigorous shaking of the flask, until the liquid appears black because of the finely dispersed palladium and until a small excess of formic acid is indicated by a strong development of gas. It is advisable to keep an empty beaker handy during this operation in case of excessive foaming. After cooling, the palladium is filtered on a Büchner funnel. When the filtrate contains colloidal palladium it can be precipitated by the addition of formic acid. The palladium is washed with water, spread on watch glasses, and dried in an evacuated desiccator over sulfuric acid.

Preparation of Platinum Catalyst (Platinic Oxide).[19]

$$H_2PtCl_6 + 6NaNO_3 \rightarrow Pt(NO_3)_4 + 6NaCl + 2HNO_3$$
$$Pt(NO_3)_4 \rightarrow PtO_2 + 4NO_2 + O_2$$
$$PtO_2 + H_2O \rightarrow PtO_2 \cdot H_2O$$

In a porcelain casserole or in a Pyrex beaker,[19a] a solution of 3.5 g. of a commercial c.p. chloroplatinic acid (e.g., Mallinckrodt Chemical Works, St. Louis, Missouri) in 10 ml. of water is prepared. To this solution is added 41 g. of c.p. potassium nitrite.[19] The mixture is evaporated to dryness by gently heating over a Bunsen burner under stirring with a glass rod. The temperature is raised to 350° to 370° within 10 minutes. Fusion takes place under development of brown nitrous oxides, and a precipitate of brown platinum oxide gradually separates out. When excessive foaming occurs, the mass is stirred more vigorously, and a second Bunsen flame is directed at the top of the reaction mixture. If the burner under the casserole is removed when foaming starts, the top of the fused contents solidifies, and material may be carried over the side of the casserole. After 15 minutes, when the temperature has reached about 400°, the evolution of gas has greatly decreased. After 20 minutes, the temperature should be 500° to 550°. At this point, the evolution of gas has almost

[18] J. Tausz and N. von Putnoky, *Ber.* **52**, 1573 (1919).
[19] R. Adams, V. Voorhees, and R. L. Shriner, *Org. Syntheses, Coll. Vol.* **1**, 463 (1948). The description of the preparation of the platinic oxide catalyst is an abstract of the method of its preparation. The original reference should be consulted for some important properties of the catalyst and for some important suggestions not directly concerned with its application to the synthesis of L-α-GPE.
[19a] W. F. Short, *J. Soc. Chem. Ind.* **55**, 14T (1936).

subsided, and a gentle evolution of gas takes place. The temperature is kept at this point with the full force of the burner directly under the casserole until the fusion is complete after 30 minutes. The temperature conditions indicated above are important for the preparation of a catalyst of optimal activity.[20] If the gas pressure is not sufficient to attain the required temperature with an ordinary Bunsen burner, a Meker burner may have to be used. After cooling, the contents of the reaction vessel are treated with 50 ml. of water. The brown precipitate settles to the bottom and can be washed by decantation once or twice. It is then filtered through a hardened filter paper or a Gooch crucible and washed practically free of nitrates. No difficulties are encountered with the filtration if the fusion has been carried out according to the directions given, but if the temperatures are too low, or if the heating is not continued for a sufficiently long time, colloidal precipitates are frequently obtained which are difficult for filter. Occasionally the precipitate tends to become colloidal near the end of the washing. In this case it is preferable to stop washing, since small contaminations with nitrate do not impair the catalytic activity. All filtrates should be checked for the presence of platinum before being discarded.[19] The oxide is dried in a desiccator and weighed out for its use. The yield is 1.57 to 1.65 g. (95 to 100% of the theoretical amount).

Catalytic Hydrogenolysis of Phosphorylated Product. The phosphorylated product (18.6 g., 0.04 moles) is dissolved in 140 ml. of 99% ethanol. The clear solution is shaken with 4.65 g. (0.0435 mole) of freshly prepared palladium black in an all-glass reduction vessel of 800- to 1000-ml. capacity in an atmosphere of pure hydrogen at room temperature at a pressure of 40 to 50 cm. of water until the absorption of hydrogen ceases. This requires approximately 1 hour during which about 1 l. of hydrogen is absorbed. The hydrogen is replaced by nitrogen, the mixture filtered, the catalyst washed with 40 ml. of 99% of ethanol, and the combined filtrates transferred to another hydrogenation vessel of the same capacity. After addition of 5.4 g. (0.022 mole) of freshly prepared platinic oxide and 4 ml. of 5 N sulfuric acid, the mixture is again shaken in an atmosphere of hydrogen until the absorption of the gas ceases. During this phase, which lasts between 1 and 2½ hours, 5000 to 6000 ml. of hydrogen is absorbed. The hydrogen is replaced by nitrogen, the catalyst is filtered off and washed with 99% ethanol, and the combined filtrates are concentrated under reduced pressure at a bath temperature of 30° to 35° to a volume of approximately 50 ml.

Deacetonation. The concentrate is mixed with 200 ml. of water, and the mixture, which has a pH of 1.7 to 1.9, is kept at room temperature

[20] A. H. Cook and R. P. Linstead, *J. Chem. Soc.* **1934,** 952.

(25°) until 6 to 7 hours have elapsed since the start of the second hydrogenation. For the removal of the sulfuric acid and of a small amount of glycerophosphoric acid, an aqueous solution of lead acetate is added cautiously until no further precipitation occurs. The mixture is centrifuged, and a stream of hydrogen sulfide is passed through the slightly opalescent supernatant. After removal of the lead sulfide, the clear solution is concentrated to a small volume under reduced pressure at a bath temperature between 35° and 40°. Foaming is prevented by the addition of caprylic alcohol whenever necessary. When a precipitate appears on the addition of lead acetate, the treatment with lead acetate and hydrogen sulfide is repeated. The solution is finally concentrated under reduced pressure at a bath temperature between 35° and 40° to a viscous oil.

The colorless oil, weighing 8.2 g., is dissolved in 140 ml. of water, and the solution is stirred with 14 g. of Amberlite IR-120 (hydrogen form) for 1.5 hours. The filtrate is concentrated to an oil under reduced pressure at a bath temperature between 35° and 40°. The oil is dissolved in 12 ml. of water. After addition of 18 ml. of 99% ethanol, the mixture is clarified by centrifugation. To the decanted supernatant, 100 ml. of 99% ethanol is added, and the crystallization of the oily precipitate is induced by scratching the walls of the flask with a glass rod. The first crystals begin to appear within half an hour and form clusters of narrow prisms which may attain considerable length (6 to 7 mm.) when the crystallization is allowed to proceed slowly and without disturbance. The crystals are filtered off by suction and are washed with 20 ml. of 95% ethanol. They are dried *in vacuo* over calcium chloride. Yield, 4 g. (approximately 40% of the theory when the calculation is based on the amount of acetone-L-α-glycerylphenylphosphoryl-N-carbobenzoxyethanolamine). The substance may be recrystallized from its concentrated aqueous solution by adding 99% ethanol in the manner just described.

Some Properties of O-(L-α-glycerylphosphoryl)ethanolamine. Melting point, 86° to 87°; $\alpha_D^{26} = -29°$ (water, C 7.6). At room temperature, L-α-GPE (monohydrate) is insoluble in acetone, ether, chloroform, ethyl acetate, benzene, petroleum ether, dioxane, anhydrous pyridine, and dimethyl formamide. It is slightly soluble in 95% ethanol and easily soluble in methanol, glacial acetic acid, and water. The water of crystallization can be removed at a vacuum of 0.2 mm. over phosphorus pentoxide during 6 hours at 100°.

L-α-GPE resembles in principle L-α-GPC in its behavior toward hydrolysis reagents. It is completely resistant toward prostatic phosphomonoesterase, and can be differentiated from phosphorylethanolamine by incubation with this enzyme. Its ethanolamine group is quantitatively transformed to free ethanolamine ions by a 20 minute-hydrolysis in N hydrochloric acid in a boiling water bath. The hydrolysis of L-α-GPE by N aqueous sodium hydroxide proceeds somewhat slower than that of L-α-GPC at 100°, and much slower at room temperature.[20a]

The identification of L-α-GPE by paper chromatography has been described by Campbell and Work,[21] its behavior during ion exchange chromatography by Tallan et al.[22]

IV. Preparation of O-Phosphorylethanolamine (Molecular Weight 141.1)

$$NH_2 \cdot CH_2 \cdot CH_2 \cdot OPO_3H_2$$

O-Phosphorylethanolamine may be obtained in crystallized form either as barium salt or as the free ester. It has been isolated from tissues by Outhouse[23] and by Colowick and Cori,[24] but its synthesis in the laboratory is the practical method for its preparation in quantity. Outhouse synthesized the ester by phosphorylation of ethanolamine with phosphorus oxychloride, Plimmer and Burch by phosphorylation of ethanolamine with a mixture of phosphoric acid and phosphorus pentoxide. Since the latter procedure is occasionally unsuccessful, perhaps owing to the quality of the sample of commercial phosphoric acid used, the procedure of Outhouse will be described in detail.

During a reinvestigation of the procedure of Plimmer and Burch, the author found recently that the phosphorylation of ethanolamine to O-phosphorylethanolamine can be accomplished in good yields with phosphoric acid when the commercial preparations of phosphoric acid are treated according to Riley[8] (see p. 348) prior to the phosphorylation, and when the phosphorylation is carried out in an oil bath of 160° to 180°.

Procedure. A mixture of 10 ml. of ethanolamine and 5.5 ml. of water is added dropwise from a buret to 15 ml. of phosphorus oxychloride in a 500-ml. suction flask immersed in cold water. After completion of the reaction, the mixture is evacuated on a suction pump for 1 hour for the purpose of removing as much as possible of the hydrochloric acid. At this stage, about 80% of the phosphorus is esterified. The mixture is taken up in 500 ml. of water, and the reaction of the solution is brought to pH 10 by the addition of a concentrated solution of barium hydroxide. The precipitated barium phosphate is removed by centrifugation, and the barium salt of ethanolamine phosphoric acid is precipitated from the combined supernatant solution and washings by the addition of 2 vols. of

[20a] G. Schmidt, M. J. Bessman, and S. J. Thannhauser, J. Biol. Chem. **203**, 849 (1953).
[21] P. N. Campbell and T. S. Work, Biochem. J. **50**, 449 (1951-1952).
[22] H. H. Tallan, S. Moore, and W. H. Stein, J. Biol. Chem. **211**, 927 (1954).
[23] E. L. Outhouse, Biochem. J. **31**, 1459 (1937).
[24] S. P. Colowick and C. F. Cori, Proc. Soc. Exptl. Biol. Med. **40**, 586 (1939).

alcohol. The precipitate is filtered, washed with alcohol and ether, dried at room temperature, and redissolved in 200 ml. of water. The solution is again precipitated with alcohol. After three reprecipitations, the barium salt is crystalline. The yield is about 50% of the theoretical.

The crystalline free ester may be prepared from the barium salt by removing the barium ions with sulfuric acid and by crystallizing the aminoethanol phosphoric acid from 80% of aqueous methanol. The melting point (corrected) of the free ester was found to be 244°.

Chargaff and Keston[25] prepared P^{32}-labeled aminoethanol phosphoric acid according to the procedure described above.

Some Properties of Ethanolamine Phosphoric Acid. Like phosphorylcholine, phosphorylethanolamine is very resistant toward acid and alkaline hydrolyzing agents. Its quantitative cleavage requires refluxing for 18 hours with 6 N aqueous hydrochloric acid at 110° (bath temperature 135° to 140°).[22] Phosphorylethanolamine is rapidly hydrolyzed by alkaline intestinal and by acid prostatic phosphatase. It is not oxidized by periodic acid, in contrast to ethanolamine.

The brucine salt of the ester crystallizes from acetone, but not from ethanol or methanol. The more stable quinine salt may be crystallized from alcohol.

Phosphorylethanolamine may be identified by paper partition chromatography or by ion exchange chromatography.[22,26-28]

[25] E. Chargaff and A. S. Keston, *J. Biol. Chem.* **134,** 515 (1940).
[26] A. H. Gordon, *Biochem J.* **45,** 99 (1949).
[27] J. Awapara, A. J. Landua, and R. Fuerst, *J. Biol. Chem.* **183,** 545 (1950).
[28] T. Astrup, G. Carlstrom, and A. Stage, *Acta Physiol. Scand.* **24,** 202 (1951).

[59] Nitrogenous Constituents of Phospholipids

By C. Artom

I. Hydrolysis of Lipid Extracts

Total Hydrolysis

Nitrogenous contaminants are present in unfractionated lipid extracts and also in the phospholipids precipitated from these extracts with acetone + $MgCl_2$. These contaminants can be largely eliminated by emulsifying the solution of lipids in chloroform with the same volume of 0.25 M $MgCl_2$, the process being repeated six or seven times (McKibbin and Taylor[1]). Other procedures for the purification of tissue lipids or

[1] J. M. McKibbin and W. E. Taylor, *J. Biol. Chem.* **178,** 17 (1949).

phospholipids have been described.[2] It seems, however, that, except for the determination of sphingosine, based on the measurement of chloroform-soluble N, the procedures described below can be applied directly to lipid extracts, even if incompletely purified.

Hydrolysis by refluxing with saturated $Ba(OH)_2$ for 2 to 5 hours is said to yield maximum liberation of choline.[1] In the writer's experience, however, partial destruction of ethanolamine and, less so, of serine may occur during hydrolysis with alkalis. When these compounds are determined, the procedure of Thannhauser et al.[3] seems preferable. The lipid extract is refluxed for 3 hours with 5 or 10 ml. of 6 N HCl in methanol. This is easily removed at 60° under slightly reduced pressure. The dry residue is taken up in a definite volume of water. An aliquot of the suspension may be used in the determination of sphingosine. The remaining suspension is chilled and filtered through well-packed asbestos with the aid of some Celite. If the liquid, while filtering slowly, is kept in a refrigerator, a perfectly clear filtrate is generally obtained which can be used for the other determinations.

Selective Hydrolysis

The procedure was suggested by Schmidt et al.[4] The lipid sample is placed in a 15-ml. graduated centrifuge tube. After evaporation of the solvent, the dry residue is dissolved in 0.3 ml. of hot ethanol and emulsified in 5 ml. of N KOH, blown vigorously from a pipet. The emulsion is left at 38° for 16 to 24 hours, with occasional shaking. One milliliter of 1% egg albumin, 1 ml. of 5 N HCl, and 1 ml. of 10% TCA are then added. After standing for 2 hours at room temperature, the emulsion is brought to a definite volume, centrifuged, and filtered through paper without suction. The filtrate contains as free choline all the choline liberated from lecithin but none from sphingomyelin.[5] On the other hand, ethanolamine is still largely present in water-soluble combinations (probably glycerol-phosphorylethanolamine). These combinations can be split by refluxing the filtrate for 2 hours with one-tenth its volume of concentrated HCl (Artom, unpublished).

[2] E. Le Breton, Bull. soc. chim. biol. **3**, 539 (1921); J. Folch and D. D. Van Slyke, Proc. Soc. Exptl. Biol. Med. **41**, 514 (1939); G. Brante, Acta Physiol. Scand. **18**, Suppl., 63 (1949); J. Folch, I. Ascoli, M. Lees, J. A. Meath, and F. N. LeBaron, J. Biol. Chem. **191**, 833 (1951).
[3] S. J. Thannhauser, J. Benotti, and H. Reinstein, J. Biol. Chem. **129**, 709 (1939).
[4] G. Schmidt, J. Benotti, B. Hershman, and S. J. Thannhauser, J. Biol. Chem. **166**, 505 (1946).
[5] M. H. Hack, J. Biol. Chem. **169**, 137 (1947).

II. Choline

Physiological, microbiological, chromatographic, and chemical techniques have been used for the determination of choline. Among the last group of procedures, the determination as the enneaiodide seems to offer the highest degree of sensitivity, flexibility, and relative specificity. The reineckate methods are somewhat shorter and have been used more extensively, however.

Determination as Enneaiodide

Principle. The procedure described here is the original Roman's method[6] with minor modifications.[7-9] Choline is precipitated as the enneaiodide; the iodine is extracted with chloroform and titrated with $Na_2S_2O_3$.[10]

Reagents

I_2-KI, 16 g. of I_2 and 20 g. of KI in 100 ml. of water.
0.1 N $Na_2S_2O_3$, from which a 0.005 N solution is freshly prepared by dilution.

Procedure. A portion of the lipid hydrolyzate is placed in a 15-ml. glass-stoppered centrifuge tube. The tube is chilled, and 0.4 ml. of I_2-KI reagent is added for each milliliter of hydrolyzate. After 30 minutes in ice, the tube is centrifuged at 2500 r.p.m. for 15 minutes, and replaced in ice. The supernatant fluid is removed with an alundum immersion filter stick, and the precipitate is washed with 1-ml. portions of iced water until the wash solution is colorless. Usually five or six washings are sufficient. Care is taken not to disturb the precipitate and to leave it covered with a thin layer of water. Then 2 to 3 ml. of chloroform is added. The walls of the tube and the filter stick are rinsed with chloroform, a little chloroform being forced also through the filter with the aid of a rubber bulb. The red or pink chloroform solution is then titrated with 0.005 N $Na_2S_2O_3$ until the color disappears. The end point is easily

---

[6] W. Roman, *Biochem. Z.* **219**, 218 (1930).

[7] E. Kirk, *J. Biol. Chem.* **123**, 623 (1938).

[8] B. N. Erickson, I. Arvin, D. M. Teague, and H. H. Williams, *J. Biol. Chem.* **135**, 671 (1940).

[9] C. Artom and W. H. Fishman, *J. Biol. Chem.* **148**, 415 (1943).

[10] The sensitivity of the procedure can be increased sixfold by converting the I_2 to IO_3^- and titrating in a water medium in the presence of an excess of KI (Erickson *et al.*[8]). In a method recently described [H. D. Appleton, B. N. LaDu, Jr., B. B. Levy, J. M. Steele, and B. B. Brodie, *J. Biol. Chem.* **205**, 803 (1953)], the enneaiodide dissolved in ethylene dichloride is determined spectrophotometrically. It is claimed that the method can estimate as little as 5 γ of choline.

detected by viewing the tube against a white background. One milliliter of $Na_2S_2O_3$ = 0.067 mg. = 0.55 micromole of choline. On samples containing between 0.3 and 2 mg. of choline, duplicate determinations agree within 2 to 3%.

Determination as Reineckate

Principle. Kapfhammer and Bischoff[11] were the first to determine choline by precipitating it with reinecke acid. The method was adapted to a colorimetric procedure by Beattie.[12] The spectrophotometric determination described below was suggested by Hack.[5]

Reagent. Saturated solution of NH_4-reineckate in 0.5 N HCl. This solution must be prepared the same day.

Procedure. To an aliquot of the filtered lipid hydrolyzate 2 ml. of the reinecke acid solution is added. After 30 minutes at room temperature, the precipitate is collected on a Pyrex M filter tube by means of gentle suction, washed twice with 2 ml. of ethanol, and dried by drawing air through the filter. It is dissolved in 3 ml. of acetone, and its optical density is determined in a Beckman spectrophotometer at 526 mμ. The concentration of choline is estimated from a standard curve, prepared by treating in exactly the same manner known amounts of choline chloride. In the range between 150 and 600 γ, duplicate determinations agree within 3%.

III. Ethanolamine and Serine

Separate determinations of the amino N (by the HNO_2 method) and of the amino acid-carboxyl C (liberated by ninhydrin) have been employed for estimating the sum of ethanolamine + serine, and serine, respectively. These procedures applied to unpurified or unfractionated lipid extracts lack specificity. In other methods, ethanolamine is first distilled, then determined in the distillate acidimetrically, colorimetrically, or from the amino N. Paper chromatography was used also for the separation of serine and ethanolamine.

Determination from NH_3-Periodate

Principle. This procedure is a slight modification of the method published by Artom.[13] Sphingosine is removed by extraction of the lipid hydrolyzate with chloroform. Ethanolamine is separated from serine by adsorption on Permutit and then eluted with NaCl. From the NH_3 liberated by periodate in the effluent and in the eluate, serine and ethanolamine, respectively, are determined.

[11] J. Kapfhammer and C. Bischoff, *Z. physiol. Chem.* **191**, 179 (1930).

[12] F. J. R. Beattie, *Biochem. J.* **30**, 1554 (1936).

[13] C. Artom, *J. Biol. Chem.* **157**, 585 (1945).

Reagents

Permutit.

NaCl, 30%.

Paraffin oil.

Alkaline borate, 8.2 g. of boric acid in 100 ml. of N NaOH.

0.2 M periodic acid ($HIO_4 + 2H_2O$).

0.1 N H_2SO_4 and 0.1 N $Na_2S_2O_3$, from which 0.01 N and 0.005 N solutions, respectively, are freshly prepared by dilution.

KI, 10%.

KIO_3, 5%.

Soluble starch, 1 g. in 100 ml. of NaCl, 30%.

Procedure. The aqueous suspension of the lipid hydrolyzate[14] is extracted in a small separatory funnel with chloroform (see below: Determination of Sphingosine). The aqueous phase is freed from traces of the organic solvent with a stream of air, filtered, and sent through a Permutit column[15] which is then washed with a few milliliters of H_2O. Ethanolamine is eluted with 5 ml. of 30% NaCl, followed by three washings with 1 to 2 ml. of H_2O.[16] The serine-containing effluent, and the ethanolamine-containing eluate, respectively, are transferred to the digestion flask of the Parnas-Wagner apparatus for micro-Kjeldahl.[17] The following are then added in the order given: 1 drop of paraffin oil, 3.5 ml. of alkaline borate, and 2 ml. of periodic acid. The mixture is steam-distilled for 7 minutes into 5 or 10 ml. of 0.01 N H_2SO_4. The receiving flask is lowered, the distillation continued for 1 minute, and the tip of the condenser rinsed with water. After cooling, 3 ml. each of KI and KIO_3 are added. The flask is stoppered, and, after 5 minutes, the liberated iodine is titrated with 0.005 N $Na_2S_2O_3$ from a microburet, graduated in intervals of 0.05 or 0.01 ml. The titration is completed after addition of 10 drops of starch solution. A blank is run on all reagents, including the methanolic HCl used in the lipid hydrolysis.

One milliliter of 0.005 N $Na_2S_2O_3$ equals 5 micromoles of ethanolamine

[14] For quantitative adsorption of ethanolamine the solution should contain only minimal amounts of other cations. The easiness with which methanolic HCl can be removed from the hydrolyzate makes this method of hydrolysis especially suitable for the subsequent separation of ethanolamine from serine.

[15] A piece of glass tubing 120 mm. long and 7 mm. in external diameter, containing 1.5 to 3 g. of Permutit.

[16] When sufficient material is available the sum of ethanolamine + serine may be determined on one fraction of the hydrolyzate. On another fraction, after adsorption of ethanolamine on Permutit, serine is estimated. Ethanolamine is calculated by difference.

[17] J. K. Parnas and R. Wagner, *Biochem. Z.* **125**, 253 (1921).

(0.300 mg.) or serine (0.525 mg.). For amounts between 20 and 400 micromoles, duplicate analyses usually agree within 3 to 5%. Recoveries of 90 to 100% were obtained with addition of pure ethanolamine and/or serine to lipid extracts. With mixtures of C^{14}-serine and inactive ethanolamine, not more than 1% of the radioactivity was found in the ethanolamine fraction.[18]

Determination as Dinitrophenyl Derivatives

Principle. The formation of colored phenyl derivatives of substances containing a free NH_2 group was first used by Robins and Lowry[19] on brain lipids. In the method of Axelrod *et al.*,[20] described below, the ethanolamine and serine derivatives are separated on the basis of their differential solubilities in organic solvents. This highly sensitive procedure has been applied to plasma and red cells only. It can probably be extended to other lipid extracts, after removal of the nitrogenous contaminants.

Reagents

DNFB, 0.1 ml. of dinitrofluorobenzene dissolved in 2 ml. of absolute alcohol, shortly before being used.

Methyl isobutyl ketone (MIK).

Standard 0.01 M ethanolamine and serine solutions, from which working standards are prepared by dilution.

NaHCO$_3$, 2.5%.

0.1 N NaOH.

5 N, 2 N, and N HCl.

Procedure. Two milliliters of the hydrolyzate of purified[21] lipids is transferred into a centrifuge tube,[22] and 0.1 ml. of DNFB reagent and 1 ml. of NaHCO$_3$ are added. The tube is stoppered and placed in a water bath at 75 to 80° for 1 hour. The tube is cooled, and 8 ml. of chloroform is added, followed by 10 minutes of shaking and centrifugation.

Two milliliters of the aqueous layer is transferred to another tube[22] containing 1 ml. of N HCl. Ten milliliters of MIK is added, followed by 10 minutes of shaking. Eight milliliters of MIK solution is added to a third tube[22] containing 4 ml. of 0.1 N NaOH. After 5 minutes of shaking, the MIK phase is removed by aspiration. Then 3 ml. of the aqueous phase is transferred to a cuvette containing 0.5 ml. of 2 N HCl, and the

[18] M. Levine and H. Tarver, *J. Biol. Chem.* **184**, 427 (1950).

[19] E. Robins and O. H. Lowry, *Federation Proc.* **10**, 238 (1951).

[20] J. Axelrod, J. Reichenthal, and B. B. Brodie, *J. Biol. Chem.* **204**, 903 (1953).

[21] The procedure of Folch and Van Slyke for the purification of plasma lipids[2] was found satisfactory.

[22] Fifteen-milliliter glass-stoppered centrifuge tubes.

optical density is determined at 420 mμ. The reading indicates the serine content of the hydrolyzate.

After removal of the aqueous phase, 6 ml. of the chloroform phase is transferred to a centrifuge tube.[23] The chloroform is evaporated to dryness. Ten milliliters of petroleum ether and 4 ml. of 5 N HCl are added, followed by 5 minutes of shaking. Then 3 ml. of the aqueous phase is transferred to a cuvette, and the optical density is read at 420 mμ. The reading indicates the ethanolamine content of the sample.

Reagent blanks are used for the zero setting. Standards of ethanolamine and serine are treated exactly as described above. Optical densities of 0.416 and 0.103 are obtained with 20 γ of ethanolamine and 20 γ of serine, respectively.

IV. Sphingosine

The procedures described below are based on the solubility of sphingosine in organic solvent, or on the resistance of sphingolipids to hydrolysis. Dihydrosphingosine and possibly other nitrogenous constituents of phospholipids, unidentified or only incompletely identified, may be included in the determinations described below.

Determination of the Chloroform-Soluble N

Principle. The method described by McKibbin and Taylor,[24] consists in a preliminary purification of the lipids, followed by hydrolysis, extraction with CHCl$_3$, and determination of total N in the CHCl$_3$ extract.

Reagents. All reagents required for a total N determination (micro-Kjeldahl with nesslerization of the distillate). See Vol. III [145].

Procedure. The lipid extract is purified by repeated emulsification with aqueous MgCl$_2$ (see above[1]). It is then hydrolyzed.[25] The hydrolyzate is transferred to a 125-ml. separatory funnel. The flask is dried at 60° and then rinsed with chloroform and warm water, the final volumes of the water and chloroform phases being about 75 and 25 ml., respectively. After shaking, the chloroform layer is withdrawn into a 50-ml. centrifuge tube, and the extraction is repeated twice with 10 to 15 ml. of chloroform. The combined chloroform extracts are cleared by centrifugation and transferred to a 50-ml. volumetric flask. Chloroform is finally added to the mark, and aliquots of the extract are used for duplicate determinations of total N. The method was tested in the range of 3 to 20 micromoles of sphingosine. An average of 93% of sphingosine, added to lipids, was recovered.

[23] Forty-milliliter glass-stoppered centrifuge tubes.

[24] J. M. McKibbin and W. E. Taylor, *J. Biol. Chem.* **178**, 29 (1949).

[25] The authors used a 5- to 7-hour hydrolysis with saturated Ba(OH)$_2$, followed by a 1¾-hour hydrolysis with concentrated HCl (1:10).

Determination from NH_3-Periodate

Principle. Sphingosine is separated from the lipid hydrolyzate by chloroform extraction, as in the method described above. The chloroform solution of sphingosine is emulsified with aqueous boric acid and periodate. The NH_3 liberated is then determined. Alternatively, sphingosine can be estimated from the difference between two NH_3-periodate determinations, on the products of total lipid hydrolysis and on the products of the selective hydrolysis of glycerolphospholipids. Both procedures can probably be applied directly to lipid extracts, without further purification (Artom, unpublished).

Reagents

Saturated solution of boric acid.
0.2 *M* sodium periodate.
N NaOH. Other reagents as listed in the determination of ethanolamine and serine by the NH_3-periodate method.

Procedure. The lipid hydrolyzate is extracted with $CHCl_3$. The chloroform extract is evaporated to 1 or 2 ml. Five milliliters of saturated boric acid and 2 ml. of Na periodate are added. The flask is shaken vigorously and repeatedly during 1 hour. It is then placed on a water bath at 60°, and a current of air is passed until the last traces of chloroform are removed. The content is transferred into the Parnas-Wagner distillation apparatus, and 3 ml. of *N* NaOH is added. The steam distillation and the iodometric titrations of the distillate are carried out as described for the determination of ethanolamine and serine. One milliliter of 0.005 *N* $Na_2S_2O_3$ equals 5 micromoles of sphingosine (1.495 mg.).

Alternative Procedure. A portion of the lipid extract is hydrolyzed with methanolic HCl which is later removed. Then 1 to 2 ml. of chloroform, 5 ml. of boric acid, and 2 ml. of Na periodate are added, and the total NH_3 liberated by periodate is determined, as described above. Another portion of the extract is hydrolyzed with normal KOH at 38° (see Selective Hydrolysis of Lipid Extracts). The filtrate is refluxed for 2 hours with one-tenth its volume of concentrated HCl. It is then neutralized with saturated NaOH, and the NH_3 liberated by periodate is determined. Sphingosine is estimated from the difference between the two determinations of NH_3-periodate.[26]

---

[26] The ultramicromethod, briefly described by Robins and Lowry,[19] is based also on the selective hydrolysis of the lipid extract.[4] The unhydrolyzed sphingolipids are extracted with octanol and hydrolyzed. Sphingosine is finally determined colorimetrically as the dinitrophenyl derivative.

V. Dimethylethanolamine

Dimethylethanolamine (DME) is not known to be a normal constituent of natural phospholipids. However, it is a likely precursor of choline, and its incorporation into tissue phospholipids can be easily demonstrated after administration of massive doses to the animals[27] and also in *in vitro* experiments on isolated tissues.[28]

Principle. In the method, briefly described by Artom and Crowder,[27] DME is steam-distilled into standard acid, and the excess acid is titrated. The method includes also certain steps by which interfering substances are eliminated.[29] Some of these steps can probably be omitted, when relatively high amounts of DME are estimated and interfering substances are present in negligible amounts (such as in hydrolyzates of purified lipid extracts).

Reagents

$NaNO_2$, 30%, prepared immediately before being used.

Glacial acetic acid.

Saturated NaOH.

Alkaline borate (6.2 g. of boric acid in 100 ml. of N NaOH).

KI, 10%.

KIO_3, 4%.

0.01 N HCl and 0.005 N $Na_2S_2O_3$, freshly prepared from the 0.1 N solutions.

Soluble starch, 1% in NaCl, 30%.

Standard $(NH_4)_2SO_4$, containing 0.05 mg. N in 1 ml.

Nessler's reagent.

Procedure. To the sample, placed in a 100-ml. beaker, 3 ml. of 30% $NaNO_2$ and 1.5 ml. of glacial acetic acid are added and stirred vigorously. After 5 minutes, the solution is made alkaline to phenolphthalein with saturated NaOH and transferred to the distilling flask of the Wagner-

---

[27] C. Artom and M. Crowder, *Federation Proc.* **8**, 180 (1949).

[28] M. Crowder and C. Artom, *Federation Proc.* **11**, 199 (1952).

[29] In a direct steam distillation the following values were obtained (in micromoles of ammonia or volatile amines per 100 micromoles of substance): trimethylamine, 97; creatinine, 27; methylethanolamine, 20 to 30; arginine, 29; asparagine, 19; glutamine, 3; urea, 2.8; guanine, 2; guanidoacetic acid, 2; ethanolamine, 2; choline, 1; histidine, 1; glutathione, 0.9; xanthine, 0.9. All other substances tested (adenine, adenylic acid, alanine, aspartic acid, betaine, cysteine, cystine, glycine, lysine, methionine, ornithine, proline, sarcosine, serine, taurine, thymine) gave values between 0 and 0.5. When the procedure was carried out as described, identical values were obtained on DME solutions in the presence or in the absence of the substances listed above.

Parnas micro-Kjeldahl apparatus. One drop of paraffin oil and 3.5 ml. of alkaline borate are added, a 125-ml. side-arm flask with a one-hole rubber stopper being substituted for the receiving flask. The side arm is connected to a vacuum pump, and NH_3-free air is passed through the system at room temperature. After 30 minutes, the side flask is removed. A 250-ml. flask containg 5 or 10 ml. of 0.01 N HCl is used as a receiver. The distilling flask is connected with the steam generator, and steam is passed through the system for 15 minutes at a rate sufficient to bring the distillate to about 120 ml. From the distillate, an aliquot is taken for the colorimetric determination of NH_3. On the remaining distillate, the free acid is titrated iodometrically (see above: Determination from NH_3-Periodate). The aliquot for NH_3 is evaporated in a beaker to about 5 ml., then transferred to a calibrated colorimeter tube. Nessler's reagent is added, and NH_3 is determined, using a 420-mμ filter. A reagent blank is run throughout the whole procedure. After correction for the blank, the combined acidity and the ammonia values are expressed in micromoles for the total volume of distillate. The difference between these two values indicates the micromoles of DME in the sample. Recoveries of 90 to 97% of the DME added to lipid extracts were obtained. Duplicate analyses agree within 3 to 4%.

[60] Chromatographic Separation of Volatile Fatty Acids

By ERNEST BUEDING

Chromatographic separation of volatile fatty acids can be carried out because of differences in their partition coefficients between water and organic solvents. Their solubility in organic solvents increases with the length of their carbon chain; it decreases as the pH of the aqueous phase is raised and as the concentration of butanol in certain organic solvents, e.g., chloroform, is lowered. Accordingly, separation of the acids is carried out by using two variables: (1) the pH of the buffer present in the non-mobile aqueous phase, and (2) the butanol concentration of the eluent.

A. Separation of Volatile Acids Containing One to Six Carbons

Quantitative separation of acids according to the length of their carbon chain is achieved by means of the procedures described below.[1] n-Butyric, n-valeric, and n-caproic acids are eluted together with their corresponding isomers.

[1] E. Bueding and H. Yale, J. Biol. Chem. **193**, 411 (1951).

Reagents

Chloroform (reagent grade), washed free of alcohol by percolation with distilled water for about 1 hour. The chloroform is separated and filtered.

Celite (Johns-Manville, No. 535) is washed in a column with peroxide-free ether (about 1 l. per 500 g. of Celite). The powder is then dried *in vacuo* over anhydrous calcium chloride. In preparing the columns 0.6 ml. of buffer is added per gram of Celite and ground with it in a mortar. Chloroform which has been equilibrated with the buffer is added to the mixture until, after mixing, a small volume of supernatant solvent is present.

Butanol (reagent grade), purified by distillation *in vacuo*.

Phosphate buffers (2 *M*) (pH 7.6, 6.5, and 3.5), prepared from stock solutions (2 *M*) of K_2HPO_4, KH_2PO_4, and H_3PO_4. The pH of these buffers is determined on appropriate dilutions (1:30). Buffers of pH 6.5 and 3.5 and the stock solution of KH_2PO_4 are kept at 37° in order to prevent crystallization. *All solvents used for each column are equilibrated with the buffer used as the nonmobile phase* (2 to 5 ml. of buffer per 100 ml. of solvent). The solvent is filtered in order to remove any water remaining in suspension.

Indicators. Solutions of bromothymol blue and thymol blue.[2]

The sizes of the columns vary according to the *total* amount of acid contained in the mixture to be analyzed.

Mixtures Containing 0.01 to 0.15 Milliequivalent of Acid. Eight grams of Celite and a glass tube (inside diameter, 12 mm.) provided with a stopcock are used. A small volume of chloroform is put into the column, and the Celite-buffer-chloroform mixture is added, about 1 g. at a time. After each addition, the Celite is stirred with a slender rod and tamped down with a glass rod which has a flattened end, the diameter of which is almost as large as that of the column. The rod is twisted as it is lowered, an even downward pressure being exerted. The column is rotated, and any floating particles of Celite are tamped down evenly. It is very important to pack the Celite into the column carefully and evenly in order to avoid the formation of vertical "channels" through which the acids flow more rapidly, a condition which results in overlapping of the acids instead of their separation. If the buffer is ground too firmly with the Celite, or if too much pressure is exerted in the packing of the column, some buffer may be squeezed out of the Celite, causing a high blank value.

The acid mixture to be analyzed is added to the Celite column in a total volume not exceeding 10 ml. of chloroform; the effluent is allowed

[2] W. M. Clark, "Determination of Hydrogen Ions," 3rd ed., p. 94, 1928.

to flow from the column at its natural or at an accelerated rate obtained by air pressure. The rate of flow appears to have no effect on the results obtained.

Mixtures Containing 0.15 to 1.0 Milliequivalent of Acid. Forty grams of Celite and glass columns which have an inside diameter of 26 mm. are employed. A perforated porcelain disk is used in addition to the glass wool as support for the Celite. The Celite-chloroform suspension is put into the column and stirred well with a rod. Excess solvent is added to within 1 inch of the top. A one-holed rubber stopper, which carries tubing connected with a source of air pressure, is inserted into the top of the column. The air pressure is allowed to reach about 400 mm. of Hg; then the stopcock is opened and the solvent allowed to flow out until its level is only slightly above the surface of the Celite. The pressure packing is repeated, and the Celite is further packed down by hand, as described above. At most, 25 ml. of solvent containing the mixture of volatile acids is placed on the column. Allowing the columns to run dry has no adverse effect on the results obtained if the period during which there has been no solvent above the Celite is of short duration (30 minutes or less). Drying over a longer period of time appears to lower the recovery of the acids.

Micromethod (mixtures containing 1 to 20 microequivalents of acid).[3,4] The procedure is the same as the one described above for 0.01 to 0.15 milliequivalent, except that 4 g. instead of 8 g. of Celite is used and titration is carried out with an alkali of lower normality (0.001 N; see below).

Preparation of Samples for Chromatographic Analysis. Steam distillates are neutralized with sodium bicarbonate (pH 7 to 8), concentrated *in vacuo* to a small volume and lyophylized. The residue is taken up in 2 ml. of pure chloroform (equilibrated with water) and made acid to pH 2 (red to thymol blue) with concentrated H_2SO_4. Anhydrous sodium sulfate is added to remove water; thereafter the mixture is centrifuged and the supernatant decanted. The sodium sulfate is washed twice with small portions of chloroform followed by one washing with 1% butanol in chloroform. A small aliquot of the combined supernatant fluids is titrated in order to determine the total acidity.

An alternative method[5] consists in dissolving the dry sodium salts of the acids in a small volume of buffer and grinding this solution thoroughly in a mortar with Celite (1 g. of Celite per 0.6 ml. of buffer). The amount of Celite should not exceed 5% of that used in preparing the columns. Chloroform equilibrated with the buffer is added to the mortar and ground with the buffer-salt-Celite mixture which is then placed and tamped down on the column.

[3] E. F. Kohlmiller, Jr. and H. Gest, *J. Bacteriol.* **61**, 269 (1951).
[4] H. Gest, personal communication.
[5] H. Rudney, personal communication.

Titration. Aliquots of the effluent are titrated after adding to the sample an equal volume of distilled water and 2 drops of a solution of bromothymol blue. CO_2-free nitrogen is passed through the mixture before and throughout the titration with an aqueous potassium hydroxide solution (0.001 to 0.01 N). The alkali required to neutralize a similar aliquot obtained from the column before any acid has been eluted is used as the blank value.

Elution of the Acids. When the buffer of the nonmobile phase has a pH of 7.6, caproic acid and its isomers are eluted with pure chloroform, and valeric acid and the lower volatile acids are held by the column. With 5% butanol in chloroform, valeric acid and its isomers are eluted, and, when the butanol concentration is increased to 20%, butyric and isobutyric acids are obtained. Propionic, acetic, and formic acids remain on the column. These acids are eluted separately by the use of buffers having a lower pH. At pH 6.5, acids containing four or more carbon atoms are eluted with pure chloroform, propionic acid with chloroform containing 3 to 5% butanol, and acetic acid with 20% butanol. Formic acid, because of its relatively low solubility in organic solvents, requires a pH of 3.5 to pass into 20% butanol in chloroform. In this case all other volatile acids are removed previously from the column, with chloroform or chloroform containing butanol in a concentration of less than 20%. Quantitative recoveries are obtained with all volatile acids except formic acid, which is recovered only to an extent of approximately 75%. For further details, the reader is referred to the table.

ELUTION OF VOLATILE ACIDS FROM CELITE COLUMNS

pH of buffer	Butanol in $CHCl_3$, %	Nature of acids eluted	Approximate volume required to elute acids at maximum capacity of each column, ml.		
			4-g. Celite column	8-g. Celite column	40-g. Celite column
7.6	0	C_6	50	80	400
7.6	5	C_5	40	70	300
7.6	20	C_4	40	70	450
6.5	0	C_4 to C_6	50	90	550
6.5	3	Propionic	40		
6.5	5	Propionic		60	400
6.5	20	Acetic	50	100	600
3.5	0	C_2 to C_6		50	400
3.5	5	Acetic		50	250
3.5	20	Formic		80	400

B. Partial Separation of Pentanoic Acids and of α,β-Unsaturated Pentenoic Acids by Partition Chromatography

Partial separation of the isomers of valeric acid and of α,β-unsaturated pentenoic acids can be carried out by partition chromatography when the pH of the buffer is 7.8 and when 1% butanol in chloroform is the eluent. A mixture not exceeding 0.01 milliequivalent of pentanoic acids is placed on a column (inside diameter, 16 mm.) consisting of 12 g. of Celite and 7.2 ml. of potassium phosphate buffer (2 M, pH 7.8). For a total load of 0.2 milliequivalent of these acids, a column with an inside diameter of 26 mm., 48 g. of Celite, and 28.8 ml. of buffer are used. Under these conditions, trimethylacetic acid is eluted first; it is followed by n-valeric acid and finally by a mixture of α-methylbutyric and isovaleric acids. The behavior of these latter two acids on partition chromatography is so similar that they cannot be separated by this method. The same procedures can be applied for the partial separation of α,β-unsaturated pentenoic acids;[6] cis-α-methylcrotonic (tiglic) acid and β,β-dimethylacrylic acid are eluted first (shortly prior to n-valeric acid). $trans$-β-Ethylacrylic acid is eluted simultaneously with α-methylbutyric acid, and α-ethylacrylic acid appears only after elution of these former two acids has been completed. $trans$-α-Methylcrotonic (angelic) acid remains on these columns but can be eluted when the concentration of butanol is raised to 20%.

C. Preparation of p-Bromophenacyl Derivatives of Volatile Acids after Their Separation by Partition Chromatography

Isolation and identification of the acids after their chromatographic separation may be carried out by preparation of their crystalline p-bromophenacyl esters. To each fraction isolated by means of the buffered Celite column, water is added (one-third of the volume of the solvent), followed by 90% of the equivalent amount of sodium hydroxide and slightly more than 10% of the equivalent amount of sodium bicarbonate required to neutralize the acid. The mixture is shaken vigorously in a separatory funnel; after the two layers have separated, the chloroform layer is drawn off and extracted a second time with one-ninth its volume of water and a small amount of sodium bicarbonate. The two aqueous extracts are combined, filtered, concentrated to a small volume in $vacuo$ (35 to 40°), and lyophilized.

The lyophilized sample is dissolved in 10 ml. of water, acidified with 2 N sulfuric acid to pH 2, and concentrated in $vacuo$. The distillate is trapped in a dry ice-acetone bath, neutralized with a solution of sodium

---

[6] E. Bueding, $J.$ $Biol.$ $Chem.$ **202**, 505 (1953).

tions, the low-temperature vacuum distillation technique of Grant[5] can be recommended. A suitable two-armed distillation apparatus with a relatively large surface and a short diffusion path is made from the bulb of a 50-ml. Florence flask and a 25×150-mm. test tube. A 1- to 3-ml. sample containing 1 to 50 micromoles of volatile acid is acidified, placed in the 50-ml. side arm, and shell-frozen by immersion in a dry ice-alcohol mixture. The apparatus is closed with a rubber stopper carrying a stopcock and is evacuated by means of a good vacuum pump. The stopcock is closed, the arm containing the sample is removed from the dry ice bath, and the other empty arm is immersed in its place. The distillation of the volatile constituents of the sample to the cold arm is allowed to proceed to completion without artificial heating (1 to 3 hours).

Separation of Fatty Acids in a Mixture

A. Formic Acid. This acid can be separated readily from other volatile acids (pK_a 4.75 to 4.85) by taking advantage of the fact that it is a stronger acid (pK_a 3.75). To a mixture of acids, an amount of alkali equivalent to the formic acid is added, and the solution is evaporated to dryness. The higher acids are volatilized while the formate is retained in the residue.

Formic acid can be quantitatively and specifically removed from a mixture of fatty acids by oxidation to carbon dioxide with mercuric sulfate.[2] To 100 ml. of steam distillate containing formic acid is added 25 g. of $MgSO_4$ and 5 ml. of a 10% (w/v) solution of $HgSO_4$ in 2 N H_2SO_4. The mixture is placed in a distilling flask, heated to boiling, refluxed gently for 5 minutes, and then distilled at the rate of 4 to 5 ml./min. The distillation is discontinued when the residue begins to crystallize. Volatile acids, other than formate, are recovered in the distillate.

Formic and acetic acids can be separated from longer chain fatty acids by azeotropic distillation from benzene solution as described by Schicktanz *et al.*[6]

B. C_2 to C_{12} Normal Fatty Acids. (1) METHODS USING SILICA GEL COLUMNS WITH INTERNAL INDICATOR. The silica gel partition chromatogram of Elsden,[7] with bromocresol green as internal indicator, will resolve a chloroform solution of acetic, propionic, *n*-butyric, and *n*-valeric into its components. One hundred micromoles of total acid can be used on a column of 7-mm. diameter.

[5] W. M. Grant, *Ind. Eng. Chem. Anal. Ed.* **18**, 729 (1946).

[6] S. T. Schicktanz, W. I. Steele, and A. C. Blaisdell, *Ind. Eng. Chem. Anal. Ed.* **12**, 320 (1940).

[7] S. Elsden, *Biochem. J.* **40**, 252 (1946).

To 3 g. of dry silica gel[8] is added approximately 1.8 ml. of a 0.2% (w/v) bromocresol green solution containing sufficient alkali or acid so that the final mixture has a light-blue color. The exact quantity of liquid and acid or alkali that must be added to give the correct consistency and color must be determined for each batch of gel. The mixture is homogenized and mixed thoroughly in a mortar. Approximately 30 ml. of water-saturated chloroform containing 1% (v/v) of butanol (CB$_1$) is slowly added, and the mixture is ground to a smooth slurry, which is poured into the chromatogram tube consisting of a 10-ml. buret (7-mm. inside diameter) with a small plug of cotton just above the stopcock. After the column has settled and packed, 1 ml. of 0.03 N n-caproic acid in CB$_1$ is run onto the column and washed through with the same solvent. The surface of the column must be smooth and must be covered with solvent at all times. The column should have a greenish-blue color. The fatty acid bands are yellow with a sharp leading edge and a diffuse rear edge.

When 2 ml. of a chloroform solution containing formic, acetic, propionic, n-butyric, and n-valeric acids is placed on the top of the column and the chromatogram is developed with CB$_1$, formic acid remains at the top of the column, and the other fatty acids pass down the column as discrete yellow bands, moving at different rates. n-Valeric acid moves fastest, and acetic acid slowest. The relative rate of movement of an acid is measured by the R value, defined as the rate of movement of a compound down the column, divided by the rate of movement of the meniscus of the developing solvent above the column. Although the R value changes somewhat with the amount of acid in the sample, it is characteristic of the compound under a given set of conditions. Elsden obtained the following R values when 20 micromoles of each acid was applied to a column: acetic 0.3, propionic 0.75, n-butyric 1.05, and n-valeric 1.3.

The fatty acids must be dissolved in chloroform when applied to the Elsden column. This is accomplished by evaporating an aqueous solution of the neutralized acids to a volume of 0.1 to 0.2 ml., adding 3 to 4 g. of finely powdered anhydrous KHSO$_4$, and extracting the resulting mass with six 4-ml. portions of dry CB$_5$. When quantitative extraction is not required, a less thorough extraction with CB$_1$ is adequate. An aliquot of the extract, not larger than 3 ml. and containing from 10 to 100 micromoles of total acid, is added to a 3-g. column which is then developed with water-saturated CB$_1$.[9]

The fatty acids, tentatively identified by their R values, are collected

---

[8] Acid silicic (precipitated), analytical reagent, supplied by Mallinckrodt Chemical Works, is more satisfactory than other similar products tested.

[9] The best concentration of butanol varies with the particular batch of gel from CB$_{0.6}$ to CB$_5$ and therefore should be determined empirically.

as they are eluted from the column. The quantity can be most conveniently determined by titration with a 0.05 M solution of KOH in methanol,[10] using a microburet. The methanolic KOH must be protected from carbon dioxide. Dry CO_2-free air is passed through the sample for 1 minute preceding and during the titration. Two drops of 0.04% (w/v) methanolic cresol red indicator and 0.5 ml. of methanol are added per 5 ml. of solution undergoing titration. The color changes very sharply from yellow to bluish-red at the end point. Because the indicator is preferentially soluble in water, the tubes in which the column eluate is collected must be dry.

Propionic, butyric, and valeric acids are readily eluted from a column with CB_1 and titrated directly. To elute acetic acid with a reasonable volume of solvent, it is necessary to use CB_5 or CB_{20}. The latter solvents also elute sufficient bromocresol green indicator to interfere with the titration of the acetic acid. However, the acid may be recovered free of indicator if necessary by neutralizing it, evaporating off the organic solvent, acidifying the residue, and subjecting it to steam distillation.

Elsden's method can be modified so as to permit the separation of the C_2 to C_7 normal fatty acids, by replacing the chloroform-butanol solvents by wet cyclohexane containing 0.6% (v/v) butanol. The C_7 and C_8 acids do not separate. Since cyclohexane has a lower specific gravity than chloroform, the former solvent passes much more slowly through the silica gel column under gravity. Therefore it is necessary to apply 2 to 5 pounds of air pressure to the top of the column to increase the rate of flow. With cyclohexane, n-heptanoic acid does not produce a visible band on the column but passes through ahead of the band caused by n-caproic acid.

A similar method employing a silica gel column with internal indicator for the separation of C_5 to C_{10} fatty acids has been described by Ramsey and Patterson.[11] In this procedure the stationary phase is methanol containing bromocresol green indicator, and the mobile phase is 2,2,4-trimethylpentane. As the acids emerge from the column, they are titrated with 0.02 N sodium ethylate. When a mixture of the C_6, C_7, C_8, and C_9 acids is passed once through the column, each separated acid is contaminated with about 4% of the next higher homolog.

2. OTHER METHODS USING PARTITION CHROMATOGRAPHY ON COLUMNS. Moyle et al.[10] have described a column method for the quantitative separation of lower fatty acids in which the stationary phase consists of heavily buffered silica gel and the moving phase consists of mixtures of chloroform and butanol. The C_2 to C_7 normal fatty acids can be separated

[10] V. Moyle, E. Baldwin, and R. Scarisbrick, *Biochem. J.* **43,** 308 (1948).
[11] L. L. Ramsey and W. I. Patterson, *J. Assoc. Offic. Agr. Chemists* **31,** 139 (1948).

in 100-micromole quantities on a 5-g. column, but the C_7 acid cannot be separated from higher homologs.

James and Martin[12] have applied gas-liquid partition chromatography to the separation of C_1 to C_{12} fatty acids in amounts of the order of 5 to 10 micromoles. This column method permits the separation of *n*- and iso-butyric acids and of the four isomeric valeric acids. The column (4 to 11 feet long) consists of Celite coated with silicone DC 550 containing 10% (w/w) of pure stearic acid. The moving phase is nitrogen gas. The jacketed column is operated at 100° or 137°, depending on the acids to be separated. The acids pass through the column at characteristic rates, formic acid moving fastest and dodecanoic acid slowest. The method is more suitable as an analytical than as a preparative procedure.

The countercurrent method of Craig and his associates[13,14] may be used for the separation of C_2 to C_{18} normal fatty acids when the required apparatus is available. With a given pair of solvents, three or four acids differing by one carbon atom can be partially separated. Although the acids generally are not separated quantitatively by this method, a sample of each acid may be obtained free from contaminating acids. The method has the advantage that the results are quantitatively reproducible, and relatively larger quantities of acid (up to 1 millimole of each of four acids) can be separated.

Paper Chromatography of C_1 to C_9 Fatty Acids

Of the several methods for paper chromatography of fatty acids that have been reported,[15-17] that of Reid and Lederer[18] appears to combine the most desirable features. This method involves ascending chromatography of the ammonium salts of fatty acids with *n*-butanol-aqueous ammonia as solvent system. The free fatty acids appear as yellow spots on a purple background after the developed chromatogram is sprayed with a bromocresol purple solution containing formaldehyde. The latter reacts with the ammonium ion to form the weak base hexamethylenetetramine and thereby accentuates the acidic spots. Fatty acids of the normal series from acetic to heptanoic can be separated, identified, and estimated with an accuracy of $\pm 5\%$ for each acid. Heptanoic acid cannot be separated from octanoic and nonanoic acids, but octanoic acid can be

---

[12] A. T. James and A. J. P. Martin, *Biochem. J.* **50**, 679 (1952).
[13] G. Sato, G. T. Barry, and L. C. Craig, *J. Biol. Chem.* **170**, 501 (1947).
[14] G. T. Barry, Y. Sato, and L. C. Craig, *J. Biol. Chem.* **188**, 299 (1951).
[15] F. Brown and L. P. Hall, *Nature* **166**, 66 (1950).
[16] E. R. Hiscox and N. J. Berridge, *Nature* **166**, 522 (1950).
[17] E. P. Kennedy and H. A. Barker, *Anal. Chem.* **23**, 1033 (1951).
[18] R. L. Reid and M. Lederer, *Biochem. J.* **50**, 60 (1951).

estimated in the absence of heptanoic and nonanoic acids. Isomeric acids cannot be separated.

Reagents

Butanol-NH_3 solvent. Redistilled *n*-butanol is saturated with an equal volume of aqueous 1.5 M NH_3.

Indicator solution. 0.04% (w/v) bromocresol purple is dissolved in a 1:5 (v/v) dilution of formalin in ethanol. The indicator solution is adjusted to pH 5 with 0.1 N NaOH.

Procedure. Sheets of Whatman No. 1 (special chromatographic) paper, 30 × 30 cm., are exposed to the atmosphere over concentrated aqueous ammonia in a closed vessel for 3 to 4 hours immediately before use. Samples of approximately 5 μl. containing 0.5 to 1.5 micromoles of the ammonium salt of each fatty acid are placed 3 to 4 cm. apart along a "starting line" 3 cm. from the bottom of the paper. Each sheet is stapled in the form of a cylinder which is stood in a 0.5-cm. layer of butanol phase of the solvent in a convenient glass jar with close-fitting lid. The aqueous solvent phase is placed in a beaker which is stood in the butanol phase. The same solvent mixture can be used for at least four chromatograms. The chromatogram is developed for 18 to 24 hours, when the solvent will have reached the top of the paper. When development is complete, the paper is removed and allowed to dry at room temperature. The paper is then sprayed with sufficient indicator solution to allow complete and even penetration of indicator to the reverse side of the paper. After the sprayed chromatogram has been held for 2 to 3 minutes in an atmosphere saturated with a 3% (v/v) dilution of aqueous ammonia (specific gravity 0.88), the acids appear as bright-yellow spots on a stable purple background. The R_f values are given in Table I.

TABLE I
R_f VALUES OF AMMONIUM SALTS OF FATTY ACIDS

Acid	Single acids	Mixtures of acids
Formic	0.10	—
Acetic	0.11	0.10–0.11
Propionic	0.19	0.18–0.20
n-Butyric	0.29	0.29–0.32
n-Valeric	0.41	—
i-Valeric	0.40	0.39–0.44
n-Hexanoic	0.53	0.53–0.59
n-Heptanoic	0.62	0.62–0.64
n-Octanoic	0.65	0.64–0.66
n-Nonanoic	0.67	—

For quantitative estimation of the fatty acids on the basis of spot size, the acid spots are traced onto graph paper and an estimate of their areas is thus obtained. Two standard solutions, containing known concentrations of the ammonium salts of the fatty acids under investigation, must be chromatographed on the same sheet with each test solution. It is advantageous to run two standards with six or seven different unknowns on each of four sheets. Calculations are then based on the mean of quadruplicate estimates, and variations from sheet to sheet become less important. The logarithm of the quantity of acid is directly proportional to the spot area.

Identification and Estimation of Volatile Acids

Duclaux Distillation. One of the most generally applicable methods for the identification of a volatile fatty acid is the so-called Duclaux distillation. The method depends on the fact that, when a dilute aqueous solution of a volatile acid is distilled under well-defined conditions, the rate of distillation of titratable acid is characteristic of the particular compound and, in general, differs markedly with different acids. When only a single volatile acid is present, it can usually be identified by its distillation rate. The method frequently can be used to determine the relative amounts of two or three known acids in a mixture.

A simple and satisfactory Duclaux distillation is carried out by distilling at a constant rate from a definite initial volume of solution and titrating three successive portions of distillate. The apparatus consists of a 150-ml. round-bottomed, short-necked flask, attached by means of standard taper (24/40) joints and a short descending connecting tube to a vertical condenser.[19] The upper portion of the flask should be covered with asbestos to prevent refluxing. Exactly 35 ml. of a steam distillate, containing one or more fatty acids, is placed in the flask along with a generous pinch of talc or pumice boiling chips. The flask is heated with a hot flame over an area about 3 cm. in diameter, the rest of the flask being shielded by an asbestos mat to prevent superheating of the vapor above the liquid level. The distillate is collected in three volumetric flasks each calibrated to hold exactly 10 ml. The three portions of distillate are titrated with CO_2-free alkali to the phenol red end point. The results are expressed as the cumulative percentages of acid in each fraction referred either to the total acid initially present in 35 ml. or to the acid distilling in the first 30 ml.

Typical data for a number of volatile acids are given in Table II. The percentages shown are readily reproducible to within ±0.2 with a given apparatus. The distillation rates increase appreciably with the concentra-

[19] A. Knetemann, *Rec. trav. chim.* **47**, 950 (1928).

tion of acid in the sample; for best results the concentration should be kept within a definite range, e.g., 10 to 30 micromoles per 35 ml.

A single unknown acid can be identified by comparison of its Duclaux values with those of known acids. Only in exceptional cases (i.e., acetic and vinyl acetic acids) are the constants of two acids sufficiently similar to cause uncertainty.

TABLE II

TYPICAL DUCLAUX VALUES FOR PURE ACIDS

Concentration: 10 to 30 μM. per 35 ml.

a = % of total vol-acid b = % of acid in 30 ml. of distillate

Fraction, ml.	Formic		Acetic		Propionic		n-Butyric		Vinylacetic	
	a	b	a	b	a	b	a	b	a	b
10	5.8	15.0	14.8	23.5	28.2	35.4	41.2	46.0	19.5	25.6
20	15.0	38.9	34.9	55.0	54.6	70.3	70.6	78.8	44.4	58.0
30	38.6	(100)	63.5	(100)	80.0	(100)	89.5	(100)	76.5	(100)

Fraction, ml.	Isobutyric		n-Valeric		n-Caproic		n-Heptanoic		n-Octanoic	
	a	b	a	b	a	b	a	b	a	b
10	54.0	57.5	55.7	57.8	69.2	70.6	81.0	82.0	90.4	90.4
20	82.3	87.5	85.0	88.0	91.5	93.8	96.0	97.0	99.0	99.0
30	94.0	(100)	96.0	(100)	97.5	(100)	99.0	(100)	100	(100)

When two acids are present in a sample, they distill independently. For a mixture of acetic and butyric acids, for example, the first 10 ml. of distillate contains 0.148 of the total acetic acid and 0.412 of the butyric acid. The following relations apply when t_{10}, t_{20}, and t_{30} represent the cumulative titrable acidity in the three 10-ml. fractions, and A and B are the total amounts acetic and butyric acids in the 35-ml. sample.

$$0.148A + 0.412B = t_{10}$$
$$0.349A + 0.706B = t_{20}$$
$$0.635A + 0.895B = t_{30}$$

The values of A and B can be calculated from any two of these equations. For example, solving the first and third equations for A and B, we get

$$A = 3.9t_{30} - 6.96t_{10}$$
$$B = 2.43t_{10} - 0.36A$$

Similarly, three acids in a mixture may be estimated by the use of three equations involving three unknowns.

Formic Acid. The presence of formic acid in a mixture of volatile acids is indicated by its ability to reduce mercuric chloride to calomel, which precipitates as fine white needles (see below). Pyruvic acid in large amounts interferes with this test, since it also reduces mercuric chloride.

The reduction of mercuric chloride is employed in the following method of Leloir and Munoz[20] for the estimation of formate. To a test tube with a glass stopper, add 5 ml. of a solution of volatile acids containing 2 to 15 micromoles of formic acid neutralized to phenol red. Add 1 ml. of a reagent containing 30 g. of $HgCl_2$, 30 g. of $NaAc \cdot 3H_2O$, and 8 g. of NaCl, made up to 100 ml. with water. Heat the tube in a 100° water bath for 2 hours. Cool and add 0.5 ml. of glacial acetic acid, 1 ml. of saturated KI, and 2 ml. of 0.03 N I_2. Shake until the precipitated Hg_2Cl_2 is completely dissolved, then back-titrate the excess iodine with 0.01 N $Na_2S_2O_3$. One milliliter of 0.01 N $Na_2S_2O_3$ is equivalent to 5 micromoles of formate.

A more specific quantitative method for formate is given by Pirie.[21] This depends on the oxidation of formate with $HgCl_2$ and estimation of the resulting carbon dioxide manometrically. See Vol. III [52].

Other methods for the estimation of formate are available.[22,23]

Acetic Acid. A highly specific identification can be made as sodium uranyl acetate.[24] A distillate containing at least 10 micromoles of acetic acid is neutralized with a suspension of uranyl carbonate, and the supernatant is concentrated to a small drop on a slide. On the addition of a small crystal of NaCl, characteristic tetrahedral crystals of sodium uranyl acetate form at or near the edge of the drop. No other fatty acid gives similar crystals. Propionic and higher acids interfere with the acetate test only when they constitute more than 40% of the total acid.

Acetic acid can be separated and estimated quantitatively by diffusion techniques[25,26] as well as by the steam distillation and chromatographic methods previously described. For enzymatic methods, see Vol. III [48].

Other Fatty Acids. The identification and estimation of small quantities of the C_3 to C_{10} fatty acids can be achieved by one or more of the techniques already described, namely, steam distillation, Duclaux distillation, column partition chromatography, and paper chromatography. The method of choice depends on the number and nature of the components

[20] L. F. Leloir and J. M. Munoz, *Biochem. J.* **33**, 737 (1939).
[21] N. Pirie, *Biochem. J.* **40**, 100 (1946).
[22] M. J. Pickett, H. L. Ley, and N. S. Zygmuntowicz, *J. Biol. Chem.* **156**, 303 (1944).
[23] W. M. Grant, *Anal. Chem.* **19**, 206 (1947).
[24] P. D. C. Kley, "Organische Mikrochemische Analyse," Voss, Leipzig, **1922.**
[25] S. Black, *Arch. Biochem.* **23**, 347 (1949). See Vol. III [47].
[26] E. J. Conway and M. Downey, *Biochem. J.* **47**, iv (1950).

in the sample. For preliminary examination of a sample the paper chromatographic method of Reid and Lederer[18] is convenient. When only a single volatile fatty acid is indicated by the chromatogram, the preliminary identification can be confirmed by Duclaux distillation and the quantity can be estimated by steam distillation of a suitable aliquot. When two fatty acids other than formic acid are present, confirmation of the identification of the components can be accomplished by separating them by one of the column chromatographic methods followed by Duclaux distillation of the individual acids. When the sample is thought to contain branched-chain fatty acids, it can be examined by application of the gas-liquid partition chromatography method of James and Martin.[12] Duclaux distillation will also distinguish between the normal and branched-chain butyric or valeric acids, provided that only one isomer is present in the sample. The estimation of the ratio of two known fatty acids (C_2 to C_6) in a mixture can be accomplished by the Duclaux distillation method or partition between two immiscible solvents.[27] These methods give more accurate results ($\pm 3\%$) when the chain lengths of the acids differ by at least two carbon atoms.

The estimation of three or more volatile acids (C_2 to C_{10}) in a mixture presents considerable difficulties. In the special case of a mixture of acetic, n-butyric, and n-caproic acids, each component can be estimated with an accuracy of $\pm 10\%$ by Duclaux distillation, provided that no one acid constitutes less than 20% of the total. With most other mixtures of three or more acids the Duclaux method has little value. To estimate the components in such mixtures the column methods of James and Martin[12] and others[7,10-12,28] or the paper chromatographic method of Reid and Lederer[18] may be used in a quantitative manner. The liquid-liquid partition methods of Sato et al.[13] and Barry et al.[14] are also applicable both for identification and estimation of fatty acids in complex mixtures when the necessary apparatus is available.

A convenient method for the estimation of n-octanoic or n-nonanoic acid in amounts of 0.4 to 7 micromoles in the presence of the C_2 to C_6 fatty acids has been described by Lehninger and Smith.[29] The method depends on the measurement of the turbidity resulting from the addition of silver nitrate to the fatty acid under special conditions.

[27] O. L. Osburn, H. G. Wood, and C. H. Werkman, *Ind. Eng. Chem. Anal. Ed.* **8,** 270 (1936).

[28] M. W. Peterson and M. J. Johnson, *J. Biol. Chem.* **174,** 775 (1948).

[29] A. L. Lehninger and S. W. Smith, *J. Biol. Chem.* **173,** 773 (1948).

[62] Molecular Distillation of Lipids and Higher Fatty Acids[1]

By EDMOND S. PERRY

Molecular distillation is particularly well suited for the distillation of the constituents of fats and oils. In fact, the more successful applications of this type of distillation have been in this field of endeavor. Molecular stills, like conventional stills, are used for purification, isolation, or concentration of components from the mixture in which they occur. The advantage of the molecular still lies in its ability to permit distillation at temperatures 50 to 150 degrees lower than those obtained in ordinary vacuum stills for the distillation of comparable materials. Therefore, substances having molecular weights up to about 1200 can be distilled without harm from thermal hazard. Distillations for the purposes mentioned above are referred to as pot or cyclic distillation in accordance with the kind of still employed. The choice of still usually depends on the separation required and on the quantity of material available for distillation. When the material is in ample supply, it is best to use a flowing-film cyclic still because of its superior performance. The cyclic stills can also be used by special technique for the characterization of a specific molecular species and for the estimation of purity or potency of constituents in mixtures. This technique is called the analytical distillation or the "elimination curve" technique. It is preferably performed in a cyclic flowing-film still because the precise management required to carefully control the operation can be done better in these stills.

For an extensive discussion of the various aspects of molecular distillation, summary publications[1a,2] are recommended. In this brief chapter the subject is limited to a description of typical apparatus and procedures which apply to fatty acids and lipids for either purification or analytical studies. The commonly used molecular stills are described in one section, and the methods are given in another. In each case an actual application is mentioned to exemplify the method. By this means it is hoped that each reader can readily select the right combination of still and method to suit his particular problem.

[1] Communication No. 197 from the Laboratories of Distillation Products Industries, Division of Eastman Kodak Co., Rochester, N.Y.

[1a] A. W. Weissberger, ed., "Distillation," Vol. IV, Chapter 4, Interscience Publishers, New York, 1951.

[2] K. Hickman, *Chem. Revs.* **34**, 51 (1944).

Stills

Pot Type. The pot-type molecular stills have been made in many forms and modifications ranging in sizes to accommodate from microgram to several hundred gram quantities. A few of the simpler varieties are shown in Fig. 1. The concentric tube styles offer simplicity as well as utility for the distillation of small samples, especially where only a few fractions are desired. Stills made from straight-walled 25-mm. tubes can handle as

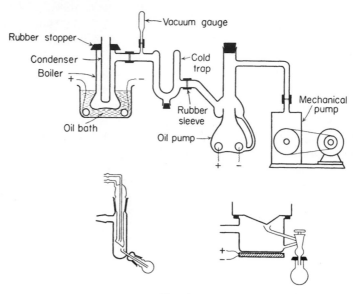

Fig. 1

little as 0.5 g., whereas the flared construction shown in Fig. 1 is suitable for larger quantities. Micro pot stills to handle smaller charges have been described.[3,4] The more elaborate designs of Fig. 1 feature distillate withdrawal tubes which provide for continuous operation.[1a,2,5] These are useful when the sample to be distilled is sufficiently large so that the hold-up on the still surfaces is not significant. Their performance, however, is no better than that of the concentric tube stills. Distillation in any kind of pot still should be done with the thinnest possible layer of distilland because the rate of distillation is directly proportional to the area of the distilling surface. Deep layers also increase the thermal hazard to the distilling molecules.

[3] C. W. Gould, G. Holzman, and C. Niemann, *Anal. Chem.* **20**, 361 (1948).

[4] J. R. Matchett and J. Levine, *Ind. Eng. Chem., Anal. Ed.* **15**, 296 (1943).

[5] B. Riegel, J. Beiswanger, and G. Lanzl, *Ind. Eng. Chem., Anal. Ed.* **15**, 417 (1943).

Flowing-Film Type. The flowing-film molecular stills are made in two versions. These are illustrated schematically in Fig. 2. Many modifications of the falling-film variety exist.[6-8] The centrifugal model[9] shown is at present the only one of its kind in general use. These stills are usually constructed to take a maximum charge of 1000 g. The lower limit may be as little as 50 g., depending on which fractions are of interest, since the

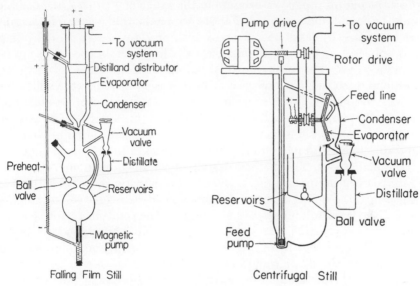

Falling Film Still   Centrifugal Still

FIG. 2

hold-up will approach 25 g. A semimicro falling-film still of novel design has been reported.[10]

The auxiliary equipment required to produce vacuum and measure vacuity for a complete distillation system will include an oil diffusion pump of the G-4 type[11] for pot stills and the GB-25 type[12] for the flowing-film stills, Pirani gauge, and mechanical pump with suitable capacity for backing the oil pump. The typical pumping system shown schematically in Fig. 1 is suitable for the stills of either Fig. 1 or 2. The larger stills, of course, demand the larger pumps.

[6] K. Hickman, *Ind. Eng. Chem.* **29,** 968 (1937).

[7] F. W. Quackenbush and H. Steinbock, *Ind. Eng. Chem., Anal. Ed.* **15,** 468 (1943).

[8] J. K. Taylor, *J. Research Natl. Bur. Standards* **37,** 173 (1946).

[9] R. Biehler, K. Hickman, and E. Perry, *Anal. Chem.* **21,** 638 (1949).

[10] I. A. Berger, *Anal. Chem.* **20,** 980 (1948).

[11] Single Stage Oil Diffusion Pump Type G-4, Consolidated Vacuum Corporation, Rochester, N. Y.

[12] Booster Diffusion Pump Type GB-25, Consolidated Vacuum Corporation, Rochester, N. Y.

Methods

Preliminary Procedure. Before undertaking the molecular distillation of any material some preliminary precautions and preparations with both the apparatus and the sample are recommended. It is always advisable to determine whether the still is free of leaks before the sample is admitted. Unless the pumps produce a vacuum in the range of 1 to 10 μ, the trouble should be found and remedied. Failure to attain this degree of vacuum may be caused by leaks and in many instances by contaminated pump fluids, solvents, or water in the sample, drift in the calibration of the pressure gauges, etc. The sample for distillation should be freed of solvents, water, and dissolved corrosive gases before it is admitted to the still. This can be done usually by evacuation under water aspirator pressures at steam bath temperatures. The sample should be admitted, whenever possible, while the still is under vacuum of the mechanical pump and the cold trap is filled with refrigerant. This facilitates degassing of the sample.

Pot Distillations. The method described for the concentric tube stills of Fig. 1 is typical for any pot-type still. The substance to be distilled is charged into the boiler, and the amount should be regulated so that the depth of the layer will not exceed 5 or 6 mm. The condenser tube is set in place and the system is evacuated. Some heat may be applied by the oil bath to aid degassing. If, during degassing, the distilland splashes onto the condenser, the latter should be replaced by a clean one before proceeding with the distillation. Sometimes it is better to do the degassing without the condenser tube, in which case the outer tube can be sealed with an inverted rubber stopper. After degassing, the condenser is filled with refrigerant (usually dry ice and acetone) and the distilling pressure is established with both pumps. The oil bath is heated quickly to the desired temperature and maintained at this level for 15 to 30 minutes, depending on the thickness of the distilland layer. The oil bath is then removed, and the still is allowed to cool to room temperature before the vacuum is broken. The distillate is contained on the outer surface of the condenser tube and can be removed by solvents or otherwise.

If two or more fractions are desired, this procedure is repeated. Usually temperature increments of 25 degrees are used between fractions. The use of two or more condenser tubes allows for faster operation. While one tube is in the still, the other is being cleaned of the previous distillate.

An example of the application of this method is the tocopherol concentration step by molecular distillation prior to the Emmerie-Engel determination for tocopherols in edible fats as used by Quaife and Harris.[13] When gram quantities of fats containing γ-tocopherols ranging from 0.06

[13] M. Quaife and P. Harris, *Ind. Eng. Chem., Anal. Ed.* **18,** 707 (1946).

to 2.0 mg./g. were distilled by the above method, the resulting distillates, which amounted to 15 to 20 mg., contained 92 to 102% of the tocopherol. This intermediate distillation step thus provides a means of determining the low level of tocopherol potencies in most natural fats and lipid extracts which otherwise are too dilute and impure for the existing analytical methods.

Cyclic Distillation. The procedure for the cyclic distillation is described for a laboratory centrifugal still as shown in Fig. 2. The method is equally applicable to a falling-film still. The sample to be distilled is prepared as already described and then charged into the still while the latter is under vacuum of the mechanical pump. The feed pump circulates the distilland over the rotor to degas it. Some heat on the rotor will hasten the degassing, but the amount should never be great enough to induce distillation. Usually one pass over the rotor is sufficient, but with difficultly degasable materials two or more may be necessary. If a considerable quantity of volatiles are collected in the cold trap, they should be removed from the trap before continuing the distillation. Otherwise some will eventually pass over to the pumps and foul the fluids. The pressure in the system is finally reduced to the distilling range, and the heat to the rotor is increased until distillation begins. For convenience, the rotor heat is adjusted so that the distilland temperature is fixed at an even 10° mark where the first fraction will be collected. At this point the ball valve in the upper reservoir is closed and the distillate stopcock is turned 180°. The distillate passes into the receiver. All the distilland now passes over the evaporator, and as the last bit leaves the lower reservoir the ball valve is opened to allow the residue in the upper reservoir to flow down to the lower one where it then becomes the distilland for the next cycle. The distillate stopcock is turned to its original position, and the distillate returns to the still. Meanwhile the next temperature level is established. It is common practice to increase the temperature 10 degrees for each fraction. In the meantime the distillate receiver is replaced with a fresh one, and as soon as the next temperature is reached the second fraction can be collected.

This procedure is repeated for as many temperature steps as desired. At each temperature a fraction is obtained, thus rendering a series of fractions containing the components of the original sample in a state of separation.

Sometimes the sample for distillation is not large enough to permit complete distillation in the flowing-film stills owing to the hold-up of these units. The use of an equal quantity of a relatively nonvolatile carrier oil (see below) will increase the volume of the sample, thus making it possible to do the distillation. The added carrier oil will have little or no effect on the separation as long as it does not distill with the components of the sample.

Farmer and Vander Heuvel[14] subjected the methyl esters of the unsaturated fatty acid components of cod liver oil to cyclic distillation for the purpose of separating the various acids in the mixture. A double distillation was performed from which seven fractions and a residue resulted. Of these, four fractions, which comprised 61% of the original charge, were each found to consist of an essentially pure fatty acid ester.

The superior performance of the cyclic falling-film still over the simple pot-type still has been shown by Hickman[6] for the distillation of the relatively heat-sensitive vitamin A from halibut liver oil. The greater thermal hazard[15] of the pot stills is readily apparent from this comparison.

Analytical Distillation. The procedure for the analytical distillation is quite similar to that for cyclic distillation with the exception that the substance being distilled must be carried in a specially prepared substrate and that the conditions of the distillation must be very closely controlled. The method is based on the fact that at each successive temperature the component under study is eliminated in greater amounts until a maximum is reached. Thereafter, the rate of its elimination decreases to zero. A plot of the concentration of the substance in each fraction versus the temperature of the fraction gives rise to a curve having the shape of a probability distribution curve. The temperature at the maximum of the peak is known as the "elimination" temperature and is a specific property of the distilling molecular species. This curve was called the "elimination" curve by Hickman,[6] and the theoretical background was shown by Embree.[16] Departure in shape of the experimental curve from that of the theoretical is evidence that more than one substance is contributing to it. A single substance will yield a sharp peak, whereas a mixture of isomers or members of a homologous series will give an irregular curve or one with a broad plateau.

The elimination curve technique gives a characteristic but relative temperature for each molecular species. It must be noted that the elimination temperature is not an absolute quantity for the substance but is also dependent on both the apparatus and the conditions of the method. In other words, two independent laboratories may not obtain the identical elimination temperatures for the same material. Standardization is possible, however, by the use of "pilot" dyes[6] which permit measurement of the difference in elimination temperatures between the pilot and the

[14] E. H. Farmer and F. A. Vander Heuvel, *J. Soc. Chem. Ind. (London)* **57**, 24T (1938).
[15] K. Hickman and N. Embree, *Ind. Eng. Chem.* **40**, 135 (1948).
[16] N. Embree, *Ind. Eng. Chem.* **29**, 975 (1937).

substance being studied. This difference is a fixed and reproducible quantity. The anthraquinone dyes have been used for this purpose.[6]

Constant Yield Oils. The special substrate mentioned above is an expedient to permit analytical work and to overcome certain difficulties in carrying out the distillation. It is a mixture of two oils, a so-called "constant yield oil" (C.Y.O.) and a relatively nonvolatile carrier oil. The C.Y.O. is a mixture of equal parts of many distillate fractions. Each fraction distills at a definite temperature, and the number of fractions in the mixture fixes the temperature range to be covered in the distillation. Thus, a mixture consisting of 11 fractions, the first of which distills at 100°, the second at 110°, the third at 120°, and so on, will cover the temperature range between 100° and 200°. The function of the C.Y.O. is to maintain uniformity in distillation rates for all cycles and to regulate the drainage from the condensers. The carrier oil is used to increase the volume of the distilland, and since it does not distill it has an equalizing effect on the feed rates from the first to the last cycle of the distillation. The utility of the C.Y.O. and carrier oils will become readily apparent as the method is studied more closely.

A preparation for triglyceride C.Y.O. has been reported,[17] in which the acids of natural oils are used. In another method[18] four saturated acids are esterified two at a time with glycerin using sulfuric acid and xylene. From the three triglyceride mixtures which result, each of different distilling range, the C.Y.O. is made. The preparation of these esters is indicated in the table.

TRIGLYCERIDE ESTERIFICATION

	Low-boiling fraction, g.	Medium-boiling fraction, g.	High-boiling fraction, g.
Caproic (C_6)	316		
Caprylic (C_8)	322	322	
Capric (C_{10})		326	326
Lauric (C_{12})			330
Glycerin	152	127	109
H_2SO_4	0.2 ml.	0.2 ml.	0.2 ml.
Xylene	1000 ml.	1000 ml.	1000 ml.

The esterification reaction goes to 95% of completion in 13 to 14 hours. The triglyceride oils are subjected to cyclic distillation in the

[17] J. G. Baxter, E. LeB. Gray, and A. O. Tischer, *Ind. Eng. Chem.* **29**, 1112 (1937).
[18] C. Robeson and C. Eddinger, Laboratory Report, March 12, 1946, Research Laboratories of Distillation Products Industries, Rochester, N. Y.

molecular still. The first 8 to 9% of the distillate is discarded, and the remaining distillates in each case are mixed together. A satisfactory C.Y.O. results from a blend of 2 parts each of the C_6–C_8 and the C_{10}–C_{12} fractions with 1 part of the C_8–C_{10} fraction. A superior C.Y.O. is produced by redistilling this blend, containing an equal weight of carrier oil, into a series of fractions at 10-degree intervals such as is done for the elimination curve and then combining equal parts of the distillates into one composite.

Triglyceride carrier oils are made from any refined vegetable oil such as corn, soy, cottonseed, or castor oils. To remove the volatile fraction the refined oil is stripped in the molecular still in three or four successive cycles at 10-degree intervals beginning at about 200°.

Hydrocarbon constant yield oils are useful because they can be prepared from automotive lubricants. The distillation of a mixture of equal parts of SAE 10 and 50 motor oils by the method of the analytical distillation will give a series of fractions and a residue from which a blend of equal parts of each fraction constitutes a fair C.Y.O. The redistillation of this blend with an equal quantity of the residue in a similar manner will produce a much superior C.Y.O. The residue from the first distillation of the 10 and 50 oils can be used as a carrier oil. Triglyceride carrier oils can also be used with the hydrocarbon C.Y.O.

Procedure. The method for the analytical distillation will be described as it applies to a falling-film cyclic molecular still of the kind shown in Fig. 2. Let it be assumed that the elimination curve for stearic acid is desired. A 10-degree temperature interval and 10-minute cycle will be used. Five grams of stearic acid is dissolved in 100 ml. of C.Y.O. (range 80 to 160°) and 300 ml. of carrier oil. The mixture is charged through a funnel which is inserted in the tubulation of the upper reservoir and degassed as described for the cyclic distillation. The system is finally evacuated to the micron or distilling range. The feed rate is then adjusted so that the entire charge is pumped over the evaporator in exactly 9 minutes, and simultaneously the heat is increased until the oil temperature is set at 80°. The reservoirs are segregated by the ball valve, and the distillate stopcock is opened to the sample bottle. At the end of 9 minutes all the distilland will have passed over the evaporator. The ball valve is opened, and distillation is allowed to continue for 1 minute more, during which time the oil in the preheat line is displaced. At the end of the tenth minute the collection of the distillate is stopped by turning the stopcock so that the distillate will return to the still. The evaporator temperature is immediately increased to the next level, 90°, and when established at this value the operation is repeated to get the second fraction. The conditions for increasing the evaporation temperatures should be predetermined so that only about 1 minute will be required to reach the next temperature. During this minute the distillate receiver is charged and the feed pump is adjusted (the adjustment required should be predetermined) so that the remaining distilland will pass over the evaporator again in 9 minutes. The distillate thus obtained constitutes the second fraction. The successive repetition of the above steps through the 150-degree cycle will give a series of eight fractions. A plot of the acid content in each fraction against the temperature of the fraction gives the elimination curve for stearic acid.

The elimination temperature for several saturated and unsaturated fatty acids has been determined by Gray and Cawley,[19] using the analytical method. For the saturated acids they found a 5-degree rise in the elimination temperature for each additional CH_2 group. A comparison of the temperature for the unsaturated acids with stearic acid showed that each nonconjugated double bond lowered the maximum 2 degrees. If the bonds are conjugated, however, each additional double bond raises the maximum 3 degrees. This information was used by these workers to settle the controversy over the structural differences between vitamins A and A_2.[19,20]

The elimination curve technique has been applied to many of the constituents of fats and oils, as, for example, the natural steroids,[21] cholesterol esters,[22] the natural vitamins A,[6,23-26] natural vitamin D,[27] and the tocopherols.[28]

[19] E. LeB. Gray and J. D. Cawley, *J. Biol. Chem.* **134**, 397 (1940).
[20] A. E. Gillam, I. M. Heilbron, W. E. Jones, and E. Lederer, *Biochem. J.* **32**, 405 (1938).
[21] K. Hickman, *Ind. Eng. Chem.* **32**, 1451 (1940).
[22] G. Fletcher, M. Insalaco, J. Cobler, and H. Hodge, *Anal. Chem.* **20**, 943 (1948).
[23] K. Hickman, *Ind. Eng. Chem.* **29**, 1107 (1937).
[24] H. Kascher, J. G. Baxter, *Ind. Eng. Chem., Anal. Ed.* **17**, 499 (1945).
[25] N. Embree, *Chem. Revs.* **29**, 317 (1941).
[26] E. LeB. Gray, K. Hickman, and E. Brown, *J. Nutrition* **19**, 39 (1940).
[27] K. Hickman and E. LeB. Gray, *Ind. Eng. Chem.* **30**, 796 (1938).
[28] M. Stern, C. Robeson, L. Weisler, and J. Baxter, *J. Am. Chem. Soc.* **69**, 869 (1947).

[63] Preparation and Assay of Cholesterol and Ergosterol

By THRESSA C. STADTMAN

Preparation of Substrates

Colloidal Cholesterol Suspension. Dissolve about 0.5 g. of cholesterol in 10 ml. of warm acetone. Pour slowly and with vigorous stirring into 100 ml. of boiling water. Filter to remove floating particles and sediment. The residual acetone may be removed by boiling *in vacuo* when it is desirable to eliminate this component from the reaction mixtures. Ethanol may be used instead of acetone if it does not interfere. An alternative procedure in which the acetone solution of cholesterol is slowly poured into a 1% solution of serum albumin or casein may also be employed. These protein-stabilized colloidal emulsions are somewhat easier to handle than the protein-free suspensions, since the latter must be added to the reaction mixtures after the enzyme rather than before.

Colloidal Ergosterol Suspension. Prepare in the same manner as the cholesterol suspensions.

Determination of Cholesterol and Ergosterol by Liebermann-Burchard Reaction

Reagents

> Absolute ethanol-acetone, 1:1.
> Chloroform, reagent grade.
> Acetic anhydride-sulfuric acid reagent. Four parts of cold reagent-grade acetic anhydride and 1 part of cold reagent-grade concentrated sulfuric acid are mixed slowly and with constant swirling in an ice-cooled flask just prior to use.
> Cholesterol standard. Dissolve 100 mg. of cholesterol (melting point 149°) in chloroform, and adjust to 100 ml. at refrigerator temperature. Dilute this stock solution 1:10 with chloroform at the same temperature for a working standard. Store both solutions in the refrigerator.
> Ergosterol standard. Dissolve 100 mg. of ergosterol in chloroform and dilute as directed for the cholesterol solutions.

Procedure. The samples to be assayed (containing 0.1 to 0.5 mg. cholesterol) are adjusted to 80% (v/v) ethanol-acetone (1:1) in order to deproteinize the reaction mixtures and extract the cholesterol. This is conveniently carried out by transferring the reaction mixture to a centrifuge tube and rinsing the reaction vessel with small aliquots of ethanol-acetone until the desired amount has been added to the tube. The supernatant solution is clarified by centrifugation and decanted into a clean test tube. The precipitate is extracted with approximately 0.5 vol. of warm ethanol-acetone for 5 to 10 minutes in a water bath (40 to 50°) to effect complete recovery of the cholesterol. The sample is again centrifuged, and the supernatant solution added to the first extract. The combined ethanol-acetone extracts of the samples are then evaporated to dryness in a water bath under air jets.

When the samples are completely dry, 5 ml. of chloroform are pipetted into each tube. Then 2 ml. of cold acetic anhydride-sulfuric acid reagent are added to each sample and mixed by swirling. The tubes are placed in a 16 to 18° water bath, and the color allowed to develop for 15 minutes. The blue-green color is measured immediately at 625 mμ and compared with a standard curve over the range of 0.1 to 0.5 mg. of cholesterol. Standards should be included with each assay.

Since cholesterol esters give a higher color yield than free cholesterol, the samples should first be subjected to alkaline hydrolysis[1] if there is any indication that esterification may have occurred.

The method is carried out in exactly the same fashion for the estimation of ergosterol (in the absence of cholesterol) or for sitosterol. Standard curves must be constructed for each sterol, since the color yield with ergosterol is about two and one-half times that of cholesterol on a molar basis, whereas sitosterol gives values slightly less than cholesterol.

Various ketosteroids also yield colored products when incubated for longer times with the concentrated sulfuric acid-acetic anhydride reagent used in the Liebermann-Burchard test. Thus, in the absence of reactive sterols, such compounds as cholestenone, testosterone, progesterone, dehydroisoandrosterone (Δ^5-androstene-3β-ol-17-one), or the acetate of the last compound can be measured by allowing maximal color development to occur (2 hours at 22°) and comparing with a standard curve for each compound over the range of 0.1 to 0.5 mg. The cholestenone color is measured at 625 mμ, whereas the hormones are measured at 420 mμ.

Determination of Cholestenone as Its 2,4-Dinitrophenylhydrazone

Reagents

    Absolute ethanol.

    0.1% 2,4-dinitrophenylhydrazine in 2 N HCl in absolute ethanol.

    Standard cholestenone solution. Dissolve 10 mg. of cholestenone (melting point 81°) in absolute ethanol, and dilute to 100 ml.

[1] G. R. Kingsley and R. R. Schaffert, *J. Biol. Chem.* **180**, 315 (1949).

with absolute ethanol at temperature of refrigerator where solution is to be stored.

Chloroform, reagent grade.

Procedure. Reaction mixtures to be assayed for cholestenone are deproteinized and extracted as described above for cholesterol except that *absolute ethanol* is used instead of the *ethanol-acetone* reagent. The combined ethanol extracts are adjusted to suitable volumes, and 2-ml. aliquots containing 10 to 200 γ of cholestenone are pipetted into test tubes. To each of these is added 1 ml. of 0.1% alcoholic 2,4-dinitrophenyl-hydrazine in 2 N HCl. The samples are mixed and heated for 5 minutes at 100°. They are then cooled, diluted to 10 ml. with chloroform, and read at 390 mμ against a reagent blank. A standard curve is constructed over the range of 10 to 200 γ of cholestenone.

Direct Determination of Cholestenone. With partially purified enzyme preparations that exhibit low absorption at 240 mμ it is possible to follow cholestenone formation from cholesterol by measuring the increase in absorption at this wavelength. For this purpose samples are adjusted to 80% ethanol (v/v) and aliquots containing 10 to 100 γ of cholestenone are diluted to 3 ml. with ethanol and read at 240 mμ in a Beckman Model DU spectrophotometer. A standard curve is constructed over the same range of concentrations.

Section III

Citric Acid Cycle

[64] Chromatographic Analyses of Organic Acids

By J. E. VARNER

Column Chromatography

Principle. The procedure is essentially that of Bulen *et al.*[1] and consists in a preliminary or survey separation, followed, if necessary, by additional separations. This survey separation is conducted by adding a mixture of organic acids to the top of a silica gel column and developing with a series of *n*-butanol-chloroform solvents. Fractions of the effluent are collected and titrated with standard alkali.

Apparatus

Microburet, 10 ml.
Intermittent siphon, to deliver about 3 ml.
A chromatographic tube, 12-mm. inside diameter, 24 cm. long, with a 14/35 standard taper outer joint at the top and a stopcock at the bottom.
A chromatographic tube, 7-mm. inside diameter, 75 cm. long, with a 14/35 standard taper outer joint at the top and a stopcock at the bottom.
A solvent reservoir consisting of a 500-ml. round-bottomed flask with a 14/35 standard taper inner joint attached to the bottom, and a three-way stopcock (Pyrex 7380, 2 mm.) attached to the neck of the flask. A funnel is attached to one arm of the Y and an atomizer bulb or some other source of pressure to the other.

Reagents

Silica gel is prepared from Mallinckrodt's silicic acid (analytical reagent) by removal of the fine particles through repeated suspension in distilled water and decantation of the slower settling particles until approximately one-third of the original material is removed. The remaining fraction containing the coarse particles is filtered, dried in an oven at 100° for 24 hours, and stored in a closed container.
Eluting solvents are prepared from U.S.P. chloroform (washed twice with water) and c.p. *n*-butanol to contain 5, 15, 20, 25, 35 and 50% (v/v) *n*-butanol in chloroform and from water-washed benzene and c.p. *n*-butanol to contain 25% (v/v) *n*-butanol in

[1] W. A. Bulen, J. E. Varner, and R. C. Burrell, *Anal. Chem.* **24,** 187 (1952).

benzene. Each solvent mixture is equilibrated against 0.5 N sulfuric acid by shaking the two phases in a separatory funnel and passing the solvent layer through dry filter paper to remove suspended water droplets.

Phenol red indicator is prepared by grinding 100 mg. of the solid with 5.7 ml. of 0.05 N sodium hydroxide and making up to 100 ml. Standard 0.01 N sodium hydroxide (carbonate-free).

Procedure. The survey column is prepared as follows: A glass-wool plug is placed in the bottom of the chromatographic tube (12 mm. $\times$ 25 cm.) to support the column. Eight grams of the prepared silica gel is mixed with 5.5 ml. of 0.5 N sulfuric acid in a mortar. The resulting free-flowing powder is slurried in 60 to 70 ml. of chloroform and added to the chromatographic tube in successive portions. A slight pressure is applied to the top of the tube to speed the removal of excess solvent, care being exercised not to allow the solvent level to fall below the top of the column. This procedure gives a uniformly packed column 14.5 cm. long. The mixture of free organic acids, containing a total of 10 to 100 mg., dissolved in 0.5 N sulfuric acid is mixed thoroughly with 1 g. of silica gel, and the resulting free-flowing powder is transferred quantitatively to the top of the column. This transfer is easily accomplished by pouring the mixture into the tube through a short-stemmed funnel which, along with the container, is rinsed with 2 to 3 ml. of chloroform. If the acids are liberated from their sodium salts, a slight excess of sulfuric acid may be used in the liberation with no undesirable effects. With a thin glass rod, the silica gel containing the sample is slurried with the chloroform used in rinsing, the excess chloroform is drained through the column, and a glass-wool plug is pressed down firmly on the surface of the sample.

The solvents are added to the top of the column through the reservoir which permits the application of pressure. The solvent schedule is 100 ml. of 5%, 135 ml. of 15%, 100 ml. of 25%, 300 ml. of 35%, and 150 ml. of 50% *n*-butanol-chloroform. The 5% solvent is added to the reservoir, and the pressure is adjusted so that the effluent is collected at the rate of 80 ml. per hour. Subsequent solvents are added just before the last of the preceding solvent passes into the column. The effluent is cut into fractions with an intermittent siphon arbitrarily adjusted to deliver 3.3 ml. Individual fractions are titrated with 0.01 N sodium hydroxide after the addition of 8 to 10 ml. of carbon dioxide-free water and 2 drops of phenol red indicator. Near the end point vigorous agitation is necessary to ensure intimate contact between the two phases.

Table I indicates the relative positions of those organic acids which have been separated on this column. Fumaric, glutaric, and formic acids are not separated from each other on the survey column but can be readily separated by a second treatment on a longer column.

The fractions containing these acids are combined in a separatory funnel, and the water layer is separated. The solvent layer is washed once with water; if necessary, sufficient 0.01 N sodium hydroxide is added to make the solution alkaline to phenol red. The water solutions are combined and evaporated to dryness on a water bath. The sodium salts obtained are dissolved in a volume not exceeding 0.5 ml. of 0.5 N sulfuric

TABLE I
The Survey Separation

		Fraction number		
Acid		Initial	Peak	Final
Acetic		14	17	20
Pyruvic		23	26	30
Glutaric		35	37	40
Fumaric	Group I	35	37	40
Formic		37	40	43
Lactic		46	49	52
Succinic	Group II	49	52	57
α-Ketoglutaric		51	57	63
trans-Aconitic		63	69	77
Oxalic		81	83	90
Glycolic	Group III	83	88	98
cis-Aconitic		82	87	95
Malic		114	118	125
Citric		143	150	159
Isocitric		161	170	183
Tartaric		204	210	223

acid. An amount of 4 N sulfuric acid equivalent to the organic acid salts present is added. This solution is mixed thoroughly with 1 g. of dry silica gel and transferred quantitatively to the top of a chromatographic column 60 cm. long and 7 mm. in diameter prepared from 12 g. of silica gel mixed with 8.5 ml. of 0.5 N sulfuric acid. This column is developed by the addition of 20% n-butanol-chloroform. Fractions are collected at the rate of 30 ml. per hour with the aid of an intermittent siphon delivering 0.9 ml.

Table II shows that fumaric, glutaric, and formic acids are completely separated by this longer column.

It can be seen from Table I that lactic, succinic, and α-ketoglutaric acids may not separate completely in the survey separation. The degree of separation does give information useful in the tentative identification of these acids. The separation of these acids on a longer silica gel column is not practical. If these acids occur together in amounts such that they are not separated from each other by the survey separation, the following procedure may be used.

TABLE II
SEPARATION OF GROUP I

	Acid	Fraction number		
		Initial	Peak	Final
Fumaric		26	28	32
Glutaric		34	38	43
Formic		52	55	61

The fractions containing these acids are combined, and the water layer is separated. This solution of the sodium salts of the organic acids is evaporated to a volume of 10 ml. on a water bath. The α-ketoglutaric acid is separated as the 2,4-dinitrophenylhydrazone by adding 2 ml. of a 1% solution of 2,4-dinitrophenylhydrazine in 10% sulfuric acid and allowing the mixture to stand for 30 minutes. The mixture is then extracted twice with 2 to 3 ml. of ether. The combined ether extracts are extracted once with 10 ml. of water, and this water solution in turn is extracted twice with 2 to 3 ml. of ether. The combined ether layers contain the α-ketoglutaric acid derivative, and the combined water layers contain lactic and succinic acids. The separation of lactic and succinic acids is effected chromatographically by a modification of the method of Neish.[2] The fractions which contain these acids are combined, and the water layer is separated and evaporated to dryness. The free acids are liberated from their salts and added to the top of a survey column as above. The column is developed by the addition of 25% n-butanol-benzene. Table III shows the separation of lactic and succinic acids by this procedure.

If the conditions of the initial extraction of the tissue are such that cis-aconitic acid is not converted to trans-aconitic acid, cis-aconitic acid is eluted from the survey column with glycolic and oxalic acids. The fractions containing these three acids are combined, the water layer separated, and the solution of the sodium salts evaporated to about 10 ml. and transferred quantitatively to a centrifuge tube. The solution is made just acid to phenol red with dilute sulfuric acid, and the oxalate is precipitated by adding a slight excess of solid calcium carbonate and digesting for 30

[2] A. C. Neish, Can. J. Research B27, 6 (1949).

minutes in a boiling water bath. The mixture of calcium oxalate and calcium carbonate is centrifuged, the supernatant fluid decanted, and the precipitate washed twice with 2 ml. of hot water. Oxalic may be recovered as such from this precipitate by acidification and extraction, or it may be determined by the standard permanganate titration procedure. The supernatant solution and washings are combined and evaporated to dryness, and 1 ml. of 0.5 N sulfuric acid is added. This acidified mixture is kept in a closed container in a water bath at 65° for 1 hour. This treatment quantitatively transforms *cis*-aconitic to *trans*-aconitic acid. The resulting

TABLE III
SEPARATION OF GROUP II

Acid	Fraction number		
	Initial	Peak	Final
Succinic	11	14	17
Lactic	18	21	26

TABLE IV
SEPARATION OF GROUP III

Acid	Fraction number		
	Initial	Peak	Final
trans-Aconitic	14	16	24
Glycolic	36	41	52

trans-aconitic acid and glycolic acid are separated by addition to the top of the survey column in silica gel as previously described and elution with 20% *n*-butanol-chloroform. Table IV shows the separation achieved by this procedure.

Any of the accepted extraction methods may be used for the preparation of samples to be separated by this procedure. In general, tissue extracts which are free of carbohydrates and amino acids are to be preferred. In some instances (tomato fruit, *Bryophyllum* leaves, citrus fruits, etc.), a 0.5- to 1.0-g. sample of the homogenized, acidified tissue can be mixed with 1 g. of silica gel and added directly to the top of the column for separation.

Recoveries of acids are, with the exception of formic and oxalic, 95 to 100%. The recovery of formic acid after elution from the two columns is 85%; the recovery of oxalic acid is also 85%.

The sensitivity of the method using columns of the size recommended permits detection and tentative identification of quantities as small as 2 to 5 micromoles of the individual acids. It should be emphasized that the column diameter recommended here is merely a convenient one for general use. Columns as small as 2-mm. inside diameter or as large as 50-cm. inside diameter can be used with equal facility by adjusting the sample size, the solvent schedule, and the fraction size according to the area of the column.

An Alternative Procedure. An extremely useful procedure is the separation of organic acids by anion exchange chromatography as described by Busch *et al.*[3] and by Owens *et al.*[4] Since the order of emergence of the acids is different, the anion exchange procedure is a valuable complementary procedure. See Vol. III [70].

Paper Chromatography

Principle. The organic acids are separated as anions on paper sheets using an alkaline-developing solvent. The procedure is adapted from that of Jones *et al.*[5]

Materials and Reagents

Chromatographic jar, 12 to 18 inches high, with cover.
Whatman No. 1 filter paper sheets.
Chlorophenol red, 0.04% in alcohol plus sufficient dilute alkali to make the solution definitely alkaline.
Ethanol-water-ammonium hydroxide (conc.), 8:1:1 (v/v).

Procedure. Ether, ether-acetone, or water solutions (2 to 20 mg./ml.) of the acid mixtures are spotted on a paper sheet 35 cm. long on a line 5 cm. from the paper-solvent contact and allowed to dry in air. The spot diameters should be uniform and 5 to 10 mm. in diameter. Satisfactory separations may be obtained by either ascending or descending solvent development after the solvent front has advanced only 20 to 30 cm. If a mixture contains several slow-moving components, a longer period of development is advantageous. When the solvent front has moved the desired distance, the solvents are removed by steam distillation by blowing steam directly at the face of the paper which is kept dry by irradiation with a battery of infrared lamps.[5] Alternatively, the paper is allowed to

[3] H. Busch, R. B. Hurlbert, and V. R. Potter, *J. Biol. Chem.* **196,** 717 (1952).
[4] H. S. Owens, A. E. Goodban, and J. B. Stark, *Anal. Chem.* **25,** 1507 (1953).
[5] A. R. Jones, E. J. Dowling, and W. J. Skraba, *Anal. Chem.* **25,** 394 (1953).

dry in a stream of air for a few minutes, then sprayed to a light dampness with water and placed in an oven at 95° for 5 minutes. For some purposes simply drying in a stream of air for a few hours is satisfactory. The dried sheets are sprayed with indicator solution until clearly defined spots are obtained. Table V shows the R values for several organic acids.

TABLE V[a]

R Values for Some Nonvolatile Organic Acids

Acid	$\%R_{malic}$[b]
Citric	41.5–42.5
Aconitic	48–52; 68–71
Malonic	82–83.5
Tartaric	89–90.5
Malic	100
Succinic	119–113
Glutaric	134.5–135.5
Fumaric	141–143
Glycolic	190–193

[a] These data are selected from a more comprehensive table published by A. R. Jones, E. J. Dowling, and W. J. Skraba, *Anal. Chem.* **25**, 394 (1953).

[b] Descending chromatograms run at 25° for 20 to 24 hours with ethanol-water-ammonium hydroxide (8:1:1).

[c] Two spots, presumably the *cis* and *trans* forms.

This procedure has a high resolving power for complex mixtures. For example, a mixture of citric, oxalic, malic, succinic, fumaric, glycolic, and acetic acids are completely separated from one another on an ascending chromatogram after the solvent has advanced only 20 to 25 cm.

For best results, the organic acid mixtures should be free of inorganic salts, sugars, and other interfering substances. Bryant and Overell[6] have developed a method involving an anion exchange resin for removing these substances from alcohol extracts and water extracts of plant tissues.

Alternative Procedures. A sensitive procedure for the separation and detection of volatile organic acids has been described by Jones *et al.*[5] A method using an acidic solvent system has been adapted for quantitative work by Bryant and Overell.[6]

[6] F. Bryant and B. T. Overell, *Biochim. et Biophys. Acta* **10**, 471 (1953).

[65] Preparation of α-Keto Acids

By ALTON MEISTER

Many of the more than sixty α-keto acids that have been described are of biochemical interest as intermediates in amino acid metabolism, as participants in the citric acid cycle, or as intermediates in other biochemical reactions. With the exception of the α-keto acid analogs of 3,5-diiodotyrosine, 5-hydroxy-L-lysine, 4-hydroxy-L-proline, and L-threonine,[1] all the α-keto acids corresponding to the amino acids known to occur in hydrolyzates of protein have been prepared. Four α-keto acids, pyruvic, oxalacetic, α-ketoglutaric, and oxalosuccinic, are intermediates in the citric acid cycle, and, with the exception of oxalosuccinic acid,[2] the α-amino analogs of these acids are widely distributed in nature.

The α-keto analogs of amino acids possessing two asymmetric carbon atoms, e.g., threonine, isoleucine, 5-hydroxylysine, would be expected to exist in two optically active forms. Thus far this has been demonstrated by actual preparation of the separate isomers only in the case of the optically active α-keto-β-methylvaleric acids derived from isoleucine.[3] Where the β-carbon atom is the second center of asymmetry, enolization is associated with racemization. α-Keto acids capable of enolization may theoretically exist in *cis*- and *trans*-enol forms, and this has been demonstrated experimentally in several instances.[4]

Methods of Preparation of α-Keto Acids by Organic Synthesis

A number of organic synthetic procedures have been devised,[5] among which the following have been frequently employed:

1. Condensation of diethyl oxalate with a fatty acid ester, followed by hydrolysis and decarboxylation of the resulting oxalo acid ester:

[1] DL-α-Keto-β-hydroxybutyric acid has been synthesized (see the table), although the isomer corresponding to L-threonine (or D-allothreonine) has not as yet been prepared.

[2] α-Aminotricarballylic acid is known [J. P. Greenstein, *J. Biol. Chem.* **109,** 529 (1935); **116,** 463 (1936); J. P. Greenstein, N. Izumiya, M. Winitz, and S. M. Birnbaum, *J. Am. Chem. Soc.* **77,** 707 (1955)] but has not as yet been found in nature.

[3] A. Meister, *J. Biol. Chem.* **190,** 269 (1951).

[4] See, for example, A. Wohl and P. Claussner, *Ber.* **40,** 2308 (1907); H. Gault and R. Weicke, *Bull. soc. chim.* [4] **31,** 867 (1922).

[5] For a review of organic synthetic procedures, see K. L. Waters, *Chem. Revs.* **41,** 585 (1947).

$$RCH_2COOC_2H_5 + COOC_2H_5 \xrightarrow{C_2H_5ONa} COOC_2H_5 \rightarrow RCH_2COCOOH$$
$$\underset{COOC_2H_5}{|} \qquad \underset{\underset{\underset{R}{|}}{COCHCOOC_2H_5}}{|}$$

2. Hydrolysis of oxime esters:

$$RCCOOR' \rightarrow RCOCOOR' \rightarrow RCOCOOH$$
$$\underset{NOH}{\|}$$

3. Hydrolysis of nitriles:

$$RCOOH \rightarrow RCOX \rightarrow RCOCN \rightarrow RCOCOOH$$

4. Hydrolysis of dehydropeptides:

$$RCH_2CCOOH \rightarrow RCH_2COCOOH$$
$$\underset{NCOR'}{\|}$$

Preparation of α-Keto Acids by Enzymatic Methods

α-Keto acids can be prepared in many instances by enzymatic oxidation of the appropriate amino acid analogs. The metabolic relationship between α-amino and α-keto acids has been recognized for many years.[6,7] In more recent studies optically specific amino acid oxidases[8–13] have been found to oxidize amino acids according to the following equations:

$$RCHCOOH + O_2 \xrightarrow{enzymatic} \left[\underset{\underset{NH}{\|}}{RCCOOH} \right] + H_2O_2 \qquad (1)$$
$$\underset{NH_2}{|}$$

$$\left[\underset{\underset{NH}{\|}}{RCCOOH} \right] + H_2O \xrightarrow{spontaneous} \underset{\underset{O}{\|}}{RCCOOH} + NH_3 \qquad (2)$$

$$\underset{\underset{O}{\|}}{RCCOOH} + H_2O_2 \xrightarrow{spontaneous} RCOOH + H_2O + CO_2 \qquad (3)$$

[6] O. Neubauer, *Deut. Arch. klin. Med.* **95**, 211 (1909).
[7] F. Knoop, *Z. physiol. Chem.* **67**, 489 (1910).
[8] H. A. Krebs, *Biochem. J.* **29**, 1620 (1935).
[9] A. E. Bender and H. A. Krebs, *Biochem. J.* **46**, 210 (1950).
[10] M. Blanchard, D. E. Green, V. Nocito, and S. Ratner, *J. Biol. Chem.* **155**, 421 (1944); **161**, 583 (1945).
[11] P. K. Stumpf and D. E. Green, *J. Biol. Chem.* **153**, 387 (1944).
[12] E. A. Zeller, *Advances in Enzymol.* **8**, 459 (1948).
[13] H. A. Krebs, "The Enzymes" (Sumner and Myrbäck, eds.), Vol. 2, Part 1, p. 499, Academic Press, New York, 1951.

In the presence of catalase, reaction 3 does not take place, and the products of the oxidation are α-keto acid and ammonia. By means of ion exchange chromatography or other purification procedures, the α-keto acids or their salts may be obtained as pure compounds in good yields. The enzymatic procedures permit preparation of a number of α-keto acids not yet available by organic synthetic methods. Other advantages include ready availability of the amino acid starting materials and enzymes, simplicity of operation, good yields, and, in the case of the α-keto-β-methylvaleric acids, preservation of the optical configuration of the β-carbon atom.

Procedure for the Preparation of α-Keto Acids Using L-Amino Acid Oxidase[14]

Starting Materials

Enzyme preparation. Dried snake venom[15] is dissolved in water and dialyzed at 5° for 6 to 24 hours against running water. A white precipitate, which may form during dialysis, is removed by centrifugation.

Crystalline catalase[16] is dialyzed for 12 to 24 hours at 5° against running water.

The L isomer (or racemate) of the appropriate amino acid.

Oxidation. Catalase solution and the venom preparation are added successively to a solution of the amino acid,[17] and the pH is adjusted to

---

[14] A. Meister, J. Biol. Chem. **197**, 309 (1952).

[15] Rattlesnake (Crotalus adamanteus), moccasin (Agkistrodon piscivorus piscivorus), and copperhead (Agkistrodon contortrix) venoms are recommended, although venoms of other snakes could probably be used. See also Vol. II [24].

[16] J. B. Sumner and A. L. Dounce, J. Biol. Chem. **121**, 417 (1937); **127**, 439 (1939). Final concentrations of 1 to 3 units/ml. are recommended. For units, see R. N. Feinstein, J. Biol. Chem. **180**, 1197 (1949). See also Vol. II [137].

[17] The relative amounts of venom preparation and amino acid to be used must be determined by pilot runs, which may be conveniently carried out in a Warburg apparatus or by determinations of ammonia. Sufficient enzyme should be used to complete the oxidation in about 1 day. In some cases shorter periods of oxidation are desirable owing to destruction of the keto acid (e.g., p-hydroxyphenylpyruvic acid); in others, oxidations lasting 2 days have not appeared to lower the yield. In general high initial concentrations (0.1 to 0.3 M) of amino acid favor more rapid oxidation, and this is especially important when less susceptible substrates are employed. Suspensions of the less soluble amino acids should be used. Several examples may be given: The author used 200 mg. of Crotalus adamanteus venom for the oxidation of 4.01 g. of DL-methionine (final volume, 100 ml.). For the oxidation of 2.0 g. of L-leucine, 300 mg. of venom in a final volume of 100 ml. was used. Eight hundred milligram of venom was employed for the oxidation of 4.0 g. of DL-valine (final volume, 75 ml.).

7.2 by addition of 2 N sodium hydroxide. The mixture is incubated at 37°, and a stream of rapidly bubbling oxygen is passed into the solution.[18] The progress of the oxidation is conveniently followed by determinations of ammonia.

Isolation. When oxidation is complete, the solution is dialyzed[19] against several changes of water[20] at 5° for 12 hours, and the combined dialyzates are evaporated *in vacuo* to a volume of about one half of the original volume used in the oxidation step.[21] The concentrated solution[22] is treated with Norit, filtered, and added to the top of a well-washed strong cation exchange resin column in the acid form.[23] Elution is carried out with water. Fractions of the effluent which are acid to Congo red paper are combined and adjusted to pH 4.5 with 2 N sodium hydroxide or barium hydroxide, and evaporated *in vacuo* to about one-tenth of the original volume used in the oxidation. The barium or sodium salt of the keto acid crystallizes on slow addition of cold acetone or ethanol. The product is recrystallized from water-ethanol or water-acetone, filtered, and dried over phosphorus pentoxide. The yields of barium salts are usually somewhat greater than those of the corresponding sodium salts.

The procedure described above has been used for the preparation of the sodium salts of α-ketobutyric, α-ketoheptylic, α-keto-ε-hydroxycaproic, α-ketophenylacetic, α-ketosuccinamic, β-cyclohexylpyruvic, α-ketoisocaproic, α-keto-γ-methiolbutyric, α-ketocaproic, α-ketovaleric, phenylpyruvic, p-hydroxyphenylpyruvic, and α-ketoisovaleric acids, and for the barium salts of α-keto-δ-carbamidovaleric acid, α-ketoglutaramic acid, the γ-ethyl ester of α-ketoglutaric acid, and β-sulfopyruvic acid. It has also been used for the preparation of S-benzyl-β-mercaptopyruvic and α-keto-δ-nitroguanidinovaleric acids. In the preparation of α-keto-δ-

---

[18] Caprylic alcohol should be added initially, and if necessary during the course of oxidation, to avoid foaming. Substitution of air for oxygen results in slower oxidative rates.

[19] In some cases, the solution may be acidified with hydrochloric acid, and the keto acid obtained by extraction with ether. Purification may be accomplished by ion exchange or by distillation.

[20] The volume of each change is ten times the volume of the original solution used in the oxidation.

[21] The enzyme may be recovered in 70 to 90% yield by lyophilization of the contents of the dialysis sack.

[22] If the racemic amino acid is used, the D isomer may precipitate at this point and may be removed by filtration.

[23] Dowex 50 (250 to 500 mesh) obtained from the Dow Chemical Company may be used. The resin is washed successively with water, 4 N sodium hydroxide, water, 4 N hydrochloric acid, and water. A column 15 cm. in height and 2.5 cm. in diameter and a flow rate of 1 to 3 ml./min. is suitable for a solution obtained in the oxidation of 0.02 mole of L-amino acid.

PREPARATION OF α-KETO ACIDS

α-Keto acid†	α-Amino acid analog†	References to preparation†	Melting point of 2,4-dinitro-phenylhydrazone (67)†
Pyruvic	Alanine	(12, 13, 14*)	216° h
α-Ketoadipamic (1)	α-Aminoadipamic acid (homoglutamine)	(15*)	(68)
α-Ketoadipic	α-Aminoadipic acid	(16*)	208° h
α-Ketobutyric	α-Aminobutyric acid	(17, 18*, 19*)	198° h
α-Ketoheptylic (2)	α-Aminoheptylic acid	(18*, 19*)	130°, e, l
α-Keto-ε-hydroxycaproic	α-Amino-ε-hydroxycaproic acid	(19*)	183° h
Mesoxalic	Aminomalonic acid	(20)	202° (uncorr.) hc
α-Ketophenylacetic	α-Aminophenylacetic acid	(21, 22, 19*)	193° h
dl-Oxalosuccinic (3)	α-Aminotricarballylic acid (11)	(23, 24, 25)	(68)
α-Keto-δ-guanidinovaleric	Arginine	(26*)	216°, 267° (69)
α-Ketosuccinamic	Asparagine	(27*)	183° e
Oxalacetic	Aspartic acid	(28, 29, 30*, 31, 32)	218° h
α-Keto-δ-carbamidovaleric	Citrulline	(19*)	190° h
β-Cyclohexylpyruvic	β-Cyclohexylalanine	(19*)	189° h
β-Sulfopyruvic	Cysteic acid	(33*)	210° a
β-Mercaptopyruvic (4)	Cysteine	(34*, 35*)	195–200° (35), 161–162° (33)
α-Keto-γ-ethiolbutyric	Ethionine	(19*)	131° h
α-Ketoglutaric	Glutamic acid	(36, 37, 38)	220° h
α-Ketoglutaric-γ-ethyl ester	Glutamic acid-γ-ethyl ester	(19*)	(68)
α-Ketoglutaramic (5)	Glutamine	(27*)	(68)
Glyoxylic	Glycine	(39, 40)	203° h
β-Imidazolylpyruvic	Histidine	(41, 41a)	190–192° hc, 240° e, l (69)
α-Keto-γ-hydroxybutyric acid (6)	Homoserine	(42, 42a, 19*)	
dl-α-Keto-β-methylvaleric (3)	dl-Isoleucine (or dl-alloisoleucine)	(43, 44, 45*)	169° h
d-α-Keto-β-methylvaleric (3)	l-Isoleucine (or d-alloisoleucine)	(46*)	176° h
l-α-Keto-β-methylvaleric (3)	d-Isoleucine (or l-alloisoleucine)	(46*)	176° h

PREPARATION OF α-KETO ACIDS (*Continued*)

α-Keto acid†	α-Amino acid analog†	References to preparation†	Melting point of 2,4-dinitrophenylhydrazone (67)†
α-Ketoisocaproic	Leucine	(47, 48*, 18*, 19*, 49*)	162° h
Trimethylpyruvic	*tert*-Leucine	(50*)	180° h
α-Keto-ε-aminocaproic (7)	Lysine	(26*)	212° h
α-Keto-γ-methiolbutyric (8)	Methionine	(51*, 19*)	150° h
α-Keto-δ-nitroguanidinovaleric	Nitroarginine	(26*)	225° ac
α-Ketocaproic	Norleucine	(52, 18*, 19*)	153° h
α-Ketovaleric	Norvaline	(53, 18*, 19*)	167° h
α-Keto-δ-aminovaleric (7)	Ornithine, proline	(26*)	232–242° (70), 219° (71), 211–212° (26)
Phenylpyruvic (8a)	Phenylalanine	(54, 55*, 19*)	162–164°, 192–194° (72)
S-Benzyl-β-mercaptopyruvic	S-Benzylcysteine	(56, 33*)	150° a
β-Hydroxypyruvic	Serine	(57*)	162° e
DL-α-Keto-β-hydroxybutyric (3)	DL-Threonine (or DL-allothreonine)	(57*, 58)	157–158° (57), 198° (58)
β-[3,5-Diiodo-4-(3′,5′-diiodo-4′-hydroxyphenoxy)phenyl]pyruvic	Thyroxine	(59)	
β-Indolylpyruvic	Tryptophan	(60, 61, 62, 63)	169° (69)
p-Hydroxyphenylpyruvic (9)	Tyrosine	(64*, 19*)	178° h
α-Ketoisovaleric (10)	Valine	(53, 65, 66*, 19*)	196° h

* References marked with an asterisk describe procedures which have been successfully repeated in the author's laboratory.
† Numbers in parentheses refer to footnotes. See pp. 410–411.

1. See (15) for N-methyl derivative.
2. See K. L. Waters [*Chem. Revs.* **41**, 585 (1947)] for references to the preparation of the straight-chain α-keto acids possessing 8, 9, 10, 11, 12, 13, 15, 16, and 18 carbon atoms, and branched-chain α-keto acids possessing 7, 10, and 11 carbon atoms.
3. Where the configuration of the second asymmetric center of the amino acid is unknown, the corresponding α-keto acid is designated *dl*, *d*, or *l*, rather than DL, D, or L, in accordance with currently accepted rules of nomenclature [H. B. Vickery, *J. Biol. Chem.* **169**, 237 (1947)].
4. For disulfide form, see J. Parrod, *Compt. rend.* **226**, 736 (1948); for α-keto-β-mercaptobutyric acid, see (34).
5. See (27) for N-methyl derivative, and (15) for α-keto-N-dimethylglutaramic acid, α-keto-*dl*-γ-methylglutaramic acid; γ-(α-ketoglutaryl)-glycine, γ-(α-ketoglutaryl)-L-alanine, γ-(α-ketoglutaryl)-β-alanine, β-oxalacetylglycine, and β-oxalacetyl-L-alanine have been prepared by a similar procedure (T. T. Otani and A. Meister, Abstracts, American Chemical Society Meeting, Cincinnati, Ohio, March, 1955).
6. See (42) and N. Hall, J. E. Hynes, and A. Lapworth, *J. Chem. Soc.* **107**, 132 (1915) for preparation of several similar α-keto acids.
7. See (26) for N-carbobenzoxy derivative, and (19) for N-chloroacetyl derivative of α-keto-ε-aminocaproic acid.
8. See (34) for α-keto-β-methiolpropionic acid and several related α-keto acids. The α-keto analogues of methionine sulfone and the methionine sulfoxides have been prepared by enzymatic methods (*cf.* (19)).
8a. The preparation of thienylpyruvic acid has been described [J. A. Jacquez, R. K. Barclay, and C. C. Stock, *J. Exptl. Med.* **96**, 499 (1952)].
9. The o-hydroxy [J. Plochl and L. Wolfrum, *Ber.* **18**, 1183 (1885)], m-hydroxy, and 2,5-dihydroxy derivatives [O. Neubauer, *Deut. Arch. klin. Med.* **95**, 211 (1909)] have also been described.
10. The α-keto analog of pantoic acid, α-keto-β,β-dimethyl-β-hydroxymethylpropionic acid, has been prepared as the γ-lactone [R. Kuhn and T. Wieland, *Ber.* **75B**, 121 (1942); S. H. Lipton and F. M. Strong, *J. Am. Chem. Soc.* **71**, 2364 (1949)].
11. Two diastereoisomeric forms exist.
12. J. J. Berzelius, *Ann. Physik* **36**, 1 (1835).
13. J. W. Howard and W. A. Fraser, *Org. Syntheses Coll. Vol.* **1**, 475 (1941) (2nd Ed.).
14. V. E. Price and L. Levintow, *Biochem. Preparations* **2**, 22 (1952).
15. A. Meister, *J. Biol. Chem.* **210**, 17 (1954).
16. H. Gault, *Bull. soc. chim.* (France) **11**, 382 (1912).
17. L. Claisen and E. Moritz, *Ber.* **13**, 2121 (1880).
18. F. Adickes and G. Andresen, *Ann.* **555**, 41 (1943).
19. A. Meister, *J. Biol. Chem.* **197**, 309 (1952).
20. R. S. Curtiss, *Am. Chem. J.* **35**, 477 (1906).
21. L. Claisen and F. H. Morley, *Ber.* **11**, 1596 (1878).
22. B. B. Corson, R. A. Dodge, S. A. Harris, and R. K. Hagen, *Org. Syntheses Coll. Vol.* **1**, 241 (1941) (2nd Ed.).
23. F. Lynen and H. Scherer, *Ann.* **560**, 163 (1948).
24. S. Ochoa, *J. Biol. Chem.* **174**, 115 (1948).
25. C. G. Baker, *Biochem. Preparations* **4**, in press.
26. A. Meister, *J. Biol. Chem.* **206**, 579 (1954).
27. A. Meister, *J. Biol. Chem.* **200**, 571 (1953).

28. W. Wislicensus, *Ann.* **246,** 306 (1888).
29. H. J. H. Fenton and M. A. Jones, *J. Chem. Soc.* **77,** 77 (1900).
30. L. O. Krampitz and C. H. Werkman, *Biochem. J.* **35,** 595 (1941).
31. J. C. Roberts, *J. Chem. Soc.* **1952,** 3315.
32. C. Heidelberger, *Biochem. Preparations* **3,** 59 (1953).
33. A. Meister, P. E. Fraser, and S. V. Tice, *J. Biol. Chem.* **206,** 561 (1954).
34. J. Parrod, *Compt. rend.* **215,** 146 (1942); *Bull. soc. chim.* (France) **14,** 109 (1947).
35. F. Schneider and E. Reinefeld, *Biochem. Z.* **318,** 507 (1948).
36. E. E. Blaise and H. Gault, *Compt. rend.* **147,** 198 (1908).
37. W. Wislicensus and M. Waldmuller, *Ber.* **44,** 1564 (1911).
38. L. Friedman and E. Kosower, *Org. Syntheses* **26,** 42 (1946).
39. S. R. Benedict, *J. Biol. Chem.* **6,** 51 (1909).
40. S. Weinhouse and B. Friedmann, *J. Biol. Chem.* **191,** 707 (1951).
41. G. Barger and C. P. Stewart, *J. Pharmacol. Exptl. Therap.* **29,** 223 (1926).
41a. H. P. Broquist and E. E. Snell, *J. Biol. Chem.* **180,** 59 (1949).
42. H. Schinz and M. Hinder, *Helv. Chim. Acta* **30,** 1349 (1947).
42a. H. Hift and H. R. Mahler, *J. Biol. Chem.* **198,** 901 (1952).
43. A. Mebus, *Monatsh.* **26,** 483 (1905).
44. L. Bouveault and R. Locquin, *Compt. rend.* **141,** 115 (1905).
45. A. Meister and J. P. Greenstein, *J. Biol. Chem.* **195,** 849 (1952).
46. A. Meister, *J. Biol. Chem.* **190,** 269 (1951).
47. R. Locquin, *Bull. soc. chim.* France (3), **31,** 1147 (1904).
48. J. P. Greenstein and V. E. Price, *J. Biol. Chem.* **178,** 695 (1949).
49. A. Meister, *Biochem. Preparations* **3,** 66 (1953).
50. C. Glücksmann, *Monatsh.* **10,** 770 (1889).
51. W. M. Cahill and G. G. Rudolph, *J. Biol. Chem.* **145,** 201 (1942).
52. K. Kondo, *Biochem. Z.* **38,** 407 (1912).
53. E. Moritz, *J. Chem. Soc.* **39,** 13 (1881).
54. J. Plochl, *Ber.* **16,** 2815 (1883).
55. R. M. Herbst and D. Shemin, *Org. Syntheses Coll. Vol.* **2,** 519 (1943).
56. J. A. Stekol, *J. Biol. Chem.* **176,** 33 (1948).
57. D. B. Sprinson and E. Chargaff, *J. Biol. Chem.* **164,** 417 (1947).
58. E. Hoff-Jorgensen, *Z. physiol. Chem.* **265,** 77 (1940).
59. A. Canzanelli, R. Guild, and C. R. Harington, *Biochem. J.* **29,** 1617 (1935).
60. A. Ellinger and Z. Matsuoka, *Z. physiol. Chem.* **109,** 259 (1920).
61. C. P. Berg, W. C. Rose, and C. S. Marvel, *J. Biol. Chem.* **85,** 219 (1929).
62. R. W. Jackson, *J. Biol. Chem.* **84,** 1 (1929).
63. L. C. Banguess and C. P. Berg, *J. Biol. Chem.* **104,** 675 (1934).
64. O. Neubauer and K. Fromberg, *Z. physiol. Chem.* **70,** 326 (1910).
65. K. Brunner, *Monatsh.* **15,** 751 (1894).
66. J. W. Cornforth, *in* "Chemistry of Penicillin" (H. T. Clarke *et al.*, eds.), p. 783, Princeton University Press, Princeton, 1949.
67. The solvents used for crystallization are designated as follows: h-water, hc-hydrochloric acid, a-alcohol, e-ethyl acetate, l-ligroin, ac-glacial acetic acid.
68. These derivatives have been found to contain variable amounts of the 2,4-dinitrophenylhydrazones of α-ketoadipic or α-ketoglutaric acids.
69. P. K. Stumpf and D. E. Green, *J. Biol. Chem.* **153,** 387 (1944).
70. H. A. Krebs, *Enzymologia* **7,** 53 (1939).
71. M. Blanchard, D. E. Green, V. Nocito, and S. Ratner, *J. Biol. Chem.* **155,** 421 (1944).
72. W. S. Fones, *J. Org. Chem.* **17,** 1534 (1952).

guanidinovaleric acid, the ion exchange step may be omitted, since the free keto acid crystallizes on evaporation of the protein-free solution. For the preparation of the α-keto analogs of lysine and ornithine, the corresponding ω-N-carbobenzoxy amino acid derivatives are oxidized. When oxidation is complete, the respective ω-N-carbobenzoxy-α-keto acids are extracted from the acidified oxidation mixture with ethyl acetate, and crystallized. Removal of the carbobenzoxy group by treatment with HBr in glacial acetic acid yields the keto acid hydrobromides.

D-Amino acid oxidase, obtained from hog kidney acetone powder, may also be used for the preparation of a number of keto acids, including some of those listed above.[24] When crude D-amino oxidase preparations are used, it is advisable to extract the keto acid from the acidified oxidation mixture and purify by distillation *in vacuo* rather than use the ion exchange procedure. In the preparation of the isomers of α-keto-β-methylvaleric acid, the pH should not exceed about 8.0 in order to avoid enolization of the keto acid and consequent racemization.[3]

Choice of Method of Preparation

The table lists a number of references to the preparation of specific α-keto acids. Although no attempt has been made to include all the methods available in the literature, a number of representative methods (including those successfully employed in the author's laboratory) have been given. The decision as to which of the several methods is preferable for the preparation of a given α-keto acid will depend in part on the available starting materials and on the quantity of product desired. Fifteen of the α-keto acids listed have thus far been prepared only by enzymatic procedures, and at least ten others may also be obtained in this manner. Where enzymatic and organic synthetic methods exist for the preparation of the same compound, such factors as yield and degree of enzymatic susceptibility must be considered. For the preparation of quantities of 1 to 10 g., the enzymatic methods are preferable to organic synthetic procedures for α-ketobutyric, α-ketoheptylic, α-ketophenylacetic, α-keto-isocaproic, α-ketocaproic, α-ketovaleric, phenylpyruvic, p-hydroxyphenylpyruvic, and α-ketoisovaleric acids. For larger quantities it may be desirable to use organic synthetic methods in cases where the enzymatic susceptibility of the corresponding amino acids is relatively low, necessitating large amounts of enzyme. Although the cost of large amounts of snake venom may be prohibitive, hog kidney D-amino acid oxidase can be prepared at much lower cost. Thus the enzymatic method, with D-amino acid oxidase, is suitable for the preparation of large quantities of α-ketoisocaproic, α-ketobutyric, α-ketovaleric, and α-ketocaproic

[24] A. Meister, *Biochem. Preparations* **3**, 66 (1953).

acids. Excellent synthetic methods exist for p-hydroxyphenylpyruvic, phenylpyruvic, and α-ketoisovaleric acids; these procedures are probably preferable for the preparation of large amounts of these keto acids.

General Methods of Characterization

Most of the α-keto acids readily form oximes, phenylhydrazones, semicarbazones, and 2,4-dinitrophenylhydrazones. Of these, the latter have been most commonly employed for keto acid identification in biochemical studies. Usually addition of an excess of warm 1% 2,4-dinitrophenylhydrazine in 2 N hydrochloric acid to a concentrated aqueous solution of the keto acid results in the precipitation of the corresponding crystalline hydrazone. In some instances this derivative can only be prepared by treating the keto acid with 2,4-dinitrophenylhydrazine dissolved in sulfuric acid and ethanol.[25] The melting points of a number of such derivatives are given in the table. In some instances, two forms of the hydrazone may be prepared by crystallization from different solvents. For example, crystallization of phenylpyruvic acid 2,4-dinitrophenylhydrazone from water or alcohol yields a product with a melting point of 192 to 193°, whereas crystallization from ethyl acetate and petroleum ether results in a derivative melting at 160 to 162°.[26] The two forms are interconvertible, and a mixture of the two forms melts at 192 to 193°. Paper and column chromatography of keto acid 2,4-dinitrophenylhydrazones has also been useful in identification.[27] Here again pure 2,4-dinitrophenylhydrazones may be resolved in some chromatographic systems yielding two forms, possibly representing *syn* and *anti* isomers.[28] Supplementary characterization may often be carried out by hydrogenation of α-keto acid hydrazones yielding the corresponding α-amino acid, which can be isolated or compared by paper chromatography with known amino acids.[29,29a] Paper chromatography of free α-keto acids has also been described,[30,31] although as yet such procedures are not as sensitive as those involving 2,4-dinitrophenylhydrozones.

[25] R. L. Shriner, and R. C. Fuson, "Systematic Identification of Organic Compounds," 3rd ed., John Wiley & Sons, New York, 1948.

[26] W. S. Fones, *J. Org. Chem.* **17**, 1534 (1952).

[27] See, for example, D. Cavallini, N. Frontali, and G. Toschi, *Nature* **163**, 568, **164**, 792 (1949); E. Kulonen, E. Carpen, and T. Ruokolainen, *Scand. J. Clin. & Lab. Invest.* **4**, 189 (1952); G. A. Le Page, *Cancer Research* **10**, 393 (1950).

[28] D. E. Metzler and E. E. Snell, *J. Biol. Chem.* **198**, 353 (1952).

[29] E. Kulonen, *Scand. J. Clin. & Lab. Invest.* **5**, 72 (1953).

[29a] G. H. N. Towers, J. F. Thompson, and F. C. Steward, *J. Am. Chem. Soc.* **76**, 2392 (1954).

[30] T. Wieland and E. Fischer, *Naturwissenschaften* **36**, 219 (1949).

[31] H. E. Umbarger and B. Magasanik, *J. Am. Chem. Soc.* **74**, 4253 (1952).

α-Keto acids are quantitatively decarboxylated by ceric sulfate[32,33] or by hydrogen peroxide,[34] and these procedures are useful in the determination of purity of a given α-keto acid preparation. Other procedures which may be valuable in certain instances include preparation of the sodium bisulfite derivative, nonenzymatic transamination to form the corresponding amino acid, and ultraviolet and infrared absorption studies. Certain α-keto acids, e.g., α-keto-γ-methiolbutyric and β-mercaptopyruvic, give specific color reactions also characteristic of the analogous amino acids. Finally, enzymatic and microbiological procedures may be useful for the characterization and quantitative determination of α-keto acids. For example, many α-keto acids are reduced by lactic dehydrogenase,[14,35] and at least five of these[36] are reduced at rates of about the same order of magnitude as pyruvate. The yeast α-keto acid decarboxylase also exhibits a wide range of substrate specificity.[14]

[32] C. Fromageot and P. Desnuelle, *Biochem. Z.* **279**, 174 (1935).
[33] H. A. Krebs and W. A. Johnson, *Enzymologia* **4**, 148 (1937).
[34] A. Meister, *J. Biol. Chem.* **200**, 571 (1953).
[35] A. Meister, *J. Biol. Chem.* **184**, 117 (1950).
[36] α-Ketobutyrate, glyoxylate, β-mercaptopyruvate, β-hydroxypyruvate, and DL-α-keto-β-hydroxybutyrate.

[66] Determination of α-Keto Acids

By THEODORE E. FRIEDEMANN

Assay Methods

Principle. The methods are based on the observations of Dakin and Dudley[1] that the *p*-nitrophenylhydrazones of keto acids are soluble in, or are extracted from organic solvents by, Na_2CO_3 solution and yield brilliant red to purple colors on addition of strong alkali. Lu[2] adapted these properties of hydrazones to the determination of microgram quantities, using reaction with 2,4-dinitrophenylhydrazine, extraction of hydrazones by ethylacetate, re-extraction by 10% Na_2CO_3 solution, addition of *N* NaOH, and photoelectric determination of color density. In the procedures of Friedemann et al.[3,4] described below, the steps are

[1] H. D. Dakin and H. W. Dudley, *J. Biol. Chem.* **15**, 127 (1913).
[2] G. D. Lu, *Biochem. J.* **33**, 249 (1939).
[3] T. E. Friedemann and G. E. Haugen, *J. Biol. Chem.* **147**, 415 (1943).
[4] T. E. Friedemann, V. M. Kinney, and P. K. Keegan, unpublished.

essentially the same; however, the separate determination of keto acids is based on differences in the rates of reaction with 2,4-dinitrophenyl-hydrazine, on the distribution of the hydrazones in the aqueous phase and various solvents, and, to some extent, on differences in light absorption of the hydrazones in alkali. Thus, at 25°, the monocarboxylic keto acids (and ketones and aldehydes, in general) react completely with the hydrazine within about 4 minutes, the dicarboxylic keto acids within about 20 minutes. The monocarboxylic acid hydrazones are preferentially extracted from acid aqueous solution by aromatic hydrocarbons; the dicarboxylic acid hydrazones by aliphatic and aromatic alcohols. Both groups of hydrazones are readily extracted by esters and ethers. Since a large part of the energy of metabolism in most living organisms is obtained from carbohydrate through the formation of pyruvic acid, this acid is the principal monocarboxylic keto acid in biological materials; also, since oxalacetic acid is unstable, yielding pyruvic acid and CO_2, α-ketoglutaric acid is the principal dicarboxylic acid, unless the sample is analyzed immediately. That these two are the predominating keto acids in blood and urine has been shown by chromatographic[5,6] and enzymatic[7,8] procedures.

Reagents

Protein precipitant, 10% TCA solution. Prepare fresh solution each month. Store in the refrigerator.

Acetic acid, 5% solution.

Lloyd's reagent.

Hydrazine reagent, 0.1% solution. Dissolve 500 mg. of 2,4-dinitro-phenylhydrazine in 500 ml. of 2 N HCl. Prepare fresh solution each month; store in the refrigerator. If the hydrazine is not readily and completely soluble, recrystallize from ethyl acetate.

Solvents. Ethyl benzene, benzyl alcohol, and ethyl acetate are the recommended solvents in methods 1, 2, and 3. Benzene, n-butyl alcohol, and dibutyl ether may be substituted in the respective methods 1, 2, and 3 if the preferred solvents are not available. All solvents should be distilled after procurement. Dibutyl ether should be redistilled at frequent intervals and stored in the refrigerator in order to prevent the accumulation of peroxides.

[5] D. Seligson and B. Shapiro, *Anal. Chem.* **24**, 754 (1952).
[6] E. Kulonen, E. Karpén, and T. Ruokolainen, *Scand. J. Clin. & Lab. Invest.* **4**, 189 (1952).
[7] H. A. Krebs, *Biochem. J.* **32**, 108 (1938).
[8] P. E. Simola and F. E. Krusius, *Z. physiol. Chem.* **261**, 209 (1939).

Sodium carbonate, 10% solution of anhydrous salt. Filter and keep in a Pyrex container.

Sodium hydroxide, 1.5 N solution.

Keto acid standards. Redistill pyruvic acid (Eastman) *in vacuo* and immediately prepare 0.1 to 0.5 N stock solution. Weigh accurately 8 to 25 g. of the acid in a weighing bottle; transfer quantitatively to a 1-l. or 500-ml. volumetric flask, using boiled, cold distilled water. Check the purity of the acid by titration with standard alkali. Solutions have been stored in the refrigerator for more than a year without apparent deterioration. α-Keto-glutaric acid may be obtained from commercial sources. Recrystallize if not pure, as determined by titration with standard alkali. Transfer 100 to 500 mg. to a volumetric flask, using 5% acetic acid solution; dilute to the mark and store in the refrigerator. Prepare working standards, using 5% acetic acid for all dilutions.

Collection of Samples and Preparation for Analysis. Draw *blood* from the vein with a minimum of stasis. Tap the syringe gently while holding it in a vertical position to remove the bubbles of gas; bring the plunger to the mark, and expel the contents through the needle into 5 vol. of TCA solution contained in a stoppered centrifuge tube. Mix the contents, and immmediately place in the refrigerator. Centrifugate, and analyze the clear supernatant solution. Collect *urine* in plastic or amber glass bottles containing enough 20 N H_2SO_4 to give 0.05 to 0.1 N acidity at the end of the collection period. Urine contains considerable quantities of interfering carbonyl compounds, apparently derived from ingested foods, which can be removed by Lloyd's reagent. For every 10 ml. of the acidified urine add 0.75 g. of Lloyd's reagent; shake vigorously and filter immediately. *Tissue* suspensions, as used in *in vitro* studies of intermediary metabolic reactions, may be treated directly with TCA solution, which precipitates the proteins and stops the reactions. Tissues from living animals are frozen in liquid nitrogen; they are crushed[9] or ground; the cold samples are weighed and added to measured volumes of cold TCA solution. Weighed, fresh *plant tissues* are ground in measured volumes of cold TCA solution in the Waring blendor. The tissue suspensions are cleared by centrifugation.

General Procedure. Transfer 3.0 ml. of sample extract to each of two 18 × 200-mm. test tubes. For each series of not more than ten tubes, prepare three reagent blanks containing 3.0 ml. of TCA or acetic acid solution as in the sample extract. Incubate all tubes 10 minutes in a bath or pan of water maintained at 25 ± 2°. At accurately timed intervals,

[9] J. B. Graeser, J. E. Ginsberg, and T. E. Friedemann, *J. Biol. Chem.* **104**, 149 (1934).

say 30 seconds, add in turn 1.0 ml. of hydrazine solution to each tube. Continue incubation at 25° for the exact time indicated below, at which time add the exact volume of appropriate solvent from a buret to each tube at the timed intervals. Mix immediately by aerating the contents for 2 minutes with a rapid stream of air or nitrogen through a Wright capillary pipet in each tube (10-mm. o.d. glass tubing; upper portion 4 inches long; lower portion 8 inches long, drawn to fine capillary with square-cut tip; both ends flamed). Remove the capillary pipets. Centrifugate at low speed to facilitate separation of phases. Ethyl benzene and ethyl acetate separate readily without centrifugation. Attach rubber bulbs to capillary pipets, and remove the aqueous reaction phase. Give a violent circular motion to the tube to dislodge any remaining solution. (If considered necessary, add 1 ml. of water, mix contents by circular motion of tube, and centrifugate.) Remove the remaining aqueous phase. Add 6.0 ml. of Na_2CO_3 solution from a pipet. Remove the rubber bulbs from the capillary pipets, and again mix the contents by rapid aeration for 2 minutes. Remove the capillary pipets; centrifugate if necessary. Insert rapidly a 5-ml. pipet through the solvent layer, and gently blow out any solvent which may have entered. Transfer 5.0 ml. of the lower carbonate extract to a colorimeter tube, held in the bath or pan of water at 25°. At definite intervals, say 30 seconds, add 5.0 ml. of NaOH solution in turn to each colorimeter tube. A red to orange-red color is produced. At the end of 5 minutes, at the same time intervals between each tube, determine the color density, after setting the galvanometer reading to 100 by means of the reagent blanks. Use a photoelectric colorimeter provided with a filter having maximal transmittance at 420 mμ; if a spectrophotometer is used, set the instrument at wavelength 435 mμ.

METHOD 1. *Determination of pyruvic acid and other monocarboxylic keto acids.* Proceed as above. Incubate the sample extract and hydrazine solution for exactly 5 minutes; then add 3.0 ml. of ethyl benzene.

METHOD 2. *Determination of α-ketoglutaric acid and other dicarboxylic keto acids.* Proceed as above. Incubate for exactly 25 minutes; then add 8.0 ml. of benzyl alcohol.

METHOD 3. *Determination of total keto acids.* Proceed as above. Incubate the sample extract and hydrazine solution for exactly 25 minutes; then add 8.0 ml. of ethyl acetate.

Calculations

The procedure for calculating the individual acids is illustrated by the following data and formulas based on analysis of standard solutions, using the Evelyn colorimeter with filter 420. Thus, the following respective calculated color densities per micromole of pyruvic and α-ketoglutaric acids were observed: by method 1, 2.037 and 0.149; by method 2, 0.181 and 1.862; by method 3, 2.119 and 1.919. If D_1, D_2, and D_3 represent the observed densities by methods 1, 2, and 3, and P and G represent micromoles of pyruvic and α-ketoglutaric acids in 3 ml. of the sample extract taken for analysis, then

$$D_1 = 2.037P + 0.149G$$
$$D_2 = 0.181P + 1.862G$$
$$D_3 = 2.119P + 1.919G$$

By using any two of the methods and by solving the two respective equations for P and G, the individual acids may be evaluated. However, the most accurate results are obtained with methods 1 and 2 which provide sharp separation of the hydrazones.

Concentrations in Blood and Urine

The methods have yielded the following results. The blood from ten moderately active subjects contained 0.86 to 1.63 mg. of pyruvic acid per 100 ml., with an average of 1.204 mg.; it contained 0.10 to 0.32 mg. of α-ketoglutaric acid, with an average of 0.19 mg. The 24-hour urine contains an average of 20 ± 8 mg. of pyruvic acid and an average of 33 ± 15 mg. of α-ketoglutaric acid.

[67] Preparation and Assay of Oxalacetic Acid

By SAMUEL P. BESSMAN

Preparation

Reactions

2 Na diethyloxalacetate + $H_2SO_4 \rightarrow$ 2 Diethyloxalacetate + Na_2SO_4
Diethyloxalacetate + 2 HOH $\rightarrow$ Oxalacetic acid + 2 ethanol

Method.[1-3] One hundred grams (0.48 mole) of sodium diethyloxalacetate is dissolved in 500 ml. of 10^{-3} M disodium ethylenediaminetetraacetate (Na_2-EDTA). When solution is complete, the mixture is placed in a liter separatory funnel and concentrated sulfuric acid is added dropwise until acid to congo red paper. Diethyloxalacetate separates as a heavy oil. Sufficient ether is added with shaking to bring all the oil into

[1] L. S. Simon, *Compt. rend.* **137**, 855 (1903).
[2] P. P. Cohen, *J. Biol. Chem.* **136**, 565 (1940).
[3] L. O. Krampitz and C. H. Werkman, *Biochem. J.* **35**, 595 (1941).

an upper organic phase, and it is then separated. The aqueous phase is extracted three more times with 50-ml. aliquots of ether. The combined ether extracts are washed with 100 ml. of 10^{-3} M Na$_2$-EDTA, separated, and dried with sodium sulfate. The ether is removed in a stream of air on the water bath in a tared 500 ml. flask with a ground-glass stopper (yield 67 to 75%). Concentrated HCl, four times the weight of the ethyl oxalacetate, is added and mixed well until solution of the ester occurs. The flask is stoppered and placed in the refrigerator for 7 to 10 days. It should be shaken two or three times daily. The thick crystalline precipitate of oxalacetic acid which forms is collected in a sintered glass funnel and washed with ice-cold HCl. The cake is pressed with a flattened glass rod, and the funnel with the cake is placed in a vacuum desiccator containing potassium hydroxide pellets. Renew the vacuum daily. When the odor of HCl has disappeared (this may require 3 to 5 days, during which the spent KOH is replaced and the crusts of KCl which form over the KOH are broken up), a small amount of fresh KOH is put in the desiccator together with a dish of P$_2$O$_5$ (with great care to prevent contact of the KOH and P$_2$O$_5$). After 1 day in the presence of P$_2$O$_5$ the crystals are removed from the funnel with a porcelain or glass spatula and placed in a flask for recrystallization (yield 10.0 to 15.8%).

A small amount of *dry* acetone is added, sufficient to dissolve the acid completely at boiling temperature. With no further heating, and with vigorous shaking, *dry* benzene is added drop by drop, until cloudiness appears, then very slowly with continuous shaking until crystals appear. Two volumes of benzene can then be added to the rapidly swirled mixture. After an hour at refrigeration temperature, the crystals can be filtered off and washed with dry benzene. This powdery crystalline material is stable at room temperature over calcium chloride. The yield is 8.5 g., or 13.4% of theoretical. The melting point is 150 to 152°. Another 2 g. can be obtained by working up the mother liquors. It is free of oxalate as shown by failure to form a precipitate with calcium chloride.

Assay of Oxalacetic Acid

The assay of this material is complicated by its rapid rate of spontaneous decomposition, which is catalyzed by a multitude of substances,[4-6] many of which might be present as tissue components. The spontaneous decomposition rate of the pure material at pH 5.0 is about 1% per minute at 30°.[6]

This marked lability of oxalacetic acid is utilized for its assay, either

[4] F. L. Breusch, *Biochem. J.* **27**, 1757 (1933).

[5] H. A. Krebs, *Biochem. J.* **36**, 303 (1942).

[6] S. P. Bessman and E. C. Layne, Jr., *Arch. Biochem.* **26**, 25 (1950).

as evolved CO_2 in the presence of a catalyst[7,8] or as the dinitrophenyl-hydrazone of pyruvic acid after catalytic decarboxylation.[9] The two methods can be combined to give an estimate of purity of the compound.

CO_2 Evolution

Oxalacetic acid → Pyruvic acid + CO_2

A Warburg flask is charged with 2 ml. of acetate buffer, pH 5.0, containing nickel sulfate, 0.010 M. In the side arm a freshly prepared, approximately 0.05 molar solution of OAA in 10^{-3} M Na$_2$ EDTA is placed. After temperature equilibration at 25° the taps are closed, and the oxalacetic acid is tipped. Reaction is complete in 20 to 30 minutes. Recrystallized material assays 90 to 95% by this method.

For assay of oxalacetic acid production in a reaction mixture, the method of Ostern as modified by Greville is used. Immediately at the end of the reaction 50% trichloroacetic acid or citric acid is added to a final concentration of about 5%, and the flasks are placed in crushed ice. Aniline citrate (equal parts of aniline and 50% citric acid) equal to 10% of the final volume is placed in the side arm. The flasks are equilibrated in a bath set at 5° if possible, since at this temperature no other substance yields CO_2 on addition of aniline. After equilibration the taps are closed and the aniline tipped. Readings are taken until no further evolution of CO_2 takes place. A small volume contraction due to the dilution of the aniline citrate can be corrected by a control reagent flask. Recoveries of oxalacetic acid by this method are highly variable, depending on the catalytic effect of tissue components on the decomposition of this labile intermediate. These factors can be controlled to some extent by suitable blanks. The use of EDTA in the deproteinizing solution might be of great value in facilitating recovery of the oxalacetic acid.

Acetoacetic, acetonedicarboxylic, and oxalsuccinic acids all yield CO_2 in the presence of aniline. Acetoacetic and acetone-dicarboxylic acids do not interfere at low temperatures but can be troublesome at 25°. A check on this interference is possible by running an acetone determination on the solution remaining in the flask and subtracting the micromoles of acetone found from the micromoles of CO_2 evolved. Since acetonedicarboxylic acid will yield two moles of CO_2 and acetoacetic acid one, the general reactions studied must be designed to include this possibility.

Colorimetric Determination. As a further check on the reliability of the estimation of oxalacetate a determination of pyruvate can be made after the decarboxylation. This is probably the most accurate determina-

[7] P. Ostern, *Z. physiol. Chem.* **218,** 160 (1933).
[8] G. D. Greville, *Biochem. J.* **33,** 718 (1939).
[9] See Vol. I [125].

tion of oxalacetate formation, although an indirect one. Aniline citrate or nickel sulfate is added to a final concentration of 5% or 0.10 M, resp., at the end of the reaction, and an estimation of pyruvate by the method of Friedemann and Haugen[10] is made after 30 minutes at room temperature. The pyruvate formed is corrected by preformed pyruvate and represents oxalacetate. This method eliminates the interference of other β-keto acids completely.

It might be noted that the pyridine-acetic anhydride method for measuring citrate, described by Saffran and Denstedt,[11] is sensitive to oxalacetic acid. This method has been employed on occasion in our laboratory and has yielded accurate results in the absence of citrate. In the presence of citrate an estimate can be obtained by performing the citrate determination before and after catalytic decarboxylation with aniline.

[10] See Vol. III [66].
[11] M. Saffran and O. F. Denstedt, *J. Biol. Chem.* **175**, 849 (1948).

[68] Preparation of Tricarboxylic Acids

By DANIEL H. DEUTSCH and ROBERT E. PHILLIPS

I. Preparation of *dl*-Isocitric Acid

Principle. Several methods for the preparation of isocitric acid have been reported.[1] The following method is a modification of the procedure of Pucher and Vickery which was based on the work of Fittig and Miller. The condensation of anhydrous sodium succinate with chloral in the presence of acetic anhydride gives trichloromethylparaconic acid. Hydrolysis of this acid yields isocitric acid. It is separated from the allo form by conversion to the lactone and crystallization from ethyl acetate. *dl*-Isocitric acid is readily prepared by hydrolysis of the pure lactone with dilute alkali.

Reagents

Sodium succinate, 100 g., anhydrous. Sodium succinate hexahydrate is dried at 110° to constant weight, rapidly powdered, and kept dry until ready for use.
Chloral, 96 g., anhydrous.
Acetic anhydride, 63 g., 95%.

[1] R. Fittig and H. E. Miller, *Ann.* **255**, 43 (1889); W. Wislicenus and M. Nassauer, *Ann.* **285**, 1 (1895); J. P. Greenstein, *J. Biol. Chem.* **116**, 463 (1936); G. W. Pucher and H. B. Vickery, *J. Biol. Chem.* **163**, 169 (1946).

is evaporated (below 25°) to approximately 20 ml. and diluted to 26 ml. with concentrated hydrochloric acid. After standing overnight at 0° any solid material is filtered off and discarded. The clear filtrate is cooled to −10° in an ice-salt bath and diluted slowly with 65 ml. of ice water with vigorous stirring. It is then neutralized (congo red paper) to pH 4 to 5 by the careful addition of the sodium hydroxide (cooled to −10°), the temperature of the reaction mixture being maintained below −2°. About 26 ml. is needed. When the temperature of the mixture has again reached −10°, the barium acetate solution (150 ml. cooled to −10°) is added rapidly, yielding a bulky white precipitate. The pH of the mixture is adjusted to 7.3 with the cold sodium hydroxide (ca. 8 ml.). Ethyl alcohol (60 ml., cooled to −10°) is added, and stirring is continued for 10 to 15 minutes. The suspension is centrifuged at 0°, and the supernatant liquid discarded. The precipitate is washed in a refrigerated centrifuge at 0° successively with 25, 50, and 100% ethyl alcohol and then ether. After drying at room temperature *in vacuo* over calcium chloride and paraffin, the barium salt weighs approximately 25 g. (25%). It is stored over calcium chloride in the refrigerator. The purity varies between 65 and 80%. α-Ketoglutaric acid accounts for 2 to 5% of the impurities; the others remain unidentified.

A generally satisfactory solution for use is prepared as follows, working at all times at 0°: A weighed amount of the salt is suspended in water and brought into solution with a few drops of 2 N hydrochloric acid. The barium is then precipitated with the calculated amount of sulfuric acid and the barium sulfate removed by centrifugation and discarded without washing. The supernatant liquid is then adjusted to the desired pH and volume. It contains approximately 70% of the calculated amount of oxalosuccinic acid.

Stability. Though oxalosuccinic acid is quite stable in concentrated hydrochloric acid, it readily decarboxylates in dilute acid. In this respect it is even less stable than oxalacetic acid. Polyvalent cations accelerate the decarboxylation. The dry barium salt at 0° is stable for several months, and solutions of the acid may be kept for several hours at 0° at neutral pH.

III. The Preparation of *cis*-Aconitic Acid and *cis*-Aconitic Anhydride

Principle. *trans*-Aconitic acid is partially converted into a mixture of *cis*-aconitic anhydride and itaconic anhydride by the action of acetic anhydride.[5] Pure *cis*-aconitic anhydride is isolated by extraction with chloroform and several crystallizations from benzene. The acid is prepared from the anhydride by hydrolysis with water at room temperature.[5]

[5] R. Malachowski and M. Maslowski, *Ber.* **61B**, 2521 (1928).

Materials

    trans-Aconitic acid, 50 g.

    Acetic anhydride, 50 g., 95%.

Procedure. The aconitic acid and the acetic anhydride are mixed and stirred mechanically for 3 hours at 40 to 45°. The black reaction mixture is then cooled to room temperature and, after the addition of 100 ml. of chloroform, filtered. The filtrate is shaken with 5 g. of decolorizing carbon to remove some of the tarry impurities, filtered, and evaporated *in vacuo.* To the black residue, 40 ml. of toluene is added, and the evaporation is repeated. Extraction of the semisolid residue with 300 ml. of boiling benzene gives a pale-yellow solution which on cooling to 10° deposits large, slightly yellow crystals of *cis*-aconitic anhydride containing one-third of a mole of benzene of crystallization. This material is very hygroscopic and must be protected from moisture. Further purification is achieved by dissolving the crude product in 250 ml. of boiling benzene, decolorizing with 5 g. of carbon, filtering, and chilling the solution to 10°. The product is rapidly filtered, washed with petroleum ether (30 to 60°), and dried at room temperature in high vacuum until the odor of benzene is no longer detectable. The yield of pure *cis*-aconitic anhydride, melting point 75 to 76°, is approximately 20 g. When free of benzene, *cis*-aconitic anhydride is not hygroscopic.

    cis-ACONITIC ACID. A solution of 10 g. of *cis*-aconitic anhydride in 20 ml. of water is allowed to stand at room temperature for 3 hours. The solution is then frozen and lyophylized to yield quantitatively pure *cis*-aconitic acid, melting point 125 to 127°.

[69] Assay of Tricarboxylic Acids

By JOSEPH R. STERN

    The interconversion of the three tricarboxylic acids—citric, *cis*-aconitic and *d*-isocitric—is catalyzed by the enzyme aconitase which occurs in bacterial, plant, and animal cells. At equilibrium[1] (pH 7.4, 25°) the amounts of each acid present are: citric 90.9%; *cis*-aconitic 2.90%; and *d*-isocitric 6.20%. Thus in practice it is usually sufficient to determine citric acid as an assay for tricarboxylic acids, assuming that the three acids are in equilibrium by virtue of aconitase activity. However, in many soluble extracts of cells and tissues aconitase activity is very low—

[1] H. A. Krebs, *Biochem. J.* **54**, 78 (1953).

particularly after dialysis—and one cannot assume equilibrium. Hence it may be necessary to determine the individual acids. At present, satisfactory methods are available for the microestimation of citric and d-isocitric acids in biological mixtures. The determination of cis-aconitic acid presents certain problems.

With the increasing application of enzymatic methods to problems of chemical analysis, one can foresee the adoption of the recently described adaptive enzyme citridesmolase,[2] which cleaves citric acid to oxalacetic and acetic acids, to the assay of tricarboxylic acids. The oxalacetic acid formed from citric acid could be followed optically either directly or by reduction through coupling with the DPN-linked malic dehydrogenase; then addition of aconitase would effect the conversion of both cis-aconitic and d-isocitric acids to oxalacetic or malic acids.

Estimation of Citric Acid

General Considerations. The colorimetric determination of citric acid is based on the oxidation of citric acid to pentabromoacetone and conversion of the latter to a colored complex. Since the application of this procedure to biological materials by Pucher et al.,[3] many variations have been proposed in the conditions of the oxidation and of the colorimetric reaction with a view to increasing the accuracy and/or the sensitivity of the method. The procedure to be described is that developed by Natelson et al.,[4] as modified slightly by Elliott.[5] In the experience of the writer, it is highly satisfactory from the standpoint of accuracy, sensitivity, ease of manipulation, freedom from interference, and stability of the color complex. The reader is also referred to the procedure of Weil-Malherbe and Bone,[6] modified by Taylor.[7] The method of Saffran and Denstedt[8] for determination of citric acid, based on the Furth-Herrmann reaction, lacks specificity and accuracy. Where these are not primary considerations, it may be useful because of its simplicity.

Reagents

Citric acid stock solution. Dissolve 50 mg. of anhydrous citric acid (analytical reagent) in 50 ml. of 1 N H_2SO_4. Store in the refrigerator. Four milliliters of stock solution diluted to 100 ml. with

[2] D. C. Gillespie and I. C. Gunsalus, *Bacteriol. Proc.* **1953**, 80.
[3] G. W. Pucher, C. C. Sherman, and H. B. Vickery, *J. Biol. Chem.* **113**, 235 (1936).
[4] S. Natelson, J. B. Pincus, and J. K. Lugovoy, *J. Biol. Chem.* **175**, 745 (1948).
[5] K. A. C. Elliott, private communication.
[6] H. Weil-Malherbe and A. D. Bone, *Biochem. J.* **45**, 377 (1949).
[7] T. G. Taylor, *Biochem. J.* **54**, 48 (1953).
[8] M. Saffran and O. F. Denstedt, *J. Biol. Chem.* **175**, 849 (1948).

1 N H_2SO_4 constitutes the standard solution of which 1.0 ml. contains 40 γ of citric acid.

18 N H_2SO_4 (analytical reagent).

1 M potassium bromide solution (reagent grade).

5% potassium permanganate solution (reagent grade).

6% hydrogen peroxide, made by diluting 10 ml. of 30% H_2O_2 (analytical reagent) to 50 ml. Store in the refrigerator.

Petroleum ether,[9] boiling point 90 to 100°.

Thiourea solution. Two grams of sodium borate (reagent grade) is dissolved in 100 ml. of 4% thiourea. If opalescent, the solution is filtered. The final pH should be 9.2.

Deproteinization. Trichloroacetic acid is mixed with the solution to be examined, final concentration 5 to 10%. The precipitate is removed by centrifugation.

Removal of Interfering Substances. A sample of the supernatant containing 10 to 100 γ of citric acid is placed in a glass-stoppered Pyrex centrifuge or ignition tube having a scratch mark indicating 1.0 ml. volume. Water is added to a final volume of 3.0 ml., followed by 0.1 ml. of 18 N H_2SO_4 and an alundum grain or Hengar granule. The tube is placed in an oil bath at 120 to 140° (mineral oil is satisfactory) and the solution evaporated to the 1.0-ml. mark. It is then cooled to room temperature. This step removes certain interfering substances such as β-keto acids, acetaldehyde, and acetone.

Conversion to Pentabromoacetone. To the cooled solution is added 5 drops of MKBr, then 10 drops of 5% $KMnO_4$, with shaking to ensure mixing. The solution is allowed to stand for 10 minutes during which excess of permanganate, indicated by the purple color, is maintained. The tube is placed in ice water and the excess permanganate decolorized by dropwise addition of 6% H_2O_2. Care should be taken not to leave an excess of peroxide.

Extraction of Pentabromoacetone. Two and one-half milliliters of heptane is added to the tube, and the tube is stoppered with a minimum of silicone grease. The pentabromoacetone is extracted by shaking mechanically for 10 minutes. Clear water and heptane layers usually separate completely on standing; otherwise they should be centrifuged lightly.

Extraction with Thiourea. Two milliliters of the heptane layer is transferred to a clean glass-stoppered tube, with a delivery pipet, and 4.0 ml. of thiourea solution is added. At this stage a blank tube containing

[9] A chemically pure heptane (boiling point 97 to 99°) available from Phillips Petroleum Co. is used.

2.0 ml. of heptane and 4.0 ml. of thiourea solution is added. The tubes are stoppered and shaken mechanically for 10 minutes. A sample of the aqueous layer is transferred to a 12 × 75-mm. Coleman cuvette, and the color is read against the aqueous layer from the blank tube in a Coleman spectrophotometer at 430 mμ. To correct for any difference in cloudiness between the unknown and the blank—usually negligible—a reading is also taken at 650 mμ and subtracted from that at 430 mμ.

One milliliter of standard citric acid solution (equivalent to 40 γ) is carried through the procedure. This gives an optical density of 0.120 ± 0.05 under the above conditions. A linear relationship is obtained between amount and color intensity over the range 10 to 100 γ of citric acid. In the range 100 to 200 γ of citric acid, the color intensity falls 5 to 15% below the expected value.

The yellow color of the thiourea-pentabromoacetone complex is quite stable at room temperature. Its intensity is affected significantly by changes in the pH of the buffer solution. If too much citric acid is used, a pink precipitate of the complex will appear at the interface between the heptane and the aqueous layers.

Special Points. The thiourea-borax solution reacts with rubber; hence glass-stoppered tubes are essential.

None of the following substances interferes with the method: aconitic acid, isocitric acid, pyruvic acid, lactic acid, fumaric acid, succinic acid, malic acid, maleic acid, tartaric acid, α-ketoglutaric acid, β-keto acids, β-hydroxy acids, saturated and unsaturated fatty acids, acetaldehyde, ethanol, glucose.

Elliott[5] has pointed out that the maximum extinction of the pentabromoacetone-thiourea complex is at $\lambda = 430$ mμ, not 445 mμ as Natelson et al.[4] reported.

Estimation of *d*-Isocitric Acid

Principle. The sole micromethod for the determination of *d*-isocitric acid in biological materials is the enzymatic one introduced by Ochoa.[10] This method, which is rapid, sensitive, and specific, displaces the older polarimetric macromethod introduced by Martius,[11] who first described *d*-isocitric acid in tissues. It utilizes the TPN-linked *d*-isocitric dehydrogenase system of pig heart which catalyzes the reversible reaction 1, and is a prime example of the modern marriage of enzymatic and optical

$$d\text{-Isocitrate}^{---} + \text{TPN}^+ \underset{\longleftarrow}{\overset{\text{Mn}^{++}}{\longrightarrow}} \alpha\text{-Ketoglutarate}^{--} + CO_2 + \text{TPNH}$$

(1)

[10] S. Ochoa, *J. Biol. Chem.* **174**, 133 (1948).
[11] C. Martius, *Z. physiol. Chem.* **257**, 29 (1938).

techniques. Because of the very high affinity of substrate for the enzyme and the loss of much of the CO_2 formed, the forward reaction proceeds essentially to completion with micrograms of d-isocitric acid when an excess of TPN$^+$ and enzyme are present. Measurement of the TPNH formed is carried out in the Beckman spectrophotometer at wavelength 340 mμ.

Reagents

0.25 M tris-HCl buffer, pH 7.4.
0.018 M MnCl$_2$.
0.00135 M TPN. Dissolve 1 mg. of TPN (pure) in 1.0 ml. of water.
Pig heart isocitric dehydrogenase.

Preparation of Enzyme. The preparation of isocitric dehydrogenase from pig heart is described in Vol. I [116]. The dialyzed initial extract of the heart acetone powder is a quite satisfactory source of the enzyme for use as reagent. It can be stored for months in the deep-freeze. A sample of this fraction containing 50 to 100 γ of protein is employed in the test.

Preparation of Sample. The solution to be examined is deproteinized by addition of 1.0 N acetic acid to bring the pH to 4 and incubated in a boiling water bath for 2 minutes. The denatured protein is removed by centrifugation, and the supernatant is neutralized. Trichloroacetic acid (5% final concentration) may also be used for deproteinization, since the small amounts present in the sample of neutralized supernatant required for assay do not inhibit isocitric dehydrogenase.

Procedure. The following reactants are placed in a Corex or silica cuvette ($d = 1.0$ cm.): 0.3 ml. of buffer, 0.1 ml. of MnCl$_2$ (1.8 μM.), 0.1 ml. of TPN$^+$ (0.135 μM.), 50 to 100 γ of heart enzyme, a volume of neutralized, deproteinized solution containing 5 to 20 γ of d-isocitrate, and water to a final volume of 3.0 ml. A cuvette containing all the reactants except TPN and sample is used as a blank. The reaction is carried out at room temperature and is started by addition of enzyme. The increase in optical density at 340 mμ (ΔE_{340}) is recorded at 0.5-minute intervals until a constant reading is obtained. After the assay, it is wise to add routinely 5 to 10 γ of known d-isocitrate to check its recovery and hence the absence of interfering factors in the material being tested.

Calculations. The molecular extinction coefficient (ϵ_{340}) of TPNH is 6220.[12] Under the assay conditions, therefore, an increase in optical density of 0.01 corresponds to the reduction of 0.0048 μM. of TPN or the oxidation of 0.92 γ of d-isocitric acid. Thus the micrograms of d-isocitric acid present in the sample are obtained by multiplying the ΔE_{340} by 92.

---

[12] B. L. Horecker and A. Kornberg, *J. Biol. Chem.* **175**, 385 (1948).

Accuracy. In the writer's experience 85 to 90% of *d*-isocitric acid (added as *dl*-isocitric acid) is recovered by this method. The discrepancy is probably accounted for more by impurities in the *dl*-isocitric acid used (melting point 125°) than by failure of the reaction to proceed to completion.

Possible Interfering Reactions and Materials. In the amounts used the extract is essentially free of aconitase, so citric acid and *cis*-aconitic acid do not interfere. Except for weak glutamic acid dehydrogenase activity, it is free of other TPN⁺-linked dehydrogenases. Thus the combined presence of both α-ketoglutarate and NH_3 will cause some interference. Pyruvic and lactic acids, and oxalacetic and malic acids, do not interfere because of the extremely slow reaction of lactic and malic dehydrogenases with TPN⁺. Excessive amounts of oxalacetic acid can be removed by boiling the sample at acid pH. Extracts of fresh heart should not be used as a source of the isocitric dehydrogenase system, for they contain the TPN⁺-linked "malic" enzyme,[13] which apparently is destroyed during preparation of the acetone powder. If large amounts of phosphate or pyrophosphate are present in the sample, Mg^{++} may be used as metal cofactor to avoid precipitation of the Mn^{++}.

Estimation of *cis*-Aconitic Acid

No specific generally applicable method is available for the determination of *cis*-aconitic acid. One or more of the procedures mentioned below may satisfy the particular requirements of the experimenter.

Nonspecific methods such as catalytic hydrogenation[14] and oxidation with permanganate[15] have been successfully adapted to the determination of *cis*-aconitic acid in mixtures of tricarboxylic acids, e.g., in the study of aconitase action. The spectrophotometric method of Racker[16] for following enzymic transformations of ethylenic acids could also be applied quantitatively in this situation. The application of the Furth-Herrmann procedure under controlled temperature conditions[8] (0°) gives a reliable micromethod for determining *cis*-aconitic acid in pure solution or in the presence of other ethylenic acids likely to be encountered in biological fluids. However, in the experience of the writer,[13] this method fails in biological samples containing tricarboxylic acids owing to serious interference by equilibrium amounts of citric acid.

---

[13] J. R. Stern, unpublished observations.
[14] W. A. Johnson, *Biochem. J.* **33**, 1046 (1939).
[15] S. R. Dickman, *Anal. Chem.* **24**, 1064 (1952).
[16] E. Racker, *Biochem. et Biophys. Acta* **4**, 211 (1950).

Separation of Tricarboxylic Acids

Partition chromatography on silica gel has been used successfully to separate and to determine tricarboxylic acids present in fruit[17] and in animal tissues.[18] cis-Aconitic acid has also been separated from other tricarboxylic acids by anion exchange chromatography,[19] but with poor recoveries.

[17] F. A. Isherwood, *Biochem. J.* **40**, 688 (1946).
[18] C. E. Frohman, J. M. Orten, and A. H. Smith, *J. Biol. Chem.* **193**, 277 (1951).
[19] H. Busch, R. B. Hurlbert, and V. R. Potter, *J. Biol. Chem.* **196**, 717 (1952).

[70] Isolation and Assay of Succinate and Fumarate

By HARRIS BUSCH

Isolation and Assay of Succinate

Principle. Mono-, di-, and tricarboxylic acids are chromatographically eluted from anion exchange resins with increasing concentrations of volatile acids.[1,2]

Preparation of the Resin. Dowex 1 (chloride form) is obtained from the Dow Chemical Company. Particles of 200 to 400 mesh size are obtained by suspension of the resin in water and decantation of the fine particles after the desired coarse particles settle. Not infrequently, the resin is contaminated by a dark-colored deposit which is generally more dense than the desired resin. Accordingly, in preparation of the resin, the mixture is suspended in water and, after a 2- to 5-minute interval, the dark particles have settled and the suspension containing both coarse and fine resin particles is decanted into another flask. After a 30-minute interval, the supernatant suspension is decanted, and the desired resin remains at the bottom of the flask. The initial supernatant suspension contains a good deal of desirable resin, and for better yields the processes of settling and decantation may be repeated a number of times. For conversion of the resin from the chloride to the acetate or formate form, a batch of resin is poured into a large glass column fitted with a one-hole rubber stopper at the bottom; the glass column rests on a metal ring. The rubber stopper is covered with glass wool above and is connected by tubing to the drain. The resin is treated with a solution of 1 M sodium acetate or

[1] H. Busch, R. B. Hurlbert, and V. R. Potter, *J. Biol. Chem.* **196**, 717 (1952).
[2] H. Busch, *Cancer Research* **13**, 789 (1953).

sodium formate until completeness of removal of chloride ion is demonstrated by failure of development of a precipitate in the effluent on addition of a strongly acidified solution of silver nitrate. Acidification of the silver nitrate is particularly necessary in the preparation of the acetate form of the resin inasmuch as the alkaline solution results in a precipitate of silver acetate and silver hydroxide. When the conversion to the desired form is completed, the column is washed with a volume of distilled water equivalent to two or three times the volume of the resin bed. Large batches of the resin may be prepared in either the acetate or formate form and stored at room temperature for three months without alteration in the chromatograms. The hydroxide form is much less stable.

Apparatus. A continuously increasing concentration of eluting acid is obtained when concentrated acid enters dropwise into a mixing flask containing distilled water; the weak acid produced passes over into a column of liquid above the resin column. The apparatus consists of a reservoir containing 4 N formic acid or 8 N acetic acid, connected by capillary tubing to a mixing flask which is stirred by a magnetic stirring bar encased in acid-resistant plastic. The mixing flask is connected to a single glass column or a manifold to which a number of columns can be joined. All connecting joints are standard taper 14/35. When constant air pressure is used to force acid from the reservoir into the mixing flask, a Mason jar or a filter flask may be used as the reservoir; when gravity is used for pressure, a separatory funnel suffices as the reservoir. The constant-pressure source is connected to the bench line and consists of a reducing valve, a manostat, and a gauge supplied by the Moore Products Company. It is useful to have an open line of rubber tubing connecting the reservoir and mixing flask to permit equalization of pressure as pressure is first being applied to the system; equalization of pressure prevents an initial rapid flow of strong acid into the mixing flask. In the absence of such a connection, 15 to 30 ml. of acid enter the mixing flask and result in a rapid increase in acidity passing through the resin column. When the flow rate from the resin column is satisfactory, the line connecting the reservoir and mixing flask is closed.

Preparation of the Columns. Glass columns which are 1 cm. in diameter and approximately 30 cm. in length are prepared to receive the resin by packing the tapered lower end with glass wool. A suspension of resin sufficient to produce a resin column 11.5 to 12.5 cm. in length is then pipetted into the glass column. After the water passes through the resin and the surface is sharply demarcated, the upper surface of the column is covered with a layer of glass wool. Particles adherent to the sides of the glass column can be washed out with distilled water. The sample for analysis is added to the column at pH 7 to 8 to ensure optimal adherence of the anions to the resin. If quantitative adherence of neutral amino acids is also desired, the pH should be 10 or above. After the sample has entered the column, 10 to 15 ml. of wash solution is added. Both the sample and the wash enter the resin column under the influence of gravity. An additional 15 ml. of distilled water is added to the column; this volume above the resin bed functions as a second pool for mixing with entering acid and retards the increase of acidity entering the resin column. The top of the glass column is a 14/35 standard taper outer end, and the lower end of the capillary tubing outflow from the mixing bowl or manifold connected to the mixing bowl is the corresponding inner end.[2] When samples are collected in automatic fraction collectors, the columns can be clamped to stationary bars in addition to being secured to the standard taper inner ends of the mixing flasks by springs. When the columns are connected to a manifold and collection of the fractions is carried out manually, the springs are sufficient to hold the columns to the manifold.[2] All the joints should be well lubricated. The eluate is collected at the rate of 0.5 ml./min., and the usual air pressure required to produce this rate is 0.5 to 2 p.s.i. The volumes of the individual fractions are approximately 2 ml.; the most satisfactory test tubes for collection of the fractions are 18 × 120-mm. soft glass tubes available at A. H. Thomas Company. When a separatory funnel is used as the reservoir and the pressure is hydrostatic, the length of the connecting tubing is approximately 3 feet.

Desiccation. For routine purposes, desiccation is carried out in vacuum desiccators 250 mm. in diameter. A mixture of 4-mesh calcium chloride (anhydrous) and purified sodium hydroxide scales in a 1:1 ratio is adequate for removal of volatile acid and water. The pressure in the desiccators is lowered to 25 to 40 mm. of mercury on a water pump. Owing to boiling of the solutions in the test tubes when the pressure is lowered rapidly, the pressure is slowly reduced over a 30-minute period. The desiccators are placed under infrared lamps suspended 2 feet above the tops of the desiccators. For optimal recovery, the source of heat is removed as soon as the samples are dry, i.e. in about 10 to 20 hours. For acids which sublime readily, evaporation of the acidic eluant is carried out at higher pressures and in the absence of a heat source. Palmer[3] has reported increased yields by placing samples in a 60° water bath and blowing an air stream over the liquid surface.

Isolation and Determination of Succinate. Succinic acid emerges from the Dowex 1 (formate) column in a volume of 8 to 10 ml. when the concentration of the formic acid reaches approximately 1 N. When the concentration of formic acid in the reservoir is 4 N and the volume in the mixing flask is 200 ml., the succinic acid emerges between fractions 25

[3] J. K. Palmer, personal communication.

and 35. With 8 N acetic acid in the reservoir and 200 ml. of distilled water in the mixing flask, succinic acid emerges from the Dowex 1 (acetate) column between fractions 65 and 75 in a volume of 15 to 20 ml. The quantity of succinic acid in the fractions is determined by titration of the acidic residues with 0.01 N NaOH after desiccation. The melting point of the recovered succinic acid varies from 181 to 189° compared with 189° for authentic samples. The method of Frohman and Orten for determination of polycarboxylic acids by a fluorometric procedure[4] can be applied to small quantities of succinic acid. The recovery of succinic acid ranges from 90 to 100% with an average of about 96% for fifty determinations.

Sublimation. Further purification of the succinic acid is achieved by sublimation. After the sample has been acidified and dried in the sublimation flask, the flask is connected to the condenser and the apparatus is placed in a sand bath at a temperature of 100 to 115°. The system is evacuated to a pressure of 2 to 10 mm. of mercury, and these conditions are maintained for 60 minutes. The melting point of the succinic acid which accumulates on the cold plate of the condenser is 188 to 189°, and recoveries of 85 to 90% are obtained.

Isolation and Assay of Fumarate

The isolation and assay of fumaric acid are based on the general principles for anion exchange chromatography of the carboxylic acids described in the section on isolation and assay of succinate. Fumaric acid emerges from the Dowex 1 (formate) column at a concentration of 3 to 3.5 N formic acid. When the concentration of formic acid in the reservoir is 6 N, fumaric acid emerges in fractions 85 to 100 in a volume of approximately 20 ml. The quantity of acid is determined by titration of the acid residues with 0.01 N NaOH after desiccation. A number of very useful procedures can also be utilized for determination of fumaric acid; these include titration of fumaric acid by $KMnO_4$,[5] fluorometric analysis,[4] and spectrophotometric assay.[6] Recovery of the fumaric acid ranges from 85 to 100%, and the melting points of the fumaric acid residues range from 280 to 290°, compared to 290° for authentic samples.

[4] C. E. Frohman and J. M. Orten, *J. Biol. Chem.* **205**, 717 (1953).
[5] F. B. Straub, *Z. physiol. Chem.* **236**, 42 (1936).
[6] E. Racker, *Biochim. et Biophys. Acta* **4**, 211 (1950).

[71] Isolation and Assay of L-Malate

By Seymour Korkes

Isolation

Principle. L-Malate is extracted into ether after deproteinization, purified if necessary by silica gel chromatography, and isolated and characterized as the cinchonine salt.[1]

Procedure. The reaction mixture (preferably containing 0.5 millimole of L-malate, if the entire procedure is to be followed) is deproteinized by boiling or addition of a nonether-extractable precipitant such as sulfuric or tungstic acid, and sufficient sulfuric acid added to bring the mixture to pH 1.0. It is clarified by filtration or centrifugation and extracted with ether in a continuous liquid-liquid extractor for 70 to 100 hours. The ether layer is evaporated,[2] and the residue taken up in solvents appropriate for silica gel chromatography.[3] The malate peak may be identified by separating the aqueous phase after titration of an aliquot, washing the organic phase with small portions of water, and evaporating the combined washings and initial aqueous layer to dryness, redissolving in a measured volume of water, and assaying the solution by the method to be described below. Alternatively, the malate may be tentatively identified by ascending paper chromatography of a suitable aliquot on Whatman No. 1 paper, using n-butanol saturated with 2 N formic acid as the developing solvent, and locating the components by indicator spray after the papers have been thoroughly air-dried to remove the formic acid.[4] In this procedure, L-malate has an R_f of 0.46 to 0.52. Effluent fractions from the chromatograph column containing the malic acid are pooled and the solvent removed by vacuum distillation. From the titration data, the total amount of malate is calculated, and the oil remaining after distillation is dissolved in sufficient acetone to give a concentration of approximately 1%. The solution is brought to its boiling point, and 1 mole of cinchonine is added for each mole of malic acid present (200 mg. of cinchonine per 100 mg. of malic acid). The mixture is refluxed for 20 to 30 minutes and allowed to cool. The precipitate can be filtered off immediately, or the mixture can be placed in an icebox overnight. The acid cinchonine salt of L-malate (the solubility in cold acetone is 5 mg./ml.) is

---

[1] H. D. Dakin, *J. Biol. Chem.* **59**, 7 (1924).

[2] If malate is the major organic acid present, preparation of the cinchonine salt may be attempted directly on the ether extract.

[3] C. S. Marvel and R. D. Rands, Jr., *J. Am. Chem. Soc.* **72**, 2642 (1950).

[4] J. W. H. Lugg and B. T. Overell, *Australian J. Sq. Research* **A1**, 98 (1948).

removed by centrifugation or filtration and recrystallized from boiling water. The solubility of this salt in water at 10° is about 2%, but it is readily soluble at 100°.

The crystals obtained in this manner melt sharply at 192°.[5] The specific rotation ($[\alpha]_D^{19}$) is $+146°$ in water at concentrations from 0.5 to 2%. Further evidence for the identity of the compound may be obtained by measurement of the enhanced rotation in the presence of uranyl acetate as follows: A solution of the cinchonine salt of known concentration is quantitatively decomposed with ammonia, and the precipitated cinchonine is removed by filtration. An aliquot of the filtrate is acidified with a drop of acetic acid, and 5 mg. of uranyl acetate is added per milligram of malic acid present. The solution is allowed to stand for 1 hour at room temperature and is then diluted to a suitable concentration (0.1 to 0.2%) and its rotation observed.

$$[\alpha]_D^{19} = -482°$$

Assay of L-Malate

Principle. L-Malate is quantitatively decarboxylated by preparations obtained from *L. arabinosus* to yield lactate and CO_2; the latter is estimated manometrically.[6]

Procedure. *L. arabinosus* 17-5 (American Type Culture Collection) is carried on slants containing 1% yeast extract, 1% glucose, and 1.5% agar. For large-scale growth, an inoculum in 1% glucose—1% yeast extract is grown at 30° for 18 hours. This is added in the proportion of 1 ml. of inoculum per 100 ml. of growth medium. The medium contains 2% glucose, 2% DL-malic acid, 1% yeast extract, 1% nutrient broth, 1% sodium acetate·$3H_2O$, 0.1% K_2HPO_4, and 5 ml. of the salts B mixtures of Wright and Skeggs[7] per liter. This latter mixture contains per ml.: 2 mg. of NaCl, 40 mg. of $MgSO_4·7H_2O$, 2 mg. of $MnSO_4·4H_2O$, and 2 mg. of $FeSO_4·7H_2O$. The organisms are grown for 18 to 24 hours at 30° and may be agitated by a stream of air or nitrogen. They are harvested by centrifugation and washed with cold distilled water.

If the sample to be assayed contains both malate and fumarate, a slow liberation of carbon dioxide from fumarate is observed. Two alternative procedures are available to circumvent the problem of the small amounts of fumarase present in the organisms.

1. The suspension of organisms is lyophilized and stored at 0°. For assay of malate, 20 mg. of lyophilized organisms in aqueous suspension is

[5] S. Ratner and A. Pappas, *J. Biol. Chem.* **179**, 1183 (1949); Dakin[1] gives 197 to 198° for this compound.

[6] M. L. Blanchard, S. Korkes, A. del Campillo, and S. Ochoa, *J. Biol. Chem.* **187**, 875 (1950).

[7] L. D. Wright and H. R. Skeggs, *Proc. Soc. Exptl. Biol. Med.* **56**, 95 (1944).

placed in the side bulb of a Warburg vessel. The main space should contain the sample of L-malate (5 to 15 micromoles, depending on the flask constant; adjusted to pH 5); 0.5 ml. of 2 M potassium acetate buffer, pH 5; and 3 micromoles of $MnCl_2$. The total volume is brought to 2 ml. The reaction is run at 38° in air. The malate should be quantitatively decarboxylated in 20 to 30 minutes. If pyruvate is known to be present in the sample, its contribution to the CO_2 evolution may be completely eliminated by addition of an equimolecular quantity of semicarbazide, adjusted to pH 5. The addition of 50 micromoles of semicarbazide is without effect on malate decarboxylation. If fumarate is present, the sum of malate and fumarate may be determined by the addition of excess fumarase[8] and running the reaction at pH 6. Alternatively, the fumarate may be chemically reduced to succinate by the following procedure.[9] To 3 ml. of sample in a 10-ml. centrifuge tube are added 0.1 g. of granulated Zn (20 to 30 mesh), 0.1 ml. of 0.1 M $CuSO_4$, and 0.1 ml. of 50% H_3PO_4. After 45 minutes at room temperature, an additional 0.1 ml. of H_3PO_4 is added. The reaction is complete in 90 minutes. The sample is brought to pH 5 with 10 N NaOH with bromocresol green as internal indicator, made up to 4 ml. with water, and centrifuged. A suitable aliquot is then analyzed manometrically as described above.

2. Activity toward fumarate and pyruvate can be completely removed by partial purification of the enzyme.[10,11] The purification need be carried only past the calcium phosphate gel treatment to yield a preparation suitable for assay.

[8] See Vol. I [122].

[9] H. A. Krebs, D. H. Smyth, and E. A. Evans, Jr., *Biochem. J.* **34**, 1041 (1940). The procedure is reported by P. M. Nossal [*Biochem. J.* **50**, 349 (1952)] to be satisfactory with biological materials assayed by this procedure.

[10] S. Korkes, A. del Campillo, and S. Ochoa, *J. Biol. Chem.* **187**, 891 (1950).

[11] S. Kaufman, S. Korkes, and A. del Campillo, *J. Biol. Chem.* **192**, 301 (1951).

[72] Itaconic Acid and Related Compounds

By HELGE LARSEN

A. Itaconic Acid

$$CH_2{=}C{-}COOH \quad \text{(Molecular Weight 130.1)}$$
$$| $$
$$CH_2{-}COOH$$

Itaconic acid is known as a dissimilation product of certain Aspergilli, in particular *A. terreus*, when grown on glucose or sucrose media.

Physical and physical-chemical data, useful in identification: melting point 161°; phenacyl ester,[1] melting point 78 to 79.5°; *p*-bromophenacyl ester,[2] melting point 117.4; benzylammonium salt,[3] melting point 132 to 133°; dibenzyl amide,[4] melting point 106°. Electrolytic dissociation constants at 25°: $K_1 = 1.5 \times 10^{-4}$, $K_2 = 2.2 \times 10^{-6}$. Infrared spectrum of dibenzyl amide.[4] Solubility data:[5] 5.9 g. at 10° and 8.3 g. at 20° in water; 20 g. at 15° in 88% ethanol. Slightly soluble in ether, very slightly soluble in chloroform, carbon disulfide, benzene, and petroleum ether. Itaconic acid is not steam-volatile. *R* values are given in the table.

Assay Methods

Qualitative Determination. In addition to the use of the characterization data given in the above paragraph, it is recommended to carry out a quantitative catalytic hydrogenation, e.g., according to the procedure given by Calam *et al.*,[6] by shaking in aqueous solution with a palladium-charcoal catalyst. Methylsuccinic acid, which is obtained as hydrogenation product, can be isolated by ether extraction and characterized by its melting point, 111 to 112°.

The *R* values for itaconic acid in various solvent systems are given in the table on p. 443. The original fermentation solution, which is to be tested for organic acids by paper chromatography, is shaken with an H^+-charged cation exchange resin, e.g. Dowex 50, and centrifuged to free the solution for metal ions before spotting it on the paper. Besides applying neutral 0.04% bromophenol blue in 95% ethanol, which reveals the acidic nature of the compound, the following sprays are recommended by Buch *et al.*[7] (1) Equal parts of 0.1 *N* $AgNO_3$ and 0.1 *N* NH_4OH mixed just prior to use. Keep the sprayed paper at room temperature away from sunlight and observe after 4 hours. Itaconic acid gives a gray spot on a light gray background. (2) Ten per cent solution (by volume) of acetic acid anhydride in pyridine. The sprayed paper is heated for 5 minutes at 100°. Itaconic acid reveals its presence by forming an orange to brown spot on a white background. The spot gives a strong fluorescence in ultraviolet light. (3) A saturated solution of ammonium vanadate in water. The paper is observed while still wet after the spraying. Itaconic acid gives a yellow spot on a white background.

[1] J. B. Rather and E. E. Reid, *J. Am. Chem. Soc.* **41**, 75 (1919).
[2] T. L. Kelly and P. A. Kleff, *J. Am. Chem. Soc.* **54**, 4444 (1932).
[3] R. Boudet, *Bull. soc. chim. France* **1948**, 390.
[4] R. W. Stafford, R. J. Francel, and J. F. Shay, *Anal. Chem.* **21**, 1454 (1949).
[5] Grams of itaconic acid per 100 ml. of solvent.
[6] C. T. Calam, A. E. Oxford, and H. Raistrick, *Biochem. J.* **33**, 1488 (1939).
[7] M. L. Buch, R. Montgomery, and W. L. Porter, *Anal. Chem.* **24**, 489 (1952).

Quantitative Determination. *Principle.* The most satisfactory method was developed by Friedkin[8] on the basis of bromine absorption according to Koppeschaar.[9] A selective absorption of bromine by the itaconic acid is ensured by buffering the solution at pH 1.2.

Reagents

Bromine, 1.0 ml.
Potassium bromide, 3.0 g.
Potassium chloride, 1.87 g.
1.0 N HCl, 48.5 ml.
Add water to 500 ml.

Dissolve the bromine and potassium bromide in a small amount of water before adding the other constituents. The resultant pH is 1.2 ± 0.1. Storage of the solution in a brown bottle in the refrigerator or in the dark slows the deterioration of the bromine.

Procedure. Itaconic acid can usually be determined directly in a fermentation solution, after the mycelium has been removed by filtration or centrifugation. If the solution is buffered against acid, it must be slightly acidified before the bromine is added. If stored in a refrigerator the temperature of the bromine reagent must be raised to room temperature (20°) before use.

Sample Containing 15 to 150 Mg. of Itaconic Acid. The sample (1 to 2 ml.) is pipetted into a 125-ml. iodine flask, and to the sample is added 50 ml. of the bromine reagent. The iodine flask stopper is water-sealed to prevent loss of bromine vapor. After 10 minutes at 20° the flask is placed in an ice bath. After 5 minutes 5 ml. of potassium iodide solution (50 g. of KI in 100 ml. of water) is poured into the well surrounding the stopper. The stopper is then lifted carefully so that the potassium iodide solution is sucked into the flask by the previous cooling. After 10 minutes the released iodine is titrated with 0.1 N thiosulfate, with starch as indicator. Fifty milliliters of the bromine reagent is treated in the same way as the sample. The titer difference between blank and sample, a, is equivalent to milliliters of 0.1 N itaconic acid; $a \times 0.1 =$ milliequivalents of itaconic acid; $a \times 0.0065 =$ grams of itaconic acid in sample taken for analysis.

Sample Containing 0.5 to 5 Mg. of Itaconic Acid. The following semimicro modification of the above procedure has been found satisfactory for biochemical studies in the author's laboratory. The sample (1 ml.) is pipetted into a 50-ml. iodine flask, and 2 ml. of the bromine reagent is added. The reaction mixture is treated exactly as described above, with

[8] M. Friedkin, *Ind. Eng. Chem. Anal. Ed.* **17**, 637 (1945).
[9] W. F. Koppeschaar, *Z. anal. Chem.* **16**, 233 (1876).

the modification that 1 ml. of potassium iodide solution is added and the released iodine titrated with 0.01 N thiosulfate. Special care must be taken to add exactly the same amounts of the bromine reagent in the flask containing the sample as in that containing the blank.

The following substances do not interfere with the itaconic acid determination: glucose, sucrose, lactic, acetic, oxalic, tartaric, malic, fumaric, succinic, citric, aconitic, isocitric acids, D-gluconolactone. Pyruvic and α-ketoglutaric acids do interfere when present in amounts greater than 1.5 mg. in a sample taken for the semimicro analysis, and 15 mg. in a sample taken for the macroanalysis.

Preparation

Biosynthetic Method. *Principle.* Fermentation of glucose or sucrose by *Aspergillus terreus*, strain NRRL 1960, in aerated, acid media. The fermentation products are: mold mycelium, carbon dioxide, itaconic acid (55 to 59% by weight of sugar consumed), and a trace of malic acid. The latter is easily separated from itaconic acid because of its greater solubility.

The mold can be obtained from the collections of the Fermentation Division, Northern Regional Research Laboratory, U.S. Department of Agriculture, Peoria, Illinois, or the Centraalbureau voor Schimmelcultures, Baarn, Holland.

A number of different media have been proposed in the literature for the fermentation of sugar to itaconic acid by *A. terreus*. The one which has proved most successful in the author's hands is that proposed by Pfeifer *et al.*[10] and has the following composition: 6.60% glucose (monohydrate), 0.27% $(NH_4)_2SO_4$, 0.08% $MgSO_4·7H_2O$, 0.18% corn steep liquor; pH adjusted to 5.0. A somewhat different medium has been worked out in the author's laboratory and has been found to excel the others with respect to its ability to give high consistent yields of itaconic acid: 10% sucrose or glucose, 0.3% $(NH_4)_2SO_4$, 0.3% $CaSO_4$ (anhydrous), 0.05% $MgSO_4·7H_2O$, 0.01% KH_2PO_4, in tap water; no adjustment of pH.

Procedure. A medium of the above composition is distributed in 100-ml. portions over a number of 500-ml. flasks. After sterilization each flask is inoculated with a large number of spores[11] (several loopfuls) from a Czapek-Dox medium:[12] 3% sucrose, 1.5% agar, 0.3% $NaNO_3$, 0.1% K_2HPO_4, 0.05% $MgSO_4·7H_2O$, 0.05% KCl, 1 mg.% $FeSO_4·7H_2O$, in tap water. The flasks are incubated at 30° in a shaking machine which

[10] V. F. Pfeifer, C. Vojnovich, and E. N. Heger, *Ind. Eng. Chem.* **44**, 2975 (1952).

[11] It is recommended to use a spore culture which is less than 3 weeks old.

[12] C. Thom and K. B. Raper, "A Manual of the Aspergilli," p. 32, The Williams & Wilkins Co., Baltimore, 1945.

ensures a good aeration of the cultures (e.g., a reciprocating shaker with an amplitude of about 4 cm. and about 180 oscillations per minute). The shaking must not stop during the fermentation, since the itaconic acid-producing enzymatic mechanism is extremely sensitive to lack of oxygen. The fermentation[13] is followed by titrating 5-ml. samples with $N/10$ NaOH. At peak titer, usually after about 120 hours of incubation, the mycelium is removed by filtration (coarse glass filter), and the filtrate evaporated on a steam bath to one-tenth to one-twelfth its original volume. After cooling, the mother liquid is sucked off from the itaconic acid crystals on a coarse glass filter. The raw itaconic acid (about 50 g. per 100 g. of sugar) is dissolved in a minimum amount of hot water; the solution is treated with activated charcoal and filtered while still hot. White crystalline itaconic acid separates out on cooling. The purity of the product is about 99%. The yield is about 43 g. of itaconic acid per 100 g. of sugar. For final purification the itaconic acid is recrystallized from hot ethyl acetate. Purity >99.5%.

Synthetic Method. *Principle.* Decarboxylation of citric acid by pyrolysis.[14,15]

Procedure. A 300-ml. Pyrex Kjeldahl flask is fitted with an outlet tube 12 mm. in diameter bent for downward distillation. It is attached to a 100-cm. water-cooled condenser having an indented Pyrex inner tube. Two 500-ml. long-necked distilling flasks are used in series as receivers, the vapors being led to the center of each flask by means of an adapter and glass tubing. Both receivers are cooled in a mixture of ice and water.

120 g. of citric acid is placed in the Kjeldahl flask and gently heated with a free flame until melted. Then the flask is heated very rapidly with a large Meker burner, and the distillation is completed as rapidly as possible (10 to 12 minutes). The distillation is stopped as soon as the vapors in the reaction flask become yellow. This procedure is repeated nine times, using 120-g. portions of citric acid and a clean Kjeldahl flask for each distillation. All the distillates are collected in the same set of receivers.

The distillate is placed in an evaporating dish, 50 ml. of water is added, and the mixture allowed to stand on a steam bath for 3 hours. On cooling it sets to a semisolid mass; this is filtered and washed with 150 ml. of water. The residue consists of 138 g. of perfectly white crystals of itaconic acid melting at 165°. By concentrating the filtrate, an additional

---

[13] The pH of the culture gradually drops to a value of 2.1 before the rate of acid production reaches its maximum.

[14] R. L. Shriner, S. G. Ford, and L. J. Roll, *Org. Syntheses* **11**, 70 (1931).

[15] C. V. Wilson and C. F. H. Allen, *Org. Syntheses* **13**, 111 (1933).

42 g. of product melting at 157 to 165° is obtained. The total yield is 26 to 27% of the theoretical amount.

B. Itatartaric Acid

CH_2OH—$C(OH)$—$COOH$ (Molecular Weight 164.1)

      |

CH_2—$COOH$

This compound has been shown by Stodola et al.[16] to accumulate, besides itaconic acid, in surface cultures of *Aspergillus terreus*, strain NRRL 265S14 (mutant), when grown on a glucose medium. Itatartaric acid is easily converted to its γ-lactone, i.e., oxyparaconic acid.

Assay Method

The following procedure has been used by Stodola et al.[16] to identify itatartaric acid. The fermentation liquor is concentrated *in vacuo* (50°) to one-twelfth its original volume, and crystals of itaconic acid removed. The filtrate is extracted with ether, and the extract concentrated to remove more itaconic acid. The new filtrate is concentrated to a gum which is dissolved in methanol and treated with an excess of diazomethane. The methyl esters are fractionated *in vacuo*, and 700 mg. of the viscous oil obtained in the fraction with the boiling range of 129 to 134° (2 to 3 mm. Hg) is heated for 2 hours at 60° with 0.5 ml. of methanol and 2.0 ml. of benzylamine. The reaction product is dissolved in ether, and the solution washed with diluted HCl. The crystals obtained by concentration of the ether are recrystallized from methanol-ether in the form of long needles of itatartaric acid dibenzylamide; melting point 103 to 104°, $[\alpha]_D^{23} = -42°$ (in 95% ethanol).

For further identification the dibenzylamide (192 mg.) is dissolved in methanol (7.5 ml.) and added to a solution of H_5IO_6 (0.5 g.) in 1 N H_2SO_4 (0.5 ml.). Needles appear and are filtered off 1 hour later after the addition of water (2 vol.); melting point 172 to 173°. The oxidation product (10 mg.) is dissolved in 1 N NaOH (2 ml.) containing a few drops of ethanol. Two per cent Hg-Na (0.6 g.) is added gradually. Crystals appear and are filtered off. By crystallization from ethanol, pure *dl*-malic acid dibenzylamide is obtained, as shown by its melting point (145 to 146°), X-ray diffraction patterns, and mix melting point with an authentic sample.[17]

The other cleavage product obtained by treating the itatartaric acid

---

[16] F. H. Stodola, M. Friedkin, A. J. Moyer, and R. D. Coghill, *J. Biol. Chem.* **161**, 739 (1945).

[17] See footnote 16 for the preparation of *dl*-malic acid dibenzylamide from *dl*-malic acid.

dibenzylamide with H_5IO_6 can be shown to be formaldehyde by dimedon precipitation and mix melting point of the crystallized condensation product on admixture with an authentic sample.

By fractionation of the methyl esters mentioned above, a fraction consisting of a viscous oil with a boiling range of 151 to 154° (2 to 3 mm. Hg) was obtained.[16] Its identity with the γ-lactone methyl ester of itatartaric acid (oxyparaconic acid methyl ester) was shown by its methoxyl content of 20.4% (calculated 21%) and by conversion to the benzylamide, melting point 102 to 103°. The latter gave no depression in melting point on admixture with the benzyl amide from the lower boiling fraction.

SOLVENT MIXTURES AND R VALUES FOR IDENTIFICATION OF ITACONIC ACID
BY PAPER CHROMATOGRAPHY

(Descending or ascending chromatograms at room temperature. Whatman No. 1 or Schleicher & Schüll No. 2043B filter paper. Optimum amount of acid about 100 γ)

Solvent mixture	Reference	$\% R_f$	$\% R_{malic}$*	Notes
1-Pentanol:5 M aq. formic acid = 1:1 by vol. (upper phase)	a, b	70–72	230–250	This system gives a very good resolution of the usual metabolic acids. However, aconitic, glutaric, and levulinic acids overlap itaconic acid.
Ethyl acetate:4.4 M aq. acetic acid = 1:2 by vol. (upper phase)	c	76–79	141–157	Fumaric and succinic acids overlap itaconic acid; aconitic acid does not!
Ether:13.1 M aq. acetic acid = 15:4 by vol. (upper phase)	b	85–88	240–265	Fumaric, glutaric, and levulinic acids overlap itaconic acid.
2-Ethyl-1-butanol:5 M aq. formic acid = 2:3 by vol. (upper phase)	b	64–65	355–370	This system has the greatest resolving power of those listed, and also gives the least variations in the R_f values of the usual metabolic acids.[b] Aconitic, glutaric, and levulinic acids overlap itaconic acid.
95% ethanol:conc. ammonium hydroxide: water = 8:1:1 by vol.	b		125–128	Basic solvent mixture! Citraconic acid overlaps itaconic acid.

* R_{malic} expresses the ratio of the distance moved by itaconic acid to that moved by pure malic acid.

[a] M. L. Buch, R. Montgomery, and W. L. Porter, *Anal. Chem.* **24**, 489 (1952).

[b] A. R. Jones, E. J. Dowling, and W. J. Skraba, *Anal. Chem.* **25**, 394 (1953).

[c] Routinely used in the author's laboratory.

Preparation

Principle. Oxidation of itaconic acid with $KMnO_4$ in alkaline solution according to Fittig and Köhl.[18] Itatartaric acid has not been obtained and characterized as the free acid. It forms a stable Ca salt ($C_5H_6O_6Ca\cdot\frac{1}{2}$ H_2O) which is only slightly soluble in water.[18]

Procedure. Five grams of itaconic acid is dissolved in 400 ml. of water, and the solution is made alkaline with an excess of Na_2CO_3. The solution is cooled to $0°$, and a 2% solution of $KMnO_4$ is added dropwise with stirring until the reaction is finished. Care is taken to keep the temperature at $0°$ during the oxidation. The reaction mixture is filtered to remove MnO_2. The filtrate is neutralized, concentrated to about one-third its original volume, and acidified with HCl while still hot. A solution of $CaCl_2$ is added at room temperature, and then an excess of NH_4OH. A precipitate of Ca oxalate is removed by filtration. By heating the filtrate to boiling the Ca salt of itatartaric acid precipitates. The precipitate is removed by filtration and dried.

[18] R. Fittig and W. Köhl, *Ann.* **305**, 41 (1899).

Section IV

Proteins and Derivatives

[73] Spectrophotometric and Turbidimetric Methods for Measuring Proteins

By Ennis Layne

I. Turbidimetric Methods

Principle. The turbidity produced when protein is mixed with low concentrations of any of the common protein precipitants can be used as an index of protein concentration.

Reagents and Procedures. The three most common procedures are listed below. Any multiple or fraction of the suggested volumes may be employed as desired. The protein sample should contain 0.5 to 1.5 mg. of protein per milliliter.

1. One milliliter of protein sample is mixed with 4.0 ml. of 1.25% trichloroacetic acid.[1]

2. One milliliter of protein sample is mixed with 4.0 ml. of 0.75% potassium ferrocyanide, and 1 drop of glacial acetic acid is added.[2]

3. One milliliter of protein sample is mixed with 4.0 ml. of 2.5% sulfosalicylic acid.[3]

The resulting turbidity is maximum after about 10 minutes and may be measured spectrophotometrically in the wavelength region of 600 mμ.

Standardization. Standardization may be effected by comparison with the turbidity produced by a suspension of a dried protein precipitate, or reference may be had to the methyl acrylate-styrene polymer of Haslam and Squirrel.[4,5]

Discussion. Heepe *et al.*[2] report that the turbidity formed by sulfosalicyclic acid and potassium ferrocyanide-acetic acid is more stable than that of trichloroacetic acid. Only potassium ferrocyanide-acetic acid, however, gives a turbidity which is independent of dispersion.

Turbidimetric techniques are rapid and convenient, but they yield different values with different proteins. They do not permit differentiation between protein and acid-insoluble compounds such as nucleic acids.

[1] M. Kunitz, *J. Gen. Physiol.* **35**, 423 (1952).
[2] F. Heepe, H. Karte, and E. Lambrecht, *Hoppe-Seyler's Z. physiol. Chem.* **286**, 207 (1951).
[3] F. Heepe, H. Karte, and E. Lambrecht, *Z. Kinderheilk* **69**, 331 (1951).
[4] E. J. King, *Biochem. J.* **48**, 50 (1951).
[5] J. Haslam and D. C. M. Squirrel, *Biochem. J.* **48**, 48 (1951).

II. Protein Estimation with the Folin-Ciocalteu Reagent

Principle. The method described below is that of Lowry et al.[6] The final color is a result of: (1) biuret reaction of protein with copper ion in alkali, and (2) reduction of the phosphomolybdic-phosphotungstic reagent by the tyrosine and tryptophan present in the treated protein.

Reagents

Reagent A, 2% Na_2CO_3 in 0.1 N NaOH.

Reagent B, 0.5% $CuSO_4 \cdot 5H_2O$ in 1% sodium or potassium tartrate.

Reagent C, alkaline copper solution. Mix 50 ml. of reagent A with 1 ml. of reagent B. Discard after 1 day.

Reagent D, carbonate-copper solution. Mix 50 ml. of 2% Na_2CO_3 with 1 ml. of reagent B. Discard after 1 day.

Reagent E, diluted Folin reagent. Dilute the Folin-Ciocalteu reagent (below) to make it 1 N in acid.

Folin-Ciocalteu reagent[7] (may be obtained commercially from Fisher Scientific Co., Will Corporation, etc.). Reflux gently for 10 hours a mixture consisting of 100 g. of sodium tungstate ($Na_2WoO_4 \cdot 2H_2O$), 25 g. of sodium molybdate ($Na_2MoO_4 \cdot 2H_2O$), 700 ml. of water, 50 ml. of 85% phosphoric acid, and 100 ml. of concentrated hydrochloric acid in a 1.5-l. flask. Add 150 g. of lithium sulfate, 50 ml. of water, and a few drops of bromine water. Boil the mixture for 15 minutes without condenser to remove excess bromine. Cool, dilute to 1 l., and filter. The reagent should have no greenish tint. (Determine the acid concentration of the reagent by titration with 1 N NaOH to a phenolpthalein end point.)

Protein Standard. Working standards may be prepared from human serum diluted 100- to 1000-fold (approximately 700 to 70 γ /ml.). These in turn may be checked by Kjeldahl nitrogen determination (see Vol. III [145]) or by comparison against a standard solution of crystalline bovine albumin (Armour and Co., Chicago); 1 γ is the equivalent of 0.97 γ of serum protein. Dilute solutions of bovine albumin show a marked tendency to undergo surface denaturation and thus are poor standards.

Procedure for Proteins in Solution or Readily Soluble in Dilute Alkali. (Directions are given for a final volume of 1.1 to 1.3 ml., but any multiple or fraction of the volumes given may be employed as desired. With final

[6] O. H. Lowry, N. J. Rosebrough, A. L. Farr, and R. J. Randall, *J. Biol. Chem.* **193**, 265 (1951).

[7] O. Folin and V. Ciocalteu, *J. Biol. Chem.* **73**, 627 (1927).

volumes of less than 0.1 ml., reaction tubes of a small diameter (3 mm.) *must* be used.)

Add up to 0.2 ml. of a sample containing 5 to 100 γ of protein and 1 ml. of reagent C to a 3- to 10-ml. test tube. Mix well, and allow to stand for 10 minutes or longer at room temperature. Add 0.10 ml. of reagent E rapidly with immediate mixing. After 30 minutes or longer, read the sample in a colorimeter or spectrophotometer. For the range 5 to 25 γ of protein per milliliter of final volume, it is desirable to make readings at or near the absorption maximum at 750 mμ. For more concentrated solution the readings may be kept in a workable range by reading near 500 mμ. Calculate the protein concentration by comparison with a standard curve. If it is known that the particular proteins being estimated give a color intensity different from the standard at equivalent concentrations, an appropriate correction may be made.

It is unnecessary to bring all the samples and standards to the same volume before the addition of the alkaline copper reagent, provided that corrections are made for small differences in final volume. The critical volumes are those of the alkaline copper and Folin reagents.

If the protein is already present in a dilute solution (less than 25 γ/ml.), 0.5 ml. may be mixed with 0.5 ml. of an exactly double-strength reagent C and otherwise treated as above.

Procedure for Insoluble Proteins, Etc. Many protein precipitates, e.g., tungstate precipitates, will dissolve readily in the alkaline copper reagent. However, after proteins have been precipitated with trichloroacetic or perchloric acid, for example, they will dissolve rather poorly in the 0.1 N alkali of this reagent.

With small amounts of precipitated protein (5 to 100 γ), thinly spread, 0.1 ml. of 1 N NaOH may be added. When the protein has dissolved (after ½ hour or more), 1 ml. of reagent D (no NaOH) is added, followed after 10 minutes by 0.1 ml. of Folin reagent E as usual.

With larger samples, or very insoluble precipitates, it may be necessary to heat for 10 minutes or more at 100° in 1 N alkali. Although this may lower the readings, they will be reproducible and can be measured with similarly treated standards.

Discussion. Few substances encountered in biological material cause serious interference with the method. Neither acid extracts nor lipid extracts of rabbit brain, kidney, liver, skeletal muscle, or heart give appreciable color.

However, tryptophan, tyrosine, most phenols (except nitrophenol), uric acid, guanine, and xanthine react with the Folin reagent to produce color. No more than a trace of color was obtained with adenosine, adenine, guanosine, hypoxanthine, cytosine, cytidine, uracil, thymine, or thymidine.

Neither color nor interference with protein color development was observed with the following substances in the given final concentrations: urea (0.5%), guanidine (0.5%), sodium tungstate (0.5%), sodium sulfate (1%), sodium nitrate (1%), perchloric acid (0.5%, neutralized), trichloroacetic acid (0.5%, neutralized), ethyl alcohol (5%), ether (5%), acetone (0.5%), zinc sulfate (0.1%), and barium hydroxide (0.5%).

Ammonium sulfate in a final concentration greater than 0.15% decreases the color development. Glycine (0.5%) decreases the color with protein by 50%. Hydrazine over 0.5 mg. % produces interference.

There are two major disadvantages of the Folin reaction. (1) The amount of color varies with different proteins. In this respect it is less constant than the biuret reaction (see p. 451) but more constant than the ultraviolet absorption at 280 mμ (see p. 454). (2) The color is not strictly proportional to concentration.

There are, however, certain advantages to the Folin reagent. It is more convenient and as sensitive as digestion and subsequent nesslerization. It is ten to twenty times as sensitive as measurement of the ultraviolet absorption at 280 mμ, and it is much more specific and less liable to inaccuracy due to turbidity. It is 100 times as sensitive as the biuret reaction.

III. Biuret Method

Principle. Substances containing two or more peptide bonds form a purple complex with copper salts in alkaline solution. The reagents and conditions described below are those of Gornall *et al.*[8] and represent modifications of methods reported by Robinson and Hogden[9] and Weichselbaum.[10]

Reagents

Biuret reagent: Dissolve 1.50 g. of cupric sulfate ($CuSO_4 \cdot 5H_2O$) and 6.0 g. of sodium potassium tartrate ($NaKC_4H_4O_6 \cdot 4H_2O$) in 500 ml. of water. Add, with constant swirling, 300 ml. of 10% sodium hydroxide (prepared from stock, carbonate-free, 65 to 75% sodium hydroxide solution). Dilute to 1 l. with water, and store in a paraffin-lined bottle. This reagent should keep indefinitely but must be discarded if, as a result of contamination or of faulty preparation, it shows signs of depositing any black

[8] A. G. Gornall, C. S. Bardawill, and M. M. David, *J. Biol. Chem.* **177,** 751 (1949).
[9] H. W. Robinson and C. G. Hogden, *J. Biol. Chem.* **135,** 707 (1940).
[10] T. E. Weichselbaum, *Am. J. Clin. Pathol. Suppl.* **10,** 40 (1946).

or reddish precipitate. (The addition of 0.1% potassium iodide may prevent excessive reduction and has no detectable effect on the rate, degree, or quality of biuret color.)

Procedure. To 1.0 ml. of a solution containing 1 to 10 mg. of protein per milliliter add 4.0 ml. of biuret reagent, mix by swirling, and allow to stand for 30 minutes at room temperature (20 to 25°). With a photoelectric colorimeter or spectrophotometer transmitting maximally in the region of 540 to 560 mμ, determine optical density or per cent transmission against a blank consisting of 4.0 ml. of biuret reagent plus 1.0 ml. of water or of an appropriate salt solution.

The concentration of protein in the sample is obtained by reference to a calibration curve established with a clear solution of serum protein which has been assayed by the Kjeldahl nitrogen method.

The presence of large amounts of lipoid material may yield a cloudy reaction mixture which can be cleared by shaking with 1.5 ml. of diethyl or petroleum ether and then centrifuging in a capped tube and reading the aqueous phase as above.

Discussion. There are practically no substances other than protein normally present in biological materials which give the biuret reaction to an extent sufficient to cause significant interference. (Bile pigments absorb light very weakly in the region of 540 to 560 mμ.) Although the color development with various proteins is not identical, deviations are encountered less frequently than with the Folin-phenol, the ultraviolet absorption, or the turbidimetric method. However, more material is required for assay by the biuret method.

The biuret method as described above cannot be used in the presence of ammonium salts. Their interference can be minimized by the analysis of the separated precipitate as described by Robinson and Hogden.[9]

IV. Protein Estimation by Ultraviolet Absorption

Principle. Most proteins exhibit a distinct ultraviolet light absorption maximum at 280 mμ, due primarily to the presence of tyrosine and tryptophan. Since the tyrosine and tryptophan content of various enzymes varies only within reasonably narrow limits, the absorption peak at 280 mμ has been used by Warburg and Christian[11] as a rapid and a fairly sensitive measure of protein concentration. Unfortunately, nucleic acid, which is apt to be present in enzyme preparations, has a strong ultraviolet absorption band at 280 mμ. Nucleic acid, however, absorbs much more strongly at 260 mμ than at 280 mμ, whereas with protein the reverse is true. Warburg and Christian have taken advantage of this

[11] O. Warburg and W. Christian, *Biochem. Z.* **310**, 384 (1941).

fact to eliminate, by calculation, the interference of nucleic acids in the estimation of protein.

Procedure. The optical density of an appropriately diluted protein solution is obtained at both 280 mμ and 260 mμ.

Calculations. The concentration of protein in the solution is calculated according to the treatment of Warburg and Christian, who have determined the extinction coefficients (β) of crystalline yeast enolase and of yeast nucleic acid. They define β as follows:

$$\beta = \frac{1}{c} \cdot \frac{1}{d} \ln \frac{i_0}{i}$$

where c is the concentration in milligrams per milliliter, d is the length of the light path in centimeters, and i_0 and i are the intensities of the incident and transmitted light beams respectively.

For pure enolase:

$$\frac{\beta_{280}}{\beta_{260}} = \frac{2.06}{1.18} = 1.75$$

For pure yeast nucleic acid (Merck):

$$\frac{\beta_{280}}{\beta_{260}} = \frac{24.8}{50.8} = 0.49$$

Warburg and Christian have arrived.by calculation at a series of β_{280} and β_{260} values for various mixtures of protein and nucleic acids. Solving the equation defining β for protein plus nucleic acid concentration (C_{p+n}) in a mixture, converting the logarithm to the base 10, and substituting optical density yields the relation

$$C_{p+n} = \frac{2.303}{\beta_{280}} \times \frac{1}{d} \times \text{optical density at 280 m}\mu$$

The concentration of protein (C_p) in the mixture is given by the equation

$$C_p = \frac{2.303}{\beta_{280}} \times \frac{\text{per cent protein}}{100} \times \frac{1}{d} \times \text{optical density at 280 m}\mu$$

The per cent protein and the β_{280} are constant for a given protein-nucleic acid mixture whose composition may be defined as a function of the ratio of the optical density at 280 mμ to that at 260 mμ ($R_{280/260}$). One may, according to the treatment of Warburg and Christian, calculate for each mixture a factor (F) defined as follows:

$$F = \frac{2.303}{\beta_{280}} \times \frac{\text{per cent protein}}{100}$$

Some values of F and $R_{280/260}$ are listed in the accompanying table.

PROTEIN ESTIMATION BY ULTRAVIOLET ABSORPTION

$R_{280/260}{}^a$	Nucleic acid, %	F^b
1.75	0.00	1.116
1.63	0.25	1.081
1.52	0.50	1.054
1.40	0.75	1.023
1.36	1.00	0.994
1.30	1.25	0.970
1.25	1.50	0.944
1.16	2.00	0.899
1.09	2.50	0.852
1.03	3.00	0.814
0.979	3.50	0.776
0.939	4.00	0.743
0.874	5.00	0.682
0.846	5.50	0.656
0.822	6.00	0.632
0.804	6.50	0.607
0.784	7.00	0.585
0.767	7.50	0.565
0.753	8.00	0.545
0.730	9.00	0.508
0.705	10.00	0.478
0.671	12.00	0.422
0.644	14.00	0.377
0.615	17.00	0.322
0.595	20.00	0.278

^{a} Ratio of optical density at 280 mμ to optical density at 260 mμ.

^{b} $F \times 1/d \times D_{280}$ = mg. protein per milliliter, where d is cuvette width in centimeters, and D_{280} is the optical density at 280 mμ.

To find the protein concentration of an unknown sample, it is not necessary that the nucleic acid concentration be known. One experimentally measures the optical density of the sample at 280 mμ and at 260 mμ, calculates their ratio ($R_{280/260}$), then determines the corresponding factor (F) from the table and substitutes in the following equation:

Protein concentration (mg./ml.) $= F \times 1/d \times$ optical density at 280 mμ

where d is. the cuvette width in centimeters. The method gives considerable error with mixtures containing more than 20% nucleic acids, or with very turbid solutions.

Alternatively, Kalckar[12] has used the following equation suggested by Lowry:

$$\text{Protein concentration (mg./ml.)} = 1.45D_{280} - 0.74D_{260}$$

where D_{280} and D_{260} are optical densities at 280 and 260 mμ, respectively.

A somewhat closer fit to the data of Warburg and Christian is afforded by the equation:

$$\text{Protein concentration (mg./ml.)} = 1.55D_{280} - 0.76D_{260}$$

Discussion. Considerable error may be present, since different proteins and nucleic acids do not always have the same absorption. Other substances such as purine and pyrimidine nucleotides may also exhibit absorption in the region from 260 to 280 mμ.

However, the estimation of protein may be carried out quickly and easily on a small quantity of material and under conditions (in the presence of ammonium sulfate or other salts) which render difficult the application of other methods.

[12] H. M. Kalckar, *J. Biol. Chem.* **167**, 461 (1947).

[74] Titrimetric Procedures for Amino Acids

(Formol, Acetone, and Alcohol Titrations)

By M. LEVY

Principles. The stoichiometric pH values[1] for the extreme anionic and cationic forms of amino acids are inaccessible as end points of titration in aqueous solutions because of the narrow zones between the pK values of the groups concerned and the solvent blank. The solvent blank may be thought of as an envelope of pH values determined by the behavior of strong acids and alkalies in the solvent. For water this envelope is defined by the assumption of complete dissociation for strong mineral acids and alkalies and by the behavior of the solvent as a proton acceptor and donor defined in terms of K_w. For any other solvent the same assumptions can be made about complete dissociation, but a new solvent constant, $K_{sol.}$ needs to be evaluated. Thus in a 90% alcohol solution p$K_{sol.}$ may be considered to be 17 or 18 instead of 14 as in water. This results partly from the

[1] A stoichiometric pH is defined as the pH conferred on a solution by addition to the solvent of a pure compound only. Solutions of glycine hydrochloride, glycine, and sodium glycineate define the three stoichiometric pH values of glycine.

decrease in concentration of water and probably also from other less well-defined changes in activity coefficients, dielectric constant, etc.

The procedures described in this section successfully separate the stoichiometric pH from the envelope of solvent titration by affecting differentially the apparent pK values of the amino acids and of the solvents compared to the values in aqueous solution. In the formol titration this results from the reaction of the uncharged amino groups with formaldehyde according to the equation[2]

$$R{-}CH\begin{array}{c} COO^- \\ \\ NH_3{}^+ \end{array} + 2CH_2O \leftrightarrow RCH\begin{array}{c} COO^- \\ \\ NH_2(CH_2O)_2 \end{array} + H^+$$

in effect decreasing the pK of the amino group. In spite of a narrowing of the solvent titration envelope because of the acidity of formaldehyde,[3] a sufficiently steep pH change occurs about the stoichiometric pH to give a useful end point with phenolphthalein[4] and a quite accurate one with electrometric end point determination.[5] The analytical application of the formol reaction was initiated by Sørensen.[4]

In the alcohol[6] and acetone[7] titrations the decrease in water content and the other conditions increase the spread within the borders of the solvent envelope. The apparent pK values of the carboxyl groups rise and the pK values of amino groups fall by undetermined amounts. The combined effect is to leave a region about the stoichiometric pH where the rate of change of pH on addition of titrant is high enough to be useful. A successful titration then depends on the choice of an indicator whose pK is in this region. Because of the nature of the solvents, electrometric methods are not ordinarily applicable.

The stoichiometry of the titrations is the relationship between the equivalents of alkali or acid used and the number of moles of amino acid. The relationship depends not only on the end-point pH but also on the initial pH established in the original aqueous solution.[8] If a pure compound is weighed for analysis, this pH will automatically be one of the stoichiometric points. If one is dealing with an unknown solution, the

[2] M. Levy, *J. Biol. Chem.* **99**, 767 (1933).
[3] M. Levy, *J. Biol. Chem.* **105**, 157 (1934).
[4] S. P. L. Sørensen, *Biochem. Z.* **7**, 45 (1907).
[5] M. S. Dunn and A. Loshakoff, *J. Biol. Chem.* **113**, 359 (1936).
[6] K. Linderstrøm-Lang, *Compt. rend. trav. lab. Carlsberg. Sér. chim.* **17** No. 4 (1927).
[7] R. Willstäter and E. Waldschmidt-Leitz, *Ber.* **54**, 2988 (1921).
[8] D. D. Van Slyke and E. Kirk, *J. Biol. Chem.* **102**, 651 (1933).

initial pH should be established by titration with acid or base as required if one wishes an absolute measure. Of course if only the extent of increase or decrease is desired, as in following enzymatic hydrolysis, a zero time blank includes the initial adjustment. The table gives a survey of the various simple possibilities of the titrations for the typical amino acids glycine, aspartic acid, and lysine.

| | Charge form | | | |
| | Aqueous solution | | At end point | |
Amino acid	Isoelectric	pH 6.5	Formol or alcohol	Acetone
Glycine	R^{+-}(pH 6)	R^{+-}	R^-	R^+
Aspartic acid	R^{+-}(pH 3)	R^{+--}	R^{--}	R^+
Lysine	R^{+-}(pH 9)	R^{++-}	R^-	R^{++}

By starting at the aqueous isoelectric pH in each case, the amount of alkali used in the formol or alcohol titrations is equivalent to the carboxyl groups, whereas in the acetone titration the amount of acid used is equivalent to the amino groups. If the original aqueous solution is adjusted to pH 6.5, then the formol and alcohol titrations measure the equivalents of amino groups, whereas the acetone titration measures the equivalents of carboxyl groups. It is of course impossible to adjust a mixture to the varied isoelectric points, but pH 6.5 is close to a stoichiometric point for all amino acids except histidine. By 'adjustment of an unknown mixture in water to pH 6.5 followed by titration in formaldehyde or alcohol with NaOH, one is measuring the amino (and imino) groups. Starting from the same pH, the acetone titration with HCl measures the carboxyl groups.

The imidazole group of histidine presents a special problem in that its pK [9] is near the pH 6.5 stoichiometric point for the other amino acids. The initial adjustment to pH 6.5 will therefore leave part of it combined with H^+, and this amount will be included in the alcohol or formol titrations. The presence of large amounts of imidazole may make the adjustment indefinite and impractical. No adequate compromise suggests itself for mixtures. If pure histidine is to be titrated, the stoichiometric points at each side of the imidazole are available as initial points.

At the usual end point in the formol titration the guanidine group of arginine[9] is still positively charged. It will not be included in the titrations although the corresponding amino group will.

Apparatus. Only the usual titrimetric apparatus, burets, pipets, and flasks are needed. Many variations in size of sample are possible.

[9] M. Levy and D. E. Silberman, *J. Biol. Chem.* **118**, 723 (1937).

Formol Titration

Reagents

> Standard NaOH (as concentrated as the instruments will allow).
> Formol solution. Commerical 37% formaldehyde is neutralized to
> pH 6.0 by addition of alkali.
> Indicator, 1% phenolphthalein in 95% alcohol.

Procedure. The unknown is adjusted to pH 6.5 by addition of acid or
alkali. Dilution is avoided by using relatively concentrated reagents. This
adjustment may be made with a pH meter or with an outside indicator.
Enough formol solution to make the final formaldehyde concentration 6 to
9% (after adding titrant) is added along with 1 or 2 drops of phenolphthalein solution. Alkali is added to the appearance of a definite pink
(pH 8.8), and the volume recorded. If the formaldehyde is initially at pH
6.5, no blank is necessary or desirable.

In place of the indicator a glass electrode pH meter setup may be used
to follow the titration in detail. The end point can be determined very
accurately by the method of increments. This involves determination of
the ratio of pH increment to titrant increment as a function of total
titrant. The ratio is maximal at the end point.

Alcohol Titration

Reagents

> Alcoholic KOH, made up in 90% ethanol and standardized daily
> against $KH(IO_3)_2$.
> Absolute alcohol.
> 1.0% thymolphthalein in alcohol.

Procedure. The unknown is adjusted to pH 6.5 in water by addition of
acid or alkali with as little dilution as possible. One drop of thymolphthalein per 0.1 ml. of solution is added, and the mixture titrated with
the KOH solution until a faint blue appears. Absolute alcohol is added,
and the color disappears. The titration is continued, and when the blue
color returns additional alcohol is added. This process is continued until
9 ml. of absolute alcohol has been added per 0.1 ml. of unknown solution.
The alcohol is added portionwise to prevent precipitation of incompletely
titrated material. A blank with water in place of the sample is titrated,
and the titration value subtracted from the experimental titration.

Acetone Titration

Reagents

HCl in 90% ethanol, standardized daily against a standard aqueous alkali with methyl red indicator. The concentration should be as high as convenient for the burets, etc., available.

Acetone, reagent grade.

1% benzeneazo-α-naphthylamine in acetone.

Procedure. The unknown is adjusted to pH 6.5 by addition of acid or alkali with as little dilution as possible. One-tenth milliliter of indicator solution is added per 0.1 ml. of unknown, and the mixture is titrated with the alcoholic HCl until the color becomes pink. Then 0.9 ml. of acetone is added per 0.1 ml. of unknown. The indicator will turn yellow. The titration is continued until a match is obtained against a solution made up with indicator, acetone, and alcoholic HCl to give 0.001 N HCl and containing a volume of water equal to that of the original sample. The difference between blank and experimental values is equivalent to the carboxyl groups in the dissolved substance.

Remarks and Interferences. The preliminary adjustments to pH 6.5 may be dispensed with if one is only interested in the increments of groups as in a hydrolysis where samples are taken at time intervals.

Interferences may be expected if the solutions contain materials which buffer in water at pH 6.5 or in the stoichiometric zones in the various methods. An instance has been cited in the imidazole group of histidine. Especially in the alcohol and acetone titrations precipitation of incompletely titrated material may occur. If this appears to occur, the solvents should be added in portions as suggested above. Formol titrations are free of this inconvenience.

[75] Gasometric Procedures for Amino Acids

By REGINALD M. ARCHIBALD

I. Determination of Aliphatic Amino Nitrogen by Nitrous Acid*

$$RNH_2 + HNO_2 \rightarrow ROH + H_2O + N_2$$

Principle. Van Slyke[1,2] evolved a manometric procedure in which 4 to 600 γ of nitrogen liberated in the above reaction can be measured to

* Determination by Ninhydrin, Arginase and Decarboxylase will be found on pp. 1041–1049.

[1] J. P. Peters and D. D. Van Slyke, "Quantitative Clinical Chemistry," Vol. 2, p. 385, Williams & Wilkins, Baltimore, 1932.

[2] D. D. Van Slyke, *J. Biol. Chem.* **83**, 425 (1929).

±0.1 γ/ml. (1% accuracy for 50 γ in a 5-ml. sample). A solution of the amine to be measured and acetic acid are placed in the Van Slyke-Neill blood gas apparatus. After they have been freed of air, a saturated solution of sodium nitrite is added. The evolved NO is removed with alkaline permanganate, and the residual N_2 is measured manometrically. The α-NH_2 nitrogen of amino acids is liberated quantitatively in 3 to 4 minutes at room temperature; other NH_2 groups of amino acids or other substances are liberated more slowly, as indicated in Table I. All amino acids obtained by acid hydrolysis of proteins react quantitatively to yield only 1 mole of N_2 except lysine and hydroxylysine which eventually yield 2 moles, and proline and hydroxyproline which yield none. All amino acids including lysine and ornithine react in a reasonably short interval with all their nitrogen except tryptophan and asparagine which react with half, histidine and citrulline with one-third, arginine with one-fourth, and proline, oxyproline, pyrrolidonecarboxylic acid and p-aminobenzoic acid, which yield no nitrogen. Only the free primary NH_2 group of peptides such as leucylleucine react. In proteins and polypeptides presumably the free amide group of glutamine, the ε-NH_2 of lysine (when not involved in a peptide bond), and any α-NH_2 group of terminal amino acids in which the α-NH_2 is not involved in a peptide bond will each yield 1 mole of N_2. The N of the peptide bond, however, does not react.

Reagents

Sodium nitrite reagent, 800 g. of sodium nitrite dissolved in 1 l. of warm water (60% by volume).

Glacial acetic acid.

Alkaline permanganate, 1 l. of 10% NaOH shaken with 50 g. of $KMnO_4$ until the solution is saturated.

Caprylic alcohol may be used to prevent foaming.

Procedure. A 1- to 8-ml. aliquot of solution of the amino acid to be analyzed is delivered from a rubber-tipped stopcock pipet into the Van Slyke-Neill blood gas apparatus,[3] followed by 1 ml. of glacial acetic acid. A drop of caprylic alcohol may be added if the unknown would otherwise cause foaming. The air in this mixture is then extracted by shaking for 2 minutes and ejected.[4,5] Two milliliters of the $NaNO_2$ solution is measured into the cup and allowed to run into the chamber. The nitrous acid thus

[3] D. D. Van Slyke, *J. Biol. Chem.* **73**, 121 (1927).

[4] Here the details of technique described by Van Slyke should be followed precisely. In this summary, because of limitation of space, important details such as the opening, closing, and sealing of stopcocks and the washing of apparatus are seldom mentioned.

[5] See p. 279 in ref. 1.

TABLE I
Usual and Anomalous Behavior of Amino Acids with Nitrous Acid

Reactant	Temperature, °C.	Short period Moles N_2 per mole	Minutes	Longer period Moles N_2 per mole
Most amino acids,[a] including methionine[b]	20–22	1.0	5	1.0 (60 min.)
Proline[a]	20–22	0.0	5	0.0
Hydroxyproline[a]	20–22	0.0	5	0.0
Tryptophan[a]	22	1.01	5	
Histidine[a]	22	1.00	5	
Arginine[a]	20	1.00	5	>1.00 if mineral acid is present[c]
Lysine[a]	20	1.70	5	2.0 (30 min.)
Ornithine[d]	25	1.86	3	2.0 (10 min.)
Citrulline[d]	25	1.16	3	1.50 (30 min.) 1.60 (1 hr.)
Asparagine[a]	22	1.00	5	>1.0 if mineral acid is present[c]
Glutamine[e]	22.5	1.80	4	
Cystine[a,f]	19	1.07	4	1.09 (30 min.)
Cysteine[f]	25(?)	1.28	?	
Glutathione[f]	25(?)	1.94	4	2.53 (10 min.)
Glycine[a,f]	19	1.03	5	
Glycylglycine[a,f]	21–22	1.24	4	1.35 (60 min.)
Thirteen different uramido acids[d]	25	0.19–0.31	3	
Argininic acid[d]	25	0.005	3	0.06 (30 min.)
Ammonia[a]	20	0.15	5	1.0 (2 hr.)
Purine and pyrimidines[a]	20	—		All NH_2 groups react in 2–5 hr.
Guanosine[a]	20	—		1.25(?)
Guanidine[a]	20	0.0	5	—
p-Aminobenzoic acid and other aromatic amines[g]	20	0.0		—
Urea[a]	20	0.14	5	1.0 (1 hr.) 2.0 (8 hr.)

[a] D. D. Van Slyke, *J. Biol. Chem.* **9**, 185 (1911).

[b] B. W. Chase and H. B. Lewis, *J. Biol. Chem.* **101**, 735 (1933).

[c] H. Gilman, "Organic Chemistry," 2nd ed., Vol. 2, p. 1092, John Wiley & Sons, New York, 1943.

[d] A. G. Gornall and A. Hunter, *Biochem. J.* **34**, 192 (1940).

[e] P. B. Hamilton, *J. Biol. Chem.* **158**, 397 (1945); H. B. Vickery, G. W. Pucher, and H. E. Clark, *Biochem. J.* **29**, 2712 (1935); H. A. Krebs, *Biochem. J.* **29**, 1960 (1935).

[f] A. B. Kendrick and M. E. Hanke [*J. Biol. Chem.* **117**, 161 (1937); **132**, 739 (1940)]

formed reacts with the amino groups and liberates N_2. One minute before the reaction time is complete, the mercury level is lowered almost to the 50-ml. mark, and the chamber is shaken, during which time the mercury level is adjusted occasionally to near the 50-ml. mark to compensate for the NO gas evolved by decomposition of nitrous acid. The lower the temperature and the greater the volume of the sample taken for analysis, the longer is the reaction time required.[4,6] When the total volume of solution in the chamber is 8 ml., the time required for the α-NH_2 nitrogen of acids is 3 minutes at 25°, 4 minutes at 20°, and 6 minutes at 15°. The N_2 and NO are then transferred to a modified Hempel pipet charged with permanganate solution.[4,7] The gases are followed by the aqueous reaction mixture up to the cock of the Hempel pipet. By shaking the pipet horizontally for 20 to 40 seconds, the NO is completely absorbed. After the rest of the reaction mixture has been ejected from the chamber, the chamber is washed.[4,8] Then 10 ml. of water is admitted to the chamber and evacuated with shaking for 1 minute to remove dissolved air which is then ejected. One milliliter of this extracted water is ejected into the cup to provide a liquid seal. The N_2 is returned from the Hempel pipet to the manometric apparatus.[4,9] With the nitrogen below a mercury-sealed cock, manometric readings are taken (p_1), with the mercury level at 2 ml. if the pressure is 100 mm. or greater, otherwise at the 0.5-ml. mark. Then gas (but not liquid) is ejected and a pressure reading (p_0) is taken at the same mark.

A similar run with water replacing the amino acid solution serves as a blank where $p_1 - p_0 = c$.

$$\text{Mg. of amino nitrogen} = (p_1 - p_0 - c) \times \text{factor}$$

Factors recorded in detail elsewhere[10] may be interpolated from a plot of data in Table II.

For determination of amino nitrogen in 5 ml. of 1 to 10 Folin-Wu filtrate of blood which contains less than 50 mg. of urea nitrogen per 100

---

[6] See p. 395 in ref. 1.

[7] See p. 390 in ref. 1.

[8] See p. 391 in ref. 1.

[9] See p. 392 in ref. 1.

[10] See Table 40 in ref. 1.

---

obtained theoretical yields with cystine and glycine by using 2% KI in the acetic acid. M. S. Dunn and I. Porush [*J. Biol. Chem.* **127**, 261 (1939)], report that potassium iodide brings cysteine also into line. However, iodide cuts the yield from tryptophan to less than half of theoretical for 1 nitrogen.

[g] L. Gattermann, "Practical Methods of Organic Chemistry," 3rd American ed., p. 237, The Macmillan Co., New York, 1928.

TABLE II
FACTORS BY WHICH MILLIMETERS OF HG PRESSURE $(p_1 - p_0 - c)$ ARE
MULTIPLIED TO GIVE MILLIGRAMS OF $NH_2 - N$ IN SAMPLE ANALYZED

Temperature, °C.	For pressures at 0.5 ml.	For pressures at 2.0 ml.
15	0.000390	0.001561
24	378	11
34	365	1459

ml. of blood, the reaction must be timed to within 10 seconds and 7% of the urea nitrogen subtracted from the nitrogen evolved with nitrous acid. When blood urea nitrogen is greater than 50 mg.%, the urea is converted to NH_3 and CO_2 by preliminary treatment of the blood with urease.[4] Extra $\frac{2}{3}$ N H_2SO_4 is required in the Folin-Wu precipitation to neutralize the ammonia which is subsequently removed by boiling the 1 to 10 filtrate with a few drops of magnesium hydroxide suspension.[4,11]

[11] See p. 396 in ref. 1.

Note: For the continuation of gasometric procedures in the determination of amino acids, see the sections on Ninhydrin, Arginase and Decarboxylase, pp. 1041–1049.

[75A] Determination of Amino Acids by Specific Bacterial Decarboxylases in the Warburg Apparatus

$$RCHNH_2COOH \rightarrow RCH_2NH_2 + CO_2$$

By VICTOR A. NAJJAR

The following L-amino acids have been measured in biological fluids and hydrolyzates by means of the corresponding bacterial decarboxylase.[1]

1. Lysine   → Cadaverine   + CO_2
2. Ornithine   → Putrescine   + CO_2
3. Arginine   → Agmatine   + CO_2
4. Histidine   → Histamine   + CO_2
5. Tyrosine   → Tyramine   + CO_2
6. Aspartic acid   → β-Alanine   + CO_2
7. Glutamic acid   → γ-Aminobutyric acid + CO_2

Assay Method

Principle. The enzymatic decarboxylation of amino acids proceeds to completion, and the amount of CO_2 produced in the reaction is a direct

[1] For a detailed account of the enzyme properties and method of culture see Vol. II [20]. These organisms can be obtained from the American Type Culture Collection, 2029 M St., N. W., Washington, D. C.

measure of the amino acid content of the sample. Such determinations can therefore be made manometrically with the Warburg technique by measuring the pressure change resulting from the liberation of CO_2 from the reaction mixture, provided the pH is 5.3 or below. Above this value the CO_2 formed during the reaction is not liberated quantitatively into the gas phase, but a considerable amount, depending on the pH, may be retained in solution as bicarbonate. Where it is found necessary or convenient to use higher pH values to obtain maximum rate of decarboxylation, an amount of acid, 0.1 to 0.2 ml. of 5 N H_2SO_4, is added from the side bulb to the reaction mixture to effect complete liberation of CO_2.

Reagents

Buffer, 0.2 M. The type of buffer chosen depends on the pH desired for optimum enzyme activity; phosphate, acetate, and phosphate-citrate are used.[2]

5 N H_2SO_4.

Enzyme.

Procedure. The procedure used to measure the amino acid content of a sample is common to all the enzymes discussed below.[1,3] The main compartment of the Warburg vessel contains the sample to be assayed and the appropriate buffer. One side bulb contains the enzyme, and the other the acid, if an acid tip is necessary. The amount of amino acid in the sample should be within the measuring range of the manometer. It is always preferable to use two different amounts of the sample or more, and each determination should be run in duplicate to ensure greater accuracy. A blank manometer containing no substrate is run simultaneously. The gas phase is room air. The bath is set at an appropriate temperature, 30 to 38°, and the reaction is started by tipping in the enzyme. When the reaction reaches completion, the corrected readings obtained remain constant, an acid tip is then made if desirable.

Preparation of the Enzymes and Specific Conditions Used

In general, stock cultures of the organism possessing the desired enzyme are grown on broth agar slants and kept well-stoppered in the icebox. Periodically, every 2 to 4 months, the organism is transferred into fresh agar slants. In this manner a strain can be maintained unchanged for years. The anaerobes are ordinarily kept in the growth medium under anaerobic conditions with occasional transfers into fresh media. In order

[2] E. F. Gale, *Biochem. J.* **39**, 46 (1945).
[3] See standard texts on Warburg manometric techniques.

to obtain one or more liters of culture of reproducible characteristics, the organism is inoculated into a few milliliters of the medium and incubated for 6 to 8 hours at 30 to 37°. This serves as the standard inoculum for the large quantity of culture desired. A smear preparation from the standard inoculum is previously checked for contamination.

Lysine Decarboxylase. Bacterium cadaveris (N.C.T.C. 6578),[1] isolated by Gale,[2] is inoculated into a medium consisting of 3% pancreatic casein digest and 2% glucose. The former supplies amino acids, cofactors, and minerals. It is incubated at 25° for 24 hours. The cells are centrifuged down, washed with water, and 5 vol. of acetone added at room temperature. The suspension is then filtered with moderate suction in a Büchner funnel, with a rapid-flow filter paper. Just before the precipitate gets dry, it is washed with 100 ml. of acetone followed by 50 to 100 ml. of ether. The acetone powder so prepared keeps in the desiccator for 4 to 6 weeks at room temperature. Ten milligrams is used for each Warburg vessel, with phosphate buffer, 0.2 M, pH 6.0. When the reaction reaches completion, acid is tipped in (see above) and a pressure reading taken from which the amount of CO_2 liberated can be calculated.[3]

Arginine Decarboxylase. E. coli[2] (N.C.T.C. 7020) is grown under the same conditions as for lysine decarboxylase. The acetone powder is prepared in like manner. Phosphate-citrate buffer, 0.2 M, pH 5.2, is used. Ten milligrams of the enzyme is used for each cup. No acid tip is needed in this reaction. The enzyme keeps in the desiccator at room temperature for a period of 4 to 6 weeks.

Tyrosine Decarboxylase. Streptococcus faecalis[2] (N.C.T.C. 6783) is grown on 3% casein digest, 1% glucose, and 0.1% marmite (or Difco yeast extract) at 37° for 16 hours. Acetone powder is prepared as above. Ten milligrams of the powder is used for each vessel. It keeps for 2 to 8 weeks at room temperature in the desiccator. The reaction is carried out in phosphate-citrate buffer, 0.2 M, pH 5.5, followed by an acid tip.

Ornithine Decarboxylase. Clostridium septicum[2] pIII (N.C.T.C. 547) is grown in 3% casein digest and 2% glucose, with pieces of heart muscle, at 37° for 16 hours. The organism is harvested by centrifugation and washed with water. A washed suspension of the organism amounting to 10 to 15 mg. dry weight is used for each vessel. The activity of the enzyme deteriorates after 24 hours. Phosphate-citrate buffer, 0.2 M, pH 5.5, is used for the reaction. An acid tip is used for complete CO_2 evolution.

Histidine Decarboxylase. Clostridium welchii BW 21[2] (N.C.T.C. 6785) is grown in 3% casein digest, 2% glucose, and heart muscle at 37° for 16 hours. Acetone powder is formed as before except that 3 vol. of acetone is added to 1 vol. of cell suspension. The powder maintains its activity for 2 to 4 months at room temperature in the desiccator. The determination is performed at pH 4.5 in 0.2 M acetate buffer, with 40 mg. of powder for each vessel.

Glutamic Acid Decarboxylase. Clostridium welchii SR 12 (N.C.T.C. 6784) is grown in 3% casein digest and 2% glucose with heart muscle in a hydrogen atmosphere at 37° for 12 hours or 30° for 16 hours. The organism is centrifuged and washed with water. Ten to fifteen milligrams dry weight of the suspension is used for each Warburg vessel. The determination is performed in acetate buffer, 0.2 M, pH 4.5. The suspension remains active for 3 to 4 days at room temperature if kept under hydrogen. Gale,[2] under these conditions, could not detect any decarboxylase activity other than that for glutamic acid. Meister *et al.*,[4] however, under slightly different conditions of growth and pH, obtained in addition aspartic decarboxylase activity (see below). Furthermore, there is an active glutaminase in the organism. In the presence of glutamine and glutamic acid, the CO_2 evolution represents both values, and the ammonia liberated corresponds to the value of glutamine present. The difference between the two values is equivalent to that of glutamic acid.[5]

Aspartic Acid Decarboxylase. Meister *et al.*[4] showed that *Clostridium welchii* SR 12 (A.T.C.C. 6784)[1] has, in addition to glutamic acid decarboxylase, an aspartic acid decarboxylase. The latter is inhibited by 0.006 M semicarbazide. The inhibition is reversed by α-keto acids such as pyruvic acid. This formed the basis for quantitative measurement of aspartic acid in the presence of glutamic acid. The organism is grown as described in Vol. II,[1] lyophilized, and stored at 5°. Under these conditions the activity is preserved for at least 6 to 8 months. One hundred milligrams of cells per milliliter of suspension is prepared in 0.2 M acetate buffer, pH 4.9. The main compartment of the vessel contains 2 ml. of the sample, 0.2 ml. of 3 M acetate buffer, pH 4.9, and 0.5 ml. of a semicarbazide hydrochloride solution prepared by dissolving 0.401 gm. in 100 ml. of an equal mixture of 0.4 M acetate buffer, pH 4.9, and 0.072 N NaOH. The reaction is started by tipping in 0.3 ml. of bacterial suspension. After glutamic acid decarboxylation is complete, the manometers are reset, and 50 micromoles of pyruvate and 30 mg. of bacteria in a convenient volume are tipped in from the other side bulb. The CO_2 evolution that follows is due to aspartic acid.

The determination of glutamine, glutamic acid, lysine, and arginine has been done with *E. coli*[6] (A.T.C.C. No. 11246) isolated by Najjar and Fisher.[7] The organism is grown as described earlier.[1,7] Washed suspen-

[4] A. Meister, H. A. Sober, and S. V. Tice, *J. Biol. Chem.* **189,** 591 (1951).
[5] H. A. Krebs, *Biochem. J.* **43,** 51 (1948).
[6] V. A. Najjar and J. Fisher, unpublished.
[7] V. A. Najjar and J. Fisher, *J. Biol. Chem.* **206,** 215 (1954).

sions decarboxylate glutamine, glutamic acid, histidine, lysine, and arginine. An acetone powder is prepared by adding 9 vol. of acetone to 1 vol. of cell suspension at room temperature and filtering as described above. Such preparations lose the ability to decarboxylate glutamine and histidine and decarboxylate only lysine, arginine, and glutamic acid at pH 5.3 in 0.2 M acetate buffer. The glutamic acid decarboxylase activity has a pH range of <4 to 5.8 with an optimum at pH 5.0. No activity is exhibited above pH 5.8. Similarly, arginine decarboxylase activity has a pH range of <4.8 to 6.1 with an optimum at 5.4 and no activity above pH 6.1. Lysine decarboxylase, on the other hand, has a wide pH range of 4.6 to >7.5 with an optimum at 6.2 and shows about one-fourth of the maximum activity at pH 7.5. Thus with acetone powder, 10 mg. per vessel, phosphate buffer at pH 6.3, and an acid tip, lysine can be measured in the presence of the other amino acids. The lysine and arginine decarboxylases are destroyed[7] by incubating a 2% suspension of the acetone-dried cells in water at 55° for 4 to 7 days, or in 0.3 M acetate buffer, pH 5.0, with a few drops of toluene at 37° for 3 to 5 days. A second acetone powder may be prepared from this material which has glutamic acid decarboxylase activity free from the other decarboxylases. This preparation is used to measure glutamic acid with 10 to 20 mg. of the enzyme and acetate buffer, 0.1 to 0.2 M, pH 5.3.

It is clear that measurements with fresh untreated acetone powder at pH 5.3 give the total value of lysine, arginine, and glutamic acid. At pH 6.3 in phosphate buffer the value for lysine is obtained, and from the treated acetone powder at pH 5.3 the glutamic acid value is obtained. The value for the three amino acids obtained with untreated acetone powder less the sum of the values of lysine and glutamic acid represents the value for arginine present in the sample.

In order to determine glutamine in the presence of the other amino acids, histidine, if present, is first determined separately as described above. The sample is then treated with fresh acetone powder in the Warburg apparatus in 0.2 M acetate buffer, pH 5.3, until CO_2 evolution stops. The manometers are then reset, and 10 to 20 mg. of washed cells is tipped in. The CO_2 evolution that follows is due to histidine and glutamine. Knowing the histidine value, one can obtain that of glutamine by the difference. Washed cell suspensions retain activity for at least 10 days in the icebox.

Our experience with this organism is limited to assay of known amino acid mixtures, casein hydrolyzates, and pancreatic digests of casein. Its use in determining amino acids in biological fluids and tissue extracts should be done with caution and checked with the other decarboxylases described above.

[76] Colorimetric Procedures for Amino Acids

By JOSEPH R. SPIES

I. Phenol Reagent Method

Principle. This spectrophotometric method, first used by Folin and Denis[1] and further studied by Folin et al.,[2,3] is based on the blue color formed by phenol reagent (phosphotungstic-phosphomolybdic acid) and tyrosine and or tryptophan. The procedure described below is based on methods described by Anson[4] for measurement of proteolytic activity of enzymes with hemoglobin as substrate and by Hull[5] for studying proteolysis of milk proteins.[6]

Reagents

Phenol reagent. One hundred grams of $Na_2WO_4 \cdot 2H_2O$, 25 g. of $Na_2MoO_4 \cdot 2H_2O$, 700 ml. of H_2O, 50 ml. of 85% H_3PO_4, and 100 ml. of concentrated HCl are boiled under reflux for 10 hours. Then 174 g. of $Li_2SO_4 \cdot H_2O$, 50 ml. of H_2O, and a few drops of Br_2 are added, and the solution is boiled for 15 minutes without the condenser. The reagents should be nitrate-free. The solution is cooled and diluted to 1 l. The solution, which should have no green color, is filtered and stored in a brown bottle. For use, this stock solution is diluted with 2 vol. of H_2O.

2.8 N sodium carbonate solution.

Tyrosine, analytically pure.

Procedure. To 5 ml. of test solution in a 50-ml. flask is added 10 ml. of sodium carbonate solution. The solutions are mixed, and 3 ml. of diluted phenol reagent is added, with shaking of the flask during the addition. After 5 minutes the transmittancy of the solution is read on a spectrophotometer at 650 mμ. A standard curve is prepared with tyrosine. Transmittancies range from 82 to 10% for 0.02 to 0.24 mg. of tyrosine per 5-ml. aliquot. Conformity to Beer's law over the entire range of con-

[1] O. Folin and W. Denis, *J. Biol. Chem.* **12**, 239 (1912).

[2] O. Folin and J. M. Looney, *J. Biol. Chem.* **51**, 421 (1922).

[3] O. Folin and V. Ciocalteu, *J. Biol. Chem.* **73**, 627 (1927).

[4] M. L. Anson, *J. Gen. Physiol.* **22**, 79 (1938).

[5] M. E. Hull, *J. Dairy Sci.* **30**, 881 (1947).

[6] In both methods, undigested protein is precipitated by trichloroacetic acid and the liberated tyrosine and tryptophan contents of the protein-free filtrates is taken as a measure of proteolytic activity.

centration is obtained. Results may be expressed as tyrosine units, a unit being 1 mg. of tyrosine per liter of sample or its equivalent.[7] Other units also are used for defining activity of specific enzymes.[4]

Discussion. The reaction is not specific for tyrosine and tryptophan, color being given also by phenols and other reducing agents. Tyrosine and tryptophan give the same amount of color on a molar basis. Since the color is not stable, readings should be made at the specified time. Color is developed only in alkaline solution, so care must be taken that final test solutions be alkaline. Test solutions were 1.5 N in sodium carbonate in Hull's method and 0.2 N in sodium hydroxide in Anson's method after neutralization of the trichloroacetic acid used to precipitate undigested proteins in their applications of the method.

II. Ninhydrin Method

Principle. This spectrophotometric method is based on the formation of a blue color by reaction of ninhydrin and compounds having free amino groups. The procedure described below is based on the methods described by Moore and Stein[8] for the analysis of effluent in chromatographic separation of amino acids and peptides.

Reagents

Ninhydrin. Ninhydrin, commercially prepared by the method of Teeters and Shriner,[9] is recrystallized within several months of use. One hundred grams of ninhydrin is dissolved in 250 ml. of hot water. The hot solution is stirred with 5 g. of decolorizing carbon and filtered. The filtrate is stored at 4° overnight. The crystals are washed on the filter five times with 20-ml. portions of cold water. The ninhydrin is air-dried and stored in dark glass.

Ninhydrin solution. 0.80 g. of reagent $SnCl_2 \cdot 2H_2O$ is dissolved in 500 ml. of the citrate buffer. This solution is added to 20 g. of ninhydrin dissolved in 500 ml. of methyl cellosolve. This reagent can be stored under nitrogen for at least a month. If the number of tests to be run does not warrant this precaution, a smaller quantity of ninhydrin solution can be freshly prepared as needed.

0.2 M citrate buffer, pH 5.0. 21.008 g. of reagent-grade citric acid monohydrate is dissolved in 200 ml. of N NaOH and diluted to 500 ml.

[7] A. B. Storrs, *J. Dairy Sci.* **30**, 885 (1947).
[8] S. Moore and W. H. Stein, *J. Biol. Chem.* **176**, 367 (1948).
[9] W. O. Teeters and R. L. Shriner, *J. Am. Chem. Soc.* **55**, 3026 (1933)

Methyl cellosolve. The methyl cellosolve should give a clear solution when mixed with an equal volume of water and should give a negative or very faint peroxide test with 10% aqueous KI.[10]

Diluent solvent. Equal volumes of water and reagent grade n-propanol are mixed.

Solutions for analysis. Solutions of amino acids of 1.6 to 2.0 mM or suitable concentrations of other compounds to give equivalent color intensities are used for analysis of 0.1-ml. samples. Acidic solutions should be neutralized to the end point of methyl red before analysis.

Procedure. Moore and Stein[8] give details for application of the method to large numbers of samples with modifications as to volume and concentration of test solutions. They also give extensive details of special techniques and equipment such as automatic and self-adjusting pipets, selection of matched photometer tubes, water bath, mechanical shaker, and automatic pipetting machine. Users of the method interested in these details should consult the original article. The basic method described below is for 0.1-ml. samples of concentrations suitable for the described standard curve. Modifications can be made to accommodate the particular needs of the worker.

The sample (0.1 ml.) is placed in an 18 × 150-mm. photometer tube, and 1.0 ml. of ninhydrin solution is added. The tube is covered with an aluminum cap, and the contents mixed. A rack of tubes is heated for 20 minutes in a boiling water bath. Five milliliters of diluent is then added to each tube, and the contents mixed. The tubes are wiped, shaken, and transferred to a dry rack. Readings are taken on a spectrophotometer at 570 mμ starting 15 minutes after the tubes have been removed from the water bath. The color is stable for at least 1 hour. Appropriate correction for blank is made either by reading blank solutions against diluent and zeroing the instrument on the blank reading so obtained or by means of an appropriate blank solution in conventional manner.

A standard curve is prepared with 0.1-ml. samples of leucine at six concentrations from 0.5 to 2.0 mM. The curve follows Beer's law through an optical density of 0.50, with a deviation of 4% at optical density of 1.0.

The method can be employed for the determination of proline and hydroxyproline by measuring the yellowish-red reaction products at 440 mμ. The color values are ¼ and ⅟₇, respectively, of those obtained with equimolar concentrations of leucine. Color development with proline and hydroxyproline is 80 to 90% complete in 20 minutes at 100°. Stand-

[10] Cellosolve stored in lacquered cans gives turbid solutions.

ard curves can be prepared for these amino acids with a 30- to 40-minute heating period. Cysteine also requires special treatment.

Discussion. Color development is obtained with compounds having NH₂ groups including amino acids, peptides, primary amines, and

TABLE I

Color Yields from Amino Acids and Other Compounds on Molar Basis Relative to Leucine

(Determined on 0.1-ml. aqueous samples of 2.0 mM solutions; heating time, 20 minutes; read at 570 mμ^a)

Compound	Color yield	Compound	Color yield
Alanine	1.01	Glutathione	0.76
Arginine	1.00	Glycine ethyl ester	1.00
Aspartic acid	0.88	Glycyltyrosine	0.88
Citrulline	1.03	Glycylphenylalanine	1.04
Glutamic acid	1.05	Glycylglycine	0.89
Glycine	1.01	Glycylleucine	1.05
Histidine	1.04	Leucylglycine	0.92
Isoleucine	1.00	Phenylalanylglycine	0.97
Leucine	1.00	Phenylalanine ethyl ester	0.98
Lysine	1.12	Histamine	0.65
Methionine	1.00	Taurine	0.97
Phenylalanine	0.88	Tyramine	0.64
Serine	0.94	Sarcosine	0.84 ca.
Threonine	0.92	Glucosamine	1.00
Tyrosine	0.88	Creatine	0.03
Valine	1.02	Creatinine	0.03
Cysteine	0.15 ca.	Dibenzylamine	0.04
Half-cystine	0.54	Glycine anhydride	0.01
Tryptophan	0.72 ca.	Urea	0.03
Proline	0.05	Adenine	0.00
Hydroxyproline	0.03	p-Aminobenzoic acid	0.00
Ammonia	0.98 ca.	Diethylbarbituric acid	0.00
Asparagine	0.94	Glucose	0.00
Glutamine	0.99	Uric acid	0.00

[a] Slight revision of the color values of the following amino acids was made by Moore and Stein [*J. Biol. Chem.* **192**, 663 (1951)] in a method using 1-ml. samples and 2 ml. of ninhydrin solution. The revised values were: aspartic acid 0.93, serine 0.97, alanine 0.99, valine 1.00, and tyrosine and phenylalanine 0.91. Most of the other values were checked and found to be the same as above.

ammonia. The generality of the reaction is no disadvantage in working with known mixtures of amino acids and peptides as in chromatographic analysis. The lack of specificity may be a disadvantage in working with unknown biological systems. Reaction with ammonia must be taken into account, and the possible uptake of atmospheric ammonia by acid test solutions must be determined by appropriate blank analyses. Color intensities of equimolar quantities of various amino acids relative to standard leucine vary. The relative color values of amino acids and some related compounds are given in Table I.[8] Moore and Stein[8] give correction factors for evaporation of solvent where water-organic solvent systems are analyzed. The accuracy of the method is 2% for individual amino acids with samples in the range of 2.5 γ of NH_2 nitrogen. No color is formed by equimolar quantities (relative to leucine) of adenine, p-aminobenzoic acid, diethylbarbituric acid, glucose, and uric acid. Creatine, creatinine, dibenzylamine, and glycine anhydride formed only 1 to 4% of the color intensity of leucine.

III. Copper Salt Method for Amino Acids and Peptides (620 mμ)

Principle. This spectrophotometric method is based on the blue color of the complex copper salts of amino acids and peptides. The copper salts of the amino acids are converted to the copper salt of alanine to eliminate the variation in color values of equimolar concentrations of copper salts of the amino acids. Peptide copper salts are also converted to the copper salt of alanine, the latter serving as a reference standard. The procedure described below is based on a method developed by Spies and Chambers[11] for studying rate and degree of protein hydrolysis.

Reagents

Copper chloride. 28 g. of reagent-grade $CuCl_2 \cdot 2H_2O$ per liter of H_2O solution.

Sodium phosphate. 68.5 g. of reagent-grade $Na_3PO_4 \cdot 12H_2O$ per liter of H_2O solution.

Sodium borate buffer, pH 9.1 to 9.2. 40.3 g. of reagent-grade anhydrous $Na_2B_4O_7$ is dissolved in 4 l. of H_2O, and the solution is filtered. Reagent-grade NaCl is dissolved in the buffer at the rate of 6 g. per 100 ml. This NaCl-containing solution is referred to as buffer solution.

Washed $Cu_3(PO_4)_2$.[12] To 40 ml. of sodium phosphate solution is added 20 ml. of $CuCl_2$ solution with swirling. The suspension is centrifuged for 5 minutes. The precipitate is washed twice by resuspension in 60 ml. of buffer solution followed by centrifuging. The washed precipitate is suspended in 100 ml. of buffer solution. $Cu_3(PO_4)_2$ suspensions in glass-stoppered bottles give maximum color after 4 days of storage, but slight decrease in color is obtained after 10 days of storage.

---

[11] J. R. Spies and D. C. Chambers, *J. Biol. Chem.* **191**, 787 (1951).
[12] W. A. Schroeder, L. M. Kay, and R. S. Mills, *Anal. Chem.* **22**, 760 (1950).

Alanine. Alanine is recrystallized with special care to prevent contamination with traces of insoluble foreign matter which might interfere with spectrophotometric analyses and to obtain analytically pure alanine for the standard curve.

Procedure. To 5.0 ml. of test solution in a 25-ml. glass-stoppered Erlenmeyer is added 5.0 ml. of the $Cu_3(PO_4)_2$ suspension. The suspension

TABLE II

RELATIVE COLOR INTENSITIES PRODUCED BY 0.0250 M SOLUTIONS OF AMINO ACIDS ALONE AND WITH ADDED ALANINE AS COMPARED TO ALANINE AS 100%

Amino acid	Alanine equivalence, %	
	Without added alanine[a]	With added alanine[b]
Alanine	100.0	100.0
Arginine monohydrochloride	103.8	98.8
Aspartic acid	89.3	99.0
Glutamic acid	104.6	101.2
Glycine	79.8	98.8
Histidine monohydrochloride	145.2	123.6
Hydroxyproline	125.9	103.6
Isoleucine	109.2	101.2
Leucine[c]	101.8	98.2
Lysine monohydrochloride	102.0	98.2
Methionine[c]	[d]	102.4
Phenylalanine[c]	101.8	99.4
Proline	123.7	100.6
Serine	91.3	99.5
Threonine	95.0	100.0
Tryptophan[c]	108.9	99.4
Tyrosine[e]	99.6	99.4
Valine	105.6	100.4

[a] Average deviation of all results $\pm 0.4\%$.

[b] Average deviation of all results $\pm 0.2\%$.

[c] Analyzed in solution containing 0.01667 M glycine and 0.00833 M amino acid.

[d] Copper salt precipitated before readings could be taken. There was no precipitation in test with added alanine.

[e] Tyrosine was added in the solid state to a water-copper phosphate suspension. The tyrosine dissolved in 1 hour, and transmittancies were determined as usual.

is shaken occasionally during 5 minutes and then centrifuged in a capped 12-ml. tube for 5 minutes. The clear blue solution is carefully decanted (9.5 ml. being obtained) into a 25-ml. flask containing 200 mg. of alanine. The alanine is dissolved, and the transmittancy of the solution is determined at 620 mμ. Blank solutions for analysis of amino acids and peptides are made by mixing 5.0 ml. of buffer solution and 5.0 ml. of water; then

200 mg. of alanine is dissolved in 9.5 ml. of this solution. Blank solutions for analysis of protein hydrolyzates are made by mixing 5.0 ml. of buffer solution with 5.0 ml. of the diluted hydrolyzate. After centrifugation, 200 mg. of alanine is dissolved in the decanted solution.

The standard curve is prepared with eight alanine solutions ranging from 5 to 40 mM, each solution differing by 5 mM. Conformity to Beer's

TABLE III

ANALYSIS OF DIPEPTIDES AND TRIPEPTIDES WITH ADDED ALANINE

Substance	Color of solution before adding alanine	Molarity of test solution	Alanine equivalence of test solution, %	Ratio of mol. wt. to equivalent wt. found	Formula of copper salt[a]
Alanine	Blue	0.0250	100.0	1.00	CuA$_2$
Dialanylcystine	Blue	0.00625	96.4 ± 0.0	3.86	Cu$_2$P
Alanylglycine	Blue	0.0125	100.0	2.00	CuP
Glycylalanine	Blue	0.0125	100.0 ± 0.4	2.00	CuP
Glycyl-α-aminoisobutyric acid	Blue	0.0125	102.4 ± 0.0	2.05	CuP
Glycyl-α-amino-n-butyric acid	Blue	0.0125	99.8 ± 0.2	2.00	CuP
Glycyldehydrovaline	Green-blue	0.0125	90.2 ± 0.0	1.80	CuP
Glycylsarcosine	Blue	0.0125	35.0 ± 0.2	0.70	CuP$_3$
Glycylserine	Blue	0.0125	104.3 ± 0.3	1.90	CuP
Sarcosylphenylalanine	Blue	0.0125	98.8 ± 0.0	1.98	CuP
Glycylalanylglycine	Purple[b]	0.0125	100.2 ± 0.2	2.00	CuP
Glycylglycylalanine	Purple	0.0125	99.6 ± 0.0	1.99	CuP
Glycylglycylleucine	Purple	0.0125	95.5 ± 0.2	1.91	CuP
Glycylleucylglycine	Purple	0.0125	99.2	1.98	CuP
Leucylglycylglycine	Purple	0.0125	99.8 ± 0.2	2.00	CuP

[a] Calculated on the basis of alanine having the formula CuA$_2$; P stands for dipeptide or tripeptide.

[b] Test solutions turned characteristic blue on addition to 200 mg. of alanine.

law over the entire range of concentrations is obtained. Acid solutions should be brought to between pH 7 to 9.1 before analysis.

Discussion. The molar color equivalence of seventeen amino acids differs from that of alanine from 0 to 3.6%, as shown in Table II. Histidine gives an alanine color equivalence value of 124%. Cystine, phenylalanine, methionine, leucine, and tryptophan form insoluble or partially soluble copper salts. These amino acids can be analyzed by forming a mixed salt with glycine. Addition of glycine in analysis of protein hydrolyzates is not necessary because of the small proportion of these refractory amino acids and the solubilizing effect of the other amino acids present. The copper salt of cystine is too insoluble to be determined even with the mixed salt method. Dipeptides react with copper as though the component amino acids are free, whereas only two of the amino acids of tripeptides react with copper as shown in Table III. In general, only those amino acids of peptides or proteins react with copper that have a free carboxyl group adjacent to peptide-linked nitrogen or with a free amino group adjacent to a peptide-linked carboxyl group. Therefore, the copper-binding capacity of protein hydrolyzates increases during hydrolysis until all the α-amino nitrogen is either free or in dipeptide form. Application of the method to the study of the rate of hydrolysis of proteins is described in the original paper.[11] Solutions (25 mM) of lactic acid, glucose, sucrose, inositol, mannitol, and ammonium chloride give no color with copper in this test. Noninterference by ammonia is an advantage of the method. The average deviation of analyses is $\pm 0.2 \%$.

IV. Copper Salt Micromethod for Amino Acids and Peptides (Ultraviolet)

Principle. The method is based on the observation that copper salts of amino acids have an absorption maximum at 230 mμ. The procedure described below is based on a method developed by Spies[13] for studying protein hydrolysis.

Reagents

0.05 M copper chloride. 8.52 g. of reagent-grade $CuCl_2 \cdot 2H_2O$ per liter of H_2O solution.

Sodium borate buffer, pH 9.1 to 9.2. 40.3 g. of reagent-grade, anhydrous $Na_2B_4O_7$ is dissolved in 4 l. of H_2O solution and filtered. For use, 6.0 g. of reagent-grade NaCl is dissolved in 100 ml. of borate solution. This NaCl-containing solution will be referred to as buffer solution.

Alanine. Analytically pure alanine.

Procedure. to 5.0 ml. of buffer solution in a 12-ml. glass-stoppered centrifuge tube[14] is added 5.0 ml. of test solution. One-tenth milliliter of $CuCl_2$ solution is added, and the tube is stoppered, shaken, and let stand for 10 minutes at a convenient room temperature. For maximum accuracy the same temperature $\pm 1°$ is maintained for all determinations. The suspension is centrifuged in the stoppered tube. The clear solution is poured into a quartz cuvette, and transmittancy is determined at 230 mμ. The ultraviolet-absorbing linkage of the copper salts is relatively unstable

[13] J. R. Spies, *J. Biol. Chem.* **195**, 65 (1952).
[14] Obtainable from Kimble Glass Co., Toledo, Ohio.

in the presence of excess copper salt; therefore centrifugation at exactly 10 minutes after addition of the copper chloride is recommended. After centrifugation, the loss of absorbing capacity of the solutions is less than 2% in 60 minutes.

TABLE IV

ALANINE EQUIVALENCE OF COPPER SALTS OF 0.0003 M SOLUTIONS OF AMINO ACIDS MEASURED AT 230 mμ

Amino acid	Wavelengths of maximum absorbency,[a] mμ	Alanine equivalence, %	
		At 230 mμ[b]	At wavelength of maximum absorbency
Alanine		100.0	
Arginine	229–230	103.3	
Aspartic acid	229–232	96.9	
Cystine[c]	232–236	80.9	83.7
Glutamic acid	229–233	103.8	
Glycine	226–227	100.2	102.7
Histidine	225–227[d]	83.5	87.1
Hydroxyproline	237–239	91.1	98.1
Isoleucine	230–232	102.7	
Leucine	228–233[d]	103.9	
Lysine	230–232	98.9	
Methionine	229–230	102.6	
Phenylalanine	230–235	91.1	
Proline	239–240	83.5	90.4
Threonine	230–233	100.7	
Tryptophan	[e]	144.1	
Tyrosine	232–233	94.7	97.7
Valine	230–231	101.5	

[a] Range of wavelength over which transmittancy varied by not more than ±0.1% at readings near 27% transmittancy.

[b] Average deviation of all results, ±0.8%.

[c] A 0.00015 M solution was used.

[d] Maximum absorbency may extend to lower wavelengths than given.

[e] Reading at 230 mμ was lowest obtainable with full slit opening.

Blank solutions are made by adding 5.0 ml. of the test solution to 5.0 ml. of buffer solution. A control analysis of a freshly prepared solution of alanine can be made with each day's tests to check for slight variations in absorbency due to distilled water or other causes because of the sensitivity of the method. This control analysis is not necessary if maximum accuracy is not required or if experience shows it to be unnecessary.

The standard curve is prepared with alanine solutions in concentrations from 50 to 500 μM. Conformity to Beer's law over the entire range of concentration is obtained. Zero concentration of alanine gives a transmittancy value of approximately 75% due to absorption of copper salt in solution.

Discussion. The molar alanine equivalence of eleven amino acids is within 5%, three within 9%, and three within 20%, of the 100% value of alanine, as shown in Table IV. Histidine gives an alanine equivalence of 83.5, and tryptophan gives an alanine equivalence of 144%. An

TABLE V

ALANINE EQUIVALENCE OF COPPER SALTS OF DIPEPTIDES AND TRIPEPTIDES
MEASURED AT 230 mμ

| | | | Alanine equivalence of test solutions,[a] % | |
Substance	Wavelength of maximum absorbency, mμ	Molarity of test solutions × 10⁴	Found[b]	Estimated if amino acids were free
Alanine		3.0	100.0	100
Dialanylcystine	217–220	0.75	77.2	100
Alanylglycine	<213	1.5	79.5	100
Glycylalanine	<213	1.5	75.2	100
Glycylaminoisobutyric acid	<213	1.5	73.5	100
Glycylamino-*n*-butyric acid	<214	1.5	75.4	100
Glycyldehydrovaline	223–224	1.5	47.1	
Glycylsarcosine	216–220	1.5	18.7	
Glycylserine	<214	1.5	69.8	100
Sarcosylphenylalanine	222	1.5	68.7	100
Glycylalanylglycine	<213	1.5	84.6	150
Glycylglycylalanine	<212	1.5	85.8	150
Glycylglycylleucine	<214	1.5	84.8	150
Glycylleucylglycine	<213	1.5	89.0	150
Leucylglycylglycine	<213	1.5	87.2	150

[a] Based on an absorbency of 0.0003 *M* alanine as 100%.
[b] Average of duplicate determinations.

advantage of this method is that the copper salts of cystine, phenylalanine, methionine, leucine, and tryptophan are all soluble enough for analysis without formation of the double salt with glycine, as was necessary with the 620-mμ method of analysis.[11] Those amino acids whose absorption differs from that of alanine can be analyzed by applying an appropriate correction or by using standard curves prepared for the amino acid being analyzed.

The absorbing capacity of the copper salts of free amino acids at 230 mμ is more than that of the copper salts of the corresponding peptide-linked amino acids, as shown in Table V. This increase in absorption provides the basis for adopting this method to study rate and degree of protein hydrolysis. Application of the method to the study of protein hydrolysis is described in the original paper, and use of the method in following proteolysis of cottonseed allergen has been described.[15]

Test solutions of 500 μM (twice the concentration of amino acid usually employed) in glucose, galactose, sucrose, mannitol, inositol, and ammonium chloride did not interfere with the test. Noninterference by ammonia is an advantage of the method. The necessity of avoiding traces of copper or amino acids in reagents, distilled water, or on glassware is obvious from consideration of the sensitivity of the method. Particular care must be taken to exclude contaminating copper from blank solutions.

The precision of the method determined from duplicate analyses of eight 300 μM alanine solutions is: standard deviation of the mean $\pm 1.1\%$, and standard deviation of a single pair of duplicates $\pm 3.2\%$.

[15] J. R. Spies, D. C. Chambers, E. J. Coulson, H. S. Bernton, and H. Stevens, *J. Allergy* **24**, 483 (1953).

[77] Microbiological Determination of Amino Acids

By Esmond E. Snell

Introduction

The use of microorganisms for quantitative estimation of amino acids was a natural outgrowth of their earlier successful use in the determination of vitamins. The general principles on which such methods are based, an account of their development, and details of many individual assay methods, have been treated in several review articles.[1-5]

The methods depend on employment of microorganisms for which individual amino acids are essential nutrients. In a basal medium lacking a single essential amino acid but complete in other respects, the growth of such an organism is, within limits, a function of the amount of the amino acid added. The concentration of amino acid in a given sample can thus be determined by comparing the growth response (measured in

[1] E. E. Snell, *Physiol. Revs.* **28**, 255 (1948); *Advances in Protein Chem.* **2**, 85 (1945); *Ann. N. Y. Acad. Sci.* **47**, 161 (1946).

[2] E. E. Snell, *in* "Vitamin Methods" (P. György, ed.), Vol. 1, p. 327, Academic Press, New York, 1950.

[3] B. S. Schweigert and E. E. Snell, *Nutrition Abstr. & Revs.* **16**, 497 (1947).

[4] E. C. Barton-Wright, "The Microbiological Assay of the Vitamin B-Complex and Amino Acids," Pitman, London, 1952.

[5] M. S. Dunn, *Physiol. Revs.* **29**, 219 (1949); *Food Technol.* **1**, 269 (1947).

any convenient fashion) to aliquots of the sample with that permitted by known amounts of the pure amino acid. A great many microorganisms that require one or more amino acids are known, and, in theory, assay methods based on any one of these could be developed. In practice, the lactic acid bacteria show many advantages and have been utilized widely with great success. Although different authors have employed different organisms of this group and slightly different basal media for determination of various amino acids (and even of a single amino acid), all the methods are fundamentally similar. Once the technique is learned for a single amino acid with a single assay organism, its application to any amino acid in any of several variations with respect to test organism or basal medium offers no difficulty.

In proposing variations in test media and test organisms for an amino acid, individual authors have not, in general, established by comparative tests the superiority of one organism or test medium over another. Rather, they have been content to develop methods that give apparently accurate and reproducible results and are conveniently operable in their own laboratory. It is probable that each of the methods of recent origin are of comparable accuracy, and there is little basis for choice among them. Some of the methods, however, can be applied to any of several amino acids without change in methodology save for the necessary change in amino acid composition of the basal medium. Some of them, too, have been used successfully in several different laboratories. These are the methods selected for presentation here.

Preparation of Samples for Assay

Many investigations of the utility of peptides as sources of amino acids for microorganisms have been made.[5,6-13] Depending on the amino acid being determined, the assay medium, and the test organism, individual peptides may exhibit activity lower than, equal to, or higher than that expected from their amino acid content. In general, organisms that utilize a given peptide for growth are able to hydrolyze it;[6,9-13] the presence or absence of peptidases for such peptides suffices to explain activities equal to or less than that expected from the composition of the

[6] H. Kihara and E. E. Snell, J. Biol. Chem. 197, 791 (1952).
[7] H. Kihara, O. Klatt, and E. E. Snell, J. Biol. Chem. 197, 801 (1952).
[8] A. I Virtanen and V. Nurmikko, Acta Chem. Scand. 5, 97, 681 (1951).
[9] J. M. Prescott V. J. Peters, and E. E. Snell, J. Biol. Chem. 202, 533 (1953).
[10] M. S. Dunn and L. E McClure, J. Biol. Chem. 184, 223 (1949).
[11] M. Klungsoyr, R. J. Sirny, and C. A. Elvehjem, J. Biol. Chem. 189, 557 (1951).
[12] V. J. Peters and E. E. Snell, J Bacteriol. 67, 69 (1954).
[13] D. Stone and H. D. Hoberman, J. Biol. Chem. 202, 203 (1953).

peptide. Tentative explanations for instances in which peptides are more active than their component amino acids also have been advanced; here, too, the organisms are generally able to hydrolyze the peptides in question.[6,7,9,13]

These results make it clear that as a general rule proteins must be hydrolyzed completely to their component amino acids before application of microbiological methods can be expected to yield valid results. In some instances, enzymatic digests or partial acid hydrolyzates have been assayed with results that check those obtained after complete acid hydrolysis.[11] Only where such a check has been obtained on the specific proteinaceous material being investigated should partial hydrolyzates be employed for assay. The variable activity of different peptides for individual assay organisms also emphasizes that the increase in amino acid content observed in a sample after acid hydrolysis is not a reliable measure of the "bound" amino acid content of the sample.

Procedure. The hydrolytic procedure necessary to release amino acids completely (for purposes of microbiological assay) from a protein-containing sample can be readily determined by sampling the hydrolyzate at intervals, assaying the several samples, and selecting a time interval at which further hydrolytic treatment fails to alter the values obtained. Studies of this nature (summarized elsewhere[3,11]) show that the time required for this purpose varies somewhat with the sample and with the test organism. For routine application, it is best to adopt a procedure known to effect release of the amino acids from the most resistant samples to an extent sufficient for complete utilization by the most demanding assay organisms.

A convenient procedure[14,15] for acid hydrolysis consists in placing the finely divided, weighed sample together with a 40-fold excess of 3 N hydrochloric acid in a test tube. The tube is sealed, then heated at 15 lb. of steam pressure (121°) in the autoclave for 5 to 10 hours. The tube is opened, and the sample neutralized before assay. Interfering materials present in the humin of such hydrolyzates of some foodstuffs can be removed by adjustment to pH 4.0 and filtration through a fritted glass filter before final adjustment to neutrality.[16]

Since tryptophan is destroyed by acid hydrolysis, it must be liberated by alkaline or enzymatic hydrolysis. Autoclaving (121°) of 0.5 g. of sample with 10 ml. of 5 N NaOH in a sealed Pyrex tube for 5 to 15 hours

[14] L. M. Henderson and E. E. Snell, *J. Biol. Chem.* **172**, 15 (1948).
[15] J. L. Stokes, M. Gunness, I. M. Dwyer, and M. C. Caswell, *J. Biol. Chem.* **160**, 35 (1945).
[16] M. J. Horn, A. E. Blum, C. E. F. Gersdorff, and H. W. Warren, *J. Biol. Chem.* **203**, 907 (1953).

has been recommended[14,15,17] for this purpose; hydrolysis with $Ba(OH)_2$ should not be used.[17] This procedure appears to racemize the peptide-bound L-tryptophan of proteins completely (it does not racemize free L-tryptophan), and, since D-tryptophan is inactive for assay organisms so far utilized, one must multiply the assay result by 2 if a standard of L-tryptophan is employed.

In carbohydrate-containing foods, acid hydrolysis also destroys much tyrosine;[18] hydrolysis by autoclaving the sample with 5 N NaOH for 5 hours in the autoclave at 15 lb. of pressure (121°) is recommended.[18] This effects complete racemization of the tyrosine.

As mentioned earlier, generally applicable enzymatic digestion procedures have not been developed. In a recommended procedure for tryptophan analysis, a suspension of the finely divided sample (200 mg.) is layered with toluene and shaken with a mixture of pancreatin (20 mg.) and hog intestinal mucosa (4 mg.) for 1 to 5 days.[14,19] Tryptophan analyses of such digests agreed well with those obtained by alkaline hydrolysis.

Amino Acid Standards. The development of highly sensitive microbiological methods, and later of paper chromatographic methods, for analysis of amino acids revealed significant and previously unsuspected contamination of certain amino acids (both natural and synthetic) with other amino acids. Awareness of this situation has led to marked improvement in the quality of products now available from chemical supply houses. Despite this improvement, the analyst should assure himself of the chemical and optical purity of the amino acids used in establishing standard dose-response curves. Criteria of purity and methods for recrystallization of the amino acids have been summarized by Dunn and Rockland.[20] The optical purity of most of the amino acids can be checked conveniently by means of D- and L-amino acid oxidases as described by Greenstein et al.[21] Special care should be taken with isoleucine, since the synthetic product may contain significant amounts of alloisoleucine, and the product isolated from proteins by older procedures is also frequently impure. Convenient procedures for purification of this amino acid are now available.[22] Many of the older microbiological values for the iso-

[17] W. A. Krehl, J. de la Huerga, and C. A. Elvehjem, *J. Biol. Chem.* **164**, 551 (1946).

[18] M. Gunness, I. M. Dwyer, and J. L. Stokes, *J. Biol. Chem.* **163**, 159 (1946).

[19] I. T. Greenhut, B. S. Schweigert, and C. A. Elvehjem, *J. Biol. Chem.* **165**, 325 (1946).

[20] M. S. Dunn and L. B. Rockland, *Advances in Protein Chem.* **3**, 295 (1947).

[21] J. P. Greenstein, S. M. Birnbaum, and M. C. Otey, *J. Biol. Chem.* **204**, 307 (1953); A. Meister, L. Levintow, R. B. Kingsley, and J. P. Greenstein, *J. Biol. Chem.* **192**, 535 (1951).

[22] J. P. Greenstein, S. M. Birnbaum, and L. Levintow, *in Biochem. Preparations* **3**, 84 (1953).

leucine content of proteins are high because of an impure standard. In general, only the L-amino acids should be taken as standards unless it has been established that the D-isomers have no activity under the test conditions.

Methods

Excellent procedures studied in connection with one or more individual amino acids are available in the literature but cannot be considered here. A particularly valuable compilation of individual procedures that permit determination of eighteen amino acids is that of Barton-Wright;[4] Dunn and co-workers[23] have summarized the several individual procedures adopted in their laboratory for twelve amino acids. The procedures to be presented below are the general ones of Henderson and Snell,[14] Baumann and co-workers,[24] and Stokes et al.[15,18] Since the methodology in all cases is similar, detailed description of one method will suffice to illustrate the procedure and certain general principles involved. Summaries of the various methods then will be presented; where clarification is desired, details will become clear from examination of the one procedure presented in full.

Assay Organisms

The assay organisms utilized with methods to be presented later are listed in Table I, together with the amino acids for assay of which they are most commonly employed. It will be observed that several different organisms can be employed for determination of certain of the amino acids. Agreement of assay results obtained with different test organisms is an excellent indication of the reliability of the values obtained for a given amino acid.[1-3]

Stock cultures of these organisms are conveniently carried as stab cultures in any of several media.[2] One of the simplest of these contains 1% yeast extract, 1.0% glucose, and 1.5% agar.[2,24] In recent years there has been a tendency to employ more complex media for the stock cultures and to increase the frequency of transfer from monthly to biweekly or even weekly. Barton-Wright[4] recommends the medium of Nymon and Gortner,[25] which contains 1 g. of glucose, 1 g. of tryptone (Difco), 0.2 g. of K_2HPO_4, 0.3 g. of $CaCO_3$, 0.5 ml. each of inorganic salts A and B,[26]

[23] M. S. Dunn, M. N. Camien, R. B. Malin, E. A. Murphy, and P. J. Reiner, *Univ. Calif. (Berkeley) Publs. Physiol.* **8,** 293 (1949).

[24] B. F. Steele, H. E. Sauberlich, M. S. Reynolds, and C. A. Baumann, *J. Biol. Chem.* **177,** 533 (1949).

[25] M. C. Nymon and W. A. Gortner, *J. Biol. Chem.* **163,** 277 (1946).

[26] Salts A contain 25 g. of KH_2PO_4 and 25 g. of K_2HPO_4 in 250 ml. of water; salts B contain 10 g. of $MgSO_4 \cdot 7H_2O$, 0.5 g. of $FeSO_4 \cdot 7H_2O$, 0.5 g. of NaCl, and 0.5 g. of $MnSO_4 \cdot 4H_2O$ in 250 ml. of water.

10 ml. of liver extract,[27] and 1.5 g. of agar per 100 ml. Henderson and Snell[14] employ their basal assay medium (Table II) modified by using glucose and sodium citrate at one-half the specified level, and adding per 100 ml. of single-strength medium 0.5 g. of tryptone, 0.5 g. of yeast extract, and 2 g. of agar. However, limited studies of Dunn and co-workers[28] indicate that the medium chosen for carrying stock cultures

TABLE I

COMMONLY USED CULTURES FOR DETERMINATION OF AMINO ACIDS

Name and strain	ATCC[a]	For determination of
Lactobacillus arabinosus 17-5[b]	8014	Glutamic acid, leucine, isoleucine, valine, phenylalanine, tryptophan
Lactobacillus delbrückii 5[c]	9595	Phenylalanine, tyrosine, serine
Leuconostoc citrovorum[d]	8081	Alanine; could also be used for most other amino acids
Leuconostoc mesenteroides P-60[e]	8042	All amino acids *except* alanine
Streptococcus faecalis R[f]	8043	Glutamic acid, histidine, lysine, arginine, leucine, isoleucine, valine, methionine, threonine, tryptophan

[a] American Type Culture Collection, 2029 M Street, N. W., Washington 6, D. C.
[b] A strain of *Lactobacillus plantarum*.
[c] Possibly identical with *Lactobacillus casei* or *Lactobacillus helveticus* [M. S. Dunn, S. Shankman, M. N. Camien, and H. Block, *J. Biol. Chem.* **168**, 1 (1947)].
[d] Recent studies indicate that this organism belongs to the genus *Pediococcus* rather than *Leuconostoc* [E. M. Jensen and H. W. Seeley, *J. Bacteriol.* **67**, 484 (1954)].
[e] Studies of C. S. McCleskey [*J. Bacteriol.* **64**, 140 (1952)] indicate that this culture is misclassified and is probably a strain of *Streptococcus equinus*.
[f] Designated in some early literature as *Streptococcus lactis* R or RglA.

is not a critical factor in success of these methods. The procedure had little effect on the subsequent quantitative response of *Lactobacillus arabinosus* to several essential amino acids.

Stab cultures prepared in the medium of choice are incubated at 37° until good growth appears in the line of the stab (usually 24 hours; and no longer than 48 hours), then refrigerated at 4 to 12° until the next transfer. Fresh transfers are prepared at intervals of 1 to 4 weeks. *Streptococcus faecalis* grows satisfactorily either in stab cultures or on agar

[27] Approximately 450 g. of fresh liver is ground and suspended in 2 l. of water. The suspension is heated for 1 hour on the steam bath, filtered through cheese cloth, the filtrate adjusted to pH 7.0, again heated for 15 minutes, filtered through paper, and stored in a brown bottle under toluene in the cold.
[28] M. S. Dunn, S. Shankman, M. N. Camien, and H. Block, *J. Biol. Chem.* **168**, 1 (1947).

TABLE II

COMPOSITION OF ASSAY MEDIA FOR AMINO ACIDS

(Milligrams per 100 ml. of double-strength medium)

Constituents	Medium		
	(1) Henderson and Snell[a]	(2) Steele, Sauberlich, Reynolds, and Baumann[b]	(3) Stokes, Gunness, Dwyer, and Caswell[c]
Glucose	4000	5000	2000
Sodium citrate	4000	—	—
Sodium acetate	200	4000	1200
Inorganic salts			
NH$_4$Cl	600	600	—
K$_2$HPO$_4$	1000	120	100
KH$_2$PO$_4$	—	120	100
MgSO$_4$·7H$_2$O	160	40	40
MnSO$_4$·4H$_2$O	32	4	2
FeSO$_4$·7H$_2$O	8	2	2
NaCl	8	2	2
Purine and pyrimidine bases			
Adenine sulfate	2	2	2
Guanine hydrochloride	2	2	2
Xanthine	2	2	—
Uracil	2	2	2
Vitamins			
p-Aminobenzoic acid	0.04	0.02	0.008
Biotin	0.002	0.0002	0.00004
Calcium pantothenate	0.2	0.1	0.04
Folic acid	0.002	0.002	0.0004
Folinic acid	—[d]	—[e]	—
Nicotinic acid	0.2	0.2	0.04
Pyridoxal·HCl	0.04	0.06	—
Pyridoxamine·2 HCl	—	0.06	0.08
Pyridoxine·HCl	—	0.2	—
Riboflavin	0.2	0.1	0.04
Thiamine	0.2	0.1	0.04
Amino acids[f,g]			
DL-Alanine	200	40	40
L-Arginine·HCl	40	48.5	40
L-Asparagine	—	80	—
L-Aspartic acid	100	20	20
L-Cysteine·HCl	—	10	—
L-Cystine	10	—	40
L-Glutamic acid	200	60	20
Glycine	10	20	40
L-Histidine·HCl	10	12.4	40

TABLE II (*Continued*)

Constituents	(1) Henderson and Snell[a]	(2) Steele, Sauberlich, Reynolds, and Baumann[b]	(3) Stokes, Gunness, Dwyer, and Caswell[c]
Amino acids (*Continued*)			
DL-Isoleucine	20	50	40
DL-Leucine	20	50	40
L-Lysine·HCl	40	50	20
DL-Methionine	20	20	40
DL-Phenylalanine	20	20	40
L-Proline	10	20	40
DL-Serine	20	10	40
DL-Threonine	20	40	40
DL-Tryptophan	20	8	80
L-Tyrosine	10	20	40
DL-Valine	20	50	40
DL-Norleucine	—	—	40

[a] *J. Biol. Chem.* **172,** 15 (1948).

[b] *J. Biol. Chem.* **177,** 533 (1949).

[c] *J. Biol. Chem.* **160,** 35 (1945).

[d] Must be added (ca. 10 γ per 100 ml.) if the medium is used with *Leuconostoc citrovorum.*

[e] When used with *L. citrovorum,* 0.08 ml. of injectable liver concentrate (reticulogen, 20 U.S.P.) is added to supply this subsequently identified growth factor. This can now be replaced by the synthetic growth factor at a convenient level (e.g., 10 γ per 100 ml.).

[f] The amino acid to be determined is omitted from the basal medium.

[g] L- or DL-Amino acids may be used interchangeably; L-amino acids are employed at one-half the concentration specified for DL-amino acids, DL-amino acids at twice that specified for L-amino acids.

slants; according to Barton-Wright,[4] better results are obtained in some assays if the organism is carried in the latter fashion.

Method of Henderson and Snell[14]

Basal Medium. The composition of the basal medium is given in Table II. Glucose, sodium citrate, and sodium acetate are added in solid form; inorganic salts, purine and pyrimidine bases, vitamins, and amino acids are conveniently combined from previously prepared stock solutions. Where many routine assays for a single amino acid are being made, it is convenient to prepare a mixture of the dry amino acids lacking the one to be determined.

This medium was developed as a single assay medium suitable for culture of any of the common assay organisms listed in Table I except *Leuconostoc citrovorum*. It becomes suitable for the latter organism on addition of folinic acid (leucovorin, 10 γ per 100 ml.), which this organism requires and which was discovered subsequent to the development of this medium.[29] The medium is suitable for assay of any of the amino acids required by such organisms. For this purpose, the single amino acid to be determined is omitted from the medium. The citrate of this medium serves only as a buffer; the above-cited members of the genus *Lactobacillus* (but not *S. faecalis*) grow equally heavily and somewhat more rapidly if the sodium citrate is omitted and the sodium acetate is increased to 4 g. per 100 ml. When this change is made, the concentration of manganese sulfate should be reduced to 8 mg. per 100 ml. Precipitation of inorganic salts is minimized by adding them after the other ingredients have been dissolved and the solution made almost to volume. For assay of most amino acids, the pH is adjusted to approximately 6.8; for assay of glutamic acid and proline, the initial pH of the medium is 6.0.[14]

Inoculum Cultures. For growing inocula, basal medium 1 (Table II) is modified by lowering the glucose and sodium citrate to one-half the level specified and adding 0.5 g. of tryptone and 0.5 g. of yeast extract per 50 ml. of the double-strength medium. The volume is adjusted to 100 ml., and separate aliquots of 5 ml. are transferred to individual tubes. These are plugged with cotton, sterilized in the autoclave at 15 lb. of pressure for 20 minutes, and held in the cold room until required. A transfer from the appropriate stock culture is made to this inoculum medium, and the subculture is incubated at 37° for 10 to 15 hours. The cells are centrifuged and resuspended in a volume of sterile 0.9% sodium chloride solution five to twenty-five times as great as that of the medium in which they were grown.[30] One drop of the resulting suspension is used to inoculate each assay tube containing 2 ml. of medium. If volumes other than 2 ml. are employed, the amount of inoculum suspension is altered correspondingly.

Procedure. Assays may be carried out in any convenient volume from 0.2 ml.[31] to 10 ml.; the *concentration* (amount per unit volume) of sample and standard required to permit growth remains constant whatever the volume. The range over which growth is a function of concentration is illustrated for L-leucine in Fig. 1 and summarized for the various amino

[29] H. E. Sauberlich and C. A. Baumann, *J. Biol. Chem.* **176,** 165 (1948).

[30] Suspensions of *Streptococcus faecalis* were diluted approximately 25-fold, those of the remaining organisms from 5- to 10-fold, depending on the density of growth. The amount of inoculum is not a critical factor in determining success of the assay.

[31] L. M. Henderson, W. L. Brickson, and E. E. Snell, *J. Biol. Chem.* **172,** 31 (1948). Special small-size test tubes are used for assays conducted on the micro scale.

acids in Table III. With few exceptions it varies only slightly from one organism to another or from one assay medium to another.

A sufficiently large total volume for convenient manipulation but one that is saving of materials is 2 ml., and the following description pertains to assays conducted on this scale. Samples and standard are dispensed to 18 × 150-mm. Pyrex culture tubes in volumes of 0.2 to 1.0 ml. The standard amino acid should be added at several different concentrations within the range shown in Table III (as illustrated for leucine in Fig. 1);

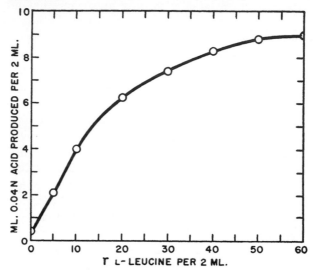

FIG. 1. A typical standard curve showing the response of *Lactobacillus arabinosus* to leucine.[14]

the sample also is added at several different concentrations estimated to supply the amino acid being determined in similar amounts. The volume in each tube is brought to 1 ml. by addition of water, and 1 ml. of the double-strength basal medium (from which the amino acid to be determined has been omitted) is added. Each tube is plugged with cotton or covered by aluminum caps or by a special cover made to fit the rack for the tubes and lined with a heavy padding of cotton, then sterilized by autoclaving at 10 to 12 lb. of pressure for 10 minutes.[32] The tubes are cooled and then inoculated with 1 drop of the inoculum suspension prepared as described in the preceding section. They are then incubated for 60 to 72 hours at 37°, after which acid production in each tube is measured

[32] Excessive autoclaving produces extensive browning of the medium (especially where the initial pH is 6.8 or higher) and should be avoided. Even shorter sterilization times can be employed successfully if care is taken to avoid contamination and the assays are conducted in a dust-free environment.

by titration with 0.04 N alkali.[33] Since the medium is heavily buffered, use of indicators gives a less satisfactory end point than does electrometric titration[14] to approximately pH 7.3. An inexpensive assembly for the latter purpose has been described.[14,31]

A standard curve is constructed by plotting acid production against amount of the standard amino acid added. Such a standard curve, showing the response of *L. arabinosus* to increasing amounts of L-leucine, is

TABLE III

CONCENTRATIONS OF VARIOUS AMINO ACIDS SUITABLE FOR ESTABLISHING A
STANDARD CURVE

Amino acid[a]	Assay range,[b] γ/ml.	Amino acid[a]	Assay range,[b] γ/ml.
Alanine	0–25	Lysine	0–20
Arginine	0–20	Methionine	0 5
Aspartic acid	0–20	Phenylalanine	0–5
Cystine	0–5	Proline	0–10
Glutamic acid	0–25	Serine	0–10
Glycine	0–15	Threonine	0–10
Histidine	0–5	Tryptophan	0–2
Isoleucine	0–15	Tyrosine	0–10
Leucine	0–15	Valine	0–15

[a] L-Isomer in all cases. In addition to setting up several points on the standard curve within this range, it is well to have one higher concentration (e.g., two to four times as great as the highest listed here) to establish a point of maximum growth.

[b] The values are taken from many papers on assay of individual amino acids and from the following summary papers: E. E. Snell, *Ann. N. Y. Acad. Sci.* **47**, 161 (1946); M. S. Dunn, S. Shankman, M. N. Camien, W. Frankl, and L. B. Rockland, *J. Biol. Chem.* **156**, 703 (1944); B. F. Steele, H. E. Sauberlich, M. S. Reynolds, and C. A. Baumann, *J. Biol. Chem.* **177**, 533 (1949).

shown in Fig. 1. Acid production in response to known amounts of sample is interpolated onto this curve to determine directly the amount of amino acid present in the sample. The amino acid content per unit weight of sample as calculated from several increasing concentrations should agree closely (e.g., within ±5%), and the values are averaged for the final result.

Assay Organisms for Individual Amino Acids and Specificity toward Optical Isomers. In the original study of this method,[14] three organisms

[33] For special purposes, turbidimetric estimation of growth after 18 to 36 hours of incubation is useful and gives results that agree closely with those obtained acidimetrically. Two-milliliter cultures may be diluted to 8 ml. with water prior to such estimation if desired.

were found most useful. *Lactobacillus arabinosus* was employed for determination of glutamic acid, leucine, phenylalanine, valine, and tryptophan; *Streptococcus faecalis* for arginine, histidine, methionine, and threonine; and *Leuconostoc mesenteroides* P-60 for aspartic acid, glycine,[34] histidine, lysine, proline, tyrosine, and isoleucine. In some instances the choice is arbitrary; for example, both *S. faecalis* and *Leuco. mesenteroides* give excellent results in tryptophan determination and, unlike *L. arabinosus*, are unable to utilize either indole or anthranilic acid in place of the intact amino acid.[35] Use of the medium for determination of alanine, serine, and cystine was not investigated; *Leuconostoc citrovorum* 8081 requires each of these amino acids[24,36,37] and might be applied to their determination in the basal medium supplemented with folinic acid, as noted earlier. *Leuconostoc mesenteroides* has been applied successfully to the determination of cystine in a somewhat different medium;[38] it appears probable that the same technique would permit utilization of this organism and the present medium for determination of this amino acid.

Certain α-keto acids and a smaller number of α-hydroxy acids can replace the corresponding α-amino acids in these test organisms.[39] In most instances, these related compounds are much less active than the amino acid itself, and their stability is such that they would not commonly occur in samples for assay. To the extent that they do occur, they must be considered as possible interfering materials. A check of the apparent amino acid content before and after extraction of the acid-hydrolyzed sample with ethyl acetate would in most cases reveal their presence.

For all the test organisms listed above, the D-isomers of the essential amino acids are essentially inactive; that is, the DL-isomers of the amino acids are 50% as effective on the weight basis as are the L-isomers.[3] Exceptions to this general rule are D-alanine, which has 100% of the activity of L-alanine for *Leuco. citrovorum* 8081;[24] D-aspartic acid, which has 100% of the activity of L-aspartic acid for *Leuco. mesenteroides* under the assay conditions recommended here;[24,40] D-methionine, which has low activity (<10% that of L-methionine) for *Leuco. mesenteroides* P-60[24] and *S. faecalis* R[14] but none for *Leuco. citrovorum*;[24] D-tyrosine, which has low

[34] Samples of DL-alanine available in 1947 were contaminated with glycine. It was necessary, therefore, to recrystallize this amino acid before satisfactory determinations of glycine could be obtained.

[35] E. E. Snell, *Arch. Biochem.* **2,** 389 (1943).

[36] H. E. Sauberlich and C. A. Baumann, *J. Biol. Chem.* **177,** 545 (1949).

[37] E. M. Jensen and H. W. Seeley, *J. Bacteriol.* **67,** 484 (1954).

[38] M. N. Camien and M. S. Dunn, *J. Biol. Chem.* **183,** 561 (1950).

[39] J. T. Holden, R. B. Wildman, and E. E. Snell, *J. Biol. Chem.* **191,** 559 (1951).

[40] M. N. Camien and M. S. Dunn, *Proc. Soc. Exptl. Biol. Med.* **75,** 74 (1950).

activity for both *Leuco. mesenteroides* and *Leuco. citrovorum*[24] but is inactive for *L. delbrückii* 5;[18] and D-glutamic acid, which has variable low activity (usually less than 5% that of L-glutamic acid) for *L. arabinosus* in the presence of L-glutamic acid.[3] In such instances, only the L-amino acids should be employed as standards.

Atypical Standard Curves and Their Significance. In the above and similar assay media an atypical sigmoidal standard curve is obtained with *L. arabinosus* in response to glutamic acid.[1,41-43] Frequently, such standard curves can be used without difficulty, since the organism responds to standard and sample alike (cf. Lewis and Olcott[41]). Where this is not true, it can be readily recognized by a "drift" in assay values calculated from different concentrations of the sample. In general, it is more satisfactory to modify the procedure in such instances to obtain a regular standard curve. With glutamic acid, for example, the tendency toward sigmoidal curves is overcome successfully in any of several ways: (1) addition of small amounts of glutamine insufficient to replace glutamic acid completely but sufficient to initiate growth,[43] or addition of small amounts of an enzymatic casein digest sufficient for the same purpose;[42] (2) use of an inoculum approximately five times as large as for other amino acids and initial adjustment of the basal medium to pII 6.0;[14,42] or (3) use of asparagine (at a concentration of 20 mg. per 100 ml. of double-strength medium) in place of aspartic acid in the medium.[44] Of these alternatives, the second is simplest and suffices to permit excellent assays for glutamic acid to be obtained. The reasons for this behavior are, however, instructive and warrant further examination. The lag in the response of *L. arabinosus* to low concentrations of glutamic acid has been shown[45] to be due to the inhibition of utilization of low levels of glutamic acid by the high level of aspartic acid (which functions as a structurally related antimetabolite of glutamic acid) in the medium. Asparagine is a much less potent inhibitor,[45] and thus its substitution for aspartic acid at a somewhat decreased concentration prevents the lag in the dose-response curve, as noted above. The process being inhibited is by-passed by supplying glutamine or appropriate peptides of glutamic acid; hence no lag in the response to these substances is observed.[42,43]

Similar antagonistic relationships between other amino acids appear to be the most common explanation for atypically shaped (sigmoidal)

[41] J. C. Lewis and H. S. Olcott, *J. Biol. Chem.* **157**, 265 (1945).
[42] L. R. Hac, E. E. Snell, and R. J. Williams, *J. Biol. Chem.* **159**, 273 (1945).
[43] C. A. Lyman, K. A. Kuiken, L. Blotter, and F. Hale, *J. Biol. Chem.* **157**, 395 (1945).
[44] W. Baumgarten, A. N. Mather, and W. Stone, *Cereal Chem.* **22**, 514 (1945).
[45] W. L. Brickson, L. M. Henderson, I. Solhjell, and C. A. Elvehjem, *J. Biol. Chem.* **176**, 517 (1948).

standard curves, which have been observed with various test organisms in several different assay media. These inhibitory relationships vary from one test organism to another, and, where such curves are observed, choice of an alternative assay organism may alleviate the difficulty. For example, *L. arabinosus* is not suited to determination of isoleucine by the procedure outlined here because its response to this amino acid is readily inhibited by leucine and valine.[45] This is not true for *Leuco. mesenteroides* or for any of four other organisms tested (*S. faecalis* R, *L. delbrückii* 3, *L. casei*, or *Leuco. mesenteroides*); one of the latter group is therefore employed for the determination.[14,45]

Accuracy of the Microbiological Procedure. Recoveries of added amino acids and repeated assays of individual proteins indicate that the experienced operator should have no difficulty in obtaining values accurate to within ±5% of the true value. In some instances, reproducibility may be considerably better than this.[1-5,14,23]

Summarized Methods

Method of Henderson and Snell[14] (see detailed presentation above)

Assay organisms:

> *L. arabinosus* 17-5: For glutamic acid, leucine, phenylalanine, tryptophan, and valine.
>
> *Leuco. mesenteroides* P-60: For aspartic acid, glycine, histidine, isoleucine, lysine, proline, and tyrosine.
>
> *S. faecalis* R: For arginine, histidine, methionine, and threonine.

Stock cultures: Carried as stabs by biweekly transfer in medium 1 (Table II) with glucose and sodium citrate reduced to one-half specified concentration and supplemented with 0.5% tryptone, 0.5% yeast extract, and 2.0% agar. Incubated for 24 to 48 hours at 37°; held at 4 to 12° between transfers.

Basal medium: Medium 1, Table II, with appropriate amino acid omitted. Initial pH 6.8 (pH 6.0 for glutamic acid and proline assay).

Inoculum medium: Medium 1, Table II, supplemented per 100 ml. (single strength) with 0.5% tryptone and 0.5% yeast extract and with glucose and sodium citrate lowered to one-half concentration specified.

Concentration of amino acids for standard curves: Approximately five different points (in duplicate) within range specified in Table III.

Sterilization: 10 minutes at 10 to 12 lb. of pressure.

Inoculum: Incubated for 8 to 12 hours at 37°. Cells centrifuged and resuspended in 5 to 25 vol.[46] of 0.9% saline. One drop of the resulting suspension is used per 2-ml. assay tube.

[46] *S. faecalis* diluted 25-fold; *L. arabinosus* and *Leuco. mesenteriodes* 5- to 10-fold. The

Incubation time and temperature: 60 to 72 hours at 37°.

Measurement of response: Acidimetric. Turbidimetric determinations at 24 to 48 hours are useful for many purposes.

Method of Steele, Sauberlich, Reynolds, and Baumann[24]

Assay organisms:

Leuco. citrovorum 8081 (*Pediococcus* sp.): For alanine, arginine, cystine, glutamic acid, glycine, histidine, isoleucine, methionine phenylalanine, proline, threonine, tyrosine, and valine.

Leuco. mesenteroides P-60 (*Streptococcus equinus*): For all amino acids except alanine.

Stock cultures: Carried as stabs with transfers at least once a month in medium containing 1% yeast extract, 1% glucose, and 1.5% agar.

Basal medium: Medium 2, Table II, with the amino acid to be determined omitted. Initial pH 6.8 to 7.0.

Inoculum medium: Medium IV [47] supplemented with 0.2% yeast extract. This is essentially identical with the basal medium used here, but with all amino acids except cysteine and tryptophan replaced with acid hydrolyzed casein (1.0 g. per 100 ml. of double-strength medium).

Concentrations of amino acids for standard curves: Approximately five different points in duplicate within range specified in Table III.

Sterilization: Not specified, but apparently about 10 minutes at 15 lb. of pressure.

Inoculum: Incubated for 20 to 24 hours at 37°, centrifuged, and diluted with sterile 0.9% NaCl solution to a standard turbidity ($G = 65$ to 70 in Evelyn tube and colorimeter, filter 660). One drop of this diluted suspension used to inoculate each 2-ml. assay tube.

Incubation time and temperature: 72 hours at 37° for acidimetric assays, 20 hours for turbidimetric assays.

Measurement of response: Electrometric titration after 72 hours, or turbidimetric estimation after 20 hours.

Method of Stokes, Gunness, Dwyer, and Caswell[15,18]

Assay organisms:

Streptococcus faecalis R: For arginine, histidine, isoleucine, leucine, lysine, methionine, threonine, tryptophan, and valine.

Lactobacillus delbrückii 5: For phenylalanine and tyrosine.

Stock cultures: Carried as stab cultures by monthly transfer in a medium containing 1 g. of glucose, 0.5 g. of Bacto peptone, 0.6 g. of anhydrous

---

amount of inoculum is not critical. For glutamic acid and proline assay, the test organisms are resuspended in an equal volume of saline.

[47] H. E. Sauberlich and C. A. Baumann, *J. Biol. Chem.* **166**, 417 (1946).

sodium acetate, 0.5 ml. of salts A,[26] 0.5 ml. of salts B,[26] and 1.5 g. of agar per 100 ml. of medium (pH 6.8). Stored in the refrigerator between transfers.

Basal medium: Medium 3, Table II, with the amino acid to be determined omitted. Initial pH 6.8.

Inoculum medium: Same as that given above for stock cultures, but with agar omitted.

Concentrations of amino acids for standard curves: Several different concentrations of the standard in the range shown in Table III.

Sterilization: Autoclaved at 15 lb. of pressure for 13 minutes.

Inoculum: Incubated for 16 to 24 hours at 37° in 8 ml. of inoculum medium, centrifuged, washed with water, suspended in 100 ml. (*S. faecalis*), or 20 ml. (*L. delbrückii*) of water. One drop of this suspension used for each assay tube of 10 ml.

Incubation time and temperature: 40 hours at 37° for *S. faecalis*, 72 hours at 37° for *L. delbrückii*.

Measurement of responses: Titration of acid produced.

[78] Measurement of Amino Acids by Column Chromatography

By EDWARD L. DUGGAN

Amino acids may be fractionated into classes or individuals by chromatography on various adsorbents.[1-3] One method, analytical in design, is superior to others in simplicity of operation and versatility of application.

The method developed by Stein and Moore originally required starch as the adsorbent.[4] The sensitivity of this fractionation to inorganic salts and the low capacity of such columns resulted in the application of ion exchange resins to the fractionation. The method in use in various laboratories during the past four years is that described by Moore and Stein.[5] Here the sulfonated polystyrene, cross-linked with 8% divinyl benzene (Dowex 50 × 8) is used. No prior desalting is required, and samples of 2 to 10 mg. of amino acids may be used. A scheme for the isolation of the

[1] S. Moore and W. H. Stein, *Ann. Rev. Biochem.* **21**, 521 (1952).

[2] E. Lederer and M. Lederer, "Chromatography," Chapter 26, Elsevier Publishing Co., Houston, Texas, 1954.

[3] P. L. Kirk and E. L. Duggan, *Anal. Chem.* **26**, 165 (1954).

[4] W. H. Stein and S. Moore, *J. Biol. Chem.* **176**, 337 (1948).

[5] S. Moore and W. H. Stein, *J. Biol. Chem.* **192**, 663 (1951).

common amino acids has also been described.[6] Potentially volatile eluents are used.

More recently, the same authors have developed a refined separation which uses Dowex 50 × 4 400-mesh resin.[7] The separation of fifty compounds was accomplished, under conditions of fractionation and analysis which possesses simplicity and fewer steps than those of the older method. These workers indicate certain difficulties in regard to mesh size and cross-linking of the Dowex 50 × 4 resin,[8] although these properties of various lots of the resin should stabilize in the near future.

The fractionation of the amino acids is abstracted in detail for the latest procedure.[7] The last section is devoted to the latest modification of the photometric determination of amino acids, using ninhydrin.[9] This common method of determination for the amino acids is implicit in the analytical scheme.

Ion Exchange Fractionation[7]

Reagents and Materials

Dowex 50 × 4 resin, 400 mesh.[8,10]

Dowex 50 × 5 resin, 400 mesh (required in smaller amount than the first resin, for the adjustment of performance of various lots of the 4% cross-linked resin).

Buffers (see Table I[7]).

Detergent, BRlJ 35.[11]

Antioxidant, thiodiglycol.[12]

Ion exchange columns, 0.9-cm. diameter, 165-cm. length from fused sintered plate, enclosed in two West-type condenser jackets, each 75 cm. long.

[6] C. H. W. Hirs, S. Moore, and W. II. Stein, *J. Biol. Chem.* **195**, 669 (1952)

[7] S. Moore and W. H. Stein, *J. Biol. Chem.* **211**, 893 (1954).

[8] Communication from S. Moore and W. H. Stein, April 1955. This note indicates that manufacturers use dry-screening for size *before sulfonation*. Certain lots of 200 to 400 mesh would not pass through a wet 200-mesh screen. "Minus 400 mesh" is the correct grade to use, and advantages in reproducibility of column performance accrue from the wet-screening through a 325-mesh screen, using the fraction passing the screen for column preparation. The authors indicate sulfonated polystyrene resins are now produced by Rohm and Haas Company, Philadelphia, Pennsylvania. (Amberlite IR-120 (XE 69) ≅ Dowex 50 × 8.   Amberlite IR-112 (XE 100) ≅ Dowex 50 × 4.)

[9] E. J. Harfenist, *J. Am. Chem. Soc.* **75**, 5528 (1953).

[10] Dowex resins are provided in washed, wet-screened form by Bio-Rad Laboratories, Berkeley, California.

[11] Atlas Powder Company, Wilmington, Delaware.

[12] Redistilled Kromfax solvent, available from Carbide and Carbon Corporation, New York, New York.

TABLE I
Buffers for Ion Exchange Analysis of Amino Acids
(S. Moore and W. H. Stein, J. Biol. Chem. 211, 893 (1954).)

Buffer	Citric acid · H$_2$O, g.	Acetic acid (glacial), g.	NaOH (97%), g.	Na acetate · 3H$_2$O, g.	Concentrated HCl, ml.	Final volume l.
I pH 2.2 ± 0.03 Na citrate (0.2 N Na$^+$)	105		42		80	5
II pH 3.1 ± 0.03 Na citrate (0.2 N Na$^+$)	714		282		393	34
III pH 5.1 ± 0.02 Na citrate-acetate (2.0 N Na$^+$)	3570	730	1600	4630		34

Notes:

1. Avoid overheating of buffer III during neutralization of the citric acid. Add the required concentrated alkali slowly.
2. Buffers I and II require 6.8 ml. of BRIJ 35 detergent solution and 5 ml. of thiodiglycol per liter of buffer, added prior to use. The BRIJ 35 solution is prepared by dissolving 50 g. of the commercial product in 150 ml. of hot water. It is ready for use after cooling.
3. Buffer III requires 1.35 ml. of the BRIJ 35 solution and 5 ml. of thiodiglycol per liter. The lowered (0.1%) concentration of detergent maintains equal drop size for this concentrated buffer as for the other eluents.
4. The buffers are filtered and stored in the cold room. This presents a problem in many laboratories with limited cold-room area. The buffers for the previous procedure used by Moore and Stein [J. Biol. Chem. 192, 663 (1951)] were saturated with thymol and stored at 4°.

Fraction collector, drop-counting, Technicon or equivalent.[13]
Gradient mixing chamber, prepared from a 500-ml. ground joint
wash bottle (Corning 1660).
Magnetic stirrer and stirring bar.
Separatory funnel, 500 ml., with pressure-safe stopcock.
Screen, 200 mesh, 8-inch diameter.[8]

Fractionation Procedure[7]

Column Preparation. The Dowex 50 × 4 resin is washed and wet-
screened by a procedure similar to that used for the Dowex 50 × 8
resin.[5] A pound of resin is washed with 4 N HCl (4 to 8 l., until filtrate is
nearly colorless) on a Büchner filter. The excess acid is removed by
several washes with distilled water. The sodium form of the resin is prepared by washing the resin with 2 N NaOH until the filtrate is alkaline.

At this point the resin is suspended in 3 vol. of N NaOH and heated
for 3 hours on the steam bath with occasional shaking. With 1-hour settling periods, the supernatant alkali solution is removed and replenished
with fresh hot N NaOH solution, for a total of five treatments. The resin
is filtered, washed free of excess alkali, and driven through a 200-mesh
screen with a strong jet of tap water. About 50% of the commercial product will pass the screen.[8]

A 100-ml. portion of settled resin is transferred to a Büchner filter
and overlaid with a filter paper disk to prevent channels. The resin is
washed with a liter of 1 N NaOH and 800 ml. of diluted pH 5 buffer
(buffer III, diluted 1:10). The resin is stirred with about 200 ml. of the
buffer to give a suspension.

The suspended resin is transferred to the 165 × 0.9-cm. chromatograph tube in six or seven portions, so that about 25-cm. sections of the
bed are formed at one time. Transfer is made by means of a funnel with its
tip bent to direct the stream against the wall of the tube. Each section is
allowed to settle under air pressure of 10 to 15 cm. Hg until no further
compaction occurs. After settling, the excess buffer is withdrawn by
suction-tubing (⅛-inch Tygon), leaving a liquid level of 10 cm. above the
resin, as the next segment is poured. The final settled resin bed is adjusted
to 150-cm. length.

If the resin column will not allow a flow rate of 5 ml./hr. at 15 cm. of
pressure, it is necessary to remove the finest 10% of the resin particles.
This is most conveniently accomplished by transfer of the suspended
resin to a graduate. After settling, the upper layer of particles may be
removed by suction into a collecting trap.

[13] Prototype was described by W. H. Stein and S. Moore, *J. Biol. Chem.* **176,** 337
(1948).

The completed resin column is connected by a microrubber stopper (Arthur H. Thomas 8823A) to a glass U tube, to an end of which is joined 3 feet of Tygon tubing (⅛ inch I.D.) terminating in a 500-ml. separatory funnel, with stopcock. The funnel is half-filled with 0.2 N NaOH (containing 5 ml. of the BRIJ 35 solution per liter). At least 100 ml. of the alkaline solution is passed through the column, under 15 to 20 cm. of pressure.

A day before the column is to be used, the alkali above the resin and in the funnel is replaced by the buffer (pH 3.1 or 2.2) with which the resin is to be equilibrated. About 200 ml. of buffer is passed overnight through the resin, and the pH of the effluent buffer is checked to determine that equilibration has occurred. The stopcock of the funnel is closed at that time, since excess treatment with buffer increases the ammonia blank obtained. The pH 3.1 buffer is commonly used in the equilibration, unless mixtures containing taurine and urea are to be analyzed.

Operation of Columns. The solution to be chromatographed is adjusted to pH 2.0 to 2.5, using test paper. One to five milliliters of sample is added to the surface of the resin column with a bent-tip pipet. Three 0.3-ml. aliquots of pH 2.2 buffer are used to wash in the sample. The pH 3.1 buffer is then added above the resin in the tube and to the separatory funnel. If the pH 2.2 buffer was used to equilibrate the resin, the first effluent shows a pH of 2.2, and the successive effluents may remain at a pH less than 3.1 until about 120 ml. of the buffer has passed through the column.[7] This volume includes the fractions corresponding to the first six compounds in Fig. 1. Stein and Moore start the column on a Friday at 3 ml./ hr., and make the temperature change and a change to 6 ml./hr. on the next Monday. The gradient elution is set up on Tuesday and the rate is brought to 8 ml./hr.

Moore and Stein use a synthetic mixture of amino acids corresponding to a serum albumin hydrolyzate.[14] A 0.1-ml. aliquot of the 10% solution of amino acids is added to 4 ml. of pH 2.2 buffer. A 1-ml. aliquot of this solution is added to the column. This aliquot contains 0.1 to 0.2 mg. of each component and 2.5 mg. of total amino acids. Up to 25 mg. of amino acids may be used, without great loss of resolution. The complex mixtures (see Fig. 1) were made by appropriate additions to the amino acid mixture.

The column is positioned over a drop-counting fraction collector. Water from a constant-temperature bath is circulated through the jackets of the column. The first temperature required is 30 ± 1°. The effluent fraction is usually 2 ml. in volume, collected at a maximal rate

[14] S. Moore and W. H. Stein, *J. Biol. Chem.* **178**, 53 (1949).

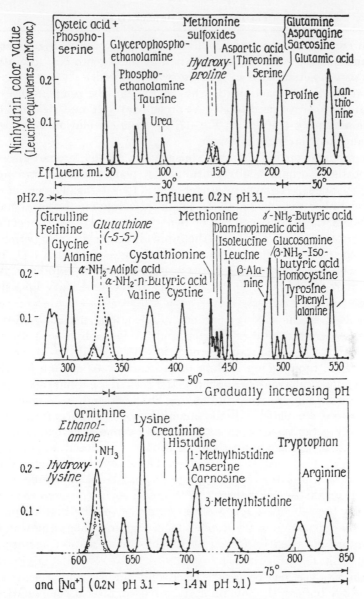

FIG. 1. Separation of amino acids and related compounds from a synthetic mixture containing fifty components. The column of Dowex 50 × 4 (150 × 0.9 cm.) was operated with sodium acetate and citrate buffers at the temperatures indicated. The amino acids were each present in 0.05- to 0.20-mg. quantities. The effluent was collected in 1-ml. fractions, except in the range from cystathionine through leucine, in which 0.5-ml. fractions were collected. The positions of hydroxyproline and glutathione, which were not included in the mixture, are indicated by dotted curves. The concentrations of hydroxylysine and ethanolamine were determined after removal of NH₃ by evaporation. The dotted base line at 600 and 725 effluent ml. indicates the positions at which there is a change in blank resulting from the elution of traces of NH₃ contained in the influent buffers. (S. Moore and W. H. Stein, *J. Biol. Chem.* **211**, 895 (1954).)

of 8 ml./hr. (12 ml./hr./cm.²). The analysis depicted in Fig. 1 required 1-ml. and 0.5-ml. fractions, as indicated in the legend. No deaeration of buffers or protection with an oil layer was required.

After the serine fraction (210 ml. in Fig. 1), the temperature of the circulating water is raised to 50°. At an effluent volume twice that of the aspartic acid peak (340 ml. in Fig. 1), the gradient elution with increase in pH and ionic strength is begun.

The gradient mixing chamber consists of a 500-ml. ground-glass joint wash bottle (Corning 1660) containing a plastic-covered stirring bar (1¼ × ¼ inches). A magnetic stirrer is used for mixing. The outlet of the mixing chamber is connected to the top of the column in the usual manner. The inlet of the chamber is cut or bent to allow vertical connection to the separatory funnel. The tip of the funnel is drawn down to permit connection to the "wash bottle" by a 3-inch length of ¼-inch I.D. Tygon tubing which tightly encloses a 1½-inch length of capillary tubing (1-mm. bore, 6-mm. O.D.). The capillary prevents density inversion of the two contacting buffers.

The flask of the mixing chamber is filled with pH 3.1 buffer. The well-greased ground joint is inserted and firmly secured with springs or rubber bands. The separatory funnel is partially filled with the same buffer. The Tygon tubing leading from chamber to column is clamped shut. Vacuum is alternately applied to the neck of the funnel and released, so that air is removed from the mixing chamber and associated tubing. Buffer should completely fill the chamber, so that the gradient will be dependably established, even as pressures change. The excess buffer is removed from the separatory funnel (above the closed stopcock), and the funnel is filled with pH 5.1 buffer. The assembly is mounted a few feet from the top of the chromatograph tube. The pH 3.1 buffer fills the volume from the funnel stopcock to the resin surface, except for the air-lock below the rubber stopper sealing the end of the chromatograph tube.

For the separation of tryptophan from arginine, the circulating water temperature is raised to 75°, after the histidine fraction has appeared (700 ml., Fig. 1).

If the analysis has been performed only for the amino acids eluted by the 0.2 N buffers, the column may be reused, after 0.2 N NaOH treatment. The complete analysis requires freshly prepared columns, since the resin bed shrinks during contact with the concentrated buffer. The resin is removed from the chromatograph tube by vacuum (water pump) into a filter flask, as water is applied to the tip of the tube. The resin is washed on a Büchner filter with the cold N NaOH and pH 5 buffer as previously described. It may then be poured to prepare a column. The treatment with 0.2 N NaOH and eluent buffer is again required.

The recoveries of amino acids in a *synthetic mixture* on three chromatograms are listed.[7] The amino acids yielding $100 \pm 1\%$ recovery are: threonine, serine, proline, glutamic acid, glycine, alanine, leucine, tyrosine, phenylalanine, lysine, and arginine. The other amino acids gave the following recoveries: aspartic acid, 102%; valine, 102%; cystine, 98%; methionine, a reproducible 90%; isoleucine, 97%; ammonia, 105%; and histidine, 98%. Cysteine may be determined as cystine after air oxidation. Tryptophan[15] must be determined on an alkaline hydrolyzate (8 hours, 90°, 5 N NaOH) which is chromatographed on a 30-cm. starch column after neutralization. The eluent was 0.1 N HCl. On known protein samples, 75% yield of tryptophan was obtained.

Sample Preparation. The recovery values for amino acid mixtures is a necessary step in development and control of a complex method. However, information regarding changes in amino acid content during hydrolysis of a protein or necessary deproteinization steps is also required.

Harfenist,[9] Smith *et al.*[16,17] and Hirs *et al.*[15] have relied on hydrolysis of protein samples for several periods of time to indicate the corrections necessary for incomplete hydrolysis or for destruction of the amino acid.

Smith *et al.*[16] found a linear decrease of amino acid content with time of hydrolysis (sealed tube, O_2 absent, 6 N HCl, 105°, 22 or 70 hours). The amino acids were destroyed at different rates. Serine, threonine, aspartic acid, and lysine were involved. Hirs *et al.*[15] using similar hydrolytic conditions, found that the decomposition of serine, threonine, cystine, and tyrosine obeyed first-order kinetics and could be represented by the relation

$$\log A_0 = \frac{t_2}{(t_2 - t_1)} \log A_1 - \frac{t_1}{(t_2 - t_1)} \log A_2$$

A_1, A_2 and A_0 are the quantities of amino acid present after t_1, t_2, and zero hours of hydrolysis. The calculated recoveries of four amino acids by the first-order relation are given in the last column of Table II.[15] The table presents the experimental recoveries after the treatment of a complete amino acid mixture similar in composition to bovine serum albumin.

Aspartic and glutamic acid were found to decompose during hydrolysis,[16,17] although Hirs *et al.*[15] could not detect destruction during heating of the amino acid mixture. The factors in destruction of arginine and lysine during hydrolysis also are unknown. Proline is also destroyed to a significant extent during hydrolysis of ribonuclease[15] or plasma com-

[15] C. H. W. Hirs, W. H. Stein, and S. Moore, *J. Biol. Chem.* **211**, 941 (1954).
[16] E. L. Smith and A. Stockell, *J. Biol. Chem.* **207**, 501 (1954).
[17] E. L. Smith, A. Stockell, and J. R. Kimmel, *J. Biol. Chem.* **207**, 551 (1954).

ponents.[18] Significant losses by decomposition occur during hydrolysis of the following amino acids: serine, threonine, cystine, tyrosine, and proline. Losses may occur also in the case of aspartic and glutamic acids, lysine, and arginine. Methionine and glutamine[18] are unstable during ion exchange, the amide yielding ammonia in large amount.

A few amino acids exist in peptides resistant to acidic hydrolysis, notably isoleucine and valine.[9,15,16] Harfenist[9] recommends 100 hours of hydrolysis, a period highly unfavorable for the labile amino acids previously mentioned. It appears that several hydrolysis periods are required

TABLE II

RECOVERIES OF AMINO ACIDS FROM STANDARD MIXTURE HEATED WITH 6 N HCl
(C. H. W. Hirs, W. H. Stein, and S. Moore, *J. Biol. Chem.* **211**, 941 (1954).)

| | Per cent recovery | | | | Per cent recovery calculated for 0 time |
| | Time of "hydrolysis" | | | | |
Amino acid	22 hr.	70 hr.	23 hr.	71 hr.	
Serine	83.7	61.0	84.9	60.3	98.6
Threonine	92.4	81.0	92.7	80.8	98.7
Cystine	96.2	83.1	98.0	84.6	104.0
Tyrosine	89.7	76.6	94.9	89.2	97.3

to enable complete liberation of isoleucine and valine, together with the amino acids which are linked in these resistant peptides. The several hydrolyses will allow correction of the content of the unstable amino acids, along the lines indicated by Hirs *et al.*[15]

The widening application of the fractionation-analysis scheme of Moore and Stein[7] to the "non-protein nitrogen" fraction of plasma[18] or of tissue cytoplasm[19] has established the number and quantities of compounds present. In addition, the application has emphasized the need of a simple protein removal step. These authors[18,19] after comparing equilibrium dialysis, ultrafiltration, and picric acid deproteinization,[20] settled on the last method as most rapid and consistent in recovery of added amino acid. Picric acid removal (using strongly basic resin, Dowex 2 × 8, chloride form) and concentration on a rotary evaporator to one-tenth the original plasma volume are the additional steps required before the sample is ready for analysis.[18] An aerobic oxidation designed to oxidize -SH groups to -S-S- is also described. Further refinements in sample preparation and hydrolytic conditions are to be expected.

[18] W. H. Stein and S. Moore, *J. Biol. Chem.* **211**, 915 (1954).
[19] H. H. Tallan, S. Moore, and W. H. Stein, *J. Biol. Chem.* **211**, 927 (1954).
[20] P. B. Hamilton and D. D. Van Slyke, *J. Biol. Chem.* **150**, 231 (1943).

Photometric Analysis of Fractions[21]

Materials and Reagents

Ninhydrin. The product of Dougherty Chemicals, New York, New York, is used without recrystallization.

Hydrindantin, prepared by reduction with ascorbic acid. Dissolve 80 g. of ninhydrin in 2 l. of water at 90°. Add while stirring 80 g. of ascorbic acid (Merck) in 400 ml. of water at 40°. Remove the heat. Crystallization begins immediately and proceeds for 30 minutes. Cool the solution by running tap water for a 1-hour period. Filter the hydrindantin crystals, wash with water, and dry over P_2O_5 in a vacuum desiccator protected from light. Store in a dark glass bottle. The small anhydrous crystals are desired for ready solubility in methyl Cellosolve. Hydrindantin is available from Dougherty Chemicals also.

Sodium acetate buffer, 4 N, pH 5.5. Dissolve 2720 g. of the reagent trihydrate in hot water (2 l.). Cool to room temperature, add 500 ml. of glacial acetic acid, and bring the solution to 5-l. volume. The pH for the buffer as prepared should be 5.51 $\pm$ 0.03. Store at 4°.

Ninhydrin reagent. Dissolve 20 g. of ninhydrin and 3 g. of hydrindantin in 750 ml. of peroxide-free methyl Cellosolve. Stir without aeration. Add 250 ml. of the pH 5.5 buffer, and transfer the resulting reddish reagent to a 1-l. brown glass reservoir for storage under nitrogen. The second and third bottles of the storage system[22] are of 2-l. size. In place of the 250-ml. dropping funnel previously used, an inlet-outlet tube has been mounted which bears a 6-mm. bore stopcock and small ball joint. The ball joint is normally covered with a socket-type stopper. Filling of the reservoir is accomplished as follows: A bent glass filling tube is attached at the ball joint. The tube dips into the beaker of fresh reagent. Air is displaced from the tube by use of positive air pressure on the water in the last bottle of the storage assembly. Vacuum applied to the third bottle draws the reagent into the reservoir. The reagent is less stable than the previous ninhydrin-stannous chloride reagent, so the preparation of fresh reagent at about weekly intervals is recommended.

Pipetting machines[23] are used. One dispenses 1-ml. aliquots of

---

[21] S. Moore and W. H. Stein, *J. Biol. Chem.* **211**, 907 (1954).

[22] S. Moore and W. H. Stein, *J. Biol. Chem.* **176**, 367 (1948).

[23] Such machines are built by Baltimore Biological Laboratory, Baltimore 2, Maryland. Similar ones are sold by Scientific Glass Apparatus Co., Bloomfield, New Jersey; or by Fisher Scientific Company, Pittsburgh, Pennsylvania.

the ninhydrin reagent (1.00 ml. ± 0.5%). The other dispenses
aliquots of the required diluent (5.00 ml. ± 0.6%).

Diluent. *n*-Propanol-water was previously used.[22] The revised
method uses 1:1 ethanol-water as a convenient diluent.

Matched collecting-photometer tubes. 800 to 1000 soft-glass test
tubes are matched for inner and outer diameters and tested for
matched photometric response, by means of a methyl red-HCl
solution.[22] Stein and Moore standardized on the following specifications: 16.25 ± 0.15 mm. I.D., 18.3 mm. O.D., optical density
(525 mμ) = 0.700 ± 0.005 unit. Tubes of slightly different diameters and wall thickness could, of course, be used; the matched
diameters and photometric response are the important factors.
Tubes are marked for correct alignment in the spectrophotometer
and numbered for identity. Covers of aluminum are used during
the heating procedure (Aloe Scientific Co.).

Metal racks. Fifty-tube aluminum racks are available from the
Technicon Company, New York, New York, or they may be
obtained from various supply sources.

Covered water bath. A bath of sufficient dimensions to allow insertion of the fifty-tube rack. Heat application should be
sufficient to provide vigorous boiling within 2 minutes of rack
insertion.

Spectrophotometer, Coleman Model 6A or equivalent.

Analytical Procedure (see also Vol. III [76]). A 1-ml. aliquot of the
ninhydrin reagent is used for either the 1-ml. or 2-ml. column fractions.
The concentrated acetate buffer of the reagent described suffices to bring
most 2-ml. fractions to the required reaction pH. The buffer capacity of
the reagent is about 0.1 meq./ml. for either acid or alkali. When this
capacity is exceeded, prior adjustment of pH is mandatory.

The capped tubes are shaken for 5 to 10 seconds and heated for exactly
15 minutes in the covered water bath. The diluent is added in 5-ml.
aliquots from the second pipetting machine, using single, double, or
triple aliquots. The tubes are kept out of sunlight.

After dilution, the tubes are wiped dry and transferred to a dry rack.
The set of tubes are cooled to <30° in front of a fan and thoroughly
shaken (30 seconds). The shaking serves to oxidize the remaining hydrindantin so that the reddish color of this compound does not increase the
absorbence of the blank mixtures. The tubes should be shaken uniformly
in the rack, rather than individually. The blank readings (at 570 mμ)
should be below 0.10 on the optical density scale, compared to water.

The tubes are read at 560 mμ against suitably chosen blank tubes.

TABLE III
COLOR YIELDS FROM AMINO ACIDS AND RELATED COMPOUNDS ON MOLAR BASIS
RELATIVE TO LEUCINE. (S. Moore and W. H. Stein, *J. Biol. Chem.* **211,** 907 (1954).)

Determined on 2-ml. samples of 0.1 mM solutions in the buffers (pH 2.2 to 5) in which the compounds emerge in the Dowex 50-X4 chromatographic procedure; heating time, 15 minutes; read at 570 mμ. (The urea and creatinine were 3.0 mM.)

Compound	Color yield	Compound	Color yield
Aspartic acid	0.94	β-NH$_2$-isobutyric acid	0.44
Threonine	0.94	Carnosine	0.93
Serine	0.95	Citrulline	1.04
Proline (440 mμ)	0.225[a]	Creatinine	0.027
Glutamic acid	0.99	Cysteic acid	0.99
Glycine	0.95	Diaminopimelic acid (per	1.24
Alanine	0.97	2 NH$_2$ groups)	
Valine	0.97	Ethanolamine	0.91
Half cystine	0.55	Fellnlne	0.05
Methionine	1.02	Glutamine	0.99
Isoleucine	1.00	Glucosamine	1.03
Leucine	1.00	Glutathione (oxidized, half)	0.93
Tyrosine	1.00	Glycerophosphoethanol-	0.50
Phenylalanine	1.00	amine	
Ammonia	0.97	Hydroxylysine	1.12
Lysine	1.10	Hydroxyproline (440 mμ)	0.077[a]
Histidine	1.02	Methionine sulfone	1.02
Tryptophan	0.94	Methionine sulfoxide	0.98
Arginine	1.01	1-Methylhistidine	0.88
α-NH$_2$-adipic acid	0.96	3-Methylhistidine	0.86
β-Alanine	0.50	Ornithine	1.12
Anserine	0.78	Phosphoethanolamine	0.43
Asparagine	0.95	Sarcosine	0.28
α-NH$_2$-n-butyric acid	1.02	Taurine	0.88
γ-NH$_2$-butyric acid	1.01	Urea	0.0314

[a] The readings taken at 440 mμ are first converted to "leucine equivalents" by using the same conversion table that is employed for the other amino acids measured at 570 mμ, and the concentrations of the imino acids are subsequently obtained by dividing by the above color yields.

Usually the blank tube is one immediately before or after a peak. Difficulties with variable blanks occur at about 590 ml. and 730 ml. (see Fig. 1), in the region where sample ammonia and contaminant ammonia are eluted.[24] The blank for methionine and the leucines is chosen after the leucine peak. The blank for ammonia is taken near the lysine peak. Histidine recovery is based on a blank reading taken just before the peak

[24] This difficulty is especially apparent when samples high in glutamine or asparagine, e.g., plasma samples, are chromatographed. See W. H. Stein and S. Moore, *J. Biol. Chem.* **211,** 915 (1954).

of this amino acid. High blank readings may be lowered by protection of collecting tubes on the fraction collector with a Lucite cover lined with cloth impregnated with citric acid. During long storage prior to analysis, the tubes are stoppered with citric acid-treated corks wrapped in aluminum foil.

Calculations. A standard curve is plotted for six to ten leucine standards in the range of 0.05 to 0.20 mM concentration. Aliquots of 1 and 2 ml. are used in the procedure described. Two direct reading tables[22] may be prepared relating optical density (in increments of 0.01 units) to micromoles of leucine present in the reaction mixture. In one table, a particular optical density value will translate to three values in micromoles, according to whether 5 ml., 10 ml., or 15 ml. of diluent is added.

The optical density value for a particular fraction is corrected for blank absorption before reference to this table of leucine equivalents. See the previous section regarding proper choice of blank tubes in a fractionation series.

The micromoles of "leucine" divided by "color yield" equals micromoles of the particular compound. The color yield values for 48 compounds are given in Table III.[21] The micromoles of an amino acid are integrated under the peak to yield total micromoles of amino acid.

[79] Identification of Amino Acids by Paper Chromatography

By WILLIAM STEPKA

Introduction

Filter paper was introduced into partition chromatography by Consden *et al.*[1] after preliminary efforts to effect a separation of the components of a mixture by liquid-liquid partition in a countercurrent machine.[2] Efficient devices of this kind are structurally complicated, and the paper chromatogram was evolved by a gradual extension of the principles to simplified systems wherein solid supports were employed to hold one of the phases.[3-5] Attempts to overcome a few of the disadvantages inherent in some of the solid supports, such as silica gel, led to successful trials with filter paper as a support for the aqueous phase.[1]

[1] R. Consden, A. H. Gordon, and A. J. P. Martin, *Biochem. J.* **38,** 224 (1944).
[2] A. J. P. Martin and R. L. M. Synge, *Biochem. J.* **35,** 91 (1941).
[3] A. H. Gordon, A. J. P. Martin, and R. L. M. Synge, *Biochem. J.* **37,** 79 (1943).
[4] A. J. P. Martin and R. L. M. Synge, *Biochem. J.* **35,** 1358 (1941).
[5] A. J. P. Martin, *Ann. N. Y. Acad. Sci.* **49,** 249 (1948).

Although initially the technique was devised for and used in the separation of amino acids and their derivatives, the authors were cognizant of its general applicability.[4] Their predictions were fulfilled soon after the introduction of the filter paper modification. Since then the technique of chromatography on paper has been extended to almost all the classes of compounds of biological interest. The list continues to expand as the technique itself reveals the presence of hitherto unknown or unsuspected compounds in diverse biological fluids and extracts.

The following description will deal primarily with the application of filter paper chromatography to the qualitative analysis of amino acids. An attempt will be made to describe the equipment, materials, and manipulations in sufficient detail to be useful to the beginner confronted with the average biochemical problems.

It is not the author's intention to review the voluminous literature on the subject. For extensive bibliographies the reader is referred to the appropriate sections in a selected number of sources.[6-12] Instead, emphasis will be placed on those systems and techniques which have produced consistent and reliable results for the author. The sources from which these have been derived are numerous and varied: most have come from the literature; some from discussions with various colleagues, particularly Dr. C. E. Dent, who guided the author through his early experiences in paper chromatography; and a few have been originated by the author. Reference citation will be kept to a minimum because it is felt that much in the practice of paper chromatography already has passed into the scientific domain. Furthermore, only in a comprehensive review of the voluminous literature would it be possible to trace the various modifications and variations of the original techniques.

In practice, qualitative paper chromatography consists of a sequence of basic operations: (1) preparation of the sample, (2) determination of optimal aliquot, (3) application of the sample to the paper sheet, (4) development of the chromatogram with selected solvents, (5) detection of

[6] J. Balston and B. E. Talbot, "A Guide to Filter Paper and Cellulose Chromatography." Reeve Angel, London, 1952.

[7] F. Cramer, "Papierchromatographie." Verlag Chemie, Weinheim, 1953.

[8] R. J. Block, E. L. Durrum, and G. Zweig, "A Manual of Paper Chromatography and Paper Electrophoresis." Academic Press, New York, 1955.

[9] R. C. Brimley and F. C. Barrett, "Practical Chromatography." Reinhold Publishing Corp., New York, 1953.

[10] E. Lederer and M. Lederer, "Chromatography: A Review of Principles and Applications." Elsevier Publishing Co., New York, 1953.

[11] W. Stepka, in Corcoran, "Methods in Medical Research." Vol. 5, p. 25. The Yearbook Publishers, Chicago, 1952.

[12] R. Consden, Brit. Med. Bull. 10, 177 (1954).

the separated compounds, and (6) identification of the spots. Special requirements frequently may add steps to the basic operations. It may be necessary, for example, to resolve overlapped spots to facilitate identification. Elution might be a step in such a process.

The operations may often be successfully carried out with simple or improvised equipment; however, for serious routine work it will most often prove to a laboratory's advantage to provide itself with permanent chromatographic equipment capable of processing sets of 8 to 16 sheets simultaneously. The full power of paper chromatography usually can be realized most readily with equipment which permits the running of duplicate and control chromatograms simultaneously in the same chamber.

Apparatus

Chambers. The major prerequisite of a suitable chamber for paper chromatography is that it be vapor-tight and capable of maintaining a saturated atmosphere with respect to the developing solvents. Unless it is possible to reserve a separate chamber for each solvent system, chambers should be constructed from materials impervious to the components of the systems. Bare wood will absorb a variety of solvents; paraffin-impregnated wood will absorb n-butanol. The vapors subsequently given off by the absorbed substances may disturb the phase relationships of other solvent systems if these are used in the contaminated chambers. Various difficulties such as streaking and variable rates of flow may result as a consequence.

Thus, if it is contemplated to utilize a chamber for more than one solvent system it would be desirable to choose one with an impervious, cleanable interior such as glass, Formica, or stainless steel. The last, it should be remembered, is subject to corrosion by HCl. Formica when properly bonded to wood with a phenolic resin glue makes a very satisfactory lining resistant to all the solvents in common use.

Commercial units are available in any of the above materials from one or more of the following: Aloe Scientific Company, St. Louis 12, Missouri; Research Equipment Corporation, Oakland, California; and Will Corporation, Rochester 3, New York. If desired, Formica-lined plywood cabinets of 16-sheet capacity may be constructed according to the specifications in ref. 11. For convenience the cabinet should be equipped with leveling devices.

Battery jars, aquaria, and various other enclosures have been proposed as containers for chromatography. Some of these are available commercially, but it is the author's opinion that none of these can compete with the operational simplicity and economy of space offered by rectangular chambers.

Trough Assembly. Troughs and associated equipment designed to fit the respective chambers are also available from the above dealers in chromatographic supplies. Although fragile, glass troughs are preferred by the author, owing to their chemical inertness. Polished stainless steel, if of the proper grade such as Type No. 316, may be used successfully in some applications[13-15] but is not so versatile as glass because it precludes the use of corrosive HCl either in the solvent or in the atmosphere of the chamber. Some stainless steels, owing to the catalytic action of heavy metals, cause excessive oxidation of phenol in phenol-containing solvents. The resulting oxidation products which remain on the paper after drying interfere with the flow of the second solvent, particularly in the region of the phenol front. Retardation of flow over this region produces an uneven front during development in the second solvent.

Manufacturers of glass equipment, such as the Scientific Glass Apparatus Company, Bloomfield, New Jersey, will fabricate glass troughs to specifications. Troughs 24 inches long made from 35-mm. (O.D.) tubing split longitudinally in half with ends turned up are a convenient size. Such troughs will hold about 200 ml. of solvent, providing ample reserve for additional solvent should it be desired to use heavy papers such as Whatman 3MM and similar grades.

Trough supports, antisiphoning rods, and weights for holding the papers in the trough in descending chromatography are available commercially. Those who wish to construct their own may find the author's previous directions[11] of some help.

Drying Chambers. In between solvents, and again at the conclusion of the run, the paper must be dried sufficiently to ensure adequate removal of the solvents. If paper chromatography is to be used as a routine tool, then this circumstance along with the added convenience justifies the expense of a specially designed drying chamber with forced draft ventilation. For small-scale sporadic work a conventional fume hood may prove adequate. Since the dimensions of commercial antisiphoning rods vary, it is not possible to present here a detailed set of directions for the construction of a drying box. General design considerations may prove useful to some, however.

Transfer of the chromatograms from the chromatography chamber to the drying cabinet is facilitated if the latter is designed to accommodate

[13] A. A. Benson, J. A. Bassham, M. Calvin, T. C. Goodale, V. A. Haas, and W. Stepka, *J. Am. Chem. Soc.* **72**, 1710 (1950).
[14] S. M. Partridge, *Biochem. J.* **42**, 238 (1948).
[15] A. J. Woiwod, *J. Gen. Microbiol.* **3**, 312 (1949).

the antisiphoning rods. Completed chromatograms may then be clipped to the rods and, with the latter serving as handles, transferred to the drying chamber where the ends of the rods fit into notched cleats. The cleats are affixed to opposite sides of the box 2 to 3 inches below the rim. At least 2 inches should be allowed between notches.

The inside width of the chamber will thus be slightly in excess of the length of the antisiphoning rod. Its length will depend on the capacity required. Since no more chromatograms can be run than can be dried, the drying chamber should accommodate the full capacity of the available chromatography chambers in order to avoid backlogs. Thus for two 8-trough chambers the minimum inside length of the drying cabinet should be 5 feet 6 inches. To the author's knowledge none of the drying chambers offered commercially have this capacity.

The overall depth should be at least 29 inches to provide vertical room for standard chromatographic papers and for a horizontal perforated diffusing baffle 4 inches above the true bottom. The space below the baffle should be connected to a blower capable of moving about 10 cabinet-volumes of air per minute and exhausting to the outside. The cover for the chamber may be constructed from two perforated sheets of thin plywood spaced about 1 inch apart. If the perforations in the two sheets are not aligned directly one above the other, but are displaced relative to one another, the cover will also serve as a dust trap minimizing contamination of the chromatograms with atmospheric dust. This arrangement in which the air flows from top to bottom reduces the hazard of contact between wet chromatograms due to the flutter sometimes induced in systems where air flows in the reverse direction.

Some workers dry their chromatograms with the aid of heat. This is not recommended, as elevated temperatures, besides having possible destructive effects on the experimental compounds,[16-19] accelerate the oxidation of phenol and thereby cause discoloration of the paper. An efficient drying cabinet or fume hood will remove phenol adequately from the chromatograms at room temperature in about 12 hours—most other solvents in considerably less time.

Auxiliary Equipment. Application onto the paper of the preparations to be chromatographed requires some volumetric device designed for the transfer of small volumes. The Kirk-type micropipets, now available from many sources, are ideally suited for this, especially when connected

[16] M. K. Brush, R. K. Boutwell, A. D. Barton, and C. Heidelberger, *Science* **113**, 4 (1951).

[17] L. Fowden, *Biochem. J.* **48**, 327 (1951).

[18] A. R. Patton, E. M. Foreman, and P. C. Wilson, *Science* **110**, 593 (1949).

[19] A. J. Woiwod, *Nature* **166**, 272 (1950).

to a syringe-type control for easy adjustment of the meniscus. A set ranging in volume from 1 to 300 μl., with duplicates of the most commonly used sizes such as 5, 10, 25, and 50 μl., is desirable.

A heat-gun such as one of the better-quality hair dryers which will withstand continuous duty over prolonged periods greatly speeds the application of the preparations to the origin when volumes in excess of 20 μl. need to be applied. The heat-gun directed over the origin is used to accelerate evaporation of the liquid phase.

For detecting colorless spots on the chromatogram, reagents which form colored derivatives with the compounds in question must be applied to the paper. This is conveniently done with the aid of a sprayer, preferably one which can be operated from a compressed air line. A satisfactory sprayer is offered by the Research Equipment Corporation and may also be selected from among the various types of nasal and perfume atomizers on the market. The essential feature is that it be capable of delivering a uniform, finely dispersed spray. From some designs the initial delivery comes in droplets larger than the paper can absorb without streaking.

For some purposes detection of the spots may also be accomplished with the aid of ultraviolet light either as fluorescent or as fluorescence-quenching areas. Amino acids, not normally fluorescent except for tryptophan and histidine, can be induced to fluoresce on paper in ultraviolet light if the chromatogram is first subjected to temperatures in the neighborhood of 80°.[18-20] The General Electric "black-light" lamp, B-H4, or any other lamp with similar emission characteristics is very satisfactory for exciting this particular fluorescence.

Many of the spray reagents in use require elevated temperatures in order to obtain the characteristic colors with the compounds it is desired to detect. For this an oven large enough to accommodate the sheet and capable of being heated to about 100° is needed. Such an oven can also be used to initiate the reaction between paper and amino acids which enables the latter to be revealed as fluorescent spots.

Paper. Much has been written on the subject of papers suitable for chromatography. Investigators who wish to become better acquainted with the properties of different papers or those who may contemplate quantitative chromatography are referred to the review of Balston and Talbot[6] and to the studies by Kowkabany and Cassidy[21] and by Müller and Clegg.[22] These references discuss in detail the influence of manufacturing processes on flow rates and of residual impurities on chromatographic properties such as influence on the solvent and on the shape and

[20] D. M. P. Phillips, *Nature* **161,** 53 (1948).
[21] G. N. Kowkabany and H. G. Cassidy, *Anal. Chem.* **22,** 817 (1950).
[22] R. H. Müller and D. L. Clegg, *Anal. Chem.* **23,** 403 (1951).

movement of the spots. On the basis of these discussions, papers may be selected to fit individual needs.

Among the available papers Whatman Nos. 1, 4, and 3MM are the most popular. Though not necessarily the best, they are suitable for routine qualitative chromatography of the amino acids. The author shares the experience of Jones[23] that amino acids tend to form rounder spots on Whatman No. 7, probably owing to the more random orientation of the fibers in this paper.[6,22] In other respects it is similar to No. 1. The foregoing Whatman papers are available in packages of 100 selected sheets marked "For Chromatography."

Schleicher and Schuell No. 507 is preferred by Bull et al.[24] and by Redfield.[25] From the work of Kowkabany and Cassidy[21] Whatman No. 3 and S. and S. 597 emerge as having the best over-all characteristics. In applications where a heavier paper is indicated the author prefers to replace Whatman No. 3 with No. 3MM because it has the advantages of a smooth surface.

It is a wise precaution against future troubles to adopt scrupulous habits in handling the stock of filter paper intended for chromatographic use. In general, excessive handling with bare hands should be avoided. Amino acids transferred to the paper from the sweat on the fingers may cause spurious spots to appear. Similarly the oil left behind as a result of handling may have a waterproofing effect and thus produce localized areas where the distribution equilibria between mobile and stationary phases will be disturbed. These difficulties can be prevented by wearing clean surgical gloves when handling the paper.

Particles of dust, apart from introducing fluorescent or quenching material, may contain catalytically active substances which may enhance the oxidation of some components of the solvent such as phenol. The possibility that such material may also interfere by reacting with the components of the mixture to be analyzed also should be kept in view.[1] Prudence therefore dictates that the papers should be protected from dust both prior to and during use.

Adsorption of ammonia is also a potential source of trouble, owing to the fact that the oxidation of phenolic compounds is accelerated in alkaline media. Again this may effect either the solvent or the compounds being chromatographed.

Where difficulties of the above sort are encountered or if prior presence in the paper of disturbing influences, such as metals, is suspected, it is possible to wash the paper with dilute acids[26] or with water containing

[23] T. S. G. Jones, *Discussions Faraday Soc.* **7**, 285 (1949).
[24] H. B. Bull, J. W. Hahn, and V. H. Baptist, *J. Am. Chem. Soc.* **71**, 550 (1949).
[25] R. R. Redfield, *Biochem. et Biophys. Acta* **10**, 344 (1953).
[26] C. S. Hanes and F. A. Isherwood, *Nature* **164**, 1107 (1949).

small amounts of a chelating agent. Whenever washing of the paper is indicated, the author prefers to use Perma Kleer-80, a broad-spectrum chelating agent, available from the Refind Products Corporation, Lyndhurst, New Jersey. The latter, in contrast to ethylenediaminetetraacetic acid, is soluble throughout the pH range and thus can be used as an acid reaction. Two milliliters of liquid Perma Kleer-80 per liter of distilled water provides an effective wash for the papers. Washing may be accomplished either chromatographically or in bulk[26] and should be followed by an adequate rinse with metal-free water.

Alternatively, the deleterious effects caused by the presence of heavy metals may be suppressed by incorporating complexing agents such as 8-hydroxyquinoline, HCN, or H_2S directly into the solvent.[1] Perma Kleer-80 or ethylenediaminetetraacetic acid may also be used in this way.

Routine Practice

Preparation of Sample. Analysis by means of paper chromatography has such wide applications that it becomes a practical impossibility to include specific directions for preparing each among the variety of possible samples. The investigator must therefore be guided by the general principles imposed by the chromatographic technique. Not every sample will yield a satisfactory chromatogram without some prior manipulation or modification.

In the simplest of situations such as investigations of free amino acids in extracts of tissues it will usually be necessary to adjust the concentration of the solute so that only small volumes (< 100 μl.) need be applied to the origin. Normally this will involve concentrating the starting material to some degree. An example of this application to animal tissues is Walker's[27] investigation of free amino acids in tissues of the rat and cow. Aliquots representing an amount of fresh tissue containing 5 mg. of nitrogen gave good chromatograms. Amino acids in plant extracts usually yield satisfactory spots from an applied aliquot representing 50 to 100 mg. fresh weight of tissue.[28,29]

Body secretions and fluids also offer the possibility of being examined without extensive pretreatment. Thus Dent[30-32] routinely obtains good chromatographic separations of amino acids from 25 μl. of urine, 125 μl. of

[27] D. M. Walker, *Biochem. J.* **52**, 679 (1952).
[28] A. A. Benson, J. A. Bassham, M. Calvin, T. C. Goodale, V. A. Haas, and W. Stepka, *J. Am. Chem. Soc.* **72**, 1710 (1950).
[29] M. A. Joslyn and W. Stepka, *Food Research* **14**, 459 (1949).
[30] C. E. Dent, *Lancet II*, 637 (1946).
[31] C. E. Dent, *Biochem. J.* **43**, 169 (1948).
[32] C. E. Dent and J. M. Walshe, *Brit. Med. Bull.* **10**, 247 (1954).

human sweat, and from 125 to 625 μl. of protein-free (ultrafiltrate) plasma or cerebrospinal fluid applied directly to the paper.

The monograph by Roberts *et al.*[33] is replete with examples of the application of paper chromatography in conjunction with radioactive tracer techniques to the study of intermediary metabolism. The work is too comprehensive to review here but it is recommended as an excellent source of information on methodology and investigational philosophy (see particularly Chapter 3).

Protein hydrolyzates do not present too much of a preparative problem, since easily removable catalysts can be chosen to effect the hydrolysis. Thus excess HCl may be removed by successively evaporating the hydrolyzate to dryness over P_2O_5 in an evacuated desiccator containing a dish of alkali pellets. The resulting amino acid hydrochlorides may then be redissolved, and aliquots representing 100 to 300 γ of the initial protein can be applied directly to the paper. Useful chromatograms may be obtained in this way, but the positions of the basic and acidic amino acids will be displaced from those normally obtained when these are applied as the free compounds. The basic members will be displaced toward the origin; the acidic away from it. To obtain a more normal pattern the hydrolyzate may be applied to the paper with the origin resting over a small dish containing a few drops of approximately 3 M ammonium hydroxide.

Alkaline hydrolysis can be carried out most conveniently with barium hydroxide, since barium ion is easily precipitated as the carbonate or sulfate.

Enzymatic hydrolysis of proteins has certain advantages in some applications, as, for example, when it is desired to examine a protein for amide content.[34] It has the further advantage that, if the buffer salts are kept low, the progress of hydrolysis can be followed simply by removing aliquots and chromatographing them directly. Controls to show that the enzyme is not contributing amino acids through its own breakdown should be incubated concurrently.

Some materials such as incubation mixtures in buffered or physiological saline solutions and growth media containing excessive amounts of salts usually need to be desalted before successful chromatograms can be obtained from them.

Electrolytic desalting may be accomplished with the apparatus described by Astrup *et al.*[35] However, the finding by Stein and Moore[36] that

[33] R. B. Roberts, P. H. Abelson, D. B. Cowie, E. T. Bolton, and R. J. Britten, *Carnegie Inst. Wash.* **607**, 31–46 (1955).

[34] W. Stepka and W. N. Takahashi, *Science* **111**, 176 (1950).

[35] T. Astrup, A. Stage, and E. Olsen, *Acta Chem. Scand.* **5**, 1343 (1951).

[36] W. H. Stein and S. Moore, *J. Biol. Chem.* **190**, 103 (1951).

arginine suffers considerable destruction and the expensive apparatus combine to make the method less desirable.

A simple and convenient ion exchange procedure using only one type of resin, Nalcite SAR (National Aluminate Corporation, Chicago, Illinois), has been described by Piez et al.[37] Here the solution is passed first through a column of the resin in the hydroxide form which retains the amino acids along with the other anions. Arginine, strongly basic, is not retained. Elution is accomplished with 1 N HCl, the eluate containing the amino acids, contaminating anions, and hydrogen ion being collected in a flask. Batchwise addition of the resin in the bicarbonate form results in an exchange of the contaminating anions for the bicarbonate. The amino acids cannot participate in the exchange because of the acid environment which, by decomposing the bicarbonate ion to CO_2, helps to drive the anion exchange to completion. Where the loss of arginine can be tolerated, the method of Piez et al. is very useful.

Roberts et al.[38] give a detailed description of a more elaborate but quantitative procedure making use of both cation and anion exchange resins. The amino acids are obtained in two groups, one containing the basic amino acids, the other the neutral and acidic ones. Their fractionation method makes available for chromatography other nonvolatile products with but little extra effort.

Determination of Optimal Aliquot. Each grade of filter paper has an upper limit for the amount of material that may be successfully resolved on it, and this limit may depend on both the qualitative and the quantitative composition of the sample. Beyond this limit the chromatographic system becomes overloaded and the spots show a pronounced tendency to streak. Ultimately the loss of resolution becomes complete, and no useful information can be gained from such chromatograms. Hence, it obviously is of some importance to know what the upper limit is for a particular sample and chromatographic system. It is likewise of practical importance to know how much material needs to be applied so that components present in low concentration will not be missed.

The optimal aliquot thus usually turns out to be a compromise between the requirements for ideal resolution and those for the lower limit of detection. The optimal aliquot can be found conveniently and quickly by unidimensional chromatography. With the quantities given in the preceding section as a guide, a graded series of spots about 2 inches apart is applied along a line near one edge of a paper sheet. This is then chromatographed unidimensionally and finally sprayed with a detecting reagent.

[37] K. A. Piez, E. B. Tooper, and L. S. Fosdick, *J. Biol. Chem.* **194,** 669 (1952).
[38] R. B. Roberts, P. H. Abelson, D. B. Cowie, E. T. Bolton, and R. J. Britten, *Carnegie Inst. Wash. Publ.* **607,** 186 (1955).

From the intensity of color and the character of the zones under each origin a clue to the appropriate aliquot may be obtained.

Application of the Sample. The paper sheet is first prepared by outlining the origin. This is done by scribing a circle of 1-cm. radius near one corner of the sheet. The point for the center is chosen so that when the sheet is hanging in the trough the origin is as close as possible to the anti-siphoning rod without touching it at any point. The object is to make the most effective use of the functional paper area and, at the same time, to prevent contact between the origin and surfaces which become wet with a film of the solvent. Instances of streaking have been traced to such contact. The location of the origin thus depends on the geometry of the apparatus in use. A simple template is an aid when many papers need to be marked. More than one sheet should be prepared for each sample to make duplicate chromatograms available.

It is advisable to displace the origin away from the second edge by about 1 cm. to allow for "fanning out" of the spots during the run in the first solvent. In this way the spots along the starting line for the second solvent are kept free of the antisiphoning rod.

The controlled micropipets described earlier facilitate the application of the optimal aliquot as determined by the procedure described in the preceding section. If the operation is carried out in a stream of warm air, such as from a hair dryer, and if the tip of the pipet is moved about the origin, an almost continuous discharge can be maintained from the pipet without spilling over the boundary of the circle. This gives a more uniform distribution of the solutes over the origin than do successive applications of small volumes to the center. In the latter the solutes have a tendency to concentrate at the periphery. Such a "doughnut" in effect presents the solvent with two origins, and slow-moving solutes may appear as double spots.

Some authors recommend that a total given volume be applied by successive spotting of very small volumes, frequently 5 μl. or less, in order to confine the origin to a tiny area. It is this author's experience that such practice has merit only when the amount of material available is so limited that it is necessary to work near the lower limits of detection. Keeping in mind that the paper can be overloaded, it is easy to see that a given aliquot just sufficient to overload an origin of area x would be suboptimal if applied to an origin of area $2x$. Owing to the effects of overloading, the spots derived from the former may be larger and more distorted than those from the latter. Here again a compromise must be made with the various factors involved.

After the sample has been applied, the sheet is prepared for the trough by making a half-inch fold along one of the edges adjacent to the origin.

Development. Development can be carried out in either one of two ways, upward or downward. The latter was used by the originators of paper chromatography. In it the solvent flows through the paper in a downward direction from a trough which also serves as a suspension for the sheets. Downward development, aided by gravity, has the advantage of being faster than the upward modification of Williams and Kirby[39] in which the bottom of the paper either dips into or stands in a tray of solvent. Here the solvent is made to creep through the paper by capillarity against the force of gravity.

Downward flow is capable of utilizing the full area of the sheet, an advantage when a large number of compounds are to be resolved. Some solvents cannot rise to the top of the sheet or do so very slowly in upward development. Hence the area available for the separations is either limited or the time required must be greatly extended. Nevertheless, situations do arise where the retarding influence of gravity in upward development may be utilized to advantage, as, for example, when it is desired to use solvents which move too rapidly in downward flow. On final analysis the present author prefers downward development for routine work; consequently the following description will be directed primarily toward the latter technique. The basic principles, however, apply to either.

When the origins have been spotted and the papers creased, the sheets may be loaded into the troughs—two sheets per trough. The trough may be loaded conveniently while resting on a flat surface large enough to accommodate the two sheets. The two folded edges are made to interlock within the trough, and then the sheets can be anchored in place by resting a weight inside the upper fold. The completed assembly is then transferred into the chamber. When the chamber has been fully loaded, solvent is poured into the troughs and the cover is tightened in place.

Here we may interject that not a few beginners fail to utilize the full capacity of their chambers. Ordinarily this stems from an overcautious desire to prevent the chromatograms from touching one another. It is essential that contact between *wet* chromatograms be prevented; but contact between dry papers when first loaded into the cabinet and before solvent is added to the troughs does no harm, particularly when it does not involve the origin. The papers will hang freely and will assume a strictly vertical position when they lose their rigidity soon after the solvent is added to the trough.

Development is normally allowed to proceed until the solvent has migrated to within approximately a centimeter from the bottom edge of the sheet. Phenol solvents require between 22 and 27 hours to travel the

[39] R. J. Williams and H. Kirby, *Science* **107**, 481 (1948).

18-inch width of a standard sheet; butanol-acetic acid-water solvents traverse the long dimension of the sheet in approximately 15 hours.

Occasionally a specific purpose may dictate that the solvent be allowed to run off the paper for varying lengths of time. Such occasions arise when it is desired to separate two overlapping compounds of low mobility. In extended development some provision must be made to maintain a steady state with respect to solvent flow rate throughout the development. This can be achieved by serrating the lower edge of the sheet, thus providing points from which the solvent may drip off.[40] The serrations prevent the formation of a "bead" of solvent on the lower edge which otherwise would retard the flow through the sheet. To maintain a steady flow over prolonged periods may also require replenishing the solvent during the run.

When the solvent has migrated the desired distance, either close to the lower margin in normal development or off the sheet in extended development, the chromatograms are transferred from the chamber to a well-ventilated hood or to a specially designed drying box (see *Drying Chambers* above) where the solvent is allowed to evaporate from the paper. Because elevated temperatures may have any one of several deleterious effects either on the solvent or on the compounds being chromatographed, it is advisable to dry the chromatograms at room temperature.

The transfer from the chromatographic chamber to the drying chamber is accomplished most easily by first clipping the sheet to its antisiphoning rod and then, by making use of the free ends of the rod, pulling the sheet out from under the anchoring weight. Clips made of stainless steel are available commercially. Satisfactory substitutes can be made easily by bending 2-inch lengths of 6-mm. glass rod into the form of a staple with the inner radius adjusted to fit snugly over the antisiphoning rod plus two thicknesses of the chromatographic paper. Two such staples will prevent the chromatogram from slipping off the antisiphoning rod during the transfer and drying operations.

The success of subsequent steps in the chromatographic procedure may depend on thorough removal of the solvent from the paper. Excessive residues of solvent components may, for example, prevent the spots from responding properly to the detecting agent. In two-dimensional chromatography they may disturb the flow or alter the properties of the second solvent. Since the rate at which the chromatograms dry depends on a number of factors, notably, the nature of the solvent to be removed, temperature, humidity, and the volume of air passing over the sheets, it is inadvisable to prescribe a definite drying time. An admittedly subjective,

[40] M. A. Jermyn and F. A. Isherwood, *Biochem. J.* **44,** 402 (1949).

but generally useful, test to determine whether or not the chromatograms
are sufficiently dry for subsequent operations can be applied, however.
It consists in gently tapping the suspended sheet with the fingers and
comparing the sound with that similarly produced by a dry control sheet.
With but little experience adequately dried papers can be easily recognized by their sound.

For two-dimensional chromatography the dried sheets are next turned
through 90 degrees with respect to the first chromatographic direction,
prepared for the troughs and chamber, and finally chromatographed in
the second solvent. On completion of the second run the sheets are removed, dried, and subsequently sprayed or otherwise treated with detecting agents to reveal the location of the scattered spots.

Solvents. Solvents of various compositions have been described for
paper chromatographic separation of amino acids. They vary in their
effectiveness; some are generally useful, and others are effective only for
the separation of the members within selected groups. Time and available
facilities generally will not allow the individual investigator to try them
all. In the long run he will profit most by selecting a pair for routine work
and by concentrating his efforts on these so as to become thoroughly
familiar with their chromatographic characteristics for the class of compounds involved. Experience soon will teach which compounds may be
resolved well and which poorly or not at all. Subsidiary solvents can then
be selected for resolving cutouts of the overlaps from the general references given below and in the introduction.

Phenol, nearly saturated with water, is a useful solvent for the separation of the amino acids. In it the amino acids span the entire R_f range
from near zero to near unity (Fig. 1). This is an extremely valuable property for qualitative analysis which overrides some of the disadvantages of
phenol, such as ease of oxidation. This disadvantage, however, is not a
serious problem if care is taken to use purified phenol and to exclude heavy
metals both from the prepared solvent and from the paper (see *Paper*
above).

Simple distillation at atmospheric pressure from an all-glass apparatus
equipped with a short condenser is effective in providing phenol suitable
for chromatography. If about 600 g. of the distillate is collected in tared
glass containers of about 1-l. capacity, subsequent preparation is greatly
simplified. In stoppered containers the redistilled phenol can be stored for
long periods if kept cold. To prepare the solvent it is only necessary to
add 39 ml. of metal-free water per 100 g. of phenol and to effect solution
by warming in a water bath. Stirring shortens the time required to dissolve the phenol crystals and also allows some control over the temperature of the solution. If, when solution is complete, the temperature

of the solvent is above the working temperature, it can be cooled quickly by shaking under the cold-water tap. Prepared thus, the solvent exists as a single phase at temperatures above 20°. It remains suitable for chromatographic use as long as it remains colorless when viewed in a clear 100-ml.

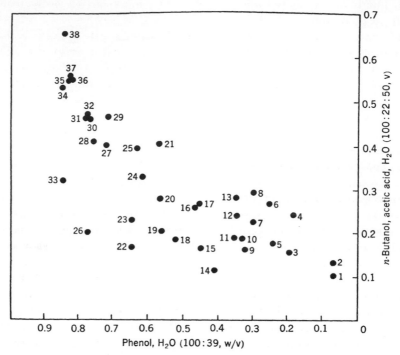

FIG. 1. Standard map of some amino acids and related compounds showing approximate R_f values on Whatman No. 1 filter paper (18¼ × 22½ inches). In compiling data for the map each solvent was allowed to migrate to the edge.

(1) Phosphoserine. (2) Cysteic acid. (3) Cystine. (4) Aspartic acid. (5) Lysine. (6) Glutamic acid. (7) Serine. (8) α-Aminoadipic acid. (9) Ornithine. (10) Asparagine. (11) Taurine. (12) Glycine. (13) Dihydroxyphenylalanine. (14) Djenkolic acid. (15) α-γ-Diaminobutyric acid. (16) Homoserine. (17) Threonine. (18) Arginine. (19) Glutamine. (20) Alanine. (21) Tyrosine. (22) Histidine. (23) Hydroxyproline. (24) β-Alanine. (25) α-Aminobutyric acid. (26) Methionine sulfoxide. (27) γ-Aminobutyric acid. (28) Kynurenine. (29) Tryptophan. (30) Valine. (31) Methionine. (32) Norvaline. (33) Proline. (34) Phenylalanine. (35) Isoleucine. (36) Leucine. (37) Norleucine. (38) α-Aminocaprylic acid.

graduated cylinder. The author makes extensive use of the above mixture as the first solvent in routine two-dimensional chromatography.

The n-butanol-acetic acid-water system of Partridge,[14] or some modification thereof,[11] is frequently employed either as the second solvent after phenol-water or as one member of other solvent pairs. Prepared in the proportions given by Partridge, the solvent gives low R_f values for the amino acids. These may be increased by using a modification derived by testing combinations from the most promising points along the critical line of a phase diagram for the ternary mixture.[11] The most favorable distribution of the amino acids (Fig. 1) was obtained with a mixture containing the following volumes of the components: n-butanol, 100 ml.; glacial acetic acid, 22 ml.; water, 50 ml. Owing to slow esterification, only freshly prepared solvent should be used.

Mixed in the given proportions, the components form a single phase at room temperature. Separation into two phases as esterification alters the proportions usually does not occur in less than 48 hours; hence the composition of the solvent remains relatively undisturbed during the 15 hours required for the run. In the author's experience with this solvent no chromatograms have been lost owing to phase separation during the run. The formula also provides a safety factor against temperature fluctuations.

Redfield[25] has described solvents of high resolving power for two-dimensional chromatography of amino acids. For the first dimension he employs a mixture containing methanol (80 ml.), water (20 ml.), and pyridine (4 ml.), and for the second run a quarternary mixture of tert-butanol (40 ml.), methyl ethyl ketone (40 ml.), water (20 ml.), and diethylamine (4 ml.). These solvents may be used as prescribed or in combination with the phenol-water or butanol-acetic acid-water systems described above. From among these four, effective pairs can be selected to meet the needs of most situations.

Although the solvents suggested in the foregoing are capable of resolving most of the amino acids present in multicomponent mixtures such as protein hydrolyzates or tissue extracts, overlaps still occur. These can best be dealt with by cutting out and rerunning unidimensionally in solvents known to resolve the compounds in question.

Several such solvents for specific overlaps may be found in the general references.[8,11] In addition to these the buffered systems described by McFarren[41] enable one to select a solvent for resolving almost any overlap likely to be encountered among the amino acids. The technique employs both buffered paper and buffered solvents. The paper is dipped in 0.066 M buffer solution of the appropriate pH and is then allowed to dry. The solvent is prepared by equilibrating the organic component(s) with the same buffer solution. With this technique it is possible to exploit differences in the properties of the compounds comprising the overlap. Thus leucine, isoleucine, and norleucine can be separated with a solvent

[41] E. F. McFarren, *Anal. Chem.* **23**, 168 (1951).

made by mixing equal volumes of benzyl and n-butyl alcohols and then saturating the resulting mixture with phosphate buffer of pH 6.2. Similarly, phenol buffered at pH 2 with $KCl:HCl$ buffer will separate threonine from alanine and valine from methionine. Other situations can be met by consulting the extensive tables provided in the original reference.[41]

Comment. Before passing on to the next topic it may be appropriate to interpose a few comments pertaining to procedural details. It is frequently suggested that additional solvent, usually the aqueous phase, be placed in trays in the bottom of the cabinet. With the solvents recommended above, no discernible benefit has ever been observed from this practice in the author's laboratory. Likewise our experience has been that it is unnecessary to equilibrate the papers in the chamber before adding the solvent to start the development. It is therefore suggested here that the individual worker first test the desirability of such procedures before adopting them as routine. It is possible that benefits accrue from such practices only under certain specific, but undefined, conditions, and where equally good results can be obtained without resorting to them a considerable amount of time and trouble can be saved.

Detection. Compounds may be detected on chromatograms by methods falling into four general categories: (1) spraying with reagents to form colored derivatives, (2) examining in ultraviolet light and observing either fluorescent or fluorescence quenching areas, (3) bioautographic means wherein the compound(s) to be detected either stimulates or inhibits the growth of selected microorganisms, and (4) radioautography for compounds which have been previously tagged with a radioactive tracer. From among these it is possible to select a detecting agent which is either general for a class of compounds or specific for individual compounds.

Of the various methods, spraying the chromatograms with reagents to reveal the compounds as colored spots is most commonly used, and for the amino acids a ninhydrin spray, owing to its sensitivity, is the most popular. With it, the amino acids appear as bluish-purple spots against a white background; the imino acids as yellow spots. Other naturally occurring substances, including non-α-amino acids and primary amines, also give the typical amino acid reaction.[31]

As originally introduced,[1] the spray reagent contained 0.1% ninhydrin in n-butanol, but subsequent investigators substituted various other solvents and some have preferred to increase the concentration of the reagent. The author employs a 0.25% solution of ninhydrin in 95% ethanol. Working in a ventilated hood, this is applied uniformly to the chromatogram in the form of a dense mist generated by a sprayer connected to the laboratory air supply. The colors develop either on prolonged standing at room temperature or within a few minutes in an oven at temperatures between 70° and 100°. For qualitative analysis rapid development is more convenient. At elevated temperatures the colors reach their final intensity more quickly if a tray of water is placed in the oven along with the sprayed papers to increase the humidity. Ninhydrin as supplied by commercial sources is usually satisfactory for qualitative purposes, but it may be recrystallized, if desired, according to the directions given by Moore and Stein.[42]

Less sensitive than the ninhydrin reagent, but useful in situations where it is desired to avoid destruction of the amino acids, is the spray reagent of Kemble and Macpherson.[43] These workers utilized the principle of the formol titration as the basis for their indicator spray which is prepared by adding 3 ml. of formalin and 0.1 ml. of 60% aqueous KOH to 20 ml. of 0.15% (w/v) solution of bromothymol blue in 95% ethanol. The reagent may be applied directly to chromatograms developed in neutral solvents whereupon the amino acids appear as yellow spots against a blue background. Chromatograms developed in solvents containing volatile, acidic, or basic components must first be steamed to remove the residual acids or bases which otherwise would control the response of the pH sensitive dye. In either case, the yellow spots must be outlined on the chromatogram for future reference, since the contrast between spots and background eventually disappears as the background fades from blue to yellow.

Occasionally, as for purposes of identification or for increased sensitivity, it is desirable to have available more or less specific spray reagents. Hence, a few of those more useful for selected amino acids will be described.

Histidine is one member of the amino acid family whose response to the ninhydrin spray is below average. The color commonly obtained, a greyed purple, is difficult to see. Histidine is readily detected by means of the Pauly diazo reagent with which it gives an intense cherry-red color. A suitable reagent can be prepared by mixing equal volumes of sulfanilic acid solution (1% in 1 N HCl, w/v) and aqueous sodium nitrite solution (0.7%, w/v). The area occupied by histidine is sprayed with the freshly prepared reagent, and the paper is allowed to lose excess moisture in a current of air for a few minutes. Finally the area is resprayed with 10% aqueous sodium carbonate. The characteristic color appears immediately if histidine is present. Other imidazoles and tyrosine also react. Residual phenol interferes by producing a background of similar color; hence chromatograms which have been run in this solvent must be freed of phenolic residues by washing with either benzene or acetone before the test. Since the reagent detects the imidazole group, it can be applied to chromatograms previously sprayed with ninhydrin.

[42] S. Moore and W. H. Stein, *J. Biol. Chem.* **176,** 367 (1948).
[43] A. R. Kemble and H. T. Macpherson, *Nature* **170,** 664 (1952).

Proline and hydroxyproline both give a yellow color with ninhydrin, but they may also be detected with an isatin reagent. A convenient formulation of the reagent, described by Smith,[44] contains 0.2% (w/v) of recrystallized isatin in acetone. This is sprayed onto the sheet over the areas occupied by the prolines. If the reaction is allowed to proceed at room temperature in the dark, only proline and hydroxyproline react and appear as blue spots. On the other hand, if, *after the acetone has evaporated*, the sprayed sheet is placed in an oven at approximately 65°, the prolines will appear as blue spots in about 15 minutes. Other amino acids give colors ranging from pink through purple, some changing to blue on standing, others fading.

Hydroxyproline may be detected specifically[45] if the isatin reaction at elevated temperatures just described is followed by a spray of Ehrlich's reagent. Under these conditions the blue color of hydroxyproline is uniquely transformed to an intense cerise. This test is both very sensitive and very selective. The Ehrlich reagent, used fresh, is prepared[44] by dissolving 1 g. of p-dimethylaminobenzaldehyde in a mixture containing 10 ml. of concentrated HCl and 90 ml. of acetone.

Tryptophan and other indole compounds may be detected directly with the Ehrlich reagent described in the preceding paragraph. Water may be substituted for the acetone, and the substitution may prevent clogging of sprayers prone to clog when used for reagents dissolved in highly volatile solvents. With Ehrlich's reagent, tryptophan gives a lavender color within a few minutes after spraying at room temperature.

Arginine responds to the Sakaguchi reagent with which it forms a red color. The sensitivity is less than with ninhydrin, but the reaction is fairly specific, since only a few other naturally occurring guanidine derivatives react. Combined arginine also reacts. Horne and Pollard[46] used the reagent to detect streptomycin on strip chromatograms. Their spraying procedure required three independent sprayings, but this may conveniently be reduced to two, as follows: spray first with a 0.2% solution of α-naphthol in 0.5 N NaOH; wait until excess moisture has evaporated and then spray with a 5% aqueous solution of sodium hypochlorite. The red color forms immediately.[47]

Amino acids normally do not fluoresce in ultraviolet light but can be induced to do so if heated in the presence of filter paper. Phillips[20] pioneered in the use of this phenomenon for the detection of amino acids

[44] I. Smith, *Nature* **171**, 43 (1953).
[45] J. B. Jepson and I. Smith, *Nature* **172**, 1100 (1953).
[46] R. E. Horne and A. L. Pollard, *J. Bacteriol.* **55**, 231 (1948).
[47] For a further discussion of colorimetric procedures for the determination of amino acids, see Vol. III [76].

on paper chromatograms. The fluorescence is apparently due to a reaction between the amino acids and free aldehydic groups in the paper.[18,19] Only a small fraction of the amino acid in each spot reacts, and the remainder is free to react with ninhydrin or other reagents.

It is commonly held that detection by fluorescence is less sensitive than with ninhydrin. It has, however, been the author's experience with chromatograms which first have been viewed under ultraviolet light and then sprayed with ninhydrin that the color developed with the latter did not extend quite to the boundaries previously outlined by fluorescence. This observation has been interpreted to indicate that, of the two methods of detection, fluorescence is the more sensitive. To induce fluorescence the chromatograms were heated for 4 to 6 hours in an oven at temperatures between 60° and 65°. They were illuminated immediately on removal with a General Electric B-H4 "black-light" lamp. The amino acids appeared as areas of blue-white fluorescence against a purplish background.

Bioautographic techniques for qualitative analysis of amino acids are not likely to find extensive use in competition with the other, more simple, methods which are available. Nevertheless they may occasionally be useful for specific requirements such as providing substantiating evidence for the identity of a given spot if the proper mutant organism is available. The studies of Long and Williams[48] on the *Lactobacillus bulgaricus* factors illustrate the use of this technique.

Radioautography constitutes a powerfully sensitive means for detecting compounds labeled with radioactive tracers on paper chromatograms. It is sufficient to place the chromatogram in intimate contact with a sheet of No-Screen X-ray film and to protect the film from light during the exposure. Depending on the quantity of radioactive material present, the exposure may be as short as a few hours or may require up to three weeks for spots containing only a few counts above background. Weak spots are extremely difficult to detect by scanning with the usual counting equipment, but the film, because it integrates the radioactivity throughout the exposure, is capable of revealing the spots clearly. After exposure the film is developed according to the manufacturer's directions.

The largest sheets of film readily available are 14 × 17 inches and are thus smaller than the standard chromatographic sheets. Consequently when radioautographs are contemplated it is convenient to outline the *chromatographic* area to be covered by the film before the sample is applied and then, in development, to keep each solvent from exceeding its marked limit. The spots will thus be confined to an area capable of being covered by the film.

The completed chromatograms are then folded to size with a sheet of

48 C. L. Long and W. L. Williams, *J. Bacteriol.* **61,** 195 (1951).

film used as a template, marked with radioactive ink, and taken to a darkroom where they are placed in contact with the film. For isotopes emitting weak radiations, such as C^{14} or S^{35}, elaborate equipment is not necessary. The cardboard sheets used for protective material in the original containers for the film provide a satisfactory substitute for a press or film holder. The marked chromatogram is slipped under a film in its individual folder, and the folder then may be sandwiched in between two of the cardboard separators. A stack can be built up with folders and separators alternating. Finally the stack is bound tightly with masking tape to ensure intimate contact between films and chromatograms and may be placed into an empty film box or wrapped in several layers of wrapping paper to protect it from light. Chromatograms of compounds labeled with more energetic isotopes such as P^{32} may be handled in similar fashion except that Duralumin separators must be substituted for the cardboard separators.

Marking the chromatograms with radioactive stamp-pad ink before exposure to the film provides a convenient means for accurately superimposing the film on the chromatogram after the film has been processed. To prepare the ink, enough nonvolatile radioactivity, such as from S^{35} or C^{14}, is added to stamp-pad ink so that the mark will produce between 100 and 200 counts per minute when counted with a thin-window Geiger-Müller tube. The marker may be any asymmetrical form carved into a small rubber stopper attached to a glass rod.

When many radioautograms are processed simultaneously, it is subsequently helpful in sorting them to have the films carry some identification. This can be accomplished in the darkroom before the chromatograms are placed in contact with the film by using a blunt pencil to mark the film with an identifying code. The film, sensitive to the pressure from the pencil point, will carry a permanent legend after being processed.

The work of Benson et al.[13] illustrates a specific application of the general principles involved.

Identification. When nonspecific methods have been used to reveal the location of the compounds on the chromatogram, it remains to identify each spot in order to fulfill the aim of qualitative analysis. This may be a formidable task, particularly if an hitherto unknown compound is involved. Nevertheless, techniques are available by which it is possible to accumulate evidence for the identity of a given spot on a chromatogram.

To begin with, the relative position occupied by a spot offers a clue for tentative identification by reference to a map plotted from data obtained by chromatographing known compounds under the same conditions. For this, relative position is a more reliable guide than absolute R_f value because the latter is subject to much greater variation.

Close observation and experience soon will enable one to recognize some obvious and some subtle characteristics exhibited by individual compounds. Among these are such qualities as speed of response to the spray reagent, color, and shape of spot. These, together with relative position, permit a good guess as to identity. Supporting evidence, however, must be obtained. For this, many approaches are possible and are limited only by the ingenuity of the investigator.

If a specific reagent for the compound in question is available, the tentative identification can be tested directly by spraying a duplicate chromatogram. If not, it is possible to test the identity by cochromatographing the unknown with an authentic sample of the known compound. For this the tentatively identified spot is cut out, eluted from the paper, and mixed with a few micrograms of the suspected known compound. Finally the chromatographic behavior of the mixture is observed. If known and unknown are identical, the mixture will yield a single spot. Even so, this does not necessarily establish identity—compelling evidence based on cochromatographic behavior requires that the mixture be tested in several diffrent solvent systems.

Cochromatography is especially valuable in testing the provisional identity of compounds labeled with radioactive isotopes. Here the chromatogram of the mixture first provides a radioautogram which later can be superimposed on the sprayed chromatogram for comparison of the correspondence between colored area on the paper and exposed area on the film. If the radioactive unknown and nonradioactive known are identical, the blackened area on the film corresponds in every detail, such as size, shape, and presence of irregularities, with the colored area. This is analogous to the isotopic ratio method of Udenfriend.[49] If the known and unknown are not identical, the distribution of radioactivity will not be uniform throughout the colored area even when known and unknown are closely overlappcd and undifferentiated by color alone. Nonidentity is thus easily and clearly recognized.

Where radioactively labeled knowns are available, the techniques of cochromatography and radioautography can be combined in reverse, i.e., to test the identity of nonradioactivc eluates.

Confirmatory evidence for identity may also be obtained by showing that an eluate of the unknown spot and a sample of the suspected known can, under identical conditions, both be converted to the same derivative as shown by comparing the chromatographic behavior of the reaction products. A typical illustration is Dent's[31] oxidation of methionine to its sulfone and of cystine to cysteic acid. Keston et al.,[50] using labeled

[49] S. Udenfriend, *J. Biol. Chem.* **187,** 65 (1950).
[50] A. S. Keston, S. Udenfriend, and M. Levy, *J. Am. Chem. Soc.* **69,** 3151 (1947).

reagents, have applied an elegant modification of the principle to test the identity of the I^{131}-pipsyl derivatives of the amino acids. They established identity either by adding the corresponding nonradioactive pipsyl derivative and demonstrating unchanged isotopic concentration after repeated recrystallization, or by adding the corresponding S^{35}-pipsyl derivative, chromatographing the mixture, and showing an unchanged isotopic ratio (I^{131}/S^{35}) throughout the resulting spot.

Enzymatic conversions may also be utilized in tests of identity and in favorable instances may offer the advantages of specificity. To illustrate, an eluate of a spot tentatively identified as glutamic acid and authentic glutamic acid can each be incubated with glutamic acid decarboxylase. The presence of γ-aminobutyric acid on parallel chromatograms of both incubation mixtures provides strong evidence in support of the tentative identification. Synge[51] and Jones[23] have described the use of D-amino acid oxidase for determining the configuration of amino acids on paper chromatograms.

Comment. Paper chromatography as a technique for qualitative analysis provides only microgram quantities of materials, and hence the classical methods of chemical identification are, in a practical sense, precluded. It is therefore necessary to rely on indirect evidence to establish the identity of any given spot. To make the identification, and the conclusions based on it, meaningful requires that all available means be exploited in support of the identification; R_f values alone do not suffice to establish identity. On the other hand, judicious use of the available identification procedures combined with the specificity inherent in the chromatographic method has provided, and can continue to provide, much valuable information.

Common Difficulties. In conclusion some of the more commonly encountered difficulties and possible remedies may be enumerated.

Slanted fronts may be caused by poorly leveled cabinets. This condition causes solvent to pile up at one end of the trough where more solvent is fed to the sheet as compared with the high end of the trough. Leveling usually eliminates the difficulty. If it does not, or if the difficulty is more noticeable during the second run, it may be suspected that decomposition products from the first solvent may play a role. These are usually concentrated near the boundary of the first solvent and may retard the movement of the second solvent over this area. The remedy depends on the nature of the first solvent. If it is phenol, the following may help: elimination of heavy metal contaminants from the paper and from the water used in liquefying the phenol; discontinuing the use of ammonia

[51] R. L. M. Synge, *Biochem. J.* **44**, 542 (1949).

vapors in the chamber; and drying at room temperature instead of at elevated temperatures.

Streaked spots may result from a variety of causes. Overloading at the origin and excessive salt concentration in the sample are two of the most commonly encountered causes. The first can be corrected by decreasing the size of the aliquot or by increasing the area of the origin; the second by desalting. Occasionally, streaking may be traced to contact between the solutes and some moist surface film such as may be formed on the antisiphoning rod. Reorienting the origin to preclude such contact will eliminate difficulties arising from this cause.

Temperature fluctuations, if of sufficient amplitude to bring about a separation of phases in the solvent during chromatography, invariably produce streaked spots. The condition can be recognized by the indistinct solvent boundary and by the water-logged appearance of the paper and sometimes by the condensate which forms on the window of the chamber. Incomplete removal from the paper of a prior solvent is another factor which similarly may influence the stability of a succeeding solvent. Some of the three- and four-component mixtures in use, being completely saturated with water, are delicately poised at the critical line and are therefore easily disturbed. An excessive amount of residual solvent from the preceding run may do so through its influence on the mutual solubilities of the components.

Thorough drying between runs will remedy the latter difficulty, while closer control of the temperature will reduce losses from that cause. A measure of protection against both disturbances can be obtained by formulating the solvent with somewhat less water than is required for saturation at the working temperature. In this way solvents capable of withstanding fluctuations of 5° can be prepared. These also will be buffered against the effects of residual solvents.

Summary. The following sequence presents in summary form the operations involved in a typical analysis. It is intended to serve as a flexible guide subject to modification as dictated by experience.

1. With unfamiliar samples, suspected to contain all or many of the amino acids, begin by concentrating the sample to contain 2.5 to 12.5 mg. of total amino acids per milliliter or the equivalent in amino nitrogen. In material, such as incubation mixtures, where the concentration of the components can be controlled, attempt to adjust the concentration so that a 20- to 50-μl. aliquot will contain 10 to 20 γ of each amino acid. Such mixtures may be sampled directly without concentration.

2. Determine the optimal aliquot by spotting the sample in a graded series of spots, ranging from 5 to 100 μl., along one edge of a standard

sheet. Chromatograph unidimensionally in phenol solvent and spray with ninhydrin reagent. Guided by the color intensity and the resolution, select the optimal aliquot.

3. Prepare sheets for two-dimensional chromatography. Apply the optimal aliquot to the origin near one corner of a filter paper sheet, working in a stream of warm air if more than 20 μl. needs to be applied.

4. Load troughs, transfer to the chromatographic chamber, and chromatograph in the first direction. Allow the solvent to migrate to within 1 cm. from the bottom edge of the sheet.

5. Clip each sheet to its antisiphoning rod and transfer to the drying cabinet. Leave until the solvent has been adequately removed.

6. Prepare the dried sheets for the second run. Follow step 4 but use the second solvent.

7. Repeat step 5.

8. Detect spots by spraying the chromatograms with ninhydrin reagent.

9. Resolve suspected overlaps by spraying duplicate chromatograms with specific reagents or by using selected solvents to rerun an eluate of the overlapped area. The latter can be located by fluorescence under ultraviolet light on heat-treated duplicate chromatograms.

10. Tentatively identify the spots observed in the two preceding steps by comparing their chromatographic pattern with a standard map and by noting individual characteristics such as hue and shape of spot.

11. Obtain confirmatory evidence by a combination of tests including cochromatography, specific reagents, derivatives, etc.

[80] Procedures for the Synthesis of Substrates for Peptidases

By EMIL L. SMITH

Many peptides suitable as substrates for the assay of peptidases have been prepared by synthetic methods. It is obviously impossible to describe all these compounds and the many satisfactory alternative methods which can be devised for their synthesis. The concern here will be with a few methods which illustrate the general principles and which, at the same time, give specific directions for preparing certain peptides which may be used for the assay of particular enzymes. An excellent survey of the field of synthetic peptide chemistry has been made by Fruton;[1] his article contains a listing of compounds prepared up to 1949.

[1] J. S. Fruton, *Advances in Protein Chem.* **5**, 1 (1949).

Method I (Fischer)

The general principle of this method involves the coupling of an α-halogen acid halide to an amino acid (or its ester), followed by amination of the product. The method is mostly employed at present for the preparation of certain glycyl derivatives because of the general availability of chloroacetyl chloride. Only a few other α-halogen acid halides or α-halogen acids are available commercially as the optically inactive compounds. The preparation of glycyl-L-leucine and of triglycine (glycylglycylglycine) by the Fischer method is given below.

The specific reactions are as follows:

(1A) Chloroacetyl chloride + L-leucine + NaOH
$$\rightarrow \text{Chloroacetyl-L-leucine} + \text{NaCl}$$

(1B) Chloroacetyl-L-leucine + $NH_3 \rightarrow$ Glycyl-L-leucine + NH_4Cl

(2A)   Diketopiperazine + $H_2O \xrightarrow{\text{alkali}}$ glycylglycine

(2B) Chloroacetyl chloride + glycylglycine + NaOH
$$\rightarrow \text{Chloroacetylglycylglycine} + \text{NaCl}$$

(2C) Chloroacetylglycylglycine + $NH_3 \rightarrow$ Diglycylglycine + NH_4Cl

The preparation of many similar compounds may be found in the volumes of Fischer's collected papers.[2]

Method II (Bergmann and Zervas)

For the preparation of more complex peptides, particularly when an optically active residue is to be in the N-terminal position, the method of choice is the carbobenzoxy method of Bergmann and Zervas.[3] In this method, the carbobenzoxy group is used to protect the amino group of the N-terminal residue during the coupling reaction. The coupling may be achieved in various ways: by acid halide, by acid azide, or, as shown more recently, by mixed acid anhydrides.[4,5] The carbobenzoxy group is subsequently removed by catalytic hydrogenation at atmospheric pressure in the presence of a palladium catalyst. The main steps for the preparation of glycyl-L-proline, carnosine nitrate (β-alanyl-L-histidine nitrate), and hydroxy-L-prolylglycine are shown, and the detailed

[2] E. Fischer, "Untersuchungen über Aminosäuren, Polypeptide and Proteine," Vol. 1, Springer, Berlin, 1906; Vol. 2, 1923.
[3] M. Bergmann and L. Zervas, *Ber.* **65**, 1192 (1932).
[4] J. R. Vaughan, Jr., *J. Am. Chem. Soc.* **73**, 3547 (1951); J. R. Vaughan, Jr., and R. L. Osato, *ibid.* **73**, 5553 (1951).
[5] R. A. Boissonnas, *Helv. Chim. Acta* **34**, 874 (1951).

description is given below. The carbobenzoxy (carbobenzoxyloxy) group ($C_6H_5CH_2OCO$—) is indicated by the abbreviation Cbz.

(3) $C_6H_5CH_2OH + ClCOCl \rightarrow C_6H_5CH_2—O—COCl + HCl$
    Benzyl alcohol + phosgene $\rightarrow$ Cbz chloride

(4)   Cbz chloride + glycine + NaOH $\rightarrow$ Cbz-glycine + NaCl

(5A)   Cbz-glycine $\xrightarrow{PCl_5}$ Cbz-glycyl chloride

(5B) Cbz-glycyl chloride + L-proline
    $\xrightarrow{NaOH}$ Cbz-glycyl-L-proline + NaCl

(5C)   Cbz-glycyl-L-proline $\xrightarrow[\text{palladium}]{H_2}$ Glycyl-L-proline + toluene + CO_2

(6A)   L-Histidine HCl $\xrightarrow[\text{HCl}]{\text{methanol}}$ L-Histidine methyl ester HCl

(6B) Cbz chloride + β-alanine + NaOH $\rightarrow$ Cbz-β-alanine + NaCl

(6C) Cbz-β-alanine + CH_3OH $\xrightarrow{HCl}$ Cbz-β-alanine methyl ester + H_2O

(6D) Cbz-β-alanine methyl ester + hydrazine
    $\rightarrow$ Cbz-β-alanine hydrazide + CH_3OH

(6E) L-Histidine methyl ester di-HCl + 2Na
    $\xrightarrow{\text{methanol}}$ L-Histidine methyl ester + 2NaCl

(6F)   Cbz-β-alanine hydrazide $\xrightarrow{HNO_2}$ Cbz-β-alanine azide

(6G) Cbz-β-alanine azide + L-histidine methyl ester
    $\rightarrow$ Cbz-β-alanyl-L-histidine methyl ester

(6H) Cbz-β-alanyl-L-histidine methyl ester
    $\xrightarrow{NaOH}$ Cbz-β-alanyl-L-histidine (Cbz-carnosine) + CH_3OH

(6I) Cbz-β-alanyl-L-histidine
    $\xrightarrow[\text{HNO}_3]{H_2,\ Pd}$ β-Alanyl-L-histidine·HNO_3 (carnosine nitrate)

(7A) Hydroxy-L-proline + methanol + HCl
    $\rightarrow$ Hydroxy-L-proline methyl ester HCl

(7B)   Cbz-glycine $\xrightarrow{PCl_5}$ Cbz-glycyl chloride

(7C) Cbz-glycyl chloride $\xrightarrow{\text{heat}}$ CO—NH—CH₂—CO (glycine
    └———O———┘
    carbonic acid anhydride) + benzyl chloride

(7D) Glycine carbonic acid anhydride + benzyl alcohol + HCl
$$\rightarrow \text{Glycine benzyl ester hydrochloride} + CO_2$$

(7E) Cbz chloride + hydroxy-L-proline methyl ester
$$\rightarrow \text{Cbz-hydroxy-L-proline methyl ester}$$

(7F) Cbz-hydroxy-L-proline methyl ester + hydrazine
$$\rightarrow \text{Cbz-hydroxy-L-proline hydrazide}$$

(7G) Cbz-hydroxy-L-proline hydrazide + HNO_2
$$\rightarrow \text{Cbz-hydroxy-L-proline azide}$$

(7H) Cbz-hydroxy-L-proline azide + glycine benzyl ester
$$\rightarrow \text{Cbz-hydroxy-L-prolylglycine benzyl ester}$$

(7I) Cbz-hydroxy-L-prolylglycine benzyl ester
$$\xrightarrow{\text{Pd, } H_2} \text{Hydroxy-L-prolylglycine}$$

Method III (Amino Acid Amides)

Certain aminopeptidases hydrolyze simple amino acid amides, and such compounds are relatively easy to prepare. Most (but not all) of the amino acid amide hydrochlorides may be obtained in satisfactory yield by direct amidation at room temperature in anhydrous methanol-ammonia of the corresponding amino acid ester hydrochloride. The method is illustrated by the synthesis of L-leucinamide hydrochloride.

(8A) L-Leucine + ethanol $\xrightarrow{0°, \text{ dry HCl}}$ L-Leucine ethyl ester HCl

(8B) L-Leucine ethyl ester HCl $\xrightarrow[\text{methanol} - NH_3]{\text{anhydrous}}$ L-Leucinamide HCl

Amide hydrochlorides of other amino acids have also been prepared by this method, including those of L-tryptophan, L-phenylalanine, L-tyrosine, L-proline, L-alanine, hydroxy-L-proline, DL-norleucine, DL-norvaline, α-amino-N-butyric acid, L-aspartic acid (as the diamide), and L-histidine. The method fails or results in poor yields with lysine and glutamic acid.

Preparation of the corresponding esters of valine, isoleucine, and alloisoleucine requires reflux conditions, and the amidation of these esters takes longer and results in poorer yields.

Glycyl-L-leucine[2]

Chloroacetyl-L-leucine (1A). L-Leucine (5 g.) is dissolved in 38.2 ml. of N NaOH and cooled to 0°. There are added simultaneously with shaking and cooling in five portions 8.6 g. of chloroacetyl chloride and 27 ml. of 5 N NaOH over a period of about 20 minutes. About 10 minutes after the final portion is added, the mixture is extracted with ethyl ether to remove excess acid chloride, and the aqueous solution is acidified to Congo red with 6 N HCl. After cooling, the crystalline product is collected and washed with cold water. The yield is about 6.7 g.; melting point, p. 136°. The compound may be recrystallized from about 11 vol. of hot water.

Glycyl-L-leucine (1B). The above chloroacetyl compound (5 g.) is finely pulverized and added to 10 vol. of 25% aqueous ammonia and allowed to stand at room temperature in a pressure bottle until complete solution occurs; this requires approximately 2 days. The mixture is then concentrated at 30 to 50° under reduced pressure. The crystalline product is repeatedly extracted with hot 90% ethanol to remove ammonium chloride. The residue is then dissolved in 10 vol. of hot water and treated with Norit. Hot absolute ethanol (2 vol.) is then added cautiously to the aqueous solution. Yield, about 3.5 g.; $[\alpha]_D^{20} = -35°$ (water).

Triglycine (Diglycylglycine)[2]

Chloroacetylglycylglycine (2A and 2B). Finely pulverized glycine anhydride (10.8 g.) is dissolved in 54 ml. of 2 N NaOH by shaking at room temperature. Fifteen minutes after complete solution has been achieved, the solution is cooled to 0° and there are added gradually over 40 minutes with cooling and stirring 12 g. of chloroacetyl chloride and 26 ml. of 5 N NaOH. The solution is then acidified to Congo red with 6 N HCl (about 23 ml.) and allowed to stand at 0°. Yield, about 11 g., although an additional 2 or 3 g. may be obtained by concentration of the mother liquor; melting point, 178 to 180°.

Triglycine (2C). The chloroacetyl compound (10 g.) is dissolved in 50 ml. of 25% aqueous ammonia and allowed to remain in a pressure bottle for 24 hours at room temperature. The solution is concentrated to dryness under reduced pressure at 30 to 50°. The residue is dissolved in 100 ml. of water, and 250 ml. of ethanol is added. The crystals are collected, and the recrystallization is done once more to remove all the ammonium chloride. Yield, about 7 g.; melting point, near 182° with decomposition. The tripeptide is an excellent substrate for aminotripeptidase (see Vol. II [9]).

Glycyl-L-proline[6]

Carbobenzoxy Chloride[3,7] (3). **This entire procedure must be performed in a well-ventilated hood!** One liter of toluene in a 2-l. Claisen flask containing boiling chips is cooled in an ice bath, and 200 g. of dry phosgene

---

[6] M. Bergmann, L. Zervas, H. Schleich, and F. Leinert, *Z. physiol. Chem.* **212,** 72 (1932).

[7] H. E. Carter, R. L. Frank, and W. H. Johnston, *Org. Synthesis* **23,** 13 (1942).

gas from a cylinder is slowly bubbled in. (The amount of phosgene absorbed is checked by weighing the flask and tubes.) The inlet tube should dip into the solution. The outlet tube should be conducted to a toluene trap at 0° and then to the exhaust flue of the hood. (It is a good precaution to keep handy a methanolic solution of ammonia for the destruction of phosgene. This should be added to the trap at the end of the experiment.) To the cold phosgene solution, 180 ml. of benzyl alcohol (reagent grade or redistilled) is added; the mixture is swirled once to mix and then allowed to remain in the ice bath for about an hour. The reaction mixture is then removed from the ice bath and permitted to stand under the hood at room temperature for several hours. When the reaction is complete, dry air or carbon dioxide is passed into the solution to remove the excess phosgene and the liberated HCl. The solution is then concentrated at reduced pressure in a water bath at 30 to 40° in order to remove the excess toluene; the temperature is finally permitted to rise to about 60°. About 240 g. of the acid chloride, which is suitable for synthetic work, is obtained. It should be stored in the refrigerator (0 to 5°) in the dark. When freshly prepared, the carbobenzoxy chloride is colorless or a slightly yellow or greenish yellow; after some months, it will finally become dark brown, but it may still be used for the carbobenzoxylation of simple amino acids, e.g., glycine, β-alanine.

N-Carbobenzoxyglycine[3] *(4)*. To a solution of 7.5 g. of glycine in 25 ml. of 4 N sodium hydroxide at 0° are added with cooling and shaking in 5 portions over a period of about 20 minutes 17 g. of carbobenzoxy chloride and an additional 25 ml. of 4 N sodium hydroxide. On acidification to Congo red with concentrated hydrochloric acid, the compound crystallizes in long prisms. The yield is about 75% of theory. The substance is recrystallized from hot chloroform or from methanol-water. Melting point, 120°.

N-Carbobenzoxyglycyl Chloride *(5B)*. To 6.3 g. of finely pulverized carbobenzoxyglycine suspended at 0° in 35 ml. of anhydrous diethyl ether (alcohol-free) is added, with cooling and shaking, 6.7 g. of finely pulverized phosphorus pentachloride. When solution is nearly complete (15 to 20 minutes), the solution is decanted or filtered to remove the excess PCl_5 and is rapidly concentrated to a thick sirup *in vacuo* with no heat applied to the distilling flask. Frost should form on the outside of the flask. If the compound crystallizes, it is washed on a Büchner funnel with cold petroleum ether (dried over P_2O_5). The compound may be crystallized by washing the residue three times by decantation with anhydrous petroleum ether at 0°, followed by chilling under dry petroleum ether in an alcohol-dry ice bath. The yield is about 5.3 g. The acid chloride must be prepared fresh and used immediately for the coupling reaction.

N-Carbobenzoxyglycyl-L-proline (5B). To a solution of 5 g. of L-proline in 25 ml. of 2 N sodium hydroxide are added in portions at 0° over a period of 20 minutes, with cooling and shaking, an additional 20 ml. of 2 N sodium hydroxide and 10 g. of carbobenzoxyglycyl chloride. The solution is acidified to Congo red with 6 N HCl. The compound is extracted into ethyl acetate; the solution is dried over Na₂SO₄ and concentrated to a small volume. After a period of standing in the refrigerator, the yield is about 10 g. After recrystallization from ethyl acetate, the melting point is 156°.

Glycyl-L-proline (5C). Hydrogen is bubbled through a solution of 3 g. of the carbobenzoxy dipeptide in 50 ml. of methanol with 1.5 ml. of water and 1.5 ml. of glacial acetic acid in the presence of palladium black.[8,9] After about 30 minutes the evolution of carbon dioxide ceases as tested with a solution of BaCl₂ in which the exit tube is dipped. The solution is filtered to remove the catalyst, diethyl ether is added, and the compound is allowed to crystallize slowly in the cold. The yield is 80 to 100% of theoretical. The dipeptide is obtained as the hemihydrate in the form of prisms (melting point, 185°). This peptide is a substrate for prolidase (see Vol. II [10]).

Carnosine (β-alanyl-L-histidine)[10-12]

Histidine Methyl Ester Dihydrochloride[13] (6A). Thirty grams of dry, finely powdered histidine monohydrochloride is covered with 450 ml. of dry methanol, the mixture is heated under reflux, and dry HCl is passed into the mixture until all the material dissolves (3 to 4 hours). The solution is then cooled to 0°, and the histidine methyl ester dihydrochloride is allowed to crystallize. This material is re-esterified, collected, and dried

---

[8] A simple method of performing such hydrogenations is to use an Erlenmeyer flask with an inlet tube, leading from the hydrogen, which nearly touches the bottom of the flask, and an exit tube at the top of the flask leading to the atmosphere. Rubber connections should be avoided because of the poisonous effect of the sulfur on the catalyst. It is preferable to select an all-glass apparatus with ground-glass seal. The contents of the flask are mixed by a magnetic stirrer while the hydrogen is passed into the reaction mixture.

[9] The catalyst may be prepared by the procedure of R. Willstätter and E. Waldschmidt-Leitz [*Ber.* **54**, 128 (1921)]. A satisfactory catalyst which is commercially available is the 84.1% palladium oxide (American Platinum Works, Newark, New Jersey).

[10] The procedure given here is essentially that of R. H. Sifferd and V. du Vigneaud [*J. Biol. Chem.* **108**, 753 (1953)]. Some modifications developed later[11,12] have been incorporated.

[11] H. T. Hanson and E. L. Smith, *J. Biol. Chem.* **179**, 789 (1949).

[12] N. C. Davis and E. L. Smith, *Biochem. Preparations* **4**, 38 (1955).

[13] E. Fischer and L. H. Cone, *Ann.* **363**, 107 (1908); see also E. Fischer.[2]

in a vacuum desiccator over solid sodium hydroxide (melting point, 197 to 198°; yield, 28 to 30 g.).

Carbobenzoxy-β-alanine (6B). Forty-five grams of β-alanine is dissolved in 250 ml. of 2 N NaOH. The solution is cooled to 0° and then, with good stirring and cooling, 93.5 g. of carbobenzoxy chloride is added dropwise from a separatory funnel simultaneously with 125 ml. of 4 N NaOH added in the same manner. The total time for the addition should be about 45 minutes. The stirring is continued at room temperature for an additional 20 minutes. The reaction mixture is then washed in a 1-l. separatory funnel with 5 × 100 ml. of ether to remove unreacted carbobenzoxy chloride. (A few drops of pyridine in methanol are added to the ether extracts to destroy the excess acid chloride. The mixture is allowed to evaporate under a hood.)

The aqueous solution is transferred to a 1-l. beaker and slowly acidified (foaming!) to Congo red by the addition of 6 N HCl. The solution is allowed to stand in the cold (5 to 10°). The carbobenzoxy-β-alanine is recrystallized by being dissolved in the minimum amount of hot methanol with hot water being added until an opalescence is noted. The mixture is cooled at room temperature and then overnight at 0°. The recrystallized material separates as white platelets (melting point, 104 to 105°; yield, 90 to 95 g.).

Carbobenzoxy-β-alanine Hydrazide[11] *(6C and 6D)*. Sixty grams of recrystallized carbobenzoxy-β-alanine is dissolved in 500 ml. of dry methanol, and the solution is cooled in an ice-salt bath. Then, with continued cooling, anhydrous HCl gas is passed in for 2 hours. The solution is protected from air by a calcium chloride tube and allowed to warm up to room temperature. It is then concentrated to a thick sirup under reduced pressure at 20 to 40°. The concentration is repeated at least three times with 50 ml. of methanol each time in order to remove all traces of HCl. The residual sirup is dissolved in 100 ml. of absolute ethanol, rinsed into a 250-ml. Erlenmeyer flask, and 47 g. (a threefold excess) of 85% (or better grade) hydrazine hydrate is added to the carbobenzoxy-β-alanine methyl ester. The solution is thoroughly mixed and allowed to stand at room temperature for 2 to 3 hours and then in the refrigerator overnight. The product usually begins to crystallize in a few minutes. The carbobenzoxy-β-alanine hydrazide is collected on a Büchner funnel; it is thoroughly washed with cold absolute ethanol and air-dried (melting point 147 to 148°; yield, 55 to 58 g.).

Coupling Procedure. The two compounds, histidine methyl ester and carbobenzoxy-β-azide, which are used for the coupling, must be freshly prepared. It is most convenient to prepare the free ester first. While it is drying over sodium sulfate (see below), the azide is prepared and dried over sodium sulfate. The two solutions may then be filtered simultaneously and mixed.

Preparation of Free Histidine Methyl Ester[13] (*6E*). Twenty-nine grams of finely powdered histidine methyl ester dihydrochloride is dissolved in 300 ml. of absolute methanol; heating and stirring are necessary. When all the material is dissolved, the flask is rapidly cooled to 10°. Before crystallization begins, a cold solution of 5.52 g. of sodium in 150 ml. of absolute methanol is added, and the mixture is thoroughly swirled. Sodium chloride begins to precipitate almost immediately. To complete the removal of the salt, 200 ml. of ether is added; the mixture is again thoroughly mixed and is kept in the cold bath for 15 to 20 minutes. The sodium chloride precipitate is finely divided and difficult to remove. The filtration is best performed by gravity filtration with Whatman No. 50 fluted filter paper. The filtrate is collected in a 1-l. round-bottomed flask and concentrated to a thick sirup under reduced pressure at 30 to 40°. Chloroform which has been purified over CaCl$_2$ (25 ml.), is added, and the solution is again concentrated. This is repeated twice more to remove the last traces of methanol. The residue is then extracted with 4 × 50 ml. of chloroform. The chloroform extracts are combined, dried over Na$_2$SO$_4$, and filtered into a 500-ml. round-bottomed flask. The filtrate is concentrated under reduced pressure at 20 to 40° to approximately 100 ml. This solution of the histidine free ester is protected from moisture by a calcium chloride tube and kept at 0° or below until the azide is prepared.

Carbobenzoxy-β-alanine Azide (*6F*). Carbobenzoxy-β-alanine hydrazide (35.5 g.) is dissolved in a mixture of 600 ml. of water, 120 ml. of glacial acetic acid, and 120 ml. of 5 N hydrochloric acid. The resulting solution is cooled to 0°, and with good stirring a cold solution of 12 g. (0.27 mole) of sodium nitrite in 50 ml. of water is added slowly over a period of 10 minutes. The mixture is then stirred for another 10 minutes, and 100 ml. of cold chloroform is added. The contents of the flask are transferred to a 1-l. separatory funnel, and the chloroform solution of the azide is separated. The flask is rinsed with a 40-ml. portion of chloroform which is used for a second extraction. The combined chloroform extracts are dried over sodium sulfate and used immediately for the condensation step.

Carbobenzoxycarnosine Methyl Ester[14] (*6G*). The chloroform solution of carbobenzoxy-β-alanine azide is filtered into the chloroform solution of L-histidine methyl ester; the solution is thoroughly mixed, protected from moisture by a calcium chloride tube, and allowed to stand under the hood

[14] The coupling procedure and work up of the compound are due to Sifferd and du Vigneaud;[10] the crystallization and characterization of the carbobenzoxycarnosine ester to Davis and Smith.[12]

at room temperature overnight. (The highly toxic HN_3 is liberated during the reaction.) The resulting solution is concentrated to a thick sirup *in vacuo* at 30 to 40°. A thick gum is precipitated by the addition of 300 ml. of dry ether. Chloroform (30 ml.) is added, and the mixture is cautiously warmed on a steam bath and stirred until the gum is completely converted to cream-colored microneedles. The flask is cooled, and the crystalline carbobenzoxy-carnosine methyl ester is collected on a Büchner funnel, washed with ether, and air-dried. Melting point, 95 to 96°; yield, 37.5 g.; $[\alpha]_D^{24} = -13.25°$ (2% in 1% HCl).

Carbobenzoxycarnosine (6H). Ten grams of the carbobenzoxycarnosine methyl ester is saponified with 30 ml. of 1 N NaOH in a mixture of 50 ml. of acetone and 50 ml. of water. After 1 hour, the solution is neutralized exactly with 5 ml. of 6 N HCl and concentrated to dryness under reduced pressure at 40 to 50°. The residue is extracted with 4 × 50 ml. of hot absolute ethanol, and the ethanol extracts are combined and concentrated under reduced pressure at 40 to 50°. The residue is extracted with 4 × 50 ml. of hot absolute ethanol, and the ethanol extracts are combined and concentrated under reduced pressure until crystallization begins. The preparation is warmed on the steam bath and then allowed to cool slowly overnight. The white microneedles are collected on a Büchner funnel and air-dried; melting point, 166 to 167°; yield, 8 g. This material is suitable for reduction without recrystallization. When recrystallized from the minimum amount of hot methanol, the product melts at 171 to 172°; $[\alpha]_D^{23} + 16.4°$ (C, 1, water).

Carnosine Nitrate (6I). The carbobenzoxycarnosine (8.1 g.) is dissolved in 90 ml. of 50% methanol containing an exact equivalent of nitric acid (1.41 ml.; specific gravity, 1.42). Hydrogen is passed into the solution in the presence of palladium catalyst[9] until the evolution of carbon dioxide ceases (3 to 4 hours).[8] The catalyst is removed by filtration on a Celite pad, and 2 vol. of absolute ethanol is added to the clear filtrate, which is then slowly cooled to 0 to 5°. Carnosine nitrate crystallizes as long colorless needles; melting point, 222 to 223° d.; yield, 5.85 g.; $[\alpha]_D^{20} + 23.0°$ (C, 5.2, water).[15] The compound is a substrate for carnosinase (see Vol. II [10]).

Hydroxy-L-prolylglycine[16]

Hydroxy-L-proline Methyl Ester Hydrochloride (7A). The amino acid is suspended in 7 vol. of methanol at 0° which is then saturated with

---

[15] If the free base is desired instead of the carnosine salt, the carbobenzoxy compound is hydrogenated in the presence of sulfuric acid and the sulfate subsequently removed with barium as described by Sifferd and du Vigneaud.[10]

[16] E. L. Smith and M. Bergmann, *J. Biol. Chem.* **153**, 627 (1944).

dry HCl. The solvent is removed *in vacuo*, and the process is repeated. The yield is about 90%. After recrystallization from methanol-ether, the melting point is 163 to 164°.

Glycine Benzyl Ester Hydrochloride (7B, 7C, 7D). Finely pulverized carbobenzoxyglycine (15 g.) is treated with 16 g. of pulverized PCl₅ in 75 ml. of anhydrous ether at room temperature until solution is nearly complete. The excess PCl₅ is removed by filtration, and the solution is concentrated *in vacuo*, and finally heated to 50° for 5 minutes. Ether is added, and the glycine carbonic acid anhydride is collected and washed with petroleum ether. The anhydride is added to 25 ml. of benzyl alcohol which has been incompletely saturated with dry HCl at 0°. The anhydride dissolves with evolution of CO_2. A first crop of ester hydrochloride is removed by filtration, and a second crop may be obtained by addition of ether. The yield is about 10 g. after drying in a desiccator over NaOH pellets. Melting point, 139 to 140°.

Carbobenzoxyhydroxy-L-proline Hydrazide (7E and 7F). To 10 g. of hydroxyproline methyl ester hydrochloride in 50 ml. of water with 100 ml. of chloroform are added in portions with cooling and shaking 3 g. of MgO and 15 g. of carbobenzoxy chloride. Twenty minutes after the last addition, 2 ml. of pyridine is added. Then 5 N HCl is added until the aqueous layer is acid to Congo red. The chloroform layer is washed successively with cold water, dilute bicarbonate, water, and dilute HCl; it is then dried over Na₂SO₄. The solution is concentrated *in vacuo* at 30 to 40° with the repeated addition of absolute ethanol. The sirup thus obtained is dissolved in 50 ml. of absolute ethanol and filtered; 3.8 g. of hydrazine hydrate is added, and the solution is allowed to stand at room temperature overnight. The slight precipitate is removed by filtration, and the solution is then concentrated *in vacuo* repeatedly with ether. Yield, 8 to 10 g. of needles which are recrystallized from ethyl acetate-ether; melting point, 149 to 149.5°.

Carbobenzoxyhydroxy-L-prolylglycine Benzyl Ester (7G and 7H). Five grams of the hydrazide is converted to the azide by being dissolved in 50 ml. of water with 1 ml. of concentrated HCl and 3 ml. of glacial acetic acid. A solution of 1.5 g. of sodium nitrite in 5 ml. of water is added with cooling and shaking. After a few minutes, the oily precipitate is extracted into ethyl acetate and washed successively with cold water, cold bicarbonate, and again with cold water. After drying over sodium sulfate, the solution of the azide is added to a dry ethyl acetate solution of glycine benzyl ester previously prepared from 5.4 g. of the hydrochloride. (The free ester is prepared by dissolving the ester hydrochloride in the minimal amount of water which is then overlayered with 50 ml. of ethyl acetate. With strong shaking and cooling in an ice-salt bath, a 10% excess of

10 N NaOH is added, followed immediately by anhydrous Na_2CO_3 added in portions until the salt clumps as a solid which retains all the water. The ethyl acetate solution is decanted, and the residue is extracted three times more with 10- to 15-ml. portions of ethyl acetate. The combined extracts are dried over Na_2SO_4.[17]) The coupling mixture is allowed to stand overnight under a hood at room temperature. The reaction mixture is washed with dilute HCl and water and then dried over sodium sulfate. The solution is then concentrated *in vacuo* to yield 4.5 to 5.0 g. of needles. The substance is recrystallized from ethyl acetate. Melting point, 153°.

Hydroxy-L-prolylglycine (7I). Three grams of the above carbobenzoxy dipeptide benzyl ester is hydrogenated[8] at atmospheric pressure in methanol containing 2 ml. of glacial acetic acid and 2 ml. of water in the presence of palladium black[9] until the evolution of CO_2 ceases. The peptide which crystallizes during the hydrogenation is dissolved by the addition of a little hot water. The solution is filtered to remove the catalyst and is then repeatedly concentrated *in vacuo* with the addition of methanol. Yield, 1.3 g. of needles. The dipeptide is recrystallized from water-methanol. $[\alpha]_D^{26} = -22.4°$ (water). The peptide is a substrate for iminodipeptidase (see Vol. II [9]).

L-Leucinamide Hydrochloride[18]

L-*Leucine Ethyl Ester Hydrochloride (8A)*. A preparation of L-leucine (preferably free of methionine) is suspended in 10 vol. of absolute methanol at 0° and kept in an ice-salt bath while dry HCl is bubbled in. When the solution is saturated with the gas, the solvent is removed *in vacuo* at 30 to 35°. The residual sirup (or crystals) is again dissolved in ethanol, and the procedure is repeated. The desired product frequently crystallizes in the concentrator or will do so when anhydrous ether is added. The product is washed with ether and dried in a desiccator over NaOH pellets. The yield is nearly quantitative. Melting point, 134°. The methyl ester hydrochloride can be prepared in the same manner; melting point, 149 to 150°.

L-*Leucinamide Hydrochloride (8B)*. A recrystallized preparation of L-leucine ethyl (or methyl) ester hydrochloride (10 g.) is allowed to stand in a pressure bottle for 2 days with 50 ml. of anhydrous methanol which has been previously saturated with ammonia gas at 0°. The solution is repeatedly concentrated *in vacuo* with methanol and the crystals are filtered and washed with ether. The substance is recrystallized from methanol-ether or methanol-ethyl acetate until free of NH_4Cl. Yield,

---

[17] This general method of preparing free esters is that of Fischer.[2]
[18] E. L. Smith and N. B. Slonim, *J. Biol. Chem.* **176,** 835 (1948).

8.5 g.; melting point, 244 to 244.5°; $[\alpha]_D = 10.0°$ (1% in water).[19] This compound is the preferred substrate for leucine aminopeptidase (see Vol. II [9]).

[19] D. H. Spackman and E. L. Smith, unpublished. Previous values for the melting point and rotation were given by Smith and Slonim.[18]

[81] Procedures for Preparation of Peptide Polymers

I. Preparation of Poly-L-Lysine

By EPHRAIM KATCHALSKI

Principle. The preparation of poly-L-lysine from L-lysine outlined below consists in the following steps: L-lysine (I) reacts with benzyl chloroformate to yield N,N'-dicarbobenzoxy-L-lysine (II). II, on treatment with phosphorus pentachloride, gives ϵ,N-carbobenzoxy-α,N-carboxy-L-lysine anhydride (III). In the presence of diethylamine, acting as a polymerization initiator, III gives poly-ϵ,N-carbobenzoxy-L-lysine (IV) with the evolution of carbon dioxide. On removal of the carbobenzoxy groups of IV by means of anhydrous hydrogen bromide, poly-L-lysine hydrobromide (V) is obtained. This synthesis is based on the method described by Katchalski *et al.*[1]

[1] E. Katchalski, I. Grossfeld, and M. Frankel, *J. Am. Chem. Soc.* **70**, 2094 (1948).

$$H-\left[-HN-CH-CO-\right]-N(C_2H_5)_2$$
$$(CH_2)_4$$
$$NH_2 \cdot HBr \quad\Big]_n$$
$$V$$

Cbz $= C_6H_5CH_2OCO-$

Benzyl Chloroformate (Carbobenzoxy Chloride, Benzyl Chlorocarbonate)

Procedure. An apparatus is assembled as shown in Fig. 1. The 2-l. three-necked round-bottomed flask, A, is equipped with a dropping funnel, C, a reflux condenser, D, and an efficient spiral condenser, B,

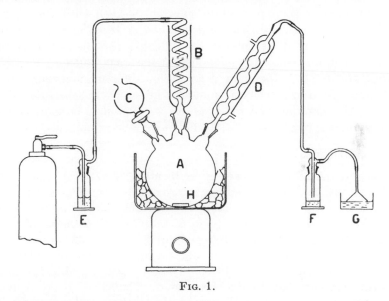

FIG. 1.

cooled with a dry ice-acetone mixture. The magnetic stirring bar, H, is enclosed in glass or Teflon.

Phosgene, conveniently withdrawn from a cylinder, is bubbled through the wash bottle, E, containing concentrated sulfuric acid, and liquefied in the spiral condenser, B. The three-necked flask is cooled with ice (Note 1), and the rate of flow of liquid phosgene is adjusted to 2 drops per second by varying the rate of flow of gas. Freshly purified benzyl alcohol (640 g.) (Note 2) is simultaneously added from the funnel at the same rate. The excess of phosgene, and the hydrogen chloride formed during the reaction, are absorbed by concentrated ammonium hydroxide kept in vessel G. A wash bottle, F, containing concentrated sulfuric acid is inserted between D and G to prevent the diffusion of ammonia into the reaction vessel, A (Note 3). When the addition of benzyl alcohol is completed, the stream of phosgene is stopped, the spiral condenser is exchanged for a gas inlet tube, and a rapid stream of dry carbon dioxide is passed through the reaction mixture for 3 to 4 hours. During the last hour the temperature is raised to 30°. The colorless benzyl chloroformate thus obtained (1000 g.) usually solidifies in the deep-freeze to a crystalline mass melting at 0°. It can be kept in glass-stoppered bottles in the cold for at least a year without deterioration.

NOTES. (1) With ice cooling, the temperature of the reaction mixture is 10 to 20°. At lower temperatures the reaction between phosgene and benzyl alcohol is slow. (2) Commercial benzyl alcohol was shaken with saturated aqueous sodium bisulfite solution, washed with water, dried over anhydrous sodium sulfate, and distilled *in vacuo* at 20 to 30 mm. The fraction boiling at 100 to 110° was collected. (3) The level of sulfuric acid in F should be below the inlet tube of the wash bottle.

Other Methods of Preparation. A somewhat different procedure for the preparation of benzyl chloroformate is given in *Organic Syntheses.*[2] However, the material prepared as above was found to be more suitable for the synthesis of crystalline N-carbobenzoxy-α-amino acids.

N,N'-Dicarbobenzoxy-L-lysine (II)

Procedure.[3] A solution of 27.4 g. (0.15 mole) of L-lysine monohydrochloride in 200 ml. of 2 N sodium hydroxide is placed in a 1-l. three-necked flask fitted with a mechanical stirrer and two dropping funnels. The flask is cooled in an ice bath, and 65 ml. (78 g., 0.45 mole) of benzyl chloroformate and 190 ml. of 4 N sodium hydroxide are added simultaneously to the vigorously stirred solution over a period of 20 minutes. The mixture is stirred for an additional 20 minutes and acidified to Congo red with concentrated hydrochloric acid. The dicarbobenzoxylysine which separates out as an oil is extracted with three 100-ml. portions of ether. The combined ethereal extracts are shaken once with 300 ml. and once with 100 ml. of 1 N potassium bicarbonate in a separating funnel. The potassium salt of the dicarbobenzoxylysine which is thus formed settles as an oil and is withdrawn from the separating funnel. It is mixed with crushed ice and acidified to Congo red with concentrated hydrochloric acid. The aqueous layer is decanted, and the semisolid material is dissolved in 200 ml. of ethyl acetate. The solution is washed twice with 50-ml. portions of water, dried over anhydrous sodium sulfate, and the solvent distilled off at 20 mm. The residue is crystallized by triturating

² *Org. Syntheses* **23**, 13 (1943).
³ Adopted from M. Bergmann, L. Zervas, and W. F. Ross, *J. Biol. Chem.* **111**, 245 (1935).

with 250 ml. of petroleum ether, which is left to cover the substance for 2 days at room temperature. It is filtered by suction, washed with petroleum ether, and dried *in vacuo* over sulfuric acid. The yield of the dicarbobenzoxy-L-lysine, melting at 55 to 65°, is 50 to 56 g. (80 to 90% of the theoretical amount). It is used without further purification for the next step.

ε,N-Carbobenzoxy-α, N-carboxy-L-lysine Anhydride (III)

Procedure. A solution of 41.4 g. (0.1 mole) of N,N'-dicarbobenzoxy-L-lysine in 500 ml. of dry benzene is introduced into a 1-l. flask, and 100 ml. of solvent is distilled off at atmospheric pressure (Note 1). The flask is cooled with ice, and 41.6 g. (0.2 mole) of pure phosphorus pentachloride is added in three portions with vigorous shaking. After 10 minutes, when most of the solid phosphorus pentachloride has disappeared, the reaction mixture is filtered rapidly with suction (Note 2) and left at room temperature for a period of 2 to 3 hours (Note 3), during which most of the anhydride crystallizes out. Then 600 ml. of dry petroleum ether is added to complete the precipitation. After 2 hours in the refrigerator the crystalline mass is filtered rapidly with suction and washed with petroleum ether. The crude anhydride is dissolved in 100 ml. of warm absolute ethyl acetate, the solution filtered, and 900 ml. of dry petroleum ether added. This purification procedure is repeated two to three times until the sample obtained is found to be free of chloride ion and gives the calculated equivalent weight on titration with sodium methoxide.[4] The final product is dried *in vacuo* (Note 4) over phosphorus pentoxide and solid potassium hydroxide. The yield of the anhydride, melting at 101° (decomp.), is 21 to 23 g. (68 to 75% of the theoretical amount).

NOTES. (1) All apparatus should be carefully dried before use. (2) If filtration is delayed, crystallization of the anhydride may prevent the removal of the excess of phosphorus pentachloride. (3) The yield of the anhydride may drop considerably if the mixture is left at room temperature for a longer period. (4) A drying tower of calcium chloride should be inserted between the desiccator and the water pump.

Poly-ε,N-carbobenzoxy-L-lysine (IV)

Procedure. In a 1-l. flask equipped with a calcium chloride drying tube, 30.6 g. (0.1 mole) of ε,N-carbobenzoxy-α,N-carboxy-L-lysine anhydride is dissolved in 300 ml. of dimethylformamide (Note 1). A solution of 0.146 g. (0.002 mole) of diethylamine (Note 2) in 2 ml. of dimethylformamide is added immediately, and the reaction mixture is left at room temperature for 48 hours. Then 600 ml. of water, containing

[4] A. Berger, M. Sela, and E. Katchalski, *Anal. Chem.* **25**, 1554 (1953).

2 ml. of concentrated hydrochloric acid, is added, and the mixture is left overnight at room temperature. The white amorphous product is filtered, washed with water until the filtrate is neutral, and dried over phosphorus pentoxide in a vacuum desiccator. The yield of polycarbobenzoxy-L-lysine is 25 g. (80% of the theoretical amount) (Note 3).

Notes. (1) Dimethylformamide (obtained from du Pont de Nemours) was purified by distillation at 25 mm. The fraction distilling at 60 to 63° was collected. If the acidity of the solvent exceeds 0.001 N, as determined by anhydrous titration with sodium methoxide,[4] it should be treated with anhydrous potassium carbonate and redistilled *in vacuo*. (2) The diethylamine is conveniently added from a stock solution, prepared by diluting 7.31 g. (10.3 ml.) of diethylamine to 100 ml. with dimethylformamide. The molar ratio between the diethylamine, acting as the initiator of the polymerization, and the ϵ,N-carbobenzoxy-α,N-carboxylysine anhydride determines the average degree of polymerization of the final product. A ratio of 1:50 was chosen in the procedure given above, where a preparation of a polycarbobenzoxy-L-lysine with an average degree of polymerization of about 50 has been described. Other amines such as ammonia, alkylamines, and esters or amides of amino acids may be selected as suitable catalysts of polymerization. (3) The number, average degree of polymerization, n, of the final product can be calculated from the total nitrogen (Dumas) and the amino nitrogen (Van Slyke[5]): $n =$ (total nitrogen)/2(amino nitrogen). Preparations obtained by following the above procedure had number average degrees of polymerization in the range of 35 to 60.

Other Methods of Preparation. Polycarbobenzoxylysine was prepared by Katchalski *et al.* by bulk polymerization. Polymerization in dioxane at elevated temperatures was carried out by Becker and Stahmann.[6] The procedure given seems preferable, as the polymer formed during polymerization stays in solution in the dimethylformamide chosen as solvent. Polymers prepared at elevated temperatures were found to contain amino as well as carboxyl end groups. The average degree of polymerization of such polymers may be calculated from the quantitative determination of both types of end groups.[7]

Poly-L-lysine Hydrobromide (V)

Procedure. Into a 250-ml. flask equipped with a calcium chloride drying tube are placed 4 g. of powdered, dry polycarbobenzoxy-L-lysine and 30 ml. of a 33% solution of hydrogen bromide in glacial acetic acid

[5] D. D. Van Slyke, *J. Biol. Chem.* **83**, 425 (1929).
[6] R. R. Becker and M. A. Stahmann, *J. Am. Chem. Soc.* **74**, 38 (1952).
[7] M. Sela and A. Berger, *J. Am. Chem. Soc.* **75**, 6350 (1953).

(Note 1). The solid goes rapidly into solution with vigorous evolution of carbon dioxide. Polylysine hydrobromide is precipitated from the clear solution after a few minutes. Precipitation is completed after an additional 30 minutes at room temperature by the addition of 150 ml. of anhydrous ether. The precipitate is centrifuged, the supernatant liquid discarded, and the centrifugate washed with 20 ml. of anhydrous ether. The polylysine hydrobromide is again separated by centrifugation, and the washing with ether repeated two to three times. The solid residue is finally dissolved in 3 ml. of water and precipitated from the clear solution by the addition of 15 ml. of absolute ethanol and 150 ml. of ether. The purified polylysine hydrobromide is centrifuged and dried in a vacuum desiccator over phosphorus pentoxide and solid potassium hydroxide. The yield of the colorless, water-soluble poly-L-lysine hydrobromide is 2.9 to 3.0 g. (91 to 94% of the theoretical amount) (Note 2).

NOTES. (1) Dry gaseous hydrogen bromide is bubbled through 67 g. of glacial acetic acid with ice cooling until the weight increases to 100 g. Moisture is rigorously excluded. The solution may be kept indefinitely in the refrigerator. The hydrogen bromide gas may be withdrawn from a steel cylinder or obtained by dropping dry bromide on tetraline warmed to 80°.[8] (2) As it has been proved[1] that no cleavage of peptide bonds occurs under the experimental conditions used, it can be assumed that the average degree of polymerization of the poly-L-lysine obtained is identical with that of the polycarbobenzoxylysine from which it was derived.

Poly-L-lysine Hydrochloride

Procedure. Poly-ε,N-carbobenzoxy-L-lysine (4.0 g.) is dissolved in 30 ml. of glacial acetic acid, and a stream of dry hydrogen chloride is passed through the solution at 70° for 2 hours. During this period polylysine hydrochloride separates out, and its precipitation is completed by the addition of 150 ml. of anhydrous ether. Further treatment is similar to that given for polylysine hydrobromide. The yield is 2.3 to 2.4 g. (92 to 95% of the theoretical amount).

Other Methods of Preparation. The above methods of preparation of hydrobromide and hydrochloride are adapted from the procedure of Ben-Ishai and Berger.[9] Poly-L-lysine hydriodide was prepared from poly-ε,N-carbobenzoxy-L-lysine by treatment with phosphonium iodide.[1] Poly-L-lysine hydrochloride was prepared from the picrate.[1]

Chemical Properties. Poly-L-lysine hydrobromide and hydrochloride dissolve very readily in water, and their aqueous solution gives a positive ninhydrin reaction and a strong biuret reaction. Poly-L-lysine forms a

[8] *Inorg. Syntheses* 1, 151 (1939).
[9] D. Ben-Ishai and A. Berger, *J. Org. Chem.* 17, 1564 (1952).

water-insoluble picrate, picrolonate, and phosphotungstate. At neutral pH and low ionic strength, polylysine forms precipitates with nucleic acids and other acidic polyelectrolytes. Poly-L-lysine hydrobromide has $[\alpha]_D^{25} = -39.5°$ (C, 5, water). Its purity may be checked by amino nitrogen (Van Slyke[5]), total nitrogen (Kjeldahl), and halogen analysis (Volhard), as well as by the quantitative yield of L-lysine on total acid hydrolysis.

Enzymatic Properties. Poly-L-lysine is readily hydrolyzed by trypsin[10,11] and acts as a pepsin inhibitor.[12] It inhibits the formation of thrombin in shed human blood[13] and antagonizes the blood clot-delaying action of heparin.[13]

II. Preparation of Poly-L-Glutamic Acid

By EPHRAIM KATCHALSKI and ARIEH BERGER

Principle. The preparation of poly-L-glutamic acid from L-glutamic acid outlined below consists in the following steps: L-glutamic acid (VI) is esterified with benzyl alcohol to yield γ-benzyl L-glutamate (VII). VII reacts with phosgene to give γ-benzyl N-carboxy-L-glutamate anhydride (VIII), which is polymerized to poly- γ-benzyl L-glutamate (IX) in the presence of diethylamine acting as a polymerization initiator. Poly-L-glutamic acid (X) is obtained from IX on treatment with anhydrous hydrogen bromide. This synthesis is based on the method described by Hanby *et al.*[14]

[10] E. Katchalski, *Advances in Protein Chem.* **6**, 123 (1951).
[11] S. G. Waley and J. Watson, *Biochem. J.* **55**, 328 (1953).
[12] E. Katchalski, A. Berger, and H. Neumann, *Nature* **173**, 998 (1954).
[13] A. De Vries, A. Schwager, and F. Katchalski, *Biochem. J.* **49**, 10 (1951).
[14] W. E. Hanby, S. G. Waley, and J. Watson, *J. Chem. Soc.* **1950**, 3239.

$$\text{H} - \left[-\text{HN} - \underset{\underset{\displaystyle \text{COOH}}{\overset{\displaystyle |}{\underset{|}{(\text{CH}_2)_2}}}}{\text{CH}} - \text{CO} - \right]_n - \text{N}(\text{C}_2\text{H}_5)_2$$

X

γ-Benzyl L-Glutamate (VII)

Procedure. Freshly distilled, constant-boiling hydriodic acid (250 ml.) and 150 g. (1.02 mole) of L-glutamic acid are added to 1 l. of benzyl alcohol, and the reaction mixture is left overnight at room temperature. The mixture is then diluted with 2 l. of ethanol and treated with 300 ml. of pyridine. After 24 hours at 0° the solid formed, consisting of a mixture of γ-benzyl L-glutamate and L-glutamic acid, is filtered and washed with ethanol and ether. The glutamic acid content of this solid is estimated by titrating a portion against alkali with bromothymol blue as indicator. The bulk of the solid is finally recrystallized from 2 l. of hot water containing 1 mole of sodium hydrogen carbonate for each mole of glutamic acid found by titration. The γ-benzyl ester crystallizes in colorless plates. It is filtered and dried *in vacuo* over phosphorus pentoxide. The yield of the ester, melting at 169 to 170°, is 96 to 110 g. (40 to 46% of the theoretical amount). $[\alpha]_D^{25} = +18.7°$ (C, 7, glacial acetic acid).

If the γ-benzyl L-glutamate obtained is contaminated with iodine, it should be recrystallized from water.

γ-Benzyl N-Carboxy-L-glutamate Anhydride (VIII)

Procedure. In a two-necked, 2-l. round-bottomed flask, equipped with a reflux condenser and a gas inlet tube extending below the surface of the reaction mixture, are placed 400 ml. of dry dioxane (Note 1) and 20 g. of γ-benzyl L-glutamate. The mixture is kept at 40° in a water bath, and a stream of dry phosgene (Note 2) is passed through the gas inlet tube for 2½ hours, whereupon all the solid goes into solution. The hydrogen chloride formed and the excess of phosgene are passed through the top of the condenser into a wash bottle containing sulfuric acid (Note 3) and finally absorbed over aqueous ammonia. The stream of phosgene is continued for an additional half-hour at a somewhat reduced rate. The supply of phosgene is then stopped, and a stream of dry carbon dioxide is blown through the solution for 3 hours to remove excess phosgene. The clear solution is concentrated *in vacuo* at 40° (Note 4), and the oily residue crystallized by rubbing with a glass rod. The crystalline mass is washed with petroleum ether and dissolved in 150 ml. of warm absolute ethyl acetate, the solution is filtered, and the γ-benzyl N-carboxy-L-glutamate anhydride is precipitated by the addition of 1.5 l. of petroleum ether. The crystals are collected, washed with petroleum ether, and dried *in vacuo* over phosphorus pentoxide. The yield of anhydride, melting at 96 to 97° (decomp.), is 15.6 to 16.6 g. (70 to 75% of the theoretical amount).

NOTES. (1) Dioxane was refluxed over solid potassium hydroxide for 12 hours, dried over sodium, and fractionated at atmospheric pressure with careful exclusion of moisture. (2) The phosgene was withdrawn from a cylinder and dried by bubbling through concentrated sulfuric acid. (3) The level of the sulfuric acid in the wash bottle must be kept below the gas inlet tube to prevent clogging. (4) The air drawn into the capillary is dried by means of a calcium chloride tube.

Other Methods of Preparations. Hanby *et al.*[14] prepared γ-benzyl L-glutamate anhydride by the action of phosphorus pentachloride on γ-benzyl N-carbobenzoxy-L-glutamate.

Poly-γ-benzyl L-glutamate (IX)

Procedure. In a 100-ml. Erlenmeyer flask, fitted with a calcium chloride tube, are placed 13.2 g. (0.05 mole) of γ-benzyl N-carboxy-L-glutamate anhydride in 34 ml. of nitrobenzene (Note 1) and 36.5 mg. (0.0005 mole) of diethylamine in 1 ml. of nitrobenzene (Note 2). The reaction mixture is kept at 30° for 5 days. Then 250 ml. of petroleum ether is added, and the colorless polymer precipitated is collected and washed with acetone. The yield of poly-γ-benzyl glutamate is 10.2 to 10.5 g. (93 to 95% of the theoretical amount) (Note 3). The polymer shows an optical rotation of $[\alpha]_D^{25} = +14°$ (C, 1, chloroform).

NOTES. (1) The nitrobenzene was dried over phosphorus pentoxide overnight, decanted, and distilled at 0.05 mm. (2) The diethylamine is conveniently added from a stock solution, prepared by diluting 3.65 g. (5.15 ml.) of diethylamine to 100 ml. with nitrobenzene. (3) A molar ratio of 100:1 between the anhydride and the polymerization initiator was chosen in order to obtain a polypeptide with an approximate chain length of one hundred glutamic acid residues. The average degree of polymerization of poly-γ-benzyl L-glutamate cannot, however, be determined by amino nitrogen analysis, as a part of the terminal amino groups disappear during polymerization owing to a specific termination reaction.[14]

Poly-L-glutamic Acid (X)

Procedure. In a wash bottle fitted with a calcium chloride tube at its outlet are placed 2.5 g. of poly-γ-benzyl L-glutamate and 70 ml. of glacial acetic acid. The reaction mixture is kept at 80° while a stream of dry hydrogen bromide[15] is passed through the solution. The solid dissolves

[15] See Note 1, p. 545.

entirely within 10 minutes, and the poly-L-glutamic acid begins to precipitate out. The stream of hydrogen bromide is continued for 20 minutes, the mixture is cooled to room temperature, and the precipitation of the polyamino acid is completed by the addition of 200 ml. of anhydrous ether. The precipitate is collected, thoroughly washed with anhydrous ether, and dried over concentrated sulfuric acid and solid potassium hydroxide. The yield of poly-L-glutamic acid is 1.45 g. (94% of the theoretical amount).

Other Methods of Preparation. The present method of debenzylation is adapted from that of Ben-Ishai and Berger.[9] Poly-L-glutamic acid was also prepared from poly-γ-benzyl glutamate by treatment with phosphonium iodide[14] and from poly-γ-ethyl L-glutamate by saponification in alcoholic potassium hydroxide.[16] Hanby *et al.*[14] report extensive racemization on alkaline hydrolysis of poly-γ-methyl L-glutamate and on reduction of poly- γ-benzyl L-glutamate with sodium in liquid ammonia.

Chemical Properties. Poly-L-glutamic acid is very sparingly soluble in water. It may be brought into solution by the addition of sodium hydroxide, sodium bicarbonate, or buffer solutions of a pH higher than pH 6. Once dissolved, polyglutamic acid remains in solution at pH values as low as pH 4. The silver and copper salts of polyglutamic acid are insoluble in neutral and alkaline environments. Poly-L-glutamic acid shows an optical rotation of $[\alpha]_D^{26} = -45$ to $-50°$ (C, 2, 4 *N* alkali). The purity of the polymer may be checked by the determination of its neutralization equivalent,[14] the yield of glutamic acid on acid hydrolysis, and by elementary analysis. Its average molecular weight cannot be determined by end-group analysis.[17]

Enzymatic Properties. Poly-L-glutamic acid is hydrolyzed by papain and carboxypeptidase but is resistant to pepsin, trypsin, and chymotrypsin.[16]

III. Preparation of Poly-L-aspartic Acid

By EPHRAIM KATCHALSKI and ARIEH BERGER

Principle. The preparation of poly-L-aspartic acid from L-aspartic acid outlined below consists in the following steps: L-aspartic acid (XI) reacts with benzyl chloroformate to yield N-carbobenzoxy-L-aspartic acid (XII), which is esterified with benzyl alcohol to dibenzyl N-carbobenzoxy-L-aspartate (XIII). Partial hydrolysis of XIII gives β-benzyl N-carbobenzoxy-L-aspartate (XIV), which, on treatment with phosphorus pentachloride, yields β-benzyl N-carboxy-L-aspartate anhydride (XV). XV is polymerized to poly-β-benzyl aspartate (XVI) in the pres-

[16] M. Green and M. A. Stahmann, *J. Biol. Chem.* **197**, 771 (1952).
[17] See Note 3, p. 548.

ence of diethylamine, acting as a polymerization initiator. The benzyl groups of XVI are removed by means of anhydrous hydrogen bromide to give poly-L-aspartic acid (XVII). This synthesis is based on the method described by Berger and Katchalski.[18]

Cbz = $C_6H_5CH_2OCO$

N-Carbobenzoxy-L-aspartic Acid (XII)

Procedure. L-Aspartic acid (24 g., 0.18 mole) is added to a suspension of magnesium hydroxide, prepared by mixing 55 g. of magnesium chloride hexahydrate with a solution of 17.5 g. of sodium hydroxide in 300 ml. of water. The mixture is cooled in an ice bath, and 38 ml. (44.5 g., 0.26 mole) of benzyl chloroformate is added in five portions with vigorous shaking. Sufficient concentrated hydrochloric acid is added to dissolve the excess of magnesium hydroxide, the solution is extracted twice with 50-ml. portions of ether, and the ethereal extracts discarded. After acidification with concentrated hydrochloric acid to Congo red, the N-carbobenzoxy-L-

[18] A. Berger and E. Katchalski, *J. Am. Chem. Soc.*, **73**, 4084 (1951).

aspartic acid is extracted from the aqueous layer with two 200-ml. portions of ether. The combined extracts are washed with 50 ml. of water, dried over anhydrous sodium sulfate, and the solvent distilled off at atmospheric pressure. The residue crystallizes on trituration with petroleum ether. The yield of N-carbobenzoxy-L-aspartic acid, melting at 110 to 114°, is 35 to 40 g. (73 to 83% of the theoretical amount). After recrystallization from benzene the melting point rises to 116°.

Dibenzyl N-Carbobenzoxy-L-aspartate (XIII)

Procedure. Thirty-six grams (0.135 mole) of N-carbobenzoxy-L-aspartic acid (melting point, 114°), 160 ml. (168 g.) of benzyl alcohol, 200 ml. of toluene, and 2 g. of *p*-toluenesulfonic acid monohydrate are placed in a 1-l. two-necked round-bottomed flask fitted with a liquid-sealed mechanical stirrer and a moisture trap, *A*, with a condenser (see Fig. 2). The reaction mixture is refluxed with stirring (Note 1) until no more water collects in the trap. With dry solvents, 8 to 9 ml. of water is collected. The contents of the flask are allowed to cool to room temperature, shaken with 4 g. of magnesium oxide for 10 minutes, and filtered by gravity. The clear solution is transferred to a Claisen flask, and the toluene and benzyl alcohol are removed *in vacuo* at 25 mm. and 1 mm., respectively. The oily residue is poured into a porcelain mortar and triturated with petroleum ether. The crystalline mass formed after a short while is crushed, repeatedly washed with small portions of petroleum ether (Note 2), and left with the same solvent overnight. The

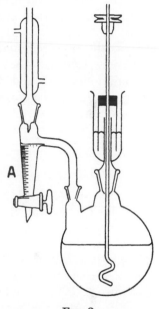

FIG. 2.

crystals are finally collected on a Büchner funnel and air-dried. The yield of the dibenzyl ester, melting at 58 to 60°, is 45 to 51 g. (75 to 85% of the theoretical amount). Recrystallization from di-*n*-butyl ether yields a product melting at 67°.

NOTES. (1) When stirring is omitted, the azeotropic removal of water is slow, and overheating may occur. The slight precipitate which appeared in some runs in the reaction mixture did not affect the purity of the final product. (2) Large amounts of petroleum ether should be avoided, as dibenzyl N-carbobenzoxy-L-aspartate is slightly soluble in this solvent.

β-Benzyl N-Carbobenzoxy-ʟ-aspartate (XIV)

Procedure. A mixture of 50 ml. of 2 N sodium hydroxide (Note 1), 240 ml. of water, and 600 ml. of dioxane (Note 2) is added to a solution of 44.7 g. (0.10 mole) of dibenzyl N-carbobenzoxy-ʟ-aspartate in 500 ml. of dioxane and 200 ml. of water. After 24 hours at room temperature the solution is brought to pH. 5.5 with concentrated hydrochloric acid and concentrated *in vacuo* at 25 mm. The residue is taken up in 120 ml. of 1 N potassium bicarbonate, and the solution is extracted with three 50-ml. portions of ether. The ethereal extracts are discarded, and the aqueous solution is acidified with 6 N hydrochloric acid to Congo red. The crystalline precipitate is filtered by suction, washed with water until the washings are neutral, and dried *in vacuo* over concentrated sulfuric acid. The yield of β-benzyl N-carbobenzoxy-ʟ-aspartate, melting at 104 to 108°, is 21.5 to 23 g. (60 to 65% of the theoretical amount). Recrystallization from benzene raises the melting point to 114°.

NOTES. (1) The exact amount of sodium hydroxide is essential for maximum yield. A standardized sodium hydroxide solution should therefore be used. (2) The dioxane used was purified by refluxing for 12 hours over sodium hydroxide and distilling at atmospheric pressure. The given amounts of dioxane and water ensure the formation of a homogeneous solution.

β-Benzyl N-Carboxy-ʟ-aspartate Anhydride (XV)

Procedure. β-Benzyl N-carbobenzoxy-ʟ-aspartate (36 g., 0.1 mole) is dissolved in 400 ml. of benzene, and 100 ml. of solvent is distilled off at atmospheric pressure (Note 1). The solution is cooled in an ice bath, and 23 g. (0.11 mole) of phosphorus pentachloride is added quickly, before crystallization sets in. The mixture is shaken for 15 minutes and decanted from the excess of phosphorus pentachloride. The clear solution is warmed to 50° for 5 minutes and left at room temperature for 4 hours (Note 2). During this time most of the product crystallizes out. Crystallization is completed by the addition of 300 ml. of anhydrous petroleum ether, and the anhydride is filtered by suction, washed with petroleum ether, and dried *in vacuo* over solid potassium hydroxide and sulfuric acid. The crude product, melting at 105 to 115°, is recrystallized from anhydrous benzene. The yield of β-benzyl N-carboxy-ʟ-aspartate anhydride, melting at 134 to 135°, is 20 to 21 g. (80 to 85% of the theoretical amount).

NOTES. (1) The distillation of benzene removes traces of water azeotropically. (2) If crystallization does not begin within 2 hours, crystals for seeding may be obtained from a small sample of the benzene solution, by adding petroleum ether, cooling, and rubbing with a glass rod.

Poly-β-benzyl L-aspartate (XVI)

Procedure. The preparation of poly-β-benzyl L-aspartate from 8.0 g. of (0.032 mole) β-benzyl N-carboxy-L-aspartate anhydride, with 23.3 mg. of diethylamine as catalyst, is analogous to the preparation of poly-γ-benzyl L-glutamate from γ-benzyl N-carboxy-L-glutamate anhydride.[19] A yield of 6.3 g. (94% of the theoretical amount) of poly-β-benzyl L-aspartate is obtained (Note 1).

NOTES. (1) The average degree of polymerization, n, can be calculated from the total nitrogen (Dumas) and the amino nitrogen (Van Slyke[5]): n = (total nitrogen)/(amino nitrogen). Various preparations had average degrees of polymerization: n = 80 to 100.

Poly-L-aspartic Acid (XVII)

Procedure. The preparation of poly-L-aspartic acid from 2.5 g. of poly-β-benzyl L-aspartate is analogous to the preparation of poly-L-glutamic acid from poly-γ-benzyl L-glutamate.[20] A yield of 1.38 g. of poly-L-aspartic acid is obtained (98% of the theoretical amount).

Chemical Properties. In its solubility and salt formation poly-L-aspartic acid closely resembles poly-L-glutamic acid.[20] Poly-L-aspartic acid shows an optical rotation of $[\alpha]_D^{25}$ = −11.2° (C, 5, water in the presence of one equivalent of sodium hydroxide). The purity of the poly-L-aspartic acid obtained may be checked as in the case of poly-L-glutamic acid.[20] The number average degree of polymerization of the preparation obtained can be calculated from the ratio of total nitrogen to amino nitrogen (Van Slyke[5]).

Enzymatic Properties. The behavior of poly-L-aspartic acid toward proteolytic enzymes has not been investigated.

Preparation of Other Water-Soluble Poly-α-amino Acids

In addition to the poly-α-amino acids described above, the preparation of the following water-soluble polyamino acids was reported in the literature: Neutral polyamino acids: poly-DL-alanine,[21] polysarcosine,[22] and poly-L-proline.[23] Basic polyamino acids: poly-DL-ornithine and poly-

[19] See p. 548.
[20] See pp. 548–549.
[21] W. T. Astbury, E. C. Dalgliesh, S. E. Darmon, and G. B. B. M. Sutherland, *Nature* **162**, 596 (1948).
[22] S. G. Waley and J. Watson, *Proc. Roy. Soc. (London)* **A199**, 499 (1949).
[23] A. Berger, J. Kurz, and E. Katchalski, *J. Am. Chem. Soc.* **76**, 5552 (1954).

DL-arginine.[24] Acidic polyamino acids: poly-DL-aminomalonic acid.[25] Poly-L-tyrosine[26] is soluble in water only at pH values above pH 8.5. Poly-p-aminophenylalanine[27] is soluble in water only at pH values lower than pH 5.

[24] E. Katchalski and P. Spitnik, *J. Am. Chem. Soc.* **73**, 3992 (1951).
[25] M. Frankel, M. Harnik, Y. Levin, and Y. Knobler, *J. Am. Chem. Soc.* **75**, 78 (1953)
[26] E. Katchalski and M. Sela, *J. Am. Chem. Soc.* **75**, 5284 (1953).
[27] M. Sela and E. Katchalski, *J. Am. Chem. Soc.* **76**, 129 (1954).

[82] Resolution of DL Mixtures of α-Amino Acids

$$\text{DL-CH}_2\text{R}'\text{CONHCHRCOOH} + \text{H}_2\text{O} \xrightarrow{\text{Acylase}} \text{L-NH}_2\text{CHRCOOH}$$
$$+ \text{CH}_2\text{R}'\text{COOH} + \text{D-CH}_2\text{R}'\text{COHNCHRCOOH} \quad (1)$$
$$\text{DL-NH}_2\text{CHRCONH}_2 + \text{H}_2\text{O} \xrightarrow{\text{Amidase}} \text{L-NH}_2\text{CHRCOOH} + \text{NH}_3$$
$$+ \text{D-NH}_2\text{CHRCONH}_2 \quad (2)$$
$$\text{D-CH}_2\text{R}'\text{CONHCHRCOOH or D-NH}_2\text{CHRCONH}_2 + \text{H}_2\text{O}$$
$$\xrightarrow[\text{Alkali}]{\text{HCl}} \text{D-NH}_2\text{CHRCOOH} + \text{CH}_2\text{R}'\text{COOH or NH}_3 \quad (3)$$

(where R = alkyl or aryl, and R' = halogen)

By JESSE P. GREENSTEIN

Principle. DL-Amino acids are converted into the corresponding acyl or amide derivatives and treated at neutral pH with the appropriate amount of acylase, carboxypeptidase, or amidase. After incubation at 38°, the reaction mixture is deproteinized and condensed *in vacuo*, and the L-amino acid separated from the D-amino acid derivative either in different solvents or by means of ion exchange resins. The D-amino acid derivatives are hydrolyzed by HCl to the D-amino acid hydrochlorides, which may be isolated as such or converted by neutralization with alkali to the free D-amino acids. The yields of each isomer range from 60 to 90%, and there is less than 0.1% contamination of each optical isomer by the other as measured by optically specific amino acids or decarboxylases. The method is a general one for α-amino acids and so far has been applied to the resolution of the amino acids listed in the table. To conserve space, however, the resolution of only those amino acids of natural or related occurrence will be described in detail. Among these is hydroxyproline, the preparation of whose isomeric forms is more conveniently effected by purely chemical means, and which is therefore described separately. The procedures are those reported from this laboratory; see refs. 1–4 for most recent publications with previous references.

Starting Materials

N-Acetylated Amino Acids. (*a*) One mole of DL-alanine, aminobutyric acid, valine, leucine, methionine, S-benzylhomocysteine, aspartic acid, or glutamic acid, suspended in a tenfold amount of boiling glacial acetic acid, is carefully treated with 1.1 mole of acetic anhydride (99%). Similarly, 1 mole of S-benzyl-L-cysteine is treated with 2.2 moles of acetic anhydride.[5] Boiling is continued for 5 minutes. The solution is allowed to cool to room temperature and evaporated *in vacuo* to dryness. The residue is dissolved in the minimum amount of acetone, chilled, and the acetyl-DL-amino acid either directly crystallized from this solvent or precipitated as crystals by addition of ether or petroleum ether. (*b*) Commercially available isoleucine occurs in two forms: A, a mixture of DL-isoleucine and DL-alloisoleucine, and B, a nearly equal mixture of L-isoleucine and D-alloisoleucine. Treatment of B as in (*a*) yields acetyl (L-isoleucine + D-alloisoleucine), melting point 150°, $[\alpha]_D^{25} = -3.0°$ (4% in ethanol),[6,7] from which of course only these two isoleucine epimers can be obtained by resolution. The four isomers can be obtained only

[1] J. P. Greenstein, S. M. Birnbaum, and M. C. Otey, *J. Biol. Chem.* **204,** 307 (1953).

[2] W. S. Fones, *J. Am. Chem. Soc.* **75,** 4865 (1953).

[3] J. P. Greenstein, S. M. Birnbaum, and L. Levintow, *in Biochem. Preparations* **3,** 84 (1953).

[4] C. G. Baker and H. A. Sober, *J. Am. Chem. Soc.* **75,** 4058 (1953).

[5] L-Cysteine is cheaper and more readily available than is DL-cysteine, and both racemization and acetylation of the S-benzyl derivative are achieved in a single step by means of excess acetic anhydride. On the other hand, DL-homocysteine is more readily available than is the L-form. The S-benzylation is accomplished by dissolving L-cysteine·HCl or DL-homocysteine in several equivalents of chilled 2 N NaOH and shaking with 2 to 3 equivalents of benzylchloride for several hours at 0°. When clear solution is achieved, acetic acid is added to the alkaline solution until a pH of 5 to 6 is reached, and the crystalline precipitate is filtered, washed, and recrystallized by dissolving in HCl and adding ammonia water to pH 5 to 6. Synthetic DL-leucine as purchased is frequently contaminated with appreciable quantities of DL-isoleucine. The impurity is carried along in the acetylation, but it is possible to remove it by several crystallizations of the acetyl-DL-leucine from 50% acetic acid. Commercially available L-leucine may contain traces of L-methionine, but combined acetylation and racemization of the L-leucine by treatment with 2.2 moles of acetic anhydride in boiling glacial acetic acid, followed by removal of the solvent, and crystallization of the residual acetyl-DL-leucine from water, serves to remove the impurity and provide the needed product.

[6] J. P. Greenstein, L. Levintow, C. G. Baker, and J. White, *J. Biol. Chem.* **188,** 647 (1951).

[7] The small rotation is due to the fact that acetyl-D-alloisoleucine has an $[\alpha]_D^{25} = -21.4°$ and acetyl-L-isoleucine has an $[\alpha]_D^{25} = +15.0°$.

from A. One mole of A in a tenfold amount of boiling glacial acetic acid is treated with 2.5 moles of acetic anhydride, the solution evaporated *in vacuo* to a sirup, and the residue treated with an equal volume of cold water. The crystals which rapidly form are nearly pure acetyl-DL-alloisoleucine.[8] They are filtered, washed with cold water, and crystallized twice from 50% acetic acid which removes any adherent acetyl-DL-isoleucine. After resolving (as below) into L- and D-alloisoleucine, each of the optical isomers is treated again with excess acetic anhydride to yield, respectively, acetyl (L-alloisoleucine + D-isoleucine), melting point 150°, $[\alpha]_D^{25} = +3.0°$ (4% in ethanol), and acetyl (L-isoleucine + D-alloisoleucine), with properties described above, and from which the four stereoisomers are subsequently separated by resolving each acetyl derivative.[3,6] (c) Racemic α,β-diaminopropionic acid·HBr[9] in aqueous solution is treated with excess silver acetate and acetic acid, the AgBr removed by filtration, and excess Ag+ removed by H_2S. The clear, H_2S-free filtrate is evaporated to dryness *in vacuo*, and the residue dissolved in glacial acetic acid. The solution is brought to a boil and treated with 2.8 moles of acetic anhydride. The solvent is removed *in vacuo*, and the residual diacetyl-α,β-diaminopropionic acid crystallized from ethanol.[10] (d) One mole of L-histidine base[11] is treated with 2 moles of acetic anhydride in hot glacial acetic acid solution, the solvent removed *in vacuo*, the process repeated, and the residual α-acetyl-DL-histidine crystallized from a little water by addition of acetone.[12] (e) One mole of L-arginine base[11] is shaken vigorously for several hours with a twentyfold amount of acetic anhydride. The resulting α-acetyl-DL-arginine is filtered and washed with ether, dried, and crystallized twice from the minimal amount of water by addition of acetone.[13]

N-Chloroacetylated Amino Acids. (a) One mole of racemic phenylalanine or tryptophan is dissolved in 1 mole of chilled NaOH and treated alternately with 1.5 moles of chloroacetyl chloride and 1.5 moles of 1 N NaOH. On acidification with 5 N HCl to pH 1.7, the chloroacetylated amino acids crystallize. The compounds are filtered, washed with cold

---

[8] Acetyl-DL-isoleucine is present in the mother liquor together with highly colored side products and unprecipitated acetyl-DL-alloisoleucine. It is not practicable at this stage to attempt an isolation of the first-mentioned fraction.

[9] This compound is readily prepared by heating available α,β-dibromopropionic acid with an excess of concentrated ammonia under pressure, and crystallizing the product several times from water-alcohol.

[10] M. Bergmann and K. Grafe, *Z. physiol. Chem.* **187,** 187 (1930).

[11] Obtained from available hydrochloride by adding 1 equivalent of LiOH followed by addition of ethanol to 80% and recrystallization from water-ethanol.

[12] M. Bergmann and L. Zervas, *Biochem. Z.* **203,** 280 (1928).

[13] M. Bergmann and H. Köster, *Z. physiol. Chem.* **159,** 179 (1926).

water, and crystallized twice from water. (b) Synthetic hydroxylysine. HCl,[14] a mixture of normal and allo forms, is treated in warm aqueous solution with an excess of copper carbonate, the excess carbonate removed by filtration, and the soluble copper salt of hydroxylysine treated with 1.5 equivalent of carbobenzoxy chloride and alkali to yield the ε-carbobenzoxy derivative. The copper is removed with H_2S, and the ε-carbobenzoxyhydroxylysines are converted to the α-chloroacetyl derivative as in (a). The acidification to pH 1.7 results in lactonization, and the product is a mixture of normal and allo-α-chloroacetyl-ε-carbobenzoxy-hydroxylysine-δ-lactones which are separated by fractional crystallization in ethyl acetate.[2] (c) One mole of racemic tyrosine, serine, homoserine, threonine, allothreonine, aminoadipic acid, α,γ-diaminobutyric acid·HCl, ornithine·2HCl, or lysine·HCl is dissolved in chilled NaOH solution equivalent to the acid groups present, i.e., HCl, COOH, and phenolic hydroxyl in tyrosine, and treated alternately with 1.5 equivalents of chloroacetyl chloride per NH_2 group and the equivalent amount of 2 N NaOH. Acidification to pH 1.7 with concentrated HCl does not result in precipitation, and the reaction mixture is therefore extracted several times with ethyl acetate. The combined extracts are dried over Na_2SO_4, evaporated to dryness, and treated with chilled dry ether or petroleum ether. Homoserine yields the chloroacetyl-γ-lactone under these conditions.[15]

N-Trifluoroacetylated Amino Acids. β-Phenylserine ethyl ester (threo or erythro)[16] is suspended in dry ether and treated with trifluoroacetic anhydride. The ethereal solution is washed successively with dilute HCl and saturated $NaHCO_3$, dried over Na_2SO_4, and the reaction product precipitated from the condensed solution by addition of petroleum ether. The trifluoroacetyl ester is saponified by addition of aqueous NaOH to its solution in ethanol, and the trifluoroacetylated β-phenylscrine extracted into ethyl acetate and crystallized by addition of petroleum ether.[17]

Proline Amide. One mole of L-proline is treated with 2 moles of acetic anhydride in boiling glacial acetic acid solution, and the solvent is removed *in vacuo*. The residual acetyl-DL-proline is refluxed with a tenfold amount of 2 N HCl for 2 hours, the solvent removed *in vacuo*, and the residual DL-proline·HCl esterified in ethanol saturated with HCl gas.

[14] G. Van Zyl, E. E. Van Tamelen, and G. D. Zuidema, *J. Am. Chem. Soc.* **73**, 1765 (1951).

[15] S. M. Birnbaum and J. P. Greenstein, *Arch. Biochem. and Biophys.* **42**, 212 (1953).

[16] The threo form is obtained by the interaction of glycine and benzaldehyde in alkaline medium, the erythro form by the interaction of ethyl glycinate and benzaldehyde.

[17] W. S. Fones, *J. Biol. Chem.* **204**, 323 (1953).

The solvent is removed *in vacuo*. The residual ester·HCl is dissolved in a little cold water, treated with excess of solid K_2CO_3, and the semifluid mass extracted several times with ether. The combined ethereal extracts are dried over Na_2SO_4, filtered, and, after removal of the solvent, distilled *in vacuo* at 2 mm. of Hg and 98°. The free DL-proline ethyl ester is converted to the amide by being dissolved in a tenfold amount of dry methanol saturated at 0° with NH_3 gas. After standing at 5° for 5 days, the solvent is removed and the residual DL-proline amide recrystallized from chloroform.[18]

The yields of the above compounds range from about 20% for proline amide to about 60% for most of the acetylated and chloroacetylated amino acids. As always when yields are considered, the economic balance between the cost of materials and the price of the investigator's time must be drawn. In all cases, the purity of these starting materials for the resolution procedures is of importance, and freedom from traces of the free DL-amino acid must, where possible, be established. With the exception of acetylhistidine and acetylarginine, the simplest procedure for this purpose is to take up the final product in dry acetone. If it forms a clear, sparkling solution, the absence of amino acid is practically assured. Otherwise, it is filtered clear with aid of Norit, the solvent is removed, and the residual product tested again. The presence of DL-amino acid in the starting materials would lead to contamination of the resolved L-isomer by the D-form from the DL-amino acid source.

The physical constants of these starting materials are collected in the table (pp. 568–569), together with those of the unnatural amino acid derivatives, as well as the hydrolytic rates by the appropriate enzyme preparations for these substrates. Although the chloroacetyl derivatives are invariably more susceptible than are the corresponding acetyl derivatives (a ratio of 4:1 for acylase and 150:1 for carboxypeptidase), and thus less enzyme would be needed for the same period of incubation, several acetyl derivatives have been recommended because their preparative procedures are simpler and more economical than those for the corresponding chloroacetyl derivatives.[19]

General Resolution Procedure

Step 1. The acylated DL-amino acid or DL-amino acid amide is dissolved or suspended in water, and the mixture brought to pH 7.0 with LiOH in the case of the former, or to pH 8.0 in the case of the latter to whose

---

[18] V. E. Price, L. Levintow, J. P. Greenstein, and R. B. Kingsley, *Arch. Biochem.* **26,** 92 (1950).

[19] In other cases, i.e., lysine, aminoadipic acid, the chloroacetyl derivatives themselves are hydrolyzed relatively slowly. In the case of the phenylserines, both acetyl and chloroacetyl derivatives are nearly resistant, and the trifluoroacetyl derivatives have had to be employed.[17]

solution MnCl₂ to about 0.01 M is also added. Water is then added to bring the final concentration of substrate to 0.10 M, and the pH checked again and adjusted if necessary.[20] In all cases but that of acetyl-S-benzyl-DL-homocysteine, complete solution is achieved.[21] No buffer is needed to maintain the pH, for there is very little change in pH as a result of the enzymatic hydrolysis of these substrates.

Step 2. The table is consulted for rates, and enough acylase I or II (see Vol. II [12]), carboxypeptidase (see Vol. II [8]), or amidase (see Vol. II [55]) is weighed to hydrolyze the susceptible L-component of the substrate in 1 to 2 hours. The enzyme is stirred to solution, and the digest placed in a water bath at 38°. At intervals, aliquots are removed for manometric ninhydrin measurements (see pp. 1041–1044), except for the β-phenylserines where the manometric nitrous acid procedure (see Vol. III [75]) is employed.[22] When the results show complete hydrolysis of the susceptible L-isomer, a small amount of the enzyme is again added, and the digestion is allowed to continue for a few hours longer. This is simply to ensure complete hydrolysis of the susceptible isomer; there is no concern with these substrates that the D-moiety of the racemate will be attacked. Glacial acetic acid is then added to bring the pH to about 5.

Isolation of the L-Isomers

Chemical Procedures. (a) Tyrosine, S-benzylcysteine, and S-benzyl-homocysteine—these L-isomers crystallize nearly quantitatively during

---

[20] Among all the amino acids whose resolution has been studied, only the chloroacetyl derivatives of nonylic, decylic, and undecylic acids must be employed at initial concentrations less than 0.10 M, i.e., 0.05 M, 0.025 M, and 0.01 M, respectively, in order to ensure complete hydrolysis of the L-isomer of the racemate. A brief time may elapse before the lactones of chloroacetyl-DL-homoserine and α-chloroacetyl-ε-carbobenzoxy-DL-hydroxylysine are opened by the added alkali. A few drops of phenol red added to the mixture may be used to follow this procedure.

[21] Acetyl-S-benzylcysteine and acetyl-S-benzylhomocysteine are brought into solution at about pH 9, and then the pH is lowered to 7 by addition of acetic acid. The former compound remains in solution at this pH, but the latter begins to slowly crystallize. For this reason, the resolution mixture of the benzylhomocysteine is gently stirred throughout the course of the enzymatic digestion.

[22] Isovaline is another amino acid whose resolution is followed by the nitrous acid procedure. During the resolution of tyrosine, S-benzylcysteine, and S-benzylhomocysteine, their L-isomers crystallize from the mixture, and homogeneous samples are difficult to acquire for analytical purposes. In such cases double or treble the usual amount of enzyme is added (see above), the digestion period is also doubled or trebled, and the mixture is subjected to frequent shaking during the digestion period. In the case of the hydroxylysines, whose α-chloroacetyl-ε-carbobenzoxy derivatives are hydrolyzed so very slowly by acylase I as to reach only about 90% completion after many days of incubation, the L-isomer will be optically pure, but the D-isomer will be appreciably contaminated with the L-form. Removal of the contaminant by lysine decarboxylase is described.[2]

the digestion. Together with protein, they are filtered, washed with cold water, warmed to 60° with a tenfold volume of 2 N HCl in the presence of Norit, and filtered clear. The filtrates are treated with 28% NH₃ water to pH 6, and the resulting preparations redissolved in HCl in the presence of Norit, filtered, and the filtrate again brought to pH 6. The protein-free L-isomers are filtered, washed with water until free of NH$_4^+$, and dried. (b) Arginine—the digest is clarified with Norit and evaporated *in vacuo* to low bulk and chilled. Then 4 N NaOH is added to pH 11, and the mixture is shaken with an excess of freshly distilled benzaldehyde in the cold for several hours. The crystalline precipitate of benzilidine-L-arginine is filtered and washed with water, alcohol, and ether. The compound is warmed to 60° with 2 N HCl, the liberated benzaldehyde extracted into ether, and the filtered acid solution evaporated *in vacuo* to small volume. Treatment with aniline to pH 6, followed by addition of hot ethanol, yields the crystalline monohydrochloride of L-arginine. (c) α,β-Diaminopropionic acid, α,γ-diaminobutyric acid, ornithine, and lysine—the digest is clarified with Norit and evaporated *in vacuo* to low bulk and treated with ethanol to 80%. The ω-acylated L-amino acids rapidly crystallize and when complete are filtered, washed with ethanol, and crystallized twice more from water-ethanol mixtures. After refluxing for 2 hours with 2 N HCl to remove the ω-acyl groups, the solutions, decolorized with Norit, are evaporated *in vacuo* to dryness. The L-ornithine residue is crystallized directly as the dihydrochloride from the dry methanol solution. The other residues are taken up in the minimum amount of water. Addition of ethanol to the solution of residual L-α,β-diaminopropionic acid leads to crystallization of the monohydrochloride. The aqueous solutions of L-α,γ-diaminobutyric acid and L-lysine residues are treated with aniline to pH 6, followed by addition of ethanol, and lead to crystallization of these compounds as the respective monohydrochlorides. (d) Hydroxylysine and allohydroxylysine—these are treated as in (c), except that the ω-acyl group, here ε-carbobenzoxy, is removed by dissolving the L-isomer in methanol-water and catalytically hydrogenating in the presence of palladium and an excess of HCl. The filtrate is evaporated *in vacuo*, and the condensed solution, brought to pH 5 with aniline, yields the L-isomers as the crystalline monohydrochlorides.[2] (e) Alanine, aminobutyric acid, valine, leucine, isoleucine, alloisoleucine, methionine, histidine, aspartic acid, glutamic acid, aminoadipic acid, serine, homoserine, threonine, allothreonine, phenylalanine, tryptophan, and β-phenylserine (threo or erythro)—the digests are clarified with Norit and evaporated *in vacuo* to small volume. Solutions of the three dicarboxylic amino acids are further reduced in pH to 3.0 by addition of dilute HCl. To each of the condensates ethanol is added to 80%. The resulting precipitate of L-amino acid is filtered and recrystallized once or twice from water-ethanol with the aid of Norit to yield the protein-free, pure product.[23] (f) Proline—the digest is clarified with Norit and evaporated in vacuo to small volume. Saturated K_2SO_3 solution is added to pH 10 to 11, and the mixture is vigorously gassed with N_2 until free of NH_3. The solution is shaken with 3 equivalents of carbobenzoxy chloride for 2 hours at 5°, and the resulting crystals of carbobenzoxy-D-proline amide are filtered, washed with water, and set aside. The filtrate is acidified to pH 1.7 with concentrated HCl, and the oily carbobenzoxy-L-proline is extracted into ether. The solvent is removed, the residual oil dissolved in methanol and water, treated with a few drops of acetic acid, and the solution shaken with palladium in presence of H_2. The filtrate is evaporated to dryness in vacuo, and the residual L-proline crystallized twice from absolute ethanol.[24]

Isolation of the D-Isomers

Chemical Procedures. (a) Tyrosine, phenylalanine, valine, leucine, alloisoleucine, S-benzylcysteine, and S-benzylhomocysteine—after removal of the L-amino acid isomers above, the filtrates are treated with concentrated HCl to pH 1.7, and, where present, alcohol is removed by evaporation in vacuo to near dryness. The crystals of acylated-D-amino acids are filtered, washed with cold water, dried, taken up in hot acetone, and filtered clear. After removal of the solvent, the residues are refluxed for 2 hours with a tenfold amount of 2 N HCl. The solutions are filtered hot with the aid of Norit and evaporated in vacuo to dryness to remove excess HCl. The residues are taken up in water and brought to pH 6 with ammonia. Where necessary, ethanol is added to 80% to complete the crystallization of the D-isomers. (b) Tryptophan—the procedure is the same as in (a) except that 2 N H_2SO_4 replaces 2 N HCl in the refluxing operation. Warm baryta solution is added to the diluted hydrolyzate in slight excess, the mixture treated with acetic acid to pH 5, and $BaSO_4$ filtered off through a layer of Norit. The clear, colorless filtrate is concentrated in vacuo to crystallization of the D-tryptophan. (c) Histidine, arginine, and α,β-diaminopropionic acid—the L-amino acid filtrates are freed from alcohol by evaporation in vacuo, and in the case of arginine from benzaldehyde by extraction with ether. The aqueous solutions are

---

[23] Substantially the same procedure is used to obtain the optical isomers of norvaline, norleucine, ethionine, pentahomoserine, and hexahomoserine.

[24] D. Hamer and J. P. Greenstein, J. Biol. Chem. 193, 81 (1951). Substantially the same procedure is used to resolve tert-leucine, except that here the final product is crystallized from water-acetone and water-ethanol mixtures; cf. N. Izumiya, S-C. J. Fu, S. M. Birnbaum, and J. P. Greenstein, J. Biol. Chem. 205, 221 (1953).

brought to pH 7, treated with *Crotalus adamanteus* (rattlesnake) venom, and aerated for 20 hours to oxidize any residual L-amino acid. The digests are acidified to pH 5, the venom protein coagulated on the water bath, and the mixtures filtered with aid of Norit. Concentrated HCl is added to 2 N; the solutions are refluxed for 2 hours and filtered hot with the aid of Norit. The water-clear filtrates are evaporated *in vacuo* to dryness, and the residues dissolved in a little water. D-α,β-Diaminopropionic acid is crystallized as the monohydrochloride by addition of ethanol to 80%. The acid histidine solution is brought to pH 7 by addition of LiOH, the D-histidine base crystallized by addition of ethanol, and the product recrystallized from a little water. The D-arginine residue is converted to the benzilidine derivative, and the pure monohydrochloride subsequently isolated as described above for the L-isomer. (d) Alanine, aminobutyric acid, isoleucine, methionine, aspartic acid, glutamic acid, aminoadipic acid, serine, threonine, allothreonine, and β-phenylserine (threo or erythro)—the L-isomer resolution filtrates after removal of ethanol by evaporation *in vacuo* are treated with concentrated HCl to pH 1.7 and extracted several times with ethyl acetate. The combined extracts are dried over Na_2SO_4, filtered, the solvent removed, the residues taken up in acetone, filtered clear, and the solvent removed. The residues are refluxed for 2 hours with a tenfold volume of 2 N HCl, except for the β-phenylserines which are warmed with 20% acetic acid on the water bath for 4 hours, and the solvent then removed *in vacuo*. In each case the residue is redissolved in a little water and brought to pH 6, except for the dicarboxylic amino acids which are brought to pH 3. On addition of ethanol to 80%, the D-amino acids separate and are recrystallized twice more from water-ethanol mixtures.[25] (e) α,γ-Diaminobutyric acid, ornithine, and lysine—the procedure is the same as in (d), except that with D-α,γ-diaminobutyric acid and D-lysine the final acid solutions are treated with aniline to pH 5 and the monohydrochlorides subsequently crystallized by addition of ethanol to 80%, whereas with D-ornithine the dry residue is crystallized as the dihydrochloride from its methanol solution. (f) Homoserine—the procedure is the same as in (d), except that the HCl hydrolysis leads to the formation of the γ-lactone. On removal of the HCl solvent, the residual D-homoserine-γ-lactone is taken up in water and crystallized by the addition of acetone. The lactone is dissolved in water and refluxed for 3 hours in the presence of an excess of pulverized ammonium carbonate. The solution is evaporated *in vacuo* to dryness, and the residue dissolved in a little water, brought to pH 5 to 6 with dilute acetic acid, clarified with Norit, and treated with excess ethanol.

[25] When ammonia is the neutralizing agent, the amino acids are crystallized until they no longer react with Nessler's reagent.

The D-homoserine is crystallized again from water-ethanol.[26] (g) Hydroxylysine and allohydroxylysine—after removal of the L-isomer, the filtrates are acidified to pH 1.7, and the D-form extracted into ethyl acetate. After removal of the solvent, the residue is dissolved in methanol-H_2O and catalytically hydrogenated in presence of Pd and HCl. By this means the ϵ-carbobenzoxy group is removed, and on filtration from the catalyst and addition of HCl to 2 N, followed by a 2-hour period of refluxing, the α-chloroacetyl group is removed. The solvent is removed $in\ vacuo$, and the residue taken up in solution and brought to pH 5. Since there is appreciable L-isomer present in the preparations,[22] they are treated with lysine decarboxylase (Vol. II [20]) to remove the contaminant, filtered from enzyme protein, and the crystalline monohydrochlorides of pure D-hydroxylysine and D-allohydroxylysine isolated by addition of excess ethanol. (h) Proline—the crystalline carbobenzoxy-D-proline amide is recrystallized from ethyl acetate (melting point 94°, $[\alpha]_D^{23} - +33.6°$ for 2% in ethanol), dissolved in methanol-H_2O, and catalytically hydrogenated in the presence of Pd and HCl. The filtrate is treated with concentrated HCl to 2 N, refluxed for 2 hours, and evaporated to dryness $in\ vacuo$.[27] The residual D-proline·HCl is freed of Cl⁻ by treatment with Ag_2CO_3, and of Ag⁺ by H_2S, and the clear filtrate is evaporated to dryness $in\ vacuo$. The residual D-proline is twice crystallized from absolute ethanol.

Hydroxyproline[28]

The optical isomers of this amino acid are most readily obtained by a chemical method starting from available hydroxy-L-proline isolated from proteins. According to the procedure of Neuberger,[29] allohydroxy-L-proline is prepared by the following steps: (a) One mole of hydroxy-L-proline is treated with 1 mole of acetic anhydride in boiling glacial acetic acid solution, the solvent removed $in\ vacuo$, and the residual acetyl-hydroxy-L-proline taken up in acetone and brought to crystallization at $-5°$ (melting point 133°, $[\alpha]_D^{25} = -118.5°$ for 2% in H_2O). (b) One mole of acetylhydroxy-L-proline is dissolved in a little dry methanol, treated with an excess of a dry ethereal solution of diazomethane, and the solvent subsequently removed, yielding a residual oil. (c) The acetylhydroxy-L-proline methyl ester is dissolved in dry pyridine and treated with recrystallized (from benzene) p-toluenesulfonyl chloride, after standing at 0° for 18 hours poured into chilled 1 N HCl, and the resulting crystals of

[26] Substantially the same procedure is followed to obtain D-pentahomoserine and D-hexahomoserine; cf. L. Berlinguet and R. Gaudry, $J.\ Biol.\ Chem.$ **198,** 765 (1952).
[27] At this stage, D-$tert$-leucine amide is hydrolyzed by refluxing for 12 hours in 5 N HCl.
[28] D. S. Robinson and J. P. Greenstein, $J.\ Biol.\ Chem.$ **195,** 383 (1952).
[29] A. Neuberger, $J.\ Chem.\ Soc.$ **1945,** 429.

N-acetyl-O-toluenesulfonylhydroxy-L-proline methyl ester crystallized from ether-alcohol (melting point 60°). (d) The previous compound is saponified by being dissolved in methanol and shaken with the calculated amount of 1 N NaOH for 18 hours, and on neutralization with the equivalent amount of 1 N HCl the crystalline acid compound appears (melting point 182°). (e) Epimerization at C_4 is effected by heating a solution of N-acetyl-O-toluenesulfonylhydroxy-L-proline containing 2 equivalents of 0.5 N NaOH at 100° for 20 minutes. After an excess of HCl is added to the cooled digest, the mixture is evaporated in vacuo to dryness, the residue extracted with hot acetone, the solvent removed, and the new residue refluxed with 2 N HCl for 2 hours. The acid solvent is removed in vacuo, the residue neutralized with LiOH, and the resulting allohydroxy-L-proline purified by several crystallizations from aqueous alcohol.

Allohydroxy-D-proline is prepared from hydroxy-L-proline by the following steps:[28] (a) One mole of hydroxy-L-proline is treated with a tenfold amount of glacial acetic acid plus 5 moles of acetic anhydride, and the mixture refluxed for 4 hours. (b) The solvent is removed in vacuo, the residual sirup dissolved in 2 N HCl and refluxed for 3 hours, the solution decolorized with Norit, and the solvent removed in vacuo. (c) The crystalline residue is an epimeric mixture of the hydrochlorides of hydroxy-L-proline (about 25%) and allohydroxy-D-proline (about 75%); it is dissolved in water and Cl^- removed with an excess of Ag_2CO_3, and the excess Ag^+ removed with H_2S. (d) The colorless filtrate is evaporated in vacuo to low bulk, ethanol added to 90%, and the resulting crystals of nearly pure allohydroxy-D-proline filtered and crystallized twice more from 90% ethanol to yield a pure product. From this preparation of allohydroxy-D-proline, hydroxy-D-proline is prepared by the procedure described above for conversion of hydroxy-L-proline to allohydroxy-L-proline.[28]

Physical characteristics of the intermediates are as follows: acetyl-allohydroxy-D-proline, melting point 145°, $[\alpha]_D^{25} = +91.0°$ for 2% in H_2O; N-acetyl-O-toluenesulfonylallohydroxy-D-proline methyl ester, melting point 143.5°; N-acetyl-O-toluenesulfonylallohydroxy-D-proline, melting point 143.5°. $[M]_D^{25}$ for hydroxy-L-proline in 5 N HCl is $-66.2°$; and for allohydroxy-L-proline, $-24.7°$. Rotation values for the corresponding D-isomers are equal in magnitude and opposite in sign. The over-all yields are about 20%.

Isolation of L- and D-Isomers by Chromatography[4]

For this purpose, Dowex 50 (alanine, methionine, valine, aspartic acid, serine, and ornithine) and Amberlite XE-64 (proline, histidine, and arginine) are used in the acid phase. The resin is washed successively with 5 N HCl, water, 1 N NaOH, water, 5 N HCl, and water. The deproteinized and condensed enzymatic digests as above are added to the top of the resin column at a rate of about 0.5 ml./min. and washed into the resin with water. The acyl-D-amino acid is then eluted from the resin by water, and as long as this component is being eluted the eluate is acid in reaction and hydrolysis with HCl leads to positive ninhydrin color. When these reactions fail, the fractions are combined, evaporated to dryness *in vacuo*, the residue dissolved in ethanol or acetone, filtered, and the filtrate evaporated to dryness *in vacuo*. The residue is hydrolyzed by refluxing for 2 hours with 2 N HCl, the solution filtered hot with aid of Norit, and the water-clear solvent removed *in vacuo*. The free D-amino acid is prepared by appropriate neutralization as described above. In preparation for the elution of the L-amino acid isomer from the column, the latter is washed with several hundred milliliters of water. Elution of the free L-amino acid is accomplished with HCl at concentrations depending on the resin and the particular amino acid involved—thus 1 N for aspartic acid and histidine, 2.5 N for alanine and serine, and 5 N for phenylalanine, ornithine, methionine, and valine.[4] When the ninhydrin reaction becomes negative in the eluate, the fractions are combined, and the free L-amino acids prepared by appropriate neutralization as described above.[30] In the case of the separation of proline from proline amide, the former is not retained by the XE-64 column used and thus is eluted by passage of water, leaving the proline amide to be subsequently eluted with relatively dilute HCl, i.e., 0.12 N.[31]

Criteria of Purity of Resolved L- and D-Isomers

These involve considerations of analytical, optical, and, in the case of amino acids with two asymmetric centers, steric purity. Analysis of the amino acid isomers for content of C, H, N, and, where appropriate, S is taken for granted as necessary routine procedure. Qualitative considerations can be met by specific color reactions where applicable, i.e., Sakaguchi (arginine), Pauly (histidine and tyrosine), isatin (proline),

---

[30] L-Isovaline is also separable from chloroacetyl-D-isovaline after action of acylase I on the chloroacetylated racemate by use of 20- to 50-mesh Dowex 50 in the acid phase, by first eluting the chloroacetyl-D-isovaline with water and later the L-isovaline with 2.5 N HCl; cf. C. G. Baker, S-C. J. Fu, S. M. Birnbaum, H. A. Sober, and J. P. Greenstein, *J. Am. Chem. Soc.* **74**, 4701 (1952).

[31] L-*tert*-Leucine is separable in the same way from D-*tert*-leucine amide on Amberlite XE-64 in the acid phase, the free L-amino acid being washed through with water, while the D-amide is subsequently eluted from the resin by means of relatively dilute HCl at 0.05 to 0.10 N; cf. N. Izumiya, S-C. J. Fu, S. M. Birnbaum, and J. P. Greenstein, *J. Biol. Chem.* **205**, 221 (1953).

platinic iodide (methionine), periodate-Nessler (hydroxyamino acids) (cf. ref. 32), and more generally by techniques of paper chromatography employing several solvent systems by which a single and known spot on the chromatogram must be obtained. Optical purity is most often determined by optical rotatory characteristics as measured in the polarimeter. These optical data are quantitative only if the analytical data reveal a chemically pure compound (elemental and chromatographic analysis), and if the rotatory characteristics of the resolved L- and D-isomers are equal in magnitude and opposite in sign. Where the L-isomer isolated from natural sources is available and meets the analytical criteria, a further basis for a quantitative criterion of purity is possible. In view of the limitations of the polarimeter, the accuracy in determining optical purity with this instrument may frequently be no better than $\pm 1\%$. A more sensitive criterion of optical purity is by means of optically specific amino acid oxidases and decarboxylases which possess an accuracy in measurement to at least 1 part in 1000.[33]

The technique is as follows: 1000 micromoles of the isomer to be tested is placed in each of four Warburg vessels, and to each of two of these flasks 1 micromole of the optical enantiomorph (usually in the form of an aliquot of a larger volume) is added. To these four flasks, as well as to two others, 1.5 to 2.5 ml. of buffer is added. In the side arms of all six vessels the appropriate enzyme solution is introduced. The flasks are tipped after a 10- to 15-minute equilibration period and read at intervals until gas evolution or consumption is complete (15 to 120 minutes). This method obviously is applicable only where the 1 micromole of added susceptible isomer is readily and quantitatively oxidized or decarboxylated in the presence of the 1000-fold amount of the resistant enantiomorph. As a rule, a considerable amount of the enzyme is employed. The criteria of purity set up demand (a) that the 1000 micromoles alone of the isomer to be tested must consume less than 1 microatom of O_2 or evolve less than 1 micromole of CO_2, while simultaneously (b) the added 1 micromole of susceptible isomer must be quantitatively oxidized or decarboxylated as shown by the increment in value over (a), all values being corrected for the enzyme blanks.

Crotalus adamanteus (rattlesnake) L-amino acid oxidase is successfully employed in determining the optical purity of the D-isomers of aminoadipic acid, histidine, β-phenylserine (erythro), aminobutyric acid, valine, leucine, isoleucine, alloisoleucine, homoserine, methionine,

---

[32] R. J. Block and D. Bolling, "The Amino Acid Composition of Foods. Analytical Methods and Results," 2nd ed., Charles C Thomas, Springfield, Ill., 1952.

[33] A. Meister, L. Levintow, R. B. Kingsley, and J. P. Greenstein, *J. Biol. Chem.* **192**, 535 (1951).

S-benzylcysteine, S-benzylhomocysteine, phenylalanine, tyrosine, and tryptophan.[1] *Bothrops jararaca* venom L-amino acid oxidase is used to determine the optical purity of D-alanine.[34] The optical purity of the D-isomers of arginine (*E. coli* 7020), lysine (*B. cadaveris* 6578), ornithine (*C. septicum* P-III plus added pyridoxal phosphate), and aspartic and glutamic acids (*C. welchii* SR 12) is tested with the mentioned decarboxylases.[33–35] Hog kidney D-amino acid oxidase is used to test the optical purity of the L-isomers of α,β-diaminopropionic acid, serine, allothreonine, proline, hydroxyproline, allohydroxyproline, alanine, aminobutyric acid, valine, leucine, isoleucine, alloisoleucine, homoserine, methionine, S-benzylcysteine, S-benzylhomocysteine, phenylalanine, tyrosine, and tryptophan.[1]

In all cases, the L- and D-isomers prepared by the resolution procedures herein described contain less than 1 part in 1000 of the enantiomorph. That the purity may even be better than this is suggested by the cases of L-alanine and L-serine which have been found to contain less than 1 part of the D-isomers in 10,000 parts of the L-forms.[33] Since the D-isomers of hydroxylysine and allohydroxylysine are treated with lysine decarboxylase as part of the preparative procedure, they, too, are of high optical purity.

These methods with the criteria stated are so far inapplicable to the D-isomers of α,β-diaminopropionic acid, serine, allothreonine, proline, hydroxyproline, and allohydroxyproline, to the L-isomers of aspartic, glutamic and aminoadipic acids, histidine, ornithine, lysine, and the erythro isomer of β-phenylserine, and to the L- and D-isomers of threonine, β-phenylserine (threo), and α,γ-diaminobutyric acid.[36]

The problem of steric purity involving isoleucine, threonine, hydroxyproline, and hydroxylysine is at present successfully approached only in the case of isoleucine. D-Isoleucine and D-alloisoleucine each treated with D-amino acid oxidase yield 2,4-dinitrophenylhydrazones of their α-keto acids with $[\alpha]_D^{25} = -16.9°$ and $+16.9°$ (2% in ethanol) respectively.[6] Similarly, L-isoleucine and L-alloisoleucine treated with rattlesnake venom L-amino acid oxidase yield hydrazones with $[\alpha]_D^{25}$ of $+17.3°$ and $-17.1°$, respectively.[37] Admixture of L-isoleucine and L-alloisoleucine,

[34] S. M. Birnbaum and J. P. Greenstein, *Arch. Biochem. and Biophys.* **39**, 108 (1952).
[35] S. M. Birnbaum, L. Levintow, R. B. Kingsley, and J. P. Greenstein, *J. Biol. Chem.* **194**, 455 (1952).
[36] They are readily applicable to optical purity determinations of a number of the unnatural amino acids, i.e., the L- and D-isomers of norvaline, norleucine, aminoheptylic acid, aminocaprylic acid, aminononylic acid, pentahomoserine, hexahomoserine, ethionine, aminophenylacetic acid, aminocyclohexylacetic acid, and aminocyclohexylpropionic acid.[1]
[37] A. Meister, *Nature* **168**, 1119 (1951).

TABLE I

ENZYMATIC RESOLUTION DATA

DL-α-Amino acid	Recommended derivative of amino acid	Molecular weight of derivative	Melting point of racemic derivative (corr.)	Enzyme used	Hydrolytic[a] rate of derivative	Molecular weight of amino acid	$[M]_D$ of[b] L-amino acid
Alanine	Acetyl	131.1	136°	Acylase I	3,200	89.1	+ 13.0
Aminobutyric acid	Acetyl	145.1	132	Acylase I	9,500	103.1	+ 21.2
Valine	Acetyl	159.2	148	Acylase I	1,660	117.1	+ 33.1
Norvaline	Acetyl	159.2	115	Acylase I	9,800	117.1	+ 29.2
Leucine	Acetyl	173.2	159	Acylase I	5,400	131.1	+ 21.0
Norleucine	Acetyl	173.2	105	Acylase I	14,400	131.1	+ 32.1
Isoleucine	Acetyl	173.2	116	Acylase I	376	131.1	+ 53.5
Alloisoleucine	Acetyl	173.2	168	Acylase I	250	131.1	+ 53.1
Isoleucine + alloisoleucine[c]	Acetyl	173.2	150	Acylase I	376	131.1	+ 53.5
Alloisoleucine + isoleucine[d]	Acetyl	173.2	150	Acylase I	250	131.1	+ 53.1
Methionine	Acetyl	191.2	112	Acylase I	24,200	149.2	+ 34.6
Ethionine	Acetyl	205.2	91	Acylase I	15,400	163.2	+ 38.7
S-Benzylcysteine	Acetyl	253.3	157	Acylase I	100	211.2	− 42.2
S-Benzylhomocysteine	Acetyl	267.3	115	Acylase I	170	225.2	+ 61.3
Histidine	Acetyl-1·H₂O	215.2	148	Acylase I	150	155.2[e]	+ 18.3[e]
Arginine	Acetyl-2·H₂O	252.2	266	Acylase I	410	174.2[e]	+ 48.1[e]
Aspartic acid	Acetyl	175.1	150	Acylase II	27	133.1	+ 33.8
Glutamic acid	Acetyl	189.2	185	Acylase I	3,080	147.1	+ 46.8
α,β-Diaminopropionic acid	Diacetyl	188.1	181	Acylase I	708	104.1[e]	+ 35.4[e]
Isovaline	Chloroacetyl	193.7	163	Acylase I	38	117.1	+ 9.7
Aminoheptylic acid	Chloroacetyl	221.7	106	Acylase I	28,200	145.2	+ 33.8
Aminocaprylic acid	Chloroacetyl	235.7	93	Acylase I	7,700	159.2	+ 36.6
Aminononylic acid	Chloroacetyl	249.8	89	Acylase I	1,600	173.2	+ 58.0[f]
Aminodecylic acid	Chloroacetyl	263.8	92	Acylase I	120	187.2	+ 58.0[f]
Aminoundecylic acid	Chloroacetyl	277.8	92	Acylase I	9	201.2	+ 58.3[f]
Serine	Chloroacetyl	181.6	123	Acylase I	11,600	105.1	+ 15.9
Homoserine	Chloroacetyl[g]	177.6	79	Acylase I	12,600	119.1	+ 21.8[h]
Pentahomoserine	Chloroacetyl	209.6	104	Acylase I	23,800	133.1	+ 38.3
Hexahomoserine	Chloroacetyl	223.7	91	Acylase I	10,000	147.1	+ 34.9

Threonine	Chloroacetyl	195.6	124	Acylase I	720	119.1	− 17.9
Allothreonine	Chloroacetyl	195.6	91	Acylase I	2,580	119.1	+ 36.3
Aminoadipic acid	Chloroacetyl	237.7	129	Acylase I	45	161.1	+ 40.3
Aminophenylacetic acid	Chloroacetyl	227.6	127	Acylase I	4,500	151.2	+254.0
Aminocyclohexylacetic acid	Chloroacetyl	233.7	175	Acylase I	4,600	157.2	+ 55.8
Aminocyclohexylpropionic acid	Chloroacetyl	247.7	142	Acylase I	350	171.2	+ 25.7
Phenylalanine	Chloroacetyl	241.7	130	Carboxypeptidase	2,750	165.1	− 7.4
Tyrosine	Chloroacetyl	257.7	158	Carboxypeptidase	2,500	181.2	− 18.1
Tryptophan	Chloroacetyl	280.7	154	Carboxypeptidase	2,000	204.1	− 69.4[f]
α,γ-Diaminobutyric acid	Dichloroacetyl	271.2	128	Acylase I	60	118.1[e]	+ 37.4[e]
Ornithine	Dichloroacetyl	285.2	105	Acylase I	304	132.2[e]	+ 37.5[e]
Lysine	Dichloroacetyl	299.2	103	Acylase I	140	146.2[e]	+ 37.9[e]
Hydroxylysine	ε-Carbobenzoxy[i] α-Chloroacetyl	354.8	152	Acylase I	<5	162.2[e]	+ 28.9[e]
Allohydroxylysine	ε-Carbobenzoxy[i] α-Chloroacetyl	354.8	145	Acylase I	<5	162.2[e]	+ 50.9[e]
β-Phenylserine (threo)	Trifluoroacetyl	277.3	161	Carboxypeptidase	1,100	181.2	− 88.1
β-Phenylserine (erythro)	Trifluoroacetyl	277.3	150	Carboxypeptidase	800	181.2	+147.3
Proline	Amide (free)	114.1	99	Amidase	7	115.1	− 69.5
tert-Leucine	Amide·HCl	166.7	337[j]	Amidase	12	131.1	+ 11.8
Cystine[k]	—	—	—	—	—	240.2	−557.4
Homocystine[k]	—	—	—	—	—	268.2	+209.2
Cysteine[k]	—	—	—	—	—	121.1	+ 7.9
Citrulline[l]	—	—	—	—	—	175.2	+ 42.4

[a] All substrates at 0.016 M concentration. Rates in terms of micromoles of the L-isomer hydrolyzed per hour per milligram of protein N at 38°. Acylated substrates at pH 7.0, amides at pH 8.0 in the presence of 0.01 M Mn++.

[b] 2 to 5% concentration, except where otherwise stated, in 5 N HCl solution; T = 24 to 26°.

[c] Epimer of L-isoleucine and D-alloisoleucine; acetyl derivative has $[\alpha]_D^{25} = -3.0°$ (4% in ethanol).

[d] Epimer of L-alloisoleucine and D-isoleucine; acetyl derivative has $[\alpha]_D^{25} = +3.0°$ (4% in ethanol).

[e] As free base.

[f] In glacial acetic acid solution.

[g] As γ-lactone.

[h] Freshly prepared solution.

[i] As δ-lactone.

[j] In sealed capillary tube.

[k] Derived from corresponding S-benzyl isomers.

[l] Derived from ornithine isomers.

or of D-isoleucine with D-alloisoleucine, results in the isolation of keto acid hydrazones with optical rotations somewhere between these extremes.[3] The sensitivity of this test is governed by the accuracy of polarimetric observations. Its inapplicability to the other amino acids in this category is due primarily to the resistance of several of the isomers to the action of the oxidases.

[83] Chemical Determination of Glutamic Acid[1-4]

By H. WAELSCH

Principle. The β-formylpropionic acid formed from glutamic acid be the action of ninhydrin is converted into the 2,4-dinitrophenylhydrazony, which is transferred into capryl alcohol and then extracted with borate buffer. The color developed after the addition of sodium hydroxide to the borate solution is measured in a colorimeter.

The synthetic 2,4-dinitrophenylhydrazone of β-formylpropionic acid is used as the standard. The stability of the hydrazone in acid and alkaline solution decreases rapidly with increasing temperature, acidity, or alkalinity, but under the conditions of the method only a negligible amount is destroyed.

Aspartic acid and cystine (or cysteine) interfere in the determination by formation of hydrazones which appear in the borate solution. Aspartic acid is removed by chromatographic separation on aluminum oxide on a microscale and elution of glutamic acid with acetic acid. The interference of cystine or cysteine is eliminated by subtraction of 12% of the amount of these amino acids present. Glutamic acid may be determined in the presence of a large excess of glutamine.

Apparent glutamine in the eluate is determined as glutamic acid by the same procedure after hydrolysis which splits the amide into glutamic acid and ammonia.

The recovery of glutamic acid from pure solutions without chromatographic analysis is quantitative if account is taken of the distribution of the synthetic hydrazone between the solvents used in the procedure. Under the conditions described, 14% of the glutamic acid present is retained in the aluminum oxide column.

[1] B. A. Prescott and H. Waelsch, *J. Biol. Chem.* **164,** 331 (1946).

[2] B. A. Prescott and H. Waelsch, *J. Biol. Chem.* **167,** 855 (1947).

[3] S. P. Bessman, J. Magnes, P. Schwerin, and H. Waelsch, *J. Biol. Chem.* **175,** 817 (1948).

[4] P. Schwerin, S. P. Bessman, and H. Waelsch, *J. Biol. Chem.* **184,** 77 (1950).

The method does not differentiate between D- or L-glutamic acid. Sensitivity of the method: 5 γ of glutamic acid $\pm 5\%$.

Apparatus

Centrifuge tubes, 15-ml. graduated and 25-ml. ungraduated.
Glass tubes, 7-mm. outer diameter, drawn out to 1-mm. tips.
Adsorption funnels (Fig. 1).
Adapters for packing the aluminum oxide in the adsorption funnels by centrifuging. Two holes corresponding to the diameter of the stem of the adsorption funnels are bored into cork plates of 3-cm. diameter and 1-cm. thickness. The adapter with the funnels in place is set on a 15-ml. centrifuge metal tube, and the assembly is fitted into a 100-ml. centrifuge metal tube.

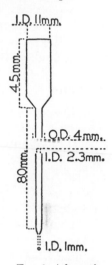

Fig. 1. Adsorption funnel.

Reagents

Ninhydrin. Eastman, recrystallized once from boiling water with the addition of Norit (yield 80%).
Acetic acid, 0.5 N.
14% guanidine carbonate solution.
12% neutral lead acetate ($+3H_2O$) solution.
5 N NaOH.
0.1% 2,4-dinitrophenylhydrazine solution in N HCl. This reagent is kept in the refrigerator and renewed at monthly intervals.
Capryl alcohol, treated with sodium and distilled.
Absolute ethanol.
Borate buffer. Twenty grams of sodium tetraborate and 5.3 g. of sodium carbonate are dissolved in 1 l. of water, and the pH is adjusted to 10.
Aluminum oxide, according to Brockmann (Merck).
N HCl.
Standard solution of 2,4-dinitrophenylhydrazone of β-formylpropionic acid. Two and one-half grams of ethyl formyl succinate is converted into β-formylpropionic acid by refluxing for 6 hours in 12.5 ml. of a 3% solution of oxalic acid.[5] The reaction mixture is cooled and added, with shaking, to 1 l. of a saturated solution of 2,4-dinitrophenylhydrazine in 2 N HCl. After several hours in the cold the hydrazone is filtered off (1.9 g. 54% yield) and

[5] E. Carrière, *Ann. chim. (Paris)* **17**, 38 (1921–1922).

recrystallized three times from 95% ethanol (melting point 202.5 to 203°, uncorrected).

From 50 to 60 mg. of the hydrazone (1 mg. = 0.521 mg. of glutamic acid) is dissolved in approximately 10 ml. of absolute ethanol with the addition of 1 ml. of pyridine. The hydrazone dissolves with difficulty, and it is necessary to shake for about 30 minutes prior to dilution to 100 ml. with absolute ethanol. From this stock solution, which may be kept in the ice box for several days, a working standard is prepared by diluting 2 ml. to 100 ml. with borate buffer.

Procedure

Determination of Glutamic Acid in Mixtures with Other Amino Acids; Chromatographic Adsorption; Preparation of Column. A suspension of 10 g. of aluminum oxide in about 50 ml. of N HCl is stirred for approximately 1 minute. The acid is decanted, and the oxide is washed with distilled water by decantation eight to ten times until the washings are neutral to litmus. "Acid" aluminum oxide so prepared can be stored under distilled water.

The outlet of the adsorption funnel is closed with a small amount of glass wool, gentle suction is applied, and about 6 cm. of the funnel stem, measured from the glass wool, is filled with oxide. After filling the funnel with water it is centrifuged in the assembly described above for 30 minutes at 2000 r.p.m.

Adsorption, Washing, and Elution. The solution containing glutamic and other amino acids is neutralized with dilute NaOH to pH 7, with the aid of a pH meter or with bromothymol blue as an internal indicator, and diluted with water to a volume such that 1 ml. contains not more than 25 γ of glutamic acid. (Concentrations of sodium chloride up to 1 M do not interfere with the adsorption and elution of glutamic.) The concentration of glutamic acid is determined in duplicate or triplicate. Two blank determinations with 2 ml. of H_2O are carried out simultaneously.

For each determination, 2 ml. of the solution is filtered by gravity through a funnel. When the surface of the fluid reaches a point about 1 mm. above the oxide, 2 ml. of water is added. The filtrate and wash water are discarded. The elution is accomplished by passing two 2-ml. portions of 0.5 N acetic acid through the column into a 15-ml. graduated centrifuge tube.

Reaction with Ninhydrin and Removal of Excess Ninhydrin. To each of the tubes containing eluate, 20 ± 0.5 mg. of ninhydrin is added and the tubes are immersed in a boiling water bath for exactly 10 minutes, after which they are transferred to an ice bath for 2 minutes. The following reagents are added in the order named, with an interval of 5 minutes after each addition: 0.4 ml. of 14% guanidine carbonate, 1 ml. of 12% lead acetate, and 0.5 ml. of 5 N NaOH. The mixtures are diluted to 6 ml. with water, well stirred, and centrifuged for 10 minutes. The supernatant solutions are decanted into small test tubes.

Formation and Extraction of Hydrazone. During the centrifuging, 3-ml. portions of a 0.1% solution of 2,4-dinitrophenylhydrazine in 1 N HCl are placed in 100-ml. test tubes and cooled in an ice bath. Glass tubes drawn to a 1-mm. tip are placed in each of the tubes, and 5-ml. portions of the supernatant fluids are added. The solutions are mixed by passing air through the tubes for about 1 second. Exactly 10 minutes later, 10 ml. of capryl alcohol is added to each tube, and the mixtures are agitated by a vigorous air stream for 1 minute. The tubes are taken from the ice bath, most of the aqueous phases are removed through the aeration tubes by suction, the turbid capryl alcohol layers are clarified by centrifuging for 5 minutes in 15-ml. centrifuge tubes, 9-ml. aliquots of the capryl alcohol are pipetted into 25-ml. centrifuge tubes, and 6-ml. portions of borate buffer are added. The mixtures are vigorously agitated by a stream of air for 1 minute, and the layers are separated by centrifuging for 20 minutes.

Development and Reading of Color. A 5-ml. portion of the borate solutions is mixed with 2 ml. of alcoholic KOH (3.5 ml. of 15 N KOH to 20 ml. with absolute ethanol) in a cuvette which is then immersed in a water bath at 25°. Exactly 2 minutes after addition of the alkali, the color is measured in the spectrophotometer at 420 mμ against a blank containing 5 ml. of borate buffer and 2 ml. of alcoholic KOH.

Preparation of Standard Curve and Calculation. Different amounts, up to 5 ml., of the working hydrazone standard solution are diluted to 5 ml. with borate buffer in the cuvettes of the spectrophotometer, and the color is developed as described in the preceding section.

The amount of glutamic acid present in the original sample is calculated from the formula

$$\text{Glutamic acid} = 1.6 \times F_c(A - B) - 0.12\ C$$

where A is the quantity read from the standard, B the quantity corresponding to the blank, and C the amount in micrograms of cystine + cysteine. The factor 1.6 is derived from the fact that three aliquots ($\frac{5}{6}$, $\frac{9}{10}$, $\frac{5}{6}$) are taken during the procedure. F_c is a constant representing the retention of a portion of the glutamic acid on the aluminum oxide column and the distribution of the hydrazone between capryl alcohol and the aqueous acid and alkaline solutions. Since each of these factors depends on definite physical relationships, the value of F_c for the conditions of the determination can be established with precision. The constants are determined in solution of known glutamic acid content.

Determination of Glutamic Acid in the Absence of Interfering Substances. When no interfering substances are present, the chromatographic adsorption is omitted and the reaction with ninhydrin is carried out in 4 ml. of 0.5 N acetic acid containing the glutamic acid. Two blanks containing 4 ml. of acetic acid are included in each set of determinations. In the calculation, instead of the constant F_c, a constant F_p, based on the distribution of the hydrazone between capryl alcohol and the aqueous solutions, is used.

Glutamine

Glutamine is determined in the filtrate from the adsorption column after hydrolysis with hydrochloric acid. The filtrate (2 ml.) of a solution containing glutamic acid and glutamine and the wash water (2 ml.) are collected in a 10-ml. volumetric flask and hydrolyzed with 2 ml. of 6 N HCl for 1 hour. The hydrolyzate is neutralized and diluted to 10 ml. as described previously. Two milliliters of the neutralized solution is taken for the glutamine determination. In pure solution and in the absence of asparagine the glutamic acid may be determined directly without further adsorption on a second column. Under such conditions the hydrolyzed solution is diluted to 10 ml. with water, and 4 ml. of this solution is treated with ninhydrin. In all determinations on tissue filtrates the hydrolyzed solution, containing glutamic acid originating from glutamine, is passed through a second column. The recovery of glutamine amounts to 95 to 105%.

Interfering Substances and Their Removal

Lactic Acid. High concentration of lactic acid interferes with the determination of glutamic acid if the "acid aluminum" is not freshly prepared.

Glutathione and Cysteine. In glutathione-containing solutions, cysteine and glutamic acid, equivalent to approximately 20% of the tripeptide when expressed as glutamic acid, are liberated under the conditions employed for glutamine hydrolysis. A nearly complete removal of glutathione or cysteine without loss of glutamic acid is accomplished by adsorbing the sulfhydryl compound on lead carbonate introduced on top of the aluminum oxide column. One milligram of lead carbonate ($2PbCO_3 \cdot Pb(OH)_2$) suspended in 0.5 ml. of water is superimposed on the aluminum oxide column under gentle suction. By this modification of the column, glutathione and any cysteine are removed to the extent of at least 99%.

Glutathione in the amounts known to occur in mammalian tissue can be successfully eliminated by this procedure.

Determination of Glutamic Acid and Glutamine in Tissues

The dicarboxylic acid and the amide may be determined in protein-free filtrates of blood, plasma, serum, and tissue. The extracts are prepared with TCA as precipitating agent. TCA is neutralized with NaOH (bromothymol blue). The final concentration of TCA is not to exceed 4%.

[83A] Determination of Aspartic Acid and Asparagine

By SAMUEL P. BESSMAN

Aspartic acid, asparagine, glutamic acid, and glutamine can be determined in the same specimen by making use of two principles. The dicarboxylic amino acids are adsorbed on acid-treated alumina and can be eluted separately, glutamic acid by acid elution, and aspartic acid by alkaline elution.[1] Neither glutamine nor asparagine is adsorbed on the alumina. The solution from which the dicarboxylic acids have been removed is then subjected to acid hydrolysis to convert glutamine and asparagine to the corresponding acids and rechromatographed on alumina.[2] Glutamic acid is determined by oxidation (ninhydrin) to succinic semialdehyde, which is converted to the dinitrophenylhydrazone, extracted selectively for an acid hydrazone, and estimated colorimetrically in alkaline solution. The aspartic acid eluate is estimated by using the ninhydrin color directly.

Neither cystine, β-alanine, nor γ-aminobutyric acid interferes with the aspartic determination, but glutathione does (approximately 15% of the aspartic acid color). Glutathione may be removed by adding 1 mg. of $PbCO_3$ to the column.[3]

Preparation of Blood or Tissue

Ten milliliters of fresh blood is placed in a tube containing 1 mg. of heparin and mixed. The plasma is removed and precipitated with ½ vol. of 15% TCA. The filtrate can be stored in the deep-freeze until used.

[1] R. Kuhn and T. Wieland, *Ber.* **73**, 962 (1940).
[2] B. Prescott and H. Waelsch, *J. Biol. Chem.* **164**, 331 (1946).
[3] S. P. Bessman, J. Magnes, P. Schwerin, and H. Waelsch, *J. Biol. Chem.* **175**, 817 (1948).

Four milliliters of filtrate is adjusted to pH 7.0 (bromothymol blue) and made to 5.0 ml. with H_2O. The TCA filtrate of tissue is prepared to a final TCA concentration of 5% and neutralized as above.

Adsorption and Elution

Preparation of Al_2O_3. Ten grams of adsorption alumina (Fisher, 150 to 200 mesh) plus 50 ml. of N HCl are stirred for 1 minute, decanted, and washed with water until the washings are neutral. The oxide, thus treated, is stored under water.

Separation of Glutamic Acid and Aspartic Acid

1. The stem tip of the funnel (bowl, 60 × 15 mm.; stem, 80 × 3 mm.) is closed with glass wool, and a suspension of Al_2O_3 is added, filling the stem with solid. (When glutathione is present, 1 mg. of $PbCO_3$ in 0.1 ml. of water is added on top of the alumina layer.)

2. Two milliliters of neutralized TCA filtrate is filtered by gravity through the funnel into a glass-stoppered tube graduated at 6.0 ml., and the funnel is washed by passing through 2.0 ml. of water.

3. The combined filtrates from step 2 are saved for glutamine-asparagine determination.[4]

4. The funnels are placed in centrifuge tubes graduated at 6.0 ml., and 4.0 ml. of 0.5 N acetic acid is passed through the funnel.

5. The eluate from step 4 is saved for glutamic acid determination. (This contains 60 to 70% of the glutamic acid and 5% of the aspartic acid. This recovery is constant for any set of determinations and is corrected for by the standards.)[4]

6. Two milliliters of N acetic acid is passed through the funnel to elute most of the remaining glutamic acid. (Using 4.0 ml. of N acetic acid in step 4 leads to simultaneous elution of aspartic acid.) The filtrate is discarded.

7. The funnel is washed with 2.0 ml. of water, and the washings discarded.

8. Four milliliters of 0.15 M Na_3PO_4 is passed through the funnels to elute aspartic acid.[4]

9. The filtrate from step 8 is saved for aspartate determination.[4]

Separation of Glutamine and Asparagine. To the filtrates from step 3 is added 0.8 ml. of concentrated HCl (approximately 12 N). The mixture is heated for 60 minutes at 100°, cooled, adjusted to pH 7.0 (bromothymol blue), and made to 6.0 ml. Two milliliters is placed on a column (as in step 1 above) and steps 1 through 9 are followed. The filtrate is discarded at step 3.

[4] These steps represent samples which contain asparagine or aspartic acid.

Glutamic Acid Estimation (See also Vol. III [83])

Reagents

Guanidine carbonate, 14%.
Lead acetate (PbAc$_2$·3H$_2$O), 12%.
NaOH, 5 N.
2,4-Dinitrophenylhydrazine, 0.1% in N HCl.
Capryl alcohol (85% pure, Rohm and Haas).
Potassium hydroxide, saturated.
Borate buffer—20 g. of Na$_2$B$_2$O$_4$·10H$_2$O, 5.3 g. of Na$_2$CO$_3$ per liter
   of water, adjusted to pH 10.0.

Procedure. The filtrate from step 5 above and 20 mg. of ninhydrin
(0.2 ml. of a 1% ninhydrin solution in citrate buffer-aspartic acid reagent
may be used) are heated for 10 minutes at 100° and cooled in ice. Then
0.4 ml. of guanidine carbonate, 1.0 ml. of PbAc$_2$, and 0.5 ml. of NaOH
are added, with stirring after each addition. The mixture is centrifuged
for 5 minutes at 2000 r.p.m. Five milliliters of supernatant is added to
3,0 ml. of 2,4-dinitrophenylhydrazine solution in ice and mixed. After
10 minutes, 10 ml. of capryl alcohol is added. Nitrogen is bubbled through
vigorously for 1 minute. The mixture is centrifuged for 1 minute at 2000
r.p.m. Then 9.0 ml. of the capryl alcohol phase is added to 6.0 ml. of
borate buffer, stirred for 1 minute with N$_2$, and centrifuged for 1 minute
at 2000 r.p.m. Five milliliters of borate extract is placed in a colorimeter
tube, and 2.0 ml. of alkali (2.3 ml. of saturated KOH made to 20 ml. with
ethanol, fresh) is added. After 2 minutes, it is read at 420 mμ.

Aspartic Acid Estimation

Reagents

Citrate buffer, pH 5.0— 42 g. of citric acid plus 200 ml. of H$_2$O
   made to pH 5.0 with saturated NaOH and diluted to 500 ml.
Ninhydrin reagent, 1% in citrate buffer.
SnCl$_2$, 400 mg. % in citrate buffer.

Procedure. The eluate from step 9 above is neutralized against phenolphthalein by addition of concentrated HCl. Two milliliters of citrate
buffer, 1.0 ml. of ninhydrin solution, and 1.0 ml. of SnCl$_2$ are added. The
mixture is heated for 15 minutes at 100° and then cooled in ice for 10
minutes in the dark. Extraction is performed with 6 ml. of isobutyl alcohol
by stirring with N$_2$. The aqueous layer is suctioned off, and the alcohol
layer read at 570 mμ.

Note. Some bromothymol blue indicator is present, but it remains in
the aqueous phase.

Comments

The procedures described above are accurate for 5 to 50 γ of glutamic acid, 10 to 80 γ of aspartic acid, 15 to 150 γ of glutamine, and 30 to 240 γ of asparagine. Duplicate determinations agree within 5%. Ninety-three to ninety-five per cent of added aspartic and 90 to 95% of added asparagine are recovered from blood. However, approximately 20% of the glutamic acid and glutamine present is also reflected in the aspartate or asparagine color, respectively. Standard solutions of aspartic and glutamic acid are chromatographed with each set of unknowns, and suitable corrections are applied. We find convenient a set of eighteen columns containing 6 unknowns, standard glutamic acid, standard aspartic acid and water blanks, all in duplicate. Three days are required for analysis of one complete set of determinations. On the first day eluates containing glutamic, aspartic, and glutamine-asparagine mixture are collected, and the amide eluate is hydrolyzed. On the next day the hydrolyzed amide eluate is chromatographed, and the glutamic and aspartic acid eluates from the previous day are assayed. Thus, with columns, one may average three complete determinations of glutamic acid, aspartic acid, and glutamine-asparagine per day.

[84] Preparation and Determination of Sulfur Amino Acids and Related Compounds

By Jakob A. Stekol

L-Cystine

$$\text{HOOC—CH—CH}_2\text{—S—S—CH}_2\text{—CH—COOH}$$
$$\quad\quad\quad | \quad\quad\quad\quad\quad\quad\quad\quad\quad\quad | $$
$$\quad\quad\quad \text{NH}_2 \quad\quad\quad\quad\quad\quad\quad\quad\quad \text{NH}_2$$

Hydrolysis of Proteins

One of the best hydrolysis media is 20% HCl in 50% formic acid. In this mixture of acids cystine of insulin is apparently spared from destruction even on prolonged digestion.[1]

Isolation from Natural Sources: Large Amounts

Human hair, cleaned from debris and defatted, is refluxed in the HCl-formic acid mixture for 12 to 18 hours. Most of the acid is removed by

[1] G. L. Miller and V. du Vigneaud, *J. Biol. Chem.* **118**, 101 (1937).

distillation *in vacuo*, and the residue is diluted with water and decolorized with carboraffin. The filtrate is adjusted with a solution of sodium acetate to pH 3 to 4, and crude cystine is allowed to crystallize in the cold. Crude cystine is removed by filtration, washed with hot water, redissolved in hot dilute HCl, and decolorized once more with carboraffin. Cystine is precipitated from the filtrate with ammonia at pH 5. For purification, the cystine is suspended in a large volume of hot water (about 1 l. per 10 g. of cystine) and dissolved by the addition of HCl. L-Cystine is precipitated from the solution with ammonia at pH 5. The process is repeated, if necessary, and the product is suspended in hot water and filtered. Further washing with water, followed by ethanol and ether, generally yields a pure product. The yields are 5 to 6% of the weight of clean hair used. L-Cystine crystallizes in hexagonal plates; $[\alpha]_D^{20} = -223°$ for a 1% solution in 1 N HCl.[2,3] The solubility of L-cystine in water at 24 to 27.5° (pH 6.4) is 0.133 g./l.[3]

Isolation of Small Amounts of L-Cystine

This is based on the procedure of Vickery and White[4] in which cystine is reduced to cysteine and the latter is precipitated by Cu_2O as the cuprous mercaptide. Lucas and Beveridge[5] and Zittle and O'Dell[6] use Cu_2O as direct reducer and precipitant of resulting cysteine, basing the procedure on the observation of Rossouw and Wilken-Jordan[7] that a preliminary reduction of cystine to cysteine is unnecessary for the formation of the mercaptide from cystine and Cu_2O, which is also true for the precipitation of glutathione with Cu_2O in acid medium.[8] Copper is removed from the mercaptide as Cu_2S by treatment with H_2S, the resulting cysteine is oxidzed to cystine by careful addition of an alcoholic solution of iodine, and the precipitation of cystine is completed by the addition of ethanol.[9]

Procedure. A hydrolyzate containing about 20 mg. of cystine is distilled *in vacuo* to complete dryness, taken up in water, and 2 ml. of N HCl is added. The solution is filtered, and the filtrate is made up to 100 ml. with water. Thirty-milliliter aliquots can be used for the precipitation of cystine as cysteine copper mercaptide, and 10-ml. aliquots can be used for

[2] H. S. Loring and du Vigneaud, *J. Biol. Chem.* **107**, 267 (1934).
[3] G. Toennies and T. F. Lavine, *J. Biol. Chem.* **89**, 153 (1930).
[4] H. B. Vickery and A. White, *J. Biol. Chem.* **99**, 701 (1932).
[5] C. C. Lucas and J. M. R. Beveridge, *Biochem. J.* **34**, 1356 (1940).
[6] C. A. Zittle and R. A. O'Dell, *J. Biol. Chem.* **139**, 753 (1941).
[7] S. D. Rossouw and T. J. Wilken-Jordan, *Biochem. J.* **29**, 219 (1935).
[8] F. G. Hopkins, *J. Biol. Chem.* **84**, 269 (1929).
[9] J. A. Stekol and K. Weiss, *J. Biol. Chem.* **179**, 67 (1949).

the quantitative determination of cystine by the Sullivan sodium 1,2-naphthoquinone-4-sulfonate method, described below, if such determinations are desired. A 30-ml. aliquot in a 50-ml. centrifuge tube is heated to boiling, and eight times the theoretical amount of Cu_2O suspension is added with stirring.[10] The oxide goes into solution, and the greenish-gray mercaptide precipitates. About a minute after the last of the reagent is added, 0.5 ml. of sodium acetate (110 g. of trihydrate in 100 ml. of water) is added with stirring. The resulting pH is 4.0. The mercaptide flocculates, and it is allowed to stand for 40 minutes at room temperature. It is then centrifuged for 5 minutes, and the clear supernatant is carefully decanted. The precipitate is washed with two or three 25-ml. portions of 95% ethanol, centrifuged each time, and the supernatant decanted. The alcohol-washed mercaptide is dissolved in 10 ml. of 1 N HCl, warmed on a water bath, and a stream of H_2S passed through the solution for 30 minutes. The resulting Cu_2S is centrifuged, and the supernatant is filtered through a fine glass filter. Cu_2S is suspended in 2 to 3 ml. of 1 N HCl, centrifuged, and the supernatant filtered through the same glass filter. From the combined filtrate H_2S is removed by a stream of air. An alcoholic solution of iodine is carefully added drop by drop to the solution, an excess being avoided (a faint yellow color should serve as an indication). The solution is next neutralized with ammonia to pH 3 to 3.5, and enough ethanol is added to make about 50% solution. After standing in the cold overnight, the crystallized product is centrifuged, suspended in a small amount of water, centrifuged, and then recrystallized from a small amount of hot dilute HCl by the addition of ammonia to pH 3 to 3.5. The product is washed with water, ethanol, then ether. The yields of chromatographically pure cystine range from 50 to 70% of theory.

D-Cystine

D-cystine is conveniently prepared by racemization of L-cystine in boiling HCl,[11] conversion of the latter to S-benzyl derivative, and resolution by means of the brucine salt of its formyl derivative.[12] Racemization of S-benzyl-L-cysteine can also be accomplished by the action of excess acetic anhydride.[12]

[10] Cu_2O suspension is prepared as follows: Fehling's solution is heated to boiling, and glucose in water is added to it. After boiling for 3 minutes, the precipitated Cu_2O is washed by decantation with water.

[11] V. du Vigneaud, R. Dorfmann, and H. S. Loring, *J. Biol. Chem.* **98**, 577 (1932).

[12] J. L. Wood and V. du Vigneaud, *J. Biol. Chem.* **130**, 109 (1939).

L- or D-Cysteine

$$HOOC—CH—CH_2—SH$$
$$\underset{NH_2}{|}$$

The most convenient method of preparation of either isomer of cysteine from either isomer of cystine is that of du Vigneaud *et al.*,[13] based on the reduction of cystine to cysteine with metallic sodium in liquid ammonia. The procedure is identical with that described below for the preparation of homocysteine, except that cystine is used instead.

Determination of Cystine and Cysteine

Comment. The choice of a method for the determination of cystine-cysteine in various materials depends largely on the nature of the material to be analyzed and the relative specificity of the method. To date, the most specific method for cystine-cysteine is that of Sullivan and co-workers,[14] based on the use of sodium-1,2-naphthoquinone-4-sulfonate. Various interfering substances were reported from time to time, but these either are not found in proteins, or are largely removed during hydrolysis, or their effect is minimized or abolished by the proper use of the procedure. The method is applicable to TCA filtrates of tissue homogenates. Amounts of about 75 γ of cystine in the volume of liquid used for analysis in the photoelectric colorimeter appear to be minimal for accurate and reproducible results. An alternate method, which requires 100 to 300 γ of cystine or cysteine for a single determination with a precision of approximately ±2% is that of Kolb and Toennies,[15] based on the use of phosphotungstic acid. This method is not recommended by the authors[15] for cystine-cysteine determination in nucleoproteins, and it does not differentiate cystine from homocystine.

Sullivan Procedure for Cystine and Cysteine[14]

Principle. Two portions of the same solution are used. In one, cystine is reduced with nascent hydrogen (Na·Hg) to cysteine, and in the other, cystine is decomposed with NaCN to NaSR and CNSR. Cysteine is determined in both solutions colorimetrically by the use of sodium-1,2-naphthoquinone-4-sulfonate in alkaline medium at 490 to 500 mμ, with cystine as the standard.

[13] V. du Vigneaud, L. F. Audrieth, and H. S. Loring, *J. Am. Chem. Soc.* **52,** 4500 (1930).
[14] M. X. Sullivan, W. C. Hess, and H. W. Howard, *J. Biol. Chem.* **145,** 621 (1942).
[15] J. J. Kolb and G. Toennies, *Anal. Chem.* **24,** 1164 (1952).

(A)   RSSR + NaCN → RSNa + RSCN (double decomposition)
(B) RSSR + H₂(Na·Hg) → 2RSH (reduction)˙

Reagents

   5 N NaOH.
   0.2% sodium amalgam. Dilute the usual 2% commercial amalgam
   with mercury to 0.2%.
   The following solutions must be prepared fresh before use:
   5% NaCN in water.
   10% Na₂SO₃ in 0.5 N NaOH.
   2% Na₂S₂O₄ (hydrosulfite, dithionite) in 0.5 N NaOH.
   1% sodium-1,2-naphthoquinone-4-sulfonate in water.

   Procedure A. To 5 ml. of solution containing cystine and cysteine add
2 ml. of 5% NaCN, mix, and let stand for 10 minutes. Add 1 ml. of 1%
sodium-1,2-naphthoquinone-4-sulfonate, shake for 10 seconds, then add
5 ml. of 10% Na₂SO₃ in 0.5 N NaOH, mix, and let stand for 30 minutes.
Add 1 ml. of 5 N NaOH and 1 ml. of 2% Na₂S₂O₄ in 0.5 N NaOH. Read
at 490–500 mμ in a photoelectric colorimeter, using cystine as the standard prepared in the same way.
   Procedure B. To 7 ml. of solution (adjusted with HCl to 0.1 N, if not
already at that acidity) add 1 ml. of 0.2% sodium amalgam, and shake
occasionally for 1 hour at room temperature. Remove 5 ml. of the supernatant, and determine its total cysteine content as described in procedure
A.
   In procedure A, the value is for free cysteine plus 1 mole of cysteine
obtained by NaCN decomposition of 1 mole of cystine. In procedure B,
the value is for free cysteine plus two moles of cysteine obtained by
reduction of one mole of cystine.

$$B - A = \text{Cystine}$$
$$B - \text{cystine} = \text{Cysteine}$$

It is advisable to conduct all manipulations by this procedure in complete
absence of exposure to direct overhead light.

Microbiological Assay of Cystine or Cysteine

   The procedure is that of Riesen et al.,[16] as modified by Toennies et al.,[17]
based on the use of oxidized peptone and *Leuconostoc mesenteroides* p-60.

[16] W. H. Riesen, H. H. Spenzler, A. R. Roblee, L. V. Hankes, and C. A. Elvehjem,
   J. Biol. Chem. **171**, 731 (1947).
[17] G. Toennies, G. D. Shockman, and J. J. Kolb, unpublished data.

The modifications of the Riesen *et al.* procedure were as follows:

1. Cystine standard solution and that containing unknown were autoclaved separately from the medium to prevent destruction of cystine.

2. Ascorbic acid was added to the medium (50 mg. per 100 ml.) to increase sensitivity.

3. Optical densities were read[18,19] to improve linearity response.

4. The media were inoculated with log phase organisms and read after 17 hours in a spectrophotometer. These modifications decrease the time needed for determination from 72 to 17 hours, and increase the sensitivity fiftyfold.

Procedure. Standards in duplicate at three levels ranging from 0.15 to 0.9 γ per 6 ml. were run simultaneously with unknowns at three levels in duplicate. The proteins were hydrolyzed in evacuated sealed tubes containing formic-hydrochloric acid mixture (3 N HCl in 50% formic acid).[20] After 17 hours of incubation subsequent to inoculation, the optical densities of the standards were plotted against micrograms of cystine and the values for the unknown were read from the graph. Riesen *et al.*[16] found that glutathione, homocystine, sodium thioglycolate, amyl mercaptan, and several other S-containing compounds could not replace cystine for growth of *Leuconostoc mesenteroides* p-60. Ergothioneine also does not replace cystine.[21] The organism does not differentiate cystine from cysteine, and it does not respond to D-isomers of either.

Cystine and Cysteine Derivatives of Biological Interest

Only brief references to the methods of preparation of these can be made.

[18] G. Toennies and D. L. Gallant, *J. Biol. Chem.* **174,** 451 (1948).

[19] G. Toennies and D. L. Gallant, *Growth* **13,** 7 (1949).

[20] Numerous preparations of nucleoproteins were hydrolyzed and assayed for cystine by this procedure. It would appear from preliminary observations that the question of optimal conditions of hydrolysis of proteins by acids is not yet satisfactorily settled for all types of proteins. In an example illustrated below, thymonuclear protein was hydrolyzed for 6 hours by employing different ratios of protein to formic-hydrochloric acid mixture, and cystine was determined by the microbiological method as modified.[17]

Weight of protein, mg.	Ml. of 3 N HCl in 50% formic acid	Per cent of cystine found
9.32	0.09	0.34 ± 0.01
7.85	0.17	0.30 ± 0.01
6.71	0.23	0.30 ± 0.01
5.27	0.25	0.27 ± 0.02

[21] J. J. Kolb and H. Heath, unpublished data.

L-*Cysteinesulfinic Acid*,[22] HOOC—CH—CH$_2$—SO$_2$H.
$$|$$
NH$_2$

2-Aminoethanesulfinic Acid,[23] H$_2$N—CH$_2$—CH$_2$—SO$_2$H. This compound has been isolated from the livers of rats which were given either cysteine or cysteinesulfinic acid,[24] and it was proved to originate from cysteine as intermediate in the formation of taurine.[25]

Taurine, H$_2$N—CH$_2$—CH$_2$—SO$_3$H. Good yields can be obtained from abalone fish[26] or synthetically.[27]

Cysteic Acid, HOOC—CH—CH$_2$—SO$_3$H. This is obtained by oxida-
$$|$$
NH$_2$
tion of cysteine with bromine.[28]

β-Thioethylamine,[29] H$_2$N—CH$_2$—CH$_2$—SH.

Lanthionine, HOOC—CH—CH$_2$—S—CH$_2$—CH—COOH. This can
$$|\qquad\qquad\qquad|$$
NH$_2$ NH$_2$
be isolated from Na$_2$CO$_3$-treated wool[30] or prepared synthetically from cysteine.[31]

Djenkolic Acid,

HOOC—CH—CH$_2$—S—CH$_2$—S—CH$_2$—CH—COOH
$$|\qquad\qquad\qquad\qquad\qquad\qquad|$$
NH$_2$ NH$_2$

This has been isolated from Djenkol beans,[32] and optically active djenkolic acid can be conveniently prepared either from cysteine and methylene dichloride,[33] or from cysteine and formaldehyde in acid medium.[34]

Sodium Thiopyruvate,[35] HOOC—C—CH$_2$—SH.
$$\|$$
O

[22] T. F. Lavine, *J. Biol. Chem.* **113**, 583 (1936).
[23] F. Chatagner and B. Bergeret, *Compt. rend.* **232**, 448 (1951).
[24] J. Awapara, *J. Biol. Chem.* **203**, 183 (1953).
[25] J. Awapara and W. J. Wingo, *J. Biol. Chem.* **203**, 189 (1953).
[26] C. L. A. Schmidt and T. Watson, *J. Biol. Chem.* **33**, 499 (1918).
[27] G. S. Marvel, C. F. Bailey, and M. S. Sparberg, *J. Am. Chem. Soc.* **49**, 1833 (1927).
[28] H. T. Clarke, *Org. Syntheses* **20**, 23 (1940).
[29] E. J. Mills and M. T. Bogert, *J. Am. Chem. Soc.* **62**, 1173 (1940).
[30] M. J. Horn, D. B. Jones, and S. J. Ringel, *J. Biol. Chem.* **138**, 141 (1941).
[31] V. du Vigneaud and G. B. Brown, *J. Biol. Chem.* **138**, 151 (1941).
[32] A. G. van Veen and A. J. Hyman, *Geneesk. Tijdschr. Ned.-Indië* **73**, 991 (1933); *Rec. trav. chim.* **54**, 493 (1935).
[33] V. du Vigneaud and W. I. Patterson, *J. Biol. Chem.* **114**, 533 (1936).
[34] M. D. Armstrong and V. du Vigneaud, *J. Biol. Chem.* **168**, 373 (1947).
[35] J. Parrod, *Bull. soc. chim.* **14**, 109 (1947); J. A. Stekol, *J. Biol. Chem.* **176**, 33 (1948).

Sodium, Potassium, or Lithium Salt of L-*Cystine*[36]
L-*Cystinedisulfoxide,*[37]

$$\underset{\underset{NH_2}{|}}{HOOC-CH}-CH_2-\underset{\underset{O}{||}}{S}-\underset{\underset{O}{||}}{S}-CH_2-\underset{\underset{NH_2}{|}}{CH-COOH}$$

Mercapturic Acids

$$\underset{\underset{NH-CO-CH_3}{|}}{R-S-CH_2-CH-COOH}$$

(R = groupings described in the table)

These S-substituted derivatives of N-acetyl-L-cysteine are formed in dogs, cats, rats, mice, swine, rabbits, and probably in man on administration of the compounds listed in the table, and are excreted in the urine.

PROPERTIES OF MERCAPTURIC ACIDS

Parent compound administered	R Grouping attached to S of N-acetyl-L-cysteine	Melting point, °C.	$[\alpha]_D$ in ethanol	Reference
Benzene	Phenyl	142	−22°	a
Bromobenzene	p-Bromophenyl	152–153	− 6.7°	b
Chlorobenzene	p-Chlorophenyl	153–154	Not reported	c
Iodobenzene	p-Iodophenyl	152–153	−10.7°	d
Fluorobenzene	p-Fluorophenyl	158–159	−19°	e
Benzyl chloride	Benzyl	147–148	−43.5°	f
p-Bromobenzyl bromide	p-Bromobenzyl	143–144ᵍ	−37°	h
Naphthalene	α-Naphthyl	170–171	−25°	i
Anthracene	α-Anthryl	193–195	−6° to −10°	j
2,3,5,6-Tetra-chloronitrobenzene	2,3,5,6-Tetrachloro-phenyl	212	+33°(±2°)	k

a S. H. Zbarsky and L. Young, *J. Biol. Chem.* **151,** 211, 487 (1943).
b E. Baumann and C. Preusse, *Ber.* **5,** 309 (1881).
c M. Jaffe, *Ber.* **12,** 1092 (1879).
d E. Baumann and P. Schmitz, *Ber.* **20,** 586 (1895).
e L. Young and S. H. Zbarsky, *J. Biol. Chem.* **154,** 385 (1944).
f J. A. Stekol, *J. Biol. Chem.* **124,** 129 (1938).
g In the original article (Stekolʰ) the melting point of the compound was reported as 118 to 119°, owing to a typographical error.
h J. A. Stekol, *J. Biol. Chem.* **138,** 225 (1941).
i M. C. Bourne and L. Young, *Biochem. J.* **28,** 803 (1934).
j E. Boyland and A. A. Levi, *Biochem. J.* **30,** 728 (1936).
k H. G. Bray, Z. Hybs, S. P. James, and W. V. Thorpe, *Biochem. J.* **53,** 266 (1953).

36 G. Toennies and T. F. Lavine, *J. Biol. Chem.* **90,** 203 (1931).
37 G. Toennies and T. F. Lavine, *J. Biol. Chem.* **113,** 571 (1936).

They are generally isolated from the urine by extraction with ether or ethyl acetate. For details of the procedures, see the references in the table. Mercapturic acids decompose on heating with alkali to the corresponding mercaptans, ammonia, and, probably, pyruvic acid. A quantitative procedure for the determination of some of these mercapturic acids in urine is based on the determination of the mercaptan, derived from them by alkaline treatment, iodimetrically or gravimetrically by precipitation of the mercuric derivative.[38]

Methionine

$$CH_3—S—CH_2—CH_2—CH—COOH$$
$$| $$
$$NH_2$$

Comment on Preparation of Methionine. None of the methods to date for the isolation of methionine from natural sources can be recommended as a preparative method for methionine. The yields are poor, and the procedures are cumbersome. One of the most convenient synthetic methods is that of Snyder et al.,[39] and one of the best chemical resolution methods for methionine is that of Windus and Marvel.[40] An excellent procedure for the isolation of methionine from biological materials as the methyl-methionine-sulfonium salt has recently been described.[41] For biological studies in which it is desired to isolate the entire molecule of methionine, particularly in tracer work, this procedure is recommended. It has been adapted in the author's laboratory, in cooperation with Dr. S. Weiss, to the isolation of methionine as the sulfonium salt from 2 to 10 g. of animal protein, and it is described in detail below.

Isolation of Methionine as Methyl-Methionine-Sulfonium Bromide

Principle.[41] Protein hydrolyzate is freed of excess acid, and the bases and free ammonia are removed by phosphotungstic acid. The filtrate containing methionine is digested in 18 N H_2SO_4 in the presence of methanol, whereby methionine is converted quantitatively to methyl-methionine-sulfonium base. The latter is precipitated as the phosphotungstate, then converted to the bromide salt. No exchange of the methyl group of methionine with that of methanol occurs. In effect, direct methylation of methionine with dimethyl sulfate takes place, in analogy with direct methylation of methionine with methyl bromide.[42]

[38] J. A. Stekol, *J. Biol. Chem.* **113,** 279 (1936).
[39] H. R. Snyder, J. H. Andreen, G. W. Cannon, and C. Peters, *J. Am. Chem. Soc.* **64,** 2082 (1942); H. R. Snyder and G. W. Cannon, *ibid.* **66,** 511 (1944).
[40] H. Windus and C. S. Marvel, *J. Am. Chem. Soc.* **53,** 3490 (1931).
[41] N. F. Floyd and T. F. Lavine, *J. Biol. Chem.* **207,** 119 (1954).
[42] G. Toennies and J. J. Kolb, *J. Am. Chem. Soc.* **67,** 849 (1945).

$$CH_3\!-\!S\!-\!R \xrightarrow[\text{methanol}]{18\ N\ H_2SO_4} \begin{array}{c} CH_3 \\ \diagdown \\ \overset{+}{S}\!-\!R \\ \diagup \\ CH_3 \end{array}$$

Procedure. Ten grams of fat and water-free protein is digested in 300 ml. of 20% HCl for 6 to 8 hours, and the acid is removed *in vacuo* by repeated distillation with water. The digest is clarified by two to three treatments with activated carbon, and the filtrate (50 ml.) is warmed on a water bath. To the warm filtrate 50 ml. of a hot solution of 30 g. of phosphotungstic acid is added. The mixture, after heating on a steam bath for 30 minutes, is kept in the dark overnight. The precipitated phosphotungstates are removed by filtration and washed with 75 ml. of a solution of 2 g. of phosphotungstic acid containing 2 ml. of 1 N HCl. The filtrate and washings (purplish) are treated with about 2 ml. of 1 N tetraethylammonium bromide to remove the excess of phosphotungstic acid. More tetraethylammonium bromide is added, if necessary, until the filtrate is completely free of phosphotungstic acid. The latter is removed by filtration through a fine glass filter, and the filtrate is concentrated *in vacuo* to about 10 ml. To the concentrate 15 ml. of 18 N H$_2$SO$_4$ and 5 ml. of methanol are added, and the precipitate, if any is formed, is removed by filtration. The filtrate and the aqueous washings are evaporated to 15 ml. *in vacuo* (foaming during evaporation), then refluxed for 0.5 hour. The brown digest is diluted with water to 200 ml. and filtered. The filtrate is warmed on a steam bath, and to it a hot solution of 10 g. of phosphotungstic acid in 12 ml. of water is added. The mixture is placed in a refrigerator for about 2 hours. The precipitate is removed by filtration through a medium glass filter and washed with water until the filtrate is free of sulfate (BaCl$_2$ test). The washed precipitate is then washed with 5 portions of 100 ml. each of 95% ethanol, which removes most of the colored material, then with ether, dried, and weighted. To the weighed phosphotungstate 4 ml. of 90% aqueous acetone is added, and the mixture is thoroughly triturated, then filtered. The trituration with dilute acetone on the undissolved precipitate is repeated with 3 additional portions of 4 ml. each, and the extracts are combined. The undissolved phosphotungstate is washed with ether, dried, and weighed. The difference in the weight of the phosphotungstate before and after trituration with acetone gives the weight of methylmethionine-sulfonium-phosphotungstate which dissolved in the acetone. For every gram of the phosphotungstate that dissolved in acetone, 1 ml. of 1 M tetraethylammonium bromide is added to the combined acetone extracts, followed by 45 ml. of water. The mixture is stirred, then centrifuged. The supernatant liquid is filtered through a fine glass filter, and the pH of the filtrate is adjusted to 5 with 1 M NH$_4$OH. The solution is then decolorized with activated carbon (in the cold), filtered through a fine glass filter, and the filtrate is evaporated *in vacuo* at 25 to 30° (inside temperature) to near dryness. Ten milliliters of methanol is added to the concentrate, and the distillation is repeated. The yellowish oil is quantitatively transferred to a 50-ml. centrifuge tube with 4 ml. of methanol, and the distillation flask is rinsed with 20 ml. of absolute ethanol. The rinse is added to the centrifuge tube containing the methanol solution of the sulfonium salt. On standing in the refrigerator, methyl-methionine-sulfonium bromide crystallizes, and it is removed by centrifugation. It is recrystallized from methanol solution by the addition of ethanol, washed with ethanol, then with ether, and dried *in vacuo* at room temperature. The yields are 30 to 40% of theory; melting point 136 to 138° with sintering (open capillary, uncorrected). The salt (derived from L-methionine) is somewhat hygroscopic, and direct exposure to moist air should be avoided. In contrast, the corresponding sulfonium salt derived from DL-methionine is not so hygroscopic. Methyl-methionine-sulfonium salts decompose in boiling water (8 hours), and much more readily in alkali (2 equivalents), quantitatively to dimethyl sulfide and homoserine:

$$R—S \overset{CH_3}{\underset{CH_3}{\diagup\diagdown}} \xrightarrow[\text{heat}]{\text{alkali}} CH_3—S—CH_3 + HO—CH_2—CH_2—\underset{NH_2}{\overset{|}{CH}}—COOH$$

Determination of Methionine

Two procedures will be described. In one, iodine, which is bound by methionine in 1:1 molar ratio, is determined either titrimetrically or spectrophotometrically. In the other, methionine reacts with nitroprusside in strong acid medium, and the resulting color is measured at 540 mμ. The iodimetric procedure is independent of methionine peptides, but not of S-substituted homocysteines or homocysteine thioacetals, homocystine, or homocysteine. Cystathionine does not react in the iodimetric procedure, whereas homolanthionine (S-γ-amino-γ-carboxyl-propylhomocysteine) does react. The colorimetric procedure responds to methionine peptides and S-substituted homocysteines, but not to homocystine, homocysteine, homocysteine thioacetals, cystathionine, or homolanthionine.

Iodimetric Procedures

Principle. An excess of iodine is added to a buffered solution containing methionine, the free iodine is removed either with Na$_2$S$_2$O$_3$ or with a mixture of isoamyl alcohol and carbon tetrachloride, the solution is acidified, and the iodine bound to methionine is liberated. The latter is either titrated or determined spectrophotometrically.

$$CH_3-S-CH_2-CH_2-CH-(COO^-)\ NH_3^+ + I_2 \underset{pH\ 1}{\overset{pH\ 7}{\rightleftarrows}}$$

$$CH_3-\overset{+}{S}-CH_2-CH_2-CH-(COO^-)\ NH + 2HI$$

A. Iodimetric Titration.[43] A single determination requires 0.2 to 5 mg. of methionine with the precision on the lower amounts of about $\pm 2\%$, employing an accurately calibrated 1-ml. buret and the necessary glassware. Generally, a 25 to 50% excess of iodine is added to the methionine solution in 1 M KI buffered at pH 7. The excess iodine is removed after 10 to 20 minutes with $Na_2S_2O_3$, the solution is acidified, and the liberated iodine is titrated with thiosulfate. It is usually advisable to run a blank, especially if other amino acids are present. In this, methionine is oxidized to the sulfoxide (which does not form dehydromethionine) by KIO_3 in 0.5 to 1.0 M HCl solution. After 10 to 20 minutes, the excess KIO_3 is converted to I_2 by KI, the solution is neutralized and buffered at pH 7, and the iodine determination is carried out as before. The difference between the amounts of iodine liberated in the two cases, expressed in milliliters of thiosulfate, is a measure of methionine present.

Milliliters $Na_2S_2O_3 \times 0.5 \times$ normality of thiosulfate

= millimoles methionine present

Reagents

Phosphate buffer, pH 7 (7 parts of M K_2HPO_4 and 3 parts of M KH_2PO_4).
0.1 N KIO_3.
5 M KI.
2 N NaOH.
0.1 N I_2, diluted to 0.01 N as needed.
0.025 N $Na_2S_2O_3$ standardized.
Starch solution as the indicator.

Blank. It is necessary that the iodine concentration correspond approximately to that in the "determination." Accordingly, the "blank" is set up first in order that after the iodate oxidation the iodine concentration may be adjusted to that of the "determination." Sufficient 6 N HCl is added to an aliquot (usually 5 ml.) of the unknown so that the

[43] T. F. Lavine, *J. Biol. Chem.* **151**, 281 (1943).

final concentration is 0.5 to 1 N HCl. Sufficient 0.1 N KIO₃ is added so
that only a slight yellow color is imparted to the solution (50% excess).
After 10 to 20 minutes the excess iodate is converted to I_2 by addition of
1 ml. of 5 M KI. About one-half of the amount of NaOH necessary for
neutralization is then added, followed by a mixture consisting of the
balance of NaOH, 1 ml. of buffer, 1 ml. of 5 M KI, and sufficient water to
make the volume the same as in the "determination." If the iodine in
the blank is in great excess, it is advisable to remove most of it before
neutralization and buffering. After 10 to 20 minutes the solution is
titrated as in the "determination."

Determination. This is set up 10 minutes after the blank is started so
that the I_2 consumption will be largely completed by the time the iodate
oxidation in the blank is finished. If the solution under consideration is
acid, the amount of alkali needed for neutralization is determined, with
methyl red as indicator. After neutralization, if necessary, of an aliquot
(usually 5 ml.) there is added sufficient buffer and 5 M KI so that there
will be 1 ml. of each for every 5 ml. of final solution, and 25 to 50% excess
of I_2, or enough to impart a definite and persistent I_2 color to the solution.
After 10 to 20 minutes the excess iodine is removed by thiosulfate, with
starch indicator. The end point should be exact, although the amount of
thiosulfate serves only as a measure of the excess iodine for comparison
with that of the "blank." For every milliliter of buffer, 1 to 1.5 ml. of
2 N HCl is added, and the liberated iodine is titrated with thiosulfate.
The iodide concentration should be at least 0.5 M at the end of the titration in order to prevent the oxidation of methionine by iodine in acid
solution. If necessary, additional 5 M KI is added after the excess I_2 is
removed and *before* the solution is acidified.

Note. Homocystine and homocysteine, if present, will interfere with
methionine determination by this procedure. Lavine[43] devised a procedure
whereby the interference by either compound is circumvented by reduction of homocystine to homocysteine and conversion of it to homocysteine
thiolactone, or by precipitation of homocystine with HgSO₄ (the report
of the procedure is in press).

B. Spectrophotometric Determination.[44] For a single determination
of approximately ±1% accuracy, two aliquots of 25 γ of methionine in
4 ml. of a neutral solution are needed. Two-tenth's to three-tenths gram
of protein and 1.2 to 2.5 ml. of 3 N HCl in 40% formic acid, sealed under
vacuum in a test tube, are heated for 13 hours at 120°. The contents of
the tube are diluted, filtered through a sintered glass filter of medium
porosity, neutralized, and made up to 100 ml.

[44] B. Bakay and G. Toennies, *J. Biol. Chem.* **188**, 1 (1950).

Reagents. Isoamyl alcohol-carbon tetrachloride, 9:1. Extract 500 ml. of isoamyl alcohol with an equal volume of 0.01% silver nitrate solution. Then extract with 10 ml. of 5 M NaI plus 1 ml. of 4 M HCl, and remove the precipitated matter by centrifugation. Extract with an equal volume of 1% stannous chloride, and wash twice with water. Distill the alcohol, mix 180 ml. of the purified product with 20 ml. of carbon tetrachloride, and extract the mixture with 10 ml. of 5 M NaI. It is advisable to perform the following final steps each day before carrying out determination. Extract the solvent with 1 to 2 vol. of 1% stannous chloride, wash twice with water, centrifuge, add 10 ml. of buffer mixture I (see below), and equilibrate by stirring for 10 minutes with bubbling nitrogen. Remove the aqueous layer, and keep the resulting solvent well stoppered.

BUFFER MIXTURE I. Mix 1.5 vol. of M KH$_2$PO$_4$ with 8.5 vol. of M K$_2$HPO$_4$ and 10 vol. of 5 M NaI. Saturate with nitrogen for 10 minutes, and, after 1 to 2 days of standing, extract the liberated iodine with a small volume of solvent. Filter, boil while the nitrogen is bubbling through the solution until the color is at a minimum, and let cool without stopping the nitrogen. Keep stoppered.

BUFFER MIXTURE II. Prepare before use by mixing 40 ml. of buffer mixture I with 10 ml. of a 0.0025 M iodine solution.

	Determination of			
	Methionine plus non-methionine		Non-methionine	
	Tube I	Tube II	Tube III	Tube IV
Aliquot, ml.	0	4	0	4
Water, ml.	5	1	4	0
Buffer mixture II, ml.	5	5		
Iodate mixture, ml.			1	1
Waiting time, min.			30	30
Buffer mixture III, ml.			5	5
Waiting time, min.	30	30	30	30
Bubbling of nitrogen, min.	10	10	10	10
Solvent mixture, ml.	10	10	10	10
Bubbling of nitrogen, min.	2	2	2	2
Centrifugation and cooling, min.	5	5	5	5
Aqueous layer, ml.	10	10	10	10
Aqueous optical density	a_0	a_4	c_0	c_4
4 N HCl, ml.	1	1	1	1
Waiting time, min.	10	10	10	10
Aqueous layer, ml.	11	11	11	11
Aqueous optical density	b_0	b_4	d_0	d_4

BUFFER MIXTURE III. Mix 80 ml. of buffer mixture I and 20 ml. of approximately 3 N NaOH. Filter if necessary. The NaOH must be adjusted so that 5 ml. of this mixture, 4 ml. of water, and 1 ml. of the iodine mixture yield a pH of 7.0.

IODATE MIXTURE. Prepare before use by mixing equal volume of 0.01 N KIO_3 and 6 N HCl.

4 N HCl AND WATER. Saturate with nitrogen before use.

Procedure. Into a series of four calibrated tubes the unknown and the reagents are added in the sequence outlined in the table on page 591. Each tube is read at 420 mμ in a Coleman spectrophotometer.

The optical density $\times$ 1000 is designated for each reading as a_0, a_4, c_0, etc. These can be expressed in micrograms of methionine by preparing a standard curve using pure methionine. Subtract a_0 from a_4 (Δa), b_0 from $b_4(\Delta b)$, c_0 from $c_4(\Delta c)$, and d_4 from $d_4(\Delta d)$.

$$\text{Methionine} = (\Delta b - \Delta a) - (\Delta d - \Delta c)$$

Colorimetric Procedure[45]

Reagents

5 N NaOH; 6 N HCl; 1% glycine in water.

10% sodium nitroprusside in water. Make up at room temperature; store in a brown bottle in a refrigerator.

Procedure. Use 0.2 to 1 mg. of methionine in 1 to 7.5 ml. of solution, and add water to 7.5 ml. Add 1.5 ml. of 5 N NaOH, 1.5 ml. of 1% glycine, and 0.3 ml. of sodium nitroprusside. Mix after each addition. Put in a water bath at 37 to 40° for 15 minutes. Chill in an ice bath for 5 to 7 minutes. Then add 3 ml. of 6 N HCl to the tubes as they are removed from the ice bath. Let the HCl run down the side of the tube, and *do not mix*. Stopper. Place all tubes in a basket, and shake together for exactly 1 minute. Let stand at room temperature for 15 minutes, and read against water using a 540-mμ filter. Prepare a reagent blank at the same time using the same volumes. If the hydrolyzate of protein is colored, add all reagents except the nitroprusside and read both blanks against water, subtracting the two from unknown. The standard curve is prepared in the range of 0.1 to 1 mg. of pure methionine.

[45] T. E. McCarthy and M. X. Sullivan, *J. Biol. Chem.* **141**, 871 (1941), as modified by D. Bolling, *in* "The Amino Acid Composition of Proteins and Foods" (R. J. Block and D. Bolling, eds.), 2nd ed., Charles C Thomas, Springfield, Ill., 1951.

Preparation of Methionine Derivatives of Biological Interest

$$\text{HOOC—CH—CH}_2\text{—CH}_2\text{—S—S—CH}_2\text{—CH}_2\text{—CH—COOH}$$
$$\overset{|}{\text{NH}_2} \qquad\qquad\qquad\qquad \overset{|}{\text{NH}_2}$$

Homocystine

$$\text{HOOC—CH—CH}_2\text{—CH}_2\text{—SH}$$
$$\overset{|}{\text{NH}_2}$$

Homocysteine

Neither homocystine nor homocysteine has as yet been found in nature, but considerable experimental evidence points to the probable biological role of these amino acids as intermediates in methionine metabolism. Both compounds can be prepared synthetically[39] or from commercial methionine. Both compounds of either optical configuration can be prepared by chemical resolution procedures.[46]

Homocystine from Methionine. This is based on the original procedure of Butz and du Vigneaud[47] with minor modifications and simplifications. One hundred grams of commercial methionine is refluxed in 400 ml. of 18 N H_2SO_4 (882 g. of concentrated H_2SO_4 and 1 l. of water) for 8 hours. The digest is poured into 1 l. of water, decolorized with activated carbon, and filtered. The filtrate is neutralized with concentrated ammonia to about pH 7 and left in a refrigerator overnight. The precipitated crude homocystine is dissolved in about 1 l. of boiling water by the addition of concentrated HCl, decolorized with carbon, filtered, and the filtrate is neutralized with ammonia. Homocystine is thoroughly washed with hot water, dilute ethanol, then with absolute ethanol and ether. The yield is about 35 to 40% of methionine used. Inactive homocystine thus obtained crystallizes in irregular hexagonal plates and decomposes at 260 to 265°. Its solubility in water at 25° is 1 g. per 5 l.

It should be noted that during the decomposition of methionine in 18 N H_2SO_4 considerable amount of methyl-methionine-sulfonium base is formed in addition to homocystine.[48] In the presence of methanol (1:1 molar ratio to methionine) no homocystine but only the sulfonium base is formed.[48] This observation was the basis for the preparative method of methyl-methionine-sulfonium bromide from methionine. Various alcohols, other than methanol, similarly form the corresponding sulfonium bases with methionine in 18 N H_2SO_4.[49]

[46] V. du Vigneaud and W. I. Patterson, *J. Biol. Chem.* **109**, 97 (1935).
[47] L. W. Butz and V. du Vigneaud, *J. Biol. Chem.* **99**, 135 (1932).
[48] T. F. Lavine and N. F. Floyd, *J. Biol. Chem.* **207**, 97 (1954).
[49] T. F. Lavine, N. F. Floyd, and M. Cammaroti, *J. Biol. Chem.* **207**, 107 (1954).

Homocysteine.[50] This can be prepared either from S-benzyl-homocysteine[46] or from homocystine of either optical form. No racemization occurs during the manipulations. Twelve grams of homocystine is added to about 500 ml. of liquid ammonia which is kept at boiling point (omit cooling mixtures, and work under a hood at room temperature). Small pieces of metallic sodium (c.p.) are added until a blue color results, which persists for about 5 to 10 minutes. Enough ammonium iodide is then added to dispel the blue color (due to excess sodium). Ammonia is allowed to evaporate spontaneously (do not warm!), and the residual ammonia is removed by evacuation in the absence of oxygen. The flask is then flushed with nitrogen, about 50 ml. of freshly boiled water is added, and the flask is cooled in an ice bath. The solution is neutralized to litmus with 45% HI, a small amount (about 2 g.) of activated carbon is added, and the solution is filtered under nitrogen. To the filtrate about 500 ml. of absolute ethanol is added, and the mixture is placed in a refrigerator overnight. The homocysteine crystallizes in large flat plates. It is removed by filtration under nitrogen and washed with absolute ethanol, then ether. Both solvents must be free of peroxide, aldehydes, and air to avoid oxidation. The yield is about 80%. The product contains about 92 to 96% of sulfhydryl. DL-Homocysteine melts at 232 to 233°. For further purification, if necessary, 3 g. of homocysteine is dissolved in 30 ml. of water, 100 ml. of absolute ethanol is added, and the homocysteine that precipitates immediately is removed by filtration. To the filtrate 400 ml. of absolute ethanol is added to obtain the second crop of homocysteine.

Homocysteine Thiolactone Hydriodide. The formation of homocysteine thiolactone from methionine in boiling 57% HI was first demonstrated by Baernstein.[51] During the process the methyl group of methionine yields quantitative amounts of methyl iodide. Later the formation of the thiolactone from homocysteine in 20% HCl was also reported.[50] One gram of methionine is boiled for 10 hours with 20 ml. of 57% HI containing 1% H_3PO_2. The digest is evaporated to dryness *in vacuo*. The residue is dissolved in hot absolute ethanol, and homocysteine thiolactone hydriodide is precipitated by the addition of 3 vol. of ether, filtered, and washed with ether. It crystallizes in hexagonal rods; melting point 204 to 206°.

Homocysteine Thiolactone from Homocysteine.[50] The hydrochlorides can be prepared by heating on a water bath for 1 hour D- or L-homo-

[50] B. Riegel and V. du Vigneaud, *J. Biol. Chem.* **112**, 149 (1935).

[51] H. D. Baernstein, *J. Biol. Chem.* **106**, 541 (1934). The demethylation of methionine in boiling HI to yield methyl iodide is the basis for a quantitative determination of methionine in natural materials. See above reference, and, by the same author, *J. Biol. Chem.* **115**, 25, 33 (1936). The method is satisfactory if special precautions indicated by the author are observed. It requires several hours for completion. Any S-substituted homocysteine or cysteine will interfere, however.

cysteine, obtained from D- or L-homocystine or D- or L-S-benzylhomo-
cysteine by the liquid ammonia technique described above, in concentrated HCl. The active isomers of homocysteine thiolactone hydrochloride
melt at 194°. The L-isomer has a specific rotation of $[\alpha]_D^{26} = +21.5°$ for a
1% solution in water, and the enantiomorph has an equal rotation of
opposite sign.

It is of interest to note here that the S-C bond of homocysteine thiolactone is that of a *thio ester*, and its reactivity, particularly in the peptide
bond formation, deserves investigation. Concerning this suggestion, it has
been reported that when homocysteine thiolactone hydriodide is freed
from the hydriodide by dilute alkali it reacts with itself to give the diketopiperazine of homocysteine.[52]

Methionine Sulfoxide[53]

$$HOOC—\overset{|}{\underset{NH_2}{CH}}—CH_2—CH_2—\overset{||}{\underset{O}{S}}—CH_3$$

Dissolve 200 mM. of DL-methionine in a mixture containing 200 mM.
of HCl (10% excess), 150 ml. of water, and 250 ml. of methanol. Add 240
mM. (about 20% excess) of hydrogen peroxide (30% solution) and water
to 500 ml. After mixing, cool under the tap, and let stand for 30 minutes.
Add 230 mM. (5% excess) of amylamine and methyl alcohol equal to
about one-half the total volume of the mixture. Filter if necessary. Add
3 vol. of acetone (about 2400 ml.), and the sulfoxide will precipitate.
The precipitate is filtered as soon as the supernatant liquid becomes clear
(about 10 minutes). Wash the sulfoxide by making about six resuspen-
sions in 500-ml. portions of acetone to free it from amylamine hydrochloride. Dry in air, then *in vacuo* at 100°. The yield is 95%; melting
point 225 to 230° with decomposition. The solubility in water at 25° is
about 66 g. per 100 ml.

The diastereoisomeric sulfoxides derived from L-methionine have also
been prepared.[54]

Dehydromethionine[55]

$$HOOC—\underset{NH—————}{CH}—CH_2—CH_2—\overset{+}{S}—CH_3$$

[52] V. du Vigneaud, W. I. Patterson, and M. Hunt, *J. Biol. Chem.* **126**, 217 (1938).
[53] G. Toennies and J. J. Kolb, *J. Biol. Chem.* **128**, 399 (1939).
[54] T. F. Lavine, *J. Biol. Chem.* **169**, 477 (1947).
[55] T. F. Lavine, *Federation Proc.* **4**, 96 (1945); U.S. Patent 2,465,461 (1949).

The formation of dehydromethionine from methionine and iodine at pH 7 is the basis for the quantitative iodimetric and spectrophotometric determination of methionine described in previous sections.

Procedure. The Ag_2O from 11.8 mM. of $AgNO_3$ and 12.5 mM. of NaOH is centrifuged and washed twice by stirring with 50 ml. of water and centrifuging. Five millimoles of methionine and 35 ml. of water are added to the Ag_2O in the centrifuge tube, and the mixture is stirrred. Ten milliliters of 0.5 M I_2 in methanol is added, and the mixture is stirred until the precipitate changes color from a tan to white. The mixture is centrifuged immediately, and the clear liquid is decanted. Then 0.1 N HCl is added to obtain pH 7, and the solution is filtered to remove any AgCl. The filtrate is evaporated in a vacuum desiccator in the presence of P_2O_5. The residue is treated with 10 to 20 ml. of methanol and filtered. Ethyl ether is added to the filtrate until turbidity develops. After some crystallization takes place, ether is again added to turbidity, and the process is repeated until the mother liquor remains clear. A total of 5 vol. of ether is usually necessary. The yield is 78%.

Dehydro-L-methionine decomposes at 207 to 210° with effervescence. $[\alpha]_D^{26} = +57$ to 61° for an 0.5 M aqueous solution. It crystallizes in tufts or clusters of quill-shaped needles, often of large size, is hygroscopic, and is soluble in methanol. The DL-compound crystallizes in rectangular plates or prisms with a tendency to cluster. It is less soluble in methanol than the L-derivative; melting point 193–197° with decomposition.

Dehydromethionine acts as an oxidizing agent toward acidified KI, cysteine, and neutral bisulfite, regenerating methionine in each instance. The compound is assayed by titrating the iodine liberated in the presence of 1 M KI (added first) and 0.5 M HCl solution. It is stable in aqueous solution (pH 7). In acid or alkaline solution methionine sulfoxide is formed. The substance shows no increased acidity in the presence of formaldehyde (free amino or imino groups are absent, $C-NH_2$ or $C=NH$).[55]

Methyl-Methionine-Sulfonium Iodide or Bromide[42]

$$\text{HOOC}-\underset{\underset{NH_2}{|}}{CH}-CH_2-CH_2-\overset{+}{S}\underset{CH_3}{\overset{CH_3}{<}}$$

A mixture of 40 mM. of methionine, 65 ml. of 89% formic acid, 20 ml. of acetic acid, and 10 ml. (160 mM.) of methyl iodide is kept in a dark place at 25° for 3 days. The solution is distilled *in vacuo* to a sirup. Digestion with 40 ml. of absolute methanol produces a granular precipitate. After filtering and washing with methanol and acetone it is dissolved in 30 ml. of warm 50% ethanol and recrystallized by the addition of 100 ml. of absolute ethanol. Yield 75%; melting point about 150° with evolution of gas. The bromide is similarly prepared with methyl bromide; meeting point 146–148°. The bromide is somewhat more soluble in methanol than the iodide.

Ethionine[56]

$$HOOC—CH—CH_2—CH_2—S—CH_2—CH_3$$
$$|$$
$$NH_2$$

The procedure is identical with that employed for the preparation of methionine from homocystine and methyl iodide by the liquid ammonia technique,[46] except that ethyl iodide is used instead of the methyl iodide. Ethionine crystallizes in large plates; melting point 272°, with effervescence at 284°. L- or D-Ethionine was similarly prepared with either L- or D-S-benzylhomocysteine or L- or D-homocystine.[57] L-Ethionine has the specific rotation of $[\alpha]_D^{23} = +22° \pm 1°$ for a 1% solution in 1 N HCl. Analogous derivatives of ethionine can be prepared by the same procedures which were described above for those of methionine.

Cystathionine

$$HOOC—CH—CH_2—CH_2—S—CH_2—CH—COOH$$
$$|\qquad\qquad\qquad\qquad\qquad\qquad |$$
$$NH_2 \qquad\qquad\qquad\qquad\qquad NH_2$$

L-Cystathionine was isolated from the media containing methionine on which a mutant of *Neurospora* was grown,[58] and it has been detected chromatographically in rat livers.[59] L-Cystathionine cleaves to cysteine in rat liver preparations.[60]

Two methods of preparation of all diastereoisomeric forms of cystathionine have been described. In one,[61] β-chloro-α-aminopropionate is condensed with homocysteine in alcohol in the presence of alkali; in the other,[62] cysteine is condensed with 3,6-bis(β-chloroethyl)-2,5-diketopiperazine[39] in liquid ammonia, and the resulting condensation product is hydrolyzed with HCl to cystathionine.

[56] H. M. Dyer, *J. Biol. Chem.* **124,** 519 (1938).

[57] J. A. Stekol and K. Weiss, *J. Biol. Chem.* **179,** 1049 (1949).

[58] N. H. Horowitz, *J. Biol. Chem.* **171,** 255 (1947).

[59] W. Hess, *Arch. Biochem. and Biophys.* **40,** 127 (1952).

[60] F. Binkley, W. P. Anslow, Jr., and V. du Vigneaud, *J. Biol. Chem.* **143,** 559 (1942); F. Binkley, *J. Biol. Chem.* **155,** 39 (1944); W. P. Anslow, Jr., and V. du Vigneaud, *J. Biol. Chem.* **170,** 245 (1947).

[61] W. P. Anslow, Jr., S. Simmonds, and V. du Vigneaud, *J. Biol. Chem.* **166,** 35 (1946).

[62] S. Weiss and J. A. Stekol, *J. Am. Chem. Soc.* **73,** 2497 (1951).

Properties[61]

L(-)-Cystathionine: $[\alpha]_D^{21.5} = +23.5°$ for a 1% solution in 1 N HCl. D(+)-Cystathionine: the specific rotation is equal and opposite in sign. L(-)-Allocystathionine: $[\alpha]_D^{21} = -25°$ for a 1% solution in 1 N HCl. The specific rotation of D(+)-allocystathionine is equal and opposite in sign. Cystathionine is stable in 25% HCl, decomposes in HI or HBr. It is insoluble in organic solvents, somewhat soluble in cold water, and much more so in hot water. Exact solubilities are not known as yet.

Homolanthionine

$$\text{HOOC—CH—CH}_2\text{—CH}_2\text{—S—CH}_2\text{—CH}_2\text{—CH—COOH}$$
$$\overset{|}{\text{NH}_2} \qquad\qquad\qquad\qquad\qquad \overset{|}{\text{NH}_2}$$

This compound has not been found in nature as yet, but its sulfur has been shown to be available for the synthesis of cystine in the rat.[63] All optical forms of homolanthionine have been synthesized.[62]

Methods of Determination

None has been reported that is specific, as far as the author is aware. In a mixture containing methionine, homocystine, homocysteine, cystine, and cysteine, homocystine and homocysteine can be determined by difference, by the methods for cystine-cysteine,[14] methionine,[45] and total disulfides and sulfhydryls present.[15] Lack of specificity of the latter method,[15] however, may cause complications in some materials.

Ergothioneine

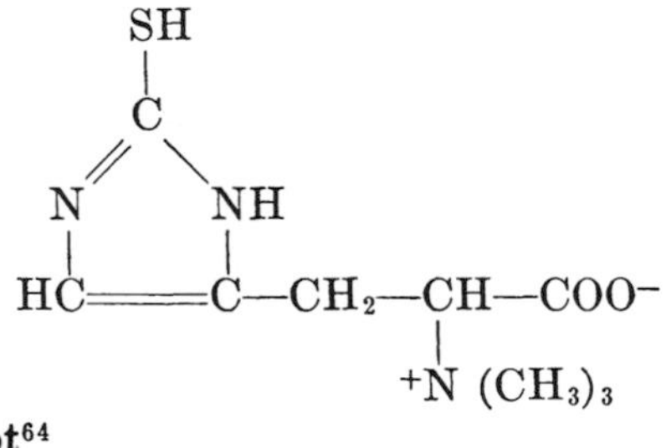

$$\text{SH}$$
$$|$$
$$\text{C}$$
$$N \diagup\!\!\diagdown NH$$
$$\text{HC} =\!\!=\!\!= \text{C—CH}_2\text{—CH—COO}^-$$
$$\overset{|}{{}^+\text{N (CH}_3)_3}$$

Isolation from Ergot[64]

Five hundred grams of finely ground ergot corns is suspended in 2 l. of water containing 1 ml. of glacial acetic acid, and the mixture is brought

---

[63] J. A. Stekol and K. Weiss, J. Biol. Chem. **175**, 405 (1948); **179**, 67 (1949).
[64] G. Hunter, G. D. Molnar, and N. J. Wight, Can. J. Research **E27**, 226 (1949).

to boil with stirring. While hot it is poured through cheesecloth, and the residue is pressed in a filter press, washed twice by bringing to boil with 1 l. of water each time, filtering, and expressing as before. The volume of filtrate and washings is 2660 ml. The filtrate is treated with a saturated aqueous solution of uranium acetate in slight excess as tested, on a tile, by a 5% aqueous solution of potassium ferrocyanide. This requires about 532 ml. of uranium acetate. The mixture is allowed to settle overnight, and the clear supernatant is decanted or preferably siphoned off. The remainder is centrifuged. The liquids are combined. The precipitate is suspended in equal volume of water, recentrifuged, and the washings are added to the main filtrate. The final volume is 4450 ml. The solution is made 0.5 N acid with H_2SO_4 and heated to 60°. An aqueous suspension of Cu_2O is added, and the mixture is stirred. Cu_2O is added until a definite excess is present. The precipitate is allowed to settle, the supernatnat is syphoned off, and the remainder is centrifuged. The copper precipitate is washed at 60° with its own volume of 0.5 N H_2SO_4, which had been brought to boil with a little Cu_2O added. The washing is repeated twice. The copper precipitate is suspended in hot water, and copper is removed with H_2S. The filtrate and washings are freed from H_2S, then neutralized to litmus with barium hydroxide. At this point the solution must be free of barium and sulfate. The filtrate is reduced *in vacuo* to 15 ml., decolorized with charcoal, and the filtrate is further reduced *in vacuo* to 5 ml. Massive crystallization sets in. The separated material is redissolved by heating over a flame, and 10 to 15 ml. of absolute ethanol is added. The crystallized material is removed after several hours of standing in the cold, washed with ethanol, and dried in air. The yield is 1.3 g. of crude ergothioneine. The crude material contains hypoxanthine. For purification, the crude material is suspended in water, and the insoluble hypoxanthine is removed by filtration. The filtrate is treated with charcoal, filtered, and pure ergothioneine is crystallized by addition of ethanol. The melting point is 290° with decomposition.

Synthesis of Ergothioneine

An efficient synthesis was reported by Heath *et al.*[65] The specific rotation of synthetic material of $[\alpha]_D = +47°$ for a 1% solution in water, and the ultraviolet absorption spectrum maximum at 2580 A., $\epsilon = 16,000$, were given. R_f was 0.87 in phenol, 0.32 in collidine. The synthetic and natural ergothioneines were identical.

[65] H. Heath, A. Lawson, and C. Rimington, *J. Chem. Soc.* **1951**, 2215.

Method of Determination of Ergothioneine

An improvement of the method of Hunter[66] for estimation of ergothioneine in simple solutions and in blood has been reported by Melville and Lubschez.[67]

[66] G. Hunter, *Can. J. Research* **E27**, 230 (1949).
[67] D. B. Melville and R. Lubschez, *J. Biol. Chem.* **200**, 275 (1953).

[85] Preparation of S-Adenosylmethionine

By G. L. CANTONI

Preparation

The structure of S-adenosylmethionine[1] is shown below:

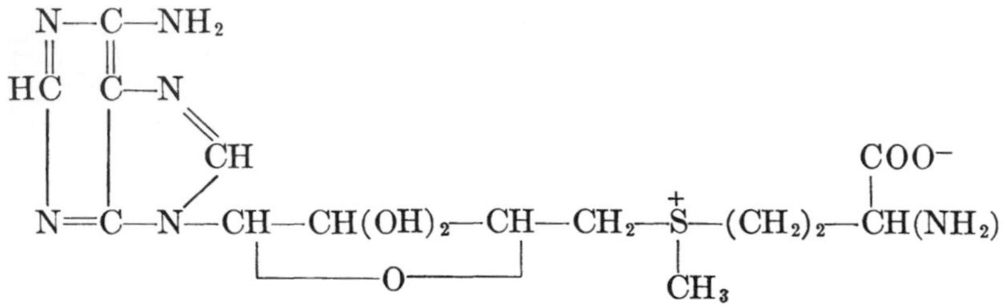

An enzyme system obtained from rabbit liver[2] catalyzes its formation according to the reaction

$$\text{L-Methionine} + \text{ATP} \xrightarrow[\text{Mg}^{++}]{\text{GSH}} \text{AMe} + 3\text{IP}$$

In acid media, sulfonium compounds form insoluble salts with reinecke acid; this property was used to good advantage to isolate AMe at a high level of purity.

[1] The following abbreviations will be used: AMe—S-adenosylmethionine, i.e., active methionine, ATP—adenosinetriphosphate; IP—orthophosphate; GSH—reduced glutathione; MAE—methionine-activating enzyme; THAM—trishydroxymethylaminomethane; TCA—trichloroacetic acid; PCA—perchloric acid; MEK—methyl ethyl ketone.
[2] See Vol. II [33].

Reagents

K₄ATP, adjusted to pH 7.4, 0.15 M.

MgCl₂, 2.0 M.

THAM buffer, pH 7.4, 1.0 M.

L-Methionine.

GSH, 25 mg./ml., neutralized with KOH immediately before use.

MAE,[2] 15 to 30 units/ml., adjusted to pH 7.4 with dilute KOH immediately before use.

TCA, 100% solution, 100 g. in 100 ml.

TCA, 10%.

Ammonium reineckate,[2a] 1.5 g. in 100 ml. of 5% TCA.

PCA, 6% solution.

Procedure

Enzymatic Formation of AMe. A reaction mixture containing 1600 μM. of ATP, 40 μM. of MgCl₂, 30 ml. of THAM buffer, 5 ml. of the GSH solution, and 3000 μM. of L-methionine (added as a solid) was warmed to 37°. By means of a glass electrode the pH was checked and if required adjusted to 7.4 to 7.6 by addition of 2 N KOH. Fifteen milliliters of the enzyme solution was then added, and the reaction mixture incubated at 37°. Aliquots were removed periodically (every 30 to 60 minutes). The choice of the deproteinizing agent for these aliquots depended on the further treatment of the sample: for determination of orthophosphate[3] addition of 2 vol. of 10% TCA was suitable, whereas for determination of AMe, by the ion exchange technique described below, deproteinization with equal volumes of 6% PCA was preferred.

During the incubation the pH of the reaction mixture became gradually more acid and was maintained at 7.4 ($\pm$0.2) by addition of KOH as required. The duration of incubation varied, depending on the activity of the enzyme; generally under the conditions described 3 to 4 hours was a satisfactory period, at which time the reaction was terminated by addition of one-tenth the volume of 100% TCA. The resulting suspension was cooled in ice and centrifuged in the cold, the residue washed once with 15 ml. of cold 10% TCA, and the extracts pooled and stored in the icebox.

The activity of the MAE, i.e., the rate of formation of AMe, can be determined by measuring the amount of orthophosphate liberated from ATP in the presence and absence of methionine.[4] This method, however, is accurate only for the determination of initial reaction rates or under

[2a] *Org. Syntheses Coll. Vol.* **2**, 555 (1943).

[3] G. L. Cantoni, *J. Biol. Chem.* **189**, 745 (1951).

[4] G. L. Cantoni, *J. Biol. Chem.* **204**, 403 (1953).

conditions where the concentration of ATP is high enough to saturate both the MAE and other contaminating ATPases. When used to determine the formation of AMe, the accuracy of the method progressively diminishes because of increasing differences in the concentration of ATP in the experimental and control samples.

A more direct method to determine the formation of AMe was as follows: A 1.0-ml. aliquot of the reaction mixture was deproteinized with 1.0 ml. of 6% perchloric acid and centrifuged; 1 ml. of the supernatant was neutralized to pH 7.8 or 8.0 and diluted to 16 ml. An aliquot containing approximately 5 μM. of adenine nucleosides was then passed through a small (10 × 20 mm.) ion exchange column, Dowex 1 resin, X10, 100- to 200-mesh chloride form, and the effluent collected in a 50-ml. graduated cylinder. The column was washed with H_2O until 50 ml. was collected. The optical density of this effluent at 260 mμ was used to estimate the AMe content ($E = 15,200$).

Other procedures (unpublished), based on the precipitation of AMe as the reineckate, are also available for the determination of AMe.

Purification of AMe. The NH_4 reineckate solution, precooled, was added to the pooled TCA extracts until precipitation was complete (2 to 3 vol. required), and the suspension was kept in ice overnight. The pink microcrystalline precipitate was collected by filtration on a sintered glass funnel (M porosity), washed twice with ice-cold ammonium reineckate (0.5% in H_2O), and dried *in vacuo* over silica gel. The whole operation was carried out in a cold room at 2°. The resulting powder (1.0 to 2.0 g.) was then stored in the cold.

To regenerate AMe from its reineckate, the salt was dissolved in a small volume of MEK, and the resulting solution centrifuged at room temperature to remove a small insoluble residue. The clear, dark-red supernatant was then transferred to a separatory funnel and 0.1 N H_2SO_4 was added in slight excess. After equilibration and separation of the two phases, the aqueous phase was re-extracted with fresh MEK until colorless. Reineckates have high ultraviolet absorption and complex spectra with a maximum at 305 mμ ($E = 15,000$); optical density at 305 mμ can therefore be used as a sensitive method for the detection of traces of reineckate ion (after removal of MEK). To remove MEK from the water phase and to concentrate AMe sulfate as desired, the aqueous layer was extracted twice with 2 vol. of ether and concentrated under reduced pressure. The resulting solution then was adjusted to pH 5.5 to 6.0 and kept in the icebox until needed.

AMe can be assayed by a variety of chromatographic chemical and enzymatic techniques.[4,5,6] The procedure described above consistently

[5] G. L. Cantoni, *J. Biol. Chem.* **209**, 647 (1954).
[6] See Vol. II [34].

yielded preparations which assayed at 92 to 98% purity level, in terms of adenine content.

Properties

AMe is quite stable in the cold at pH between 3.5 and 6.0. In more acid solutions the stability of AMe is limited primarily by the stability of the adenine-ribose linkage. In neutral or mildly alkaline solutions AMe decomposes rapidly owing to cleavage at the sulfonium center. Thiomethyladenosine and homoserine have been identified as products of this cleavage;[7] under more drastic conditions decomposition goes further and other products are formed (ethylene?). Other properties are described in ref. 4.

[7] J. Baddiley, G. L. Cantoni, and C. A. Jamieson, *J. Chem. Soc.* **1953**, 2662.

[86] Glutathione—Isolation and Determination[1]

By Konrad Bloch *and* John E. Snoke

Isolation

All isolation procedures for glutathione (GSH) are based on the ability of the tripeptide to interact with heavy metal salts to form water-insoluble mercaptides. In the first isolation of GSH from yeast by Hopkins,[2] mercuric sulfate and cuprous oxide were employed to precipitate GSH from protein-free extracts. Formation of the mercury complex is prevented by higher concentrations of chloride ion.[3] A procedure which uses cadmium chloride for the initial precipitation of GSH has been introduced by Binet and Weller[4] and, as adapted by Waelsch and Rittenberg,[5] is suitable for the isolation of small amounts of GSH from small quantities of tissues.

Reagents

10% TCA.
6% $CdCl_2 \cdot 2\frac{1}{2}$ H_2O.
10 N NaOH.

[1] The following abbreviations are used: GSH, glutathione; TCA, trichloroacetic acid; SSA, sulfosalicylic acid.
[2] F. G. Hopkins, *J. Biol. Chem.* **84**, 269 (1929).
[3] W. Stricks and I. M. Kolthoff, *J. Am. Chem. Soc.* **75**, 5673 (1953).
[4] L. Binet and G. Weller, *Compt. rend.* **198**, 1185 (1935).
[5] H. Waelsch and D. Rittenberg, *J. Biol. Chem.* **139**, 761 (1941).

Cu_2O, 1% aqueous suspension.[6]

H_2SO_4, 2 N and 0.5 N.

Procedure. To ground tissue or tissue extract is added an equal weight of TCA solution. The residue obtained on centrifugation is extracted twice more with half the original volume of TCA. To the combined extracts is added an amount of $CdCl_2$ solution equal to one-fourth the volume of the extracts. The solution is brought to pH 5 by addition of 10 M NaOH and then adjusted to pH 6.5 with bicarbonate. The precipitated Cd complex is kept at 0° for 1 hour and then washed twice with ice-cold distilled water. The precipitate is dissolved in a minimum of 2 N H_2SO_4, and then 3 ml. of 0.5 N H_2SO_4 is added for each 10 mg. of GSH expected. The solution is filtered if necessary and the amount of GSH present determined in an aliquot by one of the methods described below. The solution is warmed to 40°, and Cu_2O suspension containing 2.5 mg. of Cu_2O for each 10 mg. of GSH is added dropwise with gentle shaking. The precipitate is left at 0° for several hours, separated by centrifugation, and washed successively two times with 0.5 N H_2SO_4, three times with distilled water, and two times with methanol. If the cuprous mercaptide is discolored, it may be redissolved by the addition of an excess of Cu_2O suspension in 0.5 N H_2SO_4.[7] After filtration, the mercaptide reprecipitates on aerating the solution. The recoveries of mercaptide in this purification procedure are usually low.

For isolation of the free tripeptide the cuprous mercaptide of GSH is decomposed in aqueous suspension by H_2S, and the solution, after removal of copper sulfide, is brought to dryness by lyophilization. Free GSH has been crystallized by evaporation of aqueous solutions[8] or by precipitation with ethanol.[9] For elementary analysis, the cuprous mercaptide is a convenient derivative. It is dried to constant weight *in vacuo* at 80°. The absence of contaminants in GSH isolated by this procedure may also be checked by chromatography on paper as described by Hanes *et al.*[10]

Chemical Analysis

A number of methods are being used for routine analysis of GSH in biological material. They are based on the reactivity of the SH group in

---

[6] Although the directions given in the literature call for Cu_2O freshly precipitated from Fehling's solution, it has been the experience of this laboratory that commercial preparations are adequate. Uniform suspension of the Cu_2O is achieved by addition of a small quantity of an anionic detergent.

[7] N. W. Pirie, *Biochem. J.* **24**, 51 (1930).

[8] C. R. Harrington and T. H. Mead, *Biochem. J.* **29**, 1602 (1935).

[9] V. du Vigneaud and G. L. Miller, *J. Biol. Chem.* **116**, 469 (1936).

[10] C. S. Hanes, F. J. R. Hird, and F. A. Isherwood, *Biochem. J.* **51**, 25 (1952).

GSH and are therefore suitable only if it is established that other SH compounds are not measured under the same conditions. These methods are titration with iodine,[11] reduction of nitroprusside,[12] and polarographic analysis.[13,14]

Miller and Rockland[14] have described a method for separation of cysteine from GSH on paper chromatograms and for the subsequent assay of the tripeptide by means of ninhydrin.

Enzymatic Analysis (Glyoxalase Method)[14a]

Principle. In 1932 Lohmann[15] discovered that the conversion of methylglyoxal to lactic acid by glyoxalase (equation 1) requires the

$$CH_3COCHO + H_2O \xrightarrow{\text{glyoxalase}} CH_3CHOHCOOH \qquad (1)$$

presence of GSH. It was later shown by Woodward[16] that the rate of reaction 1 is dependent on the concentration of GSH. Racker[17] has shown that two enzymes participate in the glyoxalase reaction and by using purified enzymes has been able to develop an assay for GSH which will be preferred when greater sensitivity is required. The method of Woodward is more applicable to routine use and is therefore described below. As far as is known, GSH is the only naturally occurring compound which is active in the glyoxalase assay.[17a]

Reagents

Source of enzymes. An acetone powder prepared from baker's yeast[18] is freed before use from GSH by washing 1 g. of dried yeast three times with distilled water. The washed yeast is suspended in 4 to 5 vol. of water and used immediately.

Methylglyoxal. Commercial methylglyoxal (Commercial Solvents, 30% aqueous solution) is distilled at atmospheric pressure. Its concentration is determined as pyruvic acid by neutralizing an aliquot of the distillate, adding of neutral 5% H_2O_2 solution and

[11] A. Fujita and I. Numata, *Biochem. Z.* **299**, 249 (1938).
[12] R. R. Grunert and P. H. Philips, *Arch. Biochem. and Biophys.* **30**, 217 (1951).
[13] R. E. Benesch and R. Benesch, *Arch. Biochem. and Biophys.* **28**, 43 (1950).
[14] J. M. Miller and L. B. Rockland, *Arch. Biochem. and Biophys.* **40**, 416 (1952).
[14a] See Vol. I [70].
[15] K. Lohmann, *Biochem. Z.* **254**, 332 (1932).
[16] G. E. Woodward, *J. Biol. Chem.* **109**, 1 (1935).
[17] E. Racker, *J. Biol. Chem.* **190**, 685 (1951).
[17a] A recent discussion of Methods for Detection and Assay of Glutathione can be found in Glutathione, A Symposium, Academic Press, New York, 1954.
[18] R. Albert, E. Buchner, and R. Rapp, *Ber.* **35**, 2376 (1902).

of a known quantity of 0.1 N NaOH, and back-titrating 10 minutes later with 0.1 N HCl.

2% sulfosalicylic acid (SSA).

95% N_2 − 5% CO_2 gas mixture.

0.2 M sodium bicarbonate.

Procedure. The determination of GSH is based on the manometric measurement of the CO_2 produced in the bicarbonate buffer by the lactic acid which is formed enzymatically from methylglyoxal. A standard curve is prepared with amounts of GSH ranging from 0.025 mg. to 0.15 mg. If GSH is determined in tissue extracts which have been deproteinized by SSA, it is essential to prepare the standard curve with GSH dissolved in SSA of the same strength.[19]

Tissue extracts are prepared by deproteinization with the 2% SSA and subsequent neutralization. They should contain GSH in amounts corresponding to those used for the preparation of the standard curves. To the main compartment of the manometric flasks is added 0.15 to 0.3 ml. of yeast suspension, 3 mg. of methylglyoxal, 0.4 ml. of 0.2 M NaHCO$_3$, and water to a final volume of 2 ml. The GSH-containing extract is placed into the side arm, the flasks are gassed with the N_2-CO_2 mixture, and after temperature equilibration with shaking the GSH solution is tipped in. The amount of GSH is determined by comparing the rate of CO_2 evolution with that obtained in the standardization experiments. It is advisable to use a standard curve with at least two known concentrations of GSH (freshly prepared) with each series of determinations of unknown. In the authors' experience the accuracy of the method with samples containing 0.1 to 0.3 mg. of GSH per milliliter does not exceed 10%.

Other enzyme systems which could be adapted to the quantitative assay of GSH are nitrate reductase[20] and GSH reductase.[21]

---

[19] This is necessary because it has been observed (1) that sulfosalicylic acid inhibits the glyoxalase reaction and (2) that GSH is unstable when dissolved in this reagent.

[20] L. A. Heppel and J. R. Hilmoe, *J. Biol. Chem.* **183**, 129 (1950); see Vol. II [57].

[21] See Vol. II [126].

[87] Assay of Aromatic Amino Acids

I. Phenylalanine

By Sidney Udenfriend

Chemistry of the Reaction. The two products formed by the enzymatic decarboxylation of phenylalanine are CO_2 and phenylethylamine. Procedures are available for the assay of phenylalanine by the measurement of the CO_2 evolved after enzymatic decarboxylation.[1]

Measurement of the phenylethylamine offers certain advantages over measurement of the CO_2. The amine can still be identified structurally with the parent amino acid and has solubility characteristics which permit it to be separated from all unchanged amino acids and from many other amines. The employment of enzymatic CO_2 evolution as a specific assay method necessitates the use of highly purified enzyme preparations to remove other decarboxylases and other sources of CO_2. Formation of phenylethylamine can be due only to the presence of phenylalanine. Since the amine can be separated from unchanged amino acids and from other amines, it is possible to use unpurified extracts as sources of decarboxylase activity. A further advantage of the amine procedure is the increased sensitivity (0.01 micromole) compared with that of the CO_2 method (1.0 micromole). The method is based on decarboxylation of the amino acid by an acetone powder preparation of *Streptococcus faecalis*. The resulting phenylethylamine is extracted into chloroform and assayed by a modification of the methyl orange procedure of Brodie and Udenfriend.[2,3]

Reagents and Materials

An acetone-dried powder of *S. faecalis* was prepared by the method of Epps.[4] This preparation is stable for at least eight months when kept at $-10°$. A 15-l. carboy of medium inoculated with *S. faecalis* will yield approximately 11 g. of powder. The powder is suspended in water immediately before use, 50 mg./ml. of water.

0.7 M citrate buffer, pH 5.5.

10 N NaOH.

[1] E. F. Gale, *Biochem. J.* **41**, viii (1947).
[2] B. B. Brodie and S. Udenfriend, *J. Biol. Chem.* **158**, 705 (1945).
[3] S. Udenfriend and J. R. Cooper, *J. Biol. Chem.* **203**, 953 (1953).
[4] H. M. R. Epps, *Biochem. J.* **39**, 42 (1945).

0.00100 M L-phenylalanine. This standard solution is stable indefinitely when stored in the cold.

Chloroform containing 3.3% isoamyl alcohol. The solvents are washed successively with 1 N NaOH, 1 N HCl, and twice with water before mixing.

Methyl orange reagent. 500 mg. of methyl orange is dissolved in 100 ml. of warm H_2O. The resulting solution is washed several times with an equal volume of chloroform. The methyl orange reagent is made by diluting this solution with an equal volume of saturated boric acid solution. This dilution is made just prior to use, since the methyl orange precipitates within an hour.

Alcoholic H_2SO_4 solution. Two milliliters of concentrated H_2SO_4 in 100 ml. of absolute alcohol.

Procedure. To 2 ml. of a solution containing 0.03 to 0.30 micromole of L-phenylalanine in a 60-ml. glass-stoppered bottle, add 0.5 ml. of 0.7 M citrate buffer, pH 5.5, and 0.5 ml. of the *S. faecalis* suspension. Incubate the bottle at 37°, with shaking, for 2 hours. After cooling, add 0.3 ml. of 10 N NaOH and 15 ml. of the $CHCl_3$-isoamyl alcohol reagent, and shake the bottle for 5 minutes. Transfer the contents of the bottle to a test tube, and centrifuge for 2 minutes. Remove the upper aqueous phase by aspiration with a fine-tipped pipet. Transfer as much of the chloroform phase as possible into a 15-ml. glass-stoppered tube, carefully avoiding any of the alkaline aqueous phase which may remain on the walls of the tube. Add 0.5 ml. of the methyl orange reagent, and shake the tube for 2 minutes. Centrifuge the tube at high speed, and completely remove the excess methyl orange by aspiration with a fine-tipped pipet. Transfer 10 ml. of the solvent phase to a colorimeter tube containing 2 ml. of the alcoholic H_2SO_4 solution. Mix the contents of the tube, and determine the optical density in a colorimeter at 540 mμ. Standards are run through the entire procedure for comparison, optical density being proportional to concentration.

To determine L-phenylalanine in plasma or in tissue extracts, it is necessary to deproteinize the sample. Add 0.1 vol. of 1 N HCl, and heat in a boiling water bath for 3 minutes. Neutralize the sample by the addition of 0.1 vol. of 1 N NaOH, and remove the proteins by centrifugation. An appropriate aliquot of the supernatant fluid is then assayed by the above procedure. No phenylalanine is lost on precipitation of the proteins in this manner. In Table I are presented the recoveries of known amounts of phenylalanine added to a liver extract and to plasma.

Protein hydrolyzates can be assayed after neutralization and appropriate dilution.

Phenylalanine decarboxylase activity varies from one *S. faecalis* preparation to another. With some preparations almost complete decarboxylation occurs within 30 minutes, whereas with others only 75 to 95% decarboxylation takes place in 2 hours, regardless of the amount of acetone powder employed. The simplest explanation for these low values is the presence in the crude acetone powder of other enzymes, such as

TABLE I

RECOVERY OF L-PHENYLALANINE FROM PLASMA, LIVER HOMOGENATE, AND SIMULATED PROTEIN HYDROLYZATE

Tissue	L-Phenylalanine added, μM.	L-Phenylalanine found, μM.	Recovery, %
Liver extract[a] (rat)	0.000	0.093	
	0.100	0.201	104
	0.100	0.218	112
	0.200	0.324	110
	0.200	0.311	106
	0.400	0.476	97
	0.400	0.502	102
Plasma[b] (dog)	0.000	0.117	
	0.100	0.219	101
	0.100	0.224	103
	0.200	0.313	99
	0.200	0.320	101
Simulated protein hydrolyzate[c]	0.00	0.00	
	0.30	0.34	112
	0.30	0.31	103
	0.30	0.29	97
	0.30	0.31	103

[a] 0.2 ml. of supernatant fluid from a liver homogenate, equivalent to 40 mg. of the original wet tissue.

[b] 0.60 ml. of dog plasma was used.

[c] A solution of amino acids similar in composition to a hydrolyzate of serum albumin, without phenylalanine.

transaminases, which also act on phenylalanine. The presence of a competing transaminase was indirectly shown by demonstrating the inhibition of phenylethylamine formation by the addition of α-ketoglutarate. With each preparation, the per cent of phenylethylamine obtained from any given amount of phenylalanine remains fairly constant from day to day. The unknown samples are always calculated with respect to phenylalanine standards that are simultaneously run through the procedure.

Specificity of Method As Applied to Plasma. L-Phenylalanine and tyrosine are the only known amino acids which are decarboxylated by acetone-dried *S. faecalis preparations.* Tyramine, the product derived from tyrosine, does not react with the methyl orange reagent.

The specificity of the method for plasma was examined by the technique of countercurrent distribution. A large sample of dog plasma filtrate was incubated with the decarboxylase preparation and then extracted into chloroform containing 3.3% isoamyl alcohol, as described in the method for plasma. The apparent phenylethylamine was subjected to an eight-transfer countercurrent distribution with the solvents, chloroform containing 3.3% isoamyl alcohol, and 0.5 M borate buffer, pH 8.1. With these solvents phenylethylamine was distributed about equally between the two phases.

Except in the end funnels, which contained only a small percentage of the total material, the ratio of the amount of material in the chloroform phase to the total amount in both phases was constant and almost identical with that found for authentic phenylethylamine measured at the same time with the same batch of equilibrated solvents. A small amount of nonphenylethylamine material was found in funnels 1, 2, and 9. The theoretical amount of phenylethylamine that should be present in these funnels was calculated from the measured partition ratio of phenylethylamine by application of the binomial expansion in the manner described by Williamson and Craig.[5] In all, only about 0.11 micromole of 3.82 micromoles was nonphenylethylamine, less than 3% of the total.

A further check on the specificity of the method was obtained by paper chromatography. An aliquot of each funnel, in the countercurrent distribution outlined above, was dried on paper, and chromatograms were developed with a mixture of butanol-propionic acid-water (5:3:2). The dried chromatograms were sprayed with ninhydrin. With samples from funnels 3 to 8 a single spot appeared, having the color and R_f corresponding to phenylethylamine.

D-Phenylalanine yields no measurable amounts of phenylethylamine when carried through this procedure. The method can therefore be used to determine L-phenylalanine in the presence of its optical isomer.

II. Tyrosine

By SIDNEY UDENFRIEND

Chemistry of the Reaction. The Gerngross-Herfelt test for substituted phenols[6] involves reaction with nitrosonaphthol, in nitric acid, to yield unstable red-colored derivatives. This reaction has long served as a qualitative test for tyrosine in biological material. If one permits the

[5] B. Williamson and L. C. Craig, *J. Biol. Chem.* **168**, 687 (1947).
[6] O. Gerngross, K. Voss, and T. Herfelt, *Ber.* **66**, 435 (1933).

reaction to continue for a longer time and at an elevated temperature, the red intermediate changes to a stable yellow compound. The nitrosonaphthol, which is itself yellow, can then be extracted from the reaction mixture with ethylene dichloride or chloroform. The extracted reaction mixture containing the tyrosine-nitrosonaphthol chromophore[7] (apparently a condensation product of 1 mole equivalent each of tyrosine and nitronaphthol) can then be assayed colorimetrically. The procedure has been employed successfully in the studies on the enzymatic conversion of phenylalanine to tyrosine[8] and to determine tyrosine in protein hydrolyzates.[7]

Reagents

1-Nitroso-2-naphthol. 0.1% 1-nitroso-2-naphthol[9] in 95% alcohol.
Nitric acid reagent.[10] 1:5 nitric acid containing 0.5 mg./ml. of $NaNO_2$.

Procedure. To 1.0 ml. of plasma or tissue extract are added 3.0 ml. of water and 1.0 ml. of 30% trichloroacetic acid, and the mixture is centrifuged after 10 minutes. To 2 ml.[11] of the deproteinized plasma, tissue extract, or protein hydrolyzate containing 0.03 to 0.80 micromole of tyrosine in a glass-stoppered centrifuge tube are added 1 ml. each of the nitrosonaphthol and the nitric acid reagents. The tube is stoppered, placed in a water bath at 55° for 30 minutes, and cooled. Ten milliliters of ethylene dichloride is added, and the tube is shaken to extract the unchanged nitrosonaphthol. The tube is then centrifuged at low speed, and the supernatant aqueous layer is transferred to a cuvette. The optical density is determined at 450 mμ in a spectrophotometer or photoelectric colorimeter. No change in reading was observed after the solution remained overnight at room temperature. The optical densities are proportional to the concentration, at least up to 0.8 micromole of tyrosine.

Standards are prepared by treating known amounts of tyrosine or tyramine with nitrosonaphthol in the same manner as the unknown samples. For any particular series of determinations, the values obtained are highly reproducible. However, standards are run with each set of

---

[7] S. Udenfriend and J. R. Cooper, *J. Biol. Chem.* **196,** 227 (1952).
[8] S. Udenfriend and J. R. Cooper, *J. Biol. Chem.* **194,** 503 (1952).
[9] The practical grade of 1-nitroso-2-naphthol (Eastman Kodak Company) can be used without purification.
[10] Trichloroacetic acid filtrates inhibit the formation of the nitrosonaphthol derivative unless freshly diluted nitric acid is used. It was found that traces of nitrite overcome this inhibition.
[11] For samples smaller than 2 ml., the volume is made up to 2 ml. with distilled water.

determinations, since there is a small daily variation in the optical densities.

An optical density of about 0.200 is obtained with the Beckman spectrophotometer when 0.1 micromole of tyrosine is carried through the procedure, only about 57% of the theoretical value. The losses for the tyrosine derivative can be satisfactorily accounted for. About 27% of the tyrosine undergoes nitration, and about 19% of the nitrosonaphthol derivative is removed by the ethylene dichloride extraction.

TABLE II

RELATIVE SENSITIVITY AND SPECIFICITY OF VARIOUS PROCEDURES FOR DETERMINATION OF TYROSINE

Procedure	Sensitivity optical density, $\mu M./ml.^a \times 10^3$	Apparent tyrosine per gram of rat liver, $\mu M.$
Nitrosonaphthol	7.3	0.33
Nitration[b]	4.7	1.27
Pauly-Weiss modification[c]	3.9	4.58
Folin-Ciocalteu modification[d]	13.9	10.4
Millon[e]	3.6	0.55[f]

[a] Absorption was measured in the Beckman spectrophotometer at the wavelength suggested in the description of each of the procedures.

[b] F. D. Snell and C. T. Snell, "Colorimetric Methods of Analysis," p. 200, Van Nostrand, New York, 1937.

[c] R. J. Block and D. Bolling, "The Amino Acid Composition of Proteins and Foods," 2nd ed., p. 114, Charles C Thomas, Springfield, Ill., 1951.

[d] O. Folin and V. Ciocalteu, J. Biol. Chem. **73,** 627 (1927).

[e] F. W. Bernhart and R. W. Schneider, Am. J. Med. Sci. **205,** 636 (1943).

[f] High concentrations of trichloroacetic acid inhibit the Millon reaction. In this case, proteins were removed with tungstic acid, as suggested in the procedure.

Recovery from Tissues. Tyrosine, when added to plasma or liver homogenates in amounts greater than 0.3 micromole, is recovered quantitatively. Below 0.3 micromole, recoveries are somewhat low, presumably because of coprecipitation with proteins.

Specificity. The method does not distinguish between tyrosine and tyramine. However, the amount of tyramine normally present in tissues is negligible, if it is present at all. Tissues contain other nontyrosine material which reacts with nitrosonaphthol to yield colored derivatives. A sample of pooled human plasma was analyzed for tyrosine as described in the analytical procedure. The absorption spectrum of the nitrosonaphthol chromophore differed from that of the prepared tyrosine derivative. It is obvious from this result that the method, as applied to plasma, is not specific. However, it is more specific than other chemical procedures for tyrosine assay. A comparison of the apparent tyrosine content of a rat liver homogenate, as measured by the nitrosonaphthol method and by a number of other methods, is shown in Table II. The nitrosonaphthol procedure gave the lowest value, indicating that this method excluded much nontyrosine material measured by the other procedures.

Sensitivity. The relative sensitivities of various methods for the determination of tyrosine are presented in Table II. Only the Folin-Ciocalteu method has a higher sensitivity.

Modifications of the Millon reaction[12] have been used successfully in enzymatic studies. Many substances, including inorganic phosphate, interfere with the reaction. It is therefore necessary to run internal standards with all determinations.

III. Tryptophan

By SIDNEY UDENFRIEND and RALPH E. PETERSON

The indole nucleus of tryptophan undergoes many reactions which can be utilized for analytical purposes. The two most common analytical procedures involve condensation with aldehydes[13] and reaction with $HgSO_4$-$HgCl_2$.[14] These procedures are sensitive and fairly specific for indole compounds. However, the rate of color development and the final intensity of the color can be influenced by extraneous materials. For precise work it is therefore necessary to run internal standards in each sample to be analyzed to correct for interfering substances.

The following adaptation of the procedure by Graham *et al.*[15] has been utilized successfully to determine tryptophan in tissues and biological extracts.

Reagents

0.5% *p*-dimethylaminobenzaldehyde (DMAB) in 12 *N* HCl.
0.2% $NaNO_2$.

Assay. Transfer 0.5 ml. of a deproteinized solution containing 0.05 to 0.5 micromole of tryptophan to a colorimeter tube containing 2.5 ml. of the DMAB reagent, and allow to stand for 30 minutes at room temperature, preferably in the dark. Then add 2.5 ml. of absolute alcohol and 2 to 3 drops of 0.2 *N* $NaNO_2$, and let stand at room temperature for another 30 minutes. The optical density of the resulting blue chromophore

[12] F. W. Bernhart and R. W. Schneider, *Am. J. Med. Sci.* **205**, 636 (1943).
[13] F. B. Hopkins and S. W. Cole, *Proc. Roy. Soc. (London)* **68**, 21 (1901).
[14] J. W. H. Lugg, *Biochem. J.* **31**, 1423 (1937).
[15] C. F. Graham, E. P. Smith, S. W. Hier, and D. Klein, *J. Biol. Chem.* **168**, 711 (1947).

is measured at 620 mμ. Standards are prepared at exactly the same time by adding known amounts of tryptophan dissolved in 0.050 ml. of solution to 0.5 ml. of deproteinized tissue blank and developing the color as usual.

By means of a Beckman spectrophotometer with a 1-cm. light path, an optical density of about 0.275 is obtained for 0.1 micromole. With amounts of tryptophan up to 0.5 micromole, the response is linear.

Precipitation of Proteins. Several different protein precipitants have been used successfully with this procedure. Proteins may be precipitated by addition of 2 ml. of 7.5% trichloracetic acid to 2 ml. of an enzyme reaction mixture. Recoveries of tryptophan are quantitative unless the protein concentration is too high. Deproteinization may also be carried out with zinc acetate and NaOH as described by Knox and Mehler.[16]

Specificity. Condensation with DMAB is not specific for tryptophan, since practically all indoles react to some extent. Tryptophan can be assayed in the presence of kynurenine, 3-hydroxykynurenine, and other nonindole metabolites. The method as presented is not applicable to studies involving the conversion of tryptophan to other indole derivatives, i.e., indole, 5-hydroxytryptophan, tryptamine, unless these compounds can first be removed. In many cases removal can be readily accomplished by extraction of the interfering indole with a suitable solvent.

The only available procedures which are specific for tryptophan are those involving microbiological assay. A number of such procedures have been published.[17,18] These may be time-consuming for enzymatic studies, but they are specific for the natural isomer of tryptophan.

[16] W. E. Knox and A. H. Mehler, *J. Biol. Chem.* **187**, 419 (1950).
[17] M. S. Dunn, H. F. Schott, W. Frankl, and L. B. Rockland, *J. Biol. Chem.* **157**, 387 (1945).
[18] B. S. Schweigert, H. E. Sauberlich, and C. A. Elvehjem, *J. Biol. Chem.* **164**, 213 (1946).

[88] Preparation and Determination of Intermediates in Aromatic Ring Cleavage

By D. L. MacDonald and R. Y. Stanier

The primary substrates and cyclic intermediates in the oxidation of simple, nonnitrogenous aromatic compounds are all substances which are either commercially available or easily prepared by well-known methods. Hence their preparation does not appear to merit inclusion here. We have instead presented the methods used to prepare and determine the aliphatic products of ring oxidation—β-ketoadipic acid, β-carboxymuconic acid, cis,cis-muconic acid, and γ-carboxymethyl-Δ$^{\alpha}$-butenolide.

I. β-Ketoadipic Acid

Determination[1]

Principle. Catalytic decarboxylation to levulinic acid and measurement of the amount of CO_2 evolved.

Reagents

0.5 M acetate buffer, pH 4.2.
0.1 M 4-aminoantipyrine.

Procedure. The sample to be assayed, acidified to pH 4.0 to 4.5, is placed in the main compartment of a manometric flask, together with 2 ml. of acetate buffer. In the side arm is placed 0.4 ml. of 4-aminoantipyrine. After thermal equilibration the two solutions are mixed, and readings are continued until CO_2 production ceases.

Preparation

Principle. β-Ketoadipic acid has been prepared by the acid hydrolysis of diethyl β-ketoadipate[2] or of dimethyl β-ketoadipate.[3] The diethyl β-ketoadipate can be prepared by treatment of diethyl β-keto-α-carbethoxyadipate with β-naphthalenesulfonic acid,[2] with concentrated H_2SO_4 followed by ethanol,[4] or more simply by boiling with water.[4] The dimethyl β-ketoadipate has been prepared by the decomposition of dimethyl β-keto-α-acetyladipate using ammonia in ether.[3] For the preparation of the free acid the simplest procedure is the direct hydrolysis of diethyl β-keto-α-carbethoxyadipate, using concentrated hydrochloric acid[5] as described below.

Procedure: (a) Preparation of diethyl β-keto-α-carbethoxyadipate.[2,5] A solution of 40 g. of succinic anhydride in 45 ml. of absolute alcohol is heated on the steam bath until the alcohol ceases to reflux. The excess alcohol is removed by distillation under reduced pressure, and the residue of ethyl hydrogen succinate is refluxed for 1 hour with 45 ml. of thionyl chloride. The excess thionyl chloride is removed *in vacuo*, and the β-carbethoxypropionyl chloride is distilled, giving 60 g., boiling point 98° at 15 mm.

Magnesium malonic ester is prepared by adding a solution of 50.5 g.

[1] W. R. Sistrom and R. Y. Stanier, *J. Bacteriol.* **66,** 404 (1953).
[2] B. Riegel and W. M. Lilienfeld, *J. Am. Chem. Soc.* **67,** 1273 (1945).
[3] J. C. Bardhan, *J. Chem. Soc.* **1936,** 1848.
[4] S. F. MacDonald and R. J. Stedman, *Can. J. Chem.* **33,** 458 (1955).
[5] U. Eisner, J. A. Elvidge, and R. P. Linstead, *J. Chem. Soc.* **1950,** 2223.

of redistilled diethyl malonate in 25 ml. of absolute alcohol to a suspension of 7.9 g. of aluminum-free magnesium turnings in 8 ml. of absolute alcohol and 0.5 ml. of carbon tetrachloride. To this mixture is added 100 ml. of anhydrous ether, and the mixture is refluxed until the magnesium dissolves (about 6 hours). The ether and alcohol are removed by distillation, first at atmospheric pressure and then at reduced pressure. About 100 ml. of dry benzene is then added, and after the residue dissolves the benzene is removed at reduced pressure.

The magnesium malonic ester is dissolved in 200 ml. of dry ether, and to this is added the β-carbethoxypropionyl chloride dissolved in 50 ml. of dry ether, and the mixture is stirred under reflux for 4 hours. The mixture is cooled, and with continued stirring a cold solution of sulfuric acid (9.5 ml.) in water (160 ml.) is added slowly. The ethereal layer is separated, the aqueous layer is extracted once with ether, and the combined ethereal solution is washed with a little water and dried with Na_2SO_4. The ether is removed by distillation, finally under reduced pressure, and the product distilled. After a small forerun, there is obtained about 75 g. of diethyl β-keto-α-carbethoxyadipate distilling at 144° (0.2 mm.).

(b) *Preparation of β-ketoadipic acid.* Diethyl β-keto-α-carbethoxyadipate (10 g.) is kept with concentrated hydrochloric acid (30 ml.) at room temperature for 36 hours, during which time CO_2 is evolved. The solution is concentrated *in vacuo* at 30° to 35°, and the last traces of hydrochloric acid are removed by drying *in vacuo* over KOH. The β-ketoadipic acid (melting point 115°) is obtained in quantitative yield. The material may be recrystallized from ethyl acetate, or from a mixture of acetone and petroleum ether. The pure compound melts at 122° to 123° with decomposition.

II. β-Carboxymuconic Acid

Determination[6]

Principle. The biologically active isomer (presumably all *cis*) of β-carboxymuconic acid is converted stoichiometrically to β-ketoadipic acid by a specific inducible enzyme system. At a wavelength of 270 mμ the former compound has a strong light absorption, whereas the latter is essentially transparent. Hence the change in optical density at 270 mμ on treatment with the enzyme can be used as a measure of the amount of active β-carboxymuconic acid present. The method is particularly valuable in that it does not measure the biologically inactive isomer, which is almost invariably present as a decomposition product in preparations of the active isomer.

[6] D. L. MacDonald, R. Y. Stanier, and J. L. Ingraham, *J. Biol. Chem.* **210,** 809 (1954)

Reagents. The enzyme system responsible for converting β-carboxy-muconic acid to β-ketoadipic acid is present in cells of *Pseudomonas fluorescens,* but only after growth at the expense of a metabolic precursor (*p*-hydroxybenzoic acid). Preparations satisfactory for the assay can be obtained as a by-product of the preparation of protocatechuic acid oxidase,[7] by elution of the alumina C_γ gel employed in the final step with 0.01 *M* phosphate buffer at pH 6.5. Although highly active, such preparations still contain protocatechuic acid oxidase and hence cannot be used to determine β-carboxymuconic acid in materials that also contain proto-catechuic acid.

0.1 *M* phosphate buffer, pH 7.0.

Procedure. The optical density at 270 mμ is determined on a solution containing β-carboxymuconic acid, after appropriate dilution with phos-phate buffer, by reading against a buffer blank of the same volume. A suitable volume of the enzyme system is then added to each cuvette, and readings are continued until the optical density at 270 mμ reaches a stable value. Corrected for dilution by the enzyme, the decrement in optical density is a measure of the amount of active β-carboxymuconic acid (E_{270} at pH 7.0 is 6400).

In samples that also contain protocatechuic acid, this substance can be estimated enzymatically by conversion to β-carboxymuconic acid with protocatechuic acid oxidase, after which the total amount of β-car-boxymuconic acid can be estimated as described above, and corrected for the amount of protocatechuic acid initially present.

Preparation

The only available procedure is an enzymatic one, no chemical method being known.

Reagents

Purified protocatechuic acid oxidase.[7]
0.1 *M* Na protocatechuate.
0.1 *M* phosphate buffer, pH 7.5.
1 *N* NaOH.
5 *N* HCl.
Ether.
Methanol.
Isopropanol.

Procedure. A flask of 3-l. capacity containing 250 ml. of phosphate buffer, 25 ml. of 0.1 *M* Na protocatechuate (2.5 millimoles) and approx-

[7] See Vol. II [37].

imately 600 units of protocatechuic acid oxidase is placed at 30° and agitated mechanically to ensure good aeration. The course of the oxidation is followed by periodic measurement of optical density at 270 and 290 mμ on diluted samples of the mixture. In neutral solution, protocatechuic acid has molar extinctions of 3890 and 2730 at 290 and 270 mμ, respectively, and β-carboxymuconic acid has extinctions of 1590 and 6400 at these wavelengths. At appropriate intervals, an additional 7.5 millimoles of substrate is added in lots of 2.5 millimoles. The pH is maintained slightly above 7.0 by periodic additions of 1 N NaOH. In order to maintain the rate of the reaction at a relatively high level, it is desirable to add a further 600 units of enzyme about halfway through the oxidation.

When the reaction has reached completion, the mixture is chilled to 0°, rapidly adjusted to pH 2 with 5 N HCl, and rapidly extracted with an equal volume of ether, previously chilled to $-15°$, by vigorous shaking for 5 minutes in a separatory funnel. The ether layer is removed, and the aqueous layer is re-extracted twice more in the same fashion. The combined ether extracts are shaken with a volume of chilled 1 N NaOH slightly less than that required for the complete neutralization of the contained β-carboxymuconic acid, and the pH of the aqueous extract is adjusted to 7.0. This extract is then dried by lyophilization. The dried material should be natural β-carboxymuconic acid of about 80% purity, the principal contaminant being biologically inactive β-carboxymuconic acid formed during the extraction procedure. The natural isomer can be obtained as a sodium salt of about 95% purity by the following procedure. All operations must be conducted as close to 0° as possible. One gram of the dried powder is dissolved in 30 ml. of H_2O, and methanol (150 ml.) is added. The precipitate is removed by centrifugation and discarded. To the clear supernatant is added, portionwise, over a period of 24 hours, 280 ml. of isopropanol. The crystalline sodium salt is collected by filtration, recrystallized by the same procedure, and dried *in vacuo* at room temperature over P_2O_5 and KOH. The recovery is 80%. The stability of the crystalline sodium salt is not known, but it is probably advisable to store it at $-15°$.

III. *cis,cis*-Muconic Acid

Preparation

Principle. cis,cis-Muconic acid is prepared by the oxidation of phenol with peracetic acid[8] or perpropionic acid.[9] Perpropionic acid has been

[8] J. A. Elvidge, R. P. Linstead, B. A. Orkin, P. Sims, H. Baer, and D. B. Pattison, *J. Chem. Soc.* **1950**, 2228; J. A. Elvidge, R. P. Linstead, P. Sims, and B. A. Orkin, *ibid.* **1950**, 2235.

[9] W. R. Sistrom and R. Y. Stanier, *J. Biol. Chem.* **210**, 821 (1954).

reported to be less likely to explode than peracetic acid, but caution should be observed during distillation and handling of any peracid.

Procedure. Peracetic acid (755 g.) of approximately 13% strength is obtained by adding a mixture of 642 g. of acetic anhydride and 5 ml. of concentrated sulfuric acid during 9.5 hours to 150 g. of 30% hydrogen peroxide, while stirring and cooling (4 to 9°). After a further 3 hours, the product is distilled at water pump pressure from a bath at 30° to 35° into an ice-cooled receiver. The peracid content is determined by adding an aliquot to an excess of aqueous potassium iodide solution and titrating the iodine liberated. Phenol (22.8 g.) is dissolved in 435 g. of 13.5% peracetic acid, and the solution is kept in the dark for 15 days. Care should be taken to avoid undue exposure to light, or the application of heat. The precipitated *cis,cis*-muconic acid (12 g.) is removed by filtration and dissolved in 10% aqueous sodium bicarbonate, and the solution is washed with an equal volume of ether, stirred with charcoal, and filtered. Acidification of the filtrate with hydrochloric acid gives the acid (melting point 184°). The product can be crystallized from ethanol, with considerable loss of material, and then melts at 194° to 195° (heating at 10° per minute).

IV. γ-Carboxymethyl-Δ^α-butenolide

Preparation

Principle. Lactonization of *cis,cis*-muconic acid, either chemical or enzymatic. The chemical lactonization gives the racemic mixture, and the enzymatic lactonization gives the (+) isomer. The (−) isomer can be prepared from the racemic mixture by selective bacterial oxidation of the (+) isomer. Since both isomers are of value in enzymatic studies, their preparation will be described in addition to the preparation of the racemic mixture.

Procedure for Chemical Lactonization.[8] Fifteen grams of *cis,cis*-muconic acid is suspended in a cold mixture of 60 ml. of sulfuric acid and 20 ml. of H_2O. After 24 hours, the mixture is poured into 200 g. of crushed ice and the solution is made just acid to congo red with concentrated NH_4OH and extracted continuously with ether for 48 hours. The ether extract is evaporated and the residue crystallized from benzene-ethanol. The lactone (12 g.) separates as prisms (melting point 110.5° to 111.5°).

Procedure for Enzymatic Lactonization (Isolation of (+) Isomer).[9] The reaction mixture contains 4 millimoles of *cis,cis*-muconic acid, 1.5 millimoles of $MnCl_2$, 10 millimoles of phosphate buffer (pH 6.0), and 150 units of lactonizing enzyme[7] in a total volume of 200 ml. This is incubated at 30° for about 2 hours, the course of the reaction being followed by measuring the change in optical density of diluted samples at 230 and 260 mμ. The reaction does not proceed to completion, and the incubation should be terminated when the optical densities reach a constant value. The molar extinction coefficients of the muconic acid and the lactone are 16,900 and 40, respectively, at 260 mμ; and 10,600 and 1520, respectively, at 230 mμ.

After termination of the enzymatic reaction, the precipitate of manganous phosphate is removed by centrifugation, and the supernatant is shaken with chloroform to remove protein. It is then evaporated *in vacuo* (bath temperature 30°) to a volume of 60 ml. This solution is decolorized with charcoal, acidified to pH 2.5 with H_2SO_4, and continuously extracted for 48 hours with ether. The ether extract is evaporated, and the resulting sirup is taken up in 10 ml. of absolute alcohol, decolorized with charcoal, and dried *in vacuo* over Drierite for 12 hours. Crystallization should occur when the container is scratched. The crude product is recrystallized from ethyl acetate-benzene, and then melts at 76° to 77.5°.

Procedure for the Biological Resolution of the Racemic Lactone (*Isolation of* (−) *Isomer*).[9] A culture of *Pseudomonas fluorescens* is grown in a medium containing mandelic acid or benzoic acid as the principal carbon source.[7] The cells are harvested by centrifugation and resuspended in 0.05 M phosphate buffer (pH 7.0) at a density of 375 Klett units (red filter). To 100 ml. of this suspension in a 1-l. Erlenmeyer flask is added 500 mg. of the synthetic lactone. The mixture is placed at 30° and subjected to mechanical agitation in order to ensure aeration. The course of the oxidation is followed by periodically determining the optical density at 230 mμ on suitably diluted samples, read against a similar bacterial suspension without lactone. When no further decrement in the OD_{230} occurs, the cells are removed by centrifugation, and the residual lactone [the (−) isomer] is isolated from the supernatant by the procedure described above for the isolation of the (+) isomer. The purified lactone melts at 78° to 79°.

[89] L-Kynurenine and N^1-Formyl-L-kynurenine

By V. H. AUERBACH and W. E. KNOX

I. Preparation and Properties of L-Kynurenine

Preparation

Principle. L-Tryptophan is acetylated with acetic anhydride by the method of du Vigneaud and Sealock.[1] The N$^\alpha$-acetyl-L-tryptophan is ozonolyzed and hydrolyzed to L-kynurenine and isolated as the sulfate,

---

[1] V. du Vigneaud and R. R. Sealock, *J. Biol. Chem.* **96**, 511 (1932).

by the method of Warnell and Berg.[2] The method is equally applicable for the preparation of D- or DL-kynurenine. Kynurenine can also be made by enzymic oxidation of L-tryptophan[3] and by synthesis[4] and resolution.[5]

Reagents

L-Tryptophan (6.8 g. dissolved in 10 ml. of water plus 16.8 ml. of 2 N NaOH).

Ozone. The amount delivered in a steady stream of at least 2% is calibrated accurately with 5% KI and thiosulfate titration.[6] An exactly equimolar amount is used to oxidize the acetyltryptophan.

Procedure. The dissolved tryptophan is acetylated in an ice bath by a total of 80 ml. of 2 N NaOH and 8 ml. of acetic anhydride, each added in eight equal portions over 15 minutes with adequate shaking between additions. After the additions the mixture is allowed to stand at room temperature for 20 minutes. It is then cooled in an ice-salt bath, and 35.8 ml. of 6 N H_2SO_4 is added. The white crystals of N^α-acetyl-L-tryptophan are filtered, washed with 25 ml. of 0.2 N HCl and with water, and recrystallized from water as long flat needles. The yield is 7.4 g., melting point 187 to 188° (uncorr.), $[\alpha]_D^{25} = +31°$ (1% solution dissolved with 1 equivalent of NaOH).

N^α-Acetyl-L-tryptophan is finely powdered, and 6.75 g. (0.027 mole) is suspended in 225 ml. of glacial acetic acid. Ozone is passed into the suspension at room temperature with adequate stirring for complete absorption until exactly 1 equivalent (0.027 mole) has been added, as determined by the previous calibration and by the more rapid release of iodine from a 5% KI trap attached to the outflow. The reaction mixture is then hydrolyzed with an initial addition of 27 ml. of concentrated HCl by refluxing under a condenser fitted with a Bunsen valve. The refluxing is continued for 5½ hours, during which time three additions of 10 ml. of HCl are made. Then 1.6 ml. of concentrated H_2SO_4 is added, and the mixture is concentrated to a thick sirup under reduced pressure in N_2.

The sirup is dissolved in 125 ml. of hot 50% EtOH, treated with a small amount of Norit, and reconcentrated *in vacuo* to 10 ml. The mixture is then dissolved in 30 ml. of hot EtOH and allowed to crystallize at 0°. Recrystallization from hot 66% EtOH, after a second treatment with Norit, is facilitated by a slight excess of H_2SO_4. The yield is 3.3 g. of

[2] J. L. Warnell and C. P. Berg, *J. Am. Chem. Soc.* **76,** 1708 (1954).
[3] O. Hayaishi, *Biochem. Preparations* **3,** 108 (1953).
[4] C. E. Dalgliesh, *J. Chem. Soc.* **1952,** 137.
[5] A. Butenandt and R. Weichert, *Z. physiol. Chem.* **281,** 122 (1944).
[6] L. I. Smith, F. L. Greenwood, and O. Hudrlik, *Org. Syntheses* **26,** 63 (1946).

L-kynurenine sulfate monohydrate as nearly white needles. The free base can be prepared by addition of an equivalent amount of Ba(OH)$_2$ to an aqueous solution of the sulfate, filtration, and concentration under reduced pressure in N$_2$ to the point where crystallization begins. Two volumes of hot EtOH is then added to complete the crystallization.

Properties

Physical Properties. L-Kynurenine sulfate monohydrate melts with decomposition at 179 to 180°, and the free base at 191° (uncorr.). The specific rotations of the two compounds in 1% aqueous solutions are, respectively, $[\alpha]_D^{25} = +9.7°$ and $-30.5°$. The ultraviolet absorption curve at neutral pH is a convenient means of identification and measurement.[7] The maximum absorption at 360 mμ has a molar extinction coefficient of 4.50×10^3.

II. Preparation and Properties of N^1-Formyl-L-kynurenine

Preparation

Principle. Formic-acetic anhydride is used to formylate only the aromatic amino group of L-kynurenine, according to the method of Dalgliesh.[4] Acid hydrolysis to the starting material is more difficult to avoid if L-kynurenine sulfate is the starting material. D- and DL-Kynurenine can be used to form the corresponding formyl compounds.

Reagents

L-Kynurenine[8] (1.04 g., 0.005 mole, dissolved in 2.5 ml. of 98 to 100% formic acid).

Formic-acetic anhydride, 0.005 mole. Mix 0.5 ml. of acetic anhydride and 1.0 ml. of 98 to 100% formic acid in an ice bath, and allow the mixture to stand at room temperature 30 minutes before use.

Procedure. The L-kynurenine in formic acid is added to the solution of formic-acetic anhydride at room temperature and permitted to stand for 2 hours. The reaction mixture is then mixed into 50 vol. of anhydrous ether. After 2 hours at 0° the amorphous product is filtered, washed with ether, and immediately dried to avoid hydrolysis. The yield is nearly 100% of spectroscopically pure formylkynurenine, but it contains a little formic acid. It can be crystallized from water as a hydrate if precautions are taken to exactly neutralize the accompanying acid during solution to avoid hydrolysis.

[7] A. H. Mehler and W. E. Knox, *J. Biol. Chem.* **187**, 431 (1950).

Properties

Physical Properties. The dried compound melts with decomposition at 162°. The ultraviolet absorption curve at pH 7.0 has maxima at 260 mμ and 321 mμ with molecular extinction coefficients of 10,980 and 3750, respectively.[4,7] Kynurenine from incomplete reaction or from hydrolysis and N$^\alpha$-acylkynurenine due to use of excess acetic anhydride in the synthesis may be detected by their characteristic absorption at 360 mμ.

Enzymatic Properties. N^1-Formylkynurenine is hydrolyzed to kynurenine by the kynurenine formamidase (formylase) of liver. This reaction is about seven times as fast as the reaction of the same enzyme on formylanthranilic acid. Spontaneous hydrolysis is much more rapid than for acetyl compounds and is avoided by preparation of solutions at a neutral pH and without heating.

[90] Isolation and Determination of Histidine and Related Compounds

By HERBERT TABOR

A large number of procedures are available for the isolation and determination of each of the compounds of this group. Limitations of space do not permit the inclusion of all these methods, and, in general, those procedures have been selected which have been used in this laboratory for certain enzymatic studies. A number of other techniques would be equally suitable for these purposes, and more suitable for others; attention is therefore directed to the references covering these alternate procedures.

Preparation and Isolation

Histidine.[1-6] Although histidine is available commercially, isolation procedures are frequently employed in various laboratory experiments.

[1] The chemical syntheses of histidine are not covered here, as they are reviewed by K. Hofmann;[2] for the synthesis of α-C^{14}-histidine, see G. Wolf;[3] for the synthesis of histidine labeled with C^{14} in the C-2 position of the imidazole ring, see Borsook *et al.*;[4,5] for N^{15} histidine (labeled in the γ-nitrogen), see Tesar and Rittenberg.[6]

[2] K. Hofmann, "Imidazole and Its Derivatives," Interscience Publishers, New York, 1953.

[3] G. Wolf, *J. Biol. Chem.* **200,** 637 (1953).

[4] H. Borsook, C. L. Deasy, A. J. Haagen-Smit, G. Keighley, and P. H. Lowy, *J. Biol. Chem.* **187,** 839 (1950).

[5] M. Toporek, Atomic Energy Commission Report UR-203 (1952).

[6] C. Tesar and D. Rittenberg, *J. Biol. Chem.* **170,** 35 (1947).

Histidine is most commonly isolated from an acid hydrolyzate of protein.[7] After removal of the acid the histidine can be isolated from the solution by a number of different procedures.[8,9] A frequent method for the isolation of histidine and other imidazoles involves precipitation with mercuric sulfate in dilute sulfuric acid,[10] although other amino acids such as tryptophan and cystine are also precipitated by this reagent. Silver salts have also been used to precipitate histidine;[10,11] at an alkaline pH arginine is also precipitated, but at a neutral pH it largely remains in solution. Phosphotungstic acid[12] and electrodialysis[13] have been used to separate all the basic amino acids. The isolated histidine has been further purified by crystallization as the diflavianate,[14] the di-(3,4-dichlorobenzenesulfonate),[15] and the nitroanilinate.[16] Although these methods have given good results in many laboratories, the procedures have frequently been laborious, and the possibility of both incomplete precipitation as well as incomplete separations was frequently present. More recently, therefore, the introduction of ion exchange resins, such as Dowex 50[17,18] and Amber-

---

[7] In the presence of fat or carbohydrate materials the removal of this material and the precipitation of the protein with trichloroacetic acid is advisable prior to the hydrolysis. The protein is usually hydrolyzed by boiling in 6 N HCl or H_2SO_4 for 16 to 36 hours.

[8] The various procedures adopted by different investigators for the isolation of histidine from small quantities of protein have been summarized by Block and Bolling.[9] Three of the more recent papers containing procedures for histidine isolation are: L. Levy and M. J. Coon, *J. Biol. Chem.* **192**, 807 (1951); G. Ehrensvärd, L. Reio, and E. Saluste, *Acta Chem. Scand.* **3**, 645 (1949); G. Wolf.[3]

[9] R. J. Block and D. Bolling, "The Amino Acid Composition of Proteins and Foods," pp. 3–39, Charles C Thomas, Springfield, Ill., 1951.

[10] A. Kossel and A. J. Patten, *Z. physiol. Chem.* **38**, 39 (1903).

[11] H. B. Vickery and R. J. Block, *J. Biol. Chem.* **93**, 105 (1931); H. O. Calvery, *J. Biol. Chem.* **83**, 631 (1929); R. J. Block, *J. Biol. Chem.* **106**, 457 (1934).

[12] D. D. Van Slyke, A. Hiller, and R. T. Dillon, *J. Biol. Chem.* **146**, 137 (1942).

[13] A. H. Gordon, A. J. P. Martin, and R. L. M. Synge, *Biochem. J.* **35**, 1369 (1941); E. Sperber, *J. Biol. Chem.* **166**, 75 (1946); for other references on electrodialysis see p. 34 in ref. 9.

[14] H. B. Vickery and C. S. Leavenworth, *J. Biol. Chem.* **76**, 707 (1928).

[15] H. B. Vickery, *J. Biol. Chem.* **143**, 77 (1942); **144**, 719 (1942); H. B. Vickery and J. K. Winternitz, *J. Biol. Chem.* **156**, 211 (1944).

[16] R. J. Block, *Proc. Soc. Exptl. Biol. Med.* **37**, 580 (1937); *J. Biol. Chem.* **133**, 67 (1940).

[17] The chromatography of histidine and other amino acids on Dowex 50-H, with HCl as the eluting agent, was reported by W. H. Stein and S. Moore [*Cold Spring Harbor Symposia Quant. Biol.* **14**, 189 (1950)]. In their procedure the histidine is eluted with 4 N HCl after prior elution of a variety of other amino acids with 1.5 N HCl; the histidine is eluted by this method in a much smaller volume than in the elution procedure described here. Further modifications of the chromatographic

lite XE-64,[19] has been particularly useful. In this laboratory we have employed a combination of Dowex 50 chromatography and crystallization as the 3,4-dichlorobenzenesulfonate.

A solution containing approximately 3 millimoles of histidine in 100-ml. volume is placed on a Dowex 50 column (8 to 12% crosslinked; hydrogen form; approximately 200 mesh; height 15 cm.; diameter 2.2 cm.). The column is washed with 15 ml. of water and eluted with 2.7 N HCl. The tubes are assayed for histidine colorimetrically (see p. 630). The histidine-containing fractions (usually 230 to 430 ml.)[17] are then evaporated to dryness on the steam bath or *in vacuo*.

For many purposes this is sufficient purification. If further purification is desired,[20] the histidine-containing residue is dissolved in 10 ml. of water and treated with 1.5 g. of 3,4-dichlorobenzenesulfonic acid; solution is effected by heating. On cooling in ice, crystals of histidine di-(3,4-dichlorobenzenesulfonate) separate and are collected by filtration on a sintered-glass funnel. The crystals are then washed with several milliliters of a cold 5% solution of 3,4-dichlorobenzenesulfonic acid. If desired, the material can be recrystallized many times from a 5% solution of 3,4-dichlorobenzenesulfonic acid with very little loss of histidine. After drying over $CaCl_2$ *in vacuo* the excess reagent can be removed by several washings with ether. The 3,4-dichlorobenzenesulfonate can be converted to the dihydrochloride by treatment with Dowex 1-Cl or by repeating the chromatography on Dowex 50. Quantitative removal of the 3,4-dichlorobenzenesulfonic acid can be easily checked by measuring the optical density at 275 mμ, where this compound has a high absorption. The histidine-containing solution is then evaporated to dryness on the steam bath. The purity of the isolated histidine can best be determined by optical rotation,[21] elementary analysis, and paper chromatography (see the table).

procedure have been published more recently.[18] These involve the use of low crosslinked resins, as well as buffers of progressively changing pH and ionic strength.

[18] S. Moore and W. H. Stein, *J. Biol. Chem.* **192**, 663 (1951); W. H. Stein, *J. Biol. Chem.* **201**, 45 (1953); S. Moore and W. H. Stein, *J. Biol. Chem.* **211**, 893 (1954).

[19] J. C. Winters and R. Kunin, *Ind. Eng. Chem.* **41**, 460 (1949); C. H. Hirs, S. Moore, and W. H. Stein, *J. Biol. Chem.* **200**, 493 (1953).

[20] Occasionally the histidine solution (adjusted to pH 7) is passed through a Dowex 1-acetate column at this point and eluted with 0.1 N acetic acid or 0.1 M sodium acetate. This step is particularly valuable in separating histidine from traces of urocanic acid, which are usually not completely removed by the other procedures.

[21] $[\alpha]_D^{22.7} = +13.0°$ for L-histidine in 6 N HCl ("Handbook of Chemistry and Physics," Chemical Rubber Publishing Co.). The free base can be obtained by adjusting a solution of histidine dihydrochloride to pH 7.0 and adding sufficient ethanol to induce crystallization. $[\alpha]_D^{25}$ for the free base in water is $-39.7°$ [L. Levintow, V. E. Price, and J. P. Greenstein, *J. Biol. Chem.* **184**, 55 (1950)].

Urocanic Acid (Imidazoleacrylic Acid). Urocanic acid has been prepared chemically[22] by the action of trimethylamine on imidazolechloropropionic acid. The latter was formed by the reaction of sodium nitrite and histidine in concentrated hydrochloric acid.

Urocanic acid[23-25] has been more conveniently prepared by the action of a heated histidase preparation (*Pseudomonas*) on histidine (see Vol. II [29]). Eighty milliliters of heated *Pseudomonas* extract, 75 millimoles of histidine, 150 mg. of streptomycin, 300 mg. of sodium thioglycolate, and water (380 ml. total volume; pH 9.5) are incubated at 38° for 24 to 48 hours. The formation of urocanic acid is followed by measuring the optical density in a 1:5,000 dilution at 277 mμ (see p. 633). When the optical density reaches the maximum, the solution is adjusted to pH 4.5 with acetic acid and evaporated to small volume. White crystals of urocanic acid dihydrate separate on cooling. These are collected by filtration and recrystallized from hot water (yield 80 to 90%). This procedure has also been followed for small-scale preparations with similar yields.

The purity of the isolated urocanic acid dihydrate can be determined by its spectrum[25] at various pH values (see p. 633). The solubility of urocanic acid dihydrate in water is approximately 0.1% at 0° and 6% at 100°; the hydrochloride and sodium salts are very soluble. Urocanic acid loses its water of crystallization slowly at room temperature. The anhydrous form is more rapidly prepared by heating the dihydrate for several hours at 100°. The anhydrous compound is not noticeably hygroscopic.

Histamine. A number of chemical methods have been described for the synthesis of histamine.[2] The most convenient preparative method for laboratory use, however, is the microbiological decarboxylation of L-histidine to histamine, which usually proceeds in 80 to 90% over-all yield. Although histamine is available commercially, this procedure is still frequently employed in the laboratory for the preparation of isotopically labeled histamine. For details on the procedures for the microbiological decarboxylation of histidine, reference is made to the articles of Epps,[26] Schayer,[27] and Rodwell.[28]

[22] S. Edlbacher and H. von Bidder, *Z. physiol. Chem.* **276**, 126 (1942).
[23] The preparation of urocanic acid described here is based on the procedure reported by Mehler *et al.*[24,25]
[24] A. H. Mehler, H. Tabor, and O. Hayaishi, *Biochem. Preparations* **4**, 50 (1955).
[25] A. H. Mehler and H. Tabor, *J. Biol. Chem.* **201**, 775 (1953).
[26] H. M. R. Epps, *Biochem. J.* **39**, 42 (1945).
[27] R. W. Schayer, *J. Am. Chem. Soc.* **74**, 2440 (1952).
[28] A. W. Rodwell, *J. Gen. Microbiol.* **8**, 233 (1953).

Most histamine studies involve the purification of solutions containing small amounts of histamine prior to bioassay or colorimetric determinations.[29–32] A convenient first step in this purification, introduced by McIntire et al.,[29] depends on the extraction of histamine into butanol at an alkaline pH. An alkaline salt mixture (0.4 g. of $Na_3PO_4 \cdot H_2O$, 2.3 g. of Na_2SO_4) is added to each 10 ml. of the aqueous solution containing histamine. This solution is then extracted (usually by mechanical shaking for 15 minutes) with an equal volume of n-butanol or t-butanol (90% extraction). The butanol is then separated and evaporated in vacuo to dryness. The residue, which usually contains 0.1 to 1 micromole of histamine, is dissolved in water and placed on a column containing 1 g. of Amberlite XE-64 (H form); the histamine is eluted with 0.4 N HCl.

This stage of purification is usually adequate for colorimetric procedures and for bioassays.[33] If further purification or isolation is desirable, the procedures chosen will depend on the quantities of histamine present, as well as the probable contaminants. Histamine can be conveniently precipitated as the picrate or as the di-(3,4-dichlorobenzenesulfonate) in a manner analogous to that described for histidine.[34] After removal of the anion the free base can be purified by sublimation in vacuo (<0.1 mm., temperature approximately 90°). In most physiological experiments, however, the quantity of histamine present is too small to permit isolation of the crystalline material.

[29] F. C. McIntire, L. W. Roth, and J. L. Shaw, J. Biol. Chem. 170, 537 (1947). McIntire et al. used a column of cotton-acid succinate for the adsorption of histamine; this material was also used by Rosenthal and Tabor,[30] and by Millican et al.[31] The use of Decalso and charcoal as adsorption materials for histamine has recently been reported by M. Roberts and H. M. Adam [Brit. J. Pharm. 5, 526 (1950)]. With all these chromatographic procedures it is advisable to carry out a preliminary determination of the size of the column required for the materials being assayed. Amberlite IRC-50 columns have been used by Lubschez.[32]

[30] S. M. Rosenthal and H. Tabor, J. Pharmacol. Exptl. Therap. 92, 425 (1948).

[31] R. C. Millican, S. M. Rosenthal, and H. Tabor, J. Pharmacol. Exptl. Therap. 97, 4 (1949).

[32] R. Lubschez, J. Biol. Chem. 183, 731 (1950).

[33] An alternate procedure which has been commonly employed in preparing histamine solutions for bioassay has involved prolonged acid hydrolysis and ethanol extraction. Two examples of the procedures are the methods of Barsoum and Gaddum[33a] and of Code.[33b]

[33a] G. S. Barsoum and J. H. Gaddum, J. Physiol. 85, 1 (1935).

[33b] C. G. Code, J. Physiol. 89, 257 (1937).

[34] A. Galat and H. L. Friedman, J. Am. Chem. Soc. 71, 3976 (1949).

Acetylhistamine. To synthesize acetylhistamine[35,36] 3.3 g. of histamine base is refluxed with 15 ml. of acetic anhydride for 24 hours on the steam bath; the open end of the condenser is protected by a calcium chloride tube. The reaction mixture is then treated with 200 ml. of water and evaporated to dryness on the steam bath. The residue is extracted with absolute ethanol. After evaporation of the ethanol, crystalline acetylhistamine is obtained. This can be recrystallized from ethanol-ether and further purified by sublimation *in vacuo* (0.05 mm., about 140°).

For the isolation of acetylhistamine from urine or other solutions, 23 g. of Na_2SO_4 and 7 g. of Na_2HPO_4 are added to each 100 ml. of urine. The solution is adjusted to pH 8 [37,38] and extracted three times with 0.3 vol. of *n*-butanol. Further purification can be effected by chromatography on cotton-acid succinate or Amberlite XE-64.[39]

Imidazoleacetic Acid.[40,41] Imidazoleacetic acid is synthesized by hydrolysis of cyanomethylimidazole; the latter is prepared by the action of hypochlorite solutions on histidine. Nineteen millimoles of sodium hypochlorite (volume 27 ml.)[42] is added dropwise to 10 millimoles of

[35] R. C. Millican, unpublished modification of P. van der Merwe, *Z. physiol. Chem.* **177**, 308 (1928). A satisfactory preparation of histamine base can be obtained from the more readily available histamine dihydrochloride by dissolving 5.2 g. of the latter salt in 45 ml. of hot methanol; after cooling, 119 ml. of 0.5 N ethanolic NaOH is added. CO_2 is bubbled through the solution to neutralize any excess alkali, and the solution is evaporated to dryness. It is not necessary to separate the histamine from the inorganic salt. The free base can also be conveniently prepared from the hydrochloride by passage through an anion exchange column or by a conventional treatment with silver carbonate.

[36] H. Tabor and E. Mosettig, *J. Biol. Chem.* **180**, 703 (1949).

[37] At pH 8 the distribution coefficient for acetylhistamine is 0.85 and that for histamine is 0.1, permitting a partial separation of acetylhistamine and histamine. With more alkaline solutions the distribution coefficient for histamine increases, and both compounds are extracted into the butanol phase.

[38] R. C. Millican, *Arch. Biochem. and Biophys.* **42**, 399 (1953).

[39] Various eluting agents can be used, including dilute acetic or hydrochloric acid, sodium acetate, or sodium phosphate. With cotton-acid succinate, acetylhistamine is completely separated from histamine by a preliminary wash with dilute Na_3-$(PO_4)_2$ [30,31,38] prior to elution of the histamine with 0.4 N HCl.

[40] This presentation is based on an unpublished method of H. Bauer and H. Tabor. The use of hypochlorite solutions in the preparation of cyanomethylimidazole is based on the method of H. Dakin [*Biochem. J.* **10**, 319 (1916)] which employed chloramine-T. Cyanomethylimidazole can also be prepared by reacting chloromethylimidazole with KCN, but this procedure is somewhat less convenient. See F. C. Pyman, *J. Chem. Soc.* **99**, 668 (1911); Mehler *et al.*[41]

[41] A. H. Mehler, H. Tabor, and H. Bauer, *J. Biol. Chem.* **197**, 475 (1953).

[42] Commercial hypochlorite solutions (such as Clorox) are suitable. The concentration of active chlorine is determined by a thiosulfate titration.

histidine monohydrochloride suspended in 5 ml. of water; the temperature is maintained at 15 to 25°. After storage at room temperature overnight, the dark-brown solution is alkalinized with 2 g. of Na_2CO_3 and distilled to dryness *in vacuo*. The dry residue is refluxed three times with a total of 500 ml. of ethyl acetate. The ethyl acetate is decanted and evaporated to dryness, leaving a residue of crude cyanomethylimidazole. The latter is dissolved in 20 ml. of 0.05 N KOH, and steam is passed through until all the NH_3 is liberated. The solution is then saturated at 0° with HCl gas; the insoluble NaCl is removed by filtration through a sintered-glass funnel, and thoroughly washed with cold concentrated HCl. The filtrate is evaporated to dryness, and the imidazoleacetic acid hydrochloride is recrystallized from ethanol-ether. The over-all yield is approximately 70%.

For the isolation of imidazoleacetic acid from urine,[41] incubation mixtures, etc., the solution is passed through a Dowex 1-acetate column and eluted with 0.2 M sodium acetate. Imidazoleacetic acid, free of salt, is then obtained by passing the eluate through another Dowex 1-acetate column and eluting the imidazoleacetic acid with 1 N acetic acid; the eluate is then evaporated to dryness *in vacuo*. Further purification is effected by adsorption on Dowex 50-H and elution with HCl, crystallization from acetone-water, chromatography on Dowex 1-acetate, and sublimation at 0.025 mμ and 140 to 145°.

L-*Histidinol Dihydrochloride*. L-Histidinol dihydrochloride is synthesized by the reduction of monobenzoyl-L-histidine methyl ester with lithium aluminum hydride in a tetrahydrofuran solvent, and removal of the benzoyl group by hydrolysis. For further details see ref. 43.

Miscellaneous Compounds. A number of imidazoles of biological interest have not been covered in the above presentation; some of these are listed here with references to procedures for their preparation: Anserine,[44] carnosine,[44] methylhistidine,[45] ergothioneine,[46,47] imidazole-

---

[43] H. Bauer, E. Adams, H. Tabor [*Biochem. Preparations*, submitted], and P. Karrer, H. Suter, and P. Waser [*Helv. Chim. Acta* **32**, 1936 (1949)] prepared L-histidinol dihydrochloride by the LiAlH$_4$ reduction of dibenzoylhistidine methyl ester. Small amounts of L-histidinol were isolated from a histidine-requiring *E. coli* mutant by H. J. Vogel, B. D. Davis, and E. S. Mingioli [*J. Am. Chem. Soc.* **73**, 1897 (1951)].

[44] O. K. Behrens and V. du Vigneaud, *J. Biol. Chem.* **120**, 517 (1937); V. du Vigneaud and O. K. Behrens, *Ergeb. Physiol.* **41**, 917 (1939); R. H. Sifferd and V. du Vigneaud, *J. Biol. Chem.* **108**, 753 (1935); H. T. Hanson and E. L. Smith, *J. Biol. Chem.* **179**, 789 (1949); R. A. Turner, *J. Am. Chem. Soc.* **75**, 2388 (1953); H. Kroll and H. Hoberman, *J. Am. Chem. Soc.* **75**, 2511 (1953).

[45] W. Sakami and D. W. Wilson, *J. Biol. Chem.* **154**, 215 (1944).

[46] H. Heath, A Lawson, and C. Rimington, *J. Chem. Soc.* **1951**, 2215.

[47] G. Hunter, G. D. Molnar, and N. J. Wight, *Can. J. Research* **E27**, 226 (1949).

glycerine,[48] imidazoleacetol,[48] imidazolelactic acid,[49] imidazolepropionic acid,[50] imidazolepyruvic acid,[51] pilocarpine,[52] and spinacin.[53]

Determination

Coupling with Diazotized Amines. The most commonly used methods for the determination of imidazole compounds are modifications of the procedure introduced by Pauly[54] in 1904, which depends on the coupling of imidazole compounds with diazotized sulfanilic acid at an alkaline pH. Although these procedures have been very helpful, they all have definite limitations. The tests are rather nonspecific and give positive reactions with a large number of imidazoles, phenols, and certain other compounds; the color and extinction coefficient vary markedly with different compounds. Biological materials, furthermore, frequently contain compounds (e.g., uric acid, copper) which inhibit the color development. To assay a specific compound, therefore, chromatographic or solvent purification is required. The colors obtained are rather unstable and require rapid, carefully timed measurements or the addition of a stabilizing agent, such as an organic solvent. The available procedures cannot all be included here, and thus this presentation will be restricted to the diazotization procedure[55] used in this laboratory.

1. Preparation of the Diazo Reagent. Ten milliliters of a stock 0.1% *p*-nitroaniline solution[56] (in 0.1 *N* HCl) is cooled and treated with 1 ml. of 4% sodium nitrite. This solution of diazotized *p*-nitroaniline is kept cool and made up every 3 to 4 hours.

2. General Assay. One-half milliliter of sample plus 2 ml. of 0.4 *N* HCl are mixed in a test tube[57] and cooled in ice. One-half milliliter of the

---

[48] B. N. Ames, H. K. Mitchell, and M. B. Mitchell, *J. Am. Chem. Soc.* **75**, 1015 (1953).

[49] S. Fränkel, *Monatshefte für Chemie* **24**, 229 (1903).

[50] Although several synthetic procedures have been reported[2] imidazolepropionic acid is most conveniently prepared by the reduction of urocanic acid [G. Barger and A. J. Ewins, *J. Chem. Soc.* **99**, 2336 (1911); also ref. 25].

[51] H. P. Broquist and E. E. Snell, *J. Biol. Chem.* **180**, 59 (1949).

[52] A. R. Battersby and H. T. Openshaw, *in* "The Alkaloids," p. 201 (Manske and Holmes, eds.), Academic Press, 1953.

[53] D. Ackermann and S. Skraup, *Z. physiol. Chem.* **284**, 129 (1949).

[54] H. Pauly, *Z. physiol. Chem.* **42**, 508 (1904).

[55] These procedures are based on the diazotization procedures published by S. M. Rosenthal and H. Tabor.[30] Modifications have been reported in refs. 32 and 41.

[56] The use of diazotized *p*-nitroaniline in histidine determination was first reported by E. Gebauer-Fulnegg, *Z. physiol. Chem.* **191**, 222 (1930).

[57] If tyrosine, tyramine, or other phenols are present, interference due to these compounds can be eliminated by adding 0.5 ml. of 4% NaNO₂ at this point, and heating the test tubes for 2 minutes in boiling water. The test tubes are then thoroughly cooled in ice water.

diazo reagent is added with shaking, followed by three successive 0.1-ml. additions of 20% sodium carbonate. One milliliter of t-butanol is then added, followed by 0.2 ml. of 20% sodium carbonate and 0.12 ml. of 5 N NaOH. Standard solutions are similarly treated, and the densities of the red solutions are read at 550 mμ. The sensitivity of the test with histidine is approximately 1 to 2 γ; the accuracy of a single reading is only ±10 to 20%. A large variety of imidazoles react in this test; some imidazoles, however, such as imidazolealdehyde and imidazolecarboxylic acid, do not give a significant color at the concentrations usually measured (1 to 20 γ).

3. *Histamine Assay.*[30,58,59] In this procedure the final azo compound is extracted into methyl isobutyl ketone and is thereby separated from the azo compound formed with histidine or with other compounds having acidic groups. Two and one-half milliliters of a 10% trichloroacetic acid filtrate (or a comparable hydrochloric acid or sulfuric acid solution) is placed in a test tube[57] and cooled in ice. One-half milliliter of the diazo reagent is added, followed by 0.6 ml. of a 20% sodium carbonate solution. After the solution is thoroughly mixed, an additional 0.25 ml. of the sodium carbonate is added; after 1 minute 0.3 ml. of 5 N NaOH is also added. The final pH is 10.1 to 10.5. The solution is then extracted with 1 to 2 ml. of methyl isobutyl ketone. This is separated,[60] and the absorption measured at 550 mμ. Visual comparison of the red color with that formed with standard solutions of histamine has also given satisfactory results.

4. *Spray Reagent*[61] *for Paper Chromatograms.* A convenient procedure for visualizing imidazoles is the application of a diazotized sulfanilic acid spray, followed by a 5% sodium carbonate spray. The diazotized sulfanilic acid solution is prepared by slowly adding, with stirring, 25 ml. of a cold 5% sodium nitrite solution to 5 ml. of a cold stock sulfanilic acid solution (0.9 g. of sulfanilic acid, 9 ml. of concentrated HCl, and water to 100 ml.). This reagent can be kept for at least 5 days at 0°.

[58] Acetylhistamine, imidazole ethanol, and histidinol react in this test in addition to histamine. The sensitivity of the test is 0.5 to 1 γ; the accuracy about ±10%.

[59] A more sensitive adaptation of this procedure has been reported by Lubschez.[32] The sensitivity of the method is 0.1 γ.

[60] In the presence of ammonium salts an interfering red color appears in the methyl isobutyl ketone layer; this is eliminated by shaking the ketone layer with 2 vol. of 0.05 M barbital buffer (pH 7.7) before the spectrophotometric reading.

[61] This procedure has been reported by B. Ames and H. K. Mitchell [*J. Am. Chem. Soc.* **74**, 252 (1952)]. Similar procedures have also been reported by a number of other investigators, including R. J. Block [*Arch. Biochem. and Biophys.* **31**, 266 (1951)], who used diazotized sulfanilamide, and F. Sanger and H. Tuppy [*Biochem. J.* **49**, 363 (1951)], who used diazotized p-anisidine.

Coupling with Dinitrofluorobenzene. The coupling of amino acids with dinitrofluorobenzene to form the colored dinitrophenyl derivatives has been employed by a number of investigators for the determination of amino acids (see Vol. IV [9]). This procedure has been adapted for histamine determination by McIntire et al.[62] and by Graham et al.[63] The dinitrophenyl derivative of histamine has been separated from most interfering substances by extraction into methyl-n-hexyl ketone[63] or by counter-current extraction on a cotton-benzoate column.[62] The sensitivity of the test is 0.01 γ and thus is as sensitive as many bioassay procedures. The methods are relatively recent, and therefore the occasional occurrence of interfering substances has not been completely ruled out yet.

Bioassay. Biological assays have been most commonly used for the determination of small quantities of histamine (0.01 to 1 γ); these methods usually depend on the *in vitro* contraction of the guinea pig uterus or small intestine, the contraction of the fowl cecum, or the lowering of the blood pressure of an anesthetized cat in response to histamine.[33b,64] Although these procedures have been very useful, they are somewhat cumbersome; to obtain accurate and reproducible results considerable experience with the procedures is usually necessary. Other pharmacologically active materials, such as N-methylhistamine, acetylcholine, and potassium, interfere with the results, and consequently adequate preliminary purification is necessary. Increased specificity is also obtained by using atropinized systems and by the inhibition of the histamine response by antihistaminics. For further details reference is made to the papers of Code,[33b] Barsoum and Gaddum,[33a] Ahlmark,[64] and Adam et al.[64]

Bromination. A specific colorimetric technique for histidine depends on the formation of a violet color when a histidine solution is treated with bromine water.[65] Although some reaction is given by 1-methylhistidine and by histamine, other imidazole compounds give essentially no color in this test. Considerable difficulty has been experienced, however, in obtaining reproducible results with this test and in eliminating various inhibitory effects of other materials. Consequently a number of modifications have been reported, some of which are listed in ref. 66.

[62] F. C. McIntire, F. B. White, and M. Sproull, *Arch. Biochem.* **29**, 376 (1950).

[63] H. T. Graham, O. H. Lowry, and F. B. Harris, *J. Pharmacol. Exptl. Therap.* **101**, 15 (1951).

[64] A. Ahlmark, *Acta Physiol. Scand* **9**, Suppl. 28 (1944); M. Guggenheim, "Die biogenen Amine," S. Karger, Basel, 1951; H. M. Adam, D. C. Hardwick, K. E. V. Spencer, and W. T. S. Austin, *Brit. J. Pharmacol.* **9**, 360 (1954).

[65] F. Knoop, *Beitr. Chem. Physiol. u. Pathol.* **11**, 356 (1908).

[66] E. Racker, *Biochem. J.* **34**, 89 (1940); R. Kapeller-Adler, *Z. physiol. Chem.* **264**, 131 (1933); **271**, 206 (1934); R. J. Block.[16] Other relevant papers are cited in ref. 9.

Enzymatic Methods. An active histidase preparation (*Pseudomonas*) has been described in Vol. II [29], which quantitatively converts histidine to urocanic acid. The latter is determined by its absorption at 277 mμ (see below).[67]

Ultraviolet Absorption. Although most of the imidazole compounds have only end absorption, urocanic acid[25] and certain other imidazole

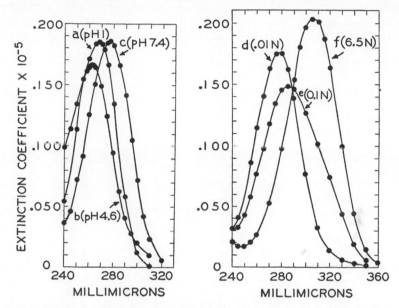

Fig. 1. Effect of pH on the ultraviolet absorption spectra of urocanic acid. The data are expressed as molar extinction coefficients. The spectra were determined with 5×10^{-5} M solutions of urocanic acid in the following solvents: 0.1 N HCl (curve a); CO$_2$-free water (curve b); 0.01 M potassium phosphate, pH 7.4 (curve c); 0.01 N NaOH (curve d); 0.1 N NaOH (curve e); and 6.5 N KOH (curve f). Reprinted from A. H. Mehler and H. Tabor [*J. Biol. Chem.* **201**, 775 (1953)] with the permission of the Editors of the *Journal of Biological Chemistry*.

compounds (such as imidazole aldehyde, imidazolecarboxylic acid,[68] and ergothioneine[47]) have well-defined absorption peaks in the ultraviolet spectrum. The molar extinction coefficient of ergothioneine is 16,000 at $\lambda_{258m\mu}$[46]; the molar extinction coefficient of urocanic acid at pH 7.4 is 18,800 at 277 mμ. The effect of pH on the absorption of urocanic acid is shown in Fig. 1.

[67] This procedure has been used by E. Adams for the assay of histidine in his studies on the enzymatic conversion of L-histidinol to histidine by a DPN-linked histidinol dehydrogenase. Adams has also used the latter enzyme for the assay of histidinol [*J. Biol. Chem.* **209**, 829 (1954)].

[68] R. A. Turner, *J. Am. Chem. Soc.* **71**, 3472 (1949).

$R_f{}^a$ VALUES OF VARIOUS IMIDAZOLE DERIVATIVES

	Diazo[b] spot color	t-Butanol, 70[c] / Formic acid, 15 / H_2O, 15	n-Propanol, 75[d] / 28% NH_4OH, 1.5 / H_2O, 23.5	n-Propanol, 75[d] / Glacial acetic, 1.5 / H_2O, 23.5	Ethanol, 77[c] / H_2O, 23
Acetylhistamine	Red	0.76	0.81	0.56	0.72
4-Amino-5-carboxamide[e] imidazole	Purple[f]	0.54	0.48	0.48	0.50
Aminomethyl-imidazole dihydrochloride	Yellow	0.32[g]	0.52	0.25	0.46[g]
Carnosine	Red	0.31	0.20	0.11	0.14
Cyanomethyl-imidazole	Red	0.69	0.79	0.69	0.75
Ergothioneine[h]	Orange[f]	0.33	0.22	0.24	0.39
Formylhistidine[i]	Red	0.56	0.29	0.18	0.32
Histamine dihydrochloride	Red	0.31	0.62	0.18[g]	[g]
Histidine monohydrochloride	Red	0.21	0.29	0.13	0.14
Histidinol dihydrochloride	Red	0.35[g]	0.62[g]	0.21[g]	0.51[g]
Hydroxyethylimidazole	Red	0.66	0.80	0.52	0.67
Hydroxymethylimidazole	Pink	0.63	0.69	0.47	0.67
Imidazole	Yellow	0.68	0.83	0.51	0.72
Imidazoleacetic acid[k]	Red	0.63	0.33	0.34	0.37
Imidazoleacetol[j]	Red	0.52	g	0.39	0.26
Imidazole formaldehyde	f	0.52	0.74	0.71	0.74
Imidazoleglycerine[j]	Yellow	0.44	0.41	0.37	0.52
Imidazolelactic acid	Red	0.49	0.32	0.20	0.29
Imidazolepropionic acid	Red	0.69	0.39	0.44	0.39
N-Ethylhistamine[k]	Red	0.49	0.82	0.30[g]	0.63
N-Methylhistamine[k]	Red	0.36	0.75	0.25[g]	0.52
Urocanic acid	Orange[f]	0.74	0.32	0.65	0.63

[a] Whatman No. 1 paper; ascending method; 25°; 19 hours.

[b] With the spray reagent of Ames and Mitchell (see p. 631); each spot contained approximately 10 γ of material. With large quantities pink spots with yellow centers are seen instead of the red spots listed.

[c] Solvent previously used by A. Meister, H. Sober, and S. Tice, J. Biol. Chem. 189, 577 (1951).

[d] The propanol solvents have been introduced by B. N. Ames and H. K. Mitchell, [J. Am. Chem. Soc. 74, 252 (1952)] for the chromatography of imidazole compounds. The results of this table are in essential agreement with the R_f values reported by Ames and Mitchell. The data in this table represent unpublished data of H. Tabor and R. C. Millican.

[e] Kindly supplied by J. Rabinowitz.

[f] Spot can be visualized by the quenching of fluorescence observed with ultraviolet irradiation (2537-A. filter).

[g] Considerable tailing.   [h] Kindly supplied by S. Spicer.

[i] Kindly supplied by H. Bond.   [j] Kindly supplied by B. Ames.

[k] Synthesized in collaboration with H. Bauer.

Microbiological Assays. *Streptococcus faecalis* R, *Leuconostoc mesenteroides* P-60, *and Lactobacillus fermenti* 36 have been commonly used as assay organisms for the microbiological assay of L-histidine. The details of the various procedures are reported in the publications listed in footnote 69. Mutant strains of *Escherichia coli* are also satisfactory for histidine assays.

Paper Chromatography. The R_f value for a number of imidazoles in four solvents are listed in the table.

[69] M. S. Dunn, M. N. Camien, S. Shankman, and L. B. Rockland, *J. Biol. Chem.* **159,** 653 (1945); M. S. Dunn, S. Shankman, and M. N. Camien, *ibid.,* **161,** 669 (1945); L. M. Henderson and E. E. Snell, *ibid.,* **172,** 15 (1948); L. M. Henderson, W. L. Brickson, and E. E. Snell, *ibid.,* **175,** 31 (1948); M. J. Horn, D. B. Jones, and A. E. Blum, *U. S. Dept. Agr. Publ.* No. 696 (1950).

[91] Glycocyamine and Creatine

By J. W. DUBNOFF

Determination of Glycocyamine (and Arginine)[1]

Principle. Glycocyamine is separated from arginine and stronger bases by passing the solution through a simple Permutit column and quantitatively determined by a modified Sakaguchi reagent.[2]

Reagents

3% sodium chloride.

0.3% sodium chloride.

Permutit according to Folin. Permutit can be regenerated after use by allowing 3% sodium chloride to percolate through and then washing with distilled water until chloride-free. This is most conveniently done in large batches on a Büchner funnel.

10% urea in water.

0.2% naphthol in absolute alcohol, diluted with 4 vol. of the 10% urea solution before use.

Hypobromite solution. 0.66 ml. of liquid bromine is added to 100 ml. of 5% sodium hydroxide. This solution should be kept cold and not longer than 1 or 2 days.

10 mg. % glycocyamine in 0.1 N hydrochloric acid. Standard solutions are made on the day on which they are to be used by diluting this stock solution with water.

[1] J. W. Dubnoff and H. Borsook, *J. Biol. Chem.* **138,** 381 (1941).
[2] C. J. Weber, *J. Biol. Chem.* **86,** 217 (1930); **88,** 353 (1930).

Adsorption Column. The Permutit is contained in the stem of a glass funnel whose dimensions are: upper part 15 mm. external diameter, 100 mm. long; stem 7 mm. external diameter, 100 mm. long. The lower end of the stem is slightly constricted. A small amount of cotton is placed above the constriction, 0.9 g. of Permutit is poured in, and the stem is tapped gently to settle the particles.

Preparation of Solutions for Analysis. For complete separation of glycocyamine and arginine the salt concentration of the solution should not be over 0.5%. If neither of these compounds is present in amounts over 2 mg. %, the salt concentration may be as high as 1%. Urine is usually diluted five to ten times with water. Blood filtrates may be prepared by deproteinizing according to Folin and Wu or by heat coagulation at pH 6 after 1:10 dilution with water. Tissue extracts are diluted to contain 1 g. of tissue (fresh weight) in 40 ml. of suspension. The pH is adjusted to 5.0, and the suspension immersed in a boiling water bath for 10 minutes, cooled, and filtered. Analyses are carried out on the filtrates.

Procedure. Five milliliters of the solution to be analyzed is passed through the Permutit column, and the small amount of glycocyamine remaining in the column is removed with 5 ml. of 0.3% sodium chloride. The combined filtrate contains all the glycocyamine. (Arginine may be eluted by passing 10 ml. of 3% sodium chloride through the column and determined on the eluate by the same colorimetric procedure.)

A 2-ml. aliquot is taken for analysis. It is first cooled in an ice bath, and then 0.5 ml. of the ice-cold naphthol-urea solution is added. After 2 minutes 0.2 ml. of ice-cold sodium hypobromite solution is added by means of a micropipet which should be emptied with sufficient force to mix rapidly and completely with the solution. This pipet is placed above the solution level and kept away from the sides of the test tube to prevent contamination by urea in the solution. The color is simultaneously developed in a series of standard solutions containing 0, 0.25, 0.5, 1.0, and 2.0 mg. % of glycocyamine. After 20 minutes the color development is complete, and it remains stable for 2 hours if the solutions are kept in an ice bath. The tubes are shaken for a few seconds to remove excess gas, warmed by immersion in water at room temperature, and the intensity of the color measured in a spectrophotometer or a colorimeter, with light of approximately 0.525 μ (yellow-green).

Comment. The addition of urea before the hypobromite obviates the careful timing required if the usual procedure is followed and prevents interference by normal amounts of ammonia, histidine, tyrosine, and tryptophan. Arginine and methyl guanidine, both of which react with the Sakaguchi reagent, do not appear in the glycocyamine fraction.

If the salt concentration of the unknown solution is kept below the limits prescribed, no difficulty is encountered with Permutit according to Folin. A more elaborate column with Amberlite IR-4 as an exchanger has been described by Sims.[3]

Preparation of Glycocyamine.[4] To 0.75 g. of glycine in 5 ml. of water is added 0.84 g. of cyanamide and 3 drops of concentrated ammonia. The mixture is kept at room temperature for 3 days. The crystalline precipitate is filtered off and recrystallized from a large volume of water. Yield, 0.75 g.

Determination of Creatine

Principle. This method[5] depends on the color reaction developed by creatine in the presence of diacetyl and α-naphthol described by Barrett.[6]

Preparation of Tissues for Analysis. Samples are homogenized or ground with sand and 5 ml. of 10% trichloroacetic acid (TCA), and the residue is re-extracted with 10 ml. of 5% TCA. The use of metaphosphoric acid has been recommended for this extraction[7] to prevent a slow ultra violet light-catalyzed reaction between TCA and the reagents.[5]

Reagents

   Stock alkali solution containing 30 g. of NaOH and 80 g. of Na_2CO_3 in 500 ml.
   Diacetyl. A 1% solution diluted 1:20 before use. 1% α-Naphthol made up just before use in the stock alkali.

Procedure. To a neutral solution containing not more than 60 γ of creatine is added 2 ml. of 1% α-naphthol in alkali followed by 1 ml. of diluted diacetyl. The solution is shaken, and the color is measured after 30 minutes at 0.525 μ.

Comment. Sulfhydryl compounds interfere with application of this method, particularly in liver. *p*-Chloromercuribenzoate (CMBA) and other thiol inhibitors have been used to overcome this difficulty.[7,8] Since these reagents interfere if added in excess, it is necessary to determine the thiol content by I_2 titration and to add equivalent amounts of the inhibitor. Amino acids and other unknown compounds may also reduce color.[5,7] Arginine, guanidine and glycocyamine produce about one-ninth the color of creatine.

Creatine and creatinine are in equilibrium. The slow interconversion

[3] E. A. H. Sims, *J. Biol. Chem.* **158,** 239 (1945).

[4] A. Strecker, *J. Prakt. Chem.* 530 (1861).

[5] P. Eggleton, S. R. Elsden, and Nancy Gough, *Biochem. J.* **37,** 526 (1943).

[6] M. M. Barrett, *J. Pathol. Bacteriol.* **42,** 441 (1936).

[7] I. Abelin and J. Raaflauk, *Biochem. Z.* **323,** 382 (1952).

[8] A. H. Enner and L. A. Stocker, *Biochem. J.* **42,** 557 (1948).

shows a complex pH dependence. For this reason creatine analyses by this method should be performed without delay.

Determination of Total Creatine[9]

Principle. Creatine is converted to creatinine, adsorbed on Lloyd's reagent, washed free of interfering substances, and reacted with the Jaffe reagent.

Reagents

Lloyd's reagent.
0.01 N HCl.
10% NaOH.
0.5% picric acid solution. Ten parts of this solution is diluted with 1 part of 10% NaOH.

Procedure. Tissues are extracted at pH 5.0 for 10 minutes in a boiling water bath. The pH is adjusted to 2.0 $\pm$ 0.2, and a 5.0-ml. aliquot is added to a 16 × 125-mm. ignition tube. The samples are autoclaved for 20 minutes at 125°. After they are cooled, approximately 50 mg. of Lloyd's reagent is added, and the tubes are shaken for 5 minutes. The tubes are centrifuged, and the Lloyd's reagent washed by resuspending one or more times with 0.01 N HCl. Three milliliters of the sodium picrate solution is added. The tubes are shaken and centrifuged. The color is measured at 0.525 μ and compared with standards ranging from 0 to 2 mg. %, which have been carried through the same procedure.

Comment. The Jaffe reagent for creatine or total creatine reacts with a great many reducing substances. However, these may be separated from creatine by the adsorption procedure. Of the bases carried along arginine does not interfere but glycocyamine produces about one-ninth the color of creatine. This method may be used for creatinine if the autoclaving step is left out. The diacetyl method is recommended for its simplicity. However, the modified Jaffe reaction is more specific and less likely to be inhibited by the usual additions in enzyme studies. For more positive identification, the bacterial digestion procedure[10] or the semiquantitative paper chromatography method[11] may be used.

Synthesis of Creatine[12]

Sarcosine (0.445 g.) is dissolved in 2.0 ml. of water containing 2 drops of concentrated ammonia. Cyanamide (0.4 g.) in 3 ml. of water is added.

[9] H. Borsook, *J. Biol. Chem.* **110**, 481 (1935).
[10] R. Dubes and B. F. Miller, *J. Biol. Chem.* **121**, 429 (1937).
[11] S. R. Ames and H. A. Risley, *Proc. Soc. Exptl. Biol. Med.* **69**, 267 (1948).
[12] K. Bloch and R. Schoenheimer, *J. Biol. Chem.* **131**, 111 (1939).

The solution is kept for 48 hours at room temperature. The creatine hydrate is filtered off and recrystallized from a small amount of water. The hydrate may be converted to creatine by heating at 105°.

Isolation of Creatine from Rat.[12] Extract the ground carcass continuously for 8 hours with 95% ethanol. Evaporate the extract to a small volume *in vacuo*. Make the residue approximately 1 N with HCl, and extract with ether to remove fats. Boil for 4 hours to convert the creatine to creatinine. Bring to dryness *in vacuo* and redissolve in 300 ml. of water. Clear by adding a solution of 1 g. of lead acetate followed by NaOH until it is just alkaline to phenolphthalein. Remove the lead with H_2S. Make the solution just acid to Congo red with HCl, and treat with 0.5 g. of picric acid and 0.5 g. of potassium picrate. The crystals formed after 2 days in the refrigerator are recrystallized from 300 ml. of a solution containing 0.12% picric acid and 0.12% potassium picrate. This purification may be repeated many times without appreciable loss of material.

[92] Isolation and Determination of Arginine and Citrulline

By EVELYN L. OGINSKY

I. Arginine

Determination

Colorimetric Method. *Principle.* The procedures described have been modified from those of Archibald,[1,2] primarily to decrease reaction volumes. Arginine in the sample is decomposed by arginase to ornithine and urea, and the latter is assayed colorimetrically.

Reagents

Arginase[3] solution, 5 to 10 units/ml.
Acid mixture (1 vol. concentrated H_2SO_4:3 vol. sirupy H_3PO_4:1 vol. H_2O).
Urea standard, 50 γ/ml., in H_2O.
α-Isonitrosopropiophenone, 4 g. in 100 ml. of 95% ethanol.

[1] R. M. Archibald, *J. Biol. Chem.* **157**, 507 (1945).
[2] R. M. Archibald, *J. Biol. Chem.* **165**, 293 (1946).
[3] See Vol. II [49].

Procedure. Step 1. Standard Curve. The standard urea curve is established by the addition of 2.5 ml. of the acid mixture and 0.4 ml. of α-isonitrosopropiophenone solution to a series of eight large test tubes containing from 0 to 150 γ of urea in 3.5 ml. of H_2O.

Step 2. Treatment of Assay Sample. The solution to be assayed may contain between 50 and 1000 γ of L-arginine per milliliter and should be adjusted to pH 9.5. Add 0.5 ml. or less of this assay solution to 1 ml. of arginase, and bring the final volume to 1.5 ml. with H_2O. Incubate for 10 minutes at room temperature; then add 1.5 ml. of the acid mixture. Remove a 2-ml. sample, and add for urea determination 2.5 ml. of H_2O, 1.5 ml. of acid mixture, and 0.4 ml. of α-isonitrosopropiophenone.

If the original sample contains large amounts of protein, add 1.5 ml. of 10% metaphosphoric acid instead of the acid mixture at the end of the incubation with arginase. Centrifuge, after allowing the tubes to stand for 15 minutes. Dilute the supernatant 1:10 with H_2O, remove a 3.5-ml. sample, and add 2.5 ml. of acid mixture and 0.4 ml. of α-isonitrosopropiophenone.

Step 3. Urea Determination. Mix the assay tubes prepared in steps 1 and 2 thoroughly, stopper, and boil them in the dark for 60 minutes. After they have cooled at room temperature for 15 minutes, read the color developed at 540 mμ. Multiply the amount of urea found in the 2-ml. assay sample by a factor of 4.35 to obtain the amount of arginine in the original sample volume. If metaphosphoric acid is used in step 2, an appropriate correction for dilution of the sample should be applied. Interfering color reactions with such compounds as alloxan and allantoin are discussed by Archibald.[2] Citrulline also forms a colored reaction product, but only at relatively high levels and can be corrected for by zero time controls in step 1.

Arginine can also be estimated by modification of the Sakaguchi reaction.[4]

Microbiological Methods. L-Arginine can also be assayed by measurement of the growth response of bacteria for which the amino acid is an essential nutrient. Assay of 0 to 100 γ with *Streptococcus faecalis* by use of a complicated synthetic medium has been reported by Stokes *et al.*:[5] growth response of the organism can be measured as increase in titratable acidity or in turbidity. An *Escherichia coli* mutant requiring arginine[6] has been employed in the author's laboratory, by measurement of growth as turbidity in a simple synthetic medium[7] supplemented with

[4] H. T. Macpherson, *Biochem. J.* **36,** 59 (1942).
[5] J. L. Stokes, M. Gunness, I. M. Dwyer, and M. C. Caswell, *J. Biol. Chem.* **160,** 35 (1945).
[6] B. D. Davis, *Experientia* **6,** 41 (1950).
[7] B. D. Davis and E. S. Mingioli, *J. Bacteriol.* **60,** 17 (1950).

0 to 25 γ of arginine per milliliter. Neither organism responds to ornithine or citrulline.

Preparation

Principle. Arginine from protein hydrolyzates, or from enzymatic synthesis, is isolated as the flavianate, which is then hydrolyzed to the hydrochloride. The procedure described is that reported by Cox[8] for the large-scale isolation of L-arginine from gelatin.

Procedure. Step 1. Isolation of the Flavianate. Hydrolyze 1 kg. of gelatin by boiling with 2.3 l. of concentrated HCl for 24 hours. Concentrate the hydrolyzate to a thick sirup under reduced pressure. Add 3 l. of H_2O, filter, and make the filtrate neutral to Congo red with concentrated NaOH solution. Precipitate the arginine by adding, with stirring, 300 g. of flavianic acid (1-naphthol-2,4-dinitro-7-sulfonic acid) in 1500 ml. of H_2O; continue stirring for 60 minutes. Store in the refrigerator for several days to complete the precipitation. Filter, wash the arginine flavianate precipitate with ice-cold H_2O until washings are free of Cl^-, and dry.

Step 2. Conversion of Flavianate to Hydrochloride. For every 100 g. of arginine flavianate, add 200 ml. of concentrated HCl, and steam for 1 to 2 hours with stirring. Cool the mixture in ice. Filter off the free flavianic acid precipitate on a porous glass filter, and wash with cold concentrated HCl until the filtrate has only pale yellow color. Concentrate the combined filtrates *in vacuo* to a thick sirup, using caprylic alcohol if necessary to prevent foaming. Dissolve the sirup in 400 ml. of 95% ethanol with gentle warming, and then place it in the refrigerator overnight. Filter off the precipitate, and add to the filtrate with stirring 50 g. of aniline to remove excess HCl. Scratch the sides of flask to induce precipitation, store in the refrigerator overnight, and then filter off the arginine hydrochloride precipitate. Wash thoroughly with cold 95% ethanol.

Step 3. Recrystallization of Arginine Hydrochloride. Dissolve the above precipitate in 400 ml. of H_2O, decolor with 10 to 15 g. of Norit, and filter. Concentrate the filtrate *in vacuo* to sirup, and after cooling to room temperature add 95% ethanol with stirring until permanent turbidity appears. Add an additional 400 ml. of ethanol, and place in the refrigerator overnight. Filter, and wash the precipitate thoroughly with cold ethanol. Dry *in vacuo*.

The melting point of L-arginine hydrochloride isolated by this procedure was reported as 222° (corrected). L-Arginine hydrochloride isolated by a less effective alternative procedure through the benzylidene derivative[9] was reported to have an $[\alpha]_D^{25}$ of $+12.2°$ ($c = 5$ in H_2O).

[8] G. J. Cox, *J. Biol. Chem.* **78,** 475 (1928).
[9] E. Brand and M. Sandberg, *Org. Syntheses* **12,** 4 (1932).

II. Citrulline

Determination

Principle. The determination of citrulline by the method of Archibald[10] is based on the formation of a colored reaction product with diacetyl monoxime in acid solution.

Procedure. To the assay sample contained in 4 ml. of aqueous solution, add 2 ml. of acid mixture (1 vol. concentrated H_2SO_4:3 vol. sirupy H_3PO_4) and 0.25 ml. of a 3% aqueous solution of diacetyl monoxime. Shake the tubes thoroughly, boil them in the dark for 15 minutes, then cool them in the dark for 10 minutes at room temperature with occasional vigorous shaking. Read the color developed at 490 mμ. The assay range is from 10 to 100 γ of citrulline, and both D and L isomers will react. Arginine does not give a color reaction. Interference of urea can be eliminated by pretreatment of the sample with urease; interference of allantoin, by resin adsorption and elution.[10]

Preparation

Principle. The L-isomer of citrulline is formed from L-arginine by arginine desimidase,[11] and is isolated via the copper salt by the method of Fox.[12]

Reagents

Arginine desimidase extract (free of citrullinase activity).[11]
L-Arginine HCl (dissolve 3 g. in 50 ml. of H_2O, and adjust to pH 68. with KOH).

Procedure. In 10 ml. of H_2O, dissolve an amount of arginine desimidase extract calculated, from prior assay, to decompose the 3 g. of arginine hydrochloride in 48 hours. Add the arginine solution, and incubate the mixture at 37°. The pH should be adjusted to 6.8 with acid as necessary. After 48 hours' incubation, heat the mixture to 100° for 10 minutes, and centrifuge. Adjust the clear supernatant to about pH 9, and aerate until the odor of ammonia disappears. Neutralize to pH 7, add 2 g. of CuO, boil for 30 minutes, and filter while hot. Suspend the CuO residue in 50 ml. of H_2O, boil for 5 minutes, and refilter. Wash the CuO with 50 ml. of boiling H_2O. Concentrate the combined filtrates in order to precipitate the copper salt of citrulline, which is separated by filtration and then dissolved in 200 ml. of H_2O. Gas the solution with H_2S until no more black

---

[10] R. M. Archibald, *J. Biol. Chem.* **156**, 121 (1944).
[11] See Vol. II [50].
[12] S. W. Fox, *J. Biol. Chem.* **123**, 687 (1938).

precipitate is formed (5 to 10 minutes), filter, and concentrate the filtrate *in vacuo* to 10 to 15 ml. Add 30 vol. of absolute alcohol, cool the mixture in ice, and place in the cold room for a few days. Filter the citrulline, wash with ice-cold absolute alcohol, and dry *in vacuo* over P_2O_5. L-Citrulline prepared by this procedure had an $[\alpha]_D^{24}$ of $+18.0°$ ($c = 5.15$ in $1 M$ HCl), and a melting point of 200 to 202° (uncorrected);[13] both values agree with those in the literature obtained after other procedures.

[13] E. L. Oginsky and R. F. Gehrig, *J. Biol. Chem.* **198**, 799 (1952).

[93] Preparation and Determination of Argininosuccinic Acid

By S. RATNER

$$
\begin{array}{ll}
\text{NH} & \text{COOH} \\
\| & | \\
\text{HN—C—NH—CH} \\
| & | \\
(\text{CH}_2)_3 & \text{CH}_2 \\
| & | \\
\text{HC—NH}_2 & \text{COOH} \\
| \\
\text{COOH}
\end{array}
$$

(Molecular weight 290.3)

Enzymatic Synthesis

Principle. The compound[1] can be prepared enzymatically with either one of the two enzyme systems which participate in the synthesis of arginine from citrulline (Vol. II [48]). It is more convenient to employ the second of the two, the splitting enzyme,[2] which catalyzes the following reversible reaction: argininosuccinic acid $\rightleftarrows$ arginine + fumaric acid. With high concentrations of arginine and fumaric acid, about 75% of the substrates is converted to argininosuccinic acid when equilibrium is reached.

Preparation of Enzyme. For high purity and yield of argininosuccinic acid, it is desirable that the splitting enzyme be free of arginase and fumarase. Pig kidney is an excellent enzyme source for this purpose. It is free of arginase at the start and is rendered free of fumarase by a heat step early in the purification procedure, thus affording an enzyme preparation which has the activity and concentration necessary to reach equilibrium in a few hours.

[1] S. Ratner, B. Petrack, and O. Rochovansky, *J. Biol. Chem.* **204**, 95 (1953).
[2] S. Ratner, W. P. Anslow, Jr., and B. Petrack, *J. Biol. Chem.* **204**, 115 (1953).

The methods of assaying enzyme activity and preparing the enzyme are identical with the procedures given for preparing the splitting enzyme of beef liver[2] (Vol. II [48]). Essentially, the tissue is dried with acetone and extracted. The extract is treated with ammonium sulfate to obtain a 0 to 30 fraction which is then refractioned; the 20 to 30 fraction is subjected to heat treatment at pH 5.1. The heat step should be carried out just prior to use. Enzyme activity closely parallels the data given for liver and frequently exceeds it. The procedure starts with 200 g. of acetone-dried tissue. Approximately two batches of this size are required to prepare about 25 g. of the barium salt. The procedure given below works equally well when scaled down to one-tenth size.

Incubation. The enzyme solution obtained after the heat step contains about 3 to 4 mg. of protein per milliliter and should have a specific activity (micromoles split per hour per milligram of protein) of 45 to 55. The pH is brought to 7.3 with 0.3 vol. of 0.1 N potassium phosphate buffer, pH 7.5. To 420 ml. of the solution, containing 65,000 to 80,000 units and about 1.4 g. of protein, are added 210 ml. each of 0.5 M potassium fumarate and L-arginine monohydrochloride which have been previously adjusted to pH 7.5. The mixture is incubated for 3 hours at 38°. The heat step reproduces well, but the precaution is taken to assay a small amount of the neutralized enzyme solution while the large-scale incubation is in progress. If the units are below the desired range, the incubation time can be somewhat prolonged. For the concentrations and scale given, equilibrium is approached in about 2 hours with 80,000 units. Not only are the yields lower if equilibrium is not reached, but contamination by fumarate is likely to be increased. Unduly prolonged incubation leads to losses through ring closure.

Isolation Procedure

Deproteinization. The mixture is chilled rapidly to 0°, treated with 50 ml. of cold 100% trichloroacetic acid, allowed to stand for 5 minutes, and centrifuged at 0° for 10 minutes at 2000 r.p.m. Rapid manipulation is facilitated by carrying out the incubation and deproteinization in 250-ml. centrifuge bottles. Trichloroacetic acid is rapidly removed from the supernatant fluid by four successive extractions in a separatory funnel with equal volumes of ether.

Removal of Phosphate and First Alcohol Precipitation. The aqueous layer is chilled to 0°, and 200 ml. of 1.0 M $BaCl_2$ is added. The pH of the solution is adjusted to 9 with saturated $Ba(OH)_2$. About 200 ml. is required. A heavy precipitate of barium phosphate appears. The procedure, starting with the heat step, should be carried to this point in a single day. The mixture may then be left to stand at 0° over one night with a slight loss. Each of the alcohol steps should be completed the day they are started.

The precipitate is removed by suction filtration on a pad of suitable filter aid such as Celite (Johns-Manville), with the filter flask kept cold. To the filtrate, now 1340 ml., is added 3 vol. of cold 95% ethyl alcohol with hand stirring. A bulky white precipitate comes down which is collected by centrifugation at 0° after standing an hour in the cold. The precipitate is dried by successive treatments with 75, 95, and 100% alcohol and finally with ether. Each drying step is carried out by suspension in 1 l., followed by centrifugation. About 20 minutes at 2000 r.p.m. in 250-ml., flat-bottomed, glass bottles is required for good packing. In order to obtain the thorough drying which is essential, the packed layer in each bottle should be no more than 2 to 3 cm. high, and about 150 ml. of the alcohol or ether per bottle is necessary at each step. After the ether has been drained off, the precipitate is spread along the walls by gentle tapping, the bottles are set on the side, and the residual ether is allowed to evaporate off completely at room temperature for about 2 hours, leaving 30 to 33 g. of a white, finely divided, light powder which assays at 1.6 to 1.7 μM. of argininosuccinic acid per milligram of barium salt.

Second Alcohol Precipitation. The powder is dissolved in 500 ml. of water, and the precipitation with alcohol repeated without pH adjustment or addition of $BaCl_2$. The solution should be faintly alkaline to phenolphthalein. A relatively small first fraction is removed 5 minutes after the addition of 100 ml. of alcohol. One liter of alcohol is added to the supernatant, and the mixture is allowed to stand for about 30 minutes. After the main fraction has been collected by centrifugation, it is dried with alcohol and ether as before and then dried *in vacuo*, over P_2O_5, at room temperature overnight; 20 to 25 g. is obtained, containing 1.9 to 2.1 μM./mg. The final yield is about 70% of theory calculated from the value expected at enzymatic equilibrium. The product keeps well when stored dry at 0°.

As substrate for assay of the splitting enzyme, the preparation must be substantially free of fumarate. Although the contamination may seem low on a weight basis, it is important that the molar ratio to argininosuccinic acid be low, since fumarate inhibits the splitting enzyme in the direction of cleavage by 30% at equimolar concentrations.

The free amino acid is prepared in amorphous form by rapidly removing barium with an exact equivalent of H_2SO_4 in the cold followed by lyophilization. It is extremely hygroscopic.

When a concentrated aqueous solution of the free amino acid is allowed to stand at 25° for 1 or 2 days, extensive ring closure occurs, and after addition of alcohol to .50% the anhydride crystallizes out slowly on further standing.

Assay-Procedure

Principle. The method of estimation,[2] which is highly specific, depends on the observation that cleavage to arginine, and then to urea, proceeds to completion in dilute solution, in the presence of excess splitting enzyme and excess arginase.

Reagents

0.5 M potassium sulfate.

1.0 M potassium phosphate buffer, pH 7.5.

Splitting enzyme from beef or kidney. The 20 to 30 fraction, obtained after the second ammonium sulfate treatment, is suitable.

Arginase (Vol. II [48]).

Preparation of Sample. A weighed sample of the barium salt (about 40 mg.), dissolved in 9.5 ml. of water, is treated with 0.5 ml. of 0.5 M potassium sulfate, and the mixture centrifuged. The quantity is chosen so that the clear supernatant will contain about 6 to 7 μM./ml.

Procedure. An aliquot containing 1 to 4 μM. is added to the incubation mixture consisting of 0.1 ml. of buffer, 100 units of splitting enzyme in 0.3 ml., 13 Van Slyke-Archibald units of arginase in 0.3 ml., and water to make a final volume of 2.0 ml. Splitting enzyme is omitted from one tube to check that the material is free of arginine. The tubes are incubated at 38° for 1 hour, 0.5 ml. of 25% trichloroacetic acid is added, and urea is then estimated colorimetrically in 0.5 ml. of the filtrate as in the assay of the splitting enzyme. The results are expressed as micromoles per milligram of the barium salt.

Purity. The purity cannot be calculated accurately without a barium analysis, since the barium content varies. After the first precipitation, the barium salt contains about 2.4 equivalents of barium per mole of argininosuccinic acid, and about 1.7 equivalents after the second precipitation. For these values, 2.21 and 2.47 μM. of argininosuccinic acid per milligram of barium salt, respectively, corresponds to 100% purity. The final product, about 85% pure, invariably contains a few per cent of barium fumarate and water or alcohol of hydration. The product remains unchanged for at least a year when stored under anhydrous conditions at 0°. The lyophilized, free amino acid is about 95% pure and contains about 4% fumaric acid.

Properties

Physical and Chemical Behavior.[1] Both the barium salt and the free amino acid are highly water-soluble. The latter has an isoionic point of 3.2. The pK values are 1.62, 2.70, and 4.26 for the three carboxyl groups, 9.58 for the α-amino group, and >12 for the guanidino group. Hot alkali causes cleavage to ornithine and aspartic acid, and hot acid causes ring closure. No color is formed with the colorimetric methods employed for estimating arginine, creatine, citrulline, or urea. The α-amino group is ninhydrin responsive.

Reversible Ring Closure.[1] The compound spontaneously undergoes reversible ring closure to form a cyclic anhydride which is inactive as substrate for the splitting enzyme. The rate is temperature- and pH-dependent; at pH 7.2 the loss is 3 to 4% in 24 hours at 0° and 40% in 30 minutes at 100°; at pH 3.2 the loss is 7% in 1 hour at 25° and 100% in 30 minutes at 100°.

The anhydride, which can be prepared in crystalline form,[1] has an isoionic point close to pH 6. In 0.1 N NaOH, it is converted to the active form at the rate of 28, 38, and 44% in 1, 2, and 3 hours. It is extensively decomposed in hot alkali but is stable to hot acid.

Chromatographic Behavior. Argininosuccinic acid and the anhydride each give well-defined ninhydrin reactive spots on paper chromatograms with R_f values[1] of 0.27 and 0.49, respectively, in 80% phenol-water; these are unaffected by NH$_3$. In the presence of an acetic acid atmosphere, the respective R_f values are 0.65 and 0.77.[3]

[3] J. B. Walker and J. Meyers, *J. Biol. Chem.* **204,** 115 (1953).

[94] Determination and Isolation of Carbamylglutamic Acid and Related Compounds

By PHILIP P. COHEN

Determination of Carbamylglutamic Acid and Related Compounds

A colorimetric method for estimation of carbamylglutamic acid (CLG) and related compounds has recently been published by Koritz and Cohen.[1]

[1] S. B. Koritz and P. P. Cohen, *J. Biol. Chem.* **209,** 145 (1954).

Reagents

    50% (v/v) H_2SO_4.

    1% aqueous sodium diphenylamine-*p*-sulfonate. Store in the dark.

    3% aqueous diacetylmonoxime.

    1% potassium persulfate. Store in the refrigerator.

Procedure. To 3 ml. of sample are added 6 ml. of 50% H_2SO_4, 0.1 ml. of the amine, and 0.25 ml. of diacetylmonoxime. The tubes are mixed, capped (one-hole alkali-treated rubber stoppers in which a short section of capillary tubing is placed are used), and placed in a boiling water bath

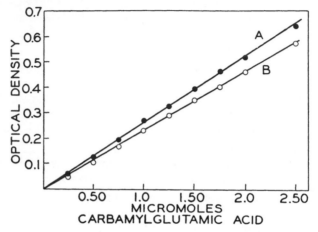

Fig. 1. Optical density values. Curve *A*, read immediately after color development; curve *B*, reaction after exposure to artificial light at room temperature for 80 minutes.

for 10 minutes. The tubes are cooled in a bath of cold tap water, and to the cold tubes 0.25 ml. of the persulfate solution is added. The contents of the tubes are rapidly mixed, capped, and placed in a boiling water bath for 1 minute. The tubes are then cooled immediately and thereafter protected from direct sunlight. They are read at 550 mμ after 10 minutes. It is convenient to use a metal rack to contain the tubes so that the various heatings and coolings may be carried out simultaneously for all the tubes.

    The absorption curve of the color formed has a broad peak with a maximum at 560 mμ, although all results reported were obtained at 550 mμ. A typical standard curve is shown in Fig. 1. The reproducibility of values for quantities above 0.5 μM. is in the range of $\pm 3\%$ of the mean, and for quantities below 0.5 μM. in the range of $\pm 6\%$. For best results known standards should be prepared at the same time, since on standing the color gradually fades. However, as may be seen from the 80-minute curve of Fig. 1, the fading is proportionally the same for all values so that the linearity of the curve is preserved. The fading of the color is slowest in the dark and in artificial light and somewhat faster in daylight out of the direct sun. Direct sunlight causes rapid and nonproportional changes in the color. In practice the tubes are protected from daylight after the addition of persulfate.

TABLE I

CHROMOGENIC EQUIVALENTS OF VARIOUS SUBSTANCES

Compound	Chromogenic equivalent[a]	Compound	Chromogenic equivalent[a]
Carbamylglutamic acid	0.337	Hydantoin	0.810
Thiocarbamylglutamic acid	0.001	Hydantoin-5-acetic acid	0.024
Phenylcarbamylglutamic acid	0.001	Hydantoin-5-propionic acid	0.029
Carbamyl-α-methylglutamic acid	0.008	Uric acid	0.016
Glutamine	0.002	Uracil	0.002
Carbamylaspartic acid	0.440	Adenine	0.001
Carbamylglycine	3.22	ATP	0.010
Carbamylalanine	0.582	Urea	2.60
Carbamylleucine	0.303	Phenylurea	5.82
Carbamylarginine	0.460	Methylurea	3.98
Arginine	0.210	Glycocyamine	0.020
Citrulline	3.90	Semicarbazide	0.036
Tryptophan	0.068	α-Ketoglutaric acid semicarbazone	0.030

[a] Chromogenic equivalent is defined as the optical density of 1.0 μM. of the substance run under conditions described for the assay and at an arbitrary cell width. Linearity between the amount of substance tested and optical density has been assumed up to or down to 1 μM. of the substance. Tryptophan gives a brownish color rather than the usual color with carbamyl compounds.

Specificity. As may be expected from the nature of the method any substance with a ureide grouping will generally give some color. This may be seen from Table I. Of various substances tested, the following gave no color at the 20-μM. level: histidine, tyrosine, glutamic acid, methionine, ornithine, proline, asparagine, N-formylglutamic acid, N-formylaspartic acid, N-formylglycine, N-formylleucine, N-butyrylglutamic acid, N-valerylglutamic acid, carbamylproline, alanine, cystine, cysteine, N-acetylglutamic acid, N-propionylglutamic acid, glyclyglycine, glycylglutamic acid, thiourea, and urethan. Of interest is the property of some substances which by themselves give no color but in the presence of CLG intensify the color produced. Histidine and ornithine act in this way, 5 μM. of either amino acid increasing the color value of 1 μM. of CLG by 23%.

Sulfur compounds, on the other hand, result in an inhibition of color formation. Cysteine, cystine, and methionine will inhibit the color to varying degrees, depending on their concentration. This inhibition for a range of concentrations of cysteine up to 5 μM. can be overcome by increasing the amount of persulfate to 0.75 ml. The linearity of the reaction for the different amounts of CLG is not destroyed under these

TABLE II
RECOVERY OF CARBAMYLGLUTAMIC ACID FROM RAT LIVER HOMOGENATES

Homogenate	Carbamyl-glutamic acid added, μM.	D	D_c	Recovery, %	Amberlite IR-100 effluent			
					Carbamyl-glutamic acid added, μM.	D	D_c	Recovery, %
+	0.3	0.328	0.107	119	0.5	0.245	0.215	131
+	0.6	0.437	0.216	120	1.0	0.462	0.432	132
+	—	0.221			—	0.030		
—	0.3	0.090			0.5	0.164		

To 6 ml. of a mixture containing 5 ml. of CLG of the desired concentration and 1 ml. of 30% (wet weight) rat liver homogenate (in isotonic KCl), 4 ml. of 5% HClO₄ was added, the protein centrifuged down, and the clear supernatant sampled. Tungstic acid and trichloroacetic acid may also be used as deproteinizing agents. D represents density of deproteinized samples. D_c represents density of deproteinized samples corrected for tissue blank.

conditions. Essentially the same results are obtained with cystine and methionine.

Uric acid has the property of an increasing chromogenic equivalent on standing, owing in all probability to the degradation of uric acid to more chromogenic urea derivatives. Adenosine triphosphate at the usual concentrations encountered does not interfere with the determination.

The recovery of CLG added to rat liver homogenates is high. The reason is unknown but may be due to the type of synergism previously noted with histidine and ornithine or to the presence of partial breakdown products of proteins. Some support is lent to the latter suggestion by the fact that tissue blanks increase with incubation time. However, recoveries of all concentrations of CLG used are fairly constant for a given run so that the concentration curve remains linear. Depending on the origin of the protein, the blank value may be quite high. Such a high

blank will make the determination of a small amount of CLG very difficult. In such cases the acidic protein filtrate may be passed through an
Amberlite IR-100 column prepared according to Archibald[2] or treated
according to the procedure outlined in the next section. The Amberlite
IR-100 column permits the CLG to pass through while retaining a large
percentage of the non-CLG chromogenic material (evidently basic in
nature) present in the filtrate (see Table II). For results relative to a
control the usual standard curve may be used, but if absolute amounts
of CLG are required the values must be read from a standard curve
prepared under the special conditions of the experiment in question.

Separation and Isolation of Carbamylamino Acids and Related Compounds by Column Chromatography[3]

A strongly basic anion exchange resin, Dowex 2-X10, 200 to 400 mesh,
is employed routinely. This resin, which is available commercially in the
form of its chloride, is initially converted into the formate form. It is
then packed in columns in the usual manner and washed with 88%
formic acid, distilled water, formic acid-ammonium formate buffer, and
distilled water.

The solution to be fractionated is adjusted to pH 8 to 9 and poured
on the column. The mixture is eluted by increasing concentrations of
ammonium formate-formic acid buffer. The apparatus used to obtain a
smooth increase in the concentration of the eluant consists of an eluant
reservoir and an eluant mixing vessel. The reservoir is connected to the
mixing vessel in such a way that a constant air pressure applied to the
reservoir forces liquid from the reservoir into the mixing vessel. The latter
is stirred magnetically and is in turn connected to the column, so that
liquid entering the mixing chamber from the reservoir forces an equal
amount of liquid out of the mixing vessel into the column. Conventional
filter flasks serve well as reservoir and mixing vessels. The rate of flow
through the column is controlled by the air pressure applied to the
reservoir. The concentration gradient is fixed by starting the experiment
with a suitable concentration of ammonium formate-formic acid buffer,
e.g., 0.600 M, in the reservoir, and 500 ml. of distilled water in the mixing
vessel. The concentration of the eluant at any given stage of the elution
may be calculated from the equation $X = x\left(1 - e - \dfrac{L}{V}\right)$, where X is
the concentration of the eluant entering the column when the concentration of the solution in the reservoir is x, the volume of liquid in the mixing

[2] R. M. Archibald, *J. Biol. Chem.* **156**, 121 (1944).
[3] J. M. Lowenstein and P. P. Cohen, unpublished studies.

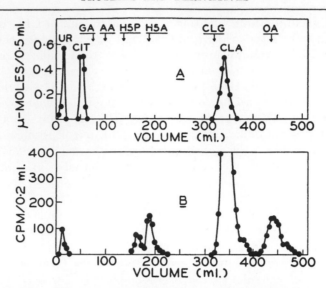

FIG. 2. Elution patterns obtained with a Dowex 2-formate column, 1.1 × 10.5 cm. Starting conditions: reservoir ammonium formate-formic acid buffer, pH 4.0, 0.600 M with respect to formate-formic acid, mixing vessel 500 ml. water. Rate of elution 0.3 ml./min., approximately.

A, artificial mixture. (The arrows indicate the positions at which certain other compounds, not included in the run, are eluted.) The following abbreviations are used: UR, urea; CIT, citrulline; GA, glutamic acid; AA, aspartic acid; H5P, hydantoin-5-propionic acid; H5A, hydantoin-5-acetic acid; CLG, carbamyl-L-glutamic acid; CLA, carbamyl-L-aspartic acid; OA, orotic acid; AMP, adenosine monophosphate.

B, reaction mixture from tissue slice experiment.

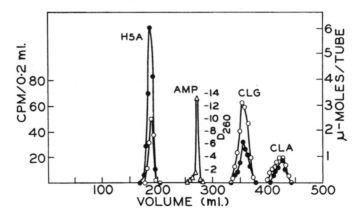

FIG. 3. Elution pattern of an artificial mixture obtained with a Dowex 2-formate column, 1.0 × 20 cm. Starting conditions as for Fig. 2, except that pH of buffer was 3.0. Methods of determining points: ○, radioactivity; ●, colorimetric method for carbamyl compounds; △, optical density at 260 mμ.[6]

vessel is V, and the volume of liquid which has passed through the mixing vessel is L. The values of x and V are fixed by the experimental conditions, and L may be determined by measuring the volume of eluate which has emerged from the column. Tables of $(1 - e^{-x})$ are to be found in many reference books, and hence the calculation of X is a simple matter.

The theoretical and practical aspects of gradient elution chromatography are discussed in papers by Lakshmanan and Lieberman[4] and Busch et al.[5]

Elution patterns of mixtures of compounds are shown in Figs. 2 and 3. The results are highly reproducible both with artificial and tissue reaction mixtures. The volume at which a particular compound is eluted can be reproduced to within $\pm 3\%$.

At the conclusion of each run the column is washed as described above and is then ready for further use.

[4] T. K. Lakshmanan and S. Lieberman, Arch. Biochem. and Biophys. 45, 235 (1953).
[5] H. Busch, R. B. Hurlbert, and V. R. Potter, J. Biol. Chem. 196, 717 (1952).

[94A] Carbamyl Phosphate

By L. Spector, M. E. Jones, and F. Lipmann

Carbamyl phosphate is formed through the reaction of KH_2PO_4 with KCNO. When the reaction is complete, potassium and most of the inorganic phosphate are removed simultaneously as the perchlorate and as the lithium salt, respectively, by the addition of lithium perchlorate. Lithium carbamyl phosphate is obtained from the filtrate by alcohol fractionation.

Procedure. In 100 ml. of water are dissolved 13.6 g. (0.10 mole) of KH_2PO_4 and 8.1 g. (0.10 mole) of KCNO. The solution is warmed at 30° for 30 minutes and then chilled. Subsequent operations (until the drying step) are best performed in the cold room. An ice-cold solution of 7.2 g. (0.30 mole) of lithium hydroxide and 16.7 ml. of 12 N perchloric acid (0.20 mole) in 83 ml. of water is added to the solution of crude potassium carbamyl phosphate. The mixture is thoroughly agitated and allowed to stand for 1 to 2 hours before removal of the precipitate[1] on a Büchner funnel. To the magnetically stirred filtrate cold absolute ethanol is added, dropwise, from a buret until 70 ml. has been added. The precipitate which

[1] This precipitate is a mixture of potassium perchlorate and lithium phosphate. It is discarded.

forms is removed by filtration and discarded.[2,3] The filtrate is treated in the same manner with an additional 50 to 55 ml. of absolute ethanol. The precipitate is collected and dried over phosphorus pentoxide in a vacuum desiccator at room temperature, the desiccant being changed once or twice a day until the evolution of volatile material from the product is complete (1 to 4 days). Yield, 4 g.[4,5]

Stability. Solutions of lithium carbamyl phosphate in water have been found to have the following half-lives: 16 hours at 0°; 2 hours at 30°; 40 to 50 minutes at 37°. In the frozen state the solutions appear to be stable indefinitely. The dried solid salt has stood at room temperature for over 3 weeks with no detectable change.

Assay

Since carbamyl phosphate is for practical purposes stable on short exposure to the conditions of the Fiske-SubbaRow colorimetric test,[6] it is most conveniently measured by differential phosphate analysis. Two tubes are required for each analysis—one for the assay of orthophosphate and another for the determination of the sum of orthophosphate and carbamyl phosphate (after hydrolytic decomposition of the latter).

Orthophosphate Assay. The Fiske and SubbaRow reagents are used, and the minor modifications made in the conventional procedure are designed to avoid hydrolysis of carbamyl phosphate during the determination. For protein precipitation, ice-cold TCA (final concentration, 3%) is used and the protein precipitate is centrifuged down in chilled metal cups (2 to 3 minutes) or in a refrigerated centrifuge. Samples are kept ice-cold until ready for analysis. A suitable aliquot is transferred to an assay tube, diluted to 80% of its final volume, the molybdate and reducing reagents[7] are added to each tube separately at timed intervals,

---

[2] Ordinarily the solid begins to separate at 65 to 70 ml. of ethanol, but if the addition has been too rapid the precipitate will sometimes fail to appear. However, continued stirring has invariably resulted in the eventual appearance of precipitate.

[3] This precipitate is a mixture of residual potassium perchlorate and lithium phosphate along with a forerun of lithium carbamyl phosphate. No attempt has been made to recover the latter from the mixture.

[4] The product prepared in this way generally gave the following analysis: 75% (by weight) lithium carbamyl phosphate; 10% lithium phosphate; 1 to 2% urea; and traces of unidentified carbon- and nitrogen-containing substances. The remainder appears to be firmly bound water.

[5] If desirable, the lithium salt can be converted to potassium carbamyl phosphate with potassium Dowex 50. All operations must be carried out at 0° to avoid hydrolysis of carbamyl phosphate.

[6] C. H. Fiske and Y. SubbaRow, *J. Biol. Chem.* **66**, 375 (1925); see Vol. III [114].

[7] The aminonaphtholsulfonic acid reagent, when mixed with standard orthophosphate solution and the molybdate reagent, should produce, within 5 minutes at room temperature, a color intensity not less than 95% of that obtained in 20 minutes. This reagent is stable when stored well-stoppered and in the refrigerator.

and the tube is then made to volume and mixed. The color is read 5 to 7 minutes after the addition of the reagents.[8,9]

Carbamyl Phosphate Hydrolysis. Carbamyl phosphate can be hydrolyzed to yield orthophosphate by heating in a boiling water bath for 2 minutes or by exposure to 0.1 N base for 10 minutes at room temperature.[10] The heat treatment is more rapid and may be preferred when there are no other heat-unstable phosphates in the sample. Alkaline hydrolysis is in general a milder, more specific procedure and is preferable in the presence of other heat- and acid-labile phosphates, such as the adenosine polyphosphates.[11] After hydrolysis of the carbamyl phosphate, the total orthophosphate is determined by the Fiske and SubbaRow method. The difference between this and the orthophosphate value represents the carbamyl phosphate content.

The term "hydrolysis" has been used in an inclusive sense to indicate inorganic phosphate release from carbamyl phosphate regardless of the detailed mechanism. It should be noted, however, that at least two types of reaction are included here. In acid solutions the reaction appears to be a true hydrolysis while in neutral or basic solution, the reaction is predominantly a splitting of carbamyl phosphate to inorganic phosphate and cyanate and not a hydrolysis. The formation of cyanate has been confirmed by a differential ammonia analysis and by the specific blue color which cyanate gives with cobalt.

[8] The use of the timed interval limits the number of tubes which may be run as a set to approximately ten. If samples are kept well chilled, however, several sets may be run in sequence.

[9] The reagents of O. H. Lowry and J. A. Lopez [*J. Biol. Chem.* **162**, 421 (1946)] may be used in place of those of Fiske and SubbaRow, but the substitution offers no particular advantage; see also Vol. III [114].

[10] The rate of hydrolysis of carbamyl phosphate is approximately the same from pH 1.5 to 9.0. In this range heating is necessary for rapid decomposition, whereas at strongly alkaline reaction decomposition occurs in a few minutes at room temperature. Therefore, the pH used for the hydrolysis is mainly dictated by the stability of any other phosphate compounds present in the sample.

[11] Although carbamyl phosphate may be determined in the presence of the adenosine polyphosphates by hydrolysis with heat, the alkaline hydrolysis is more suitable. If heat is used, however, the TCA filtrate should be diluted to bring the molarity to 0.05, for under these conditions the adenosine polyphosphates are hydrolyzed only to the extent of 3% in the 1-minute heating.

[95] Determination and Preparation of the Hydroxyamino Acids

By Warwick Sakami

Methods of Determination

Specific Chemical Procedures. Serine and threonine in proteins have been determined by estimation of the aldehydes quantitatively formed from these amino acids by periodate cleavage. Periodate reacts according to the following equations:

$$\overset{3}{C}H_2OH\overset{2}{C}H\overset{1}{C}HNH_2COOH \xrightarrow{IO_4^-} HCHO + NH_3 + \overset{3}{C}HO\overset{2}{C}O\overset{1}{O}H$$

$$\overset{4}{C}H_3\overset{3}{C}HOH\overset{2}{C}HNH_2\overset{1}{C}OOH \xrightarrow{IO_4^-} \overset{4}{C}H_3\overset{3}{C}HO + NH_3 + \overset{2}{C}HO\overset{1}{C}OOH$$

In the presence of excess periodate at weakly alkaline or weakly acid pH the glyoxylic acid is rapidly oxidized to formate and CO_2.[1] Formaldehyde and acetaldehyde have been quantitatively separated by aeration[2] or diffusion[3] and determined. In the presence of excess amino groups such as are present in protein hydrolyzates, acetaldehyde is quantitatively removed from the reaction mixture whereas formaldehyde is quantitatively retained.[2] Acetaldehyde has been trapped in bisulfite and estimated by the Clausen procedure in which aldehyde-bound bisulfite is determined iodometrically.[1,4] Two hundred micrograms of threonine may be determined by this procedure.[3] A more sensitive and specific method involves the determination of acetaldehyde by its specific color reaction with *p*-hydroxydiphenyl;[4] 30 γ of threonine may be determined by this procedure.

Formaldehyde has been determined gravimetrically by precipitation of the dimedon derivative from the reaction mixture.[5] A number of hours are required for complete precipitation of the dimedon. A generally more convenient method involves distillation of the formaldehyde and its determination by the Clausen procedure[4] or by the specific[6] color reaction

[1] D. B. Sprinson and E. Chargaff, *J. Biol. Chem.* **164,** 433 (1946).

[2] L. A. Shinn and B. H. Nicolet, *J. Biol. Chem.* **138,** 91 (1941).

[3] T. Winnick, *J. Biol. Chem.* **142,** 461 (1942).

[4] R. J. Block and D. Bolling, "Amino Acid Composition of Proteins and Foods," 2nd ed., Charles C Thomas, Springfield, Ill., 1951.

[5] B. H. Nicolet and L. A. Shinn, *J. Biol. Chem.* **139,** 687 (1941).

[6] D. A. MacFadyen, *J. Biol. Chem.* **158,** 107 (1945).

with chromotropic acid.[7] The Clausen procedure is preferred by Block and Bolling[4] for the determination of serine and threonine in proteins (method of Rees modified by Block and Bolling). Micromole quantities of serine have been determined by means of chromotropic acid added directly to the periodate reaction mixture after destruction of excess periodate with bisulfite.[8] Brief mention has been made of the possibility of determining threonine and serine in the presence of interfering carbohydrate as the difference in acetaldehyde or formaldehyde formed by periodate before and after acetylation.[4] It has been reported that carbohydrate in protein samples does not interfere with the determination of serine because they are converted during the hydrolysis procedure into substances (presumably furfural derivatives) that do not yield formaldehyde when treated with periodate.[7] On the other hand, glucosamine yields 0.1 mole of formaldehyde per mole of glucosamine after boiling for 24 hours with 6.8 N HCl.[7]

The sum of serine and threonine in protein hydrolyzates has been determined by estimation of ammonia formed by periodate oxidation.[9,10] Ammonia reacts with the aldehydes which are also formed in the reaction but is displaced from its combination and smoothly evolved when other amino acids are present.[9,10] Hydroxylysine interferes in this determination but may be removed by procedures which separate basic amino acids.[9] Phosphatidylserine[11,12] and the sum of phosphatidylserine and ethanolamine[11,12] have also been estimated by determination of periodate ammonia.

CO_2 is rapidly and quantitatively evolved from serine and threonine carboxyl carbon in the presence of excess periodate.[1] Since hydroxylysine and ethanolamine should not yield CO_2 under these conditions, periodate-CO_2 determination may be of some value in serine and threonine determination under special circumstances.

Microbiological Determination. Microbiological determination is of particular value when the determination of the amount of L-amino acid is desired. *L. mesenteroides*[13] and *S. faecalis*[14,15] are commonly employed in the determination of serine and threonine, respectively. Only the

[7] J. M. Boyd and M. A. Logan, *J. Biol. Chem.* **146,** 279 (1942).

[8] W. R. Frisell, L. A. Meech, and C. G. Mackenzie, *J. Biol. Chem.* **207,** 709 (1954).

[9] D. D. Van Slyke, A. Hiller, and D. A. MacFadyen, *J. Biol. Chem.* **141,** 681 (1941).

[10] B. H. Nicolet and L. A. Shinn, *J. Biol. Chem.* **142,** 139 (1942).

[11] C. Artom, *J. Biol. Chem.* **157,** 585 (1945).

[12] C. Artom, *J. Biol. Chem.* **157,** 595 (1945).

[13] J. Lascelles, M. J. Cross, and D. D. Woods, *J. Gen. Microbiol.* **10,** 267 (1954).

[14] M. J. Horn, D. B. Jones, and A. E. Blum, *J. Biol. Chem.* **169,** 739 (1947).

[15] S. C. Balasubramanian and M. Ramachandran, *J. Gen. Microbiol.* **7,** 44 (1952).

L-forms of the amino acids are utilized by these organisms.[16,16a,17] A microdetermination has been reported.[18] Very small quantities (0.063 to 0.25 γ) of threonine and serine have been determined with *L. fermenti 36* and *L. casei*, respectively, employing the microtechniques of Linderstrøm-Lang and Holter. It should be noted that members of the B₆ group can substitute for threonine in the nutrition of the *lactobacilli*.[19] These substances may be destroyed by exposure of the sample to ultraviolet light.[16a]

Quantitative Chromatography. Serine and threonine in protein hydrolyzates have been obtained as separate fractions by chromatography on a starch column,[20,21] or on a column of sodium-Dowex 50[22] eluted with pH 3.41 buffer, and have been determined in these fractions by a photometric ninhydrin procedure.[23] Dowex chromatography is generally preferable, since it is not sensitive to the presence of salt and is more rapid and convenient to operate. The Dowex procedure has found wide application in the determination of amino acids in urine and other biological fluids as well as in the analysis of proteins. Desalting may be accomplished by an electrolytic procedure[24,25] or by adsorption of the amino acids on a sulfonated resin in the H⁺ form and elution with ammonia[26] which requires no special equipment. Stein and Moore[24] have reported that the recovery of threonine after electrolytic desalting is slightly low. Threonine and serine are not clearly separated from one another on Dowex 50 eluted with HCl.[27]

The determination of threonine and serine by quantitative paper chromatography requires only a few micrograms and has a maximum accuracy of better than ±10%. Salt generally has an undesirable effect. With suitable precautions it is possible to obtain clean-cut reproducible

[16] The activity of D-serine was 12% of that of L-serine. This may have been due to lack of optical purity of the D-serine, since DL-serine was 50% as active as the L-form.[16a]
[16a] B. F. Steele, H. E. Sauberlich, M. S. Reynolds, and C. A. Baumann, *J. Biol. Chem.* **177**, 533 (1949).
[17] J. L. Stokes, M. Gunness, I. M. Dwyer, and M. C. Caswell, *J. Biol. Chem.* **160**, 35 (1945).
[18] L. M. Henderson, W. L. Brickson, and E. E. Snell, *J. Biol. Chem.* **172**, 31 (1948).
[19] J. L. Stokes and M. Gunness, *Science* **101**, 43 (1945).
[20] S. Moore and W. H. Stein, *J. Biol. Chem.* **178**, 53 (1949).
[21] W. H. Stein and S. Moore, *J. Biol. Chem.* **178**, 79 (1949).
[22] S. Moore and W. H. Stein, *J. Biol. Chem.* **192**, 663 (1951).
[23] S. Moore and W. H. Stein, *J. Biol. Chem.* **176**, 367 (1948).
[24] W. H. Stein and S. Moore, *J. Biol. Chem.* **190**, 103 (1951).
[25] T. Astrup, A. Stage, and E. Olsen, *Acta Chem. Scand.* **5**, 1343 (1951).
[26] M. E. Carsten, *J. Am. Chem. Soc.* **74**, 5954 (1952).
[27] W. H. Stein and S. Moore, *Cold Spring Harbor Symposia Quant. Biol.* **14**, 179 (1950).

chromatograms.[28,29] Fowden locates amino acids on the filter paper by their fluorescence after heating.[28] The lower limit of amino acid concentration required by this procedure is about 20 γ per square inch.[30] Amounts of amino acid too small to give a fluorescent spot are located (less accurately) by spraying a control chromatogram with ninhydrin. Amino acids are then extracted from the paper under conditions which serve to remove ammonia. Ammonia interferes in the subsequent photometric ninhydrin determination of amino acid. The preferred paper chromatographic separation of serine or threonine depends on the nature of the other substances present. Two-dimensional chromatography on paper buffered at pH 7 to 7.5 with phenol-water as the first solvent and collidine-lutidine as the second separates serine and threonine·from one another and from the other amino acids of protein hydrolyzates.[29]

Methods of Preparation. Serine and threonine have been isolated from protein hydrolyzates by a single chromatogram on a sodium-Dowex 50 column. Buffer has been removed from the amino acid samples by desalting on a Dowex 50 (H$^+$) column.[31] Serine and threonine have also been separated on ammonium-Dowex 50 columns by elution with ammonium formate buffers.[32] This procedure possesses the advantage over the sodium-Dowex 50 chromatography that the buffer may be removed by sublimation. However, the resolution is not as good, and in the published procedure of amino acid separation four separate chromatograms are required to obtain pure serine and threonine. Serine and threonine have also been isolated in larger amount by displacement chromatography.[33] This procedure also requires four chromatographic separations to obtain single amino acid fractions containing serine and threonine, and these samples must subsequently be freed of contaminating sodium hydroxide. Elution chromatography of amino acids on Dowex 50 with HCl solutions does not separate serine from threonine.[27] However, a fraction consisting solely of serine and threonine may be obtained by chromatography of the neutral amino acids, and this may be resolved into its components by one of several published procedures. Serine-threonine mixtures may be resolved by chromatography on starch with aqueous phenol as eluant,[34] by chromatography on ammonium-Dowex 50

[28] L. Fowden, *Biochem. J.* **48,** 327 (1951).
[29] J. F. Thompson and F. C. Steward, *Plant Physiol.* **26,** 421 (1951).
[30] D. M. Phillips, *Nature* **161,** 53 (1948).
[31] R. Swick, D. Buchanan, and A. Nakao, *J. Biol. Chem.* **203,** 55 (1953).
[32] C. H. R. Hirs, S. Moore, and W. H. Stein, *J. Biol. Chem.* **195,** 669 (1952).
[33] S. M. Partridge and R. C. Brimley, *Biochem. J.* **51,** 628 (1952).
[34] D. F. Elliott, *Biochem. J.* **45,** 429 (1949).

with ammonium-formate in ethanol, pH 3.1, as eluant,[32] or by displacement from a sulfonated polystyrene resin by sodium hydroxide in aqueous acetone.[33] A column of starch 1.8 × 25 cm. will separate 100 mg. of threonine-serine mixture.[34] Serine has also been purified from serine-threonine mixtures by precipitation as the p-hydroxyazobenzenesulfonate.[35,36] The serine salt is considerably less soluble than the threonine derivative.[35] In the routine determination of the C^{14} distribution of rat protein serine in the author's laboratory, a serine-threonine fraction is first prepared by Dowex 50 chromatography of the neutral amino acids. Serine is then precipitated as the p-hydroxyazobenzenesulfonate which is directly degraded by periodate.[37] The precipitant may be removed by treating a solution of the sulfonate with Amberlite IR-4B. Threonine has been freed of serine by heating with hydroiodic acid and red phosphorus.[38] This procedure converts serine to alanine but destroys only a small amount of threonine. Threonine was isolated from the reaction products by chromatography on Dowex 50.

[35] W. H. Stein, S. Moore, G. Stamm, C.-Y. Chou, and M. Bergmann, *J. Biol. Chem.* **143,** 121 (1942).
[36] C. Gilvarg and K. Bloch, *J. Biol. Chem.* **193,** 339 (1951).
[37] W. Sakami, *J. Biol. Chem.* **187,** 369 (1950).
[38] G. Ehrensvard, L. Reio, E. Saluste, and R. Stjernholm, *J. Biol. Chem.* **189,** 93 (1951).

[96] (−)-Carnitine

By H. E. CARTER and P. K. BHATTACHARYYA

$$(CH_3)_3\overset{+}{N}\text{-}CH_2CH(OH)CH_2C\bar{O}_2$$

Assay Method

See following paper [97] for procedure.

Isolation and Purification Procedures

Carnitine can be obtained from liver extract or meat extract. The latter is the preferable source, since it contains 2 to 3% of carnitine. The procedure described below for the preparation of carnitine is that of Carter *et al.,*[1] modified by the incorporation of an IRC-50 resin column step to remove choline.

[1] H. E. Carter, P. K. Bhattacharyya, K. R. Weidman, and G. Fraenkel, *Arch. Biochem. and Biophys.* **38,** 405 (1952).

Step 1. Preparation of a Phenol Extract. A solution of 450 g. of Wilson's beef extract in 2500 ml. of water is stirred with 1000 ml. of molten phenol, and the mixture is filtered over 200 g. of Celite. On standing overnight the phenol layer is separated and the aqueous layer is re-extracted with two 500-ml. portions of phenol. The phenol layers are washed with 500 ml. of water and then are poured into a mixture of 7 l. of ether and 1000 ml. of water in a 12-l. flask. A brown precipitate is removed by filtration, and the aqueous layer is separated. The ether-phenol layer is washed twice with 700-ml. portions of water. The aqueous layers containing the crude carnitine are combined.

Step 2. Removal of Choline on IRC-50 Resin. The aqueous layer above is passed through a 1.5-inch column containing 250 ml. of Amberlite IRC-50 buffered to pH 7 in the sodium phase at a rate of not more than 1000 ml./hr. The column is then washed with 4.0 l. of distilled water. The aqueous effluent and washings are combined and concentrated to a volume of about 1000 ml. The pH of this solution is adjusted to 1.5 with concentrated HCl. The acidified solution is decolorized by stirring with 100 g. of Norit A with heating. Three such treatments with Norit A generally give a colorless solution.

Step 3. Preparation of Second Phenol Extract. The colorless solution from the Norit treatment (total solids, about 60 g.) is made basic with concentrated NH$_4$OH and filtered through hardened filter paper. The filtrate is extracted with 400-ml. and 200-ml. portions of molten phenol, and the phenol layers are washed with the same 200-ml. portion of water. The carnitine is transferred to an aqueous phase by treatment with 1.8 l. of ether and 200 ml. of water as previously described. The aqueous layer (825 ml.; 17 g. of solid) is used in the next step.

Step 4. Purification over Alkaline Alumina. Alcoa activated alumina (80 to 200 mesh) is washed repeatedly (fifteen to twenty times) with distilled water in a beaker and then is transferred to a Büchner funnel and washed with water under suction until the solid content of the filtrate is less than 1 mg. per 25 ml. (about 20 l./kg. required). A suspension of 1200 g. of the washed alumina in 3 l. of 80% aqueous methanol is poured into a 3-inch column, and the alumina is allowed to settle overnight. The column is then washed with 1000 ml. of 80% methanol. The aqueous layer above is concentrated to 60 ml. under reduced pressure, and a crop of creatine is obtained on cooling overnight (2.0 to 3.0 g.; melting point 256 to 258° after recrystallization from water). The filtrate is evaporated to dryness and left overnight under 500 ml. of absolute ethanol. The residue (creatine; 2.5 to 3.0 g.) is removed by filtration, and the ethanolic extract is evaporated to dryness. The residue is dissolved in 50 ml. of 80% aqueous methanol and chromatographed over the alumina column. The column is developed with 80% methanol at a rate of about 5 ml./min. Fractions of 50- to 100-ml. quantities are collected. The first 900 ml. contains no solid. The next 600 ml. contains mainly creatinine, and the subsequent fractions (1000 to 1100 ml.) contains the bulk of the carnitine. This fraction on evaporation deposits about 5.0 g. of crystalline carnitine (melting point 192 to 195°, micro block). This material is essentially pure. The melting point can be raised to 196 to 198° by recrystallization from *anhydrous* ethanol-acetone mixtures. (Dissolve 1.0 g. in 50 ml. of 3:2 acetone-ethanol. Cool, and add 140 ml. of ice-cold acetone. Keep overnight at −9°, and decant the supernatant liquid. Yield of recrystallized carnitine, 0.6 g. *Anhydrous conditions must be rigorously maintained.*)

Properties

Pure (−)-carnitine is a white, extremely hygroscopic solid melting at 196 to 198° and giving a specific rotation of $[\alpha]_D^{22} = -23.5°$ (0.5% solution in water). Carnitine gives a number of crystalline salts (hydrochloride, melting point 142°; chloroaurate, melting point 153 to 155°; chloroplatinate, melting point 214 to 218°; ethyl ester hydrochloride, melting point 146°).

In the preparation of carnitine from natural sources a major problem is the separation from choline whose salts have very similar solubilities. This difficulty was nicely solved by the discovery that carnitine is not retained by IRC-50 in the sodium phase, whereas the nonamphoteric choline is held.

On vigorous treatment with strong alkali carnitine decomposes, giving an almost quantitative yield of trimethylamine. On treatment with concentrated sulfuric acid, carnitine yields crotonobetaine.

DL-Carnitine has been prepared by several procedures,[2] but all involve an extensive series of reactions. Adequate resolution procedures have not been developed. If natural (−)-carnitine is desired, the isolation procedure is preferable.

[2] H. E. Carter and P. K. Bhattacharyya, *J. Am. Chem. Soc.* **75**, 2503 (1953).

[97] The *Tenebrio* Assay for Carnitine

By G. FRAENKEL

Assay Method

Principle. The larvae of a beetle, the meal worm *Tenebrio molitor*, when grown on a semisynthetic diet consisting of casein, glucose, cholesterol, a salt mixture, and nine known members of the vitamin B complex, survive and grow at a normal rate up to the age of about five weeks. Then growth is gradually reduced, and the larvae begin to die. Most larvae are dead by the seventh week. On addition of at least 35 γ of carnitine per gram of the dry diet, they survive and continue to grow. In the routine assay for carnitine, larvae are raised in bulk on the deficient basal diet for 4 weeks, and then transferred to the experimental diets. The assay is based on both rate of growth and mortality.[1-4]

TABLE I[a]

NUMBERS SURVIVING AND AVERAGE WEIGHTS OF *Tenebrio* LARVAE ON THE
BASIC DIET AND WITH THE ADDITION OF YEAST OR GRADED
DOSES OF CARNITINE

(Twenty larvae were used in each of the control diets, and 60 larvae with each
concentration of carnitine; at 30° and 60% relative humidity.)

Weeks	Control diets		Carnitine, γ/g. of dry diet				
	No addition	2% yeast	0.19	0.375	0.75	1.5	50
	No. Mg.	No. Mg.	No. Mg.	No. Mg.	No. Mg.	No. Mg.	No. Mg.
5	9 7.4	17 9.5	45 8.3	48 8.9	48 10.7	50 10.0	50 10.2
6	6 9.7	17 15.0	42 13.3	48 14.9	46 19.3	49 19.1	50 19.4
7	5 12.2	16 23.5	40 21.3	47 22.3	46 29.6	47 29.8	50 29.2
8	4 13.2	15 31.4	38 21.0	47 31.8	46 42.2	47 41.0	48 45.6
9	4 16.2	15 39.7	36 34.6	47 42.5	45 59.5	47 57.6	48 59.2
10	3 24.5	15 50.7	34 47.5	46 56.0	45 64.6	47 72.5	48 74.6

[a] G. Fraenkel, *Biol. Bull.* **104,** 359 (1953).

Tenebrio molitor. *T. molitor* normally feeds on dried cereals in store and can usually be obtained from pet shops.[5] The general nutritional requirements have been described by Fraenkel *et al.*[6] Stock cultures are raised on whole-wheat flour with the addition of 5% dried brewer's yeast. At a temperature of 30° and humidity of 60 to 70%, the larvae become fully grown and reach a weight of about 135 mg. after approximately twelve weeks. Pupation takes place two to eight weeks later, and this stage lasts about one week. In the routine assay for carnitine it is not necessary to grow larvae to full size. Assays are as a rule terminated after ten weeks,

[1] G. Fraenkel, M. Blewett, and M. Coles, *Nature* **161,** 981 (1948).
[2] G. Fraenkel, *Arch. Biochem. and Biophys.* **34,** 457 (1951).
[3] G. Fraenkel, *Biol. Bull.* **104,** 359 (1953).
[4] G. Fraenkel, *Arch. Biochem. and Biophys.* **50,** 486 (1951).
[5] Or from biological supply houses, such as Turtox, 761–763 East 69th Place, Chicago 37, Ill.
[6] G. Fraenkel, M. Blewett, and M. Coles, *Physiol. Zool.* **23,** 92 (1950).

when larvae on the positive diet, which contains 2% of yeast, have reached a weight of 50 to 60 mg. By this time, in the absence of carnitine, most larvae have died (Table I). To grow larvae to full size and adulthood would require at least twice the time and quantity of food.

The newly formed pupae usually lie on top of the food and are placed in a separate dish on cotton, to be protected from feeding larvae. The adult beetles which emerge are then transferred to a pot which contains a layer of white flour plus 5% dried yeast. The bottom of the pot is covered by a disk of rough brown paper. On the top of the flour is placed a watch glass or petri dish with moist cotton so that beetles can mount the cotton and drink, but the flour will not be caked by the moisture. The beetles lay eggs which in part are stuck on the paper and in part lie freely in the flour. Twice a week the paper with the eggs is removed, and the remaining eggs are sifted from the flour. The young larvae which hatch from the eggs after 6 days are then grown on the basal diet (see below) for four weeks. Up to 500 larvae can be supported on 25-g. diet. By the end of this period they weigh 2 to 2.5 mg. and are then collected and placed on the experimental diets.

Twenty larvae are used in each individual test. These are divided into lots of 10 and each lot is placed into a 1-ounce wide-mouthed bottle, each containing 3 g. of food. This amount of food suffices for normal growth of 10 larvae up to a weight of 60 to 70 mg.

Stock cultures and assays are kept in a chamber with the temperature controlled at 30° and the relative humidity at 60 to 70%. Temperature and humidity control are important to obtain reproducible results. At a lower humidity and temperature, growth is correspondingly slower. At a higher humidity diets become moldy. Even under controlled conditions variations occur from experiment to experiment which makes it necessary to keep suitable positive and negative controls in each series of test.

Preparation of the Diets. The basic diet consists of:

		Vitamins (γ/g. dry diet)	
Casein, vitamin-free (Labco)	20 parts	Thiamin	25
Glucose	80 parts	Riboflavin	12.5
Cholesterol	1 part	Nicotinic acid	50
McCollum's salt mixture No. 185	2 parts	Pyridoxine	12.5
Water	10 parts	Pantothenic acid	25
		Choline chloride	500
		i-Inositol	250
		Folic acid	2.5
		Biotin	0.25

It has been found convenient to suspend the minerals in a small quantity of water and mix them thoroughly with the casein. The mixture is then dried in an oven at 80 to 100° with frequent stirring and finally ground in a Wiley mill to a fine powder. (It appears that heated "salted" casein supports better growth than the dry unheated mixture of casein and minerals.) Cholesterol and glucose are then added, and the diet is again thoroughly mixed. This stock diet is weighed into the wide-mouthed bottles in lots of 6 g. When an experiment is to be started, the vitamins are dissolved in water in amounts which will add the above-named quantities to 1 g. of the diet, employing 0.05 ml./g. If the carnitine preparation to be tested is in solution, suitable dilutions are added in amounts of 0.05 ml./g. of diet. This quantity of water is added to the diet if dry substances are to be tested for carnitine activity. The diets, with the vitamin solution and extracts added, are then mixed with a spatula and placed in the constant-temperature chamber for 2 days, at which time the moisture content of the diets has become equilibrated to that of the air and is about 10%. They are then ground in a mortar and subdivided into two bottles, each containing 3 g. Ten four-week old larvae which have been grown on the deficient diet are placed into each bottle, and these are kept in the chamber for the duration of the experiment. As a rule the larvae are counted at the end of the sixth week, and weighed in groups of the survivors of each bottle at the end of the seventh, eighth, and tenth weeks, when the test is usually terminated (Table I). The larvae are then returned into the stock culture.

Carnitine is highly soluble in water, and consequently assays are frequently performed on extracts. The material is extracted with water, either in powdered state or as a homogenate, in a boiling water bath for 30 minutes. Homogenates are prepared with the Potter-Elvehjem homogenizer, with the Waring blendor, or by grinding with sand in a mortar, according to consistency and amounts of material available. After removal of the supernatant, the solids are re-extracted in the same manner once or twice, and the combined extracts finally concentrated to the desired volume. The extracts are then pipetted into the diets in suitable dilutions, in amounts of 0.05 ml./g. Substances which can be finely powdered may be mixed into the diets dry. With extracts, a better distribution in the diet is ensured, but there sometimes remains an uncertainty as to the completeness of the extraction process. In many cases, however, tests made with both methods agreed closely.

Criteria of Assays and Calculations

The assays are based on the assumption that 0.35 γ of carnitine per gram of diet ensures survival of the larvae and growth at a near optimal level during the first ten weeks (Table I). This allows the amount of carnitine in a particular material which has been added to the basic diet to be calculated. This assay is, however, restricted to the ten-week period. Larvae which are allowed to grow to maturity on 0.35 γ of carnitine per gram of diet grow more slowly, more larvae die, the pupae are lighter, and the adults are malformed. For maximal weights (130 to 150 mg.), optimal survival, and perfect adults. 1.5 γ of carnitine is required (Table II).[7] It

TABLE II

DEVELOPMENT OF *Tenebrio* AT DIFFERENT LEVELS OF CARNITINE[a]

Amount of carnitine, γ/g. diet	No. of starting larvae[b]	No. of dead larvae	Average pupal weight, mg.	Average age until pupation, weeks	Average longevity of adults, weeks	No. of adults		
						Normal	Abnormal	Total
0.09	12	10	88.8	19.5	—	0	2	2
0.18	20	11	94.1	16.9	1	0	9	9
0.37	20	4	106.2	17.0	1	1	15	16
0.75	20	1	117.7	16.9	2	14	5	19
1.5	20	1	122.4	15.3	5.0	19	0	19
50.0	20	0	130.1	16.0	6.8	20	0	20

[a] Condensed from G. Fraenkel and P. I. Chang, *Physiol. Zool.* **27**, 40, Table 6 (1954).

[b] Weight about 90 mg.

would be impracticable to base an assay on the successful completion of the whole life cycle, not only because of the length of time and larger amounts of food involved, but also on account of the great individual variability in growth rate, and the period between attainment of full size and pupation.

Possible Use of Other Organisms

A dietary carnitine requirement has so far been demonstrated in four insect species only, all belonging to the same family of coleoptera, *Tenebrionidae*. Of these, *Tenebrio obscurus*, which closely resembles *T. molitor*, is more difficult to breed, in that larvae frequently die before pupation. The other two species are very much smaller, but their life cycle is correspondingly shorter. *Tribolium confusum*, which reaches pupation at the age of 20 days at a weight of 2 mg., responds to a carnitine deficiency solely by a failure to form perfect adults, and this phenomenon has shown some variability of expression.[8] *Palorus ratzeburgi*, which pupates after 30 days at a weight of only 1 mg., appears to react to a carnitine deficiency similarly to *T. molitor*, by an increased mortality, a reduced growth rate,

[7] G. Fraenkel and P. I. Chang, *Physiol. Zool.* **27**, 40 (1954).
[8] E. W. French and G. Fraenkel, *Nature* **173**, 173 (1954).

and reduced adult emergence.[9] However, relatively little is known about the biology of this insect, which is commonly confused with *Tribolium* species and is, by virtue of its small size, far less easy to handle than *Tenebrio*.[10]

[9] M. I. Cooper and G. Fraenkel, *Physiol. Zool.* **25,** 20 (1952).

[10] Postscript (added September 25, 1956). The above described biological testing method for carnitine, which was the basis for the recognition of the B_T deficiency of *Tenebrio* and the subsequent isolation of carnitine and which has been used in hundreds of assays has failed to work in our laboratory since the summer of 1953. Carnitine deficiency symptoms closely resembling those described above have, however, since been reported for *Tribolium* from the Department of Zoology, University of Munich, Germany (G. Frobrich, *Naturwiss.* **40,** 344, 566, 1953) and *Tenebrio* from the Department of Biochemistry, University of Liège, Belgium (J. Leclercq, *Biochem. Biophys. Acta* **13,** 160 (1953)). The observation of a diminished response to a carnitine deficiency by *Tenebrio* was subsequently also made at the University of Liège (Leclercq, personal communication). A joint investigation, carried out at the laboratories at Liège and Urbana, into the conditions under which a carnitine deficiency becomes manifest in *Tenebrio* led to the following results (Fraenkel, G. and Leclercq, J., *Arch. Intern. Physiol.* **64,** 1956, in press): (1) Most commercially obtainable "vitamin-free" caseins gave results which suggested an optimal or suboptimal content of carnitine, while other samples, especially those successfully used in the past, still gave rise to a clear-cut carnitine deficiency. (2) Different strains of *Tenebrio* reacted differently to a deficiency of carnitine in the diet. The strain grown at Urbana has over the years becomes less sensitive to the absence of carnitine. (3) The larvae required in addition to carnitine another indispensable growth factor which is present in the water-insoluble fraction of yeast (2% of the diet). In many instances, a carnitine deficiency, expressed by a high death rate of the larvae, is enhanced in the presence of this factor. The difficulties in making the *Tenebrio* assay for carnitine work again have, at the time of writing, not entirely been resolved, and it is not known by which methods of purification a casein can be produced which is totally devoid of carnitine activity and nutritionally satisfactory in every other respect. It is, however, possible to produce a typical carnitine deficiency in *Tenebrio* by the addition to the diet of γ-butyrobetaine, which acts as a specific inhibitor for carnitine (Bhattacharyya, P. K., Friedman, S., and Fraenkel, G., *Arch. Biochem. Biophys.* **54,** 424, 1955).

Section V

Nucleic Acids and Derivatives

[98] Determination of Nucleic Acids by Phosphorus Analysis

By GERHARD SCHMIDT

The methods for the determination of nucleic acids by phosphorus analysis were devised at a period preceding the development of the much more specific spectrophotometric and chromatographic procedures for nucleic acid analysis in tissues. The subsequent comparison of the results obtained by phosphorus analysis with the more recent techniques has shown that particularly the ribonucleic acid values obtained from some tissues include contaminating phosphorus compounds whose elimination is necessary, particularly in incorporation experiments with P^{32}. Little reliable information regarding the nature or the origin of these contaminants is as yet available, apart from the observation that they are of low molecular weight. Consequently the only essential modifications of the original technique which have been developed for the purpose of avoiding this error consist in steps aiming at a preliminary isolation of the nucleic acids prior to the hydrolytic degradation procedures necessary for the analysis. The application of procedures including such steps is now generally considered as necessary for turnover experiments with P^{32}. The yield in isolation of the nucleic acid is at best 70%.

These introductory remarks show that the procedures for nucleic acid determination in tissues by phosphorus analysis cannot be considered as routine techniques, but that the extent of contamination must be carefully evaluated in their application. A more detailed discussion of this problem will be given after the description of those procedures in which the isolation of the nucleic acid fraction is omitted.

I. Determination of Total Nucleic Acids According to Berenblum, Chain, and Heatley[1]

Principle. The phosphorus fraction of animal tissues remaining after complete extraction of the lipids and of the acid-soluble phosphorus compounds consists—according to present information—of the sum of nucleic acid phosphorus and phosphoprotein phosphorus. Since the amounts of the phosphoproteins in most tissues except eggs, mammary gland, and milk are very small in comparison to those of the nucleic acid phosphorus, the amounts of phosphorus obtained after extraction of the lipids and of the acid-soluble phosphorus compounds represent a good approximation

[1] I. Berenblum, E. Chain, and N. G. Heatley, *Biochem. J.* **33**, 68 (1939).

of the concentration of nucleic acid phosphorus in many animal tissues except for the materials listed above and for nervous tissue.

II. Partition of the Phosphorus Compounds Obtained According to the Berenblum-Chain-Heatley[1] Procedure into DNA, RNA, and Phosphoproteins According to Schmidt and Thannhauser[2]

Principle. The mixture of the three fractions obtained by the Beren-blum-Chain-Heatley procedure can be quantitatively partitioned into its components by selective alkaline hydrolysis. During a 16-hour incubation at room temperature in N sodium or potassium hydroxide, DNA remains still acid-insoluble, RNA is quantitatively transformed to acid-soluble nucleotides without any formation of inorganic phosphate,[3-6] and the phosphorus of the phosphoproteins is quantitatively converted to inorganic phosphate.[6]

Removal of Phospholipids and of Acid-Soluble Phosphorus Compounds

Modified Technique of Berenblum et al.[1] Amounts of tissues, usually of 1 to 2 g., are homogenized in 10 ml. of alcohol-ether mixture (3:1, v/v). The suspension is quantitatively transferred to a 200-ml. Erlenmeyer flask with an additional 15 ml. of the alcohol-ether mixture and boiled on the water bath for a few minutes. After cooling, the suspension is filtered on a Büchner funnel over Whatman No. 1 paper which had been covered with a thin layer of Hyflo Filter-Aid (Johns-Manville) and washed several times with ether. The powdery filter cake is transferred to a small round-bottomed flask and refluxed with 30 ml. of a mixture of equal volumes of chloroform and methanol. The lipid-free residue is filtered in a similar manner as described above. The amount of Filter-Aid should be kept at a minimum compatible with the quantitative removal of the filter cake from the paper.

The dry precipitate is transferred quantitatively to a 50-ml. centrifuge tube preferably equipped with a ground-glass stopper and suspended in 15 ml. of approximately 0.1 N ice-cold glycine-perchloric acid buffer of pH 2.5. The suspension is mechanically shaken in a cold room for 20 minutes, centrifuged,[7] the supernatant is discarded, and the residue is resus-

[2] G. Schmidt and S. J. Thannhauser, *J. Biol. Chem.* **161**, 83 (1945).

[3] G. Schmidt, R. Cubiles, N. Zöllner, L. Hecht, N. Strickler, K. Seraidarian, M. Seraidarian, and S. J. Thannhauser, *J. Biol. Chem.* **192**, 715 (1951).

[4] K. C. Smith and F. W. Allen, *J. Am. Chem. Soc.* **75**, 213 (1953).

[5] G. de Lamirande, C. Allard, and A. Cantero, *J. Biol. Chem.* **214**, 519 (1955).

[6] A. M. Crestfield, K. C. Smith, and F. W. Allen, *J. Biol. Chem.* **216**, 185 (1955).

[7] The supernatant solution occasionally contains floating particles resisting centrifugation. They are collected by filtration on a small Büchner funnel on Whatman No. 50 (hardened) filter paper and combined with the sediment.

pended in 15 ml. of the acid buffer. Six such extractions are carried out. The last supernatant solution must be free of phosphorus after ashing. The last residue is suspended in 10 ml. of N potassium hydroxide and treated as described below.

The extraction procedure just described has the advantage of permitting lipid-phosphorus determinations and nucleic acid determinations in the same sample. It also has the advantage of avoiding alterations of the solubility of phospholipids—a factor of some importance in experiments with nervous tissue. On the other hand, the protracted extraction with the acid buffer involves the danger of an appreciable degree of acid hydrolysis of the purine-deoxyriboside bonds of DNA.

Technique of Schmidt and Thannhauser.[2] The danger just described is largely avoided in the method of Schmidt and Thannhauser, who extract the acid-soluble phosphorus compounds with trichloroacetic acid prior to the extraction of the phospholipids.

Amounts of tissue up to 2 g. are homogenized under cooling in ice water and mixed immediately with several volumes of 10% trichloroacetic acid. After 10 minutes of standing in the cold, the suspension is filtered over a thin layer of Hyflo Filter-Aid (Johns-Manville Co.). If the presence of Filter-Aid interferes with the purposes of the experiment, the extractions and washings of the tissues may also be carried out on the centrifuge. The filtered material is washed extensively with at least twelve portions of 10% and later of 5% trichloroacetic acid. The last washing is free of acid-soluble phosphates.

The residue is quantitatively transferred from the filter to a beaker and suspended in 95% alcohol. (It is not advisable to leave the precipitate on the filter for the first alcohol washings, since the replacement of TCA by alcohol causes swelling of the particles of the residue and consequently clogging of the filter pores.) The suspension is filtered over a thin layer of Hyflo Filter-Aid and washed several times with alcohol and ether. The powder is refluxed for 10 minutes in 100 ml. of an ethanol-ether mixture (3:1, v/v), filtered, and refluxed for 2 hours with a mixture of chloroform and methanol (1:1, v/v). The extracted tissue powder is filtered, washed with ether, and dried in a vacuum desiccator until the ether is removed. The powder is suspended in 5 ml. of N potassium hydroxide[8] per gram of moist tissue (0.3 N potassium hydroxide, as

---

[8] Although the concentration of N potassium hydroxide is sufficient to transform RNA-phosphorus completely to acid-soluble phosphorus compounds, de Lamirande *et al.*[5] found that 1.5 N potassium hydroxide is required for its complete conversion to mononucleotides, at least under the conditions of tissue analysis when the presence of considerable quantities of proteins exerts a strong buffer effect. Hydrolysis of the tissue nucleic acid fraction with 0.3 N potassium hydroxide yielded only

suggested by Davidson et al.,[9,10] is probably preferable but has not been tested so far by the author) and is slowly shaken at room temperature for 16 hours. Under these conditions, the tissue powder is practically completely dissolved, and the small deposit at the end of the hydrolysis consists only of the Filter-Aid.[11] The supernatant liquid is centrifuged off, and an aliquot is set aside for the determination of the total phosphorus (sample A). Another aliquot is used for the determination of the ultraviolet absorption at 260 mμ.

For the partition of the phosphorus into DNA, RNA, and phosphoprotein phosphorus, a 5-ml. aliquot is acidified to pH 1 by the addition of a measured volume of a 15% solution of perchloric acid. The copious precipitate is centrifuged off and contains quantitatively the deoxyribonucleic acid and a large amount of protein degradation products. The supernatant, which contains quantitatively the ribonucleotides and the inorganic phosphate formed from the phosphoproteins, is set aside. The precipitate is suspended in a small amount of water and dissolved by the addition of the necessary amount of N potassium hydroxide. It is reprecipitated by adding a sufficient volume of a mixture of trichloroacetic acid and hydrochloric acid to bring the final concentration of free hydrochloric acid to approximately 0.2 N. The precipitate is centrifuged and washed two times with a mixture of 5% trichloroacetic acid and 0.2 N hydrochloric acid. In a suitable aliquot of the supernatant solution (without the washings) the total phosphorus is determined according to the method of Fiske and SubbaRow (sample B).

Another aliquot of the supernatant solution (exclusive of the wash-

---

70% of the amount of mononucleotides as compared to the yield observed with 1.5 N potassium hydroxide. Smith and Allen[4,6] similarly reported that, after hydrolysis of 200 mg. of yeast RNA with 5 ml. of N sodium hydroxide at room temperature for 24 hours, approximately 3% of the total nucleotides were present in the form of oligonucleotides. Although the presence of acid-soluble oligonucleotides does not affect the results of RNA-phosphorus determinations, complete conversion to mononucleotides is, of course, essential for experiments designed to check the result of phosphorus determination by chromatographic separation of the nucleotides. Incomplete conversion of tissue RNA to mononucleotides might be partially responsible for some discrepancies[8,9] observed by several authors between the RNA-phosphorus values and the yield of mononucleotides.

[9] J. N. Davidson and R. N. Smellie, Biochem. J. 52, 594 (1952).

[10] I. Leslie, in "The Nucleic Acids" (E. Chargaff and J. N. Davidson, eds.), Vol. 2, p. 1. Academic Press, New York, 1955.

[11] When the procedure is carried out as described in this chapter, the silicates of the Filter-Aid do not interfere with the colorimetric phosphorus determinations, although small amounts of inert silicates are extracted by the potassium hydroxide and appear as insoluble granules during the ashing. Contact of alkali-treated Filter-Aid with acid, however, must be carefully avoided. For this reason, the alkaline hydrolyzate must be centrifuged prior to acidification.

ings) is neutralized with potassium hydroxide and stirred with 1 to 3 g. of Amberlite cation exchanger IR-100H (analytical grade). By this treatment, peptones are adsorbed which would interfere with the phosphorus determination by the formation of precipitates on addition of ammonium molybdate. The ion exchanger is centrifuged, washed with water, the combined filtrate and washings are made up to a measured volume with water, and the inorganic phosphorus is determined in the combined filtrate and washings according to Fiske and SubbaRow (sample C).

When samples A, B, and C represent the values per 100 g. of tissue, the difference A − B represents the concentration of DNA-phosphorus, B − C that of RNA-phosphorus, and C that of the phosphoprotein phosphorus.

Limits of Application of the Determination of Nucleic Acids by Phosphorus Analysis

The procedure described gave very satisfactory recoveries of DNA, RNA, casein, or ovovitellin, when mixtures of these substances were added to tissue homogenates. Obviously, however, the accuracy of the figures greatly depends on the mutual proportions, particularly between DNA and RNA. (Since the phosphoprotein phosphorus is based on determinations of inorganic P, this fraction can be determined with a fair degree of accuracy even in the presence of large amounts of nucleic acids.) The proportion between DNA and RNA varies within very wide ranges from very small values (for example, in the unfertilized arbacia egg, where this ratio is extremely small, and in pancreas, where it is approximately 0.1) to the high values in thymus or in mature sperm cells or in isolated chromosomes. Since the procedure is essentially a differential method, the accuracy of the values of the two components is of similar degrees only when both their concentrations are within similar ranges.

A more serious shortcoming of the procedure is the occurrence in some tissues of unknown phosphorus-containing substances of non-nucleotide nature which are not completely removed by the extraction of the acid-soluble and the lipid-phosphorus compounds. These phosphorus compounds appear in the RNA fraction, and their presence may result in RNA-phosphorus figures which are too high. Davidson and Smellie[9,10] tested this possibility by paper electrophoresis of the ribonucleotides of the alkaline hydrolyzate of liver after injection of P[32]-phosphate. Comparison of the spots visualized by ultraviolet absorption with radioautographs of the same hydrolyzate revealed spots showing radioactivity without ultraviolet absorption. These authors estimate that the RNA-phosphorus values obtained with the Schmidt-Thannhauser procedure on rat liver are approximately 25% too high.

On the basis of a quantitative comparison between the RNA-phosphorus figures and the ultraviolet absorption of brain hydrolyzates, Folch,[12] as well as Rossiter and collaborators,[13,14] came to the conclusion that the discrepancies are much higher in analyses of nervous tissues.

It is difficult to decide at present to what extent the presence of these contaminations is due to incomplete removal of acid-soluble and lipid-bound phosphorus from the nucleic acid fraction or to the actual occurrence of unknown phosphorus compounds linked to protein. The difficulty created by the incomplete extraction—even of inorganic phosphate—from tissue is known to many investigators studying the incorporation of labeled—inorganic or organic—phosphorus into the protein fraction of homogenates. Davidson and Smellie[9,10] have convincingly demonstrated contamination of the RNA fraction with inorganic P^{32} added in the form of phosphate simultaneously with trichloroacetic acid to liver homogenates. The author has experienced a similar difficulty in incorporation studies with the protein fraction of mammary gland homogenates.

On the other hand, the almost negligible amounts of inorganic phosphorus found in the RNA fraction of all tissues except those containing appreciable amounts of phosphoproteins render it unlikely that incompleteness of the extraction of acid-soluble phosphorus compounds could cause appreciable errors in the chemical determination of RNA, although it interferes with measurements of the specific radioactivity. Of course, the possibility is not excluded that certain acid-soluble phosphoric acid esters might remain adsorbed to the protein fraction to a higher degree than does inorganic phosphate.

For the serious discrepancies between ultraviolet extinction and phosphorus' concentration encountered in the analysis of nerve tissue, Folch and Le Baron[12] advanced the explanation that certain phospholipids, in particular phosphoinositides, are not completely extracted by the lipid solvents used.

The presence of unknown phosphorus components of proteins must likewise be considered as a source of error in nucleic acid determinations. So far, however, no phosphorus-containing constituents of proteins except phosphoric acid esters of hydroxyamino acids have been reported.

In regard to Davidson and Smellie's experiments in which the phosphorus figures obtained by direct determination were compared with those calculated from the ultraviolet extinctions of the nucleotide spots

12 J. Folch and F. N. Le Baron, *Federation Proc.* **10**, 185 (1951).
13 J. E. Logan, W. A. Nannell, and R. J. Rossiter, *Biochem. J.* **51**, 470 (1953).
14 H. A. Deluca, R. J. Rossiter, and K. P. Strickland, *Biochem. J.* **55**, 193 (1953).

on paper electrophoresis strips, the question might also be raised whether some nucleotide losses might not have occurred during the 18-hour electrophoresis at pH 3.5, and whether the conversion of RNA to mononucleotides was complete.[8]

These comments show that the interpretation of the phosphorus values should always be checked by determinations of the ultraviolet absorption (see Vol. III [106]) of the final filtrate containing the RNA nucleotides.

For incorporation studies with P^{32}, this precaution is not sufficient, however, and a preliminary isolation of the nucleic acid fraction must precede its partition by alkaline hydrolysis. The yield of the several isolation techniques with gram amounts of tissue is 70% at best, and the conclusions regarding turnover rates reached on the basis of these procedures rest on the assumption that the behavior of the isolated portions is representative of that of the fractions lost during the procedure.

III. Procedures for Incorporation Studies of P^{32} in DNA and RNA of Tissues

Method of E. Hammarsten,[15] as Applied by Deluca, Rossiter and Strickland[14]

Extraction of Acid-Soluble Phosphorus Compounds. The tissue (amounts of 500 mg. or more) is suspended in approximately 20 vol. of 10% trichloroacetic acid solution at 0° and homogenized in a Potter-Elvehjem type of apparatus. After centrifugation, the residue is extracted seven times with 10-vol. portions of a solution of 0.04 M of acid potassium phosphate in 10% trichloroacetic acid and two times with 20 vol. of a 10% trichloroacetic acid solution. The radioactivity of the final extracts is negligible.

Extraction of Phospholipids. The residue is extracted two times in the cold with 10-vol. portions of 95% ethanol and four times with 10-vol. portions of Bloor's alcohol-ether mixture, at boiling temperature. The final residue is washed with 10 vol. of ether.

Extraction and Isolation of the Nucleic Acid Fraction. The residue is suspended in 2 ml. of a 10 M aqueous solution of urea per 500 mg. of fresh tissue, and the suspension is kept at room temperature for 5 minutes. After addition of 6 vol. (of the original tissue sample) of a saturated aqueous solution of sodium chloride and ammonium sulfate, the suspension is boiled for 1 minute. The suspension is centrifuged, the supernatant liquid is set aside, and the residue is extracted twice with 4 vol. of the salt-urea mixture (150 g. of urea dissolved to 1 l. with the salt mixture)

[15] E. Hammarsten, *Acta Med. Scand.* **128**, Suppl. 196, 634 (1947).

at boiling temperature. The nucleic acids are precipitated from the combined extracts by the addition of approximately 1 ml. of a saturated aqueous copper sulfate solution to 7 ml. of the extract. After standing overnight, the suspension is centrifuged and the precipitate is washed twice with a 1.8% solution of copper chloride. The copper nucleates are decomposed in the following manner: The precipitate is suspended in 0.8 ml. of a concentrated potassium acetate buffer of pH 6.4 (100 g. of potassium acetate dissolved in 100 ml. of water and titrated to pH 6.4 with glacial acetic acid). The suspension is centrifuged, and the residues are extracted three times with portions of 0.2 ml. of a 5 M urea solution. The nucleates are precipitated at $-15°$ with 10 ml. of 95% ethanol and a drop of a saturated solution of sodium chloride. The precipitate is redissolved in 1 ml. of water, and the solution is reprecipitated at 0° by the addition of 0.11 ml. of hydrochloric acid followed by addition of 10 ml. of a mixture of 9 vol.-parts of 95% ethanol and 1 vol.-part of aqueous N hydrochloric acid at $-15°$. After 2 hours of standing, the centrifuged precipitate is dissolved by neutralizing with 0.01 N sodium hydroxide, and the solution is reprecipitated with acid and acid alcohol as described before. This precipitate is used for the partition into DNA and RNA according to Schmidt and Thannhauser[2] (see below).

Method of Davidson and Smellie[9]

A somewhat simpler isolation procedure was described by Davidson and Smellie. The tissue powder obtained after extraction with trichloroacetic acid and lipid solvents is extracted three times for 1 hour each time with a solution containing 10 g. of sodium chloride in 100 ml. of solution, at 100°. The combined extracts are precipitated with 2 vol. of 95% alcohol, and the precipitate is washed with alcohol and ether and dried.

Partition of the Nucleic Acid into Ribo- and Deoxyribonucleic Acids According to Schmidt and Thannhauser. The nucleic acid precipitate is dissolved in 0.3 N potassium hydroxide (Davidson and Smellie's modification of the method of Schmidt and Thannhauser, who used N alkali). One milliliter of alkali is recommended for 5 mg. of nucleic acid. The amount of alkali must be sufficient to prevent a substantial lowering of the pH by the hydrolytic liberation of acidic groups from RNA; on the other hand, an undue excess of alkali should be avoided, since it would cause unnecessary dilution of the nucleate solution. The alkaline solution is incubated for 16 hours at 30°. During this time the degradation of ribonucleic acid to acid-soluble nucleotide is quantitative, whereas that of deoxyribonucleic acid results in the formation of products which can still be precipitated quantitatively by acidification with perchloric acid at pH 1. If the temperature does not exceed 30°, no appreciable destruction of the purine or pyrimidine moieties occurs, whereas at higher temperatures (37°, as originally recommended by Schmidt and Thannhauser) considerable deamination of cytidylic acid results. After the completion of the hydrolysis, the largest part of the potassium ions is removed by dropwise addition of 10% perchloric acid until a pH between 7 and 8 is reached. The suspension of potassium perchlorate is kept for several hours in the refrigerator for the purpose of maximal precipitation and is diluted to a measured volume with a minimal amount of water. After centrifugation, an aliquot of the supernatant is brought to pH 1 by further addition of a measured volume of 1% perchloric acid. The precipitate which contains the deoxyribonucleic acid is centrifuged off after a few minutes standing.

Deoxyribonucleic Acid Fraction. The precipitated deoxyribonucleic acid is dissolved in a small volume of dilute sodium hydroxide and reprecipitated with perchloric acid. For P^{32} determinations, it is obviously advisable to dissolve the precipitates in solutions of nonlabelled ribonucleotides and inorganic phosphates. This purification is repeated several times. The final solution is used for phosphorus and radioactivity determinations. The absence of contaminating ribonucleotides can be checked by ionophoresis and radioautography.

Remarks. The extent of purification required for specific activity determinations depends on the ratio between ribonucleic and deoxyribonucleic acids in the isolated total nucleic acid fraction. This ratio varies over a very wide range according to the analyzed tissue. It is easy to obtain sufficiently pure deoxyribonucleic acid samples from tissues like thymus or lymph glands or from sperm cells; but extensive purification is necessary in work with liver, pancreas, intestinal mucosa, and unfertilized eggs. For this reason, a detailed description of a generally valid purification procedure of the deoxyribonucleic acid fraction cannot be given. The general principle described in the preceding paragraph must be adapted to the specific requirements of the investigation.

Ribonucleotide Fraction. The supernatant contains the ribomononucleotides but might still contain traces of inorganic phosphate. These can be removed by treatment of the neutralized hydrolyzate with magnesium oxide, preferably after addition of amounts of unlabeled phosphate containing 5 to 10 mg. of phosphorus. The purity of the ribonucleotide fraction can be checked by ionophoresis and radioautography according to Davidson and Smellie.[9]

[99] Determination of Nucleic Acids in Tissues by Pentose Analysis

By WALTER C. SCHNEIDER

Principle. The procedure for the determination of nucleic acids described below is based on the finding that nucleic acids can be separated from other tissue compounds by their preferential solubility in hot trichloroacetic acid.[1] The isolated nucleic acids are then quantitated by means of colorimetric reactions involving the pentose components of the nucleic acids.

Reagents

Diphenylamine reagent. One gram of purified diphenylamine is dissolved in 100 ml. of reagent glacial acetic acid and 2.75 ml. of reagent concentrated sulfuric acid. Although it is possible that this reagent may be stored without undesirable effects, we have found it convenient to prepare it immediately before use. Commercial diphenylamine is steam distilled and/or recrystallized from boiling hexane to obtain a white crystalline product.

Orcinol reagent. One gram of purified orcinol is dissolved, immediately before use, in 100 ml. of concentrated HCl containing 0.5 g. of $FeCl_3$. Since commercial orcinol can vary in color from light pink to brick red, we have purified it by dissolving it in boiling benzene, decolorizing with charcoal, and crystallizing after adding hexane. A perfectly white crystalline product can be obtained in this manner.

Separation of Nucleic Acids from Tissue.[1] The isolation of the nucleic acids from liver will be described. Application of the method to other tissues and to tissue extracts will require modifications dictated by their nucleic acid content.

Removal of Acid-Soluble Compounds. One milliliter of a 20% rat liver homogenate is mixed with 2.5 ml. of cold 10% TCA and centrifuged. The sediment is washed once with 2.5 ml. of cold 10% TCA.

Removal of Lipoidal Compounds. The final sediment remaining after removal of the acid-soluble compounds is extracted twice with 5 ml. of 95% ethanol and recovered by centrifugation. This procedure is generally sufficient to remove the phosphorus-containing lipids.[2] In cases where it

[1] W. C. Schneider, *J. Biol. Chem.* **161**, 293 (1945).
[2] W. C. Schneider and H. L. Klug, *Cancer Research* **6**, 691 (1946).

is not, additional extractions with boiling alcohol-ether or other solvent mixtures can be made. Incomplete extraction of the lipoidal materials will be indicated by the presence in the nucleic acid fraction of phosphorus that cannot be accounted for as nucleic acid phosphorus.[1]

Removal of Nucleic Acids. The lipid-free tissue residue is suspended in 1.3 ml. of water and 1.3 ml. of 10% TCA, and the mixture is heated for 15 minutes at 90° with occasional stirring. This treatment quantitatively splits both DNA and PNA from the tissue proteins and leaves the latter as an insoluble residue which is centrifuged off and washed with 2.5 ml. of 5% TCA. The combined extracts constitute the nucleic acid fraction of the liver.

Alternate Procedure for Isolation of Nucleic Acids; Separation of DNA and PNA.[3,4] The lipid-free tissue residue is suspended in 1 N KOH (2 ml.) and incubated for 20 hours at 37°. DNA and protein are then precipitated by the addition of 0.4 ml. of 6 N HCl and 2 ml. of 5% TCA and centrifuged off, leaving the PNA breakdown products in solution. The DNA in the sediment is then brought into solution by heating it with TCA as above.

DNA and PNA can also be separated by treating the lipid-free residue with perchloric acid at 4° for 18 hours to release the PNA and then releasing the DNA by heating with perchloric acid.[5] The use of perchloric acid has been found to interfere with the determination of DNA in liver (ref. 6, confirmed by the author), since it extracts some material which gives a false test with diphenylamine.

Estimation of DNA.[7] One milliliter of the nucleic acid extract is mixed with 2 ml. of diphenylamine reagent and heated for 10 minutes in boiling water. The intensity of the blue color is read at 600 mμ, the wavelength of maximum absorption. A standard curve is prepared relating optical density to micrograms of DNA-P, with purified DNA as the standard. The practice of using the organic P content of DNA as the reference rather than DNA weight has the advantage that DNA samples of varying degrees of purity will yield the same curve when diphenylamine color intensity is plotted against DNA-P.[1]

Estimation of PNA.[8] Two-tenths milliliter of the nucleic acid extract is diluted to 1.5 ml. and heated with 1.5 ml. of orcinol reagent for 20 minutes in boiling water. The intensity of the green color is read at 660 mμ.

[3] G. Schmidt and S. J. Thannhauser, *J. Biol. Chem.* **161**, 83 (1945).
[4] W. C. Schneider, *J. Biol. Chem.* **164**, 747 (1946).
[5] M. Ogur and G. Rosen, *Arch. Biochem.* **25**, 262 (1950).
[6] V. R. Potter, R. O. Recknagel, and R. B. Hurlbert, *Federation Proc.* **10**, 646 (1951).
[7] Z. Dische, *Mikrochemie* **8**, 4 (1930).
[8] W. Mejbaum, *Z. physiol. Chem.* **258**, 117 (1939).

A standard curve is prepared relating color intensity to micrograms of PNA-P with purified yeast nucleic acid serving as the standard. Samples of PNA of different purities will yield the same calibration curve when the plot is made in this manner. Since DNA also reacts with orcinol, a calibration curve is also prepared for this material.[1]

Calculation of DNA Content.[1] The DNA content of the nucleic acid extract is given by the following equation:

$$\text{Micrograms DNA-P per ml.} = \frac{\text{Optical density at 600 m}\mu}{0.019}$$

The factor 0.019 is the optical density per microgram of DNA-P observed in the Beckman model DU spectrophotometer with 1-cm. cells. The DNA-P content per milliliter of extract would be equivalent to 40 mg. of liver tissue, since the extract was one-fifth as concentrated as the original homogenate.

Calculation of PNA Content.[1] The PNA content of the extract is given by the expression:

$$\text{Micrograms PNA-P per 0.2 ml.} =$$
$$\frac{(\text{Optical density at 660 m}\mu + 0.008) - (\text{DNA-P per 0.2 ml.} \times 0.013)}{0.116}$$

The values 0.013 and 0.116 are the optical densities per microgram of DNA-P and PNA-P, respectively, in the orcinol reaction with 1-cm. cuvettes. The factor 0.008 in the equation is due to the fact that the calibration curve does not go through the origin. If DNA and PNA are separated as in the alternate procedure, the correction for DNA in the equation above drops out. We have found, however, that in mixtures of DNA and PNA the contributions of each nucleic acid to the optical density are strictly additive. For this reason, unless the ratio of DNA to PNA is excessively high (as might be the case in the analysis of isolated nuclei), the separation of the two nucleic acids is unnecessary.

Alternate Colorimetric Reactions for DNA and PNA. DNA can also be determined by its reaction with carbazole,[7,9] cysteine,[10,11] tryptophan,[12,13] or indole.[14] Of these reactions only cysteine and indole do not react with PNA. Tryptophan gives a red color with DNA and a green color with PNA, and carbazole gives a red color with both DNA and PNA. The

[9] S. Gurin and D. B. Hood, *J. Biol. Chem.* **139**, 775 (1941); **131**, 211 (1939).
[10] Z. Dische, *Proc. Soc. Exptl. Biol. Med.* **55**, 217 (1944).
[11] P. K. Stumpf, *J. Biol. Chem.* **169**, 367 (1947).
[12] P. Thomas, *Z. physiol. Chem.* **199**, 10 (1931).
[13] S. S. Cohen, *J. Biol. Chem.* **156**, 691 (1944).
[14] G. Ceriotti, *J. Biol. Chem.* **198**, 297 (1952).

color produced in the reaction of carbazole with DNA is about ten times as intense as with the same amount of PNA. Furthermore, carbazole has the advantage that it reacts only with the pyrimidine-bound sugars of DNA, whereas diphenylamine, cysteine, and indole react only with the purine-bound deoxyribose. This difference between the carbazole and the other reactions is important because it provides a simple means of differentiating between two different portions of the DNA molecule. Tests on a number of tissues with the diphenylamine and carbazole reactions have shown that the amount of DNA present is the same, regardless of which reaction is employed for the determination. This indicates, of course, that the DNA of these tissues contains equal proportions of purine and pyrimidine nucleotides, a fact that the recent analyses of isolated nucleic acids for purine and pyrimidine bases has amply established. Of the color reactions for DNA, the carbazole and indole reactions are the most sensitive. The latter reaction is somewhat more difficult to carry out, owing to the fact that several chloroform extractions are required to remove colors produced by interfering materials.

Several colorimetric reactions have been used for the determination of PNA. The phloroglucinol reaction[15] is specific for PNA, and orcinol reacts with both DNA and PNA. The orcinol reaction has the advantage of being approximately ten times as sensitive as the phloroglucinol reaction, however, for the estimation of PNA. Another method which is specific for PNA in suitably prepared samples is to hydrolyze the pentose of PNA to furfural, distill it off, and estimate the furfural from the intensity of the color produced in its reaction with aniline acetate.[16,17]

General Comments. The determination of nucleic acids in tissues is largely a problem in identification. By means of the extraction procedures described above and the colorimetric reactions of PNA and DNA, a considerable degree of specificity is placed on the determination of these compounds. Occasionally however, false results will be obtained, owing to the presence of materials in the nucleic acid extracts that interfere with the pentose reactions. The presence of interfering materials may be obvious if the colors produced are sufficiently different from those observed with DNA and PNA. Such is the case with nucleic acid extracts of serum[18,19] and ascitic fluid.[20] To guard against such interferences, it is a wise policy to determine the nucleic acid content of the extracts by two

[15] L. Hahn and H. von Euler, *Arkiv Kemi* **22A**, No. 231 (1946).
[16] J. Brachet, *Enzymologia* **10**, 87 (1941).
[17] J. N. Davidson and C. Waymouth, *Biochem. J.* **38**, 39 (1944).
[18] T. Yamashita and M. Yamada, *Chem. Abstr.* **46**, 10362 (1954).
[19] S. Niaza and D. State, *Cancer Research* **8**, 653 (1948).
[20] E. Shelton, *J. Natl. Cancer Inst.* **15**, 49 (1954).

or more independent procedures and compare the results.[1] For example, the nucleic acid phosphorus estimated from the pentose reactions can be compared with the actual organic phosphorus content, with the ultraviolet absorption at 260 mμ, or with the purine and pyrimidine content. Determination of the N:P ratio of the nucleic acid extract is also useful in revealing whether a nucleic acid is present in the extract.[1] By means of combinations of two or more of these methods, it has been possible to show that nucleic acids are present in the extracts of tissues prepared as described above and, in many cases, account almost exclusively for the material extracted from the tissues.[1,2,4,21]

It should be emphasized that the extraction methods described above were developed for nucleic acid determinations by spectrophotometric methods. Although it was at first thought that these procedures might be directly applicable to isotopic work,[1] it has become quite clear that the separations are not sufficiently refined for such studies. The methods have served, however, as starting points for other separation procedures more suitable for isotopic work.[22]

[21] R. A. Huseby and C. P. Barnum, *Arch. Biochem.* **26**, 187 (1950).
[22] J. N. Davidson and R. M. S. Smellie, *Biochem. J.* **52**, 599 (1952).

[100] Preparation of Ribonucleic Acid from Plant Viruses

By C. A. KNIGHT

Three methods of preparing nucleic acids have been applied to plant viruses to give high yields of seemingly undegraded RNA. However, no one of these appears to be universally feasible or desirable, although the detergent method approaches this objective. In view of this situation, and the fact that each procedure has some distinct advantage, all three are given here, together with the suggestion that one of the three will almost surely prove useful for preparation of nucleic acid from any plant virus.

Heat Denaturation Method. This procedure is a modification of the method of Cohen and Stanley,[1] which has given good results with numerous strains of tobacco mosaic virus and with cucumber viruses 3 and 4.[2]

Six milliliters of 0.3 *M* sodium chloride is heated in a 15-ml. conical glass centrifuge tube in a water bath at 100°. To this is added 2 ml. of aqueous highly purified virus solution at a concentration of 30 to 80 mg.

[1] S. S. Cohen and W. M. Stanley, *J. Biol. Chem.* **144**, 589 (1942).
[2] C. A. Knight, *J. Biol. Chem.* **197**, 241 (1952).

of virus per milliliter. The mixture is stirred by being drawn up and down in a dropping pipet for about 15 seconds. By this time the mixture has reached a temperature of about 100° and heating is continued for 1 minute; the tube is then withdrawn and placed in an ice bath. (Rapid equilibration of temperature is possible with small volumes; therefore when large amounts of nucleic acid are desired, several tubes are used rather than one large vessel. With strains of tobacco mosaic virus, the appropriate number of tubes and amount of virus to be employed can be estimated on the basis of a nucleic acid content of 6% and a final yield of about 80%.) After chilling, the reaction mixture is spun at 7000 r.p.m. in an angle centrifuge in order to remove coagulated protein. The clear supernatant fluid containing the sodium nucleate is dialyzed overnight at 4° against 18 l. of flowing distilled water in a Kunitz-Simms[3] rocking dialyzer. The dialyzed nucleate is concentrated to about a twentieth of its volume by a stream of air directed against the collodion bag, and the small amount of insoluble matter appearing during concentration is removed by centrifugation at 3000 to 7000 r.p.m. for about 10 minutes. (If the equipment is available, a superior removal of impurities is achieved at this stage by centrifugation at 40,000 r.p.m. for 1 hour in one of the 40 series of rotors of the Spinco model L centrifuge. The gravitational field thus obtained will remove not only insoluble matter but also un-denatured virus and soluble virus fragments, while the nucleate remains in the supernatant fluid.) The clear, viscous solution of nucleate can be used directly, or it can be lyophilized to give a white solid which dissolves very readily in water.

RNA preparations obtained by this method should give negative tests for protein by the most sensitive tests and should have extinction coefficients per mole of P at 260 mμ of about 9000.

Ethanol Method.[4,5] At present this method has been applied successfully only to turnip yellow mosaic virus. It is probably the mildest method by which any RNA has been prepared, and, since turnip yellow mosaic virus preparations contain as much as 35% RNA, this virus constitutes a good source of nucleic acid.

A solution of virus at about 10 mg./ml. in water containing a small amount of salt is adjusted, if necessary, so that it is neutral to bromothymol blue. (The presence of salt is essential for the subsequent coagulation of protein and release of nucleic acid. However, the common method of preparing the virus[4] involves precipitation with ammonium sulfate, and, when such precipitated virus is redissolved in water, the residual salt

---

[3] M. Kunitz and H. S. Simms, *J. Gen. Physiol.* **11**, 641 (1928).
[4] R. Markham and K. M. Smith, *Parasitology* **39**, 330 (1949).
[5] R. Markham, personal communication.

is adequate without further addition.) An equal volume of ethanol, previously shaken with $CaCO_3$ to neutralize, is added, and after about 2 minutes the coagulated protein is spun down by centrifugation for 15 minutes at 10,000 r.p.m. The supernatant fluid contains the nucleate, which can be precipitated by raising the ethanol concentration to 75%.

Detergent Method.[6] This modification of the method used by Bawden and Pirie[7] appears to give higher yields. The disadvantages of the detergent method as compared with the other two are the greater number of manipulations required in the preparative procedure and the greater difficulty in separating reagent from nucleic acid-free protein when it is desired to save the latter. On the other hand, the detergent method has been applied successfully to more viruses than the other techniques, having been used in its present form to isolate nucleic acid from tomato bushy stunt, potato X, and southern bean mosaic viruses.[6] From the results of other experiments,[8] it is likely that the method could also be applied to other viruses.

To 4 ml. of aqueous solution of virus at about 10 mg./ml. in a 12-ml. glass tube is added 1 ml. of 10% Duponol C solution (purified sodium dodecyl sulfate gives similar results). The mixture is heated in a boiling water bath for 4 minutes and then chilled in an ice bath. The reaction mixture is dialyzed against several changes of distilled water (or against flowing water in the Kunitz-Simms apparatus) in order to remove the bulk of the free detergent. The dialyzed preparation is then made 1 N with respect to sodium chloride by adding an appropriate amount of 5 N sodium chloride and heated at 100° for 3 minutes. After being chilled in an ice bath, the protein precipitate is centrifuged off, washed twice with small portions of hot 1 N sodium chloride, and the combined supernatant fluid and washings dialyzed against several changes of distilled water (or against flowing water in the Kunitz-Simms apparatus). The final dialyzed preparation is concentrated to about a twentieth of its volume by a stream of air directed at the collodion bag. The concentrate is clarified by centrifugation as described for the heat denaturation method, to yield a clear, viscous solution of the nucleate which may be used directly or which may be converted to a white solid by lyophilization. When such nucleate preparations are tested for protein, as, for example, by a sensitive variation of the biuret test,[9] from essentially nothing to about 5% protein will be found, depending on the virus.[6] The yield of

[6] R. W. Dorner and C. A. Knight, *J. Biol. Chem.* **205,** 959 (1953).

[7] F. C. Bawden and N. W. Pirie, *Biochem. J.* **34,** 1278 (1940).

[8] F. C. Bawden, "Plant Viruses and Virus Diseases," 3rd ed., Chronica Botanica Co., Waltham, Mass., 1950.

[9] J. D. Smith and R. Markham, *Biochem. J.* **46,** 509 (1950).

nucleate for the three viruses listed above ranged from 35 to 80%, based on recovery of P. It is noteworthy that the concentrations of virus and detergent to be used for best results vary somewhat for different viruses and, therefore, the procedure given above, although basic, may have to be modified with respect to virus and detergent concentrations to give optimal results with each virus.

The following modification of the detergent method has provided an additional procedure for obtaining RNA from tobacco mosaic virus, and the product has been employed in experiments on the reconstitution of virus by co-polymerization of nucleic acid and protein components.[10] Equal volumes of 2% solutions of virus and sodium dodecyl sulfate, adjusted to pH 8.5, are mixed and held at 40° for 16–20 hours. Following this treatment, ammonium sulfate is added to 0.35 saturation, and the protein precipitate is separated by centrifugation. The supernatant fluid is stored overnight at 4° during which time 60 to 90% of the nucleic acid precipitates out and is obtained by centrifugation. The RNA is further purified by twice dissolving in ice water and precipitation with two volumes of cold ethanol, adding, if necessary, a few drops of 3 M acetate at pH 5. The RNA solution is finally centrifuged with refrigeration at 40,000 R.P.M. for 2 hours. The supernatant fluid, containing the RNA, should give a spectrophotometric curve characterized by a minimum at about 230 mμ, a maximum at about 258 mμ and a ratio of the maximum to minimum absorbance of about 3.0.

[10] H. Fraenkel-Conrat and R. C. Williams, *Proc. Natl. Acad. Sci. (U.S.)* **41**, 690 (1955).

[101] Preparation of Ribonucleic Acid from Yeast and Animal Tissues

By GERHARD SCHMIDT

I. RNA from Yeast

Among the cellular organisms, yeast is at present the most satisfactory starting material for the preparation of practically pure RNA in quantity because its DNA content is so small (<2% of RNA) that steps aiming at the separation of the two nucleic acid types are practically superfluous. This permits the application of suitable denaturation procedures resulting in instantaneous inactivation of nucleases as an initial step; enzymatic degradation of RNA during its isolation is thus efficiently avoided.

In the earlier methods, the separation of RNA from the proteins was accomplished by treatment of the yeast cells with alkali.[1] This principle is now completely abandoned, owing to the instability of RNA under these conditions; for a considerable period this treatment was replaced by the use of salt solutions of high ionic strength effecting dissociation of the nucleoproteins into their components.[2] Since the experiments of Sreenivasaya and Pirie[3] and of Mirsky and Pollister,[4] detergents (particularly sodium dodecyl sulfate) are increasingly used for the extraction of ribonucleates because the RNA specimens obtained are of higher molecular weights and because the yields are much better in comparison with the results of procedures involving the exclusive use of salt solutions.

The best available procedure for the preparation of yeast RNA is that of Crestfield et al.[5]

Principle. Ribonucleate is extracted from the yeast cells (*S. cerevisiae*) by short heating with an aqueous solution of sodium dodecylsulfate, the crude ribonucleate is precipitated by alcohol, and pure ribonucleate settles out from the aqueous solution of the precipitate on addition of sufficient amounts of sodium chloride to bring its concentration to normality and on subsequent standing of this solution at 0°.

Purification of Sodium Dodecylsulfate. Portions (300 g.) of Duponol C are extracted with 7 l. of boiling ethanol. The hot solution is filtered through a Büchner funnel over a pad of Filter-Aid and yields 170 g. of crystalline sodium dodecylsulfate on standing at 0°. (S, 9.8%; theoretical value, 11.1%.)

Extraction of RNA. A 4-l. beaker is covered with a plate containing a narrow slot which permits insertion of a thermometer and of the rod of a mechanical stirrer. To the beaker is added 500 ml. of an aqueous solution containing 2% sodium dodecylsulfate, 4.5% ethanol, and 0.0125 M primary and 0.0125 M secondary sodium phosphates. The solution is heated to boiling by a Fisher burner with continuous stirring. Then 150 g. of yeast cut into very fine pieces is added in one batch to the boiling mixture. Immediately after addition of the yeast the beaker is again covered. The solution, whose temperature is 83 to 87° immediately after addition of the yeast, is heated for 1 minute with the burner under stirring (temperature 92 to 94°) and then transferred for 2 minutes to a boiling-water bath in order to avoid excessive foaming. The hot solution is cooled to 4° within 4 to 8 minutes by pouring it into a 2-l. beaker which is immersed

[1] E. J. Baumann, *J. Biol. Chem.* **33**, XIV (1918).
[2] G. Clarke and S. B. Schryver, *Biochem. J.* **11**, 319 (1917).
[3] M. Sreenivasaya and N. W. Pirie, *Biochem. J.* **32**, 1707 (1938).
[4] A. E. Mirsky and A. W. Pollister, *J. Am. Chem. Soc.* **74**, 1724 (1952).
[5] A. M. Crestfield, K. C. Smith, and F. W. Allen, *J. Biol. Chem.* **216**, 185 (1955).

into a mixture of dry ice and Cellosolve, with manual stirring. The nucleate extract is centrifuged for 30 minutes at 0° at 2000 r.p.m.

Precipitation of Crude Nucleic Acid Fraction. The supernatant solution is poured into 2 vol. of cold ethanol. The suspension is centrifuged in the cold at 2000 r.p.m. for 15 minutes, and the precipitate is washed twice with 150-ml. portions of 67% alcohol. Flocculation is achieved by adding 5 to 10 drops of a 2 N sodium chloride solution to the suspension of the precipitate in the wash alcohol. Dispersion of the precipitate in the washing liquid is preferably carried out by gradual addition of the washing liquid and thorough mixing with the aid of a glass rod. The washed precipitate is suspended in 80% ethanol and left overnight in the refrigerator.

Isolation of RNA. After centrifugation, the precipitate is dissolved in 130 to 180 ml. of water. A turbid solution with a pH of approximately 8 results. The pH is brought to 7 by the addition of 1 N acetic acid. The solution is centrifuged in a preparative Spinco centrifuge for 30 minutes at 20,000 r.p.m. at 0°. Some floating particles may resist centrifugation. The supernatant solution is mixed with a sufficient amount of solid sodium chloride to bring its concentration to normality. The mixture is left standing at 0° for 30 minutes. Under these conditions the ribonucleate separates out as a gel which is collected by centrifugation at 2000 r.p.m. for 1 hour. The gel is washed with three successive portions of 150 ml. of 67% alcohol to each of which 1 ml. of a 2 M solution of sodium chloride has been added. The precipitate is dissolved by gradual addition of water under continuous stirring (pH 7). The solution is dialyzed for 36 hours against distilled water at 4°. The water is frequently changed. The dialyzed solution is filtered through Celite (Johns-Manville) on a Büchner funnel, and finally through a D-7 Steriflo asbestos pad (F. R. Hormann and Company). The clear, colorless solution is lyophilized. The yield is 1.4 g.

Some Properties of RNA Isolated According to the Procedure Described

Sodium ribonucleate obtained according to the method described above has a tough consistency and is of fibrous structure. It dissolves only slowly in water. Phosphorus, 8.2%; secondary phosphoryl groups, approximately 6.4% of the amount of total phosphoryl groups (purified commercial RNA, 17%). Intrinsic viscosity: 70 (purified commercial RNA, 7.8).[6] Extinction at 260 mμ, $E_{1\,cm.}^{1\%} = 208$ (determined in the presence of 0.01 M hydrochloric acid); extinction at 260 mμ after alkaline hydrolysis to mononucleotides, $E_{1\,cm.}^{1\%} = 284$ (determined in the presence of

---

[6] The viscosity of ribonucleate solutions prepared according to Crestfield *et al.*[5] decreases soon during aging of the solutions; it is also dependent on the extent of grinding applied to the dry preparation prior to dissolving it.

0.01 M hydrochloric acid). Sedimentation in the ultracentrifuge yields two boundaries. Freshly dissolved samples contained predominantly the slower component, whereas samples which had been stored for 1 to 2 days in the refrigerator yielded mainly the boundary of the faster component.

II. RNA from Animal Tissues

All procedures available for the preparation of RNA from animal tissues are less satisfactory than that described for its isolation from yeast. One of the reasons for this fact is the presence in animal tissue of amounts of DNA accounting for higher percentages of the total nucleic acid fraction than those found in yeast. The only known exception is the composition of the nucleic acid fraction in unfertilized arbacia eggs in which the concentration of DNA is extremely small compared to that of RNA.

So far, the removal of DNA contaminations from RNA fraction has been accomplished only on undenatured nucleoprotein mixtures under conditions which are not well defined and which are not easily reproducible. DNA-containing nucleoproteins have been shown to be selectively precipitated in guanidine chloride solutions of high ionic strength.[7] Kay and Dounce[8] used the insolubility of DNA nucleoproteins in sodium chloride solutions of low ionic strength for their separation from RNA protein complexes by centrifugation.

All these procedures, however, involve relatively protracted treatments of the tissue suspensions under conditions compatible with the activity ranges of nucleases. A certain degree of enzymatic degradation of RNA during the preparation is thus unavoidable. This difficulty is particularly serious in attempts to isolate RNA from pancreas, owing to the stability of ribonuclease activity even toward relatively drastic conditions of protein denaturation. The only possibility to obtain pancreas-RNA which is not partially degraded by ribonuclease during the isolation is the deproteinization of frozen ground pancreas tissue with trichloroacetic acid and the subsequent separation of RNA from the denatured proteins.[9] According to the experiences of the author, it is unavoidable, however, that RNA samples prepared in this way are contaminated with 10 to 20% of DNA.

So far, the best source for the preparation of animal ribonucleic acid is liver tissue, whose ribonuclease activity is relatively low. The procedure described is that of Kay and Dounce.[8] Other isolation methods have been developed by Volkin and Carter[7] and by Grinnan and Mosher.[10]

[7] E. Volkin and C. E. Carter, J. Am. Chem. Soc. **73**, 1516 (1951).
[8] E. R. M. Kay and A. L. Dounce, J. Am. Chem. Soc. **75**, 4041 (1953).
[9] J. E. Bacher and F. W. Allen, J. Biol. Chem. **183**, 641 (1950).
[10] E. L. Grinnan and W. Mosher, J. Biol. Chem. **191**, 719 (1951).

Preparation of Liver-RNA According to Kay and Dounce[8]

The tissue is cooled in an ice bath as soon as possible after the death of the animal. It may be stored at $-15°$ unless it is used immediately. Fifty grams of the frozen material is finely chopped and homogenized for 5 minutes at 0 to 3° in a Waring blendor with 200 ml. of an ice-cold 0.9% solution of sodium chloride made up in 0.01 M sodium citrate solution. The homogenate is centrifuged for 30 minutes at a temperature of 0 to 3° at 2500 r.p.m. for the removal of deoxyribonucleoproteins. The centrifugation is repeated with the decanted supernatant solution. A sufficient amount of N hydrochloric acid is added to the decanted supernatant solution to bring the pH to 4.5. After centrifugation of the heavy precipitate in the cold for 15 minutes, the sediment is suspended in 0.9% sodium chloride solution to a volume of 100 ml. Then 9 ml. of a 2% solution of sodium dodecyl sulfate is added to the suspension with stirring. The pH is brought to 7 by cautious neutralization with a 10% solution of sodium hydroxide, and the stirring is continued for 3 hours. Solid sodium chloride is added to the suspension until its concentration is 1 M. The slightly viscous solution is centrifuged in a Servall centrifuge at 14,000 r.p.m. The nucleic acid is precipitated from the decanted, clear supernatant solution by the addition of 2 vol. of alcohol. The precipitate is centrifuged, washed with alcohol and acetone, and dried in air. The treatment with dodecyl sulfate (stirring time 2 hours) and the precipitation of RNA is repeated according to the procedure just described, on the solution of the dried precipitate, which is dissolved in 100 ml. of water.

The dried nucleate precipitate is dissolved in 30 ml. of water, and the solution is cooled to 0°. The solution is brought to a sodium chloride concentration of 0.9% by the addition of the solid salt, and the pH is adjusted to 4.5 by cautious addition of N hydrochloric acid. The solution is centrifuged in the cold for 1 hour at 14,000 r.p.m. The supernatant solution is decanted into a beaker containing a sufficient quantity of sodium chloride to bring its concentration to 1 M. After the salt has been completely dissolved, the pH is adjusted to 7.0 by cautious addition of a 0.1 M solution of sodium hydroxide in the cold. The nucleate is precipitated by the addition of 2 vol. of 95% ethyl alcohol. The precipitate is centrifuged, washed with alcohol and acetone, and dried in the air.

The preparation gives negative biuret, Sakaguchi, and diphenylamine tests. It contains 7.5% phosphorus; extinction at 260 mμ, $E_{1\,cm.}^{1\%} = 191$. At 30°, a 3% aqueous solution of the ribonucleate has a relative viscosity of 2.25.

[102] Isolation of Sodium Deoxyribonucleate in Biologically Active Form from Bacteria

By ROLLIN D. HOTCHKISS

Principle and General Remarks

Bacteria are disintegrated, by techniques suited to the strain used, in the presence of sodium citrate, which reduces the Mg^{++} concentration sufficiently to minimize deoxyribonuclease destruction of the DNA.[1] The salt present is low enough so that DNA becomes soluble only as fast as it becomes dissociated from DNA-protein or other combinations. This limiting process varies in rate and completeness from strain to strain of bacteria.

Native biologically active DNA can withstand alcohol precipitation, shaking with chloroform or surface-active denaturing agents, heating in water up to 80°, and pH variation between about 4.5 and 10, but is denatured by a variety of exposures beyond these limits. Therefore it is always maintained as a Na, Ca, or other salt, and DNA here should be understood to mean the Na salt. Deproteinization procedures take advantage of the chloroform surface denaturation (method of Sevag[2]) or treatment with surface-active agents.[3] Ribonuclease is used to depolymerize RNA; the action often appears sluggish in the presence of DNA. Degraded RNA, small proteins which not readily denatured, and most polysaccharides can usually be separated by their greater solubility, as compared with DNA, on alcohol precipitation.[1] (In particular, they are not likely, even if precipitated, to gather as a fibrous "spool" removable by winding on a glass rod.)

For quick removal of deoxyribonuclease and bulk protein impurities, a prompt chloroform treatment is given. Early subsequent precipitation by alcohol aids in the denaturation of more protein and speeds the removal of the residual protein, including the difficultly denaturable ribonuclease, if the latter has been added.

Reagents

Na deoxycholate, 5%. Crude or purified deoxycholic acid is suspended in H_2O and brought into solution with small portions of 2 N to 6 N NaOH. After back neutralization to pH approx-

---

[1] M. McCarty and O. T. Avery, *J. Exptl. Med.* **83**, 97 (1946).

[2] M. G. Sevag, D. B. Lackmann, and J. Smolens, *J. Biol. Chem.* **124**, 425 (1938).

[3] E. R. M. Kay, N. S. Simmons, and A. L. Dounce, *J. Am. Chem. Soc.* **74**, 1724 (1952).

imately 8 to 9 if necessary, the solution is adjusted with H_2O to contain 50 mg. of the Na salt per milliliter.

Saline, 0.85 g. of NaCl per 100 ml. in H_2O.

Citrate-saline. The above saline is brought to about 0.1 M in Na citrate by addition of 30 mg. of crystalline trisodium salt per milliliter.

Ribonuclease. Approximately 0.5 mg./ml. of solution of crystalline ribonuclease. It is important that this preparation be known not to act on (reduce viscosity of) DNA. A qualitative test with a viscous solution of calf thymus DNA will suffice.

Meat infusion-neopeptone medium. Aqueous infusion of macerated beef hearts (450 g. of meat to 1 l.) is heated to 85°, filtered, and 5 g. of NaCl and 10 g. of neopeptone (Difco Neopeptone) are added per liter. The whole is sterilized by autoclaving, clarified if necessary, and brought to pH 7.5.

Procedure: Preparation of active transforming DNA from pneumococcus. Pneumococci are grown as unshaken cultures in 1.5-l. portions of meat infusion-neopeptone medium (refined media based on casein hydrolyzate and vitamins may be used also), contained in 2-l. flasks. The inoculum consists of 10 ml. of a fresh culture of the donor strain from which DNA is to be prepared. Sterile glucose (3 ml. of 20% solution) is added to each flask, and alkaline phosphate (15 ml. of sterile 0.5 M K_2HPO_4) to control the acidity that will be formed. After about 16 hours, growth is heavy and can be augmented by 2 to 3 hours of incubation after further additions of glucose, and phosphate or NaOH to maintain the pH near 7.5. The culture should not be allowed to reach a pH below 6 and should (with pneumococcus) be used within 1 to 2 hours after it has reached maximum growth as judged by previous experience with the medium used in similar vessels and similar conditions of aeration.

The cells are centrifuged down at an appropriate speed (e.g., 2500 to 5000 r.p.m. in the usual laboratory centrifuge),[4] and the supernatant is removed by decantation and pipetting. The cells are collected and resuspended, without washing, in citrate-saline (about 0.5% and not over 1% of the original culture volume) at about 25°. Residual culture medium may contribute to this volume, provided 30 mg. of crystalline trisodium citrate is added for each milliliter of medium remaining.

[4] It has proved convenient with pneumococcus to add a low concentration of antiserum (e.g., 2 ml. of crude therapeutic antipneumococcal horse serum or 0.5 ml. of concentrated antiserum globulin per liter) to give a partial agglutination of the cells. This treatment, followed by 30 minutes of incubation, allows much more complete separation of the cells at lower speeds and gives no interference with further purification.

Lysis is initiated by adding 1 ml. of 5% sodium deoxycholate per 25 ml. of suspension. Left at room temperature with occasional shaking, the culture becomes within 1 to 2 minutes progressively more ropy and then viscous. After about 5 minutes no further change is apparent, and $\frac{1}{2}$ to 1 vol. of chloroform and $\frac{1}{40}$ vol. of isoamyl alcohol are added. Vigorous mechanical shaking in a well-stoppered centrifuge bottle for 15 to 20 minutes leads to protein interfacial denaturation. The operation at this point and from here on (except the treatment with ribonuclease) may be conducted at lowered temperature. Efficient completion of the processing within 2 to 3 hours at room temperature appears, however, to give as satisfactory a product, and protein denaturation is favored by alcohol precipitation in this way. Overnight steps are conducted at refrigerator temperature.

Centrifugation now leads to the separation of three layers: a bottom layer of chloroform, an intermediate layer of denatured protein-chloroform emulsion, and a cloudy, yellowish, viscous supernatant containing DNA. Ordinarily, washing or processing of the emulsion does not prove worth while; if there is reason to believe that the DNA-proteins have not been completely dissociated, the emulsion may be re-extracted with deoxycholate or 1 M NaCl.

The turbid top layer is removed by a pipet and precipitated with 1.5 vol. of 95% alcohol. At this point the precipitate is stringy, but often not the coherent fibrous mass that is given by purified DNA. It is collected promptly by winding on a clean glass rod or by centrifugation, care being taken to exclude in so far as possible the slowly forming, more compact precipitate, containing sodium citrate and other contaminants from the lysate. The stringy and partly flocculent precipitate is returned (within 30 minutes for ease of solution) into saline and shaken to disperse. The appropriate volume will depend on the yield—usually about 10 ml. per 1.5 l. of original culture gives a turbid solution, freely manipulated, yet viscous enough to retard the rise of small air bubbles.

If RNA is to be removed, 0.05 mg. of crystalline ribonuclease is added per 10 ml., the solution is incubated for 15 minutes at 37°, a second portion of enzyme is added, and incubation is continued for a further 15 minutes. The solution is overlayered with 1.5 vol. of alcohol, and the two layers are mixed with a glass rod moved in a circular path, always in one direction. Fibrous precipitate will wind upon the rod; if difficulty is encountered in recovering all of it, brief centrifugation and alcohol reprecipitation of the sediment from solution in a smaller volume of saline should serve to separate most of the high-molecular DNA. The precipitate has a smaller bulk than before, since the crude material contains three to four times as much RNA as DNA. (Addition of crystalline deoxyribonuclease instead of the ribonuclease gives at this step a high-molecular preparation of RNA.) The combined fibrous precipitates are redissolved without being allowed to dry, in about 0.8 or less of the previously used volume of saline, shaken as before with chloroform and isoamyl alcohol for 15 minutes, and then centrifuged. The protein-containing emulsion layer is now usually much smaller than before, and the upper layer is nearly clear and colorless. After pipetting off, the upper layer is treated with chloroform at least once more; the material is often virtually protein-free at this point and may be considered so when the chloroform emulsion layer is only a faint scum (after the room-temperature alcohol precipitations already specified).

Two to three alcohol precipitations of the aqueous layer are used to remove the chloroform, amyl alcohol, and remaining low-molecular impurities, and to sterilize the material if it is to be used for bacteriological purposes. Precipitations can be conducted conveniently and rapidly by adding 1.2 to 1.5 vol. of alcohol and winding the fibrous fresh precipitate upon a glass rod, removing the "spool" of precipitate into a new portion of saline (sterile if desired). Much of the alcohol can be removed by squeezing the precipitate against the wall of the container while it is on the glass rod. It is important, if the glass rod is used to gather precipitate, that the solution be kept at an appropriately small volume (1 to 5 ml./ mg. of DNA); otherwise more and more material will be discarded at each precipitation as fibers too small to be collected.

The clear colorless solution contains DNA in biologically active form (see Vol. III [105]). Some of the DNA may have been denatured by careless handling or delay; a small amount of degraded RNA ("core material") may remain. Ordinary proteins, polysaccharides, and low-molecular materials have been essentially completely removed. From 1 l. of pneumococcal culture (about 30 mg. of dry bacteria) approximately 1 mg. of DNA can be recovered. From a liter of the more dense *Escher-ichia coli* cultures, after the longer extractions mentioned below, approximately 6 to 8 mg. of DNA has been recovered. Preparations are well preserved in physiological saline solutions in the cold, and, better still, as alcohol precipitates, under 70 to 90% aqueous alcohol, well stoppered.

Modifications of the Procedure. It is possible to recover DNA from uncentrifuged pneumococci by conducting lysis and alcohol precipitation directly in the citrate-treated culture. To prepare cells of other bacterial species for deoxycholate extraction, it has variously been useful to freeze and thaw, to grind, or sonically to disintegrate centrifuged cells. It has been helpful to conduct the deoxycholate treatment for longer times, or at temperatures from 37° up to 55°, if a viscous extract is not obtained at lower temperatures. *Hemophilus influenzae* DNA, which can also be tested for biological activity, has been separated from protein by Duponol treatment, and RNA has been removed by adsorption on charcoal or, together with polysaccharides, by electrophoresis[5] (see also Vol. III [103]). From staphylococci, a DNA preparation having apparently specific stimulatory action on protein synthesis was obtained[6] by a method involving fractionation with an organic quarternary base.[7]

Deoxyribonuclease from some sources is activated by Mn^{++} as well as by Mg^{++}; it should be pointed out that citrate does not effectively repress the nuclease if Mn^{++} is present. Ethylenediaminetetracetate may be used in place of citrate as complexing agent.[5]

For use in ordinary microbial genetic work, the DNA preparations often need not be freed from RNA; the latter may even serve some purpose as a buffer, a stabilizer, and a "carrier" in reprecipitation.

[5] S. Zamenhof, G. Leidy, H. E. Alexander, P. L. FitzGerald, and E. Chargaff, *Arch. Biochem. and Biophys.* **40,** 50 (1952); S. Zamenhof, H. E. Alexander, and G. Leidy, *J. Exptl. Med.* **98,** 373 (1953).

[6] E. F. Gale and J. P. Folkes, *Biochem. J.* **59,** 661 (1955).

[7] A. S. Jones, *Biochim. et Biophys. Acta* **10,** 607 (1953); S. K. Dutta, A. S. Jones, and M. Stacey, *ibid.*, p. 613.

[103] Preparation and Assay of Deoxyribonucleic Acid from Animal Tissue

By STEPHEN ZAMENHOF

Introduction

At present it is not known whether even a most carefully isolated DNA can in all respects be identical with the DNA as it existed in the living cell. However, such a carefully isolated DNA preparation may exhibit certain features which it had when it was in the living cell. The most obvious and the most important of these features is the proper biological activity. Such a DNA preparation may be called "functionally intact," and its properties are of more interest than the properties of the products of an undetermined degree of "denaturation"; in particular, the studies of the functionally intact DNA preparation may lead to correlation between the function and the structure.

One proper biological activity which can be demonstrated in certain DNA preparations is their transforming activity[1,2] (see Vol. III [105]). In

[1] O. T. Avery, C. M. MacLeod, and M. McCarty, *J. Exptl. Med.* **79,** 137 (1944).

[2] H. E. Alexander and G. Leidy, *J. Exptl. Med.* **93,** 345 (1951).

such preparations one can indeed study the physicochemical properties and the biological activity and determine the correlations between these two properties.[3]

Unfortunately, at present only a few bacterial DNA's lend themselves to the assay for transforming activity. The studies of other DNA preparations are open to the objection that the starting material was a product of degradation. Some of the properties studied, such as chemical composition, are less likely to change on mild "denaturation" (for a recent review, see ref. 4); others, such as resistance to various physical and chemical agents, will be greatly altered.[3,5]

Although a completely satisfactory study of such DNA preparations cannot be offered at present, advantage can be taken of the fact that the DNA of different species, although chemically different,[4] do exhibit similarities in their resistance to various agents. Thus, the isolation of any DNA under conditions which would not inactivate another DNA having transforming activity may lead to a "functionally intact" product. Another possibility is the addition of an active transforming principle to serve as a marker.

The above considerations are the basis of the isolation procedures described in this chapter.

Sources of DNA

Although in theory every animal tissue is a source of DNA, in practice some types of tissue are less suitable for the following reasons.

1. The tissue may exhibit deoxyribonuclease (DNase) activity difficult to inhibit and sufficiently strong to injure the DNA during the initial stages of preparation (example: pancreas).

2. The high ratio of other constituents (RNA, proteins, polysaccharides) to DNA may make the purification tedious (example: egg).

3. The strong bond between the DNA and the protein may make the extraction difficult (example: mammalian sperm[6]).

Unquestionably the best source in all the above respects is the thymus gland. The DNase content is low,[7] and its injurious action can be prevented by the steps described later on. The ratio of DNA to other cell constituents is high and can be easily increased by washing the homog-

---

[3] S. Zamenhof, H. E. Alexander, and G. Leidy, *J. Exptl. Med.* **98**, 373 (1953).

[4] S. Zamenhof, *in* "Phosphorus Metabolism" (McElroy and Glass, eds.), Vol. 2, p. 301, The Johns Hopkins Press, Baltimore, 1952.

[5] S. Zamenhof, G. Griboff, and N. Marullo, *Biochim. et Biophys. Acta* **13**, 459 (1954).

[6] S. Zamenhof, L. B. Shettles, and E. Chargaff, *Nature* **165**, 756 (1950).

[7] S. Zamenhof and E. Chargaff, *J. Biol. Chem.* **180**, 727 (1949).

enate prior to extraction. The latter can be easily achieved by strong salt solutions.[8] Calf thymus is most easily obtainable in the metropolitan area, but the thymus of other animals (pig, sheep) may be used as well.

Another recommendable tissue is the spleen; however, the DNA content and the DNase absence are here less satisfactory than in the thymus.

A convenient source is found in fowl (chicken, duck) or fish (salmon, carp) erythrocytes.[4]

The preparations from these three sources will be described below. A classical source rich in DNA is also found in fish and invertebrate spermatozoa;[4] it must, however, be emphasized that the seminal fluid may contain strong DNase.[6]

Occasionally, the nature of the problem under investigation may involve the isolation of DNA from less suitable tissues or even from the whole animal. In these cases the methods described below under Special Purification Procedures may lead to products of satisfactory purity; however, the integrity of the DNA so obtained may be lost in cases when the enzymatic activity is too strong.

Preparation of DNA

1. From Calf Thymus Glands.[5] Not more than 30 minutes after the glands have been removed from the animal they are frozen on dry ice. When so preserved, the glands can be stored at −30° for several days prior to processing; if the glands are frozen later than 2 hours after removal, the final product may already be damaged as tested by its stability to heat (see below under Assays).

Fifty grams of frozen glands are thawed, freed from fat, cut to small pieces, and washed in five 100-ml. ice-cold portions of 0.1 M aqueous sodium citrate (pH 7.4) or 0.1 M ethylenediaminetetraacetate (EDTA) (pH 7.35). The mixture is then homogenized in a cooled Potter-Elvehjem glass tissue grinder with addition of 250 ml. of ice-cold sodium citrate or EDTA as above. The homogenate is centrifuged for 30 minutes at 1800 × g, and the supernatant discarded. The residue is then stirred in 250 ml. of citrate or EDTA and centrifuged as before. This operation is repeated two more times. The final residue is extracted with 1000 ml. of 2 M aqueous NaCl solution.[8] The extraction is best accomplished by resuspending small portions of the residue in 2 M NaCl, homogenizing in a cooled glass tissue grinder for 3 minutes, and gradually adding more 2 M NaCl solution until the total volume for this portion has been used up and the residue is uniformly distributed. The mixture is left at 4° for 1 day and then centrifuged for 1 hour at 1900 × g. To the opalescent supernatant 2 vol. of absolute ethanol is added at a rate of 150 ml./hr.,

[8] A. E. Mirsky and A. W. Pollister, *Proc. Natl. Acad. Sci. (U.S.)* **28**, 344 (1942).

with constant slow swirling of the recipient flask; the resulting fibers are lifted, washed in 75% aqueous ethanol, drained, and redissolved, as described before, in a cooled tissue grinder in 1000 ml. of 0.14 M aqueous NaCl solution made 0.015 M with respect to sodium citrate (pH 7.1), hereafter called "standard buffer." To this solution is added $\frac{1}{9}$ vol. of 5% Duponol[9] in 45% aqueous ethanol,[10,11] and the mixture is stirred for 1 hour at room temperature. Solid NaCl is then added, to obtain a final concentration of 5% NaCl. The mixture is stirred for another $\frac{1}{2}$ hour at room temperature and then left overnight at 4°. The mixture which contains precipitated Duponol is then centrifuged for 1 hour at 31,000 $\times$ g; to the clear supernatant 2 vol. of ethanol is slowly added, as described before; the resulting fibers are lifted, washed in 75% aqueous ethanol, redissolved in 1000 ml. of standard buffer with $\frac{1}{9}$ vol. of 5% Duponol, and the mixture is stirred for 1 hour. Solid NaCl is then added to obtain 5% NaCl solution, and the mixture is stirred, stored, and centrifuged, as described before. To the final clear supernatant 2 vol. of absolute ethanol is slowly added, as described before, and the fibers are lifted, washed in 75% ethanol, and redissolved in standard buffer to yield a clear, very viscous solution, containing 1 to 1.5 mg. of DNA/per milliliter. Total yield, 600 to 950 mg.

The product is stored in the frozen state at $-15°$. When so stored, it does not undergo any demonstrable change for at least 1 year. Storage in nonbuffered or low ionic strength solutions as well as drying (even from the frozen state) are avoided, as they invariably lead to a degraded product.[3,5] Drying with absolute ethanol and ether is particularly harmful.[3,5]

All the steps are performed at 0 to 4°, with glass, rubber, or plastic tools and vessels. In particular, minute traces of rust ($Fe^{++} \rightarrow Fe^{+++}$) cause rapid degradation;[3,5] for this reason the use of a Waring blendor or other "stainless steel" disintegrators is avoided.

All the steps from thawing the glands until the first alcohol precipitation are performed as fast as possible; otherwise, owing to the action of thymus DNase, a product is obtained which can be shown to be degraded (see under Assays).

The above described precautions refer to DNA from all sources investigated thus far.

2. From Calf Spleen.[5] One hundred grams of spleen is reworked as described for calf thymus, but with only 250 ml. of solution for extraction

[9] A mixture of sodium lauryl sulfate and other fatty alcohol sulfates manufactured by E. I. Du Pont de Nemours & Co., Wilmington, Delaware.

[10] A. M. Marko and G. C. Butler, *J. Biol. Chem.* **190**, 165 (1951).

[11] A. L. Dounce, N. S. Simmons, and E. R. M. Kay, *Federation Proc.* **10**, 177 (1951); E. R. M. Kay, N. S. Simmons, and A. L. Dounce, *J. Am. Chem. Soc.* **74**, 1724 (1952).

or for deproteinization because of the lower DNA content in spleen tissue. The finally deproteinized material is redissolved in standard buffer to yield a clear, very viscous solution containing 1 to 1.5 mg. of DNA/per milliliter. Total yield, 100 to 300 mg.

3. *From Chicken Erythrocytes.* One hundred milliliters of fresh chicken blood (from approximately three chickens) is mixed with 24 ml. of 0.1 M sodium citrate solution. The mixture is centrifuged for $\frac{1}{2}$ hour at 1800 $\times$ g, and the plasma is discarded. The packed erythrocytes (containing some white cells) are frozen (12 hours at $-15°$) and thawed at 23°. The cells, now broken by freezing and thawing, are stirred in 60 ml. of the same citrate solution and centrifuged for $\frac{1}{2}$ hour at 1800 $\times$ g. The supernatant is discarded. This last operation is repeated three times more. The washed nuclei so obtained are suspended in 1 vol. of the same citrate solution and homogenized in a cooled glass tissue grinder as described before. The homogenate is centrifuged for 1 hour at 1800 $\times$ g, and the supernatant discarded. The sediment is resuspended in 1 vol. of the same citrate solution, again centrifuged as above, and the supernatant is discarded. The sediment is then extracted with 400 ml. of 2 M NaCl solution. Further procedure is as described for calf thymus, but with 400 ml. of standard buffer for deproteinization. Total yield, 300 to 340 mg.

Special Purification Procedures

Removal of RNA.[12] This step may be necessary when DNA is being prepared from organs rich in RNA (such as liver).

Activated charcoal (Norit A, pharmaceutical grade) is washed with running water for 18 hours, then once with 8 vol. of 2 M NaCl solution and twice with 8 vol. of standard buffer. Then $\frac{1}{15}$ to $\frac{1}{20}$ vol. of the wet charcoal is added to the DNA-RNA solution in standard buffer containing 0.5 to 1 mg. of DNA per milliliter. The mixture is shaken at 0 to 4° for 1 hour and then centrifuged at 31,000 $\times$ g for 1 hour. DNA recovery is of the order of 94%. RNA content can be reduced to less than 1%, depending on the amount originally present.

Removal of Polysaccharides. No universal chemical method for removal of high molecular weight polysaccharides has been reported. Purification by electrophoresis[13] is satisfactory, but the amounts so purified are limited.

DNA solution to be purified, containing 1 to 3.3 mg. of DNA per milliliter of standard buffer, but no impurities other than polysaccharide, is dialyzed for 12 hours against the 200-fold volume of the

---

[12] S. Zamenhof and E. Chargaff, *Nature* **168**, 604 (1951).
[13] S. Zamenhof, G. Leidy, H. E. Alexander, P. L. FitzGerald, and E. Chargaff, *Arch. Biochem. and Biophys.* **40**, 50 (1952).

same buffer. The solution is then subjected to electrophoresis at 1.5° in a Tiselius cell,[14] with automatic slow compensation, for at least 7 hours, at an electrode potential of 120 volts. At the end of this period the compensation is stopped and the still hypersharp peak (boundary) of DNA is permitted to migrate into the clean top section of the cell up to the capacity of the latter. If the polysaccharide itself exhibits anodic migration, the separation is terminated not later than when the foot of the polysaccharide peak nearly touches the dividing line between the cell sections. After the separation of the sections in the usual way, the DNA fraction is collected from the top section. Up to 25% of the DNA contained in the original solution is thus obtained. The remainder is recovered from the other cell sections and resubjected to electrophoresis two to four more times. When it becomes too diluted, its content is precipitated with 2 vol. of ethanol, redissolved in a smaller volume, dialyzed and subjected to electrophoretic separation as before, but in a smaller cell.

Assays

General Notes. The colorimetric determinations are carried out in 8-mm. I.D. Pyrex tubes and read in 3-mm.-wide microcells[15] in a Beckman spectrophotometer. The liquids are dispensed by means of a 1-ml. total capacity Gilmont microburet.[16] The viscosity determinations are carried out at 23° ± 0.1° in an Ostwald-type microviscosimeter[17] (water value 20 to 25 seconds) with exactly 0.250 ml. of liquid delivered by microburet by means of a 1-mm. O.D. plastic tube; the determinations are made not earlier than ½ hour after the sample has attained the temperature specified.

Assays for Purity of DNA. (1) *DNA content* is estimated by comparison with a "reliable" sample of DNA. The value of such an estimation is therefore limited, yet most convenient for routine work. For comparison the colorimetric micromethod based on the Dische reaction with diphenylamine[18] is recommended.

To a 0.1-ml. sample containing 10 to 25 γ of DNA, 0.2 ml. of diphenylamine reagent[18] is added, and the mixture is heated for 10 minutes in boiling water. The reading is effected at 610 and 650 mμ against a similarly

---

[14] The instrument used was model 38 of the Perkin-Elmer Corporation, Glenbrook, Connecticut. Cells of a capacity of 6 ml. can be employed; a special extension tube in the top section of the cell is advisable to permit the separation of larger volumes.

[15] O. H. Lowry and O. A. Bessey, *J. Biol. Chem.* **163**, 633 (1946); cells obtained from the Pyrocell Manufacturing Co., New York, New York.

[16] The Emil Greiner Co., New York, New York.

[17] Otto R. Greiner Co., Newark, New Jersey.

[18] Z. Dische, *Mikrochemie* **8**, 4 (1930); also personal communication; R. Steele, T. Sfortunato, and L. Ottolenghi, *J. Biol. Chem.* **177**, 231 (1949).

treated blank, and the difference is compared with the similar difference obtained for the standard.

The colorimetric method based on the indole reaction[19] is also recommended. The method as described by Ceriotti can also be scaled down ten times, thus permitting a DNA assay on as little as 0.5 to 1 γ of DNA.

(2) *RNA content* is estimated by a microquantitative modification[20] of the colorimetric reaction with orcinol suggested by Pesez.[21] In this method the deoxyribose is destroyed by strong acid so that the traces of RNA can be estimated in the presence of a large excess of DNA.

A sample containing 0.5 to 3 γ of RNA and up to one hundred times as much DNA is cautiously evaporated to dryness (desiccator). One-tenth milliliter of orcinol reagent[22] is added, and the solution is heated in boiling water for 10 minutes. The sample is cooled, 0.3 ml. of water is added, and the color is read at 660 and 600 mμ against a similarly treated blank; the difference in reading is then compared with the corresponding value for the similarly treated standard solution of similar concentration. Sensitivity (including possible error caused by DNA), 1% of the DNA content.

(3) *Protein content* can be estimated by a micromodification (scaled down ten times) of the Sakaguchi method.[23] The method assumes that the reacting arginine amounts to 20% of the protein. This assumption, true for calf thymus histone, may not be true for other proteins; for this reason a more general method based on a modification of the biuret reaction has been developed.[20]

To a sample containing 2 to 50 γ of protein and up to 500 γ of DNA in 0.2 ml. of water, 0.1 ml. of the stable biuret reagent[24] is added, and the absorption read at 310 mμ and 390 mμ against a similarly treated blank. To correct for the absorption of the DNA, the procedure is repeated with reagent in which $CuSO_4$ solution has been replaced by water. The difference between readings at 310 and 390 mμ in this last determination is subtracted from the similar difference in the first determination (complete biuret reagent), and the value so obtained is compared with the value for a similarly treated standard (commercial crystalline protein such as ribonuclease or lysozyme).

Assays for Integrity of DNA. VISCOSITY. Viscosity as such is a property of DNA not sufficient to estimate its integrity. Specific viscosity (η_{sp}), as

[19] Z. Dische, *Biochem. Z.* **204,** 431 (1929); G. Ceriotti, *J. Biol. Chem.* **198,** 297 (1952).

[20] S. Zamenhof and E. Chargaff, unpublished results.

[21] M. Pesez, *Bull. soc. chim. biol.* **32,** 701 (1950).

[22] W. Mejbaum, *Z. physiol. Chem.* **258,** 117 (1939).

[23] S. Sakaguchi, *J. Biochem. (Japan)* **5,** 25 (1925); H. Schwander and R. Signer, *Helv. Chim. Acta* **33,** 1521 (1950).

[24] Prepared by adding, drop by drop, 40 ml. of 1% aqueous $CuSO_4$ solution to 150 ml. of 40% aqueous NaOH solution, and filtering through "medium" fritted filter.

ordinarily determined in the Ostwald-Fenske viscosimeter, depends on the dimensions of the viscosimeter, the quantity of liquid, and the salt concentration and impurities; in addition, η_{sp} is a nonlinear function of the DNA concentration. Where all the above factors are identical for a sample tested and for a "reliable" reference sample, the integrity of the former can be roughly compared with the latter; however, even this method is not reliable, as will be seen below.

CONSTANCY OF VISCOSITY. Considerably degraded DNA may have a tendency to repolymerize, thus attaining high viscosity. Such samples are thixotropic;[25] in addition, the viscosity increases on standing. The viscosity of some other degraded DNA's may decrease on standing. In contrast to this, the undegraded DNA retains its η_{sp} constant[5] within $\pm 2\%$ over a period of at least 8 hours at $23° \pm 0.1°$.

STABILITY TO HEAT.[5] A sample of DNA having constant η_{sp} at $23°$ is still damaged if it has lost its stability to heat. The instability to heat has been found thus far to be the most reliable physical indicator of the injury to DNA.

The viscosity of a sample of DNA *in standard buffer* (see Preparation of DNA) containing 400 to 550 γ of DNA per milliliter is determined. The sample is then heated in a stoppered tube for 1 hour at $76° \pm 0.3°$, cooled to $23°$, well mixed to return the moisture condensed on the walls, and the viscosity determined again. The η_{sp} of an undamaged DNA will remain the same as before heating within $\pm 2\%$; the damaged DNA will show loss of η_{sp}, depending on the degree of injury.[5]

BEHAVIOR TOWARD DNASE. The undamaged DNA exhibits a characteristic lag period at the onset of the depolymerizing action of the crystalline pancreatic DNase;[3,5] this lag period is shortened or nondemonstrable in the DNase-damaged DNA. To demonstrate the lag period any viscosimetric method of DNase assay can be followed; DNase concentration should be so low as to induce an 8% drop of η_{sp} at $30°$ in not less than 4 hours; under such conditions the undamaged DNA will exhibit no drop of viscosity (lag period) for at least $1\frac{1}{2}$ hours.

BIOLOGICAL ACTIVITY. When facilities for testing the transforming activity (of bacterial DNA) are available (see Vol. III [105]), this property can be employed for testing the loss of integrity of DNA, whether during the proposed preparation or during subsequent treatments.[3,5]

Such pilot preparation (from calf thymus) is performed as described before, but washing of the homogenate is omitted and only 30 mg. of tissue is used. To the tissue, just before homogenizing (extracting), is added 100 γ of crude or purified DNA of *Hemophilus influenzae* having transforming activity; the latter serves as a marker (tracer) of integrity.

[25] S. Zamenhof and E. Chargaff, *J. Biol. Chem.* **186**, 207 (1950).

The DNA (of ox, in this case) is considered undamaged if at the end of the preparation (or any other proposed treatment) the mixture shows undiminished transforming activity toward *H. influenzae* (per microgram of DNA of *H. influenzae*).

[104] The Characterization of Deoxyribonucleic Acid by Viscosity Measurements

By C. A. THOMAS, JR.

The viscosity increment of a solution of macromolecules is a measure of the size of the dissolved particles. Since the viscosity increment of aqueous deoxyribonucleic acid (DNA) is exceptionally high, its precise determination should provide valuable information about the size and shape of this molecule in solution. In order to perform a measurement which is meaningful in terms of molecular properties, however, η_{sp}/c[1] must be measured in solutions which are dilute enough to allow independent molecular motions and then extrapolated to zero concentration. The viscosity of DNA solutions is also found to depend strongly on shear. Since the present theoretical status of shear dependence allows interpretation of the intrinsic viscosity only when Brownian motion is overwhelming, viscosity measurements must be performed over a range of shears and then extrapolated to zero shear. Thus it is necessary to make viscosity measurements in very dilute solutions of DNA and at very low rates of shear; these conditions present experimental difficulties which have only recently been overcome. We will now consider some of the measurements of the intrinsic viscosity which have successfully eliminated the effects of both concentration and shear.

Intrinsic Viscosity—Determination. The table on p. 705 summarizes the available data on the intrinsic viscosity of DNA.

Intrinsic Viscosity—Interpretation. The intrinsic viscosity determined above may be used as a qualitative measure of the molecular size and presumably of the molecular weight. Since the DNA molecule in solution is probably not a simple linear polymer but a strand composed of two polynucleotide chains, one would expect no simple relationship between the intrinsic viscosity and the extent of degradation. However, for a qualitative comparison of different preparations, the intrinsic viscosity provides a sensitive measure of differences. Moreover, since the viscosity

---

[1] $\eta_{sp} = \dfrac{\eta_{solution} - \eta_{solvent}}{\eta_{solvent}}$; $\lim\limits_{c \to 0} (\eta_{sp}/c) \equiv [\eta]$ the intrinsic viscosity.

Intrinsic Viscosity of DNA
(Neutral pH; in presence of NaCl)

Investigators	Preparation	Viscometer	$[\eta]$	NaCl conc.	DNA conc.	Gradients
Pouyet (1952)[a]	Schwander and Signer	Couette	57	10%	0.002–0.008%	0.246 sec.$^{-1}$
Conway and Butler (1954)[b]	Unpubl. enzyme method	Couette	40	0.1 M	0.001–0.01%	0.5–15 sec.$^{-1}$
Signer and Berneis (1952)[c]	?	Tilting capillary	42	?	0.139%	6–20 sec.$^{-1}$
Reichmann et al. (1954)[d]	Modified Schwander and Signer	Capillary	48	0.20 M	0.002–0.007%	60–200 sec.$^{-1}$
Reichmann et al. (1954)[d]	Simmons B	Capillary	53	0.20 M	0.002–0.008%	60–200 sec.$^{-1}$

Note: Although the gradients obtained with the capillary viscometer of Reichmann *et al.* are not so low as those of the Couette and tilting capillary, it appears that they are low enough to permit extrapolation to zero shear in 0.20 M NaCl. In all the measurements quoted above, enough NaCl is present to eliminate the large interaction effects which are present in the absence of electrolyte.

[a] J. Pouyet, *Compt. rend.* **234**, 152 (1952).

[b] B. E. Conway and J. A. V. Butler, *J. Polymer Sci.* **12**, 199 (1954).

[c] R. Signer and K. Berneis, *Makromol. Chem.* **8**, 268 (1952).

[d] M. E. Reichmann, S. Rice, C. A. Thomas, and P. Doty, *J. Am. Chem. Soc.* **76**, 3047 (1954).

is a monotonically decreasing function of the gradient, the intrinsic viscosity measured at some higher, more experimentally accessible average gradient (ca. 1000 sec.$^{-1}$) can also be utilized to qualitatively detect changes in molecular configuration. Indeed, this was the only way viscometric studies were performed until recently.

Coming now to the more quantitative interpretation in terms of molecular dimensions, we find that it is necessary to combine the intrinsic viscosity with an independent determination of the molecular weight of the solute particle. The molecular weight has been determined by light scattering for the last two entries in the table above. The intrinsic viscosity can now be interpreted in terms of two alternative models: the rigid ellipsoid of revolution (with no hydration) or the gaussian coil. Choosing the first of these possibilities, one calculates an axial ratio of 425 for an intrinsic viscosity of 50. Equating the hydrodynamic volume with the volume of the particle calculated from the molecular weight (6×10^6) and the bulk density (1.63) leads to a length (major axis) of

12,750 A. and a diameter (minor axis) of 30 A. The radius of gyration of such a particle would be 3700 A. However, an independent measure of the radius of gyration, determined from the angular intensity distribution of scattered light, yields a value which is almost a factor of 2 *smaller* than 3700 A. Thus the interpretation of the viscosity on the basis of an ellipsoid or rodlike particle is not correct.

On the other hand, interpreting the intrinsic viscosity in terms of a randomly coiled polymer by utilizing the Flory-Fox[2] relationship which has been shown to have wide applicability, we can relate the mean-square end-to-end distance of the polymer chain, $\overline{R^2}$, to the intrinsic viscosity.

$$[\eta] = \Phi \frac{(\overline{R^2})^{3/2}}{M}$$

The quantity Φ is a constant, the value of which is taken to be 2.1×10^{21}. By employing the molecular weight value of 6.85×10^6, which was determined by Reichmann *et al.*[3] for their modified Schwander and Signer preparation, they obtain 5400 A. for $\sqrt{\overline{R^2}}$. This value is in complete agreement with the same quantity determined by light scattering. This agreement substantiates not only this value for the dimension but also the choice of the random coil as being the most appropriate simple model for the DNA molecule in solution.

Instead of combining the intrinsic viscosity with an independent molecular weight determination to calculate the molecular size, it is possible to combine the intrinsic viscosity and the sedimentation constant to calculate the molecular weight. This can be done by utilizing the Flory-Mandelkern[4] relation which is somewhat analogous to the Svedberg equation and has received general confirmation. This relation is:

$$M = \left[\frac{\eta_0 N s_0 [\eta]^{1/3}}{\Phi^{1/3} P^{-1} (1 - \bar{v}\rho)} \right]^{3/2}$$

where P is a constant for sedimentation analogous to Φ, and $\bar{v}$ is the partial specific volume and ρ the density of the solution. The quantity s_0—the sedimentation constant extrapolated to zero concentration—has been determined by Simmons as 15.5×10^{-13} by extrapolation from concentrations as low as 0.03% DNA. Taking $[\eta] = 53$ from the table for the Simmons preparation leads to a molecular weight of 5.8×10^6. This value

[2] P. J. Flory and T. G. Fox, *J. Am. Chem. Soc.* **73**, 1904 (1951).

[3] M. E. Reichmann, S. Rice, C. A. Thomas, and P. Doty, *J. Am. Chem. Soc.* **76**, 3047 (1954).

[4] L. Mandelkern, W. R. Krigbaum, H. A. Sheraga, and P. J. Flory, *J. Chem. Phys.* **20**, 1392 (1952).

is to be compared with the light-scattering molecular weight of 5.85×10^6 for the same preparation of DNA showing excellent agreement.

Some Important Variables Which Affect the Viscosity of DNA. In addition to the effects of concentration and shear, there are many variables which affect the viscosity of DNA solutions. One of the most interesting is the presence or absence of added electrolyte. At moderate concentrations of DNA, the addition of small quantities of electrolyte causes a large decrease in the viscosity. Although the effect is not so large as is found in more flexible polyelectrolytes, this viscosity decrease was interpreted in terms of a contraction of the molecule, brought about by the shielding of the repelling charges on the polymer strand, by the added counter ions. However, recent measurements of very dilute DNA solutions in the absence of added salt at very low gradients by Conway and Butler[5] show that, as the concentration of DNA becomes small, the decrease in viscosity brought about by the addition of salt becomes less. These data apparently indicate that the intrinsic viscosity of DNA in the absence of added salt is approximately the same as the intrinsic viscosity measured in 0.10 M NaCl. This point requires still further investigation because there are not sufficient measurements at low enough concentrations to assure that both of these values are identical. However, it is clear that the effect in very dilute DNA solutions (ca. 0.002%) is much less than at moderate concentrations (ca. 0.01%). They conclude from these observations that the large increase in viscosity, which was formerly attributed to the expansion of the molecule on the removal of electrolyte, is predominately due to increased interparticular interaction.

Another rather exceptional change in the viscosity of DNA solutions comes about when the pH of the solution is gradually lowered. Reichmann *et al.*[6] have found that the intrinsic viscosity decreases to a value of 2.8 at pH 2.60 in 0.20 M NaCl. This decrease in viscosity is not accompanied by depolymerization, because they were able to confirm the same value of the molecular weight at pH 2.60. as found at neutral pH. This transformation can be explained in terms of the model of a gaussian polymer. Picturing the DNA molecule as being composed of two polynucleotide chains held together by hydrogen bonds along their length, we note that lowering the pH to 2.60 would titrate a large number of groups involved in hydrogen bonds, thus destroying them. If these hydrogen bonds between the chains were broken, the molecule would become more flexible. Being more flexible, the coil would contract into a more compact form because of the greater configurational entropy involved.

A decrease in viscosity similar to that found on lowering the pH is

[5] B. E. Conway and J. A. V. Butler, *J. Polymer Sci.* **12**, 199 (1954).
[6] M. E. Reichmann, B. Bunce, and P. Doty, *J. Polymer Sci.* **10**, 109 (1952).

found when DNA solutions are heated. The heat stability of DNA from various origins has been investigated by Zamenhof.[7] Light-scattering experiments have indicated that this decrease in viscosity brought about by heat treatment does not at first cause any decrease in the molecular weight.[8] Thus it appears that heat treatment is analogous to exposure to low pH values. Presumably the hydrogen bonds between the polynucleotide chains are broken by heat as well as by low pH.

Conclusion. Although the measurement of the intrinsic viscosity in a manner which successfully eliminates the effects of both concentration and shear is difficult experimentally, its determination provides a sensitive measure of molecular size. Moreover, when such a sensitive measurement is combined with an independent molecular weight determination, the absolute value of the size can be obtained. On the other hand the molecular weight itself can be calculated by combination with the sedimentation constant.

Viscosity measurements have proved especially valuable in investigating the effects of pH and heat on the DNA molecule. In both cases the molecular weight was found to remain constant whereas the size of the molecule, as reflected by the viscosity, undergoes substantial contraction. This contraction in size is thought to be the result of the destruction of hydrogen bonds between polynucleotide chains.

[7] S. Zamenhof, H. E. Alexander, G. Leidy, *J. Exp. Med.* **98**, 373–397 (1953).
[8] Paul Doty and S. A. Rice, *Biochim. et Biophys. Acta* **16**, 446–447 (1955).

[105] Methods for Characterization of Nucleic Acid[1]

By ROLLIN D. HOTCHKISS

I. Characterization of Nucleic Acids by Spectrophotometry

Principle. Both DNA and RNA absorb ultraviolet light strongly, with maximum absorption about 260 mμ, falling off to a minimum about 230 to 240 mμ and almost disappearing at 310 mμ.

High-molecular DNA and RNA can undergo a denaturation change which results in an increase of ultraviolet absorption in saline solution of about 33% at 260 mμ (and of similar magnitude along the whole absorption curve). The absorption increment from NaOH treatment has been used in the author's laboratory since 1948 to determine DNA in the presence of RNA (because the latter is usually not high molecular).

[1] For colorimetric procedures for analysis of nucleic acid constituents, see Vol. III [12, 13, 99].

After Kunitz reported that pancreatic deoxyribonuclease (DNase) hydrolysis resulted in an increase in absorption,[2] it was quickly discovered that the alkali and enzymatic effects were equivalent and not additive. Enzymatically hydrolyzed DNA does not show further increase in ultraviolet absorption when made alkaline; likewise alkali- (or acid-) denatured DNA, reneutralized, no longer gives the increase on hydrolysis with enzyme. It was also shown at the same time (with M. McCarty) that high-molecular RNA from *Pneumococcus* would show a similar 30 to 33% absorption increment when treated with *ribonuclease* (RNase) or alkali.

A number of authors have observed some of these absorption shifts in the years since, particularly with DNA. It seems clear that some have come to look on the extinction coefficient in relation to phosphorus content as some sort of criterion of nativeness in DNA,[3] but the quantitative differences between different preparations and different modes of denaturation discouraged interpretation of the absorption increment of denaturation. Beaven *et al.*[4] have recently proposed the use of alkali denaturation, without phosphorus analysis, in much the same way as described here in judging the quality of a preparation. As stated, the method has been used by the author with increasing confidence during the past several years.

It is important to make as clear as present knowledge permits what features of DNA or RNA structure can be correlated with a 33% absorption increment. In operational terms, the increment is shown by bacterial DNA of high biological activity, tissue DNA prepared by a variety of gentle means, DNA isolated while still attached to histone[5] (on alkali denaturation), and highly viscous RNA preparations from pneumococcus. The value of 33% is believed to be close to the ideal maximum, although a number of DNA and RNA preparations have been described which gave absorption increments less than this on hydrolysis and had been considered native. The author believes such preparations to be nevertheless denatured and has been able to make similar ones by irreversibly damaging intact DNA by drying, by heating, or by suitably drastic shift of pH or ionic strength. His native material, like that of a number of other workers, gives an increment close to 33%. Few RNA preparations of good

[2] M. Kunitz, *J. Gen. Physiol.* **33,** 349 (1950).

[3] G. Frick, *Biochim. et Biophys. Acta* **8,** 625 (1952); J. Shack, R. J. Jenkins, and J. M. Thompsett, *J. Biol. Chem.* **203,** 373 (1953); R. Thomas, *Biochim. et Biophys. Acta* **14,** 231 (1954); E. Chargaff, *in* "The Nucleic Acids" (Chargaff and Davidson, eds.), Vol. I, p. 336. Academic Press, New York, 1955.

[4] G. H. Beaven, E. R. Holiday, and E. A. Johnson, *in* "The Nucleic Acids" (Chargaff and Davidson, eds.), Vol. I, p. 526. Academic Press, New York, 1955.

[5] J. Shack and J. M. Thompsett, *J. Biol. Chem.* **197,** 17 (1952); *J. Natl. Cancer Inst.* **13,** 1425 (1953).

quality have been described (they are often prepared with use of alkali). In particular, it should be noted that air or solvent drying of DNA usually leads to partial denaturation, in this sense. High specific viscosity is not a reliable criterion of integrity of DNA; the temperature dependence of viscosity (see Vol. III [103]) is probably a much better one.

The ability to show an absorption increment, then, appears to depend on some not adequately understood feature of the spatial fitting or hydrogen bonding of the purine and pyrimidine bases, easily destroyed by denaturation. It may be that different nucleic acids of widely different base composition will show somewhat discrepant absorption increments. The methods which follow may therefore be somewhat liable to reinterpretation or quantitative correction as more is learned about the molecules involved.[1]

Differentiation of Native and Denatured DNA and RNA. Exactly 2.5 or 3.0 ml. of the nucleic acid solution, at appropriate concentration (about 10 to 50 γ/ml. in NaCl of 0.1 M or stronger) is placed in a quartz cell having a 1-cm. light path. The optical density is determined at 320 mμ (as indication of extraneous scattering or absorption) and at 260 mμ. Saline in the same volume serves as blank. Strong NaOH is added to each cell, 0.05 ml. of approximately 6 N NaOH per 3 ml., or an equivalent amount. The absorption of the alkaline solution, corrected back to the volume before dilution with alkali, less the initial absorption at 260 mμ, gives the "absorption increment." This increment, multiplied by 3 (that is, 1/0.33), is taken as the 260-mμ absorption value for the native DNA and RNA initially present. An optical density of 1.0 at 260 mμ (1-cm. light path) corresponds to about 45 γ of DNA or RNA per milliliter.

Freshly made NaOH, kept in Pyrex, adds only a small absorption to the blank. The quantity used corresponds to a final concentration of 0.1 M, or about a threefold excess to allow for buffering by other constituents (final pH to be more than 11.5). Certain substances, such as phenols, or proteins in twentyfold excess over nucleic acid, might give sufficient alkali response to necessitate the before and after observation of absorption at a number of wavelengths, or other blank determinations.

Specific Spectrophotometric Determination of High-Molecular DNA and RNA in Mixtures. In a similar way, small amounts of purified ribonuclease (RNase) or deoxyribonuclease (DNase) may be added to neutral solutions of nucleic acid, and the absorption increment after enzymatic reaction is complete is a measure of the corresponding native nucleic acids.[1] It is convenient to add the enzymes in succession, RNase first, since it is more likely than DNase to be free from the other enzyme; it may also reduce inhibition which RNA can exert on certain preparations

[1] See Addendum, p. 715.

of DNase. Magnesium salt is added at the beginning, since DNase will require Mg^{++} as activator, and its addition may influence the absorption.

Reagents

Mg-saline, 0.9% NaCl containing 0.005 M MgCl$_2$ (or MgSO$_4$).

RNase, crystalline ribonuclease, 0.5 mg./ml. in H$_2$O. This preparation should be sufficiently free of DNase so as not to reduce the viscosity obviously when incubated 30 minutes with an equal volume (e.g., 0.3 ml. of 0.2%) of viscous calf thymus or other DNA.

DNase, crystalline pancreatic deoxyribonuclease, 0.1 mg./ml. in 0.9% NaCl. (Commercial streptococcal DNase will also serve.)

Procedure. The nucleic acid is diluted in Mg-saline to an approximate total concentration of 10 to 50 γ/ml. and brought to pH between 6 and 7.5. A known 2.5- or 3.0-ml. amount is placed in a 1-cm. quartz cell, and the optical density measured at 260 mμ (and at 320 mμ to estimate extraneous scattering or absorption) against a Mg-saline control. A known volume of RNase (0.01 ml. or the equivalent, with micropipet) is added to the blank and the unknown with mixing. Hydrolysis requires less than 10 minutes at room temperature when active enzyme is used. When the absorption at 260 mμ becomes stable, the optical density, corrected to the volume before RNase addition, less the initial value, gives the absorption increment due to ribonucleic acid hydrolysis. The increment, times 3, is taken as the original absorption of native RNA.

After the absorption at 260 mμ is constant, a known volume of DNase (0.06 ml. of above solution, or the equivalent) is added to the blank and the unknown. The corrected absorption increment is calculated exactly as for RNA; the onset of change usually comes only after some delay, since the early stages of hydrolysis do not result in optical density change. The increment, times 3, is taken as the original absorption of native DNA (45 γ/ml. giving optical density of 1.0 per centimeter).

The initial absorption at 260 mμ, less the sum of native RNA and DNA determined as above, gives a measure of the denatured or degraded nucleic acids, plus other constituents absorbing at this wavelength.

Remarks. The absorption increment is a property the meaning of which is not yet fully clear. As indicated, it may for the moment be considered an empirical property (like the inevitable proportionality constant in a colorimetric analysis), one which seems to be closely correlated with biological and physical integrity. The method outlined above may need some quantitative revision as more information on nucleic acid structure becomes known.

For economy, the small correction due to the enzymes added can be separately determined for each nuclease and deducted, instead of eliminated by incorporation in the blank as above. Either RNase or DNase may be used alone; if the DNase preparation has RNase activity, however, it will be necessary to add pure RNase before using the DNase. Since high-molecular RNA is not usually available, the DNase preparation should be tested for RNase in some other way, e.g., by observing or measuring whether it decreases the acid precipitability of an ordinary RNA preparation.

II. Characterization of DNA by Biological Activity
(Bacterial Transformation)

Principle. DNA acts as the bearer of specific genetic properties in some bacteria, and perhaps in most cells. By introducing DNA coming from one strain into a suitable, related but different, strain of certain bacteria, heritable properties characteristic of the donor cells can be transferred to the recipient line.[6,7] The proportion of the recipient cells "transformed" is a measure of a specific biological activity of the material. This proportion is determined quantitatively by exposing a measured sample of cells from the transformed culture to a selective environment which will permit growth of only the modified cells, and counting the number of cells which have produced colonies after a suitable time.[7,8]

The response of a pneumococcal population to suitable DNA is dependent on (1) appropriate pH and concentrations of calcium and serum albumin, (2) the physiological state of the culture, (3) randomness or synchrony in the division state[8] of individual cells, (4) the total DNA concentration, and (5) the proportion of this DNA which carries the biological activity being measured. Factor 1 is provided by choosing suitable media; factors 2 and 3 have to be controlled in each experiment by comparison against a standard transforming DNA. Increasing concentration of DNA (factor 4) gives a dose-response curve showing transformants increasing linearly with DNA at low concentrations (up to about 0.2 γ/ml.), leveling off to a saturation plateau at high concentrations (0.2 to 5 γ/ml.).

DNA from a species not suitable for transformation can be tested by incorporating it as an inhibitory substance in known ratios to active DNA. The inhibitory activity of the DNA from a heterologous source is a test of its own biological intactness, since various kinds of denatured DNA do not inhibit transformation.

Materials. TRANSFORMABLE BACTERIA. A number of strains of *Pneumococcus* have been transformed, especially several derived from strain

[6] O. T. Avery, C. M. MacLeod, and M. McCarty, *J. Exptl. Med.* **79,** 137 (1944).

[7] R. D. Hotchkiss, *Cold Spring Harbor Symposia Quant. Biol.* **16,** 457 (1951).

[8] R. D. Hotchkiss, *Proc. Natl. Acad. Sci. U.S.* **40,** 49 (1954).

R36A.[6] *Hemophilus*[9] and meningococcal[10] strains have been used similarly. It is likely that other transformable strains and species will become available from time to time; this expectation is one reason for attempting to illustrate here a general method of bioassay.

DNA PREPARATION. In the example given, the quantitative marker introduced will be streptomycin resistance of *Pneumococcus*. The DNA used comes from one-step, high-level streptomycin resistant pneumococci. (*Hemophilus* or meningococcal, etc., mutants could be used, corresponding to the transformable species of bacteria available.) Preparation is carried out as described in Vol. III [102] the DNA being precipitated with alcohol to sterilize it, but never dried to a water-free state. Dilutions are made in sterile 0.9% NaCl, standard and unknown being prepared in comparable dilutions.

DNASE. Commercial pancreatic deoxyribonuclease is used in a working dilution of 0.03 mg./ml. in sterile growth medium, or saline, stored in the refrigerator.

SERUM ALBUMIN. Bovine serum albumin (Armour Plasma Fraction V) is dissolved at 4 g. per 100 ml. in distilled H_2O, adjusted to pH 7.5, and sterilized by filtration. Other purified albumin preparations from various species suffice, crude serum and other proteins do not.

GROWTH MEDIUM. This should be suited to the organism used. For pneumococcus, the medium may be 1% neopeptone (Difco), 5% fresh meat infusion (lean beef ground into an equal weight of distilled water, strained, heated to 85°, and filtered), and 5% yeast extract (baker's yeast cake slurried into an equal weight of boiling water, filtered). When brought to pH 7.6, to 0.3% concentration of NaCl, clarified, and autoclaved, this is the basal growth medium. Additional requirements for transformation are supplied by adding K_2HPO_4 to a final concentration of M/40, glucose to 0.03%, serum albumin to 0.2%, and $CaCl_2$ to 0.003% (all in per cent weight per 100-ml. volume). For *Hemophilus* and other organisms, serum albumin has not proved necessary.

Other media, such as one based on casein hydrolyzate and vitamins,[11] may be used, with appropriate addition of serum albumin.

SELECTIVE MEDIUM. In this example, streptomycin is the selective agent. It has also been used for *Hemophilus*;[9] other quantitative selection principles are described elsewhere[7,12] for *Pneumococcus*. In the author's laboratory, quantitation is customarily carried out in liquid selective media containing antiserum globulin which causes single cells to grow as

[9] H. E. Alexander and G. Leidy, *J. Exptl. Med.* **97,** 17 (1953).
[10] H. E. Alexander and W. Redman, *J. Exptl. Med.* **97,** 797 (1953).
[11] M. H. Adams and A. S. Roe, *J. Bacteriol.* **49,** 401 (1945).
[12] R. D. Hotchkiss and J. Marmur, *Proc. Natl. Acad. Sci. U.S.* **40,** 55 (1954); J. Marmur and R. D. Hotchkiss, *J. Biol. Chem.* **214,** 383 (1955).

colonies.[8] Since this method has not been worked out for other bacteria as yet, the more adaptable streptomycin-blood-agar plate scoring is described here. For many organisms nutrient agar without blood would suffice.

The basal neopeptone medium described is brought to 100°, and 15 g. of agar per liter dissolved in it. After cooling to 45°, 0.1% serum albumin, 2% sterile defibrinated whole rabbit or human blood, and streptomycin (150 γ/ml.) are added, and standard plates of about 20 ml. are poured. Such plates, made without the streptomycin, should support growth of surface colonies from single cells of the sensitive recipient strain to be used; with the drug present, only the resistant transformants or mutants should grow.

Procedure for Quantitative Transformation. Albumin-containing medium is inoculated with approximately 5000 cells/ml. (10^{-5} dilution of a fresh full-grown culture of transformable strain), and divided into 1.5-ml. portions in small culture tubes. After 4 hours of undisturbed incubation (in a water bath at 37° for accurate definition of time intervals), standard and unknown DNA's are added in a series of saline dilutions to separate cultures. Appropriate amounts are from about 1 to 2 γ/ml. (saturation level) down to 0.01 γ/ml. or less, in any desired dilution steps.

Exactly 5 (or 10) minutes after addition of DNA, a drop of DNase solution is added to each culture to stop the formation of further transformants, and incubation is continued for 75 minutes. At the end of this time, the cultures are chilled in ice and diluted as soon as possible for streaking on selective medium. There should be a countable number of transformant colonies from 0.10 ml. evenly spread on the surface of the selective medium, using 1:20 dilution (at high DNA) or undiluted culture (for low DNA concentrations).

The yield of colonies, in relation to those from a parallel culture treated with a standard DNA bearing the same marker, is a reliable measure of the transforming activity of a preparation. The slope of the linear region of the response curve indicates the active DNA concentration. The number of transformants at saturation concentration expresses the activity, or quality, level of the DNA preparation (active/total, modified by affinity constants, degree of degradation, etc.). DNA from heterologous sources can be evaluated by its ability, in known admixtures, to interfere with transformation by active specific material. The absolute yield is meaningful only when the condition and amount of inoculum, the time of growth, and the medium have been closely controlled. A larger inoculum gives a larger population, sensitive to DNA considerably earlier than the 4 hours used here, but the period of sensitivity is brief and must be rather closely outlined or it may be entirely missed.

Alternative assay methods, in which DNA is diluted to the end point at which transformation ceases to show up,[6,9,13] are possible but are quantitatively less reliable, since both statistical and physiological uncertainties produce much scatter at the dilution end point.

[13] S. Zamenhof, H. E. Alexander, and G. Leidy, *J. Exptl. Med.* **98**, 373 (1953); A. W. Ravin, *Exptl. Cell Research* **7**, 58 (1954).

ADDENDUM

The increasing evidence that many high-molecular RNA preparations tend to depolymerize spontaneously in saline suggests (May, 1956) that one should if possible determine for each RNA the denaturation increment (more or less than 33%) with reference to an arbitrarily defined native state.

[106] Methods for Characterization of Nucleic Acids by Base[1] Composition

By AARON BENDICH

I. Methods of Hydrolysis

Principle. The liberation of bases from nucleic acids depends on the cleavage of acid-labile glycosidic bonds. Except for the T-even bacteriophages of *Escherichia coli*, the DNA's of which contain the base 5-hydroxymethylcytosine[2] (which is largely destroyed during hydrolysis with $HClO_4$), hydrolysis by means of $HClO_4$ is the method of choice for both RNA and DNA. For DNA's in general, and for the above phage DNA's in particular, hydrolysis with formic acid is also valuable.

Reagents

70 to 72% $HClO_4$ (ca. 12 N). A good analytical reagent grade is supplied by the Mallinckrodt Chemical Works (catalog No. 2766).

88% formic acid reagent grade supplied by Merck & Co.

Procedure. (a) ($HClO_4$).[3] A weighed nucleic acid preparation (dry) is intimately mixed with 70 to 72% $HClO_4$ (5 mg. per 0.1 ml. of $HClO_4$)[4] in a small glass-stoppered test tube or centrifuge tube and is heated at 100° for 60 minutes with occasional agitation.[5] The mixture is cooled and

[1] The term "base" refers to the purines and pyrimidines isolable from nucleic acids, despite the lack of demonstrable basic properties in uracil and thymine.

[2] G. R. Wyatt and S. S. Cohen, *Biochem. J.* **55**, 774 (1953).

[3] Based mainly on A. Marshak and H. J. Vogel, *J. Biol. Chem.* **189**, 597 (1951).

[4] A 2.5% loss of thymine from insect virus DNA has been found by G. R. Wyatt [*J. Gen. Physiol.* **36**, 201 (1952)] with this proportion of DNA to $HClO_4$. He advises the use of half this quantity of $HClO_4$.

[5] At somewhat lower temperatures, incomplete liberation of cytosine from RNA is observed together with the appearance of a cytidylic acid [P. M. Roll and I. Welicky, *J. Biol. Chem.* **213**, 509 (1955)]. It is imperative that a temperature of 100° be maintained.

diluted (to 0.500 ml.) with water. The mixture is ground with a glass rod to produce a homogeneous suspension which is then centrifuged to separate the solution from the black particulate residue. The clear supernatant fluid is then ready for chromatographic analysis and for determination of nitrogen or phosphorus content.

(b) FORMIC ACID.[2] A specimen of DNA (0.7 mg.) or virus (1.5 mg.) containing DNA (but not RNA) is weighed into a Pyrex glass tube (6 mm. internal diameter). Formic acid (0.5 ml. of 88%) is added, and the tube is sealed in an oxygen-gas flame about 20 mm. above the surface of the liquid. The sealed tube is heated in an oven at 175° for 30 minutes. Care must be exercised in handling the tube, since pressure develops owing to the formation of *carbon monoxide* resulting from the decomposition of formic acid at elevated temperatures. The tube is cooled and thoroughly secured in a stand. The pressure is then released by directing a hot pinpoint flame to the very top of the tube. The expanding gases force open a small hole at the point of application of the flame. The top of the opened tube is cut off, and the hydrolyzate is taken to dryness below 75° under reduced pressure in a stream of nitrogen gas. The residue is redissolved in 25 μl. of N HCl, and the solution is then ready for paper chromatographic analysis (8-μl. portions) or phosphorus estimation (2-μl. portions).

II. Determination of Base Composition by Paper Chromatography

Principle. The hydrolyzates obtained as described above are applied to filter paper strips,[3,6,7] and the bases contained therein are separated by an isopropranol-HCl solvent system.[7] In addition to its high resolving power, this HCl-containing solvent system has an advantage over other neutral or ammoniacal systems, since it permits the analysis of larger quantitites of guanine. The separated bases are, in turn, extracted from the paper with dilute acid, and the base content of the extract is determined by ultraviolet spectrophotometry.[2,3,6,7]

Reagents

Isopropanol. Commercial absolute isopropanol may contain peroxides. A sample (1 to 2 ml.) of the alcohol is acidified with a few drops of dilute HCl; a small crystal of KI is added. If peroxide is present, the solution turns pale yellow-brown owing to the formation of iodine which is easily detected by starch indicator. If present, peroxide is removed from the isopropanol by contact treatment with solid ferrous sulfate until the test is negative.

[6] E. Vischer and E. Chargaff, *J. Biol. Chem.* **176**, 703 (1948).
[7] G. R. Wyatt, *Biochem. J.* **48**, 584 (1951).

The isopropanol is fractionally distilled, and the fraction boiling at 82 to 83°/760 mm. is collected and stored in the dark.

Concentrated HCl. A reagent grade is used (sp. gr. 1.19).

0.1 N HCl.

Solvent System.[7] To 65 ml. of peroxide-free absolute isopropanol is added 16.7 ml. of concentrated HCl (12.0 N). After mixing, water is added to 100 ml. The solvent is best used soon after mixing, since poor resolution of the bases results if the solvent is more than a few days old. R_f values of purines and pyrimidines in this and other solvent systems are listed in Table I. It is understood that such values serve only as a guide, since absolute R_f values vary considerably.

Procedure. (a) CHROMATOGRAPHY. Sheets of Whatman No. 1 or Schleicher and Schuell No. 597 filter paper, measuring about 15 × 50 cm., are used. Three longitudinal lines are drawn in pencil to divide the paper sheet into four equal lanes. The hydrolyzate (7 to 18 μl.) is deposited[8] in the center of two of the lanes on a transverse "starting line" drawn across the lanes about 7 cm. from the top of the sheet. An identical volume of 2 N HClO₄ or N HCl is deposited in a third lane, to serve as a blank. In the remaining lane is placed a control volume of a known mixture containing about 8 to 25 γ of each of the bases of interest made up in 2 N HClO₄ or N HCl. The sheet is dried in air or in a stream of warm (but not hot) air.

A glass cylindrical jar (about 50 cm. high with an inside diameter of about 25 cm.) is fitted with a glass trough supported about 5 to 10 cm. from the top. The isopropanol-2N HCl solvent is placed in the trough, and the jar is sealed with a glass cover to permit the vapors to saturate the atmosphere in the jar. (A small beaker containing the solvent is kept on the bottom of the jar to ensure the saturation.)

Two sheets may be run in the same apparatus. The sheets are hung by immersing their folded tops in the trough containing the solvent, and the jar is closed. Development is allowed to proceed until the solvent has moved about 35 cm. from the "starting line"; at room temperature, this takes about 15 hours. The papers are hung upside down to dry overnight in air in a well ventilated room.

(b) ELUTION.[3] The separated bases are located on the paper chromatograms as dark patches against a background of general paper fluorescence

---

[8] A Levy-type micropipet [M. Levy, *Compt. rend. trav. Lab. Carlsberg, Sér. chim.* **21**, 101 (1936)], a Scholander-type microburet [P. F. Scholander, *Science* **95**, 177 (1942)], or a Gilmont ultramicroburet (Emil Greiner Co., New York) may be used to dispense the small volumes. These volumes should contain about 5 to 30 γ of each base, although amounts up to 75 γ can still be resolved from mixtures in a 35-cm. movement of the solvent front.[7]

TABLE I

R_f VALUES OF PURINES AND PYRIMIDINES (DESCENDING CHROMATOGRAPHY)

Solvent system[a]	Filter paper		Guanine	Adenine	Hypoxanthine	Xanthine	Cytosine	5-Methylcytosine	Uracil	Thymine	Ref.
2 N HCl in: isopropanol Water	65 35	b	0.25	0.36	0.31	0.25	0.47[c]	0.55[c]	0.68	0.77	d
n-Butanol, saturated with water (NH₃ atmosphere)		b	0.0	0.45	0.18	0.01	0.26	ca. 0.31	0.35	0.54	e
n-Butanol 0.1 N NH₃	6 1	f	0.35	0.80			0.51	0.65		1.0	g
n-Butanol Diethylene glycol 0.1 N HCl	4 1 1	f	0.08	0.20	0.29	0.31	0.34		0.60	0.74	h
n-Butanol Water Formic acid	77 13 10	b	0.13	0.33	0.30	0.24	0.26		0.39	0.56	i
5% Na₂HPO₄, isoamyl alcohol	j	b	0.02	0.44	0.57	0.49	0.73		0.73	0.73	j
n-Butanol, saturated with 10% aqueous urea	j	b	0.05	0.41	0.29	0.12	0.29		0.35	0.52	j

[a] The composition of the solvent systems is given in volume-volume proportions. For other solvent systems, see R. J. Block, E. L. Durrum, and G. Zweig, "Paper Chromatography and Paper Electrophoresis," p. 206, Academic Press, New York, 1955.

[b] Whatman No. 1 filter paper.

[c] 5-Hydroxymethylcytosine has about the same R_f value as cytosine; G. R. Wyatt and S. S. Cohen, Nature 170, 1072 (1952); Biochem. J. 55, 774 (1953).

[d] G. R. Wyatt, Biochem. J. 48, 584, 581 (1951).

[e] R. D. Hotchkiss, J. Biol. Chem. 175, 315 (1948); the value 0.31 is estimated from the "epicytosine" peak.

[f] Schleicher and Schuell filter paper No. 597.

[g] These are relative values, with thymine arbitrarily taken as 1.0: E. Chargaff, R. Lipshitz, C. Green, and M. E. Hodes, J. Biol. Chem. 192, 223 (1951).

[h] E. Vischer and E. Chargaff, J. Biol. Chem. 176, 703 (1948).

[i] R. Markham and J. D. Smith, Biochem. J. 45, 294 (1949).

[j] These two solvent systems are used in turn in a two-dimensional ascending chromatographic procedure for the separation of bases, nucleosides, and nucleotides: C. E. Carter, J. Am. Chem. Soc. 72, 1466 (1950).

when viewed in a dark room under an ultraviolet lamp.[9,10] The spots are outlined in pencil,[11] and rectangles or disks containing them are cut out. Identical areas of paper are cut from corresponding positions in the blank lane. Contact with the fingers is to be avoided. Each area of paper is then cut up in small pieces, which are placed in a small, wide test tube and eluted, in the closed tube, with constant shaking for 2 hours at room temperature with 5.00 ml. of 0.1 N HCl. The eluates are decanted from the paper and centrifuged to remove shreds.

(c) 5-METHYLCYTOSINE. Since the amounts of 5-methylcytosine in many DNA's are usually small, both it and cytosine are often analyzed together and expressed as cytosine. The following procedure, developed by Wyatt,[12] permits the determination of 5-methylcytosine when it comprises as little as 0.1% of the original nucleic acid.

The hydrolyzate (6 to 8 mg. of DNA) is deposited in a band 25 cm. long across the top of a filter paper sheet which is then developed as above in the isopropanol-HCl system. A transverse strip containing *all* of both cytosine and 5-methylcytosine is cut from the dried chromatogram. One end of this strip is cut to a point, and the other end is placed in a trough containing water in a chromatographic tank. The pyrimidines migrate with the water front and are eluted quantitatively in the first 0.5 ml. to drip from the bottom of the paper strip. The eluate, collected in a small test tube, is evaporated to dryness and is redissolved in 0.1 N HCl (0.04 ml.). Measured volumes are deposited on filter paper and developed in an *n*-butanol-aqueous ammonia solvent. A mixture of 6 vol. of *n*-butanol (Mallinckrodt reagent grade) and 1 vol. of 0.1 N ammonia[13] is suitable. The separated spots are eluted as above with 0.1 N HCl and measured against corresponding blanks carried along in parallel.

Alternative procedures have been briefly described.[13,14]

(d) ULTRAVIOLET SPECTROPHOTOMETRY. The eluates are placed in 1-cm. matched silica cells, and the optical density (or extinction, E) is read against the corresponding eluates of the blank lanes in a calibrated

[9] E. R. Holiday and E. A. Johnson, *Nature* **163**, 216 (1949).

[10] A suitable ultraviolet lamp (General Electric mercury lamp No. G8T5) equipped with a Corning glass filter No. 9863 is supplied (Model MR4) by G. W. Gates and Co., Inc., Franklin Square, Long Island, New York. X-ray or any other ultraviolet absorbing goggles or glasses should be worn by the viewer for protection of the eyes.

[11] The guanine is seen as a light blue fluorescent spot, since it is present in the paper in the form of the hydrochloride. Better definition of the guanine spot is achieved, if necessary, by briefly exposing the chromatogram to ammonia vapor.

[12] G. R. Wyatt, *Biochem. J.* **48**, 581 (1951).

[13] E. Chargaff, R. Lipshitz, C. Green, and M. E. Hodes, *J. Biol. Chem.* **192**, 223 (1951).

[14] E. Chargaff, C. E. Crampton, and R. Lipshitz, *Nature* **172**, 289 (1953).

Beckman photoelectric quartz spectrophotometer. (The method of calibration is furnished in the set of directions supplied with the instrument.) Readings are recorded at 5-mμ intervals from 230 to 310 mμ, as well as at the absorption maximum, and absorption spectra may be plotted for comparison with known bases. A convenient basis for the assessment of purity and identity (in addition to the R_f value and the appearance of the original spot on the chromatogram) is the ratio of extinction values at selected wavelengths. A selection of such ratios is given in Table II.

(e) CALCULATIONS. The concentration of the purines and pyrimidines may be determined either from the absorption maxima extinctions or, preferably, by a variation of this method that may be termed the differential extinction technique.

1. From Absorption Maxima. The molecular extinction coefficients (ϵ) at the absorption maxima are listed in Table II.[15] The relationship between ϵ and extinction (E) or optical density (OD) is

$$\text{Molarity} = \frac{E}{\epsilon} = \frac{\text{OD}}{\epsilon}$$

It is convenient, for purposes of calculation, to define the *OD unit*. It is defined here as the OD (at the maximum) of the solution (or eluate) multiplied by the total volume in milliliters. For example, OD (at 249 mμ) of a guanine eluate is found to be 0.582. Since 5.00 ml. of 0.1 N HCl was used for the elution, there are therefore 0.582 × 5.00 or 2.91 OD units of guanine in the total eluate. Since there is 0.0901 micromole of guanine per OD unit (Table II), the eluate contains 2.91 × 0.0901, or 0.262 micromole of guanine.

2. The Differential Extinction Technique.[6] Since 0.1 N HCl extracts of filter paper show a low but variable absorption in the ultraviolet, it is preferable to calculate concentration from the difference (Δ) in the extinction values read at the absorption maximum and at another wavelength such as, for example, 290 or 310 mμ. For standard solutions containing

---

[15] Unfortunately, the published ultraviolet absorption data for certain purines and pyrimidines are not always in agreement. The differences, which may be as high as several per cent, arise from (a) the failure to use matched silica cells, (b) the use of "standards" the elementary composition (degree of hydration, empirical formula of a salt form, etc.) of which has not been determined or recorded, (c) the failure to demonstrate homogeneity, and (d) the failure to calibrate the spectrophotometer. A discussion of the criteria of purity of these compounds is given by A. Bendich, *in* "The Nucleic Acids" (Chargaff and Davidson, eds.), p. 81, Academic Press, New York, 1955. The spectral data selected here are in agreement (±1%) with values determined independently in this Laboratory.

TABLE II

ULTRAVIOLET ABSORPTION PROPERTIES OF PURINES AND PYRIMIDINES
(0.1 N HCl)[a,b]

	Wavelength at maximum, mμ	$\epsilon \times 10^{-3}$[c]	Ref.	Ratio of extinctions			Micromole[d] OD unit
				250/260	280/260	290/260	
Adenine	262.5	12.6	e	0.78	0.37	0.036	0.0794
Guanine	249	11.1	e	1.37	0.82	0.48	0.0901
Uracil	260	8.15	f	0.85	0.176	0.010	0.123
Thymine	265	7.95	g	0.69	0.55	0.115	0.126
Cytosine	276	10.0	h	0.48	1.50	0.77	0.100
5-Methylcytosine	283	9.79	h	0.40	2.53	2.26	0.102
5-Hydroxymethyl-cytosine	279.5	9.70	i				0.103

[a] See ref. 15.

[b] These values apply for 0.01 N HCl as well.

[c] ϵ is the molecular extinction coefficient, $\epsilon = \dfrac{OD}{molarity} = \dfrac{E}{molarity}$, where OD is optical density, and E is extinction.

[d] OD unit is defined here as optical density (or extinction) of a solution, at the maximum, multiplied by its total volume in milliliters. For example, 200 ml. of a solution of thymine in 0.1 or 0.01 N HCl is obtained from a chromatogram. The solution shows an OD of 0.240 (at 265 mμ, the maximum) and thus contains $0.240 \times 200 = 48.0$ OD units of thymine or $48.0 \times 0.126 = 0.605$ micromole of thymine.

[e] E. Vischer and E. Chargaff, *J. Biol. Chem.* **176**, 703 (1948).

[f] This Laboratory. A slightly different value[6] is used below.

[g] G. R. Wyatt, *Biochem. J.* **48**, 584 (1951).

[h] D. Shugar and J. J. Fox, *Biochim. et Biophys. Acta* **9**, 199 (1952).

[i] G. R. Wyatt and S. S. Cohen, *Biochem. J.* **55**, 774 (1953).

10 γ of base per milliliter of 0.1 N HCl solution, the following Δ values have been determined:

Adenine[6]	$E_{262.5} = 0.930;$	$E_{290} = 0.030;$	$\Delta = 0.900$
Guanine[6]	$E_{249} = 0.737;$	$E_{290} = 0.262;$	$\Delta = 0.475$
Uracil[6]	$E_{259} = 0.738;$	$E_{280} = 0.148;$	$\Delta = 0.590$
Thymine[16]	$E_{265} = 0.632;$	$E_{290} = 0.083;$	$\Delta = 0.549$
Cytosine[17]	$E_{276} = 0.910;$	$E_{300} = 0.047;$	$\Delta = 0.863$
5-Methylcytosine[17]	$E_{283} - 0.603;$	$E_{310} = 0.036;$	$\Delta = 0.567$

An example[6] of the application of this method is given. For an adenine eluate, $E_{262.5}$ is found to be 0.311, and E_{290}, 0.017. The difference, Δ_x, is therefore $0.311 - 0.017 = 0.294$. In 1 ml., therefore, there is present

[16] This Laboratory.

[17] D. Shugar and J. J. Fox, *Biochim. et Biophys. Acta* **9**, 199 (1952).

$10\Delta_x/\Delta = 2.94/0.900 = 3.27\ \gamma$ of adenine; the total eluate (5.00 ml.) thus contains 16.35 γ of adenine.

Accuracy. With practice on artificial mixtures of purine and pyrimidine standards, it is often possible to carry out routine determinations with a recovery within $\pm 2\%$, or less. However, recovery of bases from nucleic acid hydrolyzates, expressed as nitrogen recovered relative to the nitrogen content of the hydrolyzate, is usually within 2 to 4% for DNA and about 3 to 5% for RNA. When expressed as moles of base per mole of phosphorus in the hydrolyzate, the recoveries tend to be 1 to 3% lower than that based on nitrogen.

Alternative Methods of Analysis. Columns of cation and anion exchange resins[18] and of starch[19] have been used for the separation and estimation of nucleic acid bases. A useful paper chromatographic technique for the quantitative estimation of adenine and guanine and the nucleotides of uracil and cytosine in mild acid hydrolyzates of RNA has been described.[20] A differential spectrophotometric procedure for the determination of pyrimidine nucleosides and purines of RNA hydrolyzates has been developed.[21] A suitable method for the separation and estimation of the mononucleotides of DNA has been developed by enzymatic hydrolysis and anion exchange chromatography.[22]

An excellent discussion of the chromatographic separation and estimation of nucleic acid components has recently appeared.[23]

[18] W. E. Cohn, *Science* **109**, 377 (1949); *J. Cellular Comp. Physiol.* **38**, Suppl. 1, 21 (1951).
[19] M. M. Daly, V. G. Allfrey, and A. E. Mirsky, *J. Gen. Physiol.* **33**, 497 (1950).
[20] R. Markham and J. D. Smith, *Biochem. J.* **45**, 294 (1949); see also *Biochem. J.* **46**, 509 (1950).
[21] H. S. Loring, J. L. Fairley, H. W. Bortner, and H. L. Seagran, *J. Biol. Chem.* **197**, 809 (1952).
[22] R. O. Hurst, A. M. Marko, and G. C. Butler, *J. Biol. Chem.* **204**, 847 (1953).
[23] G. R. Wyatt, *in* "The Nucleic Acids" (Chargaff and Davidson, eds.), p. 243, Academic Press, New York, 1955.

[107] Methods of Isolation and Characterization of Mono- and Polynucleotides by Ion Exchange Chromatography[1]

By WALDO E. COHN

A. Preparation of Digests

Since ion exchange chromatography is markedly influenced by the amounts and kinds of ions present in all solutions used, including the digest containing the substances to be chromatographed, the digestion step itself must be ordered so as to avoid introducing unwanted ionic material or to remove it, if introduction is unavoidable. Thus high concentrations of buffers or large volumes of solution must be avoided; so must ionic species which will conflict with the subsequent chromatography (e.g., sulfate in a chloride sequence, or chloride in a formate elution) or with assay methods (e.g., TCA absorbs ultraviolet light; arsenate assays as phosphate). Hence, in choosing digestion procedures, constant reference must be made to the subsequent chromatographic separations which are described in Section B, this article.

1. Alkaline Hydrolysis of RNA (or Polynucleotides Thereof).[2] The material to be hydrolyzed is made 0.3 N (0.1 to 0.5 N is the range in the literature) in free NaOH (or KOH) and allowed to stand at 37° overnight.[3,4] Longer periods at lower temperatures (or alkalinity) or shorter periods at higher temperatures (or alkalinity) seem equally effective but have not been carefully explored. Excess alkali may cause the appearance of non-nucleotide phosphorus-containing material from tissues or further degradation of nucleotides.

For ion exchange analysis, it is desirable that the material be absorbed from solutions with molarity less than 0.02 in total anion and that excess OH^- (and CO_3^{--}) be removed. If the volume of the hydrolyzate is small, the excess alkali may be neutralized with NH_4Cl (or NH_4 acetate, formate, or sulfate, as the subsequent procedure requires) and the solution diluted to the appropriate ionic strength. Alternatively, to avoid dilution, the alkali may be removed by titration with a strong-acid cation exchanger in the acid form[5] (Dowex 50 or IR-120 in the H^+ form; large mesh material is preferable). Once the pH of the solution has dropped

[1] This contribution was prepared under Contract No. W-7405-eng-26 for the Atomic Energy Commission.

[2] W. E. Cohn and J. X. Khym, *Biochem. Preparations* **5**, in press.

[3] G. Schmidt and S. J. Thannhauser, *J. Biol. Chem.* **161**, 83 (1945).

[4] E. Volkin and W. E. Cohn, *in* "Methods in Biochemical Analysis" (Glick, ed.), Vol. 1, p. 287. Interscience, New York, 1954.

below 10 (it is safe to go to 5 or even less[5]), the solution may be buffered to the desired pH and ionic strength and absorbed on the exchange column (see Section B, this article).

2. Hydrolysis of RNA by RNAse (see Vol. II [62]). Kunitz[6] first isolated crystalline ribonuclease and determined the optimum conditions for its activity as pH 7.7 (7.0 to 8.2) and 65°. The enzyme is sufficiently active at 25°, however, and is commonly used at that temperature or at 37°. Kunitz followed its action by determining the increase in acid-soluble P. It is quite certain, however, that there is considerable hydrolysis subsequent to complete reduction of RNA to acid-soluble (or dialyzable) fragments.[6-8] Since each hydrolytic cleavage releases one secondary phosphate acid group, the action of the enzyme is conveniently followed by titration[8,9] to a constant pH (7.0 to 7.5), which also has the advantage of eliminating buffers where these would interfere with subsequent chromatography. The rates of reaction thus determined are considerably (perhaps ten times) slower than those determined by the decrease in acid insolubility. Table I summarizes some of the kinetic data in the literature in terms of the time for half completion. Since the approach to the end point is asymptotic,[8] no definite completion time can be given; a practical limit is ten times the half-time.

Magnesium ion is reported to be inhibitory.[10]

Conversion of the digest to pH 2 serves to precipitate undigested RNA and protein.[8] The supernatant is then adjusted to the proper pH and ionic strength for chromatography (see Section B).

3. Hydrolysis of DNA by DNAse (see Vol. II [63]). The kinetics of DNAse action have been examined by Kunitz,[11] subsequent to his isolation of the crystalline enzyme. Magnesium ion, in proportion to DNA but in considerable excess of it (in terms of P), is required for maximum activity, and the rate of hydrolysis is proportional to both DNA and DNAse concentrations over a wide range of each. The optimum is 6.5. The various criteria of hydrolysis (drop in viscosity, increase in dialyzability and in acid solubility, increase in secondary phosphate titration,

[5] S. E. Kerr, K. Seraidarian, and G. B. Brown, *J. Biol. Chem.* **188**, 207 (1951).

[6] M. Kunitz, *J. Gen. Physiol.* **24**, 15 (1940). See also G. Schmidt, *in* "The Nucleic Acids" (Chargaff and Davidson, eds.), p. 555, Vol. 1, Academic Press, New York, 1955.

[7] R. Markham and J. D. Smith, *Biochem. J.* **52**, 552, 558, 565 (1952).

[8] E. Volkin and W. E. Cohn, *J. Biol. Chem.* **205**, 767 (1953).

[9] E. Volkin and C. E. Carter, *J. Am. Chem. Soc.* **73**, 1516 (1951).

[10] C. Lamanna and M. F. Mallette, *Arch. Biochem.* **24**, 451 (1949).

[11] M. Kunitz, *J. Gen. Physiol.* **33**, 349, 361 (1950). See also G. Schmidt, *in* "The Nucleic Acids" (Chargaff and Davidson, eds.), p. 555, Vol. 1, Academic Press, New York, 1955.

increase in ultraviolet absorption) do not all change at the same rate. Acid solubility (dialyzability), indicating the production of smaller fragments, changes more slowly than does viscosity or ultraviolet light absorption, but more rapidly than does the number of titrable acid groups. The last

TABLE I

RATES OF HYDROLYSIS OF YEAST RNA BY CRYSTALLINE RNASE

Source	Concentration of RNA, %	Concentration of RNAse, %	Temperature, °C.	pH	Observation	Half time, minutes
a	2	0.0005	25	8.0	Acid solubility	$<10^b$
a	2	0.005	25	8.0	Acid solubility	$<10^c$
a	2	0.05	25	8.0	Acid solubility	$<10^d$
e	10	1.0	?	5.5^f	Prostatic phosphatase^g	<60
e	10	0.04	25	5.5^f	Prostatic phosphatase^g	300
h	0.7^i	0.01	37	7.2	Titration^j	40
h	0.7^i	0.1	25	7.2	Titration^j	10
h	0.7^i	0.2	25	7.2	Titration^j	5
h	$0.7^{i,k}$	0.01	25	7.2	Titration^j	75
h	0.5^l	0.04	25	7.2	Titration^j	30
m	5–9	ca. 0.1	25	6.0	Dialysis	ca. 100

[a] M. Kunitz, *J. Gen. Physiol.* **24**, 15 (1940).
[b] 73% acid-soluble in 10 minutes (maximum = 87%).
[c] 81% acid-soluble in 10 minutes (maximum = 87%).
[d] 84% acid-soluble in 10 minutes (maximum = 87%).
[e] G. Schmidt, R. Cubiles, N. Zöllner, L. Hecht, N. Strickler, K. Seraidarian, M. Seraidarian, and S. J. Thannhauser, *J. Biol. Chem.* **192**, 715 (1951).
[f] Starting pH; not buffered.
[g] Prostatic phosphatase used to determine number of phosphate monoesters produced.
[h] E. Volkin and C. E. Carter, *J. Am. Chem. Soc.* **73**, 1516 (1951).
[i] Subsequently chromatographed.
[j] End point is 0.40 to 0.45 mole of hydroxide consumed per mole of P, depending on pyrimidine content.
[k] Calf liver RNA.
[l] Subsequently hydrolyzed with intestinal phosphatase (see Table III, last entry).
[m] B. Magasanik and E. Chargaff, *Biochem. et Biophys. Acta* **7**, 396 (1951).

named, in connection with the preparation of the smaller polynucleotides, would be the property of major interest, but the intermediate poly-nucleotides may be more accessible by use of dialysis as the criterion.

Data indicating the rate of splitting of DNA by DNAse under various conditions are summarized in Table II.

The choice of conditions will depend not only on the level of degrada-

TABLE II
RATE OF HYDROLYSIS OF DNA BY CRYSTALLINE DNASE

Source	Concentration of				Temperature, °C.	pH [b]	Observation	Half time, minutes
	DNA, %	DNAse, %	Mg, M	Mg:P [a]				
c	0.5	0.0001	0.03	2	25	7.6	Titration	30 [d]
							Acid solubility	15
							Optical density	5
c	0.5	0.0002	0.03	2	25	7.5	Titration	13
c	0.016	0.000083	0.05	100	30	5.5	Acid solubility	9
							Viscosity	1
e	0.5–2.0	0.10 [f]	0.003	0.2–0.05	30	7.5	Acid solubility [g]	10
e	0.5–2.0	0.002 [f]	0.003	0.2–0.05	30	7.5	Acid solubility [g]	250
h	0.5	0.0012	0.0035	0.2	25	7.2	Titration	60 [i]
j	1.25	0.0025	0.003	0.07	37	7.4–7.7	Dialysis	ca. 300
j	0.6	0.02	0.003	0.14	37	7.4–7.7	Dialysis	250
k	0.5	0.01	0.025	1.5	37	6.5	(None)	i, l
m	0.1	0.24	0.036	10	37	6.5	(None)	n
o	0.067	0.0023	0.015	6	?	?	Titration	p

[a] Half the maximum rate of digestion (followed by optical density increase) is achieved by a magnesium ion concentration of four to six times the DNA-P concentration; the maximum rate is achieved by a 10- to 20-fold excess of Mg over P. Mg = P gives about 10% of the maximum rate.

[b] Optimum pH = 6.5. Increased ionic strength decreases the rate (50% decrease in going from 0.04 to 0.10 ionic strength). Hence, buffers and Mg must be used sparingly.

[c] M. Kunitz, *J. Gen. Physiol.* **33**, 349, 361 (1950).

[d] End point: 1 mole of OH⁻ per 3.65 moles of P (4.4 on repetition).

[e] J. A. Little and G. C. Butler, *J. Biol. Chem.* **188**, 695 (1951).

[f] Lyophilized DNAse preparation made by method of M. McCarty, *J. Gen. Physiol.* **29**, 123 (1946).

[g] Less than 20% acid-insoluble material at end point.

[h] E. Volkin, J. X. Khym, and W. E. Cohn, *J. Am. Chem. Soc.* **73**, 1533 (1951).

[i] Estimated from completion time of 7 to 8 hours (0.20 mole of OH⁻ consumed per mole of P at completion); for subsequent phosphatase treatment, see Table III, entry f.

[j] S. Zamenhof and E. Chargaff, *J. Biol. Chem.* **187**, 1 (1950).

[k] R. L. Sinsheimer and J. M. Koerner, *Science* **114**, 42 (1951).

[l] Not known. Total digestion time = 72 hours with stirring under hexane. Subsequent phosphatase digestion described in Table III, entry j.

[m] R. L. Sinsheimer and J. F. Koerner, *J. Biol. Chem.* **198**, 293 (1952).

[n] Not known, total digestion time = 48 hours with stirring under hexane. Subsequently hydrolyzed with rattlesnake diesterase to a quantitative yield of mononucleotides.

[o] R. L. Sinsheimer, *J. Biol. Chem.* **208**, 445 (1954).

[p] Not given; end point of 1 micromole/4.4 micromoles of P.

tion desired but also on the subsequent procedures (e.g., further hydrolysis by intestinal phosphatase or venom diesterase; see p. 742). The only case so far reported where the DNA-DNAse digest has been chromatographed by ion exchange without further degradation is that of Sinsheimer[12] (see Section B.3).

4. Hydrolysis by Snake Venom Diesterase (see Vol. II [89], also Section C.3.b). (*a*) RNA.[13] An amount of this enzyme preparation is used which will give over 80% splitting (as measured by the consumption of alkali to maintain the pH at 8.5 to 8.6) in relation to the total RNA-P in the substrate in a period of 6 to 8 hours without significant release of inorganic phosphate due to the residual nucleotidase activity. As prepared by Volkin,[13] from 1 to 3 ml. of the enzyme solution is used for about 20 mg. of RNA. A typical hydrolysis progressed as follows: A mixture of 16 mg. (51.5 micromoles of P) of calf liver RNA, 1.0 ml. of 0.05 M MgCl$_2$, and 1 ml. of diesterase preparation was adjusted to pH 8.6 and maintained at this pH by the intermittent addition of 0.02 M Na OH from a microburet. The temperature was 25°. At the end of 7 hours, approximately 41 micromoles (80% of total P) of OH$^-$ had been consumed, and inorganic phosphate had risen to 7 micromoles (14% of total P). The reaction was terminated by the addition of NH$_4$OH to 1 M concentration, and the mixture was diluted to 50 ml. preparatory to absorption on the ion exchange column (Section B).

(*b*) DNA-DNAse DIGEST.[14] The action of snake venom diesterase preparations on the DNA-DNAse digest is apparently much more rapid, in so far as preparations in different laboratories can be compared. Hurst and Butler[15] report 88% hydrolysis of 10 ml. of 0.5% "magnesium oligonucleotide"[16] in 2 hours at 37° by 2 ml. of diesterase preparation in 0.05 M Veronal buffer at pH 9.0 with only 5.6% release of inorganic phosphate from AMP in the same period.

Hurst *et al.*[17] found it advisable to heat the DNA before the use of DNAse and diesterase to inactivate a phosphomonoesterase activity apparently contaminating the DNA which otherwise degrades approximately 14% of the mononucleotides[13] produced by the diesterase. Their combined procedure for preparation of deoxynucleotides from DNA is as follows.

A mixture of 30 mg. of DNA in 5 ml. of H$_2$O (0.6% solution) is heated

[12] R. L. Sinsheimer, *J. Biol. Chem.* **208**, 445 (1954).

[13] W. E. Cohn and E. Volkin, *J. Biol. Chem.* **203**, 319 (1953).

[14] W. E. Cohn, E. Volkin, and J. X. Khym, *Biochem. Preparations* **5**, in press.

[15] R. O. Hurst and G. C. Butler, *J. Biol. Chem.* **193**, 91 (1951).

[16] J. A. Little and G. C. Butler, *J. Biol. Chem.* **188**, 695 (1951).

[17] R. O. Hurst, A. M. Marko, and G. C. Butler, *J. Biol. Chem.* **204**, 847 (1953).

to 100° in a water bath for 10 minutes. $MgSO_4$ to 0.025 M (Mg:P = 1.25) and 0.02 mg. of DNAse per milligram of nucleate (0.12 ml. of 0.5% solution giving a 0.012% DNAse solution) are added, and the mixture is incubated at 37°. After 1 hour, 1 to 2 ml. of diesterase, prepared from Russell's viper venom by the Hurst and Butler method[15] and containing 0.01% merthiolate, is added. This is sufficient to complete the hydrolysis to mononucleotides in 1 to 2 hours. The pH is maintained at 8.3 to 8.5 with 0.1 N NaOH. When the consumption of alkali ceases, the mixture is acidified to pH 4 with 2 N acetic acid (an acetate ion exchange system is used subsequently; otherwise, a different acid would be used).

Sinsheimer and Koerner[18] attained a quantitative yield of mononucleotides by digesting a DNA-DNAse digest (see Table II) at pH 9 with 0.1 ml. of a rattlesnake diesterase preparation per milliliter of (approximately 0.1%) digest for 6 hours at 37°.

5. *Hydrolysis with Snake Venom* (dried rattlesnake venom) (see Vol II [89]). (a) RNA.[13,19] A mixture of 20 mg. of RNA and 10 mg. of venom (*Crotalus ademanteus*), in 5 ml. of 0.1 M borate buffer (pH 8.5) containing 0.01 M $MgCl_2$, is digested overnight at 37°. (Alternatively, the buffer may be omitted and the progress of the digestion followed by titration with NaOH from a microburet, with the pH maintained at 8.5 to 8.6.) The digestion is about two-thirds completed in 3 hours. The digest is diluted to 50 ml. and made ammoniacal in preparation for absorption on the anion exchange column. (If nucleosides are not to be adsorbed, the pH may be as acid as 3 in the absorption step, since the strongly absorbed pyrimidine nucleoside diphosphates are the sole nucleotide products.)

(b) DNA-DNAse DIGEST. The action of rattlesnake venom on the DNA-DNAse digest seems to be more rapid than on RNA, according to the figures given by Hurst *et al.*[20] Two milligrams of venom digested 250 mg. of "magnesium oligonucleotide"[16] (in 30 ml. at pH 9.4) to 60% inorganic phosphate and 78% uranium-soluble phosphate in 4 hours at 37°. (Differences in rate are to be expected not only between nucleic acids and between each nucleic acid and its nuclease digest but also between different samples of crude venom on the same substrate. Each situation must be independently assayed for conformity to a predetermined standard of hydrolysis.) The digest was absorbed at pH 4.5 on a Dowex-1-acetate column (see Section B.3).

6. *Hydrolysis of Nuclease Digests to Nucleotides by Intestinal Phosphatase* (see Vol. II [90]). Although the first production of 5′ deoxy-

[18] R. L. Sinsheimer and J. F. Koerner, *J. Biol. Chem.* **198**, 293 (1952).
[19] W. E. Cohn and E. Volkin, *Arch. Biochem. and Biophys.* **35**, 465 (1952).
[20] R. O. Hurst, J. A. Little, and G. C. Butler, *J. Biol. Chem.* **188**, 705 (1951).

nucleotides from DNA [21,22] and of 5' nucleotides from RNA [23] was achieved by the hydrolysis of nuclease digests with intestinal phosphatase with arsenate to inhibit monoesterase action, this method has been largely supplanted by the use of snake venom diesterase preparations on whole RNA [13,19] or on the DNA-DNAse digest. [15,17,18,20] The presence of arsenate

TABLE III

HYDROLYSIS OF NUCLEASE DIGESTS WITH INTESTINAL PHOSPHATASE
INHIBITED WITH ARSENATE

	Concentration of			Tem-pera-ture, °C.	pH	Criterion	Time, hours
Source	DNA,[a] %	Intestinal phosphatase,[b] %	Arsenate, M				
c	0.14	0.004[c]	0.002	—	8	67% Uranium-soluble[d] 10% Inorganic P	2[e] 24[e]
f	0.5	0.2	0.005	25	8.4	100% Titration[g]	9–10[h,i]
j	0.5	0.25	0.05	37	8.5	—	20[i,k]
l	RNA 0.5	0.1	0.004	25	8.5	86% Titration[m]	1.5[i,n]

[a] Digested with DNAse.

[b] Commercially available preparation, unless otherwise specified.

[c] C. A. Zittle, *J. Biol. Chem.* **166**, 491 (1946). (Prepared own phosphatase.)

[d] About one-third of digest was uranium-soluble at the start.

[e] Composition of final digests not known.

[f] E. Volkin, J. X. Khym, and W. E. Cohn, *J. Am. Chem. Soc.* **73**, 1533 (1951).

[g] Ca. 20% owing to prior nuclease digestion (Table II, entry h).

[h] About two-thirds recovered as mononucleotides, one-tenth as bases and nucleosides (see Table II, entry h, for prior nuclease treatment).

[i] Preparation chromatographed subsequently.

[j] R. L. Sinsheimer and J. F. Koerner, *Science* **114**, 42 (1951).

[k] About one-quarter of material recovered as bases and nucleosides, half as mononucleotides.

[l] W. E. Cohn and E. Volkin, *Nature* **167**, 483 (1951).

[m] 36% owing to prior RNAse digestion (see Table 1, footnote j).

[n] 60% of P recovered as inorganic P, 25% as nucleotides. Nucleosides present (ca. 30%).

and the lack of complete suppression of monoesterase action, leading to appreciable breakdown of mononucleotide before complete hydrolysis of polynucleotides,[22,23] are among the possible explanations for the loss in popularity. However, three procedures for DNA are summarized (in Table III), two of them having led to the actual production of the deoxynucleotides and one having led to the first isolation of 5' ribonucleotides.

[21] W. Klein, *Z. physiol. Chem.* **224**, 244 (1934).

[22] E. Volkin, J. X. Khym, and W. E. Cohn, *J. Am. Chem. Soc.* **73**, 1533 (1951).

[23] W. E. Cohn and E. Volkin, *Nature* **167**, 483 (1951).

In the absence of arsenate, nucleosides are the chief, if not the only, end products.[24]

B. Ion Exchange Chromatography of Nucleic Acid Digests[2,14,25]

1. Preparation of the Anion Exchange Column (see Tompkins[26] and Samuelson[27]).[2,25] It is customary to use sintered glass disks in glass tubing to support the column, but any porous support is satisfactory. The porosity of the support should be relatively coarse so as not to impede the flow unnecessarily. The exit tube may be doubled back on itself to a height greater than the top of the column as a precaution against running dry (provided that gravity flow is used).[26] The exit tube is kept at minimum diameter to reduce the mixing of various portions of the effluent. Various schemes have been advanced for preventing disturbance of the top of the resin bed by the influent stream, but this is more often ignored. It is, however, of the greatest importance that no channeling occur.

The anion exchangers are of the polystyrene, strong-base (i.e., quaternary ammonium) type in the form of spherical beads (200 to 400 mesh is the range usually used). Dowex 1 or 2 [28] and Amberlite IR-400 or IR-410 [29] are examples of this type;[27] most methods have been developed around Dowex 1 because of its early availability in fine-mesh form.

The degree of crosslinking[27] (i.e., percentage divinylbenzene used to link the linear vinylbenzene molecules into a three-dimensional structure) specified is usually 8 to 10%, and such material may be regarded as standard. All mononucleotide separations have been carried out on material of this kind, but there is disagreement in the literature as to its suitability for polynucleotide separation. Merrifield and Woolley[30] and Sinsheimer[12] report satisfactory polynucleotide separations with it, whereas Carter and Cohn[31] report otherwise. Satisfactory separations of polynucleotides on a material of a lower degree of crosslinking (2%) are reported by Volkin and Cohn.[8] It should be noted that material of a lower degree of crosslinking may also be used for mononucleotide separations. Its only disadvantage is the mechanical one of swelling and shrinking in response to changes in the osmotic pressure of the surrounding aqueous solution.

[24] W. Andersen, C. A. Dekker, and A. R. Todd, *J. Chem. Soc.* **1952**, 2721.
[25] W. E. Cohn, *in* "The Nucleic Acids" (Chargaff and Davidson, eds.), Vol. I, p. 211, Academic Press, New York, 1955.
[26] E. R. Tompkins, *J. Chem. Educ.* **26**, 32, 92 (1949).
[27] O. Samuelson, "Ion Exchangers in Analytical Chemistry," Wiley, New York, 1953.
[28] Dow Chemical Company, Midland, Mich.
[29] Resinous Products Corporation, Philadelphia, Pa.
[30] R. B. Merrifield and D. W. Woolley, *J. Biol. Chem.* **197**, 521 (1952).
[31] C. E. Carter and W. E. Cohn, *J. Am. Chem. Soc.* **72**, 2604 (1950).

The resin may be converted into the appropriate salt form (chloride, formate, etc.) and washed batchwise (e.g., in a stoppered graduate cylinder) before being placed in the column, or it may be washed *in situ*. It has been found useful to include in the treatment of each new batch of exchanger a series of washes calculated to remove organic material from the polystyrene matrix. This may be done by the following sequence: water, acetone-water, acetone, petroleum ether, acetone, acetone-water, water. Conversion to the desired salt form may be accomplished before or after this sequence. In general, the exchanger is received in the chloride form; washing with strong HCl removes acid-soluble impurities and leaves the exchanger in the chloride form. A water wash to remove HCl is all that is required before use. Conversion to the formate or acetate form is best accomplished with the sodium salts of the appropriate acids, since the acids themselves, because of their low ionization, are very inefficient at replacing chloride. Again, a water wash to remove excess anion is necessary.

A general direction as to the amount of washing required is that the wash effluents should satisfy the conditions of minimum, constant assay in the tests which will be applied subsequently to the active effluents.

The column is formed by simply slurrying the resin into the tube, taking care to keep it covered with water at all times, and allowing it to settle. After any regeneration (elution) sequence and water wash, it may be advisable to resuspend the resin and allow it to settle; this is also the remedy for an air-bound column. Whether or not to repeat the original wash sequence between runs is a matter of judgment based on the previous use and projected use of the column.

2. The Absorption Step.[2,25] The capacity of the exchanger material is approximately 1.2 meq./ml. of apparent bed volume. In general, it is well not to chromatograph more than 1 to 2% or so of the ionic capacity of the column, if the maximum degree of resolution is desired. This corresponds to 35 to 70 mg. of nucleic acid digest on a 10-ml. (e.g., 0.80 sq. cm. $\times$ 12 cm.) column. For the preparation of larger amounts with no loss of resolution, the area of the column may be increased in proportion. For example, 2-g. amounts of the isomeric nucleotides have been separated on a 12-cm. $\times$ 33-sq. cm. column, and 0.7-g. amounts of RNA-RNAse digests on 15-cm. $\times$ 3.7-sq. cm. columns with the same precision as is afforded by the analytical (10-ml.) columns. With greater risk of overlapping of peaks, larger amounts can be handled.

The object of the absorption step is to retain all the substances of interest at the top of the column. The success of the step depends on the equilibrium between solution and exchanger (which is determined principally by the ionic strength of the medium) and the rate of exchange (which depends on flow rate and mesh size and the charge on the ions in question). Both equilibrium constant and rate of absorption depend on charge which is influenced by pH; hence, this is the most important factor in determining how successful the absorption step will be. Second in importance is probably the ionic strength of the medium; the greater the concentration of competing ions, the poorer the absorption of the nucleotides.

Since the secondary pK's of phosphate esters are in the neighborhood of 6 [25] (versus inorganic P at 7), this pH gives the nucleotides an average charge of -1.5; at 7, it is -1.9. Hence, absorption of phosphate esters is markedly increased by going to pH 6 or 7. This can be used to offset high ionic strength (e.g., above 0.02 M Cl^-). A disadvantage, however, is the larger amount of acid subsequently consumed to convert the absorbed substances to the pH of elution; if dilute (e.g., 0.002 N) acid is to be used, this disadvantage can be of major importance.

A rule of thumb for absorption is pH >6, competing ions <0.02 M. It is also well to keep the concentration of stronger anions to a minimum (e.g., sulfate in a chloride system, chloride in an acetate or formate system), since they will be preferentially absorbed and may cause difficulties during the elution sequence.

In general, it is advisable to wash the sample through the exchanger with water. If the pH of the absorption step is 7 or greater or if any substantial amounts of potentially conflicting anions are introduced, it is advisable to wash the column again with 0.01 to 0.02 N ammonium chloride (or formate or acetate, depending on the form of the exchanger) until the pH of the effluent falls to 7 or less, and until the conflicting anions are removed. This buffers the absorbed materials and also serves to remove absorbed carbonate (which, as CO_2, can spoil a column when acid is passed through) and other undesirable absorbed anions.

3. The Elution Sequence.[2,14,25] The elution of nucleotides from the column is effected by increasing the acidity and/or the ionic strength of the influent solution above that (or those) which prevailed in the absorption and washing sequences. The large differences in the ionic properties, and hence exchanger affinities, of the various mononucleotides (ignoring polynucleotides and polyphosphates for the moment) make it relatively impossible to separate all of them sequentially with a single influent. For this reason, the concentration of hydrogen ion or of competing ion or both are increased either discontinuously, in a series of steps, or continuously by a mixing device (gradient elution).[32]

The chemical principles underlying gradient elution do not differ from

[32] H. Busch, R. B. Hurlbert, and V. R. Potter, *J. Biol. Chem.* **196,** 717 (1952); R. B. Hurlbert and V. R. Potter, *J. Biol. Chem.* **209,** 1 (1954).

those of ordinary elution by a sequence of reagents. Since so much has been published in terms of the latter, only these methods will be presented.

(a) THE CHLORIDE SYSTEM. Simple elution with a series of dilute HCl solutions suffices to separate the usual mononucleotides and even some of the isomers.[2,25,33] The cytidylic and the adenylic acids are removed rapidly and slowly, respectively, by 0.002 N HCl (about 1.0 to 1.5 l. is required for a 10-ml. column). Increasing this to 0.003 N removes the uridylic and the guanylic acids, again rapidly and slowly, respectively (about 2.5 to

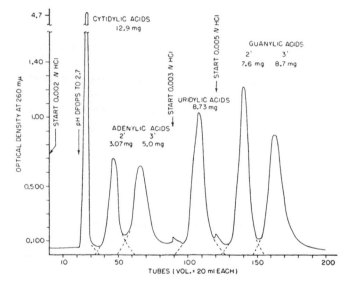

FIG. 1. Anion exchange chromatogram of alkaline hydrolyzate of about 50 mg. of rat liver RNA. Column: Dowex-1-Cl⁻, 400 mesh, 6 cm. × 0.72 sq. cm. [From E. Volkin and C. E. Carter, *J. Am. Chem. Soc.* **73**, 1516 (1951).]

3.0 l. is required for a 10-ml. column). Flow rates as high as 3 ml./sq. cm./min. have given satisfactory results, and there is no indication that this is an upper limit. With some risk of incomplete separation of cytidylic and adenylic acids, 0.003 N HCl may be used throughout (ca. 5 l. is required). A more rapid elution of the guanylic acids is afforded by 0.005 N HCl, as shown in Fig. 1.

Some of the various isomeric forms of the mononucleotide monophosphates are not well separated in the HCl system, although 5′ nucleotides (ribo or deoxy) always precede the 2′ and 3′, which appear in that order. The purine nucleotides can be separated, and the cytidylic isomers yield to 0.001 N HCl (but not to 0.002 N), but the uridylic acids have not so far been separated in the HCl system. Separation of the latter pair is best achieved in formate or acetate systems (see Section B.3.b). The 5' deoxy and 5' ribo compounds are separable with borate.[25]

At higher pH (ca. 5.6) and correspondingly higher chloride ion concentration (0.02 to 0.1 M), the pyrimidine nucleotides are eluted together and the purine nucleotides are eluted together.[33] The total volume at 0.02 M required is about 5 l.; at 0.1 M, it is 300 ml.

The pyrimidine diphosphonucleosides are not rapidly eluted by these reagents, although cytidine diphosphate tends to overlap guanylic acid, the last of the monophosphates. They are eluted with 0.007 to 0.01 N HCl (cytidine diphosphates, about 1 l. required) and 0.05 to 0.075 N HCl (thymidine and uridine diphosphates). It is apparent that the diphosphates follow the same order of elution in the acid region as do the monophosphate analogs (see Fig. 1 in Vol. III [120]).

The di-, tri-, and polynucleotides fall into sequences which can be predicted qualitatively from the behavior of the component mononucleotides. Dinucleotides behave like diphosphates; they tend to follow the mononucleotides. Trinucleotides tend to follow dinucleotides, etc. Thus, at pH 2, cytidine diphosphate precedes guanylyl cytidylic acid (GC) which in turn precedes ADP; adenylyl cytidylic (AC) precedes GC. AAC precedes GC; AGC follows it.

The separation of many of the RNA-RNAse products has been achieved[8] on columns of about the same length as described but with an anion exchange material of 2% divinylbenzene content. For a 10-ml. column, the volumes and compositions of eluting solutions and the substances (represented by the initials of the nucleotides) eluted by each were as follows:

0.005 N HCl	1.0 l. C, AC, U
0.01 N HCl	0.7 l. AAC, GC
0.01 N + 0.0125 N NaCl	1.0 l. GAC, AGC, AU
0.01 N + 0.025 N NaCl	0.3 l. (AAG)C
0.01 N + 0.05 N	1.2 l. AAU, (AAG)U, GGC, GU, (AGG)C
0.01 N + 0.05 N NaCl	0.7 l. GAU, AGU
0.01 N + 0.1 N NaCl	0.7 l. (AGG)U, Polys, GGU
0.01 N + 0.2 N NaCl	0.3 l. Polys
0.01 N + 0.3 N NaCl	0.3 l. Polys
0.01 N + 1 N NaCl	0.3 l. Polys
2 N HCl	

(Parentheses indicate that the exact sequence of the nucleotides is not known.) No systematic exploration of such variables as length of column, amounts of reagents, gradient elution, and pH, have been made.

The final separation of the polynucleotide fractions obtained in

[33] W. E. Cohn, *J. Am. Chem. Soc.* **72**, 1471 (1950).

acetate systems (see Section B.3.b) from the DNA-DNAse digest (Section A.3) has been accomplished in chloride systems buffered at pH 4.0 to 4.4 with 0.05 to 0.1 M acetate.[12] In all cases, the column (originally in the acetate form, and maintained in that form during the reabsorption of the acetate-containing fractions) was converted to the chloride form by the eluting solution so that the final separations were accomplished by stepwise increase in chloride ion; the range used was 0.02 M to 0.2 M to elute all di- and some trinucleotides.

(b) THE FORMATE (AND ACETATE) SYSTEMS. Although the dilute HCl systems are advantageous if reabsorption is contemplated, pH regulation

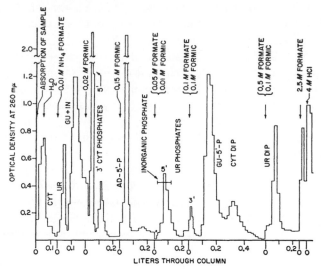

FIG. 2. Anion exchange chromatogram of snake venom diesterase digest of 16 mg. of calf liver RNA. Column: Dowex-1-formate, 400 mesh, 5.8 cm. × 0.9 sq. cm. [From W. E. Cohn and E. Volkin, *J. Biol. Chem.* **203**, 319 (1953).]

and buffer capacity are more easily maintained with weaker acids. Formate (and acetate) systems, which offer the possibility of complete removal of solvent (the acids and ammonium salts of both are volatile *in vacuo* at relatively low temperatures), have been used to advantage in the separation of mononucleotides and in the separation of mononucleotide isomers (see Fig. 2). The first three fractions are nucleosides.

The cytidylic acid isomers are removed (in the order 5′, 2′, 3′, if all are present) by 0.01 M formic acid (ca. 500 ml.), the adenylic acids by 0.1 M formic acid (ca. 700 ml.).[13,19,23] Although 1 M formic (ca. 700 ml.) will remove both guanylic acid and uridylic acid (in that order),[33] the uridylic acid isomers do not separate well, if at all. The guanylic and uridylic acids have usually been removed (and separated) by similar quantities of

0.05 M formate (NH_4 or Na) plus 0.005 to 0.01 M formic acid (uridylic) and 0.1 M formate plus 0.1 M formic acid (guanylic).[13,19] The latter brings off cytidine diphosphate on the heels of the 5' guanylic acid.[13] Uridine diphosphate can be removed with a small volume (ca. 200 ml.) of 0.5 M formate plus 0.1 M formic acid.[13]

The 5' deoxynucleotides behave exactly as do the 5' ribonucleotides in this system.[22,25]

Although acetate systems have been used to separate the cytidylic acids[34] and adenylic acids, no other acetate system experiments involving the isomeric ribonucleotides have been reported. However, the separation of (5') deoxynucleotides (produced as described in Section A.4.b) in an acetate system has been accomplished with increasing concentrations of acetate buffer at pH 4.3 as follows[35] (in milliliters per 10 ml. of resin bed):

0.04 M	ca. 1000 ml.	Deoxycytidylic (and 5-methyldeoxycytidylic)
0.08 M	ca. 1000 ml.	Thymidylic
0.15 M	ca. 1000 ml.	Deoxyadenylic
0.3 M	ca. 1000 ml.	Deoxyguanylic

A similar separation was achieved with 0.5 M acetic acid with increasing amounts of sodium acetate (from none to 0.1 M).[20]

Sinsheimer[12] has used the acetate system to fractionate the DNA-DNAse digest also (see Section A.3). Volumes of 500 to 1000 ml. (per 10 ml. of resin bed) of pH 5.5 acetate buffer at concentrations of 0.1, 0.24, 0.30, and 0.45 M were used to get four fractions; these were rechromatographed on an acetate column with 0.05 to 0.1 M acetate at pH 4.0 to 4.4 containing increasing amounts of chloride (0.015 to 0.20 M). These solutions soon convert the column to the chloride form so that the rechromatography is actually accomplished in a chloride system with the acetate buffer serving to regulate pH.

Gradient elution[32] has been applied to the formic acid (and formate) elution of various nucleotides, including those found in alkaline hydrolyzates of RNA. The concentration of the influent was automatically increased, rapidly at first, then more slowly with increasing time. A method of achieving a linear increase in concentration with volume (or time) has been reported.[36] This may have an advantage over the type reported.

4. Concentration by Ion Exchange.[2,33] The concentration of the relatively dilute solutions which are obtained in the analytical chromatographic separations described in Section B.3 may often be most easily achieved by a reabsorption–re-elution sequence on a smaller ion exchange

[34] W. E. Cohn, *J. Am. Chem. Soc.* **72**, 2811 (1950).
[35] R. L. Sinsheimer and J. F. Koerner, *Science* **114**, 42 (1951).
[36] C. W. Parr, *Biochem. J.* **56**, xxvii (1954).

column. If the elution has been accomplished with dilute acid and no salt (e.g., 0.002 to 0.01 N HCl or 0.01 N formic acid for mononucleotides), it is sufficient to add NH_4OH to an alkaline pH and reabsorb the solution on an exchanger bed of tenfold smaller volume (e.g., 1 cm. $\times$ 1 sq. cm., if a 10-cm. $\times$ 1-sq. cm. bed was originally used). Solutions containing stronger acid or salt and which, on neutralization with NH_4OH, will contain total anion in excess of 0.02 M may have to be diluted in order to effect complete absorption of the nucleotide material. Polynucleotides, with stronger affinities for the exchanger, can be reabsorbed from stronger salt solutions.

The size of the reabsorbing column can be diminished to the point of incipient breakthrough (about 8 ml. of exchanger per gram of nucleotide can be used[2]), since the object is to approach saturation of the column with the material being concentrated. Elution of this near-saturated column is then achieved with acid and/or salt of a concentration which will remove the substance in question in the volume desired.

This sequence also permits the replacement of one bulk ion with another. Thus a concentration column containing material reabsorbed from a formate solution may be converted to the chloride form without loss of material by washing, after the absorption step, with sufficient NH_4OH-NH_4Cl solution (of low ionic strength) to replace all formate by chloride. Then the removal may be effected with HCl of the desired concentration.

C. Identification and Characterization of Nucleotides in the Elution Sequence

1. Ultraviolet Absorption Spectra.[4,25,37] The development of a dependable and sensitive spectrophotometer for the determination of the ultraviolet absorption characteristics of aqueous solutions has made it possible to follow the course of an elution sequence of nucleic acid components with ease and with complete conservation of material. The purine and pyrimidine components have relatively high extinction coefficients in the ultraviolet which vary quantitatively with both the substituents and the ionic forms of the acid and basic groups in and on the rings (which makes the extinction pH sensitive).[4,37] The position of the phosphate group in the nucleotides affects to a smaller degree the spectra of the pyrimidine nucleotides but has no discernable effect on the spectra of the purine nucleotides. The sugar constituent has no measurable effect (except at pH values high enough to ionize the OH groups);[38] thus cytidine 5′-phos-

[37] G. H. Beavan, E. R. Holiday, and E. A. Johnson, *in* "The Nucleic Acids" (Chargaff and Davidson, eds.), Vol. I, p. 493, Academic Press, New York, 1955.
[38] J. J. Fox, L. F. Cavalieri, and N. Chang, *J. Am. Chem. Soc.* **75**, 4315 (1953).

phate has the same spectrum (except at very high pH) as does deoxycytidine 5'-phosphate. The spectra of polynucleotides are only approximately those of a mixture of the constituent mononucleotides; the divergence from an arithmetic mean becomes greater with increasing size of the polynucleotide, apparently reaching a maximum with whole nucleic acid. This divergence may be considered as owing to the "shadowing" of each light-absorbing unit by others in its immediate vicinity in contrast to the independence of the absorption of each in free solution.

The absolute value of the extinction at some particular wavelength, in relation to phosphorus or some other independently determined constituent of the molecule, gives the extinction coefficient (optical density per centimeter of path length in a molar solution) at that wavelength and under the conditions existing in the solution (e.g., pH). Most nucleotide extinction coefficients have been based on phosphorus determinations and have utilized either the wavelength of maximum absorption (in one or the other ionic form) for each substance or 260 mμ for all.

The ratio of the relative extinctions at two wavelengths affords a quantitative measure of the shape of the spectrum which is, in turn, a characteristic of the substance in question. The shape of the spectrum is influenced by pH,[4,37] in relation to the various pK's affecting the ionic form of the substance,[25] and mass law considerations apply. The determination of such ratios, even at a variety of pH values, is not destructive of the material and is therefore a most useful method of characterization when the amounts of material are limited. Different workers have taken different wavelengths for comparison, thus making comparisons between them difficult.

Most of the ionizable groups in the purine and pyrimidine bases have pK values in the regions 3 to 6 and 9 to 10.[25] Thus, with a few exceptions, the pH values of 2, 7, and 12 have been used to measure the extinction coefficients and ratios of purine and pyrimidine constituents. Table IV[4] is an example of the type of data which are obtained and which can be used for routine preliminary identification of nucleic acid constituents.

2. *Chemical Determinations* (see Vol. III [109]). Nucleotide components in an elution sequence can be detected, with varying degrees of accuracy and response, by phosphorus, sugar, or purine or pyrimidine base analyses (nitrogen analyses are tenuous because of the possibility of nitrogen loss from the exchanger itself).

The phosphorus of the nucleotides is not easily acid-hydrolyzable, although that of the purine 2'- or 3'-nucleotides is more labile than that of the corresponding pyrimidine nucleotides which have a stable N-glycoside linkage, or of the 5' isomers which are very stable. Only the purine nucleotides respond as readily as ribose to the orcinol reaction, owing to

TABLE IV

SPECTROPHOTOMETRIC CONSTANTS OF NUCLEIC ACID DERIVATIVES[a]

pH →	$\epsilon_{260} \times 10^{-3}$			$\epsilon_{250}/\epsilon_{260}$			$\epsilon_{280}/\epsilon_{260}$			$\epsilon_{290}/\epsilon_{260}$		
	2	7	12	2	7	12	2	7	12	2	7	12
(Orotic acid)[b]	3.45			0.57	0.57	0.8	1.9	1.68	1.60	1.65	1.35	1.55
(Uric acid)[b]				1.0	1.7	1.5	2.7	2.7	2.4	2.6	3.8	3.9
Adenosine ⎱	14.2	15.0	15.0	0.85	0.80	0.80	0.22	0.15	0.15	0.03	0.002	0.002
Adenylic acid ⎰												
Adenine	12.7	13.3	10.2	0.76	0.76	0.57	0.375	0.125	0.60	0.035	0.005	0.025
Hypoxanthine	7.7[c] , 8.1		11.0	1.40[c]	1.32	0.78[c]	0.07[c]	0.092	0.14[c]	0.005	0.010	0.015
pH 10.5		11.0			0.84			0.124				
Inosine	7.4	7.4	12.1	1.68	1.68	1.05	0.24	0.25	0.18	0.025	0.025	0.008[d]
Guanine	8.1	7.3	6.6	1.37	1.42	0.93[c]	0.84	1.04	1.15[c]	0.50	0.54	0.59
Guanylic acid ⎱	11.8	11.8	11.8	1.02[c]	1.15	0.89	0.68	0.68	0.60	0.40	0.28	0.12
Guanosine ⎰												
Xanthine	8.15	7.5[c]	4.4[c]	0.58	0.68[c]	1.20[c]	0.50[c]	0.75[c]	1.95[c]	0.08	0.20[c]	1.40[c]
pH 10		5.2[d]			1.29			1.71			0.92	
Xanthosine	8.7	7.9	7.9	0.75	1.29[c]	1.30	0.28	1.10[c]	1.13	0.03	0.58[c]	0.61
pH 8		7.65[d]			1.30			1.13			0.61	
Thymine	7.4	7.4	3.7	0.67	0.67	0.65	0.53	0.53	1.31	0.09	0.09	1.41
Thymidylic acid ⎱	8.4	8.4	6.7	0.64	0.65	0.74	0.72	0.73	0.67	0.23	0.24	0.17
Thymidine ⎰				0.65	0.65	0.75	0.72	0.72	0.67	0.235	0.235	0.16
Uridine	9.9	9.9	7.3	0.74	0.74	0.83	0.35	0.35	0.29	0.03	0.03	0.02
Uridylic acid ⎱ 2′				0.80	0.78	0.85	0.28	0.30	0.25			
⎪ 3′				0.76	0.73	0.83	0.32	0.35	0.25			
⎰ 5′				0.74	0.73	0.82	0.38	0.40	0.33			

TABLE IV (Continued)

pH→	$\epsilon_{260} \times 10^{-3}$			$\epsilon_{250}/\epsilon_{260}$			$\epsilon_{280}/\epsilon_{260}$			$\epsilon_{290}/\epsilon_{260}$		
	2	7	12	2	7	12	2	7	12	2	7	12
Uracil	8.2	8.2	4.1[c]	0.84	0.84	0.71	0.175	0.175	1.40	0.01	0.01	1.27
Cytosine	6.2	5.7	4.95[c]	0.48	0.78	0.76[c]	1.53	0.58	0.80[c]	0.78	0.08	0.32[c]
5-Methylcytosine	3.6[d]	4.45[d]	7.4	0.41	0.81	0.86	2.66	1.20		2.42	0.55	0.31[c]
Cytidine	6.2	7.4	7.4	0.45	0.86	0.86	2.10	0.93	0.95[c]	1.55	0.29	0.31[c]
Cytidylic acid 2'	6.8 }	7.6	7.6 }	0.48	0.90	0.90	1.80	0.85	C.85	1.22	0.26	0.26
Cytidylic acid 3'	6.5 }			0.45	0.86	0.86	2.00	0.93	C.93	1.43	0.30	0.30
Cytidylic acid 5'	6.2	7.4	7.4	0.46	0.84	0.84	2.10	0.99	0.99	1.55	0.33	0.33
Deoxycytidylic acid	(6.2)[c,d]	(7.4)[c,d]	(7.4)[c,d]	0.46	0.82	0.82	2.12	0.99	0.99	1.55	0.30	0.30
5-Methyl deoxycytidylic acid	(3.2)[c,d]	(6.0)[c,d]	(6.0)[d]	0.36	0.96	0.96	3.15	1.52	1.52	3.4	1.02	1.02

[a] W. E. Cohn, in Oak Ridge National Laboratory Report ORNL-889, p. 50 (1950), and unpublished data.
[b] All values tentative.
[c] Rapid shift in values with pH; unreliable without exact pH control. Values in parentheses are approximate or relative.
[d] G. H. Beaven, E. R. Holiday, and E. A. Johnson, in "The Nucleic Acids" (Chargaff and Davidson, eds.), Vol. I, p. 493, Academic Press, New York, 1955.

the acid sensitivity of the N-glycoside linkage; the uridylic acids respond very slowly, the cytidylic acids almost not at all. The purine deoxynucleotides respond to the diphenylamine reaction; the pyrimidine ones do not.

Hydrolysis to free bases can be accomplished in a variety of ways, and the bases can then be determined chemically or chromatographically.[25,39,40]

Polynucleotides of RNA can be hydrolyzed in alkali under the same conditions as have been described for whole RNA.[7,8] Each nucleotide is obtained as a mixture of 2′ and 3′ isomers except for that which holds a monoesterified phosphate group or no phosphate group at all. In the latter cases, a 3′ nucleotide or nucleoside results. The amounts of these afford measures of the size of the polynucleotide in question, and their kind specifies the nature of one end group; the 2′ and 3′ nucleotides in the digest represent the other components of the original polynucleotide.

3. *Enzymic Determinations* (see Vol. III [111]). (*a*) PHOSPHATASES. Nucleotides possessing free (i.e., monoesterified) phosphate in the 2′, 3′, or 5′ position are susceptible to the action of prostatic, intestinal, and bone phosphatases (see Vol. II [79], [80], and [82]). The 5′ nucleotidases (Vol. II [85]) are specific for 5′-linked phosphate. Since both intestinal phosphatase and snake venom contain diesterases, they cannot be used as nucleotidases on polynucleotides because they will convert the polynucleotides to free 5′ nucleotides and then hydrolyze these to inorganic phosphate. (The possible presence of diesterase or nuclease activity[41] in phosphatase preparations, and of phosphatase in DNA preparations,[17] must be considered.) The 3′,5′ (and 2′,5′) nucleoside diphosphates are resistant to the venom and bull semen 5′ nucleotidases but not to one found in potato or to intestinal phosphatase.[19]

A nucleotidase specific for 3′ phosphate groups has been found (Vol. II [86]).

The amount of phosphate removed by such phosphatases, in relation to the total present, affords a measure of the size of a polynucleotide as well as a definition of the nature of the end phosphate group. It also makes it possible to identify the end nucleoside residues of polynucleotides by subsequent alkaline or diesterase hydrolysis, as is described in this and the following subsections, respectively.

(*b*) DIESTERASES (see Vol. II [89] and [90], also p. 728, this article). Snake venom diesterase shows a variable degree of activity toward the polynucleotides derived as final products from the RNA-RNAse digest.

[39] E. Chargaff, R. Lipshitz, C. Green, and M. E. Hodes, *J. Biol. Chem.* **192**, 223 (1951).
[40] G. R. Wyatt, *Biochem. J.* **48**, 584 (1951).
[41] G. Schmidt, R. Cubiles, and S. J. Thannhauser, *Cold Spring Harbor Symposia Quant. Biol.* **12**, 161 (1947).

The smaller polynucleotides seem more resistant than the larger, and those ending in uridylic acid more resistant than those ending in cytidylic. Preliminary removal[8] of the end (monoesterified) phosphate or its preservation in cyclic form[7] markedly enhances the speed of the hydrolysis. Thus it would appear that singly esterified 3′ phosphate groups at the ends of polynucleotides inhibit the diesterase to a variable degree.[13] Since the DNA-DNAse polynucleotides do not contain such groups, they are readily hydrolyzed by venom diesterase.[17,18]

Hydrolysis of polynucleotides by diesterase liberates primarily 5′ nucleotides. In the case of a 3′ end phosphate group, a 3′, 5′ diphosphate is liberated; in the case of a nucleoside linked to the chain only through its own 3′ position, a nucleoside is liberated. Thus the end nucleosides of polynucleotide chains can be identified.

The action of the diesterase of intestinal phosphatase on RNA is not well known, since the interfering monoesterase activity of the preparation has never been eliminated completely, even with large amounts of arsenate.[42] Arsenate does seem to inhibit such activity with respect to DNA fragments, however.[43]

[42] C. A. Zittle, L. A. Wells, and W. G. Batt, *Arch. Biochem.* **13**, 395 (1947).
[43] C. A. Zittle, *Arch. Biochem.* **13**, 191 (194)..

[108] Paper Chromatography of Mononucleotides and Oligonucleotides

By Roy Markham

The paper chromatography of nucleotides has so far been used as a means of separating definite groups of substances for further treatment, such as hydrolysis by specific enzymes or by chemical methods, or separation by electrophoresis on paper, or as a means for the separation and identification of mixtures having a restricted composition. As there are now at least twenty-eight different mononucleotides obtainable from ribonucleic acid alone (including 2-thiouridylic, 8-azaguanylic, and 3-chlorocytidylic acids), and twenty-two possible ones from deoxyribonucleic acids, of which some eleven have actually been isolated (including 8-azaguanylic acid, 5-chloro-, bromo-, and iodouridylic acids, 5-methyl- and 5-hydroxymethylcytidylic acids, and 5-methylaminopurine deoxyribonucleotide), it is evident that a complete separation of all nucleotides obtainable from nucleic acids is a matter for the employment of a number of techniques, only one of which is paper chromatography. Fortunately no one nucleic acid has so far been found to have more than six bases in its nucleic acid so that the separations required are not so complex.

Detection of Spots

All nucleotides may be detected by their ultraviolet absorption visually[1,2] or by photography.[3,4] There seems little doubt that the latter method is slightly more sensitive, but it is more complex and if sufficient material is available the first method is quite adequate. Frequently the quantity of material is the limiting factor. The order of detection is about 5γ of a nucleotide, depending on the position of its absorption maximum.

Guanylic acids and particularly 8-azaguanylic acids fluoresce in the ultraviolet ($\lambda = 254$ and 265 mμ), and they and their compounds may be detected by photography.[5,6] The sensitivity of the method is some ten times that of the absorption technique, but to obtain this sensitivity a special ultraviolet light source must be used,[4] because the photographic paper is sensitive to ultraviolet light of all wavelengths. Guanylic acid fluoresces only in acid (the papers are exposed to HCl vapor), whereas 8-azaguanylic acid fluoresces better under alkaline conditions (NH_3 atmosphere).

Nucleotides may also be detected by spraying with dilute $FeCl_3$, which complexes with the phosphate groups. The spots are then detected by spraying the paper to detect free ferric ions (sulfosalicylic acid). Phosphate compounds remain colorless against a purple background.[7] Other methods for detecting phosphorus compounds may, of course, be used. Ribonucleoside 5'-phosphates may also be detected by spraying with $NaIO_4$, and then testing for aldehydes[8] or for excess $NaIO_4$.

Paper

For general use Whatman No. 3 MM paper is most suitable. Whatman No. 1 is to be preferred for quantitative work. Acid-washed papers give diffuse spots and are not to be recommended.

General Procedure

Descending chromatography is almost imperative because many solvents require to be run off the end of the paper. Very large tanks are

[1] E. R. Holiday and E. A. Johnson, *Nature* **163**, 216 (1949).
[2] C. A. Carter, *J. Am. Chem. Soc.* **72**, 1466 (1950).
[3] R. Markham and J. D. Smith, *Nature* **163**, 250 (1949).
[4] R. Markham and J. D. Smith, *Biochem. J.* **49**, 401 (1951).
[5] R. Markham and J. D. Smith, *Biochem. J.* **46**, 513 (1950).
[6] R. E. F. Matthews, *Nature* **171**, 1065 (1953).
[7] H. E. Wade and D. M. Morgan, *Nature* **171**, 529 (1953).
[8] J. G. Buchanan, C. A. Dekker, and A. G. Long, *J. Chem. Soc.* **1950**, 3162.

to be avoided if possible. Appropriate marker spots should be used as R_F values tend to be variable.

Solvents

Generally it is advisable to use solvents which do not absorb ultraviolet light. Strongly acid solvents will hydrolyze purine deoxyribonucleotides, cyclic mononucleotides, and ribo-oligonucleotides.

1. Isopropanol (680 ml.), concentrated HCl (176 ml.), and water to 1 l.[9] This mixture separates cytidylic and uridylic acids ($R_F = 0.61$ and 0.79) from the products of hydrolysis of ribonucleic acids (1 N HCl for 1 hour at 100°). Uracil and thymine will interfere if present. Markers are uridylic and cytidylic acids.

2. Isopropanol (700 ml.), water (30 ml.), and 0.35 ml. of NH$_3$ solution (0.880 specific gravity) per liter of volume.[10] This solvent is useful for separating ribomononucleotides from a mixture such as that produced by KOH hydrolysis of ribonucleic acid (1 N, 18 hours at 20°). Guanylic acid ($R_F \simeq 0.15$) is separated from adenylic, cytidylic, and uridylic acids ($R_F \simeq 0.25$). Markers are guanylic acid and adenylic acid. If NH$_3$ concentration is too low, uridylic acid may run ahead of adenylic acid. Nucleoside 5'-phosphates run very close to 3'- or 2'-phosphates. Thiouridylic acid runs near uridylic acid, azaguanylic acid slightly faster than guanylic acid. Cyclic mononucleotides run faster than nucleoside 3'-phosphates. In order of increasing R_F values, the following substances are separated:

a. Large polynucleotides and guanylic acid-containing trinucleotides.
b. Guanylic acid-containing dinucleotides and trinucleotides without guanylic acid.
c. Guanylic acid plus nonguanylic acid-containing dinucleotides.
d. Cyclic trinucleotides not containing guanylic acid.
e. Trinucleoside diphosphates plus cyclic dinucleotides containing guanylic acid.
f. Mononucleotides other than guanylic acid.
g. Cyclic dinucleotides not containing guanylic acid and dinucleoside monophosphates containing one guanine residue.
h. Dinucleoside monophosphates not containing guanine.
i. Cyclic guanylic acid.
j. Guanosine.
k. Cyclic adenylic, cytidylic, and uridylic acids.
l. Nucleosides other than guanosine ($R_F \simeq 0.6$).

[9] G. R. Wyatt, *Biochem. J.* **48**, 584 (1951).
[10] R. Markham and J. D. Smith, *Biochem. J.* **52**, 552, 558 (1952).

These groups of substances are readily resolvable by electrophoresis or by degradation methods. Azaguanylic acid compounds have somewhat higher R_F values than do the corresponding guanylic acids and are readily identified by their fluorescence. Xanthine compounds run with guanosine compounds, and hypoxanthine ones with adenine compounds. Substituted uridylic acids run close to similar uridylic acid compounds.

This solvent may also be used for deoxyribonucleic acid derivatives. The R_F values are slightly higher with these compounds. Thymine compounds all have higher R_F values than the similar adenine derivatives.

3. Saturated $(NH_4)_2SO_4$ in water (800 ml.), N sodium acetate (180 ml.), and isopropanol (20 ml.).[5] This solvent separates the isomeric adenylic and guanylic acids, and, in a long run, the isomeric cytidylic acids. The order of running (increasing R_F) is cyclic adenylic acid, adenosine 3'-phosphate, adenosine 2'-phosphate, cyclic guanylic acid, guanosine 3'-phosphate, guanosine 2'-phosphate, cyclic uridylic and cytidylic acids, cytidine 2'-phosphate, and cytidine 3'-phosphate. Uridylic acids run on or near the cytidylic acids. As a marker a complete KOH digest of yeast ribonucleic acid may be used (1 N KOH at 20°, 18 hours neutralized with 1 vol. of 1 N acetic acid). The xanthine and hypoxanthine cyclic nucleotides run with the corresponding guanine and adenine nucleotides, as do their nucleoside 2'-phosphates. The corresponding nucleoside 3'-phosphates, however, run with the 2'-phosphates apparently because they have no $-NH_2$ group.

A mixture giving a somewhat similar separation is given by Carter (5% Na_2HPO_4 in water saturated with isoamyl alcohol).[2]

4. The mixture phenol (85), isopropanol (5), formic acid (10), and water (100) separates the ribomononucleotides, in the order of increasing R_F: uridylic acid, guanylic acid, cytidylic acid, and adenylic acid.[11] It is the only single system which will do this, but unfortunately it is opaque to ultraviolet light. It might, however, be used for the first dimension of a two-dimensional chromatogram.

5. Isobutyric acid-ammonium isobutyrate separates guanylic acid plus uridylic acid from cytidylic acid and adenylic acid.[12]

Further Separation. Mixtures separated by chromatography can be separated further or resolved into their constituents by a number of methods.

1. Cyclic nucleotides may be transformed to the corresponding nucleoside 2'- and 3'-phosphates by 0.1 N HCl for 4 hours at 20°.

[11] P. Boulanger and J. Montreuil, *Bull. soc. chim. biol. Paris* **33**, 784 (1951).
[12] B. Magasanik, E. Vischer, R. Doniger, D. Elson, and E. Chargaff, *J. Biol. Chem.* **186,** 37 (1950).

2. Pyrimidine cyclic nucleotides may be turned into nucleoside 3'-phosphates by pancreatic ribonuclease.

3. Purine cyclic nucleotides may be turned into nucleoside 3'-phosphates by leaf ribonuclease.[13]

4. Nucleoside 3'-phosphates may be dephosphorylated by 3'-nucleotidase.[14]

5. Nucleoside 5'-phosphates are dephosphorylated by snake venom (rattlesnake or Russel's viper).

6. Noncyclic nucleotides are dephosphorylated by prostatic phosphomonoesterase. Cyclic ones are not.

7. Cyclic nucleotides have smaller electrophoretic mobilities at pH 7.5 than do corresponding nucleoside 2'-, 3'-, or 5'-phosphates.

8. Ribonucleoside 5'-phosphates are oxidized by $NaIO_4$.

9. Ribonucleoside 5'-phosphates form borate complexes and may be recognized by their enhanced electrophoretic mobility at pH 9.2 (0.05 M borax).

10. Dinucleotides move more rapidly on electrophoresis at pH 5.0 than do mononucleotides. Tri- and polynucleotides move even faster (rates 1:1.4:1.5).

11. Dinucleotides may be dephosphorylated to the corresponding dinucleoside phosphates. These may then be degraded by $NaIO_4$ followed by exposure to pH 10 leaving a mononucleotide.[15]

Several of the unusual nucleotides mentioned in the introductory remarks have been observed by my colleagues, Drs. R. E. F. Matthews and J. D. Smith, and Mr. D. B. Dunn (private communications).

[13] M. Holden and N. W. Pirie, *Biochem. J.* **60**, 39 (1955).
[14] L. Shuster and N. O. Kaplan, *J. Biol. Chem.* **207**, 535 (1953); Vol. II [86].
[15] P. R. Whitfeld, *Biochem. J.* **58**, 390 (1954).

[109] Chemical and Enzymatic Methods for the Identification and Structural Elucidation of Nucleic Acids and Nucleotides

By GERHARD SCHMIDT

I. Comments on the Hydrolysis of Nucleic Acids

Hydrolytic cleavage by acid or alkali reagents and by enzymes is an essential tool in the analysis of nucleic acids. Owing to the great variety of problems and conditions in the application of this tool, flexibility of techniques and adaptation of "standard" procedure to the requirements of each individual problem are very important. Rigid and detailed directions can be given only in a few isolated instances. The fact that nucleotides are composed of constituent groups of different chemical structures frequently offers the possibility to check the results of degradation procedures with independent techniques. For example, the course of hydrolysis of a nucleic acid sample may be followed by phosphorus determinations, by ultra-violet spectrophotometry of the purine- and pyrimidine-containing degradation products, by carbohydrate analysis, and by chromatographic methods. It is advisable to take full advantage of the multiple analytical approaches which are possible owing to the peculiarities of nucleic acid structure, particularly in problems involving tissue analysis.

II. Recently Suggested Symbols for the Structure of Nucleotides

The introduction of special symbols for the constituent groups of nucleic acids has become necessary to permit a concise presentation of the types of linkages involved in hydrolytic degradation procedures. The development of a suitable system of such symbols has been initiated in several recent papers.[1,2] The abbreviations used in some sections of this article are largely based on this system.

Nucleoside groups are designated by the capital letters of the respective bases (A for adenosine, C for cytidine groups, etc.); phosphoryl groups are designated by the letter p; the locations of the phosphoric ester bonds are represented by the numbers of the phosphorylated carbohydrate groups placed between the symbol p and that of the respective nucleoside groups; and 2',3'-cyclic phosphodiester bonds are indicated by the symbol p!.

For example, the symbols (1) A3p5G3p5C3p, (2) A3p5G3p5Cp!, (3) p5A3p5G3p5C, and (4) A3p5G3p5C represent (1) adenine-guanine-cytosine trinucleotide with a 3'-phosphomonoester group on the terminal cytidine group; (2) the analogous trinucleotide with a terminal cyclic 2',3'-phosphodiester group; (3) a trinucleotide of analogous sequence with a 5'-phosphomonoester group on the terminal adenosine group; and (4) a trinucleoside diphosphate of analogous sequence of constituent groups.

It is obvious that the proposed system of symbols does not cover all naturally occurring pertinent structures. For example, no provisions are made for the differentiation between ribo- and deoxyribonucleotides, for the natural cytosine analogs, or for the presentation of compounds such as apurinic acid. Extension of applicability, however, could be achieved only at the expense of clarity and simplicity of the symbols.

---

[1] L. A. Heppel, P. J. Ortiz, and S. Ochoa, *Science* **123,** 415 (1956).

[2] L. A. Heppel, P. R. Whitfeld, and R. Markham, *Biochem. J.* **60,** 8 [1955]. The author wishes to express his thanks to Dr. Heppel for permission to reproduce essential parts of his scheme in this paper.

III. Stability of Organic Groupings of Nucleic Acid toward Hydrolyzing Agents[3]

It is hardly necessary to point out that the stability of the constituent groups of nucleotides toward hydrolyzing agents might differ considerably from that of the corresponding free compounds.

A. Purines

Adenine and guanine as free compounds or as nucleotide constituents are practically quantitatively stable toward 0.5 N sulfuric, hydrochloric, and perchloric acids at 100° for 2 hours,[4] toward N sulfuric and hydrochloric acids at 100° for 1 hour,[3,5] toward 70% (12 N) perchloric acid at 100° for 1 hour,[6,7] and toward N alkali at room temperature or at 37° for 24 hours.

Appreciable losses (about 10%) of adenine occur during heating of nucleotides or nucleosides in N sulfuric acid for 2 hours. Glycine was found to be present among the products of more drastic acid degradation of adenine in amounts corresponding to 1.8% of the total nucleic acid nitrogen.[8]

B. Pyrimidine Groups

Information regarding the stability of pyrimidine groups of nucleic acids is particularly important owing to the drastic conditions required for their quantitative liberation from nucleic acid derivatives.

According to Wyatt,[6] cytosine and uracil are nearly quantitatively recovered after heating with 12 N perchloric acid at 100° for 1 hour. Considerable destruction of cytosine by its deamination to uracil occurs during heating of this base as well as during heating of its nucleic acid derivatives with N sulfuric, hydrochloric, or concentrated formic acid at temperatures between 100° and 175°.[9,10] Loring et al.[10] utilized the extensive deamination of ribocytidylic acid during treatment with N hydrochloric acid at 100° for 14 hours for the development of a convenient procedure for the preparation of a mixture of 2'- and 3'-ribo-

[3] H. S. Loring in "The Nucleic Acids" (E. Chargaff and J. N. Davidson, eds.), Vol. I, p. 191. Academic Press, New York, 1955.

[4] G. Schmidt, Z. physiol. Chem. 208, 225 (1932).

[5] J. D. Smith and R. Markham, Biochem. J. 46, 509 (1950).

[6] G. R. Wyatt, Biochem. J. 48, 584 (1951).

[7] A. Marshak and H. J. Vogel, J. Biol. Chem. 189, 597 (1951).

[8] R. D. Hotchkiss, in "Phosphorus Metabolism" (W. D. McElroy and B. Glass, eds.), Vol. 2, p. 426. Johns Hopkins Press, Baltimore, 1952.

[9] E. Vischer and E. Chargaff, J. Biol. Chem. 176, 703 (1948).

[10] H. S. Loring, J L. Fairley, H. W. Bortner, and H. L. Seagran, J. Biol. Chem. 197, 809 (1952).

uridylic acids. Ribocytidylic acid is also deaminated to a considerable extent in N sodium hydroxide during 24 hours of standing at 37°, but not at 24°.

Hydroxymethylcytosine, which is reasonably stable toward 12 N perchloric acid at 100° for 1 hour in the form of the free compound, is largely destroyed when DNA containing this pyrimidine as a constituent group is treated under these conditions.[11,12]

C. Carbohydrate Groups

D-Ribose is sufficiently stable toward N hydrochloric acid at 100° for 1 hour to permit quantitative reductometric studies of the acid hydrolysis of purine nucleosides, as well as paper chromatographic identification.[9] The great lability[13–15] of D-deoxyribose was recognized early by Levene and London.[13] Its isolation—chromatographically or by preparative methods—requires enzymatic conversion of DNA to nucleosides and hydrolysis of the latter in 0.01 N hydrochloric acid for 12 minutes. The liberation of the aldehyde groups of the deoxyribose moieties of DNA required for the Schiff test is usually carried out by hydrolysis of DNA or purine deoxyribonucleotides in N sulfuric acid at 60° for 5 to 10 minutes. The main degradation product of D-deoxyribose under more drastic conditions of hydrolysis is levulinic acid.[16]

The occurrence of intramolecular phosphoryl migration in 2'- and 3'-nucleotides or ribose phosphates in acid solution must always be considered.[17,18] No migration occurs, however, in aqueous solutions of the free phosphoric acid esters. 2'- and 3'-Mononucleotides are reasonably stable in dilute alkali at 25°. On the other hand, alkaline hydrolysis of ribopolynucleotides always results in the formation of mixtures of the two isomers, even though enzymatic hydrolysis of the same substrates to mononucleotides yields exclusively the 3'-isomers. This demonstrates that the formation of mixtures of both isomers during alkali hydrolysis does not indicate the presence of 2'-phosphoric ester linkages in ribopolynucleotide molecules but that it must be attributed to secondary phosphoryl migration occurring during an intermediary phase of the hydrolytic cleavage of the internucleotide bonds. Brown and Todd's

11 G. R. Wyatt and S. S. Cohen, *Biochem. J.* **55**, 774 (1953).

12 A. D. Hershey, J. Dixon, and M. Chase, *J. Gen. Physiol.* **36**, 777 (1953).

13 P. A. Levene and E. S. London, *J. Biol. Chem.*, **81**, 711 (1928).

14 W. Klein, *Z. physiol. Chem.* **255**, 82 (1938).

15 E. Chargaff, E. Vischer, R. Doniger, E. Green, and F. Misani, *J. Biol. Chem.* **177**, 405 (1948).

16 R. E. Deriaz, M. Stacey, E. G. Teece, and L. F. Wiggins, *J. Chem. Soc.* **1949**, 1222.

17 W. E. Cohn, *J. Am. Chem. Soc.* **72**, 2811 (1950).

18 D. M. Brown and A. R. Todd, *J. Chem. Soc.* **1952**, 44.

concept[18] of the intermediary formation of 2′,3′-cyclic phosphodiesters and their subsequent hydrolysis to mixtures of 2′- and 3′-phosphomonoesters is supported by the observation that cyclic mononucleotides were isolated from hydrolyzates of ribonucleates by refluxing them with weak alkali, for example in the presence of barium carbonate.[19]

IV. Hydrolysis of Nucleic Acids and Their Derivatives

A. Hydrolysis of Ribopolynucleotides with Alkali

Ribopolynucleotides are hydrolytically depolymerized in a sharply defined manner on standing in aqueous solutions of 0.3 N to 1.0 N sodium or potassium hydroxide for 16 hours at 24°.[20,21] Under these conditions, each interlinking phosphodiester group is converted to a phosphomonoester group (secondary phosphoryl group) by cleavage of the phosphoester bond of each 3′-mononucleotide group with the 5′-positions of the adjacent nucleotide, whereas the resulting phosphomonoester groups as well as the preformed terminal secondary phosphoryl groups remain as phosphoric ester linkages (see Table II).[22] Neither the extent nor the mechanism of this hydrolysis is influenced by the degree of polymerization.[1,20,22] Thus oligonucleotides with terminal 3′-secondary phosphoryl groups are hydrolyzed quantitatively to a mixture of 2′- and 3′-mononucleotides. (The formation of 2′-mononucleotides is caused by secondary migration of original 3′-phosphoryl groups.[18])

Incubation of riboöligonucleotides with terminal 5′-phosphoryl groups results in the quantitative conversion of all nonterminal nucleotide groups to a mixture of 2′- and 3′-mononucleotides, whereas the terminal nucleotide group containing the 5′-phosphomonoester group appears in the alkaline digest as 3′ (and 2′?),5′-diphosphonucleoside and as nucleoside, respectively[1] (see Table I).

Oligonucleotides from which the terminal secondary phosphoryl groups are removed yield mixtures of 2′- and 3′-mononucleotides and one nucleoside per molecule of oligonucleotide. The nucleoside originates from the terminal group containing the free 2′- and 3′-glycolic hydroxyl groups.[23]

Practically, it is convenient to perform the hydrolysis with potassium

[19] R. Markham and J. D. Smith, *Biochem. J.* **52**, 565 (1952).
[20] G. Schmidt, R. Cubiles, N. Zöllner, L. Hecht, N. Strickler, K. Seraidarian, M. Seraidarian, and S. J. Thannhauser, *J. Biol. Chem.* **192**, 715 (1951).
[21] J. N. Davidson and R. N. Smellie, *Biochem. J.* **52**, 594 (1952).
[22] W. E. Cohn and E. Volkin, *Nature* **167**, 483 (1951).
[23] W. E. Cohn, D. G. Doherty, and E. Volkin, *in* "Phosphorus Metabolism" (W. D. McElroy and B. Glass, eds.), Vol. 2, 339. Johns Hopkins Press, Baltimore, 1952.

hydroxide, since a large part of the potassium ions can be removed from the hydrolyzate by neutralization with perchloric acid.[21]

Refluxing of ribonucleates in buffered aqueous solutions between pH 6 and pH 8 for 8 days results in their conversion to inorganic phosphate and nucleosides.[24] The procedure may be used for the large-scale preparation of ribonucleosides. Of historical importance is the conversion of aqueous solutions of ribonucleotides to nucleosides by heating solutions of the former in dilute ammonia in an autoclave at an oil bath temperature of 175° for 3 hours.[25] According to Allen and Bacher,[26] the hydrolytic cleavage of the phosphomonoester bonds of ribonucleotides by alkali at 100° is catalyzed by the presence of lanthanum salts. These authors observed conversion of ribocytidylic and uridylic acids to the respective nucleosides to an extent of over 90% by heating their aqueous solutions for 100 minutes in the presence of 0.0115 M lanthanum nitrate and 0.031 sodium hydroxide (initial concentrations).

The great importance of alkaline hydrolysis for the analysis of ribopolynucleotides is obvious. Deoxyribopolynucleotides are depolymerized only slightly by alkali hydrolysis, and no significant information regarding their structure is obtained by treatment with alkali.

B. Hydrolysis of Polynucleotides with Acid

1. Ribopolynucleotides. The effects of strong acids on ribopolynucleotides are much more complicated than those of alkali, owing to the fact that not only a part of the phosphoester bonds but also the N-glucoside linkages of the purine-containing groups are very sensitive toward acids. All interlinking phosphodiester groups are much more acid-labile than the phosphomonoester groups formed as intermediaries. This was postulated by Jones,[27] who found that the time of the liberation of orthophosphate from ribonucleic acid was almost identical with that obtained with an equimolar mixture of the four "yeast" mononucleotides. The acid lability of phosphodiester bonds of pyrimidine nucleotide groups, in contrast to the stability of the phosphomonoester bonds of the free pyrimidine nucleotides, is demonstrated by the fact that, under suitable conditions of acid hydrolysis, almost the total amounts of the pyrimidine nucleotide groups of polynucleotides can be recovered in the form of mononucleotides.

[24] K. Holle, K. Dimroth, L. Jaenicke, and R. Hamm, Ger. pat. 824,206 (Dec. 10, 1951) (cc. 12 p. 16) [*C.A.* **48**, 7672 (1954)].

[25] P. A. Levene and W. A. Jacobs, *Ber.* **44**, 746 (1911).

[26] F. W. Allen and T. E. Bacher, *J. Biol. Chem.* **188**, 59 (1951).

[27] Walter Jones, "Nucleic Acids," p. 44. London, 1920.

The pyrimidine ribonucleotides obtained by acid hydrolysis are mixtures of 2'- and 3'-mononucleotides.[18] In this respect, the results of acid and alkali hydrolysis of ribonucleic acids are similar. Both the N-glucosidic bonds and the phosphomonoester bonds of pyrimidine nucleotides are much more stable than those of the purine ribonucleotides. Under conditions resulting in the quantitative liberation of the purine and of the phosphate groups of purine mononucleotides, 90 to 95% of the pyrimidine ribomononucleotides are still present as such in the hydrolyzates. More details regarding the behavior of ribonucleotides toward acids are described in Section V.B.5 of this chapter.

Riboöligonucleotides have been isolated from RNA hydrolyzates obtained by incubation of ribonucleic acid in 6 N hydrochloric acid for 3 minutes at room temperature.[28]

2. Deoxyribopolynucleotides. The N-glucoside bonds of the purine groups in deoxyribonucleotides are much more sensitive toward acids than those of the RNA-purine groups. Complete liberation of both purine bases from thymus deoxyribonucleate was accomplished by Tamm *et al.*[29] by incubation of DNA at pH 1.6 (0.02 N HCl) at 37° for 26 hours. At 100°, the liberation of the purines was complete after incubation in 0.1 M glycine buffer at pH 2.8 in 1 hour.[29] Hydrolysis of DNA in methanol saturated with hydrogen chloride[30] for 3 to 5 hours at 50° or with a 1.5% methanolic solution[31] of hydrogen chloride for 20 hours at 37° resulted likewise in complete release of the purines. Hydrolysis of DNA at pH 1.6 at 37° is essentially limited to the liberation of purines and to slight depolymerization of the polynucleotide chain. Except for the purines, no dialyzable degradation products appear. The purine-free degradation products were designated by Tamm *et al.*[29] as apurinic acids (corresponding to the thymic acids of the older literature which were obtained from DNA by hydrolysis under more drastic conditions).

PREPARATION OF APURINIC ACID.[29,32] Two-tenths per cent solutions of highly polymerized sodium deoxyribonucleate were mixed with approximately 0.27 vol. of 0.1 N hydrochloric acid so that the pH of the suspension was 1.6. The suspension was incubated for 26 hours at 37° neutralized to pH 7.3, and dialyzed through cellophane against 0.2 M borate buffer at 4° for 22 hours, against running tap water at 12° for 22 hours, and against frequently changed distilled water at 4° for 24 hours. The contents of the dialyzing bags were lyophilized to a white

[28] R. B. Merrifield and D. W. Woolley, *J. Biol. Chem.* **197**, 521 (1952).
[29] C. Tamm, M. E. Hodes, and E. Chargaff, *J. Biol. Chem.* **195**, 49 (1952).
[30] E. Vischer and E. Chargaff, *J. Biol. Chem.* **176**, 715 (1948).
[31] M. M. Daly, V. G. Allfrey, and A. E. Mirsky, *J. Gen. Physiol.* **33**, 497 (1950).
[32] C. Tamm, H. S. Shapiro, and E. Chargaff, *J. Biol. Chem.* **199**, 313 (1952).

fluff which was hygroscopic and easily soluble in water. The yield was almost quantitative with respect to the initial amount of pyrimidines.

The type of the pyrimidine-containing degradation products obtained by acid hydrolysis of DNA or DNA oligonucleotides (including dinucleotides) at higher temperatures (e.g., 0.01 N hydrochloric acid for 1 or 2 hours at 100°) differs essentially from that found in acid hydrolyzates of RNA or riboöligonucleotides, inasmuch as considerable amounts of deoxyribonucleoside-3,5-diphosphates[33,34] and deoxyribonucleosides are formed. Quantitative studies on the hydrolytic formation of deoxyribonucleoside diphosphoric acids are an important approach to the elucidation of the sequence of nucleotide groups in DNA.[34]

According to Brown and Todd,[18] the different behaviors of ribo- and deoxyribopolynucleotides during hydrolytic degradation in acid and alkaline media are now largely explained by the presence of a hydroxy group in vicinal position to a phosphoric acid ester bond in the individual nucleotide groups of RNA.

C. Hydrolysis of Nucleic Acid and Nucleotides by Enzymes

The action of various hydrolyzing enzymes, particularly that of phosphomono- and diesterases on nucleic acid derivatives, is discussed in Vol. II [62, 63, 79, 80, 85, 86]. In order to avoid repetition, the action of those enzymes which are used as tools in studies on the structure of polynucleotides has been summarized in the form of tables which are somewhat modified and extended presentations of a scheme given in a paper by Heppel *et al.*[2] In addition to the action of specific enzymes, some essential chemical degradation procedures which were discussed in the preceding chapters have been included in the tables.

Table I refers to representative ribotrinucleotides containing exclusively pyrimidine bases (any cytosine group could be replaced by a uracil group without affecting the validity of the table), but differing from one another by the type of the terminal group; furthermore, the behavior of the different pyrimidine mononucleotides toward the various hydrolyzing agents is presented.

Table II refers to a ribotrinucleotide containing a purine nucleotide group (adenosine or guanosine phosphate) and to the different types of purine mononucleotides. The presentation of separate schemes for pyrimidine oligonucleotides and for oligonucleotides containing both pyrimidine and purine groups is desirable because of the different behavior of both types toward acid hydrolysis and toward ribonuclease. The behavior of short-chain oligonucleotides toward hydrolyzing agents

[33] P. A. Levene and W. A. Jacobs, *J. Biol. Chem.* **12**, 411 (1912).
[34] H. S. Shapiro and E. Chargaff, *Federation Proc.* **15**, 352 (1956).

TABLE I

DEGRADATION OF PYRIMIDINE RIBOÖLIGONUCLEOTIDES AND PYRIMIDINE MONONUCLEOTIDES

Hydrolysis conditions	Degradation products						
	C3p5C3p5C3p	C3p5C3p5C5Cp!	C3p5C3p5C5C	p5C3p5C3p	C3p	p5C	C3p!
N HCl, 100°, 1 hour	3C(2),3p	3C(2),3p	2C(2),3p + C	?	C(2),3p	φφ	C(2),3p
0.1 N HCl, 20°, 4 hours	3C(2),3p	3C(2),3p	2C(2),3p + C	?	C(2),3p	φφ	C(2),3p
N NaOH, 25°, 18 hours	3C(2),3p	3C(2),3p	2C(2),3p + C	p5C(2),3p C(2),3p	C3p C2p?	φ	C(2),3p
0.1 M NaIO$_4$		φφ	Oxidized C3p5C3p	C Oxidized p5C3p5C3p	φφ	Oxidized Inorganic P, oxidation product	φφ
0.1 M NaIO$_4$, subsequent incubation at pH 10.5	φφ	φφ	Oxidized C3p5C3p		φφ	φφ	φ
Ribonuclease,[a] long incubation	3C3p	3C3p	2C3p + C	?	φφ	φφ	C3p
Intestinal polynucleotidase	Only action on RNA and DNA studied. The mononucleotides formed are exclusively 5'-mononucleotides.				φφ	φ	φ
Intestinal phosphodiesterase[b]	3C3p	Only action on cyclic nucleotides studied.		?	φφ	φφ	C3p
Spleen polynucleotidase[b]		2C3p + C			φφ	φ	φ
Spleen polynucleotidase, spleen phosphodiesterase[b]	3C3p	2C3p + C2p	2C3p + C	3C5p	φφ C	C	C2p C3p
Snake venom, crude[c]	Terminal C3p inhibitory, C, p5C(2),3p from higher polynucleotides and RNA				Φ	Φ	C3p
Snake venom phosphodiesterase[c]	p5C, p5C(2),3p from higher polynucleotides and RNA	2C5p + C	3C5p	3C5p	φ	φ	C3p
Prostate phosphomonoesterase[d]	C3p5C3p5C + p	φφ	φφ	?	C	C	φφ
5'-Nucleotidase (snake venom)[e]	?	?	?	C3p5C3p5C	C	C	?
3'-Nucleotidase[f]				?	C	φ	?

[a] See Vol. II [62].
[b] See Vol. II [90]; see also M. A. Maver and A. E. Greco, *Federation Proc.* **13**, 261 (1954).
[c] See Vol. II [89].
[d] See Vol. II [79].
[e] See Vol. II [85].
[f] See Vol. II [86].

is representative of that of the nucleic acids in many, but not in all instances. For this reason, supplementary information concerning the hydrolysis of ribo- and deoxyribonucleic acids is presented in Table III.

TABLE II

DEGRADATION OF RIBOÖLIGONUCLEOTIDES CONTAINING PYRIMIDINE AND PURINE GROUPS AND OF PURINE MONONUCLEOTIDES

Hydrolysis conditions	C3p5A3p5C3p	A3p	p5A	Ap!
1.5 N HCl, 100°, 1 hour	2C(2),3p + [A] + p	[A] + p	[A] + p[b]	[A] + p
N HCl, 100°, 1 hour	2C(2),3p + [A] + p[b]	[A] + p[b]	[A] + p[b]	[A] + p[b]
0.1 M NaIO₄	⊖	⊖	Oxidized	⊖
0.1 M NaIO₄, subsequent incubation at pH 10.5	⊖	⊖	Oxidized + p	⊖
Ribonuclease, long incubation[a]	C3p + A3p5C3p	⊖	⊖	⊖
Spleen polynucleotidase, spleen phosphodiesterase	2C3p + A3p	⊖	⊖	A2p
Snake venom, crude	See Table I, column 1	⊖	A	A3p
Snake venom diesterase	⊖[c]	⊖	⊖	A3p
Prostatic phosphomono-esterase	C3p5A3p5C + p	A + p	A + p	⊖
5′-Nucleotidase (snake venom, semen)	⊖	⊖	A + p	⊖

[a] See Vol. II [90]; see also M. A. Maver and A. E. Greco, *Federation Proc.* **13**, 261 (1954).

[b] The formation of P is not quantitative.

[c] Because of terminal C3p-grouping.

In addition to the symbols explained in Section I, the following abbreviations are used in the tables:

1. Free purine bases are indicated by the capital letters [A] for adenine and [G] for guanine (in brackets).

2. Mixtures of 2′- and 3′-mononucleotides are indicated by the term (2),3 between the symbols for the nucleoside and the phosphoryl group; e.g., C(2), 3p designates a mixture of 2′- and 3′-cytidylic acids.

3. ⊖ The symbol indicates degradation negligible.

4. A question mark indicates that the information available is insufficient.

The tables are intended to facilitate the selection of degradation procedures suitable for the approach of structural problems. It will be necessary to consult other pertinent sections of this book or the original literature regarding detailed experimental conditions (pH optima, enzyme inhibitors, duration of treatment, etc.).

Only those enzymes or chemical reagents have been included whose action has been shown to give conclusive or at least strong supporting evidence in favor of structural theories. Only the characteristic degradation products are listed. Products not pertinent to their analyses have been omitted.

TABLE III
DEGRADATION OF NUCLEIC ACIDS

Hydrolysis conditions	Typical degradation products	
	RNA	DNA
1.5 N HCl, 100°, 1 hour	Purines + inorganic P + pyrimidine (2),3 mononucleotides	Purines + some inorganic P + pyrimidine nucleotide diphosphates + pyrimidine-5′-mononucleotides + levulinic acid
0.1 N HCl, 20°, 4 hours	Pyrimidine(2),3 mononucleotides	Intermediaries between DNA and apurinic acid
0.1 N HCl, 37°, 27 hours	Pyrimidine(2),3 mononucleotides	Apurinic acid
0.1 M NaIO₄	⊖	⊖
N NaOH, 20°, 18 hours	(2),3 Mononucleotides (practically quantitative)	Acid-insoluble polynucleotides
Ribonuclease	See Tables I and II	⊖
Snake venom, crude	Nucleosides, nucleoside diphosphates	Nucleosides + inorganic P (preceding incubation with DNase I necessary)
Snake venom phosphodiesterase	5′-Mononucleotides, nucleoside diphosphates	5′-Mononucleotides[a,b,c] (yield quantitative, except with DNA of T2r+-bacteriophage[d]) (preceding incubation with DNase I necessary)
Deoxyribonuclease I	0	5′-Terminal oligonucleotides, 5′-mononucleotides
Deoxyribonuclease (*Micrococcus pyogenes*)	0	Oligonucleotides, 3′-mononucleotides[e]

[a] R. L. Sinsheimer and J. F. Koerner, *J. Biol. Chem.* **198**, 293 (1952).
[b] R. L. Sinsheimer, *Science* **120**, 551 (1954).
[c] M. Privat de Garilhe and M. Laskowski, *Biochim et Biophys. Acta* **18**, 370 (1954).
[d] Yield 62 per cent, probably because oligonucleotides containing glucosyl-hydroxy-methyl cytidylic acid groupings in certain structural positions might be resistant toward snake venom phosphodiesterase.
[e] L. Cunningham, B. W. Catlin, and M. Privat de Garilhe, *J. Am. Chem. Soc.*, in press (1956).

Although the behavior toward enzymes of the specific trinucleotides referred to in the tables is characteristic for that of other tri- and dinucleotides of respective types, it would be erroneous—with few exceptions—to extend all analogies indiscriminately to polynucleotides with longer chains. One of these exceptions is crystalline pancreas ribonuclease, whose action, after exhaustive digestion, yields the same type of end product regardless of the molecular weights of the substrates. The action of some other enzymes, however, is strongly influenced by the chain lengths of the substrates. Prostatic phosphomonoesterase hydrolyzes terminal 3′-pyrimidine nucleotide groups of di- or trinucleotides much faster than those of higher oligonucleotides, and this enzyme is at present not suitable for the quantitative determination of such terminal groups in ribonucleic acids or long-chain polynucleotides. Another example of the influence of chain length is the strong inhibitory effect of 3′-terminal phosphomonoester groups on the action of snake venom phosphodiesterase. In di- or trinucleotides, this effect is so strong that such substrates are practically resistant toward phosphodiesterase in the absence of phosphomonoesterase of suitable specificity, yet ribonucleic acid or long-chain ribopolynucleotides are slowly hydrolyzed by snake venom diesterase presumably because the inhibitory influence of 3′-terminal phosphomonoester groups decreases with increasing chain lengths.

Finally, it should be pointed out that certain enzymes listed in the tables have not as yet been obtained as enzymatically homogeneous preparations. All preparations of prostatic phosphomonoesterase contain small contaminations of ribonuclease whose interference is negligible for end group determinations on small oligonucleotides, owing to the rapid phosphomonoesterase action and the relatively large amount of terminal phosphoryl groups. On the other hand, these contaminations eliminate at present the application of prostatic phosphomonoesterase as end group reagent for ribonucleic acids because the use of high enzyme concentrations and long incubation periods would be required. The action of spleen phosphodiesterase on cyclic nucleotide groups and on internucleotide phosphodiester bonds must be attributed to two different enzymes, according to Heppel and Hilmoe.[2] This observation could not be considered in the tables, since the experimental details have not been published as yet.

V. Chemical and Enzymatic Methods for the Analysis of Phosphoryl Groups of Nucleotides

A. Introduction

Since the individual components of all known nucleotide polymers or coenzyme nucleotides are linked with each other by bonds involving phosphoryl groups, information regarding the quantitative proportions between the various types of these groups (primary, secondary, and, possibly, tertiary) in a given nucleotide is essential for the analysis of its structure. Furthermore, it is evident that chemical or enzymatic depolymerization of polynucleotides involve shifts of these proportions and may be studied by methods permitting their determination.

In a straight-chain, noncyclic polynucleotide, each phosphoryl residue has a strongly dissociating group of a pK between 1 and 2, whereas the terminal phosphoryl group has in addition a secondary dissociation with a pK_2 of approximately 6. The proportion between secondary phosphoryl groups and total phosphorus in a given nucleotide fraction thus yields the amount of "P-terminal" end groups and permits conclusions regarding the chain length of the polynucleotide. Information regarding the nature of the P-terminal nucleotide group is obviously essential for the investigation of the nucleotide sequence in a given polynucleotide. The techniques for the partition of primary and secondary phosphoryl groups in phosphoric acid esters may be divided into two groups: (1) titration and (2) specific enzymatic hydrolysis. No specific chemical reaction for this differentiation comparable to the use of fluorodinitrobenzene for the differentiation between primary amino groups and peptide bonds is so far known.

The application of both types of techniques is limited. Titration yields accurate results with nucleotides of low molecular weight when the P-terminal groups account for a considerable percentage of the total phosphoryl groups.

The use of specific phosphatases permits in principle end group determinations, even in polynucleotides of high molecular weights. The fact must be considered, however, that as a rule the rate of enzymatic hydrolysis of the end groups of high polymers is much slower than that observed with mononucleotides or short-chain oligonucleotides. Consequently, large amounts of enzyme are usually necessary to obtain degrees of hydrolysis approaching completion during reasonably short incubation periods. The use of excessive amounts of a specific enzyme, in turn, has the disadvantage that the activities of small amounts of contaminating enzymes become appreciable and interfere with interpretation of the results. This difficulty is particularly serious in the field of the phosphoesterases, since none of these enzymes has been obtained so far in crystallized form.

B. Noncyclic Mononucleotides

The amount of secondary phosphoryl groups of a nucleotide fraction containing noncyclic mononucleotides exclusively is equimolar to the amount of its total phosphorus. Equimolarity between total and secondary phosphoryl groups may be used as corroborative evidence compatible with the mononucleotide nature of a given nucleotide fractson or the completeness of hydrolysis of the interlinkages between the nucleotide groups of a polynucleotide. Independent evidence is required for the purpose of establishing the proportion of nucleoside mono- and diphosphates in a given nucleotide mixture.

1. Titration of Secondary Phosphoryl Groups of Nucleotides. Owing to the dissociation of the primary amino groups, the two-step nature of the titration curves of adenylic, cytidylic, and guanylic acids is hardly noticeable, and only uridylic and thymidylic acids titrate with a clear first inflection in the region around pH 5. The interference of the amino groups with the titration of the secondary phosphoryl groups may be overcome by titrating between ranges of high pH values in which the dissociation of the amino groups is negligible. This can be done in two ways: (1) Crestfield et al.[35] titrated between pH 6 and the second inflection point in the region of pH 8 under the assumption that the pK of secondary phosphoryl groups of nucleotides has a value of 6, regardless of the nature of the bases and of the chain length. On this basis, the molar amounts of secondary phosphoryl groups were calculated by doubling the values of amounts of the alkali consumed between pH 6 and the inflection point. (2) According to Seraidarian,[36] the Henderson-Hasselbalch equation may be transformed to permit computation of the equivalent amounts of all dissociable groups of identical pK from the slope of the titration curve between any two pH values, pH$_1$ and pH$_2$, within the range of their dissociation.

$$\text{pH}_1 = \text{p}K + \log \frac{T - U_1}{U_1} \tag{1}$$

$$\text{pH}_2 = \text{p}K + \log \frac{T - U_2}{U_2} \tag{2}$$

$$U = T - A \tag{3}$$

$$T = \frac{U_1\left(\dfrac{[\text{H}_2^+]}{[\text{H}_1^+]} - 1\right)}{\left(\dfrac{[\text{H}_2^+]}{[\text{H}_1^+]} \times \dfrac{U_1}{U_2}\right) - 1} \tag{4}$$

In these equations, U_1 represents the molar alkali consumption between pH$_1$ and the second inflection point; U_2, that between pH$_2$ and the second inflection point; T, the total molar amount of secondary phosphoryl groups.

Both titration methods are based on the assumption that differences

[35] A. M. Crestfield, K. C. Smith, and F. W. Allen, *J. Biol. Chem.* **216**, 185 (1955).
[36] M. Seraidarian, Thesis, Tufts College, 1952.

between the pK_2 values of different mononucleotides are so small that they cause only negligible errors in the interpretation of titration figures. The method of Seraidarian, however, has the advantage of checking this premise for each titration. It follows from equation 4 that the calculated T-values must be constant within a reasonably wide pH range for arbitrarily chosen pairs of pH values. This was found to be the case with solutions of pure mononucleotides.

2. *Quantitative Hydrolysis of Secondary Phosphoryl Groups by Acid Prostatic Phosphatase.* The phosphoryl groups of all mononucleotides (and many other phosphoric acid monoesters) are converted quantitatively[20] to inorganic phosphate by acid prostatic phosphatase. Phosphodiester groups are resistant toward this enzyme.[20] A mixture of nucleotides, the total phosphoryl groups of which are converted to inorganic phosphate by incubation with acid prostatic phosphatase, may be assumed to consist exclusively of mononucleotides or of mixtures of nucleoside mono- and diphosphates. For example, the phosphatase technique may be conveniently employed to select suitable conditions of alkaline hydrolysis for the complete cleavage of ribopolynucleotides to mononucleotides.

In the application of phosphatase for the quantitative determination of secondary phosphoryl groups in nucleotide mixtures, it is obviously advisable to work with a large excess of enzyme and with reasonably long incubation periods. For example, the use of 300 phosphatase units (Vol. II [79]) and an incubation period of 2 hours at 37° provide suitable conditions for the complete conversion of 1 mg. of mononucleotide phosphorus to inorganic phosphate.

When mixtures of mononucleotides and oligonucleotides are incubated with acid prostate phosphatase, the inorganic phosphate liberated includes the phosphoryl groups of the mononucleotides as well as those of the P-terminal end groups of noncyclic oligonucleotides (see below).[19,20,22,23,37,38]

The behavior of nucleoside diphosphates toward phosphatases is not sufficiently well known. Crestfield and Allen[38a] found recently that ribonucleoside diphosphates are resistant toward 5'-nucleotidase of snake venom.

3. *Differentiation of 5'-Nucleotides from Other Isomers (2'- or 3'-Nucleotides).* OXIDATION OF 5'-RIBONUCLEOTIDES WITH SODIUM PERIODATE.[20] The percentage of 5'-ribonucleotides in a mononucleotide mixture can be accurately determined by oxidation with sodium periodate. The

[37] G. Schmidt, R. Cubiles, and S. J. Thannhauser, *J. Cellular Comp. Physiol.* **38**, Suppl. 1, 61 (1950).

[38] R. Markham, R. E. F. Matthews, and J. D. Smith, *Nature* **173**, 537 (1954).

[38a] A. M. Crestfield and F. W. Allen, *J. Biol. Chem.* **219**, 103 (1956).

oxidation of the 2',3'-glycolic group of ribosides or ribotides is completed within 5 minutes at pH 5.5 under consumption of 1 mole of periodate per mole of 5'-mononucleotide. Amounts consuming several milliliters of 0.01 M periodate solution are suitable for analysis.

2'- or 3'-Ribonucleotides, deoxyribonucleotides, and deoxyribo-nucleosides do not react with periodate.

5'-Adenylic acid can be qualitatively differentiated from its 2'- and 3'-isomers according to Parnas and Klimek[38b] by its ability to form soluble copper complexes in alkaline solution. (Two volumes of a 2.5 per cent solution of sodium adenylate in 0.02 N sodium hydroxide are cautiously mixed with one volume of a 1.4 per cent solution of crystalline copper sulfate.) In presence of 5'-adenylate, a clear, deeply blue solution is obtained, whereas in presence of the other isomers, a blue precipitate forms which yields a colorless solution on filtration.

ENZYMATIC HYDROLYSIS OF 5'-NUCLEOTIDES BY 5'-NUCLEOTIDASES. Purified 5'-nucleotidase of human or bull semen[39] or of snake venom[40] is a sensitive reagent for the determination of the amounts of 5'-nucleotides in mononucleotide mixtures. It hydrolyzes specifically 5'-ribo-nucleotides as well as 5'-deoxyribonucleotides.

IDENTIFICATION OF 5'-RIBOADENYLIC ACID BY MUSCLE DEAMINASE.[41-43] 5'-Riboadenylic acid and, at slower rates, 5'-deoxyriboadenylic acid[44] are specifically deaminated by a purified deaminase of muscle. The specificity of the enzyme is so characteristic that it can be used for determination of 5'-adenylic acids in deproteinized tissue filtrates. A convenient spectrophotometric procedure for the quantitative determination of 5'-adenylic acid on this basis has been developed by Kalckar.[42] A modified description of this technique is given in Vol. II [68].

4. Enzymatic Differentiation between 2'- and 3'-Ribomononucleotides.[45] 3'-Ribomononucleotides are specifically dephosphorylated by a phosphatase found in barley by Shuster and Kaplan.[45] 2'-Ribonucleotides are hydrolyzed at negligible rates by this enzyme, 5'-nucleotides not at all.

5. Differentiation of Purine (2'- or 3'-) and Pyrimidine Ribonucleotides by Acid Hydrolysis.[20] PRINCIPLE. The phosphoryl groups of 2'- and 3'-purine ribonucleotides are quantitatively converted to inorganic phos-

[38b] I. K. Parnas and R. Klimek, *Z. physiol. Chem.* **217,** 75 (1933).

[39] L. A. Heppel and R. J. Hilmoe, *J. Biol. Chem.* **188,** 665 (1951); see also Vol. II [85].

[40] R. O. Hurst and G. C. Butler, *J. Biol. Chem.* **193,** 91 (1951); see also Vol. II [89].

[41] G. Schmidt, *Z. physiol. Chem.* **179,** 243 (1928).

[42] H. M. Kalckar, *J. Biol. Chem.* **167,** 461 (1947).

[43] G. Nikiforuk and S. P. Colowick, *J. Biol. Chem.* **219,** 119 (1956); see also Vol. II [68].

[44] C. E. Carter, *J. Am. Chem. Soc.* **73,** 1537 (1951).

[45] L. Shuster and N. O. Kaplan, *J. Biol. Chem.* **201,** 535 (1953); see also Vol. II [86].

phoric acid by heating in 1.5 N sulfuric acid in a boiling water bath for 1 hour. Only practically negligible amounts of inorganic phosphoric acid are formed under these conditions from pyrimidine nucleotides.

LIMITATIONS. 5'-Purine ribomononucleotides as well as 5'-purine deoxyribonucleotides are dephosphorylated to a considerable extent but not quantitatively under these conditions. The procedure is therefore applicable only under conditions excluding the presence or intermediary formation of 5'-phosphomonoester groups in more than negligible quantities. This is the case in alkaline or acid hydrolyzates and in ribonuclease digests of ribonucleic acids.[20,23] Owing to the small amounts of 5'-P-terminal groups in ribopolynucleotides synthesized by the polynucleotide phosphorylase of Grünberg-Manago and Ochoa,[1,46] it is likely that the procedure can be used for the analysis of such polynucleotides. Ribonucleic acid samples of animal origin are frequently contaminated with appreciable quantities of DNA (see Vol. III [101]). The hydrolysis procedure described in this section can be applied to such samples only after removal of the contaminating DNA by alkali hydrolysis and subsequent precipitation of DNA by acidification.

PROCEDURE. When 5'-nucleotides are absent (e.g., in alkali or ribonuclease digests of yeast RNA), a sample containing up to 5 mg. of nucleotide phosphorus is mixed with an equal volume of 3 N sulfuric acid and placed for exactly 1 hour in a strongly boiling water bath. The test tube containing the hydrolyzate is covered with a glass bulb. The hydrolysis mixture is then rapidly cooled to room temperature in a water bath, and the inorganic phosphorus is determined according to the method of Fiske and SubbaRow. 5'-Purine nucleotides must be removed prior to the acid hydrolysis by incubation with purified 5'-nucleotidase. The difference between the amounts of inorganic phosphate determined before and after acid hydrolysis represents the phosphorus of the 2'- and 3'-purine nucleotides with non-cyclic terminal groups.

C. Oligonucleotides

1. Nature of the Phosphoryl Groups. The nature of the phosphoryl groups in straight-chain oligonucleotides is analyzed according to the general methods described in the preceding section on mononucleotides. Any straight-chain noncyclic oligonucleotide contains one terminal secondary phosphoryl group in either the 3'- or the 5'-position. The question of the natural occurrence of terminal 2'-phosphomonoester groups is still in a controversial stage. This possibility has been discussed recently in attempts to interpret the formation of 2',5'-nucleoside diphosphates in digests of ribonucleates and ribopolynucleotides with snake

[46] M. Grünberg-Manago and S. Ochoa, *Biochim. et Biophys. Acta* 20, 269 (1956).

venom. At any rate, the existence of coenzymes (TPN) with a 2'-phospho-monoester group and the discovery of a spleen phosphodiesterase which hydrolyzes cyclic nucleotides under formation of 2'-mononucleotides seem to suggest that this possibility should be considered.

In this connection, it should be mentioned that observation of the enzymatic formation of nucleoside diphosphates from ribonucleic acids is at present the only evidence for considering the possibility of branched-chain structures for these polymers. The problem of branching in poly-nucleotide chains is still in the stage of a working hypothesis, however, and has not been included in this discussion.

2. Determination of Total Terminal Phosphomonoester Groups. The chain length is determined by the proportion of terminal phosphoryl groups to the total phosphoryl groups in the sample. Titration and deter-mination of the amount of inorganic phosphate liberated during ex-haustive digestion with phosphomonoesterases are the most suitable methods available for this purpose.

This simple correlation between chain length and relative amount of secondary phosphoryl groups holds of course only in absence of cyclic 2',3'-nucleotide groups and of nucleoside diphosphates. Cyclic pyrimidine nucleotide groups are converted to phosphomonoester groups by pro-longed incubation with large amounts of ribonuclease I. All cyclic nucleotide groups are transformed to the corresponding phosphomono-esters by a specific enzyme discovered by Heppel and Hilmoe in spleen (see p. 766). Correction of end group determinations in nucleotide mix-tures for the possible presence of nucleoside diphosphates requires at present chromatographic analysis.

TITRATION. Each phosphoryl group of a straight-chain oligonucleotide has a strong primary dissociation of a pK between 1 and 2. The true pK_1 can be determined only on deaminated oligonucleotides or on natural oligonucleotides containing exclusively uracil or thymine as bases. In oligonucleotides which contain primary amine groups, the titration range of the latter groups overlaps with that of the primary phosphoryl dissociations.

On the basis of this assumption, according to which the possibility of the occurrence of triesterified phosphoryl groups is neglected, the amounts of primary phosphoryl groups are usually not determined by direct titration but are calculated from the values of total phosphoryl groups.

The techniques for the titration of secondary phosphoryl groups have been described in Section V.B.1. The fact that the T-values (equation 4, p. 760) of yeast ribonucleic acids are constant only within the range between 5.7 and 6.3 seems to indicate that mixtures of polynucleotides might contain terminal secondary phosphoryl groups with appreciably different pK_2 dissociations. It seems preferable, therefore, to determine the approximate amounts of terminal secondary phosphoryl groups in mixed polynucleotides and nucleic acids by the technique of Crestfield et al.[35]

ENZYMATIC CLEAVAGE OF TERMINAL PHOSPHOMONOESTER (SECONDARY PHOSPHORYL) groups. Prostate phosphatase or bone phosphomono-esterase[19,37,47,48] attacks only phosphomonoester groups but leaves the phosphodiester bonds between the nucleotide groups intact. Since many samples of purified prostate phosphatase are contaminated with small amounts of ribonuclease, the use of this phosphatase should be limited to the determination of the terminal groups present in the oligonucleotide mixture obtained from ribonucleate by exhaustive digestion with pan-crease ribonuclease.

3. Determination of Terminal 3'-Phosphomonoester Groups. The specific 3'-nucleotidase of Shuster and Kaplan[45] has not been tested as yet as a tool for the determination of terminal 3'-phosphomonoester groups of oligonucleotides. In principle, the application of the enzyme for this purpose seems to be possible, despite the fact that even purified prepa-rations contain appreciable amounts of phosphodiesterase activity. Since the phosphodiesterase action, however, does not result in inter-mediary formation of 3'-phosphomonoester groups, one might expect that its presence will not interfere with the application of 3'-nucleotidase for the determination of terminal 3'-phosphomonoester groups.[49]

Up to the present, the procedures used for the actual determinations of 3'-phosphomonoester groups have been limited to the use of prostate phosphomonoesterase or bone phosphomonoesterase. The origin of the inorganic phosphate from terminal 3'-phosphomonoester groups may be ascertained by determining the increase of periodate consumption or by incubating another sample of the substrate with purified 5'-nucleo-tidase from semen or snake venom.

4. Detection of Terminal 2',3'-Cyclic Phosphodiester Groups. So far, formation of cyclic phosphodiester groups has been successfully demon-strated in di- and trinucleotides obtained by hydrolysis of higher poly-nucleotides with small amounts of ribonuclease or with barium carbonate. Their detection is based mainly on their paper electrophoretic separation and on their conversion to noncyclic nucleotides by the action of ribo-

[47] E. Volkin and W. E. Cohn, J. Biol. Chem. 205, 767 (1953).
[48] J. M. Gulland and E. M. Jackson, Biochem. J. 32, 590 (1938).
[49] Under the above conditions it has been found that approximately 7% of the total phosphate of yeast RNA is liberated by the 3'-nucleotidases (L. Shuster, thesis, Johns Hopkins University, 1954).

nuclease or spleen phosphodiesterase. The problem of the presence of terminal cyclic nucleotide groups in genuine ribonucleic acids is as yet undecided. The observation of Heppel and Hilmoe,[50] who found that spleen phosphodiesterase converts cyclic nucleotides (in contrast to internucleotide bonds) to the corresponding 2'-nucleotides, seems to offer interesting possibilities for the search of such terminal groups in higher polynucleotides.

5. *Determination of Terminal 5'-Phosphomonoester Groups.* Terminal 5'-phosphomonoester groups in oligonucleotides are determined (1) by the amount of inorganic phosphate liberated by the action of purified 5'-nucleotidase of semen[39] or snake venom;[40] (2) by the amounts of nucleoside diphosphate formed during hydrolysis of the oligonucleotide sample with N potassium hydroxide during 6 hours at 24°. The second procedure is suitable for the detection of terminal 5'-phosphomonoester groups in higher polynucleotide and nucleic acids.

6. *Stepwise Degradation of Straight-Chain Oligonucleotides with a Terminal 3'-Phosphomonoester Group with Phosphomonoesterase and Periodate According to Whitfeld*[51] *and to Brown et al.*[52] PRINCIPLE. After hydrolysis of the terminal 3'-phosphomonoester group by incubation with prostatic phosphomonoesterase, the resulting terminal 2',3'-glycol group is converted to the corresponding terminal dialdehyde by oxidation with periodate. Incubation of the oxidation product at pH 10 at 37° for 18 hours results in its cleavage under liberation of the terminal group in the form of an oxidized nucleoside and under formation of an oligo-nucleotide in which the original preterminal group is in P-terminal position.

This procedure may be repeated with the remaining oligonucleotide until finally only the nucleoside remains which originates from the de-phosphorylation of nucleotide groups on the other end of the original oligonucleotide.

VI. Remarks Concerning the Cleavage of Internucleotide Bonds

The compilation of the principal degradation procedures of nucleic acids and nucleotides in Tables I, II, and III shows that some structural features of polynucleotides may be explored by the systematic applica-tion of chemical and enzymatic hydrolysis reagents. In general, the results obtained with the different degradation methods included in the tables are remarkably consistent, and their somewhat simplified presentation in the form of tables is justifiable. On the other hand, the origin of certain types of degradation products is still under investigation, and some

50 L. A. Heppel and R. J. Hilmoe, Vol. II [90].
51 P. R. Whitfield, *Biochem. J.* **58,** 390 (1955).
52 D. M. Brown, M. Fried, and A. R. Todd, *J. Chem. Soc.* **1955,** 2206.

exceptions from the general rules have been observed. In this section some of the more complex aspects of nucleic acid structure will be briefly discussed.

A. Ribonucleic Acids

Position of the terminal phosphomonoester groups of nucleic acids and polynucleotides. The existence of enzymes suitable for the degradation of ribonucleates to 3'- as well as to 5'-mononucleotides was conclusively demonstrated by Cohn and Volkin.[53] Mononucleotides of the former type are formed by the action of crystallized pancreas ribonuclease and by spleen polynucleotidase,[50,54] mononucleotides of the latter by snake venom phosphodiesterase and intestinal polynucleotidase. In contrast to the two series of mononucleotides obtained from ribonucleic acids, only one series of riboöligonucleotides had been obtained until very recently, namely oligonucleotides with terminal 3'-phosphomonoester groups (and oligonucleotides with terminal 2',3'-cyclic phosphodiester groups).

The first evidence for the natural occurrence of ribonucleic acids with terminal 5'-phosphomonoester groups was furnished by Markham *et al.*,[38] who demonstrated that a structure of this type is characteristic for the ribonucleic acid component of tobacco mosaic virus. Recently, Heppel *et al.*[1] have shown that the polynucleotides synthesized by polynucleotide phosphorylase belong to this type of nucleotide polymers. Furthermore, they found that a ribonuclease prepared from liver nuclei hydrolyzes enzymatically synthesized polynucleotides to oligonucleotides with terminal 5'-phosphomonoester groups.

In all these cases, the 5'-nucleotide nature of the terminal phosphomonoester group has been postulated on the basis of the fact that (2')3',5'-nucleoside diphosphates were formed during exhaustive hydrolysis of these compounds with dilute alkali at room temperature.

Enzymatic formation of nucleoside diphosphates. The biological importance of nucleoside diphosphate derivatives is unquestioned. Several essential coenzymes are derivatives of ribonucleoside diphosphates (TPN,[55] coenzyme A,[56] "active sulfate"[57]). Nucleoside diphosphates have been obtained (besides 5'-mononucleotides) as products of the enzymatic degradation by snake venom phosphodiesterase of ribonucleic acids and riboöligonucleotides with terminal 3'-phosphomonoester groups.[38a,46] In contrast to the formation of nucleoside diphosphates by alkaline hydrolysis, the explanation of their enzymatic formation is not

[53] W. E. Cohn and E. Volkin, *J. Biol. Chem.* **203,** 319 (1953).
[54] M. A. Maver and A. E. Greco, *Federation Proc.* **13,** 261 (1954).
[55] A. Kornberg and W. E. Pricer, Jr., *J. Biol. Chem.* **186,** 557 (1950).
[56] T. P. Wang, L. Schuster, and N. O. Kaplan, *J. Am. Chem. Soc.* **74,** 3204 (1952).
[57] P. W. Robbins and F. Lipmann, *J. Am. Chem. Soc.* **78,** 2652 (1956).

yet clear. Three possibilities are considered: (1) Nucleoside diphosphates could arise from the hydrolysis of terminal nucleotide groups of nucleic acids with terminal 3'-phosphomonoester groups. (2) They could arise from analogous terminal hydrolysis of oligonucleotides with terminal 3'-phosphomonoester-groups. (According to this interpretation, one would have to assume that snake venom phosphodiesterase preparations contain nucleases whose action would result in the formation of such oligonucleotides from ribonucleic acids.) (3) They could arise from the cleavage of nucleic acids with branched chains by the cleavage of 2',3'-interlinkages at the branching points and subsequent cleavage of the resulting oligonucleotide with its terminal 3'-phosphomonoester group.

B. Deoxyribonucleic Acids

The effects of chemical hydrolyzing reagents on DNA differ sharply from those on RNA because of the absence of free hydroxy groups in adjacent positions to the phosphoric acid ester groups. In consequence, the phosphodiester bonds are very stable toward alkali, even at 100°. The relatively slight depolymerization of DNA by alkali has so far remained without significance for structural studies because the degradation products are polynucleotides of high molecular weight, and their separation and analysis have not yet been attempted.

The behavior of the purine and pyrimidine deoxyriboside bonds toward acid has been discussed already (see p. 753).

In regard to the effect of acids on the internucleotide bonds, their preservation under the conditions 'of the preparation of apurinic acid (see p. 753) demonstrates that the phosphodiester bonds of DNA are less labile toward acids than those of RNA. At 100°, however, these linkages are cleaved by N hydrochloric acid within an hour under formation of considerable quantities of pyrimidine deoxynucleoside diphosphates—in contrast to the absence of the corresponding nucleoside diphosphates in acid hydrolyzates of RNA.

Apart from the isolation of apurinic acid and of nucleoside diphosphates, chemical hydrolysis of DNA to degradation products of nucleotide level has not been useful so far either in the isolation of hydrolysis products or in structural considerations based on quantitative analysis of the hydrolyzates. The enzymatic cleavage of deoxyribopolynucleotides by intestinal or snake venom phosphodiesterase is at present the only method which permits their complete cleavage to 5'-deoxyribonucleotides. Both enzymes do not act on highly polymerized preparations of DNA but require a preceding depolymerization to oligonucleotides by incubation with crystallized pancreas deoxyribonuclease.[58,58a] In contrast

[58] R. L. Sinsheimer and J. F. Koerner, *J. Biol. Chem.* **198**, 293 (1952).
[58a] S. Zamenhof and E. Chargaff, *J. Biol. Chem.* **187**, 1 (1950).

to the strongly inhibitory effect of terminal 3'-phosphomonoester groups on these enzymes, the terminal 5'-phosphomonoester groups which are formed by the action of pancreas deoxyribonuclease do not interfere with the action of snake venom or intestinal phosphodiesterase. The amounts of terminal phosphomonoester groups in deoxyribooligonucleotides may be determined by titration or by specific enzymatic hydrolysis in a manner analogous to that used for this purpose on riboöligonucleotides. No observations concerning the action of the 3'-nucleotide phosphomonoesterase on the 3'-phosphomonoester groups of 3',5'-pyrimidine deoxynucleoside diphosphates have been reported as yet.

Crystallized pancreas deoxyribonuclease hydrolyzes approximately 25% of the internucleotide bonds in DNA under formation of a mixture of oligonucleotides with 5'-terminal phosphomonoester groups and 5'-deoxyribomononucleotides. Similar to pancreas ribonuclease digests of RNA, the deoxyribooligonucleotides formed by pancreas deoxyribonuclease contain higher percentages of purine nucleotide groups in comparison with those of the substrate DNA. In contrast to pancreas ribonuclease, however, the specific structural features determining the susceptibility of internucleotide bonds to pancreas deoxyribonuclease action are ast yet unknown.

Deoxyribonucleases whose action results in nucleotide and oligonucleotide mixtures whose composition differs from those obtained with pancreas deoxyribonuclease have been detected in some tissues. Some of these nucleases are capable of cleaving considerably higher percentages of the total interlinkages in comparison to pancreas deoxyribonuclease.[59,60] So far, however, the latter enzyme has been used most widely as a tool for structural studies of DNA, particularly as a tool for the preliminary depolymerization required for the action of snake venom and intestinal phosphodiesterases.

VII. Comments on the Fractionation of Oligonucleotides, Polynucleotides; and Nucleic Acid by Dialysis

Nucleates, including commercial ribonucleates, do not pass collodion or cellophane membranes regardless of the ionic strength of the solvent or of the outside liquid. Since chemical or enzymatic cleavage of the internucleotide bonds is always accompanied by the appearance of dialyzable fragments, dialysis has been frequently employed for separation of the degradation products from the undegraded or slightly degraded part of the original nucleate as well as for attempts to achieve a partition of the degradation products into fractions of high and low molecular weights. The nondialyzable degradation products have been termed

[59] M. Privat de Garilhe and M. Laskowski, *Biochim. et Biophys. Acta* **14**, 154 (1954).
[60] M. Webb, *Exptl. Cell Research* **5**, 27 (1953).

as "cores" by Zamenhof and Chargaff.[61] It must be emphasized, however, that the dialyzability of polynucleotides depends strongly on the ionic strength of the solution. Markham and Smith[19] found that large percentages of the high-molecular polynucleotide fraction which did not dialyze at low ionic strength diffused readily through cellophane in N sodium chloride solution.

VIII. Comments on the Separation of Nucleic Acid Derivatives by Precipitating Reagents

Although precipitation procedures have been replaced by chromatographic techniques for the quantitative analysis as well as for the isolation of many nucleic acid derivatives, the former are still indispensable for both purposes. In the field of quantitative analysis, the most convenient method for the separation of nucleic acids from their derivatives of low molecular weight is the fractionation of these compounds on the basis of their solubility behavior toward acids. Furthermore, the quantitative and specific precipitation of free purines as silver compounds has important analytical application. For the chromatographic isolation of nucleic acid derivatives, it is frequently necessary to use precipitation methods as preliminary purification steps (see, for example, Sanadi et al.[62]) prior to their partition on ion exchange columns or on paper.

It should be pointed out that, except for three purines, no methods are available permitting the complete precipitation of nucleotides, nucleosides, or free bases as groups of compounds. In part, this is due to the fact that the derivatives of adenine, guanine, and cytosine differ from those of uracil and thymine by the absence of a primary amino group in the latter. Further general differences preventing the development of group reagents for the precipitation of nucleic acid derivatives is the presence of glycolic hydroxyls in 5'-nucleotides, and their absence in 2'- or 3'-nucleotides and in deoxyribonucleotides.

The choice of precipitating reagent depends therefore largely on the nucleotide composition of the material to be analyzed. Some methods which yield satisfactory results with muscle extracts whose nucleotide fraction consists predominantly of 5'-adenylic acid derivatives are useless for the analysis of nucleotides of many other tissues or of nucleic acid digests.

A. Separation of Nucleic Acids from Oligo- and Mononucleotides by Precipitation with Acids

Mononucleotides and short-chain oligonucleotides are acid-soluble and can be separated from nucleic acids by precipitation of the latter

[61] S. Zamenhof and E. Chargaff, J. Biol. Chem. 178, 531 (1949).
[62] D. R. Sanadi, D. M. Gibson, P. Ayengar, and M. Jacob, J. Biol. Chem. 218, 505 (1956).

with 0.5 N hydrochloric or sulfuric acids (final concentration); 0.5% nucleate solutions are completely precipitated, 0.1% solutions with yields of over 90%. The nucleic acid precipitates which appear in very dilute solutions at first in opalescent form settle after a few minutes as sticky sediments which tend to adhere to the walls of the glass vessel and can be easily separated by centrifugation. Owing to the sticky consistency of the precipitates, thorough washing is usually difficult. It is preferable to redissolve the precipitates in small volumes of water and to reprecipitate the nucleic acids by cautious addition of hydrochloric or sulfuric acids. All precipitation should be carried out in the cold for the purpose of preventing hydrolysis. In the presence of proteins, nucleic acids may be quantitatively precipitated together with the proteins from solutions which are much more dilute than those indicated above, since they are carried down during deproteinization at 7% concentration of trichloroacetic acid. Presumably, other acid protein precipitants could be applied with similar results.

Concerning the specifity of acid precipitation of nucleic acids, it must be emphasized that polynucleotides of higher molecular weights which are present in digests of nucleates with ribonuclease I or deoxyribonuclease I, or which may occur in tissues are partially acid-insoluble. No detailed information is available concerning the quantitative correlation of molecular weight of polynucleotides with their solubility in acids.

B. Precipitation of Nucleic Acids with Uranium Salts in Acid Solution (McFadyen's Reagent[63])

Quantitative separation of nucleic acids (and polynucleotides of sufficiently high degrees of polymerization) from oligo- and mononucleotides may be achieved with an equal volume of a 0.25% solution of uranyl acetate in a 2.5% solution of perchloric acid (modified McFadyen reagent; the original McFadyen reagent contains trichloroacetic acid instead of perchloric acid[63]). Flocculent precipitates and clear filtrates are almost always obtained with this reagent.

Ribonucleates are precipitated in high yields from 10% solutions with 5 to 10 vol. of glacial acetic acid. This procedure is useful for the purification of ribonucleic acids, but separation of ribo- and deoxyribonucleic acids is not achieved, in contrast to earlier statements by Levene.[64] It is easy to demonstrate that highly polymerized deoxyribonucleates are likewise precipitated by glacial acetic acid.

Nucleic acids are precipitated at neutral or slightly alkaline reaction

[63] D. A. McFadyen, *J. Biol. Chem.* **107**, 297 (1934).
[64] P. A. Levene and L. W. Bass, "Nucleic Acids," pp. 301, 307. Chemical Catalog Company Inc. New York, 1931.

by barium, calcium, and many heavy metal salts. Except for the lanthanum salts[15,65] of nucleic acids, which were used by Hammarsten for the quantitative precipitation of ribonucleates from dilute solutions, metal salts of nucleic acids have not been employed extensively for nucleic acid analysis, owing to the difficulty of recovering nucleic acids from such precipitates.

It is important, however, to be aware of the possibility that extraction of nucleates at neutral reaction (e.g., by concentrated sodium chloride solution) from tissues might be incomplete in the presence of high concentrations of calcium salts such as occur under physiological conditions in lactating mammary glands.

Additional procedures for the precipitation of dilute nucleate solutions are discussed in Vol. III [100–103].

C. Reagents for the Precipitation of Mononucleotides

No general statements regarding the behavior of short-chain oligonucleotides can be made at present. Pure representatives of such compounds became available only during the last five years owing to the methods developed by W. E. Cohn, E. Volkin, and their collaborators. Information regarding the behavior of oligonucleotides are not as yet sufficiently extensive to justify a discussion here. It may be anticipated that the behavior of oligonucleotides toward precipitating reagents will depend on the presence in their molecules of bases with primary amino groups, particularly of guanine, and on the presence of terminal secondary phosphoryl groups and on other factors.

Mononucleotides, like other phosphoric acid monoesters, are precipitated at pH 9 in the presence of an excess of barium acetate by the addition of 2 to 4 vol. of alcohol. Guanylic acid and nucleoside polyphosphoric acids form water-insoluble barium salts in contrast to those of the other mononucleotides which are water-soluble at concentrations encountered in tissues and in ordinary enzyme experiments. *Sodium or potassium salts of nucleotides* are water-soluble in high concentrations, except secondary sodium or potassium guanylates which settle out in the refrigerator at a pH of approximately 5 from a 5% solution of guanylates.[66] The yields of precipitated guanylates are increased by the presence of 10% sodium acetate in the solution.

5'-AMP, ADP, and ATP are quantitatively precipitated by *copper sulfate and calcium hydroxide,* but 2'- or 3'-nucleotides are only partially precipitated by this reagent, perhaps because of the absence of the complex-forming glycolic group. Usually, appreciable degradation of

[65] E. Hammarsten, *Acta Med. Scand.* **128,** 634 (1947).
[66] R. Feulgen, *Z. physiol. Chem.* **106,** 249 (1919).

nucleoside polyphosphate occurs when the precipitates are exposed to the alkaline reaction of the precipitating reagent for several hours at room temperature.

Uranyl acetate, which completely precipitates 5'-adenylic acid as well as the (2'),3'-nucleotides of adenine, guanine, and cytosine from very dilute aqueous solutions, does not precipitate 2'- or 3'-uridylic acids.

Similarly, uridylates are not precipitated by a 20% solution of *mercuric acetate* or by a 20% solution of *phosphotungstic acid* in 12.5% sulfuric acid, although the nucleotides containing primary amino groups are precipitated by these reagents.

Precipitation of nucleotides as *lead salts* (either by neutral or, if necessary, by basic lead acetate) is a useful step in the preparative separation of the nucleotides from nucleosides, amino acids, sugars, and basic cell constituents, but it is not quantitative in dilute nucleotides solutions. No precipitation method is known that would permit the quantitative separation of nucleotides as a group from other organic tissue constituents.

D. Nucleosides

The nucleosides, except the guanosines, are easily soluble in water at acid, alkaline, or neutral reactions. The guanosines are only slightly soluble in water near the neutral point and usually precipitate directly during standing of neutralized hydrolyzates of nucleic acid in the refrigerator in the form of gelatinous precipitates, which consist of fine needles.

Nucleosides containing a primary amino group may be precipitated from dilute solution with *mercuric acetate* or *phosphotungstic acid*. Uridine is not precipitated under these conditions.

Lead acetate and ammonia, which is used for the precipitation of ribonucleosides, leaves large parts of deoxyribonucleosides in the supernatant solutions.[67]

E. Purines

The purine bases in nucleic acid hydrolyzates are quantitatively and fairly selectively precipitated with silver nitrate at a pH of approximately 1 and at approximately 1×10^{-2} molarity. In hydrolyzates of tissue filtrates which are free of proteins and peptones and which contain adenine and guanine as the only purine substances this precipitation is so specific that the decomposition liquid of the silver precipitates can be directly used for the quantitative spectrophotometric determination of the two purines. Quantitative and highly selective precipitation of

[67] F. Bielschowsky, *Z. physiol. Chem.* **120**, 134 (1932).

the purines as silver compounds occurs even in acid hydrolyzates of tissue homogenates; however, some contaminations absorbing light of wavelengths less than 250 mμ are often encountered and must be removed by additional purification steps (e.g., chromatography, countercurrent distribution) prior to spectrophotometric analysis. Complete separation of the purines from the pyrimidine bases is always accomplished by this procedure.

The rather unique advantage of the solubility properties of the purine silver compounds is the fact that they permit the separation of the purines as a group from almost all other organic tissue constituents and their degradation products.

Procedure for the Precipitation of the Purine Bases of Nucleates.[10,68-70] The samples containing about 1 mg. or more of each individual purine in 5 ml. are hydrolyzed in 0.5 N sulfuric acid in a boiling water bath for 2 hours; they are brought to pH 1 by dropwise addition of concentrated sodium hydroxide. The purines are precipitated at 50° by the addition of 1 ml. of an aqueous 1 M solution of silver nitrate. For quantitative recovery it is essential to leave the suspension of the silver purine precipitate in the refrigerator overnight. When the purines are precipitated from hydrolyzates which have been standing for several days, it is advisable to heat the suspensions to 50° in order to redissolve purine sulfates which might have crystallized on standing in the hydrolyzates.

If the presence of nitrate ions is undesirable in the supernatant solution containing the pyrimidine compounds the precipitation of the purines may be carried out with hot saturated aqueous silver sulfide solution. Owing to the slight solubility of silver sulfate, the use of the nitrate is preferable whenever feasible.

For the quantitative recovery of the purines the silver precipitates are washed several times with 0.1 N sulfuric acid. The purines are liberated by heating the finely suspended, washed precipitates in a boiling water bath with several successive portions of N hydrochloric acid.

Purines may be quantitatively precipitated from highly dilute solutions as cuprous complexes.[71-74] The sample containing at least 0.5 mg. of purine is transferred to a conical, 50-ml. centrifuge tube, diluted to 25 ml., and neutralized with phenolphthalein. The tube is heated in a boiling water bath, and the purine bases are precipitated by adding 0.8 ml. of a saturated solution of sodium bisulfate and 1 ml. of a 10%

[68] R. Feulgen, *Z. physiol. Chem.* **102**, 244 (1918).
[69] S. E. Kerr and K. Seraidarian, *J. Biol. Chem.* **159**, 211 (1945).
[70] G. Schmidt and P. A. Levene, *J. Biol. Chem.* **126**, 423 (1938).
[71] M. Krüger and J. Schmid, *Z. physiol. Chem.* **45**, 1 (1905).
[72] G. Schmidt, *Z. physiol. Chem.* **129**, 191 (1933).
[73] L. Graff and A. Maculla, *J. Biol. Chem.* **110**, 71 (1935).
[74] G. H. Hitchings and C. H. Fiske, *J. Biol. Chem.* **140**, 491 (1940).

solution of $CuSO_4 \cdot 5H_2O$. After 3 minutes of heating, the precipitate is collected by centrifugation and washed twice with 5-ml. portions of hot, 1% acetic acid. The precipitate is suspended in 3 ml. of 3 N hydrochloric acid and heated to boiling cautiously over a microburner flame. After addition of 15 ml. of hot water, the tube is returned to the boiling water bath and hydrogen sulfide is passed through the solution for 3 minutes. The mixture is cooled, rinsed into a 25-ml. volumetric flask, diluted to the mark, mixed, and filtered.

This precipitation method has somewhat fallen into oblivion because it cannot be used in the presence of trichloroacetate ions, and because the precipitates from tissue hydrolyzates contain nitrogenous non-purine substances such as cystine and glutathione. The purines are quantitatively precipitated from tissue hydrolyzates (obtained with 0.5 N sulfuric acid) without deproteinization. It appears possible that the interference of contaminations might be avoided by ultraviolet spectrophotometry of the purines liberated from their cuprous complexes.

Gulland and McRae have described the quantitative precipitation of purine bases with palladous chloride.[75] This reagent has been seldom applied in nucleic acid analysis (see page 777), although it would seem useful for the isolation of purines from solutions in which their precipitation as silver complexes is rendered difficult by the presence of high concentrations of chlorides.

[75] J. M. Gulland and T. F. McRae, *J. Chem. Soc.* **1932**, 2231.

[109A] Colorimetric and Enzymatic Methods for the Determination of Some Purines and Pyrimidines

By GERHARD SCHMIDT

Introduction. Although chromatographic procedures are most widely applied for the analytical determination of purines and pyrimidines, colorimetric and enzymatic methods are indicated in many instances because of the expediency of these procedures. When sufficient material is available (at least 0.5 mg. of purines in 5 ml.) it is advisable to combine the colorimetric procedures described below with a preceding separation of the purines and pyrimidines by precipitation of the free purines as silver or palladous complexes in acid solution (see Vol. III [109]).

Since the experimental details of enzymatic purine and pyrimidine determinations are described in the sections on individual enzymes, only brief discussions of these methods will be given here.

I. Colorimetric Methods

Colorimetric Estimation of Thymine According to Hunter,[1] Modified by Woodhouse,[2] and by Pircio and Cerecedo[3]

Principle. Thymine yields a red color with the diazo reagent of Koessler and Hanke[4] in alkaline solution and in the presence of a reducing agent.

Diazo Reagent

> One and one-half milliliters of a sulfanilic acid solution (4.5 g. of sulfanilic acid is dissolved in 45 ml. of concentrated hydrochloric acid, and the volume of the solution is made up to 500 ml.) and 1.5 ml. of a 5% sodium nitrite solution are mixed and cooled; after 5 minutes, 6 ml. of the sodium nitrite solution is added, and the volume is made up to 50 ml. The solution is kept in the cold and can be used for 24 hours.

Procedure. To 2.5 ml. of a 1.1% solution of sodium carbonate is added 1 ml. of the diazo reagent. After 1 minute, 0.5 ml. of a solution containing between 10 and 100 γ of thymine is added.

The mixture is kept at room temperature for 20 minutes and assumes a faint yellow color. Then 1 ml. of 3 N sodium hydroxide is added. After 1 minute 0.2 ml. of a 20% solution of hydroxylamine (w/v) is added. An intense red color rapidly develops which is stable for several hours. The extinction values obey Beer's law in concentrations between 10 and 100 γ/ml.

Specificity. None of the other nucleic acid purines or pyrimidines yields the intense red color produced by thymine, but in sufficiently large concentration they give yellow tints which interfere with the quantitative evaluation of the thymine test. The purines may be easily separated from thymine by precipitation as silver complexes in acid solution, however, and uracil does not appreciably interfere when present in concentrations equal to those of thymine. Cystine produces a color similar to thymine. Other amino acids yield yellow colors of varying intensities in concentrations of 200 γ/ml. Some aldehydes, furfural, and decomposition products produced by mild hydrolysis of nucleic acids give pink colors. Thus, the reaction cannot be applied to the direct determination of thymine in nucleic acid hydrolyzates[4] but must be separated from other

[1] G. Hunter, *Biochem. J.* **30,** 745 (1936).
[2] D. L. Woodhouse, *Biochem. J.* **44,** 185 (1949).
[3] A. Pircio and L. R. Cerecedo, *Arch. Biochem.* **26,** 209 (1950).
[4] K. K. Koessler and M. T. Hanke, *J. Biol. Chem.* **39,** 407 (1919).

products of the reaction mixture by precipitation with silver nitrate and barium hydroxide.

The modification of Hunter's procedure by Pircio and Cerecedo[3] permits the estimation of cytosine and of thymine in nucleic acid hydrolysates without preliminary isolation of the pyrimidine bases. 50 to 75 mgs.[4a] of nucleic acid are hydrolysed with 0.5 to 1 ml. of concentrated formic acid in a small sealed Pyrex bomb tube for 2 hours at 175°. The dark brown hydrolysate is evaporated to dryness *in vacuo* over calcium chloride and potassium hydroxide. The residue is extracted with anhydrous ether for the removal of formic acid, the suspension is centrifuged, and the ether is decanted. The residue is dissolved in 0.1 N hydrochloric acid, and quantitatively transferred to a small beaker with small amounts of water. The pH is adjusted to 1.4, and 12 drops of a 2 per cent aqueous solution of palladous chloride are added for the purpose of precipitating the purines. The mixture is heated to boiling, cooled, and filtered. The flocculent precipitate is washed with water, and the combined filtrate and washings are adjusted to a pH between 5.5 and 6. The solution is brought to a volume of 50 ml., filtered if necessary, and a sample of 5 ml. is evaporated to dryness in a centrifuge tube. The residue is extracted with ether for the complete removal of levulinic acid, the suspension is centrifuged, and the ether is poured off. The air-dried residue is allowed to stand overnight with 5 ml. of water. One-ml. aliquots of the filtered solution are used for the determination of thymine by the method of Hunter.

Colorimetric Determination of Uracil and Cytosine According to Soodak, et al.[5]

Principle. Brominated uracil and cytosine reduce arsenotungstate. Adenine thymine, 5-methylcytosine, 5-nitrouracil, isodialuric acid, dialuric acid, isobarbituric acid, barbituric acid, and alloxan do not react after bromination. Guanine produces a significant degree of color after bromination, but it can be easily removed by precipitation with silver salts.

The extinction of the reaction products obtained with uracil is about twice that obtained with an equimolar amount of cytosine. In mixtures containing both uracil and cytosine the sum of their extinctions is determined in an aliquot. In a second aliquot, cytosine is removed by adsorption on the cation exchanger Decalso. From the extinction of the filtrate

---

[4a] The amounts of starting materials suggested for the application of the procedures of this chapter represent those recommended in the original papers. Very likely, modifications of these techniques for considerably smaller quantities could be developed without great difficulties.

[5] M. Soodak, A. Pircio, and L. R. Cerecedo, *J. Biol. Chem.* **181**, 713 (1949).

the amount of uracil is calculated. The excess of the extinction over that of the filtrate represents the extinction caused by the concentration of cytosine in the mixture.

Reagents

Solution of lithium arsenotungstate.[6] One hundred grams of sodium tungstate is dissolved in 500 ml. of water, and 140 g. of arsenic pentoxide is added to the solution. This mixture is boiled under a reflux condenser for 1 hour. The condenser is then removed and the boiling continued until the volume of the solution is about 200 ml. This solution is slowly poured with stirring onto 100 g. of lithium chloride. The stirring is continued until all the white lithium chloride has gone into solution, and then the mixture is chilled thoroughly (at least to 10°) for about 2 hours. The lithium compound of arsenotungstic acid, by this time well settled, is filtered off on a Büchner funnel and dried as completely as possible by suction. The precipitated salts should weigh about 130 g. They are dissolved in water and made up to 500 ml.

Urea-cyanide solution. A 2.5% sodium cyanide solution containing 25% urea is added to a saturated aqueous solution of bromine.

Standard pyrimidine solution. Aqueous solutions (0.0002 or 0.0003 N) of uracil and of cytosine monohydrate are freshly prepared from 0.002 N stock solutions containing in 100 ml.

Decalso cation exchanger (The Permutite Company, New York). The Zeolite exchanger (60 to 80 mesh) is stirred or shaken for 15 minutes with four portions (each 8 to 10 vol.) of a 3% solution of acetic acid. Between the third and fourth washings, a 20-minute treatment with a 25% solution of potassium chloride is carried out. The Decalso is washed frequently with water, filtered, dried in air, and stored.

Procedure. DETERMINATION OF THE EXTINCTION OF THE SUM OF URACIL AND CYTOSINE. The unknown, containing between 0.05 and 0.3 micromole of pyrimidine in neutral aqueous solution, is measured into a test tube calibrated to 25.0 ml. The volume is brought to 2 ml. with distilled water, and 7 drops of bromine water added. After standing for exactly 5 minutes, the excess bromine is removed by aeration, and the aeration tube rinsed down with approximately 3 ml. of water. Five milliliters of the urea-cyanide solution is added, followed by 1.5 ml. of Newton's reagent. The tubes are shaken and allowed to stand for 1 hour or more for color development. After dilution to the 25-ml. mark, the solutions are read

[6] E. B. Newton, *J. Biol. Chem.* **120,** 315 (1937).

in an Evelyn photoelectric colorimeter with a 660- or 690-mμ filter. Standard solutions of the respective pyrimidine, as well as a reagent blank, are carried along with those of the unknown. All solutions should be kept at room temperature. Curves obeying the Beer-Lambert law are obtained when densities are plotted against the concentrations.

REMOVAL OF CYTOSINE BY ION EXCHANGE ADSORPTION. Five milliliters of the mixture of pyrimidines is placed in a 15-ml. centrifuge tube, 2 g. of Decalso is added, and the tube is stoppered with a clean rubber stopper and shaken for 5 minutes by hand. The zeolite is then removed by centrifugation. A suitable aliquot of the filtrate is taken for analysis. The color production is due to uracil alone.

A blank and standards for uracil and cytosine are run along with the analysis of the unknowns. From the dilution factor, the optical density of the uracil in 1 ml. of the original mixture is calculated, and this value is subtracted from that obtained in the ion exchange. The difference corresponds to the density of cytosine in 1 ml. of the original mixture.

Colorimetric Determination of Guanine, Xanthine and 2-Hydroxyadenine According to Hitchings[7]

Guanine, xanthine, and 2-hydroxyadenine are reduced by the phenol reagent of Folin.[8] The extinction of the blue solution is determined at 625 mμ.

Procedure. The solution containing between 0.7 and 2.8 micromoles of guanine or xanthine is measured into test tubes calibrated to contain 25 ml. To each tube are added 1.5 ml. of phenol reagent (containing 150 g. of lithium sulfate per liter[9]) and 8 ml. of a sodium carbonate solution saturated at room temperature. The solutions, including a standard, are diluted to the mark and placed in a water bath of 40° for 20 minutes. The reaction mixtures are cooled to room temperature in a water bath, and the extinctions are determined photometrically at 625 mμ. The molar extinctions of guanine, xanthine, and 2-hydroxyadenine are practically the same. Uric acid obviously interferes with the procedure.

Colorimetric Determination of Adenine According to Woodhouse[10]

Principle. Free adenine is reduced to a diazotizable amine by reduction with zinc dust in sulfuric or hydrochloric acid. The reaction product is determined according to the Bratton-Marshall[11] reaction.

[7] G. H. Hitchings, *J. Biol. Chem.* **139**, 843 (1941).

[8] O. Folin, "Laboratory Manual of Biological Chemistry," 5th ed., p. 339. New York and London, 1934.

[9] O. Folin and V. Ciocalteu, *J. Biol. Chem.* **73**, 629 (1927).

[10] D. L. Woodhouse, *Arch. Biochem.* **26**, 209 (1950).

[11] A. C. Bratton and E. K. Marshall, Jr., *J. Biol. Chem.* **128**, 537 (1939).

Procedure. The solution containing adenine (5 to 40 mg. in 20 to 30 ml.) in N sulfuric acid is heated with 0.1 g. of zinc dust in a water bath of 90° for 30 minutes under occasional gentle shaking. The suspension is then filtered through a small pad of cotton wool. Samples of the filtrate (e.g. 2 ml.) are pipetted into graduated glass-stoppered cylinders. After addition of 1 ml. of a 0.1% solution of sodium nitrite and standing at room temperature for 10 minutes, the solution is thoroughly mixed with 1 ml. of a 0.1% solution of sodium sulfamate. After 2 minutes, 1 ml. of freshly prepared Bratton-Marshall reagent is added. The volume is made up to 10 ml. with water, and the extinction of the orange-colored solution is determined after 10 minutes, at 505 mμ. It follows Beer's law within the range of concentrations indicated.

Adenine and the pyrimidine compounds of nucleic acids do not interfere with the reaction.

II. Enzymatic Methods

Enzymatic Determination of the Sum of Bound and Free Guanine

Principle. Guanine and guanosine are rapidly and quantitatively deaminated by aqueous extracts of rabbit or rat liver. Free adenine and free or bound cytosine are not appreciably deaminated by rabbit liver preparations. For the enzymatic determination of guanine, the sample of unknown purine compounds is hydrolyzed with 0.5 N hydrochloric acid in a boiling water bath for 2 hours. The hydrolyzate is incubated with the enzyme at pH 8.5 in 0.1 M glycylglycine buffer. The amount of guanine is calculated either from that of ammonia or that of xanthine formed by the enzyme. For technical reasons the spectrophotometric determination of xanthine is preferable whenever the optical properties of the digests are suitable for absorption measurements.

Separate determinations of free guanine and of guanosine appear to be feasible with sufficiently purified enzyme preparations according to Kalckar. Owing to the ubiquitous occurrence of adenosine deaminase, and of nucleases and nucleotidases, such a partition requires extensive control experiments concerning the specificity of the deaminase preparation, whenever the presence of adenine nucleotides or nucleosides in the unknown purine sample has to be considered. In view of the availability of chromatographic methods for the partition of bound and free purines, the most useful analytical application of guanine deaminase is the determination of total guanine groups.

Experimental procedures for the enzymatic determination of guanine may be found in Vol. II [72].

Enzymatic Determination of Riboadenosine

A procedure for the determination of adenosine on the basis of enzymatic conversion to inosine by the highly specific action of the adenosine deaminase of the mucosa of calf small intestines has been developed by Kornberg and Pricer (Vol. II [69]). The enzyme does not deaminate natural deoxyriboadenosine, riboguanosine, and ribocytidine. It converts 2,6-diaminoribofuranosyl purine to guanosine. The behavior of deoxyribosyl guanine and deoxyribosine cytosine toward the enzyme has not yet been tested.

In the presence of high concentrations of orthophosphate, the phosphatase activities present in adenosine deaminase preparations are sufficiently suppressed to prevent action of the deaminase on phosphorylated adenosine compounds.

Experimental procedures for the use of adenosine deaminase for the enzymatic assay of adenosine are described in Vol. II [70].

Enzymatic Determination of the Sum of Hypoxanthine and Xanthine with Xanthine Oxidase

Both hypoxanthine and xanthine can be quantitatively dehydrogenated to uric acid by xanthine dehydrogenase. The amount of uric acid formed is calculated from the increase of the extinction at 290 mμ. The experimental procedure is described in Vol. II [73].

Enzymatic Determination of Uric Acid

The amounts of uric acid present in unknown samples may be calculated from the decreases of the extinctions at 290 mμ observed after enzymatic conversion of uric acid to allantoin by purified uricase preparations. The decrease is 0.072 for a uric acid concentration of 1 γ/ml. in 0.1 N glycine buffer at pH 9. The preparation and the assay conditions of uricase are described in Vol. II [73].

[110] Microbiological Assay Method for Deoxynucleosides, Deoxynucleotides, and Deoxynucleic Acid

By E. HOFF-JØRGENSEN

Assay Method

Principle. The lactic acid bacterium *Thermobacterium acidophilus R 26* Orla Jensen (ATCC 11506) requires a deoxyribonucleoside as an essential growth factor. Neither vitamin B_{12} nor any other substance tested can replace the requirement for a deoxyribonucleoside. This organism, therefore, as showed by Hoff-Jørgensen,[1] can be used as a test organism for microbiological assays of deoxynucleosides and deoxynucleotides, and also of deoxynucleic acid (DNA) after depolymerization with DNAase.

Stock Cultures. Stock cultures are maintained in the following medium by weekly transfer: 0.1 g. of cysteine and 0.5 g. of yeast extract (Difco) are dissolved in 100 ml. of skimmed milk, pH 6.8. The milk medium is dispensed in 2-ml. quantities to test tubes 100 × 10 mm.; about 0.1 g. of $CaCO_3$ is added to each tube. The tubes are plugged with cotton, autoclaved at 120° for 10 minutes, inoculated with a wire loop, incubated for 24 hours at 37°, and stored in a refrigerator.

Inoculum Medium. Fifty milliliters of double-strength basal medium is mixed with 50 ml. of a solution containing 2 μM. of a deoxyriboside in 50 ml. of water (the minimum amount of peptone, e.g., Difco, which gives maximum growth can be used instead of a deoxyriboside). The medium is dispensed in 5-ml. quantities to 15-ml. centrifuge tubes, each containing a glass bead. The tubes are plugged with cotton, autoclaved at 120° for 10 minutes, and stored in a refrigerator. Fresh inoculum medium is prepared every month.

Inoculum. A small loopful of the stock milk culture is transferred to a tube containing 5 ml. of the inoculum medium. After incubation at 37° for 20 to 24 hours, the cells are centrifuged, washed once with 10 ml. of sterile saline, and resuspended in 10 ml. of sterile saline. One small drop of this suspension is used to inoculate each assay tube.

Standards. Stock solution: 10^{-4} g. mole of a deoxyriboside, e.g., 24 mg. of thymidine, is dissolved in 100 ml. of 25% ethanol. This solution is stable for at least one year. Working standard: 5×10^9 g. mole of deoxyriboside per milliliter; 50 μl. of the stock solution is diluted to 10 ml. with water.

Basal Medium, Double Strength (100 ml.). HCl-hydrolyzed casein solution, 30 ml.; papain-hydrolyzed casein solution, 10 ml.; salt A, 5 ml.; salt D, 1 ml.; Tween 80, 10%, 1 ml.; cytidylic acid solution, 1 ml.; potassium acetate solution, 5 ml.; thioglycolic acid solution, 1 ml.; adenine-guanine-thymine solution, 1 ml.; vitamin solution, 1 ml.; glucose, 3 g.; DL-tryptophan, 20 mg.; L-cysteine, 20 mg. Dissolve the glucose, tryptophan, and cysteine in the previously mixed solutions, adjust the pH to 6.7 with 1 N KOH, and add water to make 100 ml.

Prepare the various solutions as follows:

HCl-Hydrolyzed Casein. Mix 250 g. of purified casein (e.g., Labco "vitamin-free") with 1 l. of 20% HCl, and reflux the mixture for 8 to 10 hours. Remove excess HCl by

[1] E. Hoff-Jørgensen, *Biochem. J.* **50**, 400 (1951).

distillation under reduced pressure on a water bath until a thick paste remains. Add 800 ml. of water, and repeat the distillation. Dissolve the residue in water, adjust the solution to pH 3.5 with concentrated, NH_3 and dilute with water to ca. 2 l. To this solution, add 25 g. of activated charcoal (e.g., Darco G-60), stir for 30 minutes, and filter, using filter aid (e.g., Super-Cel). Dilute the filtrate to 4 l. with water. Casein hydrolyzates which have been hydrolyzed for more than 10 hours or which have been treated with too much charcoal give poor growth.

Papain-Hydrolyzed Casein. Suspend 125 g. of purified casein (e.g., Labco "vitamin-free") in ca. 1 l. of water, adjust the pH to 6.0, add a suspension of 5 g. of papain in about 100 ml. of water, shake, and incubate the mixture at 45 to 50° for 2 to 4 hours. Glacial acetic acid is now added to adjust the pH to ca. 4.0. After heating to ca. 100°C. add 25 g. of activated charcoal, stir for 30 minutes, and filter. Filtrations are aided by use of Super-Cel. The clear filtrate is diluted to 2 l., preserved under toluene, and stored in a refrigerator. If the blank is too high, repeat the treatment with charcoal.

Salt A. Dissolve 20 g. of monobasic potassium phosphate (KH_2PO_4) in water to make 100 ml.

Salt D. Dissolve 0.2 g. of sodium chloride, 0.3 g. of Mohr's salt [$Fe(NH_4)_2(SO_4)_2$ ·$6H_2O$], 0.8 g. of manganese sulfate ($MnSO_4·4H_2O$), 4 g. of magnesium sulfate ($MgSO_4$ ·$7H_2O$), and 2 ml. of 1 N HCl in water to make 100 ml.

Tween 80 Solution. Dissolve 10 g. of Tween 80 (polyoxyethylene sorbitan monoöleate) in water to make 100 ml. Store in a refrigerator.

Cytidylic Acid Solution. Dissolve 1 g. of cytidylic acid in water, adjust the pH to 7.0 with ca. 2 M sodium acetate solution, and add water to make 100 ml. Store in a refrigerator.

Potassium Acetate Solution. Dissolve 500 g. of potassium acetate in water to make 1000 ml.

Adenine-Guanine-Thymine-Uracil Solution. Dissolve 0.2 g. each of adenine sulfate, guanine hydrochloride, uracil, and thymine with the aid of heat in 10 ml. of 1 N HCl. Add water to 100 ml.

Thioglycolic Acid Solution. Dissolve 1 g. of thioglycolic acid in water to make 100 ml.

Vitamin Solution. Dissolve 0.5 mg. of folic acid and 5 mg. each of p-aminobenzoic acid, riboflavin, nicotinic acid, and calcium pantothenate in 50 ml. of water. Store under a preservative in a refrigerator. Prepare a fresh solution every month.

Procedure

The assay is carried out in lipless *uniform* test tubes (100 × 8 mm. I.D.). To each series of tubes the standard vitamin solution is added in the following amounts: 0, 0, 0.1, 0.2, 0.4, 0.6, 0.8, and 1.0 ml., each with an error of not more than 2%. Each level is set up in duplicate. The extract of the sample to be assayed is similarly added to a series of tubes in the following amounts: 0.2, 0.4, 0.6, and 0.8 ml., also in duplicate. All tubes are diluted to 1.0 ml. with distilled water, and 1 ml. of the basal medium is added to each tube. The tubes are shaken, covered with glass or aluminum caps, autoclaved at 120° for 5 minutes, cooled to room temperature, and inoculated with one drop of the immediately previously prepared inoculum suspension. To two of the four tubes containing 0 ml.

of standard no inoculum is added. These tubes are used as blanks in the turbidimetric determination of growth. All tubes are incubated at 37° for 24 to 36 hours.

Determination of Response. The tubes are shaken, and the turbidity is read in a photometer (e.g., Lumetron 402 C, Photovolt Corporation, 95 Madison Avenue, New York); λ = ca. 650 mμ. The microcuvettes are filled with a pipet and emptied with a piece of plastic tubing connected to a suction pump.

Calculation of Results. A standard dose-response curve is prepared by plotting the average of the turbidity values found at each level of the deoxynucleoside standard against the amount of deoxynucleoside present. The deoxynucleoside content of a sample is determined by interpolating the response to the known amount of the test solution onto this standard curve. The deoxynucleoside content per milliliter of the test solution is then calculated for each of the duplicate sets of tubes, and the deoxynucleoside content of the sample is calculated from the average of the values.

Preparation of Samples for Assay

Deoxynucleosides. A solution containing about 3 mμM. (or 0.5 to 1 γ) of deoxynucleoside per milliliter is prepared in water or in a not more than 0.05 M maleic acid buffer, pH 6.5.

Deoxynucleotides. Incubation of a solution of deoxynucleotides with crude intestinal phosphatase[2] is without effect on the response; it is therefore concluded that deoxynucleotides give the same response as deoxynucleosides on a molar basis.

Deoxynucleic Acid. Pure DNA has a growth effect which is less than 1% of the effect of the deoxynucleosides present in the DNA. If the DNA is depolymerized by deoxyribonuclease,[3] however, the growth response is equivalent to the effect of the calculated content of deoxyribosides in the DNA.

Samples of bacteria, yeast, or tissue may be analyzed either in the wet state or after drying with acetone. For the analysis of bacteria and yeast, the cells should be disintegrated, e.g., in "the tuning fork disintegrator" (obtained from H. Mickle, Middlesex, England).

The sample containing at least 0.2 γ of DNA-P is placed in a small test tube. An exactly measured amount of 0.5 N NaOH solution (e.g., 0.5 ml.) is added, or if the sample is a solution enough 1.0 N NaOH solution to make the final solution 0.5 N in NaOH. The tube is placed in a boiling water bath for 10 minutes. During this time the tissue is disintegrated with a glass rod. After the incubation at 100°, 5 vol. of a solution containing 0.06 g. mole of maleic acid and 0.01 g. mole of magnesium sulfate

---

[2] G. Schmidt and S. J. Thannhauser, *J. Biol. Chem.* **149**, 369 (1943); see Vol. II [80].

[3] M. Kunitz, *J. Gen. Physiol.* **33**, 349 (1950).

per liter are added for each volume of 0.5 N sodium hydroxide used above. The pH of the mixture should now be 6.3 to 7.0. In order to depolymerize the DNA, 0.1 ml. of a solution usually containing 100 γ of crystalline deoxyribonuclease (Worthington Biochemical Laboratories, Freehold, New Jersey) is added, and the mixture is incubated for 16 to 20 hours at 37°. For each new material assayed, the extraction procedure and the minimum amount of DNase which gives maximum response should be found by experiments. After incubation the mixture is diluted to contain about 3 mμM. of deoxynucleoside per milliliter and assayed (1 mole of deoxynucleoside $\sim$ca. 310 g. of DNA).

Differentiation between Purine and Pyrimidine Deoxynucleosides

As the pyrimidine deoxynucleosides are stable toward mild acid hydrolysis, whereas the purine deoxynucleosides are not, it is possible to distinguish between these two types of deoxynucleoside by assaying the depolymerized sample before and after boiling for 5 minutes at pH 1. Before assay the acid solution must be neutralized.

Specificity, Sensitivity and Accuracy

The method seems to be absolutely specific for the deoxyribonucleic linkage and allows the determinations of amounts greater than about 2 γ of deoxynucleosides, deoxynucleotides, or DNA with a standard error of about 5%.

[111] Preparation of Nucleoside Diphosphates and Triphosphates[1]

By ROBERT B. HURLBERT

The primary purpose of this article will be to provide information on the preparation and purification of an important group of nucleotides which have been only recently discovered: UDP, UTP, GDP, GTP, CDP, and CTP.[1a] Occasional mention will also be made of the related

[1] The preparation of this manuscript was supported in part by a grant (No. C-646) from the National Cancer Institute, National Institutes of Health, United States Public Health Service.

[1a] The 5'-mono-, di-, and triphosphates of adenosine, cytidine, guanosine, uridine, and inosine are abbreviated as: AMP, ADP, and ATP; CMP, CDP, and CTP; GMP, GDP, and GTP; UMP, UDP, and UTP; and IMP, IDP, and ITP, respectively. Uridine diphosphoglucose and uridine diphosphogalactose are designated as UDPG, uridine diphospho-N-acetylglucosamine as UDPAG, and guanosine diphospho-mannose as GDPM. CDP-X represents derivatives of CDP, e.g. CDP-choline.

monophosphates as well as of UDPG, UDPAG, GDPM, and the adenosine and inosine phosphates, although these compounds are more fully covered elsewhere in this volume. It should be recognized that, although development in this field is rapid, it is still in its early stages, so that the procedures described here may soon be improved. In many cases it is not possible to provide tested details of procedure; in place of these, brief reviews of available information are provided. The preparation of these nucleoside diphosphates and triphosphates by isolation from natural sources and by chemical and enzymic syntheses will be considered in addition to their purification by ion exchange chromatography. These compounds have also recently become available as commercial products from several companies.

Natural Occurrence

All or most of these nucleoside-5'-phosphates have recently been found in small amounts in extracts of rat tissues,[2-5] yeast,[6,7] rabbit muscle,[8] and guinea pig mammary gland.[9] Evidence for their presence in *Lactobacillus arabinosus*[10] has been reported. These nucleotides are apparently involved in metabolism as coenzymes,[11a,b] as precursors of coenzymes,[12-16] and as precursors of the nucleic acids.[17-19]

[2] H. Schmitz, V. R. Potter, R. B. Hurlbert, and D. M. White, *Cancer Research* **14**, 66 (1954).

[3] H. Schmitz, R. B. Hurlbert, and V. R. Potter, *J. Biol. Chem.* **209**, 41 (1954).

[4] R. B. Hurlbert, H. Schmitz, A. F. Brumm, and V. R. Potter, *J. Biol. Chem.* **209**, 23 (1954).

[5] H. Schmitz, W. Hart, and H. Ried, *Z. Krebsforsch.* **60**, 301 (1955); H. Schmitz, V. R. Potter, and R. B. Hurlbert, *ibid.* **60**, 419 (1955).

[6] S. H. Lipton, S. A. Morell, and A. Frieden, *J. Am. Chem. Soc.* **75**, 5449 (1953).

[7] H. Schmitz, *Biochem. Z.* **325**, 555 (1954).

[8] R. Bergkvist and A. Deutsch, *Acta Chem. Scand.* **8**, 1889 (1954).

[9] E. E. B. Smith and G. T. Mills, *Biochim. et Biophys. Acta* **13**, 587 (1954).

[10] J. Baddiley and A. P. Mathias, *J. Chem. Soc.* **1954**, 2723.

[11a] D. R. Sanadi, D. M. Gibson, P. Ayengar, and M. Jacob, *J. Biol. Chem.* **218**, 505 (1956); P. Ayengar, D. M. Gibson, C. H. Lee Peng, and D. R. Sanadi, *ibid.* **218**, 521 (1956).

[11b] E. B. Keller and P. C. Zamecnik, *Federation Proc.* **14**, 234 (1955).

[12] E. P. Kennedy and S. B. Weiss, *J. Am. Chem. Soc.* **77**, 250 (1955).

[13] C. E. Cardini, L. F. Leloir, and J. Chiriboga, *J. Biol. Chem.* **214**, 149 (1954).

[14] I. D. E. Storey and G. J. Dutton, *Biochem. J.* **59**, 279 (1955).

[15] E. E. B. Smith and G. T. Mills, *Biochim. et Biophys. Acta* **13**, 386 (1954).

[16] A. Munch-Petersen, *Arch. Biochem. and Biophys.* **55**, 592 (1955).

[17] E. L. Bennett and B. J. Kreuckel, *Biochim. et Biophys. Acta* **17**, 503, 515 (1955).

[18] R. B. Hurlbert and V. R. Potter, *J. Biol. Chem.* **209**, 1 (1954).

[19] M. Grunberg-Manago and S. Ochoa, *J. Am. Chem. Soc.* **77**, 3165 (1955); M. Grunberg-Manago, P. J. Ortiz, and S. Ochoa, *Science* **122**, 907 (1955).

Some information is available on the actual amounts of these nucleotides in various sources. Bergkvist and Deutsch[8,20] found in rabbit muscle (values expressed as micromoles per 100 g. of wet weight): ATP, 300 to 400; UTP, 4 to 8; CTP, 0.6 to 1.2; and GTP, 4 to 8. Rough estimates made from complete chromatograms published by Schmitz et al. indicate the following concentrations: In rat muscle:[2] ATP, 300 to 350; UTP, ca. 7; CTP, present but low; GTP, ca. 7. In yeast:[7] ATP, 17; UTP, 11; CTP 6; GTP 8. In rat liver[4] the following values were estimated: ATP, 34; UTP, 8; CTP, 2; and GTP, 8. The amounts of the mono-and diphosphates were also estimated. It may be noted that in the rat muscle the triphosphates of each series were predominant, in yeast similar amounts of the mono-, di-, and triphosphates within each series were noted, and in rat liver the mono- and diphosphates predominated. Yeast and liver, but not muscle, contain relatively large amounts of the uridine diphosphate derivatives UDPG and UDPAG. Cabib et al.[21] found a total of 7.5 to 30 micromoles of these derivatives per 100 g. (wet weight) of yeast, and in rat liver about 50 micromoles per 100 g. was found.[18] Other derivatives present in these latter sources are UDP-glucuronic acid,[14,15,18] GDPM,[22] and possibly CDP-choline.[12]

Formation of Nucleoside Di- and Triphosphates from the Monophosphates

The 5′-monophosphates of cytidine, guanosine, and uridine are readily available by chemical and enzymic preparations, as well as commercially. Relatively simple procedures for the chemical synthesis of these compounds in good yield have been described by Khorana et al. The formation of GMP,[23] UMP,[24] and CMP[24] is achieved by the action of mild phosphorylating agents on the corresponding 2′,3′-isopropylidene nucleosides. The nucleotides may also be obtained in good quantity by digestion of ribonucleic acid with snake venom diesterase (cf. Cohn[25]).

Possible preparations of the nucleotides labeled isotopically are:

1. From ribonucleic acid labeled in vivo by various labeled precursors such as adenine,[26] orotic acid,[18] inorganic phosphate-P^{32}, and $C^{14}O_2$.

2. By chemical phosphorylation of cytidine or uridine with phosphoric acid-P^{32} [24] or by phosphorylation of cytidine-4-C^{14}.[27]

[20] R. Bergkvist and A. Deutsch, Acta Chem. Scand. 8, 1880 (1954).
[21] E. Cabib, L. F. Leloir, and C. E. Cardini, J. Biol. Chem. 203, 1055 (1953).
[22] E. Cabib and L. F. Leloir, J. Biol. Chem. 206, 779 (1954).
[23] R. W. Chambers, J. G. Moffatt, and H. G. Khorana, J. Am. Chem. Soc. 77, 3416 (1955).
[24] R. H. Hall and H. G. Khorana, J. Am. Chem. Soc. 77, 1871 (1955).
[25] W. E. Cohn, Vol. III [107].
[26] S. E. Kerr, K. Seraidarian, and G. B. Brown, J. Biol. Chem. 188, 207 (1951).
[27] L. Grossman and D. W. Visser, J. Biol. Chem. 209, 447 (1954).

3. By preparation of UMP or GMP through enzymic condensation of labeled orotic acid[28,29] or guanine,[30] respectively, with 5-phosphoribosyl-1-pyrophosphate.

Chemical Syntheses of the Di- and Triphosphates. Straightforward preparations of UDP, UTP, ADP, and ATP from the corresponding nucleoside-5'-monophosphates have been reported by Hall and Khorana[31] and by Khorana.[32] In both these procedures, the monophosphate is treated at room temperature with an excess of phosphoric acid and dicyclohexyl carbodiimide in aqueous pyridine, to give approximately equimolar amounts of the di- and triphosphates with a total conversion of about 33% of the monophosphate in the case of AMP and 40 to 45% in the case of UMP. Similar procedures for the guanosine and cytidine series are reported by these authors to be in the process of development.

Enzymic Syntheses of the Di- and Triphosphates. Several systems for the enzymic phosphorylation of nucleoside monophosphates to the diphosphates and triphosphates have been described. It should be stressed that these have not been developed as large-scale preparations, so that neither optimum conditions nor the extent of contamination of the products with reactants and side products have been determined. The methods thus far best described and most readily utilized are the method of Hecht *et al.*[33] and of Herbert *et al.*,[34] which will be described here, and the method of Lieberman *et al.*[35]

The method of Lieberman *et al.*[35] has been applied to the preparation of uridine and guanosine polyphosphates. Nucleotide phosphokinase, partially purified from autolyzates of yeast, is coupled with the phosphopyruvate (pyruvate phosphokinase) system (prepared from rabbit muscle, see Vol. I [60]) as the phosphorylation source. As compared with the procedure to be described here, this method requires more time in the preparation of the enzymes and substrates but has an advantage in that the products are isolated from a relatively simple reaction mixture.

Strominger *et al.*[36] have described briefly the preparation of an "adenylate kinase" type of enzyme from calf liver acetone powder which will utilize ATP for the phosphorylation of AMP, UMP, and CMP to the

[28] I. Lieberman, A. Kornberg, and E. S. Simms, *J. Biol. Chem.* **215**, 403 (1955).
[29] R. B. Hurlbert and P. Reichard, *Acta Chem. Scand.* **9**, 251 (1955).
[30] A. Kornberg, I. Lieberman, and E. S. Simms, *J. Biol. Chem.* **215**, 417 (1955).
[31] R. H. Hall and H. G. Khorana, *J. Am. Chem. Soc.* **76**, 5056 (1954).
[32] H. G. Khorana, *J. Am. Chem. Soc.* **76**, 3517 (1954).
[33] L. I. Hecht, V. R. Potter, and E. Herbert, *Biochim. et Biophys. Acta* **15**, 134 (1954).
[34] E. Herbert, V. R. Potter, and Y. Takagi, *J. Biol. Chem.* **213**, 923 (1955).
[35] I. Lieberman, A. Kornberg, and E. S. Simms, *J. Biol. Chem.* **215**, 429 (1955).
[36] J. L. Strominger, L. A. Heppel, and E. S. Maxwell, *Arch. Biochem. and Biophys.* **52**, 488 (1954).

diphosphates. A dialyzed extract of chicken liver, metabolizing hexose diphosphate or phosphoglyceric acid, has been found to phosphorylate UMP rapidly,[29] and a yeast preparation has been reported which will catalyze the phosphorylation of UMP by ATP.[37]

A crude enzyme preparation from rabbit muscle, which will rephosphorylate ADP in systems utilizing ATP as phosphate donor, has been described by Ratner and Pappas.[38] This preparation, in the presence of ATP and phosphoglyceric acid as substrate, has been tested on a small scale as a "high-energy" phosphate source for the phosphorylation of ribose phosphate, UMP, CMP, and AMP in tissue extracts.[29,39] These experiments suggest that this system is capable of providing a simple source for the phosphorylation of the nucleoside monophosphates, in the presence of a suitable nucleotide phosphokinase.

Procedure of Hecht et al. and Herbert et al. Rat liver cytoplasm, under conditions of vigorous oxidative phosphorylation, is used for the phosphorylation of UMP, CMP, and GMP.[33,34,40] A rat liver is rapidly excised and chilled in ice-cold 0.25 M sucrose solution. (Preliminary perfusion of the liver *in situ* may be desirable.) A portion of the liver is homogenized in 9 vol. of 0.25 M sucrose solution by use of a tube and pestle homogenizer.[41] This suspension is centrifuged at 600 × g. for 10 minutes in a horizontal head (e.g., International No. 269) to sediment the nuclei. The resulting supernatant fraction, as the sole enzyme source, is added as soon as possible to the cold reaction mixture described below. All the foregoing operations are carried out at 0° to 1°. The complete reaction mixture, in a 125-ml. Erlenmeyer flask, is incubated at 30° with shaking (gas phase, air) at about 80 to 100 oscillations per minute to ensure adequate oxygenation.

The reaction mixture consists of nucleoside monophosphate (15 micromoles), ATP (0 to 1 micromoles), potassium pyruvate (75 micromoles), potassium fumarate (30 micromoles), potassium glutamate (75 micromoles), potassium phosphate buffer, pH 7.2 (75 micromoles), magnesium chloride (45 micromoles), and sucrose (870 micromoles). The total volume is 15 ml., including the cytoplasmic fraction from 750 mg. of rat liver. The ATP, fumaric acid, glutamic acid, and pyruvic acid are commercial products. The pyruvic acid is subjected to vacuum distillation before neutralization and use.

[37] A. Munch-Petersen, *Acta Chem. Scand.* **8**, 1102 (1954); *ibid.* **9**, 1523, 1537 (1955).
[38] S. Ratner and A. Pappas, *J. Biol. Chem.* **179**, 1183 (1949).
[39] R. B. Hurlbert, unpublished experiments. For detailed chromatographic separation of the nucleotides and charcoal adsorption see also R. G. Hansen, R. A. Freedland and H. M. Scott, *J. Biol. Chem.* **219**, 391 (1956).
[40] E. Herbert and V. R. Potter, *J. Biol. Chem.* **223**, 453 (1956).
[41] V. R. Potter, Vol. I [2].

The reaction is stopped by the addition of 0.1 vol. of cold 4.4 N perchloric acid (or equivalent) during chilling of the flask. The precipitated proteins are centrifuged, washed with 3 ml. of cold 0.2 N perchloric acid, and the cold extract is neutralized immediately with 2 N potassium hydroxide to pH 6 to 7, with the use of phenol red as internal indicator. Care is taken to minimize the formation of local concentrations of alkali which hydrolyze the pyrophosphate linkages. The mixture is chilled nearly to the freezing point to bring about the most complete precipitation of the potassium perchlorate, which is then centrifuged or filtered off on glass wool. The solution may be quick-frozen and stored, but it is advisable to proceed immediately with the isolation of the nucleotides, which is described later.

Pilot runs for this procedure should be checked at various time points by determination of inorganic phosphate esterified and by chromatographic analysis for the relative amounts of the products. Useful techniques for paper and simplified ion exchange chromatography are discussed later.

By this procedure it is possible to phosphorylate 10 to 12 micromoles of CMP or UMP in 10 minutes. Since the system is known to be able to maintain active oxidative phosphorylation for 40 to 60 minutes, it is apparent that higher amounts of the monophosphates could be used. The longer time periods favor formation of UDP-carbohydrate compounds and of derivatives of CDP. The relative amounts of di- and triphosphates formed will depend on the speed and time of the phosphorylation and the original level of the monophosphate.

The preparation of these di- and triphosphates labeled with P^{32} in the P_2 and P_3 positions may be accomplished by inclusion of labeled inorganic phosphate in the reaction mixture. The preparation of the di- and triphosphates labeled elsewhere in the molecule may proceed from the labeled monophosphate, obtained as discussed earlier.

Other Preparations of the Diphosphates and Triphosphates

The naturally occurring nucleoside diphosphate derivatives may occasionally be convenient sources for the preparation of the di- and triphosphates.

Park[42] has found *Staphylococcus aureus*, grown in the presence of penicillin, to accumulate as much as 500 micromoles of uridine pyrophosphate-amino sugar compounds, per 100 g. of packed cells, which may be extracted and hydrolyzed to yield UDP. Certain preparations of yeast have been reported by Leloir and co-workers to contain as much as 20 micromoles of UDPAG and 10 micromoles of UDPG per 100 g.[21] in addi-

[42] J. T. Park, *J. Biol. Chem.* **194**, 877 (1952).

tion to GDPM.[22] Hen oviduct was found by Strominger to contain large amounts of uridine nucleotides and GDPM.[43] These compounds may be hydrolyzed by 0.01 N HCl at 100° for 5 minutes to produce the nucleoside diphosphates. UDPG, UDPAG, and GDPM will also undergo pyrophosphorolysis to form the triphosphates by the action of an enzyme preparation from *Saccharomyces fragilis*, as described briefly by Munch-Petersen[37] (cf. ref. 44). Radioactive phosphate has been introduced into the triphosphates by the use of labeled pyrophosphate in this reaction.[16] Mills *et al.* have reported an enzyme, from the nuclear fraction of guinea pig liver, which is also capable of causing the pyrophosphorolysis of UDPG and UDPAG.[45]

Several enzyme systems have been described for the interconversion of the various nucleoside diphosphates and triphosphates. Strominger[46] has found the phosphopyruvate-pyruvic phosphokinase system capable of phosphorylating GDP and CDP as well as ADP and UDP.[47] Berg and Joklik[48] have described an enzyme preparation from autolyzed yeast or rabbit muscle which can catalyze the reaction

$$N_1TP + N_2DP \rightleftarrows N_1DP + N_2TP$$

for the di- and triphosphates of the nucleosides uridine, inosine, and adenosine. Sanadi *et al.*[11] describe an enzyme preparation from pork kidney mitochondria which catalyzes the above type of reaction for the adenosine, guanosine, and inosine compounds. The effectiveness of these interconversions may be increased for preparative purposes by coupling through the adenosine phosphates with phosphate accepting or generating systems, such as the hexokinase-glucose or phosphopyruvate-pyruvic phosphokinase systems.

The preparation of ITP by the action of nitrous acid on ATP has been described.[49]

Extraction and Concentration of the Nucleoside Diphosphates and Triphosphates

The isolation of these nucleotides in practical amounts from any source (natural, synthetic, or enzymic) is primarily dependent on ion

[43] J. L. Strominger, *Biochim. et Biophys. Acta* **17**, 283 (1955).

[44] H. M. Kalckar, *Science* **119**, 479 (1954).

[45] G. T. Mills, R. Ondarza, and E. E. B. Smith, *Biochim. et Biophys. Acta* **14**, 159 (1954).

[46] J. L. Strominger, *Biochim. et Biophys. Acta* **16**, 616 (1955).

[47] A. Kornberg, *in* "Phosphorus Metabolism" (McElroy and Glass, eds.), Vol. I, p. 410. The Johns Hopkins Press, Baltimore, 1951.

[48] P. Berg and W. K. Joklik, *J. Biol. Chem.* **210**, 657 (1954).

[49] N. O. Kaplan, Vol. III [123].

exchange chromatography for their final purification. It is sometimes necessary or convenient to effect a preliminary concentration of the nucleotides by precipitation as the barium salts or adsorption on charcoal, especially when the source is of excessive volume or contains large amounts of salts, pigments, organic reagents, etc. Paper chromatography (or paper electrophoresis) is highly useful as an independent analytical method to follow the purification and establish purity of the preparations.

The extraction of the nucleotides from the natural and enzymic sources depends on the usual methods for the disruption of tissues and cells, and the precipitation of proteins, nucleic acids, lipids, etc. The commonly used extracting agents are boiling water, boiling 50% ethanol, cold trichloroacetic acid, or cold perchloric acid. Three typical procedures may be cited to serve as examples for both extraction and concentration of the nucleotides (see also corresponding preparations in this volume): (1) the preparation of ATP from trichloroacetic acid extracts of rabbit muscle by precipitation with barium acetate at pH 6.8, by LePage;[50] (2) the preparation of coenzyme A from hot-water extracts of dried yeast by adsorption on charcoal, by Stadtman and Kornberg;[51] and (3) the preparation of triphosphopyridine nucleotide by adsorption on charcoal from hot-water extracts of pork liver, by LePage and Mueller.[52] The initial concentrates in these procedures, or slight modifications of them, are also likely to be good sources for the preparation of the nucleoside diphosphates and triphosphates (cf. ref. 11a).

Some information on the distribution of the nucleotides during barium salt fractionations is available. From the work of Bergkvist and Deutsch[8,20] and Smith and Mills[15] it appears that the nucleoside di- and triphosphates follow ADP and ATP in the barium precipitations. For a more general preparation of the nucleotides in example 1 above, the use of 3 vol. of ethanol in conjunction with the barium precipitation would be recommended. Such a barium precipitate contains most of the phosphate compounds of the extract, including nucleotide derivatives, inorganic phosphate, sugar phosphates, etc.[53] Subsequent fractionation of such preparations by barium salt precipitation under different conditions of pH [20] or by charcoal adsorption is a possibility.

Concentration of the nucleotides from large volumes of extracts by adsorption on activated, acid-washed charcoal appears to be relatively

[50] G. A. LePage, *Biochem. Preparations* **1**, 5 (1949).
[51] E. R. Stadtman and A. Kornberg, *J. Biol. Chem.* **203**, 47 (1953).
[52] G. A. LePage and G. C. Mueller, *J. Biol. Chem.* **180**, 975 (1949).
[53] G. A. LePage, *Cancer Research* **8**, 197 (1948); and *in* "Manometric Techniques in Tissue Metabolism" (Umbreit, Burris, and Stauffer, eds.), 3rd ed. Burgess Publishing Co., Minneapolis, in press (1956).

selective for the purines and pyrimidines and their derivatives.[9,16,17,21,54] Inorganic phosphate, sugar phosphates, salts, and proteins are not adsorbed to an interfering degree. Nearly complete adsorption of the nucleotides from acidic (or neutral) solutions may be effected by 5 to 20 mg. of charcoal per milliliter. Common agents for elution of the charcoal are 50% ethanol, 50% ethanol–0.1 to 1% concentrated ammonia, 40% acetone–0.1% concentrated ammonia, pyridine-ethanol (1:1), 3 to 20% pyridine in water, and 10% amyl alcohol. The ethanolic agents appear to be more effective for elution of the nucleotides; the pyridine solutions for the free bases and nucleosides. The elution is often not completely quantitative (80–90%), and the recovery depends on the kind of charcoal used, as well as the degree of activation.

In many situations preliminary concentration of the nucleotides is not necessary. The nucleotides in cold trichloroacetic acid extracts (extracted with ether and neutralized) and cold perchloric acid extracts (neutralized with potassium hydroxide and the potassium perchlorate precipitate removed) may often be sorbed directly on the chromatographic column. Filtrates prepared from extracts or enzymic reactions which have been heated to coagulate the proteins may often be chromatographed successfully.

It should be stressed that in the preparation of the nucleotides a number of precautions must be observed against enzymic or chemical destruction of the labile phosphate linkages. Several examples are as follows: (1) Animal tissues must be chilled or quick-frozen at death of the animal. Yeast should be in active culture, collected cold, and quick-frozen. Both should be worked up as soon as possible. (2) Alkaline conditions should be avoided. (3) Acidic conditions, when used, should be mild and cold and of short duration.

Chromatographic Separation

The ion exchange procedures using Dowex 1 (chloride) resin columns, applied originally to the separation of the adenosine nucleotides by Cohn and Carter,[55] have been used for the preparation of the uridine and guanosine phosphates as well.[6,9,15,21,22,35] The use of Dowex 1 (formate) resin columns, eluted by mixtures of formic acid and sodium formate, were, however, found by Bergkvist and Deutsch[56] to be more easily employed for good resolution of complex mixtures of uridine, guanosine, inosine, and adenosine polyphosphates. Another procedure, which will be described here, involves the separation of nucleotide mixtures by means

[54] R. K. Crane and F. Lipmann, *J. Biol. Chem.* **201**, 235 (1953).
[55] W. E. Cohn and C. E. Carter, *J. Am. Chem. Soc.* **72**, 4273 (1950).
[56] R. Bergkvist and A. Deutsch, *Acta Chem. Scand.* **8**, 1877 (1954).

of Dowex 1 (formate) columns and gradient elution techniques with formic acid and ammonium formate. This procedure has been treated in detail previously.[4]

Gradient Elution Apparatus. The term "gradient elution" refers to a procedure for the mechanical mixing of the eluting agents to provide a smooth and reproducible increase in the eluting power. The gradient elution technique, in conjunction with an automatic fraction collector, is convenient in that more detailed chromatograms may be obtained with less attention. More important, the technique is especially suitable for the resolution of complex mixtures of compounds of unknown or widely varying affinity for the exchanger. In addition, the eluted compounds are usually obtained in smaller volumes with less "tailing" (cf. refs. 4, 57).

An apparatus of the "constant mixing volume" type is used to obtain the concentration gradient. In this system, concentrated eluent is forced by air pressure or gravity from a "reservoir" flask into a "mixer" flask (which contains originally water or other dilute eluent) and, simultaneously, mixed eluent is forced over onto the chromatographic column. For the reservoir flask, a large suction flask may be used, with the air pressure applied to the side arm, and an outlet tube extending from the bottom out through a rubber stopper in the neck. The mixer flask is usually a round-bottomed flask with a side arm (ground-glass joint) to receive the outlet tube from the reservoir. The outlet tube from the mixer flask extends from the center of the flask out through the neck (rubber or ground-glass joint). Tygon plastic tubing is used to provide flexibility in the outlet tubes, and all connections are airtight. Mixing in the mixer flask is provided by a plastic-coated magnetic bar, driven by a magnet and motor outside the flask.

The concentration gradient given by this system is not linear but "convex"; that is, the concentration of the mixed solution rises most rapidly at first, then approaches asymptotically that of the reservoir solution.[58] To achieve a more nearly linear effect, it is desirable to conduct the elution over several short ranges of concentration of eluent, rather than one long range. This is done by several substitutions of a higher concentration of eluent in the reservoir.[59]

[57] R. M. Bock and N. S. Ling, *Anal. Chem.* **26**, 1543 (1954).

[58] The formula, previously incorrectly printed,[4] for the rise in concentration of the eluent is:

$$\text{Eluent concentration} = \text{Reservoir concentration} \times \frac{(\text{antilog}: X/2.3V) - 1}{\text{antilog}: X/2.3V} \text{ where}$$

X/V is the ratio of the volume of eluent passed (X) to the volume of the solution in the mixer flask $(V,$ a constant).

[59] A completely linear gradient (cf. ref. 57) theoretically should give improved resolution in the early part of a range and should be more convenient in permitting longer

To speed up the chromatography of a number of similar samples, the gradient elution technique may be utilized for the simultaneous elution of several columns. Each column should be packed in an identical way, so that flow rates are similar, and should have its own outlet tube from the mixer, which is made larger in proportion to the number of columns. Fractions are collected by an automatic fraction collector, such as the simple concentric row type which changes at timed intervals. Any number of columns may be eluted, up to the limitations of the fraction collector.

Chromatographic Procedure. Dowex 1 (chloride), $\times 8$ or $\times 10$,[60–62] 200 to 400 mesh, from which the finest particles have been removed by suspension and decantation, is washed with 3 M sodium formate until the effluent is free of chloride, then with water. To form the chromatographic column, a slurry of the resin is poured in a number of 2-cm. sections (each allowed to settle and drain before the next) into a glass column tube. A convenient support for the column is provided by crimped paper disks, cut from Soxhlet extraction thimbles, resting on constrictions in the glass tube. The resin column is then washed with 2 to 3 column volumes of 88% formic acid to remove colored impurities and rinsed with water until the pH of the effluent rises to 5 or 6.

The sample is adsorbed on the column from a dilute, neutral solution (about 0.01 to 0.05 M, pH 6 to 7) at a flow rate of 0.6 to 0.7 ml./min./sq. cm. of cross section of column. The total salt content of the sample solution should be considered in relation to the total capacity of the column used (1.2 meq./cc. of moist resin). The maximum amounts of the samples (salts plus nucleotides) which may be chromotographed have not been determined by us; however, good resolution of samples containing 1 to

---

ranges. Such an apparatus may be constructed from two identical bottles connected by a small-diameter siphon tube or bottom outlets so that a common descending liquid level is maintained as the eluent flows from the reservoir to the mixer and from the mixer to he column. The elution patterns given by the two types of apparatus are similar.

[60] Dowex 1 is a strong base anion exchanger, composed of a polystyrene resin with quaternary nitrogen substitutents on the benzene rings, and is obtainable from the Dow Chemical Co., Midland, Michigan. Dowex 2 gives chromatographic results similar to those given by Dowex 1. The $\times 8$ and $\times 10$ resins (8% and 10% divinyl benzene as cross-linking agent) are satisfactory for the separations described here; however, less highly cross-linked resins have been found to favor separation of compounds of higher molecular weight, such as coenzyme A[51] or polynucleotides[61] which require a 2% DVB resin. Improved resolution of UTP and GTP is obtained by use of the 4% DVB resin.[39]

[61] W. E. Cohn, *in* "The Nucleic Acids" (Chargaff and Davidson, eds.), Vol. I, p. 211. Academic Press, New York, 1955.

[62] See Vol. III [107].

5% of the total capacity of the column have been obtained, depending on the nature and complexity of the mixture. The column is washed with several column volumes of water, or until no more light-absorbing materials are removed, before elution is begun.

Two different elution systems will be described (see Fig. 1). In the formic acid system, mixtures of formic acid and ammonium formate in

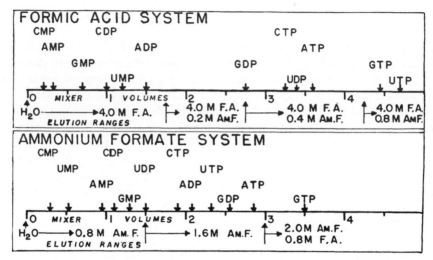

FIG. 1. Gradient elution chromatography with Dowex 1 (formate). The elution sequence and approximate elution positions of the nucleotides in the diagram apply to several combinations of resin column and mixing flask volumes (see text for details). The approximate eluent concentration at the end of each concentration range, and other compounds eluted are as follows:

Formic Acid System. First range: $0 \rightarrow 4$ M formic acid for 1.75 mixer volumes, end concentration 3.2 M formic acid, ca. pH 1.6. CDP-X and DPN are eluted in region of CMP, TPN in region of GMP, IMP and orthophosphate between CDP and UMP, and GDPM after ADP. Second range: $\rightarrow 4$ M formic acid–0.2 M ammonium formate for 1.0 mixer volume, end concentration 3.6 M formic acid–0.12 M ammonium formate, ca. pH 2. UDPAG and UDPG are eluted before GDP. Third range: $\rightarrow 4$ M formic acid–0.4 M ammonium formate for 1.5 mixer volumes, end concentration 3.9 M formic acid–0.33 M ammonium formate, ca. pH 2.3. Pyrophosphate and UDP-glucuronic acid are eluted in region of UDP. Fourth range: $\rightarrow 4$ M formic acid–0.8 M ammonium formate for 1.0 mixer volume, end concentration 4 M formic acid–0.6 M ammonium formate, ca. pH 2.6. ITP is also eluted in this range.

Ammonium Formate System. First range: $0 \rightarrow 0.8$ M ammonium formate, pH 5, for 1.5 mixer volumes, end concentration 0.6 M ammonium formate. DPN is eluted before CMP, orthophosphate and TPN before UMP; also UDPAG, IMP, and UDPG between UMP and AMP. Second range: $\rightarrow 1.6$ M ammonium formate, pH 5, for 1.5 mixer volumes, end concentration 1.35 M ammonium formate. Third range: $\rightarrow 2.0$ M ammonium formate–0.8 M formic acid, pH 4, for 1.0 mixer volume, end concentration 1.75 M ammonium formate–0.5 M formic acid, pH ca. 4.15.

the pH range 1.5 to 3 are employed to elute the nucleotides in the order C-A-G-U.[1] In the ammonium formate system, an elution order of C-U-A-G is obtained by use of solutions of ammonium formate at pH 5 (with the exception that GTP is eluted at a lower pH). Mixtures of compounds not resolved by one system are usually resolved by the other. To obtain complete purity it is often necessary to chromatograph a compound by both systems.

For a resin column of 1.2 to 1.4 × 20 cm. (column volume 20 to 30 cc.) a mixer flask of 500-ml. volume is used. For a column of 1.8 to 2.0 × 20 cm. (ca. 50 to 60 cc.) a mixer of 1000-ml. volume is used. The flow rate is 0.6 to 0.8 ml./min./sq. cm. of cross section of the resin bed, and fractions corresponding in size to one-fourth to one-third of the column volume are collected by an automatic fraction collector. To obtain the first "elution range," in the formic acid system, elution is begun with water in the mixer flask and 4 M formic acid in the reservoir. Enough fractions are collected to give a total eluate volume corresponding to 1.75 "mixer volumes" (i.e., when a 500-ml. mixer flask is used, 875 ml.). The second range is obtained by replacing the reservoir solution with 4 M formic acid–0.2 M ammonium formate, and subsequent ranges are obtained in a similar way. The contents of the mixer flask are not changed.

To give the best resolution, the column size, mixer size, and concentration ranges should be in proper proportion for the compounds to be chromatographed. Examples of the proportions suitable for the resolution of larger batches of nucleotides are given above. The capacity of the system may be scaled up or down to conform to the quantity of material to be separated by changing the volume of resin and the mixer volume in the same proportion. Greater resolving power with a given volume of resin may be obtained (provided the capacity of the resin is adequate) by increasing the ratio of length to diameter of the column and by increasing the mixer volume so the rise in concentration of the eluent is less steep. Complete diagrams of the application of such systems of greater resolving power to the chromatography of prepared mixtures and tissue extracts have been published.[2,4,7,39]

The chromatographic fractions are analyzed consecutively by measurement of their light absorption at 260 mμ and at 275 mμ by use of a Beckman spectrophotometer, Model DU. The ratios, or the consecutive *changes* in ratios, of the light absorption values (As$_{275}$/As$_{260}$) are valuable in the identification of the chromatographic peaks and estimation of the homogeneity of the peaks. Characteristic 275/260 ratios for the nucleotides in the *eluants employed here* are: A, 0.4; U, 0.6; G, 0.75; and C, 1 to 2. To speed up the analysis of the fractions a, device is sometimes used for the automatic transfer of the fraction from tube to Beckman cuvette and back again.[63] Otherwise transfers are most easily made by means of pipets provided with rubber bulbs, and several Beckman cuvettes, each for a consecutive series of fractions, may be used simultaneously to permit several readings with each setting of the instrument.

Recovery of the Nucleotides from Chromatographic Eluates. The nucleotides are eluted from the columns by solutions containing relatively large amounts of formic acid and ammonium formate. Although the di- and triphosphates are fairly stable in this pH region (1.5 to 3), they are stored in the cold when the fractions are collected. The formic acid is readily removed at low pressure by distillation, desiccation or lyophilization. The ammonium formate may be sublimed at a temperature of 35° to 40° at a pressure of less than 1 mm. of Hg. The use of special sublimation vessels, or other means of reducing the path between the warmed sample and the condensing surface, greatly speeds this procedure. It is also possible to convert the ammonium formate to formic acid by rapid passage of the eluate through an excess of ice-cold Dowex 50 (H+),[64] followed by lyophilization of the sample to remove formic acid. Since with this procedure the free acids are obtained, it is believed to be safer to neutralize the nucleotides immediately at the end of the lyophilization or in some cases to add back a small amount of sodium or ammonium formate before the solution is lyophilized.

Charcoal has been used[21,39] to adsorb the nucleotides from the acidic, salt-containing eluates. The procedure is convenient and gives recoveries of 85 to 90%.

The nucleotides may be precipitated as their sodium salts from neutral solutions containing several milligrams of nucleotide per milliliter by the addition of 3 vol. of acetone at 0°.

Stepwise Chromatographic Elution. The method of Herbert *et al.*,[34] which has been extensively used for analytical purposes, employs stepwise elution by formic acid and ammonium formate solutions of different pH to achieve a more selective elution and resolution of known mixtures of nucleotides. This system of elution (scaled upward in capacity) can be recommended for large-scale nucleotide preparations especially when the mixtures are not complex and manual collection of fractions is employed. In devising an elution scheme for a given separation problem it is helpful to consider the behavior of the compounds in question under various different pH conditions (cf. Fig. 1), (cf. refs. 34, 56, 61). Often pairs of compounds which might not be well resolved at a constant pH may be separated by use of an elution step at a different pH.

[63] The Gilson Automatic Transferator, obtainable from Gilson Medical Electronics, 714 Market Place, Madison, Wisconsin.

[64] A sulfonated polystyrene cation exchange resin obtained from the Dow Chemical Co., Midland, Michigan, washed with NaOH, HCl, and water.

TABLE I

R_f VALUES FOR THE 5'-RIBONUCLEOTIDES[a]

Ethanol-ammonium acetate	Isobutyric acid-ammonium hydroxide	Phosphate-ammonium sulfate-propanol
GTP, 0.03	CTP, 0.07	AMP, 0.29
ATP, 0.04	UTP, 0.08	ADP, 0.37
CTP, 0.04	CDP, 0.13	ATP, 0.48
UTP, 0.04	UDP, 0.13	GMP, 0.50
GDP, 0.06	GTP, 0.18	GDP, 0.58
ADP, 0.07	ATP, 0.20	GTP, 0.61
CDP, 0.07	UMP, 0.23	UMP, 0.64
UDP, 0.09	GDP, 0.24	CMP, 0.67
GMP, 0.09	CMP, 0.28	UDP, 0.71
CMP, 0.13	ADP, 0.30	CDP, 0.71
AMP, 0.15	GMP, 0.31	UTP, 0.77
UMP, 0.18	AMP, 0.45	CTP, 0.77

[a] The solvent proportions are given in the text. These data are taken from R. M. Bock, S. A. Morell, N. S. Ling, and S. H. Lipton, *Arch. Biochem. and Biophys.* in press, and from Circulars OR-7 and OR-10 of the Pabst Laboratories. Ascending chromatography on Whatman No. 1 paper was used. Since the R_f values are easily influenced by slight variations in the conditions, standardization with known compounds is essential.

Paper Chromatography. The purity of the nucleotide preparations may be checked by the following systems, preferably by the use of descending development for the first three.

1. Ethanol (75 ml.)–1 M ammonium acetate, pH 7.5 (30 ml.), Paladini and Leloir.[65]
2. Isobutyric acid (100 ml.)–1 M ammonium hydroxide (60 ml.)–0.1 M disodium Versene (1.6 ml.), Krebs and Hems.[66]
3. 1% ammonium sulfate (50 ml.)–isopropyl alcohol (100 ml.), Anand *et al.*, described by Hall and Khorana.[31]
4. 0.1 M phosphate, pH 6.8 (100 ml.)–ammonium sulfate (60 g.)–*n*-propanol (2 ml.).[66a]

These systems give good resolutions of the mono-, di-, and triphosphate of a series. Since the systems vary in the degree of separation of the various series, it is advisable to use at least two such systems for determinations of purity of the nucleotide preparations. R_f values for all the nucleotides have been reported and are summarized in Table I.

[65] A. C. Paladini and L. F. Leloir, *Biochem. J.* **51**, 426 (1952).
[66] H. A. Krebs and R. Hems, *Biochim. et Biophys. Acta* **12**, 172 (1953).
[66a] R. M. Bock, N. S. Ling, S. A. Morell, and S. H. Lipton, *Arch. Biochem. and Biophys.* **62**, 253 (1956).

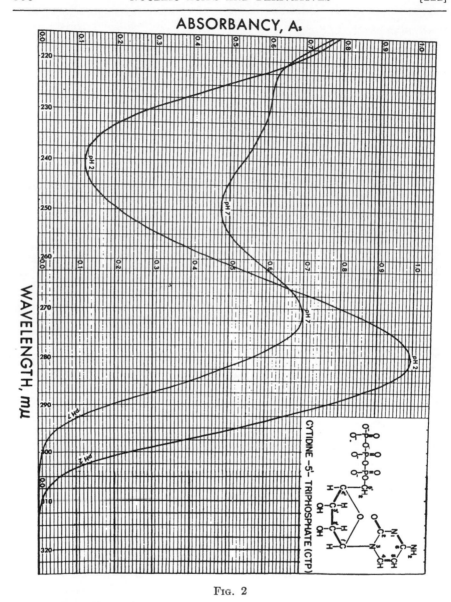

Fig. 2

Figs. 2–5. Ultraviolet absorption spectra of the nucleoside-5′-triphosphates, determined on solutions of the purified sodium salts. The spectra of UTP at pH 2, CTP at pH 11, and ATP at pH 11 are essentially identical with the corresponding spectra at pH 7. Figures reproduced from Bock et al.[66a] and from Circulars OR-7 and OR-10 of the Pabst Laboratories.

Properties of the Diphosphates and Triphosphates

In Figs. 2 to 5, spectra of CTP, ATP, GTP, and UTP at two or three different pH values are given.[66a,67] The spectra of UTP at pH 2, CTP at pH 11, and ATP at pH 11 are identical with the spectra at pH 7. These

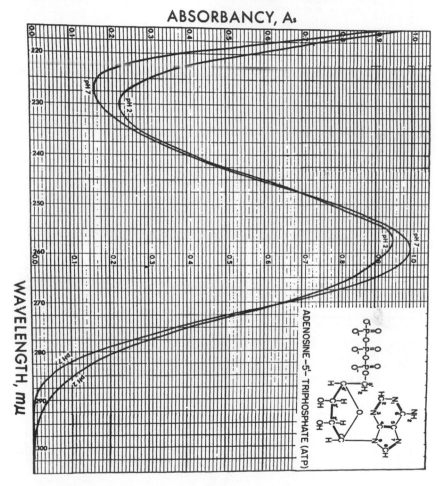

Fig. 3

[67] Taken from Circular OR-7, March, 1955 (ATP, UTP, and CTP), and Circular OR-10, January, 1956 (GTP), of the Pabst Laboratories, with the kind permission of the Pabst Brewing Co., Milwaukee, Wisconsin. More complete spectrum data are available from these circulars. The author is grateful to Dr. S. A. Morell and Dr. A. Frieden of the Pabst Laboratories and to Dr. R. M. Bock of the University of Wisconsin, for their cooperation in making these data available, some in advance of publication.

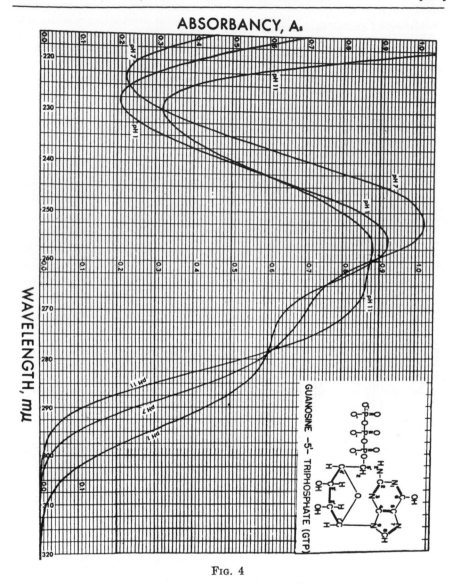

FIG. 4

spectra were recorded on solutions of purified sodium salts of the nucleotides by means of a calibrated Cary Model 11 recording spectrophotometer. The purification was carried out by ion exchange chromotography
and precipitation or crystallization of large batches of these compounds
as the sodium salts, and was checked by paper chromatography and comparison of spectra with the corresponding mononucleotides prepared by
chemical syntheses.

The spectra of these nucleoside triphosphates, as well as the diphosphates, are for practical purposes identical with those of the nucleoside-5'-phosphates from which they are derived. The spectra of the nucleoside-5'-phosphates resemble closely the spectra of the corresponding

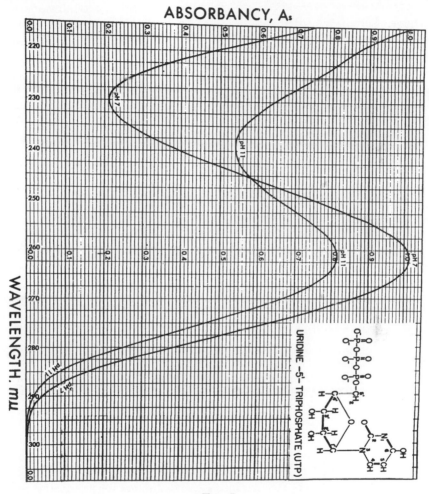

Fig. 5

nucleosides (especially under neutral conditions), and resemble less closely the spectra of the nucleoside-2' or 3'-phosphates (cf. ref. 68).

The relative absorbance data in Figs. 2 to 5 may be used to calculate molar absorptivities. Table II gives the concentrations of the sodium salts of the nucleotides used for these spectrum studies, and the composition of the salts as determined by knowledge of the precipitation conditions, titration studies, and analysis for water of hydration. From these data, the molar absorptivity values have been calculated. Similar data for the nucleosides and the mono- and di-phosphates have also been obtained, and the molar absorptivities determined for these compounds were found to be in good agreement with those for the triphosphates. Although here is presented spectrum data only for the triphosphates, the molar absorptivities recorded in Table II are averages of the values calculated for the 5'-mono-, di-, and triphosphate of each nucleoside.

TABLE II[a]

CHARACTERISTIC DATA FOR THE 5'-RIBONUCLEOTIDES

Compound[b]	Molecular[b] weight	mg./ml.[b]	pH	$\lambda_{max.}$, mμ	$\epsilon_{max.}$[c]	ϵ_{260}[c]
Na$_2$H$_2$ CTP 4H$_2$O	599	45.2	2	280	13,200	6,240
			7	272	9,250	7,520
Na$_2$H$_2$ ATP 4H$_2$O	623	39.6	2	257	15,240	14,870
			7	259	15,780	15,700
Na$_2$H$_2$ GTP 2H$_2$O	603	43.9	1	256	12,520	11,970
			7	253	13,980	11,930
Na$_3$H UTP 2H$_2$O	586	57.8	7	262	10,020	9,920
			11	262	7,900	7,870

[a] Data taken or calculated from R. M. Bock, S. A. Morell, N. S. Ling, and S. H. Lipton, *Arch. Biochem. and Biophys.* in press, and from Circulars OR-7 and OR-10 of the Pabst Laboratories.

[b] These data apply to the solutions used for determination of the spectra in Figs. 2 to 5.

[c] These values are averages of molar absorptivities calculated for the 5'-mono-, di-, and triphosphate of each nucleoside.

Other workers have determined the molar absorptivities of these nucleotides: Schmitz *et al.*[3] reported the following ($\epsilon \times 10^{-3}$ at $\lambda_{max.}$ at pH 1): 13.7 at 278 mμ for CTP; 9.9 at 260 mμ for UTP; and 13.3 at 256 mμ for GTP. Bergkvist and Deutsch[8,20] found 10.0 at 261 to 262 mμ for UTP; 10.85 at 256 mμ for GTP. By way of comparison, the values reported for the *nucleosides* by Beaven *et al.*[68] are 13.4 at 280 mμ for cytidine; 10.1 at 262 mμ for uridine; 12.2 at 257 mμ for guanosine; and 14.6 at 257 mμ for adenosine.

Highly characteristic spectrum shifts occur as the pH is changed, corresponding to changes in the proportion of the different ionic forms of the purine or pyrimidine bases. From pK data of the type given by Beaven *et al.*[68] for the nucleosides, it may be seen that pH 1, pH 7, and

[68] G. H. Beaven, E. R. Holiday, and E. A. Johnson, *in* "The Nucleic Acids" (Chargaff and Davidson, eds.), Vol. I, p. 493. Academic Press, New York, 1955.

pH 11 are values at which all the nucleosides are predominantly in one ionic form or another, with the result that the spectra are unchanged over a range of at least 1 to 2 pH units. The presence of the charged phosphate group in the nucleotides appears to change only slightly the pK values observed for the nucleosides.[66a] In the identification and estimation of purity of these compounds it is important to make measurements at at least two accurately controlled pH values corresponding to two different ionic forms. It is also important to make readings at lower wavelengths, since the presence of many nonnucleotide materials is detected by high light absorption in the region of 215 to 230 mμ.

[112] The Preparation and Assay of Cyclic Nucleotides

By Roy Markham

The cyclic nucleotides (Fig. 1) are the 2′,3′-monophosphate esters of the nucleosides. It will be noted that they are intramolecular phosphodiesters, or anhydrides of normal nucleotides, and as such belong to the class of "high-energy" phosphate compounds. These nucleotides are known to occur as intermediates in the breakdown of ribonucleic acids by enzymes and by other hydrolytic agents, and it would seem quite likely that they are concerned in reactions involving nucleotide rearrangements in polynucleotide chains. So far they have not been recognized in the free condition, but there is now ample evidence that they occur as terminal groups in polynucleotides from many sources.

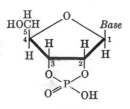

Fig. 1

Preparation of Cyclic Nucleotides

There are two methods which may be used for the preparation of cyclic nucleotides, namely, the degradation of substances containing phosphodiester structures involving ribonucleosides, and the dehydration of nucleotides (nucleoside 2′- or 3′-phosphates) chemically. In the former procedure a certain amount of choice of method and of substrate is permissible, and, in at least one instance, theoretical yields may be obtained, although on a small scale. The chemical synthesis of cyclic nucleotides, on the other hand, enables larger quantities of cyclic nucleotides to be obtained, but they contain impurities which are difficult to remove. In addition, cyclic nucleotides have a tendency to hydrolyze unless kept in a dry state in the cold, and so for some purposes it is desirable to isolate them by paper chromatographic methods shortly before they are used. This limits the usefulness of methods for producing large quantities.

Degradative Methods

Enzymic Method. (1) *Pancreatic Ribonuclease.* The cyclic nucleotides containing pyrimidine bases may be obtained in reasonable quantity from digests of ribonucleic acid treated with pancreatic ribonuclease. To a solution of yeast ribonucleic acid (500 mg. in 10 ml. adjusted to pH 7) ribonuclease is added (100 γ), and the solution is placed in a cellophane dialysis bag in a large volume of distilled water. Cyclic nucleotides diffuse out together with smaller quantities of dinucleotides and larger polynucleotides and thus escape further enzyme action. The dilute solution is concentrated by vacuum distillation for the separation of its constituents. A certain amount of trial may be necessary to establish optimum conditions for this procedure, and there is no objection to reducing the enzyme concentration. A simpler method, however, is to change the temperature. The rate of diffusion of the cyclic mononucleotides is inversely proportional to the viscosity of the water, which only doubles from 0° to 25°. The enzyme activity, however, increases at a much greater rate, so that at any enzyme concentration the yield of cyclic mononucleotides increases as the temperature is dropped.

THE CHROMATOGRAPHIC SEPARATION. The digest is chromatographed on Whatman No. 3 MM paper in a solvent consisting of isopropanol 70 parts, water 30 parts, and 0.35 ml. of NH_3 solution (specific gravity 0.880) added for each 1 l. of gas space in the tank. The quantity which may be placed on a sheet 25 cm. wide is 20 to 30 mg. of total solids in 0.5 ml. The solvent is allowed to flow until the lowest band observed nearly reaches the bottom of the sheet. For this, descending chromatography is essential. The band of cyclic nucleotides (the lowest band) runs about 1.7 times as fast as yeast adenylic acid. The product is eluted directly on to a strip of No. 3 MM paper for electrophoresis. The electrophoresis is carried out at 20 v./cm. in 0.05 M ammonium formate buffer at pH 3.5. Uridylic acid forms a band moving 16 cm. in 2 hours, whereas cytidylic acid moves about 6.5 cm. in 2 hours, both toward the positive electrode.[1] If desired, the above separation may be avoided by deaminating the cyclic cytidylic acid ($NaNO_3$ + acetic acid), but the product is then cyclic uridylic acid exclusively.

2. *Plant Ribonuclease.* The leaf ribonuclease of Holden and Pirie[2] degrades ribonucleic acids to give more or less quantitative yields of

[1] R. Markham and J. D. Smith, *Biochem. J.* **52**, 552 (1952).
[2] M. Holden and N. W. Pirie, *Biochem. J.* **60**, 39 (1955).

pyrimidine cyclic nucleotides which are themselves resistant to the enzyme. The enzyme is best used in 0.05 M KH_2PO_4, which serves to prevent phosphatase action on the purine noncyclic mononucleotides which are liberated as well. This enzyme is an ideal tool for the production of labeled cyclic pyrimidine nucleotides. Incubation may be continued indefinitely.

Leaf ribonuclease is also capable of forming purine cyclic nucleotides in large amounts, but these are subject to the further action of the enzyme and eventually end up as nucleoside 3'-phosphates. In order to avoid complications the enzyme is allowed to act on the "core" resulting from the action of pancreatic ribonuclease on yeast ribonucleic acid which has no pyrimidine nucleoside 3'-phosphate esters left in its structure. The only cyclic mononucleotides resulting are thus cyclic guanylic and cyclic adenylic acid, having R_F values of 1.35 and 1.7 times that of yeast adenylic acid. The chromatographic bands yield pure materials. Naturally, as the enzyme degrades the purine cyclic nucleotides, the reaction has to be stopped. Dialysis may also be used, but it will then be impossible to prevent some phosphatase action if the enzyme is not sufficiently pure. The resulting nucleosides will then have to be removed by electrophoresis at a pH below 8 and above 2, where they have no negative charge, and cyclic nucleotides are charged negatively and are stable.

Chemical Method. Treatment of ribonucleic acid with alkalies in water usually results in the formation of cyclic mononucleotides as transient intermediates in the hydrolysis. Of the various alkaline substances suitable for the production of cyclic mononucleotides in quantity $BaCO_3$ and $CaCO_3$ would seem to be the most suitable. $BaCO_3$ is practically insoluble in water (0.0065% at 100°), but the solution degrades ribonucleic acid readily. The solution is not alkaline enough, however, to hydrolyze cyclic nucleotides rapidly, so that appreciable yields of the latter can be obtained. This method is suitable only for the production of purine cyclic mononucleotides.

Procedure. Ribonuclease "core" is heated in cellophane bags in water with excess solid $BaCO_3$ by being immersed in a large volume of boiling water. The optimum time is about 1 hour. The liquid outside the bag is concentrated, and the cyclic adenylic and guanylic acids are isolated by chromatography on paper.[3]

Chemical Synthesis

The chemical synthesis of cyclic nucleotides starts with nucleoside 2'- or 3'-phosphates. The mechanism therefore involves the elimination of the elements of water. So far, two different methods have proved success-

[3] L. A. Heppel and P. R. Whitfeld, *Biochem. J.* **60**, 1 (1955).

ful, namely, treatment with trifluoroacetic anhydride and with dicyclo-hexylcarbodiimide (DCC). Of these the latter and more recent method would appear to be more generally useful.

Procedure with Dicyclohexylcarbodiimide. To 10 mg. of yeast adenylic acid (i.e., the mixture of adenosine 2'- and 3'-phosphates obtainable commercially) in 0.1 ml. of water plus 0.2 ml. of pyridine, 200 mg. of DCC (now commercially available) in 0.5 ml. of pyridine is added. The reaction is optimal at 15 minutes at 20°. Any precipitate is spun off, and the supernatant liquid is diluted with 2 vol. of water, extracted with ether several times to remove pyridine, and then chromatographed on paper in isopropanol-water (70:30) in an ammonia atmosphere. The product is about 70% cyclic adenylic acid, contaminated with the starting material and with adenylyl cyclohexylurea, which has an R_F of about 0.95.

The methods of Dekker and Khorana[4] and of Terner and Khorana[5] should be of general application. However, the yields of the cytidylic and uridylic acid compounds are small in aqueous pyridine, and a more generally useful solvent is dimethyl formamide to which a saturated solution of the nucleotide is added, extra dimethyl formamide being added if the mixture becomes cloudy. The reaction mixture may be applied directly to the chromatogram. Guanylic acid reacts to a negligible extent in either medium.[5a] For a review of carbodiimides, see Khorana.[6]

The other method is the earlier and original method of synthesis by Brown *et al.*[7]

Procedure with Trifluoroacetic Anhydride. Trifluoroacetic anhydride (5 ml.) is added to dry yeast adenylic acid (1 g.), and the mixture is left in a stoppered flask at 20° for 18 hours. The solution is evaporated under reduced pressure, and the residue is washed with dry ether and ground to a powder. The powder is added to a solution of NH_3 in dry ethanol at 0° (25 ml.) and left for 20 minutes. On evaporation of the ethanol and NH_3 the product contains the cyclic nucleotide. Further purification is by cellulose column chromatography or by chromatography on paper. The number of reaction products is large, but the only one likely to cause trouble is adenosine 2'- (or 3'-)phosphate ethyl ester. This contaminant originates from the ethanol used, and other less reactive alcohols may be used for dissolving the NH_3. The most promising alcohol, namely

[4] C. A. Dekker and H. G. Khorana, *J. Am. Chem. Soc.* **76**, 3522 (1954).
[5] G. M. Terner and H. G. Khorana, *J. Am. Chem. Soc.* **77**, 5349 (1955).
[5a] R. Markham, unpublished observations.
[6] H. G. Khorana, *Chem. Revs.* **53**, 145 (1953).
[7] D. M. Brown, D. I. Magrath, and A. R. Todd, *J. Chem. Soc.* **1952**, 2708.

tert-butanol, unfortunately is solid in the cold, but benzyl alcohol and isopropanol have been employed successfully. The alkali may also be varied, triethylamine, triethanolamine, and cyclohexylamine having been used. The method is a general one and may be applied to other mononucleotides.

Recognition and Estimation of Cyclic Nucleotides

The cyclic nucleotides are easily recognized because their diester structure modifies their physical properties, and also their reaction to enzymes and other agents. These characteristic properties are summarized as follows:

1. Cyclic nucleotides have higher R_F values than the corresponding noncyclic nucleotides in organic solvents such as isopropanol 70%, water 30% (v/v), with NH_3 in the atmosphere 0.3 ml./l. of gas space of NH_3 solution, 0.880 specific gravity (solvent 1), and smaller R_F values in aqueous chromatographic solvents like 80% saturated aqueous $(NH_4)_2SO_4$, 18% M sodium acetate, 2% isopropanol (v/v) (solvent 2).

2. Since there is no secondary phosphate dissociation, no increase in mobility is shown over the pH range 5.5 to 7.5 on electrophoresis on paper. Below this range the mobilities are very similar to those of the corresponding noncyclic nucleotides.

3. For the above reason prostate phosphomonoesterase has no effect on cyclic nucleotides.

4. Cyclic nucleotides are hydrolyzed quantitatively by 0.1 N HCl at 20° in 4 hours. This treatment is not sufficient to hydrolyze an internucleotide link or an ester linkage in, say, a nucleotide methyl ester, which would resemble a cyclic nucleotide in three properties listed above. After hydrolysis a mixture of nucleoside 2'- and 3'-phosphates is left. These substances may be recognized by changes in chromatographic behavior, mobility at pH 7.5, and susceptibility to phosphomonoesterase.

5. Cyclic nucleotides are hydrolyzed by enzymes in the following ways: (a) Pyrimidine cyclic nucleotides are hydrolyzed by pancreatic ribonuclease to give the corresponding nucleoside 3'-phosphates.

(b) All cyclic nucleotides are hydrolyzed by spleen phosphodiesterase to give the corresponding nucleoside 2'-phosphates.[8] This reaction may be used to detect cyclic terminations of small polynucleotides if present.

(c) Cyclic mononucleotides are hydrolyzed by snake venom (rattlesnake or Russel's viper) diesterase[9] to the nucleoside 3'-phosphates. Cyclic dinucleotides yield a nucleoside 3',5'-diphosphate under this treatment,

---

[8] P. R. Whitfeld, L. A. Heppel, and R. Markham, *Biochem. J.* **60**, 15 (1955); see also Vol. II [90].

[9] See Vol. II [89].

to which ribodinucleotides of the normal type are highly resistant. The specific phosphodiesterase from the venom gives the same result.

(d) Purine cyclic nucleotides are converted to the corresponding nucleoside 3'-phosphates by plant leaf ribonuclease.[2,10] The pyrimidine cyclic nucleotides are unchanged by this treatment.

Estimation. Cyclic mononucleotides are best estimated by paper chromatographic methods. The positions in which they run in solvent 1 are characteristic and approximate to those of the corresponding nucleosides, from which they may be separated by:

1. Hydrolysis to the corresponding nucleoside 2'- and 3'-phosphates by N KOH at 20° for 18 hours, or enzymically, and rechromatographing in solvent 1. Individual nucleotides may then be isolated by any of the usual methods.

2. Electrophoresis at pH 3.5, when the nucleosides remain still or move to the cathode. The mononucleotides move toward the anode in the order uridylic acid, guanylic acid, adenylic acid, cytidylic acid (relative mobilities 1.0, 0.9, 0.5, 0.25). Xanthylic and inosinic acids move with uridylic acid and may have to be separated from it by chromatography. Xanthylic acid, however, moves with guanylic acid in solvent 1, so the separation is simplified. This pair, either cyclic or noncyclic, is best separated at a pH of 2.5 to 3, when the guanylic acid will move behind the xanthylic acid.

Nucleotides may, of course, be hydrolyzed to the corresponding bases or nucleosides and then estimated after chromatography on paper.

In the event that a methyl or ethyl ester of a nucleotide is present as a contaminant, this may chromatograph close to the cyclic nucleotide and will have to be removed by a specific hydrolysis, as described above. The products may then be separated by chromatography in solvent 1 or by electrophoresis on paper. A mixture of a cyclic nucleotide and its methyl or ethyl ester, as obtained in solvent 1, may also be hydrolyzed by spleen phosphodiesterase, the former yielding a nucleoside 2'-phosphate and the latter a nucleoside 3'-phosphate. These may be separated by ion exchange chromatography, by paper chromatography if they are adenine or guanine nucleotides (in solvent 2), or by the use of the specific 3'-nucleotidase from plants.[11]

According to the method of separation, the extinction coefficient of the appropriate derivative (i.e., nucleotide or base) is used for the quantitative measurement. Cyclic mononucleotides are, of course, hydrolyzed by 0.1 N HCl, and the extinction coefficients of the mixed nucleoside 2'- and 3'-phosphates may be used for the spectrophotometry of the solutions.

[10] R. Markham, unpublished experiments.
[11] L. Shuster and N. O. Kaplan, *J. Biol. Chem.* **207,** 535 (1953); see Vol. II [86].

[113] Chemical Synthesis of Nucleosides and Nucleotides

By ALEXANDER R. TODD

I. Synthesis of Nucleosides

Although synthesis of all the natural ribonucleosides has been effected,[1] the methods involve relatively complex operations, and for most biochemical studies the preparation of the ribonucleosides and deoxyribonucleosides from natural sources is recommended. Should synthetic material be necessary, the original literature should be consulted.[1,2]

II. Synthesis of Nucleotides

In the preparation of nucleotides from nucleoside derivatives, dibenzyl phosphorochloridate [$(C_6H_5CH_2O)_2POCl$] is frequently employed as the phosphorylating agent. As a general reagent its preparation is therefore described before the preparation of individual nucleotides.

Dibenzyl Phosphite.[3] A solution of phosphorus trichloride (261 ml. ≡ 1 mole) in dry benzene (2.25 l.) is cooled to 0°, and to it is added, with stirring, a mixture of benzyl alcohol (616 ml. ≡ 2 moles) and diethylaniline (960 ml. ≡ 2 moles) previously cooled to 0°. The addition is carried out at such a rate that the reaction temperature remains at ca. 5° (≯ 10°) and with efficient cooling in an ice-salt mixture takes approximately 2½ hours. The mixture is stirred for a further 1½ hours with continued cooling, and then more benzyl alcohol (308 ml. ≡ 1 mole) is added during 1 hour, the temperature being kept below 10°. Stirring is continued for a further 1½ hours; then water (1 l.) is added and the phases separated.

The organic phase is washed successively with water (3 × 1 l.), aqueous ammonia (3 × 1 l. of 3 N), and water (3 × 1 l.), and dried overnight over sodium sulfate. The dried solution is then evaporated under reduced pressure with a nitrogen leak, and the residue finally stripped up to 150° (bath temperature) (oil pump vacuum, nitrogen leak). The residue sets on cooling to a crystalline mass of dibenzyl phosphite. Yield, 710 g. ≡ 90%.

Dibenzyl Phosphorochloridate.[3] Dibenzyl phosphite (39.3 g.), dissolved in dry carbon tetrachloride (200 ml.), is cooled to −15°, and a solution

---

[1] A. R. Todd *et al.*, *J. Chem. Soc.* **1947**, 1052; **1948**, 967, 1685; **1949**, 232, 1620.

[2] J. Davoll and B. A. Lowy, *J. Am. Chem. Soc.* **73**, 1650 (1951).

[3] As modified from F. R. Atherton, H. T. Openshaw, and A. R. Todd, *J. Chem. Soc.* **1945**, 382.

of chlorine in carbon tetrachloride (220 ml. of 1.4 N) is added dropwise with stirring, the temperature being kept below $-10°$. Dry nitrogen is then passed through the solution to remove chlorine and hydrogen chloride (ca. 90 minutes). For phosphorylation purposes the solution is normally used directly. Isolation of the dibenzyl phosphorochloridate is not recommended, as it is a thick oil which decomposes even on standing at room temperature. It is therefore advisable to keep dibenzyl phosphite as a stable stock reagent and to prepare as required from an appropriate quantity a carbon tetrachloride solution of the phosphorochloridate which should be used *at once* for the phosphorylation reaction. The yield of dibenzyl phosphorochloridate obtained on chlorinating dibenzyl phosphite is virtually quantitative.

A. Ribonucleotides

Adenosine-5′ Phosphate

2′,3′-O-Isopropylidene Adenosine.[4] A mixture of adenosine (30 g.; dried at 110°/1 mm. for 48 hours) and zinc chloride (80 g.) in dry acetone (800 ml.) is refluxed for 12 hours and then set aside at room temperature for 2 days, with exclusion of moisture. Acetone is removed under reduced pressure. A solution of barium hydroxide (190 g. of octahydrate in 1 l. of water) is added, and the mixture is shaken vigorously with cooling. After 5 minutes, solid carbon dioxide is added, to neutrality. The mixture is then heated to about 80° and filtered hot through a thin layer of Hyflo Super-Cel. The precipitate is washed well with boiling water and methanol, the precipitate being resuspended each time. Total filtrate and washings are then taken to ca. 500 ml., and the crystalline deposit of 2′,3′-O-isopropylidene adenosine is collected. The filtrate is taken to dryness and extracted once with boiling ethanol, ethanol removed under reduced pressure, and the residual gum crystallized from hot water. The two lots of isopropylidene adenosine are combined and given a final crystallization from hot water (24 g.), melting point 220°. (Check the absence of adenosine in the final product by running a paper chromatogram in n-butanol-water (86:14).)

2′,3′-O-Isopropylidene Adenosine-5′ Dibenzyl Phosphate.[5] A solution of isopropylidene adenosine (10 g.; dried at 110°/1 mm. for 48 hours) in dry pyridine (80 ml.) is frozen in a CO_2-acetone bath, and dibenzyl phosphorochloridate (from 20 g. of dibenzyl phosphite) is quickly added. The mixture is removed from the cooling bath and allowed to warm up till all the solid is melted, the flask being vigorously rotated. The mixture is maintained in a viscous liquid condition in the cooling bath (ca. $-30°$ to

[4] As modified from P. A. Levene and R. S. Tipson, *J. Biol. Chem.* **121**, 131 (1937).
[5] As modified from J. Baddiley and A. R. Todd, *J. Chem. Soc.* **1947**, 648.

−28°) for 6 hours, then kept at 0° overnight, with the exclusion of moisture at all times. A solution of sodium carbonate (9 g.) in water (40 ml.) is then added, and the mixture is shaken by hand. Pyridine is removed under reduced pressure (initial frothing due to CO_2 evolution), and the residue evaporated down twice with water, under reduced pressure. To the residue, water and chloroform are added, the chloroform extract washed once with dilute sodium bicarbonate solution, then water, dried (Na_2SO_4), and chloroform removed under reduced pressure. The residual gum is evaporated twice with ethanol to remove traces of chloroform, then dissolved in a small amount of ethanol (10 ml.). A mixture of 100 ml. of ether and 100 ml. of petroleum ether (boiling point 40 to 60°) is then added with vigorous shaking, and the mixture is kept at 0° for several hours. After the supernatant liquors have been decanted, the deposited gum is crystallized from ethanol-ether (ca. 14 g.), melting point 97°.

Adenosine-5′ Benzyl Phosphate.[5] 2′,3′-O-Isopropylidene ̄adenosine-5′ dibenzyl phosphate (5.4 g.) is dissolved in a mixture of dilute sulfuric acid (500 ml. of $N/50$) and alcohol (150 ml.), and the solution refluxed for 1 hour. Sulfate is removed by adding the calculated amount of barium hydroxide, and the precipitated barium sulfate is removed by filtration. The filtrate is concentrated to small bulk under reduced pressure and set aside at room temperature, giving adenosine-5′ benzyl phosphate as colorless needles (3.7 g.), melting point 234° (decomp.).

Adenosine-5′ Phosphate.[5] A solution of 2′,3′-O-isopropylidene adenosine-5′ dibenzyl phosphate (14 g.) in 50% aqueous ethanol (500 ml.) is hydrogenated with a mixture of palladized charcoal and Adam's palladium oxide, at room temperature and pressure, until hydrogen uptake ceases. Catalyst is removed, and 1 l. of 0.1 N sulfuric acid is added to the filtrate. The solution is then set aside at room temperature for 2 days. A solution of barium hydroxide (15.75 g., A.R. grade) in water (200 ml.) is added (check end point with rhodizonic acid), barium sulfate removed by filtration through Hyflo Super-Cel, and the filtrate taken to small volume (50 to 100 ml.) under reduced pressure. The filtered solution is seeded with adenosine-5′ phosphate and kept at 0°. Adenosine-5′ phosphate crystallizes as needles, melting point 190° (ca. 7 to 8 g.).

Adenosine-2′ and -3′ Phosphates

5′-O-Trityladenosine.[5a] Anhydrous adenosine (10 g.; dried at 105°/ 0.1 mm. over phosphoric oxide) is dissolved in boiling dry pyridine (300 ml.), and the solution is cooled to room temperature. Trityl chloride (11.5 g.) is added, and the solution is allowed to stand for 7 days protected from atmospheric moisture by a calcium chloride tube. The solution is

[5a] P. A. Levene and R. S. Tipson, *J. Biol. Chem.* **121**, 311 (1937).

poured into 2 l. of ice water with vigorous stirring, and the yellow gum which separates solidifies when the solution is kept overnight in the icebox. The solid product is collected, triturated with water, and dried (19 g.). The crude material is dissolved in boiling pyridine (150 ml.), and cold absolute ethanol (300 ml.) is added. When kept at 0°, 5'-O-trityl adenosine separates in colorless crystals which are collected, washed with a little ethanol, and dried (9.2 g.), melting point 250°. The mother liquors contain a further small quantity of the material together with N,O$^{5'}$-ditrityladenosine.

5'-O-Trityladenosine Dibenzyl Phosphate. 5'-O-Trityladenosine (3.1 g.) is dissolved in carefully dried pyridine (60 ml.), and the solution cooled to a temperature slightly above its melting point in an acetone-solid CO_2 bath. Dibenzyl phosphorochloridate (from 4.5 g. of dibenzyl phosphite) is run into the solution with exclusion of moisture. The solution is kept at this temperature with occasional shaking for 4 hours and then left at room temperature overnight. A solution of sodium carbonate (3.0 g.) in water (15 ml.) is added, sodium chloride removed by filtration, and the solution then evaporated to dryness under reduced pressure. Re-evaporation at reduced pressure several times with ethanol removes pyridine completely. The residual brown gum is shaken with chloroform and sodium hydrogen carbonate solution, and the chloroform layer is further washed with sodium hydrogen carbonate solution, water, and then dried over anhydrous sodium sulfate. The chloroform solution is evaporated, and the residue dissolved in ethanol (15 ml.) and set aside for 48 hours. A small precipitate of unchanged 5'-O-trityladenosine is removed by filtration, and the filtrate is evaporated to dryness, giving a solid foam. This is refluxed with dry ether (25 ml.) for 2 hours, and the hygroscopic granular product collected (4.7 g.).

Adenosine-2' and -3' Phosphates.[6] A solution of the above dibenzyl phosphate (0.41 g.) in 80% acetic acid (4 ml.) is boiled under reflux for 20 minutes. Water and chloroform are added, and the aqueous phase is separated, washed with more chloroform, and evaporated under reduced pressure. Evaporation with ethanol removes any residual acetic acid. The gum (0.22 g.) which is obtained is dissolved in water and hydrogenated at room temperature and pressure over a mixed palladium charcoal-palladium oxide catalyst. After filtering from catalyst, barium hydroxide is added to the solution, and excess barium hydroxide neutralized by carbon dioxide. The crude barium salt is obtained by evaporation of the solution and purified by precipitation from water by ethanol (3 vol.) to yield the mixed barium salts of adenosine-2' and -3' phosphates.

[6] D. M. Brown and A. R. Todd, *J. Chem. Soc.* **1952**, 44.

The above barium salt (414 mg.) is dissolved in water (90 ml.), and the solution adjusted to pH 8 to 9 and run onto a column (15 cm. × 4 sq. cm.) of Dowex 2 anion exchange resin (mesh size 200 to 400) in the formate form. The column is washed with water (150 ml.), and then elution is continued with 0.1 N formic acid at a flow rate of 3 ml./min., fractions (ca. 25 ml.) being collected. The progress of elution of the nucleotides is followed by the optical density of the fractions at 260 mμ. After the removal of the first component (adenosine-2' phosphate) is complete (volume to peak, ca. 3.5 l.), the second component (adenosine-3' phosphate) is rapidly eluted with N formic acid.

The fractions containing adenosine-2' phosphate are united and evaporated to small bulk (50 ml.) and finally freeze-dried. The product is crystallized from water, with addition of charcoal if necessary, and separates as a colorless crystalline solid (67 mg.), melting point 187° (decomp.).

The fractions containing adenosine-3' phosphate are freeze-dried as above, and the product is crystallized from water. It forms long colorless needles (unhydrated) or prismatic needles (containing 1.5 mols of water of crystallization) (120 mg.), melting point 195° (decomp.).

The two substances are distinguished, *inter alia*, by paper chromatography in the isoamyl alcohol-5% disodium hydrogen phosphate solvent with R_f values of 0.74 and 0.67 for adenosine-2' and -3' phosphate, respectively.

Adenosine-2' and -3' Benzyl Phosphates. PHENYLDIAZOMETHANE. Benzaldehyde hydrazone (2.4 g.), yellow mercuric oxide (9.0 g.), and anhydrous sodium sulfate (2.5 g.) are mixed in a flask and covered with a layer of dry ether (30 ml.). Cold saturated alcoholic potassium hydroxide solution (0.5 ml.) is added, and the mixture is shaken for 15 minutes. The deep-red solution is filtered, the residue washed with more ether, and the combined filtrates evaporated under vacuum below room temperature. The product is a red oil and should be used immediately for esterification reactions.

Adenosine-2' and -3' Benzyl Phosphates.[7] Yeast adenylic acid (0.5 g.) is suspended in dimethylformamide (5 ml.) and treated with phenyl diazomethane (from 2.4 g. of benzaldehyde hydrazone). The solution is shaken for some time, until solid material has completely dissolved. Ether, water, and barium carbonate are then added, and the solution is shaken for several hours with changes of ether. The aqueous phase is separated, filtered through Hyflo Super-Cel and evaporated under reduced pressure. The product is twice reprecipitated from water (3 to 4 ml.) by acetone and

[7] D. M. Brown, L. A. Heppel, and R. J. Hilmoe, *J. Chem. Soc.* **1954**, 40.

dried (0.44 g.). It consists of the mixed barium salts of adenosine-2′ and -3′ benzyl phosphates. These can be separated by ion exchange chromatography.

The barium salt is dissolved in water (50 ml.), brought to pH 8, and run onto a column (9 sq. cm. × 11 cm.) of Dowex 2 resin in the formate form. After washing with water (500 ml.), elution with 0.1 N formic acid is commenced. The first 5 l. containing traces of mononucleotides is discarded. Adenosine-2′ benzyl phosphate is removed slowly by the next 2 l., and the remainder rapidly eluted by changing to 0.5 N formic acid. Continued elution with this solvent then removes the adenosine-3′ benzyl phosphate. The course of fractionation is followed by observing the optical densities of the fractions at 260 mμ.

The combined fractions containing adenosine-2′ benzyl phosphate are evaporated at 20 mm., and the residue is crystallized from water. Adenosine-2′ benzyl phosphate separates as a hydrate in small irregular prisms.

The fractions containing adenosine-3′ benzyl phosphate are evaporated at 14 mm. at room temperature, with additions of water at the latter stages to reduce the formic acid concentration. The residual oil crystallizes from water in rosettes of small needles.

Adenosine-2′ and -3′ benzyl phosphates are distinguished on paper chromatograms with the butanol-acetic acid-water (4:1:5 v/v) solvent (R_f, 0.42 and 0.47 respectively), or the 5% disodium hydrogen phosphate-isoamyl alcohol solvent (R_f, 0.67 and 0.56, respectively).

Adenosine-2′,3′ Phosphate.[8] Trifluoroacetic anhydride (5 ml., 12 moles) is added to anhydrous yeast adenylic acid (mixture of 2′- and 3′-isomers; 1 g., 1 mole), and the resulting solution is set aside overnight at room temperature in a stoppered flask. The solution is evaporated under reduced pressure and the residual pale-yellow resin triturated with dry ether. The white powder so obtained is collected in a centrifuge, briefly dried under vacuum, and then added rapidly to a stirred, ice-cold saturated solution of ammonia in ethanol (25 ml.). After 30 minutes at 0°, solid material is removed by centrifuging and washed with ethanol. The combined supernatant solution and washings are evaporated under reduced pressure, and the residual gum is thoroughly washed with acetone, giving a white solid (0.63 g.) which is redissolved in a little methanol. Addition of dry ether gives the ammonium salt of the cyclic phosphate as a white, deliquescent powder.

The barium salt is prepared by heating an aqueous solution of the ammonium salt (0.1 g.) on the steam bath for 40 minutes with a slight excess of barium carbonate. The mixture is filtered, the filtrate concentrated to small bulk under reduced pressure, and excess of ethanol added.

[8] D. M. Brown, D. I. Magrath, and A. R. Todd, *J. Chem. Soc.* **1952**, 2708.

The microcrystalline precipitate of barium adenosine-2',3' phosphate (ca. 0.05 g.) is purified by reprecipitation from aqueous solution with ethanol. Light absorption in water. $\lambda_{max.}$ 260 mμ (ϵ, 13,150); $\lambda_{min.}$ 225 mμ (ϵ, 1830, for M = 397). Electrometric titration can be used to demonstrate the absence of a secondary phosphoryl dissociation (pK ca. 5.9).

The compound has an R_f value of ca. 0.43 in an isopropanol-ammonia-water system (70:5:25 v/v) and of ca. 0.57 in the 5% disodium hydrogen phosphate-isoamyl alcohol system of Carter.[9] The order of elution of the isomeric adenylic acids and cyclic product from a column of Dowex 2 resin (200 to 400 mesh; formate form), with formic acid (0.1 N) as eluant, is as follows: adenosine-5' phosphate, adenosine-2' phosphate, adenosine-3' phosphate = adenosine-2',3' phosphate (280 mμ/260 mμ, optical density ratio at peak = ca. 0.19).

The compound is very easily hydrolyzed by aqueous acid or alkali; it must be stored under anhydrous conditions.

Guanosine-5' Phosphate

2',3'-O-Isopropylidene Guanosine.[10] Guanosine (35 g.; dried for 24 hours at 110°/1 mm.) is added to a solution of anhydrous zinc chloride (85 g.) in dry acetone (600 ml.) and heated under reflux for 5 hours under anhydrous conditions, a clear solution being obtained. Acetone is removed under reduced pressure, the resulting syrup dissolved in a small amount of Cellosolve (2-ethoxyethanol), and dry ether (1500 ml.) added with shaking. The hygroscopic zinc chloride double salt is rapidly collected, washed with ether, and dried. The powdered salt is dissolved in warm Cellosolve (250 ml.), a solution of barium hydroxide (200 g. of octahydrate) in water (700 ml.) added, the mixture well shaken, and carbon dioxide passed through it until neutral to phenolphthalein. The mixture is then filtered, and the voluminous precipitate washed well with hot water (ca. 2 l.) and a small amount of hot Cellosolve. The combined filtrate and washings (ca. 3 l.), on cooling, deposits the crude product. Recrystallized twice from hot water, 2',3'-O-isopropylidene guanosine forms colorless needles (22.5 g., 64%), melting point 299° (decomp.)

Guanosine-5' Phosphate.[10] Phosphoryl chloride (1.58 ml., 1 mole) in anhydrous pyridine (20 ml.) is added dropwise during 15 minutes to a vigorously stirred solution of 2',3'-O-isopropylidene guanosine (5.5 g.; dried for 24 hours at 120°/1 mm.) in anhydrous dimethylformamide (40 ml.) and pyridine (60 ml.) at −10°, and stirring is continued for a further 2 hours. Ice-cold aqueous pyridine (50 ml., 50%) is then added during 30 minutes, followed by ice water (190 ml.) and cold 0.35 N

[9] C. E. Carter, *J. Am. Chem. Soc.* **72**, 1466 (1950).
[10] A. M. Michelson and A. R. Todd, *J. Chem. Soc.* **1949,** 2476.

barium hydroxide (245 ml.) to pH 8.7 (color change in the solution), and the mixture is evaporated to dryness. The residue is dissolved in water and filtered through Hyflo Super-Cel, the barium precipitated with sulfuric acid, and enough water and sulfuric acid added to bring the solution to a volume of 1000 ml. with an acid concentration of 0.1 N. After 2 days at room temperature the solution is neutralized with barium hydroxide and filtered hot. Lead acetate solution (35 ml., 20%) is added, and the lead salt centrifuged off, washed with water, suspended in hot water, and decomposed with hydrogen sulfide. The solution so obtained is filtered from lead sulfide, aerated, and evaporated to small volume under reduced pressure, and the nucleotide precipitated by adding acetone. The crude phosphate is then redissolved in a minimum of water, and acetone slowly added to the filtered solution. Guanosine-5' phosphate separates as a colorless mass of microcrystals (1 g., 20%), melting point 190 to 200° (decomp.).

Guanosine-2',3' Phosphate[8]

Trifluoroacetic anhydride (2.1 ml., 16 moles) is added to anhydrous yeast guanylic acid (0.35 g., 1 mole). The mixture warms slightly, and the acid gradually dissolves, giving a clear solution, which is kept overnight at room temperature in a stoppered flask and then evaporated under reduced pressure. Trituration of the residue with dry ether gives a white powder which is collected in a centrifuge and, after brief drying in a vacuum desiccator, added portionwise to a stirred ice-cold solution of ammonia in isopropanol (16 ml.). The mixture is left for 30 minutes, then centrifuged, and the solid washed with more ice-cold isopropanolic ammonia (5 ml.). The combined supernatant solution and washings are concentrated almost to dryness *in vacuo* at 20°. Methanol is added and, after recentrifugation, concentrated to very small bulk and the ammonium salt precipitated as a white powder (0.07 g.) by adding ether. It is collected in a centrifuge, washed once with a little ether containing 5% ethanol, and immediately placed in a vacuum desiccator.

The barium salt is prepared by adding a slight excess of 0.1 N barium hydroxide to an aqueous solution of the ammonium salt (0.06 g.) cooled in ice. The solution is brought to neutrality with carbon dioxide, then concentrated *in vacuo*, filtered, and finally reduced to very small bulk, and the barium guanosine-2',3' phosphate (0.05 g.) precipitated by adding acetone. Light absorption in water: $\lambda_{max.}$ 251 to 253 mμ (ϵ, 12,150 for $M = 413$). Electrometric titration can be used to demonstrate the absence of a secondary phosphoryl dissociation.

The compound has an R_f value of ca. 0.23 in the isopropanol-ammonia-water system (70:5:25 v/v), and of ca. 0.67 in the 5% disodium hydrogen phosphate-isoamyl alcohol system of Carter.[9] Elution of the isomeric guanylic acids and cyclic phosphate from a column of Dowex 2 resin (200 to 400 mesh; formate form), with 0.1 N formic acid which is 0.1 N with respect to sodium formate as eluant, follows the order: guanosine-2' phosphate, guanosine-3' phosphate, guanosine-2',3' phosphate (peak optical density ratio, 280 mμ/260 mμ = ca. 0.7).

Like other members of this class, guanosine-2',3' phosphate is very easily hydrolyzed by aqueous acids or alkalis and is stored under anhydrous conditions.

Uridine-5' Phosphate

2',3'-O-Isopropylidene Uridine.[11] Dry, finely powdered uridine (10 g.) and anhydrous copper sulfate (20 g.) are suspended in anhydrous acetone (250 ml.), concentrated sulfuric acid (0.25 ml.) added, and the suspension shaken for 48 hours at 37° with exclusion of moisture. The mixture is filtered, the residue washed with acetone (80 ml.), and the combined filtrate and washings shaken with dry calcium hydroxide (10 g.) for 1 hour. The mixture is filtered, the calcium salts washed with acetone (80 ml.), and the combined filtrate and washings evaporated to dryness under reduced pressure. The residue is recrystallized from methanol, affording fine needles (10.1 g.), melting point 163 to 165°.

5'-O-Toluene-p-sulfonyl-2',3'-O-isopropylidene Uridine.[11] Dry isopropylidene uridine (8 g.) is dissolved in anhydrous pyridine (90 ml.), toluene-p-sulfonyl chloride (12 g.) added, the mixture shaken till homogeneous, and set aside at room temperature overnight in a stoppered flask. Ice water (300 ml.) is cautiously added to the stirred solution, and the precipitate is filtered, washed with water, and recrystallized from methanol as prisms (8.2 g.), melting point 145 to 146°.

5'-Iodo-5'-deoxy-2',3'-O-isopropylidene Uridine.[11] 5'-O-Toluene-p-sulfonyl-2',3'-O-isopropylidene uridine (5 g.) and dry sodium iodide (5 g.) are dissolved in acetone (50 ml.), and the solution is heated for 2 hours at 100° in an autoclave. The resulting mixture is evaporated to dryness under reduced pressure, the residue dissolved in a mixture of chloroform and water, and the solution decolorized by the addition of a few drops of saturated sodium hydrogen sulfite solution. The two layers are separated, and the aqueous layer again extracted with chloroform. The combined chloroform solutions are shaken with water and dried over anhydrous sodium sulfate. After filtration from sodium sulfate, the filtrate is evaporated to dryness under reduced pressure, the residue dissolved in the

[11] A. Levene and R. S. Tipson, *J. Biol. Chem.* **106**, 113 (1934).

minimum quantity of methanol, n-pentane added to turbidity, and the solution set aside at 0° overnight. The colorless crystals obtained are filtered, washed with pentane, and dried, giving 5'-iodo-5'-deoxy-2',3'-O-isopropylidene uridine (4.5 g.), melting point 164°.

Dibenzyl 2',3'-O-Isopropylidene Uridine-5' Phosphate.[12] 5'-Iodo-5'-deoxy-2',3'-O-isopropylidene uridine (4.0 g.) and silver dibenzyl phosphate (4.0 g.) are suspended in benzene (50 ml.), and the solution refluxed for 1½ hours with exclusion of light and moisture. The mixture is filtered, the residue washed with benzene, and the combined filtrate and washings extracted with water (100 ml.), sodium thiosulfate solution (100 ml.), sodium hydrogen carbonate solution (100 ml.), and water (100 ml.), and dried over anhydrous sodium sulfate. The filtered solution is evaporated under reduced pressure, affording a colorless glass (5.6 g.).

Barium Uridine-5' Phosphate.[12] The above glass (5.6 g.) is dissolved in 50% aqueous ethanol (200 ml.) and shaken for 15 hours with hydrogen (1 atm.) and a mixture of palladous oxide (0.04 g.) and 10% palladized charcoal (0.04 g.). Barium uridine-5' phosphate (3.69 g.) is precipitated as a fine powder by adding ethanol (120 ml.) to the filtered solution after evaporation to 25 ml. and neutralization (pH 5 to 6) with aqueous barium hydroxide.

Benzyl 2',3'-O-Isopropylidene Uridine-5' Phosphate.[12] The above glassy dibenzyl 2',3'-O-isopropylidene uridine-5' phosphate, prepared from 5'-iodo-5'-deoxy-2',3'-isopropylidene uridine (1.7 g.) and silver dibenzyl phosphate (1.7 g.), is dissolved in anhydrous methyl cyanide (10 ml.), and the solution refluxed for 2 hours with dry potassium thiocyanate (0.43 g.). The mixture is cooled, the precipitate filtered and dissolved in dilute hydrochloric acid (5 ml. of 0.5 N), the solution rapidly extracted with chloroform (5 × 5 ml.), and the combined chloroform extracts dried over anhydrous sodium sulfate. The filtered solution is evaporated under reduced pressure, affording a hygroscopic glass which is transformed into a colorless amorphous solid (0.65 g.) by pouring a concentrated chloroform solution into a large volume of dry ether.

Benzyl Uridine-5' Hydrogen Phosphate.[13] Benzyl 2',3'-O-isopropylidene uridine-5' phosphate (0.1 g.) is dissolved in a 5% solution of hydrogen chloride in methanol (1 ml.), kept at room temperature for 15 hours, and poured into dry ether (50 ml.). The turbid solution is set aside at 0° for 6 hours, and the small rosettes of crystals (0.031 g.), melting point 129 to 132°, are separated by decantation.

[12] S. M. H. Christie, D. I. Elmore, G. W. Kenner, A. R. Todd, and F. J. Weymouth, *J. Chem. Soc.* **1953**, 2947.
[13] N. S. Corby, G. W. Kenner, and A. R. Todd, *J. Chem. Soc.* **1952**, 3669.

Uridine-2' and -3' Phosphates[14]

5'-O-Trityluridine. Dry, finely powdered uridine (10 g.) and pure dry trityl chloride (12.6 g.) are dissolved in dry redistilled pyridine (120 ml.), and the solution protected by a calcium chloride tube. After standing overnight, the solution is heated at 100° for 3 hours, cooled, and poured into water (1 l.) with vigorous stirring. The supernatant is decanted from the gummy product which is washed with water, dissolved in acetone, and the solution evaporated under diminished pressure. Residual pyridine is removed by treating the material with boiling water (2 × 200 ml.), cooling, and decanting. The solid residue is dissolved in acetone-ethanol (1:1, 250 ml.), treated with charcoal, filtered, and the solution evaporated under reduced pressure. The crude product is dried (practically quantitative yield) and crystallized by dissolving in acetone (50 ml.) and adding a little ether to the solution (yield, 9.5 g.). Evaporation of the mother liquors yields a colorless glass (10.5 g.) which can be fractionally crystallized.

Recrystallization gives the product, melting point 200°, $[\alpha]_D^{28} = +9.3°$ (in acetone).

Uridine-2' and -3' Phosphates (Uridylic Acids a and b).[10,15] Anhydrous 5'-O-trityluridine (3.1 g.) is dissolved in dry pyridine (40 cc.) and phosphorylated with dibenzyl phosphorochloridate (from 6.0 g. of dibenzyl hydrogen phosphite) as described for the adenylic acids. 5'-O-Trityluridine dibenzyl phosphate is obtained as a gum which is purified by reprecipitation from its ethanol solution by ether. The gummy product is dissolved in 50% ethanol and hydrogenated at room temperature and pressure over palladium oxide-palladium charcoal. Hydrogenation is sometimes very slow. After uptake of the theoretical amount for two benzyl groups, the catalyst is filtered off, the solution neutralized with sodium hydroxide solution, and then treated with a 20% solution of lead acetate. The lead salt is collected by centrifugation, washed with water, resuspended in water, and decomposed with hydrogen sulfide. Triphenyl carbinol and lead sulfide are filtered off, and the filtrate is evaporated to dryness. The glassy product is dissolved in methanol, and the nucleotide (1.1 g.) precipitated, by addition of acetone, as a white powder, melting point ca. 180° (decomp.). This product is a mixture of uridine-2' and -3' phosphates (ca. 40:60). Separation can be effected on the analytical scale by ion exchange chromatography.

Uridine-2',3' Phosphate.[8] This nucleoside cyclic phosphate is prepared as described for the cytidine analog. One gram of anhydrous yeast uridylic

[14] P. A. Levene and R. S. Tipson, *J. Biol. Chem.* **104**, 385 (1934).

[15] D. M. Brown, C. A. Dekker, and A. R. Todd, *J. Chem. Soc.* **1952**, 2715.

acid gives 0.61 g. of crude ammonium salt; this is free of unchanged nucleotide but contains, in addition to cyclic phosphate, some 15 to 20% of ammonium uridine ethyl phosphate. It (0.4 g.) is therefore chromatographed on a column (5.5 cm. in diameter) of powdered cellulose (240 g.), prepared as described for the corresponding cytidine derivatives. The first 500 ml. of eluate is discarded and the remainder collected in 90 fractions of approximately 20 ml. each. Paper strip chromatography shows that practically all the nucleotide ethyl ester is in fractions 14 to 34 (peak at 30), and the cyclic phosphate in fractions 44 to 85 (peak at 65).

Fractions 44 to 85 are combined and, by a procedure identical to that described for the cytidine compound, yield pure ammonium uridine-2',3' phosphate as a white, very hygroscopic powder (0.25 g.).

The barium salt is prepared as described for the cytidine derivative; from 0.1 g. of the pure ammonium salt, 0.045 g. of the barium compound is obtained by concentration of the mother liquors. Light absorption in water: λ_{max}. 258 to 259 mμ; λ_{min}. 228 to 229 mμ (ϵ, 9570, 2290, for M = 374).

The compound has an R_f of ca. 0.30 when chromatographed on paper strips in the isopropanol-ammonia-water system (70:5:25 v/v), and ca. 0.82 in the disodium hydrogen phosphate-isoamylalcohol system of Carter.[9] The order of elution of the isomeric uridylic acids and cyclic phosphate from a Dowex 2 column (200 to 400 mesh resin; formate form) with 0.01 N formic acid which is 0.05 M with respect to sodium formate as eluant is as follows: uridine-2',3' phosphate (peak optical density ratio, 280 mμ/260 mμ = 0.195), uridine-2' phosphate, uridine-3' phosphate.

The compound is easily hydrolyzed by aqueous alkali and acid and is stored under anhydrous conditions.

Uridine-2' and -3' Benzyl Phosphates.[16] Yeast uridylic acid (0.5 g.) is dissolved in dimethyl formamide (5 ml.), and phenyldiazomethane (from 1.0 g. of benzaldehyde hydrazone; see above) is added dropwise during 3 minutes, the reaction being modified by cooling. The solution is left at room temperature overnight, and then ether (40 ml.) is added. The precipitated gum is washed with more ether. The gum is dissolved in a little ethanol, and the solution diluted with an equal volume of ether. A small amount of sticky solid separates and is discarded. The mother liquors are evaporated to dryness. The residue is dissolved in water and shaken twice with chloroform. The aqueous phase is treated with saturated barium hydroxide solution (2 ml.), and then carbon dioxide is passed in, to neutrality. The solution is evaporated to 5 ml. and filtered through Hyflo Super-Cel. Evaporation under reduced pressure with additions of ethanol gives a hygroscopic solid foam consisting of the mixed barium salts of uridine-2' and -3' benzyl phosphates.

[16] D. M. Brown and A. R. Todd, *J. Chem. Soc.* **1953,** 2040.

Both isomeric uridine benzyl phosphates have the same R_f value, 0.7, in the isopropanol-ammonia-water (70:10:20 v/v) solvent. They can be separated analytically by ion exchange chromatography.

When the known crystalline uridine-3' phosphate[17] (uridylic acid *b*) is used in the above preparation, the product isolated is barium uridine-3' benzyl phosphate.

The above preparation may give a product contaminated with traces of uridylic acid and cyclic uridine-2',3' phosphate. This can readily be ascertained by paper chromatography. If this is the case, further chromatographic purification is necessary.

The barium salt (0.46 g.) is dissolved in water, and barium ion removed by passage through a column of Amberlite IRC-50 (polyacrylic acid) resin (hydrogen form). The eluate is evaporated to dryness, the residue dissolved in the isopropanol-ammonia-water (70:10:2 v/v) solvent (10 ml.) containing a little added water to effect dissolution, and the solution applied to a column (42 × 3.5 cm.) of cellulose powder (160 g. of 80 mesh). The column is developed with the isopropanol-ammonia-water solvents, and the fractions containing the product collected and evaporated. The glassy product is the ammonium salt of the uridine benzyl phosphate.

Cytidine-5' Phosphate[10]

2',3'-O-Isopropylidene Cytidine. Cytidine (4 g.; dried for 12 hours at 110°/1 mm.) is dissolved in a solution of anhydrous zinc chloride (10 g.) in dry acetone (100 ml.), and the mixture heated under reflux for 7 hours with exclusion of moisture. The clear solution is set aside overnight at room temperature, acetone then removed under reduced pressure, and dry ether (250 ml.) cautiously added to the residue, with shaking. The fine precipitate of the zinc chloride-2',3'-isopropylidene cytidine double salt is filtered off, washed with ether, dried, and added to a slight excess of warm barium hydroxide solution. The mixture is neutralized with carbon dioxide and filtered warm, the filter residue being extracted three times with hot water. The combined filtrate and extracts are evaporated to dryness under reduced pressure, and the dry solid residue extracted with hot ethanol. Removal of the solvent under reduced pressure gives a colorless glass. This is dissolved in water and filtered through Hyflo Super-Cel to remove a slight trace of semicolloidal impurity, the solution evaporated, and the residue redissolved in hot absolute ethanol, filtered, and again evaporated under reduced pressure, whereupon the isopropylidene compound is left as a glass (4.4 g., 95%).

Cytidine-5' Phosphate. Dibenzyl phosphorochloridate (from 10 g. of

[17] D. M. Brown, D. I. Magrath, and A. R. Todd, *J. Chem. Soc.* **1954**, 1442.

dibenzyl phosphite) is added to a solution of 2′,3′-isopropylidene cytidine (4.0 g.; dried for 18 hours at 60°/1 mm.) in dry pyridine (60 ml.) at −40°, and the solution maintained at −40° for 3 hours and then set aside at room temperature overnight. When the solution is worked up in the usual way, a gum is obtained which is evaporated twice with ethanol and then precipitated from concentrated ethanolic solution by ether. The resinous product is hydrogenated in aqueous ethanol (catalyst. palladium and palladized charcoal). Catalyst is removed by filtration, the filtrate concentrated to small volume under reduced pressure, and N-sulfuric acid (20 cc.) added; the mixture is kept at 70 to 75° for 1½ hours, then neutralized with barium hydroxide and barium carbonate, and filtered hot, the residue being washed with hot water. The combined filtrate and washings give on evaporation a glass which is dissolved in hot water (20 ml.) and filtered. Ethanol (30 ml.) is added to the solution and the precipitated granular *barium* salt collected, washed with ethanol and then ether, and dried (2.6 g., 38% over-all yield from cytidine). N-Sulfuric acid (ca. 3.8 ml.) is added to a solution of the barium salt (1 g.) in water till all the barium has been precipitated (rhodizonic acid). Barium sulfate is removed by filtration through Hyflo Super-Cel, and the filtrate evaporated to ca. 10 ml. under reduced pressure. To the hot solution, hot ethanol (20 ml.) is added, yielding cytidine-5′ phosphate (0.65 g., 90%) as colorless plates, melting point 233° (decomp. with vigorous effervescence).

Cytidine-2′ and -3′ Phosphates[8]

Cytidine-2′,3′ Phosphate. Trifluoroacetic anhydride (6 ml., 14 moles) is added to anhydrous yeast cytidylic acid (1 g., 1 mole). The mixture becomes warm (cool flask a little if necessary), and the acid dissolves completely within 20 minutes, giving a clear solution which is kept overnight at room temperature, then evaporated under reduced pressure; the residue is triturated with dry ether (30 ml.), falling to a white powder. The product is collected by centrifugation and washed with ether, and traces of solvent are removed in a vacuum desiccator; it is then added portionwise to stirred, ice-cold saturated ethanolic ammonia (35 ml.), and the mixture left for 30 minutes and then centrifuged. The solid is washed with ice-cold ethanol (10 ml.), and the combined supernatant liquid and washings are concentrated to small bulk *in vacuo* at 20°. The product (0.47 g.) is precipitated by ether (ca. 30 ml.), collected, washed once with ether containing ethanol (9:1), and dried in a desiccator; it is free of cytidylic acid, but in addition to the ammonium salt of the cyclic phosphate it contains an appreciable amount of ammonium cytidine ethyl phosphate. Separation of the components is achieved by means of chromatography on a cellulose column. The column (29.5 × 5.5 cm.) is prepared by allowing powdered cellulose (90 mesh Whatman; 220 g.) to settle from acetone suspension, washing with a solution of 8-hydroxyquinoline (1 mg./g. of cellulose) in acetone, then with acetone followed by isopropanol, and finally the solvent system used for chromatography, isopropanol-water-ammonia (d. 0.880)(80:18:2). The crude ammonium salt (0.36 g.) is dissolved in a small amount of solvent mixture, to which a little water has been added, and applied evenly to the top of the column, and the chromatogram is developed (8 hours) with the above solvent system. The first 500 ml. of eluate is discarded, the next 150 ml. collected in 3 × 50-ml. fractions, and subsequently 50 fractions of 20 ml. each are collected. The contents of each fraction are ascertained by paper strip chromatography. The peak of elution of the cytidine ethyl phosphate occurs in numbers 12 to 15 of the 20-ml. fractions and of the cyclic phosphate in fraction 45.

Fractions 28 to 50 (of the 20-ml. fractions) are combined and evaporated as above, and the residue is converted into a white powder (0.22 g.) by trituration with acetone. It is purified by dissolution in methanol and reprecipitation with ether; the final product is extremely hygroscopic.

The barium salt is prepared as described for the guanosine compound. In this case, however, methanol is used in place of acetone for the precipitation of the salt from concentrated aqueous solution. Light absorption in water: $\lambda_{max.}$ at 268 and 232 mμ (ϵ, 8400, 8150); $\lambda_{min.}$ at 250 and 223 mμ (ϵ, 6900, 7750) for $M = 373$. As with the other nucleoside-2',3' phosphates, electrometric titration clearly demonstrates the absence of secondary phosphoryl dissociation.

The compound has an R_f value of ca. 0.36 when chromatographed on paper strips in the isopropanol-ammonia-water system (70:5:25 v/v), and of ca. 0.81 in the 5% disodium hydrogen phosphate-isoamyl alcohol system of Carter.[9] The order of elution of the isomeric cytidylic acids and cyclic phosphate from a column of Dowex 2 resin (200 to 400 mesh; formate form), with 0.02 N formic acid as eluant, is as follows: cytidine-2' phosphate, cytidine-3' phosphate = cytidine-2',3' phosphate (peak optical density ratio, 280 mμ/260 mμ = 1.62).

The compound is easily hydrolyzed by aqueous acids or alkalis and is stored under anhydrous conditions.

Cytidine-2' and -3' Benzyl Phosphates.[16] Cytidylic acid (0.5 g. from yeast nucleic acid) is boiled under reflux with 80% acetic acid (10 ml.) for 30 minutes, in order to convert it to a mixture of the 2'- and 3'-isomers. The clear solution is evaporated to dryness under reduced pressure, and last traces of acetic acid are removed by distillation with added water. The residue is dissolved in water (5 ml.), and phenyldiazomethane (from 2.4 g. of benzaldehyde hydrazone) in dimethylformamide (7 ml.) is added dropwise with vigorous shaking. The reaction mixture is kept at room temperature during the addition by external cooling. The solution is shaken for 2 hours, excess barium carbonate and water (10 ml.) added, and further shaken. After extraction with chloroform (4 × 20 ml.), the aqueous phase is filtered through Hyflo Super-Cel and evaporated to dryness under reduced pressure (1 mm.). The residual gum is triturated with dry acetone and the solid collected, yielding the mixed barium cytidine-2' and -3' benzyl phosphates (0.51 g.), which is contaminated with some cytidine methyl phosphate. Ion exchange chromatography permits the separation of the esters.

The above barium salt (0.64 g.) is dissolved in water (50 ml.), and the solution brought to pH 8 and run onto a column (10.5 × 3 cm.) of Dowex 2 (formate). After washing the column with water, elution is commenced with 0.02 N formic acid. The first product, cytidine methyl phosphate (ca. 50 mg.), is removed (volume to peak, 400 ml.) and may be worked up by evaporation at reduced pressure and crystallization from water. Elution is continued to remove traces of cytidylic acid and cytidine-2',3' phosphate. 0.1 N Formic acid removes cytidine-2' benzyl phosphate from the column, and then cytidine-3' benzyl phosphate is eluted with 0.5 N formic acid. Progress of elution is followed by observing optical densities of the fractions (ca. 20 ml.) at 260 mμ. The fractions containing the separated products are bulked and evaporated under reduced pressure, with additions of water at the later stages to reduce the formic acid concentration.

Cytidine-2' benzyl phosphate crystallizes from water containing a little ethanol in colorless needles (50 g.), sintering at 168°, melting point 174°.

Cytidine-3' benzyl phosphate separates from water in small prisms (66 mg.) which slowly decompose from 170° with a final melting point at 203°. Both substances have the same R_f value, 0.7 in the isopropanol-ammonia-water (70:10:20 v/v) solvent; the methyl ester has $R_f = 0.5$.

B. Deoxyribonucleotides

Thymidine Phosphates[18]

5'-O-Trityl Thymidine. Triphenylmethyl chloride (3.5 g.) is added to a solution of anhydrous thymidine (2.5 g.) in dry pyridine (50 ml.), and the mixture left at room temperature for 1 week. It is then cooled to 0° and poured into ice water (500 ml.) with vigorous stirring. The precipitate is washed with water and dried *in vacuo* (P_2O_5). The product is next dissolved in acetone (5 ml.), and dry benzene (35 ml.) is added. After

[18] A. M. Michelson and A. R. Todd, *J. Chem. Soc.* **1953**, 951.

filtration, the acetone is boiled off, and the resulting solution cooled. 5'-O-Trityl thymidine separates as colorless needles (4 g., 80%), melting point 128°, $[\alpha]_D^{13} + 19.2°$ (c, 1.1 in 95% EtOH).

Thymidine-3' Phosphate. A solution of 5'-O-trityl thymidine (6.3 g.) in dry pyridine (65 ml.) is cooled to just above its melting point, and dibenzyl phosphorochloridate (from 10 g. of dibenzyl phosphite) is added. The mixture is kept at this temperature with occasional shaking during 6 hours, then left at 0° overnight. Aqueous sodium carbonate (5 g. in 30 ml. of water) is added, the mixture evaporated under reduced pressure, and the residue shaken with chloroform and aqueous sodium hydrogen carbonate. The chloroform layer is further washed with sodium hydrogen carbonate, then with water, dried (Na_2SO_4), and evaporated to a cream-colored glass.

The glass (7.6 g.) is boiled in acetic acid (50 ml. of 80%) for 7 minutes. Acetic acid is removed under reduced pressure, the residue neutralized with aqueous barium hydroxide, and triphenylmethanol removed by extraction several times with chloroform. Barium is removed from the aqueous solution by titration with sulfuric acid (rhodizonic acid) and centrifugation, and the solution concentrated to small bulk under reduced pressure and freeze-dried.

The residue is hydrogenated in aqueous ethanol (100 ml. of 75%) at room temperature at 1 atm. over a mixture of palladium and palladized charcoal catalysts. Catalyst is removed, and the filtrate concentrated to small bulk under reduced pressure, then brought to pH 7.5 with saturated aqueous barium hydroxide. The precipitate (mainly barium phosphate) is centrifuged off, washed well with water, the combined supernatant liquids concentrated to small bulk under reduced pressure, and the solution treated with aqueous lead acetate at pH 6.8. The lead salt precipitate is centrifuged off, washed well with water, and decomposed with hydrogen sulfide in the usual way. The product is converted into the barium salt, and the solution concentrated to 20 ml. under reduced pressure. The barium salt is precipitated on addition of ethanol (40 ml.) (0.5 g.). Barium thymidine-3' phosphate crystallizes slowly from concentrated aqueous solution as colorless needles, $[\alpha]_D^{20} + 7.3°$ (c, 1.5 in H_2O).

3'-O-Acetyl 5'-O-Trityl Thymidine. A solution of anhydrous 5'-O-trityl thymidine (3.75 g.) in pyridine (35 ml.) and acetic anhydride (8 ml.) is kept at room temperature for ca. 20 hours, then cooled to 0°, and poured into ice water (500 ml.) with vigorous stirring. The white amorphous precipitate is washed with water and dried (4.0 g., 98%; melting point ca. 90°). Recrystallized from benzene-light petroleum (boiling point 40 to 60°), 3'-O-acetyl 5'-O-trityl thymidine forms rosettes of needles, melting point 105°.

3'-O-Acetyl Thymidine. A solution of 3'-O-acetyl 5'-O-trityl thymidine (3.7 g.) in acetic acid (12.5 ml. of 80%) is heated under reflux for 10 minutes, cooled to room temperature, and diluted with ice water (230 ml.). The precipitate of triphenylmethanol is filtered off, and the filtrate taken to dryness under reduced pressure at <30°, giving a crystalline mass. Recrystallized from acetone or acetone-light petroleum (boiling point 40 to 60°), 3'-O-acetyl thymidine forms needles (1.8 g., 90%; melting point 176°).

Thymidine-5' Phosphate. Dibenzyl phosphorochloridate (from 5 g. of dibenzyl phosphite) is added to a solution of 3'-acetyl thymidine (2.01 g., dried at 110°/1 mm. for 12 hours) in anhydrous pyridine (25 ml.) at −30°. The mixture is kept just above its freezing point for 3 hours and then left at room temperature overnight. Water (15 ml.) and sodium carbonate (2.5 g.) are added, and the mixture is evaporated to dryness under reduced pressure. The residue is dissolved in chloroform, washed with aqueous sodium hydrogen carbonate, with water, and dried (Na_2SO_4); removal of the solvent under reduced pressure gives a gum which is evaporated twice with ethanol and dissolved in a little ethanol, and ether (200 ml.) is added to precipitate an oil. The oil is dissolved in acetone, and the solution filtered and evaporated under reduced pressure to a pale yellow glass (3.2 g.), consisting mainly of 3'-O-acetyl thymidine-5' dibenzyl phosphate.

A solution of this (1.75 g.) in aqueous ethanol (50 ml. of 75%) is hydrogenated at room temperature over a mixture of palladium and palladized charcoal catalysts. Catalyst is removed, and the solution of 3'-O-acetyl thymidine-5' phosphate brought to pH 11 with barium hydroxide and kept at 30° for 30 minutes to hydrolyze the acetyl group. The solution is then neutralized with carbon dioxide, boiled, and filtered. Lead acetate solution is added, and the gelatinous lead salt of the nucleotide is centrifuged off, washed well with water, and decomposed with hydrogen sulfide. The supernatant liquids from the precipitate of lead sulfide are concentrated under reduced pressure and finally lyophilized. The residue is dissolved in water (5 ml.), neutralized with barium hydroxide, filtered, and the barium thymidine-5' phosphate precipitated by addition of 2 vol. of ethanol. After being washed with ethanol and then ether and dried, the white amorphous solid is redissolved in water, clarified by centrifugation, and again precipitated by adding 2 vol. of ethanol; the salt is washed with ethanol, then ether, and dried (0.72 g.).

Thymidine Diphosphate.[19] A solution of thymidine (2.23 g.; dried for 12 hours at 110°/1 mm.) in dry pyridine (30 ml.) at −40° is treated with dibenzyl phosphorochloridate (from 10 g. of dibenzyl phosphite), kept

[19] C. A. Dekker, A. M. Michelson, and A. R. Todd, *J. Chem. Soc.* **1953**, 947.

just above the melting point of the mixture for 6 hours, and then left at room temperature overnight. Water (20 ml.) and sodium carbonate (5 g.) are added, and the mixture is evaporated to dryness under reduced pressure. The residue is dissolved in chloroform, washed with aqueous sodium hydrogen carbonate and then with water, and dried (Na_2SO_4); removal of the solvent under reduced pressure gives a gum which is evaporated twice with ethanol and finally dissolved in a small volume of ethanol. Addition of ether (300 ml.) gives a gummy precipitate which is washed with ether by decantation, dissolved in acetone, and evaporated to a cream-colored glass (5.6 g.) under reduced pressure.

A solution of the glass (4 g.) in aqueous ethanol (200 ml. of 50%) is hydrogenated at room temperature and atmospheric pressure, with a mixture of palladium and palladized charcoal catalysts. Catalyst is removed by filtration, and the solution is concentrated to small volume under reduced pressure. Water (100 ml.) is added, then saturated barium hydroxide solution to pH 8. After neutralization with carbon dioxide and centrifugation, the solution is evaporated to small bulk *in vacuo* and boiled for several minutes. The granular precipitate formed is filtered off, and the process repeated after concentration of the filtrate, to give a total of 1.35 g. of crude barium salt of thymidine-3',5' diphosphate. The tetra-brucine salt softens at 176° and melts at 182 to 184°.

Deoxycytidine Phosphates[20]

Deoxycytidine-3' Benzyl Phosphate. A solution of 5'-O-trityldeoxy-cytidine (10 g., see below) in dry pyridine (80 ml.) is cooled to just above its melting point, and dibenzyl phosphorochloridate (from 20 g. of di-benzyl phosphite) is added. The mixture is kept at ca. $-30°$ for 6 hours, then left at 0° overnight. Water (60 ml.) and sodium carbonate (12 g.) are added, the mixture evaporated under reduced pressure, and the residue shaken with chloroform and water. The chloroform extract is further washed with water, dried (Na_2SO_4), evaporated to a yellow glass which is dissolved in acetic acid (150 ml. of 80%), and the solution gently boiled for 7 minutes. Acetic acid is removed under reduced pressure, and water and chloroform are added to the residue together with sufficient ammonia to bring the pH to 8.5. After extraction of the aqueous layer three times with chloroform (chloroform extracts are discarded), the solution is adjusted to pH 9 and run onto a column (10 cm. $\times$ 5 sq. cm.) of Dowex 2 anion exchange resin (formate form). The column is eluted with water (deoxycytidine removed), then 0.02 M formic acid (mononucleotide material removed), and finally 0.15 formic acid. Two peaks are obtained with the last solvent: the first, corresponding to deoxycytidine-3' benzyl

[20] A. M. Michelson and A. R. Todd, *J. Chem. Soc.* **1954**, 34.

phosphate, has an optical density ratio 280 mμ/260 mμ of 1.9; the second, corresponding to cytosine-N benzyl phosphate, has an optical density ratio 280 mμ/260 mμ of 2.6 (in 0.15 M formic acid). Appropriate fractions containing deoxycytidine-3' benzyl phosphate are combined, taken to small volume under reduced pressure, and finally freeze-dried. The residue, recrystallized from water, gives deoxycytidine-3' benzyl phosphate (0.85 g.) as clusters of hydrated needles, melting point 100 to 101°; melting point after drying 150 to 151°.

Deoxycytidine-3' Phosphate. An aqueous ethanolic solution of deoxycytidine-3' benzyl phosphate (0.15 g.) is hydrogenated in the usual manner with a mixture of palladium and palladized charcoal catalysts. Deoxycytidine-3' phosphate crystallizes from aqueous ethanol in clusters of needles (0.105 g.), melting point 196 to 197° (decomp.).

5'-O-Trityldeoxycytidine. Triphenylmethyl chloride (14 g.) is added to a suspension of anhydrous deoxycytidine (5.9 g.) in dry pyridine (160 ml.), and the mixture shaken vigorously at room temperature until a clear solution is obtained (approximately 2 hours), then set aside at room temperature for 1 week, with exclusion of moisture. The solution is cooled to 0° and poured into ice water (1200 ml.) with vigorous stirring. The precipitate is collected, washed with water, and dried. The product is dissolved in acetone containing a little methanol, and the solution filtered; on cooling, the filtrate deposits 5'-O-trityldeoxycytidine as small needles (10.8 g., 89%), melting point 239°.

N,O$^{3'}$-Diacetyl-O$^{5'}$-trityldeoxycytidine. A solution of anhydrous 5'-O-trityldeoxycytidine (2.9 g.) in dry pyridine (40 ml.) and acetic anhydride (10 ml.) is kept at room temperature for ca. 20 hours, then cooled to 0°, and poured into ice water (500 ml.) with vigorous stirring. The colorless precipitate is collected, washed with water, and dried. Recrystallized from methanol, N,O$^{3'}$-diacetyl-O$^{5'}$-trityldeoxycytidine forms long needles (3.0 g., 88%), melting point 196°.

N,O$^{3'}$-Diacetyldeoxycytidine. A solution of N,O$^{3'}$-diacetyl-O$^{5'}$-trityldeoxycytidine (4.05 g.) in acetic acid (15 ml. of 80%) is heated under reflux for 5 minutes, then the acetic acid evaporated under reduced pressure below 30°. The residue is triturated with ether (100 ml.), the mixture kept at 0° for 1 hour, and the ether decanted off. The gummy solid contains two main components, corresponding to a mono- and a diacetate, as well as traces of cytosine and N-acetylcytosine. Purification is effected by countercurrent separation with ethyl acetate-water. The fractions containing the diacetate are combined and evaporated to dryness under reduced pressure. The residue crystallizes from acetone-light petroleum (boiling point 40 to 60°) as rosettes of small needles (1.04 g.), melting point 170°. Recrystallized from water, N,O$^{3'}$-diacetyldeoxycytidine forms long thin needles, melting point 171°; λ_{max}. 247, 296 mμ; λ_{min}. 227, 270 mμ; optical ratio 280/260 mμ, 0.776 in 0.015 M H·CO₂H.

Deoxycytidine-5′ Benzyl Phosphate. Dibenzyl phosphorochloridate (from 4 g. of dibenzyl phosphite) is added to a solution of anhydrous N,O³′-diacetyldeoxycytidine (0.90 g.) in anhydrous pyridine (10 ml.) at −30°. The mixture is kept just above its melting point for 6 hours and then left at 0° overnight. Water (20 ml.) and sodium carbonate (3 g.) are added, and the mixture is evaporated to dryness under reduced pressure. The residue is dissolved in chloroform, washed with aqueous sodium hydrogen carbonate, and then with water, and dried (Na₂SO₄); removal of the solvent under reduced pressure gives a thick oil (2.3 g.) which is dissolved in a mixture of dry benzene (10 ml.) and 4-methylmorpholine (20 ml.) and kept at 100° for 2 hours to effect monodebenzylation. Solvent is removed under reduced pressure, the residue dissolved in water (50 ml.), the deep-yellow solution extracted three times with chloroform, and the chloroform extracts discarded. The aqueous solution is adjusted to pH 10 with aqueous ammonia, kept at this pH at room temperature for 12 hours, then run onto a column (10 cm. × 5 sq. cm.) of Dowex 2 anion exchange resin (mesh size 200 to 400) in the formate form, and the column washed well with water. Elution is continued with 0.025 M formic acid (approximately 2 ml./min.), and the eluate collected in 20-ml. fractions in an automatically operated fraction collector. The appropriate fractions (optical density ratio 280 mμ/260 mμ = 2.1) are combined and evaporated to small volume under reduced pressure (bath temperature below 30°) and finally freeze-dried. The residue is dissolved in water and filtered, and the filtrate is evaporated to dryness under reduced pressure to give deoxycytidine-5′ benzyl phosphate as a colorless glass (0.73 g.).

Deoxycytidine-5′ Phosphate.[20] A solution of deoxycytidine-5′ benzyl phosphate (0.4 g.) in aqueous ethanol (100 ml. of 50%) is hydrogenated with a palladium catalyst at room temperature and pressure. Catalyst is removed by filtration, the filtrate concentrated to small bulk under reduced pressure and filtered, and 2 vol. of ethanol added to the filtrate. Deoxycytidine-5′ phosphate crystallizes as small needles (0.30 g., 97%), melting point 183 to 184° (decomp.).

Deoxycytidine-3′,5′ Diphosphate.[19] A solution of deoxycytidine (0.97 g.) in dry pyridine (50 ml.) is treated with dibenzyl phosphorochloridate (from 5 g. of dibenzyl phosphite) and worked up as in the phosphorylation of thymidine. After hydrogenation, the mixture is adjusted to pH 9 and run onto a Dowex 2 column (chloride form). Mononucleotide material is removed with 0.005 N hydrochloric acid; then the diphosphate is eluted with 0.009 N acid. The fractions containing deoxycytidine diphosphate are united and concentrated at room temperature under reduced pressure to small bulk (50 ml.), and the product is isolated by freeze-drying this solution. The residue is dissolved in a little water, and a solution of excess of brucine in methanol added. After evaporation to dryness under reduced pressure, the resultant crystalline mass is recrystallized twice from 80% ethanol and once from 35% ethanol, giving large hydrated needles of the tetrabrucine salt (325 mg.) of deoxycytidine-3',5' diphosphate; sinters at 180°, melts at 185°.

Section VI

Coenzymes and Related Phosphate Compounds

[114] General Procedure for Isolating and Analyzing Tissue Organic Phosphates

By CARLOS E. CARDINI and LUIS F. LELOIR

Most of the classical work on the isolation of organic phosphates has been carried out by making use of the different solubility of the barium, mercury, and lead salts of the compounds. Now that better methods are available, these classical methods still find use in the prepurification of the mixtures. The purpose of this section is to give an outline of the methods which have been used.

Extraction

This first step consists in the separation of most of the proteins. The usual procedure for preparative purposes is to introduce the material into 0.5 to 1 vol. of boiling water, maintain the temperature at about 80° for 5 to 10 minutes, cool rapidly, and filter.

An alternative is to add about 1 vol. of cold 10% trichloroacetic acid, filter, and re-extract the residue with 0.5 vol. of 5% trichloroacetic acid.

Another procedure consists in adding 1 vol. of 95% ethanol to the material, heat, then cool and filter. This method has the advantage of eliminating glycogen.

The procedure described by Warburg and Christian[1] for the extraction of FAD is very convenient because the volume of the solution is greatly reduced in one step. The method consists in adding ammonium sulfate (50 g. per 100 ml.) and extracting with phenol. The pyridine nucleotides and FAD are extracted, leaving the sugar phosphates in the water phase.

Barium Salt Fractionation

This is one of the methods which has been most widely used for analytical as well as for preparative purposes. Although it is useful, it does not give clean separations, presumably because mixed salts of the different compounds are formed. In general, the solubility of the compounds depends on the number of barium atoms in the salt. Thus, compounds with more than one phosphate group or with a carboxyl besides the phosphate give barium salts which are insoluble in water. Neutral compounds with only one phosphate group give salts which are water-soluble and may be precipitated by the addition of 2 to 4 vol. of ethanol.

The compounds giving water-insoluble barium salts at pH 7 to 8 are

[1] O. Warburg and W. Christian, *Biochem. Z.* **298**, 150 (1938).

the following: ATP, ADP, fructose diphosphate, glucose diphosphate, mono- and diphosphoglyceric acids, 2,3-dihydroxy-3-phosphoadipic acid, phytic acid, inorganic phosphate, and pyrophosphate.

Some of the compounds giving water-soluble alcohol-insoluble barium salts are: hexose, pentose, and triose monophosphates, phosphopyruvic acid, phosphocreatine, mononucleotides, DPN, and TPN.

Some other compounds such as aminoethyl phosphate and 1,2-propanediol phosphate give barium salts which are not precipitated by alcohol.

The usual procedure[2-5] is as follows: The extract is adjusted to pH 8.2 (just pink to phenolphthalein), and an excess of barium acetate is added. The precipitate contains the water-insoluble barium salts contaminated with some of the water-soluble. In order to obtain a better separation, the precipitate may be dissolved in acid, more barium acetate added, and the precipitation at pH 8.2 repeated. The supernatants are mixed, and 2 to 4 vol. of ethanol is added. If necessary, the pH is readjusted to pH 8.2. The precipitate termed the water-soluble alcohol-insoluble fraction is separated by centrifugation.

In cases where glycogen is abundant and interferes with the barium salt fractionation, it may be convenient to add 1 vol. of ethanol to the acid extract. This precipitates the glycogen, leaving the phosphate esters in solution. The supernatant is then adjusted to pH 8.2, and barium acetate is added, followed by sufficient ethanol to make a total of 2 to 4 vol. The precipitate is then dissolved in acid and again neutralized to pH 8.2. Thus, the water-soluble and water-insoluble fractions are separated.

The water-insoluble barium salts may be further fractionated by dissolving in acid and separating the precipitates formed by gradual addition of alkali. An alternative procedure is to adjust the solution to pH 3.5 and separate the precipitates formed by gradual addition of alcohol. At this pH the secondary phosphate groups are not appreciably dissociated, so that the precipitates correspond to the acid salts. Such a fractionation has been found useful in the purification of fructose diphosphate.[6,7]

Fractionation of the calcium salts has also been used with essentially the same results as with the barium salts. In cases where it is desired to prepare the barium salt of compounds which are fairly soluble in alcohol

[2] N. O. Kaplan and D. M. Greenberg, J. Biol. Chem. 156, 511 (1944).

[3] G. A. LePage, in "Methods in Medical Research" (V. R. Potter ed.), Vol. I, p. 337. Year Book Publishers, Chicago, 1948.

[4] G. A. LePage, in "Manometric Techniques and Tissue Metabolism" (Umbreit, Burris, and Stauffer, eds.), p. 185. Burgess Publishing Co., Minneapolis, 1949.

[5] J. Sacks, J. Biol. Chem. 181, 655 (1949).

[6] M. G. MacFarlane, Biochem. J. 33, 565 (1939).

[7] C. Neuberg, H. Lustig, and M. A. Rothenberg, Arch. Biochem. 3, 33 (1943).

(for instance, UDPG), it is convenient to start with a concentrated solution, add the barium as barium bromide, and precipitate with a great excess of ethanol. The advantage of barium bromide is that it is soluble in ethanol. The same is true for calcium chloride.

Mercuric Salt Fractionation

Precipitation with mercuric salts affords a method for the separation of nucleotides from the other phosphate esters. The procedure usually consists in the addition of excess mercuric nitrate or acetate to the solution which is weakly acid to Congo red paper. The nucleotides can be recovered from the precipitate by decomposition with H_2S. A considerable amount of nucleotides may remain adsorbed to the mercuric sulfide and may be eluted by exhaustive washing with 0.01 M barium acetate.

Some compounds such as phosphopyruvate[8] are rapidly decomposed by mercuric ions. In addition to nucleotides, other substances are precipitated, such as proteins and some amino acids (tyrosine, tryptophan, cysteine, and cystine). Precipitation with mercury has been often used in procedures for the analytical separation of organic phosphates. For instance, in Kaplan and Greenberg's scheme the water-soluble barium salt fraction is treated with mercuric acetate and acetic acid to about 0.03 M concentration. Adenylic acid, inosinic acid, and other nucleotides are precipitated. The filtrate contains hexose monophosphates and glycerophosphate.

Lead Salt Fractionation

Addition of neutral lead acetate leads to the precipitation of some proteins, nucleotides, and hexose diphosphates. The degree of precipitation varies with the pH. DPN and AMP remain in the supernatant. The lead salts can be decomposed with H_2S or with sodium phosphate.

Silver Salt Fractionation

This has usually been used after purification by other methods. For instance, it is used for precipitating DPN after impurities have been removed with basic lead acetate. Good separations sometimes can be obtained. For example,[1] on adding silver nitrate to a solution of FAD and FMN at pH 4, only the former is precipitated. The silver salts can be decomposed with H_2S or with HCl.

Copper Salt Fractionation

The addition of cupric ions leads to the precipitation of most of the phosphate esters. In the presence of excess alkali, adenylic acid and

[8] O. Meyerhof and K. Lohmann, *Biochem. Z.* **273**, 60 (1934).

carbohydrate esters are precipitated, while phosphocreatine remains in solution.[9]

Cuprous ions have been used as a fairly specific precipitant for SH compounds. They were used by Hopkins[10] in the purification of GSH and also by Beinert et al.[11] for the preparation of CoA. The procedure as used for GSH consists in adding CuO_2 to a solution which is 0.5 N in sulfuric acid. The CuO_2 may be prepared by adding excess glucose to Fehling's solution and heating. When the cuprous oxide has settled, the liquid is decanted, and the solid is filtered, washed, and dried.

Manganese

Inorganic phosphate can be separated from pyrophosphate by precipitation of the latter as the manganous salt. This procedure has been used for analytical purposes.[12]

Alkaloid Salts

The alkaloid salts of organic phosphates usually crystallize well and may be used for the separation and purification of some compounds. Thus, brucine has been used for the purification of hexose monophosphates,[13-15] strychnine for fructose diphosphate,[16] and a crystallized quinine salt of DPN has been prepared.[17] The alkaloid salts are usually obtained by adding a slight excess of the free alkaloid to a concentrated solution of the free ester. The free ester is prepared by treating the barium salt with the required amount of sulfuric acid, by decomposition of the lead salt with H_2S, or by passing through cation exchange resin in the hydrogen form.

The free alkaloid is then added as a concentrated solution in ethanol or methanol, the required amount being ascertained by calculation or by adding alkaloid solution until the reaction becomes slightly alkaline.

On cooling to 0°, crystallization usually takes place and is often allowed to proceed for weeks. Further crystallization may be induced by evaporation in vacuo or by the addition of acetone. Microscopical exami-

[9] C. H. Fiske and Y. SubbaRow, J. Biol. Chem. 81, 629 (1929).
[10] F. G. Hopkins, J. Biol. Chem. 84, 269 (1929).
[11] H. Beinert, R. W. von Korff, D. E. Green, D. A. Buyske, R. E. Handschumacher, H. Higgins, and F. M. Strong, J. Am. Chem. Soc. 74, 854 (1952).
[12] A. Kornberg, J. Biol. Chem. 182, 779 (1950).
[13] R. Robison and E. J. King, Biochem. J. 25, 323 (1931).
[14] H. W. Kosterlitz, Biochem. J. 33, 1087 (1939).
[15] M. L. Wolfrom, C. S. Smith, D. E. Pletcher, and A. E. Brown, J. Am. Chem. Soc. 64, 23 (1942).
[16] C. Neuberg and O. Dalmer, Biochem. Z. 131, 188 (1922).
[17] K. Wallenfels and W. Christian, Angew. Chem. 64, 419 (1952); see Vol. III [125].

nation of the crystals may give a clue as to whether more than one substance is present, and recrystallization may be carried out with water or aqueous ethanol or acetone.

An alternative procedure is to add the theoretical amount of alkaloid sulfate to the barium salt of the ester and to remove the barium sulfate by centrifugation.

The conversion of the alkaloid salt of the ester to the free acid can be carried out with a cation exchange resin. For the preparation of the potassium salt the solution of the alkaloid salt is adjusted to pH 8.4 with potassium hydroxide and the precipitated alkaloid is removed by centrifugation. The rest of alkaloid may be extracted with chloroform.

Another method is to add barium acetate and ethanol to the alkaloid salt, so that the barium salt of the ester is precipitated.

The salts of other basic compounds have been used successfully in the crystallization of some esters. Thus, the hexylamine salts of deoxyribose-1-phosphate and fructose diphosphate[18,19] and the benzylamine salt of glucuronic acid 1-phosphate[20] have been prepared.

Countercurrent Distribution

The sparing solubility of many organic phosphates in organic solvents has prevented a general use of this method. It has been applied in the purification of DPN with phenol as solvent,[21] and it should be applicable to other dinucleotides such as FAD, TPN, and uridine diphosphate compounds. The sugar phosphates are much less soluble in organic solvents, but some separation has been obtained by adding fatty acid amines in order to increase the solubility in the organic phase.[22]

Anion Exchange Resins

Following the work of Cohn on the separation of nucleotides, this method is acquiring increased importance. Cohn[23] used a sulfonated polysterene resin (Dowex 1) in the chloride form and displaced the substances with solutions of increasing chloride content. The procedure gives very good separations. Essentially the same method has been used in the fractionation of yeast nucleotides. The resin can also be used in the formate form and the displacement carried out with solutions of increased formate concentration. The advantage of this latter procedure is that the

[18] M. Friedkin, *J. Biol. Chem.* **184**, 449 (1950).
[19] R. W. McGilvery, *J. Biol. Chem.* **200**, 835 (1953).
[20] O. Touster and V. H. Reynolds, *J. Biol. Chem.* **197**, 863 (1952).
[21] G. H. Hogeboom and G. T. Barry, *J. Biol. Chem.* **176**, 935 (1948).
[22] G. W. E. Plaut, S. A. Kuby, and H. A. Lardy, *J. Biol. Chem.* **184**, 243 (1950).
[23] W. E. Cohn, *J. Am. Chem. Soc.* **72**, 1471 (1950); see Vol. III [120].

pH of the effluent is not so low as with HCl. One difficulty in these procedures is that the substances emerge from the column in very dilute solution. Concentration has been effected by neutralizing, passing through a smaller column, and then displacing with more concentrated acid. In this manner the substance can be obtained in a small volume and may even crystallize in the column. Another method is to adsorb the substance on charcoal and then elute with aqueous ethanol or acetone-containing ammonia. A useful method for the separation of sugar esters and probably other compounds consists in using the anion exchange columns in the borate form and displacing with a borate solution. The polyhydroxy compounds form complexes with boric acid which have different acid strength and a different number of boric acid residues according to the number and position of the hydroxyl groups. This gives further possibilities of separation with the anion exchange columns.[24]

Adsorbents

Activated charcoal (Norit, Nuchar, etc.) adsorbs most of the nucleotide coenzymes. It is usually used batchwise, and elution is carried out with aqueous alcohol or acetone. If the substance is not eluted, ammonia may be added to these solvents. Aqueous pyridine is also a good eluant.

Activated charcoal has also been used for analytical purposes in the separation of nucleotides from other compounds such as inorganic phosphate, sugar phosphates, and acetyl phosphate, which are not adsorbed.[25]

[24] J. X. Khym and W. E. Cohn, *J. Am. Chem. Soc.* **75**, 1153 (1953).
[25] R. K. Crane and F. Lipmann, *J. Biol. Chem.* **201**, 235 (1953).

[115] Characterization of Phosphorus Compounds by Acid Lability

By Luis F. Leloir and Carlos E. Cardini

Measurements of the rate of hydrolysis of phosphoric esters have been carried out for analyzing mixtures, as a test of homogeneity, and as a criterion of identity. The application of this procedure became general after Lohmann[1] used it for distinguishing the 6-phosphate of glucose from that of fructose.

The rate of hydrolysis of phosphoric esters in acid solutions depends on several factors. The esterification of alcohol groups yields esters which are hydrolyzed with difficulty. This is the case with the α- and β-glycero-

[1] K. Lohmann, *Biochem. Z.* **194**, 306 (1928).

phosphates, 3-phosphoglyceric acid, and phosphorylcholine. In some cases the phosphate group appears to be released not by simple hydrolysis but by decomposition of the rest of the molecule. Thus triose phosphates yield methylglyoxal instead of the corresponding triose.[2] Migration of the phosphate group under the influence of acids has been detected in several instances. Such is the case with the α- and β-glycerophosphoric acids[3] and with the 2- and 3-phosphates of purine or pyrimidine ribosides.

Among the sugar phosphates the most stable are the 6-phosphoaldoses. The pentose phosphates are less stable and are affected by the nature of the substituents. Thus, although purine ribosides hydrolyze at the same rate as ribose phosphate, the pyrimidine phosphates are more stable.

The phosphate groups at the hemiacetal OH group of sugars are all acid-labile, and in the case of ribose and deoxyribose they are still more labile. If the OH at position 2 is substituted by an amino group, the stability is increased. For instance, galactosamine-1-phosphate is much more stable than galactose-1-phosphate.[4] This is also the case for the corresponding glucose derivatives.[5] The stabilizing effect of the amino group occurs not only with the phosphates but is general for hexosaminides as compared with the glycosides.[6]

When the phosphate is bound to a carboxyl or amino group, the compounds are very labile, as for instance acetyl phosphate and phosphocreatine.

The influence of the concentration of acid has not been studied adequately. Probably the rate of hydrolysis is proportional to the H+ concentration for most compounds. However, this is not the case for glucose-6-phosphate, which hydrolyzes at about the same rate at pH 2 and in 1 N solution of acid.[7] Another exception is ethanolamine phosphate, which has the maximum rate of hydrolysis at pH 4.5,[8] whereas the rate in 1 N acid is one-third as great.

It has become usual to separate the organic phosphates into two main groups. Those which are hydrolyzed completely in 1 N acid at 100° during 7 minutes are usually called labile. Those which are not hydrolyzed under the same conditions are called stable. The conditions were selected originally for estimating the two terminal phosphate groups in ATP. Another group of compounds estimate like inorganic phosphate in the

[2] O. Meyerhof and K. Lohmann, *Biochem. Z.* **271**, 89 (1934).

[3] E. Baer and M. Kates, *J. Biol. Chem.* **175**, 79 (1940).

[4] C. E. Cardini and L. F. Leloir, *Arch. Biochem. and Biophys.* **45**, 55 (1953).

[5] D. H. Brown, *J. Biol. Chem.* **204**, 877 (1953).

[6] M. Viscontini and J. Meier, *Helv. Chim. Acta* **35**, 807 (1952).

[7] R. Robison, *Biochem. J.* **26**, 2191 (1932).

[8] E. Cherbuliez and M. Bouvier, *Helv. Chim. Acta* **36**, 1200 (1953).

usual Fiske and SubbaRow procedure and may be called the extra-labile compounds. The separation in these three groups is arbitrary, and the properties of many compounds are intermediate. However, the classification is useful. Therefore, organic phosphates may be grouped as follows.

Extra-labile. Phosphocreatine, 1,3-diphosphoglyceric acid, ribose-1-phosphate, deoxyribose-1-phosphate, acyl phosphates.

Labile.[9] Adenosinetriphosphate (67% hydrolyzed), adenosinediphosphate (50%), uridine diphosphate (35%), aldose-1-phosphates (100%), fructose-1-phosphate (70%), fructose-1,6-diphosphate (32%), glucuronic acid 1-phosphate (100%), inorganic pyrophosphate (100%).

Stable. Phosphopyruvic acid (40%), hexose-6-phosphate, pentose-3- and 5-phosphates and the corresponding mono- and dinucleotides, 6-phosphogluconic acid, glycerol and glyceric acid phosphates, inositol phosphates, phosphorylcholine, and phosphorylethanolamine.

More information on the hydrolysis of different compounds may be obtained by an examination of the table. Some of the compounds listed as labile are hydrolyzed completely under milder conditions. For instance, for aldose-1-phosphates it is sufficient to heat for a few minutes at 100° in 0.1 N acid. Other compounds such as uridine diphosphate need 30 minutes in 1 N acid at 100° to reach complete hydrolysis of the labile phosphate. Phosphopyruvic acid is intermediate between stable and labile.

Liberation of the Phosphate Group by Methods Other Than Acid Hydrolysis

Methods which are more or less specific for removing the phosphate group in certain compounds are as follows: Phosphopyruvate may be estimated by a method based on the liberation of phosphate by hypoiodite[10] or by mercuric ions.[10] Dihydroxyacetone phosphate and glyceraldehyde phosphate lose their phosphate on standing at room temperature in 1 N alkali during 20 minutes.[11] Fructose-1-phosphate, fructose-1,6-diphosphate, and glucose-2-phosphate may be estimated by a method based on the liberation of the phosphate by heating with phenylhydrazine.[12,13] Alkaline hydrolysis leads to the liberation of phosphate from the sugar esters with a free reducing group; the 1-phosphates remain unaffected.

[9] The numbers in parentheses represent the amount of phosphate hydrolyzed in 7 minutes at 100° in 1 N acid.

[10] K. Lohmann and O. Meyerhof, *Biochem. Z.* **273**, 60 (1934).

[11] K. Lohmann and O. Meyerhof, *Biochem. Z.* **273**, 413 (1934).

[12] H. J. Deuticke and S. Hollmann, *Z. physiol. Chem.* **258**, 160 (1939).

[13] A. C. Paladini and L. F. Leloir, *Biochem. J.* **51**, 426 (1952).

Method for the Estimation of Phosphate

Of the many methods available for the estimation of phosphate, that of Fiske and SubbaRow is one of the simplest and most widely used. In certain cases, however, it is convenient to use procedures in which the phosphomolybdate complex is extracted with an organic solvent such as isobutanol,[14,15] thus avoiding the interference due to colored substances, citrates, oxalates, buffers, etc.

In the Fiske and SubbaRow procedure the extra-labile compounds are estimated as inorganic phosphate because of the relatively high acid concentration (pII 0.65) and because molybdate accelerates the hydrolysis of some organic phosphates.[15-17] By measuring the color immediately after adding the reagents, however, Fiske and SubbaRow[18] were able to estimate phosphocreatine. Better results are obtained by estimating the "true" inorganic phosphate by precipitating it with magnesium mixture or with calcium salts and ethanol. In the Lowry and López method[19] the acid and molybdate concentrations are lower than in Fiske and SubbaRow's, so that some extra-labile compounds are hydrolyzed more slowly. However, certain compounds such as deoxyribose-1-phosphate are hydrolyzed even under these conditions.

Method of Fiske and SubbaRow[20]

Reagents

5 N sulfuric acid.

2.5% ammonium molybdate.

2 N nitric acid.

Reducing reagent. This may be prepared in the powdered form and dissolved before use. The solution deteriorates slowly and should not be used after more than a week. The powdered reagent is prepared by mixing thoroughly 0.2 g. of 1-amino-2-naphthol-4-sulfonic acid with 1.2 g. of sodium bisulfite and 1.2 g. of sodium sulfite. For use 0.25 g. is measured with a small spoon and dissolved in 10 ml. of water.

Standard solution. 1.3613 g. of analytically pure KH_2PO_4 is dissolved in 1000 ml. of water, a few drops of chloroform are added,

[14] I. Berenblum and E. Chain, *Biochem. J.* **32**, 295 (1938).
[15] H. Weil-Malherbe and R. H. Green, *Biochem. J.* **49**, 286 (1951).
[16] F. Lipmann, *J. Biol. Chem.* **153**, 571 (1944).
[17] H. M. Kalckar, *J. Biol. Chem.* **167**, 477 (1947).
[18] C. H. Fiske and Y. SubbaRow, *J. Biol. Chem.* **81**, 629 (1929).
[19] O. H. Lowry and J. A. López, *J. Biol. Chem.* **162**, 421 (1946).
[20] C. H. Fiske and Y. SubbaRow, *J. Biol. Chem.* **66**, 375 (1925).

and the solution is stored in the refrigerator. For use it is diluted 1:10, so that 1 ml. corresponds to 1 micromole of phosphorus.

Deproteinization. The usual procedure is to use trichloroacetic acid (final concentration, 5 to 10%) or perchloric acid (final concentration, 8%).[21] Some extra-labile compounds such as acetyl phosphate are most stable at pH 5 to 6 and are not appreciably hydrolyzed if the trichloroacetic acid solution is added cold and if the procedure is carried out rapidly.

Procedure. The standard and unknowns should contain from 0.1 to 1 micromole of phosphate. One milliliter of sulfuric acid is added, followed by 1 ml. of molybdate. After mixing, 0.1 ml. of reducing solution is added. The volume is made up to 10 ml. After mixing again, the absorbency at 660 mμ is measured after 10 minutes.

When phosphocreatine is determined, the sample is left standing with the acid and molybdate for 20 minutes before the reducing reagent is added. All the phosphocreatine is hydrolyzed under these conditions.

It is important to avoid contamination with silicates which, like phosphate, give a blue color in the Fiske and SubbaRow procedure. Silicates are usually present in alkaline reagents stored in soft glass containers and are released by glass homogenizers.

Arsenate also estimates like phosphate, and procedures have been devised in order to avoid its interference.[22]

Estimation of Labile Phosphate. After the sulfuric acid is added in the Fiske and SubbaRow procedure, water is added to complete the 5 ml., that is, to make the concentration 1 N. The tubes are heated in a boiling water bath, usually for 7 minutes, cooled, and then the procedure is continued as described previously by adding the molybdate, etc.

Estimation of Total Phosphate. One milliliter of 5 N sulfuric acid is added to the sample as described in the Fiske and SubbaRow procedure. The mixture is evaporated in the test tube over a free flame. When the contents become brown and have cooled, 1 drop of 2 N nitric acid is added and the heating is continued until white fumes appear. If the liquid does not become colorless, the addition of nitric acid is repeated. Excess nitric acid interferes with the subsequent color development. After cooling, about 1 ml. of water is added and the tube is placed in a boiling water bath for 5 minutes. After cooling, molybdate is added, etc., as described in the inorganic phosphorus procedure.

[21] C. Neuberg, E. Strauss and L. E. Lipkin, *Arch. Biochem.* **4**, 101 (1944).
[22] L. B. Pett, *Biochem. J.* **27**, 1672 (1933).

Method of Lowry and López[19]

Reagents

Ascorbic acid, 1 g. in 100 ml. of water.

Ammonium molybdate, 1 g. in 100 ml. of 0.05 N sulfuric acid.

Acetate buffer, pH 4. 0.1 N in acetic acid and 0.025 N in sodium acetate.

0.1 N sodium acetate.

Standard solution. The same as that of the Fiske and SubbaRow method may be used.

Deproteinizing agent. (1) 5% trichloroacetic acid (0.3 N), (2) 3% perchloric acid (0.3 N) (3) saturated ammonium sulfate which is 0.1 N in acetic acid and 0.025 N in sodium acetate (pH 4).

Procedure. The sample to be analyzed is deproteinized under conditions which will not hydrolyze the particular ester; e.g., ice-cold 0.3 N trichloroacetic acid or 0.3 N perchloric acid, or, particularly with very labile esters, saturated ammonium sulfate with acetic acid and sodium acetate (pH 4). If either of the acid precipitants is used, the extracts are rapidly brought to pH 4 to 4.2 by adding 4 vol. of 0.1 N sodium acetate. Most of the labile esters are reasonably stable at this pH. The extracts are diluted with acetate buffer of pH 4 until the inorganic phosphorus is 0.015 to 0.1 mM. (0.015 to 0.10 μ moles/ml.). Ammonium sulfate extracts should be diluted at least fivefold. To each volume of extract is added 0.1 vol. of 1% ascorbic acid and 0.1 vol. of 1% ammonium molybdate in 0.05 N sulfuric acid. Readings are made at 5 minutes and again at 10 minutes after the molybdate addition at a wavelength of 700 mμ. (Any wavelength between 650 and 950 mμ is satisfactory.) Simultaneous readings are made on a standard (0.05 μ moles of phosphorus per milliliter) and a blank, both of the same composition, as far as possible, as the unknown. If a difference is observed in the readings of the unknown at 5 and at 10 minutes compared to the standard, the values are extrapolated to zero time. The ascorbic acid and molybdate may be combined before addition but must then be used within 15 minutes.

In the presence of certain tissue extracts, the reaction is delayed, in which case an internal standard must be used. A standard amount of inorganic phosphate is added to a duplicate tissue aliquot, and values of the unknown are calculated from the difference between the readings of the unknown and of the unknown with added phosphate. The inhibitory effect may be partially overcome by dilution. For example, in order to avoid undue inhibition, brain and muscle extracts should be diluted to a volume 150 to 250 times that of the tissue, and liver extracts to a volume 300 to 500 times that of the original liver. In addition, the molybdate concentration may be increased to 0.15%; i.e., 0.1 vol. of 1.5% ammonium molybdate in 0.05 N sulfuric acid is added in place of the 1% solution. This results in an acceleration of color development. In studying isolated enzyme systems, this problem of inhibition is ordinarily not encountered.

The ascorbic acid concentration may be increased, if necessary, to accelerate the reaction, but with final concentrations of greater than 0.2 to 0.3% the readings of the standard increase unduly with time. The final pH may be varied, if desired, between 3.5 and 4.2.

Removal of Inorganic Phosphate. 1. CALCIUM HYDROXIDE.[18] To the neutralized sample add 0.2 vol. of 10% $CaCl_2$ saturated with $Ca(OH)_2$. Let stand for 10 minutes at room temperature, centrifuge, and wash the precipitate with a small volume of water containing the $CaCl_2$ reagent. The washed precipitate may be dissolved in dilute HCl for the estimation of inorganic phosphate.

2. CALCIUM CHLORIDE AND ALCOHOL. This method has been used by Lipmann and Tuttle[23] for the separation of inorganic phosphate from acetyl phosphate.

Reagents

3.3% solution of anhydrous $CaCl_2$ in 33% ethanol.
Neutralization solution. A mixture of 100 ml. of concentrated ammonia and 40 ml. of glacial acetic acid is made up to 1 l., and to that 100 ml. of 0.4 M bicarbonate solution is added.

Procedure. Trichloroacetic acid extract (0.5 ml.) containing about 1 to 4 micromoles is transferred to a chilled test tube containing a drop of thymol blue, and the neutralization mixture is added quickly, with local alkalinization carefully avoided, until pH 8 is reached (grayish blue color).

To the neutralized sample 2.5 ml. of alcoholic $CaCl_2$ solution is added. When the material is mixed, the color of the indicator usually turns more yellowish but should retain a bluish tinge. In order to make the precipitate bulkier, especially if only small amounts of phosphorus are present, 0.15 ml. of 0.04 M bicarbonate solution is added dropwise. The calcium precipitate (phosphate plus carbonate) is quickly centrifuged off; 1 to 2 minutes is sufficient. The supernatant is decanted carefully, the adhering fluid is washed off with 2 ml. of the alcoholic calcium chloride solution, and the precipitate is recentrifuged without stirring.

The calcium precipitate is dissolved in 0.5 ml. of 0.5 N hydrochloric acid and quantitatively brought into a volumetric flask of convenient size. Phosphorus is determined by the procedure of Fiske and SubbaRow.

[23] F. Lipmann and L. C. Tuttle, *J. Biol. Chem.* **153**, 571 (1944).

Observed Hydrolysis Constants of Some Phosphates[a]

Compound	Normality of acid	Temperature °C.	$t_{1/2}$, min.	$K \times 10^3$	Reference
Aldose-1-phosphates					
Deoxyribose-1-phosphate (pH 4)		25	12	25	b
Ribose-1-phosphate	0.5	25	2.5	120	c
Xylose-1-phosphate	0.1	36	111	2.7	d
α-D-Glucose-1-phosphate	0.1	36	158	1.9	d
α-D-Glucose-1-phosphate	0.25	37	230	1.30	e
α-D-Glucose-1-phosphate	1	33	60	5.00	f
α-D-Glucose-1-phosphate	0.95	30	200	1.15	g
α-Glucose-1-phosphate	1	100	1.05	200	h
β-D-Glucose-1-phosphate	1	33	20	15	g
α-Galactose-1-phosphate	0.25	25	333	0.90	i
α-Galactose-1-phosphate	0.25	37	50	5.9	i
α-Galactose-1-phosphate	0.1	100	~2.1	~140	j
β-Galactose-1-phosphate	0.25	37	53	5.6	k
α-Mannose-1-phosphate	0.95	30	360	0.82	g
Galactosamine-1-phosphate	1	100	4.1	73.7	i
Glucosamine-1-phosphate	1	100	~4	~75.0	l
Maltose-1-phosphate	0.1	36	214	1.4	d
Other Sugar Phosphates					
Triose phosphates	1	100	8.1	37	m
Erythrulose-1-phosphate	1	100	30	~9.7	n
Deoxyribose-5-phosphate	1	100	6.2	56	o
Ribose-3-phosphate	0.01	100	180	~1.7	p
Ribose-3-phosphate	0.25	100	66	~4.5	q
Ribose-5-phosphate	0.01	100	1000	~0.3	p
Ribose-5-phosphate	0.25	100	600	~0.5	q
Xylose-5-phosphate	1	100	90	3.3	r
Ribulose-5-phosphate	1	100	60	~5	s
D-Glucose-2-phosphate	0.1	100	136	2.18	t
Glucose-6-phosphate	0.1	100	2300	0.13	u
Glucose-6-phosphate	1	100	1300	0.22	v
6-Phosphogluconic acid	1	100		0.26–0.15	w
Fructose-1-phosphate	1	100	2.8	70	x
Fructose-1-phosphate	0.1	100	33	9	x
Fructose-6-phosphate	1	100	70	4.36	v
Mannose-6-phosphate	0.1	100	2300	0.13	v
Mannose-6-phosphate	1	100	1034	0.29	v
6-Phosphomannonic acid	1	100		0.199–0.131	w
6-Phosphomannonic acid ψ-lactone	1	100	2500	0.12	w
L-Sorbose-6-phosphate	1	100	62	4.8	v
Ketoheptose monophosphate	1	100	75	4	h

OBSERVED HYDROLYSIS CONSTANTS OF SOME PHOSPHATES[a] (*Continued*)

Compound	Normality of acid	Temperature, °C.	$t_{1/2}$, min.	$K \times 10^3$	Reference
α-Glucose-1,6-diphosphate	0.25	37	967	(a) 0.31	z
α-Glucose-1,6-diphosphate	1	30	900	(a) 0.33	aa
β-Glucose-1,6-diphosphate	1	30	220	(a) 1.37	aa
α-Mannose-1,6-diphosphate	0.95	30	1420	0.21	g
Fructose-1,6-diphosphate	1	100	5.7	(a) 52 (b) 4.2	bb
Other Phosphate Esters					
α-Glycerophosphate	1	100	3300	0.09	h
α-Glycerophosphate	2.14	127	300	0.98	cc
β-Glycerophosphate	2.035	124	300	0.97	cc
3-Phosphoglyceric acid	1	100	2140	0.14	h
Phosphorylcholine	1	100	2300	0.13	cc
Phosphorylcholine	2	124	1870	0.16	cc
Aminoethyl phosphoric acid	1	100	~940	~0.32	dd
Aminoethyl phosphoric acid	4.5	100	300	1.0	dd
Propanediol phosphate	5	100	540	0.55	ee
Enol Phosphate					
Phosphopyruvic acid	1	100	8.3	36	h
Acid Anhydrides					
Pyrophosphoric acid	1	100	1.2	250	h
Acetyl phosphate	0.5	40	11	27.8	ff
Acetyl phosphate + molybdate	0.5	25	0.86	350	ff
1,3-Diphosphoglyceric acid	Neuter	38	27	11	gg
Amide Phosphates					
Phosphocreatine	1.0	20	150	2.1	hh
Phosphocreatine	0.5	25	4	75	c
Nucleotides					
Uridine-3′-phosphate	0.1	100	720	~0.4	ii
Cytidine-3′-phosphate	0.1	100	600	~0.5	ii
Adenosine-3′-phosphate	0.1	100	60	~5	ii
Guanosine-3-phosphate	0.1	100	180	~1.60	ii
Uridine-5′-phosphate	0.1	100	3590	0.085	ii
Adenosine-5′-phosphate	0.1	100	1050	0.26	ii
Guanosine-5′-phosphate	0.1	100	1050	0.26	ii
ATP	0.1	100	8	38	kk
UDPG	0.1	100	20	~15	kk
DPN	0.1	100	780	~0.38	ll

FOOTNOTES TO TABLE ON PP. 847–848

[a] The constants are calculated with the formula $K = \frac{I}{t} \log_{10} \frac{a}{a-k}$, or more usually

$K = \frac{I}{t_2 - t_1} \log_{10} \frac{a - x_1}{a - x_2}$. The time is in minutes, and a is the initial concentration of the substance. The time for 50% hydrolysis, $t_{1/2} = 0.30/K$; the time for 98% hydrolysis, $t = 1.7/K$.

[b] M. Friedkin, H. M. Kalckar, and E. Hoff-Jørgensen, J. Biol. Chem. **178**, 527 (1949).

[c] H. M. Kalckar, J. Biol. Chem. **167**, 477 (1947).

[d] W. R. Meagher and W. Z. Hassid, J. Am. Chem. Soc. **68**, 2135 (1946).

[e] C. F. Cori, S. P. Colowick, and G. T. Cori, J. Biol. Chem. **121**, 465 (1937).

[f] M. L. Wolfrom, C. S, Smith, D. E. Pletcher, and A. E. Brown, J. Am. Chem. Soc. **64**, 23 (1942).

[g] T. Posternak and J. P. Rosselet, Helv. Chim. Acta **36**, 1614 (1953).

[h] R. Robison and M. G. MacFarlane, in "Methoden der Fermenforschung" (Bamann and Myrbäck, eds.), Vol. 1, p. 296. G. Thieme, Leipzig, 1941.

[i] H. W. Kosterlitz, Biochem. J. **33**, 1087 (1939).

[j] C. E. Cardini and L. F. Leloir, Arch. Biochem. and Biophys. **45**, 55 (1953).

[k] F. J. Reithel, J. Am. Chem. Soc. **67**, 1056 (1945).

[l] D. H. Brown, J. Biol. Chem. **204**, 877 (1953).

[m] O. Meyerhof and K. Lohmann, Biochem. Z. **271**, 79 (1934).

[n] F. C. Charalampous and G. C. Mueller, J. Biol. Chem. **201**, 161 (1953).

[o] E. Racker, J. Biol. Chem. **196**, 347 (1952).

[p] P. A. Levene and E. T. Stiller, J. Biol. Chem. **104**, 299 (1934).

[q] H. G. Albaum and W. W. Umbreit, J. Biol. Chem. **167**, 369 (1947).

[r] P. A. Levene and A. L. Raymond, J. Biol. Chem. **102**, 347 (1933).

[s] B. L. Horecker, P. Z. Smyrniotis, and J. E. Seegmiller, J. Biol. Chem. **193**, 383 (1951).

[t] K. R. Farrar, J. Chem. Soc. **1949**, 3131.

[u] R. Robison and E. J. King, Biochem. J. **25**, 323 (1931).

[v] R. Robinson, Biochem. J. **26**, 2191 (1932).

[w] V. R. Patwardhan, Biochem. J. **28**, 1854 (1934).

[x] B. Tanko and R. Robison, Biochem. J. **29**, 961 (1935).

[y] K. M. Mann and H. A. Lardy, J. Biol. Chem. **187**, 339 (1950).

[z] C. E. Cardini, A. C. Paladini, R. Caputto, L. F. Leloir, and R. E. Trucco, Arch. Biochem. **22**, 87 (1949).

[aa] T. Posternak, J. Biol. Chem. **180**, 1269 (1949).

[bb] M. MacLeod and R. Robison, Biochem. J. **27**, 286 (1933).

[cc] O. Meyerhof and W. Kiessling, Biochem. Z. **264**, 40 (1933).

[dd] E. Cherbuliez and M. Bouvier, Helv. Chim. Acta **36**, 1200 (1953).

[ee] O. N. Miller, C. G. Huggins, and K. Arai, J. Biol. Chem. **202**, 263 (1953).

[ff] F. Lipmann and L. C. Tuttle, J. Biol. Chem. **153**, 571 (1944).

[gg] E. Negelein and H. Brömel, Biochem. Z. **303**, 132 (1939).

[hh] K. Lohmann, Biochem. Z. **194**, 306 (1928).

[ii] A. M. Michelson and A. R. Todd, J. Chem. Soc. **1949**, 2476.

[jj] P. A. Levene and R. S. Tipson, J. Biol. Chem. **106**, 113 (1934).

[kk] R. Caputto, L. F. Leloir, C. E. Cardini, and A. C. Paladini, J. Biol. Chem. **184**, 333 (1950).

[ll] F. Schlenk, J. Biol. Chem. **146**, 619 (1942).

3. MAGNESIA MIXTURE

Reagent. Dissolve 5.5 g. of $MgCl_2 \cdot 6H_2O$ and 10 g. of NH_4Cl in about 50 ml. of water. Add 10 ml. of 15 M ammonia and make up to 100 ml. Filter if necessary. One milliliter of the solution corresponds theoretically to 270 micromoles of phosphoric acid.

Procedure. The sample is neutralized with 10% ammonia and excess magnesia mixture is added. The solution should be about 1.5 M with respect to NH_4OH. The solution is stored for 2 to 3 hours in the refrigerator. The filtrate can then be neutralized with HCl and used for the estimation of extra-labile esters.

[116] Determination and Preparation of N-Phosphates of Biological Origin

By A. H. ENNOR

I. Phosphocreatine

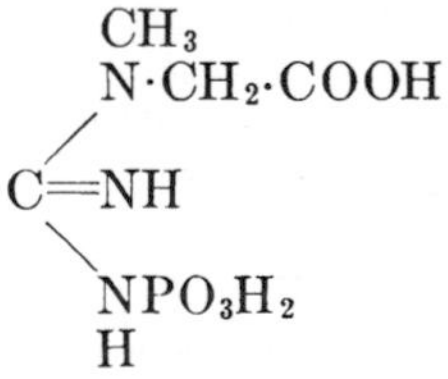

$$CH_3$$
$$N \cdot CH_2 \cdot COOH$$
$$C = NH$$
$$NPO_3H_2$$
$$H$$

Determination

Because of the ease with which the N-P bond may be broken in acid solution, the methods generally employed for the determination of phosphocreatine (PC) depend on the estimation of the P moiety. Thus a commonly employed technique is that described by LePage,[1] which depends on the determination of the P_i by the method of Fiske and SubbaRow[2] after precipitation as a Ca^{++} salt from alkaline solution. This figure is then subtracted from that obtained by a direct estimation of the P_i in the extract, since this latter is assumed to represent the sum of P_i and PC-P because of the lability of the PC in the H_2SO_4-ammonium molybdate reagent. This method lacks the sensitivity desirable if low concentrations of PC are present and has the further disadvantage that its accuracy depends on the quantitative precipitation of P_i as the Ca salt

[1] G. A. LePage *in* "Manometric Techniques and Related Methods for the Study of Tissue Metabolism" (Umbreit, Burris, and Stauffer, eds.), p. 185. Burgess Publishing Co., Minneapolis, 1951.

[2] C. H. Fiske and Y. SubbaRow, *J. Biol. Chem.* **66**, 375 (1925).

which is appreciably soluble at the recommended pH = 8.8. In addition it depends on the assumption that PC is the only organophosphate present which is hydrolyzed by acid-molybdate, and this is doubtful (cf. Ennor and Rosenberg[3]).

These disadvantages may be overcome by estimating the creatine moiety, using the method of Ennor and Stocken[4] as applied to the determination of PC by Ennor and Rosenberg.[3]

Reagents

Stock alkali. 80 g. of Na_2CO_3 + 30 g. of NaOH dissolved in water and made to 500 ml.

Diacetyl. A stock solution of approximately 1% may be prepared from dimethylglyoxime as described by Walpole.[5] When stored at 0°, it retains its activity for at least 12 months and should be diluted with water ($\frac{1}{20}$) before use.

α-Naphthol, 1% solution in stock alkali. This solution must be freshly prepared before use.

HCl, 0.4 N.

NaOH, 0.4 N.

Na p-chloromercuribenzoate, 0.05 M.

Method. The method depends on the determination of creatine before ("free creatine") and after ("total creatine" = "free creatine" + "bound creatine") acid hydrolysis. The difference between these two determinations gives "bound creatine," which is assumed to have its origin in phosphocreatine.

Tissues under examination should be dropped into liquid air or solid CO_2-ethanol immediately after excision. TCA extracts must be prepared in the cold and made to volume after adjustment to pH 7 to 7.3 by the addition of N NaOH.

Samples of 1.0 to 3.0 ml., containing 2 to 10 γ of creatine per milliliter, are pipetted into 10-ml. graduated tubes, one of which is retained for the determination of free creatine. The other is made to 3.0 ml. by the addition of the requisite volume of H_2O and placed in a water bath at 65°. After equilibration 1.0 ml. of 0.4 N HCl is added, and the contents are mixed and allowed to remain in the bath for 9 minutes.[6] Then 1.0 ml. of 0.4 N NaOH is added, and the tube and contents are rapidly cooled to room temperature by immersion in an ice bath.

[3] A. H. Ennor and H. Rosenberg, *Biochem. J.* **51**, 606 (1952).

[4] Á. H. Ennor and L. A. Stocken, *Biochem. J.* **42**, 557 (1948).

[5] G. S. Walpole, *J. Physiol. (London)* **42**, 301 (1911).

[6] These conditions minimize the conversion of PC to creatinine.

One milliliter of p-chloromercuribenzoate[7] is then added to each tube, followed by 2.0 ml. of 1% α-naphthol and 1.0 ml. of the diacetyl solution. The volume is adjusted to 10.0 ml. by the addition of H_2O, and the pink color allowed to develop at room temperature (20°) for 20 minutes[8] before determination of the color density as measured with the appropriate filter against a blank tube containing only diacetyl and α-naphthol. If a Hilger Spekker absorptiometer is employed, the color density should be determined with Ilford filter No. 604 (spectrum green). The difference between the nonhydrolyzed and hydrolyzed samples gives the creatine arising from phosphocreatine.

Under the hydrolysis conditions referred to above there is a small amount of creatinine produced directly from the phosphocreatine.[9] This, in terms of creatine, amounts to 10.8% of the total present as phosphocreatine, so that the value for the bound creatine should be corrected by this factor.

Identification

The most desirable proof of the presence of PC in any tissue is given by the separation of the compound in a form suitable for chemical characterization. However, the concentration of PC in many tissues[3] is too low to permit an economical separation in amounts large enough to allow of elementary analysis. Recourse must therefore be made to the use of other criteria[10] for its identification in protein-free tissue extracts. In general it is necessary to concentrate the PC in such extracts by a modification of the technique described by LePage[1] (Fig. 1).

Criterion 1—P: Creatine Ratio. Total P present in the solution should not be measured because of the presence of other organophosphates in addition to PC. It is thus necessary to measure the acid-molybdate-labile P, for which purpose the method of Fiske and SubbaRow[2] or that of Berenblum and Chain[11] is adequate, provided that the sample is allowed to stand with the acid-molybdate for 30 minutes at room temperature before the color density is determined in the case of the first method[2], or the aqueous solution extracted with isobutanol in the case of the second.[11]

---

[7] The addition of p-chloromercuribenzoate prevents inhibition of color development by —SH compounds present in tissue extracts, and this amount will provide an excess under most conditions.

[8] If TCA is used as the protein precipitant, the color should be allowed to develop in the dark or in a room lit by a tungsten filament lamp. This is necessary to avoid the blue color due to the reaction between TCA and α-naphthol. Such precautions are unnecessary if perchloric acid is used.

[9] H. Barker, A. H. Ennor, and K. Harcourt, *Australian J. Sci. Research* **B3**, 337 (1950).

[10] H. Barker and A. H. Ennor, *Biochim. et Biophys. Acta* **7**, 272 (1951).

[11] I. Berenblum and E. Chain, *Biochem. J.* **32**, 295 (1938).

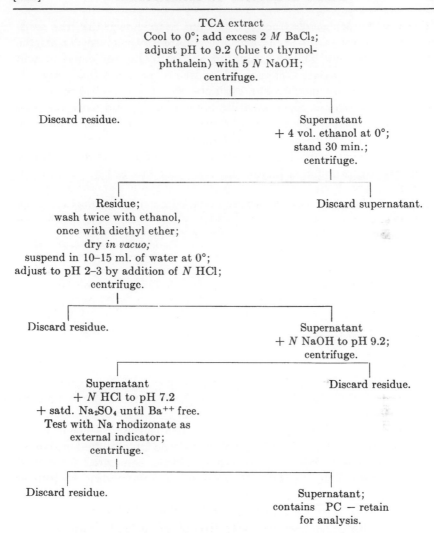

TCA extract
Cool to 0°; add excess 2 *M* BaCl₂;
adjust pH to 9.2 (blue to thymolphthalein) with 5 *N* NaOH;
centrifuge.

Discard residue.

Supernatant
+ 4 vol. ethanol at 0°;
stand 30 min.;
centrifuge.

Residue;
wash twice with ethanol,
once with diethyl ether;
dry *in vacuo;*
suspend in 10–15 ml. of water at 0°;
adjust to pH 2–3 by addition of *N* HCl;
centrifuge.

Discard supernatant.

Discard residue.

Supernatant
+ *N* NaOH to pH 9.2;
centrifuge.

Supernatant
+ *N* HCl to pH 7.2
+ satd. Na₂SO₄ until Ba⁺⁺ free.
Test with Na rhodizonate as
external indicator;
centrifuge.

Discard residue.

Discard residue.

Supernatant;
contains PC – retain
for analysis.

FIG. 1. Concentration of phosphocreatine from tissue extracts.

A simultaneous determination must be made of the small amount of P_i which will be present as a contamination, and for this purpose two methods are available.[12,13]

The creatine moiety is determined after acid hydrolysis as described above. Appropriate blank determinations should be carried out.

P : creatine ratio = 1 indicates the presence of PC.

[12] O. H. Lowry and J. A. López, *J. Biol. Chem.* **162**, 421 (1946); see also Vol. III [114].
[13] A. H. Ennor and L. A. Stocken, *Australian J. Exptl. Biol. Med. Sci.* **28**, 647 (1950).

Criterion 2—Molybdate Hydrolysis. It has been shown[9] that the products of the hydrolysis of PC in acid-molybdate are primarily creatinine and phosphoric acid. Thus a sample of authentic PC subjected to acid hydrolysis in the presence of molybdate was so changed that only 7% of the creatine equivalent to the PC appeared as such—the remainder appeared as creatinine. Since creatine is unchanged, and since no compound other than PC can give rise to creatinine under these conditions, this observation may be used as a test for the presence of PC.[10] The test should be carried out by hydrolyzing a sample of the extract in 0.1 N HCl at 65° for 9 minutes in the presence and absence of ammonium

ESTIMATION OF CREATINE AND CREATININE IN TCA EXTRACT OF GUINEA PIG LIVER BEFORE AND AFTER HYDROLYSIS IN 0.1 N HCl AT 65° FOR 9 MINUTES IN PRESENCE AND ABSENCE OF AMMONIUM MOLYBDATE ($0.5 \times 10^{-3}\ M$)
(Reproduced with the permission of the Editors of *Biochimica et Biophysica Acta*)

Treatment of extract	Creatine, γ/ml.	Creatinine, γ/ml.	Creatine $\equiv$ Creatinine, γ/ml.
Unhydrolyzed	29.3	0.0	0.0
Hydrolyzed—no molybdate	48.0	1.7	2.0
Hydrolyzed—with molybdate	30.6	16.2	18.8

molybdate ($0.5 \times 10^{-3}\ M$). Creatine and creatinine are then determined in both hydrolyzates, the former by the method described above, and the latter by the Jaffe reaction.

An example of this test (see the table) indicates the presence of PC in the extract examined, and it will be noted that the true "bound creatine" is the same within experimental error, whether calculated from acid hydrolysis ($48.0 - 29.3 + 2.0 = 20.7\ \gamma$) or acid-molybdate hydrolysis ($30.6 - 29.3 + 18.8 = 20.1\ \gamma$).

Criterion 3—Paper Chromatography. The behavior of a sample believed to contain PC is compared with that of an authentic sample of the compound. For this purpose the TCA extract of the tissue under investigation should be treated according to the scheme in Fig. 1. Both samples may then be spotted on No. 1 Whatman filter paper which has been washed with 0.1 N HCl and then with distilled H_2O until acid-free. The chromatogram is developed for 18 hours at room temperature in an alkaline medium made up of 60 ml. of *n*-propanol, 30 ml. of concentrated ammonium hydroxide, and 10 ml. of water. The paper is then dried in a current of warm air and sprayed with a mixture of 50 ml. of N HCl, 20 ml. of 1% ammonium molybdate, and 30 ml. of H_2O, and dried in an air oven at 80°. The phosphomolybdate formed by this procedure is reduced with H_2S as described by Hanes and Isherwood.[14] Under these conditions as little as 6 γ of PC may be detected about 10 cm. from the starting line after an 18-hour run. P_i, if present, appears about 2 cm. behind PC.

Preparation

The isolation of PC from skeletal muscle, as described by Fiske and SubbaRow,[15] is time-consuming and results in a low yield of the Ba salt. For these reasons the synthetic procedure described by Zeile and Fawaz,[16] as modified by Ennor and Stocken,[17] to yield the pure Na salt is to be preferred.

Step 1. Phosphorylation of Creatine. Twenty grams of a commercial sample of creatine hydrate is ground in a porcelain mortar with 60 ml. of 10 N NaOH, and 540 ml. of water is added with constant stirring. The insoluble residue is removed by filtration and the solution transferred to a round-bottomed flask immersed in an ice-salt mixture. Then 60 ml. of $POCl_3$ is added dropwise over a period of 2 hours to the vigorously stirred solution, and the alkalinity maintained by the dropwise addition of 400 ml. of 10 N NaOH over this period. Throughout these additions the temperature should be maintained at 0 to 5°. Stirring is continued for 15 minutes after the last addition.

Step 2. Isolation of NaPC. The reaction mixture is filtered and the cake of sodium phosphate re-extracted with two successive amounts of ice-cold water (200 and 100 ml.). The residue is discarded and the combined washings and filtrate brought to pH 7.6 by the addition of 5 N HCl. Then 3 vol. of ethanol is added; the precipitated NaCl and sodium phosphate are removed by filtration and discarded. The concentration of ethanol is then increased to 80%, and the precipitate consisting largely of NaCl removed by filtration. This precipitate contains appreciable amounts of PC, which may be recovered according to Fig. 2. An excess of 1% $BaCl_2$ in 80% ethanol is then added to the filtrate to precipitate BaPC and Ba phosphate. This is separated by centrifugation and the Ba salts washed with ethanol and ether and dried (fraction 1). The Ba salts (fractions 1 and 2) are then combined and extracted three times with 25 ml. of H_2O adjusted to pH 9.5 by the addition of 0.1 N NaOH. The residue is suspended in 25 ml. of H_2O, cooled to 0°, and sufficient HCl added to effect solution. Then N NaOH is added to bring the solution to pH 9.5, and the precipitate of Ba phosphate removed by filtration and discarded. The supernatant solution is combined with the original washings and Ba

[14] C. S. Hanes and F. A. Isherwood, *Nature* **164**, 1107 (1949).
[15] C. H. Fiske and Y. SubbaRow, *J. Biol. Chem.*, **81**, 629 (1929).
[16] K. Zeile and G. Fawaz, *Z. physiol. Chem.* **256**, 193 (1938).
[17] A. H. Ennor and L. A. Stocken, *Biochem. J.* **43**, 190 (1948).

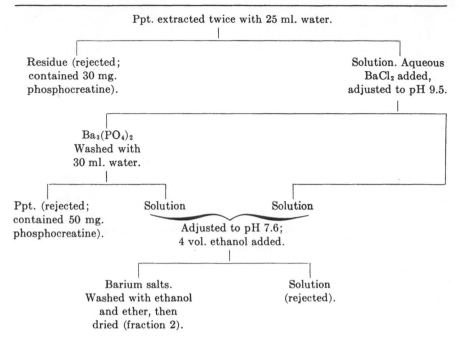

FIG. 2. Recovery of fraction 2 in synthesis of phosphocreatine. (Reproduced with the permission of the Editors of *The Biochemical Journal*.)

precipitated by the addition of 10% Na_2SO_4. The end point is judged by the use of Na rhodizonate as an external indicator. The $BaSO_4$, after separation, is given two washes with 25 ml. of H_2O, and the washes combined with the supernatant. Four volumes of ethanol is then added, and the NaPC separates immediately as a suspension of very fine oily droplets. On standing at room temperature for 12 hours crystallization is complete. The crystals, which may be either needles ($C_4H_8O_5N_3PNa_2 \cdot 4H_2O$) or six-sided platelets ($C_4H_8O_5N_3PNa_2 \cdot 6H_2O$), are separated by filtration, washed with ethanol, and dried in air. The product is completely free from P and NaCl. The yield is about 4 g., and an additional 1.5 g. of the BaPC may be obtained by the addition of an excess of 1% $BaCl_2$ in 80% ethanol-water to the mother liquor.

II. Phosphoarginine

$$\underset{\text{HOOC---CH---(CH}_2)_3\text{---N---C---NPO}_3\text{H}_2}{\overset{\overset{\displaystyle NH_2}{|}\qquad\qquad\overset{\displaystyle NH}{\underset{H}{}\;\overset{\|}{}\;H}}{}}$$

Determination

Like PC, phosphoarginine (PA) cannot be determined directly, so that either the P or the arginine moieties must be determined after acid hydrolysis.

Thus determination of P_i before and after hydrolysis for 1 minute in 0.1 N TCA at 100° has been used by Lohmann.[18] This method is subject to error, however, because of the P released from other organophosphates. More dependable results have been claimed by Mende and Chambers[19] by similar determinations before and after hydrolysis of the water-soluble ethanol-insoluble Ba salts of a tissue extract in 0.01 N HCl at 37°. Under these conditions PA is 98% hydrolyzed in 3 hours and completely hydrolyzed in 10 hours.[19] This method lacks specificity and must be correlated with simultaneous measurements of arginine[19] by the Sakaguchi reaction as modified by Weber,[20] although it should be noted that this method is not so satisfactory as that described for the estimation of creatine. A more dependable and reproducible method for the estimation of PA results from the use of the technique described for the estimation of PC (above). The method is used to estimate the arginine after its release by hydrolysis of PA in N HCl at 100° for 10 minutes. In this case increased sensitivity may be gained by increasing the concentration of α-naphthol from 1% to 5%. The subsequent procedure is then as described for the estimation of creatine and phosphocreatine except for some slight modifications.[21]

Preparation

No synthetic method is available, and attempts in this laboratory to phosphorylate the free guanidino N of arginine, even in the presence of Cu to protect the α-amino group, have not been successful. Meyerhof and

---

[18] K. Lohmann, *Biochem. Z.* **286**, 28 (1936).

[19] T. J. Mende and E. L. Chambers, *Arch. Biochem. and Biophys.* **45**, 105 (1953).

[20] C. J. Weber, *J. Biol. Chem.* **86**, 217 (1930).

[21] The reaction between α-naphthol, diacetyl, and arginine to produce a pink color does not reach an equilibrium as is the case with creatine. It is also much more temperature-dependent, so that it is desirable to run a standard amount of arginine (100 γ) with each estimation and to determine the color densities at 30 minutes in place of the usual 15 minutes.

Lohmann[22] isolated the compound from crustacean muscle by a method which depends on the solubility of the Ba salt in water and its insolubility in 60% ethanol. The yield of the pure material, which corresponded to the formula $(C_6H_{14}O_5N_4P)_2Ba \cdot 2H_2O$, is low, however. These authors obtained 0.3 to 0.4 g. from 1 kg. of minced material (50 to 100 crabs[22]).

A much simpler method which gives reproducibly high yields has recently been developed by Ennor *et al.*[23]

Step 1. Extraction of Muscle. The tail muscle[24] of live crayfish (*Jasus lalandii*) is rapidly separated from the chitinous material and dropped into liquid N_2. The hard, frozen material is then ground to a fine powder and extracted in 100-g. amounts with 450 ml. of 9% TCA in a Waring blendor for 2 minutes. The product is then filtered on a Büchner funnel (Whatman No. 541 filter paper) and the filtrate immediately made alkaline by the addition of 10 N NaOH. The filter cake is re-extracted in the blendor with 130 ml. of 5% TCA, and the suspension filtered. The filtrate, after being made alkaline, is combined with the previous filtrate and the filter cake discarded.

Step 2. Preparation of Crude Barium Phosphoarginine. The combined filtrates are adjusted to pH 9, and M barium acetate added until no further precipitation occurs. The suspension is again adjusted to pH 9 and allowed to settle overnight in the cold room. The bulk of the supernatant is syphoned off, and the remainder collected after centrifugation of the insoluble barium salts which are discarded. Then 3 vol. of 95% ethanol is added to the supernatant, and the mixture is allowed to stand overnight in the cold room. The slightly opalescent supernatant is syphoned off and discarded. The precipitate is collected by centrifugation, washed with ethanol and with ether, and finally dried *in vacuo* over $CaCl_2$. The product contains about 30% PA.

Step 3. Conversion of Crude Barium Phosphoarginine to the Sodium Salt. It is convenient, because of the bulk of the precipitates involved, to carry out this step in 5-g. lots. Five grams of the crude salts from step 2 is suspended in 150 ml. of ice-cold water in a 250-ml. centrifuge tube and sufficient 5 N HCl added to effect solution. The solution is then rapidly adjusted to pH 9.[25] The precipitate of barium phosphate is removed by centrifugation, washed once with water, and the supernatants combined and filtered.

Six milliliters of a solution[26] of $CuCl_2$ is added, and the solution is

[22] O. Meyerhof and K. Lohmann, *Biochem. Z.* **196,** 49 (1928).

[23] A. H. Ennor, J. F. Morrison, and H. Rosenberg, *Biochem. J.* **62,** 358 (1956).

[24] Two hundred to five hundred grams of muscle is obtained from each crayfish; the remainder of the animal is discarded.

[25] Blue to thymolphthalein.

[26] Fifty grams of $CuCl_2$ dissolved in 50 ml. of water.

adjusted to pH 6.5[27] by the addition of 5 N NaOH. The Cu precipitate is removed by centrifugation and the blue supernatant treated with 1 ml. of $CuCl_2$ solution. After the pH is adjusted to 6.5 the Cu salt is removed by centrifugation and the almost colorless supernatant discarded.

The Cu salts are combined, suspended in 100 ml. of water, and dissolved by the addition of 5 N HCl. Then 0.5 ml. of $CuCl_2$ solution is added, followed *immediately* by sufficient 5 N NaOH to bring the solution to approximately pH 6.5. The suspension is then carefully adjusted to this pH and centrifuged.[28] This precipitation procedure is repeated until the supernatant is free from Ba—generally five to seven precipitations.

The Cu precipitate is next suspended in 50 ml. of water, cooled in an ice-salt mixture, and the Cu removed with H_2S. The CuS is removed by centrifugation, suspended in 25 ml. of water, and again treated with H_2S. It is then recentrifuged and the CuS discarded. The supernatants are combined, adjusted to pH 5, placed in an ice bath and vigorously aerated to remove H_2S.[29] A small amount of activated charcoal is added, and the mixture shaken and filtered. The resultant solution is clear and colorless.

Step 4. Precipitation of Phosphoarginine Hydrochloride. The filtrate from step 3 is cooled to 0° and brought to pH 3.5 by the addition of 5 N HCl. Then 9 vol. of ethanol at −10° is added slowly with stirring, and the solution is carefully adjusted to pH 5.0.[30]

The mixture is allowed to stand at −10° until the precipitate has aggregated, and then it is centrifuged. The supernatant is discarded, and the precipitate of PA hydrochloride dissolved in about 30 ml. of water and sufficient 5 N NaOH added to bring the pH to about 7. The slightly opalescent solution is filtered, cooled to 0°, and adjusted to pH 3.5. Then 1.5 vol. of ethanol at −10° is added, and the solution is carefully adjusted to pH 5.0 in order to reprecipitate the hydrochloride. After recovery the material is again precipitated and if washed with ethanol and ether and dried *in vacuo* over $CaCl_2$ corresponds to the formula $C_6H_{15}O_5N_4P \cdot HCl$. The hydrochloride is unstable, however, and even in the dry condition slowly breaks down with the appearance of free arginine and P_i.

Step 5. Precipitation of Barium Phosphoarginine. The wet precipitate from step 4 is dissolved in 20 ml. of water, 5 ml. of 1 M barium acetate

[27] Just green to bromothymol blue.

[28] It is important to have the pH correct; otherwise a blue supernatant which contains PA will result. If the pH is correct, an almost colorless supernatant results in subsequent precipitations and the supernatant contains only traces of PA.

[29] Care should be taken during the aeration to maintain the solution at pH 5 to 6.

[30] Variation of this procedure is likely to produce an oil which should be centrifuged, the supernatant discarded, and the oil dissolved in water and reprecipitated with ethanol at pH 5.0.

added, and the solution adjusted to pH 9.0. If any precipitate of Ba phosphate appears it should be removed by centrifugation and the supernatant adjusted to pH 8.2. Then 4 vol. of ethanol is added slowly with stirring, at which the pH generally falls to about 7 to 7.5. The precipitate of barium PA is collected by centrifugation, washed with ethanol and ether, and dried *in vacuo* over $CaCl_2$. The product is analytically pure and corresponds to the formula

$$
\left[\begin{array}{c} \text{H} \\ | \\ \text{NPO}_3 \\ \diagup \\ \text{HN} = \text{C} \\ \diagdown \\ \text{NH} \\ | \\ (\text{CH}_2)_3 \\ | \\ \text{CH} \cdot \text{NH}_2 \\ | \\ \text{COOH} \end{array} \right] \quad \text{Ba} \cdot \text{H}_2\text{CO}_3 \cdot \text{H}_2\text{O}
$$

The yield is consistently 10 to 11 g. of the barium salt per kilogram of muscle and represents a recovery of approximately 50% of the PA originally contained therein.

Identification

The presence of PA in tissue extracts may be confirmed by methods similar to those described for PC.

Thus the molar ratio of P:arginine may be determined in the barium salts of a water-soluble ethanol-insoluble fraction of a tissue extract by the methods described above. Care should be taken, however, that arginine is the only guanidine compound present, for neither the Sakaguchi nor the α-naphtholdiacetyl reactions are specific for arginine.

Contributory evidence for the presence of PA may also be obtained by comparing the behavior of the unknown compound with that of an authentic sample. For this purpose a unidimensional ascending chromatogram run in propanol-ammonia-water (see above) is satisfactory. After drying, the chromatogram should be stood in a closed jar in an atmosphere of HCl for 10 minutes to hydrolyze the PA. It is then treated with one of the following reagents:

1. To show the presence of arginine: sprayed either with (1) a freshly prepared mixture of 1.0 ml. of 5 N NaOH and 3.0 ml. of a solution containing 2% α-naphthol and 0.05% diacetyl in 50% ethanol, in which case arginine will appear as a pink spot, or (2) a freshly prepared solution containing 0.2 ml. of 40% NaOH, 0.2 ml. of 40% urea, and 0.2 ml. of 1% α-naphthol in ethanol made to 10 ml. with water. After this treatment the paper is allowed to dry at room temperature and is then sprayed with a freshly prepared mixture of 7 ml. of water and 3 ml. of Na hypobromite (0.9 ml. of bromine in 100 ml. of 10% NaOH).[31]

2. *To show the presence of P_i:* sprayed with a solution of HCl and ammonium molybdate and reduced with H_2S as described above.

Unidimensional ascending chromatography in phenol saturated with water containing a trace of ammonia is also satisfactory. The spots may be detected by spraying with ninhydrin after a 12-hour run at 20°. A good separation of arginine from PA is obtained with the latter running about 3.5 times as fast as the former.

Other N-Phosphorylated Compounds

Although in recent years there have appeared in the literature several reports on the presence of phosphagens other than PA in the tissues of invertebrates, definite evidence for their presence has been lacking. Recently, however, Thoai et al.[32] have reported the isolation of phosphoguanidoacetic acid and phosphoguanidotaurine from *Nereis sp.* and *Arenicola sp.*, respectively. In addition Thoai and Robin[33] have described the isolation of "lombricine" (phosphoguanidoethylserylphosphate) from *Lumbricus terrestris sp.* This compound has been ascribed the following structure:

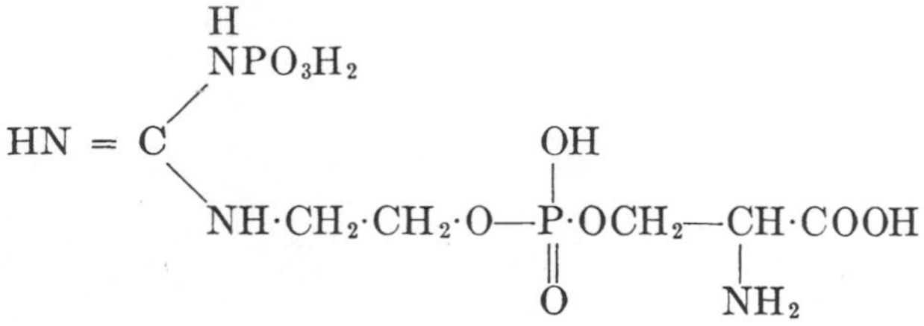

$$HN = C \underset{\underset{NH \cdot CH_2 \cdot CH_2 \cdot O - \overset{\overset{OH}{|}}{\underset{\underset{O}{\parallel}}{P}} \cdot OCH_2 - \overset{}{\underset{\underset{NH_2}{|}}{CH}} \cdot COOH}{\overset{\overset{H}{NPO_3H_2}}{\diagup}}}{}$$

Although it is suspected that these compounds play a role similar to that of PC and PA, there is as yet no definite proof of this, and little is known about their chemistry and metabolic importance.

[31] J. Roche, N. Thoai, and J. L. Hatt *Biochim. et Biophys. Acta* **14**, 71 (1954).

[32] N. Thoai, J. Roche, Y. Robin, and N. Thiem, *Biochim. et Biophys. Acta* **11**, 593 (1953).

[33] N. Thoai and Y. Robin, *Biochim. et Biophys. Acta* **14**, 76 (1954).

[117] Isolation of ATP from Muscle

By Louis Berger

The method described is based mainly on the procedures of Lohmann and Schuster[1] and Kerr[2] as modified by Dounce et al.[3] ATP is extracted from muscle with TCA and carried through a series of precipitations as the mercury and barium salt. Probably any fresh muscle may be used, but for maximum yield and highest purity rabbit muscle is recommended. The anesthesia procedure of Du Bois et al.,[4] described in step 1, is particularly recommended.

Step 1. Isolation of the Muscle. Anesthetize a large male rabbit by a series of intraperitoneal injections of magnesium sulfate solution (51% $MgSO_4 \cdot 7H_2O$). The first dose is 1 ml./kg. of body weight followed by 0.5 ml./kg. every 10 minutes until complete anesthesia is reached. This takes about 30 minutes. Then decapitate the animal, and skin, eviscerate, and chill the carcass in ice water for 15 minutes. Dissect the muscle from any part of the carcass, and pass it through a chilled meat grinder. Not more than 30 minutes should be spent in dissecting, even if some of the muscle must be sacrificed. One large rabbit (4 kg.) will yield 0.8 to 1 kg. of muscle. Unless otherwise stated, all subsequent operations are performed at temperatures as near to 0° as is practical.

Step 2. Extraction of the Muscle with TCA. Quickly weigh the muscle, and extract 200-g. portions for 2 minutes in a Waring blendor with an equal volume of 10% TCA. Squeeze the liquid through a double layer of cheesecloth and re-extract the residue as above, this time with 5% TCA. Filter the combined turbid extracts by suction through paper, recycling until a clear filtrate is obtained. Adjust the pH of the filtrate to 7 with 10% NaOH. All subsequent steps will be described for a neutralized filtrate obtained from 1 kg. of muscle. If more than one rabbit is used, it is recommended that steps 1 and 2 be performed for each rabbit individually, and the filtrates pooled.

Step 3. First Precipitation of Hg Salt of ATP. Slowly add 4 ml. of glacial acetic acid to the 2 l. of filtrate with stirring, followed by sufficient mercuric acetate reagent (20% mercuric acetate in 2% acetic acid) to

[1] K. Lohmann and P. Schuster, Biochem. Z. 282, 104 (1935).
[2] S. E. Kerr, J. Biol. Chem. 139, 121 (1941).
[3] A. L. Dounce, A. Rothstein, G. T. Beyer, R. Meier, and R. M. Freer, J. Biol. Chem. 174, 361 (1948).
[4] K. P. DuBois, H. G. Albaum, and V. R. Potter, J. Biol. Chem. 147, 699 (1943).

give complete precipitation (150 to 200 ml. required). Allow the mercury salt to settle for 1 to 3 hours, after which time most of the supernatant fluid will be clear enough to decant. Collect the precipitate at the centrifuge, and wash it once with three times its volume of a 1:40 dilution of the mercuric acetate reagent.

Step 4. First Precipitation of the Ba Salt of ATP. Suspend the washed precipitate in 400 ml. of water, and decompose the mercury salt with H_2S. This is best accomplished in a closed system by applying vacuum to the container before attaching it to the H_2S source. With continuous shaking the reaction will be complete in about 10 minutes, after which time no more H_2S will be consumed. Filter the preparation by suction; recycle if necessary to obtain a clear filtrate. Wash the HgS cake three times with 20-ml. portions of cold water. Free the combined filtrates of H_2S by passing air through the solution. About 1 hour is usually required. Adjust the pH of this solution to 7 with 10% NaOH. Add 20 ml. of 2 M barium acetate (an excess) to precipitate the barium salt of ATP. Allow the precipitate to settle for about 1 hour, after which time most of the supernatant fluid will be clear enough to decant. Collect the precipitate at the centrifuge, and wash it twice with three times its volume of cold water.

Step 5. Second Precipitation of Hg Salt of ATP. Suspend the washed barium precipitate in 500 ml. of cold water, and add the minimum amount of glacial acetic acid to effect almost complete solution. A small amount of precipitate may remain which should be filtered or centrifuged off. The pH of the solution should not be allowed to go below 3. Add one-tenth the amount of mercuric acetate reagent required in step 3 (about 20 ml.), and collect and wash the mercury precipitate as in step 3.

Step 6. Second Precipitation of Ba Salt of ATP. Decompose the washed mercury precipitate with H_2S as in step 4, and obtain an aerated filtrate as described there. Slowly add 10% NaOH until a small amount of precipitate forms, keeping the pH below 6. The precipitate is barium ATP resulting from a small amount of barium which is carried through with the mercury precipitations. If no precipitate forms when pH 6 is reached, add 1 drop of 2 M barium acetate so as to allow a small amount of precipitate to form. Filter off the precipitate, which will remove colloidal sulfides. Adjust the pH of the clear filtrate to 7 with 10% NaOH, and precipitate the barium salt of ATP with barium acetate as described in step 4. Wash the precipitate at the centrifuge successively with 100-ml. portions of the following: twice with cold water, once with 50% ethanol, once with 75% ethanol, twice with 95% ethanol, and twice with ether. Finally, air-dry the product and then place it in a vacuum desiccator over activated alumina or calcium chloride.

Two to three grams of $Ba_2ATP \cdot 4H_2O$ [5] is obtained from 1 kg. of muscle. Chromatographic analysis[6] and enzymatic assay[7] usually indicate a purity of 90 to 95%. The major impurity is the barium salt of ADP. Usually less than 1% of the total phosphorus is present as inorganic phosphorus. Preparations of ATP, virtually free of ADP, can be made from this product by chromatographic separation.[6]

[5] Commercially available from various sources as the Ba, Na, or K salt.
[6] See Vol. III [120].
[7] See Vol. III [121].

[118] Preparation of ADP from ATP

By Louis Berger

ADP is prepared from ATP by enzymatic hydrolysis using the adenosinetriphosphatase of lobster muscle.[1] Rabbit muscle[2,3] may also be used as the source of the enzyme, but it is more difficult to prepare and is more likely to contain interfering enzymes.[3]

Step 1. Preparation of the Lobster Muscle. Cut a live lobster in half, transversely, and remove the tail muscles. Cut them into strips with scissors (do not mince or grind). Suspend the strips (25 to 35 g.) in 300 ml. of cold 0.45% KCl, and mechanically stir slowly for 15 minutes. Change the solution every 15 minutes until a total of five washings have been made. The strips are now ready for use in step 3.

Step 2. Preparation of ATP Solution. Dissolve 1 g. of the sodium or potassium salt of ATP[4] in 100 ml. of water. If the barium salt of ATP[4,5] is used, it must first be freed of barium as follows: dissolve 1.1 g. in 40 ml. of cold 0.1 N HCl, and add 0.3 g. of Na_2SO_4. Centrifuge the $BaSO_4$, wash it once with 10 ml. of the cold acid, and again centrifuge. Neutralize the combined supernatant fluids with 10% NaOH to pH 7, and dilute to 100 ml. with water. If the Na or Ba salt of ATP was used to prepare the solution, add solid KCl to make the solution 0.1 M in K ions.

Step 3. Enzymatic Conversion of ATP to ADP. Add the washed muscle strips of step 1 to the 100 ml. of 1% ATP solution of step 2. Slowly stir

[1] K. Lohmann, *Biochem. Z.* **282**, 109 (1935).
[2] K. Bailey, *Biochem. J.* **36**, 121 (1942).
[3] I. Green, J. R. C. Brown, and W. F. H. M. Mommaerts, *J. Biol. Chem.* **205**, 493 (1953).
[4] ADP is now commercially available from various sources.
[5] See Vol. III [117].

the mixture for 20 minutes at 20°. Filter off the strips, and determine the decrease of acid-hydrolyzable P [6] in the filtrate. If this has not decreased 50%, a further incubation with the strips is indicated. Several 100-ml. batches of ATP solution may be treated with the same muscle strips and then pooled for subsequent processing. The rest of the procedure will be described for 100 ml. of filtrate.

Step 4. First Precipitation of Ba Salt of ADP. Add 2 to 3 ml. of 2 *M* barium acetate (an excess) to the filtrate. Cool the suspension in an ice bath. Unless otherwise indicated, all subsequent operations are performed as near to 0° as is practical. Centrifuge off the precipitate which is a mixture of the barium salt of ADP and inorganic orthophosphate. Wash the precipitate once with 3 vol. of 0.2 *M* barium acetate.

Step 5. First Precipitation of Hg Salt of ADP. Dissolve the precipitate in a minimum of 0.1 *N* HCl, and add sufficient mercuric acetate reagent (20% mercuric acetate in 2% acetic acid) to give complete precipitation. Centrifuge, and wash the precipitate with 25 ml. of a 1:4 dilution of the mercuric acetate reagent. Suspend the precipitate in 60 ml. of 0.05 *N* HNO_3, decompose the mercury salt with H_2S (see step 4, Vol. III [117]), filter, and wash the HgS cake twice with 10-ml. portions of 0.05 *N* HNO_3. Free the combined filtrates of H_2S by aeration, and adjust to pH 7 with 10% NaOH.

Step 6. Second Precipitation of Ba Salt of ADP. Add an equal volume of 95% ethanol to the neutralized filtrate. Add sufficient 2 *M* barium acetate to give complete precipitation of the barium salt of ADP. Collect the precipitate at the centrifuge, and wash it successively with 25-ml. portions of the following: twice with 50% ethanol, once with 75% ethanol, twice with 95% ethanol, and twice with ether. After air-drying, place it in a vacuum desiccator over activated alumina or calcium chloride.

The yield will be 0.25 to 0.3 g. of approximately 95% pure $Ba_3(ADP)_2 \cdot 4H_2O$.[4] The major impurity is the barium salt of inorganic orthophosphate, which can be largely removed by dissolving the product in 0.1 *N* HCl and reprecipitating the barium salt of ADP as described in step 6. If the treatment with the muscle strips in step 3 was adequate, the product will be free of ATP as determined by chromatographic analysis[7] and enzymatic assay.[8]

[6] See Vol. III [115].
[7] See Vol. III [120].
[8] See Vol. III [121].

[119] Preparation of AMP from ATP

By LOUIS BERGER

AMP is prepared from ATP by alkaline hydrolysis with Ba(OH)$_2$, essentially as described by Kerr.[1]

Step 1. Preparation of ATP Solution. Dissolve 1.5 g. of the sodium salt of ATP[2] in 100 ml. of water. If the barium salt of ATP[2,3] is used, dissolve 1.7 g. in the minimum amount of cold 0.1 N HCl, adjust to pH 5 with 10% NaOH, and dilute to 100 ml.

Step 2. Alkaline Hydrolysis of ATP to AMP. Add 2 or 3 drops of 1% phenolphthalein to the ATP solution, and make the solution pink by the addition of saturated Ba(OH)$_2$ solution. At this point most of the ATP will appear as a fine suspension of the barium salt. Immerse the preparation in a boiling water bath, and keep the suspension pink by the dropwise addition of the Ba(OH)$_2$ solution, stirring constantly. After 30 minutes or so, no further addition of Ba(OH)$_2$ will be required to keep the suspension pink. After an additional 15 minutes, remove the preparation from the boiling water bath and allow it to remain hot for an additional 30 minutes before cooling in cold water to 25°.

Step 3. Removal of Ba Salt of Inorganic PP. Add 3 N HCl to dissolve the precipitate, which is a mixture of the barium salts of AMP and PP, and dilute with water to 250 ml. While stirring vigorously, adjust the pH to 9 with 10% NaOH. This causes the precipitation of Ba PP without much Ba AMP. Filter off the precipitate, and wash it twice with 25-ml. portions of dilute NaOH (pH 9).

Step 4. Precipitation of Hg Salt of AMP. Add 0.6 ml. of glacial acetic acid to the 300 ml. of combined filtrates, followed by 50 ml. of mercuric acetate reagent (20% mercuric acetate in 2% acetic acid). Place the preparation in an ice bath, and allow it to remain there for at least 1 hour before centrifuging off the Hg salt of AMP. Wash the precipitate twice with cold 20-ml. portions of a 1:4 dilution of the mercuric acetate reagent. Suspend the washed precipitate in 30 ml. of 0.01 N H$_2$SO$_4$, decompose the mercury salt with H$_2$S (see step 4, Vol. III [117]), and wash the HgS cake twice with 5-ml. portions of water. Free the combined filtrates of H$_2$S by aeration.

Step 5. Precipitation of AMP with Ethanol. Add 4 vol. of 95% ethanol to the filtrate, and store it at −15° to −20° for 18 to 24 hours. At least 2

[1] S. E. Kerr, *J. Biol. Chem.* **139**, 131 (1941).
[2] Now commercially available.
[3] See Vol. III [117].

hours before filtering off the crystals of AMP by suction, stir the preparation to disturb any areas of supersaturation.

Step 6. Recrystallization of AMP from Water. Transfer the crystals of step 5 without washing or drying to a small beaker, and dissolve them in a minimum of water at 50° to 60° (10 to 15 ml. required). Filter while warm, by suction through a small fritted-glass funnel. Rinse the beaker and funnel with 1 or 2 ml. of warm water. Transfer the combined filtrates to an ice chest. After 18 to 24 hours, filter off the crystals by suction on Whatman No. 50 paper, and wash them with 2 to 3 ml. of cold water. Dry the product in a vacuum desiccator over activated alumina or calcium chloride.

The yield will be 0.4 to 0.5 g. of the free acid of AMP[2] and will be 98 to 100% pure by chromatographic analysis[4] and enzymatic assay.[5]

[4] See Vol. III [120].
[5] See Vol. III [121].

[120] Chromatographic Separation of ATP, ADP, and AMP[1]

By WALDO E. COHN

The separation of AMP, ADP, and ATP by ion exchange chromatography is accomplished by selective elution from strong-base anion exchangers in either the chloride[2-4] or formate[5] form (presumably, any other form would work as well, but only these two have been exploited). A typical chloride separation is as follows (details of column preparation are given in Vol. III [107]; see Fig. 1).

The mixture of AMP, ADP, and ATP (about 1 to 10 mg. of each per 10 ml. of exchanger-bed volume can be handled efficiently), at a pH of 6 to 9 and in the presence of less than 0.01 M salt (preferably chloride), is absorbed on the column and washed through with 0.01 M NH$_4$Cl. A volume of 0.01 M NH$_4$Cl is used which will reduce the pH to 7 or less and remove any undesirable anions which may have been absorbed previously [ca. 5 column volumes (C.V.)]. AMP (and inorganic orthophosphate, if

[1] Contribution prepared under Contract No. W-7405-eng-26 for the Atomic Energy Commission.
[2] W. E. Cohn and C. E. Carter, *J. Am. Chem. Soc.* **72,** 4273 (1951).
[3] J. X. Khym and W. E. Cohn, *J. Am. Chem. Soc.* **75,** 1153 (1953).
[4] W. E. Cohn, *in* "The Nucleic Acids" (Chargaff and Davidson, eds.), Vol. 1, p. 211, Academic Press, New York, 1955.
[5] R. B. Hurlbert and V. R. Potter, *J. Biol. Chem.* **209,** 1 (1954); H. Schmitz, R. B. Hurlbert, and V. R. Potter, *ibid.* **209,** 41 (1954).

present) is removed by 0.005 N HCl (a volume of up to 50 C.V. is required). ADP is removed slowly (about 120 C.V.) by 0.01 N HCl or more rapidly by 0.01 N HCl plus 0.04 N NaCl (up to 50 C.V.). ATP is removed slowly by the last-named reagent (ca. 200 C.V.) or rapidly by 0.01 N HCl plus 0.2 N NaCl.

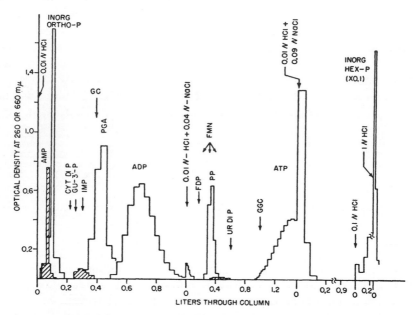

Fig. 1. Anion exchange separation of various phosphates [from W. E. Cohn, *in* "The Nucleic Acids" (Chargaff and Davidson, eds.), Vol. 1, Academic Press, New York, 1955]. Absorbed: 2 to 5 mg. of each substance (those with peak positions shown by arrows were examined in separate tests). Column: Dowex-1-Cl⁻, 200 to 400 mesh, 10 cm. × 0.76 sq. cm. Abbreviations: CYT = cytidine; GU = guanosine; GC = guanylyl cytidylic acid; UR = uridine; GGC = guanylyl guanylyl cytidylic acid; HEX-P = hexametaphosphate.

The removal of each substance may be followed by spectrophotometric observation of the eluates, and the most efficient conditions for each substance, in relation to possible contaminants, selected by varying the volumes and the chloride concentrations. The lower the chloride concentration, the larger are the volumes required. Orthophosphate can be removed by preliminary washing with 0.025 N NH₄Cl. Sugar phosphates can be removed separately by preliminary use of borate buffer, as shown by Khym and Cohn.[3] The separation of phosphoglyceric acid, inorganic pyrophosphate, and fructose diphosphate from ADP requires more careful chromatography; in this case removal of ADP with 0.01 N HCl

(100 C.V.) and a slower removal of ATP with 0.01 N HCl plus 0.04 N NaCl (200 C.V.) may be recommended.[4]

Higher concentrations of formate-formic acid mixtures serve to elute these substances separately from formate columns.[5]

The demonstration of uridylic acid derivatives[5] containing more than one phosphate group indicates the need for careful chromatography of extracts which may contain these substances. The spectra of uridine and adenosine are not too dissimilar.[6]

The nucleotides, once separated, may be concentrated[7] by neutralization with NH_4OH followed by reabsorption on smaller columns and re-elution with stronger salt at the most acid pH deemed advisable. Recovery without salt is best achieved by conversion to a formate or acetate system, followed by volatilization, or by precipitation as barium salts.

[6] See Vol. III [107], Table IV.
[7] See Vol. III [107], B, 4.

[121] Determination of ATP and ADP in Tissue Filtrates

By AGNETE MUNCH-PETERSEN and HERMAN M. KALCKAR

Principle. The method is based on the stepwise conversion of ATP and ADP to 5-AMP and subsequent deamination of the 5-AMP formed to 5-IMP by means of 5-adenylic acid deaminase, a reaction which can be followed spectrophotometrically at 265 mμ.[1] (The deaminase assay system is described in the article on 5-AMP deaminase, Vol. II [68].) The reactions which take place are the following:

$$AMP \rightarrow IMP \tag{1}$$
$$ADP \rightarrow \tfrac{1}{2}AMP + \tfrac{1}{2}ATP \tag{2}$$
$$ATP \rightarrow AMP + 2P \tag{3}$$

Reaction 1 is the deamination of 5-AMP to 5-IMP, catalyzed by the 5-AMP deaminase.[2,3] In reaction 2 the enzyme adenylate kinase, also called myokinase,[4,5] causes a conversion of ADP into ATP and 5-AMP.

[1] H. M. Kalckar, *J. Biol. Chem.* **167**, 445 (1947).

[2] G. Schmidt, *Z. physiol. Chem.* **179**, 243 (1928).

[3] H. M. Kalckar, *J. Biol. Chem.* **167**, 461 (1947); see Vol. II [96] for preparation of apyrase.

[4] H. M. Kalckar, *J. Biol. Chem.* **148**, 127 (1943).

[5] S. P. Colowick and H. M. Kalckar, *J. Biol. Chem.* **148**, 117 (1943); see Vol. II [99] for preparation of the adenylate kinase.

Reaction 3 is catalyzed by the potato apyrase[3,6] which will dephosphorylate ATP (as well as ADP) to 5-AMP.

The assay starts with the addition of deaminase to determine possible amounts of 5-AMP present. Myokinase is then added, whereby 50% of the ADP will be converted into 5-AMP which is deaminated to form 5-IMP. A subsequent addition of potato apyrase will bring about dephosphorylation of ATP to form 5-AMP, the latter being converted to 5-IMP. The presence of the 5-AMP deaminase will make reactions 2 and 3 go to completion, and the processes can be followed by the stepwise decrease in optical density at 265 mμ, caused by the deamination reaction.

Reagents

    0.3 M succinate buffer, pH 6.1, 2.5 ml.
    1 M MgCl₂ (required by the myokinase), 25 μl.
    Neutralized tissue filtrate, 0.5 ml.
    5-AMP deaminase.
    Myokinase.
    Potato apyrase.

Procedure. The mixture of buffer, MgCl₂, and tissue filtrate is placed in a 3-ml. quartz cell. Deaminase (10 γ/ml.) is added first, and deamination of the 5-AMP present will then cause a decrease in extinction at 265 mμ. After readings are constant, myokinase is added (20 γ of protein per milliliter), and readings are taken until they are constant again. At this stage the ADP has been converted into half ATP and half 5-AMP, the latter being subsequently deaminated to 5-IMP. By addition of apyrase the total amount of ATP present in the reaction mixture, including the amount which appeared through the myokinase reaction, will be converted through 5-AMP to 5-IMP, accompanied by a new decrease in extinction at 265 mμ.

The three successive optical changes measured at 265 mμ during the whole procedure will indicate the concentration of 5-AMP, ½ADP, and finally ½ADP + ATP. According to Kalckar,[1] conversion of 1 γ of 5-AMP per milliliter to 5-IMP causes a decrease in extinction of 0.019. On the basis of this the amount of the three separate adenine nucleotides in the mixture may be calculated.

In this way it is possible to analyze a mixture of the three different adenine nucleotides in pure solutions as well as in tissue and plasma filtrates. In pure solutions a final concentration of 2 γ/ml. of each adenine component will suffice for analysis. It must be added, however, that in tissue and plasma filtrates the sensitivity of the method is somewhat

[6] H. M. Kalckar *J. Biol. Chem.* **153**, 355 (1944).

decreased, as such extracts often contain substances which exert an inhibitory effect on the deaminase, particularly at low substrate concentrations.[1] In this case the substrate concentrations should not be less than 4 γ/ml. in the final solution. The perchlorate concentration in the final solution should not exceed 0.7 to 0.8% as higher concentrations tend to diminish the values obtained by the deaminase reaction.

It is important that the 5-AMP deaminase and the myokinase do not contain any appreciable adenosinetriphosphatase activity to cause a breakdown of ATP before the addition of apyrase. Evidence has been presented[7] that the 5-AMP-deaminase cannot be separated from myosin, but the adenosinetriphosphatase activity of the purified product is extremely low compared with the deaminase activity. This presence of traces of myosin implies, however, that in cases where the amount of ADP in relation to ATP is very small the method is not suited for quantitative measurements of ADP.[8]

[7] V. Sz. Hermann and G. Josepovits, *Nature* **164,** 845 (1949).
[8] E. L. Chambers and Th. J. Mende, *Arch. Biochem. and Biophys.* **44,** 46 (1950).

[122] Assay of Adenosine Triphosphate

By BERNARD L. STREHLER and W. D. McELROY

$$\text{ATP} + \text{Mg}^{++} + \text{O}_2 + \text{LH}_2 + \text{Luciferase} \rightarrow \text{Light}$$

Assay Method

Principle. This method of ATP analysis is based on the linear luminescence response of firefly extracts to added ATP when all other factors are present in excess.[1-3] Light is usually measured with a photomultiplier, the commercial Farrand photofluorometer being satisfactory. Qualitative estimations may be made with the naked eye.

Reagents. The enzyme can be prepared as described in Vol. II [150]. It is not necessary to use the purified luciferase, since the crude extracts have also been used successfully. The first ammonium sulfate precipitate is preferable, since it also has the luciferin bound to the enzyme. The only other reagents required are 0.1 M MgSO$_4$, 0.1 M Na$_2$HAsO$_4$, pH 7.4, and 0.05 M glycine, pH 7.4.

[1] W. D. McElroy, *Proc. Natl. Acad. Sci. U.S.* **33,** 342 (1947).
[2] W. D. McElroy, J. W. Hastings, J. Coulombre, and V. Sonnenfeld, *Arch. Biochem. and Biophys.* **46,** 399 (1953).
[3] B. L. Strehler and J. R. Totter, *Arch. Biochem. and Biophys.* **40,** 28 (1952).

One hundred firefly lanterns will suffice for about 500 assays. Large numbers of fireflies can readily be collected during the summer months in most temperate or tropical regions of the world, and a few hours' collection will be adequate for several thousand assays. The fireflies can be dried in a vacuum desiccator over $CaCl_2$ and maintained indefinitely in a deep-freeze. The crude enzyme may be stored in a frozen condition for long periods of time (6 months to a year) without appreciable loss of activity. Aliquots of the enzyme are usually stored frozen in conveniently sized vials, since repeated thawing and freezing does lead to some inactivation.

Procedure. Two-tenths milliliter of a firefly enzyme is mixed with 0.1 ml. of arsenate plus 0.1 ml. of $MgSO_4$ and 1.0 ml. of buffer. Just prior to light measurement, 0.2 ml. of the unknown sample is added. Readings are made at a constant time after the enzyme and unknown are mixed at room temperature. The results are compared to a standard curve relating light intensity to ATP concentration.

Sensitivity and Accuracy. With a commercial Farrand light detector, quantities of ATP as low as 1 γ of ATP per milliliter of crude extract can be measured with about 5% accuracy. With a quantum counter the sensitivity is increased by a factor of 1000 with the same accuracy.

Specificity. The assay is specific for ATP, but precursors to ATP will also produce light provided the appropriate enzymes for ATP formation are present. The purified enzyme preparation responds only to ATP. ADP, creatine phosphate, acetyl phosphate, inosine triphosphate, uridine triphosphate, guanosine triphosphate, and inorganic polyphosphates are completely inactive.

Application of Method to Crude Tissue Preparations. This method has been applied to a variety of crude, boiled tissue extracts. Ten to twenty minutes of boiling followed by immediate cooling to 0° until assay is usually sufficient. With this amount of boiling it has been found unnecessary to centrifuge the extracts prior to assay. With tissue such as brain, muscle, and kidney, 3 to 10 mg. is dropped into 3 ml. of boiling water and extracted for 10 minutes. Assay of these extracts has given excellent results when compared to other procedures. TCA extracts cannot be used directly because of the effect of TCA on the firefly enzyme.

Precautions. Among the possible sources of error are the following: inhibitory effects of substances in the extracts, turbidity, precipitation of buffer or activator, anaerobicity, other sources of luminescence. Neither ATP standards nor tissue extracts should be made up more than 1 week before use because of the slow production of pyrophosphate. All the above sources of error except the latter can be controlled by the use of appropriate internal standards, i.e., addition of known amounts of ATP to the unknown. Because of the sensitivity of the method the extracts can usually be diluted to an extent that no inhibitory effects are observed.

Other Applications. A variety of other enzymes, coenzymes, and substrates can be assayed by this method. These include myokinase, hexokinase, ATPase, apyrase, creatine-adenylic transphosphorylase, ADP, AMP, phosphocreatine, CoA, and DPN pyrophosphatase. These substances have been tested by the above procedures. It should be possible to measure quantitatively other compounds which directly or indirectly show a stoichiometric relationship to ATP.

[123] Preparation of Deaminated Adenine Derivatives

I. Preparation and Properties of Inosine Triphosphate (ITP)

By NATHAN O. KAPLAN

Preparation

Principle. The procedure outlined below is based on the deamination of ATP by nitrous acid with isolation of the ITP as a barium salt in a manner described by Kleinzeller[1] and Kaplan *et al.*[2] The deamination may also be carried out enzymatically by means of takadiastase deaminase.[2]

Reagents

ATP, 1 g. free acid or K or Na salts in 50 ml. of 2 N HAc.
NaNO$_2$, 8 g. dissolved in 30 ml. of H$_2$O immediately before use.

Procedure. The NaNO$_2$ solution is added dropwise to the ATP solution. Immediately after addition of the nitrite, a suitable aliquot is removed, diluted to 3 ml., and read at 265 mμ. The reaction mixture is then kept at room temperature, and aliquots are removed every 30 minutes and read at 265 mμ. When there is no further decrease in the reading at 265 mμ (2 to 3 hours), the mixture is neutralized with NaOH to phenol red. Then 5.0 ml. of 25% barium acetate is added along with 1 vol. of ethanol. The barium salt of the ITP is then collected by centrifugation and washed with 50, 95, and 100% alcohol and finally with ether. After drying *in vacuo* over P$_2$O$_5$, the yield of barium salt is about 1.2 g.

Properties

Chemical Properties. The ITP has a maximum absorption at 249 mμ, the millimolar extinction coefficient being 13.9. The properties of the

[1] A. Kleinzeller, *Biochem. J.* **36**, 729 (1924).
[2] N. O. Kaplan, S. P. Colowick, and M. M. Ciotti, *J. Biol. Chem.* **194**, 579 (1952).

compound are very similar to ATP, and acid hydrolysis liberates phosphate from ITP at the same rate as from ATP. Purity can be ascertained by comparing the labile phosphate in the ITP with the absorption at 249 mμ. The barium salt analyzes 85% pure.

Enzymatic Properties. ITP can be distinguished from ATP enzymatically. ITP reacts with yeast hexokinase, but at a slower rate than with ATP. IDP does not react with muscle adenylic acid kinase, and this property can be used to determine whether ITP or ATP is present, as illustrated in the accompanying table. ITP can also be distinguished

REACTIONS OF ITP AND ATP WITH HEXOKINASE AND MUSCLE ADENYLIC ACID KINASE

(2.5 μmoles of ITP and ATP is incubated with 0.01 M MgCl$_2$, 0.05 M NaHCO$_2$, 0.05 M glucose, 13 units of yeast hexokinase,[a] and with and without 0.05 ml. of muscle adenylic acid kinase.[b] Time of incubation is 60 minutes, and the temperature is 37°. After treatment with TCA, the mixtures are analyzed for labile phosphate.[c] The results of a typical experiment are summarized below.)

Additions	ATP	ITP
	(μmoles, 7 min., P left)	(μmoles, 7 min., P left)
None	5.0	5.0
Hexokinase	2.9	3.1
Hexokinase and myokinase	0.5	3.1

[a] See Volume I [32].
[b] See Volume II [99].
[c] See Volume III [114].

from ATP in the firefly luminescence system which does not react with the ITP.[3]

II. Preparation and Properties of Deamino-DPN

Preparation[2]

Principle. The deamination is carried out as in the preparation of ITP. The deaminated coenzyme is then precipitated with acid acetone.

Reagents

DPN, 1 g., 90% pure, dissolved in 50 ml. of 2 N HAC.
NaNO$_2$, 8 g., dissolved in 30 ml. of H$_2$O immediately before using.

Procedure. The NaNO$_2$ is added dropwise with constant stirring to the DPN solution. By measuring the decrease at 265 mμ, the course of the deamination can be followed. The reaction usually is completed within 2 hours. To precipitate the deaminated compound, the mixture is made

[3] See Vol. III [122].

acid to congo red with HNO_3. Four volumes of cold acetone are then added, and the mixture is left overnight at 4°. The resulting precipitate is quite oily and adheres to the walls of the flask. The supernatant from the precipitate can generally be decanted; the precipitate is dissolved in H_2O and reprecipitated with 4 vol. of cold acetone after acidification with HNO_3. This precipitate can be washed with acetone and ether and then dried. From 1 g. of DPN, 800 mg. of deamino-DPN is obtained. No nitrite can be detected in the preparation. Deamino-DPN can also be prepared by using the takadiastase deaminase.[4]

Properties

Chemical Properties. Deamino-DPN has an absorption maximum at 248 mμ. The compound gives an identical reaction with cyanide as does DPN (see Vol. III [128]). From the cyanide spectrum, the compound has been estimated to be approximately 88% pure.

Enzymatic Properties. The deaminated DPN reacts equally as DPN with some dehydrogenase, whereas with other enzymes it is either inactive or reacts at a slower rate than the natural coenzyme.[5] *Neurospora* DPNase attacks deamino-DPN at 2% of the rate of DPN. This difference can be used as a basis for differentiating between the two pyridine nucleotides. Because it can be distinguished from DPN enzymatically, the deamino-DPN has been used in studies dealing with the mechanism of dehydrogenase action[6] and also in studies involving pyridine nucleotide transhydrogenase.[7,8]

III. Preparation and Properties of Inosinic Acids[9]

Preparation

The deamination of the 2'-, 3'-, and 5'-adenylic acid isomers to form the corresponding inosinic acids is carried out in the same manner as is the preparation of ITP and deamino-DPN. After the reaction is complete (as indicated by no further decrease at 265 mμ), phenol red indicator is added, and the solution brought to a pale pink with phenol red; 10 ml. of 25% barium acetate are added, and the nucleotides are precipitated by the addition of 95% ethanol. The precipitates, after centrifugation, are washed with 66% ethanol, 100% ethanol, and anhydrous ether. The compounds are then placed in a desiccator over phosphorous pentoxide.

[4] See Vol. II [70].
[5] M. E. Pullman, S. P. Colowick, and N. O. Kaplan, *J. Biol. Chem.* **194**, 593 (1952).
[6] E. Adams, *J. Biol. Chem.* **217**, 325 (1955).
[7] N. O. Kaplan, S. P. Colowick, and E. F. Neufeld, *J. Biol. Chem.* **195**, 107 (1952)
[8] N. O. Kaplan, S. P. Colowick, and E. F. Neufeld, *J. Biol. Chem.* **205**, 1 (1953).
[9] L. Shuster and N. O. Kaplan, *J. Biol. Chem.* **201**, 535 (1953).

From 1 g. of adenine nucleotide, an average of 1.2 g. of the barium salts of the different inosinic acids are obtained.

Properties

The inosinic acids can be characterized by their absorption maximum. The 2'- and 3'-isomers can be distinguished from the 5'-isomer by their acid lability.[10] Through the use of the specific 3'-nucleotidase[11] the 3'-isomer can be identified. The inosinic acids in the free acid form are somewhat more insoluble than the adenine mononucleotides.[12]

[10] See Vol. III [114].
[11] See Vol. II [86].
[12] Uridine nucleotides can be prepared from cytosine nucleotides with nitrous acid in the same manner as the hypoxanthine nucleotides are obtained from the adenine derivatives (see footnote #9).

[124] Isolation of Diphosphopyridine Nucleotide and Triphosphopyridine Nucleotide

I. Diphosphopyridine Nucleotide

(DPN, Cozymase, Coenzyme I)

By Arthur Kornberg

Principle. DPN is extracted from yeast with hot water. After removal of other substances from the extract by basic lead acetate, DPN is precipitated as a silver salt. This is decomposed with hydrogen sulfide, and the DPN is obtained as the free acid by precipitation with acetone. Further purification is achieved by anion exchange chromatography.

Method of Preparation. The preparation of DPN from a number of source materials by various methods has been reviewed.[1] The present method[2] is a modification of the procedure of Williamson and Green.[3] Ion exchange chromatography of DPN has been employed independently by others.[4]

Purification Procedure

Starting Material. Starch-free bakers' yeast is obtained from an Anheuser-Busch yeast plant. It is shipped in bulk (not compressed) in an insulated carton containing dry ice and is used 2 days after shipping.

[1] G. A. LePage, *Biochem. Preparations* **1**, 28 (1949).
[2] A. Kornberg and W. E. Pricer, Jr., *Biochem. Preparations* **3**, 20 (1953).
[3] S. Williamson and D. E. Green, *J. Biol. Chem.* **135**, 345 (1940).
[4] J. B. Neilands and Å. Åkeson, *J. Biol. Chem.* **188**, 307 (1951).

The yeast must be starch-free, since even the small amounts of starch present in compressed yeast interfere with the filtration of the hot extract and reduce the yield of the product.

DPN of 76 to 83% Purity after Precipitation by Silver. Fifty pounds of yeast is worked up in 5-pound batches. Five pounds of crumbled yeast is added to 3 l. of boiling water in a stainless-steel pot heated over three or four Meker burners. The temperature drops to about 70° but with active stirring reaches 90° in 4 to 7 minutes. At this time a 400-ml. beakerful of Super-Cel (Johns-Manville) is stirred in, and the suspension is heated 1 to 2 minutes longer until the temperature reaches 94°. The hot suspension is filtered with suction on two large (24- or 32-cm.) Büchner funnels prepared with a single layer of coarse gauze, a Whatman No. 1 filter paper, and a 0.5-cm. layer of Super-Cel. The clear, golden filtrate is collected at room temperature and accumulated in the cold room. To the 30 l. of combined filtrates is added 900 g. of basic lead acetate, freshly suspended in 3 l. of cold water. After about 30 minutes the precipitate is filtered at room temperature on Büchner funnels (prepared as above) and discarded. Capryl alcohol is added in the event of foaming. The pH is adjusted to 6.5 with 1 N acetic acid; about 500 ml. is required. To 29 l. of combined filtrates is added 2.5 l. of 25% (w/v) silver nitrate. The preparation is shielded from light and allowed to settle in the cold room overnight.

The supernatant solution is siphoned off and discarded. The precipitate is collected in the centrifuge and washed three times with cold water. It is well suspended in 200 to 300 ml. of cold water and acidified with 2 N nitric acid. About 100 to 150 ml. is required to turn Congo red paper to a royal blue color. A few drops of capryl alcohol are added, and hydrogen sulfide is passed through the chilled suspension for about 1.5 hours with slight agitation of the flask. The silver sulfide is filtered off in the cold and washed two or three times with 100-ml. portions of 0.02 N nitric acid. The filtrate (about 800 ml.) is aerated in the cold for 15 to 30 minutes, and the DPN is precipitated by adding 5 vol. of $-10°$ acetone slowly with constant agitation. The preparation is left overnight at $-10°$, or it may be processed within a few hours. The precipitate is collected in the centrifuge, washed three times with pure acetone, and promptly dried in vacuum over calcium chloride and wax shavings. The yield is 5.5 to 6.2 g. of a white powder which is 76 to 83% DPN (corrected for moisture content of 4%).

Further Purification by Ion Exchange Chromatography. Dowex 1 resin (200 to 400 mesh, Dow Chemical Company) is washed six to seven times with 3 N hydrochloric acid until the washings are essentially free of absorption at 260 mμ, two to three times with water, and then four to six times with 2 M sodium formate until the washings show only faint traces of chloride ion. Excess formate is removed by washing two to three times with water. (Five pounds of resin is washed in a 40-l. cylinder with stirring for 8 hours. The supernatant fluid and fine particles are decanted after settling overnight.)

One gram of 76% pure DPN (1082 μM.) is dissolved in water, adjusted to pH 8.0 with sodium hydroxide, and then diluted to a final volume of 500 ml. The solution is adsorbed on a column 14 cm. long and 6.6 cm. in diameter at a rate of 25 ml./min. and washed with an equal volume of water. Elution is with 0.1 N formic acid (Merck, 88%; 4.3 ml./l.) under a hydrostatic head of 2 m. which gives a flow rate of 20 ml./min. All connections in contact with the formic acid solution are made with Tygon tubing, since rubber is attacked by these solutions. Consecutive 250-ml. fractions are collected and examined at 260 mμ. Almost all (95.3%, 1031 μM.) of the DPN appears in fractions 5 (12.6%), 6 (65.5%), 7 (12.8%), and 8 (4.4%). (Enzymatic assays of these fractions check the ultraviolet absorption values within 2%.) These four fractions are combined and acidified to pH 2.0 with 6 ml. of 2 N hydrochloric acid, and then 5.5 l. of cold acetone is added slowly with constant agitation. The preparation is left overnight in the cold room, and the precipitate is collected on the centrifuge, washed three times with pure acetone, and dried in vacuum over calcium chloride and wax shavings. The yield is 0.65 g. of a white powder which is 95% DPN (corrected for moisture content of 6%). DPN preparations of lower purity (5 to 35%) may be purified to a level of 75 to 90% by the chromatographic procedure.

Properties and Purity of Product

DPN is stable in a desiccator over calcium chloride at room temperature and is relatively stable in cold neutral solutions for weeks. It is less stable at more acid pH's and is very labile toward alkali.[5] It forms a complex with alkaline cyanide by virtue of its nicotinamide-riboside moiety, which may be used for its estimation.[6] It also forms a fluorescent condensation product with acetone.[7] DPN is readily reduced by hydrosulfite. (See Vol. III [128] for details of assay methods.)

The purity is determined by measurement of the absorption at 340 mμ of the enzymatically reduced product. The extinction coefficient of reduced DPN at 340 mμ is 6.22×10^6 cm.^{2} per mole[8] (0.1 μM. of DPNH in a volume of 3.0 ml. with a light path of 1 cm. has an optical density of 0.207). Reduction is achieved by the use of ethanol and yeast alcohol

[5] F. Schlenk, "A Symposium on Respiratory Enzymes," University of Wisconsin Press, Madison, Wis., 1942.
[6] S. P. Colowick, N. O. Kaplan, and M. M. Ciotti, *J. Biol. Chem.* **191**, 447 (1951).
[7] J. W. Huff and W. A. Perlzweig, *J. Biol. Chem.* **167**, 157 (1947).
[8] B. L. Horecker and A. Kornberg, *J. Biol. Chem.* **175**, 385 (1948).

dehydrogenase.[9,10] This system will also determine deaminated DPN (resulting from hydrolysis of the 6-amino group on the purine ring).[11] The extinction coefficient of oxidized DPN at 260 mμ is 18.0×10^6 cm.^{2} per mole.

II. Isolation of Triphosphopyridine Nucleotide

(TPN, Coenzyme II)

By B. L. HORECKER and A. KORNBERG

Triphosphopyridine nucleotide was first isolated from horse erythrocytes by Warburg *et al.*[12] by a procedure which included fractional precipitation of mercury, lead, and barium salts, as well as fractional extraction with methanol. Warburg and his coworkers later adapted this method to liver as a starting material. The details of this modified procedure, which has not been published, were made available through the courtesy of E. Haas and form the basis for the first part of the present procedure.[13] The anion exchange method is based on the technique for mononucleotide separation of Cohn[14] and the gradient elution procedure of Busch *et al.*[15]

TPN has also been obtained from hog liver by LePage and Mueller[16] by charcoal adsorption and purification by charcoal chromatography.

Assay Method

Principle. The concentration of TPN is determined by measurement of the absorption at 340 mμ of the enzymatically reduced product. The extinction coefficient of reduced TPN at this wavelength is 6.22×10^6 cm.^{2} per mole.[17] (See Volume III [128] for other assay procedures.)

Reagents

Glucose 6-phosphate (0.02 *M*).[18]
0.1 *M* MgCl$_2$.
0.04 *M* glycylglycine buffer, pH 7.5.
Glucose-6-phosphate dehydrogenase.[18] Sufficient lyophilized powder was dissolved in water to give a solution containing at least 20 units/mg.

---

[9] E. Racker, *J. Biol. Chem.* **184,** 313 (1950).
[10] A. Kornberg, *J. Biol. Chem.* **182,** 779 (1950).
[11] See Vol. III [129].
[12] O. Warburg, W. Christian, and A. Griese, *Biochem. Z.* **282,** 157 (1935).
[13] B. L. Horecker and A. Kornberg, *Biochem. Preparations* **3,** 24 (1953).
[14] W. E. Cohn, *J. Am. Chem. Soc.* **72,** 1471 (1950).
[15] H. Busch, R. B. Hurlbert, and V. R. Potter, *J. Biol. Chem.* **196,** 717 (1952).
[16] G. A. LePage and G. C. Mueller, *J. Biol. Chem.* **180,** 775 (1949).
[17] B. L. Horecker and A. Kornberg, *J. Biol. Chem.* **175,** 385 (1948).
[18] See Vol. I [42].

Procedure. To 1.0 ml. of water in a quartz or Corex cell having a light path of 1 cm., add 0.25 ml. of buffer, 0.2 ml. of $MgCl_2$, and TPN sample containing from 0.02 to 0.10 μM. After an initial optical density reading at 340 mμ, add 0.02 ml. of the glucose-6-phosphate solution. Read at about 1-minute intervals until the end point is reached. From the density increment calculate the concentration of TPN

$$\left(\text{moles per liter} = \frac{d_{\text{final}} - d_{\text{initial}}}{(6.22 \times 10^6) \times 1.0} \right)$$

Purification Procedure

Starting Material. Sheep livers are obtained at the slaughterhouse as the animals are killed. The livers are cut into several slices and are thoroughly iced and processed in the laboratory within 2 hours after slaughtering. Calf, beef, and hog livers may be used, but the yields based on fresh weight of tissue are poorer. The purity of the crude preparations is lower, but the purity of the product of ion exchange chromatography is not affected.

TPN of 8 to 13% Purity after Precipitation by Mercury. Six kilograms of liver is ground in a chilled mechanical grinder and added at once with manual stirring to 9 l. of boiling water heated over several Meker burners. Stirring is continued for about 9 minutes, until the temperature rises to 85°. The mixture is rapidly cooled to 30° in a large ice bath and strained through four layers of coarse gauze. The residue is again extracted, with stirring, with 9 l. of water for 5 to 7 minutes at 78 to 83°, cooled, and strained. Subsequent steps are carried out in a cold room. The combined extracts from 12 kg. of liver (36.3 l.) are treated with 3.0 l. of 90% trichloroacetic acid (w/v) and filtered on fluted filters (Schleicher and Schuell No. 588 paper). The filtrate (36.9 l.) is neutralized to bromothymol blue with 50% potassium hydroxide (1.75 l.). The solution (38.8 l.) is treated with 583 ml. of a freshly prepared 10% (w/v) solution of mercuric acetate and filtered on fluted filters. It is necessary to pour the filtrate back once or twice to clarify it. The filtrate (38.0 l.) is treated with 7030 ml. of a freshly prepared 10% (w/v) solution of mercuric acetate. The pH is then adjusted with 50% (w/v) potassium hydroxide to a value between 5 and 6; about 240 ml. is required.

The precipitate settles out overnight in the cold room. The clear supernatant liquid is siphoned and discarded; the precipitate is collected on two large Büchner funnels. The filter cake is washed once by suspending it in 8 l. of water and filtering. The washed cake is well suspended in 2 l. of cold water and treated with 4 N nitric acid (about 30 ml.) until a royal blue color is obtained with Congo red paper. The chilled suspension is then treated with hydrogen sulfide with gentle agitation for about 1.5 hours. The solution is centrifuged, and the residue is washed twice with 500-ml. portions of 0.2 N nitric acid. The combined supernatant liquid and washings (about 3.0 l.) are aerated in the cold for 15 to 30 minutes; 14 l. of cold acetone ($-10°$) is added. After remaining overnight at $2°$ the precipitate is collected in the centrifuge, washed three times with pure acetone, and promptly dried in vacuum over calcium chloride and potassium hydroxide. The yield is 5.1 to 8.5 g. of gray powder which contains 7 to 13% TPN (uncorrected for moisture content). This powder can be stored in vacuum at $2°$ without loss for 2 years or more.

Further Purification by Ion Exchange Chromatography. Eight and one-tenth grams of crude TPN (760 μM.) is dissolved in 80 ml. of water and adjusted to pH 6.4 with 2 M NH$_4$OH (39.0 ml.). The solution is centrifuged, the residue discarded, and the supernatant liquid adjusted to pH 7.7 with 11.0 ml. of 2 N NH$_4$OH. This solution is placed on a Dowex 1-formate (10% cross-linked) column 16.5 cm. in length and 3.5 cm. in diameter and adsorbed at the rate of 5 ml./min. The column is then washed with 100 ml. of water. Elution is begun with 1 l. of water in a closed mixing bottle equipped with a magnetic stirrer. As water is withdrawn from this vessel, it is replaced from a reservoir containing a solution which is 0.2 M with respect to formic acid and 0.2 M with respect to sodium formate. All connections in contact with the eluting solution are made with Tygon tubing, since rubber is attacked by formic acid solutions. Fractions of 24 ml. are collected at a rate of about 4 ml./min. These fractions are examined for ultraviolet absorption at 260 mμ and assayed enzymatically for TPN. With the gradient elution technique, TPN is eluted in a narrow fraction with a peak concentration of about 2 μM./ml.

Fractions 59 to 87, containing 650 μM. of TPN, together with a 25-ml. rinse are pooled and acidified with 5.0 ml. of concentrated nitric acid until royal blue to Congo red paper. The TPN is precipitated by the addition of 3.0 l. of cold acetone.

The flocculent precipitate is allowed to settle overnight in the cold room, and the clear supernatant solution is siphoned off and discarded. The precipitate is collected by centrifugation, washed with cold acetone, and dried in vacuum. If the precipitate fails to flocculate, the whole suspension must be centrifuged. Precipitate that adheres to the walls of the vessel and the centrifuge bottles is recovered by rinsing them with several small volumes of water. Any insoluble residue which is present is removed by centrifugation, and the TPN is reprecipitated from the supernatant solution with 4 vol. of cold acetone. This precipitation is carried out directly in the centrifuge bottle in which the precipitate can be collected, washed with cold acetone, and dried in vacuum. The yield is 527 mg. (78%) of a grayish-white powder which contains 88% pure TPN (corrected for a moisture content of 5%). Significant amounts of TPN may adhere to the walls of the centrifuge bottle and can be recovered in solution with a water rinse. This preparation is stable indefinitely when stored in vacuum at 2°.

A procedure for the large-scale preparation of TPN by the use of the DPN kinase from pigeon liver[19] has recently been described by Wang et al.[20]

[19] See Vol. II [111].
[20] T. P. Wang, N. O. Kaplan, and F. E. Stolzenbach, *J. Biol. Chem.* **211**, 465 (1954).

[125] Crystallization of DPN as the Quinine Salt and Preparation of Pure DPN

By Kurt Wallenfels and Walter Christian

Preparation of DPN-Quinine Salt

Principle. To a concentrated aqueous solution of crude DPN is added excess quinine and a mixture of ethanol and benzene. DPN crystallizes as the quinine salt, which may be decomposed by means of an ion exchange resin, silver nitrate, or mercuric acetate. From the salt 100% pure free DPN is obtained.[1]

Reagents

DPN, 60 to 70% pure.
Mixture of ethanol and benzene (300 ml. of absolute ethanol mixed with 700 ml. of thiophene-free benzene).
Quinine base, pure.
Weakly basic anion exchanger in acetate form (Amberlite IR-4B).
Silver nitrate solution, 20%.
Mercuric acetate, 20% solution.
Formic acid, 8.5% solution.
Acetic acid, 1 N and 0.1 N.

Procedure. Two grams of 60 to 75% pure DPN is dissolved in 20 ml. of water, traces of insoluble matter are centrifuged off, and a solution of 2.2 g. of quinine base in 17 ml. of absolute ethanol is added, giving a clear solution. The ethanol-benzene mixture is added with vigorous stirring at

[1] K. Wallenfels and W. Christian, *Angew. Chem.* **64**, 419 (1952), D. B. P. no. 913406 (1954), B. P 730.444 (1955).

room temperature until, after addition of 400 to 450 ml., the two phases just mix. The homogeneous mixture is allowed to stand at room temperature for several hours with occasional rubbing of the vessel with the glass rod. The quinine salt of DPN precipitates in fine needles which often aggregate in spherical clusters. Crystallization is complete after 48 to 72 hours. The crystals are centrifuged, washed with the ethanol-benzene mixture, and dried *in vacuo* over paraffin and P_2O_5. The yield is 2.9 g. of crystalline quinine salt. For recrystallization it is dissolved slowly with gentle warming in ten times its weight of 50% ethanol. The quinine salt crystallizes from the clear solution on addition of about 10 vol. of ethanol-benzene mixture until the phases again just mix.

Properties of DPN-Quinine Salt

Crystals are thin needles, often combined in clusters. The loss of weight on drying *in vacuo* at 58° is 5.1%. The melting point is not sharp, about 162 to 170°. The quinine salt is difficultly soluble in water, but once dissolved by warming it does not crystallize from the solution. It is easily dissolved by adding acid (acetic, formic, mineral). Mole ratio, DPN:quinine::2:3.

Decomposition of Quinine Salt and Preparation of Pure DPN

Method 1. Decomposition with Ion Exchange Resin. Two grams of quinine salt is dissolved in 20 ml. of 1 N acetic acid at 0°, and 50 ml. of ice water is added. The cold solution is placed on a column of 20 ml. of anion exchange resin IR-4B in acetate form kept in a cold room at 0 to 2°. The exchanger adsorbs DPN, but quinine is not retained and can be completely washed from the column with 0.1 N acetic acid. DPN is eluted with ice-cold 8.5% formic acid and collected in 5-ml. fractions. To each fraction 5 vol. of cold acetone is added to precipitate the DPN. The precipitate is washed with acetone and dried. The DPN content as determined with yeast alcohol dehydrogenase shows the first fractions to be 100% pure, the last 90 to 75%. If the decomposition is carried out at room temperature, the greatest portion of the resultant DPN is 80 to 85% pure.

Method 2. Decomposition with Mercuric Acetate. One gram of quinine salt is dissolved in 50 ml. of 0.5 N acetic acid. To this is added 2 ml. of 20% mercuric acetate solution, and the mixture is centrifuged. Half the volume of ethanol is added to the clear supernatant fluid. The resultant precipitate is washed with 1% mercuric acetate solution in 50% aqueous ethanol and centrifuged at high speed. The residue is suspended in 3 ml. of water at 0°, and H_2S is introduced with cooling. Mercury sulfide is centrifuged and washed with a few drops of water. The combined solutions are freed of H_2S by evacuation, and 0.1 vol. of 85% formic acid is added in the cold. DPN is precipitated with 5 vol. of cold acetone, centrifuged, washed with cold acetone, and dried. The yield of DPN is 95 to 98% pure.

Method 3. Decomposition with Silver Nitrate. One gram of quinine salt is dissolved in 25 ml. of 0.1 N HNO_3, and 1.25 ml. of 20% silver nitrate solution is added. The silver salt is precipitated with 25 ml. of ethanol, washed with ethanol and ether, and dried. It is then suspended in the minimum volume of water and just dissolved by adding 2 N HNO_3. H_2S is introduced with cooling, and the precipitate is centrifuged. To the clear solution 5 vol. of acetone is added, and the precipitate is washed with acetone and dried. The purity of the resulting DPN is 92 to 96%.

Test for Homogeneity. In addition to the enzymatic test, paper electrophoresis is eminently suitable for testing the homogeneity of the preparations. For this 0.02 ml. of DPN solution containing 0.4 mg. of activity is applied to a strip of Whatman No. 1 paper 5 cm. wide. 0.025 M phosphate buffer pH 7.6 and 10 volts per centimeter are used. At the end of the run the strip is dried and photographed in light of wavelength 273.7 mμ. The starting material shows three heavy bands, the quinine salt two bands, and the DPN prepared from the quinine salt only one band. With the given concentration as little as 5% of impurity absorbing at 260 mμ is clearly detectable. If after electrophoresis the dry paper strip is sprayed with pyrophosphate buffer, pH 8.6, containing semicarbazide, ethanol, and alcohol dehydrogenase, the DPN is reduced on paper to DPNH. The DPNH band is visualized in light of 366-mμ wavelength. In preparing DPN from the quinine salt rapidly and with effective cooling, a product is obtained with extinction coefficient 9.4 cm.^{2}/mg. [2] After a short period of storage in the desiccator at 0° this value falls to 8.3 cm.^{2}/mg. and then remains constant for some time. In the electrophoretic test preparations of 8.3 cm.^{2}/mg. behave identically with those of 9.4 cm.^{2}/mg. The quinine salt retains its properties and activity in the enzymatic assay during storage for over two years at 38°.

[2] B. L. Horecker and A. Kornberg, *J. Biol. Chem.* **175,** 385 (1948).

[126] Preparation of Reduced DPN (Chemical Method)

By ALBERT L. LEHNINGER

DPNH has often been prepared by chemical reduction in solution and such solutions used directly in enzyme experiments after oxidation of excess hydrosulfite (the reducing agent) by O_2.[1] Such solutions of course contain components other than DPNH, such as sulfite and sulfate. These may be inhibitory to enzymes or may cause undesirable side reactions. For certain types of work it is therefore necessary to have available DPNH in a solid form of high purity and reasonable stability, free of such undesirable contaminants.

There are two types of such preparations of DPNH which have gained some use: the disodium salt described by Ohlmeyer[2] and the barium salt described by Lehninger.[3] In both cases, the DPN is reduced chemically with hydrosulfite. In the former procedure, the disodium salt is isolated subsequent to the removal of sodium sulfite and sodium sulfate formed from the hydrosulfite by fractional crystallization from the aqueous solution at very low temperatures. Ohlmeyer obtained a disodium salt of apparently very high purity; Drabkin, using this method, was also able to obtain preparations of apparently high purity.[4] In this laboratory, however, experience has been much less satisfactory, despite many trials of the Ohlmeyer method as it was published or with a variety of modifications. Although the procedure is simple and the product useful for many purposes, it was not found possible to remove sulfite and sulfate completely. Since these were extremely undesirable contaminants in certain situations, it was necessary to devise another method of isolation of DPNH from the reduction medium. The preparation described below[3,5] was found to yield a highly purified barium salt free of these contaminants.

Procedure. In a flask, 500 mg. of DPN (purity = 86%, total of 662 micromoles) was dissolved in 41.0 ml. of 1.3% $NaHCO_3$, and 250 mg. of sodium hydrosulfite (1444 micromoles) was quickly added with swirling. The contents were then gassed with 95% N_2–5% CO_2. The vessel was closed off and kept at 25° for 2 hours. At the end of this time the contents were vigorously gassed with oxygen for 15 minutes to oxidize excess hydrosulfite. The slightly yellow solution contained 650 micro-

[1] O. Warburg and W. Christian, *Biochem. Z.* **297**, 66 (1938).
[2] P. Ohlmeyer, *Biochem. Z.* **297**, 66 (1938).
[3] A. L. Lehninger, *J. Biol. Chem.* **190**, 345 (1951).
[4] D. L. Drabkin, *J. Biol. Chem.* **157**, 563 (1945).
[5] A. L. Lehninger, *Biochem. Preparations* **2**, 92 (1952).

moles of DPNH as determined by light absorption at 340 mμ, using the molar extinction coefficient 6.22×10^6 cm.^{2} per mole. The solution was then chilled to 0°, and to it was added dropwise with stirring 4.0 ml. of 2 M barium thiocyanate. After addition was complete, the pH of the suspension was brought to 7.5 by the addition of about 0.4 ml. of 2.5 N NaOH. The precipitated barium salts were removed by centrifugation and discarded. (This precipitate contains the bulk of the sulfite and sulfate.) Barium thiocyanate was chosen, since this barium salt is quite soluble in ethanol-water mixtures and the presence of contaminating thiocyanate in the final product can be easily detected colorimetrically as the ferric complex. Barium iodide or bromide is also appropriate. To the clear supernatant was added an equal volume of cold ethanol. After standing for 20 minutes, the slight flocculent precipitate was centrifuged off and discarded. The clear, slightly yellow supernatant solution contained 579 micromoles of DPNH. An additional 5 vol. of cold ethanol was added to precipitate the barium salt of DPNH, which was recovered at the centrifuge and washed and dried with absolute ethanol and finally with ether.

A total of 522 micromoles of DPNH was recovered in the form of a slightly yellow powder weighing 535 mg., indicating a purity of 85% on the basis of the composition $C_{21}H_{27}O_{14}N_7P_2Ba\cdot4H_2O$. This material was found to give a faint test for thiocyanate and contained considerable barium carbonate. It was dissolved in 12.0 ml. of ice-cold carbon dioxide-free water, and some insoluble material was removed by centrifugation. Then 13.0 ml. of water was added, followed by 25.0 ml. of ethanol, and the solution was clarified by centrifugation in the cold. The barium salt of DPNH was then precipitated by the slow addition of 140 ml. of cold ethanol and washed and dried at the centrifuge with ethanol and ether. The material was dried over calcium chloride. It weighed 476 mg. and contained 501 micromoles of BaDPNH$\cdot4H_2O$, indicating a purity of 92% and an over-all yield of 75%. Yields in eight other preparations were between 65 and 69%. A checker in another laboratory reported a product of high purity but a yield of only 29%.

For use in enzyme experiments the barium salt is dissolved in water, and the barium is removed with an excess of either sodium sulfate or disodium phosphate, the choice of precipitant depending on the enzyme reaction being studied. There is no measurable adsorption of DPNH on BaSO$_4$ or Ba$_3$(PO$_4$)$_2$ precipitates. Stock solutions prepared in this way are stable for a day or two at pH 7.8 in the frozen state.

Properties and Purity of Product. The barium salt of reduced DPN is faintly yellow and dissolves readily in water to give a clear solution. On drying in a high vacuum over phosphorus pentoxide at 60° for 4 hours, the above preparation lost 7.45% of its weight, slightly less than calculated for 4 moles of water per mole of DPNH. Drying at 100° caused loss of weight corresponding to nearly 7 moles of water. The material contained less than 0.004 mole of —SCN⁻ per mole of DPNH, estimated colorimetrically as the ferric complex. The material contained carbonate or bicarbonate as measured manometrically in a Warburg vessel by tipping acid into an aqueous solution of the material in carbon dioxide-free water. The evolved gas was completely absorbable by sodium hydroxide. Calculated as barium carbonate, this impurity amounted to 3.9%.

Analysis. Found: N, 11.4%; P, 7.01% (on basis of desiccator-dry weight as given above). Calculated for $C_{21}H_{27}O_{14}N_7P_2Ba \cdot 4H_2O$: N, 11.23%; P, 7.10%. From the analytical and spectrophotometric data presented, it appears permissible to conclude that the preparation was about 92% pure and that about half of the impurity was $BaCO_3$. It is possible that the rest may be moisture not removed at 60° but labile at 100°.

Such preparations of DPNH were stored in the cold over calcium chloride. Preparations held in a desiccator for a year at room temperature showed a decline of the 340 mμ : 260 mμ ratio to about 0.26 and a decline of enzymatic assay values to about 70% purity.

The method of isolation of DPNH described has been found not to yield any purification of DPN; final products are generally no purer than the DPN used as starting material. For the preparation of enzymatically reduced DPN, see the following article, Vol. III [127].

Preparation of TPNH. In the author's knowledge, there is no report in the literature of the isolation of TPNH as either the disodium or barium salt. However, a single preparation of the barium salt of TPNH has been made in the author's laboratory by exactly the same procedure described above. The starting material was 49% pure, and the final product was estimated to be about 35% pure. There were some indications that TPNH is somewhat less stable than DPNH and that the reduction and isolation procedures were less satisfactory, since the yield was only 46%.

[127] Enzymatic Preparation of DPNH and TPNH

By GALE W. RAFTER and SIDNEY P. COLOWICK

Procedure

DPNH. A typical preparation is as follows: 200 mg. of DPN is dissolved in 20 ml. of unadjusted 0.5 M tris(hydroxymethyl)aminomethane containing 0.5 M ethyl alcohol. Approximately 1 mg. of crystalline yeast alcohol dehydrogenase is added (see Vol. I [79]). After 30 minutes of incubation at 25°, a 0.05-ml. aliquot is removed from the reaction mixture and added to 3.0 ml. of 0.1 M tris(hydroxymethyl)aminomethane contained in a 1-cm. Beckman cuvette. The optical density is read at 340 mμ. If the expected optical density is obtained, 1.0 ml. of 25% barium acetate followed by 10 ml. of 95% ethyl alcohol is added to the reaction mixture. The precipitate that forms is removed by centrifugation and discarded. To the clear supernatant an additional 100 ml. of 95% ethyl alcohol is added and the solution is allowed to stand for 2 hours at 0°. The precipitated barium-DPNH is collected by centrifugation and washed with absolute ethanol, 1:1 absolute ethanol-diethyl ether, and finally with diethyl ether. The product is dried in a vacuum desiccator containing calcium chloride. It is important to dry the material rapidly, since prolonged contact of the moist precipitate with air gives a yellow gumlike material rather than the slightly yellow powder obtained by rapid drying *in vacuo*. For use, the barium salt of DPNH is dissolved in water and a slight excess of sodium sulfate is added. The precipitated barium sulfate is removed by centrifugation.

The purity of the DPNH is equivalent to the purity of the DPN used as a starting material. Over-all yields of barium DPNH of approximately 70% are readily obtained. The product contains essentially no DPN. Both the solid barium salt and the DPNH in solution are stored at −15°.

The above procedure is essentially the same as one described earlier by Pullman *et al.*[1] Boiling the solution to inactivate the enzyme, as described by Pullman *et al.*, has been avoided in the present procedure. Nygaard and Theorell[2] describe the same modification. The boiling step is likely to result in some oxidation of the DPNH if the solution is not strongly alkaline.

TPNH. The following description is that of Evans and Nason:[3] "40 mg. of TPN were dissolved in 15.4 ml. of 0.1 M phosphate buffer, at pH 7.5, and 0.8 ml. of 0.1 M MgCl$_2$, and 3 ml. of 0.05 M D-isocitrate were added. The reaction was started by the addition of 0.8 ml. of a phosphate buffer (pH 7.5) extract of washed, acetone-dried, pig heart, containing isocitric dehydrogenase.[4] The reaction was allowed to proceed at room temperature for 30 to 45 minutes, and at intervals during this period 0.05-ml. aliquots were removed, diluted to 3.0 ml. with 0.1 M phosphate buffer at pH 7.5, and optical densities determined in a Beckman spectrophotometer at 340 mμ. The reaction was considered complete when a maximum increase in optical density was obtained. At the end of the reaction, the TPNH preparation was adjusted to pH 9.0 to 9.5,

[1] M. E. Pullman, S. P. Colowick, and N. O. Kaplan, *J. Biol. Chem.* **194,** 593 (1951).
[2] A. P. Nygaard and H. Theorell, *Acta Chem. Scand.* **9,** 1300 (1955).
[3] H. J. Evans and A. Nason, *Plant Physiol.* **28,** 233 (1953).
[4] See Vol. I [116].

placed in a boiling water bath for three minutes, centrifuged for five minutes, and the supernatant solution was used as a source of TPNH. This solution was shown to be active enzymatically by preliminary tests with partially purified glutathione reductase from peas.[5] With this system the absorption decreased to the original starting value."

No satisfactory procedure has been developed for isolating TPNH free of contaminating substrates after enzymatic reduction. Isolation as the barium salt, as described above for DPNH, is not feasible in this case because of the presence of di- and tricarboxylic acids having alcohol-insoluble barium salts.

Properties

The following statements are made with specific reference to DPNH, but they apply also to TPNH.

Extinction Coefficient. DPNH has an extinction coefficient of 16×10^6 cm.^{2}/mole at 260 mμ, attributable to the adenine moiety,[6] and of 6.25×10^6 cm.^{2}/mole at 340 mμ, attributable to the reduced nicotinamide moiety. The latter value is the average of the following three independent determinations. Ohlmeyer[7] has obtained a value of 6.28×10^6 cm.^{2}/mole for highly purified DPNH, isolated after chemical reduction. Wallenfels and Christian[8] have reported, for DPNH derived from the crystalline quinine salt of DPN, a value of 9.4 cm.^{2}/mg., which, on the basis of a molecular weight of 665, corresponds to 6.25×10^6 cm.^{2}/mole. Horecker and Kornberg,[9] using a method which is independent of the purity of the DPNH sample but depends instead on the purity of the oxidant, reported a value of 6.22×10^6 cm.^{2}/mole.

Effect of Acid and Alkali. DPNH is extremely stable in alkali (e.g., 0.1 N NaOH at 100° for 10 minutes has no effect) but is readily destroyed at pH values of 4 or lower, even at room temperature. The product formed by acid shows no absorption at 340 mμ but high absorption at 290 mμ [10] and is similar to, but not identical with, the product (DPNH-X) formed by the action of triosephosphate dehydrogenase on DPNH.[11] The acid product undergoes further transformation, with loss of 290-mμ absorption, when stronger acid is used.[11]

DPNH prepared by dithionite reduction (and thereby containing sulfite) is considerably more acid-labile than that prepared enzymatically. Whereas the latter may be used at pH values as low as 5 without appreci-

[5] See Vol. II [126].
[6] S. P. Colowick, N. O. Kaplan, and M. M. Ciotti, *J. Biol. Chem.* **191**, 447 (1951).
[7] P. Ohlmeyer, *Biochem. Z.* **297**, 66 (1938).
[8] See Vol. III [125].
[9] B. L. Horecker and A. Kornberg, *J. Biol. Chem.* **175**, 385 (1948).
[10] E. Haas, *Biochem. Z.* **288**, 123 (1936).
[11] G. W. Rafter, S. Chaykin, and E. G. Krebs, *J. Biol. Chem.* **208**, 799 (1954).

able destruction, the former undergoes decomposition even at pH 7. Sulfite has the additional property of stabilizing the 290-mμ absorbing material derived from acidified DPNH, so that it resists decomposition by stronger acid.[10]

Reaction with Redox Dyes. DPNH does not react appreciably with most redox dyes, e.g., 2,6-dichlorophenolindophenol or methylene blue, at pH values of 7 or higher. Those dehydrogenase assays described in the literature in which the reduction of a dye is taken as a measure of DPNH formation are not valid, since a flavoprotein (diaphorase) is required for reduction of the usual dyes by DPNH. Only at pH values below 7 does the rate of nonenzymatic reaction of DPNH with dyes become appreciable.[12]

[12] G. W. Rafter and S. P. Colowick, *Federation Proc.* **14**, 267 (1955).

[128] Procedures for Determination of Pyridine Nucleotides

By MARGARET M. CIOTTI and NATHAN O. KAPLAN

The general procedures for the analysis of oxidized and reduced forms of DPN and TPN are given in the accompanying table.

SUMMARY OF PROCEDURES FOR PYRIDINE NUCLEOTIDE ASSAYS

Total DPN and TPN	Cyanide addition, methyl ethyl ketone, or strong alkali fluorescence before and after treatment with *Neurospora* DPNase (see procedures I-A, II-A, and II-B)
DPN	Yeast alcohol dehydrogenase and ethanol (see procedures I-B and II-C-1)
TPN	Pig heart isocitric dehydrogenase (see procedures I-C and II-C-2)
Total DPNH and TPNH	*Clostridium kluyveri* oxidase (see procedures I-D and II-D-1)
DPNH	Yeast alcohol dehydrogenase and acetaldehyde (see procedures I-E and II-D-2)
TPNH	TPNH cytochrome c reductase or TPNH glutathione reductase (see procedures I-F, I-F-2, and II-D-3)

I. Spectrophotometric Determinations

A. DPN and TPN by the Cyanide Addition Reaction

Principle. When cyanide is added to DPN or TPN, an addition product is formed.[1] These products show a new absorption peak at 325 mμ. The millimolar extinction coefficient of both the DPN-CN and TPN-CN complexes at 325 mμ is 6.3.

Reagents

1.0 *M* potassium cyanide.

[1] S. P. Colowick, N. O. Kaplan, and M. M. Ciotti, *J. Biol. Chem.* **191**, 447 (1951).

Procedure. An aliquot of 0.1 to 0.5 ml. containing 0.1 to 0.8 micromole of DPN or TPN is added to a cuvette. A final volume of 3 ml. is obtained by the addition of 1.0 M KCN. After 1 minute the optical density at 325 mμ is measured. It is necessary to read the unknown against a cyanide blank.

Discussion. Other N-substituted nicotinamide derivatives give a similar spectrum in the presence of cyanide.[1] The millimolar extinction coefficient of the nicotinamide mononucleotide and nicotinamide riboside cyanide complexes is identical with that of DPN and TPN. The millimolar extinction coefficient of N-methyl nicotinamide-CN complex (in aqueous solution) is much lower.[1] In order to make the assay specific for DPN and TPN, samples are treated with cyanide before and after reaction with *Neurospora* DPNase (see below). Reduced DPN and TPN do not react with cyanide. The quatenary nitrogen of the nicotinamide ribose link is essential for the reaction with cyanide. The cyanide reaction is an equilibrium reaction which is dependent on hydrogen ion concentration. Lowering the pH or cyanide concentration will affect the level of the complex formed.[2]

Neurospora DPNase Assay. The sample to be treated with *Neurospora* DPNase[3] is added to an acetate or phosphate buffer solution (pH range 3 to 9). Ten to eighty units (see Vol. II [114]) of enzyme are added, and the mixture is incubated at 37° for 7.5 minutes. The difference between the reading of this sample and the untreated sample is a measure of DPN and TPN present. The decrease in cyanide complex represents cleavage at the nicotinamide riboside bond. Other enzymes split DPN at the pyrophosphate linkage. This latter type of cleavage does not produce a change in the cyanide readings.[1]

B. DPN Determination by Alcohol Dehydrogenase (ADH)

Principle. DPN is reduced in the presence of ethanol and yeast alcohol dehydrogenase. The reduced DPN is measured by its absorption at 340 mμ.

Reagents[1,4]

0.5 M ethanol in 0.1 M tris(hydroxymethyl)aminomethane (pH 10.1) or in 0.1 M sodium pyrophosphate.

Yeast ADH (see Vol. I [79]).[5]

[2] N. O. Kaplan, *Record Chem. Progr.* (*Kresge-Hooker Sci. Lib.*) **16**, 177 (1955).
[3] N. O. Kaplan, S. P. Colowick, and A. Nason, *J. Biol. Chem.* **191**, 473 (1951).
[4] E. Racker, *J. Biol. Chem.* **184**, 313 (1950).
[5] The commercial preparation obtained from Worthington Biochemical Company is suitable for this type of assay.

Procedure. To 2.5 ml. of the ethanol-tris (or sodium pyrophosphate) mixture 0.1 to 0.8 micromole of DPN is added, and the volume is brought to 3 ml. with H_2O. The optical density is taken at 340 mμ. From 10 to 50 γ of yeast ADH is then added. The final reading is taken after 1 to 3 minutes. The amount of DPN present is determined by the optical density change and can be calculated by using the millimolar extinction coefficient for DPNH (6.3).

Discussion. Pyrophosphate appears to be a better reagent than Tris, particularly when heavy metals are present in the reaction mixture. This procedure is quite specific for DPN. The reduction of TPN is about 1/1500 that of DPN.

Several analogs of DPN also react at a slower rate with the yeast ADH.[6] Nicotinamide mononucleotide and nicotinamide riboside are inactive.

Since the reduction of DPN is quite unfavorable, hydrogen ion and ethanol concentrations are important factors in an accurate determination of DPN. This assay is more specific than the cyanide addition method; however, it will not distinguish the cleavage of the nicotinamide riboside linkage from the hydrolysis of the pyrophosphate grouping of the coenzyme.

C. TPN Determination by Isocitric Dehydrogenase

Principle. Isocitrate is oxidized by isocitric dehydrogenase with TPN as the oxidizing agent. The TPNH is measured by its extinction at 340 mμ.

Reagents

> 0.05 M sodium isocitrate.
> 0.1 M potassium phosphate buffer, pH 7.5.
> 0.1 M magnesium chloride.
> Isocitric dehydrogenase (see Vol. I [116]).

Procedure. A solution containing 0.1 to 0.8 micromole of TPN is added to a cuvette. The following additions are then made: 0.1 ml. of 0.1 M $MgCl_2$, 0.1 ml. of 0.05 M Na isocitrate, and 0.1 M potassium phosphate, pH 7.5, to a final volume of 3 ml. The optical density of the sample is read at 340 mμ. To start the reaction, 0.1 ml. of isocitric dehydrogenase is added. After about 2 minutes the sample is read again. The amount of TPNH can be calculated from the increase in optical density using a millimolar extinction coefficient of 6.3.

[6] N. O. Kaplan, M. M. Ciotti, and F. E. Stolzenbach, *J. Biol. Chem.* **221**, 833 (1956).

Discussion. The crude extract from a pig heart acetone powder contains both a TPN and a DPN isocitric dehydrogenase.[7,8] Although ammonium sulfate fractionation of the crude extract does not result in a great deal of purification, it does tend to separate the two enzymes. The DPN-specific enzyme usually comes down in the 40 to 50% fraction and the TPN system in the 50 to 60% fraction. It is wise, however, to test all the fractions which are to be used for TPN determinations for DPN isocitric dehydrogenase activity.

Magnesium or manganese ions are essential for the TPN isocitric dehydrogenase. Care should be taken that adequate levels of metal are present when assaying for TPN in solutions which may contain chelating agents.

D. Determination of TPNH and DPNH

Principle. The enzyme from *Clostridium kluyveri* oxidizes reduced DPN and TPN in the presence of oxygen. The decrease at 340 mμ represents the level of the total reduced coenzymes.

Reagents

0.1 M potassium phosphate buffer, pH 7.5.
Dialyzed crude extract of *Clostridium kluyveri*.[9]

Procedure. An aliquot containing 0.1 to 0.8 micromole of DPNH or TPNH is added to a cuvette, and 0.1 M potassium phosphate buffer (pH 7.5) is added to bring the volume to 3 ml. An initial reading is taken at 340 mμ. The reaction is started by adding 0.1 ml. of the *Clostridium kluyveri* extract. The decrease in optical density indicates the amount of DPNH or TPNH that is present in the sample. The *kluyveri* extract should be dialyzed, since the extract usually contains a high concentration of pyridine nucleotides.

E. DPNH Determination

Principle. DPNH is readily oxidized by yeast alcohol dehydrogenase when acetaldehyde is present.

Reagents

1/40 dilution of yeast alcohol dehydrogenase in 0.1 M K$_2$HPO$_4$.
0.5 M acetaldehyde.
0.1 M potassium phosphate buffer, pH 7.5.

---

[7] See Vol. I [119].
[8] N. O. Kaplan, S. P. Colowick, and E. F. Neufeld, *J. Biol. Chem.* **205**, 1 (1953).
[9] M. M. Weber and N. O. Kaplan, *Bact. Proc.* **1954**, 96. See Vol. I [84] for the media for growing *Clostridium kluyveri* and the method of making the extract.

Procedure. DPNH (0.1 to 0.8 micromole) is mixed with 0.02 ml. of 0.5 M acetaldehyde and 0.1 M potassium phosphate buffer (pH 7.5) to a final volume of 3 ml. After the optical density is read at 340 mμ, 0.1 ml. of a 1/40 dilution of yeast alcohol dehydrogenase is introduced to initiate the reaction. The decrease in optical density determines the amount of DPNH present. The reaction should be complete within 1 minute.

Discussion. TPNH will also be oxidized by the ADH at a slow rate. Hence it is important to make a proper dilution of the enzyme so that no oxidiation of TPNH occurs under the conditions of the assay.

F. TPNH Determinations

1. TPNH Cytochrome C Reductase[10]

Principle. The cytochrome c reductase isolated from liver specifically requires TPNH as the reducing agent, and therefore can be used for the assay of TPNH.

Reagents

> 5% cytochrome c.
> 0.1 M potassium phosphate buffer, pH 7.8.
> TPNH cytochrome c reductase from liver (see Vol. II [123]).

Procedure.[11] The assay mixture is as follows: 0.05 ml. of a 5% solution of cytochrome c, 0.01 to 0.04 micromole of TPNH, and 0.1 M potassium phosphate buffer (pH 7.8) to a final volume of 3 ml. An initial reading is taken at 550 mμ. With optimal conditions the reaction is complete in 15 to 30 minutes. The increase in optical density after the addition of the cytochrome c reductase is a measure of the amount of TPNH present in the system.

Discussion. It is always advisable to run a control—that is, a sample without enzyme. Many tissue extracts will directly reduce cytochrome c and dyes; therefore this procedure is somewhat limited.

2. TPNH Glutathione Reductase

Principle. TPNH acts as the reducing agent in the presence of oxidized glutathione when an enzyme from either peas[12] or wheat germ[13] is added.

Reagents

> 0.1 M potassium phosphate, pH 7.5.
> 0.1 M glutathione (oxidized).
> Glutathione reductase.

[10] B. L. Horecker, *J. Biol. Chem.* **183,** 593 (1950).
[11] N. O. Kaplan, S. P. Colowick, and E. F. Neufeld, *J. Biol. Chem.* **195,** 107 (1952).
[12] L. W. Mapson and D. R. Goddard, *Biochem. J.* **49,** 592 (1951).
[13] See Vol. II [126].

Preparation of the Pea Enzyme.[8] Mechanically powdered dried peas are extracted with 10 vol. of cold 0.1 M phosphate buffer (pH 6.7). The extract is fractionated with ammonium sulfate. The precipitates are dissolved in 0.1 M Tris (pH 7.5). Most of the enzyme activity is present in the 40 to 50% and 50 to 60% $(NH_4)_2SO_2$ fractions. The fractions contain little if any DPNH glutathione reductase activity.

Procedure. Ten micromoles of glutathione and 0.1 to 0.8 micromole of TPNH are brought to a volume of 3 ml. with 0.1 M buffer (pH 7.5). Then 0.1 ml. of the glutathione reductase is used to start the reaction. The decrease in optical density at 340 mμ from the initial to the final reading is an indication of the amount of TPNH present. It is also possible to follow this reaction by the appearance of SH groups (see Vol. II [126]).

Discussion. This method is particularly useful in assaying tissue extracts for TPNH. It is difficult to employ the cytochrome c reductase or the TPNH diaphorase present in spinach,[14,15] since there is usually a rapid endogenous reduction of dye or cytochrome c. In the glutathione system, the oxidation of TPNH is determined. Endogenous reduction of glutathione does not become a factor, since an excess of the oxidized glutathione is present.

II. Fluorimetric Methods

A. DPN and TPN by Alkali Addition

Principle.[16] The oxidized pyridine nucleotides form a fluorescent compound on the addition of strong alkali.

Reagents

5.0 N sodium or potassium hydroxide.

Procedure. A sample of not more than 1 ml. is brought to a final volume of 8 ml. with 5 N NaOH or KOH. The solution is placed in a boiling-water bath for 5 minutes. The samples are cooled and read on a Coleman fluorometer with thiamine filters (B_1 and PC_1). The assay is accurate for 4 to 20 γ of pyridine nucleotides (DPN or TPN). By using the Farrand fluorometer, the method can be made much more sensitive.[17] Ultraviolet light destroys the fluorescence produced by the reaction of DPN with strong alkali.[16] By carrying out the reaction in 5 N NaOH (i.e., 1 ml.) and then diluting to 8 ml. with H_2O, however, fluorescence which is equal in magnitude and stable to ultraviolet light can be obtained.[17]

---

[14] The TPNH diaphorase from spinach has been described and purified by Drs. Jagendorf and Avron of the McCollum-Pratt Institute (in preparation).

[15] K. B. Jacobson and N. O. Kaplan, in preparation.

[16] N. O. Kaplan, S. P. Colowick, and C. C. Barnes, *J. Biol. Chem.* **191**, 461 (1951).

[17] O. H. Lowry, N. R. Roberts, J. L. Kapplahn, and C. Lewis, *Federation Proc.* **15**, 304 (1956).

Discussion. After the 5-minute boiling, no additional fluorescence will develop. The samples are stable for hours. Nicotinamide riboside and nicotinamide mononucleotide form fluorescent products identical to DPN and TPN. The alkaline product formed with N-methyl nicotinamide gives a low fluorescence (1/60 of the DPN). It is possible to make this assay specific for DPN and TPN by the use of *Neurospora* DPNase (see Section I-A).

B. Methyl Ethyl Ketone

Principle. The oxidized forms of DPN and TPN react with methyl ethyl ketone in alkaline solution followed by acidification to yield a fluorescent product.[18]

Reagents

Methyl ethyl ketone (technical grade).
0.1 M manganese chloride.[19]
3.5 N sodium hydroxide.
0.4 N hydrochloric acid.

Procedure. The sample to be assayed is mixed with 0.2 ml. of a 1/500 dilution of 0.1 M MnCl$_2$ in methyl ethyl ketone and 0.6 ml. of 3.5 N NaOH. After 5 minutes of incubation at room temperature, 0.4 N HCl is added to a final volume of 8 ml. The mixture is then heated in a boiling water bath for 5 minutes. After cooling, the sample is read in a Coleman fluorometer with thiamine filters.[20] This method is accurate for 1 to 10 γ of pyridine nucleotides.

Discussion. The order of addition of the reagents is very important. If the alkali is added before the methyl ethyl ketone, all the oxidized pyridine nucleotides will be destroyed. Before boiling it is advisable to check the pH of the sample. If it is any higher than 3, the fluorescent product will not be formed. Within the range of 1 to 10 γ of DPN this method has been very valuable in assaying tissue homogenates.[20] Nicotinamide riboside, nicotinamide mononucleotide, and N-methyl nicotinamide[21] also react in this assay, yielding the same amount of fluorescence as DPN and TPN. In order to make an accurate determination for DPN and TPN, one sample is pretreated with *Neurospora* DPNase (see Section 1-A) and then assayed with the methyl ethyl ketone reagents.

[18] K. J. Carpenter and E. Kodicek, *Biochem. J.* **46**, 421 (1950).
[19] The use of manganese was suggested by Drs. O. H. Lowry and H. Burch.
[20] The method can be made more sensitive by using the Farrand fluorometer, see footnote #17.
[21] It is of interest that the N-methyl nicotinamide gives a low fluorescence with strong alkali whereas with the methyl ethyl ketone method it produces an equimolar fluorescence as does DPN.

C. Determination of DPN or TPN by Fluorescence[22]

Principle. The specific enzymes discussed below are added to the sample to be assayed. After incubation, this sample and a control sample are treated according to the methyl ethyl ketone procedure. The fluorescence in the incubated sample will be decreased by the amount of reduced coenzyme formed (which will not react under these conditions). The reduced nucleotides are actually destroyed in the methyl ethyl ketone procedure.

1. *DPN Determination* (for reagents, see Section I-B). In using the methyl ethyl ketone method specifically for DPN, two samples are assayed. The first sample is treated directly; the second is first incubated with ethanol-Tris and ADH.

2. *TPN Determination* (for reagents, see Section I-C). A sample to be analyzed for TPN is incubated with 5 micromoles of isocitrate, 0.1 M phosphate buffer (pH 7.5), 0.01 M MgCl, and isocitric dehydrogenase. This sample is compared with an identical aliquot that has not been acted on by the enzyme.

D. Determination of DPNH and TPNH [22a]

The method can be adapted to determine reduced coenzymes. The samples are assayed before and after enzymatic reactions resulting in the oxidation of the pyridine nucleotides. The increased fluorescence produced by the oxidized coenzyme is a measure of the reduced DPN and TPN originally present.

1. *Total DPNH and TPNH* (for reagents, see Section I-D). An aliquot that is to be assayed for reduced coenzymes is incubated with the dialyzed *Clostridium kluyveri* enzyme in phosphate buffer (pH 7.5). After incubation, the treated sample and an equivalent untreated sample are analyzed by the methyl ethyl ketone procedure.

2. *DPNH Assay* (for reagents, see Section I-E). If a sample is incubated with buffer (pH 7.5), acetaldehyde, and ADH, all the DPNH present will be oxidized. As discussed above, this sample will have an increased amount of fluorescence when compared to a non-treated sample.

3. *TPNH Assay* (for reagents, see Section I-F). The first sample is analyzed directly; the second is incubated with the glutathione reductase

[22] The principle of the fluorometric procedures for the determination of oxidized and reduced coenzymes has been developed by Drs. L. Astrachan and K. B. Jacobson of this laboratory.

[22a] The levels of enzyme used in the specific assay of the various pyridine nucleotides depend on the activity and possible contamination of the enzyme preparation. The enzyme concentration and incubation time can be controlled by carrying out a determination first with known levels of the coenzymes.

from peas. The two samples are then compared to obtain the value for the TPNH present. The TPNH-specific diaphorase from spinach has been used with some success to assay small amounts of TPNH.[15] This has been achieved by adding FMN which promotes oxidation of TPNH by the enzyme. The method is not too satisfactory, as FMN will cause a nonspecific oxidation of both TPNH and DPNH. By proper controls, however, TPNH can be determined by this procedure with an accuracy of 15%.[23]

III. Extraction of DPN, TPN, DPNH, and TPNH from Tissues[24]

A. DPN and TPN [25]

A sample of tissue (up to 700 mg.) is disintegrated in a homogenizer with 5 vol. of cold 5% trichloroacetic acid. The tissue must be homogenized immediately after removal from the animal. The denatured protein is removed by centrifugation. The trichloroacetic acid can be removed by five extractions with ether. This is not necessary for the total DPN and TPN determination described above, however. In order to ensure accurate determination in the specific assays for DPN and TPN, controls should be included containing added trichloroacetic acid extract. It is important that proper neutralization of the extract be carried out before the reagents of the enzyme assay are added.

B. DPNH and TPNH

The tissue (up to 700 mg.) is immediately placed in a homogenizer containing a volume of 0.1 M sodium carbonate which is five times the weight of tissue (pH 10). The sodium carbonate solution is initially placed in a boiling bath for 3 to 5 minutes before the tissue is added. The tissue is allowed to stand in the hot carbonate for 30 seconds, and gentle homogenizing is carried out. The homogenizer is attached to a motor for rapid grinding for 30 seconds and then returned to the boiling bath for 60 sec-

---

[23] The glutathione reductase procedure appears to give much more consistent results in assaying for TPNH than the TPNH diaphorase system. Proper controls, however, should always be carried out with the pea system in order to ensure that there is no DPNH reaction with glutathione.

[24] Recently G. E. Glock and P. McLean [*Biochem. J.* **61**, 388 (1955)] have developed a system of analysis for determining tissue pyridine nucleotides which is based on the rate of reaction with yeast alcohol dehydrogenase and TPNH cytochrome c reductase. The only objection to such a method is that in measuring rates the problem of inhibitors may become a factor. In our system of analysis, where extent is measured, the question of inhibitors may not be so serious. Our data on liver pyridine nucleotides, however, is in fairly good agreement with that of Glock and McLean.

[25] The quantities of pyridine nucleotides in most tissues are usually of a magnitude which requires fluorescent procedures.

onds. After chilling, the sample is centrifuged to remove the denatured protein. The carbonate extracts are then neutralized, and the usual fluorometric or spectrophotometric assays carried out.

C. Extraction of DPN and TPN from Bacteria

A weighed sample of bacteria is sonicated with 5 vol. of trichloroacetic acid. The extract contains all the DPN and TPN present. As yet no attempt has been made to extract the reduced forms of the coenzymes from bacteria.

[129] Preparation of DPN Derivatives and Analogs

By NATHAN O. KAPLAN and FRANCIS E. STOLZENBACH

I. Nicotinamide Mononucleotide (NMN)

Assay Method

Nicotinamide mononucleotide reacts with cyanide to give a complex which gives a maximum absorption that is identical with the DPN-CN complex. NMN also gives the same molar fluorescence with methyl ethyl ketone as does DPN. However, NMN can be distinguished from DPN by its inactivity with yeast alcohol dehydrogenase and its resistance to *Neurospora* DPNase.

We have used the following method to determine NMN. A sample, depending on concentration, is assayed for total cyanide reaction or for total fluorescence produced with methyl ethyl ketone. A second sample is then treated with *Neurospora* DPNase; the loss in cyanide or methyl ethyl ketone reaction (see Vol. III [128]) represents the amount of DPN. A third sample is treated with a nicotinamide ribosidase from *L. arabinosus*;[1] this preparation will attack only nicotinamide riboside (NR). The residual cyanide or methyl ethyl ketone reaction after the combined DPNase and ribosidase treatment represents the NMN present in the sample. To ensure that the residual material is NMN, the sample can be treated with prostatic phosphatase and the ribosidase; and then re-assayed for reaction with either cyanide or methyl ethyl ketone.

NMN can also be distinguished from NR and DPN by paper chroma-

---

[1] The nicotinamide ribosidase can be obtained by sonic-oscillating fresh cells of *L. arabinosus*. The resulting extracts contain no DPNase activity and usually are devoid of 5'-nucleotidase action. These crude extracts can be used in the assay of the nicotinamide derivatives, since only nicotinamide riboside is split by the preparation.

tography, using a solvent system of equal parts of ethanol. and 0.1 N acetic acid. The various nicotinamide derivatives can also be separated by paper electrophoresis.

Preparation of NMN

Principle. DPN is split by the snake venom pyrophosphatase to form 5'-AMP and NMN. The NMN is separated from the 5'-AMP on a Dowex-formate column.

Preparation of Snake Venom Pyrophosphatase. The method is essentially that of Butler (Vol. II [89]). One hundred milligrams of *Crotalus adamanteus* venom is dissolved in 100 ml. of H_2O, and the insoluble material is removed by centrifugation. The supernatant is placed on a column composed of 12.5 g. shredded Whatman No. 5 filter paper, and the effluent is taken off in one fraction. The column is then washed with 100 ml. of H_2O, and the washings collected in *one* 60-ml. portion and *two* 20-ml. portions.

An equal volume (100 ml.) of a 0.1% NaCl solution is then added to the column, followed by an equal amount of 1% NaCl. The effluents are collected in 20-ml. portions. The pyrophosphatase activity is usually present in the second to fifth 0.1% NaCl fractions, and in the first 1% NaCl fraction. These fractions are almost completely devoid of 5'-nucleotidase activity. The active fractions are usually lyophylized; the dehydrated powder is quite stable when stored in the cold and kept dry.

Reagents

DPN.
$M/1$ NaHCO$_3$.
$M/1$ MgCl$_2$.
Lyophylized snake venom pyrophosphatase.

Procedure. One and one-half grams of DPN are dissolved in a volume of 20 ml. and neutralized with NaOH to pH 8.0. Seven and one-half milliliters of $M/1$ NaHCO$_3$ and 2.5 ml. of 0.3 M MgCl$_2$ are then added; the pH of the mixture should be about pH 8.2. One hundred milligrams of the lyophylized snake venom preparation is then added, and the mixture incubated at 37°. The destruction of DPN is followed by the alcohol dehydrogenase method (see Vol. III [128]). During the course of the reaction, the pH is checked and, when necessary, NaOH is added to bring the solution back to pH 8.2. It is possible that addition of further enzyme may be essential to bring the reaction to completion; the incubation period for complete cleavage of DPN may be from 18 to 36 hours.

At the conclusion of the reaction, the mixture is placed on a Dowex

1-formate column; for 1.5 g. of DPN, a column of 35 × 80 mm. is necessary to adsorb the 5'-AMP. The effluents are collected in 10-ml. aliquots and then washed with water. The water effluents are also collected in 10-ml. portions. The column is treated with H_2O until the NMN is completely washed off. Recovery of NMN from the column can be followed spectrophotometrically at 325 mμ by the cyanide reaction.

The effluents containing the NMN can be stored in the frozen state.[2] We have found it more convenient, however, and perhaps safer to lyophylize the effluents. From the initial 1.5 g. of DPN, 1.5 g. of lyophylized powder is obtained. The preparation contains 1 micromole of NMN per milligram of powder. Most of the impurities can be accounted for as salts; 10% of the total NMN is the α-isomer. This is approximately the concentration of the α-isomer of DPN[3] in the starting DPN preparation.

II. Nicotinamide Riboside (NR)

Assay

The method is outlined above in the description of the determination of NMN.

Preparation of NR

Principle. NMN is hydrolyzed with the prostatic monoesterase, and the inorganic phosphate is adsorbed on a Dowex 1-formate column.

Reagents

NMN.
Sodium acetate.
2 N HCl.
10% NH_4OH.
Prostatic phosphatase (see Vol. II [79]).

Procedure. Five hundred micromoles of the lyophylized NMN (0.5 g.) is dissolved in 20 ml. of H_2O, and sufficient solid sodium acetate is introduced to make the final concentration 0.1 M. HCl is then added to bring the pH to 5.3. After addition of approximately 400 Schmidt units of prostatic phosphatase, the mixture is incubated at 37°. The amount of enzyme added should be sufficient to complete the hydrolysis of the NMN in less than 12 hours. The course of the reaction can be followed by determining the appearance of inorganic phosphate by the

---

[2] A barium salt of the NMN can also be prepared by precipitation with 4 vol. of alcohol.

[3] N. O. Kaplan, M. M. Ciotti, F. E. Stolzenbach, and N. R. Bachur, *J. Am. Chem. Soc.* **77,** 815 (1955).

Fiske-SubbaRow procedure (see Vol. III [114]). Completeness of the reaction can be determining by comparing the inorganic phosphate liberated to the original NMN total P.

After the conclusion of the incubation, the mixture is heated at 100° for several minutes to inactivate the phosphatase. Any material that is precipitated is removed by centrifugation. The mixture is brought to pH 8.0 by addition of 10% NH$_4$OH and placed on a Dowex 1-formate column (20 × 100 mm.). Effluents are collected before and after washing with water, as in the preparation of the NMN. The collected NR can be stored in the frozen state, or it can be lyophylized. This procedure gives almost a quantitative yield of NR from NMN.

III. Analogs of DPN

A number of analogs of DPN have been prepared containing a component other than nicotinamide in the molecule. These analogs have been prepared by the pig brain DPNase exchange reactions (transglycosidase) outlined in Vol. II [113]. Preparation of the isonicotinic acid hydrazide analog of DPN (INH DPN) and the 3-acetylpyridine analog (AP DPN) will be described below as examples of the general procedure.

Reagents

DPN neutralized to pH 7.5 with NaOH.
INH or 3-AP.
$M/1$ phosphate buffer, pH 7.5.
Pig brain DPNase (Vol. II [113]).

Preparation of INH DPN [3,4]

The reaction mixture contains 15.0 millimoles of INH (final concentration, 0.1 M), 2 g. of DPN, 6.25 ml. of $M/1$ potassium phosphate (pH 7.5), 900 units of pig brain enzyme, and water to 150 ml.

To follow the reaction, aliquots are removed, diluted with 0.1 N NaOH, and read at 385 mμ. The INH analog shows a marked absorption at this wavelength.[4] When the increase at 385 mμ reaches a maximum (6 to 8 hours), the mixture is made 5% with trichloroacetic acid. After removal of denatured protein, the analog is precipitated by the addition of 5 vol. of cold acetone. The precipitate is washed several times with acetone, and finally with ether. The yield is 1.25 g. No free INH can be detected in the preparation. The preparation, however, still contains approximately 10% DPN. This can be removed if desired by the addition of *Neurospora* DPNase, which attacks only the natural coenzyme and

---

[4] L. S. Zatman, N. O. Kaplan, S. P. Colowick, and M. M. Ciotti, *J. Biol. Chem.* **209**, 467 (1954).

not the analog. The analog can then be repurified after removal of the enzyme with trichloroacetic acid by reprecipitation with acetone. Attempts to purify the INH analog of DPN by column chromatography have not always yielded reproducible results.

Properties of INH DPN. INH DPN, as mentioned above, can be identified by its characteristic reaction with alkali. The analog is split by the pig brain enzyme, liberating free INH. The pig brain DPNase can also promote an exchange between the analog and the nicotinamide to form DPN. The analog has been found to be inactive in replacing DPN in a number of dehydrogenases.

Preparation of AP DPN

The reaction mixture is the same as that used in preparing INH DPN.[5,6] AP DPN can be assayed with yeast ADH;[6] the analog can be distinguished from DPN, since its reduced spectrum has a maximum at 365 mμ as contrasted to the 340-mμ peak of reduced DPN. Acetyl pyridine increases the rate of DPN disappearance, and therefore AP DPN formation reaches a maximum in a considerably shorter period than occurs in INH DPN synthesis. AP DPN splitting by the pig brain DPNase is not inhibited by free acetyl pyridine, and the reaction should be stopped with trichloroacetic acid as soon as maximum synthesis is approached. The theoretical value can be obtained from a comparison of the 365/340 ratio for DPNH and AP DPNH. The ratio for DPNH is 0.69, whereas that of AP DPNH is 1.36.

After the removal of the denatured protein, the AP DPN is precipitated with 5 vol. of cold acetone. After washing with acetone and ether, the material is dried and then dissolved in water. If significant DPN is present it can be removed with the *Neurospora* DPNase, which does not act on AP DPN. The analog can be further purified by adsorption on a Dowex 1-formate column and then eluted with a 0.1 M formic acid–0.1 M sodium formate mixture.[7] The compound is then precipitated with acid acetone and washed and dried. The yield from 2 g. of DPN is usually 1 g. of analog. Purity as estimated by 260 mμ absorption and from the reaction with yeast alcohol dehydrogenase is 90 to 95%.

Properties of AP DPN.[8] AP DPN is reduced by a number of dehydrogenases. The potential of the analog is 0.08 volt more positive than DPN. The reduced spectrum of AP DPN has a maximum at 365 mμ with a millimolar extinction of 7.8. AP DPN will also react with cyanide

[5] N. O. Kaplan and M. M. Ciotti, *J. Am. Chem. Soc.* **76**, 1713 (1954).

[6] N. O. Kaplan and M. M. Ciotti, *J. Biol. Chem.* **221**, 823 (1956).

[7] See Vol. III [124].

[8] N. O. Kaplan, M. M. Ciotti, and F. E. Stolzenbach, *J. Biol. Chem.* **221**, 833 (1956).

(see Vol. III [128]) to form a complex with a maximum at 340 mμ. AP DPN can be distinguished from DPN by its difference in the reduced spectrum, its resistance to *Neurospora* DPNase, and its failure to give a fluorescent product with strong alkali (Vol. III [128]).

IV. Adenosine Diphosphate Ribose (ADPR)

Assay Method[9]

There is no chemical or enzymatic method for the determination of ADPR. This nucleotide can be determined by its adenine, phosphate, and ribose content, and by its free reducing group. The compound can also be identified by the products formed from the action of snake venom pyrophosphatase, namely 5'-adenylic acid and ribose-5'-phosphate.

Preparation

Principle.[9] DPN is hydrolyzed with *Neurospora* DPNase to yield nicotinamide and ADPR. The ADPR is then precipitated with acetone.

Reagents
DPN.
NaAc.
4 N HNO$_3$,
Neurospora DPNase (Vol. II [114]).

Procedure. One gram of 90% DPN is dissolved in 20 ml. of H$_2$O; the solution is made 0.05 M with solid NaAc. The mixture is brought to pH 5.5 with NaOH, and 8000 units of the *Neurospora* enzyme is added. The mixture is then incubated at 37°. The course of the DPN destruction is followed by assays with yeast alcohol dehydrogenase. The reaction is complete within 4 hours. After washing once with a mixture of 4 parts of absolute alcohol and 1 part of ether, the preparation is washed with anhydrous ether and dried. The yield is 760 mg. No free or bound nicotinamide can be detected in the preparation.

V. α-Isomer of DPN

Preparation

Principle. By treatment with the *Neurospora* DPNase, all the β-DPN can be hydrolyzed to ADPR, leaving only the α-isomer, which can then be isolated by the general procedures used for DPN preparation.

Procedure.[3] One gram of 90% DPN is dissolved in 20 ml. of H$_2$O, and the solution is made 0.05 M with the addition of solid NaAc. The mixture

[9] N. O. Kaplan, S. P. Colowick, and A. Nason, *J. Biol. Chem.* **192**, 473 (1951).

is adjusted to pH 5.5 with NaOH, and 8000 units of *Neurospora* enzyme is added. The course of the reaction is determined by yeast alcohol dehydrogenase assays (Vol. III [128]). When the reaction is complete, there is no detectable reaction with the yeast enzyme. Complete splitting of the β-isomer of DPN takes place within 3 to 4 hours. The cyanide complex of the α-isomer has a maximum millimolar extinction coefficient of approximately 5.0 at 332 mμ as contrasted to the maximum of 6.3 at 325 mμ for the β-isomer. Hence from the extinction at 332 mμ in cyanide, the amount of α-isomer can be determined, once all the β-isomer is destroyed.

After the conclusion of the DPNase reaction, the mixture is brought to pH 8.0 with 10% NH₄OH and placed on a Dowex 1 column (35 × 100 mm.) sufficient to adsorb all the ADPR. The effluent is collected in 10-ml. samples. A mixture of M/10 formic acid and M/10 sodium formate is then added to the column (see Vol. III [124]). Ten-milliliter samples are again collected until all the α-isomer has been eluted from the column. The fractions containing the compound are then collected and adjusted with 4 N HNO₃ (blue to congo red). To the mixture is added 6 vol. of cold acetone; after standing overnight at 4°, the resulting precipitate is removed by centrifugation, washed with acetone and ether, and dried *in vacuo*. The yield of α-isomer is 60 mg. from 1 g. of 90% DPN.

Properties. The α-isomer of DPN is present in most DPN preparations to an extent of 8 to 12%. It can be detected because of its inactivity in the yeast alcohol dehydrogenase reaction and by its resistance to the *Neurospora* DPNase. The isomer can be reduced by hydrosulfite, forming a compound with a maximum absorption at 348 mμ. The α-isomer also reacts with cyanide, yielding a complex with a 332-mμ maximum. The α-linkage is present only in the nicotinamide riboside part of the molecule; the adenylic acid moiety is a β-derivative.

[130] Preparation and Assay of 2',5'-Diphosphoadenosine Nucleotide

By T. P. WANG

$$\text{TPN} \xrightarrow[\text{pyrophosphatase}]{\text{nucleotide}} \text{2',5'-Diphosphoadenosine} + \text{NMN}$$

Procedure.[1] Fifty milligrams of TPN is incubated with 2 ml. of snake venom nucleotide pyrophosphatase,[2] 20 micromoles of MgCl₂,

[1] T. P. Wang, L. Shuster, and N. O. Kaplan, *J. Biol. Chem.* **206**, 299 (1954).
[2] See Vol. II [89].

and 300 micromoles of Tris (pH 9.5) in a total volume of 3 ml. at 37°. The splitting of the TPN is followed by using the TPN-specific isocitric dehydrogenase[3] from pig heart. The formation of NMN and 2'5'-diphosphoadenosine is complete in 2½ hours. The reaction mixture is then placed on a Dowex 1 anion exchanger (in formate form) with a column size of 5 cm. × 0.8 cm.[1]

The NMN is not adsorbed on this exchanger, whereas the diphosphoadenosine is removed by this column. After being washed with 3 ml. of water twice, the column is eluted with a mixture of 0.02 M HNO$_3$ and 0.02 M NaNO$_3$. Approximately 60 ml. of the mixture is required before the diphosphoadenosine appears in the eluate. The appearance of this compound is followed by its adsorption at 260 mμ. The eluates containing the compound are combined, and 1 ml. of 20% HgAc$_2$ (in 0.1 N HAc) is added. A white precipitate of the Hg salt of the diphosphoadenosine forms immediately. After standing in the icebox for 4 hours, the precipitate is collected by centrifugation, washed twice with water and resuspended in 2 ml. of water. The Hg salt is then decomposed with a stream of H$_2$S gas. After removal of the HgS precipitate, the solution is aerated to eliminate the excess H$_2$S. About 35 to 40 micromoles of the diphosphoadenosine compound is obtained by this procedure.

Analysis. Analyses of adenine, total phosphate, and ribose have been made on this compound and give a ratio of 1 adenine:1 ribose:2 phosphates.

Assay Method. The 2',5'-diphosphoadenosine is incubated with the prostate acid phosphatase.[4] The product will be adenosine, which can be identified by the specific adenosine deaminase of intestinal mucosa.[5] Under identical conditions, the 3',5'-diphosphoadenosine will also yield adenosine as the product. However, the 2',5'-diphospho compound can be distinguished from the 3',5'-diphospho compound by its inertness toward 3-nucleotidase.[1]

---

[3] See Vol. I [116].

[4] See Vol. II [79].

[5] A. Kornberg and W. D. Pricer, Jr., *J. Biol. Chem.* **193**, 481 (1951); see also Vol. II [69].

[131] Preparation of Coenzyme A

By Arthur Kornberg and E. R. Stadtman

Molecular Weight 767 $(C_{21}H_{36}O_{16}N_7P_3S)$

Principle

In the method described, which has also been published elsewhere,[1] CoA is extracted from yeast with hot water, chromatographed on charcoal, and precipitated with acetone. The crude preparation is purified by ion exchange chromatography, concentrated on charcoal, and precipitated with acetone, giving a product of 50 to 65% purity in a yield of 50 to 55%.

Two other methods have also been published whereby CoA of reasonably high purity can be prepared.[2,3] One of these methods[3] involves relatively few steps, but it is expensive, since large amounts of glutathione are required.

Starting Material. A crude yeast preparation suitable for CoA purification by means of ion exchange chromatography is obtained by passing over a charcoal column a water extract of dried yeast (strain G, Anheuser-Busch, Inc.). An Armour Company CoA concentrate from hog liver (13 units of CoA per milligram), and a Sigma Chemical Co. TPN concentrate (10% TPN, 6.5 units of CoA per milligram) may be subjected to ion exchange chromatography, but they give products of lower purity (ca. 150 units/mg.) and in smaller yield (30 to 35%). An actinomycetes CoA con-

---

[1] E. R. Stadtman and Arthur Kornberg, *J. Biol. Chem.* **203,** 47 (1953).

[2] J. D. Gregory, G. D. Novelli, and F. Lipmann, *J. Am. Chem. Soc.* **74,** 854 (1952).

[3] H. Beinert, R. W. von Korff, D. E. Green, D. A. Buyske, R. E. Handschumacher, H. Higgins, and F. M. Strong, *J. Am. Chem. Soc.* **74,** 854 (1952).

centrate (6.0 units/mg.) is unsuitable for ion exchange chromatography. A crude CoA concentrate from yeast supplied by Pabst and containing only 1.3 units of CoA per milligram gives similar results to those obtained with the more purified material prepared by us. However, the capacity of the resin column, in terms of the units of CoA being processed, is only about one-seventh as great as when the more purified material is used. A preliminary charcoal purification of the Pabst CoA concentrate gives a CoA preparation (10 units/mg.) which for ion exchange chromatography is as good as our crude concentrate. Thus, charcoal treatment of other commercially available crude CoA concentrates may provide suitable starting materials for ion exchange chromatography.

Procedure[4]

Preparation of Yeast Extract. Three kilograms of yeast is dropped into 15 l. of boiling water. The suspension is stirred and boiled vigorously for 5 minutes. Twenty-five pounds of cracked ice is added, and the cold suspension is centrifuged at ca. $2000 \times g$ for 30 minutes. The slightly turbid supernatant (19 l.) should contain $100,000 \pm 30,000$ units of CoA.[5]

Charcoal Chromatography of the Yeast Extract.[6] Nineteen liters of yeast extract is adjusted to pH 3.0 with $6\ N$ HCl (200 to 250 ml.). The extract

---

[4] All operations are carried out at room temperature unless otherwise indicated.

[5] All purification steps are followed by direct CoA analysis using the phosphotransacetylase assay system (E. R. Stadtman, G. D. Novelli, and F. Lipmann, *J. Biol. Chem.* **191**, 365 (1951); see also Vol. III [132]) with a slight modification to permit more rapid determinations. The method is as follows: water to give a final volume of 1.0 ml.; tris(hydroxymethyl)aminomethane·HCl buffer (M/1, pH 8.0), 0.1 ml.; dilithium acetyl phosphate (Ac~P), 6 micromoles; cysteine·HCl, 10 micromoles; an aliquot of the test solution containing 0.5 to 3.0 units of CoA; and phosphotransacetylase [E. R. Stadtman, *J. Biol. Chem.* **196**, 527 (1952)], 8 units; added in the indicated order. The reaction mixture is incubated at 28° for 5 minutes, and then 0.1 ml. of potassium arsenate (0.5 M, pH 8.0) is added. After 10 minutes, the residual Ac~P is estimated by the hydroxamic acid method [F. Lipmann and L. C. Tuttle, *J. Biol. Chem.* **159**, 21 (1945)]. Under these conditions the amount of Ac~P decomposed is proportional to the amount of CoA present in the reaction mixture. Up to twelve samples can be conveniently examined in a single assay. For reference, a standard tube containing 2 to 2.5 units of CoA and a control sample containing no CoA are included in each assay. All CoA samples must be adjusted to pH 7.5 to 8.0 prior to testing.

The reference standard used in the study was an acetyl-CoA sample prepared by the enzymatic acetylation of CoA and isolated by paper chromatography [E. R. Stadtman, *J. Biol. Chem.* **196**, 535 (1952)]. The acetyl CoA/adenine ratio of the preparation was 0.97, and it was assumed to contain 316 units of CoA per micromole [J. D. Gregory, G. D. Novelli, and F. Lipmann, *J. Am. Chem. Soc.* **74**, 835 (1952)].

[6] This charcoal step is a modification of a method developed by M. Soodak and F. Lipmann (private communication).

(which becomes turbid on acidification) is passed through a charcoal[7] column (10 cm. $\times$ 29 cm.) at a rate of 1 l. per 3 minutes. The milky-white effluent is discarded. The column is washed with 10 to 15 l. of distilled water, with 2 to 4 l. of 40% aqueous acetone, and then with 40% aqueous acetone containing 1.0 ml. of concentrated ammonium hydroxide solution (28%) per liter. The eluate is collected in 2-l. fractions. The pH of successive fractions increases gradually from 3.3 to 5.0 and then increases sharply to pH 7.5 to 9.0. In most runs the CoA is concentrated in the

TABLE I

CHARCOAL CHROMATOGRAPHY OF CoA IN CRUDE YEAST EXTRACT

Fraction number (2 l.)	Color, description	pH	CoA, % of initial
1	Straw, turbid	3.4	0
2	Straw	3.6	0
3	Straw, clear	3.9	7
4	Straw-clear	4.0	26
5	Yellow, clear	4.2	28
6	Deep amber	5.0	26
7	Deep amber	8.5	7
8	Amber	9.4	0
Total			94

eluates having a pH of 3.8 to 7.0; occasionally, however, the fraction on either side of these limits will also contain some CoA. Data from a typical experiment showing the distribution of CoA in the eluates from a charcoal column are given in Table I.

The fractions containing CoA are pooled, the pH is adjusted to 1.7 with 6 N HCl (any precipitate which forms is removed by filtration through a fluted Schleicher and Schuell No. 588 filter paper), and the CoA is precipitated by the addition of 5.5 vol. of cold (2°) acetone. After 1 to 2 hours the precipitate is recovered by filtration through a Büchner funnel (10-cm. diameter). The precipitate is washed with acetone, then with ether, and dried in a vacuum desiccator over P_2O_5; 5 to 6.5 g. of dry

---

[7] Acid-treated, degassed charcoal is prepared by suspending ca. 400 g. of unground Nuchar C charcoal in 4 l. of 6 N HCl. The suspension, in two 4-l. flasks, is kept under vacuum overnight, during which time the charcoal settles, leaving a clear supernatant. Sufficient charcoal is transferred to a large glass column (10 cm. in diameter) to give a bed height of 29 cm. The glass column is fitted at the bottom with a stainless-steel screen (80 mesh) which is packed around the edges with glass wool. The charcoal is washed with distilled water until the pH of the effluent is 3.5 to 4.0.

acetone powder is obtained containing 10 to 18 units of CoA per milligram. The over-all recovery based on the boiled yeast extract is about 80 to 95%.

Chromatography on Dowex 1 Resin.[8] Fifteen grams of acetone powder is dissolved in water, the pH is adjusted to 8.1 with 2 N KOH, and the volume to 200 ml. Any material which does not dissolve is removed by centrifugation. The crude CoA solution is placed on a Dowex-1, 2% cross-linked formate[9] column (13 cm. × 15 cm.²). Invariably the colored front progresses unevenly, owing presumably to severe shrinkage of the resin caused by the high salt concentration of the CoA solution. Therefore, when the farthest edge of the front has progressed approximately 1 cm., the upper 1-cm. layer of resin is gently stirred up with the overlying solution. After this treatment the progress of the front appears uniform. After the CoA has been applied, the column is washed with two 100-ml. portions of water and is eluted with a solution containing 25.8 ml. of formic acid (88%) per liter (0.6 M) and 18.9 g. of ammonium formate per liter (0.3 M). The rate of flow (attained with a hydrostatic pressure of 2 to 5 feet) is about 450 ml./hr. Fractions of 500 ml. are collected and assayed for CoA and for adenine (optical density at 260 mμ).

Results showing the development of a typical chromatogram are given in Table II. Elution of the CoA begins when 20 to 23 column volumes (8 or 9 fractions) of eluate have been collected. The CoA is completely eluted

---

[8] The ion exchange chromatography of CoA concentrates has also been used on a scale one-third, one-half, and two times that described. In these instances the size of the resin column was varied proportionally by changing the cross-sectional area of the column while maintaining the height of the resin bed constant. The rate of elution and volumes of the fractions collected were varied correspondingly. The elution pattern, yields of CoA, and purity of the product obtained were similar, except that a somewhat poorer over-all yield (40%) was obtained in a single experiment at two times the scale described.

Usually 14 to 16 hours is required to elute the CoA from the Dowex column. In the absence of an automatic fraction collector which permits uninterrupted elution overnight, the elution can be discontinued and finished on the following day.

The resin (200 to 400 mesh) is washed by suspending 5 pounds in 30 l. of 3 N HCl. The suspension is stirred mechanically for 8 hours, allowed to settle for 2 days, and the supernatant decanted. This step is repeated three to four times until the ultraviolet absorption of the supernatant at 260 mμ is 0.080 or less in a 1-cm. cell. The resin is washed with 30-l. batches of water (two to three times) until the test for chloride ion in the wash water is faint. The resin is transferred to a glass column fitted with a porous sintered glass base and is washed with 3 M sodium formate until the chloride ion test on the effluent is faint. Finally, it is washed with water until the effluent is practically salt-free (20 p.p.m. on a Barnstead purity meter, referred to NaCl).

Other anion exchange resins tested, including Duolite A-3 chloride, Dowex 2 formate, and Dowex 1 chloride and formate (the latter two 10% cross-linked) proved unsatisfactory.

when an additional 10 to 12 column volumes (fractions 9 to 15) are collected. In the experiment described, the CoA/adenine ratio of fractions 9 to 15 varied from 0.16 to 0.76. In other experiments ratios as high as 1.0 have been obtained. The over-all recovery of CoA from the chromatograms is generally 80 to 95%; however, in some instances recoveries as low as 70% have been observed.

TABLE II

ION EXCHANGE CHROMATOGRAPHY OF CoA

Fraction number (500 ml.)	Adenine,[a] micromoles	CoA,[b] micromoles	Micromoles CoA/Micromoles adenine
0[c]	165	0	0
1	9900	0	0
2	8500	0	0
3	1500	0	0
4	660	6	0.009
5	435	5	0.011
6	520	0	0
7	490	0	0
8	275	0	0
9	151	25	0.16
10	113	75	0.66
11	123	86	0.70
12	117	89	0.76
13	107	75	0.66
14	85	49	0.58
15	69	19	0.28
Total recovered	23,210	429	
Total, initial	25,900	450	

[a] Calculated from optical density at 260 mμ on the basis of a molecular extinction coefficient of 15.9×10^6 cm.^{2} per mole.

[b] Calculated on an assumed value of 316 units of CoA per micromole [J. D. Gregory, G. D. Novelli, and F. Lipmann, *J. Am. Chem. Soc.* **74**, 854 (1952)].

[c] Filtrate collected during the adsorption and washing.

Concentration of the Eluate. Since the CoA from the Dowex column is distributed in a large volume (2500 to 3000 ml.) of strong formate buffer, it is desirable to concentrate it and remove the excess salt before attempting to recover the CoA by acetone precipitation. This is accomplished by readsorbing the CoA on a small charcoal column and eluting with ammoniacal acetone. To obtain CoA of highest purity, the eluates from the ion exchange chromatogram in which the CoA/adenine ratio is 0.6 or greater are pooled; the pH is adjusted to 2.0 with 6 N HCl (about 100 ml.

per 2000 ml. of eluate), and the solution is passed through a small charcoal column (2.5-cm. diameter × 12 cm.) at a rate of 1.5 l./hr. (The optical
density of the effluent at 260 mμ should not exceed 0.020.) The column is
washed first with 300 ml. of water, then with 300 ml. of 40% acetone, and
the CoA is finally eluted with 40% acetone containing 1 ml. of concentrated ammonium hydroxide per liter. The flow rate is adjusted to
about 3.0 ml./min. Fractions of 50 ml. are collected. The alkaline fractions
are neutralized with 4 N nitric acid. The CoA is generally all eluted when
the pH of the effluent is alkaline.

The CoA in these fractions is sufficiently concentrated to permit detection by the nitroprusside test.[10] The fractions containing CoA (usually
fractions 3 to 6) are pooled (final volume, 150 to 200 ml.). The pooled
sample is acidified to pH 1.7 with 4 N nitric acid, and the CoA is precipitated by the addition of 7 vol. of cold acetone. The precipitate is allowed
to settle overnight, the supernatant is decanted, and the CoA is transferred to a centrifuge tube. The precipitate is washed two times with
acetone (room temperature), once with ether, and dried in a vacuum
desiccator over P_2O_5.

From 15 g. of CoA concentrate (10 units/mg.) 300 to 400 mg. of
purified material is obtained containing 200 to 270 units of CoA per
milligram. The over-all yield is generally 50 to 55%. Occasionally yields as
low as 40% and as high as 70% have been obtained. In addition to this
CoA of relatively high purity, CoA preparations containing 150 to 200
units/mg. can be obtained as a by-product by working up those eluates
from the Dowex column having a CoA/adenine ratio of less than 0.6.

Properties and Purity of Product

The composition of some purified preparations are given in Table III.
For comparison, analytical data on a sample of Pabst CoA are also
presented. The contents of phosphorus, adenine, and enzymatically
active CoA are 65 to 67% of theory. The rather good stoichiometry between these three components precludes the presence of other adenine
nucleotides as major contaminants.

It should be pointed out that CoA samples of lesser purity prepared
from the fractions just preceding the peak fractions from the Dowex column (i.e., the material with a CoA/adenine ratio of less than 0.6) do contain significant amounts of ATP.

---

[10] Drops of the various fractions are placed on a strip of No. 3 Whatman filter paper
and dried. The paper is dipped first into reagent I and then into reagent II of
Toennies and Kolb [*Anal. Chem.* **64**, 59 (1952)]. The appearance of a red spot, which
is greatly intensified by shaking the paper strip in ethyl ether, indicates the presence
of CoA (see Vol. III [13]).

Since no reductive steps are employed in the above chromatographic procedure, the CoA isolated should be present mainly as the mixed disulfide derivatives of other sulfhydryl compounds.[2,11,12] Paper chromatography of the isolated material in an 80% phenol–20% water solvent shows that the CoA moves to a spot identical with a ninhydrin-reactive

TABLE III

ANALYSIS OF COENZYME A PREPARATIONS

Sample	Total P,[a] micromoles/ mg.	Adenine,[b] micromoles/ mg.	CoA,[c] micromoles/ mg.	Micromoles P ——— Micromoles CoA	Micromoles adenine ——— Micromoles CoA
VIIa	2.56	0.82	0.77	3.32	1.06
XIII	2.55	0.91	0.83	3.07	1.10
Pabst (401-27)	2.66	0.87	0.85	3.13	1.02
Pure CoA	3.9	1.3	1.3	3.00	1.00

[a] Determined by the method of Fiske and SubbaRow [*J. Biol. Chem.* **66**, 375 (1925)].

[b] Calculated from optical density at 260 mμ on the basis of an assumed molecular extinction coefficient of 15.9×10^6 cm.^{2} per mole.

[c] Determined by arsenolysis of Ac∼P with transacetylase; see also Vol. III [132].

material ($R_f = 0.35$). After reduction of the CoA with hydrogen sulfide it moves with an R_f of 0.51; the amino compound has an R_f of 0.46, which is identical with that found for glutathione. These observations suggest that the CoA isolated may be present largely as the mixed disulfide derivative of glutathione.

[11] W. H. DeVries, W. M. Govier, J. S. Evans, J. D. Gregory, G. D. Novelli, M. Soodak, and F. Lipmann, *J. Am. Chem. Soc.* **72**, 4838 (1950).

[12] G. M. Brown and E. E. Snell, *J. Biol. Chem.* **198**, 375 (1952).

[132] Assay of Coenzyme A

By G. DAVID NOVELLI

I. Sulfanilamide Acetylation—(Range, 0.5 to 2.0 units CoA)

Principle. This method, originally described by Kaplan and Lipmann,[1] is based on the fact that a bicarbonate extract of acetone-dried pigeon liver catalyzes the acetylation of sulfanilamide in the presence

[1] N. O. Kaplan and F. Lipmann, *J. Biol. Chem.* **174**, 37 (1948).

of ATP, CoA, and Ac. The reaction is measured by noting the disappearance of sulfanilamide as measured by the diazotization method of Bratton and Marshall.[2]

Reagents

> Assay mixture—25 ml. of 0.2 M sodium citrate, 5.6 ml. of 0.02 M sulfanilamide, 6.25 ml. of 1 M sodium acetate, 20 ml. of 0.05 M ATP, 69.4 ml. of distilled water. Dispense in 10-ml. quantities and store at −10°.
>
> Cysteine-HCl, 0.2 M, store at −10°.
>
> Tris buffer, 1 M, pH 8.0 to 8.4.
>
> Reagents for determination of sulfanilamide (see Vol. I [101]).
>
> Standard CoA, 20 units/ml.
>
> Bicarbonate extract of acetone-dried pigeon liver acetone powder of pigeon liver prepared as in Vol. I [101]. Crude extract is prepared by making a 10% suspension of the acetone powder by careful hand homogenization in ice-cold 0.02 M potassium bicarbonate until a smooth suspension is obtained. The suspension is centrifuged for 20 minutes at 2000 r.p.m. in the cold. The supernatant is decanted and frozen overnight in the deep-freeze. On the next day the supernatant is allowed to thaw and then to stand for 4 hours at room temperature. This aging procedure destroys indigenous CoA. The solution is then centrifuged at 5000 r.p.m. for 1.5 hours in the cold or alternatively for 30 minutes at 30,000 r.p.m. in the Spinco preparative centrifuge. The enzyme is dispensed in plastic tubes and stored in the deep-freeze.

Assay Procedure

Step 1. Standardization of Coenzyme A Solution. In the absence of a standardized preparation, CoA must be standardized in the following manner:

a. Ten tubes (13 × 100 mm.) are set up as follows: (1) No CoA, no enzyme. (2) No CoA, three to ten varying increments of a CoA solution containing approximately 20 units of CoA per milliliter to cover the range 0.2 to 6.0 units.

b. To each tube is added 0.15 ml. of Tris buffer, 0.05 ml. of 0.2 M cysteine HCl, and 0.5 ml. of assay mixture.

c. To tubes 2 through 10 is added enzyme (0.1 to 0.3 ml.) and distilled water to bring to a common final volume of from 1.0 to 1.5 ml.

d. The tubes are incubated in a water bath at 37° for 1.5 hours, and the reaction is stopped by adding 4.0 ml. of 5% TCA.

[2] A. C. Bratton and E. K. Marshall, Jr., *J. Biol. Chem.* **128**, 537 (1939).

e. After centrifugation to remove the coagulated proteins, a 1.0-ml. aliquot is removed and residual sulfanilamide is determined by the method of Bratton and Marshall.[2]

f. The difference in the amount of sulfanilamide present in tubes 1 and 2 indicates the amount of residual CoA in the enzyme. If more than 1% of the sulfanilamide is acetylated in the absence of added CoA, the enzyme should be aged at room temperature until the residual CoA is destroyed.

g. The amount of sulfanilamide acetylated in tubes 2 to 10 is plotted against the amount of CoA added to the various tubes. One unit of CoA is defined as the amount of CoA required to produce one-half maximum acetylation. For purposes of calculation, effective saturation is taken as the point where the curve begins to turn off abruptly from its linear portion.

Step 2. Determination of CoA Content of Unknown Materials

a. Materials to be assayed for CoA, particularly animal tissues, should be heated in a boiling water bath for 5 minutes to inactivate any enzymes which destroy CoA.

b. The assay is conducted as described above, except that the following tubes are included: tube 1, no CoA, no enzyme; tube 2, no CoA; tube 3, 0.75 unit of CoA; tube 4, 1.5 units of CoA. Enough additional tubes containing various dilutions of the unknown samples as is convenient are included. The rest of the procedure is identical to step 1-*b* through 1-*f*.

Step 3. Calculations. The amount of sulfanilamide (SAM) acetylated by the unknown is divided by the amount acetylated in the presence of 1 unit of CoA.

$$\frac{\gamma \text{ SAM acetylated by unknown}}{\gamma \text{ acetylated by 1 unit CoA}} = \text{units CoA in unknown}$$

Remarks

1. 0.75 and 1.5 units of CoA are chosen as standards, since these values are generally on the linear portion of the curve.

2. The quantity of sample should be such that the acetylation value will also fall on the linear portion of the curve.

3. PABA may be substituted for sulfanilamide to obtain a longer range of linearity.

II. CoA Assay by Phosphotransacetylase

This assay is carried out as described in Vol. I [98] except that the enzyme is maintained constant at 5 units and the coenzyme A is varied. Standards containing 1 and 2 units of CoA are run with each assay. The

CoA content of the unknown samples is calculated as follows:

$$\frac{\Delta \text{ Ac-ph by unknown}}{\Delta \text{ Ac-ph by 1 unit CoA}} = \text{units CoA in unknown}$$

III. CoA Assay by Ketoglutarate Oxidation

This assay method, originally described by von Korff,[3] is based on the requirement for both DPN and CoA in the oxidation of α-ketoglutaric acid. The reaction is measured by following the reduction of DPN spectrophotometrically. The principle, reaction, and preparation of enzymes is described in Vol. I [120].

Assay Method

Reagents

> Cysteine, 0.1 M, pH 7.0.
> CoA, 5 to 15 units/ml., pH 7.
> Glycine buffer, 1 M, pH 9.0.
> α-Ketoglutarate, 0.2 M, pH 7.0.
> Deacylase.
> α-Ketoglutarate oxidase.

a. One-tenth milliliter of each of the above solutions is mixed (except that only 0.01 ml. of the oxidase is used), and the volume brought to 2.9 ml.

b. With each pair of unknown samples, a reference sample of CoA as an internal standard is treated in the same way.

c. The solutions are mixed, incubated for 5 minutes at 30°, and then transferred to Beckman cells. The optical density at 340 mg. is read against distilled water.

d. The reaction is started by adding 0.1 ml. of a 1% solution of DPN to each tube. The optical density at 340 mμ is recorded at 1-minute intervals for 3 to 5 minutes.

e. The average rate of increase in optical density per minute is calculated for each sample. The CoA content of the unknown sample is obtained by simple proportion.

IV. Pantothenic Acid Release

Principle. This method, described by Novelli *et al.*,[4] is based on the fact that *Lactobacillus arabinosus* is unable to utilize CoA or any of its enzymatic degradation products, other than pantothenic acid, to satisfy

---

[3] R. W. von Korff, *J. Biol. Chem.* **200**, 401 (1953).

[4] G. D. Novelli, N. O. Kaplan, and F. Lipmann, *J. Biol. Chem.* **177**, 97 (1949).

its nutritional requirement for the vitamin. In this procedure, pantothenic acid is released from CoA by treatment of the latter with two enzymes, and the amount of pantothenic acid thus liberated is measured microbiologically with *L. arabinosus* as test organism.

Reagents

A peptidase-like enzyme prepared from acetone-dried pigeon liver as described in Vol. I [101], except that the aging step, i.e., incubation at room temperature for 4 hours, is omitted. Instead, contaminating CoA is removed by treatment with Dowex 1. In the absence of pigeons, the enzyme may be prepared from chicken liver or hog kidney. Each preparation of this enzyme must be standardized as detailed below.

Intestinal phosphatase, prepared according to the directions of Schmidt and Thannhauser.[5]

Tris buffer, 1 *M*, pH 8.0.

Standard CoA.

A microbiological assay procedure for pantothenic acid using either *L. arabinosus*[6] or a pantothenic acid-less mutant of *Escherichia coli*.[7]

Standardization of Pigeon Liver Enzyme

Each preparation of this enzyme must be standardized by measuring its ability to liberate pantothenic acid from a given amount of CoA in the presence of an excess of intestinal phosphatase.

To a series of tubes containing 20 units of CoA, 2 to 4 Schmidt units of intestinal phosphatase[5] and 0.1 *M* Tris buffer, pH 8.5, are added increasing amounts (0.05 to 0.3 ml.) of pigeon liver extract. The volume is made up to 1.0 ml., and the tubes are incubated at 37° for 3 hours. The free pantothenate is determined microbiologically on suitably diluted aliquots. Since 1 unit of CoA contains 0.6 γ of pantothenate, the smallest amount of enzyme which liberates 12 γ of pantothenate from 20 units of CoA is taken as the standard enzyme activity.

Assay Procedure

Step 1. Preparation of Samples. Since most tissues contain enzymes which autolyze CoA, it is necessary to prepare a boiled extract of tissues as soon as possible after removal of the sample from the organism.

[5] G. Schmidt and S. J. Thannhauser, *J. Biol. Chem.* **149,** 369 (1943).
[6] H. R. Skeggs and L. D. Wright, *J. Biol. Chem.* **156,** 21 (1944).
[7] W. K. Maas and B. D. Davis, *J. Bacteriol.* **60,** 733 (1950).

Step 2. Several aliquots of the boiled extract are treated with the previously determined quantity of pigeon liver enzyme and intestinal phosphatase. The volume is made up to 1.0 or 2.0 ml. in a 0.1 M (final concentration) Tris buffer, pH 8.5.

Step 3. The tubes are incubated in a 37° water bath for 3 hours, boiled to inactivate the enzyme, and the pantothenate is determined microbiologically. At the same time aliquots of the boiled extract are tested directly for free pantothenate.

Comments and Calculations

1. The enzymatically treated samples give the value for *total pantothenate.* This value includes free pantothenate and bound pantothenate as well as any pantothenate contaminating the enzymes.

2. Enzyme control tubes, containing both enzymes with test substance, gives the pantothenate blank in the enzymes which must be subtracted from the value for total pantothenate.

3. The pantothenate value found in the aliquots without enzyme treatment gives the quantity of free pantothenate which also must be subtracted from the total pantothenate.

4. After applying these corrections, the remaining value corresponds to "*bound* pantothenate." This represents all pantothenate bound in a form which is not available to the microorganism used for assay and may include 4'-phosphopantothenate pantetheine, 4'-phosphopantetheine, dephospho-CoA, desamino-CoA and CoA. In so far as the question has been examined bound forms of pantothenate other than the intact coenzyme A represent less than 10% of the bound pantothenate occurring in fresh tissue which has not been allowed to undergo autolysis. Therefore, it appears safe to assume that the bound pantothenate represents CoA. If this value is divided by 0.6 γ it will give the CoA value in units, or if divided by 219 will give the CoA value in micromoles.

[133] Preparation and Assay of Pantethine

By ESMOND E. SNELL and EUGENE L. WITTLE

Preparation

Introduction. Work leading to the isolation, characterization, and synthesis of pantethine and its reduction product pantetheine, and clarification of the relationship of these products to other naturally occurring conjugates of pantothenic acid, has been summarized in a recent review.[1] Since the initial characterization and synthesis of this product,[2] a total of seven synthetic procedures have appeared.[2-10] Several of these are variations of similar processes. Two of the methods employ condensation of pantothenic acid esters[2,6] or pantothenyl azide[6] with β-mercaptoethylamine; these give either low yields or products difficult to purify. The process of Wieland and Bokelmann,[4] in which a mixed anhydride formed between free pantothenic acid and ethylchlorocarbonate is condensed with β-mercaptoethylamine in aqueous solution, is reportedly good,[7] although complete details have not been disclosed. It has been modified by Schwyzer[7] by use of ethylenimine in place of β-mercaptoethylamine, the substituted ethylenimine ring being subsequently opened with thiobenzoic acid to yield S-benzoylpantetheine, a crystalline product. Again details are not complete. The remaining methods[3,5,6,8-10] use the condensation of D-pantolactone with N-(β-alanyl)-2-aminoethanethiol or its disulfide. In preparing the latter intermediate, either the carbobenzoxy or phthaloyl group may be used to protect the free amino group of β-alanine; the methods are otherwise similar. Details of the carbobenzoxy procedure are described adequately in the literature.[3,5,6] Disadvantages of this process are the use of phosgene when operating on a large scale, and the fact that removal of the carbobenzoxy group by sodium and ammonia reduction also reduces pantethine to pantetheine. Both N-(β-alanyl)-2-aminoethanethiol and pantetheine oxidize on handling in air to the corresponding disulfides; this makes purification more troublesome and introduces an additional oxidative step. Use of the phthaloyl procedure avoids these disadvantages and is described below; however, certain of the crystalline intermediates formed here are not so readily purified as might be desired.

β-Phthalimidopropionyl Chloride.[11] In a 5-l. flask is placed 1 kg. (6.6 moles) of phthalic anhydride. This is melted, and 600 g. (6.6 moles) of

[1] E. E. Snell and G. M. Brown, *Advances in Enzymology* **14**, 49 (1953).

[2] E. E. Snell, G. M. Brown, V. J. Peters, J. A. Craig, E. L. Wittle, J. A. Moore, V. M. McGlohon, and O. D. Bird, *J. Am. Chem. Soc.* **72**, 5349 (1950).

[3] J. Baddiley and E. M. Thain, *J. Chem. Soc.* **1952**, 800.

[4] T. Wieland and E. Bokelmann, *Naturwissenschaften* **38**, 384 (1951).

[5] T. E. King, C. J. Stewart, and V. H. Cheldelin, *J. Am. Chem. Soc.* **75**, 1290 (1953).

[6] E. L. Wittle, J. A. Moore, R. W. Stipek, F. E. Peterson, V. M. McGlohon, O. D. Bird, G. M. Brown, and E. E. Snell, *J. Am. Chem. Soc.* **75**, 1694 (1953).

[7] R. Schwyzer, *Helv. Chim. Acta* **35**, 1903 (1952); *Experientia* **10**, 61 (1954).

[8] E. Walton, A. N. Wilson, F. W. Holly, and K. Folkers, *J. Am. Chem. Soc.* **76**, 1146 (1954).

[9] M. Viscontini, K. Adank, N. Merckling, K. Ehrhardt, and P. Karrer, *Helv. Chim. Acta* **37**, 375 (1954).

[10] R. E. Bowman and J. F. Cavalla, *J. Chem. Soc.* **1954**, 1171.

[11] S. Gabriel, *Ber.* **38**, 633 (1905); **41**, 243 (1908).

β-alanine is added in small portions over a period of 20 minutes; vigorous evolution of water results. The temperature of the melt is then raised (from about 120°) to 160° and held there for 30 to 40 minutes. The molten phthalimidopropionic acid is' poured out into a metal pan to crystallize (if allowed to solidify in the flask, the hard mass is impossible to remove). The acid is broken up and put back into the flask with 1500 ml. of benzene and 540 ml. of thionyl chloride. The slurry is heated for 3 hours, and the resulting solution is then refluxed gently for 1.5 hours. The solution is cooled and shaken while crystallizing, then slurried with anhydrous ether to facilitate filtering and the solid filtered with suction and dried in a moisture-free atmosphere. Yield, 1460 g.; melting point 104 to 106°; 92%.

Bis(2-aminoethyl) Disulfide.[12] A cooled solution of 91.1 g. (2.1 moles) of ethylenimine (Monomer-Polymer Inc.) in 819 ml. of absolute ethanol is added dropwise over a 3- to 4-hour period to 200 ml. of chilled absolute ethanol surrounded by an ice pack, while a current of hydrogen sulfide is passed through the well-stirred mixture. The solution is then concentrated under reduced pressure to 50 to 75 ml. in the absence of air to avoid oxidation of the mercaptan, and the concentrate is well cooled. β-Mercapto-ethylamine separates as a white solid which is washed with a small amount of ethyl ether and then petroleum ether and air-dried for a short time. Melting point, 97 to 99°; yield, 139 g. (85.5%).

The β-mercaptoethylamine (38.5 g., 0.5 mole) is dissolved in 100 ml. of water, and 100 ml. of dioxane is added. The solution is cooled to −6° in a salt-ice bath, and with efficient stirring a solution of 26 ml. (0.5 mole) of 30% hydrogen peroxide in 50 ml. of water is added dropwise at a rate such that the temperature remains between −4° and 0°. The reaction is exothermic, and caution must be taken to avoid a sudden temperature rise. After about 80% of the total amount of hydrogen peroxide has been added the temperature drops quite suddenly and a foam appears on the solution. A drop of the solution at this point gives a negative sulfhydryl test with nitroprusside solution. The action of the peroxide is immediately discontinued, and the disulfide solution is used directly in the next step. Alternatively, an excess of concentrated hydrochloric acid can be added and the solution evaporated in vacuum. The hydrochloride of the disulfide separates and is recrystallized from alcohol. Melting point, 212 to 216°; yield, 85%. The reactions involved in these preparations are:

$$\underset{\substack{CH_2\ \ CH_2}}{\overset{NH}{\diagup\diagdown}} \xrightarrow{\ H_2S\ } H_2NCH_2CH_2SH \xrightarrow{\ H_2O_2\ } (H_2NCH_2CH_2S{-})_2$$

12 E. J. Mills and M. T. Bogert, *J. Am. Chem. Soc.* **62,** 1173 (1940).

Bis[N-(β-phthalimidopropionyl)-2-aminoethyl] Disulfide.[10] To one-fifth of the above solution of bis(2-aminoethyl) disulfide (0.05 mole of the disulfide, equivalent to 0.1 mole of β-mercaptoethylamine) diluted to 100 ml. with water is added magnesium oxide (6 g., 0.15 mole), and the stirred suspension is treated at 0° dropwise over 90 minutes with a solution of β-phthalimidopropionyl chloride (24 g., 0.1 mole) in dry peroxide-free dioxane (180 ml.). The mixture is stirred at 0° for 30 minutes, and then 80 ml. of 1 N hydrochloric acid is added slowly. The precipitated solid is filtered, washed with water, and dried. Yield, 22 g. (80%); melting point, 193 to 200°. Recrystallization from chloroform-ethanol (1:4) gives white needles of bis[N-(β-phthalimidopropionyl)-2-aminoethyl] disulfide; melting point, 211 to 213°. Purification of this material is wasteful and troublesome, and the crude solid is usually used directly for the next step.

This reaction may also be carried out with a dilute solution of sodium hydroxide instead of the magnesium oxide above. The sodium hydroxide solution is added dropwise simultaneously with the acid chloride solution, an excess of base being avoided, and at a temperature of 8 to 15°. Yields are 70 to 90%. The reactions involved are:

Bis[N-(β-alanyl)-2-aminoethyl] Disulfide.[10] To 19 g. of the crude phthalimido derivative described above in 500 ml. of ethanol is added 40 ml. of a solution of 3.4 g. of hydrazine hydrate (3.84 ml. of 85% hydrazine hydrate) in ethanol. The mixture is refluxed for 3 hours, then concentrated in vacuum, and 50 ml. of N hydrochloric acid added slowly with shaking. The suspension is cooled, and the phthalhydrazide filtered off. The filtrate is concentrated in vacuum to an oil, the residue dissolved in 50 ml. of water, filtered, warmed to 50°, and diluted with ethanol until turbid. On standing the product crystallizes, 4 g.; melting point 204 to 207°. Concentration of the mother liquor gives additional less pure product, 5 g.; melting point 202 to 206° (total yield, 72%). Recrystallization from aqueous ethanol gives pure bis[N-(β-alanyl)-2-aminoethyl] disulfide dihydrochloride; melting point 221 to 222°. This reaction has also been run in methanol in 50 to 70% yield; melting point 212 to 218°. The reaction is

Pantethine.[6] Sodium (2.3 g., 0.1 mole) is dissolved in 150 ml. of anhydrous methanol, and to the solution is added 18.4 g. (0.05 mole) of bis(N-β-alanyl-2-aminoethyl) disulfide dihydrochloride. The solution is warmed to 40° and stirred for 1 hour, during which sodium chloride precipitates. D-Pantolactone (15 g., 0.115 mole) is then added, and the mixture is refluxed with stirring for 2 hours. After cooling, the sodium chloride is filtered out and rinsed with absolute ethanol. The filtrate is evaporated to dryness in vacuum, and the oily residue heated with stirring for 1 hour on the steam bath. It is then dissolved in a small amount of absolute ethanol, filtered from a small amount of salt, diluted with 10 vol. of anhydrous ethyl ether, mixed well, and the oil allowed to settle for several hours. The ether solution is then decanted. The oil is dissolved in 100 ml. of water, and the solution is stirred with 24 ml. of Amberlite IRC-50 resin until the pH of the solution is shifted to 6. The resin is filtered out, rinsed with water, and the filtrate evaporated to dryness in vacuum. The oil is dissolved in absolute ethanol which is then removed by vacuum distillation. The residue is then dissolved in a small amount of absolute ethanol. Ether is slowly added to incipient turbidity, and the solution is filtered by gravity from a small amount of insoluble solid and diluted with a large volume of anhydrous ether to precipitate pantethine as a clear, colorless oil. The ether solution is decanted, and the oil dried in a vacuum desiccator over phosphorous pentoxide; yield, 22 g. (80%), assaying 23,000 to 25,000 units/mg.; $[\alpha]_D^{26} = 16° \pm 1°$. The reaction is as follows:

$$2(CH_3)_2\overset{\underset{\displaystyle CH_2-O}{|}}{C}CHOHC{=}O + \left(H_2NCH_2CH_2\overset{\underset{}{O}}{\overset{\|}{C}}-NHCH_2CH_2S-\right)_2 \rightarrow$$

$$\left((CH_3)_2\overset{\underset{\displaystyle CH_2OH}{|}}{C}-CHOH\overset{\underset{}{O}}{\overset{\|}{C}}-NHCH_2CH_2\overset{\underset{}{O}}{\overset{\|}{C}}-NHCH_2CH_2S-\right)_2$$

Pantetheine is best prepared in solution as required from pantethine. If desired in bulk, catalytic reduction with hydrogen over palladium catalyst in butanol[6] or with sodium in liquid ammonia[6] are the procedures of choice. For many biological experiments, reduction in aqueous solution through interaction with an excess of some other —SH compounds (e.g., Na_2S, cysteine[13,14]) is applicable.

Assay Method

Different microorganisms vary remarkably in their ability to utilize various combined forms of pantothenic acid.[1,14,15] For one group of lactic acid bacteria (e.g., all strains tested of *Lactobacillus acidophilus*, and *L. bulgaricus* and most strains of *L. delbrückii* and *L. helveticus*) pantethine is far more active in promoting growth than is pantothenic acid. The procedure described below is essentially that of Craig and Snell,[15] which employs *Lactobacillus helveticus* 80 as test organism and was employed extensively during isolation and characterization of pantethine.[1]

Other possible procedures for estimation of pantethine, none of which has been subjected to detailed or critical investigation, consist in its hydrolysis with a liver enzyme to pantothenic acid, followed by estimation of the latter product,[5,16] or hydrolysis with acid to yield β-alanine which may then be estimated by yeast growth or other procedures.

[13] G. M. Brown and E. E. Snell, *J. Biol. Chem.* **198**, 375 (1952).
[14] G. M. Brown and E. E. Snell, *J. Bacteriol.* **67**, 465 (1954).
[15] J. A. Craig and E. E. Snell, *J. Bacteriol.* **61**, 283 (1951).
[16] G. M. Brown, J. A. Craig, and E. E. Snell, *Arch. Biochem.* **27**, 473 (1950).

Alternatively, pantethine may be converted to coenzyme A by use of a series of enzymatic reactions, and the coenzyme A formed determined enzymatically.[17]

Test Organism. *L. helveticus* 80, for which pantethine, pantetheine, or various mixed disulfides of pantetheine are approximately one hundred times as active in supporting growth as pantothenic acid,[15] is the test organism. Stock cultures may be carried by biweekly transfer in litmus milk supplemented with 0.5% yeast extract and 0.5% glucose. Such transfers are incubated at 37° for 24 hours, then held at about 4 to 12° for the remainder of the 2-week period. Alternatively, the stock culture may be maintained as stabs in the basal medium (see below) supplemented with pantethine (0.25 γ/ml.) and 0.1% of yeast extract and solidified with 1.5% agar.

Basal Medium. Composition of the double-strength basal medium is given in the table on page 925. It is convenient to prepare comparatively concentrated stock solutions of the trace ingredients (e.g., vitamins) and mix these in the proper proportions in preparing the final medium.

Inoculum. A transfer from the stock culture to 6 to 10 ml. of single-strength medium supplemented with an excess of pantethine (e.g., 0.25 γ/ml.) is made and incubated at 37 to 39° for 18 hours. Cells are centrifuged out aseptically and resuspended in an equal volume of 0.9% NaCl solution just before use.

Procedure. Reference standard and samples are distributed to assay tubes (16 × 180 mm.) in amounts sufficient to supply 0, 25, 50, 100, 150, 200, and 250 mγ of pantethine in volumes of 5 ml. or less. All tubes are diluted to 5 ml. with water, 5 ml. of the double-strength basal medium is added, and the tubes are autoclaved at 12 to 15 pounds of steam pressure for 3 to 6 minutes. After cooling in water, each tube is inoculated with one drop of the inoculum suspension. Incubation is at 37 to 39°. Growth is estimated turbidimetrically with a photoelectric colorimeter and a red filter. Satisfactory standard curves are obtainable in as little as 18 hours; heavier growth at the higher concentrations of pantethine is obtained with longer periods of incubation up to 36 to 40 hours.[18]

Standards. No entirely satisfactory reference standard of pantethine is yet available. During purification, potencies were estimated in terms of

[17] L. Levintow and G. D. Novelli, *J. Biol. Chem.* **207,** 761 (1954).
[18] For a more detailed presentation of the general technique of microbiological assay, see, e.g., E. E. Snell, *in* "Vitamin Methods" (György, ed.), Vol. 1, pp. 327–505, Academic Press, New York, 1950; or E. C. Barton-Wright "Microbiological Assay of the Vitamin B Complex and Amino Acids," Sir Isaac Pitman and Sons, London, 1952.

a single standard yeast extract;[1] different yeast extracts vary, however, in activity. Synthetic pantethine is now available; its physical state (sirup), however, makes its purification and use as a standard difficult. Crystalline S-benzoylpantetheine has full microbiological activity,[1] and might be a satisfactory standard.

COMPOSITION OF THE ASSAY MEDIUM

Component	Amount per 100 ml. of double-strength medium	Component	Amount per 100 ml. of double-strength medium
Glucose	2 g.	L-Cystine	20 mg.
Sodium acetate	2 g.	L-Cysteine·HCl	20 mg.
Sodium phosphate, dibasic	0.25 g.	Asparagin	20 mg.
		Tween 40[d]	200 mg.
Casein, acid-hydrolyzed	1.0 g.[a,b]	Oleic acid[d]	2 mg.
Potassium chloride	0.6 g.[a]	Pyridoxal·HCl	80 γ
Salts solution	1.0 ml.[c]	Thiamine·HCl	80 γ
Adenine	2 mg.	Riboflavin	160 γ
Guanine	2 mg.	Nicotinic acid	160 γ
Uracil	12 mg.	p-Aminobenzoic acid	80 γ
DL-Tryptophan	20 mg.	Folic acid	3.2 γ[a]
		Biotin	0.8 γ[a]

[a] O. D. Bird and V. McGlohon (unpublished) employ Difco Casamino Acids, 4 g. per 100 ml., in place of acid-hydrolyzed casein prepared in the laboratory. They also use 0.25 g. of KCl, and 6.0 γ of folic acid in place of the concentrations listed.

[b] Casein may be hydrolyzed either with sulfuric acid [cf. E. E. Snell and L. D. Wright, *J. Biol. Chem.* **139**, 675 (1941)] or with hydrochloric acid [cf. E. E. Snell, B. M. Guirard, and R. J. Williams, *J. Biol. Chem.* **143**, 519 (1942)].

[c] A stock solution is prepared containing, per 250 ml.: $MgSO_4·7H_2O$, 10 g.; NaCl, 0.5 g.; $FeSO_4·7H_2O$, 0.5 g.; and $MnSO_4·4H_2O$, 0.5 g. Precipitation of salts is prevented by slight acidification (e.g., addition of 2 drops of concentrated HCl).

[d] Two grams of Tween 40 (polyoxyethylenesorbitan monopalmitate) and 20 mg. of oleic acid are mixed together and diluted with water to 20 ml. Two milliliters of this solution (which can be preserved in the refrigerator) is used per 100 ml. of medium.

Specificity. Coenzyme A and pantothenylcysteine are essentially inactive for the test organism. Pantothenic acid is only about 1% as active as pantethine but is utilized more efficiently than this when pantethine also is present. Thus, contamination of samples by substantial amounts of free pantothenic acid renders the assay results meaningless. This interference can be partially corrected by including small amounts of pantothenic acid in the assay medium. Under these conditions the assay becomes much more sensitive to pantethine, but also more erratic.[15]

[134] Preparation and Assay of 4'-Phosphopantetheine and 4'-Phosphopantothenate

By G. DAVID NOVELLI

I. 4'-Phosphopantetheine

Preparation

The chemical synthesis described here was devised by Baddiley and Thain.[1] Pantetheine (3.8 g.) dissolved in 60 ml. of pyridine is oxidized to pantethine by passing a stream of oxygen through the solution with a sintered-glass bubbler until the sample no longer gives a positive nitro-prusside test for SH. Approximately 60 ml. of anhydrous benzene is added, and the solution is evaporated to a sirup under reduced pressure. The residue is dissolved in 50 ml. of anhydrous pyridine and cooled to −40°.

Dibenzyl phosphorochloridate (1.0 ml.) is added, and the solution is kept at −40° for 15 minutes. After standing at room temperature for 3 to 4 hours, the pyridine is removed by distillation under reduced pressure. The residue is dissolved in chloroform, and the solution is washed with 1 N H_2SO_4, then with H_2O, and then with $NaHCO_3$. The solution is next dried with solid $MgSO_4$ and evaporated. The residual sirup is washed three times with benzene, during which time it changes to a hard resin. The resin is dissolved in alcohol, and any remaining undissolved material is removed by centrifugation. Most of the alcohol is removed by evaporation, leaving a quite viscous solution. The viscous solution is added to 50 ml. of liquid ammonia, and small pieces of sodium are added until a transient blue color is observed throughout the solution. The solution is brought to pH 8.0 with barium hydroxide, and the precipitation barium phosphate is removed by centrifugation. The solution is reduced to a small volume of evaporation, and the barium salt of phosphopantetheine is precipitated with acetone and washed with ether and dried. The yield is 2.0 g.

The product may be purified through the silver salt, as follows: 1.2 g. of the barium salt is dissolved in 50 ml. of water, and an excess of silver nitrate solution is added. The silver salt is precipitated by adjusting to pH 7.0 with dilute aqueous sodium hydroxide. The silver salt is collected by centrifugation and washed with water until the supernatant is free of barium ions. This usually takes four to five washings. The silver salt is suspended in water and decomposed with hydrogen sulfide. The excess hydrogen sulfide is removed by aeration, and the solution is adjusted to

[1] J. Baddiley and E. M. Thain, *J. Chem. Soc.* **1953**, 1610.

pH 8.0 with barium hydroxide. The silver sulfide is removed by centrifugation, and the supernatant solution containing the barium salt of phosphopantetheine is saved. Additional material can be recovered by suspending the silver sulfide precipitate in water and treating it again with hydrogen sulfide as above. The precipitate of silver sulfide is again removed by centrifugation and discarded. The supernatant solution is pooled with the first supernatant and reduced to a small volume by evaporation under reduced pressure. The barium salt of 4'-phosphopantetheine is precipitated by the addition of acetone, washed with ether, and dried. The yield is 0.9 g.

Assay

4'-Phosphopantetheine is assayed by converting it enzymatically to coenzyme A as described in Vol. II [106].

II. 4'-Phosphopantothenate

Preparation

4'-Phosphopantothenate can be prepared by mild alkaline hydrolysis of 4'-phosphopantetheine as follows: 25 mg. of the barium salt of 4'-phosphopantetheine is dissolved in 10 ml. of 0.5 N NaOH, and the solution is placed in a boiling water bath for 2 hours. During this time the amide bond between β-alanine and 2-mercaptoethylamine will be completely hydrolyzed, yielding 4'-phosphopantothenate with only about 8% hydrolysis of the latter.

4'-Phosphopantothenate can also be synthesized either by the method of King and Strong[2] or by that of Baddiley and Thain.[3] A résumé of the latter method is given below.

Pantothenic acid 2'-benzyl ether is prepared as follows:[1] 2.8 g. of finely powdered β-alanine is dissolved in a solution containing 6.95 g. of pantolactone benzyl ether (Baddiley and Thain[3]) and 2.5 g. dimethylamine in 50 ml. of anhydrous methanol by refluxing for 4 hours. After refluxing for an additional 4 hours, the solvent is removed by evaporation under reduced pressure, to yield a sirup. The sirup is dissolved in water, acidified with dilute sulfuric acid, extracted three times with ethyl acetate, and washed with water. The solvent is again removed under reduced pressure, yielding a sirup. The sirup is dissolved in sodium bicarbonate, washed three times with ether, acidified with dilute sulfuric acid, and extracted three times with ethyl acetate. The organic layer is washed with water, dried with solid NA_2SO_4, and evaporated under reduced pressure to yield a clear sirup of pantothenic acid 2'-benzyl ether.

[2] T. E. King and F. M. Strong, *Science* **112**, 562 (1950).
[3] J. Baddiley and E. M. Thain, *J. Chem. Soc.* **1951**, 246.

The benzyl ether (3.8 g.) is dissolved in 50 ml. of anhydrous pyridine and cooled to $-20°$. Diphenylphosphorochloridate (3.63 g.) is added, and the solution is set aside for 20 hours at room temperature. Then 3.0 ml. of H_2O is added, and after standing at room temperature for 30 minutes the solvent is removed by distillation under reduced pressure. The residue is dissolved in chloroform, washed with 1.0 N H_2SO_4, then two times with water, and dried with solid $MgSO_4$. The solvent is again removed under reduced pressure, and the sirup evaporated twice with benzene to remove chloroform. The sirup is then dissolved in 100 ml. of 2 N NaOH and refluxed for 1.5 hours. After cooling, sodium ions are removed by passing over a column of Amberlite IR-120 (H form). The eluate is brought to pH 8.0 with barium hydroxide and kept at $0°$ for 12 hours, filtered, and concentrated to about 25 ml. under reduced pressure. The solution is again passed over IR-120 and again neutralized to pH 8.0 with barium hydroxide. The volume is reduced to a few milliliters under reduced pressure, and acetone is added to a concentration of 70%. A small sticky precipitate is removed by centrifugation. The compound is precipitated by adding 200 ml. of acetone. A further yield is obtained by evaporation of the mother liquors. The product is washed with acetone and dried over P_2O_5. The yield is 2.45 g. of the barium salt of O^2 benzyl pantothenic acid 4'-phosphate. The benzyl group is removed by suspending the product in 50 ml. of liquid ammonia and carefully adding small pieces of sodium. The insoluble barium salt is rubbed continuously during the reduction. The solvent is removed by evaporation, and the solid residue is passed through a small column of IR-120 resin. The product is 4'-phosphopantothenic acid.

Assay

4'-Phosphopantothenate is assayed for its content of pantothenic acid, before and after dephosphorylation with alkaline phosphatase, by using the microbiological assay for free pantothenate described by Maas and Davis.[4] See also Vol. III [132] for determination of pantothenic acid.

[4] W. K. Maas and B. D. Davis, *J. Bacteriol.* **60**, 733 (1950).

[135] Preparation and Assay of 3',5'-Diphosphoadenosine Nucleotide

By T. P. WANG

$$\text{CoA} \xrightarrow[\text{pyrophosphatase}]{\text{nucleotide}} \text{3',5'-Diphosphoadenosine} + \text{Phosphopantetheine}$$

Procedure.[1] Fifty milligrams of CoA (270 units/mg., Pabst Laboratories) is incubated with the snake venom nucleotide pyrophosphatase[2] in the same manner as in the preparation of the 2',5'-diphosphoadenosine.[3] The reaction is followed by the phosphotransacetylase assay. It is somewhat slower than the TPN splitting, and complete cleavage of CoA occurs after about 5 hours of incubation. The reaction mixture is then placed on a Dowex 1 column, washed with water, etc., in the same way as with the TPN reaction mixture. The 3',5'-diphosphoadenosine is also eluted with the 0.02 M HNO_3 and 0.02 M $NaNO_3$ mixture. About the same volume of HNO_3-$NaNO_3$ is needed to remove the 3',5'-diphospho compound off the column.

To the combined eluates containing the compound is added 0.5 ml. of 25% solution of basic lead acetate. After standing in icebox for 4 hours, the lead salt is centrifuged, washed, and decomposed with a stream of H_2S. The PbS is removed and the solution aerated. About 22 micromoles of this compound is recovered.

Analysis. Analysis of the compound gives a ratio of 1 adenine: 1 ribose:2 phosphate.

Assay Method. The 3',5'-diphosphoadenosine is dephosphorylated by incubating it with the 3'-nucleotidase.[4] The resulting 5-AMP is then demonstrated by the specific 5-AMP deaminase from muscle.[5,6]

---

[1] T. P. Wang, L. Shuster, and N. O. Kaplan, *J. Biol. Chem.* **206,** 299 (1954).

[2] See Vol. II [89].

[3] See Vol. III [130].

[4] L. Shuster and N. O. Kaplan, *J. Biol. Chem.* **201,** 535 (1953); see Vol. II [86].

[5] See Vol. II [68].

[6] 3',5'-Diphosphoadenosine has recently been identified as coenzyme for sulfate transfer reactions. It is possible that the 3',5'-diphosphoadenosine can also be assayed by its sulfate acceptor properties. (P. W. Robbins and F. Lipmann, *J. Am. Chem. Soc.* **78,** 2652 (1956).)

[136] Preparation and Assay of Adenine Pantetheine Dinucleotide (DPCoA)

By T. P. WANG

$$\text{CoA} \xrightarrow{\text{3'-nucleotidase}} \text{DPCoA} + \text{P}$$

Procedure.[1] Forty milligrams of CoA (270 units/mg., obtainable from the Pabst Laboratories), dissolved in 70 ml. of water (pH about 3.5), is neutralized with solid $KHCO_3$ to pH 7.5. With occasional stirring or shaking, the solution is kept at room temperature until the characteristic odor of —SH compounds disappears. CoA in the —SH form is not split by the 3'-nucleosidase.[2] This is due to the inhibition by the —SH group, as it has been shown that cysteine and reduced glutathione are strong

ANALYSIS OF DEPHOSPHORYLATED CoA

	Adenine[a]	Ribose[b]	Pantothenic acid[c]	Phosphate
μM./mg.	0.78	0.75	0.73	1.43
Ratio to ribose	1.04	1	0.97	1.91

[a] Adenine was estimated by UV absorption at 260 mu based on a millimolar absorption coefficient of 16,000.

[b] Ribose was determined by the orcinol method with adenylic acid as the standard (N. O. Kaplan, S. P. Colowick, and A. Nason, *J. Biol. Chem.* **191**, 473 (1951); see also Vol. III [12, 13]).

[c] Determined by the procedure of Novelli, Kaplan, and Lipmann [*J. Biol. Chem.* **177**, 97 (1949)].

inhibitors of the enzyme. Five milliliters of 3'-nucleotidase is added to the solution, which is then incubated at 37°. Small samples (corresponding to 10 units of CoA) are taken after 2 hours of incubation for the arsenolysis test.[3] As soon as all the CoA is dephosphorylated (indicated by negative arsenolysis test), the solution is evaporated under reduced pressure to a small volume (about 8 ml.). Some precipitate formed at this stage is removed by centrifugation. The clear solution is then acidified to pH 4 with 4 *N* HNO_3, and 6 vol. of cold acetone ($-15°$) are added to the acidified mixture. After standing in a deep-freeze ($-15°$) overnight, the

[1] T. P. Wang, L. Shuster, and N. O. Kaplan, *J. Biol. Chem.* **206**, 299 (1954).
[2] L. Shuster and N. O. Kaplan, *J. Biol. Chem.* **201**, 535 (1953); see Vol. II [86].
[3] E. R. Stadtman, *J. Biol. Chem.* **196**, 527 (1952), see Vol. I [98].

precipitate is collected by centrifugation. It is then redissolved in 5 ml. of water and reprecipitated with acetone as before. Any insoluble materials are removed by centrifugation or filtration before the acetone is added. After being washed once with ether, the precipitate is dried over P_2O_5 and chopped paraffin, in a vacuum desiccator. Forty milligrams of DPCoA of 50% purity is obtained.

Assay Method. See assay method for the DPCoA kinase[4] (Vol. II [110]).

Analysis. Analyses have been made for adenine, ribose, pantothenate, and total phosphate for the above preparation of DPCoA and are summarized in the table.

[4] T. P. Wang and N. O. Kaplan, *J. Biol. Chem.* **206,** 311 (1954).

[137] Preparation and Assay of Acyl Coenzyme A and Other Thiol Esters; Use of Hydroxylamine

By E. R. STADTMAN

Two general methods for the synthesis of acyl coenzyme A and other thiol esters will be described. The first is based on the procedure of Simon and Shemin,[1] in which an acid anhydride is allowed to react with the mercaptan in cold aqueous solution at pH 7.5. The second method, which has been extensively developed by Wieland *et al.,*[2-4] makes use of thiol ester interchange reactions of the type

$$RCOSR' + R''SH \leftrightarrows RCOSR'' + R'SH \qquad (1)$$

Synthesis by Reaction with Acid Anhydrides

Ten micromoles of CoASH [on basis of free —SH as determined by quantitative analysis (see section on auxiliary methods)] is dissolved in 1.0 ml. of water. The solution is cooled to 0°; 0.2 ml. of 1 M KHCO₃ is added, and the solution is neutralized to pH 7.5. Then 0.13 to 0.15 ml. of a freshly prepared ice-cold solution of acid anhydride (0.1 M) is added, and, after mixing, the solution is allowed to stand at 0° for 4 to 5 minutes.

If a drop of the reaction mixture still gives a positive —SH test with the nitroprusside reagent (see section on auxiliary methods), more acid

[1] E. J. Simon and D. Shemin, *J. Am. Chem. Soc.* **75,** 2520 (1953).
[2] T. Wieland and H. Köppe, *Ann.* **588,** 15 (1954).
[3] T. Wieland and H. Köppe, *Ann.* **581,** 1 (1953).
[4] T. Wieland and W. Schäfer, *Ann. Chem.* **576,** 104 (1952).

anhydride solution is added until the —SH test is negative.[5] The pH is finally adjusted to approximately 6.0 with HCl.

Reaction with Mixed Anhydrides of Ethyl Hydrogen Carbonate

When the acid anhydride is not readily available, a mixed anhydride of ethyl hydrogen carbonate and the acid may be used.[6] To prepare the mixed anhydride, 1.0 equivalent of the organic acid and 1.0 equivalent of pyridine are dissolved in anhydrous ethyl ether (0°). Then 1.0 equivalent of ethyl chloroformate is added dropwise with stirring. After standing an hour at 0°, the insoluble pyridine hydrochloride is removed by filtration or centrifugation. To determine the content of the mixed anhydride, an aliquot of the ether solution is added to 0.5 ml. of 2.0 N hydroxylamine solution (∼pH 7.0). The mixture is shaken. After 5 minutes a suitable aliquot is used for quantitative measurement of the hydroxamic acid by the method of Lipmann and Tuttle.[7] (Details are given below.)

To form acyl SCoA derivatives, 1.2 to 1.3 equivalents of the mixed anhydride is added dropwise with shaking to a cold (0°) aqueous solution containing 1.0 equivalent of CoASH and 0.2 M KHCO$_3$ (adjusted to pH 7.5). In those instances where the thiol ester test is applicable,[5] additional anhydride is added as necessary until the reaction mixture gives a negative —SH test with the nitroprusside reagent. The pH is adjusted to 6.0 with 1 M HCl, and the aqueous layer is extracted three times with an equal volume of ether. Finally, the last traces of ether are removed from the aqueous solution by bubbling nitrogen through it at room temperature.

Comments. In the synthesis of simple thiol esters by the anhydride and mixed anhydride procedures, the yields are normally 80 to 100%, based on the free —SH. The formation of thiol ester derivatives of α,β-unsaturated fatty acid is much poorer—15 to 50%. These low yields are due to the occurrence of side reactions in which the —SH groups add to the double bonds of the anhydride and thiol esters.

Synthesis by Ester Interchange Reaction

Although many thiol esters may serve as acyl donors for the nonenzymatic synthesis of acyl SCoA compounds, Wieland et al.[2-4] have found that S-acylthiophenol compounds are particularly well suited for this purpose.

---

[5] With certain thiol esters a negative —SH reaction is never obtained (see comments in section on qualitative —SH and thiol ester tests).

[6] T. Wieland and L. Rueff, *Angew. Chem.* **65**, 186 (1953).

[7] F. Lipmann and L. C. Tuttle, *J. Biol. Chem.* **159**, 21 (1945).

Preparation of S-Acylthiophenols. It is, of course, impossible to describe a single procedure that is universally applicable for the synthesis of all S-acyl derivatives of thiophenol. The following procedure, which has been described by Wieland and Köppe[3] for the synthesis of certain S-acyl esters of thiophenol, is presented here as a model procedure which in principle may be generally useful but which may have to be modified for the synthesis of some acyl derivatives.

Procedure. To an ether (or tetrahydrofuran) solution of the acid anhydride (or the mixed anhydride), 1 equivalent of thiophenol is added. After 24 hours at room temperature the solvent is evaporated off. The residue is taken up in water, and the oil which separates out is dissolved in ether. The solution is dried with anhydrous $NaSO_4$. After filtration, the ether is evaporated off and the acylthiophenol is contained in the residue.

Preparation of S-Acylglutathione. Three-tenths gram of glutathione (0.001 mole) is dissolved in a minimum amount of water. A methanolic solution containing 1.8 g. of acylthiophenol (0.01 mole) is added, and more methanol is added as needed to give a homogeneous solution. After 3 days at room temperature any precipitate which forms is removed by centrifugation and the supernatant solution is evaporated *in vacuo* to get rid of the methanol. The aqueous residue is extracted with ether to remove the thiophenol. The acylglutathione is precipitated by the addition of 20 vol. of acetone, and the precipitate is washed with anhydrous acetone and dried in a desiccator under vacuum. The yield of acylglutathione is about 60% based on the glutathione added.

Other Synthetic Methods

In addition to the above general methods, thiol esters have been prepared by reaction of mercaptans with acyl chlorides and thiol acids.[8] A method for the synthesis of acetoacetyl thiol esters by reaction of mercaptans with ketene has been presented by Lynen and Wieland.[9]

Specific details for the synthesis of S-lactoyl- and S-β-hydroxybutyryl-thiophenol and glutathione are reported by Wieland and Köppe.[3] The synthesis of various acyl chlorides and their use in the synthesis of S-glycolyl-, S-lactoyl-, S-mandelyl-, S-pyruvyl-, S-β-alanylthiophenol, and glutathione derivatives has also been reported by these investigators.[2] Details for the synthesis of various amino acid acyl chlorides and thiol esters are presented by Wieland and Schäfer.[4] The use of S-acetyl thio-

---

[8] I. B. Wilson, *J. Am. Chem. Soc.* **74**, 3205 (1952); E. R. Stadtman, *J. Biol. Chem.* **203**, 501 (1953).

[9] F. Lynen and O. Wieland, Vol. I [94].

phenol in the synthesis of S-acetylglutathione was reported by Wieland and Bokelmann.[10]

Properties

Most thiol esters react readily with hydroxylamine at neutral pH to give the corresponding hydroxamic acid derivatives and free mercaptans. Exceptions are found with thiol ester derivatives of β-keto acids which react with hydroxylamine to give isoxazolones and with hydrazine to give pyrazolones. Thiol ester derivatives of α-amino mercaptans viz., cysteine and cysteamine, are unstable and spontaneously rearrange to the N-acyl and (or) thiazoline derivatives. Although most thiol esters are fairly stable at neutral to acid pH, they are readily hydrolyzed by alkali. The half-life of S-acyl derivatives of saturated fatty acids with two to eight carbon atoms is only 1 to 2 minutes in 0.1 N sodium hydroxide at 30° (F. Lynen, personal communication). The S-crotonyl derivative is slightly more stable. In general it may be stated that the lower the pKa of the acid obtained from the S-acyl moiety, the more unstable is the thiol ester in alkali (T. Wieland, personal communication). Heating at 100° for 10 minutes at pH 3.0 results in little or no detectable decomposition of thiol esters of saturated fatty acids and aliphatic mercaptans. These will keep for several months at $-10°$ without significant change. Although precise stability information is not available for most other thiol esters, it appears that derivatives of α- and β-keto acids and of succinic and fluoroacetic acids are much more unstable. Thiol ester derivatives of α-amino acids are particularly unstable in the presence of bicarbonate ion.[11]

At neutral to alkaline pH, thiol esters undergo nonenzymatic ester interchange reactions of the type described by reaction 1.

Thiol esters of α,β-unsaturated acids react spontaneously with mercaptans to form addition compounds as follows:

$$R\!-\!CH\!=\!CH\!-\!COSR + R'SH \rightarrow R\!-\!\underset{\underset{SR'}{|}}{CH}CH_2COSR \qquad (2)$$

Nearly all thiol esters have characteristic light absorption bands in the ultraviolet region. Optical properties of some thiol esters are tabulated in the accompanying table. As can be seen from the table, the molecular extinction of S-acetoacetyl thiol esters is markedly influenced by pH and by the presence of Mg^{++}.

[10] T. Wieland and E. Bokelmann, *Angew. Chem.* **64**, 59 (1952).
[11] T. Wieland, R. Lambert, H. U. Lang, and G. Schramm, *Ann.* **597**, 181 (1955).

Enzymatic Assay of Ac-SCoA

Principle. In this method measurement is made of the decrease in optical density at 232 to 240 mμ that is associated with the hydrolysis of Ac-SCoA.[12-14] The complete hydrolysis of Ac-SCoA is brought about in a few seconds by incubation in the presence of phosphotransacetylase and arsenate.[12,15]

Reagents

Potassium arsenate, 0.5 M, pH 7.0.
Phosphotransacetylase, 500 units/ml.

Procedure. In a 1.0-ml. quartz cuvette (1-cm. light path) place 0.05 ml. of potassium arsenate (0.5 M), the test solution containing 0.02 to 0.1 micromole of Ac-SCoA, and water to a total volume of 1.0 ml. The reference cuvette should contain the arsenate buffer but no Ac-SCoA. Read the optical density at 232 mμ (or 240 mμ), and then add 0.01 ml. of the phosphotransacetylase preparation to the test solution and also to the reference cuvette. Wait 1 to 2 minutes (or until there is no further change in optical density), and read the optical density again. The total decrease in optical density is a measure of the Ac-SCoA. The difference in molecular extinction coefficients of Ac-SCoA and its hydrolysis products (ΔE_{232}) is 4.5×10^6 cm^2/mole. The ΔE_{240} is 3.3×10^6 cm^2/mole. Therefore, under the above conditions the changes in optical density due to 0.1 micromole of Ac-SCoA are 0.45 and 0.33 at 232 and 240 mμ, respectively.

Comments. If sodium arsenate buffer is used in place of potassium arsenate, the reaction mixture must be supplemented with 100 micromoles of potassium or ammonium chlorides.[15] If crude bacterial extracts or only partially purified preparations are used in place of highly purified phosphotransacetylase preparations, it is desirable to use fewer enzyme units and a longer incubation time.

Specificity. Of the compounds thus far examined, the method is relatively specific for acetyl and propionyl thiol ester derivatives of CoASH and pantetheine. Butyryl thiol esters are also hydrolyzed but at very much lower rates. In the presence of CoAS-transphorase[12,16] and

[12] E. R. Stadtman, *J. Cellular Comp. Physiol.* **41**, 89 (1953).
[13] E. R. Stadtman, *in* "The Mechanism of Enzyme Action" (W. D. McElroy and B. Glass, eds.) p. 581. Johns Hopkins Press, Baltimore, 1954.
[14] F. Lynen, *Federation Proc.* **12**, 683 (1953); F. Lynen and S. Ochoa, *Biochim. et Biophys. Acta* **12**, 299 (1953).
[15] E. R. Stadtman, *J. Biol. Chem.* **196**, 527 (1952).
[16] For preparation of CoA-transphorase from *C. kluyveri*, see Vol. I [99].

OPTICAL PROPERTIES OF SOME THIOL ESTERS

S-Acyl Group	Thiol Group	λ_{max}, mμ	ϵ, $\times 10^6$ cm.²/mole	$\Delta\epsilon$, $\times 10^6$ cm.²/mole*	Reference
Saturated fatty acids with 4 to 8 carbon atoms	N-Acetylcysteamine	233	4.58	4.48	a
	CoASH	260; 232†	16.4; 8.7†	0.02; 4.5	a
CH_3CO—	N-Acetylcysteamine	232	4.6	4.5	b, c
	Glutathione	232	4.5	4.5	d
	Thioglycolate	232	4.5	4.5	e
	Butyl mercaptan	233	4.7	‡	f
	Butyl mercaptan	233	4.8	‡	g
	ω-Amino-decyl thiol·HCl	232	5.7	‡	h
	Dimethylcysteamine	228	4.7	‡	h
	Thiocholine·HCl	228	5.4	‡	h
	Thiophenol	§	§	‡	h
$CH_3(CH_2)_4CO$—	CoASH	260; 232†	16.4; 9.2	0.2; 5.0†	a
CH_2OHCO—	Glutathione	235	‡		i
$CH_3CHOHCO$—	N-Acetylcysteamine	232	‡	4.2	j
	Glutathione	235	‡	4.0	k
$CH_3CHOHCH_2CO$—	Glutathione	233	‡	3.7	j
$CH_2{=}CHCO$—	Pantetheine	265	‡	6.7	l
$CH_3CH{=}CHCO$—	N-Acetylcysteamine	263; 225	6.7; 10.6	6.7; 10.0	a, c
	CoASH	260; 225†	22.6; 16.3†	6.4; 8.4	a
Octenoyl-	N-Succinylcysteamine	263; 228	7.8; 11.8	7.8; 11.3	a
	CoASH	260; 228†	24.0; 15.1†	7.8; 10. †	a
	N-Acetylcysteamine	237 (pH 7.0)	4.7	4.5	a

Acyl group	Thiol component	λ (mμ)			Ref.
CH₃COCH₂CO—	N-Acetylcysteamine	237; 302 (pH 8.0)	4.2; 5.5	4.0; 5.5	a
	N-Acetylcysteamine	237; 300 (pH 9.0)	4.0; 18.3	3.8; 18.3	a
	N-Acetylcysteamine	239; 303 (pH 9.0, 0.5 M Mg^{++})	4.5; 25.9	4. ; 25.9	a
	CoASH	260 (pH 7.0)	16.0	‡	a
	CoASH	262; 303 (pH 7.55, 0.2 M Mg^{++})	16.8; 17.0	‡; 17.0	a
HOOCCH₂CH₂CO—	CoASH	260; 232†	‡	4.5	m
NH₂CH₂CH₂CO—	Pantetheine	223	‡	4.5	e
Valyl-N-alanyl-	Glutathione	234	3.5	‡	h
Valyl-N-glycyl-	Cysteamine	236	4.75	‡	h

* Refers to change in molecular extinction obtained by hydrolysis of the thiol ester bond.
† Refers to wavelength at which there is a maximum decrease in optical density upon hydrolysis.
‡ Not determined or calculated.
§ End absorption only.
a F. Lynen, personal communication.
b E. R. Stadtman, in "The Mechanism of Enzyme Action" (W. D. McElroy and B. Glass, eds.), p. 581. Johns Hopkins Press, Baltimore, 1954.
c F. Lynen, Federation Proc. 12, 683 (1953); F. Lynen and S. Ochoa, Biochim. et Biophys. Acta 12, 299 (1953).
d W. W. Kielley, personal communication.
e E. R. Stadtman, unpublished results.
f I. B. Wilson, J. Am. Chem. Soc. 74, 3205 (1952); E. R. Stadtman, J. Biol. Chem. 203, 501 (1953).
g L. H. Noda, S. A. Kuby, and H. A. Lardy, J. Am. Chem. Soc. 75, 915 (1953).
h T. Wieland, personal communication.
i T. Wieland and H. Köppe, Ann. 588, 15 (1954).
j T. Wieland and H. Köppe, Ann. 581, 1 (1953).
k E. Racker, J. Biol. Chem. 190, 685 (1951).
l J. R. Stern, Vol. I [95].
m S. Kaufmann, personal communication.

thiol transacylases,[17] as are found in crude extracts of *Cl. kluyveri*, other thiol esters may also be measured by this assay.

Thiol Ester Derivatives of α,β-Unsaturated Fatty Acids and β-Keto Acids

Specific spectrophotometric methods for the estimation of thiol ester derivatives of α,β-unsaturated acids[18] and β-keto acids[19] have been described elsewhere.

Determination of Thiol Esters by Hydroxamic Acid Procedure

Thiol esters react quantitatively with neutral hydroxylamine to form hydroxamic acids which are estimated by the method of Lipmann and Tuttle.[7]

Reagents. The reagents are the same as those used in the procedure for estimation of acyl phosphates.[20]

Procedure: To 1.0 ml. of sample containing 0.5 to 2.5 micromoles of thiol ester, add 0.5 ml. of neutralized hydroxylamine reagent. After 10 minutes at room temperature add 1.5 ml. of the ferric chloride reagent. Centrifuge to sediment the protein, and transfer the supernatant solution to a micro-Klett colorimeter tube. Make optical density measurements immediately in a Klett photoelectric colorimeter using a No. 54 filter. Set the instrument against a blank containing all reagents except the thiol ester.

Auxiliary Methods

Qualitative —SH and Thiol Ester Tests. In most instances it is possible to determine when the acylation of sulfhydryl compounds is complete by testing a drop of the reaction mixture with the nitroprusside reagent of Toennies and Kolb.[21] With slight modification, this test can be used to detect thiol ester spots on paper chromatograms.

Reagents

Nitroprusside reagent. Dissolve 1.5 g. of sodium nitroprusside in 5.0 ml. of 2 N H_2SO_4, and then add 95 ml. of absolute methanol and 10 ml. of a concentrated NH_4OH solution (28% NH_4OH). Filter off the copious white precipitate, and keep the clear orange filtrate. This reagent is stable for weeks if kept stoppered at 0°.

Methanolic sodium hydroxide. Dissolve 2.0 g. of NaOH in 5.0 ml. of water, and add 95 ml. of absolute methanol.

Ethyl ether, analytical grade.

[17] R. O. Brady and E. R. Stadtman, *J. Biol. Chem.* **211**, 621 (1954).
[18] See Vol. I [93].
[19] See Vol. I [95].
[20] See Vol. III [39].
[21] G. Toennies and J. J. Kolb, *Anal. Chem.* **23**, 823 (1951).

Tests for Free —SH. Place a drop of the test solution on a narrow strip of Whatman No. 3 filter paper and dip the paper into the nitroprusside reagent. The immediate appearance of a red spot indicates the presence of free —SH groups. The very faint colors produced with very small quantities of —SH are made readily detectable by shaking the filter paper strip in ethyl ether. If the above test is negative, the presence of thiol esters can be demonstrated by following the nitroprusside dip with a dip in the methanolic sodium hydroxide solution. Under these conditions, the thiol esters are hydrolyzed and liberation of a free SH group produces a red color as above. To detect free mercaptans and thiol esters on paper chromatograms, a 1-cm. cross-sectional strip of the chromatogram is dipped first into the nitroprusside reagent and then into the methanolic sodium hydroxide solution. Red spots appearing after the nitroprusside dip are due to free —SH, and those formed only after a subsequent dip into alkali may be due to thiol esters. When weak spots are observed, these may be intensified by dipping the paper strip in ethyl ether.

Comments. The thiol ester test has been used successfully for the detection of various thiol esters of ordinary saturated and unsaturated fatty acids, of hydroxy acids, and of β-alanine. However, the test is not applicable to thiol esters that are readily hydrolyzed by the nitroprusside reagent alone, as, for example, thiol ester derivatives of fluoroacetate.[22] The appearance of a red color after the alkali treatment is only presumptive evidence for the presence of thiol esters, since other compounds, viz., methyl ketones, give red and blue colors with strongly alkaline nitroprusside reagents. More positive identification is made by spectrophotometric analysis of the compound after elution from paper.

Quantitative Assay of Free —SH. *The Nitroprusside Method.* This method is modified after the procedure described by Grunert and Phillips.[23]

Reagents

$(NH_4)_2SO_4$, saturated solution.

Sodium nitroprusside, 2.0 g. per 100 ml.

Cyanide reagent. Dissolve 20.7 g. of K_2SO_3 and 0.33 g. of NaCN in 100 ml.

Reference standard: Dissolve 30.7 mg. of crystalline glutathione in 100 ml. of water. One milliliter of this solution contains 1.0 microequivalent of —SH.

Procedure. In a 1.0-ml. cuvette (1-cm. light path) place 0.4 ml. of sample containing 0.03 to 0.2 microequivalent of —SH. Then, in the

[22] R. O. Brady, *J. Biol. Chem.* **217**, 213 (1955).
[23] R. R. Grunert and P. H. Phillips, *Arch. Biochem.* **30**, 217 (1951).

indicated order, add 0.4 ml. of saturated $(NH_4)_2SO_4$ solution, 0.2 ml. of 2% nitroprusside reagent, and 0.2 ml. of cyanide reagent. Mix immediately, and read at 540 mμ exactly 30 seconds after mixing. The reference cuvette should contain all reagents except the sulfhydryl compound. The optical density obtained with 0.1 microequivalent of —SH is about 0.33.

The *p-Chloromercuribenzoate Method*. This assay procedure is based on the method of Boyer,[24] which depends on the fact that *p*-chloromercuribenzoate reacts stoichiometrically with —SH groups to form mercaptides which can be quantitated by measurement of their strong light absorption at 355 mμ.

Reagents

Sodium acetate buffer, 1.0 M, pH 4.5.

p-Chloromercuribenzoate solution. Dissolve 36 mg. of *p*-chloromercuribenzoate in 2.0 ml. of 0.2 N KOH and dilute to 100 ml. Adjust pH to 7.5 with 1.0 M HCl.

Procedure. To a 1.0-ml. cuvette with a 1-cm. light path, add 0.3 ml. of 1 M acetate buffer (pH 4.5), 0.15 ml. of the *p*-chloromercuribenzoate solution, and 0.55 ml. of water. Read the optical density at 255 mμ, and then add 0.02 ml. of experimental sample containing 0.02 to 0.1 microequivalent of —SH compound. After 5 minutes read the optical density at 255 mμ again. The increase in optical density due to 0.1 microequivalent of —SH is 0.62 to 0.67. Crystalline glutathione may be used as a reference standard.

Purification by Paper Chromatography

For many enzyme studies the crude reaction mixtures obtained by the previously described synthetic methods may be used directly without purification. When necessary the thiol esters can be separated from other products by paper chromatography. The reaction mixture containing about 10 to 20 micromoles of thiol ester is adjusted to approximately pH 5.0 with 0.1 N HCl. It is streaked along the bottom edge of a large sheet of Whatman No. 3 filter paper and chromatographed in a suitable solvent system. Acyl SCoA derivatives of fatty acids with two to six carbon atoms have been successfully chromatographed in a solvent system made by mixing equal volumes of ethanol and 0.1 M sodium acetate buffer (pH 4.5). Another suitable system is prepared by mixing equal volumes of isopropyl alcohol, pyridine, and water. To locate the acyl SCoA derivatives on the paper chromatogram, a 1-cm. cross-sectional strip of the chromatogram is dipped into the nitroprusside reagents as described in the section on qualitative tests for —SH compounds and

[24] P. D. Boyer, *J. Am. Chem. Soc.* **76**, 4331 (1954).

thiol esters. The thiol esters may also be detected as "quenching" bands when the chromatogram is viewed under an ultraviolet lamp. The acyl SCoA bands are then cut from the chromatogram and eluted with distilled water. The over-all recovery is not good—30 to 65%—but compounds in which the ratio of thiol ester to adenine is 0.8 to 1.0 are readily obtained. For reasons not yet understood, two acyl SCoA bands are usually obtained upon chromatographs in the above solvent systems. In the case of acetyl SCoA, the two bands isolated in this way are undistinguishable on the basis of enzymatic assay with phosphotransacetylase.

[138] Preparation and Assay of Lipoic Acid and Derivatives

By I. C. GUNSALUS and W. E. RAZZELL

Assay Method

Principle. Lipoic acid can be detected and assayed manometrically, at levels of 1 to 10 mγ, by measuring oxygen utilization of a pyruvate dehydrogenase apoenzyme system in *Streptococcus faecalis*, strain 10Cl. The apoenzyme is prepared by growing the cells in a synthetic medium deficient in lipoic acid.[1] This assay can be performed with cell suspensions[2] or more conveniently with vacuum-dried cells[3] in which the necessary enzymes are stable for a period ranging up to five years. A curve relating the rates of oxygen consumption to the amounts of pure lipoic acid serves as standard.

Catalytic activity in this system is restricted to several lipoic acid derivatives including the dextrorototary isomer of α-lipoic acid (5-(dithiolane-3)-pentanoic acid); reduced lipoic acid, or dihydrolipoic acid[4] (6,8-dithioloctanoic acid); and a sulfoxide, β-lipoic acid, characterized as a S-benzylthiouronium salt.[5] Monoacetyl dihydrolipoic acid (6-acetyl-6,8-dithioloctanoic acid) has been prepared enzymatically (transacetylase)[4] but its catalytic activity not measured.

Lipoic acid activity is present in biological material in several bound forms, the structures of which have not been determined.[6,7]

[1] D. J. O'Kane and I. C. Gunsalus, *J. Bacteriol.* **56**, 499 (1948).

[2] I. C. Gunsalus, M. I. Dolin, and L. Struglia, *J. Biol. Chem.* **194**, 849 (1952).

[3] W. E. Razzell, G. H. F. Schnakenberg, and I. C. Gunsalus, *J. Biol. Chem.* in press.

[4] I. C. Gunsalus, L. S. Barton, and W. Gruber, *J. Am. Chem. Soc.* in press.

[5] E. L. Patterson, J. A. Brockman, Jr., F. P. Day, J. V. Pierce, M. E. Macchi, C. E. Hoffman, C. T. O. Fong, E. L. R. Stockstad, and T. H. Jukes, *J. Am. Chem. Soc.* **73**, 5919 (1951).

[6] I. C. Gunsalus, L. Struglia, and D. J. O'Kane, *J. Biol. Chem.* **194**, 859 (1952).

[7] L. J. Reed, B. G. De Busk, P. M. Johnston, and M. E. Getzendaner, *J. Biol. Chem.* **192**, 851 (1951).

Reagents

Potassium phosphate buffer, 1 M, pH 6.5.

Thiamine hydrochloride, 10 mg. per 100 ml.; store at 4°.

Adenosine, 200 mg.; plus riboflavin, 20 mg. per 100 ml. Dissolve the adenosine in 8 ml. of $M/40$ acetic acid, add the riboflavin in 70 ml. of water, adjust to pH 6.5 with KOH, dilute to 100 ml., and store away from light at 4°.

$MgSO_4$ $M/10$.

Potassium pyruvate, 0.6 M, pH 6.5.

Glutathione, 30 mg./ml., adjusted to pH 6.5 with KOH. *Fresh daily.*

$MnSO_4$, $M/50$.

$KHCO_3$, 2%.

KOH, 20%.

Lipoic acid, crystalline ($\mp$)-α-lipoic acid,[8] or (+)-α-lipoic acid.[9]

Dried apopyruvate dehydrogenase cells, 25 mg./ml. (see Enzyme Preparation, below).

Procedure. STANDARD LIPOIC ACID SOLUTION. Dissolve a known amount of crystalline lipoic acid (about 2.00 mg.) in 2.00 ml. of reagent-grade benzene, transfer 0.10 ml. to 9.90 ml. of 2% $KHCO_3$, and shake well. Dilute the resulting bicarbonate-buffered lipoic acid solution to contain 10 mγ of lipoic acid per milliliter. For example, transfer 0.10 ml. of the bicarbonate solution of lipoic acid to 9.90 ml. of water, or whatever volume is necessary to produce a final concentration of 100 mγ of lipoic acid per milliliter, and transfer 1.00 ml. of this to 9.00 ml. of water. The last two solutions should be used within an hour. The benzene solution may be stored indefinitely at 0°.

STANDARD CURVE. Transfer to a series of Warburg flasks the following reagents: 0.1 ml. of phosphate buffer, 0.4 ml. of supplements (prepared by mixing equal volumes of the thiamine, adenosine-riboflavin, and $MgSO_4$ reagents), and sufficient water to bring the volume to 1.9 ml. after the glutathione, lipoic acid, and cells are added. Place in the side arm 0.10 ml. of pyruvate, and in the center well 0.15 ml. of KOH. When cups have been prepared, add 0.10 ml. of glutathione, the lipoic acid sample

[8] M. W. Bullock, J. A. Brockman, Jr., E. L. Patterson, J. V. Pierce, and E. L. R. Stockstad, *J. Am. Chem. Soc.* **74,** 1868 (1952); C. S. Hornberger, Jr., R. F. Heitmiller, I. C. Gunsalus, G. H. F. Schnackenberg, and L. J. Reed, *J. Am. Chem. Soc.* **75,** 1273 (1953).

[9] E. Walton, A. F. Wagner, L. H. Peterson, F. W. Holly, and K. Folkers, *J. Am. Chem. Soc.* **76,** 4748 (1954).

(in 0.8 ml. or less), 0.10 ml. of $MnSO_4$, and 0.20 ml. of cell suspension to the main compartment in the order given.

Mount the cups, and equilibrate at 37° for 5 minutes. Close the stopcocks, shake for another 5 minutes to ensure against leaks or lack of equilibration, read the manometers, and tip in the pyruvate. Read every 5 minutes for 40 minutes, and calculate the oxidation rate from the linear portion of the curve. Unknown samples of lipoic acid are assayed at two or three levels by the same procedure.

Five milligrams of a typical dried cell apopyruvate dehydrogenase will take up 30 to 70 μl. of O_2 per hour as a blank, without added lipoic acid. Graded amounts of lipoic acid from 0.5 to 5 mγ of (+)-lipoic acid give a reasonably linear response to about 300 μl. of O_2 per hour at 5 mγ. Above 5 mγ the curve departs from a near-linear relationship with increasing lipoic acid but is usable to 10 to 15 mγ of (+)-lipoic acid; rate, 400+ μl. of O_2 per hour. This curve indicates a K_m for (+)-lipoic acid of about 1.7×10^{-7} moles per liter. The supplements—glutathione, thiamine, riboflavin, adenosine, and metals—will not activate the enzyme in the absence of lipoic acid, but in the presence of lipoic acid all are rate-limiting. The amounts added are in excess of the actual requirement.[3]

Lipoic acid bound in natural material is released by several means, including enzymatic and acid or alkaline hydrolysis. For routine assay two procedures have been applied: (1) heating in water at 100° for 10 minutes; (2) autoclaving in 6 N HCl at 120° for 2 hours, followed by partial evaporation *in vacuo* over KOH to reduce the anion concentration. In both cases the pH is adjusted to 6.5 with KOH and the solutions (or suspensions) made to a known volume before assay. Solutions of lipoic acid, free of extraneous material, usually assay higher after hot-water treatment than after acid hydrolysis; the reverse is true for lipoic acid bound to protein.[3]

The quantitative liberation of lipoic acid from natural material in a form soluble in benzene from aqueous acid medium has not been completely worked out. Available data indicate some loss during hydrolysis, with the 6N HCl for 2 hours at 120°, giving least destruction.

Enzyme Preparation (Pyruvate Apodehydrogenase). Streptococcus faecalis, strain 10Cl (ATCC 11,700), from a stock agar deep is inoculated into 10 ml. of AC broth,[2] incubated at 37° for 10 hours, and 1 ml. transferred to 500 ml. of the synthetic medium (see below). After 10 hours' further incubation, the 500-ml. culture is added to 5 l. of the same medium. Cells are harvested after a further 12-hour incubation, with a Sharples supercentrifuge, resuspended in 100 ml. of 0.03 M potassium phosphate buffer, pH 6.5, plus 2.0 g. of glutathione, and are dried from the frozen state *in vacuo*. The yield of dried material is about 4 g. The cells are stored over desiccant at 0°. The Q_{O_2} with pyruvate in the absence of lipoic acid is usually less than 20, and with excess lipoic acid about 100.

TABLE I

SYNTHETIC MEDIUM FOR APO CELLS[a]

Medium	Per liter	
Acid-hydrolyzed casein (H_2SO_4)	10	g.
Enzymatic casein hydrolyzate[b]	7.5	g.
Glucose	3	g.
K_2HPO_4	5	g.
Na thioglycolate	100	mg.
DL-Tryptophan	200	mg.
L-Cystine	200	mg.
Adenine, guanine, uracil (each)	25	mg.
Nicotinic acid	5	mg.
Riboflavin	1	mg.
Pyridoxine·HCl	1	mg.
Thiamine·HCl	1	mg.
Ca pantothenate	1	mg.
Folic acid	10	γ
Biotin	1	γ
Salts B[c]	5	ml.

Final pH 7.0 to 7.3; autoclave for 15 minutes at 15 pounds.

[a] I. C. Gunsalus, M. I. Dolin, and L. Struglia, *J. Biol. Chem.* **194**, 849 (1952).
[b] E. C. Roberts and E. E. Snell, *J. Biol. Chem.* **163**, 499 (1946).
[c] Salts B per 250 ml. contain 10 g. of $MgSO_4\cdot7H_2O$, 0.5 g. of NaCl, 0.5 g. of $FeSO_4\cdot7H_2O$, 0.5 g. of $MnSO_4\cdot4H_2O$, and 0.5 g. of ascorbic acid.

Preparation of Lipoic Acid and Derivatives

1. ($\mp$)-α-*Lipoic acid* (5-(dithiolane-3)-pentanoic acid) has been synthesized by several routes and characterized as a yellow crystalline solid, melting point 59 to 60°.[8] The racemate possesses 50% of the activity of the isolated dextrorotatory isomer in the catalytic pyruvate oxidation factor (POF) and enzymatic acetylation[4] systems, and 100% of the activity of the natural material for lipoic dehydrogenase.[10] Because of its availability, it serves as a convenient standard for assay. (+)-α-Lipoic acid, melting point 49 to 50°, has been synthesized[9] and shown to equal in catalytic activity the isolated substance. As the supply of this isomer increases, it should serve as a convenient standard.

2. *Dihydrolipoic acid*, reduced lipoic acid, 6,8-dithioloctanoic acid (DMO), is prepared chemically from lipoic acid—the disulfide—by chemical reduction[4] as follows: To a solution of ($\mp$)-α-lipoic acid (6.0 g.)

[10] L. P. Hager and I. C. Gunsalus, *J. Am. Chem. Soc.* **75**, 5767 (1953).

in 117 ml. of 0.25 N sodium bicarbonate, a total of 1.2 g. of sodium boro-
hydride is added portionwise; the mixture is well stirred and kept below
5°. After 30 minutes, 100 ml. of cold benzene is added and the colorless
reaction mixture acidified to pH 1 slowly with ice-cold 5 N hydrochloric

TABLE II
Biological Activities of Lipoic Acid Derivatives

| Compound | POF assay[a] | | Lipoic transacetylase |
	Catalytic (+)-lipoic = 100	Substrate O_2/mole	Substrate,[b] Ac-S/mole
(+)-α-Lipoic acid	100	0	0
($\mp$)-α-Lipoic acid	50	0	0
(−)-α-Lipoic acid	0[c]	0	0
Methyl-($\mp$)-lipoate	50	0	0
Methyl-(+)-lipoate	(Ca 100)[d]	—	—
(+)-β-Lipoic acid	(Ca 100)[d]	—	—
(−)-Dihydrolipoic acid (DMO)	100	0.5	1.0
($\mp$)-Dihydrolipoic acid	50	0.5	0.5
(+)-Dihydrolipoic acid	0[c]	0.5	0

[a] Catalytic = assay as outlined here; substrate = standard assay without pyruvate and with 10 μM. of lipoic acid derivative.

[b] Complete enzymatic acetylation, in the presence of coenzyme A, excess phosphotransacetylase, acetyl phosphate, and lipoic transacetylase. (See I. C. Gunsalus,
L. S. Barton, and W. Gruber, *J. Am. Chem. Soc.* in press.)

[c] Unnatural isomer. Synthetic sample [E. Walton *et al.*, *J. Am. Chem. Soc.* **76**,
4748 (1954)][9] showed only trace activity. We wish to thank Dr. Karl Folkers,
Merck and Company, Inc., Research Laboratories, for the (+)- and (−)-lipoic
acid samples.

[d] During isolation, conversion of α-lipoic acid to the methyl ester, with diazomethane, or to β-lipoic acid, with H_2O_2 or permanganate, was essentially quantitative and without loss of units as measured by POF assay. Methylation of
crystalline ($\mp$)-α-lipoic acid with diazomethane, followed by purification by
distillation, occurred without loss of POF activity (Gunsalus *et al.*, in press).[4]
Calculations from *Tetrahymena* data of the amounts required for half-maximum
growth [E. L. Patterson *et al.*, *J. Am. Chem. Soc.* **73**, 5919 (1951);[5] M. W. Bullock,
J. A. Brockman, Jr., E. L. Patterson, J. V. Pierce, and E. L. R. Stockstad,
ibid. **74**, 3455 (1952)] also indicate approximate equivalence of α and β forms.

acid. The solvent is evaporated in a nitrogen atmosphere under diminished pressure and the remaining oil distilled *in vacuo:* boiling point
(1 mm. Hg) 169 to 172°, yield 5.50 g. (91%).

In the same way the optical isomers are prepared; the final step of
purification—vacuum distillation—has to be omitted, since it causes

considerable racemization. $(+)$-Lipoic acid $([\alpha]_D^{20} = +113^9)$ yields $(-)$-dihydrolipoic acid $([\alpha]_D^{20} = -14.5;\ c = 0.598,$ benzene). (This levorotatory dithiol acid is the biologically active isomer; thus the reduction proceeds without inversion.)[4]

3. *Acetyl dihydrolipoic acid*, chemically, $(+)$-6-acetyl-6,8-dithiol-octanoic acid (6-Ac-DMO), is prepared from $(-)$-dihydrolipoic acid by enzymatic acetylation.[4] This derivative appears to possess biological acetylating activity but has not been separated from $(-)$-dihydrolipoic acid in sufficient quantity as the free acid to obtain valid measurements of its catalytic activity in the POF assay.

The acetyl derivative of the primary thiol group of dihydrolipoic acid has been synthesized chemically,[4] i.e., 8-acetyl-6,8-dithioloctanoic acid. This acid does not possess biological activity in the transacylations so far tested.

The catalytic and substrate activities of lipoic acid derivatives are shown in Table II.

[139] Fluorimetric Assay of Cocarboxylase and Derivatives[1]

By HELEN B. BURCH

The method depends on (1) the extraction of the sample with TCA, (2) enzymatic hydrolysis of the phosphates to thiamine, (3) conversion of thiamine to thiochrome by oxidation with $K_3Fe(CN)_6$ in alkaline solution,[2] (4) extraction of the thiochrome into *n*-hexyl alcohol, (5) measurement of total fluorescence with a sensitive fluorimeter,[3] such as the Farrand, (6) destruction of the thiochrome with ultraviolet light, and (7) measurement of the nonthiochrome-fluorescing materials which are unavoidably present. Conversion of thiamine phosphates into thiochrome phosphates may also be accomplished by ferricyanide oxidation. These can be distinguished from thiochrome by their relative insolubility in *n*-hexyl alcohol. In human blood thiamine occurs chiefly as thiamine phosphates. It is determined as total thiamine after hydrolysis by acid phosphatase. Free thiamine may be determined in rat blood or serum by omission of enzyme hydrolysis. Free thiamine is so low in human blood and serum, however, that it cannot be measured precisely by this method.

[1] H. B. Burch, O. A. Bessey, R. H. Love, and O. H. Lowry, *J. Biol. Chem.* **198,** 477 (1952).
[2] B. C. P. Jansen, *Rec. trav. chim.* **55,** 1046 (1936).
[3] O. H. Lowry, *J. Biol. Chem.* **173,** 677 (1948).

Reagents. All reagents must be as pure and as free as possible from fluorescence. Water is glass-redistilled. Glassware is cleaned by boiling in half-concentrated nitric acid, rinsing and boiling in distilled water, and finally rinsing in redistilled water. Reagents and glassware are protected from contaminating dusts during storage and use.

5 and 10% TCA, prepared from glass-redistilled acid.

4 M potassium acetate.

7.5 M NaOH, prepared from a saturated solution.

0.059 M $K_3Fe(CN)_6$.

5.5 M $NaH_2 PO_4$.

30% H_2O_2, Merck (Superoxol).

Oxidation reagent, 10 ml. of 7.5 M NaOH + 0.65 ml. of 0.059 M $K_3Fe(CN)_6$, prepared within an hour of use.

Reducing reagent, 10 ml. of 5.5 M NaH_2PO_4 + 10 μl. of Superoxol.

n-Hexyl Alcohol, redistilled. Use the fraction having the lowest fluorescence, usually the middle one, and saturate it with redistilled water.

Thiamine hydrochloride standard, 50 γ/ml. in 0.1 M HCL. A convenient working standard, 1 γ/ml., is prepared as needed.

Quinine sulfate, set standards for the fluorimeter. Stock solution: 100 mg. of quinine sulfate per liter of 0.1 N H_2SO_4. Dilute standards containing 1.33 and 0.7 mγ of quinine sulfate per milliliter in 0.1 N H_2SO_4. These are approximately equivalent in fluorescence to 5 and 2.5 mγ of thiamine hydrochloride per millimeter.

Acid phosphatase (see also Vol. II [79]). One volume of fresh or frozen human seminal fluid plasma is diluted with 5 vol. of 0.1 M KAc, pH 5, and solid $(NH_4)_2SO_4$ is added to make the solution 1.8 M. The mixture is centrifuged, and the supernatant fluid containing the enzyme is removed. Enough $(NH_4)_2SO_4$ is added to make the concentration 2.8 M, and the precipitated enzyme is collected after centrifugation. The precipitate is made up to one-half the original volume of seminal fluid with 0.1 M KAc buffer, pH 5. It is tested for acid phosphatase activity[1,4] and diluted with water to an activity equivalent to 15 to 20 moles of p-nitrophenyl phosphate hydrolyzable per liter per hour at 25°. To test, the enzyme is diluted 200-fold, and 10 μl. of the enzyme solution is incubated for 5 minutes at 25° with 200 μl. of p-nitrophenylphosphate reagent (0.4% disodium p-nitrophenylphosphate solution of the 70% product from Sigma Chemical Co., St. Louis, plus an equal volume of 0.1 M KAc buffer pH 5, added just before use).

[4] O. A. Bessey, O. H. Lowry, and M. J. Brock, *J. Biol. Chem.* **164**, 321 (1946).

The reaction is stopped by adding 2 ml. of 0.02 M NaOH. A reagent blank with 10 μl. of water is carried through the same procedure. After correction of the reading at 410 mμ for the blank reading (Beckman spectrophotometer), the activity of the undiluted sample in moles per liter per hour is $(2.21/0.01) \times 200 \times (60/5) \times$ optical density$/17,000$, or optical density $\times 0.031$, where the ϵ for p-nitrophenol in alkali is 17,000. It is essential for the complete hydrolysis of thiamine phosphates in blood filtrates to have a highly active preparation.

Special Apparatus

Fluorimeter (Farrand Optical Co., Inc., Bronx Blvd. and E. 238th St., New York, New York).

Carefully selected and matched 3-ml. (10 $\times$ 75-mm.) fluorimeter Pyrex tubes which just fit the holder without too much play. These are marked at the top to facilitate placing them in the same position for each reading.

Syringe-pipets and constriction pipets[4] of appropriate sizes. Constriction pipets of the Levy-Lang type may be purchased from Arthur H. Thomas Co., Philadelphia, Pennsylvania.

Buzzer for mixing.[5]

Semicircular test tube racks for irradiation.[5] General Electric mercury lamp, Model A-H5, and transformer mounted as previously described.[5]

Procedure: Thiamine in Whole Blood. Sixty microliters of blood collected directly into a constriction pipet is delivered into a fermentation tube containing 275 μl. of 5% TCA. Mixing is accomplished by buzzing the tube at once. After 30 minutes at room temperature, the sample is centrifuged, and 250 μl. of clear supernatant fluid is transferred to a 3-ml. tube, brought to pH 4.5 to 4.8 by addition of 30 μl. of 4 M KAc, and hydrolyzed with 10 μl. of acid phosphatase at 25° for at least 4 hours and preferably overnight. Clean rubber vial caps are used to cover the tubes during incubation, and when open they are protected from dust at all times.

Oxidation of the thiamine to thiochrome is accomplished by rapid addition (preferably from a syringe-pipet) of 120 μl. of the reagent, followed by immediate and vigorous mixing. After 15 seconds 120 μl. of the reducing agent is added. The thiochrome is extracted into 1.2 ml. of n-hexyl alcohol by vigorous mixing with the buzzer. After brief centrif-

[5] O. A. Bessey, O. H. Lowry, M. J. Brock, and J. A. Lopez, *J. Biol. Chem.* **166**, 177 (1946).

ugation (covered with Parafilm), the alcohol layer is transferred to a fluorimeter tube. One milliliter is needed for the fluorescence measurement in the Farrand, if the regular-size tubes are employed. Smaller tubes can be used with an appropriate adapter for the fluorimeter which permits a decrease in the volumes used and therefore in the quantity of thiamine required for an accurate reading. The aqueous phase should be completely avoided, since it would cause turbidity and a large error in measurement. The fluorescence is measured in the fluorimeter (R_1) after setting in the proper range with a quinine standard. The tubes, capped with rubber, are then irradiated in semicircular racks 3 inches from an A-H5 mercury lamp for 30 minutes to destroy thiochrome. The time should be periodically checked to ensure complete destruction. A fan keeps the tubes cooled from the heat of the lamp. The fluorescence is again measured (R_2). For blanks and thiamine hydrochloride standards, 60 μl. of H_2O or of standard, added in place of blood, is carried through the entire procedure (see also Vol. IV$_i$[17]). The primary filter is Corning glass 587. The secondary filters are Corning 3389 toward the phototube, a Wratten gelatin filter No. 2A (Eastman Kodak Co.) and Corning 4308.

$R_1 - R_2 = A$ = Change in fluorescence of sample on irradiation
$S_1 - S_2 = B$ = Change in fluorescence of standard on irradiation
$Bl_1 - Bl_2 = C$ = Change in fluorescence of blank on irradiation

γ thiamine per ml. of standard $\times \dfrac{A - C}{B - C} \times 100$

$$= \gamma \% \text{ thiamine in blood}$$

Since most of the thiamine in blood is contained in red cells, the whole blood analysis should be accompanied by an hematocrit to avoid confusion of change in amount of red cells with change in thiamine content.

Thiamine in Red Blood Cells. The red blood cells are isolated from oxalated or heparinized blood. The cells are measured in a constriction pipet calibrated to contain 25 μl. and rinsed into 130 μl. of 0.85% NaCl solution. After mixing, 130 μl. of 10% TCA is added and the contents are mixed vigorously on the buzzer. The procedure from this point is continued as described for whole blood.

Thiamine and Thiamine Phosphates in Rat Serum. To 120 μl. of serum in a 3-ml. test tube 660 μl. of 5% TCA is added, mixed, and treated as above. The supernatant (250 μl.) fluid is treated exactly as described for whole blood. The total thiamine, both free and phosphorylated, is obtained in this way. The free thiamine may be measured on a second 250 μl. of the supernatant fluid by simply neutralizing the filtrate and oxidizing the sample directly without enzyme hydrolysis. The thiochrome mono- and pyrophosphates resulting from the oxidation are not appreciably extracted into *n*-hexyl alcohol. Calculations are made for each aliquot as described for whole blood. The difference between the two represents the phosphorylated thiamine.

Reliability of the Methods. The specificity of this procedure for thiamine has been shown by two types of observations. The rate of destruction of the fluorescence in blood filtrate by ultraviolet light follows closely the rate of destruction of thiochrome prepared from pure thiamine when irradiated under the same conditions. The changes of blood thiamine in rats and human beings[1,6] on different levels of thiamine can be followed by means of this method.

Recovery of thiamine or thiamine pyrophosphate added to blood or blood cells is 97 to 108%.

Average values and the pooled standard deviations for replicate analyses for blood thiamine of well-nourished human beings is 4.7 $\pm$ 0.2 γ %; of red cells 8.0 $\pm$ 0.44 γ %.

The reproducibility of the method (7 to 10%) is favorable when compared to that of other available methods of analyses for thiamine (15 to 18%). In addition, it is simple and rapid to perform and easily applicable to large numbers of samples.

Whole blood and red cell thiamine is stable for 48 hours at 4°, and the TCA extracts are stable for several months when stored frozen at -20 to $-30°$ in tightly stoppered tubes.

This method has not been applied as yet to the measurement of thiamine and its phosphates in animal tissues, but its adaption to such determinations would be relatively simple.

[6] H. B. Burch, J. Salcedo, Jr., E. O. Carrasco, and C. L. Intengan, *J. Nutrition* **46**, 239 (1952).

[140] Preparation and Enzymatic Assay of FAD and FMN

By F. M. HUENNEKENS and S. P. FELTON

I. Preparation of FAD

Principle. FAD is extracted from pig liver with cold trichloroacetic acid and adsorbed from the acid solution onto a column of Florisil.[1] By elution of the column, in turn, with 5% acetic acid, water, and 0.5% pyridine, other nucleotides are removed. FAD is then eluted with 5%

[1] A magnesium silicate adsorbent obtained from the Floridin Company, Tallahassee, Florida.

pyridine and lyophilized to yield a crude concentrate ($P \cong 0.1$).[2] Further purification is accomplished by large-scale paper chromatography with t-butanol-water as the developing solvent, followed by precipitation as the uranyl salt to give a final product with $P > 0.9$.

Reagents

Trichloroacetic acid, 50% w/v.
Florisil, 30/60 mesh.
Acetic acid, 5% v/v.
Pyridine, 0.5% v/v.
Pyridine, 5% v/v.
Chloroforn.
t-Butanol-water (60:40).
Uranyl acetate, $M/50$, pH 6.0.
Ethanol.
Diethyl ether, peroxide-free.

Procedure. Pig or beef liver is the best starting material, owing to the relatively high concentration of FAD (*ca.* 20 mg./kg. wet weight of tissue[3]) and their ready availability from meat-packing plants. Kidney and heart tissues also may be used. Baker's yeast, although as rich in FAD as liver, is less suited as a source, owing to the relatively larger amount of non-nucleotide impurities present in the original extract.[3]

The fresh tissue is packed in ice and brought to the laboratory for immediate use or else stored in the frozen state. One kilogram[4] of liver is chopped into small pieces and extracted with trichloroacetic acid in the

---

[2] P = dry weight purity. A weighed sample of FAD in the free acid form is dissolved in 10^{-2} M phosphate buffer, pH 7.0, and the light absorption (E_{450}) measured at 450 mμ against a buffer blank.

$$P = \frac{\dfrac{E_{450}}{11.3 \times 10^3} \times \text{volume (ml.)} \times 7.86 \times 10^2}{\text{Weight of sample (mg.)}}$$

The extinction coefficient for FAD (pH 7) at 450 mμ is

$$\epsilon = 11.3 \times 10^6 \text{ cm.}^2 \times \text{mole}^{-1}$$

and the molecular weight of the free acid is 786. Because purified FAD is hygroscopic, it should be dried in the dark *in vacuo* over P_2O_5 at 50 to 60° before being weighed.

[3] E. Dimant, D. R. Sanadi, and F. M. Huennekens, *J. Am. Chem. Soc.* **74**, 5440 (1952).

[4] The amount of tissue that can be processed conveniently at one time is regulated, of course, by the equipment available—especially the capacity of the centrifuge. The present amount of tissue can be processed conveniently by one person in the average laboratory.

following manner: 150 g. of the tissue and 450 ml. of cold water are ground in a Waring blendor for 1 minute. Seventy milliliters of trichloro-acetic acid is added, and the blendorization is continued for an additional 30 seconds. The deproteinized slurry is centrifuged for 5 minutes at *ca.* 2300 × *g*; the opalescent,[5] yellow supernatant fluid is removed by decantation and filtered through fluted paper in order to remove fat and finely divided tissue debris. The instability of FAD in acid solutions makes it imperative that the extraction and subsequent chromatography on Florisil be performed in the cold and as rapidly as possible. Furthermore, all the steps in this procedure should be carried out with a minimum exposure of flavins to light.

A chromatographic tube, about 30 × 3 cm. is charged with dry Florisil to a height of approximately 20 cm.; the adsorbent is not tamped. A small plug of glass wool is placed on top of the adsorbent. Water is passed through the column to settle the adsorbent, followed by the trichloro-acetic acid extract.[6] The flavins, largely FAD, are adsorbed as a yellow band at the top of the column. After the entire extract has been passed through the column,[7] the column is washed, in turn, with 2 l. of acetic acid, 2 l. of water, and 2 l. of 0.5% pyridine. The water and pyridine cause the yellow flavin band to diffuse and move slowly down the column. The pyridine wash should be discontinued if appreciable amounts of the flavin appear in the effluent.

The appearance of a pale-yellow color, due to an unidentified material, in the water fraction should not be mistaken, however, for the premature elution of flavin. Finally, about 400 ml. of 5% pyridine is used to elute the flavins, leaving the column colorless except for a pale-yellow area at the top which is neglected. The flavin eluate is adjusted to pH 8, the pyridine largely removed by four extractions with 3 vol. of chloroform,[8] and the chloroform removed by gently warming the solution while the flask is connected to a water aspirator. The solution is lyophilized to dryness to yield a light-yellow fluffy powder. This is dissolved in 10 ml. of water,

---

[5] The opalescence is due to glycogen extractable by trichloroacetic acid from liver; kidney and heart extracts are sparkling clear.

[6] A large separatory funnel may be attached to the top of the column to contain the acid extract and subsequent eluting solutions. This avoids inadvertent channeling of the column due to frequent refilling and also increases the rate of flow. Air pressure or suction should not be used with these columns as they cause packing and clogging of the adsorbent.

[7] The adsorbed flavin zone should not extend below the top third of the column.

[8] If the columns have been well-washed with water and 0.5% pyridine prior to removal of the flavins, the chloroform extraction will proceed cleanly. When large amounts of contaminants are present, troublesome emulsions develop which must be broken by centrifugation.

adjusted to pH 3 with hydrochloric acid, the insoluble residue removed by centrifugation and discarded, and the supernatant fluid again lyophilized to dryness. At this stage the product consists of about 100 mg. of a granular yellow-orange powder, with $P = 0.15 - 0.20$. This material represents an almost quantitative recovery of FAD from the tissue and can be used without further purification for most purposes.[9] The flavin consists almost entirely of FAD, as judged by paper chromatography (see below), and there is relatively little contamination by other nucleotides. The R value[10] ranges between 4.0 and 7.0.

Further purification of FAD has been accomplished by a variety of methods, including extraction into phenolic solvents,[3,11-13] precipitation as heavy metal complexes,[11-13] adsorption chromatography on dicalcium phosphate,[3] partition chromatography,[3,11,12] and ion exchange chromatography.[12] In all cases the yields decrease sharply, owing to the persistence of impurities which are removed only by multiple operations and to the lability of FAD to acid, base, and light. For these reasons we have preferred the following method[14] because of its relative simplicity and larger yields.

FAD ($P \cong 0.2$) is dissolved in a minimum amount of water, and the solution applied evenly as a band at the starting line of a sheet of Munktell, Cremer-Tiselius filter paper (50 × 70 cm.).[15] About 100 mg. of FAD may be applied per sheet. The papers are run descending in a chromatocab,[16] or similar developing chamber, with t-butanol-water as the solvent. About 36 hours is required for the solvent to migrate 50 to 60 cm. After the paper has been removed from the tank and dried, the bands are located by their intense yellow fluorescence when examined under ultraviolet light. Care should be taken not to irradiate the flavins unnecessarily. FAD is the principal band, running with an R_F value of about 0.50. Faint bands, corresponding to FMN ($R_F \cong 0.65$) and FAD-X [17] ($R_F \cong 0.90$), also may be seen on the paper. The FAD band is cut out

[9] A commercial preparation of FAD of similar purity is supplied by the Sigma Chemical Co., 4648 Easton Ave., St. Louis 13, Missouri.

[10] R = ratio of light absorption at 260 and 450 mμ. For a discussion of the significance of this quantity, see ref. 3, 11, and 12.

[11] L. G. Whitby, *Biochem. J.* **54**, 437 (1953); *Biochim. et Biophys. Acta* **15**, 148 (1954).

[12] N. Siliprandi and P. Bianchi, *Biochim. et Biophys. Acta* **16**, 424 (1955).

[13] O. Warburg and W. Christian, *Biochem. Z.* **298**, 150 (1938).

[14] We are indebted to Dr. H. R. Mahler, who originally suggested this method of purifying FAD.

[15] Supplied by E. H. Sargent and Co., 4647 West Foster Ave., Chicago 30, Illinois.

[16] Research Equipment Corp., 1135 Third Street, Oakland, California.

[17] F. M. Huennekens, D. R. Sanadi, E. Dimant, and A. I. Schepartz, *J. Am. Chem. Soc.* **75**, 3611 (1953).

of the paper, and the material eluted with a minimum volume of water by means of a trough arrangement similar to that suggested by Isherwood and Hanes.[18] When dried by lyophilization the product is an orange powder with a purity of $P = 0.5 - 0.6$. The recovery is about 60% in this step, or higher if the "fringes" and "tailing" of the FAD band are included in the cut-out strip. Repetition of this process leads to a product with $P = 0.7 - 0.9$.

Crystallization of highly purified FAD from water has been accomplished by Whitby,[11] and the barium salt was crystallized earlier by Warburg and Christian.[13] Because of the high solubility of FAD, or its barium salt, in water, these methods are accompanied by some loss in material. An alternate method is to make use of the water-insoluble uranyl complex of FAD. Ten milligrams of FAD ($P \cong 0.9$) is dissolved in 5 ml. of water, and the pH adjusted to 6.0. One milliliter of uranyl acetate (about a 25% excess on a molal basis) is added dropwise over a 10-minute period with mechanical stirring. After precipitation is completed, the mixture is allowed to stand overnight in the cold. The semicrystalline[19] uranyl-FAD complex is recovered by centrifugation and washed twice with 1-ml. portions of cold ethanol and twice with 10-ml. portions of cold, peroxide-free ether. The orange solid is dried in the dark *in vacuo* over P_2O_5 at 50 to 60° and may be stored conveniently in this form. To recover FAD, the uranyl complex is suspended in the desired volume of water, sufficient 0.01 M NaOH is added to adjust the pH to 7.0, and the small amount of white precipitate (uranyl hydroxide) is removed by centrifugation. The recovery of FAD after formation and decomposition of the uranyl complex is about 70 to 80%.

Pure FAD may be obtained also by methods of chemical synthesis. Christie *et al.*[20] utilized the condensation of the mono-silver salt of FMN with 2',3'-isopropylidine adenosine-5'-benzylphosphochloridate, followed by removal of the protective groups, to prepare FAD. Recently, a simple and direct synthesis of FAD from FMN and AMP, with di-*p*-tolyl carbodiimide as the catalyst, has been achieved.[21] In both cases, however, the yield is $< 10\%$, and the FAD must be freed from other flavin impurities.

The purity and authenticity of FAD may be judged by three criteria: (1) it should run as a single, homogeneous spot when subjected to paper

[18] F. A. Isherwood and C. S. Hanes, *Biochem. J.* **55**, 824 (1953).

[19] When viewed under a low-power microscope, the uranyl-FAD complex appears as clusters of regular, orange, beadlike bodies. Similarly, on attempting to crystallize free FAD from water, we have obtained only clumped "crystalline" spheres, rather than needles, as reported by Whitby.[11]

[20] S. M. H. Christie, G. W. Kenner, and A. R. Todd, *J. Chem. Soc.* **1954**, 46.

[21] F. M. Huennekens and G. L. Kilgour, submitted to *J. Am. Chem. Soc.*

chromatography (see Part II); (2) it should have an R value equal to 3.25,[11] and (3) it should be fully active as a coenzyme in the D-amino acid apo-oxidase system (see Part II).

II. Assay of FAD

Principle. D-Amino acid oxidase is obtained from pig kidney in such a manner that the bound FAD is lost during isolation. By measuring oxygen uptake in Warburg manometers with D-alanine as substrate, the dissociated enzyme is shown to be inactive in the absence of added FAD, but it is restored to full activity as graded amounts of FAD are added to the system. This method originated with Warburg and Christian[13] and with Straub;[22] the present method represents a modification wherein the enzyme is prepared by a simple procedure. In addition, the assay of FAD by methods of paper chromatography is described.

Reagents

> Pig kidney, acetone powder.
> 1.0 N acetic acid.
> Saturated solution of ammonium sulfate.
> 1.0 N sodium hydroxide.
> Bovine serum albumin, crystalline (Armour).
> Enzyme. Dissolve 250 mg. of the lyophilized D-amino acid apo-oxidase in 10 ml. of 0.05 M pyrophosphate buffer, pH 8.3.
> 0.05 M pyrophosphate buffer, pH 8.3.
> 1.0 M DL-alanine, pH 8.3.
> FAD standard solution, 0.75 γ/ml. Concentration determined spectrophotometrically (*cf.* ref. 2).
> 6.0 N sodium hydroxide.

Procedure. Although the preparation of highly purified D-amino acid apooxidase has been described by Negelein and Brömel,[23] the following procedure is recommended for its simplicity. Thirteen grams of pig kidney acetone powder[24] is stirred with 130 ml. of water at room temperature for 20 minutes. The mixture is centrifuged at 2000 to 3000 $\times$ g for 20 minutes at 5°. The orange-brown supernatant fluid is filtered through glass wool to remove floating debris and adjusted to pH 4.5 with 1 N acetic acid. One-half volume of saturated ammonium sulfate is added, and the solution is kept at 0 to 5° for 15 minutes. The precipitate is recovered by centrifugation and resuspended in 70 ml. of ice-cold

[22] F. B. Straub, *Biochem. J.* **33**, 787 (1939).
[23] E. Negelein and H. Brömel, *Biochem. Z.* **300**, 225 (1939); see also Vol. II [23].
[24] Acetone powders of pig kidney may be prepared by the method of Morton (Vol. I [6], p. 34) or may be purchased from the Viobin Corporation, Monticello, Illinois.

water. Thirty-five milliliters of saturated ammonium sulfate is added, and the solution is kept at 0 to 5° for 15 minutes. The precipitate is recovered by centrifugation, taken up in 50 ml. of cold water, and the precipitation with ammonium sulfate (25 ml.) is repeated once again. The final precipitate is well-packed by centrifugation, drained free from supernatant fluid by inverting the centrifuge cups for a few minutes over a piece of filter paper, and dissolved in 20 ml. of cold water. The pH is adjusted to 6.8 with 1 N sodium hydroxide (about 5 to 10 drops). Two hundred milligrams of crystalline bovine serum albumin is added to stabilize the enzyme, and the solution is lyophilized to yield 950 mg. of a fluffy white powder which is stable for many months when stored in a desiccator at 0 to 5°.

Assay System. The following reagents are added, in order, to a Warburg manometer cup: 0.5 ml. of pyrophosphate buffer, 0.1 ml. of DL-alanine, 1.0 ml. of standard FAD solution, 0.2 ml. of 6.0 N sodium hydroxide (center well), 0.5 ml. of enzyme, and water to make 3.0 ml. Two additional "blank" cups are prepared in which the FAD and the DL-alanine, respectively, are omitted. In a manometer bath at 30° the cups are gassed for 3 minutes with oxygen and then allowed to equilibrate for 5 minutes with the stopcocks open. On closing the stopcocks, the oxygen uptake values are recorded for successive 10-minute intervals over the period of 1 hour. The value for the cup containing the complete system should be corrected for the two blanks.[25] The values for successive 10-minute intervals should agree to within ±5% and are averaged. Under these conditions, FAD at a final concentration of 0.25 γ/ml. will produce an oxygen uptake of about 40 μl. in 10 minutes. Concentrations of FAD in the range 0 to 0.25 γ/ml. represent the linear portion of the Michaelis curve ($K_m = 3.3 \times 10^{-7}\ M$), and several levels of FAD within this range should be run in order to establish a linear plot of activity *vs.* concentration.

In order to assay an unknown quantity of FAD, the sample replaces standard FAD in the above system and should be run at several levels and with appropriate blanks, as before. For estimating the content of FAD bound to purified enzymes or to crude tissue preparations, an aliquot of the material should be pipetted first into the manometer cup, heat-deproteinized by immersing the cup in a boiling water bath for 5 minutes, and the other reagents added to the denatured coagulum in the cup.

---

[25] The blank without added FAD should be less than 20% of the complete system. If it is higher, the enzyme has not been split adequately, relative to the bound FAD, and should be subjected again to the acid-ammonium sulfate precipitation step as described in the Procedure section.

Paper Chromatography of FAD. An additional, and very sensitive, method of assaying FAD is by means of paper chromatography in different solvent systems. The table lists the R_F values of FAD, and several

PAPER CHROMATOGRAPHY OF FLAVINS[a]

Flavin	R_F value in solvent system[b]		
	A	B	C
FAD	0.35	0.05	0.23
FAD-X	0.35	0.05	0.47
FMN	0.48	0.13	0.17
cyc-FMN	0.40	0.15	0.25
Rb	0.30	0.30	0.80

[a] Ascending, Whatman No. 1 paper.
[b] Solvent systems: A, 5% disodium hydrogen phosphate in water; B, 4:1:5 n-butanol—acetic acid—water (top phase); C, 160 g. phenol (Merck): 30 ml. n-butanol—100 ml. water (lower phase).

other flavins, in three of the most useful solvent systems. Pure FAD should migrate as a single, homogenous spot (viewed under an ultraviolet light) with the indicated R_F values. This technique will detect as little as 0.01 γ of flavin impurity. Unless extreme care is taken to apply the flavins to the paper in dim light and to run the papergrams in darkened tanks, however, there will inevitably be traces of the photodegradation products lumichrome and lumiflavin (R_F values of 0.05 and 0.15, respectively, in solvent system A). Further information about the paper chromatography of flavins is given elsewhere.[26]

III. Preparation of FMN

Principle. FMN is prepared by controlled chemical hydrolysis of FAD and is isolated essentially by the method described above for FAD.

Reagent

1.0 N hydrochloric acid.

Procedure. Twenty-five milligrams of FAD ($P > 0.25$) is dissolved in 50 ml. of hydrochloric acid, and the solution is kept in the dark at 35 to 40° for 48 hours. The solution is then poured through a Florisil column to adsorb the flavins, and the isolation procedure continued as for FAD, with the exception that the FMN band at R_F 0.65 rather than the FAD is collected in the step involving paper chromatography in t-butanol-water.

[26] F. M. Huennekens, S. P. Felton, and G. L. Kilgour, submitted to *J. Am. Chem. Soc.*

FMN has been synthesized chemically by the direct phosphorylation of riboflavin with phosphoryl chloride[27] or monochlorophosphoric acid[28] and is available commercially.[29]

IV. Assay of FMN

Principle. TPNH cytochrome c reductase is partially purified from brewer's yeast and then treated to dissociate the bound FMN. The apoenzyme may be used then to estimate added FMN. Both the preparation of the enzyme and the assay are slight modifications of the procedure originated by Haas *et al.*[30]

Reagents

Brewer's yeast, air-dried.
Ammonium sulfate.
10.0 N acetic acid.
Ammonium sulfate solution (31% saturated).
1.0 N sulfuric acid.
Congo red test paper.
0.03 M phosphate buffer, pH 7.3.
0.1 M phosphate buffer, pH 7.5.
Cytochrome c, 10 mg./ml.
TPNH, 0.4 mg./ml.
FMN, standard solution, 0.9 γ/ml.[31]

Procedure (see also Vol. III [122]). Two hundred and fifty grams of brewer's yeast is suspended in 875 ml. of water at room temperature for 42 hours. The mixture is centrifuged for 10 minutes at 2000 to 3000 $\times$ g, and the residue washed by suspension and centrifugation in an additional 400 ml. of water. The combined supernatant fraction (600 ml.) is cooled to 0°, and 183 g. of ammonium sulfate and 8.4 ml. of acetic acid are added with stirring. After the mixture has been allowed to stand for 15 minutes, the precipitate is recovered by centrifugation and dissolved in 44 ml. of water. To this solution 188 ml. of ammonium sulfate solution (31% saturated) is added with stirring, and the precipitate is removed by centrifugation and discarded. The supernatant solution (287 ml.) is treated with 34 g. of solid ammonium sulfate, and the precipitate is recovered by

[27] H. S. Forrest and A. R. Todd, *J. Chem. Soc.* **1950**, 3295.
[28] L. A. Flexser and W. G. Farkas, U. S. Patent 2,610,179.
[29] Hoffmann-La Roche, Inc., Nutley, New Jersey, and Sigma Chemical Co. (*cf.* ref. 9).
[30] E. Haas, B. L. Horecker, and T. Hogness, *J. Biol. Chem.* **136**, 747 (1940).
[31] Standardized spectrophotometrically, assuming that $\epsilon = 11.3 \times 10^6$ cm.^{2} $\times$ mole^{-1} at 450 mμ for FMN at pH 7. The molecular weight of the free acid is 456.

centrifugation and dissolved in 110 ml. of cold water. The resulting solution is dialyzed for about 18 hours against a large volume (*ca.* 20 to 30 l.) of cold, distilled water, and any precipitate is removed by centrifugation and discarded.

The partially purified reductase is now split with regard to FMN in the following manner: The well-dialyzed solution above (170 ml.) is treated with 42.5 g. of solid ammonium sulfate (final concentration 35%), and the solution is made acid with sulfuric acid to the Congo red end point. The solution is allowed to stand with stirring for 10 minutes at room temperature. The precipitate is recovered by centrifugation and dissolved in 50 ml. of phosphate buffer. The enzyme is again exposed to acid-ammonium sulfate conditions, and the precipitate (35% saturation with respect to ammonium sulfate) is dissolved in 30 ml. of buffer. The apo-reductase is stable when stored in the frozen state.

Assay System. The following components are added, in order, to a Corex cuvette having a light path of 1 cm.: 0.2 ml. of phosphate buffer (0.1 *M*), 0.1 ml. of cytochrome c, 0.1 ml. of enzyme, 0.1 ml. of FMN solution, and water to make 3.0 ml. After the solution has been allowed to stand for 5 minutes at room temperature, 0.1 ml. of TPNH is added and the optical density at 550 mμ measured at 1-minute intervals against a blank cell without cytochrome c. An additional blank should be run with FMN omitted, and the value for the experimental cell corrected accordingly. The net change in optical density is linear over a 5-minute period and should be equal to about 0.058 under the above conditions when the concentration of added FMN is 0.03 γ/ml. In the range 0 to 0.03 γ/ml. of FMN, the activity is linear with FMN concentration, and several concentrations should be used to construct the standard curve. For the determination of FMN in an unknown, the sample replaces standard FMN in the assay system. Protein-bound FMN is estimated in heat-coagulated samples, as with FAD, or on dilute (<1%) trichloroacetic acid extracts. The low value of the Michaelis constant for this enzyme ($K_m \cong 5.5 \times 10^{-8}$ *M*)[32] makes it possible to estimate very small quantities of FMN with accuracy.

FMN can be determined also by means of paper chromatography, as outlined above in Section II.

---

[32] A value of 3.0×10^{-9} has been reported by Haas *et al.*[30] for the enzyme at a higher stage of purity.

[141] Fluorimetric Assay of FAD, FMN, and Riboflavin

By HELEN B. BURCH

Principle. The assay procedure is adapted from methods described by Bessey *et al.*[1] and by Burch *et al.*[2] The methods are based on the fact that an increase in fluorescence occurs when FAD is split to FMN, and that FMN can be distinguished from riboflavin on the basis of its distribution coefficient between benzyl alcohol and aqueous solutions. The procedure for tissue analysis utilizes a sensitive fluorimeter[3] (such as that of the Farrand Optical Co., New York, New York) which permits measurements in very dilute solutions (see also Vol. IV [17]).

Reagents. All reagents are prepared with redistilled water, redistilled reagents, and glassware cleaned by boiling in half-concentrated nitric acid and subsequently rinsed, then boiled in distilled water, and finally rinsed with redistilled water.

11% TCA prepared from a 100% solution.

0.2 M K_2HPO_4.

0.05 M NaAc-0.05 M HAc buffer.

Riboflavin standards are prepared in 0.01 M HCl from a stock solution of 20 γ/ml. in 0.01 M HCl; 0.1 and 0.4 γ/ml. are appropriate for muscle and brain determinations; 1.0 and 4.0 γ/ml. for liver and kidney; and 0.5 and 2.0 γ/ml. for other tissues. The low standards are suitable for free riboflavin plus FMN.

10% sodium hydrosulfite ($Na_2S_2O_4$), prepared just before use in 5% $NaHCO_3$ solution.

Benzyl alcohol, redistilled, washed, and saturated with redistilled water.

Toluene, redistilled, washed, and saturated with redistilled water.

Volumes less than 1 ml. are measured with constriction pipets.[4] The Levy-Lang type may be purchased from Arthur H. Thomas Co., Philadelphia, Pennsylvania.

Procedure. Extraction of the Flavin Compounds. The fresh, weighed tissue sample is finely ground in a glass homogenizer at 0 to 4° with 10 ml. of water per gram of tissue (to give a dilution of 1 g. per 10.94 ml. with average tissue of specific gravity 1.06). Three-tenths milliliter of the cold

---

[1] O. A. Bessey, O. H. Lowry, and R. H. Love, *J. Biol. Chem.* **180,** 755 (1949).

[2] H. B. Burch, O. A. Bessey, and O. H. Lowry, *J. Biol. Chem.* **175,** 457 (1948).

[3] O. H. Lowry, *J. Biol. Chem.* **173,** 677 (1948).

[4] O. A. Bessey, O. H. Lowry, and M. J. Brock, *J. Biol. Chem.* **164,** 321 (1946).

suspension is mixed with 3 ml. of 11% TCA. After 15 minutes the sample is centrifuged at 4°. At the same time 0.3-ml. aliquots of riboflavin standards are carried through the entire procedure. After extraction, 0.2-ml. aliquots of the supernatant fluid are immediately removed into calibrated matched fluorimeter tubes (3 ml., 10 × 75 mm.) containing 1 ml. of 0.2 M K$_2$HPO$_4$ (final pH 6.8), and mixed. Until neutralized, the sample is kept as cold as possible to prevent hydrolysis of FAD. Other 0.2-ml. aliquots are placed in fluorimeter tubes, tightly covered with clean vial caps, and stored in the dark at 38° overnight or at room temperature for 2 days to complete the hydrolysis of FAD to FMN. The hydrolyzed samples, with suitable standards, are neutralized in exactly the same manner as the initial samples. After neutralization, special care is taken to protect the samples from the light, since both riboflavin and FMN are more sensitive to destruction by light in this concentrated salt solution than in more dilute salt solutions.

Measurement of FAD. The fluorescence of the two neutralized extracts is measured in a sensitive fluorimeter using fluorescein secondary standards for setting the instrument. These are prepared in 40% alcohol and are equivalent in fluorescence to 1 to 3 γ of riboflavin per milliliter. They can be stored in tightly stoppered tubes for two to three months. The readings are made initially (F_1) and after reduction with 10 μl. of 10% sodium hydrosulfite in 5% NaHCO$_3$ (F_2). An internal riboflavin standard can be omitted, since the high dilution prevents interference from other substances present. The reduced reading is usually so low that it is ordinarily unnecessary to correct for the volume of added reducing agent. The apparent riboflavin of the initial sample (R_i) and hydrolyzed sample (R_t) in micrograms per gram of tissue is as follows:

$$R_i = 10.94 \times \gamma \text{ riboflavin per ml. in standard} \times \frac{(F_1 - F_2)_{\text{sample}}}{(F_1 - F_2)_{\text{standard}}}$$

$$R_t = 10.94 \times \gamma \text{ riboflavin per ml. in standard} \times \frac{(F_1 - F_2)_{\text{sample}}}{(F_1 - F_2)_{\text{standard}}}$$

Under the conditions of these measurements, FAD (calculated as riboflavin) has a fluorescence equal to 15% of riboflavin, whereas FMN (calculated as riboflavin) and riboflavin are equal in fluorescence, so that FAD = $(R_t - R_i)/0.85$. The balance of the flavin consists of FMN plus free riboflavin:

$$R_t = \text{FAD} + \text{non-FAD riboflavin (FMN + free riboflavin)}$$

Measurement of Free Riboflavin. Since there is very little free riboflavin in most tissues, the above measurements will suffice for usual purposes. If it is desirable to distinguish between FMN and riboflavin, 2 ml. of the cold tissue extract is immediately neutralized with 0.5 ml. of 4 M K_2HPO_4 in a glass-stoppered centrifuge tube, thoroughly shaken with an equal volume (2.5 ml.) of benzyl alcohol saturated with water, and centrifuged briefly to clear the solution. Low concentrations of free riboflavin in normal tissues make it desirable to drive the riboflavin back into an aqueous solution before measurement of fluorescence. This is accomplished by shaking 1 vol. of the benzyl alcohol extract in a glass-stoppered vessel with 15 vol. of toluene and 1 vol. of a buffer which is 0.05 M in both sodium acetate and acetic acid. All the flavins extracted by benzyl alcohol are quantitatively driven into the aqueous layer. Standards are carried through exactly the same procedure. Fluorescence is measured as before on a 1-ml. aliquot and calculated as R_{Bz} = the apparent micrograms of riboflavin per gram of tissue extracted into benzyl alcohol. Since the partition coefficients for riboflavin, FAD, and FMN between benzyl alcohol and 10% TCA, neutralized as described above, are 4.1, 0.032, and 0.02, respectively, the apparent riboflavin in the benzyl alcohol extract driven back into water

$$R_{Bz} = 4.1/5.1 \text{ free riboflavin} + 0.032/1.032 \text{ FMN}$$
$$+ 0.02/1.02 \times 0.15 \text{ FAD}$$

By letting $R_{non\text{-}FAD}$ = FMN + free riboflavin and solving, free riboflavin in micrograms per gram of tissue = $1.3R_{Bz} - 0.04R_{non\text{-}FAD} - 0.003$ FAD.

On less-sensitive fluorimeters more concentrated extracts of tissues are measured. An aliquot of the TCA supernatant is neutralized with one-fourth its volume of 4 M K_2HPO_4 instead of five times its volume of 0.2 M. Since interfering substances are not diluted out, an internal riboflavin standard should be measured for each sample, approximately equal to it in fluorescence, followed by reduction with hydrosulfite. Both the procedure and the calculation of the apparent riboflavin in each measured sample must be altered accordingly.

Low levels of riboflavin, such as the free riboflavin in tissues or flavins in blood, cannot be measured with precision on fluorimeters of ordinary sensitivity. A fluorimeter of the Farrand type is sufficiently sensitive to measure 0.1 mγ of riboflavin per milliliter with a precision of 5%.

Reliability of the Method. The recovery of riboflavin, FMN, and FAD added to minced tissue averaged 95 to 100%. Values are quite reproducible, usually within 1% or less, except for very low levels where 1.5 to 2% may be expected. The fluorimetric measurement of FAD gives as consistent values for this coenzyme in various tissues as its measurement in the D-amino acid oxidase assay system. These sensitive procedures are easily applied to purified enzyme preparations. This is of particular advantage when small quantities of material are available.

[142] Assay and Preparation of Pyridoxal Phosphate

By I. C. Gunsalus and Roberts A. Smith

Assay Method

Principle. *Streptococcus faecalis*, strain R, grown in a vitamin B_6-deficient medium and dried *in vacuo*, contains tyrosine apodecarboxylase.[1] This tyrosine decarboxylase is activated by codecarboxylase,[2] i.e., pyridoxal phosphate (B_6-al-P). A plot of the rate of CO_2 release vs. pyridoxal phosphate—over the range 2 to 25 mγ—may be used as a standard to estimate the amount of this coenzyme in natural or synthetic material.[3] Pyridoxal can be measured with the same cells if ATP is preincubated for 10 minutes with the sample and with cells before adding tyrosine for decarboxylase measurement. In the presence of ATP, a kinase in the cells forms B_6-al-P.[4,5]

Preparation of Assay Enzyme (Tyrosine Decarboxylase). The organism *S. faecalis*, strain R (ATCC 8043), is maintained at refrigerator temperature in agar deeps of a medium composed of 1% tryptone, 0.3% K_2HPO_4, 0.3% $CaCO_3$, 0.1% glucose, 1.5% agar, and 10% by volume of liver extract (prepared by heating 1 pound of ground liver in 2 l. of water in flowing stream for 1 hour and straining off the solids). Before use, the culture is transferred to AC broth[6] containing 1.0% tryptone, 1.0% yeast extract, 0.5% K_2HPO_4, and 0.1% glucose, and incubated at 35°. A medium satisfactory for the production of tyrosine decarboxylase apoenzyme is given in the table. This medium is autoclaved for 15 minutes at 15 pounds pressure, cooled, inoculated with 0.02% of a 12- to 24-hour broth culture, and incubated for 15 to 20 hours at 37°. The cells are harvested by centrifuge, washed in saline, and resuspended in distilled water. About 10 ml. of this cell suspension is placed in a Petri dish half in a 250-mm. desiccator containing 2 to 3 pounds of 4- to 8-mesh Drierite (anhydrous calcium sulfate), and a good vacuum is immediately applied (1 mm. or less). Within 10 to 15 hours the cells dry from the frozen state to a fluffy powder, which retains the apodecarboxylase activity indefinitely if kept over Drierite.

[1] W. D. Bellany and I. C. Gunsalus, *J. Bacteriol.* **50**, 95 (1945).
[2] E. F. Gale and H. M. R. Epps, *Biochem. J.* **38**, 250 (1944).
[3] W. W. Umbreit and I. C. Gunsalus, *J. Biol. Chem.* **179**, 279 (1949).
[4] W. W. Umbreit, W. D. Bellamy, and I. C. Gunsalus, *Arch. Biochem.* **7**, 185 (1945).
[5] See Vol. II [109].
[6] A. J. Wood and I. C. Gunsalus, *J. Bacteriol.* **44**, 333 (1942).

Reagents

Tyrosine suspension 0.03 M.
Acetate buffer, 0.2 M, pH 5.5.
Enzyme (dried *S. faecalis* R), 25 mg. per 10 ml.
ATP, 10 mg./ml.

MEDIUM FOR TYROSINE APODECARBOXYLASE PRODUCTION[a]

Ingredient	Per liter
Acid-hydrolyzed casein	10 g.
Glucose	10 g.
K_2HPO_4	5 g.
Sodium acetate	2 g.
DL-Alanine	0.2 g.
L-Cystine	0.2 g.
L-Tryptophan	0.1 g.
Sodium thioglycollate	0.1 g.
Salts B[b]	5 ml.
Adenine sulfate	5 mg.
Guanine hydrochloride	5 mg.
Uracil	5 mg.
Nicotinic acid	5 mg.
Riboflavin	1 mg.
Calcium pantothenate	1 mg.
Folic acid	2 γ
Biotin	1 γ
pH	7.2–7.3

[a] W. D. Bellamy and I. C. Gunsalus, *J. Bacteriol.* **50**, 95 (1945).
[b] Salts B = $MgSO_4 \cdot 7H_2O$, 10 g.; NaCl, 0.5 g.; $FeSO_4 \cdot 7H_2O$, 0.5 g.; $MnSO_4 \cdot 4H_2O$, 0.5 g.; water, 250 ml.

Procedure.[3,4] To each Warburg flask is added, in the side arm, 0.5 ml. of 0.03 M tyrosine; in the main compartment, 1.0 ml. of 0.2 M acetate buffer, the sample adjusted to pH 5.5, water to bring the fluid volume to 2.6 ml., and 0.4 ml. (1 mg.) of dried cell preparation. The flasks are shaken at 28° for 10 minutes, and readings begun. After equilibration, the tyrosine is tipped into the main compartment, and the CO_2 evolved during five successive 5-minute intervals recorded. The average microliters of CO_2 evolved per 5 minutes multiplied by 12 equals the Q_{CO_2}.

Pyridoxal can be assayed as above, except that a 10-minute preincubation of the sample, dried cell preparation, and 0.1 ml. (1 mg.) of ATP is necessary for the conversion of pyridoxal to B_6-al-P. In assaying pyridoxal, the standard curve should be prepared with graded amounts of pyridoxal + 1 mg. of ATP.[4]

Standard Curve. For B_6-al-P, estimation may be based on a sample of known purity as determined by spectral analysis[3] or physical properties. At present, the crystalline inner salt of B_6-al-P[7] should serve as a convenient standard. A plot of microliters of CO_2 released per hour vs. micromoles of standard B_6-al-P may be used to estimate the codecarboxylase content of unknown samples. The data, from measuring CO_2 release with graded amounts of B_6-al-P, may be replotted according to the Lineweaver-Burk[8] method, and B_6-al-P estimated from the straight line given by the equation $1/v = 1/c \cdot K/V_{max.} + 1/V_{max.}$

Data for activation of a given apodecarboxylase, using both plots, are given by Umbreit *et al.* (Table II and Fig. 2 in ref. 4). With known samples, the precision of this method is about 10%. The method is specific for B_6-al-P. Pyridoxamine and pyridoxine-5-phosphates are inactive. The useful assay is 2 to 25 mγ of pyridoxal phosphate.[3] In the presence of added ATP, pyridoxal can be assayed over a range of 20 to 300 mγ by the procedure outlined.[4]

Preparation of Pyridoxal Phosphate

Principle. The following procedure is the one adopted by Wilson and Harris.[9] Pyridoxamine is phosphorylated with anhydrous phosphoric acid, oxidized with manganese dioxide, and isolated as the calcium salt. The reactions are illustrated in the following equations:

Procedure. Two hundred and twenty-five grams of phosphorus pentoxide is dissolved in 300 g. of 85% orthophosphoric acid to give an-

[7] E. A. Peterson, H. A. Sober, and A. Meister, *Biochem. Preparations* **3**, 34 (1953).

[8] H. Lineweaver and D. Burk, *J. Am. Chem. Soc.* **56**, 658 (1938).

[9] A. N. Wilson and S. A. Harris, *J. Am. Chem. Soc.* **73**, 4693 (1951).

hydrous phosphoric acid, as outlined by Ferrel *et al.*[10] Twenty-five grams of pyridoxamine dihydrochloride is dissolved in 250 g. of the anhydrous phosphoric acid. As the solid dissolves, a vigorous evolution of hydrogen chloride occurs and is dissipated by vigorous stirring. When most of the gas has been evolved, the clear viscous solution is allowed to stand in a desiccator over P_2O_5 for 3 to 6 days and the mixture poured onto 2 to 3 kg. of crushed ice and water. Then 30% NaOH is added slowly until the reaction mixture is about pH 6.0.

Manganese dioxide (10 g.) is added, and the mixture is heated at 60° for 20 minutes with frequent shaking. The manganese dioxide is oxidized to a light-colored inorganic solid, and the colorless solution becomes yellow-brown owing to the presence of the aldehyde phosphate. The solution is cooled, filtered, acidified to Congo red with phosphoric acid, and 450 to 500 g. of acid-washed charcoal (Darco G-60) added to adsorb the phosphorylated compound. The charcoal is filtered by suction and washed with water, resuspended in water, mixed with 1 to 2 vol. of Polycel—or similar cellulose material—and placed in a column. The column is washed with 2% HCl to remove the remaining phosphoric acid, and then with water to remove the HCl. Both washings should be thorough, since these do not remove the phosphorylated compound. When the column effluent is virtually Cl-free, the coenzyme is eluted with 2% ammonia. The eluate is vivid yellow, owing to the presence of the ammonium salt of B_6-al-P. About 8.5 to 9 l. of the eluate is required to release all the B_6-al-P. The eluate is concentrated under reduced pressure to about 1 l., the concentrate acidified to pH 4.0 with dilute acetic acid, and a solution of 18.5 g. of calcium acetate added. More acetic acid may be added if necessary to maintain a clear solution. A small amount of insoluble amorphous precipitate may form; if so, it is removed by filtration through Super-Cel. The clear filtrate is diluted with 3 vol. of ethyl alcohol, and the mixture allowed to stand overnight in the refrigerator. The precipitate is collected by centrifugation, washed with a 3:1 alcohol-water mixture, with alcohol-ether, and finally with anhydrous ether, then dried in a vacuum oven at 40 to 45°. The yield of calcium B_6-al-P is usually about 29 g.; yield 97%; purity about 80%.

Crystalline pyridoxal phosphate has been prepared by MnO_2 oxidation of pyridoxamine phosphate, followed by chromatography on a weak cation exchange resin, and separation from contaminating polymeric substances by dialysis. The crystalline B_6-al-P was obtained as the free inner salt, the monohydrate by lyophilizing the partially concentrated dialyzate.[7]

[10] R. E. Ferrel, H. S. Olcott, and H. L. Fraenkel-Conrat, *J. Am. Chem. Soc.* **70,** 2101 (1948).

Properties. The absorption spectrum of pyridoxal phosphate serves to characterize and to differentiate it from other phosphorylated and free forms of Vitamin B_6, including pyridoxal, as well as for its quantitative estimation in pure form and in mixtures.[11,12,13] Common to all members of the Vitamin B_6 group and to their phosphates is a band with a maximum near 285 mμ in acid solution (pH 1 to 3), which is common to all beta pyridones. If the beta (-2-) phenolic group is unsubstituted, neutralization is accompanied by the disappearance of the 285 absorption maximum and to the appearance at alkaline reaction (pH 11 to 13) of maxima near 240 and 305 mμ. In alkaline solution, pyridoxal and its 5-phosphate possess an additional yellow color, associated with the 4-aldehyde, with a maximum of 385 to 390 mμ. This band, however, can be used to differentiate the two because of its greater intensity for pyridoxal phosphate at neutral reaction. The most reliable extinction coefficients for pyridoxal phosphate were reported for the crystalline monohydrate;[7,13] E_m at 388 mμ, pH 7 = 4900; in N/10 NaOH = 6550, compared to pyridoxal hydrochloride E_m at 388 mμ, pH 7 = 200; in N/10 NaOH = 1700. Pyridoxal phosphate differs from all members of the Vitamin B_6 group in the intensity of its absorption maxima at 305 mμ in alkaline reaction (N/10 NaOH); E_m = 1100; for pyridoxal E_m = 5800; for pyridoxamine, pyridoxine, and their 5 phosphates E_m = 7000 to 8000.

Pyridoxal phosphate can thus be estimated spectrophotometrically by the increase in absorption at 388 mμ on neutralization from pH 1 to 7. Free pyridoxal interferes but slightly (E_m 200 as compared to E_m 4900 for the phosphate) especially when present as a minor impurity.

Another spectrophotometric estimation of pyridoxal phosphate has been based on loss of the (yellow) 375-mμ maximum and intensification of the 327-mμ maximum on coupling with *m*-hydroxy propadrine.[14]

[11] W. W. Umbreit, D. J. O'Kane, and I. C. Gunsalus, *J. Biol. Chem.* **176,** 629 (1948).
[12] D. Heyl, E. Luz, S. A. Harris, and K. Folkers, *J. Am. Chem. Soc.* **73,** 3430 (1951).
[13] E. A. Peterson and H. A. Sober, *J. Am. Chem. Soc.* **76,** 169 (1954).
[14] H. F. Schott and W. G. Clark, *J. Biol. Chem.* **196,** 449 (1952).

[143] Preparation and Assay of UDPG and Related Compounds[1,2]

$$\text{Gal-1-P} + \text{UDPG} \rightleftarrows \text{G-1-P} + \text{UDPGal}$$
$$\text{UDPGal} \rightleftarrows \text{UDPG}$$

By Luis F. Leloir and Alejandro C. Paladini

Assay Method

Principle. The method of estimation of UDPG is similar to that described for galactowaldenase. The reaction mixture contains Gal-1-P and a maceration juice of *Saccharomyces fragilis* plus variable amounts of a standard preparation of UDPG. Glucose-1,6-diphosphate is also added, so that the velocity of the over-all reaction is limited by the rate of production of G-1-P. Under the conditions of the test it is possible to measure from 0.001 to 0.01 micromole of UDPG.[3]

An alternative method[4] consists in using a pyrophosphorylase and phosphoglucomutase and measuring the G-6-P formed spectrophotometrically with Zwischenferment and TPN. The reactions are as follows:

$$\text{UDPG} + \text{PP} \rightleftarrows \text{G-1-P} + \text{UTP}$$
$$\text{G-1-P} \rightleftarrows \text{G-6-P}$$

With the galactowaldenase method both UDPG and UDPGal are estimated, whereas with the pyrophosphorylase method only UDPG is measured.

Reagents

Gal-1-P, 20 μM./ml. Sodium salt.
0.1 M magnesium chloride.
Glucose-1,6-diphosphate, 1 μM./ml. Sodium salt.
UDPG, 0.1 μM./ml. Sodium salt.
Enzyme. A crude extract from *S. fragilis* (see Vol. I [35]) diluted one-third with water is used.

---

[1] See Vol. I [35].

[2] The following abbreviations are used: Gal-1-P for galactose-1-phosphate and UDPGal for uridine diphosphate galactose.

[3] R. Caputto, L. F. Leloir, C. E. Cardini, and A. C. Paladini, *J. Biol. Chem.* **184**, 333 (1950).

[4] H. M. Kalckar, *Biochim. et Biophys. Acta* **12**, 250 (1953). See Vol. II [118].

Procedure. Mix the following components: 0.1 ml. of Gal-1-P, 0.02 ml. of magnesium chloride, 0.01 ml. of glucose diphosphate, the unknown preparation of UDPG in a volume of about 0.02 ml., 0.01 ml. of enzyme, and water to complete 0.2 ml. Usually two or three different amounts of the unknown are compared with a curve obtained in the same series using two or three samples of the standard solution of UDPG. Blanks are run at the same time in which the reaction is stopped at $t = 0$.

The reaction is started by adding the enzyme, and, after 20 minutes at 37°, 1.5 ml. of Somogyi copper reagent is added. The mixture is heated for 10 minutes at 100°, cooled, and 1.5 ml. of Nelson reagent and water to 7.5 ml. are added (see Vol. III [12]).

Preparation of UDPG

The treatment of baker's yeast with toluene increases the yield of UDPG. When yeast intimately mixed with 10% of its weight of toluene is incubated at 35 to 37°, there occurs an increase in UDPG which attains a maximum after about 40 minutes, followed by a slow decrease. The incubation in the absence of toluene produces no changes. The increases produced by toluene in different experiments varied from 50 to 400%, and the maximum concentration attained never exceeded 2.5 micromoles of UDPG per gram of yeast.

Procedure 1. *Step 1. Extraction.* Ten kilograms of baker's yeast is spread out in a layer about 5 cm. deep and heated in an incubator. When the temperature attains 35 to 36°, the yeast is intimately mixed with 1000 ml. of warm toluene. After 40 minutes at 35 to 36°, 10 l. of 95% ethanol is added, and the mixture is heated in a water bath until it boils. On the following day it is filtered through a 32-cm. Büchner funnel with a filter aid (Celite Hyflo Super-Cel).

Step 2. Precipitation with Mercuric Salts. The filtrate is acidified with 5 N nitric acid until acid to Congo red paper. Then 30 ml. of mercuric acetate per liter is added. After mixing, the preparation is left overnight in the icebox. The suspension is filtered through a Büchner funnel until the precipitate is nearly dry. This precipitate is then blended in 1200 ml. of 1 M ammonium acetate and left at room temperature for 2 hours. The suspension is filtered, and the filtrate is acidified to Congo red paper with nitric acid. Then 60 ml. of mercuric acetate and 1 vol. of ethanol are added, and the mixture is left overnight in the cold. After filtration the precipitate is blended in 1200 ml. of water and decomposed with hydrogen sulfide in the cold. The mercuric sulfide is filtered off, washed with 200 ml. of water, and the combined filtrates aerated and neutralized to pH 7.

REMARKS. The mercuric acetate solution is prepared by heating over a boiling water bath 13.5 g. of yellow mercuric oxide, 9.2 ml. of glacial acetic acid, and about 20 ml. of water. After cooling, the solution is made up to 100 ml.

The optimum amount of mercury reagent to be added is ascertained in preliminary trials and is sufficient to precipitate nearly all the activity. Addition of more mercuric reagent gives a precipitate of inactive substances.

The extraction of the mercuric precipitate with ammonium acetate for a second time removes more activity from the precipitate, but also larger amounts of impurities.

The decomposition with hydrogen sulfide is carried out in an ice bath and is complete in about 4 hours. The step is rather critical, since UDPG is very sensitive to acid.

Step 3. Purification with Charcoal. One-milliliter portions of extract are treated with charcoal to find the amount necessary for 90% adsorption of the material with absorbency at 260 mμ. This corresponds to nearly complete adsorption of the active substance. In the preparation described, 100 g. of Pfanstiehl's Norit A was used for the total amount of extract. After 15 minutes at room temperature with occasional shaking, the suspension is filtered. The charcoal is washed with 300 ml. of water and then suspended in 400 ml. of 50% ethanol. After 15 minutes the suspension is filtered and the filtrate concentrated at about 45° under reduced pressure to one-third of its original volume.

REMARKS. A second eluate of the charcoal still contains activity, but the purity is inferior. Contaminating substances such as glucose diphosphate are not adsorbed by charcoal, whereas others, like diphosphopyridine nucleotide, are adsorbed but not eluted appreciably with aqueous ethanol.

At this stage the substance is nearly pure, as judged by the ratio of activity to absorbency at 260 mμ, but it still contains considerable amounts of substances giving a positive ninhydrin reaction.

Step 4. Treatment with Cation Exchange Resin. The extract from step 3 is cooled, acidified to pH 3, and passed through a column (9 × 2.5 cm.) of Dowex 50 in the hydrogen form at a rate of 1 ml./min., approximately. The effluent is neutralized.

Step 5. Second Adsorption with Charcoal. The liquid is treated with 26 g. of Norit. The necessary amount is ascertained as previously described. The suspension is filtered and washed with 100 ml. of water. It is then eluted by adding 50% ethanol to the Büchner funnel which has been used for filtering, and collecting the effluents in 10- to 15-ml. fractions.

The results of a preparation which was not one of the best are shown in the Summary of Purification Procedure.

SUMMARY OF PURIFICATION PROCEDURE

Step	Volume of extract, ml.	Concentration of UDPG,[a] μM./ml.	UDPG/ total P ratio	Ratio, UDPG/ absorbancy at 260 mμ
1. Yeast extract	11,000	0.65	0.072	0.016
2. After Hg^{++} and H$_2$S	1,500	4	0.375	0.043
3. First charcoal elution	140	25	0.62	0.105
4. After resin	320	11	0.73	0.125
5. Second charcoal eluates				
Third fraction[b]	14.4	7.3	0.51	0.107
Fourth fraction	12	26	0.57	0.104
Fifth fraction	12.6	30	0.40	0.101
Sixth fraction	10.5	10	0.36	0.081
Seventh fraction	16.2	6.5	0.41	0.096

[a] Measured by enzymatic method.
[b] First and second fractions discarded.

Procedure 2. The nucleotide mixture obtained at the end of step 2 of procedure 1 is run through an anion exchange resin and eluted with solutions of decreasing pH and increasing chloride concentration, following the procedure described by Cohn.[5] The fractions corresponding to the UDPG peak are passed through a small charcoal column, and the UDPG is subsequently eluted with ethanol ammonia.[6]

The strong base Dowex 1 anionic resin (Dow Chemical Corp., Midland, Michigan), finely ground (200 to 400 mesh), is employed. The resin is converted to the chloride form with 1 N hydrochloric acid and freed from fines by six or more decantations from water. The remaining slurry is poured into a glass column of the conventional type used in chromatography (50 cm. high and 4.5 cm. inner diameter) and allowed to settle. The column of resin, which is about 30 cm. high, is washed with 1 N hydrochloric acid until the absorbency of the effluent drops to a value of 0.03 to 0.04, followed by water until the pH of the effluent is around 5.

The solution of nucleotides obtained according to procedure 1 up to the end of step 2 (approximately 1500 ml. containing about 8000 micromoles calculated as uridine from the absorbency at 260 mμ) is allowed to drain through the column at a rate of 6 to 8 ml./min. After a washing

[5] W. E. Cohn, *J. Am. Chem. Soc.* **72,** 1471 (1950).
[6] E. Cabib, L. F. Leloir, and C. E. Cardini, *J. Biol. Chem.* **203,** 1055 (1953).

with water (500 ml.), the following solutions are successively run through at the same rate: 0.01 N hydrochloric acid, 0.01 N sodium chloride in 0.01 N hydrochloric acid, 0.02 N sodium chloride in 0.01 N hydrochloric acid. Each eluent is replaced by the next after the absorbency at 260 mμ of five to ten fractions (2.5 to 5 l.) has remained under 0.1. Finally, uridine diphosphoacetylglucosamine (UDPAG)[6] and UDPG can be eluted separately, in this order, using a solution of 0.025 N sodium chloride in 0.01 N hydrochloric acid.

It is likely that the elutions with 0.01 N hydrochloric acid and with 0.01 N sodium chloride in 0.01 N hydrochloric acid are not necessary, but this procedure has not been tested.

After each run the column is regenerated with 1 N hydrochloric acid followed by water. The same column can be used several times, except for a small layer at the top which darkens during the run and is replaced each time by fresh resin.

The fractions belonging to each peak are pooled and adsorbed on a small charcoal column (3 g. of Norit A in a fritted glass funnel 4 cm. in diameter). Elution from the charcoal is carried out with a water-ethanol-ammonia mixture (40 ml. of 95% alcohol plus 1 ml. of concentrated ammonia, made up to 100 ml. with water).

Each fraction of 3 to 5 ml. is collected in the cold and immediately adjusted to pH 5 to 6 with hydrochloric acid. The fractions of high absorbency are then pooled. It is thus possible to concentrate the solutions from several liters to 20 to 40 ml. The yield of this step is 70 to 80%.

These solutions are further concentrated by evaporation under reduced pressure, and if desired the nucleotides can be precipitated as the calcium salt by addition of some drops of a saturated solution of calcium chloride in ethanol, followed by several volumes of ethanol until no more precipitation occurs.

Properties of UDPG

The theoretical molecular weight of the free acid is 566.

The liberation of glucose from UDPG by acid is practically complete in 5 to 10 minutes at 100° in acid 0.01 N. The rate of hydrolysis is about six times as high as that of G-1-P and about forty times as high as that of the labile phosphate of UDPG. These differences in rates of hydrolysis allow the isolation of UDP and UMP from UDPG.

The phosphate group connecting the glucose and UMP moieties in the UDPG molecule is acid-labile and can be hydrolyzed in 15 minutes in 1 N acid at 100°. Comparison with other known compounds reveals that the rate of hydrolysis is slower than that of the labile phosphate of ATP and of about the same order as that of 3-AMP.

UDPG loses its catalytic activity after a mild treatment with alkali, e.g., heating for 2 minutes at 100° at pH 8.5 or 30 minutes at 0° when dissolved in concentrated ammonia. This alkaline treatment of UDPG leads to the formation of UMP and glucose-1,2-monophosphate.[7]

Biological Activities of UDPG. In addition to its activity as the coenzyme of the galactowaldenase system, it has been proved that UDPG acts as a glucosyl donor in the synthesis of trehalose phosphate in yeast according to the following reaction:

$$\text{UDPG} + \text{G-6-P} \rightarrow \text{Trehalose phosphate} + \text{UDP}^{\,8}$$

It has also been found that plant enzymes catalyze a reaction between UDPG and fructose with the formation of free sucrose:

$$\text{UDPG} + \text{fructose} \rightleftarrows \text{UDP} + \text{sucrose}^9$$

Compounds Related to UDPG

Park and Johnson[10] observed the accumulation of a uridine-containing substance in *Staphylococcus aureus* cells grown in the presence of penicillin, and Park[11] isolated three uridine-5′-pyrophosphate compounds in which this residue is attached to an unidentified amino sugar. In addition to the uridine-5′-pyrophosphate amino sugar group (compound 1), which is common to all three compounds, compound 2 contains L-alanine and compound 3 contains a peptide which is composed of L-lysine, D-glutamic acid, and three alanine residues.

The labile phosphate compounds accumulate rapidly in *S. aureus* after addition of penicillin. Apparently, the growing cells are able to synthesize the compounds only for the short period (30 to 45 minutes) wherein they remain viable after the addition of penicillin. Glucose must be added in order to obtain rapid synthesis. The maximum yield is obtained when growth is allowed to proceed until about one-half the maximum population is reached before addition of penicillin. Thirty minutes after the addition of penicillin, the cells are harvested with a Sharples centrifuge.

The purification procedure used consists in four steps: (1) extraction of the packed cells with TCA; (2) fractionation of the barium salts with ethanol; (3) removal of basic impurities with a cation exchange resin (Amberlite IR-100 or IR-105, Rohm and Haas); and (4) separation of

[7] A. C. Paladini and L. F. Leloir, *Biochem. J.* **51**, 426 (1952).

[8] L. F. Leloir and E. Cabib, *J. Am. Chem. Soc.* **75**, 5445 (1953).

[9] L. F. Leloir and C. E. Cardini, *J. Am. Chem. Soc.* **74**, 6419 (1953).

[10] J. T. Park and M. J. Johnson, *J. Biol. Chem.* **179**, 585 (1949).

[11] J. T. Park, *J. Biol. Chem.* **194**, 877, 885, 897 (1952).

the three components by partition chromatography (column of Celite 545, Johns-Manville; solvent system: 0.1 M H_2SO_4 adjusted to pH 2 with NaOH, shaken with 5 vol. of 75% phenol. Lower phase used).

Studies on partially purified preparations of UDPG [7] obtained from yeast revealed the presence of a similar compound containing acetyl-glucosamine instead of glucose.[6] This compound is closely related to the compounds described by Park.

The method of preparation of uridine diphosphoacetylglucosamine (UDPAG) is similar to procedure 2 used to obtain UDPG. The yeast extract used in the chromatography is prepared as follows.

To 10 kg. of baker's yeast, 10 l. of 95% ethanol is added and the mixture is heated with continuous stirring until it boils. On the following day it is filtered through a 32-cm. Büchner funnel with a filter aid. The filtrate is acidified with 5 N nitric acid until acid to Congo red paper. Then 30 ml. of mercuric acetate (see p. 970) per liter is added. After mixing, the preparation is left overnight in the refrigerator. The suspension is filtered through a Büchner funnel. The precipitate is dried as much as possible by suction, then blended with 1200 ml. of water, and decomposed with hydrogen sulfide in the cold. The mercuric sulfide is filtered off and washed with 100 ml. of water, and the combined filtrates are aerated and neutralized to pH 6. Prior to chromatography, the solution is brought to pH 7.5 by addition of concentrated ammonia.

The chromatography in Dowex 1 and the concentration of the fraction are the same as described in connection with the preparation of UDPG.

[143A] Determination of UDPG and UTP by Means of UDPG Dehydrogenase

I. Enzymatic Determination of UDPG

By J. L. STROMINGER, ELIZABETH S. MAXWELL, and HERMAN M. KALCKAR

$$UDPG + 2DPN \rightarrow UDPGA + 2DPNH$$

Principle. UDPG is oxidized to UDP-glucuronic acid (UDPGA) in the presence of DPN and the enzyme UDPG dehydrogenase (see below). Two moles of DPN is reduced per mole of UDPG oxidized.[1] The enzyme is highly specific (for instance, it does not attack UDPGal, α-glucose-1-P, glucose, etc.), is stable, and can be purified to a rather high degree. For

[1] J. L. Strominger, H. M. Kalckar, J. Axelrod and E. S. Maxwell, *J. Am. Chem. Soc.* **76**, 6411 (1954).

the earlier method of determining UDPG with UDPG pyrophosphorylase, see Vol. II [118].

Reagents

1 *M* glycine buffer, pH 8.7.
DPN, 50 μM./ml.
UDPG dehydrogenase.

Procedure. In a quartz microcuvette having a light path of 1 cm. add 0.1 ml. of glycine buffer, 20 μl. of DPN, the UDPG sample (0.01 to 0.10 micromole), water to a final volume of 1 ml., and UDPG dehydrogenase (50 γ of protein). Take optical density readings at 340 mμ at 1-minute intervals immediately after the addition of the enzyme, and continue until no further reaction is detected. The change in optical density at 340 mμ is 12.0 per micromole per milliliter.[2]

Preparation of UDPG Dehydrogenase.[3] Calf liver acetone powder is extracted for 30 minutes with 20 vol. of chilled distilled water and spun. The supernatant is fractionated with ammonium sulfate. (The saturation percentage is based on solubility at 0°.) The precipitate formed before 40% of saturation is discarded. The most active fraction precipitates in ammonium sulfate concentrations between 40 and 55% of saturation. This precipitate is dissolved in distilled water (volume one-tenth of the original extract). The ammonium sulfate concentration in this solution is measured by a conductivity meter (Barnstead Purity Meter, Barnstead Company, Boston, Mass.). Ammonium sulfate is added as crystals until a saturation level of 35% is reached. After 10 minutes at 0° the precipitate is spun off and discarded. To the supernatant is added more ammonium sulfate until 55% of saturation. This precipitate contains the activity. It is dissolved in distilled water and can be stored at −20°. Further purification of this fraction is obtained by removing impurities by heating at pH 4.9. The pH is adjusted by means of 1 *M* acetic acid, and the temperature rapidly brought to 50°. The preparation is kept at this temperature for 1½ minutes and then rapidly cooled to 2 to 5°, and the pH is adjusted to 4.1. After 10 minutes the precipitate which is inactive is spun off and the supernatant is adjusted to pH 8.1 with 2 *N* ammonium hydroxide. The ammonium sulfate concentration is determined on the conductivity meter, and a solution of alkaline ammonium sulfate (pH 8) is added to make the mixture 35% saturated. The precipitate formed is discarded. Addition of more alkaline ammonium sulfate to the

---

[2] For reasons which are not clear, this optical density is somewhat lower than that calculated from the extinction coefficient of DPNH (see Vol. III [127]).

[3] E. S. Maxwell, J. L. Strominger, and H. M. Kalckar, excerpts from unpublished work.

supernatant to 55% of saturation yields a precipitate which contains the activity. The precipitate is dissolved in a minimum amount of distilled water and the solution immediately adjusted to pH 5.9 by means of 1 M acetic acid. Since the dehydrogenase is unstable at pH 8, the solution should never be stored until it has been adjusted to pH 5.9. The solution is then dialyzed for about 5 hours against 0.02 M acetate buffer, pH 5.9, at 0°. The dialyzed enzyme preparation is diluted five- to sixfold with cold distilled water and adjusted to a concentration of 15% acetone (temperature −5°). After standing at −5 to −6° for 30 minutes, this precipitate, which contains the activity, is collected and dissolved in a small amount of distilled water (usually about one-hundredth of the volume of the original extract); a few drops of saturated ammonium sulfate are usually added to stabilize the enzyme preparation. The latter preparation has a dehydrogenase activity per milligram of protein which is 130- to 140-fold that of the original extract, and in many cases yields as high as 30 to 40% are obtained.

II. Enzymatic Determination of UTP

By HERMAN M. KALCKAR and ELIZABETH P. ANDERSON

$$\text{UTP} + \text{G-1-P} \rightleftarrows \text{UDPG} + \text{PP}$$
$$\text{UDPG} + 2\text{DPN} \rightarrow \text{UDPGA} + 2\text{DPNH}$$

Principle. UTP, in the presence of yeast UDPG pyrophosphorylase (PP-uridyl transferase) (see Vol. II [118]), will react with α-glucose-1-phosphate to give UDPG. If DPN and UDPG dehydrogenase (see previous section) are also added to the digest, the transformation of UTP to UDPGlycosyl compound (UDP-glucuronic acid, UDPGA) is carried to completion. Moreover the reaction can be followed spectrophotometrically at 340 mμ, owing to the accompanying DPN reduction.[4] If glucose-1-phosphate is in excess, the amount of UDPG formed will be governed by the amount of UTP present. ATP is practically inactive in this reaction.

Reagents

α-G-1-P, 30 μmoles/ml. Potassium salt.
DPN, 50 μmoles/ml.
1 M-tris(hydroxymethyl)aminomethane, pH 8.1.

Procedure. To a micro silica cuvette having a light path of 1 cm., add, in a final volume of 1 ml., 0.1 ml. Tris buffer, 20 μl. of G-1-P, 20 μl. of

[4] E. P. Anderson and H. M. Kalckar, *Abstr. Biol. Bull.* (1955); E. P. Anderson, H. M. Kalckar, and A. Munch-Peterson, *Pubbl. staz. zool. Napoli*, in press.

DPN, UTP sample (0.02 to 0.1 micromole), and amounts of the enzymes corresponding to 30 γ of protein in the case of the pyrophosphorylase and 50 γ of protein for the dehydrogenase. The dehydrogenase is added as the final component, and optical density is read at 340 mμ until no further DPN reduction is detected. An increase in optical density of 1.2 corresponds to 0.1 micromole of UTP.[2]

Section VII

Determination of Inorganic Compounds

[144] Determination of Nitrate and Nitrite

By D. J. DONALD NICHOLAS and ALVIN NASON

Phenol-Disulfonic Acid Method for Nitrate

Principle. The nitration of phenol-2,4-disulfonic acid[1-3] yields an orange-brown solution. This can be measured on a colorimeter with a blue filter.

Reagents

> Phenol-disulfonic acid solution, approximately 25% w/v in concentrated sulfuric acid is made as follows: Dissolve 25 g. of phenol (c.p.) and 158 ml. of concentrated sulfuric acid (nitrogen-free). Add 67 ml. of fuming sulfuric acid containing about 20% sulfur trioxide and heat the mixture on a boiling bath for 2 hours.
> Ammonium hydroxide (c.p.) containing 28% w/w NH₃.

Procedure. The preliminary treatment of the test material will vary according to its nature. Chlorides, most forms of organic matter, and high concentrations of dissolved salts must be absent.

Place a suitable quantity of solution to be examined in a small porcelain dish, and evaporate to dryness on a boiling-water bath. To the cooled residue add 1 ml. of the phenol-disulfonic acid solution, taking care that the reagent makes contact with the whole of the residue. Allow to stand for 10 minutes. Add 10 ml. of water, cool the mixture, and add 10 ml. of ammonia solution. Again cool and dilute with water to 25 ml. Measure on a colorimeter with a blue filter.

At the same time prepare a "blank" solution and nitrate N standards of 5, 10, 20, . . . , 100 γ, which are tested as described above.

Chlorides when present in more than 2 p.p.m. may interfere with the above test. They may be removed by adding 1 ml. of glacial acetic acid to 10 ml. of the sample, followed by 0.1 g. of solid silver sulfate (nitrate-free), and, after shaking, filtering through a No. 32 Whatman paper. The test for nitrate is then applied to 5 ml. or suitable aliquot of the filtrate. The effective range is 5 to 100 γ of nitrate N.

[1] H. Sprengel, *Pogg. Ann.* **121,** 188 (1864).
[2] J. E. Eastol and A. G. Pollard, *J. Sci. Food Agr.* **2,** 266 (1950).
[3] D. J. D. Nicholas, "Chemical Tissue Tests for Determining the Mineral Status of Plants," Tintometer Ltd., Salisbury, England, 1953.

Brucine Method for Nitrate

Principle. Nitration of brucine[3] yields an orange-brown solution which can be determined on a colorimeter with a blue filter.

Reagents

Brucine, 4% w/v solution in chloroform filtered before use.
Sulfuric acid (concentrated), specific gravity 1.84.

Procedure. To an appropriate aliquot part (5 ml.) add 2 ml. of brucine from a dropping pipet. Add 5 ml. of sulfuric acid slowly from a burette. The speed of adding the acid is critical. If too fast, boiling will occur; if too slow, the chloroform is not driven off. Shake the tube continuously during this operation, and, as great heat is evolved, wear an asbestos glove. Should chloroform remain in solution, place the tubes in a water bath at 90° for about 10 minutes until the solvent is removed. Cool the tubes in air. Standard solutions are prepared as described above. Measure the color intensity with a blue filter. The effective range is 10 to 100 γ of nitrate N.

Xylen-1-ol Method for Nitrate[4,5]

Principle. The test depends on the formation of 5-nitro-2,4-xylen-1-ol, which is volatile in steam and is distilled into dilute sodium hydroxide with which it forms a red salt. This can be determined colorimetrically with a green filter.

Reagents

Alumina cream. The alumina cream is prepared by adding a slight excess of strong ammonia solution to a saturated aqueous solution of aluminum potassium sulfate (c.p.) and then adding more alum solution, slowly, until the mixture is just acid to litmus.
Lead acetate (basic) solution, 25% w/w.
Sodium hydroxide solution, 2 N.
Sulfuric acid, 85% w/w (nitrogen-free).
2,4-Xylen-1-ol solution, 1% w/v in glacial acetic acid.
Silver sulfate (nitrogen-free).

Procedure. Put 1 g. of minced tissue in a 100-ml. volumetric flask containing 20 ml. of distilled water, immerse the flask in a boiling-water bath for 15 minutes, and periodically shake the contents. Cool, and

[4] J. Blom and C. Treschow, *Z. Pflanzenernähr Düng. Bodenk.* **13a**, 159 (1929).
[5] "A Handbook of Colorimetric Chemical Analytical Methods," Tintometer Ltd., Salisbury, England, 1953.

render just acid with sulfuric acid, with bromocresol green as indicator. Oxidize any nitrites present by adding 0.2 N potassium permanganate until a faint pink color persists for 1 minute. Add 5 ml. of a solution of basic lead acetate, shake vigorously for 1 minute, then add 5 ml. of alumina cream, and again agitate the solution for 1 minute. Dilute with distilled water to 100 ml., shake again, and filter. Mix an appropriate amount of the filtrate, not exceeding 20 ml. (usually from 1 to 5 ml. is adequate), with three times its volume of sulfuric acid, adjust the temperature of the mixture to 35°, add 1 ml. of xylenol solution, and maintain at 35° for half an hour. Then dilute with 100 ml. of distilled water, and transfer to an ordinary distillation apparatus. Distill the mixture, and collect 50 ml. in a receiver containing 10 ml. of sodium hydroxide. Dilute the distillate to an appropriate volume. Measure the color intensity with a green filter.

Appropriate nitrate N standards can be put through at the same time. The effective range is 2 to 25 γ of nitrate N.

Nitrite Determination by Diazotization and Coupling Reactions

Principle. The measurement for nitrite is based on the formation of a red AZO compound. This involves, first, the reaction in acid solution of a primary amine such as sulfanilic acid or sulfanilamide with nitrite to form a diazonium salt. The latter is then coupled to an aromatic amine to yield the red AZO dye whose concentration can be determined in a colorimeter.

Reagents

1% sulfanilamide. Dissolve 1 g. of the solid in 100 ml. of an acid solution made up with 75 ml. of distilled H_2O and 25 ml. of concentrated HCl. Store in an amber reagent bottle.

0.02% N-(1 napthyl)ethylenediamine hydrochloride. Dissolve 20 mg. of the solid in 100 ml. of distilled water. Store in an amber reagent bottle.

Sodium nitrite standard. Dissolve 69.0 mg. of sodium nitrite in 1000 ml. of distilled water to give a $10^{-3} M$ solution. Dilute 3.0 ml. to 100 ml. to give a $3 \times 10^{-5} M$ solution which therefore contains 30 millimicromoles of nitrite per milliliter. The liberation of unstable nitrous oxide by any carbon dioxide present can be prevented by the addition of approximately 25 mg. of sodium hydroxide per 100 ml. of solution. The addition of 1.0 ml. of chloroform will prevent bacterial growth.

Procedure. To an appropriate aliquot (not larger than 1.3 ml.) containing 2 to 35 millimicromoles of nitrite add 0.5 ml. of sulfanilamide reagent followed by 0.5 ml. of the naphthylenethylenediamine reagent to develop the color. Add sufficient water to give a final volume of 2.3 ml. After 10 minutes determine the density of the color with a colorimeter employing a green filter. With a Klett-Summerson colorimeter and a green (540-mμ) filter, a reading of 10 colorimeter units is equivalent to 1 millimicromole of nitrite under the above conditions. Appropriate standards can be prepared ranging from colorimeter readings of 20 to 350, and therefore from 2 to 35 millimicromoles. Larger final volumes may be used with corresponding standards to suit the needs of individual laboratories. The color is stable for 2 to 4 hours.

Remarks. A variety of diazotizing and coupling reagents may be used for determining nitrite. For example, sulfanilic acid (4-aminobenzene-sulfonic acid) and α-naphthylamine or dimethyl-α-naphthylamine are also employed.[6] The advantage of the method presented above, in addition to its extreme sensitivity, is the fact that only a few minutes are necessary for full development of stable color. Although acetic or hydrochloric acid may be used in making up the sulfanilamide reagent, hydrochloric acid has been found to be more desirable, especially when reduced pyridine nucleotides are also present in the solution to be tested. The nonenzymatic disappearance of nitrite in the presence of DPNH and sulfanilamide reagent occurs to a much larger extent when acetic acid instead of hydrochloric acid is used in the sulfanilamide solution.

[6] F. D. Snell and C. T. Snell, "Colorimetric Methods of Analysis," 3rd ed., p. 804. Van Nostrand, New York, 1949.

[145] Determination of Total Nitrogen and Ammonia[1]

By ROBERT BALLENTINE

Total nitrogen in biological materials and in enzymatic reaction mixtures is determined by the Kjeldahl nitrogen method. The analysis consists essentially in three steps represented in the following equations:

$$\text{Organic-N} \xrightarrow[\text{catalysts}]{H_2SO_4} CO_2 + H_2O + NH_4HSO_4 \tag{1}$$

$$NH_4HSO_4 + 2NaOH \rightarrow NH_3 + Na_2SO_4 + H_2O \tag{2}$$

$$NH_3 + KH(IO_4)_2 \rightarrow KIO_4 + NH_4IO_4 \tag{3}$$

[1] The preparation of this manuscript was supported in part by Atomic Energy Commission contracts AT(30-1)-933 and AT(30-1)-1822.

The nature and limitations of each of these steps have received considerable attention in review articles.[2] In theory and practice the last two steps present few analytical problems, but the conversion of organic nitrogen to ammonia has evoked numerous researches and much research and discussion. Failure to recover the nitrogen quantitatively in an analytical sample may be ascribed to losses arising from the following sources. Examination of the first of the above equations reveals that one source of error lies in either over- or under-oxidation. Organic material must be oxidized to carbon dioxide; the nitrogen, however, must be liberated in a reduced form as ammonia. This requires a nice balance between the oxidation and the reduction potentialities of the digestion mixture. The organic material during its oxidation produces considerable sulfite which, along with the nascent hydrogen produced from hydrogen-rich compounds, functions to reduce the organic nitrogen and retain it as ammonia. Numerous oxidants have been employed to promote the rapid and complete oxidation of organic material in the Kjeldahl digestion; among these have been perchlorate, persulfate, hydrogen peroxide, and permanganate. Conflicting reports as to the efficiency and advisability of using any of these strong oxidants have appeared in the literature.[2d] Even hydrogen peroxide, which has been used with the greatest success, may give satisfactory results only through a balancing of errors.[3] Too vigorous oxidation leads, as in the case of perchloric acid, to the loss of ammonia as nitrogen gas[4] or as highly stable organic amines.[5] When nitrogen exists in oxidized organic bindings, failure to provide a sufficient reducing action in the digestion mixture also leads to low results. For example, aromatic amines, hydroxylamine, hydrazine, substituted hydrazones, osazones, and nitroso, nitro, diazo, and azo compounds all require special reductive treatment if their nitrogen is to be quantitatively recovered.[2d] Even in the case of proteins, where many of these systems do not occur, stable aromatic and heterocyclic ring systems can be formed during digestion, and therefore a reductive hydrolysis prior to the digestion is advisable if quantitative recoveries are to be assured.[6] In aromatic amines the low hydrogen content usually results in only small amounts of nascent hydrogen being produced, with concomitant

[2] (a) R. B. Bradstreet, Chem. Revs. 27, 331 (1940); (b) Anal. Chem. 26, 185 (1954); (c) P. L. Kirk, Advances in Protein Chem. 3, 142 (1947); (d) Anal. Chem. 22, 354 (1950); (e) C. O. Willits and C. L. Ogg, J. Assoc. Offic. Agr. Chemists 32, 118 (1949); (f) Ibid. 33, 179 (1950).
[3] P. L. Kirk, Mikrochemie 16, 13 (1934).
[4] J. P. Peters and D. D. Van Slyke, "Quantitative Clinical Chemical Methods," p. 519. Williams & Wilkins, Baltimore, 1932.
[5] L. E. Wicks and H. I. Firminger, Ind. Eng. Chem., Anal. Ed. 14, 760 (1942).
[6] R. Jonnard, Ind. Eng. Chem., Anal. Ed. 17, 246 (1945).

losses of nitrogen. A special case of loss of nitrogen from an oxidized form is that in which part of the nitrogen in the sample is either inorganic nitrate or nitrite; here special procedures are essential if quantitative recovery is to be achieved. A wide variety of reducing agents and procedures have been used, among them organic compounds such as carbohydrates[7] (sucrose, glucose, cigarette paper), hydriodic acid,[8] hypophosphorous acid,[9] and coupling nitrates with salicylic acid followed by reduction with zinc or thiosulfate.[9a] All that can be said for these methods is that there is not yet a universal digestion procedure for the determination of all types of organic and inorganic nitrogen compounds. Finally, nitrogen can be lost from the digestion mixture by volatilization. Such losses are promoted when excess neutral salt (such as potassium sulfate) is present, resulting in excessive digestion temperatures, they are also brought about by prolonged digestion periods. Losses of 0.01 mg. of nitrogen per hour per milligram of nitrogen being digested have been reported.[10] This nitrogen can frequently be recovered by condensation of the vapors issuing from the digestion flask. For many purposes, however, this loss may be unimportant and may be corrected for by blank analysis on known ammonium sulfate standards. Modern tendencies in the Kjeldahl digestion are toward the use of sealed tube methods, especially for ultramicrodeterminations.[11]

Separation of the ammonia from the digestion mixture causes little difficulty. Adequate apparatus is described in the literature and is now standardized by the Committee on Micro-Analytical Apparatus.[12] Of the numerous methods utilized for the separation of ammonia from the digest mixture, aeration, distillation, steam distillation, and isothermal diffusion may be mentioned. Also of value is direct colorimetric determination in the digest, suitable for rapid control analysis where high precision is not required.

Numerous methods for the titration of the ammonia after its separation have been used, several of which will be presented in the methods section.

[7] A. Elek and H. Sabotka, J. Am. Chem. Soc. **48**, 501 (1926).

[8] A. Friedrich, E. Kühaas, and R. Schnürch, Z. physiol. Chem. **216**, 68 (1933).

[9] S. M. Woods, D. Scheirer, and E. C. Wagner, Anal. Chem. **25**, 837 (1953).

[9a] F. Sutton and A. D. Mitchell, "Volumetric Analysis," 12th ed., p. 83. Blakiston, Philadelphia, 1935.

[10] G. R. Tristram, in "The Proteins" (Neurath and Bailey, eds.), Vol. I, Part A, p. 186. Academic Press, New York, 1953.

[11] (a) L. M. White and M. C. Long, Anal. Chem. **23**, 363 (1951); (b) B. W. Grunbaum, F. L. Schaffer, and P. L. Kirk, ibid. **24**, 1487 (1952).

[12] A. Steyermark, H. K. Alber, V. A. Alvise, E. W. D. Huffman, J. A. Kuck, J. J. Moran, and C. O. Willitts, Anal. Chem. **23**, 523 (1951).

I. Total Nitrogen—Microvolumetric Method

Digestion

Method A. The following digestion is satisfactory for most protein-containing biological samples and for many purine and pyrimidine derivatives.[13]

Reagents

Digestion mixture. Mix 3 vol. of concentrated H_2SO_4 and 1 vol. of 85% H_3PO_4.

Sucrose, c.p.

Potassium persulfate, reagent for Kjeldahl.

Procedure. Weigh or pipet a sample containing 1.0 to 0.1 mg. of nitrogen into a 10-ml. Kjeldahl flask. Add 1 ml. of the digestion mixture and a clean Pyrex bead. If heterocyclic or aromatic nitrogen is present, add 100 mg. of sucrose. Drive off excess water over a microflame or in an oven at 125°. After the flask has cooled add about 100 mg. of potassium persulfate along with 5 to 8 drops of distilled water. Again heat the flask over a microburner until white fumes start coming off. If the digest is not clear at this point, repeat the treatment with persulfate. Finally add 50 mg. of potassium persulfate along with 5 to 8 drops of distilled water, and place the flask in a sand bath at approximately 375°. After 5 hours remove the flask and cool to room temperature. *All these operations should be carried out in a hood!*

Method B. The following Kjeldahl-Gunning-Dyer[14] modification may be applied to practically all classes of animal and vegetable material, pyridine and quinoline derivatives, purines, pyrimidines, amines, amides oximes, and other substituted nitrogen compounds such as carbazole or hydrazobenzene. If hydrazine, osazones, nitro and nitroso, azo, or diazo compounds are present, a prior reduction must be made.

Reagents

Sucrose, mercuric oxide, potassium sulfate, concentrated sulfuric acid—all analytical grade.

Procedure. To the sample containing about 10 mg. of organic material and at least 0.1 mg. of nitrogen, add 100 mg. of sucrose, 40 mg. of mercuric oxide, 500 mg. of potassium sulfate, and 1.5 ml. of sulfuric acid.

[13] (a) S. Y. Wong, *J. Biol. Chem.* **55**, 427 (1923); (b) R. Ballentine and J. R. Gregg, *Anal. Chem.* **19**, 281 (1947).

[14] E. P. Clark, "Semi-Micro Quantitative Organic Analysis," p. 42. Academic Press, New York, 1943.

Start the digestion by heating the mixture gently until the frothing ceases, and then heat more vigorously until it is boiling strongly. The total time for digestion is about 1 hour, and the mixture should be colorless during the latter half of the oxidation period.

Method C. When inorganic nitrogen is present as nitrates or nitrites, direct digestion with sulfuric acid is not applicable, since nitrogen in these forms is partially lost as nitrogen oxides. For such mixtures, which are frequently encountered in plant biochemistry, the Kjeldahl-Gunning-Jodlbauer process is satisfactory.

Reagents

Sulfuric–salicylic acid. For each series of determinations freshly mix 30 ml. of sulfuric acid and 1 g. of salicylic acid.

Zinc dust, c.p.

Procedure. Weigh or pipet the sample into a 30-ml. Kjeldahl flask, and remove any water by drying in an oven. Cool the flask in an ice bath, and add 1.5 ml. of the sulfuric-salicylic reagent. Allow to stand for at least 30 minutes; then add with shaking and in small portions 100 mg. of zinc dust. Complete the digestion according to method B above.

Method D. When highly oxidized nitrogen (i.e., substituted hydrazones, osazones, nitro and azo compounds) is present in organic binding, a modification of the digestion involving a prior reduction is necessary. The method of Friedrich is perhaps the most universal method for the determination of the total organic nitrogen (but not inorganic nitrates or nitrites).[8]

Reagents

Hydriodic acid, c.p. for Kjeldahl.

Glucose, c.p.

Procedure. Weigh or pipet a 10-mg. sample containing at least 0.1 mg. of nitrogen into a 30-ml. Kjeldahl flask. Add 100 mg. of glucose or sucrose and 1 ml. of constant boiling hydriodic acid. Reflux the mixture for 45 minutes. At the end of this time increase the heat until approximately 0.7 ml. of the hydriodic acid has distilled off. Then add 500 mg. of potassium sulfate, 1 ml. of water, and 1.5 ml. of sulfuric acid. Heat the mixture until most of the water is removed, thereby steam-distilling the iodine. Repeat the distillation with water, adding several milliliters, until no further iodine is distilled off. After the digest has cooled slightly add 40 mg. of mercuric oxide and complete the digestion according to method B above.

Distillation and Determination of Ammonia

The steam distillation of ammonia from an alkalized digest is the most rapid and quantitative method for the recovery of the ammonia. A special one-piece steam distillation apparatus is recommended by the Committee for the Standardization of Micro-Analytical Apparatus.[12, 15] The use of an electrically heated steam generator greatly increases the control and smoothness of operation.

Method A. When digestion method A has been used (a method which is suitable for most biological samples), the following titration, based on an iodometric method which is both simple and direct, can be used.[13b]

Reagents

Sodium hydroxide, 10%. Dissolve 50 g. of sodium hydroxide in 500 ml. of distilled water.

Sodium thiosulfate, ca. 0.011 M. Dissolve 2.75 g. of sodium thiosulfate pentahydrate in 1 l. of distilled water, and add 3 drops of chloroform.

Potassium biniodate, 0.05 M. Dissolve 19.500 g. of potassium biniodate in distilled water, and make to 1 l. Dilute exactly fivefold for the working solution, i.e., 0.01 M (0.01 N vs. H+).

Starch indicator, 1%. Make a paste of 500 mg. of Lintner's soluble starch in a little distilled water. Pour with constant stirring into 50 ml. of boiling 20% (w/v) sodium chloride.

Potassium iodide, c.p.

Distillation. Accurately pipet 10 ml. of the 0.01 M potassium biniodate solution into a 50-ml. Erlenmeyer flask, and arrange the flask so as to immerse the delivery tip of the still in the biniodate solution. Wash the cooled digest into the still with five 1-ml. portions of distilled water containing a little phenolphthalein indicator. Follow this by sufficient 10% sodium hydroxide to neutralize the digest and make it highly alkaline. With the recommended one-piece steam distillation system, it is essential that both stopcocks be open and the steam generator be turned on during all these additions.[16] After the alkali has completely run into the distillation chamber, close both stopcocks and allow the steam distillation to proceed for 6 minutes. About 15 ml. of water should distill over during this time. Without interrupting the flow of steam, lower the flask containing the biniodate and distillate so that the tip of the condenser is about 2 cm. above the level of the liquid. Wash off the tip with a fine stream of

[15] Available as No. 7497 from Arthur H. Thomas Co., Philadelphia, Pennsylvania.

[16] Detailed description of the use of this apparatus is given in A. Steyermark "Quantitative Organic Microanalysis," p. 148. Blakiston, Philadelphia, 1951.

distilled water; then continue the distillation for 2 additional minutes. Remove the receiver.

Cleaning the Apparatus. A simple method for cleaning the apparatus after the completion of a distillation is as follows: The steam generator is switched off with both stopcocks closed, the digest thereby being sucked back into the outer jacket. It is then run out through the drain-cock. The drain is then closed, and a beaker of distilled water is brought up so as to immerse the tip of the condenser. Cooling of the steam genera-tor sucks the water back through the whole system, thoroughly flushing it out. The wash is then rejected through the draincock, the steam generator being turned on in preparation for the next sample. The drain-cock is left open at this point.

Titration. Add a small crystal (ca. 50 mg.) of potassium iodide to the receiver contents. Titrate the liberated iodine with the 0.011 *M* sodium thiosulfate. Near the end point add 1 drop of the starch indicator. In a similar fashion, run a blank titration on a 10-ml. aliquot of the potassium biniodate.

$$\frac{a - b}{a} \times 10 \, N \times 14.008 = \text{mg. N in sample}$$

where a = milliliters of thiosulfate in blank titration.

b = milliliters of thiosulfate in experimental titration.

N = normality with respect to H^+ of the potassium biniodate solution.

Method B. When the digestion has been done by any of the other methods (B, C, D) the mercury used as a catalyst will cause loss in the recovery of the ammonia if the mercury ammonium complex is not decom-posed. Sodium thiosulfate is therefore added to the alkali, thus forming mercury sulfide. It is important that the digest be made strongly alkaline but that a large excess of alkali be avoided; otherwise the mercury sulfide is decomposed and mercury appears in the distillate.[14] Since traces of hydriodic acid, iodine, and hydrogen sulfide interfere in the iodometric titration, the ammonia is trapped in boric acid and titrated with potas-sium biniodate.[17]

Reagents

Sodium hydroxide (10% w/v) + sodium thiosulfate (5% w/v).
Boric acid, 4% (w/v).
Indicator A (BCG-MR). Mix 5 parts of 0.2% bromocresol green with 1 part of 0.02% methyl red solution, both in 95% ethanol.
Indicator B (MR-MB). Mix 2 parts of 0.02% methyl red solution with 1 part of 0.2% methylene blue, both in 95% ethanol.
Potassium biniodate, the same as in method A.

[17] Modified from ref. 16, p. 135.

Procedure. Pipet into a 50-ml. Erlenmeyer flask 5 ml. of 4% boric acid solution, and add 4 drops of the indicator. The transfer of the digest and its neutralization are carried out as given under method A except that sodium hydroxide-sodium thiosulfate solution replaces plain sodium hydroxide.

When the distillation is complete, remove the flask containing the boric acid solution. Titrate the distillate with 0.01 N potassium biniodate. The end point is best observed by comparison with a blank composed of 5 ml. of boric acid solution, 4 drops of indicator, distilled water equal to the amount of distillate, and a few drops of biniodate solution.

Each milliliter of 0.01 M potassium biniodate equals 0.14008 mg. of nitrogen.

II. Total Nitrogen—Submicrospectrophotometric Method (Routine Control)

Frequently large numbers of control analyses for total nitrogen must be run. The above methods are designed for high precision (i.e., $\pm 0.2\%$). Often such precision may be sacrificed in favor of a more rapid and routine procedure. The following method has been used in the author's laboratory[18] in connection with column chromatography where a large number of samples had to be processed. By this method 5 γ of nitrogen may be determined with a precision of 2 to 3%, or 1 γ between 4 and 5%.[19]

Reagents

Digestion mixture. Mix 10 vol. of concentrated sulfuric acid and 1 vol. of 85% phosphoric acid.

Potassium persulfate, reagent for Kjeldahl.

Standard nitrogen solution. Dissolve 47 mg. of ammonium sulfate in a few milliliters of water, and make to 100 ml. in a volumetric flask. This will give a solution containing 10 γ of nitrogen per milliliter.

Nessler's reagent.[20] Dissolve 22 g. of iodine and 30 g. of potassium iodide in 20 ml. of water. After solution is complete, remove 1 ml. to a separate test tube. To the remaining solution add 30 g. of pure mercury and shake vigorously while keeping the solution cool under running cold water. Shake until all iodine color has disappeared. When a drop of the supernatant is added to a

[18] The author is indebted to Mr. Dale Cheever for working out the details of this method.

[19] See also (a) G. L. Miller and E. E. Miller, *Anal. Chem.* **20**, 481 (1948); (b) J. F. Thompson and G. R. Morrison, *ibid.* **23**, 1153 (1951).

[20] F. C. Kock and T. L. McMeekin, *J. Am. Chem. Soc.* **46**, 2066 (1924). This reagent is also available commercially as No. 2634 from Hartman-Leddon Co., Philadelphia, Pennsylvania.

milliliter of a 1% starch solution, a faint blue color should appear. If the test is negative, add dropwise more of the initial solution which was reserved, until a faint blue color is obtained. Dilute the whole solution to 200 ml., and decant from the mercury. Add this solution to 975 ml. of 10% sodium hydroxide (w/v), mix thoroughly, and allow to clear by standing for several weeks.

Digestion. Pipet a sample containing between 1 and 5 γ of organic nitrogen into a 10-ml. Pyrex test tube. Drive off the water in a vacuum desiccator or by heating in an 100° oven overnight. High concentrations of inorganic ammonia frequently present as a result of the use of ammonium buffers in column chromatography or in protein fractionation must be removed. Add sufficient 1 M sodium hydroxide to the sample to make it alkaline, and take to dryness in a vacuum desiccator over concentrated sulfuric acid. To be sure that all ammonia has been removed, remoisten the residue with water and redry. To the dried sample add 0.3 ml. of the acid digestion mixture along with a clean Pyrex bead, 30 mg. of potassium persulfate, and 2 drops of distilled water. Heat the tube over a microburner until frothing ceases and white fumes are given off. Cool, and add 10 mg. additional of potassium persulfate. Place the tubes in a sand bath at 375°. After heating for 7 hours in the same bath, remove the tubes and cool to room temperature.

Color Development. Variations in the age and mode of preparation of the primary Nessler's stock solution may require different degrees of dilution in making up the working solution. Dilute the stock Nessler's solution with 1 to 2 parts of water. Prepare a 0.05% solution of methyl cellulose, suspending the solid in boiling water and filtering through a coarse sintered-glass funnel. Chill 10 N sodium hydroxide thoroughly in an ice bath. Mix *in order* 10 ml. of diluted Nessler's solution, 2.5 ml. of methyl cellulose solution, and 10 ml. of 10 N sodium hydroxide. Immediately centrifuge at medium speed for 5 minutes to remove particles of methyl cellulose and decomposed reagent, and then filter through a coarse sintered-glass funnel. This reagent is stable for 1 hour, but for best results it should be used within 20 minutes. Without methyl cellulose the reagent becomes badly decomposed within 30 minutes and sometimes within 5 minutes. It is usually necessary to experiment with a new stock solution to determine the degree of dilution for making a stable working solution.

After the digested samples are cool, add 0.7 ml. of water and 1 ml. of ice-cold 10 N sodium hydroxide solution. When all the samples have been neutralized, add 3 ml. of the above Nessler's reagent. This can best and most reproducibly be achieved by squirting in the Nessler's reagent from a 10-ml. Luer syringe equipped with a No. 18 gage hypodermic needle. It is advisable to neutralize and add the Nessler's reagent to only about fifteen samples at a time. The time for optimum color development will vary from batch to batch of the primary stock Nessler's solution between 12 and 30 minutes. Read the color in the Beckman spectrophotometer in 1-cm. cuvettes at 425 mμ against water as a blank.

Construct a standard curve using the standard ammonium sulfate solution described under Reagents, carrying it through the complete procedure, including the acid digestion steps.[21]

III. Total Nitrogen—Ultramicromethod

The ultramicroprocedure for the determination of total nitrogen is as accurate as the micro and semimicromethods and is simpler and more rapid. Digestion is performed with simple sulfuric acid in a sealed tube.[11b] Ammonia, generated on alkalizing the digest mixture, is transferred by isothermal distillation into boric acid for direct titration. The isothermal distallation of the ammonia is readily carried out in a Conway vessel.[22]

Reagents

Sulfuric acid, concentrated.

Boric acid, 2% solution containing 0.0025% bromocresol green and 0.005% methyl red.

Potassium biniodate. To 100 ml. of 0.05 M potassium biniodate in a 500-ml. volumetric flask add methyl red and bromocresol green to give final concentrations of 0.005% and 0.0025%, respectively, and dilute to volume.

Potassium metaborate, saturated solution.

Digestion. Pyrex glass tubes of 7 mm. O.D. are thoroughly cleaned with chromic acid cleaning solution and are then cut into sections 45 mm. in length. Each section is sealed at one end in an oxygen-gas flame, great care being taken not to touch the open end of the tube with the fingers.

---

[21] The formation of color with Nessler's reagent is due to the formation of a mercury ammonium iodide precipitate. As a result, the stability of this sol depends on temperature, mode of addition of reagents, stabilizers, and neutral salt concentration. It has been found that the reagent and the color degenerate very rapidly if there are particles in the system to act as precipitation centers. These can be contributed by either methyl cellulose or by decomposed reagent (mercuric iodide). For reproducible results it is necessary to standardize the procedure carefully, and we have found that it is difficult to add the color-forming reagent in a reproducible fashion other than by means of a syringe and needle. The reagent must be added rapidly but without expelling solution from the test tube.

[22] The author is indebted to Dr. Martin Larrabee and Mr. William Stekiel for helpful and critical discussion of the distillation technique.

A small muffle furnace equipped with a thermocouple is used for heating the tubes. Inside the furnace a brass block with a number of holes drilled into the metal serves to hold the glass tubes and to protect both the furnace and tubes in case of an explosion.

With a standard capillary measuring pipet,[23] transfer a sample containing between 1 and 15 γ of nitrogen into the tube. This transfer will be more quantitative if the pipet is coated with Desicote. Place the tubes in a vacuum desiccator, and dry them (this requires 10 to 15 minutes for a 25-μl. volume). Put 10 μl. of concentrated sulfuric acid on the wall of the tube near the sample, heat the open end of the tube in an oxygen flame until it is white hot, and seal by pinching the edges together with a hot forceps. Anneal the sealed end briefly in the flame. Carefully mix the sample and the sulfuric acid by briefly centrifuging the tube. Place the tubes in the protective block in the muffle furnace, preheated to between 450 and 470°, *but definitely not above 470°*. After 30 minutes of heating, carefully withdraw the tubes with a forceps, place in a brass centrifuge tube, and briefly centrifuge. This forces the sulfuric acid to the bottom. Carefully scratch the tubes with a glass knife or diamond pencil and break into two equal portions. Heat them in a drying oven at 100° for 5 minutes to remove the sulfur dioxide and carbon dioxide which were forced into solution during the digestion. This step is essential for the quantitative recovery of ammonia.

Isothermal Distillation. Place a small disk of either Teflon or Polyethylene in the center well of a Standard Micro Conway vessel[24] and pipet onto it 50 μl. of 4% boric acid. Transfer the digest to the outer portion of the Conway cell with a suitable transfer pipet,[23] and wash out the digestion tube with 50 μl. of water. Fix the lid of the unit (made of a sheet of Lucite) in position with Dow-Corning high vacuum silicone grease. Alkalize the digest by running in 0.5 ml. of saturated potassium metaborate, immediately replacing the lid; then carefully mix the contents of the Conway vessel. Place the unit in an incubator at 37°, weighing down the lid with a 100-g. brass weight, and allow the isothermal distillation of ammonia to proceed for 2 hours. At the end of this time, open the Conway vessel and titrate the boric acid solution with 0.01 M potassium biniodate delivered from a Gilmont microburet.[25] The calculation is the same as in the macroprocedure. The sharpness of the end point is increased by using a comparison drop.

[23] Available from Micro Chemical Specialties Co., Berkeley 3, California.
[24] No. 4472-F, Arthur H. Thomas Co., Philadelphia, Pennsylvania.
[25] Available from Emil Greiner, New York City, as No. G-15391 B, capacity 0.1 ml. in 0.0001-ml. divisions.

IV. Ammonia Determination[26]

The determination of ammonia is relatively simple if highly labile amides (glutamine, for instance) are absent. If such amides are present, some modification of Archibald's method is essential.[27] For routine laboratory determination of ammonia a modification of the Conway microprocedure is most convenient.[28] The procedure for the determination of ammonia is identical to that for the isothermal distillation and titration of ammonia described in the section Ultramicrodetermination of Total Nitrogen, with the following exceptions. The solution for generating the ammonia from the sample is a borate buffer prepared by adjusting a saturated solution of sodium tetraborate to pH 10. The sample size should be such that it contains between 1 and 15 γ of ammonia.

[26] A systematic determination of ammonia, amide, nitrite, nitrate, and total nitrogen on a single sample is given by J. E. Varner, W. A. Bulen, S. Vonecko, and R. C. Burrell, *Anal. Chem.* **25**, 1528 (1953).

[27] R. M. Archibald, *J. Biol. Chem.* **151**, 141 (1943); see also Vol. III [83, 83A].

[28] E. J. Conway, "Microdiffusion Analysis and Volumetric Error," 2nd ed., p. 92. Crosby Lockwood, London, 1947.

[146] Determination of Inorganic Sulfur Compounds

By JUDITH LANGE and HAROLD TARVER

Determination of Total Sulfur

Principle. The sample is digested with nitric acid and perchloric acid which results in the conversion of all the sulfur compounds to sulfate.[1–3] Sulfate is then precipitated either as barium sulfate or as benzidine sulfate. In the former case the sulfate is determined gravimetrically, in the latter volumetrically by titration with standard base (or colorimetrically). If the material contains significant amounts of phosphorus, relative to the sulfur, it is necessary to resort to preprecipitation of the phosphate either as magnesium ammonium phosphate or as uranyl phosphate; otherwise benzidine phosphate may be precipitated along

[1] N. W. Pirie, *Biochem. J.* **26**, 2044 (1932).

[2] M. Masters, *Biochem. J.* **33**, 1313 (1939).

[3] R. J. Evans and J. L. St. John, *Ind. Eng. Chem. Anal. Ed.* **16**, 630 (1944).

with the sulfate.[4] Also, the completeness of precipitation of the benzidine sulfate should be checked when the salt concentration, e.g., NaCl, in the unknown is high.

Macro Reagents

   Concentrated HNO_3, c.p.
   $HClO_4$, 72% (sulfate-free).
   $BaCl_2 \cdot 2H_2O$, 10%.

Macroprocedure.[3] Place a measured sample of material on a Kjeldahl flask (500-ml. size or less), the amount being sufficient to give $BaSO_4$ weighable with the desired accuracy. Digest with at least 20 ml. of concentrated HNO_3 per sample in a steam or sand bath (the latter at low temperature) until solution is attained, add 5 ml. of 72% $HClO_4$ per gram of sample, and continue heating for 24 hours.[5] Raise the temperature to remove any residual HNO_3, and continue digestion with the remaining $HClO_4$ for an appropriate time, which may be as much as 15 hours.[6] Add

---

[4] The magnitude of the errors which may arise due to the coprecipitation of phosphate with the sulfate are shown by the data in the following table.

Added sulfur, μeq.[a]	Added phosphorus, μeq.	pH	Found sulfur, μeq.	Found phosphorus, μeq.[b]
—	323	2.1	—	0.5
—	323	2.3	—	32
—	323	2.5	—	50
52.0	193	2.1	52.6	0.9
52.0	323	2.1	52.0	1.2
52.0	645	2.1	52.0	1.0

[a] 1 microequivalent (μeq.) = 0.5 micromole.
[b] Determined by a modified Fiske and SubbaRow method after the titrated benzidine precipitate is evaporated down and the organic material removed by digestion with concentrated H_2SO_4 and peroxide. It is seen that there is less possibility of phosphorus contamination at low pH. With still larger precipitates, more phosphate is brought down.

[5] Add more HNO_3 as necessary if a sand bath is used and much acid is lost.
[6] When predigestion with HNO_3 at about 100° is employed, explosions with the perchloric acid are avoided and no difficulties arise. In order to ensure proper digestion, it is best to use an electrically heated Kjeldahl rack, and one with a glass manifold. It is also advisable either to determine the time required for complete oxidation or to employ some other method, such as the Carius, involving the use of fuming HNO_3 in a bomb tube, Na_2O_2 oxidation (Parr bomb), or the oxygen bomb in order to check the method.

more HClO₄ as necessary. When digestion is complete, cool the contents of the flask, dilute with water, and filter through a retentive paper into a beaker. Wash the paper thoroughly to make certain of a quantitative transfer. Then precipitate the sulfate with excess $BaCl_2$ under the usual conditions, from a volume of about 300 ml. at the boiling point, after neutralizing the HClO₄ and making to pH 2 to 3 with HCl. Allow the precipitate to stand for 1 day at room temperature, filter off in a Gooch crucible, ignite, and weigh.

Micro Reagents and Apparatus

Pirie's reagent. Solution A = 3 vol. of concentrated HNO_3, 1 vol. of 72% HClO₄; solution B = solution A saturated with $Cu(NO_3)_2 \cdot 3H_2O$. Mix 1 vol. of solution B with 3 vol. of solution A.

HCl, approximately 1 N and 6 N.

Benzidine reagent. Dissolve 12 g. of recrystallized benzidine (di)hydrochloride in 1 l. of water, containing 15 ml. of concentrated HCl, with the aid of heat. Filter after any dark oxidation products have sedimented out.[7]

NaOH, 0.01 N, CO_2-free.

Methanol, reagent-grade, 96%.

Indicators. Methyl red and phenol red.

A simple filtration apparatus has been described by Fiske.[8,9] Alternatively, an apparatus of the type shown in Fig. 1 is convenient. A modified type with a grid support for larger papers may be used to collect the precipitate for radioactivity determination, after which both precipitate and filter paper may be transferred to a volumetric flask and titrated as described below.

Microprocedure. Digest the sample, which should contain from 0.01 to 0.15 meq. of S, with 15 ml. of the reagent, using a 30-ml. Kjeldahl flask and starting at a low temperature as in the macroprocedure.[6] When most of the HNO₃ has been lost, add 10 ml. more of the reagent and continue the heating until all the HNO₃ has boiled off, taking in all about 2 hours.

[7] Deeply colored samples of benzidine hydrochloride may be purified as follows: Dissolve 50 g. in 1.8 l. of water containing 0.33 l. of 6 N HCl, warming as necessary. Filter, and reprecipitate the hydrochloride by adding 0.2 l. of concentrated HCl with mechanical stirring. Refrigerate for several hours, and wash the well-drained precipitate on a sintered-glass filter with 95% ethanol and ether to remove the excess acid. Store in a brown bottle when the ether has dried off.

[8] C. H. Fiske, J. Biol. Chem. 47, 59 (1921).

[9] P. B. Hawk, B. L. O. Oser, and W. H. Summerson, "Practical Physiological Chemistry," 12th ed., The Blakiston Co., Philadelphia, 1947. The filter is illustrated, and there is additional commentary on the method.

Raise the temperature of the sand bath, and digest with the $HClO_4$ at the boiling point for 10 hours or as long as necessary to get complete oxidation, finally increasing the heat to boil off the acid.[6] Remove the flasks from the rack, and *complete* the acid removal with the aid of a Meeker burner (that is, a burner with a large flame), but heating only enough to dry the copper salts to a brownish-black film on the surface of the bulb. Do not fail to evaporate the acid off the neck of the flask. Cool, add about 1 ml. of 6 N HCl, and dry after dissolving all the film. This drying is carried only sufficiently far that the salts which remain will still just redissolve in water.[10] Dissolve the contents of the flask in 10 ml. of water by adding 1 to 2 drops of methyl red indicator and sufficient 1 N HCl to make the reaction acid. Then slowly add 3 ml. of the benzidine reagent[11] and 10 ml. of methanol. Refrigerate (about 5°) for at least 2 hours. Filter off the precipitate, and wash at least four times with methanol. If the modified filtration apparatus is used, be sure the acid solution is washed out of any exposed part of the paper. Transfer the precipitate and filter plug (Fisk apparatus) or filter paper to a 125-ml. Erlenmeyer flask with the aid of about 50 ml. of water, add phenol red indicator, boil the flask contents, and titrate with the standard base to a definitely pink end point at the boiling point, being sure that all the precipitate has dissolved in the process.

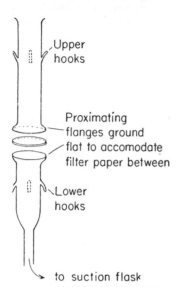

Upper hooks

Proximating flanges ground flat to accomodate filter paper between

Lower hooks

to suction flask

Fig. 1. Filter for benzidine sulfate or other small precipitates. The glass sections are held together by means of rubber bands passing between the upper and lower hooks. A convenient size of apparatus is made from 12 to 15 mm. i.d. tubing, but the same type of filter with a supporting grid and with a larger diameter is convenient for bigger samples. In this case if the precipitate is radioactive its activity may be determined first, then its titer immediately afterward.

It is advisable to check the recovery by carrying through the precipitation procedure with a known volume of a standard sulfate solution, and the blank by carrying the same standard through the whole procedure; 1 ml. of 0.01 N NaOH = 0.160 mg. of S.

[10] The removal of the excess $HClO_4$ and HCl is a critical part of the procedure. The flask obviously must not be overheated so that the contents are fused into the glass, yet the acid must be removed.

[11] The pH at this point should be between 2.0 and 2.5.

Determination of Free Sulfate (Method I)

Principle. Free sulfate may be determined either by precipitation as the barium or benzidine salts, as already indicated under the previous heading. When ethereal sulfates are present, however, as in urine, these may be hydrolyzed to a considerable extent under the conditions described for sulfate precipitation, i.e., in hot acidic solution. Consequently it is advisable to precipitate the free sulfate as the barium salt in the cold. When it is necessary to deal with very small amounts of free sulfate, the sulfate may be precipitated as benzidine sulfate and determined colorimetrically. The procedure of Letonoff and Reinhold[12] for the determination of sulfate in blood serum is described below.

Reagents

Uranium acetate, 0.4% in water.

Benzidine solution. Make a 1% solution of the free base in acetone. Allow to stand, and then filter. Store in the refrigerator, and discard when the solution becomes colored.

Standard benzidine hydrochloride solution. Dissolve 0.803 g. of purified benzidine hydrochloride in warm water, and make to 100 ml. when cool. Working standard: dilute 10 ml. to 100 ml. with water; 1 ml. of working standard = 0.01 mg. of S. Benzidine hydrochloride should be purified as indicated previously.[7]

Sodium hydroxide-sodium borate solution. Make a 1% solution of sodium borate in 0.1 N sodium hydroxide. Keep in a Pyrex, waxed, or polyethylene bottle.

Sodium-β-naphthoquinone-4-sulfonate. Dissolve 0.15 g. of the pure salt in 100 ml. of water. Maintain in the cold. Discard after 2 weeks.

Procedure. To 6 ml. of the uranium acetate reagent in a 15-ml. centrifuge tube add 2 ml. of serum. Mix by inversion, and centrifuge down the precipitated protein. Measure 4 ml. of the centrifugate into a 15-ml. centrifuge tube; add 1 ml. of acetic acid, glacial, and 9 ml. of the benzidine reagent. Allow to precipitate in the refrigerator or in an ice bath for at least 30 minutes. Centrifuge down the precipitate, and allow the fluid to drain from the inverted tube for 3 minutes. Add 14 ml. of acetone, and resuspend the precipitate. Recentrifuge and drain.

Wipe the lip of the centrifuge tube, add 1 ml. of the alkaline borate solution, and stir the precipitate until it is dissolved. Add 10 ml. of water,

---

[12] T. V. Letonoff and J. G. Reinhold, *J. Biol. Chem.* **114**, 147 (1936).

1 ml. of the naphthoquinone reagent, allow to stand for 5 minutes, and then follow with 2 ml. of acetone. Prepare simultaneously standards containing, respectively, 2 and 5 ml. of the standard benzidine hydrochloride solution, together with borate, water, color reagent, and acetone as described to make 14 ml. total volume.

Compare the colors either in a visual colorimeter or photometrically at 490 mμ. In the latter case a reagent blank should also be employed. The method described may also be used to determine sulfate in urine and other biological fluids.

Determination of Free Sulfate (Method II)[13]

Principle. The sulfate-containing solution is treated with a small excess of standard barium chloride solution, after it is freed from carbonate and other interfering ions such as those of calcium and magnesium. The solution is then buffered at pH 10 with an ammonia buffer containing the magnesium salt of ethylenediaminetetraacetate (EDTA). Subsequently the excess barium ion is titrated with a standard solution of EDTA in the form of the disodium salt. As indicator Eriochrome Black T is employed, this indicator changes from red to violet at the end point, since methyl red is also added to provide a contrast. The EDTA chelates with the excess barium or magnesium ion present, eventually removing even that which is originally chelated with the indicator, resulting in the change in color noted.

Reagents

Buffer solution. Dissolve 8.25 g. of ammonium chloride and 3.9 g. of dipotassium magnesium EDTA (or in place of the latter 3.72 g. of disodium EDTA dihydrate and 2.03 g. of magnesium chloride hexahydrate) in water along with 113 ml. of concentrated ammonium hydroxide to make 1 l.

Standard barium chloride, 0.01 M. Dissolve 2.44 g. of barium chloride dihydrate in water to make 1 l.

Standard EDTA, 0.01 M. Dissolve 3.72 g. of disodium EDTA dihydrate in water to make 1 l.

Indicator. Mix 1 vol. of 0.05% Eriochrome Black T (Superchrome Black TS) and 4.5% hydroxylamine hydrochloride in methanol with 2 vol. of 0.05% methyl red in 95% ethanol. This mixture is not too stable and should be kept under refrigeration.

---

[13] Thanks are due to Dr. E. L. Duggan for suggesting the inclusion of this method, and for his advice.

Procedure.[14-17] Acidify the unknown with hydrochloric acid, and boil to expel dissolved carbon dioxide. If calcium or magnesium is present, either remove by the use of a suitable ion exchange resin or complex with an equivalent amount of EDTA.[14] Then add a small excess of the standard barium chloride solution, accurately measured, and continue boiling for a few minutes by which time the precipitate of barium sulfate should be evident.[18] Cool the solution, and add 5 ml. of the buffer[16] together with water to make a convenient volume (25 ml.). The solution should then be at pH 10. Add 4 drops of the mixed indicator,[15] and titrate the red solution to a pure violet color with the standard EDTA solution (*T* ml. of EDTA).

Titrate a corresponding blank containing the barium ion and buffer only without the unknown (*B* ml. of EDTA), and a buffer blank without the barium ion present (*C* ml. of EDTA). The difference in titer of $B - C$ should demonstrate the equivalence of the EDTA and the barium reagents. The difference in titer of $B - T$ represents sulfate in terms of the standard EDTA.

In case of lack of correspondence between the standards, it is advisable to check the method against a standard sodium sulfate solution prepared fresh from the anhydrous salt.

Phosphate interferes with this method as with the benzidine method.

Determination of Ethereal Sulfate

Principle. Ethereal sulfate is determined as the difference between sulfate precipitated under conditions which do not lead to hydrolysis of ethereal sulfate (see under previous headings) and of total sulfate determined after the hydrolysis of the ethereal sulfates.

Procedure. Hydrolyze the ethereal sulfates in the sample by making to 0.5 N with hydrochloric acid and heating for 15 minutes at the boiling point. Proceed with sulfate determination as described in the micro-procedure by precipitating with benzidine.

[14] J. R. Munger, R. W. Nippler, and R. S. Ingols, *Anal. Chem.* **22**, 1455 (1950).
[15] B. Rehell, *Scand. J. Clin. & Lab. Invest.* **6**, 335 (1954).
[16] R. Sijderius, *Anal. Chem. Acta* **10**, 517 (1954); **11**, 28 (1954) [*C.A.* **49**, 6769, 6780 (1955)].
[17] R. D. Bond, *Chemistry & Industry* **1955**, 941.
[18] If low results are obtained, the boiling may have been insufficient.

[147] Determination of Phosphorus

(This subject, originally intended for inclusion under this number, has been incorporated into article [115] of this volume.)

[148] Determination of Metals
(Na, K, Mg, Ca, Mn, Fe, Co, Cu, Zn)*

By ROBERT BALLENTINE and DOROTHY D. BURFORD

The varieties of conditions that are encountered in enzymatic experiments place a considerable stringency on the selection of analytical methods for the metal ions. Such methods must be of sufficiently general application that they will handle not only the analyses of purified protein solutions but of tissues as well. Those that are presented here have been selected with these requirements in mind, and consequently classical and standard methods have more often been chosen since their wide applicability to a variety of tissues and experimental circumstances has been demonstrated. It has also been presumed that occasional rather than routine use will be made of the methods presented. This supposition eliminates immediately those methods which involve relatively complex instrumental techniques—such as polarography, X-ray fluorescence, and emission spectroscopy—since the equipment for them is rarely found in the average enzymatic laboratory. Indeed, in the same line of restriction, methods requiring unusual reagents have been rejected on this basis alone, rather than because of any specific shortcomings of the methods themselves. Finally, those methods which involve a large number of precise and carefully controlled manipulations have been set aside in favor of more simple ones, even when the latter may not display as high a degree of accuracy and sensitivity. In view of these restrictive criteria, it would be advisable that anyone engaging in a large-scale study of metal ions review the more recent literature with the idea of adapting some of the more facile methods of analytical instrumentation.

The attempt has been made to present three basic methods of analysis for each element. First, a rapid, sensitive colorimetric procedure is given that can be readily utilized for routine analyses of medium precision. A volumetric method is included for requirements of precision of an order of magnitude greater than that available by colorimetry, and gravimetric methods are presented because they are applicable to radioactive tracers. In a number of cases the selection of three such methods is not necessary, since one may have little advantage over another. For example, the volumetric procedure for manganese offers little increase in precision and

* The preparation of this manuscript was supported in part by the AEC contract AT(30-1)-933. The authors wish to thank Dr. Joseph R. Riden, Jr., for helpful discussions in the preparation of this manuscript.

requires considerably greater time than the colorimetric procedure. Both are based on the same empirical relationship.

One special instrumental method of analysis has been included as a separate section in view of its wide usage and applicability to elements that are otherwise difficult to determine, namely, the flame photometry of the alkali metals and alkaline earths. Flame attachments for spectrophotometers have become so widespread that it is rare to find a laboratory which does not somewhere have a flame attachment available. Certainly, when the equipment is available, this becomes the method of choice for the alkali metals and alkaline earths.

Preparation of Sample

Wet Ashing. Place an aliquot of enzyme solution or sample of tissue in a 100 ml. Pyrex Kjeldahl flask. If alcohol or any other organic solvent has been used in the preliminary preparation, place the flask overnight in an oven at 110° to remove the solvent *completely*. Then, add 30 ml. of concentrated HNO_3, followed by 5 ml. of 1:1 $HClO_4$ (72% $HClO_4$ diluted with an equal volume of water). Add a glass bead, and boil the sample until the liquid remaining in the flask gives off dense white fumes of $HClO_4$. The digest should be at most pale yellow in color. If the color of the digest is brown, add more of the HNO_3 and $HClO_4$ and repeat the oxidation. This amount of acid is sufficient for 1 g. of dry tissue.

Dry Oxidation. For the analysis of some elements it is desirable to avoid the large amount of free acid which is present when wet oxidation methods are used. Place the sample of tissue or enzyme solution in a platinum crucible (Vycor glass or fused silica crucibles may be used but are far less satisfactory, particularly when analyzing for the alkali metals, alkaline earths, iron, or cobalt, since these elements all tend to form fusion products with the silica, thus leading to low analytical recoveries). To the sample in the crucible add 2 ml. of concentrated HCl, and place the crucible in an oven at 110 to 120° overnight. If this is not done, excessive frothing will occur when the sample is first heated in the muffle furnace. Introduce the sample into a cold muffle furnace, and over a period of 1 to 2 hours bring the temperature up to 550 to 600°. Watch the sample during this initial heating period so that material is not lost owing to excessive foaming. It may be necessary to remove the crucibles from the furnace periodically to check frothing. If the muffle is already heated, introduce the samples into the heated furnace by placing them on the opened door and slowly moving them into the hottest region. This is a risky procedure, however, which may result in loss of sample. Hold the samples at 550 to 600° until at the most only a few specks of carbon remain. Depending on the amount of organic material and its nature, this may require heating periods of from 1 to 24 hours. At the end of the ignition time, it is advisable to cool the crucibles in a desiccator over silica gel. The treatment of the ash will be described under the various procedures in which dry ashing is the method of choice.

Trichloroacetic Acid Filtrates. When there is occasion to analyze a protein solution containing little material, the use of TCA may be permissible. However, the application of TCA precipitation should always be made with considerable caution. Many organic ions, such as oxalate and citrate, remain in the solution to inhibit the precipitation and the colorimetric reactions for some of the elements. Whenever there is any question, a complete destruction of the organic material is advisable. For other elements, such as copper, zinc, and manganese, one must not overlook the possibility that ionic binding of the trace element to the protein may occur, and hence considerable loss of the element will result when the protein is precipitated with TCA. For those cases in which this method is applicable, bring the aliquot for analysis to 10% (w/v) TCA by adding an aqueous 25% (w/v) solution of TCA. After letting it stand for 20 minutes, separate the material by centrifuging at 3600 r.p.m. Remove an aliquot to a small beaker with a volumetric pipet, add enough water to dilute the TCA concentration to 5%, and place the beaker in an oven at 95 to 100° overnight. This procedure results in the destruction of most of the TCA and obviates the necessity in most of the analytical procedures for a large volume of neutralizing base. The residue is usually taken up by treating with 1 ml. of N HCl and washing the beaker out with five 1-ml. portions of distilled water. The solution is finally brought to a known volume in a volumetric flask prior to analysis.

I. Sodium

Since the introduction of the complex sodium uranyl salts, sodium-zinc-uranyl acetate and sodium-magnesium-uranyl acetate, methods based on the separation of the sodium ion as the triple salt have been generally adopted. They are general and sensitive and show little interference except by those substances which precipitate uranium (phosphate, oxalate, and citrates). In biological systems this requirement for the absence of phosphates introduces some complexities and disadvantages. The sensitivity of all the methods, whether gravimetric or spectrophotometric, is limited by the *solubility* of the sodium triple salt, since essentially a gravimetric separation precedes all the various methods of determination of sodium.

Spectrophotometric Method (Range, 0.2 to 50 Mg.)

Principle. The method to be described is based on the removal of interfering phosphates as magnesium ammonium phosphate followed by precipitation of sodium as the triple salt with magnesium-uranyl acetate.[1] The final colorimetric determination is then carried out by utilizing the chromogenicity of the uranyl ion in the presence of hydrogen peroxide. This colorimetric procedure is that of Stone and Goldzieher.[2]

Reagents

Magnesium-uranyl acetate reagent. Add distilled water to 60 ml. of glacial acetic acid and 90 g. of uranium acetate to make 1 l.; warm to 70°, and stir until solution is complete. In a separate flask place 600 g. of Mg acetate and 60 ml. of glacial acetic acid with sufficient water to make 1 l. Warm this solution to 70°, and also stir until solution is complete. Mix these two solutions together, and cool to 20°. Allow to stand for at least 2 hours, and filter through a clean sintered-glass funnel. Preserve the solution in an amber bottle away from the action of direct sunlight. It is usually advisable to refilter the reagent just prior to use.

Wash liquids. Prepare a small amount of the triple salt sodium-magnesium-uranyl acetate by adding 100 ml. of the precipitating reagent to about 25 mg. of NaCl dissolved in a few milliliters of water. After it has stood for several hours, filter off the precipitate which forms, and wash on the filter, first with water and then with 95% alcohol. Dry in an oven at 90 to 100° for an hour. Prepare the wash liquid by saturating 95% ethyl alcohol at room temperature with a small portion of the dry triple salt.

$MgCl_2$, 4%.

$(NH_4)_2CO_3$, saturated aqueous solution.

H_2O_2, 30%.

Procedure. The sample for sodium analysis should be subjected to some treatment resulting in the destruction of organic material, and for this purpose dry ashing is usually preferable, although wet ashing with $HClO_4$-HNO_3 may also be used. The use of TCA filtrates must be considered with great care because of the possible interference due to the precipitation of uranium by phosphates, silicates, oxalates, and citrates. Make the sample to approximately 10 ml. in volume, and to the solution add 1 ml. of the $MgCl_2$ solution. Then bring the solution to approximate neutrality with concentrated NH_4OH, and add 1 ml. in excess. Shake the tubes vigorously, allow to stand at room temperature for 2 hours, then centrifuge for 10 minutes at 3500 r.p.m. Remove an aliquot of the supernatant to a 100-ml. beaker for analysis. Evaporate it on a steam bath to 5 ml. If the salts start to separate out, add sufficient water to keep them

[1] E. R. Caley and C. W. Foulk, *J. Am. Chem. Soc.* **51**, 1664 (1929).
[2] G. C. H. Stone and J. W. Goldzieher, *J. Biol. Chem.* **181**, 511 (1949).

in solution. Add 100 ml. of the magnesium-uranyl acetate reagent, and hold the mixture at room temperature for 45 minutes. Filter the material though a *fine* sintered-glass funnel, using the mother liquor to transfer quantitatively the precipitate to the funnel. Wash the precipitate with 5-ml. portions of the wash liquid until the wash runs through colorless (about three to five washes are usually necessary). Dissolve the precipitate from the funnel in 1 ml. of boiling water, followed by a 6-ml. wash with saturated $(NH_4)_2CO_3$ solution. Just prior to reading the color intensity in the spectrophotometer, add 1 ml. of 30% H_2O_2, and make the volume to 15 ml. with distilled water. Read in 1-cm. cuvette in the Beckman spectrophotometer at 480 mμ.[3]

Calibration curves should be prepared by suitable dilutions of a solution of the triple salt. Weighed samples may be dissolved and used for this purpose. It is essential that this calibration curve be run in considerable detail in view of the fact that Beer's law is not strictly followed at this wavelength.

Gravimetric Method

It is obvious from the above procedure that after the triple salt has been filtered and washed it may be utilized for a gravimetric determination. The use of radioactive sodium as a tracer is becoming more widespread, and the separation as the sodium-magnesium-uranyl acetate seems to be the method of choice for sample preparation.

II. Potassium

When the instrument is available, the flame photometer method for potassium is easily the favored method. All the other methods in the literature depend on an initial separation of potassium as an insoluble complex salt, the classical one being precipitation as the chloroplatinate complex. More recently, gravimetric, volumetric, and colorimetric analyses based on precipitation of potassium as either the sodium or silver cobalti-nitrite have found wide application. These methods, unfortunately, like the method utilizing platinum, all suffer from interference by ammonium ion. Inasmuch as this is a common constituent of any wet-ashed preparation, one is more or less limited either to special methods for the removal of ammonium salt (such as the controlled ex-

---

[3] It is advisable when handling more than a couple of analyses to add the peroxide and dilute to volume only in three tubes at a time in order to avoid the formation of gas bubbles which interfere in the spectrophotometry. Greater sensitivity from a spectrophotometric point of view can be obtained by reading the optical density at its point of maximum absorption, 415 mμ. No real advantage is thereby achieved, however, since the sensitivity of the method is limited by the solubility of the triple salt and the requirements for its quantitative separation.

plosion of ammonium perchlorate) or to the use of dry-ashing techniques. In the latter, losses of potassium occur both through the volatilization of the chloride and the interaction of the melt with the crucible if it is made of either silica or glass. Almost all the salts used to precipitate potassium ion show moderate solubilities of the potassium salts, and this limits the sensitivity of the methods. This limitation applies equally to indirect determinations such as the measurement of excess reagent remaining after the precipitation of potassium.

Volumetric Method (Range, 0.08 to 1.0 Mg.)

Principle. The method developed by Van Slyke and Rieben[4] utilizes the precipitation of potassium as the phospho-12-tungstate followed by titration of this material with dilute base. One advantage is the large equivalent factor of 10.3 base equivalents per potassium ion in the precipitate. The method has been modified by Folch and Lauren[5] to avoid interferences by phosphates, iron, and ammonia. The method also has the advantage of using a wet-ashing procedure, thus avoiding the various losses of potassium discussed above.

Reagents

Superoxol, 30% H_2O_2.

$Ca(OH)_2$. Wash Reagent $Ca(OH)_2$ on a filter six times with distilled water, and then dry in an oven. This procedure removes small amounts of potassium.

Standard 0.04 N NaOH, CO_2-free. Remove the CO_2 from water by adding 0.25 ml. of 0.1 N HCl to 400 ml. of water. Bring it to a boil, and allow it to cool under protection of soda-lime. To the CO_2-free water add 0.9 ml. of saturated NaOH. Mix the solution, and titrate against 0.005 N potassium biiodate (see volumetric procedure for copper), using phenolphthalein as the indicator.

Phospho-12-tungstic acid. Dissolve commercial phosphotungstic acid in an equal weight of water in a separatory funnel. Add a volume of ether equal to the volume of solution in the funnel, and shake. The heavy oil consisting of a mixture of the phospho-12-tungstic acid and ether settles to the bottom. Draw this oil off into a clean separatory funnel. Wash it three times with a volume of water equal to the volume of oil. Then dry the washed solution in a vacuum desiccator at room temperature protected from light. Store in a brown bottle. The yield runs from 30 to 80% of the

[4] D. D. Van Slyke and W. K. Rieben, *J. Biol. Chem.* **156,** 743 (1944).
[5] J. Folch and M. Lauren, *J. Biol. Chem.* **169,** 539 (1947).

initial phosphotungstic acid. The phospho-12-tungstic acid used
should be white and 1 g. may be dissolved in 5 ml. of distilled
water without heating with the production of no turbidity.[6]

Phospho-12-tungstic acid solution, 4% (w/v). Keep this solution
in a brown bottle in a refrigerator, and just prior to use centrifuge
to remove any slight turbidity which may have formed. The
solution is good for several months.

Thymol blue-phenolphthalein indicator (TBP). Dissolve 50 mg. of
thymol blue in 21.5 ml. of 0.01 N NaOH, and dilute to 50 ml.
Mix this solution with a solution of 50 mg. of phenolphthalin in
absolute alcohol.

Procedure. DIGESTION. Place the tissue or aliquot of solution to be
analyzed for potassium in a 10- or 25-ml. Kjeldahl flask. The digestion
proceeds more smoothly if the water is removed by drying overnight in
an oven at 100 to 110°. To the dried material add 0.1 ml. of concentrated
H_2SO_4, and heat in a bath so that the temperature reaches 150° about
10 minutes after starting at room temperature.[7] To promote smooth
boiling during this and subsequent procedures, it is desirable to add either
a glass bead or an Alundum chip. After the sample has blackened and
liquefied under the action of the hot H_2SO_4, add 4 drops of Superoxol, one
at a time, waiting until the blackening reappears after each addition.
Continue the Superoxol additions, 3 to 4 drops at a time, until the solution appears to blacken at a very much slower rate. About ten additions
are required. At the end of this time put the tubes on a digestion rack over
small microflames and continue the heating until dense white fumes
appear. Usually deep blackening also occurs at this time. Add Superoxol
again, dropwise, until heating for 5 minutes causes no reappearance of
the blackening. The whole procedure requires about 45 minutes.

ELIMINATION OF PHOSPHATE. To the sample in the Kjeldahl flask
add 5 ml. of water and 200 mg. of $Ca(OH)_2$. With a small stirring rod
carefully stir continuously for 5 minutes. Transfer the suspension to a
graduated centrifuge tube, make the volume to 5 ml., and centrifuge at
2000 r.p.m. for 15 minutes. Transfer 3 ml. of the above solution to a
Pyrex evaporating dish of 30-ml. capacity placed on a vigorously boiling
water bath. Allow the solution to evaporate completely to dryness, thus
eliminating any ammonia present. Remove the dish from the bath, let it

---

[6] The purified material may be purchased from Anachemia Co., 70 E. 45th St.,
New York, New York.

[7] Although a sulfuric acid bath may be used for this purpose, a safer and more desirable bath can be made from Dow Corning 550 Silicone Fluid.

cool, and add 1 ml. of 1 N HCl. Complete solution of the residue is essential and may require slight warming to bring this about. Next add a 4% solution of phospho-12-tungstic acid dropwise, 0.5 ml. being added for amounts of potassium between 80 and 800 γ. After the solutions are mixed, again evaporate to dryness on the steam bath until no detectible odor of HCl is present.

After the residue has been dried, suspend it in 2 ml. of water, and pour the suspension into a 15-ml. centrifuge tube. Transfer as much of the precipitate as possible by washing the remainder into the tube with 4 ml. of water delivered from a pipet. If some of the precipitate clings to the glass surface, transfer it with the aid of three successive 2-ml. portions of water, freeing the precipitate from the glass by means of a rubber policeman. A total of 12 ml. of water is used for the transfer. Centrifuge at 3000 r.p.m. for 15 minutes. Remove the supernatant solution with the same type of capillary tip as is used in the analysis for calcium ion, taking care not to allow the surface layer which usually contains some precipitate to come down to the level of the capillary tube. This usually leaves a few tenths of a milliliter of solution in the tube. Wash down any precipitate adhering to the capillary tube or to the walls of the centrifuge tube, and thoroughly stir up all the precipitate. Bring the total volume to 7 ml. with water. Centrifuge off the precipitate as above, and remove the supernatant. Repeat this procedure twice. Care must be taken to duplicate the washing procedure exactly, the amount of wash water used being measured accurately.

TITRATION OF POTASSIUM PHOSPHOTUNGSTATE. Add a piece of Alundum to the washed precipitate, and run in 1 ml. of 0.04 N NaOH from the alkali burette. Heat the contents of the tube by holding over a micro-burner until boiling begins. Add more NaOH to the hot solution until the precipitate is nearly dissolved, and then add 1 drop of the TBP indicator, followed by more alkali until the blue-violet color becomes permanent. Add 0.3 to 0.5 ml. more of the alkali, and boil for at least 30 seconds. (If the color changes, more alkali should be added, the 30-second boiling being repeated.) From the acid burette add enough 0.04 N H$_2$SO$_4$ to turn the indicator yellow. Then add a few more drops of the acid, and bring the solution to boiling for 30 seconds to expel CO$_2$. Finally, without waiting for the solution to cool, titrate with 0.04 N NaOH to change the color back to the violet end point, stirring the solution with a stream of CO$_2$-free air or nitrogen. The titration should be continued until the distinct violet end point appears and stays constant for 1 minute. The end point is sharp to about 0.005 ml. of the 0.04 N alkali. The reading on the alkali burette minus the reading on the acid burette, corrected for normality to 0.04 N, gives the volume of 0.04 N NaOH required to titrate the precipitate. A blank analysis should be run in which water replaces the sample.

CALCULATIONS. Milligrams of potassium in the sample = 1.677 [0.156 $(T - B)$ + 0.014], where T stands for milliliters of 0.04 N NaOH used minus milliliters of 0.04 N H_2SO_4 used, and B is this value for the blank.

Spectrophotometric Method

Principle. The above precipitate of potassium-12-phosphotungstate may be determined spectrophotometrically. The precipitate is treated precisely as the magnesium ammonium phosphate is treated in the analysis for magnesium.

Procedure. Dissolve the precipitate as above by heating with a slight excess of 0.04 N NaOH. Then add 1 ml. of acid vanadate solution (see under magnesium), followed by 7 ml. of water and 1 ml. of NH_4 molybdate. After allowing color to develop for 5 minutes, read it in the spectrophotometer at 420 mμ. Construct a standard curve, using the standard phosphate solution employed for the magnesium determination; 100 γ of phosphate is equivalent to 104.6 γ of potassium.

Gravimetric Method

There is little requirement for a gravimetric potassium method in connection with the use of radioactive tracers. Natural potassium, of course, is radioactive. Although this has been utilized for the determination of the potassium content of rocks and soil samples, the low level of natural potassium activity is such that it is not a particularly attractive method. The artificially produced isotopes of potassium are all of too short a half-life to be useful as tracers. Should a gravimetric method be required, the precipitate of potassium phospho-12-tungstate may be filtered on a fine-porosity glass filter and washed as described under the volumetric method. The papers of Rieben and Van Slyke[8] and Folch and Lauren[5] should be consulted, since the composition of the potassium phospho-12-tungstate differs, depending on the conditions of precipitation.

III. Magnesium

Spectrophotometric Method (Range 0.01 to 0.1 Mg.)

Principle. Three general procedures for the determination of magnesium in biological samples have been prevalent in the literature: (1)

[8] W. K. Rieben and D. D. Van Slyke, *J. Biol. Chem.* **156**, 765 (1944).

the precipitation as the ammonium magnesium phosphate complex; (2) the precipitation and colorimetric determination as the 8-hydroxyquinolate; and (3) the formation of magnesium lakes with various dyestuffs. The last procedure is primarily recommended on the basis of its sensitivity. It requires a high degree of control of the experimental conditions to obtain consistent analyses, however, and at best is rather low in precision in comparison to the other methods. Since temperature, pH, concentration of reagents, and concentrations of interfering elements all enter into the actual color value, these factors, as well as the time of addition of reagents and spectrophotometric determinations, must be carefully regulated. The use of 8-hydroxyquinoline is of less manipulative difficulty than the magnesium lake method but suffers from gross interferences by virtually any other ion present. For this reason, the method to be used here is based on the precipitation as the ammonium magnesium phosphate, with the subsequent determination of the phosphate concentration, as described by Simonsen et al.[9]

Reagents

Reagents for calcium removal. See section on calcium.

KH_2PO_4, 2% (w/v) aqueous.

Dilute HNO_3. Dilute 1 vol. of concentrated HNO_3 with 2 vol. of glass-distilled water.

Acid NH_4 vanadate. Dissolve 1.25 g. of NH_4 vanadate in 2200 ml. of the above dilute HNO_3. Heat until solution is complete, and after cooling to room temperature dilute to 500 ml. with the same HNO_3 concentration.

NH_4 molybdate. Make a 5% solution, being careful not to heat above 50°.

Mg wash reagent. Add 50 ml. of NH_4OH to 200 ml. of redistilled 95% alcohol, and dilute to 1000 ml. with water.

PO_4 standard. Dissolve 0.560 g. of KH_2PO_4, dried to constant weight, in 1 l. of water in a volumetric flask. One milliliter of this standard is equivalent to 100 γ of magnesium. Suitable dilute standards may be prepared by diluting this solution to give 10 and 80 γ of magnesium equivalents per milliliter.

Procedure. Since calcium ions interfere in this procedure, calcium must be removed prior to the determination of magnesium. To accomplish this the ash prepared by dry ashing (or a TCA filtrate) is treated exactly as for the determination of calcium by precipitation with oxalate. (See section on calcium, volumetric method.) The only change to be made in

[9] D. G. Simonsen, L. M. Westover, and M. Wertman, *J. Biol. Chem.* **169,** 39 (1947)

this procedure is to substitute ammonium washes for the WAE washes. Place the pooled supernatant and washes from the calcium precipitation in a centrifuge tube, add 1 ml. of the KH_2PO_4 solution, and mix. Then add 1 ml. of concentrated NH_4OH, and shake the tube vigorously for 30 seconds. Allow it to stand at room temperature for 2 hours. Centrifuge for 10 minutes at 3500 r.p.m., and carefully decant the supernatant; (follow the procedure for calcium analysis). Wash the precipitate with 5 ml. of the magnesium wash, running the wash solution down the wall of the tube, and recollect the precipitate by centrifuging. Repeat this washing procedure, and dry at 100° for 1 hour to remove any alcohol. Determine the phosphate in the precipitate by adding 1 ml. of the acid vanadate reagent and agitating until the precipitate is completely dissolved. Then add 8 ml. of water, followed by 1 ml. of 5% NH_4 molybdate, and, after allowing to stand for 5 minutes, determine the absorption in a 1-cm. cuvette at 420 mμ. As a blank, carry through the precipitation and color development procedure on a sample of water. The standard curve can be constructed by preparing suitable dilutions of the PO_4 standard solution. To suitable aliquots of these diluted PO_4 standard solutions, add 1 ml. of acid vanadate, followed by 1 ml. of NH_4 molybdate, and dilute to a final volume of 10 ml. with water.

Volumetric and Gravimetric Methods

No adequate volumetric analysis for magnesium is presented, inasmuch as all such procedures for magnesium ion require a prior quasigravimetric separation. Since the radioactive isotopes of magnesium have too short a half-life for tracer applications, probably little need for a gravimetric magnesium method will be encountered. Should such a method be desirable, the ammonium-magnesium phosphate complex may be transferred quantitatively onto a tared, fine-porosity sintered *porcelain* crucible, washed with the magnesium wash mixture, and the precipitate ignited in the muffle furnace at about 400° for 1 hour to convert the ammonium magnesium phosphate to magnesium pyrophosphate. It is weighed as such. This ignition is necessary, owing to the uncertainty of the degree of hydration of the magnesium ammonium phosphate.

IV. Calcium

Volumetric Method (Range, 0.2 to 2.0 Mg.)

Principle. The described volumetric procedure, according to Sendroy,[10] is based on the initial precipitation of calcium by oxalate from a solution

[10] J. Sendroy, *J. Biol. Chem.* **152**, 539 (1944).

of the ashed sample. The oxalate is determined by titration with perchlorato-cerate according to Smith and Getz.[11] Magnesium interferes only when present in large amounts. The reaction is stoichiometric:

$$(COOH)_2 + 2Ce^{++++} \rightarrow 2CO_2 + H_2O + Ce^{+++}$$
1 ml. 0.01 M cerate = 0.2003 mg. calcium

Reagents

HCl, 2 M.

Na acetate, 20% (w/v), made fresh.

NH$_4$ oxalate. Saturate 100 ml. of water at room temperature with reagent NH$_4$ oxalate; about 3.5 g. is required.

NH$_4$OH. Dilute 2 ml. of 28% reagent to 100 ml.

Water-alcohol-ether mixture (WEA). Mix equal volumes of water, absolute alcohol, and ethyl ether, all of which are freshly glass-distilled.

Bromcresol green indicator (BCG). 0.04% solution made by mixing in a mortar 0.1 g. of dry indicator with 14.3 ml. of 0.01 M NaOH and diluting to 250 ml. with water.

HClO$_4$, 2 M, 133 g. 72% acid diluted to 500 ml.

Ferroin indicator, 0.025 M.[12]

Perchlorato-cerate, 0.0100 M. Standardize 0.1 M perchlorato-cerate[12] by titrating an accurately weighed sample of approximately 100 mg. of sodium oxalate dissolved in 25 ml. of 2 M perchloric acid. Dilute with 2 M perchloric acid to give a 0.0100 M solution.

Procedure. PRECIPITATION. Take up the sample, dry-ashed or as TCA filtrate, in 2 ml. of 2 M HCl, and transfer quantitatively to a 15-ml. centrifuge tube with the aid of three 1-ml. water washes. In order, add 1 ml. of 20% Na acetate, 1 ml. of saturated NH$_4$ oxalate, 1 drop of BCG, and concentrated NH$_4$OH to a pH ca. 4.5. Allow to stand at room temperature overnight (10 to 16 hours). Centrifuge for 5 minutes at 2600 r.p.m.; then, using a tube drawn out to a capillary tip with the end bent into an upturned J, slowly draw off the supernatant until only 0.2 ml. remains. Wash down the walls with 3 ml. of 2% NH$_4$ solution, centrifuge, and remove the supernatant as above. Repeat twice with 3 ml. of WAE mixture, allowing the first milliliter to stir up the precipitate *gently.* Dry the tube in a 100 to 110° oven for 1 hour.

[11] G. F. Smith and C. A. Getz, *Ind. Eng. Chem., Anal. Ed.* 10, 304 (1938).

[12] Ferroin indicator (1,10[*ortho*]-phenanthroline ferrous sulfate) and perchlorato-cerate and *o*-phenanthroline are available from G. Frederick Smith Chemical Co., 867 McKinley Avenue, P. O. Box 1611, Columbus, Ohio.

Procedure. TITRATION. Dissolve the precipitate in 3 ml. of 2 M HClO₄, add 1 drop of Ferroin, and titrate with 0.01 N cerate solution, allowing 15 seconds for the first drop to react.

Spectrophotometric Method (Range, 0.002 to 0.1 Mg.)

Since colorimetric methods in general depend on initial separation of calcium as an insoluble salt (usually oxalate, picrolonate, or phosphate) little added sensitivity is gained over the volumetric procedure, and less stable solutions and more empirical calibration curves are required. Sendroy,[13] by scaling down the above precipitation procedure to a volume of 0.3 ml., has been able to determine levels of 2 γ of calcium. This small-scale procedure is not feasible with ashed samples. The determination of oxalate in such procedures by the reaction with 2,7-dihydroxynaphthalene is preferable[14] to the ceric method of Sendroy.[13]

Gravimetric Method (Range, 0.2 to 2.0 Mg.)

The isotope Ca⁴⁵ is useful for tracer applications. The precipitation method given above may be applied by filtering the precipitate on a *fine* sintered-glass filter crucible and washing as directed.

V. Manganese

The literature contains a wide variety of papers on the determination of manganese both by spectrophotometric and volumetric methods. Almost universally, however, these methods are based on the oxidation of manganese to permanganate by some oxidizing agent, the one most commonly used being periodate. In the spectrophotometric methods, the actual determination of manganese may be carried out by directly estimating the permanganate color. The sensitivity of the method may be increased by including some organic substance (such as 4,4'-tetramethyldiaminodiphenylmethane), which will react with the permanganate. This introduces numerous factors which influence the reaction, however, such as the concentrations of acids at which the reaction occurs, the strength of the periodate, the temperature, the time of the reaction, and the relation between the concentrations of the color-forming reagent and permanganate. Therefore, in the method to be given, the direct determination of the permanganate color is preferred. The method still remains a strictly empirical calibration method, however. If added sensitivity is needed, the paper of Gates and Ellis[15] may be consulted.

[13] J. Sendroy, *J. Biol. Chem.* **144**, 243 (1942).
[14] V. P. Calkins, *Ind. Eng. Chem., Anal. Ed.* **15**, 762 (1943).
[15] E. M. Gates and G. H. Ellis, *J. Biol. Chem.* **168**, 537 (1947).

Spectrophotometric Method (Range, 0.025 to 0.25 Mg.)

Principle. The spectrophotometric method presented here is based on the contribution of Willard and Greathouse.[16] This method is reasonably independent of interferences by other elements. Substances which are either strong oxidizing or reducing agents will interfere, however, and hence it is mandatory to ash the sample to remove organic materials and oxidizable inorganic substances such as ferrous iron, sulfite, and nitrite. This is most readily done by using the $HClO_4$-HNO_3 wet-ashing procedure. Chloride should not be present in any excessive amount; fortunately, it is removed by the wet-ashing procedure. The only major interferences are from chromium and cerium, two substances not normally encountered in biological materials.

Reagents

H_2SO_4, 2 N.

K or Na periodate.

Standard Mn solution. Method A: Weigh out accurately 500 mg. of electrolytic Mn metal. Dissolve in dilute HNO_3, and boil until all the oxides of nitrogen are expelled. Dilute to 500 ml. with redistilled water. Finally, dilute this stock solution to give a working standard containing 0.1 mg. of manganese per milliliter. Method B: Redistill water from an all-glass still, adding a little $KMnO_4$ to be certain that all organic reducing agents are removed. Prepare an approximately 0.1 N solution of $KMnO_4$ by dissolving 3.2 g. of $KMnO_4$ in 1 l. of redistilled water. Determine precisely the permanganate concentration by any of the standard methods such as by titration of Na oxalate. The authors prefer the use of KI as a primary standard, the reacting weight being 82.99. Weigh out accurately 80 mg. of KI, dissolve it in 40 ml. of N HCl, and add 6 to 10 ml. of 0.5 M KCN. *Hood!* Titrate with the permanganate solution until a very faint pink color persists for 20 seconds. The end point may be sharpened by adding a starch solution and titrating until colorless. One milliliter of precisely 0.100 N $KMnO_4$ contains 1.099 mg. of manganese per milliliter. Take enough of the standardized $KMnO_4$ solution to contain precisely 10 mg. of manganese. To this solution add a few milliliters of 5 M H_2SO_4 and several drops of saturated $NaHSO_3$. Then boil to remove the SO_2 and dilute the cooled solution to precisely 100 ml. in a volumetric flask.

[16] H. H. Willard and L. H. Greathouse, *J. Am. Chem. Soc.* **39**, 2366 (1917).

Procedure. Ash a sample containing 0.025 to $\pm$0.25 mg. of manganese by the $HClO_4$-HNO_3 method. Dilute the $HClO_4$ digest to 10 ml. with 2 N H_2SO_4. Then add 1 ml. of 85% H_3PO_4 and 50 to 75 mg. of K metaperiodate. Heat the solution to boiling, and keep the temperature just below the boiling point for 5 to 10 minutes. Cool, and make the volume to 25 ml. in a volumetric flask. Determine the optical density in the Beckman spectrophotometer at 525 mμ. The amount of manganese should be obtainable by reading from a calibration curve prepared simultaneously from the standard Mn solution described above. This is a rather empirical method, and it is necessary, for the best in accuracy, to run a minimum of five calibration points simultaneously with each set of determinations.

Volumetric Method

Almost all the volumetric methods are based on the oxidation of manganese to permanganate, destruction of the excess of oxidizing agent, and the determination of the permanganate present by one of the standard methods. The precision of all of these methods and their accuracy is determined (1) by the empirical nature of the oxidation of manganese to the permanganic state, a degree of oxidation which is not precisely stoichiometric owing to various side reactions, and (2) by the necessity for the precise removal of the oxidizing agent without affecting the amount of permanganate present. Accuracy in the volumetric procedures may be obtained only by carrying out careful calibration experiments with known amounts of manganese to obtain an empirical correction factor for over- and underoxidation as well as the destruction of the permanganate. It seems that these methods offer little advantage over the direct determination of the amount of permanganate present by the spectrophotometric method. Therefore, no volumetric methods have been included.

Gravimetric Methods

Procedure. Dilute the $HClO_4$-HNO_3 digest to approximately 10 ml. Add to the cold solution 1 ml. of a 10% solution of $KBrO_4$. This will cause precipitation of the manganous ion present in the digest as manganese dioxide. This may be filtered off on a sintered-glass disk and washed with a small amount of water. The radioisotopes of manganese are all of relatively short half-life (a matter of a few hours), and probably little application of the gravimetric method will be made. This method is reasonably sensitive, however, and can be quickly carried out on small samples. No figures are readily available for calculating the sensitivity of the method.

VI. Iron

Spectrophotometric Method (Range, 0.01 to 0.2 Mg.)

Principle. Owing to its very wide distribution and importance both in biological and nonbiological systems, there are more methods for the analysis of iron than virtually any other element. The method chosen for the spectrophotometric determination of iron is the *o*-phenanthroline method, selected primarily because of its sensitivity and relative freedom from interferences by other elements, a difficulty found in many other colorimetric iron analyses. It has a rather wide range of conditions under which the amount of color is proportional to the iron concentration, and the color formed is relatively stable, following Beer's law closely. The method is based on the original contributions of Fortune and Mellon.[17] The major interferences that occur are elements unlikely to be found in biological systems, such as cadmium, silver, and bismuth. It is, however, necessary to have not more than 10 p.p.m. of copper present; nickel and cobalt, which also interfere, should be present in less than 10 p.p.m. Phosphate, fluoride, pyrophosphate, chloride, sulfate, oxalate, citrate, and tartrate do not interfere unless present in extremely high amounts. This is a great advantage, as many of the other methods for colorimetric determination of iron are affected by the presence of those ions which complex iron. A final advantage is the use of the method in acid where precipitation of hydroxide does not occur. The method described here is basically that presented by Sandell.[18]

Reagents

o-Phenanthroline, 0.5% solution of the monohydrate in water. It is necessary to warm the solution to completely dissolve the reagent.[12]

Na acetate, 2 *M*.

Na acetate, 0.2 *M*.

Hydroxylamine hydrochloride, 10% aqueous solution. Make fresh each week.

Standard Fe solution. Dissolve 0.1000 g. of electrolytic Fe or Fe wire in 50 ml. of 1:3 HNO_3. Boil to expel the oxides of nitrogen, and dilute to 1 l. with Fe-free water. This solution contains 100 γ Fe per milliliter. Suitable working solutions may be prepared by dilution of this stock standard solution.

[17] W. B. Fortune and M. G. Mellon, *Ind. Eng. Chem., Anal. Ed.* **10,** 60 (1938).

[18] E. B. Sandell, Colorimetric Determination of Traces of Metals, Interscience Publishers, New York, 1944.

Procedure. The $HClO_4$-HNO_3 wet digestion is the method of choice for iron analysis. TCA precipitation is not suitable, since most commercial preparations of TCA are contaminated with traces of iron leading to high blank values. Dry ashing gives an ash usually difficult to dissolve, thus causing loss of iron in the ashing procedure. Dilute a sample containing 10 to 200 γ of Fe to 10 ml. with Fe-free water. Add 1 ml. of the hydroxylamine hydrochloride solution, and adjust the pH to between 3.0 and 6.0 with Na acetate solutions. A pH meter is advisable for this adjustment. The pH chosen will depend in part on the various interferences anticipated. For example, in the presence of pyrophosphate or fluoride, the pH must be kept as close to 6.0 as possible. In the case of maximum amounts of copper or cobalt, the pH should be kept as close to 3.0 as possible. After the adjustment of the pH, add 1 ml. of the *o*-phenanthroline reagent, mix, and dilute to 25 ml. After 5 to 10 minutes (or an hour in the presence of interference by pyrophosphate) determine the extinction at 500 mμ in the Beckman in 1-cm. cuvettes. A standard curve constructed by carrying samples of the standard Fe solution through the procedure will relate the concentration of iron to color intensity. The color intensity is usually relatively independent of the hydrogen ion concentration within the range 3.0 to 6.0, and the color formed is sufficiently reproducible that it is not necessary usually to run standards with each set of analyses.

Volumetric Method (Range, 1 to 10 Mg.)

Principle. Most volumetric procedures for iron are based on the reduction of iron to the ferrous state and back-titration with an oxidizing agent. It is unquestionable that this classical method of determination provides high accuracy, and with proper indicators, especially the use of potentiometric titration, can be made highly precise. Almost any textbook or treatise on analysis will provide a variety of such methods for iron analysis. The method to be presented here is a newer method and offers certain advantages over the classical titration in that it appears to be free from the effects of extraneous contaminating heavy metals, such as copper and zinc, etc. It is based on the complexing of iron as a chelate with ethylenediaminetetraacetic acid. The method below is taken from the paper of Cheng *et al.*[19] The method is based on the fortuitous circumstance that ferric complexes of EDTA are stable at acid pH, whereas the complexes of most other elements are stable only in the neutral and alkaline region.

[19] K. L. Cheng, R. H. Bray, and T. Kurtz, *Anal. Chem.* **25**, 347 (1953).

Reagents

Ethylenediaminetetraacetate (Versenate) solution. Dissolve 4 g. of disodium dihydrogen ethylenediaminetetraacetate in 1 l. of water.

Indicator. Dissolve 1 g. of salicylic acid in 100 ml. of ethyl alcohol.

Sodium acetate, 26% (w/v).

Standard Fe solution. The standard Fe solution described under the colorimetric procedure may be used. However, a more suitable standard is formed by dissolving 1 g. of electrolytic Fe wire in 10 ml. of 6 N HCl and diluting to 100 ml. with water. Warm on the steam bath until all the iron has dissolved, and then dilute to 1 l. This solution contains 1 mg. of Fe per milliliter.

Procedure. The wet-ashing procedure with $HClO_4$-HNO_3 is mandatory for this type of analysis. The sample should contain from 1 to 10 mg. of Fe. Dilute the digest to approximately 10 to 15 ml. in a small flask, and adjust the pH to between 2.0 and 3.0 by adding Na acetate and acetic acid. Titrate the solution with the Versenate reagent after adding 5 drops of the indicator. Be sure to titrate to the end point when the purplish-red tint has just disappeared. The solution becomes yellow or colorless. Standardize the Versenate solution against the standard Fe solution. Take an aliquot of the standard Fe containing approximately 10 mg., adjust the pH to between 2.0 and 3.0 with sodium acetate and acetic acid, and make the final volume approximately 25 to 30 ml. Add 5 drops of the indicator solution and a few crystals of NH_4 persulfate to oxidize any ferrous iron present; then titrate with the Versenate solution until the indicator turns either pale yellow or colorless.

Gravimetric Method

Principle. The utilization of iron radiotracers and the widespread importance of iron in enzymatic and metabolic systems makes some gravimetric method for analysis and radioassay essential. The best procedure seems to be the extraction of iron from the ash followed by electroplating. A system very similar to the one described for cobalt is that of Petersen.[20]

Reagents

Cupferron, 5% aqueous solution.

HCl, 1 M.

Caprylic alcohol.

NH_4 oxalate, saturated aqueous solution.

[20] R. E. Petersen, *Anal. Chem.* **24**, 1850 (1952).

Procedure. Ash with $HClO_4$-HNO_3, and dilute to approximately 25 ml. with 1 M HCl. Cool to below 25°, and add a solution of 5% cupferron with constant stirring. Usually about 2.5 ml. of solution will be required. Allow to stand at room temperature for 10 to 20 minutes, transfer to a separatory funnel, and extract with 20 ml. of chloroform. The beaker in which the precipitation has been carried out can be washed with the chloroform to transfer any precipitate remaining in it to the separatory funnel. After vigorous shaking for a few seconds, transfer the clear chloroform layer to a 100-ml. porcelain crucible. Re-extract the solution with 15 to 20 ml. of chloroform. Allow the combined chloroform extracts to evaporate in a stream of air and finally in an oven at 45 to 50° for 1 to 3 hours, thus bringing about initial decomposition of the cupferrate. To the gummy residue in the bottom of the crucible, add 0.5 ml. of capryl alcohol, cover the crucible, and ash over a gas burner or in a muffle furnace. Raise the temperature gradually to prevent any spattering of the cupferrate. The initial decomposition is usually accompanied with considerable gas evolution. Add a few milliliters of concentrated HCl to the iron oxide in the bottom of the crucible, and evaporate just to dryness on a hot plate at 100°. Add an additional 0.1 to 0.2 ml. of concentrated HCl to dissolve the iron, and dilute with 10 ml. of NH_4 oxalate solution. Transfer the solution to the electrolysis cell, rinsing out the crucible with small portions of the oxalate solution. The total volume should not exceed 17 ml. The electrolysis is carried out at 8 volts at a current density of 17 to 20 ma./cm.^{2}. Petersen electroplates on copper planchets having a total area of 2 cm.^{2}. The progress of the electrolysis is followed by removing approximately 0.5 ml. of the electrolyte to a test tube, adding 1 to 2 drops of thioglycolic acid, and 0.5 ml. of concentrated NH_4OH. A pink color indicates that the electrolysis is incomplete. After obtaining a negative test for iron, discard the oxalate solution and wash the planchet, first with water and then with acetone. Keep the planchets in a desiccator until ready to be counted. If a determination of the amount of iron is required as well as a sample for radioassay, the planchets may be made from thin platinum foil and weighed on a microbalance.

VI. Cobalt

Cobalt has figured in enzymatic studies largely as an activating ion, and the concentration of cobalt can usually be calculated. In some experiments, however, such as dialysis-equilibrium experiments, it may be necessary to know the amount of free cobalt present. In these experiments the amounts of cobalt being dealt with are usually rather low, and consequently spectrophotometric methods are a requirement if adequate sensitivity is to be obtained. Since quantities of cobalt suitable for volumetric procedures will rarely be available, no such method of analysis will be given for this element. On the other hand, utilization of Co^{60} as a radio-active tracer in cobalt metabolism and enzymatic experiments makes a gravimetric method for cobalt determination and for radioassay highly desirable. The authors' procedure seems to date to be the most satisfactory.

Spectrophotometric Method (Range, 0.005 to 0.5 Mg.)

Principle. Numerous methods for the spectrophotometric determination of cobalt have been described, using either the nitroso-R salt, or the nitrosonaphthols. Since reagents form color complexes with cobalt which are soluble in halogenated hydrocarbons, such as carbon tetrachloride, they suffer like many of the metal-complexing reagents from the fact that they react with only relative specificity with the element in question; iron, copper, nickel, and in some cases manganese, all interfere. The spectrophotometric method to be presented is based on the procedure by Almond.[21] The basis of the procedure is the extraction of cobalt and other heavy metals into chloroform solution by 2-nitrosonaphthol and the subsequent removal of the interfering elements by an extraction with KCN. The cobalt concentration is then determined spectrophotometrically.

Reagents

NH₄ citrate, 10% solution (w/v).

Borate buffer. Dissolve 19 g. of Na tetraborate hydrate in 800 ml. of water, and add 10 ml. of concentrated NH₄OH. Dilute to 1 l.

Phenol red indicator (PR), 0.02% (w/v). Add 0.1 g. of phenol red to 0.3 ml. of 1 N NaOH, and dilute with water to 500 ml.

2-Nitroso-1-naphthol reagent. Add 2 drops of 1 N NaOH to 10 mg. of 2-nitroso-1-naphthol in a 250-ml. beaker, and then add just enough water to wet the reagent. Stir, and then add water dropwise until about 2 ml. has been added and all the reagent is in solution. Dilute to 100 ml. with water. This solution may be filtered through sintered glass if necessary.

KCN, 10%.

Water. The precautions for cleanliness of glassware and the reagents discussed in connection with the determination of copper and zinc must be followed rigorously in the following cobalt analysis.

Standard Co solution. Dissolve Co metal in HCl with the aid of heat. Evaporate just to dryness on a steam bath, and take up in 500 ml. of water. The solution should contain 100 γ of Co per milliliter. If Co metal is not available, $CoCl_2·6H_2O$ may be used by dissolving 0.04 g. in water, adding 1 ml. of HCl, and diluting

21 H. Almond, *Anal. Chem.* **25,** 166 (1953).

to 100 ml. More dilute working solutions are prepared from the primary stock by suitable dilutions.

Procedure. Analysis for cobalt can best be carried out on a $HClO_4$-HNO_3 ashed sample. Dry ashing is quite unsuitable, particularly if silica, Vycor, or Pyrex crucibles are used. Platinum crucibles may be used, but the wet-ashing method is still preferable. Dilute the digest to 10 ml. with water. Add 2 drops of a phenol red solution and concentrated NH_4OH until the indicator is a faint pink (about pH 6.5). Then add 5 ml. of the NH_4 citrate solution, and titrate with the borate buffer, again to a faint pink of about pH 6.5. Add 3 ml. of the 2-nitroso-1-naphthol solution and 1 ml. of CCl_4. Shake for 60 seconds, and drain the CCl_4 into a clean tube containing 10 ml. of water and 1 drop of KCN solution. Repeat the extraction of the sample with an additional 1 ml. of CCl_4. Shake and add this to the tube containing the KCN solution. Shake the tube vigorously for 10 seconds, and then with a dry pipet carefully withdraw 1 ml. of the sample. Dilute this in a standard Beckman cuvette with an additional 2 ml. of CCl_4. Determine the extinction at 550 mμ. A standard curve should be constructed by carrying aliquots of the standard solution through the same procedure.

Gravimetric Method (Range, 0.01 to 10 Mg.)

Principle. The following method of determining Co^{60} in biological materials is according to the procedure of Ballentine and Burford.[22] The sample is wet-ashed with $HClO_4$-HNO_3; and cobalt is precipitated as cobaltic hydroxide and then electroplated as the metal from a fluoborate buffer.

Cobalt is precipitated quantitatively from an alkaline perborate solution, but several other elements (iron, silica, and the alkaline earths) also precipitate under these conditions. The alkaline earths and silica are eliminated during the electrolytic plating. Should sufficient iron be present to constitute a serious contaminant, it can be readily removed by dissolving the $Co(OH)_3$ precipitate in 6 N HCl and extracting with either isopropyl or diethyl ether.

High current densities at an alkaline pH lead to a local alkaline reaction at the cathode which causes a copious precipitate of cobaltic hydroxide with concomitant severe losses. However, the cobalt plate redissolves to some extent in the acid fluoroborate buffer. To balance these effects, therefore, the plating is done in two steps. The major part of the cobalt is deposited at a pH of 3.1 and at a low current density. The pH is then raised to 8.2 (pH obtained when 1 ml. of 1 N NaOH is added to 10 ml. of

[22] R. Ballentine and D. Burford, *Anal. Chem.* **26**, 1031 (1954).

the plating solution at pH 3.1) and the plating continued for an equivalent length of time.

Reagents

Sodium hydroxide, 9 M, carbonate-free.

Sodium perborate tetrahydrate, reagent.

Aerosol OT, 0.01% (Fisher Scientific Co.).

Hydrochloric acid, 6 N.

Cupric fluoborate solution, purified (Baker and Adamson).

Potassium fluoborate buffer. Fluoboric acid (0 5 M) is made by diluting Baker and Adamson's fluoboric acid (42 to 45%) according to the acid content on the label. Add potassium hydroxide (5 M) until the pH measured with the glass electrode is 3.1. Let stand for 30 minutes; then filter the sizable precipitate off in a sintered-glass filter funnel of medium porosity. The reagent keeps indefinitely, but after several weeks it may require refiltering.

Sodium hydroxide, 1.5 N.

Procedure. PRECIPITATION. Wet ash the sample with $HClO_4$-HNO_3. Then dilute it to 50 or 60 ml. with distilled water and bring it carefully to boiling with frequent agitation to prevent bumping. Using 9 M sodium hydroxide, make the solution just alkaline to phenolphthalein, and add 5 or 6 drops of the sodium hydroxide to give a pH of about 13. Add about 75 mg. of sodium perborate tetrahydrate immediately, shake the sample well to mix, bring again to boiling, and add another 25 mg. of sodium perborate. Then boil the sample for about 30 seconds, and allow to cool.

Transfer the sample to a 100-ml. centrifuge tube, and rinse the Kjeldahl flask with 0.01% Aerosol OT solution and then with water. Rinse down the inside of the centrifuge tube with a small amount of the Aerosol OT solution, and centrifuge the sample for 15 minutes. Remove the supernatant liquid by suction, using a glass filter stick (Pyrex No. 39535-10F), wash down the inside of the tube with about 10 ml. of Aerosol OT solution to wash out any remaining soluble salts, centrifuge the sample again for 15 minutes, and remove the supernatant liquid, using the same filter stick. In handling the centrifuged samples, great care must be taken to prevent jarring the tube, as the precipitate is very easily stirred up.

Leave the filter stick in the centrifuge tube, and wash the tube down with 2 ml. of concentrated nitric acid to dissolve the precipitate, a few drops being run into the inside of the filter stick. Then place the centrifuge tube in boiling water, and allow to stand until cool, during which time add 1 ml. of 6 N hydrochloric acid to the sample. When cool, blow the acid out of the filter stick into the centrifuge tube with compressed air, fill the filter stick with water, blow it out again, and finally rinse it off and remove. Wash the inside of the tube with about 10 ml. of Aerosol OT solution, and place the sample in an oven at 95 to 100° to dry (too high a drying temperature produces an insoluble precipitate). When thoroughly dry, add about 5 ml. of water to aid in the removal of any residual acid, and take the sample to dryness again. It can then be stored indefinitely before proceeding with the analysis.

PLATING. Rinse down the centrifuge tube containing the sample to be analyzed with 5 ml. of saturated K fluoborate buffer, of pH 3.1. Place the tube in boiling water for a few minutes, then in cold water. Pour the solution into the plating vessel, and rinse the tube with another 5 ml. of K fluoborate buffer. Centrifuge the tube briefly to collect all drops of solution hanging on the sides of the tube, and add the solution to that in the plating vessel.

The plating is carried out at a maximum voltage of 3.5 volts. (Overvoltage results in production of hydrogen and destruction of the plate.) The duration of plating is determined by the current density. At 3.5 volts, one requires a current density of 2.1 to 4.2 ma. and a total current between 40 and 80 ma. in order to complete the first part of the electrolysis in 90 minutes. Other values of current density and surface area will require adjustment as to the duration of the plating period.

At the end of the first period of plating (under the above conditions, 90 minutes), add 1 ml. of 1.5 N NaOH and continue the plating for another period of equal duration. With the current still flowing, remove the vessel from the plating set-up, pour out the solution quickly, rinse the cup with 95% alcohol, and drain dry. It is then ready to be weighed and counted.

VIII. Copper

Spectrophotometric Method (Range, 0.01 to 5 γ)

Principle. The procedure is a modification of the dithizone method given by Sandell[18] and by Bendix and Grabenstetter.[23] The spectral properties and Beer's law dependence is based on the observation of Liebhafsky and Winslow.[24] Manganese, cobalt, zinc, and ferrous iron do not interfere. The interference by ferric iron is avoided by using a reducing agent. The determination is based on a calibration curve.

[23] G. H. Bendix and D. Grabenstetter, *Ind. Eng. Chem., Anal. Ed.* **15**, 649 (1943).
[24] H. A. Liebhafsky and E. H. Winslow, *J. Am. Chem. Soc.* **59**, 1966 (1937).

Reagents

Dithizone. A 0.001 to 0.0012% (w/v) of dithizone (diphenylthio-carbazone) in *reagent* CCl₄ is made and filtered through a fine-porosity sintered-glass filter with *pressure*. The reagent must be protected from both strong light and heat. If necessary, the reagent solution (100 ml.) may be purified before use by extracting twice with 50 ml. of 1:100 NH₄OH. The aqueous layers are filtered through a coarse filter paper and then made acid with HCl. The reagent is extracted into 100 ml. of CCl₄.

Water; all glass-distilled.

Standard Cu solution. Rinse electrolytic Cu foil with 1:1 HNO₃, water, and absolute alcohol, then dry with cellulose tissue. Dissolve 100 mg. in 5 ml. of 1:1 HNO₃, dilute to 100 ml., and boil to expel oxides of nitrogen. After cooling, dilute to 1 l. For use, dilute aliquots of the stock Cu solution to give two working standards containing 1 γ/ml. and 0.1 γ/ml., respectively.

Hydroxylamine hydrochloride, 10% (w/v) in water. Use polyethylene bottles for all aqueous solutions.

Procedure. Organic material is destroyed by the standard HClO₄-HNO₃ digestion. Dilute the digest with 10 to 15 ml. of water, and add a drop of bromothymol blue. Then neutralize with NH₄OH, and adjust the volume to 25 ml.

Transfer an aliquot of the neutralized digest, containing not more than 5γ of copper, to a small separatory funnel and dilute to 25 ml. Then add 0.4 ml. of concentrated HCl, 1 ml. of hydroxylamine solution, and 5.00 ml. of dithizone reagent, in order. Shake the mixture vigorously and allow it to separate.

If the color of the CCl₄ layer is a red-violet, add an additional 5 ml. of dithizone and repeat the shaking. If still no tint of green persists, owing to excess reagent, discard the mixture and take a smaller aliquot of the digest. An excess of reagent must be present at the end of the extraction.

Waste 1 ml. of the CCl₄ through the stopcock of the funnel, and then run 3 ml. into a 1-cm. cuvette. Read the optical density at 503 mμ against distilled water.

A reagent blank must be run through the whole procedure, including ashing. Subtract the density of the reagent blank from the unknown. A standard curve is constructed, using aliquots of the standard Cu solution in place of the digests. Plot the density difference between the test samples and reagent blank versus micrograms of Cu per milliliter of CCl₄ solution. At least two check points on the standard curve should be run with each series of analyses.

Volumetric Method (Range, 0.05 to 1 Mg.)

Principle. This method is based on the modification of Scott's method by Meites.[25] It is free of interferences by calcium, magnesium, manganese, and zinc. Cobalt interferes only if the color masks the end point. The interference by iron is nullified by the addition of fluoride. The reaction is stoichiometric:

$$2Cu^{++} + 4I^- \rightarrow 2CuI + I_2$$
$$I_2 + 2S_2O_3 = S_4O_6 + 2I^-$$
$$S_2O_3 = I = Cu$$

1 ml. of 0.005 N thiosulfate = 0.3178 mg. of Cu

Reagents

>   Standard biiodate. Dissolve 1.9497 g. of K biiodate [KH(IO₃)₂] in 1 l. of boiled water to give a 0.005000 M solution.
>
>   Thiosulfate, stock ca. 0.1 M. Dissolve 24.8 g. of $Na_2S_2O_3 \cdot 5H_2O$ in 1 l. of boiled water. Add 1 drop of N NaOH and 1 ml. of $CHCl_3$ as preservatives. Store in dark bottle. Dilute an aliquot just prior to use to give an approximately 0.005 M solution. Standardize by titrating in triplicate 10-ml. aliquots of standard biiodate to which 1 to 2 g. of KI is added just before titrating.
>
>   Starch indicator. Dissolve 1 g. of Lintner soluble starch (Merck) in 100 ml. of water containing 20 g. of NaCl by gently boiling for 2 minutes. No preservative is needed, and the indicator keeps for years.

Procedure. Ash the sample with $HClO_4$-HNO_3, and continue the heating carefully until all the nitric acid is dispelled. Dilute the sample to 25 ml. with water. Add 1 g. of Na acetate trihydrate, 4 g. of NaF, and 25 g. of KI. Titrate with standardized 0.005 M thiosulfate from a 1- or 2-ml. buret, adding starch indicator just before the end point.

IX. Zinc

Of the many methods for the determination of zinc, those based on its precipitation as the insoluble ferrocyanide have been preferred in volumetric and gravimetric analyses. Second only to this, gravimetrically speaking, is precipitation of zinc as sulfide from an acidic solution. In the spectrophotometric field, however, dithizone (diphenylthiocarbazone) has been utilized almost to the exclusion of other colorimetric reagents.

[25] L. Meites, *Anal. Chem.* **24**, 1618 (1952).

One of the reasons for this is that interference by other elements can to a large extent be obviated by the formation of the zinc-dithizone complex in the presence of buffers of anions such as tartrates, which form complexes with other metals. The method given below is taken almost intact from the method of Vallee and Gibson.[26]

Spectrophotometric Method (Range, 1.0 to 30 γ)

Reagents

Dithizone. The dithizone solution used is the same as that described for copper. Water and other reagents should all be glass-distilled with precautions to avoid contamination by heavy metal ions.

Buffer solution. Dissolve 556 g. of Na thiosulfate, 90 g. of Na acetate, and 10 g. of KCN in 1 l. of Zn-free water. Adjust the solution with N acetic acid to approximately pH 5.5, using methyl red as an indicator. Then adjust the pH exactly, using a pH meter. Make the solution to 2 l. with Zn-free distilled water. To remove any traces of zinc present in the reagent and water, extract the buffer with some of the dithizone reagent, repeating the extraction until the dithizone remains a *clear green*.

Tartrate solution. Make up a 20% (w/v) solution of NaK tartrate tetrahydrate in Zn-free water. Extract with dithizone solution to remove traces of zinc as described above.

Concentrated NH_4OH, c.p. reagent or preferably NH_4OH freshly prepared from zinc-free water and gaseous NH_3.

NH_4OH, 0.1 M.

Methyl red indicator, 1:100 (w/v) alcoholic solution.

Glassware. Only Pyrex glassware should be used throughout the following procedure. Particular precautions are necessary to avoid introduction of contaminating amounts of zinc. Wash all the glassware first with soap and water, rinse with distilled water, and immerse in 2 N HNO_3 for at least 6 hours. On removal from the acid, wash the glassware with Zn-free triple-glass-distilled water. In view of the fact that separatory funnels are particularly difficult to free of any contaminating traces of zinc, after they have been washed as described above, place 20 ml. of the buffer and 5 ml. of the dithizone solution in the separatory funnel and shake, repeating this treatment until the dithizone remains a clear green. Finally, wash all the glassware with 0.01 N NH_4OH to ensure that no residual acidity is present.

[26] B. L. Vallee and J. G. Gibson, *J. Biol. Chem.* **176**, 435 (1948).

Standard zinc. Make a solution using reagent-grade 30-mesh Zn, dissolving a weighed amount in a slight excess of HCl and diluting to volume in a volumetric flask. The primary stock solution should contain 1 mg. of Zn per milliliter in approximately 0.1 N HCl. From this stock solution prepare a working Zn standard containing 100 γ per 10 ml.

Procedure. Sample preparation may best be carried out by dry ashing in platinum crucibles. For the purposes of this analysis, Pyrex, Vycor, and quartz crucibles are quite unsatisfactory. Wet ashing is also reasonably unsatisfactory, since it is difficult to obtain both $HClO_4$ and HNO_3 completely Zn-free. By careful redistillation from glass or silica vessels, these two acids may be prepared in a reasonably Zn-free state. However, where possible the dry-ashing procedure probably leads to less error. Dissolve the dry ash by warming with 13 to 30 ml. of 2 N HCl. Then evaporate to approximately 5 ml. on a steam bath and transfer to a 125-ml. separatory funnel by repeated washings with small portions of hot, Zn-free water. Transfer wet-ashed samples directly to the separatory funnel with small portions of Zn-free water. A final volume of 25 to 50 ml. of aqueous solution is optimal. To the solution in the separatory funnel add 2 ml. of the tartrate solution, along with 2 drops of methyl red indicator. Then titrate with NH_4OH and H_2SO_4 as necessay to a pH of 5.5. Add 50 ml. of the buffer solution, and allow to stand until all color has completely faded. Extract with 10 ml. of the dithizone reagent, shaking vigorously for about 2 minutes. After the dithizone in CCl_4 has collected in the bottom of the funnel, draw off the CCl_4 solution into a 50-ml. volumetric flask. Repeat this procedure until 40 ml. of dithizone reagent has been used. At this point—the last extraction—the dithizone solution should remain a clear green. Make the volume to 50 ml. Read the optical density in the Beckman spectrophotometer in a 1-cm. cuvette at 520 mμ. Read the zinc concentration from a standard curve obtained by carrying through the above procedure, with the exception of the ashing, on samples of a standard Zn solution.

Volumetric Method (Range, 1.0 to 50 Mg.)

Principle. In general a volumetric method designed primarily for precision rather than sensitivity would be carried out on protein preparations in which one does not expect to have severe interferences from other elements present. In the following procedure, however, a method is given for the separation of zinc from contaminating amounts of copper, manganese, and iron. The method is based on the precipitation of zinc as the insoluble ferrocyanide according to the procedure of Bodansky.[27]

[27] M. Bodansky, *J. Ind. Eng. Chem.* **13**, 696 (1921).

Reagents

Standard Zn solution. Dissolve 0.100 g. of Zn, reagent, 30 mesh, in 10 ml. of concentrated HCl, and dilute to 1 l.; 1 ml. is equal to 0.1 mg. of Zn.

Standard K ferrocyanide. Dissolve 7 g. of K ferrocyanide in 1 l. of water.

K Ferricyanide. Dissolve 1 g. of ferricyanide in 100 ml. of water.

Diphenylamine indicator. Dissolve 1 g. of diphenylamine in 100 ml. of concentrated H_2SO_4. This reagent should be kept in glass-stoppered bottles, out of the light, and reasonably cool.

NH_4 thiocyanate solution, 2% (w/v).

Citric acid, 50% (w/v).

Procedure. Select a sample to contain between 1 and 50 mg. of zinc. The ashing may be carried out either by dry ashing or wet ashing with $HClO_4$-HNO_3. If the dry ash is used, dissolve the ash in distilled water on a steam bath, adding 1 ml. of concentrated H_2SO_4. In the wet ash, transfer the sample to an Erlenmeyer flask with water, and add 1 ml. of concentrated H_2SO_4. In either case, heat the sample until white fumes of SO_3 are being copiously liberated. After cooling, add 25 ml. of distilled water. Precipitate the heavy metals by vigorously bubbling in H_2S. Filter off the heavy metals, and wash the precipitate with small portions of hot water. Boil the filtrate to remove H_2S, cool, and neutralize with concentrated NH_4OH. Then add 10 ml. of 50% citric acid solution. Again heat the solution to boiling, and if no Ca citrate separates, add small quantities of $CaCO_3$ until a precipitate of about 1 g. of Ca citrate is formed. Remove from the heat, and pass in a very rapid stream of H_2S until the solution has cooled. Filter through a small sintered-glass funnel, and wash the solution with 2% NH_4 thiocyanate. Finally, dissolve the precipitate in 2 ml. of concentrated HCl, and with 10 ml. of water, wash into a small Erlenmeyer flask. Dilute with water to approximately 50 ml., heat to 60°, and add 2 drops of the diphenylamine indicator, followed by 2 drops of the K ferricyanide solution. Titrate with the standard K ferrocyanide until the blue color changes to yellowish-green. The K ferrocyanide solution should be standardized against the Zn standard described above. Take an aliquot of the standard Zn solution containing approximately 25 mg. of zinc, add 2 drops of the ferricyanide and 2 drops of the indicator solution, and titrate to the same end point with K ferrocyanide. The reaction is stoichiometric.

Gravimetric Method

Suitable radioisotopes of zinc are available for tracer experiments. For the gravimetric determination of zinc, the element may be precipitated either as the ferrocyanide or the sulfide, both of which occur at different stages in the above volumetric analysis. The sulfide precipitation, involving fewer manipulations, is preferable, provided that precautions are taken to obtain a precipitate of the maximum density and crystallinity. The solution should contain zinc as the sulfate rather than as the chloride, the latter usually giving amorphous precipitates. Second, the precipitation should be initiated in a warm (60°) solution with a rapid stream of H_2S. When the precipitation is effected by a slow addition of H_2S or in a cold solution, the precipitation is not quantitative and the zinc sulfide is difficult to filter.

X. Flame Spectrophotometric Analysis for Sodium, Potassium, Calcium, and Magnesium

The flame spectrophotometer offers an accurate and convenient method of determining the sodium, potassium, calcium, and magnesium contents of tissues or enzymatic preparations. When a dilute solution of certain elements is sprayed into a flame, each element emits light of a characteristic wavelength, the intensity of which is proportional to the concentration, and which may be measured by a spectrophotometer. Because of intrinsic differences in the designs of the various flame spectrophotometers currently available, no single method is directly applicable to all of them. Consequently, the following discussion has been arbitrarily limited to the Beckman instruments, although the method can be easily adapted to other instruments.

A number of factors must be borne in mind when planning flame spectrophotometric analyses, particularly the several kinds of interferences which are encountered.[28,29] Some are intrinsic in the method of analysis; others result from the presence of other substances which affect the light emission and must either be compensated for or eliminated from the sample.

1. Band-width interference occurs when light from one metal is augmented by light from another metal at an adjacent wavelength, emitted either as monochromatic light or as a diffuse band. Such interference can be predicted by consulting spectral charts[30] showing characteristic wavelengths for various metals. This effect can usually be avoided by (1) reducing the slit width as much as possible, while still retaining adequate sensitivity; (2) choosing another wavelength for the metal being measured; (3) measuring the emission of the interfering substance and sub-

[28] *Beckman Bulletin No. 259*, Beckman Instruments, Inc.

[29] K. K. Kendall, *Beckman Bulletin No. 12*, p. 6 (1953).

[30] P. T. Gilbert, Jr., Flame Photometry—New Precision in Elemental Analysis, *Beckman Reprint R-56*, p. 6.

tracting the value obtained from that for the metal being determined; or, if necessary (4) adding the proper concentrations of the interfering substances to the standards and blank, under which conditions the sample and standards will be equally affected and the interferences thereby nullified.

2. The background radiation may be increased at all wavelengths by the continuous spectrum of substances such as proteins and aromatic hydrocarbons or by a high concentration of sodium. This type of interference can be measured by the use of a blank containing the proper amounts of the interfering substances but none of the metal being determined, and corrections made accordingly. However, as background interference varies with the square of the slit width, and the intensity of emission lines increases with the first power of the slit width,[31] this effect can usually be sufficiently reduced by using narrow slits. A photomultiplier tube further reduces such interference.

3. The emission lines produced by high concentrations of some metals, such as sodium, are of such intensity that they cause stray light which interferes at wavelengths other than those of their specific emission lines. This light, when due to sodium, can be removed by using a didymium filter;[31] other filters are available for removing light from other substances.

4. A fourth type is radiation interference, in which another substance causes a metal to have an increased or decreased light emission.[32] To avoid this type of interference, as in the case of band-width interferences, the proper concentrations of the interfering substances can be added to standards and blanks, or chemical methods may be used to remove them from a sample being analyzed. A simpler treatment which is applicable to many biological materials when determining sodium, potassium, calcium, and magnesium is that of diluting the sample sufficiently so the effect of the interfering substances becomes negligible while sufficient sensitivity is still retained for accurate analysis for the metal in question. The "detection limit" of a metal is defined as that concentration which

---

[31] *Beckman Instruction Manual No. 334-A for Models DU and B Flame Photometer,* p. 22.
[32] The light emission of sodium, calcium, potassium, or magnesium is affected by numerous substances. For instance, various organic solvents have been shown to increase the intensity of emission [G. W. Curtis, H. E. Knauer, and L. E. Hunter, Symposium on Flame Photometry, *Am. Soc. Testing Materials, Spec. Tech. Publ. No. 116,* p. 67 (1951); G. R. Kingsley and R. R. Schaffert, *J. Biol. Chem.* **206,** 807 (1954)], whereas such salts and acids as HCl, H_3PO_4, NH_4Cl, $MgSO_4$ and $CuCl_2$ are known to depress both sodium and potassium readings [J. W. Berry, D. G. Chappell, and R. B. Barnes, *Ind. Eng. Chem., Anal. Ed.* **18,** 19 (1946)]. It is also reported that 0.003% of phosphate has a depressing effect on calcium [D. F. Kuemmel and H. L. Karl, *Anal. Chem.* **26,** No. 2, 386 (1954)], and sodium, potassium, and calcium all have an effect on one another.

produces a light emission which is 1% above background or where the error becomes ±100%. (For detection limits for various elements expressed in part per million, see *Beckman Detection Table DS-2*.[33])

5. Viscosity and temperature also constitute sources of interference. The effect of viscosity can be compensated for by adding to the standards the substances which are responsible for the variation of viscosity. However, this effect is not too often a problem with aqueous solutions.

A method to determine whether or not interferences are present is given subsequently.

Principle. Various procedures for determining sodium, potassium, calcium, and magnesium by means of the flame spectrophotometer are currently in use. Samples can be prepared for analysis in a number of ways. Biological fluids such as urine, plasma, and serum may be treated with 10% TCA to remove protein,[34] or simply diluted.[35] Tissues, plant materials, and whole blood are prepared by wet ashing with $HClO_4$-HNO_3;[36,37] by dry ashing;[38] by electrolyzing an HCl extract of plant tissue with a mercury cathode;[39] or by water extraction of a frozen-dried preparation.[40]

The following method of analysis has been adapted from procedures reported by Kapuscinski *et al.*[41] and Kingsley and Schaffert.[35] Wet ashing of the sample with $HClO_4$-HNO_3, the use of a mixed standard,

[33] Detection Limits for Beckman Flame Spectrophotometers, *Beckman Detection Table DS-2.*

[34] J. W. Severinghaus and J. W. Ferrebee, *J. Biol. Chem.* **187**, 621 (1950).

[35] G. R. Kingsley and R. R. Schaffert, *Anal. Chem.* **25**, 1738 (1953).

[36] S. J. Toth, A. L. Prince, A. Wallace, and D. S. Mikkelsen, *Soil Sci.* **66**, 459 (1948).

[37] J. Q. Snyder, *Proc. Oklahoma Acad. Sci.* **31**, 134 (1950).

[38] C. E. Bills, F. G. McDonald, W. Niedermeier, and M. C. Schwartz, *Anal. Chem.* **21**, 1076 (1949).

[39] H. J. Noebels, *Beckman Reprint R-60.*

[40] A. K. Parpart, personal communication. A small piece of fresh tissue not more than 2 mm. in thickness weighing about 500 mg., is placed on ash-free filter paper in a small beaker, and the beaker is immersed in liquid nitrogen or liquid air. When the tissue is frozen, the beaker is removed from the liquid nitrogen and put directly into the freeze-dry system while a small amount of liquid nitrogen still remains on the tissue. The freeze-dry system should have a free path of only a few centimeters from tissue to drying agent (Drierite), and a high pumping speed is essential. The drying process must be complete before the vacuum is broken. When properly dried most tissues are snow-white throughout and powder easily in a mortar. The powdered tissue is mixed with 5 ml. of distilled water. Aliquots of the suspension are taken for dry weight determination. The remainder is centrifuged at 3500 rpm until a clear supernatant is obtained (about 10 minutes). The volume of the supernatant which is now ready for analysis in the flame spectrophotometer is measured.

[41] V. Kapuscinski, N. Moss, B. Zak, and A. J. Boyle, *Am. J. Clin. Pathol.* **22**, 687 (1952).

and very low ranges of concentration for the respective metals minimize interferences. Sodium and potassium, which have intense, monochromatic emission lines, are satisfactorily determined with either gas or hydrogen as fuel. Calcium and magnesium, on the other hand, have their main emissions from oxide bands. As a consequence, higher flame temperatures are necessary for analytical sensitivity. It is advisable, therefore, when determining the latter two metals to use the Beckman spectrophotometer with a photomultiplier tube equipped with a flame attachment which utilizes hydrogen as the fuel.

Reagents. Aqueous solutions used in flame spectrophotometry must be prepared, handled and stored so as to prevent inadvertent contamination with the metals being determined. Important quantities of Na and K can be contributed or removed by glass containers, filter paper, or contact with fingers, and contamination can also result from dust from clothing, high dust content of air, soap powder, or tobacco smoke. It is advisable to use glass-distilled water in preparing standard solutions and samples. Rinse glassware with HNO_3 and distilled water to remove any traces of alkali or alkaline earth elements. Store stock solutions in polyethylene bottles, and avoid the use of filter paper wherever possible.

Water, glass-distilled.

$CaCl_2$. Prepare $CaCl_2$ from Baker's $CaCO_3$ (low in alkali) by dissolving in a minimum of 1:1 HCl. Use in standard solutions.

NaCl, Baker's c.p. or Merck's reagent grade. Dry at 100° before weighing. Use in standard solutions.

$MgCl_2$. Dissolve MgO, Merck's reagent grade (99% after ignition), in a minimum amount of 1:1 HCl. Use in standard solutions.

KCl, Baker's c.p. or Merck's reagent grade. Dry at 100° before weighing. Use in standard solutions.

Procedure. PREPARATION OF SAMPLE. Select a sample of such a size that the following concentrations can be obtained by appropriate dilutions of the digest: Na, 1 to 6 p.p.m.; K, 1 to 10 p.p.m.; Ca, 1 to 10 p.p.m.; and Mg, 1 to 40 p.p.m. Digest with $HClO_4$-HNO_3. (See section on sample preparation.) Transfer the digest to a 100-ml. beaker with a small amount of hot water, removing any insoluble material if present by filtering through a fine-porosity sintered-glass funnel or by centrifugation. Evaporate just to dryness, and dilute to at least 10 ml. with distilled water. If necessary, dilute portions of this solution further so that the concentration of the element being determined falls in the middle range of the standard curve. (Biological fluids such as serum, plasma, and urine, and also purified enzyme digests, may not require the above treatment in the analysis for Na, K, and Ca but may often be prepared just by diluting the sample.)

Procedure. PREPARATION OF STANDARD SOLUTIONS AND STANDARD CURVES. Run a preliminary rough analysis by flame spectrophotometer to determine the approximate composition of the sample and also to determine whether or not interferences are present. Interferences can be detected in the following way. Determine a curve (concentration versus luminosity) using the data obtained on dilutions of a standard solution containing only the element being determined. At the same wavelength and under identical conditions, read the unknown and find the location of its reading on the standard curve. Dilute the unknown exactly by some factor, for example 1:1, and again take a reading, locating this point on the standard curve. If the two points on the curve in terms of concentration have exactly the same ratio as the dilution factor, there probably is no interference. If the ratios are different the interference should be handled in one of the afore-mentioned ways. (In the range of 1 to 10 p.p.m. Na, K, Ca, and Mg have a greatly diminished effect upon one another. In many instances, therefore, as pointed out above, simple dilution will give adequate control of interferences, and the analysis can be made using standard solutions containing only the element being determined. More often, however, the inclusion of interfering substances in the standard solutions is required.)

Prepare a mixed standard solution containing Na, K, Ca, and Mg, and any other interfering substances; such as phosphate or sulfate in the approximate ratios in which they appear in the sample, making the stock solution 100 $\times$ the concentration of the working standard. If the sample has been prepared by the $HClO_4$-HNO_3 digestion, add enough perchloric acid to compensate for the perchlorate ion in the ashed sample. (For instance, when ashing 2 ml. of serum, 3 ml. of c.p. 72% $HClO_4$ is added per liter of standard solution.) Determine standard curves for each metal at the beginning of each run by using appropriate dilutions of the stock solution to give as 100% points the following concentrations: Na, 6 p.p.m., read at 589 mμ; K, 10 p.p.m., read at 767 mμ; Ca, 10 p.p.m., read at 554 mμ; and Mg, 40 p.p.m., read at 383 mμ. Quantitative dilutions of these solutions are made to give 75%, 50%, and 25% values. The concentrations of the metals in the unknowns are determined by comparison of the luminosities with the standard curves.

OPERATING TECHNIQUES. The wavelengths and slit widths used, as well as the conditions of operation, vary according to the instrument. The proper operating conditions must therefore be determined for each instrument. Detailed instructions for operating the various models of the Beckman flame spectrophotometer are given in the instruction manuals which should be consulted before the analysis is planned. Since the line positions of the spectrophotometer shift with changing conditions, these must be checked daily and also whenever the wavelength setting is changed. The 100% standard solution is used to set the instrument for maximum brightness. It is necessary to standardize the flame spectrophotometer before each run and to check a known point on the standard curve frequently during the determination. After every two or three readings the 50% standard solution is checked again to detect any change in operating conditions of the instrument. The flame background must be subtracted from all readings; this is determined by means of a distilled water blank and should be checked frequently.

This procedure is repeated for each substance being determined, the instrument being reset and the proper solutions used to standardize the spectrophotometer in each case.

[149] Microbiological Methods for Determining Magnesium, Iron, Copper, Zinc, Manganese, and Molybdenum

By D. J. D. NICHOLAS

Microbiological methods for determining trace metals have the advantages of specificity, reproducibility, and sensitivity at low levels of metal over chemical or physical methods of analysis. Disadvantages include the necessity for rigorous pure culture techniques and the frequent use of standard growth curves to check possible mutation of the organisms.

This article describes methods for determining Mg, Fe, Cu, Zn, Mn, and Mo in biological materials, using the mold *Aspergillus niger* (Mulder strain). Other strains of *A. niger* may also be used for assays of metals.[1,2]

Principle

A. niger (M. strain) requires N, P, S, K, Mg, and the trace metals Fe, Cu, Zn, Mn, and Mo for optimum growth.[1-6] In the absence of one of

[1] D. J. D. Nicholas, *J. Sci. Food Agri.* **1**, 339 (1950).
[2] D. J. D. Nicholas, *Analyst* **77**, 629 (1952).
[3] E. G. Mulder, *Arch. Microbiol.* **10**, 72 (1939).
[4] D. J. D. Nicholas and A. H. Fielding, *Ann. Rept. Long Ashton Research Station* (Bristol, England) **126** (1947).
[5] D. J. D. Nicholas and A. H. Fielding, *J. Hort. Sci.* **26**, 125 (1951).
[6] R. A. Steinberg, *Botan. Rev.* **5**, 327 (1939).

these essential elements the growth of the fungus is retarded, as shown by reduction in dry weight yields and sporulation. Moreover, an increase in an essential element from deficiency to sufficiency levels when all others are present at optimum amounts results in a specific and quantitative increase in growth. In this way a standard growth series is prepared for any of the essential mineral nutrients and assayed by taking dry weight yields of the pads. For the bioassay of test material a known amount of it is added to a culture solution that contains all the essential elements other than the one to be determined. The growth of the fungus under these conditions depends on the amount of test element that it derives from the material added. The growth measured by dry weight yields of the mats is referred to a standard series for the element, grown under the same conditions.

Experimental Methods

Water Supply. Distilled water is prepared by being distilled twice from Pyrex glass to reduce the metal content. The minimum metal in micrograms per milliliter (p.p.m.) detected by the bioassay method is Cu, 0.001, Zn 0.005, Fe 0.002, Mn 0.002, Mo 0.00001. Water from a tinned-copper still is therefore unsuitable for assay of metals other than Mg and Zn. The metal content of water is markedly reduced by distillation either once or twice from Pyrex glass. There is no further reduction by a third distillation. Water of low conductivity prepared from standard ion exchange resins is also suitable for trace metal assays, but the Mo content may be too high for the sensitive *A. niger* assay for the element.

Glassware. Hard-glass 500-ml. Erlenmeyer flasks are thoroughly cleaned by washing several times with hot water and an organic detergent and then with concentrated nitric acid followed by single-distilled and finally with twice-distilled water from Pyrex glass. Alternatively a 0.5% w/v solution of disodium ethylenediaminetetraacetic acid is used instead of the acid wash, and flasks containing this solution are steamed in an autoclave to remove metals. Glassware and water supply may be checked for metals by shaking with 0.01% w/v solution of dithizone (diphenylthiocarbazone) in redistilled CCl_4. Metal dithizonates are indicated by a change from green of dithizone to red.

Preparation of Culture Solutions Free from Trace Metals. A suitable culture solution (1 l.) for *A. niger* is as follows: Macronutrients—dextrose, 50 g., KNO_3, 5 g., KH_2PO_4, 2.5 g., $MgSO_4 \cdot 7H_2O$, 1 g., and $Ca(NO_3)_2$, 0.5 g.; micronutrients—$FeCl_3 \cdot 6H_2O$, 20 mg., $ZnSO_4 \cdot 7H_2O$, 20 mg., $CuSO_4 \cdot 5H_2O$, 1 mg., $MnSO_4 \cdot 5H_2O$, 3 mg., $Na_2MoO_4 \cdot 2H_2O$, 1 mg.; glass-distilled water to 1 l.

The pH of the basal culture solution is 3.8; 50 ml. is used for each culture.

The macronutrients contain sufficient of the micronutrients to provide for optimum growth of the mold. Methods used to remove them from a solution of macronutrients in 200 ml. of distilled water are summarized in Table I. There is no need for purification when Mg is to be assayed.

TABLE I

REMOVAL OF TRACE METALS FROM A SOLUTION OF INORGANIC MACRONUTRIENTS AND DEXTROSE

Trace metal removed	Method used for removal	Residual metal after purification, γ/50 ml.
Copper	Add 2 ml. 10% w/v copper sulfate to 200 ml. culture solution, adjusted with hydrochloric acid to pH 3, and precipitate as a sulfide with hydrogen sulfide (15 minutes) from a Kipp's apparatus or a tank or cylinder. Let precipitate stand for 15 minutes, and filter through Whatman No. 42 paper into a clean Erlenmeyer flask. Eliminate hydrogen sulfide by boiling and agitating solution for about 20 minutes, and check removal with lead acetate paper.	0.05
Zinc and iron	Adjust 200 ml. solution to pH 5.5 with 5% w/v sodium hydroxide, and shake four times with 30-ml. portions of 5% w/v 8-hydroxyquinoline in chloroform in a liter Pyrex glass separating funnel. Extract excess quinolates with three 30-ml. portions of redistilled chloroform and then with similar quantities of redistilled ether. The latter removes chloroform. Eliminate the ether by boiling the solution on an electric hot plate. Zinc and iron are also removed at pH values above 5.5, but in alkaline solution phosphates are precipitated.	0.01
Molybdenum	The same as the coprecipitation method for the removal of copper.	0.0005
Manganese	Coprecipitation of manganese and copper diethyldithiocarbamates at pH 5.5. Add 10 ml. 0.1% w/v sodium diethyldithiocarbamate in water and 5 ml. 5% w/v copper sulfate solution. Set aside for 30 minutes, filter or centrifuge precipitate, and adjust filtrate to pH 3 with HCl. Add a further 5 ml. 5% w/v copper sulfate, precipitate copper sulfide with hydrogen sulfide from a cylinder, and proceed as described for copper.	0.01

Only microgram quantities of micronutrients are added to the culture solution so that CP or AR grade materials are usually satisfactory. Should purification prove to be necessary, as it may be for Fe, Zn, and Cu, the following procedures are used.

Extract a standard $CuSO_4 \cdot 5H_2O$ solution as a Cu dithizonate at pH 2.8, using 0.01% w/v dithizone in redistilled CCl_4. Transfer the copper dithizonate and excess dithizone into a 200-ml. beaker, and evaporate the CCl_4 on an electric hot plate. Add 10 ml. of redistilled nitric acid to digest the residual copper dithizonate, and evaporate to dryness. Add triple-distilled water, boil for 15 minutes, cool, and make

TABLE II

METAL ASSAY SERIES FOR *A. niger*

	Metal, γ/50 ml. basal culture solution	Standard deviation
Magnesium	0, 50,[a] 100, 150, 200, 300, 400,[a] 500	± 25
Iron	0, 0.1,[a] 0.25, 0.5, 1, 2.5, 5,[a] 10	± 0.1
Copper	0, 0.05.[a] 0.10, 0.20, 0.40, 0.6, 0.8,[a] 1.0, 1.5, 2, 4	± 0.05
Zinc	0, 0.25,[a] 0.5, 1, 2, 3, 4, 5,[a] 10	± 0.25
Manganese	0, 0.01,[a] 0.025, 0.05, 0.1, 0.5, 1,[a] 5	± 0.01
Molybdenum	0, 0.0005,[a] 0.001, 0.002, 0.003, 0.004, 0.005, 0.0075,[a] 0.01	± 0.005

[a] Effective range for assay, coinciding with the maximum differences in mycelial characters including spore cover and dry weights of mats.

to volume. This provides a pure copper nitrate solution free from other trace metals.

Extract a standard solution of $ZnSO_4 \cdot 7H_2O$ with 0.01% dithizone in redistilled CCl_4 at pH 7 in the presence of 1 ml. of 0.1% w/v sodium diethyldithiocarbamate (chelates Pb) per milligram of $ZnSO_4$.

$FeCl_3 \cdot 6H_2O$ is purified by extraction into redistilled ether in the presence of redistilled HCl at pH3. The ether is removed by heating on an electric hot plate. The iron salt is taken up and made to volume in triple-distilled water.

Preparation of Spore Inoculum. A. niger is readily subcultured on nutrient agar slants by a simple spore transfer from the original culture. This procedure minimizes the risk of mutation. The spores from one slant are carefully collected with a sterile platinum wire to avoid contamination and transferred aseptically to 10 ml. of sterile glass-distilled water. A drop of organic detergent may be included to disperse the spores. *A. niger* is grown at 30° for 4 days or at 25° for 5 days.

Preparation of Standard Series for the Metals. Fifty milliliters of basal culture solution is transferred into 500-ml. Erlenmeyer flasks. The necks of the flasks are closed with inverted beakers and the annular spaces plugged with nonabsorbent cotton wool. This overcomes the risk of contamination from cotton wool plugs. Trace metal standards are added in solution to the purified cultures. It is preferable to contain the added standards in 1 ml. so as to have consistent volume additions. All *A. niger* standard cultures are prepared in duplicate and are randomized in the

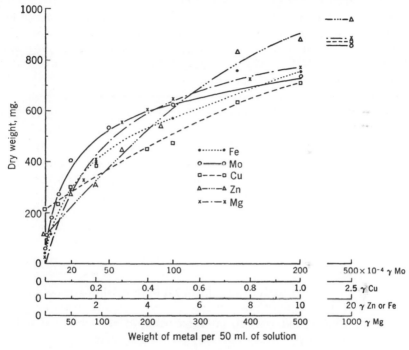

Fig. 1

incubator according to a statistical pattern. The standard errors for the assays are determined periodically as a routine check on the response of the fungus to the metals. A standard series for a metal is always included when an assay for it is being made on test materials. The effective ranges of assay for Mg, Fe, Cu, Zn, Mn, and Mo are shown in Table II and illustrated in Fig. 1.

The growth series may be assessed visually by taking into account the development of mycelia and intensity of sporulation and quantitatively by dry weight yields. The pads are filtered into a Büchner funnel, washed free from the culture solution, and weighed in 100-ml. glass beakers, which are then placed in a well-aerated oven at 90° for 12 hours when constant weight is usually reached. They are then cooled in a desiccator, weighed and redried in the oven for another 3 hours, cooled, and reweighed.

Copper has a specific effect on spore color which varies from yellow at a low copper level through brown to black at the higher levels. This color change facilitates the assay for the element.

When growth curves become erratic it is usually due to residual chelating reagents in the culture solutions.

Assay of Metals in Biological Samples. The microbiological method is particularly useful when only small amounts of tissues from normal or metal-deficient plant, animal, or bacterial sources are available. Under these circumstances the trace metal levels to be determined would be below the limits of chemical detection.

Weighed amounts of plant or animal tissues are digested in redistilled nitric acid. To 2 g. of fresh tissue or 0.25 g. of dried tissue is added 20 ml. of redistilled HNO_3. Digestion is continued for 45 minutes on an electric hot plate in a hood; after cooling, 0.5 ml. of pure 70% w/w $HClO_4$ is added, and digestion is continued until the solution is clear. The solution is evaporated to incipient dryness and, after cooling, 10 ml. of triple-distilled water is added. The solution is boiled for 15 minutes to hydrolyze the lower phosphates. The digest is made to 10 ml. with triple-distilled water, and suitable aliquots in the ratio of 1:2:3 are taken for assay. It is often unnecessary and undesirable to use $HClO_4$ to digest tissues as HNO_3 releases the metals in a soluble form suitable for assay from most biological samples.

Another suitable procedure is to burn milligram quantities of plant or animal material in a platinum crucible in an electric muffle furnace at about 400°. The cooled residue is transferred to the prepared culture solutions for the assays of Mg, Fe, Cu, Zn, Mn, or Mo.

The use of the *A. niger* method for determining small amounts of molybdenum in cauliflower leaves is shown in Table III.

TABLE III

THE MOLYBDENUM CONTENT OF WATER EXTRACTS AND ASH OF CAULIFLOWER LEAVES

Mo in water extracts[a] of leaves, γ/g. fresh weight		Mo in ash, γ/g. dry weight	
Normal	Mo-deficient	Normal	Mo-deficient
0.1 to 0.2	0.001 to 0.01	0.5 to 1	0.03 to 0.1
$SD \pm 0.002$	$SD \pm 0.001$	$SD \pm 0.005$	$SD \pm 0.002$

[a] Water homogenates of leaves prepared in a glass macerater.

It is sometimes desirable to determine the trace metal content of protein fractions during enzyme purification. The protein fractions are digested with HNO_3 before assay, as described previously. A good example of this is a study of the relation between nitrate reductase activity and molybdenum status of fractions of the enzyme from *Neurospora crassa* and soybean.[7] It was found that the amount of the micronutrient increased with increased specific activity of the enzyme.[7,8] Chemical methods would not have detected the Mo changes in the various protein fractions.

[7] D. J. D. Nicholas and A. Nason, *J. Biol. Chem.* **207**, 353 (1954); see also Vol. II [57].
[8] D. J. D. Nicholas, A. Nason, and W. D. McElroy, *J. Biol. Chem.* **207**, 341 (1954).

(ADDENDUM)

[75] Gasometric Procedures for Amino Acids

(See [75], p. 458 for Section I.)

II. Gasometric Ninhydrin Determination of Free Amino Acids

$$RCH(NH_2)COOH \rightarrow RCHO + NH_3 + CO_2$$

Principle. The reaction requires the presence, in an unconjugated state, of both the carboxyl group and a primary or secondary α-amino group.[1] Glycine, sarcosine, proline, and hydroxyproline yield CO_2 but no aldehyde. The latter three yield no ammonia. One mole of aspartic acid yields 2 moles of CO_2. β-Alanine yields only 0.16 mole of CO_2 at pH 4.7, however, and none at pH 2.5. One mole of glutamic acid likewise yields but 1 mole of CO_2 at pH 2.5. No CO_2 is evolved from amino acids in which one of the hydrogens of NH_2 is replaced by COR as in peptides and in acetylated or benzoylated amino acids. Except for glutamyl and aspartyl peptides, which have a —$CH(NH_2)COOH$ group free (e.g., as in glutathione), peptides do not give CO_2, even from their free carboxyl groups. Hence the ninhydrin CO_2 reaction differentiates amino acids from peptides more sharply than does the above-mentioned nitrous acid method. No CO_2 is given off if both hydrogens of the amide group are substituted, or if the carboxyl group is replaced by an amide or an ester group. Amines, amides, ammonia, and glucosamine give no CO_2 when heated with ninhydrin. Urea gives 0.01 mole of CO_2 in 5 minutes at 100° in the absence of ninhydrin but much less in its presence. At pH 2.5 and for a 7-minute heating period, 1 mole of CO_2 is given per mole of alanine,

[1] D. D. Van Slyke, R. T. Dillon, D. A. MacFadyen, and P. Hamilton, *J. Biol. Chem.* **141**, 627 (1941).

valine, leucine, serine, threonine, phenylalanine, tyrosine, proline, hydroxyproline, methionine, arginine, histidine, glutamic acid, hydroxy-glutamic acid, or homocystine. Molar yields of CO_2 per mole of other amino acids under these conditions are as follows: glycine 0.95, trypto-phan 0.90, aspartic acid 2.0, cystine 1.89, lysine 1.05, hydroxylysine 1.02, ornithine 1.01, β-alanine 0.0. The effects of other pH's and heating periods on molar yields of CO_2 are recorded elsewhere.[1]

Reagents

Ninhydrin.

Citrate buffer, pH 2.5. Trisodium citrate (2.05 g.) and of citric acid (19.15 g.) are ground separately in an agate mortar, then mixed. The resulting cake is ground again. The powdered buffer is con-veniently dispended from a glass spoon calibrated to deliver 50 mg.

0.5 N NaOH, with minimal CO_2 content, is best prepared by dilut-ing an aliquot of an old settled solution of 17 N NaOH with a 25% solution of NaCl.[2] It is stored to protect it from atmos-pheric CO_2.

5 N NaOH.

2 N lactic acid, 1 to 5 dilution of concentrated lactic acid in 25% NaCl.

Procedure. A 1- to 5-ml. aliquot of unknown containing 0.04 to 0.7 mg. of carboxyl carbon is placed in the reaction vessel.[2] Fifty milligrams of powdered citrate buffer is added if the aliquot is 2 ml. or less; 100 mg. is used for a 3- to 5-ml. aliquot. A few pieces of alundum and, if foam-producing materials are present, a drop of caprylic alcohol are added. To remove dissolved CO_2 the reaction vessel is vigorously boiled over a small flame for 30 seconds (or as much longer as is necessary to decompose unstable organic compounds such as α-keto acids if these are present). A stopper greased with Nevastane XX is placed on the reaction vessel which is then placed in ice water for 3 minutes. The stopper is momen-tarily removed for the addition of 50 mg. of ninhydrin if the aliquot is 2 ml. or less, or 100 mg. if it is 3 to 5 ml. The side arm of each vessel is then connected in turn to a good water pump or to a Van Slyke-Neill blood gas apparatus for evacuation to 20 to 30 mm. pressure. After the stopper of each vessel is turned and the clamp on the tubing is closed tightly, the reaction vessels are removed in turn to a rack and immersed to the level of their clamps in an actively boiling water bath for 20 min-utes. They are shaken slightly after the first minute of heating to dis-tribute the ninhydrin. The reaction vessels are then placed in a 40° bath,

[2] P. B. Hamilton and D. D. Van Slyke, *J. Biol. Chem.* **150**, 231 (1943).

and 2 ml. of the 0.5 N NaOH with NaCl is measured into the Van Slyke-Neill blood gas apparatus (#).[3] The reaction vessel is then attached to the side arm of the chamber. The clamp, the stopper of the reaction vessel, and the cock at the top of the chamber are opened to permit the gaseous contents of the reaction vessel to be transferred to the chamber (#). The mercury is caused to flow between the 50- and 2-ml. marks five times, the reaction vessel being shaken actively each time the mercury is lowered. After the chamber cock is closed, the reaction vessel is removed. Nearly all the unabsorbed gas in the chamber is expelled after this (#). One milliliter of 2 N lactic acid with NaCl is admitted to the chamber from a stopcock pipet. CO_2 is then liberated from the acidified mixture by extracting and shaking for about 30 seconds. After the mercury is adjusted again to the 50-ml. mark, the chamber is shaken for another 3.0 minutes. The pressure, p_1, is then read at the 2-ml. mark when it is greater than 100 mm. Otherwise, it is read at the 0.5-ml. mark (#). A 0.5-ml. aliquot of 5 N NaOH solution is then admitted to the chamber, and the CO_2 is reabsorbed with the aid of three excursions of the mercury one-quarter of the way down the largest bulb of the chamber. The pressure, p_2, is read at the same mark. The blank analysis for correction, c, is conducted with 2 ml. of 0.5 N NaOH, just as indicated above, except that the side arm remains sealed, and no reaction vessel is used.

$$\text{Mg. carboxyl carbon or carboxyl nitrogen} = (p_1 - p_2 - c) \times \text{factor}$$

The factors[1] can be interpolated from a plot of data given in the table.

FACTORS[a] BY WHICH THE VALUES FOR $(p_1 - p_2 - c)$ IN MM. HG ARE MULTIPLIED
TO GIVE MG. CARBOXYL CARBON OR CARBOXYL NITROGEN IN ALIQUOT TESTED

Temperature, °C.	Carboxyl carbon		Carboxyl nitrogen	
	Pressures read at:		Pressures read at:	
	0.5 ml.	2.0 ml.	0.5 ml.	2.0 ml.
15	0.0003447	0.001374	0.0004020	0.001603
25	3310	320	3861	540
35	3187	271	3717	482

[a] The factors for carboxyl nitrogen are given in detail by D. A. MacFadyen [*J. Biol. Chem.* **145**, 395 (1942)]. The factors for carboxyl carbon are obtained by multiplying those for carboxyl nitrogen by 0.85737.

[3] Here, as in other places marked (#), the details of technique described in the original article should be followed precisely. In this summary, because of limitations of space, important details such as the opening, closing, and sealing of stopcocks and the washing of apparatus are seldom mentioned.

Determination of α-Amino Nitrogen in Plasma. Freshly drawn heparinized blood is immediately centrifuged, and 10 ml. of 1% 0.0437 N picric acid solution is added to 2 ml. of plasma. The mixture is shaken vigorously for a few seconds, then centrifuged, and the supernatant fluid decanted through a funnel plugged with a small piece of cotton. Duplicate aliquots of 5 ml. of picric acid filtrate are then placed in reaction vessels, and the procedure is carried out as outlined above, except that no buffer need be added and the heating period to remove labile CO_2 is exactly 1.5 minutes. The heating period with ninhydrin is exactly 20 minutes (#).[2]

Determination of α-Amino Nitrogen in Whole Blood. To obtain the corresponding value for whole blood, 1 ml. of whole blood is mixed with 1 ml. of water; then 10 ml. of 1% picric acid solution is added. The mixture is shaken vigorously for 1 minute, then treated as above.

Determination of α-Amino Nitrogen in Cells. For determination in cells, 1 ml. of cells is mixed with 2 ml. of water. A 2-ml. aliquot of this mixture is then treated with 10 ml. of 1% picric acid and shaken as indicated above. Special precautions are required when the blood level of urea is abnormally high (#).[2]

Determination of Glutamine in Plasma. Because glutamine, which liberates CO_2 with ninhydrin, is converted by heating for 90 minutes at 100° in 0.08 M phosphate, pH 6.5, to pyrrolidone carboxylate, which does not liberate CO_2 with ninhydrin, the difference in CO_2 yields with ninhydrin before and after such heating has served as a means of measuring the glutamine content of solutions and tissues (#).[4]

Determination of α-Imino Nitrogen in Mixtures of Amino Acids. This determination can be achieved by application of another modification which involves preliminary destruction of the amino groups of amino acids with nitrous acid, removal of excess nitrite with sulfamate, and measurement of the CO_2 released from proline and hydroxyproline by ninhydrin (#).[5]

[4] P. B. Hamilton, *J. Biol. Chem.* **158**, 375 (1945); **158**, 397 (1945).
[5] P. B. Hamilton and P. J. Ortiz, *J. Biol. Chem.* **187**, 733 (1950).

III. Manometric Arginase Determination of Arginine

Arginine → Ornithine + Urea

$$CO(NH_2)_2 + H_2O \rightarrow 2NH_3 + CO_2$$

Principle. Hunter and Pettigrew's method[1] permits simultaneous action of arginase and urease. Subsequently, the evolved CO_2 is measured in the Van Slyke-Neill blood gas apparatus. About 98.5% of the theoretical yield of CO_2 is obtained.

[1] A. Hunter and J. B. Pettigrew, *Enzymologia* **1**, 341 (1937).

Reagents

Crude arginase extract.[2] For each 100 g. of minced ox, calf, or dog liver, 75 ml. of glycerol and 25 ml. of 8% aqueous solution of $MnCl_2 \cdot 4H_2O$ are added[3,4] to a Waring blendor and blended for 5 minutes. The mixture is then placed in a large flask. This is heated in a 62° to 65° bath with constant rotation of the flask until its contents reach a uniform temperature at 58°. After 15 minutes the mixture is poured on a large folded filter paper in an icebox. Twelve to eighteen hours later, 5 N NaOH of minimal CO_2 content is added to the filtrate until its pH is reduced to 7. Arginase activity of the preparation is determined,[2,3,5] and the preparation is stored, protected from CO_2, in an icebox or in a deep-freeze.

Urease. A uniform paste is made with 10 g. of Squibb's double-strength urease and water to 25 ml. The paste is dialyzed at 0° against distilled water for 6 to 8 hours to remove canavanine[6] on which the enzyme mixture would otherwise subsequently react slowly to liberate urea,[7] then CO_2. The contents of the dialysis sack are mixed with an equal volume of glycerol and stored in the icebox or deep-freeze.

Arginase-urease mixture. To 10 ml. of arginase extract is added 1 ml. of urease solution. The preparation is stored, protected from CO_2, in an icebox or in a deep-freeze. Hunter and Pettigrew,[1] who did not use dialyzed urease, found that arginine recoveries were 0.7% higher if their enzyme mixture was incubated at room temperature for 24 hours before use.

Lactic acid, 2 N.

Phosphoric acid, 2 M.[8]

18 N NaOH, carbonate-free.[8]

2 N NaOH, minimal CO_2 content.[8]

Bromothymol blue, 0.4% solution.[8]

Octyl alcohol.[8]

Procedure. A 1- to 3-ml. aliquot of unknown solution (containing 1 to 40 mg. of arginine) is pipetted into a 25-ml. volumetric flask. An

[2] A. Hunter and J. A. Dauphinee, *J. Biol. Chem.* **85**, 627 (1930).

[3] D. D. Van Slyke and R. M. Archibald, *Federation Proc.* **1**, 139 (1942).

[4] L. Hellerman and M. E. Perkins, *J. Biol. Chem.* **112**, 175 (1935).

[5] D. D. Van Slyke and R. M. Archibald, *J. Biol. Chem.* **165**, 293 (1946).

[6] R. M. Archibald and P. B. Hamilton, *J. Biol. Chem.* **150**, 155 (1943).

[7] M. Kitagawa and A. Eguchi, *J. Agr. Chem. Soc. Japan* **14**, 525 (1938).

[8] D. D. Van Slyke, *J. Biol. Chem.* **73**, 695 (1927).

equal volume of water is pipetted into a similar control flask. To each are added 0.25 ml. of 2 M H_3PO_4 and 3 drops of bromothymol blue solution. The flasks are then twice aerated according to the directions of Van Slyke to remove preformed CO_2. Water is then added until the volume is approximately 20 ml., and the H_3PO_4 is neutralized with 2 N NaOH until the indicator is pure blue (14 to 18 drops). By rapid action, minimal absorption of CO_2 before and loss of CO_2 after addition of 2 ml. of the enzyme mixture is achieved. Water is then added to the enzyme-substrate mixture to bring the volume to 25 ml. Each flask is stoppered with a one-hole rubber stopper plugged with a Vaselined glass rod. The contents are thoroughly mixed and allowed to stand at room temperature for at least 3 hours, or overnight if inhibitory substances are present.[9] A drop of octyl alcohol is caused to pass from the cup to the side arm of the Van Slyke-Neill apparatus. Then 0.2 ml. of 18 N NaOH is added to each flask during momentary removal of the plugs in the stoppers. After mixing and 5 minutes of further incubation, a 2- to 5-ml. aliquot of the reaction mixture is introduced into the chamber, followed by 1 ml. of water, with a minimum of air. The air is expelled. Then 0.5 ml. of 2 N lactic acid for a 2-ml. sample or 1-ml. of 2 N lactic acid for a 5-ml. aliquot of unknown is drawn into the chamber. The cock is sealed with mercury, the chamber evacuated, and the CO_2 extracted. The pressure, p_1, exerted by it in a 2-ml. volume is noted and recorded together with the temperature. If p_1 is less than 100 mm. Hg, the CO_2 is re-extracted and the pressure, p_1, exerted by it in a 0.5-ml. volume is recorded. The pressure, p_0, of CO_2 from the blank solution is measured in the same manner and at the same volume and temperature.

Mg. arginine per aliquot analyzed in gas machine $= (p_1 - p_0) \times$ factor

Factors can be interpolated from a plot of the data given in the table.

FACTORS[a] FOR CALCULATION OF MG. ARGININE IN ALIQUOT IN GAS MACHINE

Temperature, °C.	For 2-ml. aliquot		For 5-ml. aliquot	
	Pressures read at:		Pressures read at:	
	0.5 ml.	2 ml.	0.5 ml.	2.0 ml.
15	0.005508	0.02161	0.005987	0.02349
24	0.005250	0.02060	0.005618	0.02203
30	0.005101	0.02002	0.005412	0.02124

[a] These factors include a $+1\%$ correction factor which A. Hunter and J. B. Pettigrew [*Enzymologia* **1**, 341 (1937)] found necessary.

[9] A. Hunter and C. E. Downs, *J. Biol. Chem.* **157**, 427 (1945).

The author has made measurements of arginine in protein hydrolyzates after performing both the incubation with enzymes and the extraction and measurement of the CO_2 within the Van Slyke-Neill apparatus. By working in summer, by heating the room, or by using the apparatus in a walk-in incubator, it is possible to cut the incubation time for at least some hydrolyzates to 2 hours.[10] If this procedure is used, it is wise to neutralize the hydrolyzate with CO_2-free 5 N NaOH until the pH (by indicator) is between 8.5 and 9.3. The solution must then be protected from CO_2 until the aliquots required have been placed in the Van Slyke-Neill apparatus. Each 2-ml. aliquot (0.5 to 8.0 mg. of arginine) is transferred to the apparatus with a stopcock pipet. This is followed by 1 ml. of enzyme mixture. After the incubation, 0.5 ml. of 2 N lactic acid is added and the CO_2 is extracted and measured. In the case of the control the addition of lactic acid precedes that of the enzyme mixture, and no incubation occurs. This type of control is not satisfactory unless canavanine is previously removed from the urease to prevent liberation of CO_2 from the enzyme during incubation. Specific conditions for analysis will depend on the circumstances on hand. To make sure that all the arginine present is measured, it is wise to see if a run with a longer incubation period indicates a higher yield. If the unknown is highly buffered, it may be necessary to use 5 N in place of 2 N lactic acid to acidify the reaction mixture sufficiently to permit extraction of CO_2. It may then be necessary to use different factors in the calculation because of the decreased solubility of CO_2 in solutions which contain much salt. Some workers may prefer to reabsorb CO_2 with 5 N NaOH after reading p_1 and to read p_2 as described in the previous method with ninhydrin (p. 1043). In this case $p_1 - p_2$ for the blank equals c, and the calculation becomes

$$\text{Mg. arginine in sample} = (p_1 - p_2 - c) \times \text{factor}$$

[10] R. M. Archibald, *Ann. of N.Y. Acad. Sci.* **47**, 181 (1946).

IV. Determination of Amino Acids by Specific Bacterial Decarboxylases

$$RNH_2COOH \rightarrow RHNH_2 + CO_2$$

Principle. A number of microorganisms when grown in acidic media generate one or more amino acid decarboxylases. Preparations of *B. cadaveris*[1] have been described which are specific for L(+)-lysine, except for a very slow decomposition of hydroxylysine.[2] Arginine decarboxylase present in this preparation can be destroyed by drying the organisms

[1] E. F. Gale and H. M. R. Epps, *Biochem. J.* **38**, 232 (1944).
[2] E. F. Gale and H. M. R. Epps, *Nature* **152**, 327 (1943).

with acetone.[3] Epps[4] prepared from *Streptococcus faecalis* a decarboxylase specific for L(−)-tyrosine and L-3,4-dihydroxyphenylalanine, and from *Clostridium welchii* a preparation for L(−)histidine.[5] Another strain of this organism was found by Gale[6] to decarboxylate only L(+)-glutamic acid and β-hydroxyglutamic acid. A washed suspension of *Clostridium septicum* decarboxylated only L(+)-ornithine. Manometric measurement of amino acids in protein hydrolyzates by these decarboxylases has been conducted, for the most part, in the Warburg manometer by techniques described by these authors. Presumably the same preparations could be used just as specifically in the Van Slyke-Neill apparatus, use of which may allow even better recoveries, as smaller gas solubility correction factors are required here.[7] Indeed, rapid, specific, and accurate methods for determination of lysine, tyrosine, and glutamic acid in the Van Slyke-Neill apparatus have been developed.[8]

Procedure for Lysine. The strain of *B. cadaveris* employed by Gale and Epps when grown in glucose-citrate media, pH 6.1, is rendered free of arginine decarboxylase by drying with acetone.[3] With 1 ml. of 10% suspension of dried organisms in the Van Slyke-Neill blood gas apparatus, 2-ml. aliquots of hydrolyzates of proteins neutralized with CO_2-free NaOH gave maximum yields of CO_2 in 4 to 10 minutes at room temperature. The hydrolyzate served as an adequate buffer. Then 0.5 ml. of $5 N$ H_2SO_4 was added, the CO_2 extracted, and its pressure, p_1, and temperature were measured as outlined in the previous section (p. 1047). For determinations of blank pressure, p_0, addition of the lactic acid preceded that of the decarboxylase.

$$\text{Mg. } \alpha\text{-NH}_2\text{-N of amino acid decarboxylated} = (p_1 - p_0) \times \text{factor}$$

Factors can be interpolated from a plot of data recorded in the table.

Procedure for Glutamic Acid. Weisiger[9] obtained a very satisfactory preparation of decarboxylase for the manometric measurement of glutamic acid by growing a strain of *E. coli* in an acidic media rich in Hurona seasoning or sodium glutamate. Organisms were harvested in a Sharples centrifuge. To about 20 g. wet weight of organisms suspended in 1 l., 60 mg. of 8-amino-quinoline is added, and the mixture is heated at pH 4.5 to 40° until the arginine decarboxylase activity is reduced by

[3] C. A. Zittle and N. R. Eldred, *J. Biol. Chem.* **156**, 401 (1944).
[4] H. M. R. Epps, *Biochem. J.* **38**, 242 (1944).
[5] H. M. R. Epps, *Biochem. J.* **39**, 42 (1945).
[6] E. F. Gale, *Biochem. J.* **39**, 46 (1945).
[7] R. M. Archibald, *Ann. N.Y. Acad. Sci.*, **47**, 181 (1946).
[8] R. M. Archibald and J. R. Weisiger, *Proc. Am. Federation Clin. Research* **3**, 96 (1947).
[9] J. R. Weisiger. Unpublished results.

90%. Packed organisms are then dried in the frozen state. One-tenth gram of frozen-dried organisms is suspended in 1 ml. of 1 M guanidine hydrochloride. After 5 minutes 1 ml. of 1 M NaF, 0.3 ml. of 0.1 M iodoacetate, 1 drop of octyl alcohol, and water to bring the volume to 10 ml. are added. A 1-ml. aliquot of sample to be analyzed, 1 ml. of 0.05 M potassium phthalate buffer (pH 3.8), 1 drop of octyl alcohol, and 1 ml. of the above preparation of glutamic acid decarboxylase are admitted

FACTORS[a] BY WHICH PRESSURE OF CO_2 IN MM. HG ARE MULTIPLIED TO GIVE MG. OF α-NH_2-N DECARBOXYLATED IN ALIQUOT

Temperature, °C.	For pressures at 0.5 ml.	For pressures at 2.0 ml.
15	0.0004386	0.001721
24	4182	640
34	3988	566

[a] These figures are the product of 0.31829 and those given in Table IX of D. D. Van Slyke and J. Sendroy, *J. Biol. Chem.* **73**, 141 (1927).

to the Van Slyke-Neill apparatus, mixed, and allowed to react for 8 minutes. Then 0.5 ml. of 2 N H_2SO_4 is admitted and the CO_2 extracted and measured in the usual manner.[10] Guanidine hydrochloride serves to inactivate preferentially arginine decarboxylase. Iodoacetate inhibits the destruction of glutamic acid decarboxylase and, like fluoride, stabilizes the blank at a low value. Recovery of glutamic acid is 99.3% complete. Presumably the specific preparation of L(+)-glutamic acid decarboxylase of Najjar and Fisher[11] could be adapted to the manometric determination of glutamic acid.

[10] D. D. Van Slyke, *J. Biol. Chem.* **73**, 695 (1927).
[11] V. A. Najjar and J. Fisher, *J. Biol. Chem.* **206**, 215 (1954); see Vol. III [75A].

List of Abbreviations

(Selected from Volumes I, II and III)

A

Ac, acetate
ACF, anhydrocitrovorum factor
Ac-SCoA, acetyl coenzyme A
ADH, alcohol dehydrogenase
ADP, adenosine diphosphate
ADPR, adenosine diphosphate ribose
AMe, S-adenosylmethionine
2'-AMP, 2'-adenylic acid (a adenylic acid)
3'-AMP, 3'-adenylic acid (b adenylic acid)
5'-AMP-5'-adenylic acid (muscle adenylic acid)
AP DPN, acetylpyridine analog of diphosphopyridine nucleotide
AR, adenosine
ATP, adenosine triphosphate
ATPase, adenosine triphosphatase

B

BAL, British anti-lewisite
BCG, bacillus of Calmette and Guerin (strain of *M. tuberculosis*)

C

CDP, cytidine diphosphate
CDP-choline, cytidine diphospho-choline
CDR, cytosine deoxyriboside
CF, citrovorum factor
CHOFAH₄, N^{10}-formyltetrahydrofolic acid
CMP, cytidine monophosphate (cytidylic acid)

CoA, coenzyme A
CoASH, coenzyme A, reduced
Cr, creatine
CR, cytidine
CSA, chondroitin sulfuric acid
CTP, cytidine triphosphate

D

DAP, dihydroxyacetone phosphate
DFP, diisopropyl fluorophosphate
DNA, deoxyribonucleic acid
DNase, deoxyribonuclease
DNFB, dinitrofluorobenzene
DNP, 2,4-dinitrophenol
DOPA, dihydroxyphenylalanine
DP-enzymes, diisopropyl phosphate enzymes
DPGA, diphosphoglyceric acid
DPN, diphosphopyridine nucleotide
DPNase, DPN nucleosidase
DPNH, diphosphopyridine nucleotide, reduced
DPCoA, dephospho-coenzyme A

E

EDTA, ethylenediaminetetraacetate (Versene)

F

FAD, flavin adenine dinucleotide
FAH₄, tetrahydrofolic acid
FAH₄CHO, N^{10}-formyltetrahydrofolic acid
FDP, fructose-1,6-diphosphate
FMN, flavin mononucleotide

1051

F-1-P, fructose-1-phosphate
F-6-P, fructose-6-phosphate

G

GA, glyceraldehyde
Gal-1-P, galactose-1-phosphate
GAP, glyceraldehyde-3-phosphate
GDP, guanosine diphosphate
GDPM, guanosine diphospho-
    mannose
GMP, guanosine monophosphate
    (guanylic acid)
GPC, glycerophosphorylcholine
GPE, glycerophosphorylethanol-
    amine
G-1-P, glucose-1-phosphate
G-6-P, glucose-6-phosphate
GR, guanosine
GSH, glutathione
GSSG, glutathione, oxidized
GTP, guanosine triphosphate

H

HDP, hexose diphosphate (fruc-
    tose-1,6-diphosphate)
HxR, inosine

I

I, 5(4)amino-4(5)-imidazolecar-
    boxamide
IAA, iodoacetate
IDP, inosine diphosphate
5'-IMP, 5'-inosinic acid
INH, isonicotinic acid hydrazide
INH DPN, isonicotinylhydrazine
    analog of diphosphopyridine
    nucleotide
IR, 5(4)-amino-4(5)-imidazolecar-
    boxamide riboside

IRMP, 5(4)-amino-4(5)-imida-
    zolecarboxamide ribotide
ITP, inosine triphosphate

K

KG, α-ketoglutarate

L

LTPP, lipothiamide pyrophos-
    phate

M

MB, methylene blue
M-6-P, mannose-6-phosphate

N

NMeN, N^1-methylnicotinamide
NMN, nicotinamide mononu-
    cleotide
NR, nicotinamide riboside
NTZ, neotetrazolium

O

OAA, oxalacetate

P

PEP, phosphoenolpyruvic acid
2-PGA, 2-phosphoglyceric acid
3-PGA, 3-phosphoglyceric acid
PGA-P, 1,3-diphosphoglyceric acid
PNA, pentose nucleic acid (ribo-
    nucleic acid)
PNPA, p-nitrophenol acetate
P, orthophosphate, inorganic

POF, pyruvate oxidation factor (lipoic acid)

PP, pyrophosphate, inorganic

PuR, purine riboside

PyR, pyrimidine riboside

R

RNA, ribonucleic acid

RNase, ribonuclease

R-1-P, ribose-1-phosphate

R-5-P, ribose-5-phosphate

T

TCA, trichloroacetic acid

THAM, tris(hydroxymethyl) aminomethane

TPNH, triphosphopyridine nucleotide, reduced

TPP, thiamine pyrophosphate

TPN, triphosphopyridine nucleotide

Tris, tris(hydroxymethyl) aminomethane

TTZ, 2,3,5-triphenyltetrazolium

U

UDP, uridine diphosphate

UDPAG, uridine diphospho-N-acetylglucosamine

UDPG, uridinediphosphoglucose

UDPGA, uridine diphospho-glucuronic acid

UDPGal, uridinediphosphogalactose

UDR, uracil deoxyriboside

UMP, uridine monophosphate (uridylic acid)

UR, uridine

UTP, uridine triphosphate

X

XR, xanthosine

Author Index

The numbers in parentheses are footnote numbers and are inserted to enable the reader to locate a cross reference when the author's name does not appear at the point of reference in the text.

A

Abelin, I., 637
Abelson, P. H., 512, 513
*Ackermann, D., 630
Adam, H. M., 632
Adams, E., 629, 633, 875
Adams, M. H., 713
Adams, R., 200, 354, 355(19)
Adank, K., 919
Adickes, F., 410
Aebi, H., 373
Ahlmark, A., 632
Åkeson, Å., 876
Albaum, H. G., 87, 88, 847(q), 849, 862
Alber, H. K., 986, 989(12)
Albert, R., 605
Alexander, H. E., 696, 697, 699(3), 700, 703(3), 708, 713, 715(9)
Allard, C., 672, 673(5 see 8), 677(5 see 8)
Allen, C. F. H., 441
Allen, D. M., 326, 327(32 see 36)
Allen, F. W., 672, 674(see 8), 677(see 8), 688, 689(6), 690, 752, 753, 757(31a), 760, 765(35), 767(31a)
Allen, R. J. L., 140
Allerton, R., 100, 186
Allfrey, V. G., 723, 753
Almond, H., 1021
Alving, A. S., 78
Alvise, V. A., 986, 989(12)
Ames, B. N., 630, 631, 634
Ames, S. R., 638
Aminoff, D., 26, 97
Andersen, W., 731
Anderson, E. P., 976
Anderson, L., 27
Anderson, W. E., 309
Andreen, J. H., 586, 593(39), 597(39)
Andresen, G., 410
Anlyan, A. J., 57
Anslow, W. P., Jr., 597, 598(61), 643, 644(2), 646(2)
Anson, M. L., 467, 468(4)
Appleton, H. D., 360
Aquist, S., 22
Arai, K., 848(ee), 849

*Abraham, S., 64(9)

Archibald, R. M., 639, 640, 642, 651, 995, 1045, 1047, 1048
Ariyama, 202
Armstrong, M. D., 584
Arnstein, H. R. V., 239, 240
Aronoff, S., 213
Artom, C., 306, 329, 360, 361, 366, 657
Arvin, I., 360
Ascoli, I., 300, 310(6), 359
Asnis, R. E., 248
Astbury, W. T., 553
Astrup, T., 358, 512, 658
Atherton, F. R., 811
Atno, J., 80, 81
Audrieth, L. F., 581
Austin, W. T. S., 632
Avery, O. T., 692, 696, 712, 713(6), 715(6)
Avison, A. W. D., 229
Avrin, I., 342
Awapara, J., 358, 584
Axelrod, B., 113, 116, 117(4), 118, 119, 172, 213, 214, 228
Axelrod, J., 363, 974
Ayengar, P., 770, 786, 791(11), 792(11a)

B

Bacher, J. E., 690, 752
Bachur, N. R., 901, 902(3)
Baddiley, J., 603, 786, 812, 813(5), 919, 926, 927(1)
Baer, E., 176, 177(6), 198, 199(4), 201, 202, 207, 217, 223, 224, 228, 335, 346, 347, 350, 841
Baer, H., 618, 619(8)
Baernstein, H. D., 594
Bailey, C. F., 584
Bailey, K., 864
Bakay, B., 590
Baker, C. G., 410, 555, 556(6), 564(4), 565, 567(6)
Baker, R. H., 262
Balasubramanian, S. C., 657
Baldwin, E., 376, 382(10)
Ball, E. G., 181
Ballentine, R., 987, 989(13b), 1022
Ballou, C. E., 198, 204, 217

1055

Jackson, R. W., 411
Jacob, M., 770, 786, 791(11), 792(11a)
Jacobs, B., 287
Jacobs, W. A., 752, 754
Jacobson, K. B., 895, 898(15)
Jacquez, J. A., 410
Jaenicke, L., 752
Jaffe, M., 585
James, A. T., 377, 382
James, S. P., 585
Jamieson, C. A., 603
Jansen, B. C. P., 946
Jayko, M. E., 333
Jeans, A., 43
Jefferson, H., 21, 22(4)
Jenkins, R. J., 709
Jensen, B. N., 86
Jensen, D., 304
Jensen, E. M., 482, 488
Jensen, F. W., 234
Jephcott, C. M., 155
Jepson, J. B., 522
Jerchel, D., 114
Jermyn, M. A., 63, 64, 516
Johnson, C. M., 29
Johnson, E. A., 709, 720, 738, 739(37),
    740(d), 741, 744, 803(68), 804
Johnson, M. J., 34, 35(7), 37(7), 38(7),
    40(7), 54, 86, 161, 286, 382, 973
Johnson, W. A., 414, 430
Johnston, H. W., 352, 354(14), 532
Johnston, J. P., 96
Johnston, P. M., 941
Joklik, W. K., 791
Jones, A. R., 402, 403, 443
Jones, A. S., 606
Jones, D. B., 584, 635, 657
Jones, J. K. N., 41, 44, 45(16, 29), 46(16),
    47(16), 48, 59, 62, 63, 71(2), 236,
    238(18)
Jones, M. A., 411
Jones, M. E., 269, 391
Jones, T. S. G., 510, 526
Jones, W., 223, 752
Jonnard, R., 985
Jorpes, E., 21, 93
Josepovits, G., 871
Joslyn, M. A., 511
Jukes, T. H., 941

K

Kabat, E. A., 26
Kagan, B. M., 326, 327(38 see 36)
Kaiser, E., 326, 327(38 see 36)
Kalckar, H. M., 181, 454, 762, 791, 843,
    847(b, c), 848(c), 849, 869, 870(3),
    871(1), 968, 974, 976
Kapeller-Adler, R., 632
Kapfhammer, J., 361
Kaplan, A., 299, 300(2), 307, 312, 313(36)
Kaplan, N. O., 107, 254, 255, 266, 268,
    747, 762, 767, 791, 810, 836, 873,
    874(2), 875, 878, 882, 888, 889, 890,
    891(1), 892, 893, 894, 895(8), 898
    (15), 901, 902(3), 903, 904, 905,
    906(1), 913, 916, 929, 930, 931
Kapplahn, J. L., 895
Kapuscinski, V., 1032
Karl, H. L., 1031
Karpén, E., 413, 415
Karrer, P., 629, 919
Karte, H., 447
Kascher, H., 391
Kasdon, S. C., 57
Kass, B. M., 265
Katchalski, E., 540, 543, 544(4), 545(1),
    546, 550, 553, 554
Kates, M., 335, 346, 347, 841
Kaucher, M., 307
Kaufman, S., 437
Kay, E. R. M., 690, 691, 692, 699
Kay, L. M., 471
Keegan, P. K., 414
Keighley, G., 623
Keilin, D., 107, 108
Keller, E. B., 786, 791(11)
Keller, P. J., 51
Kelly, T. L., 438
Kemble, A. R., 521
Kemp, I., 345
Kendall, K. K., 1030
Kendrick, A. B., 460
Kennedy, E. P., 265, 377, 786, 787(12)
Kenner, G. W., 820, 954
Kerby, G. P., 22
Kerr, S. E., 156, 724(5), 725, 774, 787,
    862, 866
Kertesz, Z. I., 27, 28(2), 29(2), 30(2)
Keston, A. S., 358, 525

Riesen, W. H., 582
Riley, R. F., 348, 357
Rimington, C., 599, 629, 633(46)
Ringel, S. J., 584
Risley, H. A., 638
Rittenberg, D., 603, 623
Ritter, G. J., 32
Rivers, T. M., 101
Robbins, P. W., 767, 929
Roberts, D., 184, 185, 186
Roberts, E. C., 944
Roberts, J. C., 411
Roberts, N. R., 206, 895
Roberts, R. B., 512, 513
Robeson, C., 389, 391
Robin, Y., 861
Robins, E., 363, 365(see 26)
Robinson, D. S., 563, 564(28)
Robinson, H. W., 450, 451
Robinson, R., 155, 169, 170, 171(22), 172,
    220, 838, 841, 847(h, u, v, x), 848
    (bb, h), 849
Roblee, A. R., 582
Roche, J., 861
Rochovansky, O., 643, 647(1)
Rockland, L. B., 480, 487, 605, 614, 635
Redwell, A. W., 626
Roe, A. S., 713
Roe, E. T., 235, 236(13), 237(13), 238(13)
Roe, J. H., 70, 75, 76(8), 87, 145, 157, 168
Rolf, I. P., 331
Roll, L. J., 441
Roll, P. M., 715
Roman, W., 360
Rose, I. A., 269
Rose, W. G., 352
Rosebrough, N. J., 448
Roseman, S., 21, 26
Rosen, G., 681
Rosen, G. U., 101
Rosenberg, H., 851, 852(3), 858
Rosenthal, S. M., 627, 630(see 55), 631
    (30)
Ross, W. F., 542
Rosselet, J. P., 134, 136, 155, 847(g), 848
    (g), 849
Rossiter, R. J., 676, 677
Roth, L. W., 627

Rothenberg, M. A., 153, 164, 165(4), 167,
    836
Rothenfusser, S., 78
Rothstein, A., 862
Roussouw, S. D., 579
Ruben, S., 300, 316(5)
Rubin, J., 78
Rudney, H., 369
Rudolph, G. G., 411
Rudy, H., 336
Rueff, L., 932
Rundle, R. E., 13
Ruokolainen, T., 413, 415

S

Sabatay, S., 163, 165
Sable, H. Z., 153, 188, 233
Sabotka, H., 986
Sacks, J., 836
Saffran, M., 421, 426, 430(8)
St. John, J. L., 995, 996(3)
Saito, K., 240
Sakaguchi, S., 702
Sakami, W., 629, 660
Salcedo, J., Jr., 950
Saltman, P., 116
Saluste, E., 624, 660
Samuelson, O., 731
Sanadi, D. R., 770, 786, 791(11), 951,
    953(3)
Sandberg, M., 641
Sandell, E. B., 1017, 1024
Sanger, F., 631
Sato, Y., 377, 382(14)
Sattler, L., 84
Sauberlich, H. E., 481, 483, 484, 485, 487,
    488, 489(24), 491, 614, 658
Sax, K. B., 101
Scarisbrick, R., 376, 382(10)
Schade, A., 206
Schäfer, W., 931, 932(4), 933
Schaffer, F. L., 35, 37(10), 986, 993(11b)
Schaffert, R. R., 393, 1031, 1032
Schayer, R. W., 626
Scheirer, D., 986
Schepartz, 953
Scherer, H., 410
Schicktanz, S. T., 374
Schild, K. T., 172

Subject Index

A

Acetaldehyde,
deoxyribose-5-phosphate formation
from, 186
formation of from lactic acid, 242
hydrazones of, 284
p-hydroxydiphenyl reaction with, 244
interference of in formate determination, 289
as oxidant in yeast fermentation, 209–
210
separation of from formaldehyde, 656
Acetal phosphatides (acetal phospholipids),
glycerylphosphorylethanolamine preparation from, 349–350
hydrolysis of, 343, 346
Acetic acid, see also Acids, volatile,
colorimetric test for, 265
column chromatography of on Celite,
370
on silica gel, 374–376, 399
Duclaux distillation of, 382
value for, 380
enzymatic micromethod for, 266–269
by acetokinase reaction, 269
by DPN reduction, 269
sodium inhibition of, 269
by sulfanilamide acetylation, 266–
269
enzyme for, 268
ether-water distribution of, 265
microdiffusion method for, 263–266
paper chromatography of, 265, 377–379
separation of by azeotropic distillation, 374
steam distillation of, 373
uranyl acetate method for, 381
Acetic acid, glacial,
purification of, 200
Acetic anhydride,
acetyl phosphate preparation with,
229–230
Acetoacetic acid,
acid-catalyzed breakdown of, 269
decarboxylation of by aniline, 420
determination of, 283–285
as acetone dinitrophenylhydrazone,
283–284

by other methods, 284–285
colorimetric, 285
as Denigès complex, 285
fluorometric, 285
manometric, 284
interference by in acetate assay, 268–
269
S-Acetoacetyl thiol compounds,
optical properties of, 937
Acetobacter suboxydans,
5-ketogluconic acid from, 236–237
α-Acetohalogenoaldoses,
preparation of, 130–131
Acetohydroxamic acid,
acetate assay as, 269
as reference standard, 231
Acetoin,
chromatography of on Celite, 247–248,
249, 261
determination of, 278–279
optical rotation of (−) form of, 278
oxidation of by dichromate, 259
by $FeCl_3$-$FeSO_4$ reagent, 279, 280
preparation of (−) form of, 277–278
purification of dimer of racemic, 277
Acetokinase,
acetate assay with, 269
α-Acetolactic acid,
decarboxylation of, 282, 283
determination of, 282–283
colorimetrically, 282
gasometrically, 283
after chemical decarboxylation,
283
after enzymatic decarboxylation,
283
optical isomers of, enzymatic decarboxylation rates for, 283
preparation of, 281–282
α-Acetolactic decarboxylase,
preparation from *A. aerogenes,* 283
Acetone, see also under Acetoacetic acid,
chromatography of on Celite, 261
determination of, 283–285
distillation of, 258
as lipid solvent, 300, 313–314
titration of amino acids in presence of,
454–456, 458
Acetonedicarboxylic acid,
decarboxylation of by aniline, 420

nitrate determination with, 982
purification of hexose monophosphates
as salts of, 838
Buret, micrometer, 327
2,3-Butanediol,
chromatography of on Celite, 247–248
oxidation of by dichromate, 256
n-Butanol,
chromatography of on Celite, 261
distillation of, 258
oxidation of by dichromate, 256, 259,
261
stoichiometry of, 259, 261
Butyl mercaptan,
optical properties of S-acyl-, 936
Butyric acid,
chromatography on Celite of isobutyric
and, 370
chromatography on silica gel of n-, 374–
376
differentiation of normal and branched
chain, 382
Duclaux distillation of n-, 382
Duclaux value for n-, 380
ether-water distribution of, 265
interference of in acetate assay, 268
paper chromatography of n-, 377–379
Butyryl coenzyme A,
enzymatic cleavage of butyryl pantetheine and, 935

C

Cadaverine,
formation of, 462
Cadmium chloride,
sphingomyelin complex with, 345
Calcium,
flame spectrophotometric analysis for,
1030–1035
gravimetric method for, 1014
precipitation of by Na_2WO_4, 265
spectrophotometric method for, 1014
volumetric method for, 1012–1014
Calcium salts,
fractionation of organic phosphates as,
836–837
phosphate removal with, 846
Canavanine,
removal of from urease, 1045

Caproic acid,
ether-water distribution of, 265
Duclaux distillation of n-, 382
Duclaux value for n-, 380
and isomers of, chromatography of on
Celite, 370
Caprylic acid,
ether-water distribution of, 265
Carbamylglutamic acid (CLG),
recovery of, 650
and related compounds, determination
of, 647–651
separation and isolation of, 651–652
elution diagrams for, 652
Carbamyl phosphate,
assay of, 654–655
hydrolysis of, 655
preparation of, 653–654
Carbazole reaction,
determination of DNA by, 103
of DNA and PNA by, 682–683
of pentose by, 87
of total hexose by, 80–81
of uronic acid by, 27, 93–95
pentose and hexose interference
in, 94
Carbobenzoxychloride (benzyl chloro-
carbonate, benzyl chloroformate),
preparation of, 352, 532–533, 541–542
Carbobenzoxyethanolamine,
preparation of, 352
Carbobenzoxy group,
hydrogenolysis of, 353–355
Carbohydrate(s), see also Sugars and
Polysaccharides,
carbazole reaction for, 80–81
content of in phosphatides, 339
diphenylamine reaction for, 78–79
indole reaction for, 79–80, 98
molybdate complexes with, 213
orcinol reaction for, 88
reaction in Elson and Morgan procedure, 97
Carbon[14],
radioactivity measurement of on paper,
115
Carbon, activated, see Charcoal
Carbon column,
hyaluronic acid purification on, 23

application to carbamyl phosphate, 654–655

Flame spectrophotometric analysis, procedure, 1034–1035
sources of interference, 1030–1032

Flavinadenine dinucleotide (FAD), acid hydrolysis of, 957
chemical synthesis of, 954
crystallization of, 954
enzymatic assay of, 955–957
extinction coefficient for, 951
extraction of with phenol, 835
fluorimetric assay of, 960–962
paper chromatography of, 953–954, 957
partition coefficient for, 962
precipitation of as silver salt, 837
preparation of from liver, 950–955
various procedures for purification of, 953

Flavinadenine dinucleotide derivative (FAD-X),
paper chromatography of, 957

Flavin mononucleotide (FMN), assay of, enzymatic, 958–959
fluorimetric, 960–962
ion exchange chromatography of, 868
paper chromatography of free and cyclic, 957
partition coefficient for, 962
preparation of, 957–958
by chemical synthesis, 958
by hydrolysis of FAD, 957

Fluoride,
removal of, 373

Fluoroacetate,
thiol ester derivatives of, 939

Folin-Ciocalteu reagent (phenol reagent),
amino acid determination with, 467–468
preparation of, 448, 467
protein determination with, 448–450
purine determination with, 779
for tyrosine, 612, 613

Formaldehyde (formalin),
acetylacetone reaction with, 288–289
chromotropic acid reaction with, 247–248, 288–289
cysteine-carbazole reaction with, 77–78
formate reduction to, 287–288
formation of from glycerol, 247
hydrazones of, 284
p-hydroxydiphenyl reaction with, 244
reaction of with ammonia, 377
removal of with phenylhydrazine, 289
separation of from acetaldehyde, 656
as spray reagent for amino acids, 521

Formamide,
paper chromatography of sugar phosphates with, 116

Formic acid,
column chromatography of on Celite, 370
on silica gel, 374–376, 399, 400, 401
determination of, 286–292, 381
biological methods, 289–292
conversion to CO_2 and H_2, 289–290
inhibition of, 290
oxidation to CO_2 and H_2O, 290–292
chemical macromethods for, 286–287
oxidation with mercuric acetate, 287
oxidation with mercuric oxide, 286–287
chemical micromethods for, 287–289, 381
colorimetric, 287–289
with acetylacetone, 288–289
with chromotropic acid, 288–289
manometric, with ceric sulfate and palladium, 287
with $HgCl_2$, 381
titrimetric, with $HgCl_2$, 381
Duclaux value for, 380
formation of from glycerol, 247
nucleic acid hydrolysis by, 716
paper chromatography of, 377–379
pK_a value for, 285
production of from starch or glycogen by periodate oxidation, 47–50
separation of by alkali, 374
by chromatography, 285–286
by distillation, 285–286, 374
by oxidation with $HgSO_4$, 374
steam distillation of, 372, 373

with dicyclohexylcarbodiimide, 808

with trifluoroacetic anhydride, 808–809

by degradation of RNA, 806–807

by alkaline hydrolysis, 807

with pancreatic ribonuclease, 806

with plant ribonuclease, 806–807

Nucleotide phosphokinase,

synthesis of uridine and guanosine polyphosphates with, 788

Nucleotide pyrophosphatase, see Pyrophosphatase, snake venom

O

n-Octanoic acid,

Duclaux value for, 380

paper chromatography of, 377–379

turbidimetric method for, 382

S-Octenoyl thiol compounds,

optical properties of, 936

Oligonucleotides,

analysis of phosphoryl groups of, 763–766

chain length determination in, 764–'765

enzymatically with phosphatases, 765

by titration, 764–765

degradation of by various agents, 755–756

determination of end groups in deoxyribo-, 769

dializability of, 769–770

dependence on ionic strength, 770

paper chromatography of, 743–747

separation from nucleic acids by acid precipitation, 770–771

stepwise degradation with phosphatase and periodate, 766

terminal 2′,3′-cyclic phosphodiester groups in, 765–766

terminal 2′-phosphomonoester groups in, 763

terminal 3′-phosphomonoester groups in, 765

terminal 5′-phosphomonoester groups in, 766

Oligosaccharides, see also Disaccharides, Trisaccharides,

paper chromatography for quantitative analysis of reducing, 71

Orcinol,

detection of sugars on paper with, 64

of sugar phosphates on paper with, 114

determination of pentose with, 87–90, 92

effect of heating time on, 88

interference of tetroses with, 92

determination of pentoses and keto-heptoses with, 70–71

of phosphate esters with, 107

of PNA with, 681–682, 683

of RNA in DNA with, 702

differentiation of aldose and ketose with, 75

of pentose and heptulose with, 105–107

DNA reaction with, 683

nucleotide reactions with, 739, 742

purification of, 680

Ornithine,

N-chloroacetylation of DL-, 557

column chromatography of, 497

decarboxylase for L(+)- in *Cl. septicum*, 464, 1048

α-keto acid preparation from, 409, 412

ninhydrin color for, 503

nitrous acid reaction with, 459, 460

optical rotation of L-, 569

polymer of DL-, 553

reaction of in carbamylamino acid assay, 650

resolution of DL-, 560, 562, 564–565, 569

isolation of D-, 562, 564–565

of L-, 560, 564–565

R_f values for, 518

Orotic acid,

column chromatography of, 652

nucleotide synthesis from labeled, 787–788

spectrophotometric constants of, 740

Orthophosphate, see Phosphate, inorganic

Ovomucoid,

carbazole reaction for, 81

Vitamin(s) A,
  molecular distillation of, 388, 391
Vitreous humor,
  hyaluronic acid from, 23
Voges-Proskauer reaction,
  acetoin determination by, 278–279

W

Waring Blendor,
  lipid extraction with, 304–305
Water,
  purification of for metal assays, 1036
West reagent, 52

X

Xanthine,
  colorimetric determination of, 779
  enzymatic determination of hypoxan-
    thine plus, 781
  paper chromatography of, 716–723
    of compounds of, 746
  reduction of Folin-Ciocalteu reagent
    by, 449
  spectrophotometric constants of, 740
  volatile amines from, 366
Xanthine oxidase,
  role in inosine phosphorolysis, 181
Xanthosine,
  spectrophotometric constants of, 740
Xylan,
  content of in hemicellulose, 32
2,4-Xylen-1-ol,
  nitrate determination with, 982–983
Xylose,
  cysteine-H_2SO_4 reaction for, 87, 91
  glucose oxidase action on, 109
  R_f values for, 67
Xylose-1-phosphate,
  hydrolysis constant for, 847

Xylose-5-phosphate,
  hydrolysis constant for, 847
  R_f values for in borate, 121
Xylulose,
  orcinol reaction with, 107
  R_f values for, 67
Xylulose-5-phosphate,
  analysis of, 195
  enzymatic synthesis of, 193–195

Y

Yeast,
  fructose diphosphate preparation with,
    162–163
  D(−)-3-phosphoglyceric acid prepa-
    ration with, 209–210
  preparation of glucose-1,6-diphosphate
    from, 144–147
  spectrophotometric measurement of
    fermentation by, 255

Z

Zinc,
  gravimetric method for, 1029–1030
  microbiological methods for, 1035–
    1041
    standard curve for, 1038–1040
  purification of, 1038
  removal of from culture media, 1037
  spectrophotometric method for, 1027–
    1028
  volumetric method for, 1028–1029
Zinc hydroxide,
  protein removal by, 22, 257
Zinc sulfate,
  protein precipitation with, 278
Zwischenferment, see Glucose-6-phos-
    phate dehydrogenase

Printed and bound by CPI Group (UK) Ltd, Croydon, CR0 4YY

03/10/2024

01040508-0001